The Nucleic Acid Protocols Handbook

The Nucleic Acid Protocols Handbook

Edited by

Ralph Rapley

University of Hertfordshire
Hatfield
UK

Humana Press ✱ Totowa, New Jersey

© 2000 Humana Press Inc.
999 Riverview Drive, Suite 208
Totowa, New Jersey 07512

All rights reserved. No part of this book may be reproduced, stored in a retrieval system, or transmitted in any form or by any means, electronic, mechanical, photocopying, microfilming, recording, or otherwise without written permission from the Publisher. Methods in Molecular Biology™ is a trademark of The Humana Press Inc.

The content and opinions expressed in this book are the sole work of the authors and editors, who have warranted due diligence in the creation and issuance of their work. The publisher, editors, and authors are not responsible for errors or omissions or for any consequences arising from the information or opinions presented in this book and make no warranty, express or implied, with respect to its contents.

This publication is printed on acid-free paper. ∞

ANSI Z39.48-1984 (American Standards Institute) Permanence of Paper for Printed Library Materials.

Cover design by Patricia F. Cleary.

For additional copies, pricing for bulk purchases, and/or information about other Humana titles, contact Humana at the above address or at any of the following numbers: Tel: 973-256-1699; Fax: 973-256-8341; E-mail: humana@humanapr.com, or visit our Website at www.humanapress.com

Photocopy Authorization Policy:

Authorization to photocopy items for internal or personal use, or the internal or personal use of specific clients, is granted by Humana Press Inc., provided that the base fee of US $10.00 per copy, plus US $00.25 per page, is paid directly to the Copyright Clearance Center at 222 Rosewood Drive, Danvers, MA 01923. For those organizations that have been granted a photocopy license from the CCC, a separate system of payment has been arranged and is acceptable to Humana Press Inc. The fee code for users of the Transactional Reporting Service is: [0-89603-459-3/00 $10.00 + $00.25].

Printed in the United States of America. 10 9 8 7 6 5 4 3 2 1

Library of Congress Cataloging in Publication Data

The nucleic acid protocols handbook / edited by Ralph Rapley.
    p. cm.
  Includes index.
  ISBN 0-89603-459-3 (alk. paper) (hardcover), ISBN 0-89603-841-6 (paper)
  1. Nucleic acids—Laboratory manuals. 2. Polymerase chain reaction—Laboratory manuals. 3. Gene mapping—Laboratory manuals. 4. Molecular genetics—Laboratory manuals.
I. Rapley, Ralph.
QP620.N7987 1999
572.8—dc21   98-43385
                                                                                 CIP

Preface

There can be no doubt that some of the most spectacular advances made in science over the past few decades have been in the isolation, analysis, and manipulation of nucleic acids. This has led to a much greater understanding of mechanisms and processes across many fields of bioscience, such as biochemistry, microbiology, physiology, pharmacology, and the medical sciences to name a few. It has also led to the growth of the biotechnology industry, which seeks to develop and commercialize many of these important processes and methods. Much of this has come about because of the development of numerous molecular biology and genetic manipulation techniques. The discovery of restriction enzymes and the development of cloning vectors in the early 1970s opened the door to ways of isolating and manipulating nucleic acids that had never been thought possible. Gene probe labeling and hybridization were developed and refined to provide powerful methods of analysis. These—together with the development of DNA sequencing methods, protein engineering techniques, and PCR—have all continued to contribute substantially to the understanding of biological processes at the molecular level. The protocols for these important methods are the focus of *The Nucleic Acid Protocols Handbook*, whose aim is to provide a comprehensive set of techniques in one volume that will enable the isolation, analysis, and manipulation of nucleic acids to be readily undertaken.

The Nucleic Acid Protocols Handbook is divided into 10 parts; within each there are approximately 10 chapters. The first four parts follow one another logically: nucleic acid extraction (Part I), basic separation and analysis of DNA (II), through probe design and labeling (III), and RNA analysis techniques (IV). The following three sections deal with gene library construction and screening (V), DNA sequencing (VI), and the polymerase chain reaction (VII). Part VIII deals with the analysis of genes, mutations, and protein interactions and is followed by Part IX, on mutagenesis, transcription, and translation in vitro. This is followed finally by Part X, on gene localization and mapping *in situ*. In compiling this volume a number of techniques have been drawn and updated from versions appearing in earlier volumes of Humana Press' *Methods in Molecular Biology* series. These highly successful books have provided numerous laboratories with the techniques needed to undertake modern laboratory molecular biology success fully. As such, their format has been followed in *The Nucleic Acid Protocols Handbook*. Thus a short introduction to the basic theory of the technique is followed by a complete listing of all materials and reagents needed before a particular protocol is presented. Step-by-step instructions are then provided in the Methods section. In addition, Notes are cited throughout the Methods and appear at the end of the chapter, providing valuable and highly useful information not found in traditional scientific literature. This essential information in many cases may mean the difference between the success or failure of a particular technique and is one of the recognized key points of the Humana *Methods in Molecular Biology* series.

It is inevitable that a degree of overlap occurs between some of the chapters. Indeed, the use of the polymerase chain reaction is now so widespread that it is a key element of many of the protocols. These have been cross referenced where possible, although most of the protocols are self-contained and can be attempted without the need to read further chapters. For those new or unfamiliar to laboratory molecular biology, the compilation of protocols in *The Nucleic Acid Protocols Handbook* also provides the ability to attempt protocols confidently. The intent was not to list all protocols in molecular biology (within one volume, this is an impossible task), and certainly more advanced protocols may be found in a number of excellent texts including many in the Humana *Methods in Molecular Biology* series. It was, however, the aim to provide the most commonly used protocols and alternatives in one volume at a level accessible to most laboratories, which we believe has been achieved. In such a large compilation, much credit must go to the authors, who have devoted valuable time and effort to write and update these protocols; to Prof. John M. Walker, the series editor, for his helpful advice and guidance; and to the staff at Humana Press for their substantial efforts in the production of the volume.

Ralph Rapley

Contents

Preface v
Contributors xv

PART I NUCLEIC ACID EXTRACTION

 1 Isolation of High-Molecular-Weight DNA from Animal Cells 3
    Ian Garner
 2 Isolation of mRNA by Affinity Chromatography 9
    Sian Bryant and David L. Manning
 3 Isolation and Purification of DNA from Plants 13
    Justin Stacey and Peter G. Isaac
 4 Purification of Uncontaminated, Intact Plant RNA 17
    Shu-Hua Cheng, Brandon D. Moore, and Jeffrey R. Seemann
 5 An Improved Method to Isolate Mitochondrial RNA
    from Green Plant Tissue 23
    Fei Ye and Ralf Reski
 6 Isolating Chromosomal DNA from Bacteria 29
    Elisabeth Chachaty and Patrick Saulnier
 7 Bacterial DNA Extraction for Polymerase Chain Reaction
    and Pulsed-Field Gel Electrophoresis 33
    Elisabeth Chachaty and Patrick Saulnier
 8 Isolation of Fungal Nucleic Acids 37
    Surapareddy Sreenivasaprasad
 9 Total RNA Isolation from Bacteria 47
    John Heptinstall
10 Simultaneous RNA and DNA Extraction
    from Biopsy Material, Culture Cells, Plants, and Bacteria 53
    Udo Döbbeling
11 Spectrophotometric Analysis of Nucleic Acids 57
    John Heptinstall and Ralph Rapley

PART II BASIC SEPARATION AND ANALYSIS OF DNA

12 Restriction Endonuclease Digestion of DNA 63
    Duncan R. Smith
13 Agarose Gel Electrophoresis of Nucleic Acids 67
    D. Ross Williams and Ralph Rapley

14 Preparation of RNA Dot Blots 71
    Rachel Hodge

15 Native Polyacrylamide Gel Electrophoresis 73
    Adrian J. Harwood

16 Southern Blotting of Agarose Gels by Capillary Transfer 77
    Ralph Rapley and Jane Davenport-Jones

17 Pulsed-Field Gel Electrophoresis 81
    John Maule

18 HPLC of DNA and PCR Products 105
    Elena D. Katz

PART III PROBE DESIGN, SYNTHESIS, AND LABELING

19 End-Labeling of DNA Fragments 117
    Adrian J. Harwood

20 Nick Translation and Random Hexamer Labeling of DNA 123
    Jane Davenport-Jones

21 Generation of Labeled Probes by Polymerase Chain Reaction 127
    Y. M. Dennis Lo and Shu F. An

22 Nonradioactive Oligonucleotide Probe Labeling 135
    Sue Fowler and Ian Durrant

23 Preparation of Direct, Enzyme-Labeled DNA Probes 145
    Ian Durrant and Timothy Stone

24 Random Prime Labeling of DNA Probes
    with Fluorescein-Tagged Nucleotides 149
    Bronwen M. Harvey, Claire B. Wheeler, and Martin W. Cunningham

25 Hybridization and Detection of Fluorescein-Labeled DNA Probes
    Using Chemiluminescence 153
    Claire B. Wheeler, Bronwen M. Harvey, and Martin W. Cunningham

26 Hybridization of Enzyme-Labeled Probes
    and Detection by Chemiluminescence 157
    Timothy Stone and Ian Durrant

27 Hybridization and Competition Hybridization of Southern Blots 163
    Rosemary Kelsell

28 Autoradiography and Fluorography 169
    Eric Quéméneur

PART IV RNA ANALYSIS TECHNIQUES

29 Formaldehyde Gel Electrophoresis of Total RNA 177
    Sian Bryant and David L. Manning

30 RNA Probes for the Analysis of Gene Expression 181
    Dominique Belin

31 Primer Extension Analysis of mRNA 195
    Maggie Walmsley, Mark Leonard, and Roger Patient

Contents

32 S1 Mapping Using Single-Stranded DNA Probes 201
    Stéphane Viville and Roberto Mantovani

33 Measurements of Rate of Transcription in Isolated Nuclei
    by Nuclear "Run-Off" Assay 207
    Rai Ajit K. Srivastava and Gustav Schonfeld

34 One-Tube RT-PCR with Sequence-Specific RT Primers 213
    Ulrich Pfeffer and Paola Ferro

35 Characterization of RNA Using Continuous RT-PCR
    Coupled with ELOSA 219
    François Mallet

36 Quantitative Analysis of RNA Species by Polymerase Chain
    Reaction and Solid-Phase Minisequencing 229
    Anu Suomalainen and Ann-Christine Syvänen

37 Nonradioactive Northern Blotting of RNA 239
    Rainer Löw

38 Analysis of RNA by Northern Blotting Using Riboprobes 249
    Rai Ajit K. Srivastava

PART V GENE LIBRARY CONSTRUCTION AND SCREENING

39 Production of Double-Stranded cDNA for Gene Library Synthesis 261
    Jane Kirk and Steve Mayall

40 Using Rapid Amplification of cDNA Ends (RACE)
    to Obtain Full-Length cDNAs 267
    Yue Zhang and Michael A. Frohman

41 cDNA Library Construction
    Using Streptavidin-Paramagnetic Beads and PCR 289
    Kris N. Lambert and Valerie M. Williamson

42 Rapid (Ligase-Free) Subcloning
    of Polymerase Chain Reaction Products 295
    Alan R. Shuldiner and Keith Tanner

43 Subtraction Hybridization cDNA Libraries 305
    *Clifford W. Schweinfest, Peter S. Nelson, Michael W. Graber,
    Rita I. Demopoulos, and Takis S. Papas*

44 Cloning Polymerase Chain Reaction Products
    Utilizing the T/A Overhang and a Kit 319
    Melissa Lail-Trecker

45 Extraction and Purification of Plasmid DNA 327
    Craig Winstanley and Ralph Rapley

46 Biotinylated Probes in Colony Hybridization 333
    Michael J. Haas

47 Cloning Long Polymerase Chain Reaction Products 339
    Songrong Ren and J. Michael Ruppert

48 Cloning DNA Fragments in M13 Vectors 347
    David Walsh

49 cDNA Library Construction for the Lambda ZAP®-Based Vectors *355*
   Marjory A. Snead, Michelle A. Alting-Mees, and Jay M. Short

50 Expression and Preparation of Fusion Proteins
   from Recombinant λgt11 Phages *367*
   Sheng-He Huang and Ambrose Jong

51 Antibody Screening of Bacteriophage λgt11 DNA
   Expression Libraries *373*
   Peter Jones

52 Screening cDNA Libraries by Hybridization
   with Double-Stranded DNA Probes and Oligonucleotides *381*
   Caroline A. Austin

53 cDNA Library Screening with the Tetramethylammonium Chloride (TMAC)
   Technique Using Highly Degenerate Oligoonucleotide Probes *389*
   Bent Honoré and Peder Madsen

54 Screening Recombinant Libraries by Polymerase Chain Reaction *397*
   Michael W. King

55 Construction and Screening of Cosmid Libraries *405*
   Jens Hanke and Jörg D. Hoheisel

56 Generation of Large Insert YAC Libraries *415*
   Zoia Larin, Anthony P. Monaco, and Hans Lehrach

57 YAC Library Storage and Transport *425*
   John E. Collins, Sheila Hassock, and Ian Dunham

58 YAC Library Screening: *Preparation of Hybridization Filters
   and Polymerase Chain Reaction Pools* *431*
   Charlotte G. Cole, John E. Collins, and Ian Dunham

59 YAC Library Screening: *Hybridization
   and PCR-Based Screening Protocols* *437*
   Charlotte G. Cole, John E. Collins, and Ian Dunham

60 Phage-Display Libraries of Murine
   and Human Antibody Fab Fragments *449*
   *Jan Engberg, Lene K. Johansen, Michelle Westengaard-Hildinge,
   Erik S. Riise, and Bjarne Albrechtsen*

PART VI DNA SEQUENCING

61 Preparation and Analysis of DNA Sequencing Gels *481*
   Bimal D. M. Theophilus

62 DNA Sequencing of Plasmids *489*
   George Murphy

63 Sequencing DNA Fragments Cloned into M13
   and Phagemid Vectors *493*
   Neil Brewis

64 Direct cDNA Sequencing Using Sequential
   Linear/Asymmetric Polymerase Chain Reaction *499*
   Ivor J. Mason

65	Purification and Enzymatic Sequencing of Polymerase Chain Reaction Products 505
	Frank C. Brosius III, Lawrence B. Holzman, and Xinan Cao
66	Direct Polymerase Chain Reaction Sequencing with Denaturants 515
	Wei Zhang and Albert B. Deisseroth
67	Direct DNA Sequencing of Polymerase Chain Reaction Products Using Magnetic Beads 523
	Joakim Lundeberg, Bertil Pettersson, and Mathias Uhlén
68	Polymerase Chain Reaction Cycle Sequencing with Degenerate Primers 533
	Zhiyuan Shen, Jingmei Liu, Robert L. Wells, and Mortimer M. Elkind
69	Direct Automated Cycle Sequencing of Polymerase Chain Reaction Products 541
	Susan E. Daniels
70	Affinity-Capture and Solid-Phase Sequencing of Biotinylated Polymerase Chain Reaction Products 547
	Anu Suomalainen and Ann-Christine Syvänen
71	DNA Sequencing by the Chemical Method 553
	Eran Pichersky
72	One-Step One-Lane Chemical Sequencing of DNA 557
	Giovanna Costanzo, Ernesto Di Mauro, and Rodolfo Negri

PART VII BASIC POLYMERASE CHAIN REACTION METHODS

73	Polymerase Chain Reaction: Basic Principles and Routine Practice 569
	Lori A. Kolmodin and J. Fenton Williams
74	Primer Selection and Design for Polymerase Chain Reaction 581
	Wojciech Rychlik
75	One-Step Optimization Using Touchdown and Stepdown Polymerase Chain Reaction 589
	Kenneth H. Roux and Karl H. Hecker
76	Cloning Gene Family Members Using Polymerase Chain Reaction with Degenerate Oligonucleotide Primers 595
	Gregory M. Preston
77	Construction of Synthetic Genes by Polymerase Chain Reaction 609
	Patrick J. Dillon and Craig A. Rosen
78	Rapid Amplification of cDNA Ends 613
	David Bertioli
79	Multiplex Polymerase Chain Reaction 619
	Jerald Radich
80	Inverse Polymerase Chain Reaction 625
	Sheng-He Huang

81 Long Range Polymerase Chain Reaction 633
   William Waggott

PART VIII   ANALYZING GENES, MUTATIONS, AND PROTEIN INTERACTIONS

82 Nonradioactive Differential Display of Messenger RNA 645
   Thomas C. G. Bosch and Jan U. Lohmann

83 Gene Isolation by Exon Trapping 653
   David B. Krizman

84 DNA Rescue by the Vectorette Method 667
   Marcia A. McAleer, Alison Coffey, and Ian Dunham

85 Random Amplified Polymorphic DNA (RAPDs) 675
   Scott Tingey

86 Restriction Fragment Length Polymorphism 679
   Mohammad S. Enayat

87 Detection of Mutations in DNA and RNA by Chemical Cleavage 685
   Richard G. H. Cotton

88 Mutation Screening Using PCR-SSCP:
   Silver Staining and Isotopic Protocols 695
   Philip J. Saker

89 Detecting Point Mutations
   by Denaturing-Gradient Gel Electrophoresis 705
   *Stephen R. Dlouhy, Patricia Wheeler, James A. Trofatter,
   Peter J. Stambrook, and Jay A. Tischfield*

90 Analysis of Nucleotide Sequence Variation
   by Solid-Phase Minisequencing 717
   Anu Suomalainen and Ann-Christine Syvänen

91 The Amplification Refractory Mutation System 723
   John M. Old

92 DNase I Footprinting 729
   Benoît Leblanc and Tom Moss

93 Identification of Protein–DNA Contacts with Dimethyl Sulfate:
   Methylation Protection and Methylation Interference 737
   Peter E. Shaw and A. Francis Stewart

94 The Gel Shift Assay for the Analysis of DNA–Protein Interactions 745
   John D. Taylor, Alison J. Ackroyd, and Stephen E. Halford

95 Yeast Two-Hybrid Library Screening 757
   Ian G. Cowell

96 The Southwestern Assay 773
   Jacques Philippe

97 Nonradioactive Methods for the Detection
   of RNA-Protien Interaction 783
   *Asier Echarri, María Eugenia González, Iván Ventoso,
   and Luis Carrasco*

Contents

98 Tanscriptional Activation Analysis by the Chloramphenicol Acetyl Transferase (CAT) Enzyme Assay 793
David R. Hodge, Delores M. Thompson, Alexandra Panayiotakis, and Arun Seth

PART IX MUTAGENESIS, TRANSCRIPTION, AND TRANSLATION IN VITRO

99 Generating Nested Deletions with Exonuclease III 807
George Murphy

100 Primer-Directed Site-Specific Mutagenesis 815
Michael J. O'Donohue and G. Geoff Kneale

101 Site-Directed Mutagenesis Using a Uracil-Containing Phagemid Template 827
Christian Hagemeier

102 Site-Directed Mutagenesis Using Double-Stranded Plasmid DNA Templates 835
Jeffrey Bramam, Carol Papworth, and Alan Greener

103 Site-Directed Mutagenesis with LA-PCR™ Technology 845
Atsushi Shimada and Osamu Takeda

104 Recombination and Mutagenesis by Overlap Extension PCR 857
Robert J. Pogulis, Abbe N. Vallejo, and Larry R. Pease

105 Site-Directed Mutagenesis and Gene Fusion by Megaprimer PCR 865
Sailen Barik

106 Transcription In Vitro Using Bacteriophage RNA Polymerases 875
Elaine T. Schenborn

107 In Vitro Translation of mRNA in a Rabbit Reticuloctye Lysate Cell-Free System 885
Louise Olliver and Charles D. Boyd

108 In Vitro Translation of mRNA in a Wheat Germ Extract Cell-Free System 891
Louise Olliver, Anne Grobler-Rabie, and Charles D. Boyd

109 The *Xenopus* Egg Extract Translation System 895
Glenn M. Matthews and Alan Colman

110 Manipulation of Baculovirus Vectors 907
Claire L. Merrington, Mark J. Bailey, and Robert D. Possee

111 Procedures for the Analysis and Purification of His-Tagged Proteins 921
Richard E. Kneusel, Joanne Crowe, Melanie Wulbeck, and Joachim Ribbe

112 Detection and Immobilization of Proteins Containing the 6xHis Tag 935
Richard E. Kneusel, Melanie Wulbeck, and Joachim Ribbe

113 Expression and Purification of Recombinant Proteins
Using the pET System 947
**Robert C. Mierendorf, Barbara B. Morris, Beth Hammer,
and Robert E. Novy**

PART X GENE LOCALIZATION, MAPPING IN SITU, AND BIOINFORMATICS

114 Preparation of Tissue Sections and Slides
for mRNA Hybridization 981
Giorgio Terenghi

115 Use of Digoxigenin-Labeled Probes on Tissue Sections 985
Giorgio Terenghi

116 Gene Mapping by FISH 991
Rafael Espinosa III and Michelle M. Le Beau

117 Oligonucleotide PRINS DNA Synthesis 1011
John R. Gosden and Diane Lawson

118 Chromosome-Specific PRINS 1017
Jean-Paul Charlieu and Frank Pellestor

119 In Situ PCR Amplification of Intracellular mRNA 1023
Raymond H. Chen and Susan V. Fuggle

120 An Introduction to Bioinformatics 1031
Henry Brzeski

Index 1045

Contributors

ALISON J. ACKROYD • *Department of Biochemistry, University of Texas Southwestern Medical Center, Dallas, TX*
BJARNE ALBRECHTSEN • *Department of Biology, Royal Danish School of Pharmacy, Copenhagen, Denmark*
MICHELLE A. ALTING-MEES • *Stratagene Corporation, La Jolla, CA*
SHU F. AN • *Institute of Neurology, University of London, London, UK*
CAROLINE A. AUSTIN • *Department of Biochemistry and Genetics, The Medical School, The University of Newcastle upon Tyne, Newcastle, UK*
MARK J. BAILEY • *Institute of Virology and Environmental Microbiology, Oxford, UK*
SAILEN BARIK • *Department of Biochemistry and Molecular Biology, College of Medicine, University of South Alabama, Mobile, AL*
DOMINIQUE BELIN • *Department of Pathology, University of Geneva Medical Center, Geneva, Switzerland*
DAVID BERTIOLI • *Institute of Arable Crop Research, Rothampstead, Harpenden, UK*
THOMAS C. G. BOSCH • *Institut für Spezielle Zoologie, Universität Jena, Germany*
CHARLES D. BOYD • *University of Medicine and Dentistry, New Brunswick, NJ*
JEFFREY BRAMAN • *Stratagene, La Jolla, CA*
NEIL BREWIS • *Marie Curie Research Institute, Oxted, Surrey, UK*
FRANK C. BROSIUS III • *Department of Internal Medicine, Division of Nephrology, University of Michigan, Ann Arbor, MI*
SIAN BRYANT • *Tenovus Cancer Research Centre, University of Wales College of Medicine, Cardiff, Wales*
HENRY BRZESKI • *Department of Biosciences, University of Hertfordshire, Hatfield, UK*
XINAN CAO • *Department of Internal Medicine, Division of Nephrology, University of Michigan, Ann Arbor, MI*
LUIS CARRASCO • *Universidad Autonoma, Madrid, Spain*
ELISABETH CHACHATY • *Service de Microbiologie Médical, Institut Gustave-Roussy, Villejuif, France*
JEAN-PAUL CHARLIEU • *Medical Research Council, Human Genetics Unit, Western General Hospital, Edinburgh, UK*
RAYMOND H. CHEN • *Department of Cardiac Surgery, Brigham and Women's Hospital, Boston, MA*

SHU-HUA CHENG • *Department of Molecular Biology, Massachusetts General Hospital, Boston, MA*
ALISON COFFEY • *The Sanger Center, Wellcome Trust Genome Campus, Cambridge, UK*
CHARLOTTE G. COLE • *The Sanger Centre, Wellcome Trust Genome Campus, Cambridge, UK*
ALAN COLMAN • *Department of Surgery, Queen Elizabeth Medical Center, Birmingham, UK*
JOHN E. COLLINS • *The Sanger Centre, Wellcome Trust Genome Campus, Cambridge, UK*
GIOVANNA COSTANZO • *Instituto Pasteur, Cenci Bolognetti, Università La Sapienza, Rome, Italy*
RICHARD G. H. COTTON • *Mutation Research Centre, St. Vincent's Hospital, Melbourne, Australia*
IAN G. COWELL • *Division of Integrative Biology, Roslin Institute (Edinburgh), Midlothian, Scotland, UK*
JOANNE CROWE • *Qiagen, Hilden, Germany*
MARTIN W. CUNNINGHAM • *Research and Development, Amersham Pharmacia Biotech, Ltd., Buckinghamshire, UK*
SUSAN E. DANIELS • *Sequana, La Jolla, San Diego, CA*
JANE DAVENPORT-JONES • *Department of Biosciences, University of Hertfordshire, Hatfield, UK*
RITA I. DEMOPOULOS • *Department of Pathology, New York University Medical Center, New York, NY*
ALBERT B. DEISSEROTH • *School of Medicine, Yale University, New Haven, CT*
PATRICK J. DILLON • *Research and Development, Invitrogen, Carlsbad, CA*
ERNESTO DI MAURO • *Instituto Pasteur, Cenci Bolognetti, Università La Sapienza, Rome, Italy*
STEPHEN R. DLOUHY • *Department of Medical Genetics, Indiana University School of Medicine, Indianapolis, IN*
UDO DÖBBELING • *Department of Dermatology, University Hospital Zurich, Zurich, Switzerland*
IAN DUNHAM • *The Sanger Centre, Wellcome Trust Genome Campus, Cambridge, UK*
IAN DURRANT • *Research and Development, Amersham Pharmacia Biotech, Ltd., Buckinghamshire, UK*
ASIER ECHARRI • *Molecular Biology Department, University of Madrid, Madrid, Spain*
MORTIMER M. ELKIND • *Department of Radiological Health Sciences, Colorado State University, Fort Collins, CO*
MOHAMMAD S. ENAYAT • *Department of Haematology, Birmingham Children's Hospital NHS Trust, Birmingham, UK*
JAN ENGBERG • *Department of Biology, Royal Danish School of Pharmacy, Copenhagen, Denmark*

Rafael Espinosa III • *Section of Hematology/Oncology, Department of Medicine, University of Chicago, Chicago, IL*
Paola Ferro • *Laboratory of Molecular Biology, National Cancer Institute Genoa, Genoa, Italy*
Sue Fowler • *Research and Development, Amersham Pharmacia Biotech, Ltd., Buckinghamshire, UK*
Michael A. Frohman • *Department of Pharmacology, University Medical Center at Stony Brook, Stony Brook, NY*
Susan V. Fuggle • *Nuffield Department of Surgery, John Radcliffe Hospital, Oxford, UK*
Ian Garner • *PPL Ltd., Edinburgh, Scotland*
María Eugenia González • *Universidad Autonoma, Madrid, Spain*
John R. Gosden • *MRC Human Genetics Unit, Western General Hospital, Edinburgh, UK (Retired)*
Michael W. Graber • *Department of Pathology and Laboratory Medicine, Albany Medical Center Hospital, Albany, NY*
Alan Greener • *Stratagene, La Jolla, CA*
Anne Grobler-Rabie • *University of Medicine and Dentistry, New Brunswick, NJ*
Michael J. Haas • *Eastern Regional Research Center, U.S. Department of Agriculture, Philadelphia, PA*
Christian Hagemeier • *Laboratory of Molecular Biology and Pediatric Disease, Humboldt University, Charité Berlin, Germany*
Stephen E. Halford • *Centre for Molecular Recognition, University of Bristol, Bristol, UK*
Beth Hammer • *Novagen, Madison, WI*
Jens Hanke • *Functional Genome Analysis, Deutsches Krebsforschungszentrum, Im Neuenheimer Feld 506, Heidelberg, Germany*
Bronwen M. Harvey • *Research and Development, Amersham Pharmacia Biotech, Ltd., Buckinghamshire, UK*
Adrian J. Harwood • *MRC Laboratory for Molecular Cell Biology, University College London, London, UK*
Sheila Hassock • *Paediatric Research Unit, United Medical and Dental Schools of Guy's and St. Thomas' Hospitals, London, UK*
Karl H. Hecker • *Department of Biological Sciences, Florida State University, Tallahassee, FL*
John Heptinstall • *NES-Biosciences, Coventry University, Coventry, UK*
David R. Hodge • *Laboratory of Molecular Immunoregulation, Frederick Cancer Research and Development Center, National Cancer Institute, Frederick, MD*
Rachel Hodge • *Department of Botany, University of Leicester, Leicester, UK*
Jörg D. Hoheisel • *Functional Genome Analysis, Deutsches Krebsforschungszentrum, Im Neuenheimer Feld 506, Heidelberg, Germany*
Lawrence B. Holzman • *Department of Internal Medicine, Division of Nephrology, University of Michigan, Ann Arbor, MI*
Bent Honoré • *Department of Medical Biochemistry, University of Aarhus, Aarhus, Denmark*

SHENG-HE HUANG • *Division of Infectious Disease, Department of Pediatrics, University of Southern California, Children's Hospital of Los Angeles, Los Angeles, CA*
PETER G. ISAAC • *Nickersen-Biochem, Cambridge, UK*
LENE K. JOHANSEN • *Department of Biology, Royal Danish School of Pharmacy, Copenhagen, Denmark*
PETER JONES • *School of Biomedical Sciences, University of Nottingham, Queen's Medical Centre, Nottingham, UK*
AMBROSE JONG • *Division of Infectious Disease, Department of Pediatrics, University of Southern California, Children's Hospital of Los Angeles, Los Angeles, CA*
ELENA D. KATZ • *Department of Biotechnology, Perkin Elmer, Norwalk, CT*
ROSEMARY KELSELL • *Formerly Department of Molecular Genetics, Institute of Ophthalmology, London, UK*
MICHAEL W. KING • *Department of Biochemistry and Molecular Biology, School of Medicine, Indiana University, Terre Haute, IN*
JANE KIRK • *ICRF, Clare Hall Laboratories, South Mimms, Herts, UK*
G. GEOFF KNEALE • *Biophysics Laboratories, University of Portsmouth, Portsmouth, UK*
RICHARD E. KNEUSEL • *Qiagen, Hilden, Germany*
LORI A. KOLMODIN • *Roche Molecular Systems, Alameda, CA*
DAVID B. KRIZMAN • *Laboratory of Pathology, National Insitutes of Health, Bethesda, MD*
MELISSA LAIL-TRECKER • *Department of Anatomy, The School of Medicine, University of Connecticut Health Center, Farmington, CT*
KRIS N. LAMBERT • *Department of Hematology, University of California, Davis, CA*
ZOIA LARIN • *Institute of Molecular Medicine, University of Oxford, Oxford, UK*
DIANE LAWSON • *MRC Human Genetics Unit, Western General Hospital, Edinburgh, UK*
MICHELLE M. LE BEAU • *Section of Haematology/Oncology, Department of Medicine, University of Chicago, Chicago, IL*
BENOÎT LEBLANC • *Centre de Recherche en Cancérologie et Département de Biologie Médicale de l'Université Laval, Centre Hopital Universitaire de Québec (CHUQ), Pavillon Hôtel-Dieu de Québec, Québec, Canada*
HANS LEHRACH • *Max Planck Institut für Moleculare Genetik, Berlin, Germany*
MARK LEONARD • *Developmental Biology Research Centre, Kings College, University of London, London, UK*
JINGMEI LIU • *Department of Radiological Health Sciences, Colorado State University, Fort Collins, CO*
Y. M. DENNIS LO • *Department of Chemical Pathology, Prince of Wales Hospital, The Chinese University of Hong Kong, Hong Kong, People's Republic of China*
JAN U. LOHMANN • *Salk Institute, La Jolla, CA*
RAINER LÖW • *Zentrum Moleculare Biologie Heidelberg (ZMBH), Heidelberg, Germany*

JOAKIM LUNDEBERG • *Department of Biochemistry, Kungliga Tekniska Hogskolan, Royal Institute of Technology, Stockholm, Sweden*
PEDER MADSEN • *Department of Medical Biochemistry, University of Aarhus, Aarhus, Denmark*
FRANÇOIS MALLET • *CNRS-BioMérieux, Lyon, France*
DAVID L. MANNING • *Tenovus Cancer Research Centre, University of Wales College of Medicine, Cardiff, Wales, UK*
ROBERTO MANTOVANI • *Departimento de Genetica e di Biologia dei Microrganismi, Universita de Milano, Milan, Italy*
IVOR J. MASON • *Department of Developmental Neurobiology, King's College London, Guy's Hospital Campus, London, UK*
GLENN M. MATTHEWS • *Department of Surgery, Queen Elizabeth Medical Centre, Birmingham, UK*
JOHN MAULE • *MRC Human Genetics Unit, Molecular Genetics Section, Western General Hospital, Edinburgh, Scotland, UK*
STEVE MAYALL • *ICRF, Clare Hall Laboratories, South Mimms, Herts, UK*
MARCIA A. MCALEER • *Yamanouchi Research Institute, Littlemore Park, Oxford, UK*
CLAIRE L. MERRINGTON • *Institute of Virology and Environmental Microbiology, Oxford, UK*
ROBERT C. MIERENDORF • *Novagen, Madison, WI*
ANTHONY P. MONACO • *Wellcome Trust Centre for Human Genetics, University of Oxford, Oxford, UK*
BRANDON D. MOORE • *Department of Biochemistry, University of Nevada, Reno, NV*
BARBARA B. MORRIS • *Novagen, Madison, WI*
TOM MOSS • *Centre de Recherche en Cancérologie et Département de Biologie Médicale de l'Université Laval, Centre Hopital Universitaire de Québec (CHUQ), Pavillon Hôtel-Dieu de Québec, Québec, Canada*
GEORGE MURPHY • *John Innis Centre for Plant Science Research, Norwich, UK*
RODOLFO NEGRI • *Centro Acidi Nucleici, CNR, Rome, Italy*
PETER S. NELSON • *Department of Molecular Biotechnology, University of Washington, Seattle, WA*
ROBERT E. NOVY • *Novagen, Madison, WI*
MICHAEL J. O'DONOHUE • *UBPB-Fractionnement Enzymatique, INRA-Lille/Reims, Reims, France*
JOHN M. OLD • *Institute of Molecular Medicine, John Radcliffe Hospital, Headington, Oxford, UK*
LOUISE OLLIVER • *University of Medicine and Dentistry, New Brunswick, NJ*
ALEXANDRA PANAYIOTAKIS • *Laboratory of Molecular Oncology National Cancer Institute, Frederick, MD*
TAKIS S. PAPAS • *Center for Molecular and Structural Biology, Medical University of South Carolina, Charleston, SC*
CAROL PAPWORTH • *Stratagene, La Jolla, CA*
ROGER PATIENT • *Developmental Biology Research Centre, Kings College, University of London, London*

LARRY R. PEASE • *Department of Immunology, Mayo Clinic and Mayo Foundation, Rochester, MN*
FRANK PELLESTOR • *Medical Research Council, Human Genetics Unit, Western General Hospital, Edinburgh, UK*
BERTIL PETTERSSON • *Department of Biochemistry, Kungliga Tekniska Hogskolan, Royal Institute of Technology, Stockholm, Sweden*
JACQUES PHILIPPE • *Diabetes Unit, Department of Medicine, Centre Medical Universitaire, Geneva, Switzerland*
ULRICH PFEFFER • *Laboratory of Molecular Biology, National Cancer Institute Genoa, Genoa, Italy*
ERAN PICHERSKY • *Department of Biology, Natural Sciences Building, University of Michigan, Ann Arbor, MI*
ROBERT J. POGULIS • *Department of Immunology, Mayo Clinic and Mayo Foundation, Rochester, MN*
ROBERT D. POSSEE • *Institute of Virology and Environmental Microbiology, Oxford, UK*
GREGORY M. PRESTON • *Department of Molecular Sciences, Central Research Division, Pfizer, Inc., Groton, CT*
ERIC QUÉMÉNEUR • *Département d'Ingénierie et d'Etudes des Protéines, CEA, Saclay, France*
JERALD RADICH • *Division of Clinical Research, Fred Hutchinson Cancer Center, Seattle, WA*
RALPH RAPLEY • *Department of Biosciences, University of Hertfordshire, Hatfield, UK*
SONGRONG REN • *Department of Medicine, University of Alabama, Birmingham, AL*
RALF RESKI • *Freiburg University Institute Biology II, Frieburg/Breisgua, Germany OR Institute for General Botany, Hamburg, Germany*
JOACHIM RIBBE • *Qiagen, Hilden, Germany*
ERIK S. RIISE • *Department of Biology, Royal Danish School of Pharmacy, Copenhagen, Denmark*
CRAIG A. ROSEN • *Research and Development, Human Genome Sciences, Rockville, MD*
KENNETH H. ROUX • *Department of Biological Sciences, Florida State University, Tallahassee, FL*
J. MICHAEL RUPPERT • *Department of Medicine, University of Alabama, Birmingham, AL*
WOJCIECH RYCHLIK • *National Biosciences, Plymouth, MN*
PHILIP J. SAKER • *Complex Traits Analysis Group, Imperial College School of Medicine, Hammersmith Hospital, London, UK*
PATRICK SAULNIER • *Service de Microbiologie Médical, Institut Gustave-Roussy, Villejuif, France*
ELAINE T. SCHENBORN • *Promega Corporation, Madison, WI*
GUSTAV SCHONFELD • *Department of Internal Medicine, Washington University School of Medicine, St. Louis, MO*

Contributors

CLIFFORD W. SCHWEINFEST • *Center for Molecular and Structural Biology, Medical University of South Carolina, Charleston, SC*
JEFFREY R. SEEMANN • *Department of Biochemistry, University of Nevada, Reno, NV*
ARUN SETH • *Laboratory of Molecular Oncology National Cancer Institute, Frederick, MD*
PETER E. SHAW • *School of Biomedical Sciences, University of Nottingham, Queen's Medical Centre, Nottingham, UK*
ZHIYUAN SHEN • *Department of Molecular Genetics, University of Illinois at Chicago, Chicago, IL*
ATSUSHI SHIMADA • *Shuzo Co., Ltd., Otsu, Shiga, Japan*
JAY M. SHORT • *Diversa Corporation, San Diego, CA*
ALAN R. SHULDINER • *Johns Hopkins University School of Medicine, Baltimore, MD*
DUNCAN R. SMITH • *Molecular Biology Laboratory, Tan Tock Seng Hospital, Singapore*
MARJORY A. SNEAD • *Diversa Corporation, San Diego, CA*
SURAPAREDDY SREENIVASAPRASAD • *Department of Plant Pathology and Microbiology, Horticulture Research International, Wellesbourne, Warwickshire, UK*
RAI AJIT K. SRIVASTAVA • *Department of Internal Medicine, Washington University School of Medicine, St. Louis, MO*
JUSTIN STACEY • *Biogemma, UK, Ltd., Cambridge, UK*
PETER J. STAMBROOK • *Department of Medical Genetics, Indiana School of Medicine, Indianapolis, IN*
A. FRANCIS STEWART • *School of Biomedical Sciences, University of Nottingham, Queen's Medical Centre, Nottingham, UK*
TIMOTHY STONE • *Resarch and Development, Amersham Pharmacia Biotech, Ltd., Buckinghamshire, UK*
ANU SUOMALAINEN • *National Public Health Institute, Department of Human Molecular Genetics, Helsinki, Finland*
ANN-CHRISTINE SYVÄNEN • *Department of Medical Science and Molecular Medicine, Uppsala University Hospital, Uppsala, Sweden*
OSAMU TAKEDA • *Shuzo Co., Ltd., Kusatsu, Japan*
KEITH TANNER • *Johns Hopkins University School of Medicine, Baltimore, MD*
JOHN D. TAYLOR • *Department of Biochemistry, University of Texas Southwestern Medical Center, Dallas, TX*
GIORGIO TERENGHI • *Blond McIndoe Laboratories, University Department of Surgery, Royal Free and University College Medical School, University College London, London, UK*
DELORES M. THOMPSON • *Laboratory of Molecular Oncology, National Cancer Institute, Frederick, MD*
BIMAL D. M. THEOPHILUS • *Department of Haematology, Birmingham Children's Hospital NHS Trust, Birmingham, UK*
SCOTT TINGEY • *DuPont Agricultural Products, Wilmington, DE*
JAY A. TISCHFIELD • *Department of Medical Genetics, Indiana School of Medicine, Indianapolis, IN*

JAMES A. TROFATTER • *Department of Medical Genetics, Indiana School of Medicine, Indianapolis, IN*

MATHIAS UHLÉN • *Department of Biochemistry, Kungliga Tekniska Hogskolan, Royal Institute of Technology, Stockholm, Sweden*

ABBE N. VALLEJO • *Department of Immunology, Mayo Clinic and Mayo Foundation, Rochester, MN*

IVÁN VENTOSO • *Universidad Autonoma, Madrid, Spain*

STÉPHANE VIVILLE • *Institut de Génétique et de Biologie Moleculaire et Cellulaire, CNRS/INSERM/ULP, Strasbourg, France*

WILLIAM WAGGOTT • *Department of Cellular Sciences, John Radcliffe Hospital, University of Oxford, UK*

MAGGIE WALMSLEY • *Developmental Biology Research Centre, Kings College, University of London, London, UK*

DAVID WALSH • *Scottish Agricultural College, Edinburgh, Scotland*

ROBERT L. WELLS • *Department of Radiological Health Sciences, Colorado State University, Fort Collins, CO*

MICHELLE WESTENGAARD-HILDINGE • *Department of Biology, Royal Danish School of Pharmacy, Copenhagen, Denmark*

CLAIRE B. WHEELER • *Research and Development, Amersham Pharmacia Biotech, Ltd., Buckinghamshire, UK*

PATRICIA WHEELER • *Department of Medical Genetics, Indiana School of Medicine, Indianapolis, IN*

J. FENTON WILLIAMS • *PE Biosystems, Foster City, CA*

D. ROSS WILLIAMS • *Department of Biosciences, University of Hertfordshire, Hatfield, UK*

VALERIE M. WILLIAMSON • *Department of Hematology, University of California, Davis, CA*

CRAIG WINSTANLEY • *Department of Medical Microbiology, University of Liverpool, Liverpool, UK*

MELANIE WULBECK • *Qiagen, Hilden, Germany*

FEI YE • *Department of Biology, Massachusetts Institute of Technology, Cambridge, MA*

WEI ZHANG • *Departments of Pathology and Neuro-Oncology, The University of Texas M. D. Anderson Cancer Center, Houston, TX*

YUE ZHANG • *Department of Pharmacology, University Medical Center at Stony Brook, Stony Brook, NY*

I

NUCLEIC ACID EXTRACTION

1

Isolation of High-Molecular-Weight DNA from Animal Cells

Ian Garner

1. Introduction

Mammalian chromosomes are of the order of 12–60 times the size of that of *Escherichia coli* (4×10^3 kilobase pairs [kbp]) *(1)*. The choice of method used when purifying DNA from mammalian cells may be dictated by the use to which the product will be put as it will influence the average size of the material purified. For example, methods incorporating many aggressive manipulations will tend to shear the DNA into molecules of relatively low-mol-wt (< 50 kbp). This may be suitable for polymerase chain reaction (PCR) *(2,3)* analysis and in some cases Southern blotting *(4)* but will be unsuitable for other more demanding purposes, e.g., genomic library constructions. When performed with care, methods involving minimal manipulations will yield DNA in excess of 200 kbp, suitable for most purposes. In general, it is prudent to utilize such methods for all preparations of DNA from mammalian cells. The first three methods described below are derived from that of Blin and Stafford *(5)* and should yield high-mol-wt (HMW) DNA from solid tissues, blood, or cells in culture suitable for most purposes including cloning, PCR/RFLP analysis, and Southern blotting. The final method described is that of Lahiri and Nurnberger *(6)* and is a rapid approach that eliminates the use of solvents and enzymes, making it easier to process large numbers of samples. The material produced by this method should be approx 50 kbp and is suitable for PCR and RFLP analysis.

2. Materials

2.1. Preparation of HMW DNA from Solid Tissues

If possible, all materials should be sterilized prior to use.

1. Liquid nitrogen.
2. Porcelain pestle and mortar prechilled to –20°C.
3. Selection of spatulas.
4. 600-mL Pyrex beakers (or similar wide-based vessel).
5. TE: 10 m*M* Tris-HCl, pH 8.0, 1 m*M* EDTA.
6. Phenol saturated with TE (*see* **Note 1**).

7. Extraction buffer: 0.1M EDTA, 0.2M NaCl, 0.05M Tris-HCl, pH 8.0, 0.5% SDS, 50 µg/mL DNase-free RNase (*see* **Note 2**).
8. Proteinase K: 20 mg/mL in sterile distilled water.
9. Dialysis tubing (wide bore): Preboiled in 1 mM EDTA and rinsed with sterile distilled water.
10. 3M Sodium acetate adjusted to pH 6.0 with acetic acid.
11. Absolute ethanol.
12. 70% Ethanol.
13. Pasteur pipets with sealed hooked ends.
14. Pasteur pipets attachable to a vacuum line.

2.2. Preparation of HMW DNA from Blood

Materials as for **Subheading 2.1.** with the addition of the following:

1. Heparinized Vacutainers.
2. Hanks' buffered saline (HBS) (from Sigma, St. Louis, MO).
3. Histopaque (from Sigma).

2.3. Preparation of HMW DNA from Cells in Culture

Materials as for **Subheading 2.1.** with the addition of the following:

1. Ca^{2+}/Mg^{2+} free phosphate-buffered saline (PBS) from Gibco (Grand Island, NY) or similar supplier.
2. Rubber policemen or similar cell scrapers.

2.4. Preparation of HMW DNA from Blood Without the Use of Solvents or Enzymes

Materials 5,11,12, and 13 from **Subheading 2.1.** with the addition of the following:

1. Low salt buffer: 10 mM Tris-HCl, pH 7.6, 10 mM KCl, 10 mM $MgCl_2$, 2 mM EDTA.
2. High salt buffer: 10 mM Tris-HCl, pH 7.6, 10 mM KCl, 10 mM $MgCl_2$, 0.4M NaCl, 2 mM EDTA.
3. Nonidet P-40.
4. 10% SDS.
5. 6M NaCl.

3. Methods

3.1. Preparation of HMW DNA from Solid Tissues

1. Freshly excised tissues should be dropped immediately into liquid nitrogen. Large organs should be cut into smaller, more manageable pieces (<1 cm^3) as this will ease freezing, storage, and subsequent manipulations. Any organ can be taken, but should liver be required, a 24-h starvation period prior to sacrifice will improve DNA quality. Tissues harvested in this way can be stored at –70°C for several years prior to use.
2. Pour liquid nitrogen into the precooled mortar, add the tissue of choice (up to 1 cm^3), and grind to a fine powder with the pestle. It may be necessary to break up the tissue into smaller pieces to facilitate grinding (e.g., by wrapping in tin foil and hitting with a hammer). Add more liquid nitrogen as required to keep the sample cold.
3. Once the sample has been ground, allow the liquid nitrogen to evaporate and use a spatula to transfer the powdered tissue to the surface of 20 mL of extraction buffer in a 600-mL beaker at room temperature. Sprinkle the powder evenly over the surface of the liquid and gently swirl the beaker to submerge the material.

4. Add Proteinase K to 100 µg/mL (100 µL of stock) and gently swirl the beaker to mix the components. Incubate the beaker at 37°C for at least 3 h, preferably overnight, with gentle agitation. This can be achieved using a shaking water bath or by occasional swirling by hand. The solution should be reasonably clear and viscous at the end of the incubation. More Proteinase K may be added to achieve this (*see* **Note 3**).
5. Add 20 mL of equilibrated phenol (*see* **Note 1**) and seal the beaker with parafilm. Gently swirl by hand for 10–15 min to mix the two phases. The larger the surface area available, the easier this will be. Ideally you should generate an emulsion at this stage. It may be necessary to transfer the mixture to a larger container to achieve this.
6. Transfer the mixture to a 50-mL disposable plastic tube and centrifuge at 1500g for 10 min at room temperature to separate the two phases.
7. Remove the lower phenol phase by gentle aspiration through a Pasteur pipet attached to a vacuum line through a side arm flask. The pipet should be lowered into the lower phenol phase with the vacuum line clamped until the thread of viscous DNA has detached from the pipet tip. Slowly unclamp the vacuum line and allow the phenol phase to run into the flask. Once all of the phenol has been removed, the vacuum line is again clamped and the pipet is removed. Alternatively, the aqueous phase can be removed with a wide-bore pipet. However, care must be taken not to disturb the interface and when the DNA is very viscous this is hard to achieve (*see* **Note 4**).
8. **Steps 5–7** should be repeated to accomplish at least three extractions. The aqueous phase should be clear at this point.
9. At this stage two routes to recovery of HMW DNA are available (*see* **Note 5**).
    a. Dialyze the aqueous phase against 1000 vol of TE. This should be performed for 30 min at room temperature to prevent SDS precipitation in the sample followed by overnight at 4°C. Allow room for expansion in the dialysis bag.
    b. Transfer the aqueous phase to a fresh beaker and add sodium acetate to 0.3M. Mix by gentle swirling. Add 2 vol of absolute ethanol and mix by gentle swirling. The DNA will begin to precipitate almost immediately in a strandy complex. Initially this will be glass-like but it will begin to attain a white appearance as the precipitation proceeds. Hook out the DNA strands using a Pasteur pipet with a sealed U-shaped end before they attain too white an appearance (*see* **Note 6**). Dip the DNA in 70% ethanol for a few seconds and allow to air dry for a few minutes. Transfer the DNA to 1–3 mL TE. Gently wet the DNA in the liquid and allow it to fall off the pipet tip onto the surface of the liquid. Do not shake violently to achieve this. Leave to dissolve overnight at 4°C. If this proves difficult, incubate the tube overnight at room temperature on a gently rocking table or rotating wheel (*see* **Note 4**)
10. The absorbance of the DNA at 260 nm and 280 nm should be measured using quartz cuvets. The 260/280 ratio should be >1.8. If this is not the case, repeat steps 4–9 adding additional SDS to 1%. An A_{260} of 1.0 in a 1-cm light path is equivalent to a DNA concentration of 50 µg/mL. Store the DNA at 4°C. (*See* Chapter 11.)
11. An aliquot of the DNA should be analyzed by electrophoresis through a 0.3% agarose gel. Multimers of bacteriophage lambda generated by ligation or commercially available DNAs can serve as mol-wt markers. The prepared DNA is normally at least 100 kbp and preferably exceeds 200 kbp (*see* **Notes 7–10**).

3.2. Preparation of HMW DNA from Blood

1. Blood should be collected into heparinized vacutainers. Ideally, it should be processed immediately but can be stored overnight at 4°C.
2. Dilute 10 mL of blood with 10 mL of HBS.

3. Layer this over 5 mL Histopaque and centrifuge in a 15-mL disposable plastic centrifuge tube for 15 min at room temperature at 2000g.
4. A white band containing peripheral lymphocytes should be visible in each tube. Remove and discard the sample above this and transfer the white band to a fresh 15-mL centrifuge tube.
5. Wash the cells by adding 10 mL of HBS, mix thoroughly, and recover the cells by centrifugation for 10 min at room temperature at 2000g.
6. Discard the supernatant and resuspend the cell pellet in 20 mL extraction buffer. Continue as from **Subheading 3.1.**, **step 4**.

3.3. Preparation of HMW DNA from Cells in Culture

1. Cells ($\sim 10^8$) should be grown as a monolayer or in suspension as required (*see* **Note 9**).
2. For monolayers, decant the medium and rinse the cells twice with PBS. Recover the cells by scraping with a rubber policeman and centrifuge at 500g for 10 min in a 15-mL plastic disposable centrifuge tube at room temperature. For cell suspensions, transfer to 15-mL plastic disposable centrifuge tubes and recover the cells by centrifugation for 10 min at room temperature at 500g. Resuspend the cell pellet(s) by gentle pipeting in 10 mL PBS and recover by centrifugation as above.
3. Resuspend the cell pellet in 20 mL extraction buffer and continue as from **Subheading 3.1.**, **step 4**.

3.4. Preparation of HMW DNA from Blood Without the Use of Solvents or Enzymes

1. Collect blood as in **Subheading 3.2.**, **step 1**.
2. Transfer 5 mL of blood to a 15-mL plastic centrifuge tube. Add 5 mL of low salt buffer and 125 µL of Nonidet P-40. Mix by gentle inversion to dissolve cell membrane.
3. Centrifuge for 10 min at 1000g at room temperature to recover nuclei.
4. Discard the supernatant and wash the nuclei by gently resuspending them in 5 mL low salt buffer and centrifuging again.
5. Gently resuspend the nuclei in 0.8 mL high salt buffer and transfer to a 1.5-mL microcentrifuge tube. Add 50 µl of 10% SDS and mix thoroughly by inversion. Incubate for 10 min at 55°C.
6. Add 0.3 mL of 6M NaCl and mix by gentle inversion. Centrifuge for 5 min at maximum speed in a microcentrifuge at 4°C.
7. Recover the supernatant and add 2 vol of absolute ethanol at room temperature. Mix by gentle inversion. Recover the DNA as in **Subheading 3.1.**, **step 9b**, transferring finally to 1 mL of TE.
8. Check the quantity and quality of DNA as described in **Subheading 3.1.**, **steps 10** and **11** (*see* **Notes 7**, **8**, and **10**).

4. Notes

1. Phenol should be saturated with several changes of TE until the pH of the upper TE layer remains at 8.0. Care should be taken when handling phenol: Wear gloves, safety spectacles, and preferably work in a fume hood.
2. RNase solutions should be boiled for 5–10 min prior to use.
3. Following Proteinase K digestion, the material should be viscous before proceeding to next step. If it is not, repeat or add 100 µL more Proteinase K stock to sample and continue incubation.
4. HMW DNA should have a high viscosity because of the large size of the molecules. This will be reflected in its strandy consistency when pipeted. Always take care when pipeting and use wide bore pipets. Never vortex.

5. DNA from **Subheading 3.1.**, **step 9a** should be of superior quality to that produced at **step 9b**.
6. Care must be taken not to overprecipitate or overdry the pellet at **Subheading 3.1.**, **step 9b** as it will be very difficult or impossible to resuspend the DNA afterwards. Precipitated DNA should be hooked out when the majority of it is still relatively clear.
7. If RNA is evident when DNA is analyzed by electrophoresis (a DNase or RNase treatment can help here) use fluorimetry or ethidium bromide staining in gels to estimate DNA concentration. If enough material is present to warrant recovery, treat with DNase-free RNase and ethanol precipitate again as in **Subheading 3.1.**, **step 9b**.
8. If the DNA is degraded, repeat with freshly prepared RNase taking greater care with all manipulations. If problems persist, remove the RNase.
9. The yield from approx 10^8 cells should be approx 300 µg. The yield from 10 mL of blood should be approx 100 µg by **Subheading 3.2.** and may be up to double this by **Subheading 3.4.**
10. If yields are low, check the pH of solutions and verify that the DNA has resuspended. Material should be viscous at **Subheading 3.1.**, **step 4** before continuing on to **step 5**. Repeat if necessary (*see also* **Note 3**).

References

1. Kornberg, A. (1980) *DNA Replication.* Freeman, San Francisco, CA.
2. Saiki, R. K., Scharf, S., Faloona, F., Mullis, K. B., Horn, G. T., Erlich, H. A., and Arnheim, N. (1985) Enzymatic amplification of β-globin genomic sequences and restriction site analysis for diagnosis of sickle cell anaemia. *Science* **230,** 1350–1354.
3. Scharf, S., Horn, G. T., and Erlich, H. A. (1986) Direct cloning and sequence analysis of enzymatically amplified gene sequences. *Science* **233,** 1076–1078.
4. Southern, E. M. (1975) Detection of specific sequences among DNA fragments separated by gel electrophoresis. *J. Mol. Biol.* **98,** 503–517.
5. Blin, N. and Stafford, D. W. (1976) A general method for isolation of high molecular weight DNA from eukaryotes. *Nucleic Acids Res.* **3,** 2303–2308.
6. Lahiri, D. K. and Nurnberger, J. I., Jr. (1991) A rapid non-enzymatic method for the preparation of HMW DNA from blood for RFLP studies. *Nucleic Acids Res.* **19,** 5444.

2

Isolation of mRNA by Affinity Chromatography

Sian Bryant and David L. Manning

1. Introduction

Messenger RNA (mRNA) comprises approximately 1–5% of total cellular RNA. Although the actual amount depends on the type of cell and its physiological state, at any one time approximately 12,000 genes are being transcribed with approximately 500,000 mRNA molecules present in each mammalian cell.

Eukaryotic mRNAs are heterogeneous in size (ranging from 0.5 kb to over 20 kb) and abundance (from fewer than 15 copies to over 20,000 copies per cell). The presence of a terminal stretch of approximately 200 adenosine residues (the polyA tail) on most eukaryotic mRNAs and its absence in ribosomal and transfer RNAs has important practical consequences, as it allows polyadenylated species (messenger RNAs) to be separated from their nonpolyadenylated counterparts (ribosomal and transfer RNAs, which account for over 90% of total cellular RNA).

High-quality mRNA is needed for a number of molecular biology techniques, including cDNA library construction *(1,2)*. Not surprisingly, numerous mRNA extraction kits are now commercially available. All use the same basic principle, described in **Subheading 2.**, which involves the affinity selection of polyadenylated mRNA using oligodeoxthymidylate (oligo (dT)).

2. Materials

All materials used in this procedure should be sterile and of molecular biology grade. All Tris-containing solutions are prepared using RNase-free water and autoclaved. All other solutions, unless otherwise stated, should be treated directly with diethyl pyrocarbonate (DEPC) and autoclaved. DEPC is an efficient, nonspecific inhibitor of RNase activity. It is, however, a carcinogen and should be handled in a fume hood with extreme care. Hands are a major source of RNase activity. Because of this, gloves should be worn for all procedures.

1. RNase-free water: Add 0.1% DEPC to water. Allow to stand overnight at 37°C and autoclave to destroy residual DEPC activity. All solutions except Tris, which inactivates DEPC, can be treated in the same way.
2. SDS (sodium dodecyl sulphate): SDS is dangerous if inhaled and should be weighed in a fume hood. A 10% stock solution is normally prepared. This solution is unstable if

autoclaved, however any residual RNase activity can be destroyed by heating the solution at 65°C for 2 h.
3. Oligodeoxthymidylate-cellulose (oligo(dT)): Oligo (dT) cellulose is available commercially. Although the binding capacity of oligo(dT) cellulose varies between different suppliers, a general rule is to use 25 mg of oligo(dT) for each 1 mg of total RNA. Suspend oligo (dT) cellulose in loading buffer at a concentration of 5 mg per 1 mL loading buffer. Oligo (dT) is insoluble and should be resuspended by gentle tapping or inversion. Do not put it in a vortex. It can be stored either dry at 4°C or in suspended in loading buffer at −20°C.
4. RNase-free glass wool and Pasteur pipets: Wrap both the glass wool and pipets in aluminium foil and bake at 200°C for 2–4 h to remove any RNase activity.
5. 5 M NaCl: Store at room temperature.
6. 3 M Sodium acetate pH6: Store at room temperature.
7. Absolute alcohol: Store at −20°C.
8. 70% ethanol: Prepare this solution using DEPC-treated water. Store at 4°C.
9. Loading buffer: 0.5 M NaCl in 0.5% SDS, 1 mM EDTA, 10 mM Tris-HCl, pH 7.5 (*see* **Note 1**). Store at room temperature.
10. Elution buffer: 1 mM EDTA, 10 mM Tris-HCl, pH 7.5. The buffer can be stored at room temperature but should be preheated to 65°C prior to use.
11. Recycling buffer: 0.1 M NaOH, which should be prepared immediately before use and used fresh.

3. Methods
3.1. Preparing an Oligo (dT) Column

Oligo (dT) columns are available commercially or can be prepared by using a 1–3 mL syringe. Preparing your own columns is both easy and cheap.

1. Remove the plunger from the syringe and plug the base with glass wool.
2. Add oligo (dT) cellulose to the syringe using a sterile RNase-free Pasteur pipet. The oligo (dT) cellulose will collect, as a column, above the glass wool. The loading buffer will escape through the glass wool and can be discarded. To ensure that the oligo (dT) cellulose is packed and free from air locks, add 3 vol of loading buffer using a pipette and allow the solution to run through the column. The column is now ready for immediate use and should not be allowed to run dry.

3.2. Isolation of Poly(A+) RNA

1. Resuspend the RNA pellet in loading buffer or, if the buffer is in solution, add 1/10th vol of 5 M NaCl (*see* **Note 1**).
2. Heat denature RNA and immediately load it onto the column (*see* **Note 2**) and apply 3 vol of loading buffer.
3. Reapply the eluate to the column (*see* **Note 3**).
4. Wash with 3 vol of loading buffer (*see* **Note 4**). Discard eluate.
5. Recover the bound poly(A+) mRNA by adding 3 vol elution buffer. Collect the mRNA in a sterile tube on ice (*see* **Note 5**).
6. The mRNA is precipitated by adding 1/10th vol of 3 M sodium acetate and 2 vol of ice-cold absolute ethanol. An overnight precipitation at −20°C maximizes the precipitation of RNA.
7. Centrifuge at 15,000g for 15 min to pellet the RNA. Discard the supernatant.
8. Wash the RNA pellet in ice-cold 70% ethanol (*see* **Note 6**). Centrifuge at 15,000g for 5 min to repellet the RNA which may have been disturbed by washing. Discard the supernatant.

9. Dry the RNA pellet. Once it is dry, resuspend it in DEPC-treated water.
10. Assess the purity and integrity of mRNA (*see* **Note 7** and Chapter 11).

4. Notes

1. SDS can be omitted from the loading buffer as it may precipitate in cold or air-conditioned laboratories and clog the column. Residual SDS may coprecipitate with the RNA and interfere with other procedures such as reverse transcription.
2. RNA can be linearized or denatured by heating to 80–90°C for 5 min, cooling quickly on ice (taking care not to cause precipitation of the loading-buffer components) and immediately applying the solution to the column. This reduces the RNA secondary structure and aids binding of the poly (A+)-tailed mRNA to the oligo (dT) column.
3. This step increases the yield of poly(A+) RNA.
4. The poly (A+) tail anneals to oligo (dT) in the presence of high salt (NaCl) concentrations. The further addition of 3 vol of loading buffer ensures nonpolyadenylated RNA species are washed from the column.
5. The oligo (dT) column can be regenerated by washing with 10 vol of recycling buffer followed by re-equilibration with 3 vol of loading buffer. Oligo (dT) can be stored dry at 4°C or resuspended in loading buffer at –20°C until required.
6. This step removes any contaminating salt that may have coprecipitated with the mRNA.
7. Purity of mRNA is measured by its absorbance at 260 nm. (An absorbance of 1.0 at 260 nm is equivalent to 40 mg mRNA). An additional reading at 280 nm allows the A260:A280 ratio to be calculated. A ratio of 2.0 should be expected using this protocol. The integrity of the mRNA can be assessed by gel electrophoresis as described in Chapter 29, Formaldehyde Gel Electrophoresis of Total RNA. The mRNA should appear on ethidium bromide-stained gels as a smear ranging from 200 bp to greater than 10 kb with no detectable ribosomal RNA. If small amounts of mRNA are added to the gel, visualization by ethidium bromide may be impossible. To circumvent this problem, set up a Northern blot and hybridize using a labeled oligo(dT) primer. (*See* Chapters 37, 38.) Autoradiography should reveal mRNA fragments ranging in size.

References

1. Manning, D. L., Daly, R. J., Lord, P. G., Kelly, K. F., and Green, C. D. (1988) Effects of estrogen on the expression of a 4.4kb mRNA in the ZR-75-1 human breast cancer cell line. *Mol. Cell. Endocrinol.* **59,** 205–212.
2. Manning, D. L., Archibald, L. H., and Ow, K.T. (1990) Cloning of estrogen-responsive mRNAs in the T-47D human breast cancer cell line. *Cancer Res.* **50,** 4098–4104.

3

Isolation and Purification of DNA from Plants

Justin Stacey and Peter G. Isaac

1. Introduction

This chapter describes a DNA extraction method that can be used both on freeze-dried leaves and on fresh leaves, and is based on the method of Saghai-Maroof et al. *(1)*, modified by David Hoisington and Jack Gardiner at the University of Missouri at Columbia (personal communication). The scale of the extraction is dependent on the amount of starting material; 300–400 mg freeze-dried material requires 9 mL extraction buffer and should yield 250 µg–1 mg DNA. For high throughput screening, scaling the procedure down tenfold results in a miniprep method that is very suitable for extracting small quantities (25–100 µg) of DNA from young, fresh leaves.

The DNA is not free from contaminants such as carbohydrates, but it is of a suitable grade for enzyme digestion, Southern blotting *(2; see* Chapter 16*)*, and analysis by polymerase chain reaction (PCR), such as the random amplified polymorphic DNA (RAPD) technique *(3; see* Chapter 85*)*. We have used this procedure on wheat, barley, maize, oilseed rape, vegetable Brassicas, peas, and onions. A benefit of the method is that it requires little "hands-on" time by the operator, and can therefore be used to process large numbers of samples on a daily basis.

The procedure described in **Subheading 3.** is for 300–400 mg of freeze-dried leaf material. A detergent (cetyltriethylammonium bromide, CTAB) is used to break open plant cells and solubilize the contents. Chlorophyll and some denatured proteins are removed from the green plant tissue in an organic chloroform/octanol step, and the organic phase is separated by a brief centrifugation. At this point, the extract contains RNA and DNA, the former is removed by incubating with RNAse A. The DNA is precipitated and washed in organic solvents before redissolving in aqueous solution. The concentration of the DNA is then estimated by spectrophotometry and agarose gel electrophoresis. (*See* Chapters 11 and 13.)

2. Materials

1. 1 M Tris-HCl, pH 8.0: Filter and autoclave.
2. 5.0 M NaCl: Filter and autoclave.
3. 0.5 M EDTA (disodium ethylenediaminetetraacetic acid): Weigh out an appropriate amount of EDTA and add to stirring distilled water (about 75% of the final volume). Add sodium hydroxide pellets slowly until the solution begins to clear. Monitor the pH and add

NaOH until the EDTA has dissolved and the pH reaches between 7 and 8. Make up to the final volume, filter through a paper filter, and autoclave. Store the solution in a fridge.

4. CTAB extraction buffer: For 100 mL (enough for 10 isolations) mix 73 mL deionized water, 10 mL 1 M Tris-HCl, pH 7.5, 14 mL 5 M NaCl, 2 mL 0.5 M EDTA, pH 8.0. This solution should be filtered and autoclaved. The solution can be stored on the bench at room temperature. Immediately prior to use add 1 mL of β-mercaptoethanol and 1 g CTAB (*see* **Note 1**). Preheat the solution to 65°C.
5. Chloroform: Octanol (24:1). Store in dark at room temperature. Make up and dispense this solution in a fume cupboard.
6. Preboiled RNase A (10 mg/mL): Dissolve RNase A in water, place tube in boiling water bath for 10 min, and allow to cool on bench. Store at –20°C.
7. Isopropanol.
8. 3 M Sodium acetate, pH 6.0: Adjust the pH with acetic acid before making to the final volume. Filter and autoclave this solution and store at room temperature.
9. 76% Ethanol, 0.2 M sodium acetate: For 100 mL (10 isolations) mix 76 mL absolute ethanol, 6.7 mL 3 M sodium acetate, pH 6.0, 17.3 mL of autoclaved deionized water. Store at 4°C until ready for use.
10. 70% Ethanol. Store at –20°C.
11. 1 M Tris-HCl, pH 8.0.
12. Tris-EDTA (TE) buffer: For 100 mL mix 98.8 mL deionized water, 1 mL 1 M Tris-HCl, pH 8.0, 0.2 mL 0.5 M EDTA, pH 8.0. Filter and autoclave solution.
13. Freeze-dryer bags: We use bags made from micropore bread bag cellophane (Cryovac Packaging/Marketing Systems, St. Neots, Cambridgeshire). Bags with an open end of size 10 × 30 cm are made using a heat-sealing freezer bag sealer.
14. Miracloth (Calbiochem, San Diego, CA).
15. Glass hooks are made from Pasteur pipets by placing about 5–10 mm of the fine end of the pipet horizontally in a bunsen flame, so that the end becomes sealed. The end of the pipet will slowly droop under gravity. Remove the pipet from the flame and hold it pointing vertically (droopy end upward). The molten glass will form a hook.
16. A sample mill for grinding material, e.g., a Tecator cyclotec 1093 (Foss Tecator, Hogavas, Sweden), fitted with the finest sample mesh. Alternatively a pestle and mortar plus quartz sand can be used.

3. Method

1. Place freshly harvested plant leaf samples in labeled freeze-dryer bags. Close the bags with paper clips then place the samples in a –80°C freezer (*see* **Note 2**), and leave until frozen (longer than 6 h). Transfer the bags to a freeze dryer, evacuate the chamber, and freeze dry overnight or until the samples are dry (*see* **Notes 3** and **4**). At this point the leaves should be uniformly brittle.
2. After removing pieces of stem and leaf midribs (*see* **Note 5**), mill the samples using either a sample mill or pestle and mortar with grinding sand. Use a fresh pestle and mortar for each sample, or, if using a mill, thoroughly clean the apparatus (using a brush and vacuum cleaner) before processing the next sample (*see* **Note 6**).
3. To 300–400 mg lyophilized ground tissue in a sterile, disposable 16-mL polypropylene centrifuge tube, add 9 mL prewarmed CTAB extraction buffer (*see* **Note 7**). Mix gently by inversion.
4. Incubate the samples for 60–90 min, with occasional inversion at 65°C.
5. Allow the samples to cool by standing the tubes in a trough of water at room temperature for 5 min.
6. Add 5 mL chloroform:octanol (24:1). Rock the tubes gently (or rotate them on a tube roller) to mix for 5 min.

7. Spin the samples in a bench-top centrifuge for 2 min at 850*g* and room temperature.
8. Pour off the top (aqueous) layer into a fresh 16-mL tube (*see* **Note 8**) and add 50 µL of preboiled RNAse A (10 mg/mL). Mix the samples gently by inversion and incubate for 30 min at room temperature.
9. Add 6 mL isopropanol to each tube. Mix the samples gently by inversion until a white fluffy DNA precipitate appears (it should appear within about 1 min, *see* **Note 9**).
10. After 2–3 min, remove the precipitated DNA with a glass hook (*see* **Note 9**) and transfer to a fresh 16-mL tube containing 8 mL of cold 76% ethanol, 0.2 M sodium acetate. Leave the DNA on the hook in the tube for 20 min.
11. Transfer the DNA to a fresh 16-mL tube containing cold 8 mL 70% ethanol for a few seconds then transfer the DNA to a fresh 16-mL tube containing 1 mL TE.
12. Rock gently to disperse DNA. Once the DNA has detached from the glass hook, the hook can be removed from the tube. Leave the samples at 4°C overnight to allow the DNA to dissolve (*see* **Note 10**).
13. Calculate the DNA concentration by measuring the optical density at 260 nm of a small aliquot of the sample in quartz cuvettes in a UV spectrophotometer. A solution of 50 µg/mL has an optical density of 1 in a 1-cm cuvette (*see* Chapter 11).
14. The integrity of the DNA can be visualized by running a 10-µL aliquot of the sample on a low percentage (<0.7%) agarose gel stained with ethidium bromide. The DNA should appear as a high molecular weight band running with, or slower than, a 20-kbp size marker (such as the largest *Hind*III fragment of bacteriophage λ). See **Notes 11–14**; Chapter 13.

4. Notes

1. Some grades of CTAB do not appear to work as well as others. We use a preparation called mixed alkyltrimethylammonium bromide, available from Sigma Chemical Co. (St. Louis, MO). Poor yields of DNA have resulted when another preparation was used.
2. Samples can also be placed in layers of dry ice (solid CO_2) pellets. This is a particularly useful method when samples are being collected from the field.
3. The chamber temperature on the freeze dryer can be left at ambient (in fact, cooling the sample chamber increases the drying time).
4. Freeze drying is not the only method that can be used to dry plant material, although the integrity of the DNA may be compromised. Other methods include air drying at 65°C in an incubator with forced-air circulation or drying under vacuum. In each case, the samples should be dried to constant weight; in the case of air drying this is usually overnight.
5. Some plant species, for instance, maize and lettuce, have very pronounced leaf midribs, and these should be removed from the material before grinding. Removal of the midrib is not important where it is small or nonexistent, e.g., in very young leaves or in small-grain cereals. The reason why the midrib must be removed in some cases is that it is a major source of carbohydrate contamination.
6. Other methods of disrupting the plant tissue are available. For some species, e.g., rice and other cereals a domestic coffee mill *(4)* can also be used, provided that a reasonable amount of dried leaf is to be ground. Alternatively, the sample can be placed in a 50-mL polypropylene centrifuge tube with glass beads. The sample can then be vigorously agitated using a paint mixer for 0.5–3 min *(4)*.
7. DNA can be extracted from fresh plant tissue by grinding a leaf or leaf disc in a small amount of extraction buffer. If you are extracting from leaf discs, a scaled-down miniprep isolation procedure can be done as follows. Grind about 1 cm^2 of fresh leaf in 0.5 mL of extraction buffer using a glass rod in a small polystyrene chemical weighing boat. Pour this into a 1.5-mL centrifuge tube, wash the weighing boat out with a further 0.5 mL of extraction buffer, and pool with the first extraction. Continue from step 4, dividing the reagent volumes by 10.

8. Occasionally, leaf debris is not packed tightly enough into a solid plug separating the organic and aqueous layers. When this occurs, two things can happen. First, the whole contents of the tube can slop into the fresh tube. If this occurs, then the sample must be respun, and the aqueous phase drawn slowly up through a 1-mL Gilson-type pipet tip that has had the fine end trimmed off to make it into a wider bore; this aqueous phase should be transferred to a fresh tube with fresh RNase A. Second, a few small pieces of leaf can contaminate the new tube. If this occurs, pour the solution through Miracloth into a fresh tube.
9. After isopropanol is added, the DNA may not form a clot. Instead, it forms several smaller fragments that are very difficult to remove using a glass hook (this is usually the case with the miniprep scale). In this instance, centrifuge at 850g (or microcentrifuge for the miniprep) for 5 min to pellet the DNA. Wash the pellet with the solutions described in steps 10 and 11, repelleting after each wash. Finally resuspend the DNA in TE.
10. If the DNA solution appears turbid after standing overnight at 4°C, try heating the sample to 65°C for 10 min, inverting the tube every 3 min. Insoluble material that remains after this treatment can be removed by centrifugation at 850g for 5 min, and the cleared supernatant can be removed to a fresh tube. The pellet can be discarded (it is not DNA).
11. When checking the integrity of the DNA on an agarose gel an estimate of the relative concentrations of the samples can be made by viewing the intensity of the DNA bands, particularly if DNA molecular weight markers of known concentration are run on the same gel. This, combined with the measurement of the optical density of the DNA in a spectrophotometer, gives a more reliable estimate of the concentration. For restriction fragment length polymorphism and polymerase chain reaction analysis it is important that each sample is at the same concentration.
12. Should the DNA appear to be degraded (i.e., as a smear running down the gel), an isolation made from fresh plant tissue may yield intact DNA. When harvesting plant material for freeze drying, ensure that the tissue is immediately frozen, as this reduces DNA degradation. In addition, making fresh solutions, particularly RNase A, may cure the problem. Finally, DNA is a large molecule that can be broken by shear forces if treated with too much violence. Therefore, care should be taken to mix samples gently, never vortex the DNA.
13. Occasionally, despite having an optical density at 260 nm, there appears to be no DNA on the gel, but instead the sample well glows brightly. The presence of DNA-protein aggregates often prevents the DNA moving into the gel (*see* **Note 14**). Digesting the sample with proteinase will remove the protein, releasing the DNA. The sample should be subsequently re-extracted by performing steps 3–14, or it can be phenol extracted.
14. The DNA should not be allowed to dry at any stage during the preparation, as this hinders resuspension and solubilization in TE. This may be because the DNA and residual denatured proteins form an insoluble mass.

References

1. Saghai-Maroof, M. A., Soliman, K. M., Jorgensen, R. A., and Allard, R. W. (1984) Ribosomal DNA spacer-length polymorphisms in barley: Mendelian inheritance, chromosomal location and population dynamics. *Proc. Natl. Acad. Sci. USA* **81,** 8014–8018.
2. Southern, E. M. (1975) Detection of specific sequences among DNA fragments separated by gel electrophoresis. *J. Mol. Biol.* **98,** 503–517.
3. Williams, J. G. K., Kubelik, A. R., Livak, K. J., Rafalski, J. A., and Tingey, S. V. (1990) DNA polymorphisms amplified by arbitrary primers are useful as genetic markers. *Nucleic Acids Res.* **18,** 6531–6535.
4. Tai, T. H. and Tanksley, S. D. (1990) A rapid and inexpensive method for isolation of total DNA from dehydrated plant tissue. *Plant Mol. Biol. Reptr.* **8,** 297–303.

4

Purification of Uncontaminated, Intact Plant RNA

Shu-Hua Cheng, Brandon D. Moore, and Jeffrey R. Seemann

1. Introduction

The isolation of uncontaminated, intact RNA is essential for analyzing gene expression and for cloning genes. The presence of a large quantity of naturally occurring carbohydrates makes plant tissues one of the most difficult materials from which to isolate high-quality RNA with good yield. Such difficulty is due to copurification of carbohydrates during RNA isolation. These compounds form complexes with nucleic acids during tissue extraction and coprecipitate during subsequent alcohol precipitation steps *(1–5)*. The resulting alcohol precipitates can be gelatinous and difficult to dissolve. An RNA solution contaminated with carbohydrates is viscous and absorbs strongly at 230 nm. This ultraviolet absorption prevents an accurate quantitation of RNA concentration by a measurement of A_{260} (*see* Chapter 11). Furthermore, the contaminated RNA is not suitable for cDNA synthesis, reverse-transcription/polymerase chain reaction (RT-PCR) amplification, in vitro translation, or Northern blot analysis *(6,7)*.

As the result of industrial activity, rising atmospheric CO_2 will substantially impact plant growth and productivity, in part through effects on photosynthesis. Plants grown at elevated CO_2 typically have a lower photosynthetic capacity relative to plants grown at ambient CO_2, and show reduced levels of certain photosynthetic proteins *(8)* and mRNAs *(9)*. In order to understand the effects of elevated CO_2 on the expression and regulation of photosynthetic genes, it is essential to analyze leaf RNA levels by Northern blot hybridization in plants grown at ambient and elevated CO_2. Exposure of plants to elevated CO_2 results in a several-fold increase of total leaf carbohydrates, particularly starch (**Table 1** and **ref. 9**) and soluble polysaccharides (unpublished data; D. Gibeaut, G. R. Cramer, and J. R. Seemann), relative to ambient grown plants. Thus, the problem of carbohydrate contamination during RNA isolation is even more pronounced in plants grown at elevated CO_2. We have recently developed a simple, reliable, and inexpensive method to isolate clean leaf RNA with high yield without using time-consuming techniques such as sedimentation in cesium chloride gradients *(10)*. Plant tissue is ground in buffered guanidinium thiocyanate as described by Chomczynski and Sacchi *(11)*. After tissue extraction, the homogenates are centrifuged at a moderate *g* force to remove insoluble polysaccharides. The supernatant is then extracted using

Table 1
Qualitative and Quantitative Evaluation of Isolated Total Leaf RNA from Plants Grown at 1000 ppm of CO_2

Species	Starch[a] μmol Hexose equiv/g FW	Absorbance ratios		Yield μg RNA/g FW
		A_{260}/A_{230}	A_{260}/A_{280}	
Arabidopsis	199 (40)	2.33	1.98	915 (972)
Common bugle	198 (57)	2.02	2.14	509 (650)
Snapdragon	36 (16)	2.23	2.22	929 (636)
Parsley	53 (29)	2.20	2.13	651 (655)
Plantain	62 (13)	2.17	2.19	642 (653)
Spinach	126 (5)	2.20	2.10	733 (714)
Tobacco	346 (153)	2.10	1.90	607 (717)

Numbers in parentheses were values obtained from plants grown at ambient (360 ppm) CO_2.
[a]Starch content was determined by the method of Schulze et al. *(13)*.

acid phenol/chloroform:RNA partitions to the aqueous phase, whereas DNA and proteins are present in the interphase and the phenol phase. Most polysaccharides that remain in the aqueous phase are then selectively precipitated by potassium acetate *(12)* and the RNA is purified from residual contaminants by lithium chloride precipitation.

2. Materials
2.1. Solutions

1. 0.75 *M* sodium citrate, pH 7.0 (with HCl). 0.1% (v/v) diethyl pyrocarbonate (DEPC)-treated and autoclaved.
2. 10% (w/v) *N*-lauroylsarcosine.
3. Guanidinium thiocyanate buffer: Dissolve 250 g of guanidinium thiocyanate in the manufacturer's bottle (without weighing) with 293 mL sterile deionized water, 17.6 mL 0.75 *M* sodium citrate buffer, and 26.4 mL 10% *N*-lauroylsarcosine at 65°C (*see* **Note 1**).
4. Extraction buffer: Add 36 μL of β-mercaptoethanol(concentrated) per 5 mL of guanidinium thiocyanate buffer just before use. Polyvinylpolypyrrolidone (insoluble, 20% w/w) can be added to the extraction buffer if the tissue to be extracted contains a substantial level of polyphenols.
5. Chloroform/isoamyl alcohol 49:1 (v/v).
6. 2 *M* sodium acetate, pH 4.0 (with acetic acid), 0.1% DEPC-treated and autoclaved.
7. Acid phenol (*see* **Note 2**).
8. 100% isopropanol.
9. 70% and 100% ethanol.
10. Deionized H_2O, 0.1% DEPC-treated and autoclaved.
11. 2 *M* potassium acetate, pH 4.8, 0.1% DEPC-treated and autoclaved (*see* **Note 3**).
12. 10 *M* LiCl, 0.1% DEPC-treated and autoclaved.
13. TNE buffer: 10 m*M* Tris-HCl, pH 7.5, room temperature, 150 m*M* NaCl, 1 m*M* EDTA.
14. TE buffer: 10 m*M* Tris-HCl, pH 7.5, room temperature, 1 m*M* EDTA.

2.2. Equipment

1. Mortar and pestle.
2. Liquid nitrogen.

3. 10-mL and 30-mL Nalgene Oak Ridge polypropylene tubes, autoclaved (Nalgene, Rochester, NY).
4. Preparative centrifuge and microfuge.
5. Glass Pasteur pipet, baked at 200°C for at least 3 h.
6. Vortex.
7. 15-mL Falcon snap-capped polypropylene tube.
8. 1.5-mL Microfuge tube.

3. Method

1. Grind 0.2–0.4 g (fresh weight) of tissue in a mortar and pestle to a fine powder in liquid nitrogen. Do not allow the tissue to thaw (*see* **Note 4**).
2. Add 3.5 mL of extraction buffer and grind thoroughly while the tissue thaws. Transfer the homogenate to a 10-mL Oak Ridge tube. Rinse the mortar and pestle with 1 mL of extraction buffer and then transfer to the tube.
3. Centrifuge at 23,000g for 20 min at 4°C in a swinging bucket rotor (*see* **Note 5**).
4. Transfer the supernatant to a 10-mL Oak Ridge tube using a glass Pasteur pipet (*see* **Note 6**).
5. Add 0.4 mL of 2 *M* sodium acetate and mix by vortexing. Add 4 mL of acid phenol and vortex. Add 0.8 mL of chloroform/isoamyl alcohol and vortex.
6. Incubate the tube on ice for 15 min (*see* **Note 7**).
7. Centrifuge as in **step 3**.
8. Transfer the supernatant to a 30-mL Oak Ridge tube using a glass Pasteur pipet and add an equal volume of 2 *M* potassium acetate; mix by vortexing.
9. Incubate the tube on ice for at least 30 min (*see* **Note 8**).
10. Centrifuge at 44,000g for 20 min at 4°C in a fixed angle rotor.
11. Transfer the supernatant to a 15-mL Falcon tube and add 0.6 vol of 100%, ice-cold isopropanol; mix by vortexing. Incubate the tube at –20°C for 45 min (*see* **Note 9**).
12. Centrifuge at 2,700g for 20 min at 4°C in a swinging bucket rotor.
13. Decant the supernatant. Wash the pellet with 1 mL of 70% ethanol; spin as in **step 12** for 3 min and pour off as much of the ethanol as possible.
14. Dissolve the pellet in 400 µL of DEPC-treated water and transfer to a 1.5-mL microfuge tube.
15. Add 100 µL of 10 *M* LiCl and incubate at 4°C for at least 2 h.
16. Microfuge at 12,000g for 20 min at 4°C (*see* **Note 10**).
17. Wash the pellet 2× in 1 mL of 70% ethanol. Microfuge at 12,000g for 30 s at 4°C after each wash.
18. Resuspend the pellet in 200 µL of TNE and add 500 µL of 100% ethanol. Incubate at –20°C for at least 15 min. This alcohol precipitation step is to remove residual LiCl, which is known to inhibit cDNA synthesis reactions. If RNA of interest is used only for Northern analysis and not for synthesis of cDNA, dissolve the pellet in 200–400 µL of TE and proceed to **step 22**.
19. Microfuge at 12,000g for 5 min at 4°C.
20. Wash the pellet 2× with 1 mL of 70% ethanol. Microfuge at 12,000g for 30 s at 4°C after each wash.
21. Resuspend the pellet in 200–400 µL of TE, depending on the size of the pellet.
22. Make a 30- to 50-fold dilution of each sample and measure the absorbance at 230, 260, and 280 nm (*see* **Note 11**).

4. Notes

1. Since guanidinium thiocyanate is hazardous, it is best to prepare this solution in the manufacturer's bottle without weighing to minimize handling. When making a smaller

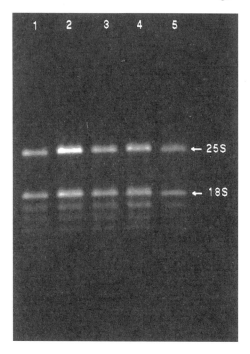

Fig. 1. Electrophoretic analysis of RNA isolated from various species grown at 1000 ppm CO_2. Two micrograms of total RNA isolated from leaves of (1) common bugle, (2) parsley, (3) plantain, (4) snapdragon, and (5) tobacco were electrophoresed on a nondenaturing agarose gel (1.4% agarose in TBE buffer containing 0.5 µg/mL ethidium bromide).

quantity of the solution, wear gloves when weighing. This solution can be stored at least 3 mo at room temperature.
2. To minimize handling, melt 500 g crystal phenol in the manufacturer's bottle at 65°C. Store in 50-mL aliquots in a –20°C freezer. Just before use, melt the aliquots at 65°C and saturate with DEPC-treated H_2O. For routine use, this solution can be kept at 4°C for up to 1 mo.
3. Dissolve 19.63 g potassium acetate in 25 mL of deionized H_2O and add glacial acetic acid until the pH is 4.8. Make up to 100 mL.
4. Use less tissue (e.g., 0.2 g) if samples contain particularly high levels of carbohydrates.
5. This initial spin removes the majority of insoluble polysaccharides, such as starch grains, by pelleting. The tissue debris forms a dark green pellet in the bottom of the tube and insoluble polysaccharides form a whitish gel-like layer on top of the tissue debris.
6. When pipetting the supernatant, care should be taken not to disturb the gel-like pellet, as it is very soft. The volume of the supernatant should be approx 4 mL. If there is a significant volume loss due to a large pellet, then compensate with extraction buffer. In our experience, there can be up to a 1-mL loss of volume in tissue extracts that contain a large quantity of starch.
7. A longer incubation period (up to 30 min total) improves the quality of RNA, particularly when there is a problem of protein contamination.
8. Polysaccharides form a whitish gel-like pellet. If the tissue used contains a high level of polysaccharides, a longer incubation period (up to 60 min) would help to precipitate more polysaccharides.

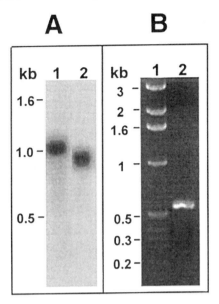

Fig. 2. (A) Northern blot analysis of total RNA isolated from tobacco leaves. One microgram of RNA was fractionated on a MOPS-formaldehyde agarose gel, blotted to a nylon membrane, and hybridized with ^{32}P-labeled cDNA fragments of spinach *psbA* (lane 1) or tobacco *rbcS* (lane 2). (B) RT-PCR of total RNA from spinach. RNA was reverse transcribed according to the standard procedure *(14)* and the first strand cDNA was used as a template to amplify a 550-bp fragment of sucrose phosphate synthase cDNA. A standard PCR reaction mixture (100 µL) was first denatured at 94°C, 4 min, followed by 35 cycles of 94°C (30 s), 50°C (30 s) and 72°C (30 s), and then extended at 72°C for 7 min. The primers used were 5'-TGCAGAGAGCAG CGGTGCTTA-3' (forward) and 5'-TCCATGCCAGGAGGAATT-3' (reverse). DNA standards (lane 1) and 5 µL of the amplified product (lane 2) were electrophoresed on an 1% agarose gel containing 0.5 µg/mL ethidium bromide.

9. Upon complete mixing of isopropanol, no precipitate should be visible. Any visible precipitate indicates the presence of a significant quantity of polysaccharides. In our experience, incubation longer than 60 min results in precipitation of polysaccharides.
10. The RNA pellet should be white. The presence of an off-white, gel-like pellet indicates contamination by polysaccharides. In such an event, repeat precipitation with potassium acetate and alcohol. Resuspend the pellet in 200 µL TE, add 1 vol of 2 *M* potassium acetate, and incubate on ice for 30 min. Spin the 1.5-mL microfuge tube at 12,000*g* at 4°C for 20 min. Transfer the supernatant to a new 1.5-mL microfuge tube, and precipitate the RNA with 2.5 vol of 100% ethanol at –20°C for 15 min. Proceed to **step 16** in **Subheading 3.**
11. The success of an RNA isolation procedure may be judged by the quantity, quality, and integrity of the RNA recovered. The RNA quality and quantity can be evaluated by measuring spectrophotometric absorbance at 230, 260, and 280 nm. An A_{260}/A_{230} ratio lower than 2 indicates contamination with polysaccharides and/or polyphenols, and an A_{260}/A_{280} ratio below 1.7 indicates contamination with proteins (*see* Chapter 11). The integrity of RNA can be assessed by the intactness of the 25S and 18S ribosomal RNA bands in an agarose gel (*see* Chapter 13).
12. Typical preparation: Using this protocol, we have successfully isolated uncontaminated, intact leaf RNA with high yield from a variety of plants grown at elevated CO_2 (Table 1).

The RNA preparation was free of DNA contamination (a common problem with many other protocols, **Fig. 1**). There was no apparent degradation of RNA as judged by the clarity and intactness of ribosomal RNA bands (**Fig. 1**). Also, the values of A_{260}/A_{230} and A_{260}/A_{280} typically were about 2, indicating little or no contamination by carbohydrates or protein (**Table 1**). The average RNA yield was greater than 500 µg/g fresh weight of leaf tissue, enough for many experiments. The yields of RNA in high CO_2 grown plants were very comparable to those from ambient grown plants and in some cases were higher. This protocol is easily scaled up or down, and can be finished within 24 h. RNA prepared by this method is suitable for poly(A^+) selection (data not shown), Northern analysis (**Fig. 2A**), cDNA synthesis, and RT/PCR amplification (**Fig. 2B**). The protocol described here is very simple, reliable, and inexpensive and allows simultaneous extraction of multiple samples (up to 12 samples can be processed at one time). Our success in isolating high-quality RNA from carbohydrate-enriched leaf tissue of diverse species demonstrates the general applicability of the method.

References

1. Lopez-Gomez, R. and Gomez-Lim, M. A. (1992) A method for extracting intact RNA from fruits rich in polysaccharides using ripe Mango mesocarp. *HortScience* **27**, 440–442.
2. Mitra, D. and Kootstra, A. (1993) Isolation of RNA from apple skin. *Plant Mol. Biol. Reptr.* **11**, 326–332.
3. Newbury, H. J. and Possingham, J. V. (1977) Factors affecting the extraction of intact ribonucleic acid from plant tissues containing interfering phenolic compounds. *Plant Physiol.* **60**, 543–547.
4. Schultz, D. J., Craig, R., Cox-Foster, D. L., Mumma, R. O., and Medford, J. I. (1994) RNA isolation from recalcitrant plant tissue. *Plant Mol. Biol. Reptr.* **12**, 310–316.
5. Wang, C.-S. and Vodkin L. O. (1994) Extraction of RNA from tissues containing high levels of procyanidins that bind RNA. *Plant Mol. Biol. Reptr.* **12**, 132–145.
6. Lay-Yee, M., DellaPenna, D., and Ross, G. S. (1990) Changes in mRNA and protein during ripening of apple fruit (*Malus domestica* Borkh. cv Golden Delicious). *Plant Physiol.* **94**, 850–853.
7. Tesniere, C. and Vayda, M. E. (1991) Method for the isolation of high-quality RNA from grape berry tissues without contaminating tannins or carbohydrates. *Plant Mol. Biol. Reptr.* **9**, 242–251.
8. Sage, R. F., Sharkey, T. D., and Seemann, J. R. (1989) Acclimation of photosynthesis to elevated CO_2 in five C_3 species. *Plant Physiol.* **89**, 590–596.
9. Van Oosten, J.-J., Wilkins, D., and Bestford, R. T. (1994) Regulation of the expression of photosynthetic nuclear genes by CO_2 is mimicked by regulation by carbohydrates: a mechanism for the acclimation of photosynthesis to high CO_2. *Plant, Cell Environ.* **17**, 913–923.
10. Glisin, V., Crkvenjakov, R., and Byus, C. (1974) Ribonucleic acid isolated by cesium chloride centrifugation. *Biochemistry* **13**, 2633–2637.
11. Chomczynski, P. and Sacchi, N. (1987) Single-step methods of RNA isolation by acid guanidinium thiocyanate-phenol-chloroform extraction. *Anal. Biochem.* **162**, 156–159.
12. Ainsworth, C. (1994) Isolation of RNA from floral tissue of *Rumex acetosa* (sorrel). *Plant Mol. Biol. Reptr.* **12**, 198–203.
13. Schulze, W., Stitt, M., Schulze, E.-D., Neuhaus, H. E., and Fichtner, K. (1991) A quantification of the significance of assimilatory starch for growth of *Arabidopsis thaliana* L. Heynh. *Plant Physiol.* **95**, 890–895.
14. Sambrook, J., Fritsch, E. F., and Maniatis, T. (1989) *Molecular Cloning: A Laboratory Manual*, Cold Spring Harbor Lab. Press, Cold Spring Harbor, NY, 141–145.

5

An Improved Method to Isolate Mitochondrial RNA from Green Plant Tissue

Fei Ye and Ralf Reski

1. Introduction

In plant cells, mitochondrial RNA (mtRNA) constitutes about only 1% of the total RNA. From this, most RNAs are ribosomal RNAs. Thus, isolation of high purified mtRNA is necessary, not only for construction of a mitochondrial cDNA library, but also for the analysis of plant mitochondrial transcription. Several methods have been frequently used for isolation of plant mtRNA *(1–3)*. However, these mtRNA preparations may be heavily contaminated by chloroplast RNA (cpRNA), especially when mtRNA is isolated from green leaves *(1,4)*. It is believed that the cpRNA sticks to the mitochondrial membrane and therefore persists after gradient purification of mitochondria. Although micrococcal nuclease would be the enzyme to remove the non-mtRNA from mitochondrial membranes prior to lysis of mitochondria, treatments with micrococcal nuclease for the mtRNA isolation from green leaves have not been effective *(4)*.

We report here a modified procedure of mtRNA isolation based on the combination of RNase A/guanidine thiocyanate/CsCl centrifugation. In our procedure, mitochondria are first separated from other subcellular components such as nuclei and plastids by differential centrifugation of leaf homogenates. The crude mitochondria are further purified by sucrose gradient centrifugation. To eliminate cpRNA, the purified mitochondria are treated with RNase A. Subsequently, RNase A is inactivated and mitochondria are lysed by adding guanidine thiocyanate in high concentration. As a strong protein denaturant, guanidine thiocyanate can inactivate nucleases very efficiently *(5)*. Mitochondrial RNA is pelleted through a CsCl gradient. Finally, coprecipitated, single-stranded DNA in the CsCl gradient can be removed from mtRNA by LiCl precipitation *(6)*.

2. Materials

1. 5.7 M CsCl solution: in 10 mM ethylenediaminetetraacetic acid (EDTA), pH 7.5, DEPC treated (*see* **Notes 1** and **2**).
2. Denaturation buffer: 50% formamide, 12% formaldehyde, 1X 4-morpholine propane sulfonic acid (MOPS) buffer (40 mM MOPS, 10 mM sodium acetate, 1 mM Na$_2$-EDTA, pH 7.0), freshly mixed before use.
3. 1 mg/mL Ethidium bromide: Diethyl pyrocarbonate (DEPC)-treated, storage at –20°C.

4. Extraction buffer: 0.35 M sorbitol, 50 mM Tris-HCl, pH 8.0, 5 mM EDTA, 0.1% bovine serum albumin (BSA), 0.25 mg/mL each spermine and spermidine, store at 4°C; add 2-mercaptoethanol to 0.2% (final concentration) just before use.
5. 4 M Guanidinium thiocyanate: in 100 mM Tris-HCl, pH 7.5 (store at 4°C), add 2-mercaptoethanol to 1% (final concentration) just before use. Store at 4°C.
6. 7.5 M Guanidinium-HCl: 10 mM DTT, pH 7.5 (adjusted with NaOH), filtrate, store at 4°C.
7. 2 M and 4 M LiCl: DEPC treated, store at 4°C.
8. Loading buffer: 50% glycerol, 0.25% bromophenol blue, 1 mM EDTA, DEPC treated, store at –20°C.
9. 10x MOPS buffer: 0.4 M MOPS, 0.1 M sodium acetate, 10 mM Na$_2$-EDTA, pH 7.0, DEPC treated.
10. 2 M Potassium acetate: pH 5.5, DEPC treated.
11. 2 M Sodium acetate: pH 7.0, DEPC treated.
12. 5% Sodium lauryl sarcosinate.
13. Tris-EDTA (TE) buffer: 10 mM Tris-HCl pH 8.0, 1 mM EDTA, DEPC treated.
14. Wash buffer: 350 mM sorbitol, 50 mM Tris-HCl, pH 8.0, 20 mM EDTA.

3. Method
3.1. Isolating of Mitochondria

All steps must be carried out at 4°C in a cold room. Solutions, bottles, and the like should be kept in wet ice.

1. Harvest 20 g of 4–6-wk-old fresh green leaves from rapeseed or other plants, cut into small segments, and chill in 200 mL ice-cold extraction buffer.
2. Homogenize leaf tissue in a blender at high speed three times (each time 5 s and with 10 s breaks in between). Filter the homogenate through two layers of cheese cloth, grade 50 (VWR), into 250 mL cold centrifuge bottles.
3. Centrifuge the filtrate at 2,000g for 10 min in a swing-out rotor. Carefully transfer the supernatant to new bottles and centrifuge at 10,000g for 20 min in a swing-out rotor.
4. Resuspend pellet in 100 mL extraction buffer and repeat **step 3** once again.
5. Resuspended the mitochondrial pellet in 20 mL ice-cold wash buffer.
6. Carefully layer each 10 mL mitochondrial suspension on top of a sucrose step gradient (9 m 0.9 M/11 mL 1.5 M/9 mL 1.75 M in wash buffer) and centrifuge for 60 min at 80,000g in a swing-out rotor (*see* **Note 3**). Collect the mitochondria from the 0.9 M/1.5 M sucrose interface (yellow band) with wide-bore pipets, then dilute with 5 vol of wash buffer over a period of 15–20 min (*see* **Note 4**).
7. Pellet the mitochondria by centrifugation at 10,000g for 20 min in a swing-out rotor and resuspend in 1 mL ice-cold wash buffer.

3.2. Isolating mtRNA

1. Coincubate mitochondria with 20 µg/mL RNase A for 60 min on the ice.
2. Add 5 vol of 4 M guanidine thiocyanate solution to the mitochondria, add 0.5 vol 5% sodium lauryl sarcosinate after 60 s at room temperature, and mix by vortexing. Centrifuge the mixture at 5,000g for 5 min in a swing-out rotor to remove insoluble debris.
3. Layer each 3.2 mL mixture onto a 1.1-mL cushion of DEPC-treated 5.7 M CsCl solution.
4. Carry out ultracentrifugation at 22,000g for 14 h in a swing-out rotor.
5. Carefully aspirate the supernatant solution and cut off the top part of the centrifuge tube that was in contact with the homogenate (All steps should avoid contamination with the finger RNase).

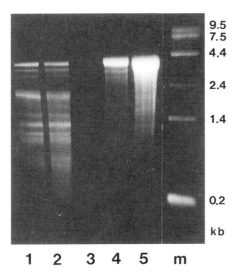

Fig. 1. Analysis of mtRNA preparation by electrophoresis in a 1.25% agarose-6% formaldehyde gel. *Lane 1:* mtRNA (5 µg) isolated with RNase treatment. *Lane 2:* mtRNA (5 µg) isolated without RNase treatment. *Lane 3:* MS2 phage RNA (10 µg) with RNase treatment. *Lane 4:* MS2 phage RNA (10 µg) without RNase treatment but using the conditions of mtRNA isolation. *Lane 5:* 10 µg MS2 phage RNA was directly loaded onto the gel. *Lane m:* RNA-ladder (BRL-Gibco, Gaithersburg, MD).

6. Dissolve the RNA pellet by extensive vortexing in 1 mL 7.5 M guanidinium-HCl solution.
7. Add 0.05 vol of 2 M potassium acetate (pH 5.5) and 0.5 vol of ethanol to the mixture.
8. Incubate at –20°C for 4 h and precipitate the mtRNA at 5000g for 10 min in a swing-out rotor.
9. Precipitate the recovered mtRNA by adding 0.1 vol of 2 M sodium acetate (pH 7.0) and 2.5 vol ethanol. Store at –20°C overnight. Centrifuge for 30 min at 10,000g.
10. Wash the mtRNA pellet with 70% ethanol, vacuum dry, and dissolve in 0.5 mL TE.
11. To obtain mtRNA free from single-stranded DNA, add an equal volume of 4 M LiCl to dissolved RNA, incubate at 4°C overnight, and collect the mtRNA by centrifugation at 10,000g for 20 min in a microcentrifuge. Wash once with 2 M LiCl and two times with 70% ethanol.
12. Vacuum dry the mtRNA and dissolve in 50 µL DEPC-treated water. Estimate the yield of mtRNA by measuring the absorbance at 260 nm.
13. For long-term storage, add 0.3 vol sodium acetate and 2.5 vol ethanol to the mtRNA and store at –70°C. Precipitate the RNA just before use.

This mtRNA preparation procedure will yield 0.3–0.5 µg mtRNA per gram fresh leaves. As a control, MS2 phage RNA (Boehringer, Mannheim, Germany) may be treated using the same conditions as mtRNA isolation. We have tested that the isolated RNA is intact (**Fig. 1**) and the contaminating cpRNA is totally eliminated (**Fig. 2**).

3.3. Formaldehyde Gel Electrophoresis

1. Melt 3.75 g agarose in 220 mL DEPC-H$_2$O plus 30 mL 10X MOPS buffer. After the agarose is cooled to 60°C, add 50 mL of 37% (12.3 M) formaldehyde solution and pour the gel in (*see* **Notes 6** and **7**).

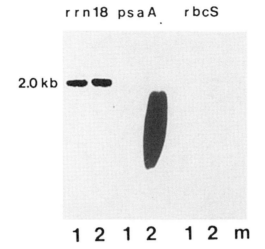

Fig. 2. Northern blot analysis of plant mtRNA preparations. The blots were probed with mitochondrial specific (*rrn*18), chloroplast specific (*psa*A), and nucleus specific (*rbc*S) gene probes. *Lane 1:* mtRNA (5 µg) isolated with RNase treatment. *Lane 2:* mtRNA (5 µg) isolated without RNase treatment. *Lane m:* RNA-ladder (BRL).

2. Denature 9 µL RNA (5–10 µg) by adding an equal volume of RNA denaturation buffer and incubate at 65°C for 5 min.
3. Add 1 µL of 1 mg/mL ethidium bromide, incubate at 65°C for another 5 min, and place samples on ice for 5 min.
4. Add 2 µL of loading buffer to each sample and load on the prepared gel.
5. Carry out electrophoresis at 5 V/cm for 3–4 h. Soak the gel in DEPC-treated H_2O for 20 min to remove the formaldehyde, and photograph the gel.
6. Further Northern blot analysis can be carried out according standard methods *(7)*. *See* Chapters 37 and 38.

4. Notes

1. Wear disposable plastic or latex gloves during work. All reagents should be used for RNA work only and kept free of ribonuclease. All glassware should be baked at 160°C for 4 h. All plasticware to be used after guanidinium thiocyanate treatment should be soaked in 0.2% DEPC for 12 h and autoclaved. All solutions to be used after guanidinium thiocyanate treatment should be treated with 0.2% DEPC for 12 h and autoclaved.
2. DEPC is a carcinogen. DEPC treatment of solutions and plasticware should be done in a chemical hood.
3. For sucrose gradient centrifugation, the prepared gradient should be allowed to equilibrate at 4°C overnight. After gradient centrifugation, wash buffer should be slowly added to collected mitochondria (over 15–20 min); this can minimize the osmotic shock.
4. For CsCl centrifugation, when different ultracentrifuge rotors are used, pay attention to maximum rotor speed and maximum run time.
5. The bulk of the DNA is removed using CsCl centrifugation. Because single-stranded DNA coprecipitates with the RNA in the CsCl gradient, 2 M LiCl precipitations are necessary to obtain pure RNA preparations.
6. Formaldehyde is very toxic. Preparation and running of formaldehyde gels should be done in a chemical hood.

7. Reagent-grade formamide can be used directly. However, if any yellow color is present, formamide should be deionized by stirring it for 1 h with 5% (w/v) resin 501-X8 (D) (Bio-Rad, Richmond, CA). After its filtration through Whatman No. 1 paper, deionized formamide should be stored in small aliquots at –70°C.

References

1. Stern, D. B. and Newton, K. J. (1986) Isolation of plant mitochondrial RNA. *Methods Enzymol.* **118,** 488–496.
2. Schuster, A. M. and Sisco, P. H. (1986) Isolation and characterization of single-stranded and double-stranded RNAs in mitochondria. *Methods Enzymol.* **118,** 488–496.
3. Schuster, W., Hiesel, R., Wissinger, B., Schobel, W., and Brennicke, A. (1988) Isolation and analysis of plant mitochondria and their genomes, in *Plant Molecular Biology* (Shaw, C. H., ed.), pp. 79–102, IRL Press, Oxford, UK, and Washington, DC.
4. Makaroff, C. A. and Palmer, J. D. (1987) Extensive mitochondrial specific transcription of the *Brassica campestris* mitochondrial genome. *Nucleic Acids Res.* **5,** 5141–5156.
5. Han, J. H., Stratowa, C., and Rutter, W. J. (1987) Isolation of full-length putative rat lysophospholipase cDNA using improved methods for mRNA isolation and cDNA cloning. *Biochemistry* **26,** 1617–1625.
6. Ye, F., Albaum, M., Markmann-Mulisch, U., and Abel, W. O. (1993) Improved method for the isolation of mitochondrial RNA from green leaves. *BioTechniques* **14,** 184.
7. Sambrook, J., Fritsch, E. F., and Maniatis, T. (1989) *Molecular Cloning: A Laboratory Manual*, 2nd ed., Cold Spring Harbor Lab. Press, Cold Spring Harbor, NY.

6

Isolating Chromosomal DNA from Bacteria

Elisabeth Chachaty and Patrick Saulnier

1. Introduction

Many methods have been described for isolating DNA from prokaryotic cells. The choice of method depends on the degree of purity of the DNA required for the analysis to be performed. Some DNA analyses (e.g., those using restriction enzymes) require DNA of high purity in relatively large amounts. This DNA can be obtained using protocols that include steps to purify DNA once released from cells, as described in this chapter. In contrast, analyses based on polymerase chain reaction (PCR) only require very small amounts of DNA whose quality can be crude. Simple, rapid methods allowing DNA to be released from bacterial cells, such as that described in Chapter 7, are sufficient for most PCR applications. Bacterial DNA can be prepared using extraction kits marketed by several manufacturers. Most of the kits use resins or membranes without organic extraction and/or the alcohol precipitation step to purify DNA. In general, techniques with extraction kits are easily and rapidly performed but are more expensive than their in-house-developed counterparts.

Plasmid and phage DNA preparation techniques have been omitted in this chapter, as they are described in Chapter 45. Rarely, large amounts of DNA (>500 μg) need to be prepared except for the construction of genomic libraries because minute amounts of specific genetic material are readily accessible to PCR analysis. Therefore, large-scale preparation of DNA is not presented in this chapter. Finally, large genomic DNA fragments need to be protected from risks of shearing when submitted to pulsed-field gel analysis. Particular extraction techniques, performed on cells previously immobilized in agarose gel are described in Chapter 7.

2. Materials

The Kate Wilson procedure *(1)*, described in the following list, produces digestible chromosomal DNA from Gram-negative bacteria efficiently, including those belonging to the genera *Pseudomonas, Agrobacterium, Rhizobium, Myxococcus,* and *Bradyrhizobium,* all of which normally produce large amounts of polysaccharides able to impede cell lysis. This procedure can be adapted for Gram-positive bacteria by adding a lysis step that includes a mixture of lysozyme and of lysostaphin.

From: *The Nucleic Acid Protocols Handbook*
Edited by: R. Rapley © Humana Press Inc., Totowa, NJ

1. Tris-EDTA (TE) buffer: 10 mM Tris-HCl, 1 mM ethylenediaminetetraacetic acid (EDTA), pH 8.0.
2. Lysostaphin: 2.5 mg/mL (stored in small, single-use aliquots at –20°C).
3. Lysozyme: 10 mg/mL (freshly prepared in TE buffer).
4. Sodium dodecyl sulfate (SDS): 10% in water (usually autoclaved and stored at room temperature).
5. Proteinase K: 20 mg/mL in sterile water or in TE buffer (stored in small, single-use aliquots at –20°C).
6. NaCl: 5 M (stored at room temperature).
7. CTAB (hexadecyltrimethyl ammonium bromide)/NaCl solution: 10% CTAB in 0.7 M NaCl. CTAB may be difficult to dissolve. Add 10 g CTAB in 80 mL 0.7 M NaCl while heating (about to 60°C) and stirring. Adjust final volume to 100 mL with 0.7 M NaCl.
8. Chloroform/isoamyl alcohol ratio: 24:1.
9. Phenol/chloroform/isoamyl alcohol: 25:24:1. Phenol is highly corrosive and can cause severe burns. Wear gloves, protective clothing, and safety glasses when handling phenol. All handling should be carried out in a chemical hood. Before use, phenol should be equilibrated to a pH greater than 7.8 because DNA will partition into the organic phase at acid pH. The preparation of phenol has been well described previously *(2)*.
10. Isopropanol.
11. Ethanol: 70%.

3. Methods

Bacteria are lysed and proteins are removed by digestion with proteinase K. Cell wall debris, polysaccharides, and residual proteins are removed by selective precipitation with cetyltrimethylammonium bromide (CTAB), and purification is completed by a phenol/chloroform extraction. Finally, DNA is recovered from the resulting supernatant by isopropanol precipitation.

1. Spin 1.5 mL of the culture (*see* **Note 1**) in a microcentrifuge for 2 min at high speed, or until a compact pellet forms. Discard the supernatant with a tip.
2. For Gram-positive bacteria, resuspend the pellet in 547 µL TE buffer. Add 10 µL of lysostaphin (2.5 mg/mL) and 10 µL of lysozyme (10 mg/mL). Mix thoroughly and incubate for 40 min at 37°C (*see* **Note 2**). Add 30 µL of SDS and 3 µL of proteinase K. Mix thoroughly and incubate for 1 h at 37°C.

    For Gram-negative bacteria, resuspend the pellet in 567 µL TE buffer. Add 30 µL of SDS and 3 µL of proteinase K. Mix thoroughly and incubate for 1 h at 37°C (*see* **Note 3**).
3. Add 100 µL of 5 M NaCl and mix thoroughly (*see* **Note 4**).
4. Add 80 µL of CTAB/NaCl solution, mix thoroughly, and incubate for 10 min at 65°C (*see* **Note 5**).
5. Add about 750 µL of chloroform/isoamyl alcohol, mix thoroughly (minimum 10 s/tube), and spin 5–10 min at high speed at room temperature (*see* **Note 6**).
6. Collect the viscous aqueous supernatant carefully in a fresh microcentrifuge tube and leave behind the interface (*see* **Note 7**).
7. Add an approximately equal volume (400–600 µL) of phenol/chloroform/isoamyl alcohol, mix thoroughly (minimum 10 s/tube), and spin as in **step 5** (*see* **Note 8**).
8. Collect the supernatant carefully in a fresh microcentrifuge tube. Add 0.6 vol isopropanol to precipitate the nucleic acids (*see* **Note 9**). Mix from top to bottom until a stringy whitish DNA precipitate is clearly visible (*see* **Note 10**).

9. Spin 10 min at high speed in a microfuge at room temperature and eliminate the supernatant (*see* **Note 11**).
10. Wash the DNA with 500 µL 70% ethanol, mix thoroughly, and spin as in the previous step.
11. Carefully remove the supernatant with a pipet tip and dry the pellet in a Speed Vac (Savant, Bioblock Scientific, Illkirch, France) (*see* **Note 12**).
12. Redissolve the DNA pellet in 20–100 µL TE buffer. Dissolution can be facilitated by heating at 60°C for 10–30 min (*see* **Note 13**).
13. Store DNA until use in small aliquots either at 4°C for a short period (24–48 h) or at –20°C or –80°C for a longer period.

4. Notes

1. Cells are cultured, under appropriate conditions either in liquid or on agar media. Cells are grown until the end of the exponential growth phase or until saturation for culture in broth. The optimal amount of the bacterial culture that should be harvested from the surface of agar plate media corresponds to the volume of a grain of rice.
2. For some Gram-positive strains, incubation time must be extended to 2 h until lightening of the bacterial suspension.
3. The solution becomes viscous as the detergent (SDS) lyses the bacterial cells' walls.
4. This step is very important because a CTAB-nucleic acid precipitate can form at room temperature when the salt concentration is lower than 0.5 *M (3)*. The purpose here is to remove cell wall debris, denatured protein, and polysaccharides complexed to CTAB while retaining the nucleic acids in solution.
5. The use of microtubes with a sure-lock system avoids loss of DNA if the cap is opened inadvertently during the incubation.
6. After centrifugation, a white interface, i.e., the CTAB-protein/polysaccharide complexes to be eliminated, should be visible.
7. With some bacterial strains, the interface formed after chloroform extraction is not compact enough to allow easy removal of the supernatant. In such cases, most of the interface can be fished out with a pipet tip and supernatant removed after spinning for 2 min.
8. This phenol/chloroform extraction step completes the elimination of proteins and CTAB precipitates.
9. There is no need to add salt because the NaCl concentration is already high.
10. Absence of visible DNA precipitate during this step may be due to shearing of DNA in to small fragments during the previous steps or to a low concentration of DNA in the solution. Additional incubation for 30 min on ice with mixing every 5 min can increase precipitation efficiency.
11. Verify the position of the DNA pellet before removing the supernatant.
12. The drying step can lead to the loss of the DNA pellet because of vacuum or to the contamination of the preparation if Speed Vac is used. To avoid using Speed Vac, remove the residual ethanol with a small pipet tip after spinning.
13. The volume of TE buffer depends on the size of the DNA pellet. For some applications, RNA should be eliminated with 1–10 units of DNase-free RNase for 30 min at 37°C. DNA concentration and purity should be assayed by spectrophotometry (ratio 260/280 nm) and controlled by gel electrophoresis, before digestion by restriction enzymes (*see* Chapters 11 and 13). If DNA is not digested, verify the amount of enzyme, enzyme buffer, and the digestion conditions used. With this DNA isolation method, NaCl concentration is high enough to interfere with some enzymatic activities. If this occurs, the DNA washing step (**Subheading 3.10.**, **step 10**) should be repeated.

References

1. Wilson, K. (1987) Preparation of genomic DNA from bacteria, in *Current Protocols in Molecular Biology* (Ausubel, F. M., Brent, R., Kingston, R. E., Moore, D. D., Seidman, J. G., Smith, J. A., et al.), Wiley, New York, pp. 2.4.1.–2.4.5.
2. Sambrook, J., Fritsch, E. F., and Maniatis, T. (1989) *Molecular Cloning: A Laboratory Manual*, 2d ed., Cold Spring Harbor Lab. Press, Cold Spring Harbor, NY, p. 134.
3. Murray, M. G. and Thompson, W. F. (1980) Rapid isolation of high-molecular-weight plant DNA. *Nucleic Acids Res.* **8,** 4321–4325.

7

Bacterial DNA Extraction for Polymerase Chain Reaction and Pulsed-Field Gel Electrophoresis

Elisabeth Chachaty and Patrick Saulnier

1. Introduction

Some polymerase chain reaction (PCR) applications such as gene detection or typing (1,2) require little purified DNA and may be performed with crude bacterial extracts. Many methods have been described for this procedure (Chapter 6). Some of them are straightforward and consist of simply boiling bacterial cells in water. However, PCR results obtained with such preparations may be erratic. The method described in this chapter (adapted from **ref. 2**) works well with Gram-positive bacteria such as staphylococci, enterococci, and *Clostridium difficile*.

Pulsed-field gel electrophoresis (PFGE) is a method used to separate large DNA fragments such as those obtained after digestion with restriction endonucleases that cut infrequently. To avoid possible risks of shearing bacterial DNA during the extraction and digestion steps, bacterial cells are immobilized prior to processing by incorporation in agarose gel. DNA extraction for PFGE is characterized by the need to prolong contact between agarose plugs and the lysis solution that must be distributed throughout the gel. However, the duration of DNA preparation has been shortened since the initial description of the method (3). The protocol described in this chapter works well with Gram-positive (4) and Gram-negative rods of clinical interest.

2. Materials

2.1. Extraction of DNA for PCR

1. Tris-EDTA (TE) buffer: 10 mM Tris-HCl, 1 mM ethylenediaminetetraacetic acid (EDTA), pH 8.
2. Lysis buffer: 50 mM Tris-HCl, 25 mM EDTA, 250 U of lysozyme per milliliter, pH 7.5. Add 15 U of lysostaphine per milliliter for staphylococci.
3. Digestion solution: 0.6% Nonidet P40 (Sigma Chemical, St. Louis, MO), 0.6% Tween 20, 0.6 µg of proteinase K per milliliter.

2.2. Extraction of Bacterial DNA for PFGE

1. TE buffer: 10 mM Tris-HCl, 1 mM EDTA, pH 8.
2. Lysis buffer: 6 mM Tris-HCl, 1 M NaCl, 0.1 M EDTA, 0.5% sodium N-lauroyl-sarcosine, pH 8 (*see* **Note 1**).

From: *The Nucleic Acid Protocols Handbook*
Edited by: R. Rapley © Humana Press Inc., Totowa, NJ

3. Lyzozyme solution: 85 mg/mL in sterile water.
   Lysostaphine solution: 1.000 U/mL in sterile water. The stock solution is stored in aliquots frozen at –20°C.
4. Low melting agarose: 2% in sterile water.
5. Sterile petri dishes 55 mm in diameter.
6. Digestion solution: EDTA 0.5 M, 1% sodium N-lauroyl sarcosine, proteinase K 2 mg/mL, pH 8.
7. PMSF (phenylmethysulfonyl-fluoride): 40 mg/mL in isopropanol. PMSF is harmful, so avoid contact with mucous membranes and eyes. It should not be inhaled and the solution must be handled in a chemical hood.

3. Method
3.1. Extraction of Bacterial DNA for PCR

1. Grow bacteria on appropriate agar or liquid medium.
2. Harvest cells from the culture medium and resuspend them in TE buffer to obtain an optical density of 2.0 at 580 nm.
3. Transfer 1 mL of this suspension into a microtube and wash the cells twice with TE buffer by centrifugation (2000g).
4. Resuspend the pellet in 500 µL of lysis buffer and incubate at 37°C for 1 h.
5. Add 500 µL of digestion solution and incubate at 56°C for 1 h.
6. Inactivate proteinase K by heating the microtube at 95°C for 10 min.
7. Store the DNA preparation frozen at –20°C in aliquots.

The volume of DNA preparation added to the PCR mixture should not exceed 1/20 of the final volume because at a lower ratio, the EDTA present in the preparation may inhibit DNA polymerase activity. The appropriate ratio is dependent on the bacterial species and PCR applications and thus should be optimized.

3.2. Extraction of Bacterial DNA for PFGE

Quantities are for the preparation of five plugs of 200 µL each.

1. Grow bacteria in appropriate liquid medium overnight and shake if necessary.
2. Pellet cells by centrifugation at 2000g and wash them twice with TE buffer.
   The amount of cells required is equivalent to that found in 5 mL of broth culture with an optical density of 1.0 at 600 nm.
3. Resuspend cells in 500 µL of TE buffer.
4. Add 75 µL of lyzozyme solution. For staphylococci, also add 15 µL of lysostaphine.
5. Add 750 µL of melted (55°C) 2% low melting agarose.
6. Mix thoroughly and pipet the agarose-cell suspension into the wells of a plug mold. Allow agarose plugs to solidify for at least 10 min at 4°C.
7. Gently remove the plugs from the mold and transfer them into a sterile petri dish containing 10 mL of lysis buffer. Incubate for 2 h at 37°C for Gram-negative bacteria and for 4 h for Gram-positive bacteria.
8. Remove the lysis buffer and replace it with 10 mL of the digestion solution. Incubate the petri dishes at 50°C overnight. Close them firmly with parafilm to avoid evaporation of the solution (*see* **Note 2**).
9. Cool the petri dish at 4°C. Transfer the agarose plugs into a new petri dish containing 10 mL of TE buffer. Wash the plugs three times in 10 mL of TE buffer with gentle shaking.
10. Remove the last 10 mL of washing buffer, add 5 mL of TE buffer.

11. Add 60 µL of PMSF isopropanol solution twice 30 min apart and incubate at 50°C (*see* **Note 3**). Rinse plugs three times in TE buffer.
12. Store plugs in 0.2 *M* EDTA at 4°C.

DNA prepared and stored as described in **Subheading 3.2.** is stable for several months. Plugs have to be rinsed three times for 10 min each in 10 mL of TE buffer to eliminate EDTA before digestion of DNA with restriction enzymes (*see* **Note 4**).

4. Notes

1. Other detergents such as sodium deoxycholate (0.2%) and Brij 58 (0.5%) (Sigma Chemicals, St. Louis, MO) may be added *(5)*.
2. Rapid protocols that include a proteinase K digestion step *(6)* or that include this step, but with a very short duration, have been proposed *(7)*. However, this may lead to inconsistent results *(7)*. On the other hand, incubation with proteinase K should be prolonged to 48 h to complete cellular protein digestion in some instances.
3. PMSF is used to inhibit residual proteinase K activity. It is insoluble in aqueous solution where its half-life is 60 min. This step is not included in many protocols which use lower proteinase K concentrations.
4. After completion of enzyme digestion, DNA electrophoresis, and ethidium bromide gel staining, some disappointments may occur:
   a. The banding pattern is absent or faint:
      i. Brightness of the agarose plug in the well of the gel is weak suggesting that its DNA content is too low (the amount of cells embedded in the agarose plug is too low).
      ii. Agarose plug in the well of the gel is bright suggesting that DNA is present but not digested (the amount of the restriction enzyme used is too low or the buffer enzyme was inappropriate).
         (a) Plugs have been insufficiently rinsed before the restriction step and residual traces of EDTA have inhibited restriction enzyme activity.
         (b) Extraction step has been unsuccessful. Increase the duration of incubation in the lysis buffer and/or digestion solution. Note that the efficiency of DNA extraction may be dependent on bacterial species, and on strains within species.
   b. Banding patterns are replaced by smears:
      i. Search for possible contamination of the preparation or of the electrophoresis buffer by exogenous DNAse.
      ii. Try to increase the duration of incubation in proteinase K digestion solution in order to destroy bacterial endonucleases. Growth conditions can influence bacterial endonuclease production. In some instances, changing the growth medium or the stage of growth (i.e., exponential versus stationary phase) at which cells are harvested can remedy the problem, but some strains constantly produce smears whatever the extraction conditions used *(4)*.

References

1. Saulnier, P., Chachaty, E., Hilali, F., and Andremont, A. (1997) Single step polymerase chain reaction for combined gene detection and epidemiological typing in three bacterial models. *FEMS Microbiol. Lett.* **50**, 311–316.
2. Goh, S. H., Byrne, S. K., Zhang, J. L., and Chow, A. A. (1992) Molecular typing of *Staphylococcus aureus* on the basis of coagulase gene polymorphisms. *J. Clin. Microbiol.* **30**, 1642–1645.
3. Schwartz, D. C. and Cantor, C. R. (1984) Separation of yeast chromosome-sized DNAs by pulsed field gradient gel electrophoresis. *Cell* **37**, 67–75.

4. Chachaty, E., Saulnier, P., Martin, A., Mario, N., and Andremont, A. (1994) Comparison of ribotyping, pulsed-field gel electrophoresis and random amplified polymorphic DNA for typing *Clostridium difficile* strains. *FEMS Microbiol. Lett.* **122,** 61–68.
5. Murray, B. E., Singh, K. V., Heath, J. D., Sharma, B. R., and Weinstock, G. M. (1990) Comparison of genomic of different enterococcal isolates using restriction endonucleases with infrequent recognition sites. *J. Clin. Microbiol.* **28,** 2059–2063.
6. Goering, R. V. and Winters, M. A. (1992) Rapid method for epidemiological evaluation of gram-positive cocci by field inversion gel electrophoresis. *J. Clin. Microbiol.* **30,** 577–580.
7. Matushek, M. G., Bonten, M. J. M., and Hayden, M. K. (1996) Rapid preparation of bacterial DNA for pulsed-field gel electrophoresis. *J. Clin. Microbiol.* **34,** 2598–2600.

8

Isolation of Fungal Nucleic Acids

Surapareddy Sreenivasaprasad

1. Introduction

The last 10 years have witnessed a growing interest in fungal molecular biology and, in a number of systems, both DNA and RNA technologies have been applied to address various aspects of fungal biology. Consequently, considerable progress has been made in the use of molecular markers, i.e., restriction fragment length polymorphisms *(1)*, fingerprints *(2)*, polymerase chain reaction (PCR) amplification profiles *(3)*, amplified fragment length polymorphisms *(4)*, and sequence data *(5)* for fungal systematics, phylogeny, ecology, epidemiology and population dynamics *(6,7)*. This has led to downstream applications such as the use of probes and primers for diagnosis of pathogenic and symbiotic fungi. Moreover, from a number of fungi, genes regulating development and differentiation processes have been isolated and are being characterized *(8,9)*. Transformation—the ability to introduce the DNA of interest into a cell—has also been achieved with a number of fungi *(10)*. These developments have opened up the possibility of genetic manipulations in fungi of agricultural and biotechnological importance.

The ability to isolate high-quality, undegraded DNA and RNA free of contaminants such as proteins, polysaccharides and the like, is crucial for their successful use in various applications. For example, DNA contaminated with phenolics and polysaccharides cannot be subjected to restriction enzyme digestion properly; degraded DNA will generate poor restriction profiles and fingerprints. These DNA samples are also unlikely to be amenable to PCR amplifications. DNA quality is also important in achieving reliable and reproducible results in PCR. Similarly, in the case of RNA, contamination with other components inhibits reverse-transcription and translation reactions, and degradation results in poor size of the cDNAs being synthesized.

In this chapter, methods for the isolation of DNA and RNA from fungi are described and these methods have been used extensively, in our laboratories and elsewhere, with various fungi. The materials required and step-by-step protocols are presented along with some practical tips to achieve better results. It is recommended that the reader goes through the entire chapter and understands the notes and the safety precautions (strictly adhering to local health and safety regulations) before starting to use any of the protocols. Good laboratory practice, e.g., wearing a laboratory coat and clean gloves, maintaining clean working bench and pipets, avoiding aerosols while pipeting, and

2. Materials

1. Extraction buffer: 200 mM Tris, 25 mM NaCl, 25 mM ethylenediaminetetraacetic acid (EDTA) and 0.5% sodium dodecyl sulfate (SDS), pH 8.5.
2. Phenol: Equilibrate 400 mL of phenol containing 0.1% 8-hydroxyquinoline, with an equal volume of 1 M Tris, pH 8.5, twice and once with an equal volume of extraction buffer (*see* **step 1**). Store the phenol at 4°C in amber bottles under a layer of extraction buffer. **Phenol is highly toxic and appropriate health and safety regulations such as wearing gloves, goggles, lab coat, and working in a fume hood should be followed while handling.**
3. RNase solution: Dissolve Ribonuclease I-A (Sigma-Aldrich Co. Ltd., Dorset, UK) 20 mg/mL in TN buffer (*see* **step 4**), boil for 15 min to make it DNase free and allow to cool at room temperature for 1 h to permit renaturation of RNase. Store the RNase solution at –20°C.
4. TN buffer: 10 mM Tris and 10 mM NaCl, pH 7.5.
5. Tris-EDTA (TE) 8: 10 mM Tris and 1 mM EDTA, pH 8.0.
6. CTAB buffer: 2% CTAB (hexadecyltrimethylammonium bromide), 100 mM Tris, 10 mM EDTA, and 700 mM NaCl. Add β-mercaptoethanol (1% by volume) just before use (**mercaptoethanol is toxic and smells horribly; work in hood, wear gloves**).
7. CHISAM: 24:1 (v/v) chloroform/ isoamyl alcohol mixture (**volatile; work in hood, wear gloves**).
8. Lysis buffer: 200 mM Tris, pH 7.5, 250 mM NaCl, 1 mM EDTA, and 1% SDS.
9. *Bis*-benzimide: Dissolve 10 mg/mL of double distilled water.
10. 20X standard saline citrate (SSC): 3 M NaCl and 0.3 M sodium citrate, pH 7.0.
11. 20X SSC saturated isopropanol: Saturate isopropanol with 20X SSC until the salt precipitates.
12. Ethidium bromide: Use a 10-mg/mL stock (**ethidium bromide is a carcinogen; purchase of stock solutions is recommended; adequate care should be taken in handling and disposal**).
13. RNA extraction buffer: 200 mM sodium acetate, 10 mM EDTA, 1% SDS. Add β-mercaptoethanol (0.5% by volume) just before use.
14. Diethyl pyrocarbonate (DEPC)-treated water: Add DEPC (0.1% by volume) to double distilled water. After overnight incubation in a shaker, autoclave the water to remove the DEPC. Incubate the water at 50°C for 1–2 h to remove traces of DEPC (**DEPC is a suspected carcinogen; take due care; open bottles containing DEPC in a fume hood, as pressure often builds up as the DEPC degrades to CO_2 and water during storage**).
15. 10X MOPS buffer: 400 mM MOPS (3-N-morpholino proanesulfonic acid), 100 mM sodium acetate, and 10 mM EDTA, adjust to pH 7.0 using 1 M NaOH.
16. RNA loading buffer: 50% glycerol, 1 mM EDTA, and 0.4% (w/v) bromophenol blue.

3. Methods

3.1. Genomic DNA Extraction

1. Grow fungal cultures in a suitable liquid medium for 3–7 d (*see* **step 1**, **Subheading 4.1.**).
2. Harvest mycelium by filtration through Whatman filter paper no. 1 (Whatman International Ltd., Kent, UK) over a Buchner funnel using suction.
3. Wrap the mycelium in aluminium foil and freeze immediately in liquid nitrogen (**follow local health and safety precautions**).

4. Pulverize the mycelium into as small pieces as possible, transfer to sterile disposable containers, and freeze dry for 24–48 h (*see* **step 2, Subheading 4.1.**).
5. Grind the freeze-dried mycelium, using a sterile mortar and pestle, to a fine powder and store in a freezer (*see* **step 3, Subheading 4.1.**).

3.1.1. Method 1

1. For DNA extraction, mix approx 300 mg of powdered mycelium with 3 mL of extraction buffer in a suitable centrifuge tube; sterile disposable needles can be used to aid the mixing; incubate at 65°C for 15 min (*see* **steps 1** and **2, Subheading 4.2.**) (*11*).
2. Add 2.1 mL phenol and 0.9 mL chloroform; mix well by inverting the tube several times (*see* **steps 4** and **5, Subheading 4.1.**).
3. Centrifuge at 13,000–18,000g for 30 min, up to 1 h at a lower speed.
4. Transfer the upper aqueous phase, using sterile disposable Pasteur pipets, taking care not to disturb the interphase, into a fresh tube (*see* **step 3, Subheading 4.2.**).
5. Add 100 µL of RNase solution and incubate at 37°C.
6. Add 2.1 mL phenol and 0.9 mL chloroform and mix well; centrifuge as in **step 3**.
7. Remove aqueous phase and repeat the phenol:chloroform extraction as in **step 6**.
8. Remove the aqueous phase, add 3 mL chloroform and mix well; centrifuge as in **step 3**.
9. Remove the aqueous phase into a graduated tube (e.g., 15 mL blood tubes), add 0.54 vol isopropanol and mix well to spool out DNA as white threads (*see* **step 5, Subheading 4.2.**).
10. Centrifuge for 2 min at 3000g to pellet the DNA.
11. Discard the supernatant, wash the pellet with 1 mL of 70% ethanol; centrifuge as in **step 10**.
12. Discard the ethanol, vacuum dry the pellet briefly to remove traces of ethanol (*see* **step 6, Subheading 4.2.**).
13. Add appropriate volume of TE 8 (500–700 µL) and leave the tubes at room temperature with gentle agitation (e.g., rotawheel/platform shaker) to dissolve the DNA.
14. Electrophorese 5 µL of the DNA sample on a 0.7% agarose gel with a known quantity of λ-DNA to quantify, based on comparative ethidium bromide fluorescence. Alternatively, DNA concentration can be determined by spectrophotometric estimation as described in **Subheading 3.4., step 2**, using 50 as the conversion factor.

3.1.2. Method 2

1. Mix approx 300 mg mycelial powder with a 4-mL cetyltrimethylammonium bromide buffer prewarmed to 65°C in a suitable centrifuge tube (*2*). Sterile disposable needles can be used to aid in mixing. Add 40 µL β-mercaptoethanol in the hood.
2. Incubate the tubes at 65°C for 30 min; mix once every 10 min.
3. Add 4 mL CHISAM, stand tubes uncapped for 2 min, and mix by incubating on a platform shaker for 15 min (*see* **steps 4** and **5, Subheading 4.1.**).
4. Centrifuge for 30 min at 13,000–18,000g.
5. Transfer the upper aqueous phase into a fresh centrifuge tube, add equal volume of isopropanol, and mix.
6. Centrifuge for 10 min at 10,000g to pellet DNA.
7. Discard the supernatant and vacuum dry the pellet briefly to remove traces of isopropanol.
8. Dissolve the DNA pellet in 600 µL sterile distilled water and add 100 µL RNase (20 mg/mL stock); incubate at 37°C for 30 min.
9. Transfer DNA to a 2.0-mL microfuge tube and add an equal volume of phenol:chloroform:isoamyl alcohol (25:24:1) and mix well by inversion.
10. Centrifuge for 15 min at 12,000–14,000g.

11. Collect the aqueous phase, add an equal volume of CHISAM, and incubate on a shaker for 15 min.
12. Centrifuge as in **step 10** and collect the aqueous phase.
13. To precipitate DNA, add 1/10 vol of 3 M sodium acetate, pH 5.2, and 2.5 vol of cold 100% ethanol (stored at –20°C), mix well and incubate at –20°C for 15 min.
14. Pellet the DNA by centrifugation for 5 min as in step 10.
15. Discard the supernatant, wash the pellet with 1 mL cold 70% ethanol (stored at –20°C) and centrifuge for 2 min as in **step 10**.
16. Vacuum dry the pellet briefly to remove traces of ethanol.
17. Dissolve the pellet in 300 µL of TE 8 by incubating on a rotawheel/platform shaker; electrophorese a 5-µL aliquot (*see* **step 14, Subheading 3.1.1.**).

3.1.3. Method 3

1. Pick a small amount of mycelium (approx equivalent to a 3–5-mm-diameter grain) from an agar culture or a liquid culture and place it into a microfuge tube (*see* **step 8, Subheading 4.2.**) *(12)*.
2. Freeze the mycelium by dipping the tube in liquid nitrogen.
3. Crush the mycelium thoroughly, as it is thawing, using a micropestle.
4. Add 100 µL of lysis buffer to the tube and mix well by vortexing.
5. Subject samples to three alternate cycles of freezing in liquid nitrogen and incubation (for 1 min) at 100°C, followed by a final incubation for 10 min at 100°C.
6. Centrifuge the extracts at 12,000–14,000g for 3 min and carefully collect the supernatant.
7. Purify the DNA using the QIAquick columns supplied by Qiagen (West Sussex, UK), according to the manufacturer's instructions.
8. Elute the DNA from the columns using 50 µL of 1 mM Tris, pH 8.5, prewarmed to 50°C.
9. Use 5–10 µL of this preparation to set up a 50–100 µL PCR.

3.2. Mitochondrial DNA Isolation

1. Extract total nucleic acids following the method described in **Subheading 3.1.1.**, using 6 g of freeze-dried mycelial powder *(13)*.
2. Use 24 mL extraction buffer, 18 mL phenol, and 6 mL chloroform:isoamyl alcohol (24:1). **Centrifuge tubes of 50–80 mL capacity will be required**.
3. Dissolve DNA in 8 mL of TE 8 in preparation for the cesium chloride/*bis*-benzimide gradient centrifugation.
4. Dissolve 12.25 g $CsCl_2$ in the (8 mL) DNA solution (final concentration 0.9423 g/mL giving a density of 1.68 g/mL); add 156 µL *bis*-benzimide (10 mg/mL stock) to achieve a final concentration of 120 µg/mL.
5. Adjust the final volume of the DNA/$CsCl_2$ solution to 13 mL with TE 8; fill a 13-mL quick-seal tube (Beckman or equivalent) and seal the tube according to the instructions of the centrifuge/tube manufacturer (Beckman or equivalent). **The volume of TE8 used for dissolving DNA and the quantities of $CsCl_2$ and *bis*-benzimide used will vary according to the capacity of the quick-seal tube used.**
6. Centrifuge at 40,000 rpm (e.g., Beckman Ti70 rotor) for 60 h at 25°C. **At higher speeds, the centrifugation time can be reduced, e.g., 48 h at 60,000 rpm.**
7. Visualize the bands under long-wave ultraviolet light (UV). **Ultraviolet light is harmful to eyes and skin; adequate eye protection must be worn, e.g., UV-resistant face mask. Handle tubes extremely carefully while transferring to the UV room and/or viewing, as any sharp movement will disturb the bands in the gradient.**

Isolation of Fungal Nucleic Acids

8. Nuclear DNA is visualized as a dense, bright, bluish-white fluorescent band in the lower half of the tube. Generally, mitochondrial DNA (mtDNA) is observed as a thin, compact, yellowish fluorescent band above the nuclear band (*see* **steps 1**, **2**, and **3**, **Subheading 4.3.**).
9. Puncture the tube at the top to release the pressure using a 21-gauge needle; do not remove the needle.
10. Puncture the tube just below the band using a 21-gauge needle on a syringe and pull the fraction containing the band slowly and carefully; collect the fraction into a 15-mL tube (e.g., blood tubes).
11. Add 10 mL isopropanol to the fraction, mix well, and let the fractions separate. Check under UV light to see the *bis*-benzimide removed by the upper isopropanol phase; remove the isopropanol and discard according to local regulations. Repeat the process at least five times or until the *bis*-benzimide is completely removed.
12. Dilute the $CsCl_2$ in the sample by adding 3 vol of sterile water. To precipitate the DNA, add 2 vol of absolute alcohol, mix well, and incubate overnight at –20°C.
13. Pellet the DNA by centrifugation at 13,000–18,000*g* for 30 min.
14. Wash the pellet with up to 20 mL 70% ethanol three times (to ensure complete removal of $CsCl_2$).
15. Vacuum dry the pellet briefly and dissolve the DNA in an appropriate volume (100–500 µL) of TE 8.
16. Subject the mtDNA fraction to restriction enzyme digestion at 37°C for 3 h or overnight e.g., mix 2 µg of DNA with 1 µL (10 U) *Eco*RI and 2 µL *Eco*RI buffer and adjust the volume to 20 µL with sterile water.
17. Electrophorese on a 0.7% agarose gel containing ethidium bromide (5 µg/mL) at 60 V for 3–5 h.
18. Digested mtDNA is visualized as discrete fragments (up to 10 generally) compared to a smear of restriction fragments from nuclear DNA.

3.3. Total RNA Isolation

1. Grind 2–5 g of fungal material (e.g., mycelium or mushroom fruit body) to a fine powder under liquid nitrogen (*see* **steps 1** and **2**, **Subheading 4.4.**) *(14)*.
2. Transfer the grindate to a Falcon tube containing 3 mL of extraction buffer per gram of fungal material and an equal volume of phenol:chloroform:isoamyl alcohol (25:24:1) preheated to 65°C (**add the grindate slowly and carefully, in a hood, to avoid loss of sample due to frothing**).
3. Mix thoroughly by inversion and /or vortexing (*see* **steps 4** and **5**, **Subheading 4.1.**).
4. Transfer the mixture to a RNase-free centrifuge tube and centrifuge at 13,000–18,000*g* for 30 min at 4°C.
5. Collect the upper aqueous phase, add an equal volume of chloroform:isoamyl alcohol (25:24:1) and mix well by inversion.
6. Centrifuge at 13,000–18,000*g* for 30 min at 4°C.
7. Collect the aqueous phase and add 12 *M* lithium chloride to a final concentration of more than 2 *M*.
8. Mix well and incubate overnight at 4°C to precipitate the RNA.
9. Pellet the precipitated RNA by centrifugation at 13,000–18,000*g* for 30 min at 4°C.
10. Discard the supernatant, wash the pellet with 5 mL 3.0 *M* sodium acetate, pH 5.2, and centrifuge as in **step 9** for 15 min.
11. Discard the supernatant, wash the pellet with 20 mL 70% ethanol, and centrifuge as in **step 9** for 15 min.
12. Repeat the 70% ethanol wash.

13. Briefly vacuum dry the pellet to remove traces of ethanol.
14. Dissolve the pellet in an appropriate volume (250–1000 µL) of DEPC-treated water (*see* **step 3**, **Subheading 4.4.**).
15. Store RNA at –70°C.

3.4. Quantification of RNA

1. Check optical density (OD) of appropriately diluted (up to 100-fold) RNA at 260, 280, and 230 nm using quartz cuvets with a UV light source.
2. RNA concentration (µg/mL) is calculated as 40 × optical density × dilution factor.
3. Ratios of 1.7–2.0 between 260 and 280, and 2 or above between 260 and 230 indicate good-quality RNA.

3.5. Formaldehyde Electrophoresis of RNA

1. In an RNase-free flask, weigh 0.6 g agarose and add 52.5 mL DEPC-treated water *(15)*.
2. Melt the agarose, let cool down to 65°C, and add 1 mL formaldehyde and 6 mL 10× 4-morpholinopropanesulfonic acid (MOPS).
3. Pour the gel into an RNase-free gel caster and allow the agarose to polymerize for an hour in a fume hood.
4. Just before electrophoresis, mix 7 µL of RNA (approx 15 µg) with 2 µL 10X MOPS, 3.5 µL formaldehyde, and 10 µL formamide; denature by incubating at 55°C for 15 min and snap cooling on ice for 5 min; Similarly denature RNA markers to be used as size standards.
5. If the RNA samples are not to be subjected to Northern analysis, mix 2.5 µL of ethidium bromide (400 µg/mL stock) to the denaturation mix.
6. Add 5 µL of loading buffer and load the samples.
7. Electrophorese in 1X MOPS buffer containing 1.8% formaldehyde at 3–4 V/cm for approximately 3 h.
8. The electrophoresed RNA can be transferred onto a nylon membrane directly by capillary transfer without pretreating the gel.
9. Size standards and RNA samples immobilized on nylon membranes can be stained using methylene blue (0.5% in 0.5 M sodium acetate, pH 5.0).

4. Notes
4.1. General

1. Avoid excess agar with the inoculum while setting up a liquid culture for DNA extraction; alternatively use a suspension as the inoculum.
2. Break the frozen mycelium into as small pieces as possible to increase the efficiency of freeze drying. Do not allow the frozen mycelium to thaw.
3. Some freeze-dried mycelial samples do not grind well; using liquid nitrogen (**take adequate precautions while handling**) while grinding helps the process.
4. Using a pipet for aliquoting/adding chloroform damages the rubber O rings; it is preferable to use a Pasteur pipet.
5. After adding chloroform or CHISAM to the homogenate, wait a couple of minutes before closing the cap and mixing, otherwise the caps pop off. With hot chloroform, the tubes can even explode; loosen the cap to release the pressure in between mixing.

4.2. Genomic DNA Extraction

1. While taking powdered mycelium for DNA extraction, leave out agar pieces, if any.
2. Completely mixing the mycelial powder and extraction buffer into a homogenate increases the efficiency of DNA extraction.

3. While taking the aqueous layer, leave a clear layer of the interphase; purity of the extracted DNA depends on avoiding these contaminants.
4. Use cut pipet tips or sterile transfer pipets while removing the aqueous phase to avoid shearing the DNA.
5. If the DNA does not spool out, do not discard; centrifugation helps to recover the DNA.
6. Do not overdry DNA pellets; doing so leads to problems dissolving the DNA. Leaving the tubes on a rotawheel for several hours at room temperature, or warming at 37–55°C for 15–30 min, is useful, but do not pipet the solution up and down, as this shears the DNA. Undissolved residues, if any, can be removed by brief centrifugation and collecting the supernatant into a clean tube.
7. If the extracted DNA is not sufficiently pure (i.e., not amenable to restriction enzyme digestion and/or PCR amplification), repeat the phenol:chloroform extraction and ethanol precipitation. Alternatively, Qiagen columns (QIAquick) can be used to purify 5–10 µg DNA, following the manufacturer's instructions.
8. Do not take too much mycelium and avoid agar as much as possible (it is preferable to pick the aerial mycelium), while setting up mini-DNA extraction following Method 3.
9. The DNA extraction Method 1 works well with a number of fungi, e.g., *Colletotrichum* spp., *Cylindrocarpon* spp., *Trichoderma* spp., and *Verticillium* spp. However, with some fungi, possibly because of higher level of polysaccharides, Method 2, using a CTAB buffer, is recommended (e.g., *Magnaporthe* spp. and *Phytophthora* spp.). Both Methods 1 and 2, using a starting material of around 300 mg freeze-dried powder, will yield 100–200 µg of DNA.
10. Performing Methods 1 and 2 yield microgram quantities of high-quality DNA but involve a number of extraction and precipitation steps. For PCR-based analyses (where only nanogram quantities are required), including cycle sequencing of amplified products, Method 3 can be used rapidly on a large number of fungal isolates, even from agar cultures. This method is also suitable in situations where there is only a limited amount of fungal material available (e.g., endomycorrhizal spores and slow-growing ectomycorrhizal cultures). We have used this method with a number of fungi, e.g., *Colletotrichum, Glomus, Pythium, Suillus, Trichoderma*, and *Verticillium*.

4.3. Mitochondrial DNA Isolation

1. Isolating mtDNA from genomic DNA using gradient centrifugation should result in two clear bands. However, because of a number of reasons (e.g., separation of repeat elements from nuclear DNA, improper formation of the gradient), more than two bands are frequently observed (the nuclear DNA band would still be recognizable by its dense fluorescence). In such situations, it is desirable to collect all the bands for further analyses.
2. It is also important to collect each band as pure as possible, avoiding contamination by other bands. If necessary, a single band can be collected as two fractions in order to ensure the availability of sufficient quantity as well as quality for subsequent analyses.
3. In certain instances, two bands may appear very close to each other and it may not be possible to collect them separately. The whole fraction containing the two bands can be collected together and subjected to a second round of gradient centrifugation. Adequate care should be taken to maintain the overall concentration of the $CsCl_2$ as described previously.

4.4. RNA Extraction

1. The worst enemy of RNA is RNase, both endogenous and external. Perform all steps quickly and hold the samples/tubes on ice as much as possible to keep the endogenous RNases inactive.

2. External RNase contamination is best avoided by following good hygiene, e.g., DEPC-treated water/buffers (Tris buffers cannot be DEPC treated but can be made in DEPC-treated water) and clean spatulas and glassware (e.g., bake the glassware at 180°C for 4 h). Use fresh gloves as often as possible. Use several pipets for RNA work only.
3. Incubating the tubes at 65°C (5–15 min) aids in dissolving RNA. If there are undissolved particles after this stage, centrifuge them at 13,000–18,000g for 2 min and collect the supernatant; the undissolved fraction has little or no RNA.
4. RNA extracted from certain samples, e.g., mycelium from a colonized compost, could be brown and viscous because of coprecipitation of other components (e.g., polysaccharides and phenols) but may still be of sufficient purity for undertaking Northern analysis. However, for other downstream applications, such as reverse transcription, purifying the RNA on Sephadex G50 columns or isolating and using mRNAs may be necessary.
5. Fungal messenger RNAs (mRNAs) usually contain a 3' poly A tail and are isolated using oligo-dT cellulose columns. A number of commercial kits are available and in our laboratory a Pharmacia kit (Amersham Pharmacia Biotech, Hertfordshire, UK) is used routinely to isolate mRNA samples, which have been used successfully to construct cDNA libraries, and for RT-PCR.

Acknowledgments

The author thanks Dr. S. Muthumeenakshi, Mr. Jack Chipili, and Dr. C. S. Kingsnorth for sharing their experiences with some of the protocols described.

References

1. Sreenivasaprasad, S. (1992) DNA sequence variation and interrelationships among *Colletotrichum* species causing strawberry anthracnose. *Physiol. Mol. Plant Pathol.* **41,** 265–281.
2. Levy, M., Romao, J., Marchetti, M. A., and Hamer, J. E. (1991) DNA fingerprinting with a dispersed repeated sequence resolves pathotype diversity in the rice blast fungus. *Plant Cell* **3,** 95–102.
3. Mills, P. R., Sreenivasaprasad, S., and Brown, A. E. (1992) Detection and differentiation of *Colletotrichum gloeosporioides* isolates using PCR. *FEMS Microbiol. Lett.* **98,** 137–144.
4. Majer, D., Mithen, R., Lewis, B. G., Vos, P., and Oliver, R. P. (1996) The use of AFLP fingerprinting for the detection of genetic variation in fungi. *Mycol. Res.* **100,** 1107–1111.
5. Sreenivasaprasad, S., Mills, P. R., Meehan, B. M., and Brown, A. E. (1996) Phylogeny and systematics of 18 *Colletotrichum* species based on ribosomal DNA spacer sequences. *Genome* **39,** 499–512.
6. McDonald, B. A. (1997) The population genetics of fungi: tools and techniques. *Phytopathology* **87,** 448–453.
7. Bourdon, J. J. and Silk, J. (1997) Sources and patterns of diversity in plant-pathogenic fungi. *Phytopathology* **87,** 664–669.
8. Tagu, D., Python, M., Cretin, C., and Martin, F. (1993) Cloning symbiosis-related cDNAs from eucalyptus ectomycorrhiza by PCR-assisted differential screening. *New Phytol.* **125,** 339–343.
9. Xu, J. R., Urban, M., Sweigard, J. A., and Hamer, J. E. (1997) The *CPKA* gene of *Magnaporthe grisea* is essential for appressorial penetration. *Mol. Pl.-Microbe Inter.* **10,** 187–194.
10. Goosen, T., Bos, C. J., and van den Broek, H. (1992) Transformation and gene manipulation in filamentous fungi: an overview, in *Handbook of Applied Mycology—Fungal Biotechnology* (Arora, D. K., Elander, R. P., and Mukerji, K. G., eds.), Marcel Dekker, New York, pp. 151–195.

11. Raeder, U. and Broda, P. (1985) Rapid preparation of DNA from filamentous fungi. *Lett. Appl. Microbiol.* **1,** 17–20.
12. Manian, S., Sreenivasaprasad, S., and Mills, P. R. (1999) DNA extraction method for PCR in mycorrhizal fungi. *FEMS Microbiol. Lett.* (in press).
13. Garber, R. C. and Yoder, O. C. (1983) Isolation of DNA from filamentous fungi and separation into nuclear, mitochondrial, ribosomal and plasmid components. *Anal. Biochem.* **135,** 416–422.
14. Hoge, J. H. C., Springer, J., Zantinge, B., and Wessels, J. G. H. (1982) Absence of differences in polysomal RNAs from vegetative monokaryon cells of the fungus *Schizophyllum commune. Exp. Mycol.* **6,** 225–232.
15. Kingsnorth, C. S., Woodhouse, C. S., and Burton, K. S. (1999) Cloning and postharvest regulation of serine proteinase transcripts from the fruit body of the cultivated mushroom *Agaricus bisporus. Microbiol.* (in press).

9

Total RNA Isolation from Bacteria

John Heptinstall

1. Introduction

The susceptibility of RNA to degradation by exogenous and endogenous RNase activity following cell lysis has been well documented *(1,2)*. Moreover RNA usually occurs complexed with protein from which it must be released. Precautions to be taken against exogenous RNase include the use of plastic gloves, autoclaving solutions after adding 0.1% (v/v) diethyl pyrocarbonate (DEPC; except Tris, which reacts), and baking glassware, spatulas, and so forth at 180°C overnight *(3)*.

The endogenous RNase level varies with cell type, thus necessary precautions will vary. These may include the use of guanidinium thiocyanate (GuSCN), phenol, a thiol reagent (β-mercaptoethanol, dithiothreitol), proteinase K, a detergent (sodium dodecyl [lauryl] sulphate, *N*-dodecyl sarkosine [sarkosyl]), placental RNase inhibitor (a protein found in placenta *[4]* and other tissues, and is sold under a variety of trade names), and vanadyl ribonucleoside complexes *(5)*. Some of the reagents (phenol, detergent, proteinase K, GuSCN) will also simultaneously deproteinize RNA.

In order to release nucleic acid from a bacterial cell, the membranes and peptidoglycan of the envelope must be disrupted. Commonly this is brought about by lysozyme/EDTA treatment *(6)*, with EDTA leading to a loss of lipopolysaccharide from the outer membrane of Gram-negative bacteria to allow access of lysozyme to peptidoglycan. The conditions are not always of sufficient duration to produce spheroplasts, which would completely lyse in the hypotonic solutions employed, unless genomic DNA is required. SDS is subsequently added, to inhibit RNase, to remove the cytoplasmic membrane, and also the protein that is complexed with RNA. GuSCN may be added at this stage, particularly if RNase activity is high *(7)*.

Following the addition of GuSCN, RNA may be separated from protein and DNA using phenol prior to precipitation with ethanol or isopropanol, although in extracting rRNA we have not found phenol treatment to be necessary. The pH must not be alkaline in view of the lability of RNA, and is normally 7.0 ± 0.2. Phenol treatment may be with buffer-saturated phenol alone, or with phenol-chloroform-isoamyl alcohol (25:24:1) in which subsequent phase separation is easier but both yield RNA in the upper, aqueous layer. Chloroform alone, without phenol, has been recommended for bacterial RNA extraction *(8)*. We have found this to give reasonably pure RNA, as judged by A_{260}/A_{280}, but very variable yields.

From: *The Nucleic Acid Protocols Handbook*
Edited by: R. Rapley © Humana Press Inc., Totowa, NJ

The RNA is precipitated from solution by adding 0.1 vol 3 M ammonium acetate, pH 5.2, and either 2.5 vol ethanol or 1.0 vol isopropanol at $-20°C$. Usually, following cell lysis, particularly if lysis is prolonged or vigorous, there will be contamination by DNA which will also be precipitated at this stage. Removal is effected with RNase-free DNase 1 (*see* **Note 1**) after which further phenol treatment may be necessary to remove the DNase.

If lysozyme was used for cell lysis it would be inactive in the presence of SDS, however proteinase K is not, even at 50°C. Therefore a rapid method of disrupting bacterial cell envelopes, inactivating RNase, and freeing RNA, is to use proteinase K and SDS simultaneously *(9)*. This treatment may be followed by GuSCN, if residual RNase is a problem, or phenol treatment, or both. If phenol treatment is used, proteinase K will result in a reduction of the white protein precipitate at the aqueous/organic interface and higher RNA yields *(10)*. The one-step acid GuSCN phenol chloroform method for eukaryotic cells *(11)* uses GuSCN and mercaptoethanol and sarkosyl for cell lysis at pH 7.0, followed by sodium acetate at pH 4.0 (to 0.2 M) and water saturated (i.e., acidic) phenol. We have found with bacterial cells that cell disruption can be carried out directly by sonication in GuSCN solution. This is rapid but not applicable to lots of samples simultaneously and may lead to shearing of high-mol-mass RNA.

Polyethylene glycol is used to precipitate intact virus particles *(12)* and plasmid DNA *(13)* during plasmid or viral DNA preparation. It has been found that, following cell disruption, the cell debris and DNA can be removed by the addition of polyethylene glycol and salt, leaving the RNA in solution *(14)*. Subsequently, an aqueous biphasic system *(15)* may be produced by the addition of more salt (*see* **Note 2**), in which RNA selectively partitions into the lower, salt phase. Whether this is done or not, the RNA may then be precipitated with or without prior phenol treatment.

2. Materials

All solutions are prepared in water that has been previously autoclaved with 0.1% (v/v) DEPC. The latter is suspected of being a carcinogen and should be handled with care. Note: Care should be exercised when handling GuSCN and SDS, particularly in the solid state.

1. 5% (w/v) Sodium dodecyl sulphate (SDS).
2. 6 M Guanidinium thiocyanate (GuSCN). Filter and store up to 1 mo at room temperature.
3. 4 M Guanidinium thiocyanate, 25 mM sodium citrate, pH 7.0, 0.1 M β-mercapto–ethanol. Filter and store up to 1 mo at room temperature.
4. Proteinase K: 0.5 mg/mL. Aliquot and store at $-20°C$.
5. Physiological saline, 0.85% (w/v) NaCl. Autoclave.
6. Polyethylene glycol 6000 (PEG) 20% (w/v), 0.75% SDS (w/v) in 7.5% (w/v) potassium phosphate, pH 7.2. Autoclave.
7. Polyethylene glycol 6000 (PEG): 12.5% (w/v) in 1% (w/v) potassium phosphate, pH 7.2. Autoclave.
8. 0.12 M Sodium phosphate buffer, pH 7.2. Autoclave.
9. Phenol, saturated with 0.12 M phosphate buffer, pH 7.2. To phenol (either freshly redistilled at 165–180°C, or provided for use in molecular biology) in a previously unopened bottle, add phosphate buffer until the bottle is full. Mix gently. Add 0.05% (w/w of phenol) 8-hydroxyquinoline as an antioxidant and allow phase separation at 4°C.

Remove and discard the upper, aqueous layer and repeat the buffer addition, mixing and decanting. Store the phenol at –20°C. Note: Phenol burns the skin. Should this occur, wash with 20% (w/v) PEG in 50% (v/v) industrial methylated spirits.
10. Chloroform (AnalaR grade).
11. 3 M Sodium acetate, pH 5.2. Autoclave.
12. TBE buffer: Tris borate EDTA buffer. 90 mM Tris(hydroxymethyl)aminomethane, 90 mM boric acid, 2.5 mM EDTA, pH 8.3. Prepare a stock 10X solution by dissolving 108 g Trizma base, 55 g boric acid, and 9.5 g disodium EDTA in water, to 1 L. Autoclave. Dilute 10X for use.
13. Agarose, 1.1% (w/v) in TBE. Autoclave. Agarose should be low or medium electroendosmosis (EEO).
14. TE buffer: 10 mM Tris-HCl, pH 8.0, 1 mM EDTA. Autoclave.
15. 10% (w/v) Sodium dodecyl sarcosinate (Sarkosyl). Autoclave.
16. L-broth. 10 g/L Tryptone, 5 g/L yeast extract, 5 g/L NaCl, 1 g/L glucose. Autoclave glucose separately.

3. Methods
3.1. Bacterial Culture

The protocol is written for *E. coli* B but is applicable to all *E. coli* strains tested. Amendments may be necessary for other bacteria (*see* **Note 3**).

1. Inoculate 50 mL of L-broth from a slope and grow overnight, 37°C.
2. Use 5 mL of this culture to inoculate 50 mL of L-broth and grow for 2.5 h at 37°C, to an absorbance (600 nm) of 0.45–0.60.
3. Centrifuge a 1.5 mL sample in an Eppendorf tube at 12,000g for 10 min. Discard the supernatant.
4. Wash in saline by resuspending the pellet in 300 µL of saline, centrifuge at 12,000g for 10 min and discard the supernatant.
5. Resuspend the pellet in 400 µL of 0.12 M sodium phosphate buffer, pH 7.2.

3.2. Lysis and Extraction with PEG 6000

1. To the resuspended pellet add 50 µL of 5% SDS and 50 µL of proteinase K (0.5 mg/mL). Vortex and incubate at 37°C for 20 min.
2. Add 500 µL of PEG 6000 (12.5% w/v) in 1% potassium phosphate, pH 7.2. Vortex and centrifuge at 12,000g for 10 min.
3. Carefully remove most of the supernatant, measuring the volume, into a sterile tube for precipitation of the RNA (*see* **Note 2**). If necessary, phenol extraction may be carried out at this stage (*see* **Subheading 3.2.7.**).
4. Add 0.1 vol 3 M sodium acetate, pH 5.2, followed by 2.5 vol ethanol (or 1 vol isopropanol) at –20°C to precipitate the RNA. Leave for at least 1 h.
5. Centrifuge at 12,000g for 15 min, discard the supernatant, drain the tube by carefully inverting onto tissue paper. Wash the pellet (which may not be visible) with 70% ethanol by resuspending and centrifuging at 12,000g for 10 min. Decant and drain as before.
6. Dissolve the pellet in TE at room temperature for at least 30 min. The yield is about 15 µg total RNA/mL of culture (**Fig. 1**). The ratio A_{260}/A_{280} is in the range 1.90–2.05.
7. Should protein removal by phenol be required (**Subheading 3.2.3.**), add, sequentially, 0.1 vol 3 M sodium acetate, pH 5.2 and 1 vol buffer-saturated phenol. Vortex and centrifuge at 12,000g for 10 min. Remove the upper, aqueous layer, repeat the extraction of this layer with 1 vol phenol, vortex, centrifuge and retain the upper layer. The RNA may be precipitated (**Subheading 3.2.4.**) without adding more sodium acetate (*see* **Note 4**).

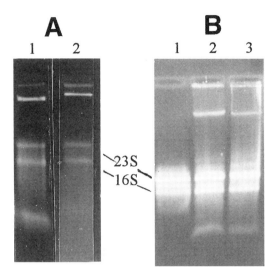

Fig. 1. Nondenaturing agarose gels of total RNA extracts from *E. coli* B. Precipitated with ethanol, but not phenol-treated. (**A**) 1.1% Agarose, about 2 μg nucleic acid per lane; lane 1, PEG; lane 2, GuSCN added *before* lysis. (**B**) 0.7% Agarose, about 6 μg nucleic acid per lane; lane 1, standard 23S and 16S rRNA; lane 2, PEG; lane 3, GuSCN added *after* lysis. A DNA band (not RNase-sensitive) is apparent between the 23S rRNA and the well in all cases shown here.

We have not found endogenous RNase to be a problem with *E. coli* and some other organisms (*Pseudomonas aeruginosa, Klebsiella aerogenes*); however, should suppression of activity be required, then GuSCN should be incorporated into the extraction, either before or after cell lysis (*see* **Note 5**).

3.3. Lysis and Extraction with GuSCN

3.3.1. Addition After Cell Lysis

1. Cells are prepared and lysed, as in **Subheading 3.2.1.**
2. Add 750 μL of 6 *M* GuSCN, vortex and centrifuge at 12,000*g* for 10 min. Transfer the supernatant to a fresh tube.
3. RNA can be precipitated directly at this stage, without significant protein contamination, by the addition of 0.1 vol sodium acetate and 2.5 vol cold ethanol, as **Subheading 3.2.4.** (**Fig. 1**).
4. Alternatively, phenol extraction may be carried out prior to precipitation (**Subheading 3.2.7.**).

3.3.2. Addition Before Cell Lysis

Bacteria can be lysed by brief sonication in the presence of GuSCN. The procedure subsequently is then the one-step acid GuSCN-phenol method *(11)*, except that the detergent is added after sonication to avoid frothing.

1. Resuspend the bacterial pellet (**Subheading 3.1.4.**) in 1 mL of 4 *M* GuSCN + sodium citrate + β-mercaptoethanol and sonicate for 20 s.
2. Add 50 μL of 10% sarkosyl, vortex and centrifuge at 12,000*g* for 10 min.
3. The RNA may be precipitated at this stage by the addition of 0.1 vol 2 *M* sodium acetate, pH 4.0 and either 2.5 vol ethanol or 1 vol isopropanol at –20°C followed by centrifugation

at 12,000g for 10 min (**Fig. 1**), or phenol-extracted in the presence of acidic sodium acetate and acidic phenol *(11)*.
4. Add, sequentially, 0.1 vol 2 *M* sodium acetate, pH 4.0, 1 vol of water-saturated phenol and 0.2 vol chloroform isoamyl alcohol (49:1, v/v), vortex and centrifuge at 12,000g for 10 min.
5. The RNA in the upper, aqueous phase is precipitated by the addition of 2.5 vol ethanol or 1 vol isopropanol at –20°C (*see* **Subheading 3.2.4.**).

4. Notes

1. RNase-free DNase from some commercial suppliers may be less than perfect *(16)*. Alternatively, various selective precipitation methods have been described for RNA, including 3 *M* sodium acetate *(10)*, 2 *M* LiCl *(10),* or 0.5 vol ethanol *(7)*.
2. At this stage, with the RNA in PEG/salt solution, more PEG and salt can be added to produce an aqueous biphasic system *(15)*, with the RNA predominantly in the lower, salt phase. Thus, add an equal volume of 20% PEG 6000 + 0.75% SDS in 7.5% potassium phosphate, pH 7.2, mix for 30 s and centrifuge for 5 min at 12,000g. The upper, PEG-rich phase will occupy some 80% of the total volume and the salt-rich, lower phase 20%, but the latter contains about 80% of the total RNA.
3. Some organisms (*P. aeruginosa, K. aerogenes*) may be extracted under the conditions given, whereas others that have been tried (*Salmonella typyhimurium, Proteus mirabilis, Serratia marcescens*) will require changes in some concentrations of reagents, in particular SDS and/or PEG may be increased.
4. If the RNA is not to be precipitated and washed, and should small amounts of phenol interfere with subsequent treatment of the RNA, then phenol can be removed by washing with chloroform. Add about an equal volume of chloroform, mix and centrifuge very briefly, discard the lower, chloroform layer. Repeat at least four times. Phenol may affect enzyme activity and certainly will give spuriously high A_{260}/A_{280} ratios since it gives a peak at 270 nm (*see* Chapter 11).
5. The efficacy of GuSCN as an inhibitor of RNase is dependent on the concentrations of both inhibitor and enzyme. Chaotropic effects are not apparent below 3 *M* GuSCN, in some cases 5 *M* may be required for sufficient inhibition *(17)*.

References

1. Barnard, E. A. (1964) The unfolding and refolding of ribonuclease in urea solutions 1. Rates and extents of physical changes. *J. Mol. Biol.* **10,** 235–262.
2. Aviv, H. and Leder, P. (1972) Purification of biologically active messenger RNA by chromatography on oligothymidylic acidcellulose. *Proc. Natl. Acad. Sci. USA* **69,** 1408–1412.
3. Sambrook, J., Fritsch, E. F., and Maniatis, T. (1989) *Molecular Cloning. A Laboratory Manual.* 2nd ed., Cold Spring Harbor Laboratory Press, Cold Spring Harbor, NY, pp. 7.3–7.5.
4. Blackburn, P., Wilson, G., and Moore, S. (1977). Ribonuclease inhibitor from human placenta. Purification and properties. *J. Biol. Chem.* **252,** 5904–5910.
5. Berger, S. L. and Birkenmeier, C. S. (1979) Inhibition of intractable nucleases with ribonucleoside-vanadyl complexes: isolation of messenger ribonucleic acid from resting lymphocytes. *Biochemistry* **18,** 5143–5149.
6. Jones, P., Qiu, J., and Rickwood, D. (1994) RNA isolation and analysis. BIOS Scientific Publishers, Oxford, UK, pp. 15–28.
7. Chirgwin, J. M., Przybyla, A. E., MacDonald, R. J., and Rutter, W. J. (1979) Isolation of biologically active ribonucleic acid from sources enriched in ribonuclease. *Biochemistry* **18,** 5294–5299.

8. MacDonell, M. T., Hansen, J. N., and Ortiz-Conde, B. A. (1987) Isolation, purification and enzymatic sequencing of RNA. *Methods in Microbiology* **19,** 357–404.
9. Hilz, H., Wiegers, U., and Adamietz, P. (1975) Stimulation of proteinase K action by denaturing agents. Application to the isolation of nucleic acids and the degradation of masked proteins. *Eur. J. Biochem.* **56,** 103–108.
10. Lizardi, P. M. (1983) Methods for the preparation of messenger RNA. *Methods Enzymol.* **96**, 24–38.
11. Chomczynski, P. and Sacchi, N. (1987) Single-step method of RNA isolation by acid guanidinium thiocyanatephenolchloroform extraction. *Anal. Biochem.* **162,** 156–159.
12. Yamamoto, K. R., Alberts, B. M., Benzinger, R., Lawhorne, L., and Treiber, G. (1970) Rapid bacteriophage sedimentation in the presence of polyethylene glycol and its application to large-scale virus purification. *Virology* **40,** 734–744.
13. Sambrook, J., Fritsch, E. F., and Maniatis, T. (1989) *Molecular Cloning. A Laboratory Manual.* 2nd ed., Cold Spring Harbor Laboratory Press, Cold Spring Habor, NY, pp. 1.23–1.41.
14. Jenné, S., Miczka, G., and Heptinstall, J. (1993) Rapid extraction of bacterial ribosomal RNA with polyethylene glycol. Sixth European Congress on Biotechnology, Firenze, Italy, **3,** WE022.
15. Albertsson, P.-A. (1971) *Partition of Cell Particles and Macromolecules.* 2nd ed., Wiley-Interscience, New York.
16. Hengen, P. N. (1996) Methods and Reagents. *Trends Biochem. Sci.* **21,** 112,113.
17. Gillespie, D. H., Cuddy, K. K., Kolbe, T., and Marks, D. I. (1994) Dissolve and capture: a strategy for analysing mRNA in blood. *Nature* **367,** 390,391.

10

Simultaneous RNA and DNA Extraction from Biopsy Material, Culture Cells, Plants, and Bacteria

Udo Döbbeling

1. Introduction

The analysis of RNA and DNA from clinical biopsy material for diagnostic and research purposes has become more and more important. Currently, available methods and kits are focusing on the extraction of only one kind of nucleic acid, but, as biopsy material is often limited, a method for the simultaneous isolation of both kinds of nucleic acids from one sample is desirable. In contrast to DNA, RNA is rapidly degraded in the biopsy material which often cannot be immediately conserved during the extraction. It was found that it is necessary to break up the tissue totally to get RNA from the inner parts of the biopsy where RNA degradation has not yet proceeded so far, as bodylike conditions have been retained in this region for a longer time. Some tissues, especially skin, are very difficult to break up and conventional methods (e.g., guanidine thiocyanate), *(1)* yield too little and strongly degraded RNA (20–30 ng/mg tissue) of total RNA with an average size of 0.2–0.4 kb).

Phenol/chloroform is a strong denaturant that can be used to break up tissue efficiently *(2)*. It was also found that phenol/chloroform can be combined with a conventional RNA extraction buffer and that the method can be designed in a way that a simultaneous isolation of high molecular weight DNA and RNA from the same sample is possible *(3)*. For skin tissue, the yield of RNA was approximately 10 times and for DNA 5 times higher than that of conventional methods (50–100 ng/mg tissue and 200–400 ng/mL, respectively).

The method can be used for fresh and cryopreserved biopsy material but also for culture cells, plant material, and bacteria *(3)*.

2. Materials

1. RNA extraction buffer: 7 M urea, 2% sodium dodecyl sulfate (SDS), 5 mM ethylenediaminetetraacetic acid (EDTA) (pH 8).
2. Phenol equilibration buffer: 0.25 M Tris (pH 8.5), 1.25 M NaCl.
   Equilibrate phenol two times with an equal volume of equilibration buffer. Add 0.1% 8-hydroxyquinoline to the phenol (yellow).
3. Phenol/chloroform: Equal volumes of equilibrated phenol and chloroform.

4. 1X TE: 10 m*M* Tris-HCl (pH 8), 1 m*M* EDTA (pH 8).
5. CsCl cushion: 5.7 *M* CsCl, 10 m*M* EDTA (pH 7.5).
6. 2X SET: 300 m*M* NaCl, 10 m*M* EDTA (pH 8), 100 m*M* Tris-HCl (pH 8).
7. Further reagents: 20% sodium dodecyl sulfate (stock), RNAse A, ethanol, isopropanol (propanol-2).

3. Methods
3.1. Tissue Disruption and RNA Isolation (see Note 1) (2–5)

1. Store biopsy material (50–200 mg of tissue) in 1 mL RNA extraction buffer (**Subheading 2.**, **step 1**) immediately after surgery (*see* **Note 2**).
2. Cut the biopsy material to small pieces in the RNA extraction buffer using a scalpel.
3. Transfer the biopsy material and RNA extraction buffer into a vial of a homogenizer on ice (e.g., Potter S or Polytron, Wimigen AG, Wohlen, Switzerland), add 1 mL phenol/chloroform (**Subheading 2.**, **step 3**) and grind either mechanically or by hand.
4. Transfer the liquified material into two Eppendorf vials. Break uncrushed material with another 1 mL RNA extraction buffer + 1 mL phenol/chloroform.
5. Centrifuge the dissolved material 10 min at full speed in an Eppendorf centrifuge.
6. Transfer the aqueous upper phase into another Eppendorf vial and extract with an equal volume of chloroform.
7. Transfer the upper phase to another Eppendorf tube containing an equal volume of isopropanol (propanol-2). Shake the vial gently by inverting until threads of high molecular DNA have contracted themselves into a small clump (*see* **Notes 3** and **4**).
8. Remove the DNA clump and keep the remaining solution of the DNA precipitation for 1 h at –20°C. Purify DNA from the DNA clump as described in **step 2**.
9. Centrifuge for 10 min at full speed in an Eppendorf centrifuge. Dissolve the pellet containing total RNA and some lower molecular weight DNA in 100 µL TE (10 m*M* Tris and 1 m*M* EDTA, pH 8). This RNA preparation can be used directly for reverse transcriptase/polymerase chain reaction (RT-PCR). Contaminations by PCR products from the residual DNA can be avoided when primers are employed that map to different exons of the mRNA to be investigated.
10. To obtain pure RNA, the solutions containing the dissolved nucleic acids from **step 8** (1 mL) or **step 9** (100 µL) are loaded onto a CsCl cushion (*5*, **Subheading 2.**, **step 5**). Use either 14 × 89 mm polyallomer tubes (Beckman Instruments, Palo Alto, CA, e.g., 1 mL nucleic acids + 9 mL CsCl cushion) or 14 × 48 mm polyallomer tubes (e.g., 0.1 mL nucleic acids + 4.9 mL CsCl cushion). Try to avoid any mixing between the nucleic acids phase and the CsCl cushion.
11. Centrifuge in a SW 40 rotor (Beckman Instruments) overnight at 35,000 rpm. Because of its lower specific weight, the remaining DNA is held back by the CsCl cushion and it can be recovered from the layer above the CsCl cushion if the cell number was so low that DNA clumps did not form.
12. Decant the CsCl solution carefully. Cut off the bottom of the tube (approx 1 cm above the bottom) and place the tube bottom upside down on filter paper to allow the remaining fluid to drain away.
13. Wash the RNA pellet on the bottom of the polyallomer tube once with 0.5 mL 80% ethanol and once with 0.5 mL 96% ethanol.
14. Dissolve the dried RNA pellet in 100 µL TE.
15. Check the integrity of the RNA by loading 1–5 µL of the RNA preparation on a 1% Tris-Acetate EDTA (TAE) agarose gel. The largest RNA molecules should be bigger than 2 kb. When the rRNA bands (28S and 18S RNAs for eukaryotic cells or 23S and 16S for prokary-

otic cells) are visible, the RNA preparation can be used for RNA protection assays and similar experiments.

3.2. DNA Isolation

1. Transfer the DNA clump, from **step 8** in **Subheading 3.1.**, into an Eppendorf vial containing 0.5 mL of 2X SET (**Subheading 2., step 6**) and treat it with 50 µg/mL RNase A for 30 min *(3,6)*.
2. Stop the RNAse A digest by adding SDS to a final concentration of 0.5%.
3. Treat the DNA with proteinase K (1 mg/mL final concentration, Boehringer Mannheim, Mannheim, Germany), until the DNA clump has dissolved and becomes transparent.
4. Extract the DNA with an equal volume of phenol/chloroform (**Subheading 2., step 3**). Centrifuge at full speed in an Eppendorf centrifuge and transfer the DNA-containing upper aqueous phase to another Eppendorf vial.
5. Extract the DNA with an equal volume chloroform, centrifuge, and precipitate the upper phase in another Eppendorf vial with 2.5 vol ethanol for 1 h at –20°C.
6. Centrifuge 10 min at full speed in an Eppendorf centrifuge. Wash the pellet with 500 µL 80% ethanol and dry.
7. For concentration measurement, dissolve the DNA in 500 µL TE (pH 8) and shear it mildly (e.g., with QiaShredder, Qiagen, Hilden, Germany).

4. Notes

1. The foregoing protocol describes the RNA/DNA isolation from biopsy material. It can also be used for plant material, culture cells, and bacteria. For plant tissue, one can start from **step 2** in **Subheading 3.1.** and for culture cells, from **step 3** in **Subheading 3.1.** RNA is much more delicate than DNA. It is hydrolyzed at alkaline pH (>10) and quickly degraded by endogenous and exogenous RNAses. The action of endogenous RNAses can be blocked by rapid lysis of the tissues and cells by RNA lysis buffer and phenol/chloroform. Phenol/chloroform should not be used in the surgery room. The biopsies should be processed as fast as possible; any delay between surgery and cell lysis will yield additional RNA degradation. DNA is generally much less affected. Frozen material (cryopreserved material) should be thawed in RNA extraction buffer. Paraffin-embedded material generally does not yield useful RNA. DNA is also strongly degraded in paraffin-fixed material, however fragments of 200–500 bp can often be amplified by PCR.
2. Degradation by exogenous RNAses can be largely avoided by wearing gloves, washing glass and plastic material with DEPC water (0.1% diethyl pyrocarbonate in water) for 2 h before use, and autoclaving all material used.
3. The DNA clump should be removed even if only RNA is to be prepared, as the DNA on top of the CsCl cushion could form a layer that hinders a large amount of RNA from sedimenting to the bottom of the polyallomer vial. A DNA contamination of the RNA may occur when the density of the CsCl cushion is too low. If no DNA clump forms, proceed as described in **Subheading 3.1., step 11**. Precipitate the DNA with an equal volume of isopropanol 1 h at room temperature. To avoid the precipitation of CsCl, do not precipitate at –20°C.
4. During the RNA preparation (**Subheading 3.1., steps 6** and **7**) a large amount of the DNA may adhere to protein in the interphase. The interphase can be transferred in this case also, as these proteins will be digested by proteinase K during the DNA preparation (**Subheading 3.2., step 3**).

References

1. Sambrook, J., Fritsch, E. F., and Maniatis, T. (1989) Extraction, purification, and analysis of messenger RNA from eukaryotic cells, in *Molecular Cloning: A Laboratory Manual*, Cold Spring Harbor Lab. Press, Cold Spring Harbor, NY, pp. 7.19–7.25.
2. Graham, D. F. (1978) The isolation of high molecular weight DNA from whole organisms or large tissue masses. *Anal. Biochem.* **85,** 609–613.
3. Döbbeling, U., Böni, R., Häffner, A., Dummer, A., and Burg, G. (1997) Method for simultaneous RNA and DNA isolation from biopsy material, culture cells, plants and bacteria. *BioTechniques* **22,** 88–90.
4. Gough, N. M. (1988) Rapid and quantitative preparation of cytoplasmatic RNA from small numbers of cells. *Anal. Biochem.* **173,** 93–95.
5. Glisin, V., Crkvenjakov, R., and Byus, C. (1974) Ribonucleic acid isolated by cesium chloride centrifugation. *Biochemistry* **13,** 2633–2637.
6. Schorpp, M., Döbbeling, U., Wagner, U., and Ryffel, G. U. (1988) 5' Flanking and 5' proximal exon regions of the two Xenopus albumin genes. Deletion analysis of constitutive promoter function. *J. Mol. Biol.* **199,** 83–93.

11

Spectrophotometric Analysis of Nucleic Acids

John Heptinstall and Ralph Rapley

1. Introduction

The ability to quantify nucleic acids accurately and rapidly is a prerequisite for many of the methods used in biochemistry and molecular biology. In the majority of situations this is carried out using spectrophotometry, which is nondestructive and allows the sample to be recovered for further analysis or manipulation. Spectrophotometry uses the fact that there is a relationship between the absorption of ultraviolet light by DNA/RNA and its concentration in a sample. The absorption maximum of DNA/RNA is approx 260 nm. This figure is an average of the absorption of the individual nucleotides that vary between 256 and 281 nm. In the case of RNA, the concentration of a sample containing RNA may be calculated from Eq. 1, where as for DNA Eq. 2 is used:

$$40 \times OD260 \text{ of the sample} = \text{concentration of RNA (µg/mL)} \quad (1)$$

$$50 \times OD260 \text{ of the sample} = \text{concentration of DNA (µg/mL)} \quad (2)$$

That is, when the OD260 of the sample is 1 the concentration of RNA will be approx 40 µg/mL (50 µg/mL for DNA). It is also possible to assess the degree of purity of the nucleic acids by examining the absorption at other wavelengths in which protein and polysaccharides have known absorption maxima. Proteins are known to absorb strongly at 280 nm and polysaccharides may be identified by their maximum at 230 nm. Therefore, the ratio of measurements of these three wavelengths 230, 260, and 280, may indicate the degree of purity of the nucleic acid sample. For example, a sample containing only RNA following an extraction method is judged as being uncontaminated if the ratio is 1:2:1 (DNA 1:1.8:1). If there is significant deviation from this, then it is evident that contaminants are present and that further purification of the sample is necessary. In many cases, the purity and the concentration may be further obscured by the presence of reagents that are used in the extraction process itself. Some of these have characteristics that are evident on a spectrophotometric scan that includes the three wavelengths indicated. Therefore, when using spectrophotometry in the analysis of DNA or RNA it is necessary to be aware of the potential problems that may result in misleading figures *(1)*.

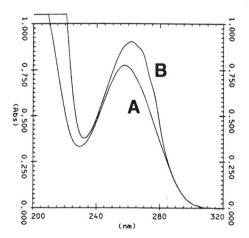

Fig. 1. Results of a spectrophotometric scan from 200–320 nm of a solution containing: (**A**) 33 µg/mL of tRNA alone (OD_{260}:OD_{280}) 2.1:1; (**B**) also contains the same concentration of tRNA, however, 16 µL/mL of phenol is present representing a $1.6 \times 10^{-3}\%$ solution. (OD_{260}:OD_{280}) 2.03:1.

2. Materials (*see* Notes 1 and 2)

1. Spectrophotometer capable of scanning in the ultraviolet wavelength range.
2. Optically matched quartz cuvets.
3. Sterile distilled water (nuclease free). This may be made in 0.1% diethyl-pyrocarbonate for the analysis of RNA.
4. 70% Alcohol.

3. Methods (*see* Notes 3 and 4)

3.1. Performing Spectrophotometry

1. For analysis of nucleic acids, clean the cuvets in 70% alcohol and leave to dry. For RNA in particular prepare the cuvets by soaking them in 0.1% diethylpyrocarbonate for at least 15 min.
2. Carry out baseline correction with H_2O or a non-UV-absorbing buffer.
3. Dilute the DNA or RNA sample with H_2O or a non-UV-absorbing buffer and carry out analysis.
4. Record absorption for 230, 260, and 280 nm and obtain a scan of absorption between 200 and 320 nm.

3.2. Analysis of RNA Spectrophotometric Scans

Figures 1–3 show typical scans of samples containing RNA in addition to some of the reagents commonly used to extract them (*2*). Very similar results are obtained with DNA. It is evident that in some cases it is necessary to scan samples to provide an indication of the degree of contamination rather than merely taking fixed readings at 260 and 280 nm. **Figure 1** shows a trace of 33 µg/mL of yeast tRNA (**Fig. 1A**) overlaid by a second trace (**Fig. 1B**) in which the sample also contains 16 µL/mL of phenol. It is clear that trace amounts of phenol remaining after an extraction can affect the OD260:280 ratio and the subsequent calculation to obtain the concentration of the RNA or DNA. In this case, the OD260:280 for **Fig. 1A** is 2.1:1, whereas for **Fig. 1B** the ratio

Spectrophotometric Analysis of Nucleic Acids

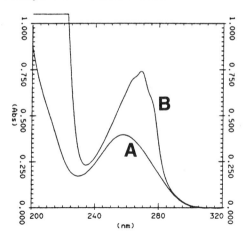

Fig. 2. Results of a spectrophotometric scan from 200–320 nm of a solution containing: (**A**) 16 µg/mL of tRNA ($OD_{260}:OD_{280}$) 2.16:1; (**B**) contains the same concentration of RNA, however, 32 µL/mL of phenol is present. ($OD_{260}:OD_{280}$) 1.87:1.

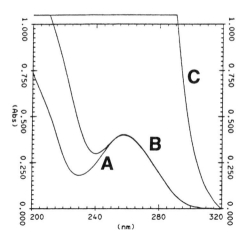

Fig. 3. Results of a spectrophotometric scan from 200–320 nm of a solution containing: (**A**) 16 µg/mL of tRNA ($OD_{260}:OD_{280}$) 2.3:1; (**B**) contains the same concentration of RNA, however, 132 µM of guanidinium thiocyanate is present ($OD_{260}:OD_{280}$) 2.1:1; (**C**) contains 2.4 M guanidinium thiocyanate and the ($OD_{260}:OD_{280}$) cannot be measured.

is 2. This phenomenon is exaggerated if the DNA/RNA concentration is lower and/or the phenol concentration is increased as evident in **Fig. 2**, where **Fig. 2A** is the pure RNA scan (16 µg/mL) and **Fig. 2B** has an additional 32 µL/mL of phenol. In this case, as with **Fig. 1**, the OD260:280 for **Fig. 2B** is misleading (1.87:1), indicating acceptably pure RNA. **Figure 3A** shows a trace of 16 µg/mL of RNA, and **Fig. 3B** contains an additional 132 µM guanidinium thiocyanate. There appears to be little difference in scans and the OD260:280 ratios are identical (2.1:1). However, increasing the guanidinium thiocyanate concentration from millimolar to molar amounts as used in a typical extraction and indicated in **Fig. 3C** has a drastic effect.

From this information it is clear to see that when analyzing ratios and concentrations of DNA or RNA spectrophotometrically it is necessary not only to derive readings at 280, 260, and 230 nm but also to scan throughout the range 200–320 nm. Trace amounts of reagents used in the extraction process can influence adversely and provide misleading data that may affect any subsequent manipulation.

4. Notes

1. It is advisable to refer to the manufacturer's instruction for the particular type of spectrophotometer used. A number of newer machines are linked to PCs and give readings at the required wavelengths.
2. Gloves should be worn throughout the procedure to minimize the effect of any exogenous nuclease activity.
3. It is always advisable when extracting and analyzing RNA to perform gel electrophoresis in addition to spectrophotometry. This gives an indication of the integrity of an extraction, as, in many cases, it is possible to visualize a major RNA species such as 16/23S rRNA for prokaryotes and 18/28S for eukaryotes.
4. It is also possible to determine approximate DNA/RNA concentrations using spot assessments with ethidium bromide. Cover a transilluminator with plastic film such as Saran Wrap. Pipet 2–5 μL of RNA samples of known concentration (0–30 μg/mL) directly onto Saran Wrap. Next to these, add the unknown RNA sample. To all samples add ethidium bromide (2 μg/mL in standard TE buffer (10 mM Tris-HCl, 1 mM ethylenediaminetetraacetic acid [pH 7.6]). It is possible to visualize the DNA/RNA and make an estimation of the concentration or alternatively record the results on a polaroid camera.

References

1. Manchester, K. L. (1995) Value of A260/A280 ratios for measurement of purity of nucleic acids. *Biotechniques* **19,** 208–210.
2. Heptinstall, J. (1997) Isolation of bacterial RNA, in *RNA Isolation and Characterization Protocols*, vol. 86 (Rapley, R. and Manning, D. L., eds.), Humana, Totowa, NJ, 47–53.

II

BASIC SEPARATION AND ANALYSIS OF DNA

12

Restriction Endonuclease Digestion of DNA

Duncan R. Smith

1. Introduction

The ability to cleave DNA at specific sites is one of the cornerstones of today's methods of DNA manipulation. Restriction endonucleases are bacterial enzymes that cleave duplex DNA at specific target sequences with the production of defined fragments. These enzymes can be purchased from the many manufacturers of biotechnology products. The nomenclature of enzymes is based on a simple system, proposed by Smith and Nathans (1). The name of the enzyme (such as *Bam*HI, *Eco*RI, and so on) tells us about the origin of the enzyme but does not give us any information about the specificity of cleavage (*see* **Note 1**). This has to be determined for each individual enzyme. The recognition site for most of the commonly used enzymes is a short palindromic sequence, usually either 4, 5, or 6 bp in length, such as AGCT (for *Alu*I), GAATTC (for *Eco*RI), and so on. Each enzyme cuts the palindrome at a particular site, and two different enzymes may have the same recognition sequence but cleave the DNA at different points within that sequence. The cleavage sites fall into three different categories, either flush (or blunt) in which the recognition site is cut in the middle, or with either 5'- or 3'-overhangs, in which case unpaired bases will be produced on both ends of the fragment. For a comprehensive review of restriction endonucleases, *see* Fuchs and Blakesley (2).

2. Materials

1. 10X stock of the appropriate restriction enzyme buffer (*see* **Note 2**).
2. DNA to be digested (*see* **Notes 3** and **4**) in either water or TE (10 mM Tris-HCl pH 8.3, 1 mM ethylenediaminetetraacetic acid [EDTA]).
3. Bovine serum albumin (BSA) at a concentration of 1 mg/mL (*see* **Note 5**).
4. Sterile distilled water (*see* **Note 6**).
5. The correct enzyme for the digest (*see* **Note 7**).
6. 5X Loading buffer: 50% (v/v) glycerol, 100 mM Na$_2$EDTA, pH 8, 0.125% (w/v) bromophenol blue, 0.125% (w/v) xylene cyanol.
7. 100 mM Spermidine (*see* **Note 8**).

3. Methods

1. Thaw all solutions, with the exception of the enzyme, and then place on ice.

2. Decide on a final volume for the digest, usually between 10 and 50 µL (*see* **Note 9**), and then into a sterile Eppendorf tube, add 1/10 vol of reaction buffer, 1/10 vol BSA, between 0.5 and 1 µg of the DNA to be digested (*see* **Note 3**), and sterile distilled water to the final volume.
3. Take the restriction enzyme stock directly from the –20°C freezer, and remove the desired units of enzyme (*see* **Notes 7** and **10**) with a clean sterile pipet tip. Immediately add the enzyme to the reaction and mix (*see* **Note 11**).
4. Incubate the tube at the correct temperature (*see* **Note 12**) for approx 1 h. Genomic DNA can be digested overnight.
5. An aliquot of the reaction (usually 1–2 µL) may be mixed with a 5X concentrated loading buffer and analyzed by gel electrophoresis (*see* Chapter 13).

4. Notes

1. Enzymes are named according to the system proposed by Smith and Nathans *(1)*, in which enzymes are named according to the bacteria from which they are first purified. Therefore, for example, a restriction enzyme purified from *Providencia stuartii*, would be identified by the first letter of the genus name (in this case, *Providencia* and hence *P*) and the first two letters of the specific epithet (in this case, *stuartii* and hence *st*) joined together to form a three-letter abbreviation: *Pst*. The first restriction enzyme isolated from this source of bacteria would therefore be called *Pst*I (with the number in Roman numerals), and the second *Pst*II, and so on. Note, however, that the name of the enzyme gives no information about the specificity of cleavage, which must be determined from one of the numerous lists of enzymes and cleavage specificities (the catalog of most suppliers of restriction enzymes will provide extensive information about restriction enzymes, such as specificity of cleavage, optimal reaction conditions, number of cleavage sites in common DNA templates, and so on, and these catalogs should be treated as valuable sources of information).
2. Each enzyme has an optimal reaction buffer. The recommended reaction conditions are normally to be found on the manufacturer's assay sheet. In practice, many enzymes share common conditions, and it is possible to make up reaction buffers that are suitable for a number of enzymes. The vast majority of enzymes will work in one of three buffers, either a high-, low-, or medium-salt buffer, recipes for which are given as follows. These buffers are normally made as a 10X stock and then 1/10 final volume is added to each digest. Great care must be taken in matching the buffer to the enzyme, since the wrong buffer can give either a dramatically reduced activity, altered specificity, or no activity at all. Several manufacturers of restriction enzymes now provide the correct buffer with their enzymes as an added benefit, and it is recommended that where these buffers are provided, they should be used.
    a. High-salt buffer (1X): 100 mM NaCl, 50 mM Tris-HCl, pH 7.5, 10 mM MgCl$_2$, 1 mM dithiothreitol (DTT).
    b. Medium-salt buffer (1X): 50 mM NaCl, 10 mM Tris-HCl, pH 7.5, 10 mM MgCl$_2$, 1 mM DTT.
    c. Low-salt buffer (1X): 10 mM Tris-HCl, pH 7.5, 10 mM MgCl$_2$, 1 mM DTT.

    In addition, two "universal buffers" are occasionally used, which are buffers in which all restriction enzymes have activity, although in some cases, activity can be reduced to only 20% of optimal activity. These are the potassium-glutamate *(3)* and potassium-acetate *(4)* buffers. These buffers can be particularly useful when a piece of DNA must be digested by two enzymes having very different optimal buffers.
3. The amount of DNA to be digested depends on subsequent steps. A reasonable amount for a plasmid digestion to confirm the presence of an insertion would be 500 ng–1 µg,

depending on the size of the insert. The smaller the insert, the more DNA should be digested to enable visualization of the insert after agarose gel analysis.
4. The DNA to be digested should be relatively pure and free from reagents, such as phenol, chloroform, alcohols, salts, detergents, and chelating agents. Any trace amounts of these chemicals will inhibit or inactivate the restriction endonuclease.
5. BSA is routinely included in restriction digests to stabilize low protein concentrations and to protect against factors that cause denaturation.
6. Good-quality sterile distilled water should be used in restriction digests. Water should be free of ions and organic compounds, and must be detergent free.
7. An enzyme unit is defined as the amount of enzyme required to digest 1 µg of a standard DNA in 1 h under optimal temperature and buffer conditions. The standard DNA used is normally λ DNA. Hence, for *Eco*RI, e.g., there are five sites for this enzyme in λ. If one is digesting PBR322, which has one site, with 1 U of enzyme for 1 h, this is actually a fivefold overdigestion.
8. Digests of genomic DNA are dramatically improved by the inclusion of spermidine in the digest mixture to a final concentration of 1 m*M* since the polycationic spermidine binds negatively charged contaminants.
    Note that spermidine can cause precipitation of DNA at low temperatures, so it should not be added while the reaction is kept on ice.
9. The smallest practical volume in which to undertake a restriction digest is 10 µL. Below this, pipeting errors can introduce significant errors in the reaction conditions. This volume also allows the entire digest to be loaded onto a small agarose gel after the addition of the stop/loading buffer. If the stock DNA concentration is too dilute to give 0.5–1 µg in 5–6 µL, then the reaction can be scaled up to 20–50 µL. If double digestion is to be undertaken (i.e., digestion with two different enzymes), then 20 µL is the recommended minimum volume, 1 µL of each enzyme can be added, and the glycerol concentration is kept low (*see* **Note 10**).
10. Many enzymes are susceptible to the presence of glycerol. The majority of stock enzymes are provided in approx 50% (v/v) glycerol. A reaction digest in which more than approx 10% (v/v) glycerol is present can give cleavage at different sites from the normal (the so-called star activity). For this reason, it is advisable to keep the enzyme total reaction volume ratio at 1:10 or lower. Similar star activity can result from incorrect salt concentrations.
11. Stock restriction enzymes are very heat labile and so should be removed from –20°C storage for as short a time as possible and placed on ice.
12. Note that the incubation temperature for the vast majority of restriction endonucleases is 37°C but that this is not true for all enzymes. Other enzymes, such as *Taq*I and *Sma*I, require different optimal temperatures (in this case 65 and 25°C, respectively). It is wise, therefore, to check new or unfamiliar enzymes before use.
13. If large-scale preparative digests are to be undertaken (100–500 µL reaction mixes), then the reaction is scaled up accordingly. However, care must be taken to ensure that the reaction components are fully mixed, especially with regard to the viscous constituents, such as DNA solutions and stock restriction enzymes. For all volume digests, vortexing should be avoided, as this can significantly reduce the activity of the enzyme. For small volumes, mixing can be achieved by tapping or gently flicking the tube with a finger (often followed by a brief 1–5 s spin in an Eppendorf centrifuge to deposit the reaction at the bottom of the tube). For larger volumes, mixing can be achieved by gentle pipeting, taking liquid from the bottom of the reaction volume, and mixing at the top of the reaction volume until a homogenous solution is obtained.

References

1. Smith, H. O. and Nathans, D. (1973) A suggested nomenclature for bacterial host modification and restriction systems and their enzymes. *J. Mol. Biol.* **81,** 419–423.
2. Fuchs, R. and Blakesley, R. (1983) Guide to the use of Type 11 restriction endonucleases. *Methods Enzymol.* **100,** 3–38.
3. McClelland, M., Hanish, J., Nelson, M., and Patel, Y. (1988) KGB: a single buffer for all restriction endonucleases. *Nucleic Acids Res.* **16,** 364.
4. O'Farrell, P. H., Kutter, E., and Nakanishe, M. (1980) A restriction map of the bacteriophage T4 genome. *Mol. Gen. Genet.* **170,** 411–435.

13

Agarose Gel Electrophoresis of Nucleic Acids

D. Ross Williams and Ralph Rapley

1. Introduction

DNA fragments may be separated by gel electrophoresis in a gel composed of agarose. This allows DNA fragments to be resolved on the basis of their molecular weight. The percentage of agarose used depends on the size of fragments to be resolved. In general a 0.8–1% gel may be used for effective separation of DNA fragments of 100–1500 base pairs *(1)*.

There are a wide variety of gel tanks and gel buffer systems in use. The mini-submarine gels are very popular because of their ease of operation, minimal use of reagents, and rapidity. The amount of DNA that may be visualized in a single band with ethidium bromide may be as little as 20 ng, although this is under ideal conditions.

As an alternative to agarose gels, DNA fragments may also be separated in polyacrylamide. This has a greater resolving power but involves a longer time for preparation than agarose gels.

2. Materials

1. Agarose (molecular biology grade) (*see* **Note 1**)
2. Running buffer (*see* **Note 2**). There are two common types of buffer systems used in agarose gel electrophoresis, Tris-borate EDTA (TBE) and Tris-acetate EDTA (TAE). To prepare stock solutions:
   10X TBE buffer: 545 g Tris, 278 g boric acid, 46.5 g EDTA in 5 L of sterile distilled water.
   50X TAE buffer: 242 g Tris, 57.1 mL glacial acetic acid, 100 mL 0.5M EDTA, pH 8.0, in 1 L of sterile distilled water.
3. Sterile distilled water.
4. Microwave oven and boiling water bath or steamer.
5. Loading buffer 50% (v/v) glycerol, 50 mM EDTA, pH 8.0, 0.125% (w/v) bromophenol blue, 1.125% (w/v) xylene cyanol (*see* **Note 3**).
6. Molecular weight size marker: Numerous commercial DNA size markers are available, either as base pair ladders (e.g., 123 bp multimer ladders) or predigested DNA (e.g., λ *Hin*dIII marker).
7. Ethidium bromide: 10 mg/mL dissolved in H_2O. Store at 4°C in a container wrapped in tin foil.
8. UV transilluminator (300 nm)
9. Polaroid camera or gel documentation system.

3. Method

1. The precise instructions for producing a gel depends on the gel-forming apparatus used. However, it is essential to make sure the gel casting stand is leak proof and sealed correctly (e.g., with end plates or waterproof tape).
2. For a 1% agarose gel in TBE running buffer: Weigh an appropriate amount of powdered agarose into a conical flask. As an example, for a 1% agarose gel, add 1g of agarose and 10 mL 10X TBE running buffer and distilled water to the final volume of 100 mL, mix thoroughly by swirling.
3. Heat the gel mix in a microwave oven (650 W) at full power for 1 min or less.
4. Remove the gel mix and allow to cool to approximately 50°C or until just cool enough to hold (*see* **Note 5**).
5. Pour the gel solution into a gel-forming tray, insert the comb template, and allow to set. This usually takes approx 25 min.
6. Remove the comb and tape or end plates and add sufficient 1X running buffer to the tank to cover the gel and electrodes.
7. To 5–10 mL of the DNA sample add 0.2 vol of loading buffer. Carefully add the sample to the well in one smooth pipetting motion (*see* **Note 4**).
8. Add an appropriate marker sample such as a λ *Hin*dIII digest or a 123 bp ladder to the end wells of the gel.
9. Carry out the electrophoresis at 100 V for approx 2 h or until the bromophenol blue dye has travelled two-thirds of the way down the gel.
10. Dismantle the gel apparatus and carefully place the gel in a tray. Add 100 mL of sterile distilled water containing 5 µL of 10 mg/mL ethidium bromide (0.5 mg/mL) and allow to stain for 15 min.
11. Destain the gel for 5 min by replacing the solution with fresh water (*see* **Note 6**).
12. Following electrophoresis remove the gel from the electrophoresis apparatus and view on a UV transilluminator.
13. Photograph the gel using a polaroid camera or make a record using a gel documentation system (*see* **Notes 7** and **8**).

4. Notes

1. Many different types of agarose for gel electrophoresis exist for general purposes; high-sieving agarose (such as Metaphor, Flowgen, FMC Corp., Litchfield, UK, which is able to resolve small base deletions); higher strength agarose for separating large fragments and low melting point gels for recovery of DNA fragments following separation. The exact concentration of gel depends on the DNA being separated and on the particular application.
2. TAE gel buffer systems are preferred to TBE systems when post-separation methods such as extraction with solid matrices (e.g., Geneclean) are to be used. However, there is little difference between the two systems for general-purpose separation of DNA.
3. Loading buffer is at 6X concentration.
4. If the samples tend to float out of the well add an additional 1–2 vol gel loading buffer and reload the sample.
5. At this point it is possible to add 2.5 µL of ethidium bromide (10 mg/mL) per 100 mL of the cooled gel solution and mix gently. This obviates the need to stain the gel following electrophoresis and saves time. However, many workers prefer to contain the ethidium bromide.
6. Discard the solution containing ethidium bromide in accordance with appropriate health and safety regulations. Remember to wear gloves when handling any solution containing ethidium bromide, which is a potent mutagen and potential carcinogen.

7. It is possible to estimate more precisely the length of a DNA fragment resolved in an agarose gel by using semi-log plots. Here the distances of all the molecular weight markers are measured relative to the front edge of the wells. Each of the distances travelled by each marker is then plotted on semilogarithmic graph paper against relative migration and a straight line drawn through the points. The length of the unknown fragment may then be estimated.
8. One common feature that may sometimes be overlooked is that larger fragments of DNA separated by gel electrophoresis bind more ethidium bromide than smaller fragments. Thus, with a restriction digest separated in an agarose gel, the intensity of the staining decreases toward the bottom of the gel. Smaller fragments may be difficult to visualize, in which case it is advisable to restain the gel for a longer period of time.

4.1. Recovery of DNA Fragments from Agarose Gels

In a number of procedures, such as subcloning or the preparation of DNA probes, fragments need to be purified from bands in agarose gels. Older methods involved some of the following approaches:

1. Running the DNA into a well cut in the gel in front of the band of interest, followed by phenol/chloroform extraction to remove contaminating agarose.
2. Binding of the DNA band to a piece of diethylaminoethyl (DEAE) cellulose filter paper inserted into the gel, followed by elution with phosphate buffer.
3. Solubilization of a block of low melting point agarose containing the desired band by heating to 60°C (or digestion of the gel with agarase) followed by phenol/chloroform extraction.
4. Electroelution of the DNA from a gel fragment into a small dialysis bag using the gel apparatus, again followed by phenol/chloroform extraction.

These kinds of procedures have largely been superseded by proprietary kits that involve binding the DNA to a solid matrix, washing away contaminants, and finally eluting the purified DNA off the matrix, which is usually retained by a small filter system. Such methods do not require solvent extractions, but do not give much better yields. Kits which work in this way include Geneclean (BIO 101, Vista, CA), Glassmax (Gibco-BRL), and a number of others, from Roche Molecular Biochemicals (Lewes, UK), Qiagen GmBH (Hilden, Germany), and Amersham Pharmacia Biotech (Amersham, UK), which vary depending on the matrix and elution technique.

4.2. Extensions of the Technique

With the addition of a few extra steps, standard agarose gels can be adapted to look for structural features in DNA molecules. (These adaptations can also be used with polyacrylamide gels.)

1. **Gels to Analyze DNA Geometry.** Regions of DNA including static curves or bends are often important in gene regulation. Such regions can be identified in ladders of DNA fragments because of their lower mobility at low temperature. If a standard gel is run at 4°C, turned through 90°, and run at 55°C, then fragments including bends will appear as bands above the diagonal of ordinary fragments produced. Bent DNA may also be detected by its effect on the mobility of fragments of equal length *(2)*.
2. **Gel Retardation.** If a DNA-binding protein is preincubated with DNA fragments, some of which contain the specific binding site for the protein, then a DNA/protein complex

will form that has a lower mobility than the unbound DNA fragment in a gel. As a result, on a gel run with adjacent lanes containing fragments in the presence or absence of protein, a band retarded when the protein is present represents a fragment that incorporates its binding site *(3)*.

References

1. Boffey, S. A. (1984) Isolation of high molecular weight DNA, in *Methods in Molecular Biology*, vol. 2: *Nucleic Acids* (Walker, J. M., ed.), Humana, Totowa, NJ, 333–341.
2. Harrington, R. E. (1993) Studies of DNA bending and flexibility using gel-electrophoresis. *Electrophoresis* **14**, 732–746.
3. Lane, D., Prentki, P., and Chandler, M. (1992) Use of gel retardation to analyse protein-nucleic acid interactions. *Microbiological Reviews* **56**, 509–528.

14

Preparation of RNA Dot Blots

Rachel Hodge

1. Introduction

RNA dot hybridizations were first described by Kafatos et al. *(1)*. These hybridizations allow rapid analysis of mRNA expression and are particularly useful in the initial characterization of clones derived from differentially expressed genes. Where accurate quantification of transcription is necessary, or many samples have to be handled, filtration manifold systems are available such as the Millipore MilliBlot system (Bedford, MA), which uses a vacuum source to transfer nucleic acid to filter.

Although it is possible to use a pure nitrocellulose membrane matrix, the nylon-based, nitrocellulose-coated membranes currently available are much easier to handle. In addition, nitrocellulose membranes are unsuitable for use with many of the nonradioactive systems now used extensively. This protocol is a slightly modified version of that supplied with Hybond N (Amersham International, Piscataway, NJ).

2. Materials

All chemicals used should be analytical reagent grade. Solutions should be treated with diethyl pyrocarbonate (DEPC, Merck Ltd, Poole, UK) prior to autoclaving (*see* **Note 1**), to destroy ribonuclease activity. Gloves should be worn at all times, particularly when handling membranes and RNA solutions.

1. Membrane: Hybond N (Amersham).
2. Sterile DEPC-treated water.
3. 10X standard sodium citrate (SSC): 1.5M Sodium chloride, 0.15M tri-sodium citrate.
4. RNA: Total RNA can be prepared by a number of methods and should be checked for degradation by electrophoresis. An RNA concentration of 10 mg/mL or above is ideal (*see* **Note 2**).
5. 10X MOPS: 0.2M 3-[*N*-morpholino] propanesulphonic acid (MOPS) sodium salt, 90 mM sodium acetate, 10 mM ethylenediaminetetraacetic acid (EDTA) disodium salt, pH to 7.0 with sodium hydroxide. Store in the dark at 4°C.
6. RNA incubation solution: To prepare 1 mL, mix 657 µL formamide, 210 µL 37% formaldehyde solution (37% w/v as supplied), and 133 µL 10x MOPS (*see* **Note 3**). This solution may be prepared fresh or stored at –20°C indefinitely. Both formamide and formaldehyde are toxic and should be handled with care in a fume hood.
7. Hair dryer (optional).

3. Method

1. Membrane should be cut to a suitable size (allow 1 cm² per RNA dot) and marked (with a pencil) to show orientation and positions for sample loading (*see* **Note 4**).
2. Wet the membrane by laying it on the surface of distilled water and then wash briefly in 10X SSC. Air dry thoroughly.
3. Thaw the RNA samples on ice and transfer appropriate amounts (*see* **Notes 5** and **6**) to fresh tubes containing a 3X volume of RNA incubation solution. Mix thoroughly.
4. Heat the RNA samples at 65°C for 5 min to denature RNA secondary structure. Cool on ice.
5. Add an equal volume of ice-cold 20X SSC and mix thoroughly.
6. Dot the RNA solution onto the membrane in 2 μL aliquots using a Gilson Pipetman (Anachem Ltd, Luton, UK). Dry the membrane between each loading. To speed up the process it is possible to use a hair dryer to dry the membrane between loadings.
7. After loading the last sample, dry the membrane thoroughly, wrap in Saran wrap (Dow Chemical Co., Midland, MI) and crosslink the RNA to the membrane using an ultraviolet cross-linker. Alternatively, the membrane can be dried, wrapped in Saran wrap (Dow), and cross-linked on an ultraviolet transilluminator (302 nm)—RNA side down—for 90 s (*see* **Note 7**).
8. The membrane can then be used in hybridization experiments immediately or stored, wrapped in Saran wrap at room temperature for up to 1 mo.

4. Notes

1. Before treating solutions with DEPC, add 0.1% diethyl pyrocarbonate, shake to disperse, and incubate at 37°C for 2 h. The solution should then be autoclaved to destroy the DEPC, which, if present, may carboxymethylate purine residues in the RNA. DEPC is highly flammable and should be handled in a fume hood. It is also suspected to be a carcinogen and so should be treated with respect.
2. Because of the subsequent eightfold dilution of the sample adding incubation solution and 10X SSC, it is highly advisable to have RNA preparations at a concentration of 10 mg/mL or above. Preparations of less than 10 mg/mL should be ethanol precipitated and redissolved at higher concentrations.
3. MOPS buffer is used where formaldehyde is a component of the incubation mixture. Tris buffers are not suitable due to the reactive amine group.
4. Great care should taken when handling membranes. Pick up the membranes by the corners with forceps and avoid touching the membranes even while wearing gloves.
5. The amount of RNA loaded per dot is obviously dependent on the abundance of the transcript in the RNA population. For a cDNA clone identified by differential screening (and hence of high abundance in the mRNA pool), a loading of 5 μg total RNA per dot should be sufficient. For medium- and low-abundance transcripts it may be necessary to load more total RNA or isolate polyA⁺ RNA for efficient detection of transcript.
6. When duplicate filters are being prepared, master mixes of RNA, incubation solution, and 10× SSC should be made for each RNA sample.
7. The appropriate exposure time for cross-linking varies with the wavelength and age of the ultraviolet bulbs in the transilluminator. To establish the optimum exposure time, a number of duplicate filters should be prepared and exposed for different lengths of time (20 s–5 min). Subsequently, filters should be hybridized, washed, and developed together. The filter giving the strongest signal indicates the optimum exposure time for cross-linking.

Reference

1. Kafatos, F. C., Jones, C. W., and Efstratiadis, A. (1979) Determination of nucleic acid sequence homologies and relative concentrations by a dot hybridization procedure. *Nucleic Acid Res.* **7,** 1541–1552.

15

Native Polyacrylamide Gel Electrophoresis

Adrian J. Harwood

1. Introduction

Agarose gel electrophoresis is generally adequate for resolving nucleic acid fragments in the size range of 100 nucleotides to around 10–15 kb (*see* Chapter 13). Below this range, fragments are both difficult to separate and hard to visualize because of diffusion within the gel matrix. These problems are solved by native polyacrylamide gel electrophoresis (PAGE). Using native PAGE, fragments as small as 10 bp and up to 1 kb can be separated with a resolution of as little as 1 bp.

Native PAGE also has a number of other advantages. It has a high loading capacity; up to 10 µg of DNA can be loaded into a single well (1 cm × 1 mm) without significant loss of resolution. Polyacrylamide contains few inhibitors of enzymatic reactions, and so PAGE is an ideal gel system from which to isolate DNA fragments for subcloning and other molecular biological techniques. It has two disadvantages: (i) The mobility of the fragments can be affected by base composition making accurate sizing of bands a problem and (ii) polyacrylamide quenches fluorescence, making bands containing less than 25 ng difficult to visualize with ethidium bromide staining; alternative means of visualizing DNA fragments are discussed in **Note 1**.

In this chapter, the preparation and use of native PAGE gels is described, as well as a method for gel purification from polyacrylamide gel slices (*see* **Note 2**).

2. Materials

1. Gel apparatus: Many designs of apparatus are commercially available. The gel is poured between two vertical plates held apart by spacers (*see* **Note 3**).
2. Deionized H$_2$O: Autoclaved water is not necessary for the gel mix or running buffer, but it should be used for diluting samples and purification from gel slices.
3. 10x TBE: 108 g of Trizma base (Tris) (Sigma Chemicals, St. Louis, MO), 55 g of boric acid, and 9.3 g of ethylenediaminetetraacetic acid (EDTA) (disodium salt). Make up to 1 L solution with deionized H$_2$O, which should be discarded when a precipitate forms.
4. Acrylamide stock: 30% acrylamide, 1% *N,N*′-methylene bisacrylamide. Store at 4°C. This is available commercially, or it can be made by dissolving acrylamide and bisacrylamide in water, which should be filtered. Acrylamide is a neurotoxin and therefore must be handled carefully. Gloves and a mask must be worn when weighing out.
5. APS: 10% Ammonium persulphate (w/v). This can be stored at 4°C for 1–2 mo.

From: *The Nucleic Acid Protocols Handbook*
Edited by: R. Rapley © Humana Press Inc., Totowa, NJ

Table 1
Gel Mixes for 3.5, 5, and 12% Polyacrylamide Gels

Acrylamide concentration	3.5%	5%	12%
10× TBE	2.5 mL	2.5 mL	2.5 mL
Acrylamide stock	5.8 mL	8.3 mL	20.0 mL
dH$_2$O	41.3 mL	38.8 mL	27.1 mL
APS	350.0 µL	350.0 µL	350.0 µL
Effective range of separation	100–2000 bp	80–500 bp	40–200 bp

6. TEMED: N,N,N',N'-tetramethyl-1,2-diaminoethane. Store at 4°C.
7. 5X sample buffer: 15% Ficoll solution, 2.5X TBE, 0.25% (w/v) xylene cyanol and 0.025% (w/v) bromophenol blue.
8. Ethidium bromide: A 10-mg/mL solution. Ethidium bromide is a potent mutagen and should be handled with care. Store at 4°C in the dark.

2.1. Purification of DNA Fragments

9. Elution buffer: 0.5 M Ammonium acetate; 1 mM EDTA, pH 8.
10. TE: 10 mM Tris-HCl, pH 7.5, 1 mM EDTA. Sterilize by autoclaving.
11. Sodium acetate: 3 M Sodium acetate, pH 5.2.

3. Methods
3.1. Separation of DNA Fragments

1. For 50 mL, enough for a $18 \times 14 \times 0.15$ cm gel, mix $10 \times$ TBE, acrylamide, H$_2$O, and APS as described in **Table 1**.
2. Just prior to pouring, add 50 µL of TEMED and mix by swirling.
3. Immediately pour the gel mix between the gel plates and insert the gel comb. Leave to set; this takes about 30 min.
4. Fill the gel apparatus with 0.5X TBE and remove the comb. Use a syringe to wash out the wells, this may take multiple washes. It is important to remove as much unpolymerized acrylamide as possible because this impairs the running in of the samples (*see* **Note 4**).
5. Add 0.2 volume of 5× sample buffer to each sample, usually in 10–20 µL of TE, water, or enzyme buffer. Mix and spin the contents to the bottom of the tube (*see* **Note 5**).
6. Load the samples on the gel and run at 200–300 V (approximately 10 V/cm) until the bromophenol blue band is two-thirds of the way down the gel; this takes about 2.5 h (*see* **Note 6**).
7. Disassemble the gel apparatus and place the gel to stain in 1 mg/mL of ethidium bromide for approximately 30 min. View the stained gel on a transilluminator. Alternative visualization methods are described in **Notes 1** and **7** and Chapter 19.

3.2. Purification of DNA Fragments

Several methods are available to extract DNA from polyacrylamide gel slices. A simple and effective way of purifying DNA from both native and denaturing polyacrylamide gels is by using the "crush and soak" method (*1*).

1. Cut the DNA band from the gel using a new scalpel blade and place in a 1.5-mL tube.
2. Break up the polyacrylamide gel slice with a yellow pipet tip and add between 1–2 vol of elution buffer.

3. Seal the tube with Parafilm® (American National Can™), and incubate at 4°C on a rotating wheel for 3–4 h for fragments smaller than 500 bp or overnight for larger fragments.
4. Spin out the lumps of polyacrylamide by centrifugation at 10,000g for 10 min and carefully remove the supernatant. Add an additional 0.5 vol of elution buffer to the pellet, vortex briefly, and recentrifuge. Combine the two supernatants.
5. Add 2 vol of ethanol and place on ice for 10 min. Centrifuge at 12,000g for 15 min.
6. Redissolve the pellet in 200 µL of TE and add 25 µL of 3 M sodium acetate. Add 400 µL of ethanol and precipitate as in **step 5**.
7. Rinse the pellet once with 70% ethanol and air dry.
8. Resuspend in an appropriate volume of TE and check the yield by gel electrophoresis.

4. Notes

1. High concentrations of DNA, e.g., as experienced when purifying oligonucleotides, can be visualized without ethidium bromide staining by ultraviolet (UV) shadowing. This is carried out by wrapping the gel in a UV-transparent plastic film, such as Saran Wrap, and then placing it onto a thin-layer chromatography plate that contains a UV fluorescent indicator (Merck 60F254, cat. no. 5554, Rahway, NJ). Long-wave UV light is shone through the gel onto the chromatography plate, causing it to glow. Regions of high DNA concentration leave a "shadow" on the plate as the transmitted UV is absorbed by the DNA. The position of the DNA can be marked on the Saran Wrap with a fiber-tip pen and then cut from the gel.
2. Recently, a meltable acrylamide substitute has been developed that potentially simplifies gel purification (available as Oligoprep, product no. EC-857 from National Diagnostics Inc., Atlanta, GA, USA). At present, this product has not been available long enough to fully assess its utility.
3. Grease and dirt on the plates can cause bubbles to form while the gel is poured. The plates, therefore, should be cleaned thoroughly and then wiped with ethanol. To help ensure that the gel only sticks to one plate when the apparatus is disassembled, apply silicon to one of the gel plates. This is easily done by wiping the plate with a tissue soaked in dimethyl dichlorosilane solution and then washing the plate in distilled water followed by ethanol. If the plates are baked at 100°C for 30 min, the siliconization will last four to five gel runs.
4. If you are separating very small fragments, e.g., less than 50 bp, the gel should be prerun for 30 min, as this elevates the resolution problem experienced with fragments running close to the electrophoresis front.
5. High-salt buffers (above 50 mM NaCl) will affect sample mobility and tend to make bands collapse. In this case, salt should be removed by ethanol precipitation.
6. Do not run the gel faster than 10 V/cm, as this will cause the gel to overheat, affecting the resolution. The gel can be run more slowly, e.g., 75 V will run overnight.
7. If the samples are radiolabeled (*see* Chapter 19), the gel should be fixed in 10% acetic acid, transferred to 3 MM paper (Whatman Int'l Ltd., Kent, UK) and dried. The dried gel is autoradiographed. If a band is to be isolated from the gel, it can be wrapped in plastic film and autoradiographed wet. The autoradiograph is then aligned with the gel and the band cut out and purified, as in **Subheading 3.2.**

Reference

1. Maxam, A. M. and Gilbert, W. (1977) A new method for sequencing DNA. *Proc. Natl. Acad. Sci. USA* **73**, 668–671.

16

Southern Blotting of Agarose Gels by Capillary Transfer

Ralph Rapley and Jane Davenport-Jones

1. Introduction

The detection of specific nucleic acid species following electrophoretic separation of a complex sample may be undertaken by the use of Southern blotting *(1)*. This technique immobilizes the separated DNA following its digestion with restriction enzymes. The separation is usually carried out using agarose gel electrophoresis (*see* Chapter 13).

The DNA is then partially cleaved by depurination, which facilitates the transfer of larger DNA fragments and is treated with alkali by sequential soaking of the gel in solutions containing HCl and NaOH, respectively. This denatures the double-stranded DNA to produce single strands that may be subsequently probed with an appropriately labeled single-stranded DNA fragment *(2)*.

Traditionally, the DNA is transferred to nitrocellulose, although now the membrane is usually constructed of nylon. Nylon has an improved capacity for DNA binding and is more robust, allowing reprobing to be undertaken. The simplest and cheapest method of transfer is by capillary action, where a solution is drawn through the gel that carries the nucleic acid fragments with it, and therefore provides a membrane with immobilized DNA, which represents the fragments' original positions on the agarose gel. There are alternative methods of transfer, such as vacuum blotting or electroblotting, which may provide a more even transfer and reduce the time but which are also more expensive. Following transfer, the DNA is fixed to the membrane by illumination with UV irradiation, after which the blot may be stored or probed.

2. Materials

1. Suitable apparatus for blotting, two buffer tanks, paper towels, Whatman 3MM paper, Maidstone, Kent, UK (*see* **Fig. 1**).
2. Nylon hybridization membrane (e.g., Hybond N+, Amersham International, Amersham, UK).
3. Depurination buffer: 0.25 *M* HCl.
4. Denaturation buffer: 1.5 *M* NaCl, 0.5 *M* NaOH.
5. Transfer buffer: 1.5 *M* NaCl, 0.25 *M* NaOH.
6. 20X standard saline citrate (SSC): 3 *M* NaCl, 0.3 *M* trisodium citrate, pH 7.0.
7. Ultraviolet light transilluminator, 312 nm output.
8. Fixing solution: 0.4 *M* NaOH.
9. Rinsing solution: 5X SSC.

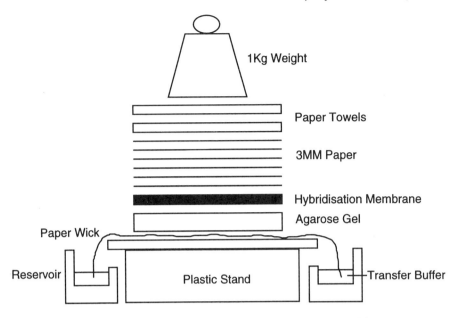

Fig. 1. A typical set-up for capillary action Southern blot.

3. Methods

1. Isolate DNA with an appropriate extraction technique and digest with a restriction endonuclease (*see* Chapters 1 and 12).
2. Separate the digested DNA by agarose gel electrophoresis (*see* Chapter 13 [*see* **Notes 1** and **2**]).
3. Clean up the gel by removing areas of the gel with a clean scalpel blade (e.g., the wells).
4. Incubate the gel in approx 3 gel vol of depurination buffer and agitate gently on an orbital shaker at room temperature for approx 30 min (*see* **Note 3**).
5. Rinse the gel in distilled water.
6. Place the gel in 3 vol of denaturation buffer. Incubate the mixture with gentle agitation at room temperature for 30 min.
7. Decant the denaturation buffer, and replace with 3 gel vol of transfer buffer. Equilibrate the gel with gentle agitation at room temperature for 30 min.
8. Place the gel on the platform of a capillary transfer system filled with transfer buffer (*see* **Fig. 1**). Place the transfer buffer in the reservoir.
9. Soak the Whatman 3MM paper in the transfer buffer.
10. Make up a wick from three sheets of 3MM paper, and place this over the platform. All air bubbles must be removed from the wick, which may be done by gently smoothing over the paper. The width and length of the platform must correspond to the size of the gel, and the wick is cut to the same width. The platform and reservoir are made to the same height.
11. Cut a piece of nylon membrane to the same size as the gel. Be careful to hold the membrane at the edges and wear gloves.
12. Wet the membrane by floating it on distilled water and then rinse it in transfer buffer. Place the membrane on the gel, and smooth out any air bubbles (*see* **Note 4**).
13. Place three sheets of Whatman 3MM paper cut to size and wetted in transfer buffer on top of the membrane.
14. Place a stack of absorbent paper on top of the Whatman 3MM paper and place a 1-kg weight on top, allowing the transfer to proceed for at least 12 h (*see* **Note 5**).

15. Disassemble the gel and membrane from the transfer system. Before separating the gel and the membrane, mark the position of the gel slots with a pencil. This will allow orientation later, as the marks will appear on the resulting autoradiograph (*see* **Note 6**).
16. Rinse the filter in 2X SSC.
17. Covalently cross-link the DNA to the matrix by exposure to a 312-nm ultraviolet light transilluminator. Place the filter, DNA side down, on a piece of clingfilm, and expose for 2–3 min (*see* **Notes 7** and **8**). The filter can be used immediately or stored dry until required.

4. Notes

1. The resolution of the separation may be improved by performing prolonged gel runs (e.g., overnight) at low voltages.
2. It is possible to determine the size of the hybridizing band following autoradiography by comparing it with standard or marker DNA (e.g., 1 DNA/*Hin*dIII, 1 kb marker, 123 bp marker). This needs to be end labeled with a radioactive or nonradioactive marker (*see* Chapter 21).
3. When the xylene cyanol loading dye changes color to a greenish color, or the bromophenol blue turns yellow, the depurination buffer may be removed.
4. Avoid trapping air bubbles in the sandwich, as this gives an uneven transfer of the DNA.
5. It is sometimes useful to put a glass plate on top of the stack in order to distribute the weight evenly. This promotes a more even transfer of the DNA.
6. An alternative to the capillary system includes vacuum blotting or electroblotting. There are a number of manufacturers that produce equipment for this purpose, and although they are more expensive, they reduce the transfer process to as little as 1 h. In some cases a more even transfer takes place.
7. For neutral nylon membranes (e.g., Hybond N, Amersham International) cross-linking is necessary. However, for positively charged membranes (e.g., Amersham Hybond N$^+$) cross-linking may be undertaken by placing the membrane in 0.4*N* NaOH for 30 min and rinsing in 5X SSC with gentle agitation for 1 min. If you are using nitrocellulose it is necessary to bake the filter at 80°C for 20–60 min.
8. Efficient crosslinking of DNA to nylon filters is achieved with an optimal amount of exposure to UV light. Some manufacturers (e.g., Stratagene, La Jolla, CA) produce UV crosslinkers (Stratalinker, Stratagene) that exposes the filter to the radiation for the optimal amount of time. It is useful if no equipment such as this is available to calibrate a UV source before use. This can be done by exposing filters with an identical amounts of DNA on each to UV for different lengths of time. Hybridization to the same probe will reveal the strongest signal that may be used to establish the optimal time for exposure. With a standard UV transilluminator, regular recalibration is required.

References

1. Southern, E. M. (1975) Detection of specific sequences among DNA fragments separated by gel electrophoresis. *J. Mol. Biol.* **98,** 503–517.
2. Evans, M. R., Bertera, A. L., and Harris, D. W. (1994) The Southern blot: an update. *Mol. Biotechnol.* **1,** 1–12.

17

Pulsed-Field Gel Electrophoresis

John Maule

1. Introduction

DNA undergoing electrophoresis in agarose assumes a conformation that only permits the movement of molecules up to about 20 kb in size. Beyond this limit, mobility rapidly decreases as the molecules become trapped in the agarose matrix. A reduction in agarose concentration to 0.5% and the application of a reduced voltage permits the resolution to be increased to a least 50 kb *(1)*, but gels of this concentration and less are too fragile and difficult to manipulate and run times become excessively long. Since the introduction of pulsed-field gel electrophoresis (PFGE) in 1983 *(2)*, this technique has been used to separate molecules as large as 12 Mb *(3)*. The technique relies on a direct current (D.C.) electric field that periodically changes direction and/or intensity relative to the agarose gel. The time interval during which the field is in any one direction is called the pulse time, and its duration is the single most important factor in determining the molecular size range over which separation is possible. In simplistic terms, large DNA molecules are able to migrate through the agarose matrix by zigzagging in response to changes in the electric field and this process has been confirmed by direct visualization *(4–6)*. Short pulse times, resulting in rapid changes in field direction, allow smaller molecules to migrate, whereas larger molecules cannot respond quickly enough and become trapped in the matrix. As the pulse time is increased, progressively larger molecules are able to migrate, but the resolution of smaller molecules is decreased. The size window of separation is influenced primarily by the choice of pulse time.

PFGE was initially used to electrophoretically karyotype a variety of prokaryotes *(7)*, but more recently the technique has been refined to become a standard molecular biology tool in a broad range of investigations (reviewed in *(8)*). Physically mapping eukaryotic genomes by the creation of large insert libraries *(9,10)* and their subsequent analysis by long-range restriction mapping *(11)* is only possible by using PFGE. This technique has also been used to study basic biological phenomena, including the elucidation of different DNA topological forms and the mapping of origins of replication in chromosomal material. The preparation and analysis of large DNA molecules is described in this chapter.

2. Materials
2.1. General Reagents

1. Yeast peptone dextrose (YPD) medium: 20 g bactopeptone (Difco, Surrey, UK), 10 g yeast extract (Difco), 1 L distilled water. Autoclave and when cool, add 50 mL 40% sterile glucose. For YPD agar, add 1.5% agar to the medium.
2. Luria broth (LB) medium: 10 g bactotryptone (Difco), 5 g yeast extract (Difco), 5 g sodium chloride. Adjust pH to 7.2 (with 5 M sodium hydroxide solution) and finally make up the volume to 1 L. Autoclave. For LB agar, add 1.5% agar to the medium.
3. 1% NDS solution: 0.45 M ethylenediaminetetraacetic acid (EDTA), 10 mM Tris-HCl, pH 9, 1% sodium N-lauroyl sarcosine (Sigma, St. Louis, MO). Mix solid EDTA and Tris with 29 g sodium hydroxide pellets in 900 mL water. Cool to room temperature and adjust pH to 9 by adding 5 M sodium hydroxide solution and finally make up to 1 L with distilled water. Autoclave. Add sodium N-lauroyl sarcosine to 1%.
4. Proteinase K (Boehringer Mannheim, Mannheim, Germany): Make up at 20 mg/mL in 1% NDS solution. Store at –20°C.
5. Low-melting-temperature agarose: Several different brands of agarose are suitable for plug formation, but make sure the sulfate content is <0.15%. Boehringer Mannheim LM-MP agarose has given consistently good results. FMC Incert (Rockland, ME) agarose is designed for plug formation but is considerably more expensive than the Boehringer product.
6. Zymolyase 100T: This is an enzyme used to digest yeast cell walls and is available from ICN Biomedicals (High Wycombe, UK). Other suitable enzymes are lyticase L5263 (Sigma), lyticase (Boehringer Mannheim), and yeast lytic enzyme (ICN Biomedicals). The unit definition differs among these various products.
7. Nunc (Roskilde, Denmark): 10 µL plastic disposable inoculating loops are ideal for maneuvering agarose plugs.
8. Fine-tip Pasteur pipets (Pastets, Alpha Laboratories, Surrey, UK): Available, in sterile packs, from several manufacturers specializing in disposable plastic laboratory products. The tip orifice is comparable in diameter to a standard 10–200 µL Eppendorf plastic tip and is, therefore, ideal for aspirating liquid surrounding an agarose plug. It avoids the possibility of sucking the plug up into the pipet.
9. Hemocytometer: Improved Neubauer design (Philip Harris Scientific, Staffordshire, UK), counting cell depth: 0.1 mm.
10. Tris-EDTA (TE): 10 mM Tris-HCl, 1 mM EDTA, pH 8. Make up as 10× stock and autoclave.
11. Bio-Rad (Richmond, CA) supplies a perforated screened cap (part no. 170-3711), that screws on to the top of a 50-mL Falcon tube in place of the normal cap. It is used to retain agarose plugs when decanting liquid from the tube.
12. Size markers, which include the size range below 100 kb, can be purchased from New England Biolabs (Beverly, MA). Midrange markers I and II cover sizes ranging from 15–291 and 24–291 kb, respectively. Bio-Rad supply concatemers of pBR328 ranging in size from 4.9–120 kb.

2.2. Saccharomyces cerevisiae and Hansenula wingei *Chromosome Preparation*

1. 0.125 M EDTA, pH 7.5: Dilute 1/2.5 to give 50 mM solution.
2. Cell wall digestion solution: 2 mL Sorbitol citrate EDTA (SCE) (1 M sorbitol, 0.1 M trisodium citrate, 60 mM EDTA pH 7—autoclave), 0.1 mL 2-mercaptoethanol, 2 mg zymolyase 100T. Store at –20°C.

3. Tris-EDTA-mercaptoethanol (TEM): 10 mM Tris-HCl, pH 8, 0.45 M EDTA, 7.5% 2-mercaptoethanol. Autoclave and add the last compound just before use.
4. Yeast nitrogen base: 10X solution is 6.7 g bacto yeast nitrogen base without amino acids (Difco), 100 mL distilled water. Filter to sterilize and store at 4°C.
5. Adeninehemisulfate casamino (AHC) medium: 10 g casamino acids (Difco), 50 mg adenine hemisulfate (Sigma). Adjust pH to 5.8 with 5 M hydrochloric acid and make up to 850 mL with distilled water. Autoclave. When medium is cool, add 50 mL sterile 40% glucose and 100 mL 10X yeast nitrogen base.
6. Hogness freezing medium: 10X stock is 36 mM K_2HPO_4, 13 mM KH_2PO_4, 20 mM trisodium citrate, 10 mM $MgSO_4$, 44% glycerol.

2.3. Schizosaccharomyces pombe *Chromosome Preparation*

1. 0.125 M EDTA, pH 7.5: Dilute to 50 mM for washing cell pellets.
2. Citrate phosphate EDTA sorbitol (CPES): 40 mM citric acid, 120 mM Na_2HPO_4, 20 mM EDTA, 1.2 M sorbitol, 5 mM dithiothreitol, pH 6. Autoclave and add the latter reagent just before use.
3. Novozym 234 (Calbiochem, La Jolla, CA): This is a cell-wall–digesting enzyme.
4. 0.1% Sodium dodecyl sulfate (SDS).
5. 0.125 M EDTA, 0.9 M sorbitol, pH 7.5—autoclave.

2.4. Lambda Concatemers Preparation

1. Lysogen N1323(λ) *(12)* carries a temperature-sensitive repressor (*cIts*), and thus can be induced at the nonpermissive temperature of 43°C. The *S* gene product is required for cell lysis, and this strain carries an amber mutation in this gene, which is not suppressed. This results in the accumulation of several hundred phage per cell, following induction and subsequent incubation. The cells can then be concentrated by centrifugation prior to lysis with chloroform; *lop*8 is a ligase-overproducing mutation that may be beneficial in the repair of single-strand breaks following induction and prolonged incubation.
2. RNase A (Sigma): Make up at 10 mg/mL in 10 mM Tris-HCl, pH 7.5, 15 mM NaCl. Heat to 100°C for 15 min. Allow to cool to room temperature, then freeze.
3. DNase 1 (Sigma): Make up at 20 mg/mL. Store at –20°C.
4. EDTA pH 8: Make up at 0.5 M and autoclave. Dilute to 0.1 M as per **Subheading 3.3., step 17**.

2.5. Mammalian Chromosome Preparation

Phosphate-buffered saline (PBS): Dissolve 0.2 g KCl, 0.2 g KH_2PO_4, and 8 g NaCl in 800 mL water. Adjust pH to 7.2 and make up to 1 L with water.

2.6. Agarose Plug Preparation

1. A plug mold can be made from a rigid polystyrene flat-bottomed 96-well microtiter plate (ICN Flow, catalog no. 76-307-05) by milling off the base or drilling out the individual wells. A Titertek plate sealing strip (ICN Flow, catalog no. 77-400-05) is secured across the top and the plate is inverted so that the base is now uppermost.
2. The mixture of cells and molten agarose is poured into a reagent trough (ICN Flow, catalog no. 77-824-01) prior to dispensing.

2.7. Agarose Bead Preparation

1. Polystyrene-dyed microparticles (Polysciences, Warrington, PA) 0.2 µm colored red (catalog no. 15705), blue (catalog no. 15706), and yellow (catalog no. 15707) (*see* **Note 1**). Microparticles are prepared as inert suspensions.

2. Liquid paraffin (The Boots Company, Nottingham, UK).
3. LiDS solution: 1% Lithium dodecyl sulfate (Sigma), 100 mM EDTA, 10 mM Tris-HCl pH 8. Dissolve EDTA and Tris and adjust the pH. Autoclave. Add lithium dodecyl sulfate (LiDS) to 1% from a filter-sterilized 20% stock solution.
4. 50 mM EDTA pH 8: Dilute sterile 0.5 M stock solution 1/10.

2.8. Digestion of Agarose-Embedded DNA

1. Phenylmethylsufonyl fluoride (PMSF) is a potent protease inhibitor and is used to destroy residual proteinase K. Great care must be exercised in handling this substance. The solid becomes electrostatically charged, so use a wooden spatula while weighing out and wear gloves throughout. Add 1 mL of propan-2-ol to 20 mg of the solid and dissolve by incubating at 50°C for 3 min. This solution is stable and may be stored frozen at −20°C (*see* **Note 2**).
2. Triton X-100: Prepare as a 10% stock solution and autoclave.
3. Bovine serum albumin (BSA): Special molecular biology grade, supplied at 20 mg in 1 mL (Boehringer Mannheim).
4. Stop buffer: 0.5× TBE, 10 mM EDTA, 2 mg/mL Orange G (Sigma).
5. Restriction enzyme buffers: Because considerable volumes are required for washing and equilibrating plugs, it is wise to prepare 10X stock solutions and autoclave.
6. Tris-EDTA-Triton X (TEX): 10 mM Tris-HCl, 1 mM EDTA pH 8. Autoclave and then add sterile Triton X-100 to 0.01%.

2.9. Separating DNA Using Contour-Clamped Homogeneous Electric Fields (CHEF)

1. Tris-boric acid-EDTA (TBE): Prepare as 5X TBE containing 450 mM Tris, 450 mM boric acid, and 10 mM EDTA.
2. Tris-acetate-EDTA (TAE): Prepare as 20X TAE containing 800 mM Tris, 400 mM sodium acetate, and 20 mM EDTA. Adjust the pH to 8.2 with acetic acid.
3. Agarose: The standard medium electroendosmosis (EEO) agarose used for conventional gels can be used for PFGE. Special PFGE agaroses are available that exhibit high gel strength and low EEO and, therefore, can be used at low concentrations. Under these conditions, faster separations times are achievable. The following PFGE agaroses are available: Fastlane and Seakem Gold (FMC), Chromosomal-grade and pulsed-field certified agarose (Bio-Rad), Boehringer multipurpose agarose, Pulsed field grade agarose (Stratagene, La Jolla CA), Kilorose and Megarose agaroses (Clontech, Palo Alto, CA) and Rapid agarose (Gibco-BRL, Gaithersburg, MD). These can be more expensive than standard-grade agarose.
4. Ethidium bromide is a mutagen and should be handled with care. Always wear gloves. Bio-Rad sells ethidium bromide tablets (each tablet makes 11 mL of 1 mg/mL solution), which reduces the danger of spillage of the solid during preparation of the solution.
5. Commercially available CHEF apparatus is available from Bio-Rad (CHEF DR-111 and CHEF Mapper) and Pharmacia (Brussels, Belgium) (Gene Navigator System) (*see* **Note 3**).
6. Details of how to construct a large gel format homemade CHEF apparatus are available from the author *(13)*. The Bio-Rad cooling module and variable speed circulating pump can be used with this apparatus.

2.10. Southern Transfer

1. HCl: Make up as 5 M (86 mL of concentrated HCl made up to 200 mL with water) and dilute 1/20 prior to use.

2. Denaturant: 0.5 M NaOH, 1.5 M NaCl.
3. Neutralizer: 1 M Tris, 2 M NaCl. Adjust pH to 5.5 with HCl.
4. Standard saline citrate (SSC): 20X stock is 3 M NaCl, 0.3 M trisodium citrate, pH 7.4.
5. Stratalinker (Stratagene) is ideal for ultraviolet crosslinking membranes.

2.11. Hybridization and Autoradiography

1. High prime DNA labeling kit (Boehringer).
2. Redivue deoxycytidine 5'-[α-^{32}P] triphosphate (≅3000 Ci/mmol) Amersham Pharmacia Biotech (Bucks, UK) catalog number AA0005.
3. Whatman (Maidstone, UK) GF/B 2.4 cm filter circles.
4. Trichloroacetic acid (TCA): Make up as 50% stock solution and dilute 1/10.
5. NICK columns (Pharmacia): prepacked with Sephadex G-50.
6. 10 × TNE: 100 mM Tris-HCl, pH 8, 10 mM EDTA, 2 M NaCl. Autoclave.
7. Sonicated salmon sperm DNA (Sigma) (D 1626): Make up at 10 mg/mL in water by stirring overnight at 4°C. Sonicate sufficiently to achieve a size of 600 bp. Store at –20°C. This reagent is used as a blocking agent to reduce nonspecific hybridization.
8. Filter hybridization mix: 5X Denhardt's solution, 5X SSC, 0.1% disodium pyrophosphate, 0.5% SDS, 10% sodium dextran sulfate (Pharmacia). Denhardt's 20X stock contains, per 100 mL, 0.4 g Ficoll 400 (Pharmacia), 0.4 g polyvinylpyrrolidone, 0.4 g BSA (Sigma fraction V). 20X SSC: *see* **Subheading 2.10.**, **item 4**. Store at 4°C.
9. SSPE: 5X stock is 750 mM NaCl, 50 mM NaH$_2$P0$_4$, 5 mM EDTA. Adjust pH to 7.4 with NaOH, and add SDS to 0.1%.
10. SSC washes contain 0.1% SDS and 0.1% disodium pyrophosphate. Maintain at 68°C until required.
11. Autoradiography film: Use Kodak XAR-5.

3. Methods

3.1. Preparation of Chromosomal DNA from S. cerevisiae and H. wingei

Yeast chromosomal DNA provides useful size markers for PFGE. Wild-type *S. cerevisiae* strains have 16 chromosomes ranging in size from 240–2200 kb, although the strain YP148 has an additional 90-kb chromosome. *H. wingei* has 7 chromosomes ranging in size from 1030–3300 kb *(14)*. This method should also be followed for the production of yeast artificial chromosome (YAC) DNA plugs.

1. Pick a single colony from a freshly grown culture, streaked on a YPD plate, and inoculate 100 mL of YPD medium in a 500-mL flask (*see* **Note 4**). YACs should be grown in AHC medium (*see* **Notes 5** and **6**).
2. Shake for 24 h at 33°C, at approx 200 rpm.
3. Dilute an aliquot 1/10 in YPD and count the number of cells using a hemocytometer. Cell count should be approx 1×10^8 cells/mL.
4. Chill the culture on ice for 15 min, and then harvest the cells by spinning at 2000g for 10 min at 4°C.
5. Discard the supernatant and gently disrupt the pellet with a sterile loop before adding 50 mL of chilled 50 mM EDTA, pH 7.5. Make sure the cells are thoroughly dispersed.
6. Spin at 2000g for 5 min at 4°C.
7. Repeat **steps 5** and **6**.
8. Finally, discard the supernatant and take up the pellet in 3 mL of ice-cold 50 mM EDTA, pH 7.5 (gives a final vol of approx 3.5 mL).

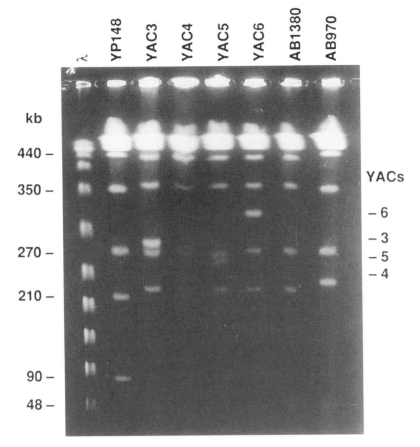

Fig. 1. Separation of yeast chromosomes, YACs, and λ concatemers by CHEF gel PFGE. The samples were loaded on to a 1% agarose gel in 0.5 × TBE. The gel was run at 6 V/cm with a 25-s pulse time for 36 h at 14°C. The positions of the four YACs are indicated. AB1380 is the S. cerevisiae host strain in which the YACs are maintained. The unresolved DNA above the 440-kb marker is called the compression zone (CZ).

9. Transfer the cells to a 20-mL universal container, with a fine-tip sterile Pastet (to disrupt any cells that are clumped), and warm to 37°C.
10. Add 6 mL of 1% low melting temperature agarose (in 0.125 M EDTA pH 7.5), which has been cooled to 50°C.
11. Finally, add 1.2 mL of cell wall digestion solution.
12. Immediately, mix thoroughly and dispense into plug molds (*see* **Subheading 3.5.**).
13. Eject the plugs into a 50-mL Falcon tube containing 25 mL of TEM solution, and incubate overnight at 37°C in a water bath.
14. Replace TEM with 20 mL 1% NDS containing 1 mg/mL proteinase K (*see* **Subheading 3.5.**).
15. Finally store plugs in 20 mL 1% NDS at 4°C.
16. Plugs should be equilibrated for at least 1 h in gel running buffer before use.

Figures 1 and **2** show the separation of some yeast chromosomes from *S. cerevisiae* and *H. wingei* (**Fig. 2**) and the resolution of six YACs ranging in size from 230–1500 kb. Yeast chromosome sizes are presented in **Table 1**.

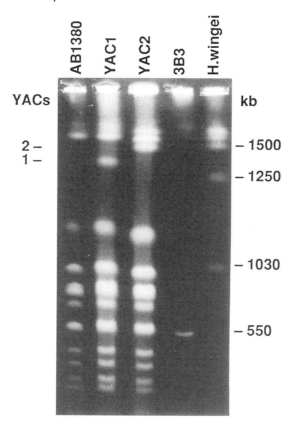

Fig. 2. The separation of large yeast chromosomes and YACs by PFGE. Samples were run on a 1% agarose CHEF gel in 0.5 × TBE at 14°C for 36 h at 6 V/cm with a pulse time of 115 s. The positions of the two YACs (sized at 1450 and 1350 kb) are indicated. *S. pombe* strain 3B3 has a minichromosome of 550 kb. Note that under these run conditions, only the smallest three *H. wingei* chromosomes are separated and the resolution of yeast chromosomes below 1000 kb is less than optimal.

3.2. Preparation of Chromosomal DNA from S. pombe

Wild-type *S. pombe* has three chromosomes ranging in size from 3.5–5.7 Mb. The strain 3B3 has an additional minichromosome of 550 kb. This yeast provides useful size markers in the megabase (Mb) size range.

1. Inoculate 5 mL of YPD medium from a single colony taken from a freshly grown plate (*see* **Notes 7** and **8**).
2. Shake at 30°C in a universal container, overnight.
3. On the next day, add the overnight culture to 100 mL of YPD in a 500-mL flask. Shake at 30°C for 24 h.
4. Dilute an aliquot of the culture 1/10 in YPD and count the number of cells in a hemocytometer. The count should be $3–5 \times 10^7$ cells/mL.
5. Chill the culture on ice for 15 min (*see* **Note 9**).
6. Spin at 2000*g* for 10 min at 4°C.
7. Discard the supernatant and gently disrupt the pellet with a sterile loop before adding 50 mL ice-cold 50 m*M* EDTA, pH 7.5.

Table 1
Chromosome Sizes (kb)

S. cerevisiae AB970	S. cerevisiae YP148	H. wingei	S. pombe 3B3
2200	2200	3300	5700
1640	1640	2900	4600
1130	1125	2600	3500
1120	1030	1800	550
955	1000	1500	
930	920	1250	
830	830	1030	
790	790		
750	750		
690	700		
585	600		
585	550		
445	440		
350	350		
285	270		
240	210		
	90		

8. Repeat **steps 6, 7,** and **6**.
9. Resuspend the pellet in 2 mL CPES, containing 0.6 mg (60 U) zymolyase 100T and 2.5 mg Novozym 234 (*see* **Note 10**).
10. Incubate at 37°C for 2 h.
11. Check for successful cell-wall digestion by mixing equal volumes of cell culture and 0.1% SDS and viewing under the microscope. Cells without cell walls are lysed by SDS. Normally >50% of the cells are in this condition (*see* **Note 11**).
12. Mix the cells (preheated to 37°C) with an equal volume of 1% low melting temperature agarose (in 0.125 M EDTA, 0.9 M sorbitol, pH 7.5), which has been maintained at 50°C.
13. Dispense into plug molds and incubate the plugs in 1% NDS + proteinase K (*see* **Subheading 3.5.**).
14. Store the plugs in 1% NDS at 4°C.
15. Soak the plugs for at least an hour in gel running buffer before use.

S. pombe chromosomes are shown separated by PFGE in **Fig. 3**. Chromosome sizes are listed in **Table 1**.

3.3. Preparation of Bacteriophage λ Concatemers

Lambda concatemers provide size markers based on a monomer of 48.5 kb. They are useful markers intermediate in size between *S. cerevisiae* chromosomes (*see* **Notes 12** and **13**).

1. Inoculate from a lysogen of N1323(λ)-*lop*8, *cts*I857, *Sam*7 into 25 mL of LB.
2. Incubate overnight at 33°C.
3. Add the overnight culture to 500 mL of prewarmed LB in a 2-L flask.
4. Shake at 33°C until OD_{600} = 0.45 (approx 3 h).

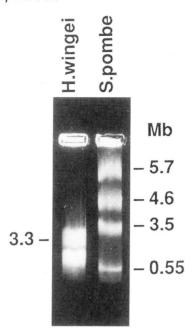

Fig. 3. *S. pombe* strain 3B3 and *H. wingei* chromosomes separated by CHEF PFGE. Samples were run at 2 V/cm, with a pulse time of 60 min for 120 h at 14°C on a 1% agarose gel in 0.5 × TBE. Note that only the largest *H. wingei* chromosome is clearly resolved, the lower band representing the other six chromosomes.

5. Induce for 15 min at 43°C.
6. Incubate for 2 h at 39°C with vigorous shaking (*see* **Note 14**).
7. Test for successful induction by adding a few drops of chloroform to a 2-mL aliquot of the culture in a glass test tube, put into a vortex, and incubate at 37°C for 10 min without shaking. The culture should clear after a few minutes. Use a culture without chloroform as a comparison.
8. Centrifuge the cells at 4000g for 10 min at 4°C.
9. Gently disrupt the pellet with a sterile plastic loop, and take up in a total volume of 200 mL of ice-cold TE.
10. Spin as in **step 8**.
11. Take up the cells in a total volume of 23 mL of ice-cold TE, having disrupted the pellet as in **step 9**.
12. Add 0.5 mL chloroform and shake gently at 37°C for 15 min.
13. Add 20 μL each of RNaseA and DNase 1 and shake gently at 37°C for 15 min.
14. Spin at 8000g for 15 min at 4°C.
15. Warm the supernatant to 37°C and mix with an equal vol of 1% low melting temperature agarose (in TE), cooled to 50°C, and dispense into plug molds (*see* **Subheading 3.5.**).
16. Incubate the plugs in 1% NDS + proteinase K (*see* **Subheading 3.5.**).
17. Rinse the plugs in 0.1 M EDTA, pH 8, and incubate in the same for 48 h at 50°C.
18. Rinse the plugs in 0.5 M EDTA pH 8, and store in the same at 4°C.
19. Plugs should be cut into small pieces and soaked in gel-running buffer for at least 1 h before loading on to the gel.

Lambda concatemers are shown separated by PFGE in **Fig. 1**.

3.4. Preparation of Chromosomal DNA from Cultured Mammalian Cells

1. Harvest the cells from culture flasks and transfer to 50 mL Falcon tubes (*see* **Note 15**).
2. Spin at 2000g for 10 min at 4°C.
3. Discard the supernatant and resuspend the cells in residual medium by vigorously flicking the base of the tube.
4. Add 25 mL of PBS to each tube.
5. Spin as in **step 2** and discard the supernatant.
6. Resuspend the cells as in **step 3** and add 10 mL of PBS to each tube. At this point, the contents of the tubes can be combined into one or more 50-mL Falcon tubes. Transfer the contents by using a fine-tip Pastet, which helps to break up any cell clumps (*see* **Note 16**).
7. Count an aliquot of the cell suspension using a hemocytometer and calculate the total cell number.
8. Repeat **steps 2** and **3**.
9. Resuspend the cells in a sufficient volume of PBS to give a cell density of 2×10^7 cells/mL (*see* **Note 17**).
10. Transfer the cells to a universal container, once again using a fine-tip Pastet.
11. Warm the cells to 37°C and mix with an equal volume of 1% low melting temperature agarose (in PBS), which has been maintained at 50°C.
12. Dispense into plug molds (*see* **Subheading 3.5.**).
13. Incubate the plugs in 1% NDS + proteinase K (*see* **Subheading 3.5.**)
14. Store the plugs in 1% NDS at 4°C.

3.5. Preparing Agarose Plugs

Both agarose plugs and beads provide an environment that protects large DNA molecules from shearing forces. This protocol allows plugs to be conveniently formed in large numbers using a mold made from a 96-well microtiter plate *(15)* (*see* **Note 18**).

1. Spray the mold with 70% ethanol and air dry prior to use.
2. Cover the base of the plate with a Titertek sealing strip and place the plate on ice.
3. Pour the molten mixture of cells and agarose into a trough and, using an 8- or 12-channel micropipet, dispense 100 µL aliquots into the mold, one row after another.
4. Allow the plugs to set on ice for 20 min.
5. Remove the sealing strip and, using a sterile plastic yellow micropipet tip, eject the plugs from the mold into a 50-mL Falcon tube. Care should be taken not to poke holes in the plugs during ejection. This can be avoided by running the tip down the wall of each well, making contact with the circumference of the plug, and then rapidly pushing downward until the plug drops from the mold.
6. The mold can now be cleaned by soaking in 0.1 *M* HCl and then thoroughly rinsing in distilled water.
7. The plugs are now incubated in detergent + proteinase K solution, although yeast plugs require a cell wall digestion step before reaching this stage (*see* **Subheading 3.1., step 13**): Up to 96 plugs are incubated in 20 mL of 1% NDS + 1 mg/mL proteinase K for 48 h at 50°C in a water bath. The solution is changed after 24 h.
8. Finally, the plugs are rinsed in the storage solution and stored in the same at 4°C. Plugs generally remain in good condition for at least a year under these conditions.

3.6. Preparing Agarose Beads

Agarose beads, although more difficult to prepare and manipulate than plugs, provide a larger surface area-to-volume ratio than do plugs. This property can be crucial during certain enzyme digestion protocols, particularly where time course procedures are required.

1. Harvest cells and wash twice in PBS.
2. Count cells in a hemocytometer and resuspend in PBS at a density of 1.25×10^7 cells/mL.
3. Add three drops of dyed microparticles per 5 mL of cell suspension and warm at 37°C.
4. Make up 2.5% low melting temperature agarose in PBS and cool to 50°C.
5. Warm some liquid paraffin to 50°C.
6. Equilibrate a 100-mL round-bottomed flask at 37°C.
7. Mix 1 vol of cells with 0.25 vol of agarose and 2.5 vol of paraffin in the flask and cover the top with Parafilm (American National Can, Greenwich, CT).
8. Immediately agitate mixture violently for 30 s in a flask shaker and then transfer the flask's contents to an ice water mixture and swirl for 10 min.
9. Pour the contents into a 50-mL Falcon tube and rinse out the flask with sufficient PBS to fill the tube.
10. Spin at 4000g for 45 s.
11. Stir the layer of beads trapped at the aqueous paraffin interface and respin for 1 min at 4000g.
12. Remove the supernatant without disturbing the pellet.
13. Resuspend the pellet in 50 mL PBS and repeat **step 10**.
14. Repeat **step 13** four more times, removing traces of paraffin from the tube with a tissue or transferring the beads to new Falcon tubes.
15. Resuspend the beads in LiDS solution to give a final vol of 50 mL.
16. Leave to stand at room temperature for 2 min.
17. Pellet the beads at 4000g for 5 min, remove the supernatant, and replace it with fresh LiDS, resuspending by vortexing if necessary.
18. Leave at room temperature for 20 min.
19. Repeat **steps 17** and **18** five more times.
20. Wash the beads extensively in 50 mM EDTA, pH 8, by repeating **steps 17** and **18** until no trace of detergent remains when the tube is shaken.
21. Store the beads in 50 mL of 50 mM EDTA, pH 8, at 4°C.

3.7. Restriction Endonuclease Digestion of DNA Embedded in Agarose Plugs

1. Soak the plugs for 10 min in a large excess of sterile TE, at room temperature. Typically, up to 10 plugs are immersed in 20 mL of TE.
2. Invert the tube frequently to achieve efficient mixing or use a tube roller.
3. Immerse the plugs in 5 mL TE containing (PMSF) at 40 µg/mL.
4. Incubate at 50°C for 30 min.
5. Repeat **steps 3** and **4**.
6. Soak the plugs for 2 h in 10 vol of 1X restriction enzyme buffer at room temperature, inverting the tube frequently to achieve efficient mixing. This preequilibration step can be reduced to 1 h, if the PMSF treatment is performed in 1X restriction buffer rather than in TE.
7. The plugs are now ready for digestion, which is performed in 1.5 mL microcentrifuge tubes—1 plug/tube.

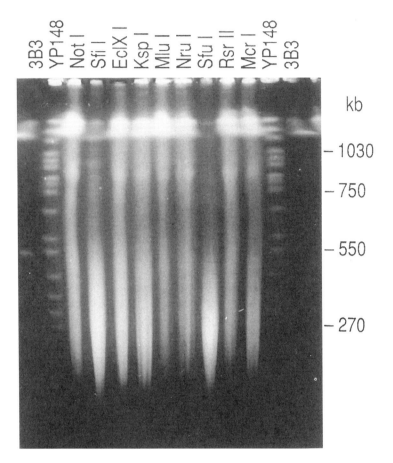

Fig. 4. Human DNA digested with rare cutter restriction enzymes and run out on a CHEF-pulsed field gel. The run conditions were 1% agarose gel in 0.5 × TAE, run at 4.5 V/cm with 70 s pulses for 27 h, followed by 120 s pulses for 21 h. Each track contains 5 μg of DNA. The gel was stained with ethidium bromide before being photographed.

8. Mix in each tube, on ice, 1X restriction buffer containing 0.1% Triton X-100 and 200 μg/mL BSA. Add 20 U of restriction enzyme (*see* **Note 19**). The final volume should be 100 μL.
9. Transfer the plug to the tube, checking that it is completely immersed in liquid and that no air bubbles are trapped around the plug. A 5-s spin in a microcentrifuge can be beneficial.
10. Incubate overnight, in a water bath, at the recommended temperature (*see* **Note 20**).
11. Double digestions can be carried out with both enzymes simultaneously if the buffer and incubation temperature are compatible. If not, the digests must be carried out sequentially and, if a buffer change is necessary, the plug must be soaked for at least 1 h in the new buffer.
12. On the next day, cool the tube on ice for 10 min and add 1 mL of ice cold TE/tube, invert once and remove TE (*see* **Note 21**).
13. Add 200 μL of stop buffer/tube and maintain on ice for 20 min (*see* **Note 22**).
14. The plug is now ready for loading into the gel.

Genomic DNA digested in plugs and separated on a CHEF apparatus is shown in **Fig. 4**. A complete list of rare cutter restriction enzymes appears in **Table 2**.

Table 2
Rare Cutter Restriction Enzymes

Enzyme	Recognition sequence[a]
Not-1	GC/GGCCGC
Asc-1	GG/CGCGCC
Fse-1	GGCCGG/CC
Srf-1	GCCC/GGGC
BssH-11	G/CGCGC
Xma-111, *Eag*-1, *EclX*-1	C/GGCCG
Sst-11, *Sac*-11, *Ksp*-1	CCGC/GG
Nae-1	GCC/GGC
Nar-1	GG/CGCC
Sma-1	CCC/GGG
Sse-83871	CCTGCA/GG
SanD-1	GG/wCCC
Rsr-11	CG/GwCCG
SgrA-1	Cr/CCGGyG
Mlu-1	A/CGCGT
Pvu-1	CGAT/CG
Nru-1	TCG/CGA
Aat-11	GACGT/C
Sal-1	G/TCGAC
BsiW-1, *Spl*1	C/GTACG
SnaB-1	TAC/GTA
Sfi-1	GGCCNNNN/NGGCC
Xho-1	C/TCGAG
BseA-1, *Acc*-111, *Mro*-1	T/CCGGA
Cla-1	AT/CGAT
Sfu-1, *Asu*-11	TT/CGAA

[a]w = A or T, r = A or G, y = C or T, and N = A, C, G, or T.

3.8. Restriction Endonuclease Digestion of DNA Embedded in Agarose Beads

1. Spin stored beads at 4000g for 5 min.
2. Withdraw an appropriate volume of beads from the pellet, using a "cutoff" micropipet tip (for n digests withdraw $n \times 100$ µL of beads) and transfer to a microcentrifuge tube. Spin for 1 min and check that the pellet contains a sufficient volume of beads. Transfer the beads to a 10-mL Falcon tube and add 10 mL of sterile 0.1% Triton X-100.
3. Mix on a tube rotator for 20 min.
4. Spin at 2000g for 5 min and discard supernatant.
5. Resuspend the beads in 10 mL sterile 0.1% Triton X-100.
6. Repeat **steps 3**, **4**, and **5** twice more and finally spin as in **step 4**.
7. Resuspend the beads to $n \times 100$ µL with water, where n is the number of digests.
8. Add n µL of 10% Triton X-100 and $n/2$ µL of BSA and mix.
9. Aliquot into microcentrifuge tubes containing 11 µL of 10X restriction enzyme buffer, using a "cutoff" tip. Add 20 U of restriction enzyme per tube and mix thoroughly.
10. Incubate for 2 h at the recommended temperature.

11. Transfer the tubes to ice and add 1 mL TEX/tube and mix.
12. Spin for 1 min and add 10 μL TE to the pellet and mix.
13. The beads are now ready for loading into the gel slots.

3.9. Resolution of Large DNA Molecules Using CHEF

There are several different instruments capable of separating DNA by PFGE (reviewed in *(8)*), but the CHEF method has proved to be the most popular *(16)*. The electric field switches through an angle of 120° and the separation pattern is uniform across the width of the gel, providing reliable lane-to-lane comparisons.

1. Select a suitable buffer for use in the CHEF:0.5X TBE gives the best results for separations up to 2 Mb (*see* **Note 23**, sentence 4).
2. Prepare a sufficient volume of the buffer to provide enough for the gel and the running buffer. This ensures that the gel and surrounding buffer are in ionic equilibrium.
3. Fill the gel tank with the correct volume of buffer and precool to the desired temperature (14°C is an optimal temperature).
4. Meanwhile, prepare the gel, having selected the most suitable type and concentration of agarose. Normally, a 1% medium EEO agarose is a suitable starting gel. Dissolve the agarose by either microwaving or heating on a hotplate. Using a conical flask 2.5× the volume of the agarose solution, preweigh the flask and contents and replenish with distilled water after the agarose has dissolved. Check that the agarose has dissolved by holding the flask up to the light, swirling the contents and checking to ensure there are no translucent lumps of solid left undissolved.
5. Allow the agarose to cool to hand hot (45°C) before pouring into the gel former. Immediately remove any air bubbles with a Pastet. It is essential that gels are cast on a level surface—if in doubt, check with a spirit level (*see* **Notes 24** and **25**).
6. If there is likely to be any delay between casting and running the gel, cover the surface with cling wrap to prevent evaporation.
7. Prior to loading, gently remove the comb and fill the slots with running buffer. This will help prevent the formation of air pockets when loading the sample plugs.
8. Plugs should be maneuvered into the gel slots by using sterile disposable inoculating loops and fine-tip Pastets. If the plugs require cutting, stand them on edge (supported by an inoculating loop) and slice downward with a sterile scalpel blade.
9. Beads are loaded by using a cutoff micropipet tip. The gel slots should not be prefilled with buffer.
10. Gently blot the gel to remove any buffer displaced from the wells, taking care not to disturb the plugs.
11. Seal the plugs/beads into position by dripping cool, molten, 1% low melting temperature agarose over the slots and allow to set.
12. Load the gel into the CHEF bath, checking that it is immersed in the correct depth of buffer. It is essential that the gel bath is level during the run.
13. Select an appropriate pulse time, voltage, and run time to separate the molecular-size range of interest and commence the run (*see* **Notes 26** and **27**).
14. A visual check to ascertain which electrodes are fizzing will confirm that the apparatus is operating correctly. Electrodes on the sides marked E and A will fizz alternatively as the field direction changes from one pulse to the next (*see* **Fig. 5**).
15. At the termination of the run, remove the gel, and stain it in distilled water containing 1 μg/mL ethidium bromide for 20 min, with gentle agitation.
16. Photograph the gel under ultraviolet light, with a ruler running down the side of the gel.

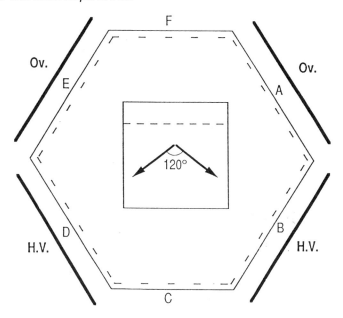

Fig. 5. Schematic diagram of the CHEF apparatus. The field switches through an angle of 120° from A–D to E–B and then back again. DNA migrates toward C.

3.10. Southern Transfer from Pulsed-Field Gels

1. The large DNA molecules separated by PFGE must be fragmented to allow efficient transfer to the hybridization membrane: Immerse the gel in 2.5× its volume of 0.25 M HCl and agitate gently for 20 min at room temperature. Do not exceed this time. This process leads to the partial depurination of the DNA.
2. Rinse the gel in distilled water.
3. Immerse the gel in 2.5× its volume of denaturant and agitate gently for 20 min. Repeat with fresh denaturant (*see* **Note 28**).
4. Repeat **step 2**.
5. Immerse the gel in 2.5× its volume of neutralizer and agitate gently for 40 min. Do not exceed this time.
6. Set up a device for Southern transfer by filling a plastic tray to the brim with 20× SSC. A sheet of 5 mm thick plastic rests across the top of the tray lengthwise, with a gap on each side to allow two sheets of Whatman 17Chr paper, placed on top of the plastic sheet, to dip into the buffer on both sides. This forms a wick that draws buffer up from the tray. Allow the paper to become saturated before the next stage.
7. Lay the gel on the paper wick, checking that there are no air bubbles trapped between the gel and the paper.
8. Cover the surrounding paper with cling wrap to prevent evaporation of the buffer.
9. Lay the hybridization membrane on to the gel, following the manufacturer's instructions.
10. The membrane should be gently rolled with a glass pipet to squeeze out any trapped air bubbles.
11. Mark the position of the slots on the membrane with a ballpoint pen so that the membrane can be orientated relative to the autoradiogram.
12. Lay on a sheet of Whatman 17Chr paper, slightly larger than the gel.
13. Lay on paper towels to cover the sheet of Whatman paper, stacked to a height of 5 cm.

14. Finally, lightly compress the paper towels with a 1-kg weight, resting on a sheet of plastic or glass to distribute the weight over the whole area of the gel.
15. After 24 h, discard any wet towels and replace with dry ones.
16. After a further 24 h, remove the membrane and wash in 2× SSC for 5 min.
17. The gel may be stained in ethidium bromide to check the efficiency of the transfer (*see* **Note 29**).
18. Crosslink the DNA to the membrane by treating with ultraviolet light or baking or both. Follow the manufacturer's instructions.

3.11. Preparation of Dried Pulsed Field Gels

There are inherent difficulties in transferring DNA from pulsed field gels on to filters by Southern blotting. Depurination or treatment with ultraviolet light is required to reduce the size of the molecules, and if this process is inefficient then transfer will be reduced. Hybridization to dried gels requires no DNA transfer, although this method is not available when using nonradioactively labeled probes.

1. After photographing the gel, follow **steps 3**, **4**, and **5** in **Subheading 3.10.**
2. Cut off one corner of the gel (for orientation purposes) and transfer it to two sheets of Whatman 3MM paper. Cover the gel in cling wrap and place on a gel drier.
3. Dry the gel under vacuum only for 20 min and then under vacuum + heat (60°C) for a further 20 min. These times are merely guidelines and may need to be adjusted to suite individual gel dryers and vacuum sources. At the end of the process, the gel should be the thickness of X-ray film but not as thin as cling wrap.
4. Wrap the gel (still stuck to the paper) in cling wrap and store at 4°C until required.

3.12. Hybridization of Radiolabeled Probes to Pulsed-Field Gel Filters

Hybridization procedures using filters carrying DNA from pulsed-field gels are not essentially any different from those methods used for filters from conventional gels, but the detection of single copy sequences does seem to be more difficult. Every effort must be made, therefore, to optimize labeling and hybridization conditions in order to produce a clear, positive result.

3.12.1. Preparing and Labeling Probes

1. Single-copy genomic probes should be prepared by isolating the insert away from the vector sequence, by digestion with appropriate restriction enzymes and separating the products on a preparative gel using low melting temperature agarose.
2. The insert band is excised from the gel (after staining with ethidium bromide and viewing with a midrange, 302-nm ultraviolet source). Because the DNA band may not occupy the whole depth of the gel, it is worth turning the gel slice on its side and trimming off excess agarose.
3. Melt the gel slice at 68°C for 5 min and add an equal volume of sterile water. Aliquot up to 13 µL for labeling and denature at 100°C for 10 min. Add 4 µL of High Prime DNA labeling system (Boehringer, Mannheim, Germany) and 30 µCi [α^{32}P] CTP in a total volume of 20 µL. Incubate at 37°C for at least 1 h.
4. Check the percentage of incorporation by spotting 1 µL from the labeling reaction on to a Whatman GF/B filter circle and counting in a scintillation counter. Incorporated label can be measured by passing 20 mL 5% TCA solution through the filter and counting it again. Incorporation should be at least 50%.

5. Unincorporated label should be separated from the labeled DNA by passage through a NICK column: Rinse the column with 10 mL TNE. Add the labeling reaction mix, made up to 100 µL with TNE. Allow to run through, then rinse the column with 300 µL TNE and finally elute with 400 µL TNE, collecting the eluate in a microcentrifuge tube.
6. Add sonicated salmon sperm DNA so that the final concentration in the hybridization mix will be 100 µg/mL.
7. Pierce the lid of the tube with a pin and denature the probe by heating at 100°C for 10 min.
8. Add directly to the hybridization mix or store on ice until required.

3.12.2. Hybridization and Autoradiography

1. Prehybridize the filter, sealed in a polythene bag, by incubating in a shaking water bath at 68°C for at least 2 h. The bag should contain 5 mL of hybridization mix/100 cm^2 of filter. Try to exclude all air bubbles from the bag (*see* **Note 30**).
2. Make an incision in the bag and add the probe, reseal the bag and incubate overnight at 68°C in a shaking water bath (*see* **Note 31**).
3. The next day, remove the filter from the bag, rinse in 2X SSC wash and then incubate in the same for 20 min at 68°C on a rotary platform.
4. Repeat, using progressively lower concentrations of SSC wash if necessary (e.g., 0.5X, 0.1X) until a reasonable signal:background level has been established, as monitored by a Geiger counter.
5. Drain the filter, blot dry with Whatman 3MM paper, enclose in cling wrap and place in a film cassette fitted with an intensifying screen (*see* **Note 32**).
6. Place an autoradiography film that has been preexposed by a short flash of light (approx 1 ms) (from a photographic flash gun fitted with a Kodak Wratten 22A filter) next to the filter, and close the cassette. Store at –70°C for an appropriate length of time and then develop the film.

Figure 6 shows an autoradiograph depicting the hybridization of a single-copy probe to a Southern blot of the pulsed field gel shown in **Fig. 4**.

3.13. Hybridization of Radiolabeled Probes to Dried Pulsed-Field Gels

1. Prehybridize the dried gel in 10 mL 5X SSPE (for up to 200 cm^2 gel) at 55°C for at least 2 h in a bag (*see* **Notes 33** and **34**).
2. Label and purify the probe as in **Subheading 3.12.1.**, **steps 1–5**.
3. Add 1 µg of sonicated salmon sperm DNA and 1/10 vol of M HCl to the probe and incubate at 37°C for 40 min.
4. Add 1/2 vol of M Tris pH 7.5 and count a 10-µL aliquot in a scintillation counter.
5. Add 5 µL of sonicated salmon sperm DNA to the probe and denature by heating at 100°C for 10 min.
6. Meanwhile, drain the prehybridization mix and refill the bag with 5 mL of 5X SSPE (for up to 200 cm^2 filter).
7. Add the probe to the bag at 2×10^6 cpm/mL of hybridization mix and incubate overnight at 55°C.
8. Remove the gel to a dark-colored tray and wash four times for 5 min at room temperature in 2X SSC.
9. Wash twice for 30 min at 55°C in 2X SSC.
10. Transfer the gel to a polythene sheet, cover with cling wrap, and autoradiograph at –70°C (*see* **Note 35**).

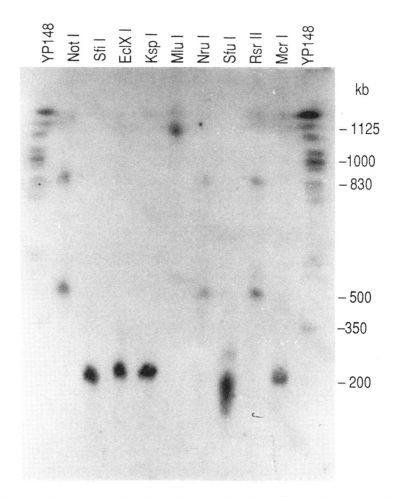

Fig. 6. Autoradiogram resulting from the hybridization of the gel shown in **Fig. 4** with a single copy probe from chromosome 11. The gel was blotted on to Hybond N (Amersham Pharmacia Biotech). The hybridized filter was washed to a stringency of 2 × SSC at 68°C and then exposed to Kodak XAR-5 film for 4 d at –70°C. Note the presence of a band at approx 200 kb in five of the digests, suggesting cleavage at a "CpG island." The nonisland cutter *Mlu*-1, which is susceptible to variable methylation, has produced very large fragments. A probe carrying the yeast repetitive element *Ty*-1 was also included and shows hybridization to the *YP*148 chromosomes.

3.14. Interpretation of Results

1. Align the developed autoradiogram with the filter and mark on the positions of the wells.
2. Bands on an autoradiogram can be sized by first measuring the distance from the sample well and then determining where the bands would appear on the photo of the ethidium-bromide–stained gel, by reference to the ruler running down the side of the gel. If appropriate size markers have been used, then it should be possible to determine the size of the band (*see* **Table 1** for details of yeast chromosome sizes). Faint bands can be enhanced by wiping the autoradiogram with dilute sodium hypochlorite solution to remove the background *(17)*.
3. If the signal on the autoradiogram is a diffuse area rather than a discrete band, this can be caused by overloading, thus next time slice up the plug and load less on the gel.

4. Occasionally, the filter may not represent a true reflection of the position of DNA on the original gel. The exact reason for this discrepancy is not known, but by hybridizing the filter with probes that light up the size markers, accurate sizing can be accomplished without reference to the original gel.
5. Multiple hybridizations to a single filter tend to give more accurate comparisons between probes than the use of filters from different gels. The subtle differences between electrophoretic runs, such as sample loadings or salt concentrations, can affect the relative mobilities of DNA molecules.
6. Ascertain that the signal has been effectively removed from the filter between hybridizations. Autoradiograph for a similar length of time to the previous exposure to make sure that all the label has been removed.
7. Always keep an exact record of the order of probes used in consecutive hybridizations, so that if unexpected bands appear on the autoradiogram, then it can be established whether they are remnants from a previous hybridization.
8. When performing double digests, always carry out single digests as well to check that the individual enzymes are active. Group the single and double digests together on the gel to make lane-to-lane comparisons easier. Check that the size of double-digest products add up to the single-digest size. Double digests revealing different hybridization patterns help to confirm that two identically sized bands are, in fact, different.
9. The appearance of more than one hybridizing band in a track often indicates a partial digest (*see* **Fig. 6**—*Not*-1, *Nru*-1 and *Rsr*-1I digests). Deliberately created partial digests can provide useful information in regions of the genome devoid of markers, especially if the probe hybridizes to near the end of a chromosome. Several methods can be used to create partial digest conditions in agarose plugs or beads *(18–20)*.
10. The total absence of a signal from a digest, but hybridization to the compression zone, indicates that the fragment has not been resolved. An increase in the pulse time, sometimes accompanied by a reduction in voltage and agarose concentration, may resolve the fragment as the window of separation is shifted to a larger size range. A comparison between **Figs. 1** and **2** indicates that a change in pulse time from 25–115 s has resulted in the maximum size resolved increasing from 440–1500 kb.
11. The appearance of a smear resulting from hybridization to a genomic digest may suggest the presence of repeated sequences in the probe. Suppression hybridization is rarely totally successful with PFGE blots, and attempts should be made to subclone out unique sequences from the probe. A smear can also appear as a result of DNA degradation. For this reason, it is prudent to include a lane of uncut DNA on the gel. Degradation can be caused by many factors, including faulty sample preparation or contaminating nucleases in the running buffer. Always change the buffer between runs and consider autoclaving the buffer if the problem persists.
12. Some restriction enzymes exhibit site preferences in that some recognition sequences are cut more readily than others. Among the rare cutter enzymes, *Nar*-1, *Nae*-1 and *Sac*-1I exhibit this phenomenon and, in fact, these enzymes require two copies of the recognition sequence before DNA cleavage can occur *(21)*. Two isoschizomers of *Sac*-1I (*Sst*-1I and *Ksp*-1) are commercially available and there is no evidence that they exhibit site preference.

4. Notes

1. Colored microparticles allow the clear visualization of both agarose plugs and beads. Mixing equal volumes of microparticles can create additional colors: blue and yellow to give green, red and yellow to produce orange, and blue and red to give mauve *(22)*.

2. Readers concerned about the toxicity of PMSF can use Pefabloc SC (Boehringer) as a substitute, although it is considerably more expensive.
3. Field inversion gel electrophoresis (FIGE) can be a substitute for CHEF PFGE. Commercially available apparatus include FIGE Mapper (Bio-Rad), for separations up to 200 kb, and Autobase (Q-Life Systems Inc., Kingston, Ontario). The latter system is supplied with run programs that control voltage and pulse time on ROM cards.
4. *S. cerevisiae* strains are best preserved by storage in Hogness freezing medium at −70°C *(23)*.
5. AHC medium selects for YAC arms because the acid hydrolysis of the casamino acids destroys tryptophan; the medium also lacks uracil.
6. This procedure is based on the published method by Carle and Olson *(24)*.
7. *S. pombe* strains are preserved in 30% glycerol + YPD at −70°C.
8. This protocol is based on the published method of Smith et al. *(25)*.
9. It is important to use ice-cold reagents throughout this protocol. Failure to do so results in extensive degradation of the DNA.
10. Spheroplast (cells without cell walls) formation in *S. pombe* is more difficult than in *S. cerevisiae*. Novozym 234 is therefore added to increase the yield.
11. Spheroplasts must be maintained in an osmotically stabilized medium by including sorbitol during spheroplast formation. Spheroplasts are sensitive to hypotonic conditions and traces of detergent, thus cell-wall digestion can be monitored by the addition of 0.1% SDS.
12. λ DNA has 12-base, single-stranded, complementary (cohesive) ends, which join under appropriate conditions and between a range of DNA concentrations *(26)*. λ DNA, in solution at 10 mg/mL, will spontaneously undergo limited concatemerization. Within agarose plugs, concatemerization is probably a very efficient process because λ DNA molecules, even at modest concentrations, are maintained in close proximity to each other, thus providing a beneficial environment for intermolecular association. The upper limit of concatemerization is probably influenced by termination resulting from damaged, single-stranded ends.
13. Wild-type λ has a monomer size of 48.5 kb, but concatemers can be formed based on other smaller λ genomes, e.g., λvir (42.5 kb) and λgt11 (43.7 kb). Bacteriophages P2 and P4 have 19-base single-strand cohesive ends, can form concatemers, and have monomer sizes of 31.8 and 11.6 kb, respectively.
14. A good yield of phage is only produced if adequate aeration is achieved at **Subheading 3.3.**, **step 6**, so vigorous shaking is important. An incubation temperature of 39°C may be beneficial in improving the phage yield.
15. Cells grown in culture should be harvested when just confluent; growth beyond this stage is characterized by the overproduction of mitochondrial DNA and the degradation of genomic DNA.
16. Some cells in culture readily form clumps, and it is essential that these clumps are disrupted before plug formation. If passage through a fine-tip Pastet fails to disperse the cells, then ejection from a syringe fitted with a 19-gauge or even a 21-gauge needle should be considered.
17. Each plug contains 1×10^6 cells in 100 µL of agarose, which is equivalent to about 10 µg of DNA. The plug may be cut in half if this quantity of DNA overloads the gel.
18. This protocol allows the rapid production of several hundred plugs at a time. The grid numbering system of microtiter plates allows different plugs to be formed within the same mold and yet preserve their identity. The plugs are round, which is beneficial for maintaining them intact during repeated manipulation.
19. The amount of restriction enzyme that needs to be added can be adjusted from experience. Some enzymes do not remain active throughout the incubation period, in which

case further enzyme can be added the next day and incubation continued for a few more hours *(27)*.
20. Some enzymes are unstable at the recommended incubation temperature. This may present a problem for the digestion of DNA in agarose, since by the time the enzyme has diffused into the plug, its activity has diminished. A preincubation period on ice may be beneficial, particularly since some enzymes are stabilized by their substrates.
21. It is important to wash the plug as in **Subheading 3.7., step 12**. During extended incubation, some DNA seeps out of the plug into the surrounding buffer and would be sheared during subsequent manipulations.
22. The inclusion of tracking dye in the stop mix colors the plugs and makes loading easier. It also allows a visual check on the correct migration of the sample during the early stages of electrophoresis.
23. The choice of buffer can influence the velocity of DNA migration. Using the standard medium EEO agarose, as well as the low EEO agaroses, the migration rate is faster in lower ionic strength buffers, such as 1X TAE, compared with the higher ionic strength TBE *(28–30)*. The effect is even more pronounced if the TAE concentration is dropped to 0.5X. TBE has a higher buffering capacity than TAE and may be the buffer of choice for prolonged electrophoretic runs as well as providing better resolution when separating molecules below 2 Mb.
24. Gels may be cast and run on glass plates when using homemade equipment. The gel thickness should be 5 mm. The glass plate does not distort the electric field *(13)*.
25. Gel combs should be of minimal thickness to provide slots tailored to the size of the plugs and should be positioned to give a gap of 0.5 mm between the bottom of the teeth and the gel support.
26. Pulse time: The most important single variable in determining the size range of molecules separated is the pulse time—the interval between the electric field switching from one direction to another. Smaller molecules, which are capable of rapid responses to changes in field direction, are separated preferentially by short pulse times. As the pulse time is increased, progressively larger molecules are separated, but the window of good resolution changes, such that molecules at the smaller end of the range are less well resolved. Pulsed field gels run at single-pulse times exhibit distinct regions of separation: Toward the top of the gel there is a region, called the compression zone (CZ), in which all molecules greater than a certain size comigrate. Below this is a region of maximum resolution in which bands are well separated, and below this is a region characterized by poorer resolution, in which mobility is linear relative to size *(31)*. Thus, maximum resolution is achieved by using the minimum pulse time capable of resolving the largest molecule of interest. Multiple consecutive pulse time regimes can be used to achieve good separation over a wide size range on a single gel. An increase in linearity of size relative to mobility can be achieved using a pulse-time ramp and this approach is particularly effective if nonlinear time ramps are employed *(13)*.
27. Field strengths of up to 10 V/cm (the distance being measured between opposite electrodes) can be used to separate molecules up to about 1.5 Mb. The higher the voltage, the faster the run time, although lower voltages tend to give better resolution, but over a narrower size range. Separation of molecules greater than about 1.5 Mb can only be achieved at reduced voltages, e.g., 3 V/cm. For separation over a given size range, or window of resolution, W, the product of the voltage gradient, V (expressed as V/cm), and pulse time, P, is roughly constant or more accurately, according to Gunderson and Chu *(32)*:

$$W = V^{1.4} \times P$$

This means in practice that, for separation in a given size range, an increase in voltage must be accompanied by an appropriate decrease in pulse time and vice versa. By solving this equation for *P*, it is possible to specify a pulse time for carrying out separation in the same size range, using different apparatus, as long as the voltage and distance, *D*, between opposite electrodes are known. Typical values for *D* are 33.5 cm for Bio-Rad CHEFs and 28 cm for the Pulsaphor hexagon system (Pharmacia, Brussels, Belgium). When the experimenter wishes to separate molecules up to a certain size limit and has no clues as to the most appropriate conditions under which to run the gel, mathematical expressions are now available that allow the various parameters to be established. An example of such a relationship (according to Smith *(33)*) is:

$$P = (R/V \times A^{-0.5} \times 5.25)^{1.25}$$

where *P* = pulse time, *R* = maximum size of molecule to be resolved, *A* = % agarose concentration, and *V* = V/cm.

28. It is important to change the denaturant halfway through **step 3**, **Subheading 3.10.**, since the HCl present in the gel will reduce the pH of the solution.
29. Alkali transfer is an alternative method worth considering *(34)*.
30. Hybridization ovens, using roller bottles, can be used as an alternative to the polythene bag method. They are intrinsically safer and use smaller volumes of reagents. Care must be taken, however, in opening the bottles and in checking the temperature inside the oven with a thermometer.
31. Filters from pulsed field gels carrying size markers can be hybridized with radiolabeled markers at the same time as the probe or separately if crosshybridization is a problem. λ DNA can be labeled and hybridized to concatemers. *S. cerevisiae* chromosomes can be hybridized to the *Ty*-1 repetitive element, which is present to various extents in all yeast chromosomes. The 90- and 1030-kb chromosomes of *YP*148 also crosshybridize with pBR322 sequences. *S. pombe* chromosomes can be hybridized to a centromeric probe, such as pSS166, which is derived from the *dg*11a region *(35)*. Minichromosomes, carried by some *S. pombe* strains, tend to crosshybridize with *Ty*-1 and pBR322 sequences.
32. The filter can be aligned accurately with the photographic film by using luminescent stickers (Glogos 11 Autorad markers, Stratagene), which are stuck to the cling wrap and produce a signal during autoradiography.
33. This protocol is based on the published method of Stoye, Frankel, and Coffin *(36)*.
34. The dried gel can be separated from the filter paper, on to which it was dried down, by briefly immersing it in water and peeling the gel off the paper. Dried gels are reasonably strong and can be maneuvered easily with care. Gels can also be hybridized in bottles. To remove the gel, reverse the orientation of the bottle and rotate in the oven briefly so that the gel rolls up.
35. Probes can be removed from dried gels by gently agitating in 0.5 *M* NaOH for 30 min followed by 30 min in a neutralizer. Several reprobings are possible without the gel disintegrating.

Acknowledgment

Thanks go to Sandy Bruce for preparing the figures used in this chapter.

References

1. Fangman, W. (1978) Separation of very large DNA molecules by gel electrophoresis. *Nucleic Acids Res.* **5,** 653–665.
2. Schwartz, D. C., Saffran, W., Welsh, J., Haas, R., Goldenburg, M., and Cantor, C. R. (1983) New techniques for purifying large DNAs and studying their properties and packaging. *Cold Spring Harbor Symp. Quant.* Biol. **47,** 189–195.

3. Orbach, M. J., Vollrath, D., Davis, W., and Yanofsky, C. (1988) An electrophoretic karyotype of *Neurospora crassa*. *Mol. Cell Biol.* **8,** 1469–1473.
4. Gurrieri, S., Rizzarelli, E., Beach, D., and Bustamante, C. (1990) Imaging of kinked configurations of DNA molecules undergoing orthogonal field alternating gel electrophoresis by fluorescence microscopy. *Biochemistry* **29,** 3396–3401.
5. Bustamante, C. Gurrieri, S., and Smith, S. B. (1990) Observation of single DNA molecules during pulsed field gel electrophoresis by fluorescence microscopy. *Methods* **1,** 151–159.
6. Gurrieri, S., Smith, S. B., Wells, K. S., Johnson, I. D., and Bustamante, C. (1996) Real-time imaging of the reorientation mechanisms of YOYO-labelled DNA molecules during 90° and 120° pulsed field gel electrophoresis *Nucleic Acids Res.* **24,** 4759–4767.
7. Maule, J. C. (1994) Electrophoretic karyotype analysis-pulsed field gel electrophoresis, in *Chromosome Analysis Protocols* (Gosden, J. R., ed.), Humana, Totowa, NJ, pp. 221–252.
8. Maule, J. C. (1998) Pulsed field gel electrophoresis. *Molecular Biotechnology* **9,** 107–126.
9. Burke, D. T., Carle, G. F., and Olson, M. V. (1987) Cloning of large segments of exogenous DNA into yeast by means of artificial chromosome vectors. *Science* **236,** 806–812.
10. Monaco, A. P. and Larin Z. (1994) YACs, BACs, PACs and MACs: artificial chromosomes as research tools. *Trends Biotech.* **12,** 280–286.
11. Bickmore, W. A. and Bird, A. P. (1992) Use of restriction enzymes to detect and isolate genes from mammalian cells. *Methods Enzymol.* **216,** 224–245.
12. Arker, W., Enquist, L., Hohn, B., Murray, N. E., and Murray, K. (1983) Experimental methods for use with lambda, in *Lambda II* (Hendrix, R. W., ed.) Cold Spring Harbor Laboratory, Cold Spring Harbor, NY, pp. 433–466.
13. Maule, J. C. and Green, D. K. (1990) Semiconductor-controlled contour-clamped homogeneous electric field apparatus. *Anal. Biochem.* **191,** 390–395.
14. Jones, C. P., Janson, M., and Nordenskjold, M. (1989) Separation of yeast chromosomes in the megabase range suitable as size markers for pulsed-field gel electrophoresis. *Technique* **1,** 90–95.
15. Porteous, D. J. and Maule, J. C. (1990) Casting multiple aliquots of agarose embedded cells for PFGE analysis. *Trends Genet.* **6,** 346.
16. Chu, G., Vollrath, D., and Davis, R. W. (1986) Separation of large DNA molecules by contour-clamped homogeneous electric fields. *Science* **234,** 1582–1585.
17. Guido, E. C. and Abhay, K. (1994) Simple method to reduce background on autoradiographs. *BioTechniques* **17,** 294.
18. Albertsen, H. M., Paslier, D. Le, Abderrahim, H., Dausset, J., Cann, H., and Cohen, D. (1989) Improved control of partial DNA restriction enzyme digest in agarose using limiting concentrations of Mg^{++} *Nucleic Acids Res.* **17,** 808.
19. Barlow, D. P. and Lehrach, H. (1990) Partial Not1 digests, generated by low enzyme concentration or the presence of ethidium bromide, can be used to extend the range of pulsed-field gel mapping. *Technique* **2,** 79–87.
20. Wilson, W. W. and Hoffman, R. M. (1990) Methylation of intact chromosomes by bacterial methylases in agarose plugs suitable for pulsed field electrophoresis. *Anal. Biochem.* **191,** 370–375.
21. Topal, M. D., Thresher, R. J., Conrad, M., and Griffith, J. (1991) Nae1 endonuclease binding to pBR322 DNA induces looping. *Biochemistry* **30,** 2006–2010.
22. Maule, J. C. (1995) Colored microparticles for clear visualization of agarose beads and plugs. *Trends Genet.* **11,** 127.
23. Werner, E., Holder, A. A., and Hoheisel, J. D. (1997) Growth and storage of YAC clones in Hogness freezing medium. *Nucleic Acids Res.* **25,** 1467–1468.
24. Carle, G. F. and Olson, M. V. (1985) An electrophoretic karyotype for yeast. *Proc. Natl. Acad. Sci. USA* **82,** 3756–3760.

25. Smith, C. L., Klco, S. R., and Cantor, C. R. (1988) Pulsed-field gel electrophoresis and the technology of large DNA molecules, in *Genome Analysis* (Davies, K. E., ed.) IRL, Oxford, pp. 41–71.
26. Mathew, M. K., Smith, C. L., and Cantor, C. R. (1988) High-resolution separation and accurate size determination in pulsed-field gel electrophoresis: DNA size standards and the effect of agarose and temperature. *Biochemistry* **27,** 9204–9210.
27. *New England Biolabs* Catalog (1998/99), Beverly, MA, p. 257.
28. White, H. W. (1992) Rapid separation of DNA molecules by agarose gel electrophoresis: use of a new agarose matrix and a survey of running buffer effects. *BioTechniques* **12,** 574–579.
29. Birren, B. W., Lai, E., Clark, S. M., Hood, L., and Simon, M. I. (1988) Optimized conditions for pulsed field gel electrophoretic separations of DNA. *Nucleic Acids Res.* **16,** 7563–7582.
30. Birren, B. W., Hood, L., and Lai, E. (1989) Pulsed field gel electrophoresis studies of DNA migration made with the PACE electrophoresis system. *Electrophoresis* **10,** 302–309.
31. Vollrath, D. and Davis, R. W. (1987) Resolution of DNA molecules greater than 5 megabases by contour-clamped homogeneous electric fields. *Nucleic Acids Res.* **15,** 7865–7876.
32. Gunderson, K. and Chu, G. (1991) Pulsed-field electrophoresis of megabase-sized DNA. *Mol. Cell. Biol.* **11,** 3348–3354.
33. Smith, D. R. (1990) Genomic long-range restriction mapping. *Methods* **1,** 195–203.
34. Reed, K. C. and Mann, D. A. (1985) Rapid transfer of DNA from agarose gels to nylon membranes. *Nucleic Acids Res.* **13,** 7207–7221.
35. Chikashige, Y., Kinoshita, N., Nakaseko, Y., Matsumoto, T., Murakami, S., Niwa, O., and Yanagida, M. (1989) Composite motifs and repeat symmetry in *S. pombe* centromeres. *Cell* **57,** 739–751.
36. Stoye, J. P., Frankel, W. N., and Coffin, J. M. (1991) DNA hybridization in dried gels with fragmented probes: an improvement over blotting techniques. *Technique* **3,** 123–128.

18

HPLC of DNA and PCR Products

Elena D. Katz

1. Introduction

The polymerase chain reaction (PCR) has rapidly become a standard laboratory technique. With the continuous development of PCR technology, there has been a growing need for PCR product quantitation. High-performance liquid chromatography (HPLC) is well accepted as a quantitative technique in many diverse applications areas because the technique can provide reliable, precise, and sensitive sample detection, and wide dynamic range.

During the last two decades, column liquid chromatography has been employed for the separation, purification, and detection of nucleic acids. Different modes of chromatography of nucleic acids have been described previously (1). Of the different chromatography methods, anion-exchange chromatography has been most commonly employed for the isolation and purification of not only oligonucleotides but also large double-stranded DNA. In anion-exchange chromatography, DNA retention is dependent on electrostatic interactions between the negatively charged phosphate groups of DNA and cationic sites of the chromatographic matrix. This process is schematically depicted in **Fig. 1**. DNA can be eluted from the anion-exchange column by altering the ionic strength of the buffer solution. In the presence of a buffer of increasing ionic strength, DNA retention is generally a function of the number of negative charges associated with the phosphate groups.

One of the most common anion-exchange materials is diethylamino ethyl (DEAE)-bonded support, and it has been extensively used to modify original soft-gel supports, porous microparticular silica, and polymer-based materials. Availability of porous microparticular silica and polymer-based supports has led to enhanced column resolution and faster analysis times. However, the columns packed with porous particles have been used only with limited success in the analysis of large double-stranded DNA fragments. HPLC columns based on nonporous, small-particle resins are available, and very fast and efficient separations of large double-stranded DNA have been demonstrated (2). The success of these columns is based on the fact that nonporous, small-particle size material with which they are packed provides improved efficiency and short analysis time. This is the result of a much faster DNA transfer between the solid

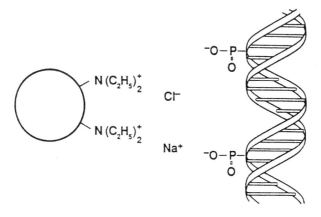

Fig. 1. Anion-exchange chromatography of DNA.

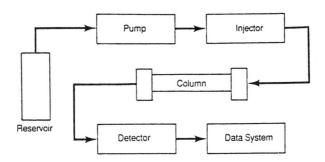

Fig. 2. Basic components of HPLC system.

and liquid column phases than is possible for conventional porous support. In addition, these columns offer high DNA recoveries because of the absence of pores.

Katz et al. have demonstrated that HPLC can be successfully employed for post-PCR analyses *(3,4)*. A typical HPLC system consists of an HPLC pump, sample injector, column, ultraviolet (UV) absorbance detector, and data-handling device, as is schematically represented in **Fig. 2**. A manual sample injector can be replaced by an autosampler; a fraction collector can also be added. Thus, the HPLC system can easily be automated and employed for the separation, quantitation, and purification of PCR products in a single step. In this chapter, the automated HPLC method for the rapid quantitation and purification of PCR products is discussed.

2. Materials

1. pBR322 DNA-*Hae*III digest.
2. 500-bp PCR product (*see* **Note 1**).
3. 115-bp HIV PCR product (*see* **Note 2**).
4. High-performance liquid chromatography (HPLC)-grade water.
5. HPLC buffer A: 1 M NaCl, 25 mM Tris-HCl, pH 9.0.
6. HPLC buffer B: 25 mM Tris-HCl, pH 9.0.
7. 0.2 N NaOH.
8. 30% Acetonitrile in HPLC-grade water.

9. An automated biocompatible HPLC system: A binary pump, a UV-VIS detector, an autosampler, a data-handling device, a liquid chromatography (LC) column oven (in this work, all instruments were from Perkin-Elmer/Cetus, Norwalk, CT).
10. Anion-exchange analytical column: a Perkin-Elmer TSK DEAE-NPR column (35 mm long and 4.6 mm inner diameter) packed with 2.5 µM particles of hydrophilic resin, bonded with DEAE groups.
11. Guard column: A PE guard column, 5 mm long and 4.6 mm inner diameter, packed with 5 µM of DEAE-NPR material.
12. Ultraviolet spectrophotometer.
13. TE buffer: 10 mM Tris-HCl, pH 8, 1 mM ethylenediaminetetraacetic acid (EDTA).

3. Methods
3.1. HPLC Separation of DNA Fragments

1. Prepare an HPLC system for PCR analysis according to the manufacturer's manual.
2. Make buffers A and B using HPLC-grade water.
3. Place a DEAE-NPR guard column followed by the analytical DEAE-NPR column between the injector valve and the UV detector.
4. Program an appropriate gradient profile according to the manufacturer's HPLC pump manual. A gradient protocol is recommended as follows:
   a. Step gradient from 44–55% A in B, for 0.1 min;
   b. Linear gradient from 55–61% A in B, for 3.5 min;
   c. Linear gradient from 61–100% A in B, for 0.5 min;
   d. Hold at 100% A for 1 min;
   e. Linear gradient from 100–44% A in B, for 0.1 min.
5. Select an appropriate starting buffer composition such as the one described in **step 4** (44% A in B). The starting buffer composition can be changed, depending on the size of the DNA fragments. Start with a lower molar concentration of NaCl (25–30% A) if the separation of PCR primers is desirable.
6. Equilibrate the column with the starting buffer composition for 15–30 min at 1 mL/min.
7. Obtain a blank chromatogram by injecting 10–20 µL of buffer A to establish a flat detector baseline. This will confirm that the total system is clean (potential sources of contamination, besides the column, could be the injector and connecting tubes).
8. Establish the performance of a new HPLC column by injecting an appropriate DNA standard. An example of the HPLC separation of an *Hae*III digest of pBR322 DNA employing the DEAE-NPR column is shown in **Fig. 3**.
9. Inject 10–50 µL of a PCR sample for PCR product separation and detection (*see* **Note 3**). Use the same operating conditions as in **step 8** for the purpose of subsequent product identification.
10. Identify a PCR product by matching its time of elution from the column with that of an appropriate DNA fragment in the chromatogram obtained in **step 8**.

3.2. HPLC Purification of PCR Products

1. Inject 100–200 µL of a given PCR sample onto the column.
2. Collect, manually or automatically, a fraction of the column buffer containing the PCR sample at the detector exit line during the chromatographic development at the appropriate time (*see* **Note 4**).
3. Desalt the purified PCR products using an established procedure *(5)*, if further PCR sample manipulations such as sequencing, cloning, or reamplification require a lower salt concentration (*see* **Note 5**).

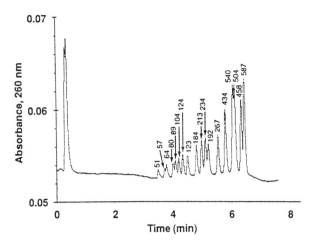

Fig. 3. Separation of *Hae*III digest of PBR322 DNA, operating conditions as described in the text. Sample concentration: 50 µg/mL in TE buffer. Injection size was 10 µL.

3.3. HPLC Quantitation of PCR Products

1. Ensure that the performance of the HPLC system is reproducible (*see* **Note 6**).
2. Inject successively two 150 µL vol of the amplified 500-bp fragment.
3. Collect the eluted products at appropriate elution times (*see* **Note 4**).
4. Note their peak areas.
5. Concentrate the samples using a SpeedVac (Savant Instruments, Inc., Farmingdale, NY) and reconstitute in a known volume of TE buffer.
6. Take an absorbance reading of the PCR sample, obtained in **step 18**, using a UV spectrophotometer, set at 260 mm (*6*).
7. Calculate the concentration of the PCR product (1 outer diameter–50 µg/mL) and its amount in the buffer volume used in **step 5**.
8. Calculate the ratio of the amount of PCR product to its peak area.
9. Use this ratio to calculate the concentration of PCR product to be quantitated (*see* **Note 7**).

4. Notes

1. 500-bp Product amplification: Amplifications were carried out using the Perkin-Elmer/Cetus GeneAmp PCR Reagent Kit. A 500-bp segment of bacteriophage λ DNA (nucleotides 7131–7630) was used as a target, the initial concentration of which was 10 pg/100 µL or $3 \times 10^{-13} M$. Using the Perkin-Elmer/Cetus GeneAmp PCR System 9600, a two-temperature step cycle PCR was carried out for 15 s at 95°C and 90 s at 68°C for 25 cycles, the final extension step was performed at 68°C for 6 min.
2. 115-bp Product amplification: A 115-bp HIV product (PCR sample courtesy of Will Bloch of Cetus Corporation) was amplified using GeneAmplimer HIV-I Control Reagents, the DNA Thermal Cycler (both Perkin-Elmer/Cetus), and two-temperature step cycle PCR as follows: during the first two cycles, the target was denatured for 1 min at 98°C and annealed/extended for 2 min at 60°C. The subsequent cycles were carried out for 1 min at 94°C and for 1 min at 60°C for 38 cycles, the final extension step was performed for 10 min at 72°C. The initial concentration of this target was 10 copies in a 100-µL reaction vol. The 115-bp HIV product was amplified in the presence of a high genomic DNA background (10 µg crude human placental DNA). A hot-start technique (*7*) was used to enhance the PCR specificity.

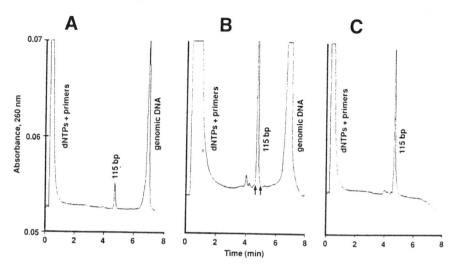

Fig. 4. **(A)** Separation of 115-bp HIV-1 product amplified as described in text. HPLC conditions as in **Fig. 2**. Injection size: 10 µL. **(B)** Separation of the same 115-bp product. Injection size: 100 µL. **(C)** Separation of 115-bp product reamplified as described in text. Injection size: 10 µL.

3. Sample volume: This volume is dependent on PCR product yield after amplification. Examples of PCR product separation are shown in **Figs. 4A** and **5A**. A 10-µL injection vol was sufficient to clearly detect both amplified products. In some cases, injection of 50 µL may be necessary. For purification purposes, 100–200 µL can be injected without column overloading (*see* **Figs. 4B** and **5B**).
4. Purification: Note in **Fig. 4B** that some nonspecific products, albeit at a very low level, were detected. However, these did not interfere with sample collection because nonspecific fragments were well separated from the 500-bp peak. During the development of this chromatogram, the 500-bp product was collected between 5.69 and 6.13 min (shown by the arrows in the figure). The 115-bp fragment was collected between 4.64 and 4.90 min (**Fig. 5B**).
5. Reamplification: The HPLC-purified, 115-bp product was desalted using a Centricon R-30 microconcentrator (Amicon, Beverly, MA). A 10-µL sample vol was then amplified using the GeneAmplimer HIV-I reagents and the GeneAmp PCR System 9600. The two-temperature step cycle PCR protocol consisted of denaturing for 15 s at 95°C and annealing/extending for 60 s at 60°C for 35 cycles. The final product was extended at 60°C for 9 min. The results of the HIV product reamplification are shown in **Fig. 5C**.
6. Reproducibility of the HPLC system: The HPLC performance can be monitored by periodically injecting an appropriate DNA standard such as an *Hae*III digest of pBR322 DNA. The reproducibility data for the peak area and retention time for a 434-bp fragment of this digest are given in **Table 1**. **Table 1** demonstrates that both the intraday and interday retention time reproducibility can be obtained within 1% relative standard deviation (SD). The interday peak area reproducibility can be obtained with 4% relative SD, whereas the intraday precision can be within 2%.
7. Quantitation of PCR products: In the HPLC quantitation of PCR products, an area under a given PCR-amplified DNA peak is directly related to the DNA concentration. The PCR product quantitation can be performed using two different approaches. One approach is to generate an LC calibration curve employing an appropriate DNA standard quantitated by UV absorbance *(4)*. The LC calibration yields a straight line when the UV absorbance

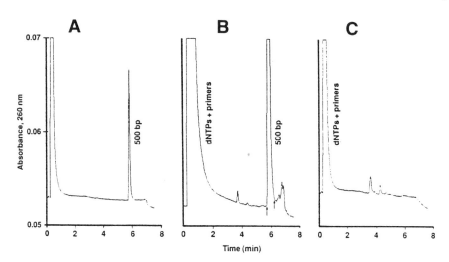

Fig. 5. (**A**) Separation of 500-bp product amplified as described in text. HPLC conditions as in **Fig. 3**. Injection size: 10 µL. (**B**) Separation of the same 500-bp product. Injection size: 100 µL. (**C**) Chromatogram of column eluent collected as described in text and amplified as in A. Injection size 10 µL.

response is plotted against the mass or concentration of DNA injected onto the column. The second approach does not involve the generation of the calibration curve and is based on the measurement of UV absorbance of a PCR product collected during a chromatographic run. To quantitate the concentration of the collected PCR product using a UV spectrophotometer, a reading is taken at 260 nm (1 outer diameter = 50 µg/mL of ds DNA), and the concentration values obtained are then related to the HPLC peak area corresponding to the PCR-product peak. This procedure can be recommended as follows. Two 150-µL portions of the amplified fragment are injected successively and the eluted products are collected in a volume of approx 400 µL. The samples are then concentrated using a SpeedVac and reconstituted in a known volume of 10 mM Tris-HCl, pH 9, 1 mM EDTA (TE) buffer. The amount of the PCR product is quantitated using a UV spectrophotometer. In this work, this protocol was employed to quantitate HPLC-purified, 500-bp product in 300-µL TE buffer. The same procedure was repeated for two 100-µL injections of the same PCR product, but the concentrated sample was reconstituted in 100-µL TE buffer. The results relating the DNA amount in the purified sample to the HPLC peak area are demonstrated in **Table 2**. The mean value of the 0.139-ng/area, which was a constant as expected, could now be used to quantitate any separated PCR-amplified DNA fragment. For example, the area of the 500-bp peak (**Fig. 4A**) was equal to 855 counts (arbitrary units). Multiplying the value of 0.14 ng/area by the peak area of 855 counts, a value of 119 ng is obtained. If an injection size is 10 µL, the concentration of the 500-bp fragment is equal to 11.9 ng/µL or, if the sample was amplified in 100 µL, to 1190 ng in 100 µL.

8. HPLC sample cross-contamination: The exquisite sensitivity of PCR may lead to a serious problem of sample cross-contamination. To check the sample cross-contamination problem, immediately after the injection of a PCR fragment, the column should be washed by injecting 100 µL of 1 M NaCl under the same gradient conditions. A volume of the eluting mobile phase can be collected during an appropriate time as indicated by the arrows of **Fig. 4B**. The resulting eluent should then be desalted using the Centricon-30 microconcentrator, and 10 µL of the concentrate amplified following an appropriate

Table 1
Peak Area and Retention Time Reproducibility

Day	Retention time (min)	Area (counts; arbitrary U)
1	5.900	276
	5.890	277
2	5.853	289
	5.903	286
	5.913	292
	5.910	281
	5.867	290
3	5.730	287
	5.730	291
4	5.810	293
	5.823	285
	5.853	299
	5.910	295
	5.833	292
	5.853	301
5	5.873	302
	5.813	305
	5.947	304
6	5.950	285
	5.930	290
	5.927	287
7	5.837	303
	5.830	306
	5.840	309
8	5.860	316
	5.863	315
	5.860	316
	5.880	313
	5.887	314
9	5.853	309
	5.887	307
	5.913	294
Mean	5.867	297
Relative SD (%)	0.9	3.9
4	5.810	293
	5.823	285
	5.853	299
	5.910	295
	5.853	301
Mean	5.847	294
Relative SD (%)	0.6	1.9

Table 2
Purified Product Quantitation

Injection size, µL	Peak area, arbitrary U	Absorbance, OD	Concentration, µg/mL or ng/µL	Amount, ng	Amount/area, ng/area
			in 300 µL		
150	14,705	0.130	6.51	1953	0.133
150	14,831	0.132	6.62	1986	0.134
			in 100 µL		
100		0.321	16.06	1606	0.157
100		0.249	12.45	1245	0.133

protocol. The sample cross-contamination was checked for the 500-bp product amplification. The results are shown in **Fig. 4C**, which clearly shows no detectable sample cross-contamination. It also shows that this amplification yielded nonspecific products, which was expected, as no DNA template was present in the PCR mixture.

9. System contamination: Presence of additional peaks that cannot be attributed to PCR product being analyzed will indicate that the HPLC system has been contaminated. In general, sources of contamination could be DNA sample carryover from one injection to another and mobile phase impurities. To prevent sample carryover, the syringe (in the case of manual injections) and the injector loop should be rinsed with a strong solvent, e.g., 1 M NaCl, prior to each sample injection. To prevent the system contamination by the mobile phase impurities, high-purity buffers and salts and HPLC-grade water must be used. To ensure that the HPLC system is clean, blank chromatograms demonstrating a flat detector baseline must be periodically obtained by injecting 10–50 µL of 1 M NaCl.

10. High-inlet column pressure: Column inlet pressure is primarily dependent on column length, particle size, and flow rate. A normal operating pressure for DEAE-NPR columns employed in this work, should be in the range of 1000–1500 psi at a flow rate of 1 mL/min at room temperature. The column pressure can increase during the post-PCR analysis owing to (1) partial clogging of connecting tubing that, in turn, can be caused by particulates from the mobile phase or injection devices, and (2) column frit clogging. To prevent the connecting tubing from clogging, in-line solvent filters should be installed in the pump outlet line before the injector. A column inlet filter and/or guard column will retain any particulate matter resulting from the rotor seal wear of the injector. HPLC columns packed with sub-5-µm particles contain low-porosity (I Jim) column frits that are more prone to clogging than the larger frits in conventional columns. Therefore, guard columns should be employed to protect the analytical columns.

11. Column maintenance: DEAE-NPR columns should be washed daily to maintain column performance during routine use. Using the injector valve, 1–2 mL of 0.2 N NaOH (by repeat injections) should be injected. If a PCR mixture that contains hydrophobic material such as mineral oil is repeatedly injected into the column, increases in column pressure can be observed. Injecting 1–2 mL of pure organic solvent (e.g., acetonitrile, methanol, or ethanol) helps keep the column pressure stable. The column should be rinsed thoroughly with HPLC-grade water and filled with 20–30% organic solvent (e.g., acetonitrile) in water before removing it from the system. Regular cleaning and thermostatting the column will ensure reproducible column performance during the post-PCR analysis.

References

1. McLaughlin, L. W. (1989) Mixed-mode chromatography of nucleic acids. *Chem. Rev.* **89,** 309–319.
2. Kato, Y., Yainasaki, Y., Onaka, A., Kitamura, T., Hashimoto, T., Murotsu, T., Fukushige, S., and Matsubara, K. (1989) Separation of DNA restriction fragments by high-performance ion-exchange chromatography on a nonporous ion exchanger. *J. Chromatogr.* **478,** 264–268.
3. Katz, E. D., Eksteen, R., and Haff, L. A. (1990) Rapid separation, quantitation and purification of products of polymerase chain reaction by liquid chromatography. *J. Chromatogr.* **512,** 433–444.
4. Katz, E. D. and Dong, M. W. (1990) Rapid analysis and purification of polymerase chain reaction products by high-performance liquid chromatography. *BioTechniques* **8,** 546–555.
5. Allard, M. W., Ellsworth, D. L., and Honeycutt, R. L. (1991) Uses for Centricon 30 microconcentrators. *Lab News* (Amicon) (Spring), 4,5.
6. Maniatis, T., Fritsch, E. F., and Sainbrook, J. (1989) *Molecular Cloning. A Laboratory Manual* (2nd ed.), Cold Spring Harbor Laboratory, Cold Spring Harbor, NY.
7. Erlich, H. A., Gelfand, D., and Sninsky, J. J. (1991) Recent advances in the polymerase chain reaction. *Science* **252,** 1643–1651.

III

PROBE DESIGN, SYNTHESIS, AND LABELING

19

End-Labeling of DNA Fragments

Adrian J. Harwood

1. Introduction

End-labeling is a rapid and sensitive method for radioactively, or nonisotopically, labeling DNA fragments and is useful in visualizing small amounts of DNA. End-labeling can also be used to label fragments at one end. All the enzymes employed are specific to either the 3' or 5' termini of DNA and will consequently only incorporate a label once per DNA strand. If double-stranded DNA is used, both ends are labeled, but single-end labeled fragments can be produced by further restriction enzyme digestion. This works well with DNA fragments cloned into polylinkers as one labeled end can be removed as a tiny DNA fragment, making subsequent purification easier. Such single-end-labeled molecules can be used to order restriction enzyme fragments and are a prerequisite for Maxam-Gilbert DNA sequencing *(1)*. End-labeled synthetic oligonucleotides have numerous applications, including sequence-specific probes *(2)*; gel retardation, and Southwestern assays *(3)* and sequencing polymerase chain reaction (PCR) products *(4)*.

There are two common methods of end-labeling, the "fill-in" reaction and the "kinase" reaction. The fill-in reaction uses the Klenow fragment of *Escherichia coli* DNA polymerase *(5)* and labels DNA fragments that have been digested with a restriction enzyme to create a 5' overhang. Klenow extends the 3' recessed end of one DNA strand by using the 5' overhang of the other strand as a template (**Fig. 1A**). This is the method of choice for double-stranded DNA fragments because of its ease. When suitable restriction enzyme sites are unavailable, or when the substrate is single stranded, the kinase reaction is used. The kinase reaction uses T4 polynucleotide kinase (T4 kinase) to transfer labeled phosphate to the 5' end of the DNA molecule (*6*; **Fig. 1B**). This method is ideal for labeling oligonucleotides, which are normally synthesized without a 5' phosphate. To label restriction-enzyme–digested DNA fragments, the terminal phosphate must first be removed by using a phosphatase, such as calf intestinal alkaline phosphatase (CIP) (obtained from Boehringer-Mannheim, Mannheim, Germany). All of these reactions can be used without labeled nucleotides to modify the DNA fragments for further recombinant DNA manipulations.

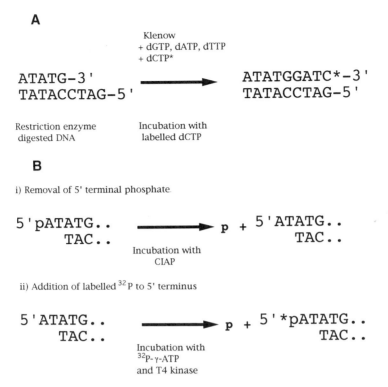

Fig. 1. End-labeling DNA. (**A**) The Fill-in reaction. (**B**) The kinase reaction.

2. Materials

Molecular biology-grade reagents should be used whenever possible. Manipulations are performed in 1.5-mL disposable, sterile polypropylene tubes, with screw tops to prevent leakage of radioactivity. Local safety precautions must be obeyed when using radioactivity.

2.1. End-Labeling with Klenow

1. 10X Klenow buffer: 200 mM Tris-HCl, pH 7.6, 100 mM MgCl$_2$, 15 mM β-mercaptoethanol, 25 mM dithiothreitol.
2. Labeled nucleotide: ^{32}P-α-dNTP (deoxyribonucleoside triphosphate), most commonly supplied as deoxyadenosine triphosphate (dATP) or deoxycitidine triphosphate (dCTP), but deoxyguanosine triphosphate (dGTP) and deoxythymidine triphosphate (dTTP) are available. It is also possible to substitute a nonisotopic label such as fluoroscein-11-dUTP (deoxyuridine triphosphate) and dioxigenin-11-dUTP.
3. Unlabeled dNTPs:
   a. dNTP mix; a mixture of 0.25 mM of each unlabeled dNTP, excluding that which corresponds to the labeled nucleotide (*see* **Note 1**).
   b. dNTP chase; 0.25 mM dNTP corresponding to the labeled nucleotide (*see* **Note 1**).
4. Klenow: The Klenow (large) fragment buffer of DNA polymerase I at 1 U/μL. Store at −20°C.
5. TE: 10 mM Tris-HCl (pH 7.5), 1 mM ethylenediaminetetraacetic acid (EDTA). Autoclave and store at room temperature.

6. Phenol: Tris-HCl equilibrated phenol containing 0.1% hydroxyquinoline (as an antioxidant). Use Ultrapure, redistilled phenol. Extract repeatedly with 0.5 M Tris-HCl (pH 8.0) until the aqueous phase is 8.0 and then extract once with 0.1 M Tris-HCl (pH 8.0). Can be stored at 4°C for at least 2 mo. Phenol is both caustic and toxic and should be handled with care.
7. Chloroform.
8. Phenol: chloroform mixture: A 1:1 mixture was made by adding an equal volume of chloroform to 0.1 M Tris-HCl, pH 8.0 equilibrated phenol. Can be stored at 4°C for at least 2 mo.
9. Ethanol and 70% ethanol (v/v in water).
10. 5 M Ammonium acetate, pH7.5: Store at room temperature.

2.2. End-Labeling with T4 Kinase

1. 10X CIP buffer: 10 mM ZnCl$_2$, 10 mM MgCl$_2$, 100 mM Tris-HCl, pH 8.3.
2. CIP: At 1 U/µL. Store at 4°C.
3. 10X kinase buffer: 700 mM Tris-HCl, pH 7.6, 100 mM MgCl$_2$, 50 mM dithiothreitol.
4. ^{32}P-γ-ATP: Specific activity > 3000 ci/mmol.
5. T4 kinase: T4 polynucleotide kinase at 1 U/µL. Store at –20°C.
6. Cold ATP: 1.0 mM ATP (freshly made from 20 mM stock).

3. Methods

3.1. End-Labeling with Klenow

1. Resuspend 1–1000 ng of DNA in 42 µL of dH$_2$O (see **Note 2**). Add 5 µL of 10X Klenow buffer, 1 µL of ^{32}P-α-dNTP, 1 µL of dNTP mix, and 1 µL of Klenow. Incubate at room temperature for 15 min (see **Note 3**).
2. Add 1 µL of dNTP chase. Incubate at room temperature for a further 15 min (see **Notes 1** and **4**).
3. Add 50 µL of TE followed by 100 µL of phenol:chloroform. Vortex briefly and separate by centrifugation at 12,000g in a microfuge (see **Note 5**).
4. Remove the aqueous (top) phase to a fresh tube and add 100 µL of chloroform. Separate the layers as in **step 3** and remove the aqueous phase to a fresh tube. Care must be taken, as the discarded reagents are contaminated with unincorporated ^{32}P-α-dNTP.
5. Add 60 µL (0.6 vol) of 5 M ammonium acetate and 200 µL (2 vol) of ethanol (see **Note 6**) and place on ice for 5 min. Centrifuge at 12,000g for 15 min. Carefully remove the supernatant (remember that it is radioactive) and wash the pellet in 70% ethanol.
6. Air dry the pellet for 10 min and resuspend in the required amount of TE (10–100 µL).

The labeled DNA can be either immediately separated by gel electrophoresis and detected by autoradiography (see **Note 7**), or digested further with a second restriction enzyme. In either case, it is a good idea to count a 1-µL sample in a scintillation counter; between 5000 and 10,000 counts are required to detect the fragment by autoradiography. Possible causes of poor labeling and possible solutions are discussed in **Notes 8–10**.

3.2. End-Labeling with T4 Kinase

1. Dissolve 1–2 µg of restriction enzyme digested DNA in 44 µL of dH$_2$O. Add 5 µL of 10X CIP buffer and 0.05–1 U of CIP (see **Note 11**). Incubate for 30 min at 37°C (see **Notes 12** and **13**).
2. Heat inactivate at 60°C for 10 min. Phenol extract and precipitate as in **Subheading 3.1.**, **steps 3–5** (see **Notes 14** and **15**).

3. Resuspend the DNA in 17.5 µL of dH$_2$O. Add 2.5 µL of 10X kinase buffer, 5 µL of ^{32}P-γ-ATP, and 1 µL of T4 kinase. Incubate at 37°C for 30 min.
4. Add 1 µL of cold ATP and incubate for a further 30 min (*see* **Note 16**).
5. Phenol extract and precipitate as in **Subheading 3.1.**, **steps 3–6** (*see* **Note 17**).

4. Notes

1. Unlabeled dNTPs are required for two reasons. First, the labeled nucleotide may not correspond to the first nucleotide to be filled within the restriction enzyme site. In the example shown in **Fig. 1A**, which is a *Bam*HI site, the labeled nucleotide, dCTP*, corresponds to the fourth nucleotide, therefore the other three nucleotides must be filled with cold dNTPs before the label is incorporated. For convenience, a general 7.5 m*M* mix of the unlabeled dNTPs can be used regardless of the actual composition of the restriction enzyme site. Second, a "chase" is required to generate molecules with flush ends as the polymerase stalls in the limited concentrations of labeled nucleotide. This step may be omitted in cases in which the heterogeneous-sized termini are not a problem, for example, when labeling large DNA fragments for separation by agarose gel electrophoresis.
2. The fill-in reaction is very robust, and provided that Mg^{2+} is present, can be carried out in almost any buffer. This means that it is possible to carry out the reaction by simply adding the labeled dNTP, unlabeled dNTPs, and Klenow directly to restriction enzyme mix at the end of digestion.
3. Because only a small region of DNA is labeled in this reaction, it proceeds very quickly. Incubation at room temperature is sufficient, unless ^{35}S-labeled dNTP is used when labeling should be carried out at 37°C. Prolonged incubation can result in degradation of the DNA ends.
4. The labeled DNA may be used for gel electrophoresis at this point, but it must be remembered that unincorporated ^{32}P-α-dNTP will be present in the DNA solution. This may increase the exposure of the operator and increase the risk of contamination when carrying out gel electrophoresis.
5. An alternative purification is to pass the DNA through a Sephadex G50 spin column.
6. If only very small amounts of DNA are present, it may be necessary to add carrier such as 10 µg of tRNA or glycogen.
7. The gel should be fixed in 10% acetic acid or trichloroacetic acid (TCA) before drying to prevent contamination of the gel dryer.
8. Klenow is rarely affected by inhibitors, but it rapidly loses its activity if it is warmed in the absence of a substrate. It can be one of the first enzymes to be lost from the general enzyme stock. If the activity of the enzyme is in doubt, carry out a test reaction by labeling control DNA. Generally, DNA markers are good for this, but check the structure of the ends before proceeding.
9. The structure of the end is important, as the enzyme can only "fill-in" those bases present in the site. Recheck the sequence of the single-strand end produced by restriction enzyme digestion. It may be possible to exchange the ^{32}P-α-dNTP for another radionucleotide that has a higher specific activity.
10. The Klenow fill-in reaction incorporates only a small number of ^{32}P-labeled nucleotides per DNA molecule. If higher levels of incorporation are required, T4 DNA polymerase may be used. T4 DNA polymerase has 200-fold higher 3'–5' exonuclease activity than does Klenow. If the DNA fragments are incubated in the absence of dNTPs, this enzyme will produce a region of single-stranded DNA, which can be subsequently labeled with higher incorporation by adding ^{32}P-α-dNTP and cold dNTPs to the mix *(7)*.
11. One unit of CIP dephosphorylates 50 pmol of ends in 1 h (for a 5-kb fragment, 1 pmol of ends is approx 2 µg).

12. The efficiency of dephosphorylation of blunt and 5' recessed ends is improved by incubating the reaction at 55°C.
13. The phosphatase reaction can be carried out in restriction enzyme buffer by adding 0.1 vol of 500 mM Tris-HCl, pH 8.9, 1 mM EDTA, and the required amount of enzyme.
14. It is important to remove all phosphatase in order to prevent removal of the newly incorporated labeled phosphate.
15. The T4 kinase reaction is very sensitive to inhibitors such as those found in agarose. Care should be taken to ensure that the DNA is inhibitor free. In addition T4 kinase will readily phosphorylate RNA molecules, therefore the presence of RNA should be avoided, as this will severely reduce the incorporation of labeled ^{32}P into the DNA.
16. The labeling reaction is only approximately 10% efficient. To get all molecules phosphorylated, it is necessary to chase the reaction with excess cold ATP.
17. This is a poor way to purify oligonucleotides. A Sephadex G25 spin column should be used instead.

References

1. Pickersky, E. (1996) Terminal labelling for Maxam-Gilbert sequencing, in *Basic DNA and RNA Protocols*, Methods in Molecular Biology, vol. 58 (Harwood, A. J., ed.), Humana, Totowa, NJ, pp. 441–446.
2. Wallace, R. B., Shaffer, J., Murphy, R. F., Bonner, J., Hirose, T., and Itakura, K. (1979) Hybridisation of synthetic oligodeoxyribonucleotides to ϕχ174 DNA: the effect of single base pair mismatch. *Nucleic Acid Res.* **6,** 3543–3557.
3. Scott, V., Clarke, A. R., and Docherty, K. (1994) *Protocols for Gene Analysis*, Methods in Molecular Biology, vol. 31 (Harwood, A. J., ed.), Humana, Totowa, NJ, pp. 339–345.
4. Harwood, A. J. and Phear, G. A. (1996) Direct sequencing of PCR products. *Basic DNA and RNA Protocols*, Methods in Molecular Biology, vol. 58 (Harwood, A. J., ed.), Humana, Totowa, NJ, pp. 403–412.
5. Klenow, H., Overgaard-Hansen, K., and Patkar, S. A. (1971) Proteolytic cleavage of native DNA polymerase into two different catalytic fragments. *Eur. J. Biochem.* **22,** 371–381.
6. Challberg, M. D. and Englund, P. T. (1980) Specific labelling of 3' termini with T4 DNA polymerase. *Methods Enzymol.* **65,** 39–43.

20

Nick Translation and Random Hexamer Labeling of DNA

Jane Davenport-Jones

1. Introduction

Labeled nucleotides (radioactive or fluorescent) can be incorporated efficiently into double-stranded DNA by nick translation. Nick translation works by using DNase and DNA polymerase I enzymes. DNase cuts one strand of the DNA, exposing 5'-phosphoryl and 3'-hydroxyl (OH) termini DNA. Polymerase I adds deoxyribonucleoside triphosphate (dNTPs), including labeled dNTPs to the exposed 3'-OH strand and at the same time, the polymerase exonuclease activity digests from the exposed 5' end. In this way, a new complementary strand, including labeled dNTPs is produced (*1*). It is also possible to incorporate radioactive nucleotides into DNA using a enzymatic primer extension technique (*2*). In this method, random hexanucleotides are annealed to the denatured DNA to be used as the probe. These are used as a primer for enzymatic extension in the presence of the four dNTPs, one of which is radiolabeled.

2. Materials (*see* Note 1)

2.1. Nick Translation of DNA

1. 10X Nick-translation buffer: 0.5 M Tris-HCl (pH 7.5), 0.1 M MgSO$_4$, 1 mM dithiothreitol, 500 mg/mL bovine serum albumin (optional).
2. DNase I: 10 ng/mL.
3. DNA polymerase I: 0.5 U/µL.
4. Unlabeled dNTP: 2 mM each of dATP, dGTP, and dTTP.
5. Radiolabeled dCTP: 10 mCi/mL [α-^{32}P]dCTP, specific activity approx 3000Ci/mmol (*see* **Note 2**). This is stored at –20°C and should be removed from the freezer approx 20 min before setting up the reaction.
6. Stop solution: 0.5 M ethylenediaminetetraacetic acid (EDTA) (pH 8.0).
7. Sephadex (Amersham Pharmacia Biotech, St. Albans, UK) separation spin column (*see* **Note 3**).

2.2. Random Hexamer Labeling of DNA

1. DNA probe to be labeled in TE buffer (10 mM Tris-HCl, 1 mM EDTA, pH 8.0).
2. Hexamer mix: 0.043 M each dCTP, dTTP, dGTP, 0.43 M HEPES, pH 7.0, 12 U/mL random hexanucleotides (Amersham Pharmacia Biotech, St. Albans, UK).

3. [α-^{32}P]dATP, specific activity 6000 Ci/mM (Amersham Pharmacia Biotech, Buckinghamshire, UK). This is stored at −20°C and should be removed from the freezer approx 20 min before setting up the reaction.
4. Stop solution: 0.5 M EDTA (pH 8.0).
5. DNA polymerase (e.g., Klenow fragment) (6 U/μL).

3. Methods

3.1. Nick Translation of DNA

1. Dilute DNA to be labeled to 20–200 ng/mL with sterile distilled H$_2$O and add 1 mg to a sterile microcentrifuge tube.
2. Add to the tube the following:
    a. 10 μL 10X nick translation buffer
    b. 10 μL 20 nM unlabeled dNTPs
    c. 10 μL 30 pmol labeled [α-^{32}P]dCTP
3. Add 1 ng/mL DNase (10 μL) and 2.5 U DNA polymerase I (5 μL). Gently mix by pipeting solution up and down.
4. Add water to ensure a final volume of 100 μL.
5. Incubate for 2 h at 15°C.
6. Stop the reaction by adding 10 μL stop solution.
7. The probe is now ready for hybridization. However, it may be necessary to remove any unincorporated nucleotides, using Sephadex spin columns (*see* **Notes 3** and **4**).

3.2. Random Hexamer Labeling of DNA

1. Take 25–100 ng of DNA to be labeled and adjust the volume of TE to 11 μL.
2. Denature the DNA by boiling for 5 min and transfer immediately to an ice bucket.
3. Add 11 μL of the primer mix, 2 μL of the [α-^{32}P]dATP and 3 U Klenow polymerase (0.5 μL).
4. Incubate the mix at room temperature for approx 4 h.
5. Add 5 μL of stop mix to terminate the reaction.
6. At this point the probe may be purified from free nucleotides by use of Sephadex spin columns (*see* **Notes 3** and **4**).
7. Following recovery of the labeled DNA, it must be rendered single stranded by boiling before it may be used in hybridization experiments.

4. Notes

1. Enzymes and buffers are now available in kit forms (Amersham [Amersham Pharmacia Biotech, Buckinghamshire, UK], Promega [Promega, Southhampton, UK]); however, slight variations exist in concentrations of enzymes and buffer ingredients.
2. Nick translation can also be used to label DNA with nonradioactive markers, including incorporation of Cy3-dCTP and fluorescein, or rhodamine-dUTP into DNA. However, radiolabeled probes are more sensitive markers for low quantities of DNA. It is also possible to label more than one dNTP, if higher specific activity is required for hybridizing low amounts of DNA. However this increases nonspecific hybridization.
3. To remove unincorporated labeled dNTPs, the probe can be purified by passing the solution through a Sephadex spin column or push column. Unincorporated dNTPs are trapped inside the Sephadex beads, whereas DNA is too big to enter the beads and passes straight through the column.
4. Percentage incorporation and the specific activity of the probe can be calculated as follows by measuring the radioactivity in the mixture before and after separation.

$$\text{percentage incorporation} = \frac{\text{cpm incorporated} \times 100}{\text{total cpm}}$$

$$\text{specific activity (cpm/mg DNA)} = \frac{\text{cpm incorporated} \times \text{dilution} \times 100}{\text{mg input DNA}}$$

References

1. Rigby, P. W. J., Dieckmann, M., Rhodes, C., and Berg, P. (1977) Labelling deoxyribonucleic acid to a high specific activity in vitro by nick translation with DNA polymerase I. *J. Mol. Biol.* **113,** 237–251.
2. Feinberg, A. P. and Vogelstein, B. (1983) A technique for radiolabelling DNA restriction endonuclease fragments to a high specific activity. *Anal. Biochem.* **132,** 6–13.

21

Generation of Labeled Probes by Polymerase Chain Reaction

Y. M. Dennis Lo and Shu F. An

1. Introduction

Nucleic acid probes are an important tool in molecular diagnosis. To facilitate the detection of hybridized probes, they are labeled with a reporter molecule, which is usually a radioisotope. For diagnostic techniques carried out in a clinical laboratory, radioisotopes are hazardous and, thus, recently there is a move to use nonisotopic labels, such as biotin and digoxigenin. Nonisotopic probes also have the advantage of much better stability over time compared with isotopic probes that have limited half-lives.

Conventional techniques using cloned DNA for probe production are time-consuming, involving the preparation of plasmid DNA with the probe DNA as an insert, restriction enzyme digestion followed by preparative gel electrophoresis to separate the probe DNA from the plasmid, and then finally the carrying out of the labeling reaction (typically nick translation *[1]* or random priming *[2]*).

The development of the polymerase chain reaction (PCR) allows an alternative, more efficient method in which probe DNA is amplified and labeled in a single step by the amplification reaction. The most important advance being the demonstration that the *Thermus aquaticus (Taq)* DNA polymerase commonly used in PCR is able to incorporate nucleotide analogs during DNA amplification *(3)*. Nucleotide analogs such as biotin-11-dUTP *(3)* or digoxigenin-11-dUTP *(4)* have been successfully used as substrates for PCR labeling. In this chapter, we present the protocols for using the PCR to generate biotin- and digoxigenin-labeled probes.

2. Materials

1. Reagent buffer, dATP, dCTP, dGTP, dTTP and *Taq* polymerase for PCR from Perkin-Elmer (Foster City, CA).
2. Biotin-11-dUTP in powder form from Sigma (St. Louis, MO), to be dissolved in 100 mM Tris-HCl, pH 7.5, 0.1 mM edetic acid (EDTA) to make a 0.3 mM solution before use.
3. Digoxigenin-11-dUTP (dig-11-dUTP) from Boehringer Mannheim (East Sussex, UK).
4. DNA thermocycler, such as the Biometra TRIO thermoblock.
5. PCR primers from British Biotechnology Ltd. (Abingdon, UK). Primer sequences are listed in **Table 1**.
6. Paraffin oil (optional) to prevent evaporation during PCR (BDH, Merck Ltd., Dorset, UK).

Table 1
Sequence of PCR Primers

Sequence name	Sequence
PCR1	5'GATTGAGATCTTCTGCGACGC3'
PCR2	5'GAGTGTGGATTCGCACTCCTC3'
BB1660	5'CTTGTTGACAAGAATCCTCAC3'
BB1661	5'GATGGGATGGGAATACA3'
BB1666	5'CAGAGTCTAGACTCGTGG3'
BB1667	5'ACAAACGGGCAACATACCTTG3'
BB1671	5'GACATACTTTCCAATCAATAG3'
T3 promoter	5'ATTAACCCTCACTAAAGGGA3'
SP6 promoter	5'GATTTAGGTGACACTATAG3'
T7 promoter	5'TAATACGACTCACTATAGGG3'

7. 1XTBS: 0.15 M NaCl, 0.015 M Tris-HCl, pH 7.2.
8. Blocking buffer: 0.1 M Tris-HCl, 0.1 M NaCl, 3 mM MgCl$_2$, 0.5% Tween-20, pH 7.5.
9. AP7.5: 0.1 M Tris-HCl, 0.1 M NaCl, 3 mM MgCl$_2$, pH 7.5.
10. AP9.0: 0.1 M Tris-HCl, 0.1 M NaCl, 0.1 M MgCl$_2$, pH 9.0.
11. 1X SSC: 0.3 M NaCl, 0.03 M sodium citrate.
12. Phosphate-buffered saline (PBS): 0.012 M Na$_2$HPO$_4$, 0.04 M KH$_2$PO$_4$, 0.15 M NaCl, pH 7.2.
13. 1 X SET: 0.15 M NaCl, 0.02 M Tris-HCl, 1 mM EDTA, pH 7.8.

3. Methods

3.1. Incorporation of Biotin-dUTP by PCR

1. Prepare PCR mix. Use dTTP and biotin-11-dUTP at a ratio of 3:1, i.e., at 150 µM and 50 µM, respectively (*see* **Note 6**). As a model system, we use PCR1 and PCR2 to amplify a 185-bp fragment from the hepatitis B virus (HBV) genome.
2. Add target DNA. For the production of vector-free probe from plasmid DNA, the amount of plasmid DNA should be kept low, at 0.2 fmol (approx 1 ng of an 8-kb plasmid) (*see* **Note 7**).
3. Carry out thermal cycling. For PCR1 and PCR2, the thermal profile is: 94°C for 10 min, followed by 25 cycles of 94°C for 2 min, 55°C for 2 min, and 72°C for 3 min.
4. Remove as much mineral oil as possible.
5. Add 1 µL of glycogen (20 mg/mL) (Boehringer Mannheim, East Sussex, UK), 10 µL of 2.5 M sodium acetate, pH 5.2, and 220 µL of ethanol. Leave the mixture at –20°C overnight or –70°C for 1 h.
6. Spin down the precipitated labeled PCR product by centrifugation at 11,600g on a Micro Centaur (MSE, Sussex, UK) for 15 min. Discard the supernatant and vacuum desiccate.
7. Dissolve the precipitated probe in 100 µL of 10 mM Tris-HCl, 1 mM EDTA, pH 8.0.
8. Electrophorese 5 µL on an agarose gel. The biotinylated PCR product runs slightly slower compared with the corresponding nonbiotinylated PCR product (**Fig. 1**).
9. The efficiency of the labeling may be checked by running 1 µL of the PCR product on an alkaline denaturing agarose gel, Southern blotting to a nitrocellulose filter (BA 85, Schleicher and Schuell, Keene, NH), and detecting the biotin label by using a streptavidin/alkaline phosphatase system *(5)* (*see* **Subheading 3.2.**) (**Fig. 1**).
10. Store labeled probe at –20°C or –70°C in aliquots.

Generation of Labeled Probes

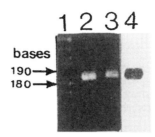

Fig. 1. Synthesis of an 185 bp biotinylated HBV probe using PCR labeling. Lane 1, pBR322 DNA digested with *Msp*I (marker); lane 2, nonbiotinylated PCR product; lane 3, biotinylated PCR product; lane 4, Southern blot of biotinylated PCR product (alkaline denaturing gel). Reprinted with permission from **ref. 3**.

3.2. Filter Hybridization Using PCR-Labeled Biotinylated Probe

As a guideline, in the example using the 185-bp biotinylated PCR probe, 50 ng of the probe in 2 mL of hybridization mix is used for every 50 cm^2 of nitrocellulose filter.

1. Following standard hybridization procedures (5), wash the filters for 5 min using 2X Tris buffered saline (TBS), 0.1% (w/v) sodium dodecyl sulfate (SDS) at 22°C. Repeat the washing step twice more, making a total of three times.
2. Wash three times, each for 5 min with 0.5X TBS, 0.1% SDS at 22°C.
3. Wash three times, each for 15 min with 0.5X TBS, 0.1% SDS at 60°C. Prior heating of the washing solution to 60°C is essential.
4. Block filters with blocking buffer for 60–90 min.
5. Dilute streptavidin to 2 µg/mL in incubation buffer AP7.5, 0.05% Tween-20.
6. Incubate filters with diluted streptavidin for 10 min with gentle rocking. As a guideline, use 3–4 mL of diluted streptavidin per 100 cm^2 filter area.
7. Wash filters three times, each with blocking buffer for 5 min.
8. Make up biotinylated alkaline phosphatase to 1 µg/mL in incubation buffer.
9. Incubate filters with biotinylated alkaline phosphatase for 10 min with gentle rocking.
10. Repeat the wash three times, each with blocking buffer for 3 min.
11. Wash filters twice, each with substrate buffer AP 9.0 for 3 min.
12. Make up nitro blue tetrazolium (NBT) (Sigma) and 5-bromo-4-chloro-3-indoly phosphate (BCIP) (Sigma) stock solutions by dissolving 1 mg NBT in 40 µL 70% dimethylformamide and 2 µg BCIP in 40 µL 100% dimethylformamide.
13. Make up substrate solution by adding 36 µL and 40 µL stock solutions of NBT and BCIP, respectively, to 12 µL of substrate buffer at 22°C.
14. Incubate filters with substrate solution for up to 20 h in plastic Petri dishes or sealed plastic hybridization bags in the dark. Monitor the color development periodically. Under optimal conditions, 1 pg of target DNA could be detected by Southern blotting using PCR-labeled biotinylated probe (**Fig. 2**).
15. Following color development, wash filters extensively with water.

3.3. Incorporation of Digoxigenin-11-dUTP

1. Set up the PCR reagent mix with 200 µ*M* each of dATP, dGTP, and dCTP. Use dTTP and digoxigenin-11-dUTP at a ratio of 3:1, i.e., at 60 µ*M* and 20 µ*M*, respectively (*see* **Note 6**).
2. Add target DNA. For the production of vector-free probe from plasmid DNA, the amount of plasmid should be kept low at 0.2 fmol (approx 1 ng of an 8-kb plasmid).

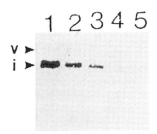

Fig. 2. Sensitivity and insert-specificity of the 185 bp PCR-generated biotinylated probe using serial dilutions of pHBV130 restricted with XhoI to release insert. v and i mark the positions of vector and insert, respectively. The amounts of insert in the lanes are: lane 1, 1 ng; lane 2, 100 pg; lane 3, 10 pg; lane 4, 1 pg; lane 5, 0.1 pg. Reprinted with permission from **ref. 3**.

Table 2
Product Size and Genomic Location of Primer Combinations

Primer combination	PCR product size (bp)	Location in HBV genome
PCR1/PCR2	185	Core-polymerase region
BB1666/BB1667	233	Surface antigen gene
BB1660/BB1661	401	Surface antigen gene
BB1666/BB1671	749	Surface-polymerase region
BB1660/BB1671	777	Surface-polymerase region

3. Carry out thermal cycling. The thermal profile is: 94°C for 10 min, followed by 30 cycles consisting of 94°C for 1.5 min, 60°C for 1.5 min, and 72°C for 3 min. This profile is found to work well with the primer combinations listed in **Table 2**.
4. Remove as much mineral oil as is possible.
5. Electrophorese 10 µL of the PCR product on an agarose gel. The digoxigenin-labeled PCR product should exhibit decreased electrophoretic mobility compared with a nondigoxigenin-labeled PCR product.
6. The efficiency of labeling may be checked by electrophoresing 1 µL of the PCR product on an alkaline denaturing agarose gel, Southern blotting to a nitrocellulose filter (BA 85, Schleicher and Schuell). Detection of the digoxigenin label is detailed below.
7. Store probe at either –20°C or –70°C in aliquots.

3.4. Production of Single-Stranded Digoxigenin-Labeled Probe

For the production of single-stranded probes, two rounds of PCR are performed: the first round without the label and a second round of asymmetric PCR for label incorporation (see **Notes 1, 2, and 10**).

1. Set up the first round PCR using 200 µM dTTP without dig-11-dUTP.
2. Following the first round, centrifuge the product through a Centricon-30 membrane (Anachem, Luton, UK) to remove the free primers.
3. Set up a second round of linear amplification using only one primer (either one of the primers used in the first round or an internal one), 20 µM dig-11-dUTP and 60 µM TTP. Sixty nanograms of the first round product is used as the template for the second round. Thirty cycles (94°C for 1.5 min, 60°C for 1.5 min, and 72°C for 3 min) are performed.

3.5. Use of Digoxigenin-Labeled Probe in Filter Hybridization

As a guideline, digoxigenin-labeled PCR probes are used at a concentration of 50 ng/µL of hybridization mix.

1. Following standard hybridization procedures (6), wash filters twice, each for 5 min at 22°C with 2X SSC, 0.1% (w/v) SDS.
2. Wash twice, each for 15 min at 65°C with 0.1X SSC, 0.1% SDS.
3. Block filters for 1 h at 22°C with 1X AP7.5, 0.5% Tween-20.
4. Dilute antidigoxigenin alkaline phosphatase Fab fragment (Boehringer Mannheim) to 150 mU/mL (1: 5000) with 1X AP7.5, 0.05% Tween-20, 2% bovine albumin (Sigma).
5. Incubate filters with antidigoxigenin for 30 min at 22°C.
6. Wash twice, each for 5 min with 1X AP7.5, 0.05% Tween-20.
7. Wash twice, each for 5 min with 1X AP7.5.
8. Wash twice, each for 5 min with 1X AP9.0.
9. Make up nitro blue tetrazolium (NBT) (Sigma) and 5-bromo-4-chloro-3-indoly phosphate (BCIP) (Sigma) stock solutions by dissolving 1 mg NBT in 40 µL 70% dimethylformamide and 2 mg BCIP in 40 µL 100% dimethylformamide.
10. Develop filters with substrate solution (36 µL NBT solution and 40 µL BCIP solution / 12 mL of 1X AP9.0). Under optimal conditions, between 1 and 10 pg of target DNA could be detected on Southern blotting using a PCR-generated digoxigenin probe.

3.6. In Situ Hybridization Using PCR-Generated Digoxigenin Probes

1. Put 5-µm sections onto 1% silane treated slides using sterile distilled water and dry overnight at 22°C.
2. Heat section for 10 min at 80°C. Then dewax in Citroclear (HD Supplies, Aylesbury, UK) three times, each for 5 min, followed by industrial methylated spirits three times, each for 5 min. Hydrate in sterile distilled water.
3. Protease digest to access nucleic acid target. This step is crucial–too much digestion results in loss of morphology and nucleic acid (especially mRNA), too little results in reduced sensitivity and higher background. Incubate slide in Protease 8 (Sigma) in PBS at 37°C with gentle shaking. Concentration and time depend on tissue type and fixation. Typically, we use a concentration of 50 mg/mL for 20 min at 22°C for DNA target and a concentration of 5 mg/mL for 20 min at 22°C for RNA.
4. Rinse slide in TBS.
5. For DNA target, denature slide/section in sterile distilled water at 95°C for 20 min. For RNA, denature slide/section in 70% formamide in 2X SET at 37°C for 20 min.
6. Rinse in ice cold water 3 min, air dry.
7. Prepare mix: 100 µL sterile distilled water, 100 µL 0.1% SDS, 100 µL 10 mg/mL polyvinyl pyrrolidone, 400 µL 50% dextran sulphate in 10X SET, 1000 µL deionized formamide. Mix and filter through 0.2 µm filter. Divide into 170 µL aliquots and store at –20°C. Before use add 10 µL 10 mg/mL salmon sperm DNA and 20 µL probe at 10 mg/mL.
8. Denature probe/mix at 95°C for 20 min. As a guideline, we use PCR-generated digoxigenin-labeled probes at a concentration of 10 ng/section for *in situ* hybridization.
9. Apply 7 to 16 µL of denatured probe/mix to coverslip and invert section onto probe/mix and seal with silicone grease.
10. For DNA, denature probe/section again by placing on a hot plate at 95°C for 10 min.
11. Hybridize at 37°C for 2–24 h in a moist Petri dish. Time depends on amount of target and concentration of probe (higher requires less time).

12. Remove coverslips in TBS/0.5% Triton X100 and rinse three times, each for 5 min at 37°C in same. Then rinse three times, each for 5 min in 0.5X TBS at 65°C.
13. Block with 15% dried milk in AP7.5/0.5% Triton X100 for 20 min at 37°C.
14. Rinse and apply alkaline phosphatase conjugated antidigoxigenin (Boehringer Mannheim) 1 to 750 in 2% BSA/AP7.5, 0.5% Triton X100 for 30 min at 37°C.
15. Rinse in AP7.5/0.5% Triton X100 three times, each for 5 min at 22°C, then in AP9.0 three times, each for 5 min at 22°C.
16. Develop in NBT/BCIP as for filters (typically overnight).

4. Notes

1. The specificity of PCR-generated single strand probes can be confirmed by demonstrating that hybridization only occurs to mRNA with antisense strand probes, but not with sense strand probes.
2. When used for *in situ* hybridization, single strand probes are found to be more sensitive than double strand ones.
3. Longer probes are generally found to be more sensitive than shorter ones. When two or more small probes that have overlapping sequences are used concurrently, an additive effect in terms of sensitivity is seen.
4. The recent development of long range PCR technology *(7)* has provided the possibility of producing longer probes than previously achievable.
5. It is also possible to generate labeled probes by using primers with 5' end labels, e.g., biotin. However, this approach has lower sensitivity than the approach outlined in this chapter because only a single label molecule is available subsequently for detection using the labeled-primer approach. However, the labeled-primer approach does have the advantage that, compared with internally labeled probes, steric hindrance is less likely to be a problem. It has been demonstrated that internally labeled biotinylated probes have reduced T_m *(8)*. This point would need to be taken into consideration when using PCR-labeled probes in hybridization experiments.
6. The ratio of the labeled-deoxynucleotide triphosphate and dTTP should be carefully controlled. In our laboratory, we use 150 µM dTTP to 50 µM biotin-11-dUTP and 60 µM dTTP to 20 µM digoxigenin-11-dUTP for the biotin and digoxigenin-labeled system, respectively.
7. When the source of probe material is plasmid DNA and it is necessary to produce vector-free probe, it is important to use a relatively small amount of plasmid DNA (approx 0.2 fmol or 1 ng of an 8 kb plasmid). The use of larger amounts of starting plasmid DNA will result in a significant amount of labeled vector sequences.
8. For probe sequences that have been cloned into expression vectors, it is possible to use universal primers to the RNA polymerase promoter sites as amplification primers. Primer sequences to the T_3, T_7, and SP6 promoter sites are listed in **Table 1**. For example, the T_3 and SP6 primers can be used as a pair in PCR for insert flanked by the T_3 promoter on one side and the SP6 promoter on the other.
9. PCR can also be used to produce labeled probes from genomic DNA. Due to the sequence complexity of DNA, we suggest that an initial round of PCR using conventional deoxynucleotide triphosphates, without the label, is first performed. The PCR product is then gel-purified and then a second round of PCR is carried out by including the labeled deoxynucleotide triphosphate to label the product. High background may result if this two-step approach is not used.
10. Single-strand PCR-labeled probes are at least twice as sensitive as their double strand counterparts of the same size, possibly due to the absence of reannealing. Single strand

probes also have the advantage of providing strand specificity. We find that heating prior to hybridization seems to improve the sensitivity of experiments carried out using single strand probes, probably as a result of the denaturation of the secondary structure of these single strand molecules.
11. Incorporation of biotin labels during PCR has also been used as a means of quantifying PCR products *(9)*.

References

1. Rigby, P. W. J., Dieckmann, M., Rhodes, C., and Berg, P. (1977) Labelling deoxyribonucleic acid to high specific activity in vitro by nick translation with DNA polymerase. *J. Mol. Biol.* **113**, 237–247.
2. Feinberg, A. P. and Vogelstein, B. (1983) A technique for radiolabeling DNA restriction endonuclease fragments to high specific activity. *Anal. Biochem.* **132**, 6–13.
3. Lo, Y. M. D., Mehal, W. Z., and Fleming, K. A. (1988) Rapid production of vector-free biotinylated probes using the polymerase chain reaction. *Nucleic Acids Res.* **16**, 8719.
4. An, S. F., Franklin, D., and Fleming, K. A. (1992) Generation of digoxigenin-labeled double-stranded and single-stranded probes using the polymerase chain reaction. *Mol. Cell. Probes* **6**, 193–200.
5. Chan, V. T. W., Fleming, K. A., and McGee, J. O. (1985) Detection of subpicogram quantities of specific DNA sequences on blot hybridisation with biotinylated probes. *Nucleic Acids Res.* **13**, 8083–8091.
6. Boehringer Mannheim. (1989) *Biochemica-Applications Manual: DNA Labelling and Nonradioactive Detection.* Boehringer Mannheim GmbH Biochemica, East Sussex, UK.
7. Barnes, W. M. (1994) PCR amplification of up to 35 kb DNA with high fidelity and high yield from Lambda bacteriophage templates. *Proc. Natl. Acad. Sci. USA* **91**, 2216–2220.
8. Weir, H. U. G., Segraves, R., Pinkel, D., and Gray, J. W. (1990) Synthesis of Y chromosome-specific labelled DNA probes by in vitro DNA amplification. *J. Histochem. Cytochem.* **38**, 421–426.
9. Duplaa, C., Couffinhal, T., Labat, L., Moreau, C., Lamaziere, J., and Bonnet, J. (1993) Quantitative analysis of polymerase chain reaction products using biotinylated dUTP incorporation. *Anal. Biochem.* **212**, 229–236.

22

Nonradioactive Oligonucleotide Probe Labeling

Sue Fowler and Ian Durrant

1. Introduction

Nonradioactive oligonucleotide probes are increasingly being used in various applications, for example, polymerase chain reaction (PCR) product detection, *in situ* hybridization and colony and plaque screening *(1)*. Oligonucleotide probes are short, defined sequences, 15–50 bases long, that are relatively easy to produce. As the base composition of oligonucleotide probes dictates the hybridization properties *(2)*, probes can be designed to discriminate between perfectly matched and mismatched target sequences.

There are three basic approaches available for nonradioactive labeling of oligonucleotide probes (*see* **Fig. 1**). Biotin phosphoramidite compounds can be used to add biotin directly to the oligonucleotide; some of these compounds can be used in a repetitive manner on an oligonucleotide synthesizer to form a tail of biotin residues *(3)*. The extremely high coupling efficiency of these compounds (>94%) allows the labeled probe to be used with the minimum of purification. These hapten-based approaches rely on the detection of the hapten with enzyme-linked immunological molecules. Alternatively, the enzyme can be linked to the oligonucleotide through an amino or thiol group attached during synthesis *(4,5)*. The amino link route has been demonstrated with alkaline phosphatase *(6)*, but the chemistry is complex and requires a number of purification processes. However, the reaction of a fivefold molar excess of suitably derivatized horseradish peroxidase (HRP) with a thiol-linked oligonucleotide *(7)* is simple and rapid and achieves a high labeling efficiency. Further probe purification is not required, as the remaining free HRP does not interfere with the subsequent hybridization and detection processes. Finally, oligonucleotides can be labeled using an enzyme catalyzed reaction. A hapten-modified nucleotide is used in conjunction with terminal deoxynucleotidyl transferase to produce a 3'-end tail on the probe sequence *(8)*. The tailing reaction is controlled to produce a short tail of 4–10 nucleotides long to give the maximum sensitivity without compromising the stringency of the probe.

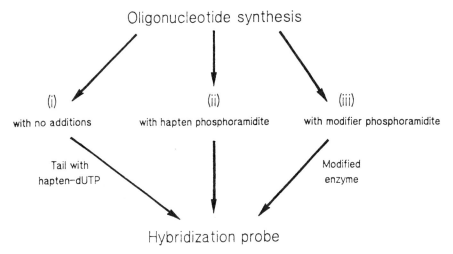

Fig. 1. Outline of oligonucleotide labeling methods.

2. Materials

2.1. 3'-End Labeling

1. Fluorescein-labeled nucleotide.
2. Cacodylate buffer: 10X stock solution containing sodium cacodylate pH 7.4, dithiothreitol, and cobalt chloride. *Note: Cacodylate is an arsenic-based compound and is classified as highly toxic. Avoid contact with skin and do not swallow. It is a possible carcinogen and there are dangers of cumulative effects. This compound, like all others, should be used within the principles of good laboratory practice. Wear appropriate protection, such as overalls, gloves, and safety glasses.*
3. Terminal transferase: 8 U/µL stock solution.

The foregoing materials are available commercially in an optimized format (ECL 3'oligo-labeling system; RPN 2130, Amersham Pharmacia Biotech, Amersham, UK).

2.2. Biotin Labeling

4. Biotin phosphoramidite: Available commercially (e.g., Amersham Pharmacia Biotech, RPN 2012).
5. Oligonucleotide synthesizer: Various commercial systems are available (e.g., Applied Biosystems, Foster City, CA).

2.3. Enzyme Labeling

6. Thiol modifier: Available commercially (e.g., Applied Biosystems).
7. Oligonucleotide synthesizer: as before.
8. Derivatized HRP: Horseradish peroxidase derivatized with a bifunctional crosslinking agent that leaves a thiol reactive group attached to the enzyme (e.g., using SMCC* [Pierce, Rockford, IL]).

2.4. Hybridization

9. 20X standard saline citrate (SSC) stock: 3 M NaCl, 0.3 M trisodium citrate.

*Succinimidyl 4-(N-maleimidomethyl) cyclohexane-1-carbonate

Oligonucleotide Probe Labeling

10. Blocking agent: Optimized blocking agent available commercially (Gene Images liquid block: RPN 3601, Amersham Pharmacia Biotech).
11. Sodium dodecylsulfate (SDS) stock: 20% (w/v) in distilled water.
12. Hybridization buffer: 5X SSC, 0.1% (w/v) SDS, 5% (w/v) dextran sulphate (e.g., Amersham Pharmacia Biotech product no. US70796), 1/20 volume liquid block reagent. Only gentle heating should be required to dissolve the components. The buffer can be made up in a large volume and stored in suitable aliquots at $-20°C$ for at least 3 mo.

3. Methods

3.1. 3'-End Labeling

1. Add the labeling reaction components (*see* **Notes 1–3**) in the following order: 100 pmol of oligonucleotide, 2.5 µL of fluorescein-labeled nucleotide, 8 µL of cacodylate buffer, sterile water to a volume of 76 µL. Then add 4 µL of terminal transferase and incubate for 90 min at 37°C.
2. Store the labeled probe on ice for immediate use or store at $-20°C$ for long-term storage of up to 6 mo.

3.2. Enzyme Labeling

1. Prepare thiol-modified oligonucleotide as detailed by the thiol modifier supplier and in compliance with the operating instructions of the oligonucleotide synthesis machine (*see* **Note 4**).
2. Purify thiol-modified oligonucleotide by a recommended method. Optimal results will be achieved by high-performance liquid chromatography (HPLC) (*see* **Note 5**; Chapter 18). Thiol-labeled oligonucleotides can be stored at $-20°C$ for several months.
3. Desalt 5 µg of oligonucleotide (*see* **Note 6**) in a volume of 50 µL on a Sephadex G25 spin column, or equivalent, equilibrated in water *(9)*.
4. Immediately add the desalted oligonucleotide to a tube of lyophilized, derivatized HRP (*see* **Note 7**). Mix gently in the pipet tip to redissolve the HRP. Incubate at room temperature for 1 h (*see* **Note 8**).
5. Store labeled probe at a concentration of 1 µg/100 µL in 50% glycerol for up to 6 mo.

3.3. Biotin Labeling

1. Use biotin phosphoramidite as detailed by the supplier (*see* **Note 9**) to perform a standard oligonucleotide synthesis under normal operating conditions (*see* **Note 10**), specifying biotin at the desired positions; usually at the 5'-end, in any number (*see* **Note 11**).
2. Purify the labeled oligonucleotide, preferably by HPLC (*see* **Note 12**; Chapter 18), and store labeled sequences at $-20°C$ for at least 6 mo.

3.4. Hybridization

1. The probes generated from all three labeling systems can be used with a simple hybridization buffer (*see* **Note 13**).
2. Prehybridize blots in the hybridization buffer for 15 min at the hybridization temperature to be used.
3. Hybridize for 1–2 h at the desired temperature, typically 42°C, at a probe concentration of 10 ng/mL (*see* **Note 14**).
4. Perform posthybridization washes at the required stringency, depending on the melting temperature of the oligonucleotide probe (*see* **Note 15**). Stringency is controlled by a combination of salt concentration and temperature, as traditionally performed with radio-

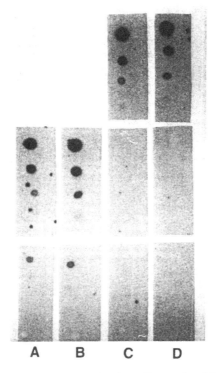

Fig. 2. Demonstration of stringency control using oligonucleotides carrying a 3'-tail of fluorescein-dUTP. Columns (**A**) and (**C**): Probes labeled with Fl-dUTP at the 3'-end and ^{32}P at the 5'-end. Columns (**B**) and (**D**): Probes labeled at the 5'-end only, with ^{32}P. Top row: 17-base probe perfectly matched to the target. Middle row: 17-base probe with one mismatch for the target. Bottom row: 17-base probe with two mismatches for the target. All filters were hybridized at 42°C for 2 h with a probe concentration of 5 ng/mL. Stringency washes were performed in 1X SSC, 0.1% SDS at room temperature (columns [A] and [B]) or at 50°C (columns [C] and [D]). Autoradiographed overnight at –70°C.

actively labeled probes. Even for probes directly linked to HRP, the temperature can be raised to 60°C for the short stringency wash incubations (*see* **Note 16**). A typical wash protocol would be as follows. (a) Wash in 2–5X SSC, 0.1% SDS for 2X 5 min at room temperature. (b) Wash in 1–3X SSC, 0.1% SDS for 2X 20 min at the desired stringency temperature.

3.5. Detection

After hybridization, the various labels used on the probes have to be visualized in some way. There are a number of methods available, principally involving HRP and alkaline phosphatase, for the production of colored precipitates or light as a final endpoint *(10)*.

1. A particularly simple, rapid, and sensitive endpoint that can be applied to all three oligonucleotide-labeling systems is the light-producing reaction, enhanced chemiluminescence, catalyzed by HRP *(11)*. Blots hybridized with probes directly labeled with HRP are placed in the ECL (Amersham Pharmacia Biotech, Amersham, UK) substrate immediately after

Oligonucleotide Probe Labeling

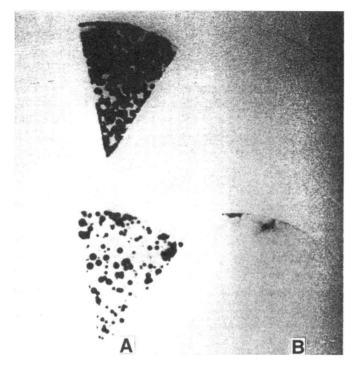

Fig. 3. Plaque screening with 3'-end labeled probes. M13 mp8 plaques containing human N-*ras* proto-oncogene sequences were lifted onto Hybond-N⁺. Probed with 5 ng/mL of (Column [**A**]) a 25-base antisense, fluorescein-dUTP tailed, probe; (Column [**B**]) a 25-base sense, fluorescein-dUTP tailed, probe. Hybridized at 42°C for 2 h; stringency wash of 1X SSC, 0.1% SDS at 42°C; detected with antifluorescein HRP and ECL; 10 min exposure.

the stringency washes. After 1 min incubation, the blots are exposed to X-ray film for the desired period (typically 10–30 min) (*see* **Note 18**) to achieve a maximum sensitivity of 25 amol.

2. The other two indirect systems require incubation of the blots in immunological reagents (e.g., antifluorescein-HRP conjugate or streptavidin-HRP complex for fluorescein and biotin, respectively). The blots are first blocked (using the blocking agent as in the hybridization buffer) for 30–60 min and then incubated in the appropriate dilution of antibody for a similar length of time. After four brief washes to remove unbound antibodies, the blots are processed with the ECL reagents as for directly labeled probes. Exposure times are again around 10–30 min for most applications (*see* **Note 18** and **Figs. 2–6**) with an absolute sensitivity of 10–20 amol for both biotin- and fluorescein-labeled probes.

4. Notes

1. The labeling reaction can be performed using as little as 25 pmol and as much as 250 pmol with equal success. The volumes of the various components should be scaled accordingly.
2. One hundred picomoles is equivalent to 560 ng of a 17-base sequence and 1000 ng of a 30-base sequence.
3. Oligonucleotide concentration is typically calculated by reading the absorbance at 260 nm and assuming that a concentration of 33 μg/mL is equivalent to 1 A_{260} U. This is only an approximation. For optimum results with this system and others, it is necessary to

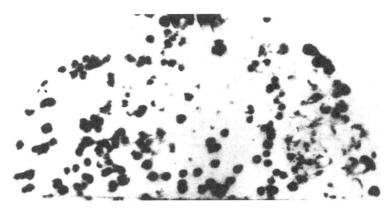

Fig. 4. Colony screening with enzyme labeled probes. *E. coli* colonies transformed with pSP65 containing human N-*ras* proto-oncogene sequences lifted onto Hybond-N$^+$. Hybridized with 5 ng/mL of an HRP-labeled 25-base probe at 42°C for 1 h; stringency wash of 3X SSC, 0.1% SDS at 42°C; ECL detection; 15 min exposure.

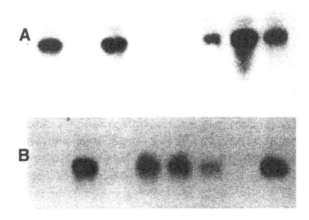

Fig. 5. PCR product detection with 5' hapten-labeled probes. Samples of human genomic DNA from normal, cystic fibrosis carrier and affected patients were subjected to PCR to amplify a 112-base fragment covering the 508 deletion locus. PCR samples were Southern blotted onto Hybond-N$^+$ and hybridized with **(A)** a 21-base, 1-biotin labeled, probe specific for the normal allele; **(B)** a 21-base, 1-biotin labeled, probe specific for the CF deletion allele. Hybridized at 42°C for 1 h; stringency wash of 1X SSC, 0.1% SDS at 42°C; detected with streptavidin-HRP and ECL; 1 min exposure.

assess accurately the concentration of the oligonucleotide probe. The relationship between concentration and absorbance is given by the following equation *(12)*:

$$E = (A \times 15{,}200) + (G \times 12{,}010) + (C \times 7050) + (T \times 8400)$$

where *A*, *G*, *C*, and *T* represent the number of times each base occurs in the probe sequence and *E* is the molar extinction coefficient (the absorbance of a 1 *M* solution).

4. The thiol modifier is dissolved in anhydrous acetonitrile and placed on the synthesizer. The thiol modifier phosphoramidite is programmed into the sequence to be placed at the 5'-end and a normal synthesis scheme followed; i.e., 15–30 s coupling time to achieve a

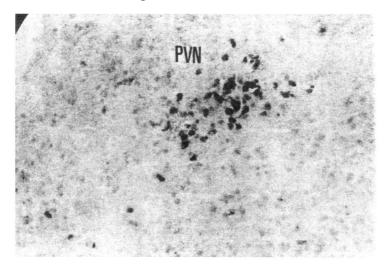

Fig. 6. *In situ* hybridization with 5' hapten-labeled probes. Cryostat sections (15 µm) were cut from the rat brain (PVN = paraventricular nucleus) and fixed with paraformaldehyde. Hybridized with 160 ng/mL of a 30-base, 10 biotin-labeled, probe specific for oxytocin mRNA for 17 h at 37°C; stringency wash of 1X SSC at 55°C; detected with streptavidin-alkaline phosphatase (Amersham Pharmacia Biotech) and NBT/BCIP color reagents: overnight color development.

coupling efficiency of >95%. Increasing the coupling time or decreasing the volume of acetonitrile used to dissolve the compound may raise the coupling efficiency to >98%.
5. Thiol modifier suppliers will probably recommend the relevant purification procedures to be followed when dealing with thiol-linked probes. During synthesis, the thiol group is protected by a trityl group and this can be used as a lipophillic tag to purify the thiol oligonucleotides from failure sequences using reverse-phase HPLC. The trityl-protecting group is subsequently removed using silver nitrate and dithiothreitol (DTT).
6. Thiol-modified oligonucleotides are stored after synthesis and purification in DTT to stop the thiol groups dimerizing. The DTT has to be removed immediately prior to coupling to the modified HRP and this is best achieved by use of a spin column. The desalted oligonucleotide should be added to the modified HRP within 5 min of preparation. Note that if commercially available spin columns are used, they will first have to be equilibrated with water. Failure to do this will alter the reaction buffer for the coupling step leading to significantly lower levels of labeling.
7. The use of spin columns (*see* **Note 6**) and the use of lyophilized, derivatized HRP ensures that a high concentration of reactants is achieved in the labeling process. It is possible to use freshly prepared, derivatized HRP in solution or alternative methods of desalting the oligonucleotide. However, the volume may increase with either of these options and the reaction may not reach maximum efficiency.
8. The labeling reaction couples >85% of the oligonucleotide to the HRP in a 1-h incubation at room temperature. However, if more convenient, it is possible to leave the coupling reaction to proceed overnight at 4°C without any difference in the final result.
9. The biotin phosphoramidite, as supplied in a solid form, is stable at room temperature for up to 1 mo. For longer term storage, store at –20°C. Only make it up in acetonitrile on the day it will be first used and, for maximum coupling efficiency, use the entire contents within 3–4 d of placing on the machine.

10. The 0.12-g bottle of biotin phosphoramidite (Amersham Pharmacia Biotech, RPN 2012) will give 10 biotinylation reactions when used in conjunction with a standard small-scale synthesis (0.2 µmol) program. The coupling of each biotin group exceeds 94% and will frequently exceed 98% when the synthesizer is set up well.
11. If it is not possible to use the entire contents of the bottle in the specified period, it is possible to remove the bottle from the machine and store the biotin phosphoramidite in acetonitrile at –20°C for at least 1 wk.
12. Biotin-labeled oligonucleotides should always be synthesized with the final trityl group left on. This is done to avoid a possible chemical reaction involving elimination of the final biotin during deprotection of the bases of the probe in a hot ammonia solution. The trityl group can then be removed separately either before or after reverse-phase HPLC purification. In many cases, because of the very high coupling efficiency achieved, it may not prove necessary to purify the probes after deprotection and trityl removal.
13. A ready-to-use buffer can also be used for the fluorescein- and biotin-labeled probes, for example, Rapid-hyb hybridization buffer (RPN 1635, Amersham Pharmacia Biotech).
14. The temperature of hybridization depends, to a large extent, on the length and base composition of the probe and independent of the type of method used for labeling. For many probes, a temperature of 42°C will be found to be adequate, although this will have to be determined empirically for each system, particularly for probes greater than 20 bases. The probe concentration recommended will work well in most applications of nonradioactive oligonucleotide probes. In some cases, it may prove necessary to alter the probe concentration, but this should only be done in the range of 5–20 ng/mL.
15. The melting temperature of an oligonucleotide probe approximates to the following formula *(13)*:

$$T_m = (4\times \text{ number of G + C bases}) + (2\times \text{ number of A + T bases})$$

Stringency washes are generally performed at 3–5°C below the melting temperature if it is necessary to distinguish between perfectly matched and mismatched probe:target hybrids. It is not possible to give exact conditions for all probes, but it should be noted that in general the stringency washes previously used with radioactive labels are equally effective with nonradioactive systems as described here. No significant effect on melting temperature has been observed on the introduction of any of the three labels discussed.

16. With HRP-labeled probes, care must be taken in the stringency washes to avoid denaturation of the enzyme label. Normal stringency washes would be performed for 2×15 min at temperatures below 42°C. At higher temperatures, up to 60°C maximum, the washes should be reduced to 2×5 min.
17. The ECL detection system is based on the HRP-catalyzed oxidation of luminol to produce blue light. The enhancement system stimulates the peak light output and sustains the light output for a number of hours *(14)*. The light can be collected on X-ray film, producing a hard-copy image similar to that traditionally obtained with radioactive labels. The system is equally applicable to directly labeled probes and the detection of haptens mediated by HRP-labeled antibodies *(15)*.
18. All three systems can be used in a variety of applications as illustrated (**Figs. 2–6**). The direct-labeled probes are particularly suitable for high-throughput systems where bulk labeling of the probes may be an advantage. The fluorescein-labeled probes offer slightly higher sensitivity and a more flexible approach and can be used in conjunction with alternative detection systems, for example, mediated through alkaline phosphatase-antibody conjugates. The biotin-labeled probes are useful in *in situ* hybridization and have a separate role in capture-based techniques, such as PCR-based sequencing and genomic walking.

References

1. Keller, G. H. and Manak, M. M. (1989) Hybridization formats and detection procedures, in *DNA Probes*. Stockton, NY, pp. 149–214.
2. Thein, S. L. and Wallace, R. B. (1986) The use of synthetic oligonucleotides as specific hybridization probes in the diagnosis of genetic disorders, in *Human Genetic Disease: A Practical Approach* (Davies, K. E., ed.), IRL, Oxford, pp. 33–50.
3. Misiura, K., Durrant, I., Evans, M. R., and Gait, M. J. (1990) Biotinyl and phosphotyrosinyl phosphoramidite derivatives useful in the incorporation of multiple reporter groups on synthetic oligonucleotides. *Nucleic Acids Res.* **18,** 4345–4354.
4. Connoly, B. A. and Rider, P. (1985) Chemical synthesis of oligonucleotides containing a free sulphydryl group and subsequent attachment of thiol specific probes. *Nucleic Acids Res.* **13,** 4485–4502.
5. Connolly, B. A. (1987) The synthesis of oligonucleotides containing a primary amino group at the 5' terminus. *Nucleic Acids Res.* **15,** 3131–3139.
6. Murakami, A., Tada, J., Yamagata, K., and Takano, J. (1989) Highly sensitive detection of DNA using enzyme-linked DNA probe. 1. Colourimetric and fluorometric detection. *Nucleic Acids Res.* **17,** 5587–5595.
7. Fowler, S. J., Harding, E. R., and Evans, M. R. (1990) Labeling of oligonucleotide with horseradish peroxidase and detection using enhanced chemiluminescence. *Technique* **2,** 261–267.
8. Durrant, I. (1991) Stringency control with 3'-tailed oligonucleotide probes. *Highlights* **2,** 6–7.
9. Sambrook, J., Fritsch, E. F., and Maniatis, T. (1982) Spun column procedures, in *Molecular Cloning: A Laboratory Manual*, Cold Spring Harbor Laboratory, Cold Spring Harbor, NY, pp. 466–467.
10. Pollard-Knight, D. (1990) Current methods in nonradioactive nucleic acid labelling and detection. *Technique* **2,** 113–132.
11. Durrant, I. (1990) Light based detection of biomolecules. *Nature (London)* **346,** 297–298.
12. Wallace, R. B. and Miyada, C. G. (1990) Oligonucleotide probes. *Methods Enzymol.* **152,** 438.
13. Wallace, R. B., Shaffer, J., Murphy, R. F., Bonner, J., Hirose, T., and Itakura, K. (1979) Hybridization of synthetic oligodeoxyribonucleotides to ϕX174: the effect of single base pair mismatch. *Nucleic Acids Res.* **6,** 3543–3356.
14. Thorpe, G. H. G. and Kricka, L. J. (1987) Enhanced chemiluminescent assays for horseradish peroxidase: characteristics and applications, in *Bioluminescence and Chemiluminescence: New Perspectives* (Scholmerich, J., Andreesen, R., Kapp, A., Ernst, M., and Woods, W. G., eds.), Wiley, NY, pp. 199–208.
15. Cunningham, M. (1991) Nucleic acid detection with light. *Life Sci. News* **6,** 2–5.

23

Preparation of Direct, Enzyme-Labeled DNA Probes

Ian Durrant and Timothy Stone

1. Introduction

The direct labeling of nucleic acid probes with horseradish peroxidase (HRP) was first described by Renz at European Molecular Biology Laboratory (EMBL) in 1984 *(1)*. The methodology was combined with enhanced chemiluminescence *(2)* (a light-producing HRP catalyzed reaction based on luminol) (*see* Chapter 26) allowing the detection of specific hybrids on membranes *(3)*. Further development led to the availability of the first light-based nucleic acid detection system; this was capable of detecting 2.5 pg of target nucleic acid on genomic Southern blots *(4)*. Subsequent protocol and reagent improvements now enable researchers to reliably detect 0.5 pg of target nucleic acid. Recent advances have now elucidated an analogous system for the labeling of DNA probes with alkaline phosphatase (AP), offering even higher sensitivity when combined with the new dioxetane based chemiluminescence substrates (*see* Chapter 26).

The HRP method (*see* **Fig. 1**) involves the labeling of a single-stranded nucleic acid probe with a positively charged HRP-parabenzoquinone-polyethyleneimine complex (labeling reagent). This complex initially associates with denatured, negatively charged nucleic acid, and is then covalently crosslinked, for example using glutaraldehyde. Probes from 50 bases to 50 kilobases in length have been successfully labeled in this way. The whole labeling protocol involves a few simple steps and takes around 30 min. It is inherently reliable, as it is a chemical labeling reaction. The scale of probe labeling can be changed easily to suit the needs of the researcher. Once a probe has been labeled, it can be stored for many months. The associated hybridization and posthybridization protocols are relatively simple, which makes probes labeled directly with HRP or AP particularly suitable for large-scale screening, where tens or hundreds of blots are processed weekly.

Probes labeled directly with HRP have been used in many membrane hybridization applications *(5,6)*, including colony and plaque screening, polymerase chain reaction (PCR) product detection/identification, yeast artificial chromosome (YAC clone) screening and restriction fragment length polymorphism (RFLP) analysis. Probes labeled directly with AP can be used in more sensitive applications, such as Southern blots of human genomic DNA and for Northern blots.

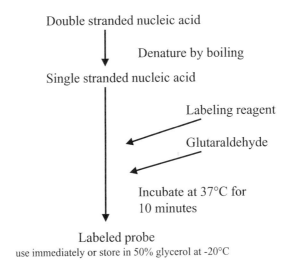

Fig. 1. Outline of the scheme for the production of horseradish peroxidase-labeled probes.

2. Materials

These materials are available in kit form, including the reagents for hybridization and chemiluminescent signal generation (ECL Direct system, RPN 3300, Amersham Pharmacia Biotech, Amersham, UK). The labeling reaction materials are stable for at least 3 mo stored at 2–8°C. The labeling reagent is light sensitive and should be protected from intense light or protracted periods of exposure to light.

1. Nucleic acid labeling reagent: Charge-modified HRP (see **Note 1**).
2. Crosslinker: Glutaraldehyde solution.

3. Method

1. Dilute the DNA to be labeled to a concentration of 10 ng/µL (see **Note 2**) with water (see **Note 3**) in a microcentrifuge tube.
2. Seal the tube and boil for 5 min in a vigorously boiling water bath (see **Note 4**).
3. Cool the DNA on ice for 5 min.
4. Centrifuge the tube briefly (10 s) to settle the liquid at the bottom of the tube.
5. Add an equal volume of DNA-labeling reagent. Mix briefly by pipet.
6. Add a volume of crosslinker equal to that of the labeling reagent. Mix briefly by pipet.
7. Incubate at 37°C for 10 min.
8. Keep the labeled probe on ice before use (see **Note 5**).

4. Notes

1. A labeling reagent based on AP is also available. The labeling protocol is similar to that described for HRP.
2. A quantity of 10 µL (100 ng) to 150 µL (1.5 µg) of probe can be labeled in each microcentrifuge tube. Poor labeling will result if a volume of DNA greater than 150 µL (1.5 µg) is present in each tube.
3. The DNA must be in a low-salt buffer (<10 mM NaCl), as ions interfere with the interactions between DNA and labeling reagent, thus reducing labeling efficiency.

4. A vigorously boiling water bath must be used to ensure complete denaturation of double-stranded probes; heating blocks appear not to denature the DNA completely. This denaturation step is not required for single-stranded DNA or RNA.
5. Probes, once labeled, can be used immediately in any membrane hybridization application or may have sterile glycerol added to a final concentration of 50% (v/v) and stored at –20°C for up to 6 mo. Glycerol of the highest quality must be used to avoid inhibition or degradation of the HRP. The recommended probe concentration is 10 ng/mL in hybridizations, although this may be reduced to 2–5 ng/mL for high target applications such as colony and plaque screening.

References

1. Renz, M. and Kurz, C. (1984) A colorimetric method for DNA hybridization. *Nucleic Acids Res.* **12,** 3435–3444.
2. Whitehead, T. P., Thorpe, G. H. G., Carter, C. J. N., Groucutt, C., and Kricka, L. J. (1983) Enhanced chemiluminescence procedure for sensitive determination of peroxidase-labelled conjugates in immunoassay. *Nature* **305,** 158–159.
3. Pollard-Knight, D., Read, C. A., Downes, M. J., Howard, L. A., Leadbetter, M. R., Pheby, S. A., McNaughton, E., Syms, A., and Brady, M. A. W. (1990) Non-radioactive nucleic acid detection by enhanced chemiluminescence using probes directly labeled with horseradish peroxidase. *Anal. Biochem.* **185,** 84–89.
4. Durrant, I. (1990) Light based detection of biomolecules. *Nature* **346,** 297–298.
5. Durrant, I., Benge, L. C. A., Sturrock, C., Devenish, A. T., Howe, R., Roe, S., Moore, M., Scozzafava, G., Proudfoot, L. M. F., Richardson, T. C., and McFarthing, K. G. (1990) The application of enhanced chemiluminescence to membrane based nucleic acid detection. *Biotechniques* **8,** 564–570.
6. Stone, T. and Durrant, I. (1991) Enhanced chemiluminescence for the detection of membrane bound nucleic acid sequences. *Genet. Anal.: Tech. Appl.* **8,** 230–237.

24

Random Prime Labeling of DNA Probes with Fluorescein-Tagged Nucleotides

Bronwen M. Harvey, Claire B. Wheeler, and Martin W. Cunningham

1. Introduction

The selection of an appropriate labeling reagent for a particular experiment, for the most part, depends on the sensitivity and resolution required. For maximum sensitivity in membrane hybridization, a radioactive label is often the label of choice.

The trend, in recent years, to the use of nonradioactive labels, such as biotin, fluorescein, and digoxygenin, continues because of the disposal and safety advantages these systems offer. Recent advances in light-based detection systems *(1,2)* mean that the sensitivity of such systems is beginning to approach that of phosphorus-32 in a number of applications.

Nucleic acids can be labeled in several different ways, resulting in the reporter molecules being incorporated directly into the nucleic acid either uniformly along its length or only at one end. Incorporation can be achieved using various polymerase or modifying enzymes or with the use of crosslinking reagents. Some reporter molecules, e.g., radioactive or fluorescent tags, can be directly detected, whereas others, e.g., biotin or digoxygenin, are detected indirectly using enzyme reaction products *(3)*. The enzyme, usually a peroxidase or phosphatase, is conjugated to a secondary molecule or ligand that ideally has a high affinity for the reporter, e.g., the biotin/streptavidin system *(4)*.

Amersham Pharmacia Biotech Ltd. (Amersham, UK) has a range of Gene Images products based on direct and indirect reporter molecules that are compatible with a number of nonradioactive detection approaches including chemiluminescence, chemifluorescence, and color *(5)*. This chapter focuses on the use of an indirect reporter molecule, a fluorescein-labeled nucleotide, which can be incorporated into nucleic acid molecules, using a suitable DNA polymerase. Details of a direct labeling system using peroxidase *(6,7)* or alkaline phosphatase may be found in Chapter 23.

The fluorescein-labeled nucleotide used in the Gene Images system is a chemically stable molecule that may be incorporated into a nucleic acid molecule using Klenow polymerase in a random prime reaction *(8)*. Subsequently, the fluorescein can be detected using a combination of an antifluorescein antibody conjugated to either horseradish peroxidase or alkaline phosphatase and an appropriate chemiluminescent substrate, e.g., ECL or CDP-*Star* (Amersham Pharmacia Biotech) *(9)*, respectively. The use of a fluorescein-labeled nucleotide, which can act as both a direct and indirect reporter molecule, has allowed the development of a unique rapid semiquantitative assay for probe labeling. Whereas other hapten-based systems require the use of a prolonged dot blot procedure to determine incorporation efficiency, this assay makes direct use of the physical properties of the fluorescein hapten. After washing away unincorporated fluorescein nucleotide, DNA tagged with fluorescein can be visualized under UV illumination. By referring to a set of standards, a semiquantitative estimate of labeling efficiency can be made.

2. Materials

Reagents 1–6 are available, fully optimized, from Amersham Pharmacia Biotech (Amersham) as the Gene Images Random Prime labeling module (RPN 3540). The system is extensively tested and necessary controls are also provided. The reagents are stable for at least 3 mo.

1. Nucleotide mix (Amersham Pharmacia Biotech): 5× stock solution containing standard nucleotides and a fluorescein-labeled nucleotide at optimized concentrations, in 10 mM Tris-HCl pH 8.0, 1 mM ethylenediaminetetraacetic acid (EDTA). Store in the dark at –20°C.
2. Reaction buffer: 5× stock containing 250 mM Tris-HCl, pH 7.5, 50 mM β-mercaptoethanol and 25 mM magnesium chloride.
3. Primers: random nonamer primers, 50 OD/mL, in an aqueous solution. Store at –20°C.
4. Enzyme solution: 4 U/μL DNA polymerase I Klenow fragment (cloned) in 50 mM potassium phosphate, pH 6.5, 10 mM β-mercaptoethanol, and 50% (v/v) glycerol. Store at –20°C.
5. Water: deionized ultrafiltered water.
6. Tris-EDTA (TE) buffer: 10 mM Tris-HCl, pH 8.0, 1 mM EDTA.
7. Positively charged nylon membrane for example Hybond N+ (Amersham Pharmacia Biotech).
8. 0.5 M EDTA, pH 8.0: The pH should be adjusted with NaOH, the EDTA will not fully dissolve unless NaOH is added.
9. Wash buffer: 2X SSC, 0.03 M trisodium citrate, 0.3 M NaCl, pH 7.0.
10. DNA fragment to be labeled.

3. Method

3.1. Labeling

1. Dilute the template DNA to be labeled to a concentration of 2–25 ng/μL in either water or TE buffer (*see* **Notes 1** and **2**).
2. Place the required tubes of labeling components, with the exception of the enzyme, in an ice bath to thaw. The enzyme should be left at –20°C until required, and returned to the freezer immediately after use.
3. Denature the template DNA by heating for 5 min in a boiling water bath, then chill immediately on ice. It is strongly advised to denature the template in a volume of at least 20 μL.
4. To a 1.5-mL microcentrifuge tube, placed in an ice bath, add the appropriate volume of each reagent in the following order: water to ensure a final reaction volume of 50 μL, 10 μL nucleotide mix, 10 μL reaction buffer, 5 μL of primers, 50 ng of denatured DNA (minimum), and 1 μL (4 U) of enzyme (Klenow fragment) (*see* **Note 3**).
5. Mix gently by pipeting up and down, and cap the tube. Spin briefly in a microcentrifuge to collect the contents at the bottom of the tube (*see* **Note 4**).
6. Incubate the reaction mix at 37°C for 1 h or room temperature overnight (*see* **Note 5**). Stop the reaction by adding 2 μL 0.5 M EDTA.

7. The labeled probe is now ready for use in hybridization (*see* **Note 6**). Further purification of the labeled probe is not required, saving loss of valuable probe and time. Probes can be stored in the dark at −20°C for at least 3 mo, avoiding frequent labeling reactions.
8. If desired the reaction can be checked using the rapid labeling assay (*see* **Subheading 3.2.**).

3.2. Monitoring Incorporation Using a Rapid Labeling Assay

1. Cut a sheet of nylon membrane allowing 1 cm between each sample. Use a pencil to indicate the sample positions.
2. Prepare a negative control consisting of a 1:5 dilution, in TE buffer, of the nucleotide mix.
3. With the nylon membrane on a clean nonabsorbent surface, carefully apply 2 µL of each labeling reaction in the appropriate position. In addition apply 2 µL of the negative control.
4. Allow the samples to absorb for 1 min and then immerse the membrane in 100 mL of prewarmed wash buffer and incubate at 60°C with gentle shaking for 15–20 min (*see* **Note 7**).
5. Remove the membrane with clean forceps and lay it on a sheet of absorbent filter paper to remove excess buffer.
6. Visualize the membrane DNA side down on a UV transilluminator (*see* **Note 8**). The fluorescein-labeled probe should be visible as a yellow/green fluorescent spot. Little or no fluorescence should be observed on the negative control (*see* **Note 9**). A semiquantitative estimate of labeling efficiency can also be obtained (*see* **Note 10**).

4. Notes

1. A wide range of DNA concentrations can be used in the labeling reaction. There can be considerable net synthesis of probe in contrast to a radioactive random prime reaction, as none of the nucleotides is present at a limiting concentration. With the Klenow polymerase, the newly synthesized DNA strand can be progressively displaced by an adjacent growing strand so that more than a single copy of the initial template can be made. This can result in a several-fold increase in the amount of DNA present as the reaction progresses. Synthesis is most efficient for lower levels of template, although net synthesis tends to increase with the amount of template.
2. A wide variety of DNA can be labeled using the Gene Images Random Prime labeling module (Amersham Pharmacia Biotech). Closed, circular, double-stranded DNA can be used as a template. The reaction yield will, however, be improved if the DNA is first linearized, as renaturation will be inhibited. Single-stranded DNA, miniprep DNA, and inserts in low-melting-point agarose can also be effectively labeled, although in some cases a longer labeling reaction may be necessary.
3. The size of the probe is determined by the ratio of the primer to template and by the nucleotide concentration. These have been fully optimized to generate probes with sizes (100–300 bp) suitable for filter hybridizations.
4. Vigorous mixing must be avoided; the resulting shearing forces will result in a loss of enzyme activity.
5. Temperature of incubation and reaction time can be chosen for convenience as the reaction reaches a plateau and does not decline significantly overnight at room temperature However the rate of reaction does depend to some extent on DNA purity. For example, with miniprep DNA or DNA in agarose, a longer incubation period may be required. With purified DNA, a slight increase in probe yield can be obtained with longer reaction times.
6. In general, the final concentration of probe in the reaction mix after labeling is between 5 and 7 ng/µL when starting with 50 ng of template. Reaction yields generally reflect the size of the template; larger templates will tend to produce higher yields.

7. Because the DNA has not been fixed to the membrane, it is necessary to avoid the use of SDS in the wash buffer, as this will strip off the labeled probe as well as the free nucleotide. More rapid removal of the unincorporated nucleotide can be obtained by use of a higher salt concentration, e.g., 5X SSC.
8. Optimum contrast is obtained using a short wavelength (254 nm) transilluminator.
9. If the negative control retains significant levels of fluorescence it may make interpretation difficult. It is possible to wash the filters for a further 15 min if this occurs, but it may also indicate that the filters were left too long after applying the samples and that the material has dried on the filter. If this is the case, the test should be repeated.
10. To obtain a semiquantitative estimate of labeling efficiency, the fluorescein can be compared to that of a dilution series of the nucleotide mix. Prepare 1/25, 1/50, 1/100, 1/250, 1/500, and 1/1000 dilutions in TE buffer, and apply 2 µL of each onto a sheet of nylon membrane (as described in **Subheading 3.2.**). Visualize, without washing, on the transilluminator alongside the prepared reaction samples. The labeling reaction has worked acceptably if the intensity is between that of the 1/25 and the 1/250 dilution. It may be possible, however, to successfully use fluorescein-labeled probes in filter hybridizations that lie outside this range. The reference filter can be stored for a week at $-20°C$ in the dark.
11. Correspondence concerning this protocol should be addressed to C. B. Wheeler.

References

1. Whitehead, T. P., Thorpe, G. H. G., Carter, C. J. N., Groucutt, C., and Kricka, L. J. (1983) Enhanced luminescence procedure for sensitive determination of peroxidase-labelled conjugates in immunoassay. *Nature* **305,** 1158–1159.
2. Bronstein, I., Edwards, B., and Voyta, J. C. (1989) 1,2-Dioxetanes: novel chemiluminescent enzyme substrates; applications to immunoassays. *J. Biolum. Chemilum.* **4,** 99–111.
3. Akhavan-Tafti, H., De Silva, R., Arghavani, Z., Eickholt, R. A., Handley, R. S., and Schaap, A. P. (1994) Lumigen PS: new chemiluminescent substrate for the detection of horseradish peroxidase. *Bioluminescence and Chemiluminescence: Fundamentals and Applied Aspects* (Campbell, A. K., Kricka, L. J., and Stanley, P. E., eds.), Wiley, Chichester, UK, pp. 199–202.
4. Langer, P. R., Waldrop, A. A., and Ward, D. C. (1981) Enzymatic synthesis of biotin labelled polynucleotides: novel nucleic acid affinity probes. *Proc. Natl. Acad. Sci. USA* **78,** 6633–6637.
5. Simmonds, A. C., Cunningham, M., Durrant, I., Fowler, S. J., and Evans, M. R. (1991) Enhanced chemiluminescence in filter based DNA detection. *Clin. Chem.* **37,** 1527–1528.
6. Renz, M. and Kurz, C. (1984) A colorimetric method for DNA hybridization. *Nucleic Acids Res.* **12,** 3435–3444.
7. Pollard-Knight, D., Read, C. A., Downes, M. J., Howard, L. A., Leadbetter, M. R., Pheby, S. A., McNaughton, E., Syms, A., and Brady, M. A. W. (1990) Non-radioactive nucleic acid detection by enhanced chemiluminescence using probes directly labelled with horseradish peroxidase. *Anal. Biochem.* **185,** 84–89.
8. Feinberg, A. P. and Vogelstein, B. (1983) A technique for radiolabelling DNA restriction endonuclease fragments to high specific activity. *Anal. Biochem.* **132,** 6–13.
9. Bronstein, I., Olesen, C. E. M., Martin, C. S., Schneider, G., Edwards, B., Sparks, A., and Voyta, J. C. (1994) Chemiluminescent detection of DNA and protein with CDP and CDP-STAR 1,2-dioxetane substrates. *Bioluminescence and Chemiluminescence: Fundamentals and Applied Aspects* (Campbell, A. K., Kricka, L. J., and Stanley, P. E., eds.), Wiley, Chichester, UK, pp. 269–272.

25

Hybridization and Detection of Fluorescein-Labeled DNA Probes Using Chemiluminescence

Claire B. Wheeler, Bronwen M. Harvey, and Martin Cunningham

1. Introduction

DNA probes can be labeled using a fluorescein-labeled nucleotide and subsequently detected with a dioxetane-based chemiluminescence system using an antifluorescein antibody conjugated to alkaline phosphatase *(1,2)*. The use of chemiluminescence avoids the fading of results associated with some colorimetric detection procedures used with some nonradioactive labeling and detection systems. Hard-copy results showing excellent resolution are produced.

The labeling procedure, based on established random primer technology *(3)* can be used with DNA of various types and purity (*see* Chapter 20). The resulting fluorescein-labeled probes can be stored for several months, thus avoiding frequent labeling reactions. Valuable probe is not lost since a further purification step before hybridization is not required.

The hybridization and wash conditions given in the following protocol are appropriate for a majority of probes labeled using the Gene Images Random Prime labeling system (Amersham Pharmacia Biotech). They allow the detection of single-copy mammalian genes without significant cross-hybridization to nonhomologous sequences. However, if these conditions are found to be insufficiently stringent for a particular probe, the stringency can be further controlled in hybridization and washing by changing the temperature and/or sodium chloride concentration.

2. Materials

Reagents for labeling DNA with fluorescein are available from Amersham Pharmacia Biotech Ltd. (Amersham, UK) as the Gene Images Random Prime module (RPN 3540). Reagents for the hybridization and detection of these probes are available from Amersham Pharmacia Biotech Ltd. (Amersham) as the Gene Images CDP-*Star* detection module (RPN 3510). Both kits have been fully optimized and extensively tested and necessary controls are provided. The reagents are stable for at least 3 mo.

1. Fluorescein-labeled probe DNA (*see* Chapter 24).
2. The following hybridization buffer can be used: 5× standard saline citrate (SSC), 1/10 dilution of Gene Images Liquid Block (RPN 3601 and supplied with RPN 3510; Amersham Pharmacia Biotech), 0.1% (w/v) sodium dodecyl sulfate (SDS), 5% (w/v) dextran sulfate, (e.g., product no. US70796; Amersham Pharmacia Biotech). Dissolve the dextran sulfate, in about three quarters of the final volume of SSC; add the dextran sulfate a little at a time while stirring and continue to stir until dissolved, which takes about 60 min. Add the liquid block and SDS. Make up to final volume with SSC. The hybridization buffer can be made up in a large volume and stored in suitable aliquots at –20°C (*see* **Note 1**).
3. 20X SSC: 0.3 M trisodium citrate, 3 M NaCl, pH 7.0.
4. Stringency wash solution 1: 1X SSC, 0.1% (w/v) SDS.
5. Stringency wash solution 2: 0.5X SSC, 0.1% (w/v) SDS.
6. Buffer A: sterile 100 mM Tris-HCl, pH 9.5, 300 mM NaCl (*see* **Note 6**).
7. Gene Images Liquid Block, supplied with the detection module.
8. Bovine serum albumin (BSA) fraction V (e.g., Sigma (St. Louis, MO) A-2153).
9. Tween 20 (e.g., product no. US20605; Amersham Pharmacia Biotech).
10. Antifluorescein antibody/alkaline phosphatase conjugate. Supplied with detection module (*see* **Note 7**).
11. CDP-*Star* detection reagent, supplied with the detection module (*see* **Note 11**).
12. Autoradiography film, e.g., Hyperfilm ECL from Amersham Pharmacia Biotech.
13. X-ray film cassette, e.g., Hypercassette from Amersham Pharmacia Biotech.
14. Gene Images Detection bags. Available separately as RPN 3609 (Amersham Pharmacia Biotech).

3. Methods
3.1. Hybridization and Stringent Washes
1. Preheat the required volume of hybridization buffer to 60°C. It is recommended that at least 0.125 mL/cm^2 of membrane be used (*see* **Notes 1** and **2**).
2. Prehybridize the blots in the hybridization buffer for at least 30 min at 60°C with constant agitation.
3. Remove the required amount of probe to a clean microcentrifuge tube (*see* **Note 3**). If the volume is <20 µL make up to this volume with water or Tris-EDTA (TE) buffer. Denature the probe by boiling for 5 min and snap cool on ice.
4. Centrifuge the denatured probe briefly, then add to the prehybridization buffer. Avoid placing the probe directly on the membrane and mix gently. Alternatively, some of the buffer can be withdrawn for mixing with the probe before addition to the bulk of the buffer. Continue the incubation overnight at 60°C.
5. Preheat stringency wash solutions 1 and 2 to 60°C. These are used in excess, e.g., 2–5 mL/cm^2 of membrane. Carefully transfer the blots, using a pair of blunt forceps, to stringency solution 1 and wash at 60°C, for a minimum of 15 min with gentle agitation. Carry out a further wash in stringency wash 2 for 15 min at 60°C (*see* **Note 4**).
6. Briefly rinse the blots in sterile buffer A at room temperature to remove any SDS.

3.2. Blocking, Antibody Incubation, and Washes
All containers used for the following steps should be rinsed with absolute ethanol before use to remove any alkaline phosphatase contamination (*see* **Note 6**).

1. Dilute liquid block 1:10 in sterile buffer A (*see* **Notes 5** and **6**). Incubate the blots with gentle agitation for 60 min at room temperature, in approx 0.75–1.0 mL of diluted liquid block per cm^2 of membrane.

2. Dilute the antifluorescein/alkaline phosphatase conjugate 1 in 5000 in freshly prepared 0.5 % (w/v) BSA in buffer A. Incubate the blots in diluted conjugate using 0.3 mL/cm^2 of membrane, with gentle agitation, at room temperature for 60 min (*see* **Note 7**).
3. Remove unbound conjugate by washing for 3×10 min in 0.3%(v/v) Tween-20 in buffer A at room temperature with agitation. An excess volume should be used (2–5 mL/cm^2) for each wash.

3.3. Signal Generation and Detection

All steps can be carried out in the darkroom if desired: it is only necessary to switch off the light after **step 4**. Wear powder-free gloves or rinse gloved hands with water before use, as the powder may interfere with the detection reaction.

1. Prepare the detection bag for use (*see* **Note 8**).
2. Drain the excess buffer from the washed blot(s) and place onto a clean, non-absorbent, flat surface. A polythene film such as Saran Wrap is suitable.
3. Pipet the CDP-*Star* detection reagent directly onto the blot(s) (30–40 µL/cm^2). Incubate for 2–5 min at room temperature (*see* **Note 9**).
4. Drain off excess detection buffer by touching the edge of the blot to a tissue. Working as quickly as possible to minimize the delay between incubating the blots in substrate and exposure to film, place blots in a prepared detection bag (*see* **Note 10**).
5. Place the blots DNA side up in the film cassette. In the dark, place a sheet of autoradiography film on top of the blots. Close the cassette (*see* **Note 11**).
6. Expose initially for 30–60 min. Remove film and develop. Multiple subsequent exposures can be made to acquire an appropriate image (*see* **Note 12**).

4. Notes

1. Some particulate material may be found in the buffer after thawing, but this should redissolve during hybridization at 60°C.
2. Hybridization can be carried out in either boxes, bags, or hybridization bottles, provided there is sufficient buffer. As a general rule, 0.25 mL/cm^2 is recommended for small blots hybridized in boxes and 0.125 mL/cm^2 for large blots hybridized in plastic bags or hybridization bottles. When hybridizing large blots, prewetting in 5X SSC is advised to minimize the hybridization volume. It is possible to hybridize several blots in the same solution, providing there is enough buffer to allow free movement of the blots. Agitation during hybridization is strongly recommended; a speed of 60–100 strokes/min is adequate.
3. For overnight hybridization to Southern blots of human genomic DNA, using the dextran hybridization buffer, a probe concentration of 10 ng/mL of labeled purified plasmid insert is usually sufficient to detect a single-copy gene in a 0.1 µg loading: representing 50 fg of specific target.
4. To increase the stringency, washes of 0.2 or 0.1X SSC with 0.1% (w/v) SDS at 60°C can be used. Alternatively, the hybridization and/or stringency washes can be carried out at 65°C.
5. This is a critical step in the detection procedure. Efficient blocking of the membrane against nonspecific binding of the antibody conjugate is essential in minimizing background. This is particularly important in colony/plaque screening.
6. The most frequent problem associated with this highly sensitive detection system is spotty backgrounds. This is almost always due to bacterial contamination of the system, leading to nonspecific detection of bacterial alkaline phosphatase (AP). Therefore, it is very important to guard against endogenous AP contamination. The most likely source of contamination is from buffer A, so this must be as sterile as possible. It should be prepared

from fresh ingredients and autoclaved immediately for 15 min at 105 kPa (15 psi). A fresh unopened bottle should be used for each experiment and any buffer remaining at the end of an experiment should be discarded.
7. The diluted conjugate should be used immediately. Several blots can be incubated together, but it is important that there is free access of the solution to the blot.
8. Cut a section from a detection bag of the appropriate size: at least 2 cm in all directions greater than the dimensions of the blot(s). Open out the two halves of the detection bag on to a clean, flat surface.
9. To avoid contamination of the detection reagent, it is recommended that a suitable aliquot is aseptically removed from the bulk solution to a separate container before use.
10. Blots can also be wrapped in SaranWrap for exposure to film. Any air pockets created in wrapping the blots should be gently smoothed out.
11. Probe bound alkaline phosphatase is used to catalyze the light production by enzymic breakdown of the stabilized dioxetane substrate CDP-*Star*. This light output is detected on light-sensitive film, providing a permanent hard-copy result. Ensure that there is no free detection reagent in the film cassette; avoid getting the film wet. The detection bag may be heat sealed to prevent the blots drying out during long exposures.
12. The optimum exposure time depends on the application. For very high target applications, a significantly shorter initial exposure may be desirable. The light output will plateau after a few hours, which makes it easier to judge the optimum exposure time and, as the signal lasts for several days, there is plenty of opportunity to obtain the optimum presentation of the result. If the initial exposure is too dark (it may even appear completely black) then a shorter exposure may still give an excellent result.
13. Correspondence concerning this protocol should be addressed to C. B. Wheeler.

References

1. Bronstein, I., Edwards, B., and Voyta, J. C. (1989) 1,2-Dioxetanes: novel chemiluminescent enzyme substrates; applications to immunoassays. *J. Biolum. Chemilum.* **4,** 99–111.
2. Bronstein, I., Olesen, C. E. M., Martin, C. S., Schneider, G., Edwards, B., Sparks, A., and Voyta, J. C. (1994) Chemiluminescent detection of DNA and protein with CDP™ and CDP-STAR™ 1,2-dioxetane substrates, in *Bioluminescence and Chemiluminescence: Fundamentals and Applied Aspects* (Campbell, A. K., Kricka, L. J., and Stanley, P. E., eds.), Wiley, Chichester, UK, pp. 269–272.
3. Feinberg, A. P. and Vogelstein, B. (1983) A technique for radiolabeling DNA restriction endonuclease fragments to high specific activity. *Anal. Biochem.* **132,** 6–13.

26

Hybridization of Enzyme-Labeled Probes and Detection by Chemiluminescence

Timothy Stone and Ian Durrant

1. Introduction

The use of enzyme-labeled probes, in conjunction with chemiluminescence, allows a flexible approach to hybridization and detection procedures. This is shown by the diversity of applications in which the system can be used, including Southern blots *(1,2)*, Northern blots *(3)*, colony and plaque screening *(4)*, yeast artificial chromosome clone screening *(5)*, and polymerase chain reaction product detection *(6,7)*.

Probes labeled directly with horseradish peroxidase (HRP) are hybridized in 6 M urea, a denaturing buffer *(8)*. This allows overnight hybridizations to be performed at 42°C and balances the maximum rate of hybridization *(9)* with maintenance of enzyme stability. The composition of the buffer has been optimized *(10)* and contains a novel rate enhancer and blocking agent, which are both necessary to prevent the labeled probe binding nonspecifically to membranes. Stringency can be controlled using the salt concentration in the primary wash buffer. Detection of HRP involves the oxidation of luminol, which decays to the ground state via a light-emitting pathway. In the presence of an enhancer molecule, there is at least 1000-fold more light produced *(11)*, giving fast and sensitive results which are recorded as a hard copy on film with excellent resolution. The light output peaks after 1–5 min and then decays slowly with a 2-h half-life *(10)*. Once the light output has decayed, largely because of irreversible HRP inactivation, blots can be immediately reprobed, without stripping. The probe concentration, length of hybridization, and the exposure time can be altered to suit the needs of a particular application.

Alkaline phosphatase (AP)-labeled probes use a similar urea-based hybridization buffer. The stringency can be controlled by both salt concentration and temperature by the use of a thermostable AP. The chemiluminescent procedure for AP enables higher sensitivity applications to be performed and multiple exposure to be taken, as the light output lasts for a number of days (*see* Chapter 24).

2. Materials

The hybridization buffer, blocking agent, and detection reagents are available in an optimized kit form, including the reagents for labeling probes directly with HRP (ECL Direct system, RPN 3000) (Amersham Pharmacia Biotech, Amersham, UK).

1. Hybridization buffer and blocking agent (Amersham Pharmacia Biotech): The hybridization buffer is a complex mixture of components that have been optimized to ensure efficient hybridization and thermal stability of the HRP label. It is based on 6 M urea as a denaturant and also contains a novel rate enhancement system. This system causes greater rates of hybridization and also aids in the stabilization of the HRP during the detection process. At room temperature, take the hybridization buffer and add sufficient solid NaCl to make a 0.5 M final concentration. Add 5% (w/v) blocking agent (this is necessary for both nylon and nitrocellulose membranes). Quickly mix the blocking agent into the buffer to ensure that no clumps of undissolved blocking agent form. Mix thoroughly, using a magnetic stirrer or roller mixer, for 1 h. The blocking agent is not fully dissolved at this stage, and heating to 42°C for 30–60 min is necessary to complete the process. A few undissolved particles of blocking agent will not affect the hybridization. Hybridization buffer and blocking agent, if stored separately, are stable at room temperature. However, when formulated as described, the hybridization buffer must be stored at –20°C, where it is stable for at least 6 mo.
2. 20X standard saline citrate (SSC): 0.3 M trisodium citrate, 3.0 M sodium chloride. Adjust to pH 7 with HCl.
3. Primary wash buffer: 360 g Urea, 4 g sodium dodecyl sulfate (SDS), 25 mL 20X SSC. Make up to 1 L. This can be kept for up to 3 mo in a refrigerator at 2–8°C. Stringency may be increased by using a lower final SSC concentration, e.g., 0.1X SSC instead of 0.5X SSC.
4. Secondary wash buffer: 2X SSC, this can be kept for up to 3 mo in a refrigerator at 2–8°C.
5. Saran Wrap (Dow Chemical Co., Midland, MI).
6. Blotting membranes: The system gives excellent results with Hybond N$^+$ (nylon) (Amersham Pharmacia Biotech). Other membrane types and membranes supplied by other manufacturers may also give good results.
7. Detection reagents 1 and 2 (Amersham Pharmacia Biotech): Detection reagent 1 contains a peracid salt as the enzyme substrate. Detection reagent 2 contains luminol and an optimized enhancer molecule. An equal volume of each of these should be mixed. The final volume required for ECL detection is 0.125 mL/cm^2 of membrane. The detection reagents are stable for at least 3 mo when stored separately at 2–8°C.
8. X-ray film cassette (e.g., Hypercassettes, Amersham Pharmacia Biotech).
9. Blue-light sensitive film (e.g., Hyperfilm ECL, Amersham Pharmacia Biotech).
10. HRP-labeled probe (*see* Chapter 23): The labeling procedure produces a probe in single-stranded form; it must not be denatured further. AP-labeled probes can also be used (*see* **Note 1**).

3. Method
3.1. Hybridization

1. Place the blot onto the surface of the fully prepared hybridization buffer, prewarmed to 42°C. Allow the blot to saturate fully before submerging it in the buffer (*see* **Note 2**).
2. Prehybridize at 42°C for 15–60 min in a shaking water bath (*see* **Note 3**).
3. Add the single-stranded probe, which has been labeled directly with HRP (*see* Chapter 23), to the hybridization buffer (*see* **Note 4**), avoiding direct addition onto the membrane (*see* **Note 5**).

Hybridization of Enzyme-Labeled Probes

4. Continue incubation at 42°C with shaking for the required length of time (from 2 h to overnight) (*see* **Note 4**).

3.2. Stringency Washes

1. Preheat the primary wash buffer to the required temperature, typically 42°C (*see* **Note 6**).
2. Pour at least 2 mL primary wash buffer/cm^2 of membrane into a clean container.
3. Remove the blots from the hybridization container and place them into the primary wash buffer. Incubate at the desired temperature in a shaking water bath for 20 min.
4. Discard the wash buffer and replace with an equivalent volume of the fresh primary wash buffer. Incubate with agitation for a further 20 min at the same temperature.
5. Discard wash buffer, place blots in a fresh container, and add an excess of secondary wash buffer (at least 2 mL of buffer/cm^2 of membrane). Incubate with agitation for 5 min at room temperature.
6. Discard the wash buffer and replace with an equivalent volume of fresh secondary wash buffer. Incubate for a further 5 min at room temperature.
7. The blots may be stored temporarily at this stage if required (*see* **Note 7**).

3.3. Detection

1. Lay a piece of Saran Wrap (Dow), with a large enough area for all the blots, onto the bench.
2. Using forceps, remove the blots from the secondary wash buffer. Drain off excess wash buffer by touching the bottom edge of the blot onto a paper towel and place the blots, DNA side up, onto the Saran Wrap.
3. Cover the blots with a volume of freshly mixed detection reagents 1 and 2 equivalent to 0.125 mL/cm^2 of membrane.
4. Leave the blots immersed for 1 min.
5. Drain off excess detection reagents by touching the bottom edge of the blot onto a paper towel and place the blots, DNA side down, onto a fresh piece of Saran Wrap.
6. Fold over the Saran Wrap to make a "parcel" of the blots. Only one layer of Saran Wrap should cover the DNA side of the blots.
7. In a darkroom, with a red safelight, place the blot parcel DNA side up in the base of an X-ray film cassette. Place a piece of blue-light sensitive film on top, close the cassette and expose the film for 1 min.
8. Remove the film and immediately replace with another sheet of unexposed film, reclose the cassette, and start a timer. Develop the first film and use result as a guide for the length of the second exposure (*see* **Note 8**).
9. After a suitable interval, normally 30–60 min for genomic Southern blots, develop the second film and interpret the results.

4. Notes

1. AP-labeled probes are used in a similar manner to that described for HRP-labeled probes except that the hybridization buffer and primary stringent wash are based on 2 *M* urea and the temperature for these steps can be varied in the region 50–65°C. The detection procedure, although still chemiluminescent, is based on a dioxetane substrate, e.g., CDP-*Star* (Tropix, Inc., Bedford, MA).
2. If hybridization is performed in a box, then the blot must be placed on top of the buffer and allowed to saturate fully before being submerged. This prevents the appearance of white patches on detection. A volume of buffer equivalent to 0.25 mL/cm^2 of membrane should be used. If the box is significantly larger than the blot, then the volume used should correspond to the area of the bottom of the box. Hybridizations may also be carried out in

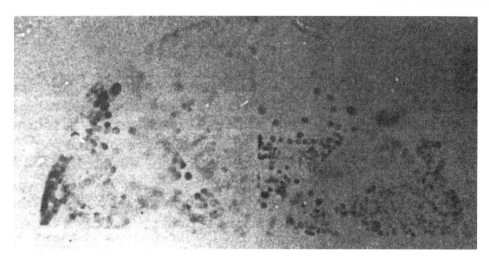

Fig. 1. Colony screening. *E. coli.* colonies transformed with a plasmid carrying a portion of the proto-oncogene Nras hybridized with a 1.5-kbp HRP-labeled Nras probe as described in the text. 1-min exposure on Hyperfilm-MP.

bags or in hybridization ovens. In these cases, the blot should be put into the container first. Buffer is added at a volume equivalent to 0.125 mL/cm² of membrane.
3. Shaking water baths should be set to 60–100 strokes/min. Blots should be allowed to move freely in boxes to prevent the appearance of white or dark patches on detection.
4. Routinely, 10 ng/mL final probe concentration and an overnight hybridization is used. This gives the maximum sensitivity of the system and is a requirement for high-sensitivity work, e.g., the detection of a single-copy gene on Southern blots of genomic DNA digests *(9)*. However, for high-target applications, such as colony or plaque screening, it may be possible to use lower probe concentrations of (2–5 ng/mL) in conjunction with a 2-h hybridization time *(10)*.
5. Probe added directly onto the membrane may result in high patchy background on detection around the site of the probe addition.
6. Stringency is controlled in the primary washes by the SSC concentration in a similar manner to posthybridization stringency washes with ^{32}P-labeled probes. The primary wash routinely contains 6 M urea and is performed at 42°C. However, it is possible to achieve similar results using a primary wash without urea with incubations performed at 55°C. Using a temperature greater than those recommended here, for either system, will result in either a very low signal because of the stringency being too great, or no signal at all because of inactivation of the peroxidase. Each of the primary washes must be carried out for no longer than 20 min for similar reasons.
7. Blots may be stored for up to 24 h at this stage, either at 4°C, wetted in secondary wash buffer, and wrapped in Saran Wrap, or at –20°C, immersed in a solution of 50% glycerol. After storage, blots should be rinsed briefly in fresh secondary wash buffer just before they are detected.
8. The length of the second exposure can be judged by assessment of the signal and background levels observed on the first film. As a guideline, typical exposure times would be 5–10 min for high-target systems (e.g., colony and plaque screening) and 30–60 min for low-target levels, such as those found in single-copy gene detection (*see* **Figs. 1** and **2**).

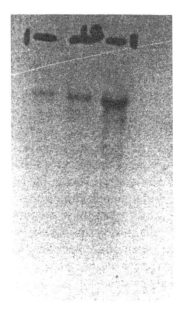

Fig. 2. Single copy gene detection. 1, 2, and 5 µg loadings of an *Eco*RI restriction enzyme digest of human genomic DNA immobilized on Hybond-N+. Hybridized with a 1.5-kbp HRP-labeled Nras proto-oncogene probe as described in the text. 30-min exposure on Hyperfilm-MP.

References

1. Heslop-Harrison, J. S. (1990) Gene expression and parental dominance in hybrid plants. *Development* **(suppl.),** 21–28.
2. Tonjes, R. R. (1991) Detecting transgene sequences by Southern blot analysis using the ECL gene detection system. *Life Sci. News* **4,** 7–8.
3. Einspanier, R. and Scham, D. (1991) Determination of insulin-like growth factor I and II in bovine ovarian follicles and cysts. *Acta Endocrinol.* **124 (suppl. 1),** 19.
4. Durrant, I., Benge, L. C. A., Sturrock, C., Devenish, A. T., Howe, R., Roe, S., Moore, M., Scozzafava, G., Proudfoot, L. M. F., Richardson, T. C., and McFarthing, K. G. (1990) The application of enhanced chemiluminescence to membrane based nucleic acid detection. *Biotechniques* **8,** 564–570.
5. Stone, T. and Durrant, I. (1992) Enhanced chemiluminescence for the detection of bound nucleic acid sequences. *Genet. Anal.: Tech. Appl.* **8,** 230–237.
6. Sorg, R., Enczmann, J., Sorg, U., Kogler, G., Schneider, E. M., and Wernet, P. (1990) Specific non-radioactive detection of PCR-amplified sequences with enhanced chemiluminescence labeling. *Life Sci. News* **2,** 3–4.
7. Tucker, S. J. (1991) ECL direct system: identification of the cystic fibrosis gene of Xenopus laevis. *Highlights* **2,** 4.
8. Hutton, J. R. (1977) Renaturation kinetics and thermal stability of DNA in aqueous solutions of formamide and urea. *Nucleic Acids Res.* **4,** 3537–3555.
9. Britten, R. J. and Davidson, E. H. (1985) Hybridization strategy, in *Nucleic Acid Hybridization: A Practical Approach* (Hames, B. D. and Higgins, S. J., eds.), IRL Press, Oxford, pp. 3–15.
10. Stone, T. (1992) ECL direct system. An analysis of filter hybridization kinetics. *Life Sci. News* **7,** 8.
11. Durrant, I. (1990) Light based detection of biomolecules. *Nature (Lond.)* **346,** 297–298.

27

Hybridization and Competition Hybridization of Southern Blots

Rosemary Kelsell

1. Introduction

This chapter describes the detection of specific DNA sequences by hybridization to a labeled probe of complementary sequence. This method is suitable for the detection of a wide range of DNA concentrations down to single-copy genes within mammalian genomic DNA (little more than 1 pg of hybridizing DNA in a total of 10 µg). In principle, hybridization consists of the annealing of a single-stranded, labeled nucleic acid probe to denatured DNA fixed to the filter. In practice, this method is designed to provide a balance between maximizing the specific signal and minimizing the nonspecific background.

One specific problem of background is in the hybridization of genomic DNA with a probe that contains repetitive DNA elements. These elements are dispersed throughout the genomes of most eukaryotic organisms so that long probes prepared from genomic DNA, such as whole λ or cosmid clones, may hybridize to many genomic locations and produce a smear that masks the signal from single-copy sequences. Competition hybridization offers a direct way to remove these repetitive elements. This technique requires prereassociation of the labeled probe in solution to sheared genomic DNA. Consequently, all the repetitive element sequences in the probe are annealed to the competitor DNA, leaving only unique sequences free to hybridize to DNA on the Southern blot. Various conditions for prereassociation have been developed *(1–3)*. This chapter describes a method based on the technique of Sealey et al. *(4)*. The protocol presented in **Subheading 3.2.** has been used successfully to compete repeats from a variety of λ clones for probing blots prepared from both pulsed-field, as well as ordinary agarose gels.

2. Materials

All general molecular biology grade reagents may be obtained from BDH Chemicals Ltd. (Poole, Dorset, UK) or Fisons (Loughborough, Leicestershire, UK). All solutions are made up using sterile distilled water and procedures required for molecular biology.

2.1. Hybridization (see Note 1)

1. 20X SSC: 3 M NaCl, 0.3 M sodium citrate, pH 7.0. Store at room temperature.
2. 100X Denhardt's solution *(5)*: 2% (w/v) bovine serum albumin (Fraction V, Sigma Chemical Co., Poole, Dorset, UK); 2% (w/v) Ficoll (Sigma); 2% (w/v) polyvinylpyrrolidone (Sigma). Store at –20°C
3. 10% SDS: Store at room temperature.
4. 10 mg/mL denatured herring sperm DNA (Sigma): Dissolve in water, shear by passing 12 times through a 17-gauge syringe needle or by sonication, and then denature by boiling for 15 min. Store at –20°C.
5. Prehybridization solution: 3X SSC; 10X Denhardt's solution; 0.1% (w/v) SDS; and 50 µg/mL herring sperm DNA. Prehybridization solution can be made as a stock and stored at –20°C.
6. Labeled probe (*see* Chapters 19, 20, or 21): 100 ng of labeled probe (*see* **Notes 2–4**). Purify in a Sephadex G50 (Pharmacia, St. Albans, Hertfordshire) spin column equilibrated in 3X SSC (*see* **Note 5**).
7. Hybridization solution: Prehybridization solution plus 10% (w/v) dextran sulphate *(6)* (Pharmacia). This is made by dissolving dextran sulphate in the prehybridization solution at 65°C, with occasional vigorous agitation. Make the hybridization solution several hours before use to allow adequate time for the dextran sulphate to dissolve; undissolved particles of dextran sulphate result in background spots. Hybridization solution can be made as a stock and stored at –20°C.
8. Hybridization oven and tubes: This author finds tubes the most convenient for hybridization. In this system, the filters are spread around the inner wall of a glass tube, which is then placed on a rotating drum in an oven. Using this method, it is easy to change solutions and the hybridization volume can be kept very small. These are commercially available (e.g., Hybaid Ltd., Teddington, Middlesex, UK and Techne, Cambridgeshire, UK) or can be homemade. Alternative systems work just as well but can be harder to set up. A common alternative is to hybridize in sealed bags immersed in a shaking water bath (*see* **Note 6**).
9. Wash solutions: Make 2 L of 3X SSC, 0.1% (w/v) SDS and 1 L of 0.1X SSC, 0.1% (w/v) SDS by dilution from stocks.
10. Suitable autoradiographic film: For example, KodakX-OMAT AR (Rochester, NY).
11. Stripping solutions: 500 mL 0.4 M NaOH and 500 mL 0.1X SSC, 0.5% SDS, 0.2 M Tris-HCl pH 7.5 (*see* **Note 7**).

2.2. Competition Hybridization

1. Oligolabeling stop buffer: 20 mM NaCl, 20 mM Tris-HCl pH 7.5, 2 mM ethylenediaminetetraacetic acid (EDTA) pH 8.0, 0.25% (w/v) SDS, 1 µM deoxycytidine triphosphate.
2. Competitor DNA: Sheared genomic DNA (of the appropriate species or DNA type) at 20 mg/mL in Tris-EDTA. This can be prepared by sonicating total genomic DNA in a suitable volume with 20–30 pulses of 5 s duration. The DNA should be of an average size of 500 bp, and an aliquot is checked on an agarose gel. The DNA is then phenol/chloroform extracted, chloroform extracted, and concentrated by ethanol precipitation. The concentration is checked by a spectrophotometer and adjusted to 20 mg/mL.

3. Methods

3.1. Hybridization

1. Rinse the filter in 2X SSC and place in a suitable container (e.g., a hybridization tube) with approx 10 mL of prehybridization solution (*see* **Note 8**). Incubate at 65°C for 3–4 h while rotating or shaking the solution (*see* **Note 9**).

2. Just prior to the hybridization step, denature the probe by boiling for 5–10 min (*see* **Note 4**) and add it directly to approx 10 mL of hybridization solution, prewarmed to 65°C (*see* **Note 10**).
3. Remove the prehybridization solution from the filter and add the completed hybridization solution. If hybridization tubes are used, this is easily done by pouring off the prehybridization solution and pouring in the hybridization solution. Incubate with rotating or shaking overnight at 65°C.
4. At the end of the hybridization, remove the filter (*see* **Note 11**) and place in approx 300 mL of 3X SSC, 0.1% SDS at room temperature for a minute to wash off the unbound probe. Repeat wash twice.
5. Wash the filter in approx 300 mL of preheated 3X SSC, 0.1% SDS at 65°C for 10 min. Repeat wash twice (*see* **Note 12**).
6. Wash the filter in 500 mL of preheated 0.1X SSC, 0.1% SDS at 65°C for 15 min. Repeat wash once (*see* **Note 13**).
7. Take the filter from the wash solution and place it wet in polythene film or a polythene bag and, if possible, seal the edges. Do not let the filter dry out or it will become extremely difficult to remove the probe for further washing or reprobing.
8. If a ^{32}P-labeled probe is being used, autoradiograph the filter for 2–24 h (*see* **Note 14**). If a longer exposure is required, the filter can be laid down for up to a further 2 wk.
9. The same filter can be reprobed after the first probe has been removed from it. Probe stripping is done by placing the filter(s) in 500 mL of 0.4 M NaOH at 42°C for 30 min and then transferring it to 500 mL of 0.1X SSC, 0.5% SDS, 0.2 M Tris-HCl pH 7.5 at 42°C for a further 30 min (*see* **Note 7**). The filter may then be reexposed to X-ray film to ensure that all the probe has been removed. Finally, it is stored in prehybridization solution at 4°C until ready for use. This author usually adds fresh prehybridization solution the next time the filter is used.

3.2. Competition Hybridization of Probe

1. Make the labeled probe up to a volume of 200 µL with an oligolabeling stop buffer. The probe is usually recovered in a volume of approx 50 µL after centrifugation through a Sephadex G50 column (*see* **Note 5**), requiring the addition of 150 µL of oligolabeling stop buffer. Add 50 µL of a 20-mg/mL solution of competitor DNA and 50 µL of 20X SSC.
2. Denature the mixture by boiling for 10 min, and then plunge into ice for 1 min.
3. Place the reaction at 65°C for 10 min (*see* **Note 15** and **Fig. 1**).
4. Add this competed probe to the hybridization solution and proceed as in **Subheading 3.1., step 3** (*see* **Notes 16** and **17**).

4. Notes

1. Other solutions work equally well for hybridization. For example, 50 mM phosphate buffer pH 6.8, 0.15% sodium pyrophosphate, 4X SSC, 5X Denhardt's solution, 0.3% (w/v) SDS, and 75 µg/mL denatured salmon sperm DNA (supplemented with 10% dextran sulphate in the hybridization mixture: Stephanie Halford, personal communication). Alternatively, 10% (w/v) SDS, 7% (w/v) PEG 6000, and 50 µg/mL denatured herring sperm DNA may be used for nylon membranes (the hybridization mixture does not require supplementation with dextran sulphate: David Kelsell, personal communication). This author's filters are made on good-quality nylon membranes (e.g., Genescreen or Genescreen plus, NEN DuPont, Boston, MA), allowing the probes to be removed and the filters to be reused with an array of probes.

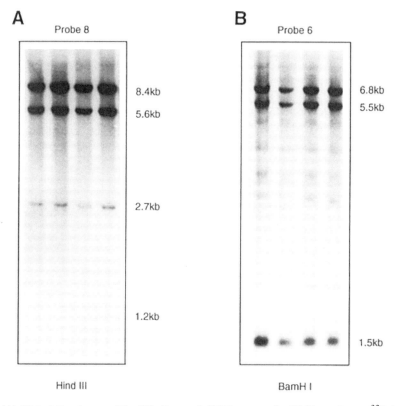

Fig. 1. **(A)** Hybridization to *Hind*III digested CHO genomic DNAs using a ^{32}P-labeled λ clone approx 15 kb long and treated using the competition hybridization protocol described in **Subheading 3.2.** There are at least four *Alu*-equivalent repeats situated within the region of DNA covered by this probe (probe 8, Davis and Meuth; *see* **ref. *11***). **(B)** Hybridization to *Bam*HI-digested CHO genomic DNAs using a different ^{32}P-labeled λ clone (probe 6, Davis and Meuth; *see* **ref. *11***). In this instance, the repeats have not been competed out of the probe as successfully as in the previous example.

2. The technique of random priming is the most efficient way of obtaining double-stranded probes labeled to a specific activity of >10^9 cpm/μg of DNA (*7*).
3. It is usually unnecessary to remove vector sequences from the DNA used for the probe, provided that crosshybridizing marker lanes are cut off the gel before capillary blotting. For example, the arms of λ clones hybridize to λ size markers, and plasmid sequences hybridize to some of the marker bands in the 1-kb ladder (Gibco-BRL, Paisley, Scotland).
4. Care should be taken if using radioactive labeled probes. This author uses screw-capped 1.5 mL tubes for making radiolabeled probes to avoid the lids popping open during the denaturation steps.
5. Commercial spin columns are available, but Sephadex G50 spin columns are cheap and simple to make. Add Sephadex G50 (Pharmacia) to 3X SSC and allow it to stand at room temperature overnight. Plug a 1-mL disposable syringe with polymer wool, fill it with the Sephadex G50, and allow the 3X SSC to drain out (the syringe should be full to the top with Sephadex G50 before it is spun). Put the syringe into a centrifuge tube (a 15-mL Falcon tube 2095 will do—Greiner-Stonehouse, Gloucestershire, UK). Spin for 3 min at 200*g* (1000 rpm in a bench centrifuge). The column should pack to a 1-mL volume. Load

the sample on top of the column and spin it under the same conditions again, collecting the sample in an uncapped Eppendorf tube. Columns should not be allowed to dry out and must be made just prior to their use.

6. To set up hybridization in bags, seal the filter in a bag, but make a funnel shape at the top of the bag with the bag sealer. The hybridization mixture containing the probe is then added to the bag through this funnel. It is relatively easy to remove air bubbles by rolling them to the top of the bag with a disposable 10-mL pipet and allowing the air to escape through the funnel. In this way, it is possible to fill bags with a minimum of spillages. This maneuver can be practiced with water in an empty bag!
7. Alternatively, filters may be stripped by agitating them in a 500-mL solution of boiling 0.1% (w/v) SDS until it reaches room temperature.
8. As a general rule, use at least 0.2 mL of prehybridization solution for every square centimeter of filter.
9. Filters can be prehybridized overnight, but this can weaken the filter and may also result in a diminished signal.
10. A probe concentration of 1–10 ng/mL gives the best signal to background ratio. If bad background is experienced with a probe at 10 ng/mL, it should be diluted tenfold and tried again.
11. The probe can be stored and reused. For radiolabeled probes, the storage time is regulated by the isotope half-life, 2 wk for ^{32}P. This is not a problem with nonradioactive probes. As the probe reanneals during hybridization, the hybridization solution should be boiled for 15 min before reuse.
12. This is a low-stringency wash and will leave the probe bound to similar (homologous) DNA sequences, as well as those that are identical. The filter can be autoradiographed at this point in order to detect these related sequences. The stringency of the wash can be increased by sequential washes at lower salt concentrations.
13. This is a high-stringency wash and should only detect identical sequences.
14. The autoradiograph sensitivity depends on the conditions under which the filter is exposed. Preflashing the photographic film with a single flash (<1 ms) of light from a flash gun with an orange filter increases the sensitivity of the film twofold and gives it a linear response to intensity of radioactivity. Additionally, placing an intensification screen behind the film and putting the assembly at –70°C gives a total tenfold increase in the sensitivity of the film but at the expense of the resolution *(8,9)*.
15. Ten minutes are adequate for the prereassociation step. However, if background repetitive sequences are still visible on the autoradiograph (*see* **Fig. 1B**), check that enough sheared genomic DNA was used. An overestimation of the concentration of the sheared genomic DNA is the most common cause of problems. If this is not the case, extend the prereassociation time for up to an hour.
16. This author and others *(10)* have found that competed probes appear to generate stronger signals than single-copy probes on Southern blots, possibly because the probes are generally longer.
17. This author has visualized fragments as small as approx 600–700 bp in size with this method.

References

1. Ardeshir, F., Giulotto, E., Zieg, J., Brison, O., Liao, W. S. L., and Stark, G. R. (1983) Structure of amplified DNA in different Syrian hamster cell lines resistant to *N*(phosphonacetyl)-L-aspartate. *Mol. Cell Biol.* **3,** 2076–2088.
2. Litt, M. and White, R. L. (1985) A highly polymorphic locus in human DNA revealed by cosmid-derived probes. *Proc. Natl. Acad. Sci. USA* **82,** 6206–6210.

3. Djabali, M., Nguyen, C., Roux, D., Demengeot, J., Yang, H. M., and Jordan, B. R. (1990) A simple method for the direct use of total cosmid clones as hybridization probes. *Nucleic Acids Res.* **18,** 6166.
4. Sealey, P. G., Whittaker, P. A., and Southern, E. M. (1985) Removal of repeated sequences from hybridisation probes. *Nucleic Acids Res.* **13,** 1905–1922.
5. Denhardt, D. T. (1966) A membrane filter technique for the detection of complementary DNA. *Biochem. Biophys. Res. Commun.* **23,** 641–646.
6. Wahl, G. M., Stern, M., and Stark, G. R. (1979) Efficient transfer of large DNA fragments from agarose gels to diazobenzyloxymethyl-paper and rapid hybridisation by dextran sulphate. *Proc. Natl. Acad. Sci. USA* **76,** 3683–3687.
7. Feinberg, A. P. and Vogelstein, B. (1993) A technique for radiolabelling DNA restriction endonuclease fragments to high specific activity. *Anal. Biochem.* **132,** 6–13.
8. Laskey, R. A. and Mills, A. D. (1975) Quantitative film detection of ^{3}H and ^{14}C in polyacrylamide gels by fluorography. *Eur. J. Biochem.* **56,** 335–341.
9. Laskey, R. A. and Mills, A. D. (1977) Enhanced autoradiographic detection of ^{32}P and ^{125}I using intensifying screens and hypersensitive film. *FEBS Lett.* **82,** 314–316.
10. Blonden, L. A. J., den Dunnen, J. T., van Paassen, H. M. B., Wapenaar, M. C., Grootscholten, P. M., Ginjaar, H. B., et al. (1989) High resolution deletion breakpoint mapping in the DMD gene by whole cosmid hybridization. *Nucleic Acids Res.* **17,** 5611–5621.
11. Davis, R. and Meuth, M. (1994) Molecular characterization of multilocus deletions at a diploid locus in CHO cells: association with an intracisternal-A particle gene. *Somat. Cell Mol. Genet.* **20,** 287–300.

28

Autoradiography and Fluorography

Eric Quéméneur

1. Introduction

Autoradiography enters its second century (Henri Becquerel, 1896) while remaining a basic method in molecular biology and despite the increasing number of phosphor imagers in laboratories and progress in nonradioactive detection methods. Film image might be preferred to numerized image because it is transparent and is exactly the same size as the sample, and this facilitates their alignment. In this chapter, we discuss the detection of radiolabeled molecules after their separation by electrophoresis or thin-layer chromatography. Obtaining photographic images implies that a few principles are known (*1*). Radioisotopes commonly used for biomolecules labeling differ in their physical properties (**Table 1**) and the energy, i.e., the reach, of the emitted radiations will dictate the practical procedure for detection and will determine the priority of resolution versus sensitivity (= speed). Two general processes have to be distinguished (*3*): direct exposure of film by β- or γ-rays or, when radiations do not efficiently reach the film, detection of secondary photons generated by excitation of a scintillator ("fluor," "phosphor") (**Fig. 1**). Low-energy isotopes detection will benefit from removing the emitters from the quenching medium, such as gel matrix. Membrane blotting is the most convenient way to achieve this prior to exposure. Collision of β- or γ-particles with silver halide crystals of the thick photographic layer covering X-ray films results in the formation of silver atom specks and generation of a latent image that is not stable and has to be irreversibly fixed by the development step. A single β- or X-particle is sufficient to create several metallic silver grains and to give a visible point. Several photon hits are required to form the necessary 3–6 atoms per grain. Therefore, any process that would stabilize the latent image and thus increase the probability of multiple hits will enhance the detection. That is why exposures involving fluorography or intensifying screens have to be performed at low temperature (–70/–80°C) or using presensitized films in which latent image formation is initiated. It is noteworthy that such preflashed films also have to be used for quantitation, as the intensity of the image on the film becomes then proportional to the amount of radioactivity in the sample.

2. Materials

1. Dried gel, membrane, or any other two-dimensional support containing radiolabeled compounds (*see* **Notes 1** and **2**).

From: *The Nucleic Acid Protocols Handbook*
Edited by: R. Rapley © Humana Press Inc., Totowa, NJ

Table 1
Characteristics of Most Frequently Used Radioisotopes

Isotope	Half-life	Radiation	Energy (MeV) E_{max}	Energy (MeV) E_{mean}	Detection method and sensitivity limit[a]
^{3}H	12.4 yr	β^-	0.018	0.0055	DA (80×10^6), F (8000)
^{14}C	5730 yr	β^-	0.156	0.050	DA (6000), F (400)
^{32}P	14.3 d	β^-	1.710	0.70	DA (500), IA (50)
^{33}P	25.4 d	β^-	0.249		DA, IA, F
^{35}S	88 d	β^-	0.167	0.0492	DA (6000), F (400)
^{45}Ca	165 d	β^-	0.256	0.077	DA, IA, F
^{125}I	60 d	γ	0.035		DA (1600), IA (100)
		X	0.027		
		Auger e^-	0.030		
^{131}I	8.04 d	γ	0.364		DA
		β^-	0.61		

[a]Abbreviations used: DA, direct autoradiography; IA, indirect autoradiography using intensifying screen; F, fluorography: Sensitivity limits are dpm/cm^2 required for a detectable image ($A_{545nm} = 0.02$ above 0.15 background) with a preflashed film in 24 h (data from **ref. 2**).

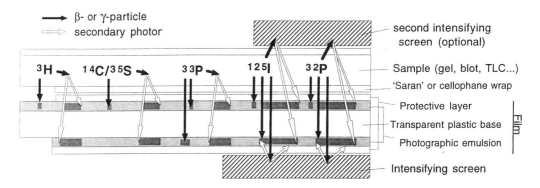

Fig. 1. Paths of radiations emitted by various radioisotopes and their ability to form direct or indirect images in the photographic emulsion layer of a X-ray film.

2. Dark room with inactinic light.
3. Appropriate X-ray films (*see* **Note 3**).
4. Film cassette.
5. Chemicals and installation for manual or automated film development.
6. Electronic flash unit able to generate <1 ms flash + orange filter to absorb blue light.
7. Intensifying screen.
8. Commercial scintillator available as solution, spray, or meltable wax (*see* **Note 4**). Alternatively, 20% (w/v) 2,5-diphenyl-oxazole (PPO) in dimethyl sulfoxide (DMSO) can be prepared.
9. Freezer (–70°/–80°C).
10. Radioactive ink: 100 µL Indian ink or equivalent is mixed with 1–5 µL of the radioactive solution used in the labeling reaction.

11. Strips for quantitative analysis of ^{3}H, ^{14}C, and ^{125}I-labeled samples are available commercially. For example, Microscales can be purchased from Amersham Pharmacia Biotech Inc. (Piscataway, NJ) in the range 3.7–4.048 Bq/mg, 3.7–31,890 Bq/g, and 44–23,900 Bq/mg for ^{3}H, ^{14}C, and ^{125}I respectively.

3. Methods
3.1. Preflashing Films

Preflashing films increases their sensitivity in photon detection-based approaches and their linear response range in quantitative analysis. It is not necessary for any other purpose.

1. In the darkroom, install the flash + filter unit at 50 cm above the film area.
2. Cut five film strips (just large enough to be processed) and cover them with a sheet of Whatman paper no. 1 (Maidstone, Kent, UK).
3. Make five 1-ms flashes, removing a strip between each in order to obtain a set of strips ranging from 1–5 ms exposure times.
4. Develop films and cut them to fit a spectrophotometer cuvette holder. Prepare an identical sample with a nonpreflashed film that will be used as reference. Read absorbances at 545 nm.
5. Choose an exposure time that gives an A_{545nm} in the range 0.10–0.20 to preflash the film to be exposed with the radiolabeled sample. If the film is not used immediately, store it at –70/–80°C. Do not preflash films that will not be used in the forthcoming days.

3.2. Direct Autoradiography (see Note 5)

1. Put the sample in the film cassette and if necessary, label using the radioactive ink. Avoid getting moisture or glove powder on the sample.
2. For samples containing low-energy isotopes, remove any barrier such as Saran Wrap or cellophane that may quench emitted radiations.
3. In the darkroom, place the film in close contact with the sample. When using single-coated film (e.g., Kodak SB, Amersham Hyperfilm βMax or ^{3}H…) or a preflashed film, the sensitive face of the film should face the gel.
4. Close the cassette and let stand at room temperature for the necessary exposure time.
5. Remove the film in the dark and process it according to the manufacturer's instructions.

3.3. Indirect Autoradiography Using Intensifying Screens (see Note 6)

Using screens is effective to detect γ- and high-energy β-emitters (^{32}P), where the isotope emissions can pass the through the film and reach the screen. Thus, screens are not efficient in enhancing ^{3}H, ^{14}C, ^{33}P, or ^{35}S.

1. In the darkroom, place the screen in the cassette and cover it with the film. If preflashed film is used, the flashed side of the film should face the screen.
2. Put the sample in the cassette after labeling with the radioactive ink, if necessary. Avoid getting moisture or glove powder on the sample, the film, or the screen.
3. When maximal sensitivity is needed, a second screen can be added over the sample, but this will decrease resolution. The opacity of some supports (e.g., blotting membranes), which absorb light, do not enable the use of two screens. Membranes can be rendered translucent by impregnating them with mineral oil *(4)*.
4. Exposure should be performed at low temperature (–70°/–80°C) and care should be taken to minimize moisture in the cassette during exposure.
5. Allow the cassette to warm to room temperature before developing the film.

3.4. Fluorography of Blotted Samples

1. Hybridization membranes can be used without further treatment except extensive washing of free radioactivity. Virtually any method or apparatus can be used to transfer labeled molecules from electrophoresis gels to membranes, provided quantitative transfer to the blot is achieved.
2. Impregnate the blot with a scintillator. (a) Spray freely to ensure uniform covering of the sample. (b) Commercial solution: Conform to supplier's instructions. (c) Homemade PPO solution (*see* **Subheading 2.**): Soak the membrane for 2 min and wash two times for 1 min in water in order to remove the DMSO and allow the PPO to precipitate. Incomplete DMSO washing would impede complete membrane drying and render it sticky to the film. (d) MeltiLex (Wallac, Wellesley, MA) is melted on the membrane surface by heating it at 75°C on a hot plate or in an oven.
3. The extent and uniformity of the scintillator impregnation can be monitored by viewing under ultraviolet light.
4. In the darkroom, place the sample and the film in the cassette so that the impregnated side of the sample faces the film. When single-coated film is used (e.g., Kodak SB), the emulsion should face the fluor. When preflashed films are used, the flashed side of the film should face the fluorographic layer.
5. Exposure should be performed at low temperature (–70°/–80°C) and care should be taken to minimize moisture in the cassette during exposure.
6. Allow the cassette to warm to room temperature before developing.

4. Notes

1. Damp samples can be used, provided moisture is prevented from contacting film by using Saran Wrap. Membranes may stick firmly to the film and tend to break or make sparks when separated. When using membranes, thorough drying before exposure is necessary.
2. Soaking gels in 50% polyethyleneglycol (PEG) 2000 prior to film exposure was shown to be an easy way to improve both autoradiographic and fluorographic detection *(5)*.
3. Most commercial films might prove suitable for any detection mode, but special care should be taken for choosing them when optimal results are requested. For example, a double-coated film is not necessary when dealing with low-energy radioisotopes, and using a single-coated film might benefit resolution even when detecting ^{32}P or ^{125}I. Direct visualization of ^3H is much more sensitive when a film that does not have a protective antiscratch layer is used. Some films are less sensitive to photons than others and might be avoided when very low activities have to be detected.
4. Commercial fluors use either PPO or sodium salicylate as scintillators with maximal emission at 388 and 420 nm, respectively. A better performance by PPO over salicylate was reported *(6)*. Care should be taken when using these compounds. The use of the meltable solid scintillator MeltiLex (Wallac, Wellesley, MA) was described as an rapid and safe alternative to liquid scintillant *(7)*.
5. Direct autoradiography is the method of choice when resolution is required (e.g., DNA sequencing). It is noteworthy that the thinner the sample, the more resolved the autoradiograph. Blotting ^3H-, ^{14}C-, or ^{35}S-labeled compounds from electrophoresis gel matrices to membranes to get rid of quenching effects benefits their detection and lower activity can be visualized. A reduction by a factor 5–10 in exposure times is also allowed, compared with dried gels.
6. Intensifying phosphor screens are made of either calcium tungstate or barium strontium sulfate, which differ slightly in their emission spectra on irradiation. However, maxima are both in the ultraviolet-blue light, respectively 370–380 and 410–450 nm, and corre-

sponding photons are well detected using most films. Screens also absorb white light and present an 'afterglow' effect that increases film contrast. The screens must be kept in the dark for several hours before exposure is started.

References

1. Hahn, E. J. (1983) Autoradiography: a review of basic principles. *Am. Lab.* **15,** 64–71.
2. Laskey, R. A. (1993) Efficient detection of biomolecules by autoradiography, fluorography or chemiluminescence. *Review 23* (RPN 1706) Amersham Pharmacia Biotech, Piscataway, NJ.
3. Laskey, R. A. (1980) The use of intensifying screens or organic scintillators for visualizing radioactive molecules resolved by gel electrophoresis. *Methods Enzymol.* **65,** 363–371.
4. Rust, S., Funke, H., and Assman, G. (1987) Mineral oil enhances the autoradiographic detection of ^{32}P-labeled nucleic acids bound to nitrocellulose membranes. *Anal. Biochem.* **163,** 196–199.
5. Mohamed, M. A., Lerro, K. A., and Prestwich, G. D. (1989) Polyacrylamide gel miniaturization improves protein visualization and autoradiographic detection. *Anal. Biochem.* **177,** 287–290.
6. Roberts, P. L. (1985) Comparison of fluorographic methods for detecting radioactivity in polyacrylamide gels or on nitrocellulose filters. *Anal. Biochem.* **147,** 521–524.
7. Quéméneur, E. and Simonnet, F. (1995) Increased sensitivity of autoradiography and fluorography by membrane blotting. *Biotechniques* **18,** 100–103.

IV

RNA ANALYSIS TECHNIQUES

29

Formaldehyde Gel Electrophoresis of Total RNA

Sian Bryant and David L. Manning

1. Introduction

RNA has the tendency to form both secondary and tertiary structures that can impede its separation by electrophoresis. As such, identical species of RNA exhibiting varying degrees of intramolecular base pairing migrate at different rates and result in the smearing of distinct RNA molecules. Consequently, the electrophoresis of RNA needs to be performed under denaturing conditions. Heat denaturing the RNA sample prior to electrophoresis is insufficient, as secondary structures will simply reform unless a denaturing system is used.

Successful electrophoresis of RNA is therefore accomplished in two steps:

1. RNA is heat denatured prior to electrophoresis.
2. During electrophoresis, conditions are be established that maintain the RNA in a denatured state.

The methodology, described in this chapter, involves the use of formaldehyde as a denaturant within the agarose gel. In addition, both formaldehyde and formamide are added to the sample before electrophoresis to aid the denaturation of the RNA sample.

For procedures such as Northern analysis, in which RNA is transferred or blotted from the gel to a solid matrix for subsequent hybridization, the optimal balance between electrophoretic resolution and efficiency of transfer is achieved with a 1–1.2% agarose gel (*see* **refs. *1–3***).

2. Materials

It is very important that all reagents used are of molecular biology grade and free from RNase contamination. As in all molecular biological procedures, gloves should be worn throughout.

1. Formaldehyde: Formaldehyde is a suspected nose, nasopharynx, and liver carcinogen, it is toxic both through inhalation and ingestion. Its use should therefore be restricted to a ventilated fume hood. Formaldehyde is routinely supplied as a 37% (v/v) stock solution. It should be stored at room temperature and out of direct sunlight to prevent oxidation.
2. Formamide: The use of this chemical should be restricted to the fume hood, as it is a suspected teratogen and is irritating to the eyes, skin, and respiratory system. Formamide should be stored in the dark and at room temperature to prevent oxidation.

3. RNase-free water: This can be achieved by treating the water first with diethyl pyrocarbonate (DEPC). Add DEPC to the water to a final concentration of 0.1%. Incubate overnight at 37°C and autoclave to destroy any residual DEPC. DEPC is an efficient, nonspecific inhibitor of RNase; however, it is carcinogenic and should be handled in a fume hood with extreme care.
4. 10X MOPS buffer: (0.2 M MOPS (3-(N-morpholino)propanesulfonic acid) pH 7.0; 50 mM sodium acetate; 10 mM ethylenediaminetetraacetic acid (EDTA), pH 8.0). This solution is prepared using RNase-free water. When autoclaved, it assumes a characteristic golden color and can be stored at room temperature.
5. Agarose (molecular biology grade): Store at room temperature.
6. Ethidium bromide solution (10 mg/mL): This is a powerful mutagen, and extreme care must be taken when using this substance. Store at 4°C.
7. A horizontal electrophoresis tank and a 11 × 14 cm casting tray.
8. RNA loading buffer: Contains 50% glycerol, 1 mM EDTA, and 0.4% bromophenol blue, made up to the required volume with DEPC-treated water.

3. Method

1. Denature the RNA in a sterile RNase-free microcentrifuge tube by mixing the following: 10 mg of RNA (in a final volume of 5 mL with DEPC-treated water), 2 mL of 10X MOPS solution, 3.5 mL of formaldehyde, and 10 mL of formamide. Incubate the RNA solution at 65°C for 15 min in a fume hood and transfer immediately to ice.
2. For a 1% agarose gel using an 11 × 14 cm casting tray, mix 0.8 g of agarose with 57.5 mL of sterile RNase-free water and boil to dissolve the agarose. Cool to 60°C, then add 8 mL of 10X MOPS buffer, 14.5 mL of 37% (12.3 M) formaldehyde, and finally 8 mL of ethidium bromide solution (at a concentration of 10 mg/mL). Mix the components of the gel by gently rotating the bottle, being careful not to introduce any air bubbles. Pour the gel into the prepared casting tray (in a fume hood), insert the comb, and let it set for a least 60 min at room temperature.
3. After cooling the denatured RNA solution in ice, add 2 mL of sterile RNA loading buffer.
4. Fill the buffer reservoirs and cover the gel with 1X MOPS buffer.
5. Pre-electrophorese the gel at 50 V for 10 min.
6. Load the heat-denatured samples into the wells and run the gel at 50 V until the bromophenol blue has moved approximately half to three quarters of the way along the gel (*see* **Note 1**).
7. Visualize RNA using a 254-nm short-wave ultraviolet transilluminator (*see* **Notes 2–7**). Gels can be photographed using a Polaroid film with a red plus yellow filter (or orange) at f4.5 for 1/2–1 s.

4. Notes

1. To avoid overheating and a smile effect on the migration of RNA and bromophenol blue dye, we routinely run gels at 4 V/cm length of gel (i.e., for a 14-cm-long gel, electrophoresis is performed at 50–60 V for approx 4 h).
2. RNA bands very broad or trailing. This problem usually occurs as a result of a fault in the sample preparation (RNA isolation) or in the loading of the gel. The salt composition (introduced with the sample) of the loading solution may be too high, too much RNA may have been loaded into the well (slight band broadening may occur with volumes above 30 mL or more than 50 mg of RNA per well) and applying higher voltages when running the gel may cause broadening and trailing probably due to excess heat. Electrophoresis may be carried out quite successfully at room temperature.

3. Little or no RNA detected in the gel after electrophoresis. If RNA is contaminated with proteins the RNA will tend to aggregate at the top of the gel, close to the well. Normally, however, this is prevented, as most procedures for RNA extraction and purification involve the use of proteolytic enzymes or deproteinizing mixtures such as phenol:chloroform.

    RNA may be degraded into fragments that are so small they pass straight through the gel. If this is the case, run the RNA on a gel containing a higher concentration of agarose, or preferably run the gel for a shorter period of time. If there is a problem with RNA degradation, then the extraction and purification methods need to be examined. Degradation of the RNA could be a result of ribonucleases in the electrophoresis buffer or tank, therefore care must be taken to ensure that the equipment is clean and that both this and the buffer are nuclease free. The gel concentration may be too high, preventing the RNA from entering it, although this is very unlikely as the concentration used in this methodology is not very high.

4. There may be problems with ethidium bromide staining in formaldehyde gels. Ethidium bromide can retard the running of the nucleic acid on the gel by up to 15%, but more importantly it can reduce the transfer efficiency of the RNA to a solid support. Ethidium bromide can be removed by soaking the gel in running buffer or transfer buffer, with several changes, for 1 h.

5. An aliquot of total cellular RNA should electrophoretically resolve only two very distinct bands, the 28s and 18s rRNAs. The appearance of these two bands gives an idea of the integrity of the RNA. The mRNA component manifests itself as a significantly lighter-stained smear above, between, and below the rRNA bands. The 5s rRNA, 5.8s rRNA, and the tRNAs all migrate close to the leading edge of the gel. Totally degraded samples appear as heavily localized smears below the level at which the 18s rRNA appears in intact samples. Heavy smears that appear along the length of the lane may be indicative of degradation or may simply mean that the sample was incompletely denatured. Fluorescence within the well suggests that genomic DNA is present within the sample.

6. The positioning of the 18S and 28S bands should be noted, as this information may be required to estimate the size of a particular RNA of interest. (The 18S rRNAs range in size from 1.8–2 kb, 28S rRNAs range in size from 4.6–5.2 kb).

7. Prolonged exposure to UV light can damage the RNA. Formaldehyde can be removed from the gel by placing it in either 1X MOPS buffer or DEPC-treated water for about 15 min, with several changes of buffer.

References

1. Boedtker, H. (1971) Conformation-independent molecular determinations of RNA by gel electrophoresis. *Biochim. Biophys. Acta* **240**, 448.
2. Rave, N., Crkvenjakov, R., and Boedtker, H. (1979) Identification of procollagen mRNAs transferred to diazobenzyloxymethyl paper from formaldehyde gels. *Nucleic Acids Res.* **6**, 3559.
3. Manning, D. L., McClelland, R. A., Gee, J. M., Chan, C. M., Green, C. D., Blamey, R. W., et al. (1993) The role of four oestrogen responsive genes pLIV1, pS2, PSYD3 and pSYD8 in predicting responsiveness to therapy in primary breast cancer. *Eur. J. Cancer* **29a**, 1462–1468.

30

RNA Probes for the Analysis of Gene Expression

Dominique Belin

1. Introduction

The isolation and characterization of RNA polymerases from the *Salmonella* phage SP6 and the *E. coli* phages T7 and T3 has revolutionized all aspects of the study of RNA metabolism *(1–6)*. Indeed, it is now possible to generate unlimited quantities of virtually any RNA molecule in a chemically pure form. This technology is based on a number of properties of the viral transcription units. First, and in contrast to their cellular counterparts, the enzymes are single-chain proteins that were easily purified from phage-infected cells and are now produced by recombinant DNA technology. Second, they very specifically recognize their own promoters (*7* and references therein), which are contiguous 17–20-bp-long sequences rarely encountered in bacterial, plasmid, or eukaryotic sequences. Third, the enzymes are highly processive, allowing the efficient synthesis of very long transcripts from DNA templates. In this chapter, the preparation of the DNA templates, the transcription from the templates of labeled synthetic RNA molecules, commonly called riboprobes, and their use in Northern and RNase protection assays are discussed.

2. Materials

These protocols require the use of standard molecular biology materials and methods for carrying out subcloning, polymerase chain reactions (PCR), and gel electrophoresis, in addition to those listed below. More rigorous precautions are required for working with RNA than are commonly used for most DNA studies because it is important to avoid RNase contamination. The two major sources of unwanted RNases are the skin of investigators and microbial contamination of solutions. Most RNases do not require divalent cations and are not irreversibly denatured by autoclaving. Gloves should be worn and frequently changed. Sterile plasticware should be used, although some mechanically manufactured tubes and pipet tips have been used successfully without sterilization. Glassware should be incubated at 180°C in a baking oven for several hours.

In addition, divalent cations (such as Mg^{2+} or Ca^{2+}) accelerate base-mediated RNA hydrolysis. RNA can be stored in sterile water but is most stable when stored in 1 mM KOAc pH 5, 0.1 mM ethylenediaminetetraacetic acid (EDTA).

2.1. Preparation of Riboprobes

1. Water: Although a number of protocols recommend treatment of the water used for all solutions with diethyl pyrocarbonate, it is unnecessary. Double-distilled water is used for the preparation of stock solutions that can be autoclaved (121°C, 15–30 min), and sterilized water is used otherwise.
2. TE: 10 mM Tris-HCl, pH 8.1, and 1 mM EDTA.
3. 10X TB: 0.4 M Tris-HCl, pH 7.4, 0.2 M NaCl, 60 mM MgCl$_2$, and 20 mM spermidine. If ribonucleotides are used at concentrations >0.5 mM each, MgCl$_2$ concentration should be increased to provide a free magnesium concentration of 4 mM.
4. 0.2 M dithiothreitol (DTT): the solution is stored in small aliquots at –20°C. Aliquots are used only once. EDTA can be included at 0.5 mM to stabilize DTT solutions.
5. Ribonucleotides: Neutralized solutions of ribonucleotides are commercially available, or can be made up from dry powder (*see* **Note 1**). Ribonucleotide solutions can be stored at –20°C for several months.
6. RNA polymerase stocks: The three RNA polymerases, SP6, T3, and T7, are available commercially. Store at –20°C.
7. RNA polymerase dilution buffer: 50 mM Tris-HCl, pH 8.1, 1 mM DTT, 0.1 mM EDTA, 500 µg/mL bovine serum albumin (BSA) and 5% glycerol. The diluted enzyme is unstable and should be stored on ice for no more than a few hours.
8. Stop-mix: 1% sodium dodecyl sulfate, 10 mM EDTA, and 1 mg/mL tRNA. The tRNA may be omitted.
9. TEN: 10 mM Tris-HCl, pH 8.1, 1 mM EDTA, and 100 mM NaCl.
10. Sample buffer (polyacrylamide/urea gels): 80% deionized formamide (*see* **Note 2**), 2 M urea, 0.1X TBE (8.9 mM Tris, 8.9 mM boric acid, 0.2 mM EDTA) and 0.01% each of xylene cyanol and bromophenol blue. Use 1–2 µL per µL of RNA in aqueous solution.

2.2. Northern Blot Hybridization

1. Formamide: pure formamide is slowly hydrolyzed by water vapor to ammonium formate and therefore must be deionized (*see* **Note 2**). Store at –20°C.
2. Glyoxal: A 30% glyoxal solution (6 M) is deionized by several incubations at room temperature with a mixed bed resin (AG501-X8, Bio-rad, Richmond, CA), until the pH is between 6 and 7 and the conductivity of 3% glyoxal in water is below 30 µS. Store at –20°C in small aliquots.
3. Denaturation buffer: 1 M glyoxal in 50% dimethylsulfoxide and 10 mM Na$_2$HPO$_4$, pH 6.8. Ethidium bromide can be included at a concentration of 50 µg/mL to visualize the rRNAs during electrophoresis.
4. 5X RSB: 50% glycerol, 10 mM Na$_2$HPO$_4$, pH 6.8, and 0.4% bromophenol blue. Autoclave and store at –20°C in small aliquots.
5. Transfer membrane: Nitrocellulose (Schleicher and Schuell, Keene, NH) and nylon membranes (Hybond N, Amersham Pharmacia Biotech, Arlington Heights, IL; or Biodyne A, Pall, Glen Cove, NY) have been used successfully.
6. 50X Denhardt's solution: Dissolve 2 g BSA (fraction V) in 80 mL sterile water. Bring the pH to 3.0 with 2 N HCl, boil for 15 min and cool on ice for 10 min. Bring the pH to 7 with 2 N NaOH and add sterile water to 100 mL. Autoclave a 100-mL solution of 2% polyvinylpyrrolidone (K90, Fluka, Buchs, Switzerland) and 2% Ficoll 400 (Pharmacia, Piscataway, NJ), and add to the 2% albumin solution.
7. Hybridization solution: 50% deionized formamide, 0.8 M NaCl, 50 mM Na-piperazine-N,N'-*bis*(2-ethanesulfonic acid), pH 6.8, 2 mM EDTA, 0.1% SDS, 2.5X Denhardt's solution, and 0.1 mg/mL sonicated and heat-denatured salmon sperm DNA. The hybridization

solution is heated, filtered over 0.45 μm Nalgene sterilization units (Nalgene, Rochester, NY) or Millipore nitrocellulose filters (Millipore, Bedford, MA) and kept at the hybridization temperature during prehybridization. The filtration step decreases nonspecific attachment of the probe to the membrane and degasses the solution.

8. 20X standard sodium citrate: 3 M NaCl and 0.3 M Na/citrate, pH 7.

2.3. RNase Protection

1. Hybridization mixture for RNase protection: 80% deionized formamide (*see* **Note 2**), 0.4 M NaCl, 40 mM Na-PIPES, pH 6.8, and 1 mM EDTA.
2. RNase digestion buffer: 300 mM NaCl, 10 mM Tris-HCl, pH 7.4, and 4 mM EDTA.
3. Pancreatic RNase: make a 10-mg/mL solution in TE containing 10 mM NaCl. Vials containing lyophilized RNases should be carefully opened in a ventilated hood to avoid contamination. Boil for 15 min and slowly cool to room temperature. Store in aliquots at −20°C.
4. T1 RNase: make a 1-mg/mL solution in TE. Adjust the pH to 7. Store in aliquots at −20°C.
5. Proteinase K: dissolve the enzyme at 20 mg/mL in water and store in aliquots at −20°C.

3. Methods
3.1. Preparation of Riboprobes
3.1.1. Linearized Plasmid Templates for Runoff Transcription

1. Subclone the desired gene fragment in a transcription vector (*see* **Note 3**).
2. Isolate plasmid DNA by alkaline lysis from a 30–100-mL saturated culture. Plasmid DNAs are purified by CsCl/ethidium bromide centrifugation or by precipitation with polyethyleneglycol (*see* **Note 4**).
3. Linearize 2–20 μg plasmid DNA with an appropriate restriction enzyme (*see* **Note 5**). Verify the extent of digestion by electrophoresis of an aliquot (0.2–0.5 μg DNA) on an agarose minigel in the presence of ethidium bromide (0.5 μg/mL) (*see* **Note 6**).
4. Purify the restricted DNA by two extractions with phenol/chloroform/isoamyl alcohol (25/24/1 by volume), and remove residual phenol by one extraction with chloroform (24/1 by volume). Precipitate the DNA with ethanol. If little DNA is present (below 5 μg), add 10 μg of glycogen (Boehringer Mannheim, Mannheim, Germany) as a carrier; this carrier has no adverse effect in the transcription reactions. After the ethanol pellet is washed, the DNA is dried in air and resuspended in TE at 1 μg/μL.

3.1.2. Synthetic and PCR-Derived Templates

The major limitation in using restriction enzymes to clone inserts and to linearize plasmid templates is that appropriate sites are not always available. Furthermore, the transcripts will almost always contain 5' and 3' portions that differ from those of endogenous RNA. One possibility in circumventing these difficulties is based on the transcription of small DNA fragments obtained by annealing of synthetic oligodeoxynucleotides *(6,8)*. An alternative and more general approach generates the transcription templates via PCR amplification of plasmid DNA. In theory, such templates could direct the synthesis of virtually any RNA sequence.

1. Design the 5' primer, which has a composite sequence: Its 5' portion is constituted by a minimal T7 promoter, and its 3' portion corresponds to the beginning of the transcript (*see* **Note 7**). Six to 10 nucleotides are usually sufficient to prime DNA synthesis on the plasmid template.

2. Design the 3' primer, which is usually 17–20 nt long and defines the 3' end of the transcript (see **Note 8**).
3. PCR amplify 2–50 ng of plasmid DNA in a total volume of 100 µL. Verify that a DNA fragment of the expected size has been amplified, and estimate the amount of DNA by comparison with known standards.
4. Purify the DNA as described in **Subheading 3.1.1., step 4** and resuspend in TE at the appropriate concentration. The PCR-derived templates are transcribed at lower DNA concentrations than plasmids to maintain the molar ratio of enzyme to promoter; a good choice is 3 µg/mL for a 100-bp fragment.

3.1.3. Basic Transcription Protocol for Radioactive Probes

1. Assemble the transcription mixture to a total volume of 10 µL by adding in the following order: water (as required), 1 µL 10X TB, 0.5 µL BSA (2 mg/mL), 0.5 µL 0.2 M DTT, 0.25 µL placental RNase inhibitor (40 U/µL), 1 µL 5 mM solution of each ribonucleotide (i.e., ATP, GTP, CTP), 5 µL α-[^{32}P]-UTP (400 Ci/mmole, 10 mCi/mL), 1 µL of restricted plasmid DNA template and 0.25–0.75 µL of RNA polymerase (see **Notes 9** and **10**).
2. Incubate 40 min at 37°C for T7 and T3 polymerases or 40°C for SP6 polymerase. After adding the same number of units of enzyme (see **Note 10**), incubate for a further 40 min.
3. Degrade the template with RNase-free DNase (1 U/µg of DNA) for 20 min at 37°C.
4. Stop the reaction by adding 40 µL of stop-mix.
5. After two extractions with phenol/chloroform, in which the organic phases are back-extracted with 50 µL of TEN, the combined aqueous phases are purified from unincorporated nucleotides by spun-column centrifugation. The spun column is prepared by filling disposable columns (QS-Q, Isolab, Hörth, Germany) with a sterile 50% slurry of G-50 Sephadex (Pharmacia, Piscataway, NJ) in TEN followed by centrifugation for 5 min at 200g_{av}. The samples are carefully deposited on top of the dried resin, and the column is placed in a sterile conical centrifuge tube. After one centrifugation for 5 min at 200g_{av}, 200 µL TEN is deposited on top of the resin and the column recentrifuged. The eluted RNA (200–400 µL) is ethanol precipitated and resuspended in water (see **Note 11**).
6. Measure the incorporation efficiency by counting an aliquot (see **Note 12** for calculations and for modifications of the basic protocol).
7. The size of the transcript may be verified by electrophoresis in polyacrylamide/urea gels (see **Note 13**).

3.2. Northern Blot Hybridization

The use of riboprobes in RNA blot hybridizations follows the same general principles as that for DNA probes. The major disadvantage in using double-stranded DNA probes results from self-annealing, which decreases the availability of DNA probes to bind to the immobilized target; this is particularly critical with heterologous probes, where the reannealed probe may displace incompletely matched hybrids. Self-annealing, of course, does not occur with single-stranded RNA probes. Most difficulties encountered with riboprobes stem from the increased thermal stability of RNA:RNA hybrids. Thus, crosshybridization of GC-rich probes to rRNAs can generate unacceptable backgrounds. This problem is often solved by increasing the stringency of hybridization, as illustrated in **Fig. 1B**. Alternatively, the template may have to be shortened to remove GC-rich regions from the probe.

1. Denature the sample in 8 µL of denaturation buffer for 15–30 min at 50°C. Add 2 µL of 5X RSB Sample Buffer and electrophorese in 0.7–2% agarose gels in 10 mM Na$_2$HPO$_4$,

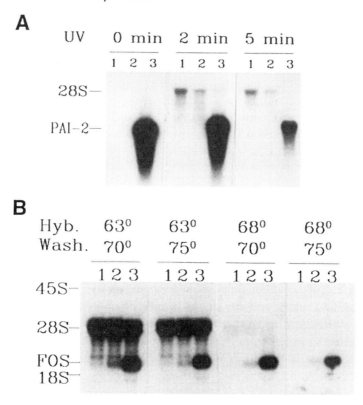

Fig. 1. Northern blot hybridization with riboprobes. (**A**) Effect of UV crosslinking. Northern blot hybridization of PAI-2 mRNA in murine total cellular RNA with an homologous cRNA probe. Lanes 1 and 2: 5 µg of placental RNA (15.5 and 18.5 d gestation) that do not contain detectable levels of PAI-2 mRNA. Lane 3: 1 µg of LPS-induced macrophage RNA, an abundant source of PAI-2 mRNA (*29*). All samples were electrophoresed and transferred together. After the membrane was cut, each filter was UV treated as described. The filters were hybridized at 58°C, washed at 70°C, and exposed together. Crosshybridization of the probe to 28S rRNA is more pronounced after UV irradiation, and specific hybridization is decreased after 5 min of UV exposure. (**B**) Effect of hybridization temperature. Northern blot hybridization of c-*fos* mRNA in rat total cellular RNA with a murine v-*fos* cRNA probe. Lane 1: uninduced cells. Lane 2: partially induced cells. Lane 3: fully induced cells (M. Prentki and D. Belin, unpublished). All samples were electrophoresed and transferred together. The filters, which were not UV crosslinked, were hybridized and washed in parallel at the indicated temperatures. The four filters were exposed together. Crosshybridization of the probe to 28S rRNA was essentially abolished by hybridizing at 68°C. Some specific signal was lost with the 75°C stringency wash.

pH 6.8. The buffer should be circularized with a peristaltic pump so that the pH near the electrodes remains neutral.
2. Transfer the RNAs by capillarity onto a membrane (*see* **Note 14**).
3. Fix the RNA by incubating the blots at 80°C under vacuum. This step is essential in removing glyoxal covalently fixed to guanine residues in the RNA.
4. Ultraviolet crosslinking is often used to improve RNA retention, although irradiation may increase background hybridization (*see* **Fig. 1A** and **Note 15**).

5. Prehybridize the blots for 4–12 h in hybridization solution. Use 200 μL/cm² of the membrane.
6. Dilute the probe in hybridization solution (25–50 μL/cm² of membrane), then add to the membrane. I frequently hybridize two to three filters per bag. Hybridize at the appropriate temperature (58–68°C, *see* **Fig. 1B** and **Note 16**), usually for 12–18 h (*see* **Note 17**).
7. Wash the membranes twice for 10–20 min at the hybridization temperature with 100 μL/cm² of 3X SSC and 2X Denhardt's solution, and then three times with 0.2X SSC, 0.1% SDS and 0.1% Na pyrophosphate at the appropriate temperature (*see* **Fig. 1B** and **Note 17**).
8. Expose the membranes to autoradiographic film. As long as the membranes are not allowed to dry, they can be further washed at increased stringencies to reduce background.

3.3. RNase Protection

This assay is based on solution hybridization and on the resistance of RNA:RNA hybrids against single-stranded specific RNases. A ^{32}P-labeled probe is synthesized that is partially complementary to a portion of the target RNA. It is hybridized in excess to the target so that all complementary sequences are driven into the labeled RNA:RNA hybrid. Unhybridized probe and any single-stranded region of the hybridized probe are then removed by RNase digestion. The "protected" probe is then detected and quantitated on a denaturing polyacrylamide gel. The probe can be used to map the ends of RNA molecules or exon-intron boundaries. It also provides an attractive and highly sensitive alternative to Northern blot hybridization for the quantitative determination of mRNA abundance.

RNase protection has a number of advantages. First, solution hybridization tolerates high RNA input (up to 60 μg of total RNA) and is not affected by the efficiency of transfer on membranes or by the availability of membrane-bound RNAs. Second, the signal-to-noise ratio is much more favorable, as crosshybridizing RNAs yield only short protected fragments. Third, a significant fraction of mRNAs is often partially degraded during RNA isolation; in Northern blots, this generates a trail of shorter hybridizing species, which reduces the sensitivity of detection. Finally, the detection of hybridized probes on sequencing gels is much more sensitive because the width of the bands is less than a tenth of those of intact RNAs in agarose gels. Only two features of Northern blots are lost in RNase protection assays: complete size determination of target RNAs and multiple use of each sample.

1. Linearize the plasmid DNA template as described in **Subheading 3.1.1.** (*see* **Note 18**).
2. Transcribe the template as described in **Subheading 3.1.3.** The amount of labeled ribonucleotide may be varied (*see* **Note 19**).
3. An optional step is to purify the full-length transcripts by electrophoresis (*see* **Note 20**). Separate the transcript on a preparative 5–6% polyacrylamide/urea gel (gel thickness: 0.4–1.5 mm). Cover the wet gel within Saran Wrap and expose for 30 s–5 min at room temperature to localize the full-length transcript. Cut the exposed band on the film with a razor blade. After aligning the cut film on the gel, excise the gel band with a sterile blade. The cut gel should be reexposed to verify that the correct band has been excised.
4. Elute the RNA from the gel either by diffusion or electroelution.
   a. Incubate the gel fragment in an Eppendorf tube in 500 μL of 0.5 *M* ammonium acetate, 1% SDS, and 20 μg/mL tRNA for 1–3 h at 37°C, or overnight at 4°C. The eluate and residual gel can be counted to ensure that more than 60% of the

RNA is eluted. After two extractions with phenol/chloroform, recover the eluted RNA by ethanol precipitation.

b. Electroelute for 1–2 h at 30 V/cm in 0.1X TBE in a sterile dialysis bag, after which invert the polarity for 30 s to detach the eluted RNA from the membrane. Purify the eluate by two phenol/chloroform extractions and ethanol precipitation with an known amount of tRNA carrier. This procedure is very sensitive to RNase degradation.

5. Resuspend the probe in water at 1–2 ng/μL. Add 1 μL of the probe to 29 μL of the hybridization mixture for each assay. The exact amount of probe is not critical, as it is in excess over its specific target (*see* **Notes 19** and **21**).

6. Lyophilize or ethanol precipitate the sample RNAs (*see* **Note 22**). Resuspend in 30 μL of complete hybridization mixture including probe, heat for 2 min at 90°C, and incubate overnight, usually at 45°C (*see* **Notes 23** and **24**).

7. Cool the samples on ice and add 300 μL of RNase digestion buffer. Digest for 1 h at 25°C with pancreatic RNase, which cleaves after uracil and cytosine residues, with T1 RNase, which cleaves after guanine residues, or with both RNases (*see* **Notes 25** and **26**).

8. Add 20 μL of 10% SDS, and degrade the enzyme(s) with 0.5 μL (10 μg) of proteinase K for 10–20 min at 37°C. Extract twice with phenol/chloroform and precipitate the RNAs with ethanol with 10 μg of carrier tRNA (*see* **Note 27**).

9. Resuspend the RNAs in sample buffer, denature the hybrids for 2 min at 90°C and electrophorese in polyacrylamide/urea sequencing gels. Alternatively, the hybrids may be analyzed on nondenaturing polyacrylamide gels. Fix the gels with 20% ethanol and 10% acetic acid to remove the urea, dry and autoradiograph (*see* **Note 28**).

4. Notes

1. Powdered ribonucleotides should be resuspended in water, neutralized to pH 7 with 1 M NaOH or HCl, and adjusted to the desired concentration by measuring the UV absorbance of appropriate dilutions:

    100 mM ATP: 1540 absorbance units at 259 nm
    100 mM GTP: 1370 absorbance units at 253 nm
    100 mM CTP: 910 absorbance units at 271 nm
    100 mM UTP: 1000 absorbance units at 262 nm

    The integrity of ribonucleotide triphosphate solutions can be verified by thin-layer chromatography on polyethyleneimine-cellulose (PEI-CEL300). The resin is first washed with water by ascending chromatography to remove residual UV-absorbing material and dried. Ten to 30 nmol of ribonucleotides are deposited on the resin, which is then resolved by ascending chromatography with 0.5 M KH_2PO_4, adjusted to pH 3.5 with H_3PO_4. After being dried, the ribonucleotides are detected by UV-shadowing at 254 nm.

2. To deionize formamide, incubate it at –80°C until 75–90% of the solution has crystallized. Discard the liquid phase, thaw, and incubate at 4°C for several hours with a mixed-bed resin (AG501-X8, Bio-rad). Use a Teflon-covered magnet that has been freed of RNase by treatment with 0.1 M NaOH for 10 min and rinsed with water and with crude formamide. Check the conductivity, which should be below 20 μS. Filter the solution onto sterile paper (Whatman, LS-14) over a sterile funnel. The absorbance at 270 nm should be below 0.2.

3. All the vectors that are commercially available consist of high-copy number *E. coli* plasmids derived from ColE1. The original plasmids (pSP64 and pSP65, Promega, Madison, WI) contained only one SP6 promoter located upstream of multiple cloning sites (MCS) *(2)*. In the second generation of plasmids, two promoters in opposite orientation flank the MCS to allow transcription of both strands of inserted DNA fragments. The pGEM series

(Promega) contain SP6 and T7 promoters *(5)*, whereas the pBluescript series (Stratagene, La Jolla, CA) contain T7 and T3 promoters.

The choice of plasmid is mostly a matter of personal preference, although it can be influenced by the properties of individual sequences. For instance, researchers have frequently observed premature termination with SP6 transcripts. The nature of the termination signals is not completely understood *(4,9)*, and their efficiency can be more pronounced when one ribonucleotide is present at suboptimal concentration (*see* **Note 13**). The problem has been sometimes solved by recloning the inserts in front of a T7 or T3 promoter. The partial recognition of T7 (T3) promoters by T3 (T7) polymerases may result in the transcription of both strands when the ratio of enzyme to promoter is not carefully controlled. This can be a source of artefacts, particularly when the templates are linearized inside the cloned inserts.

4. It is possible to use plasmid DNA from "minipreps," although transcription efficiency can be reduced, particularly with SP6 polymerase. The RNA present in the minipreps is digested with pancreatic RNase (20 µg/mL), which is removed during purification of the linearized templates. Spun-column centrifugation of the digested minipreps can improve transcription efficiency.

5. Because RNA polymerases can initiate transcription unspecifically from 3' protruding ends *(4,10)*, restriction enzymes that generate 5' protruding or blunt ends are usually preferred. If the only available site generates 3' protruding ends, the DNA can be blunt ended by exonucleolytic digestion with T4 DNA polymerase or with the Klenow fragment of DNA polymerase I. Restriction with enzymes that cut the plasmids more than once may also be used, provided that the promoter is not separated from the insert.

6. The plasmids must be linearized as extensively as possible. Because circular plasmids are efficient templates, their transcription by the highly processive enzymes may yield RNA molecules that can be up to 20 kb long and thus incorporate a significant portion of the limiting ribonucleotide.

7. There are constraints on the 5' sequence of transcripts, as the sequence immediately downstream of the start site is necessary for the transition from an abortive initiating cycling mode to the elongation mode. The first 6 nt have a strong influence on promoter efficiency; in particular, the presence of uracil residues are usually detrimental *(6,8)*. It may be necessary, therefore, to include in the 5' end of the transcripts five to six bases that differ from those present in natural RNAs. My colleagues and I have used a number of composite T7 promoters, whose efficiency is summarized in the following table. In addition to the promoter sequence (−17 to −1:5'-TAATACGACTCACTATA) at the 5' end, the first 6 nt of the templates are

Efficient promoters	Inefficient promoters
GGGAGA (T7 consensus)	GTTGGG (5% efficiency)
GGGCGA (pBS plasmids)	GCTTTG (1% efficiency)
GCCGAA	

Composite 5' primers with the SP6 promoter sequence have also been used successfully. There are also constraints on the 5' sequence of the transcripts, but the optimal sequences (GAATA, GAACA, and GAAGA) are different than those with T7 polymerase *(11–13)*.

8. The 3' end of the transcripts should be exactly defined by the 5' end of the downstream primer. However, template-independent addition of 1–2 nt during transcription usually generates populations of RNAs with different 3' ends. The proportion of each residue at the 3' end may depend on context and is influenced by the relative concentration of each ribonucleotide, limiting ribonucleotide being less frequently incorporated *(2,6,8,14)*.

9. The order of addition of components may be changed. Remember that dilutions of RNasin are very unstable in the absence of DTT, and that DNA should not be added to undiluted 10X TB. The temperature of all components must be at least at 25°C, to avoid precipitation of the DNA:spermidine complex. Because the radioactive nucleotides, which constitute half of the reactions, are provided in well-insulated vials, they can take up to 10–15 min after thawing to reach an acceptable temperature. I routinely incubate all components, except the RNasin and the polymerase, at 30°C for 15–20 min. When more than one probe is to be synthesized, a reaction mixture with all components is added to the DNA template. It is possible to reduce the total volume to 8 µL in order to conserve materials; however, because the enzymes are very sensitive to surface denaturation, these incubations are done in 400-µL vials.
10. For SP6 polymerase, use 5–10 U/µg plasmid DNA (size: 3–4 kbp). For T7 and T3 polymerases, use 10–20 U/µg plasmid DNA.
11. The purification step (**Subheading 3.1.3.**, **step 4**) may not be required, and some investigators use the transcription mixtures directly in hybridization assays. However, low backgrounds, consistency, and quantitation of the newly synthesized RNAs probably justify the additional time and effort.
12. More than 50% of the labeled ribonucleotide is routinely incorporated, and often greater than 80%. An input of 50 µCi may, therefore, yield up to 4×10^7 Cerenkov-cpm RNA. This represents 100 pmoles of UMP and, assuming no sequence bias (i.e., 25% of residues are uracil residues), 130 ng of RNA (specific activity: 3×10^8 Cerenkov-cpm/µg).

    Similar results are obtained with labeled CTP and GTP, although GTP rapidly loses its incorporation efficiency on storage. ATP is not routinely used because of its higher apparent K_m for SP6 polymerase on linear templates *(2,15)*. UTP is particularly stable and can be used even after several weeks once radioactive decay is taken into account. With each labeled ribonucleotide, the initial concentration must be >12.5 µ*M* to ensure efficient incorporation and to prevent polymerase pausing. Transcripts destined to be translated must incorporate a 5'–7 mG cap structure. This is usually achieved by performing the transcription in the presence of 500 µ*M* of a cap analogue dinucleotide and 50 µ*M* GTP. Because the cap analogue can also be used during elongation, labeled GTP should not be used to calculate incorporation efficiency.

    A variety of other labeled or unlabeled probes may also be made for use in *in situ* hybridization (*see* **refs. 16–20**).

    For the synthesis of large quantities of RNA (more than 10 RNA transcripts per DNA template molecule), each ribonucleotide is added at a final concentration of 0.5–1 m*M*; trace amounts of labeled UTP should be included to calculate transcription efficiency and to verify the size of the transcript. Total volume is increased to reduce the concentration of plasmid DNA to 30–50 µg/mL. Additional modifications of the standard protocol have been described and include the use of HEPES-KOH pH 7.5 at 120 m*M* (SP6 polymerase), 200 m*M* (T7 polymerase), or 300 m*M* (T3 polymerase) *(21)*. Furthermore, the T4 gene *32* protein can increase transcription efficiency when added at 10 µg/µg of template DNA (D. Caput, personal communication). It has been recently reported that a reduction of premature termination and an increased synthesis of large full-length transcripts (size >1 kb) can be obtained by performing the transcription at 30°C or at room temperature *(4,22)*.
13. Large transcripts that do not enter the polyacrylamide gel are diagnostic of incompletely linearized templates. Small transcripts are indicative of extensive pausing or premature termination. In addition to the natural T7 termination signal, a site located in the coding region of the *PTH* gene has been extensively characterized *(24,25)*. Mutant T7 RNA

14. Hybridization of dot-blots is often used to quantitate mRNA levels with large number of samples. To accurately quantitate specific hybrids, it is necessary to include as negative control total RNA from cells that do not express the transcript under examination.
15. The intensity provided by transilluminators, a common source of UV light, varies with time, and excessive crosslinking can severely reduce hybridization efficiency (**Fig. 1A**). Furthermore, UV crosslinking can increase crosshybridization to rRNAs, possibly by inducing covalent crosslinking with the probe. Ultraviolet irradiation should thus be limited to membranes subjected to multiple rounds of hybridization. An apparatus commercially available from Stratagene provides a means to control the dose of UV light. The appropriate dose of UV light is probably different for different membranes and has a broader range for wet than for dry membranes.
16. The major variables in optimizing signal-to-noise ratio are the temperature of hybridization and the temperature of the stringency washes (**Fig. 1B**). Many probes can be hybridized at 58°C, although the temperature must be increased to 68°C for certain pairs of probe and target RNAs. Stringency washes are usually performed at 70 or 75°C. It is difficult to increase stringency much further, as most waterbaths cannot maintain accurate temperatures at or above 80°C. For the detection of DNA targets, hybridization is usually performed at 42°C and the stringency washes at 65°C.
17. RNA probes are usually 200–800 nt long. With 600-nt-long RNA probes, a hybridization plateau is achieved in 20 h at 58°C with 2.5 ng/mL of the probe. This represents an input of 7.5×10^5 Cerenkov-cpm/mL. The probe concentration can be increased to 10 ng/mL and the hybridization time decreased to 5 h without increasing the background.
18. The probes should be 100–400 nt long, and should include at least 10 nt which are not complementary to the target RNA. Residual template DNA or antisense RNA can produce traces of full-length protected probe that must be distinguishable from the fragment protected by the target RNA (**Fig. 2**).
19. It is often useful to decrease the specific activity of the probe: more RNA is synthesized at the resulting higher ribonucleotide concentration, the probes are less susceptible to radiolysis, and less radioactivity is used. The following guideline can be used to alter the specific activity of the probes according to the sensitivity required:

Target RNA abundance	Unlabeled UTP	[32P]-UTP	Probe/sample[a]
High	100 μM	2.5 μM, 10 μCi	6–12 kcpm
Moderate	10 μM	2.5 μM, 10 μCi	60–120 kcpm
Low	—	12.5 μM, 50 μCi	300–600 kcpm

[a]The amount of probe required per sample for the detection of the target (*see* **Note 21**).

20. Purification of is often necessary for maximal sensitivity or for mapping purposes. The transcription reaction can be directly loaded on the preparative gel after DNase digestion of the template, provided that enough EDTA is present in the sample buffer to chelate all the magnesium. Omitting the DNase digestion of the template results in higher amounts of fully protected probe in the assay.
21. The probe must be in excess over the target RNA (*see* **Fig. 2A**). An input of 1–2 ng of a 300-nt probe in a total volume of 30 μL will drive the hybridization of target RNA to completion (4–8 $T_{1/2}$) in approximately 16 h. Shorter hybridizations can be performed but require higher probe input (R_0) to achieve the same extent of saturation, i.e., to maintain the $R_0 \times T_{1/2}$ value.

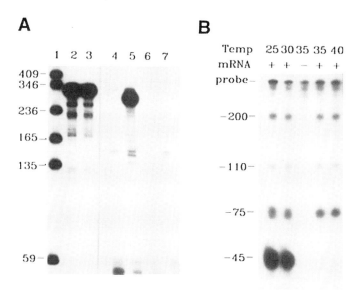

Fig. 2. RNase protection assay. (**A**) Discrimination between target-specific signal and complete probe protection by residual DNA. Detection of PN-I mRNA in total RNA from murine tissues (30). The probe (310 nt long) was gel purified. Lane 1: size markers. Lane 2: purified probe. Lane 3: probe hybridized and processed without RNase digestion. This part of the gel was autoradiographed for 6 h. Lane 4: control hybridization with 10 µg of tRNA; traces of fully protected probe are visible. Lane 5: 10 µg of RNA from seminal vesicles, an abundant source of PN-I mRNA; the specific protected fragment is 260 nt long. Lane 6: 10 µg of liver RNA, which does not contain detectable levels of PN-I mRNA. Lane 7: 10 µg of testis RNA, which contains trace levels of PN-I mRNA. This part of the gel was autoradiographed for 24 h. (**B**) Effect of hybridization temperature on the detection of short complementary RNAs. The 5' ends of phage T4 gene 32 transcripts in total RNA from bacteria carrying a gene 32 expression cassette were mapped by hybridization to a cRNA probe containing 400 nt of gene 32 upstream sequences (23). The probe was not gel purified, and hybridization was performed at the indicated temperatures. Fully protected probe results from incomplete DNase digestion of the template and is also visible in the control hybridization without target RNA. The 44-nt protected fragment is no longer detected above 30°C.

22. To facilitate the RNase digestion step, each sample should contain the same amount of total RNA. For very low abundance target RNA, the amount of sample RNA may be increased up to 60 µg. Inequalities should be eliminated by addition of tRNA. A negative control sample, containing only tRNA, is always included (**Fig. 2A**, lane 4, **Fig. 2B**, lane 3).
23. The temperature of hybridization must be reduced to detect small or very AU-rich protected fragments. For instance, a 44-nt fragment of a phage T4 gene 32 transcript (containing 35 A/U and 9 C/G) was only protected by performing the hybridization at 25–30°C (23) (**Fig. 2B**).
24. Alternatively, the probe and target RNAs may be coprecipitated with ethanol, resuspended in 9 µL TE, heated for 2 min at 90°C, and chilled on ice. After the addition of 1 µL of 3 M NaCl, 0.2 M Tris/HCl pH 7.4, 20 mM EDTA, the hybridization is carried out for 30–60 min at 70°C. Because the probe concentration is higher, there is no formamide and the incubation temperature is higher, the hybridization is driven to

completion more rapidly. The RNase digestion can be performed in 50–100 µL (*26*; J. Curran, personal communication).

25. The amount of RNase is determined by the total amount of RNA present in the samples, including that contributed by the probe. I usually add 0.5 µg of pancreatic RNase and/or 0.25 µg of T1 RNase per microgram of RNA. In most cases, digestion with pancreatic RNase alone is sufficient. When the probe and the target RNAs are from different species, the extent of homology can be sufficient to generate discrete protected fragments, particularly if digestion is performed with RNase T1 only. The temperature of digestion can be increased to 30–37°C, although this often leads to partial cleavages within the RNA:RNA hybrids.
26. To ensure that the probe remains intact during hybridization, it may be useful to include a parallel control that is hybridized and processed without RNase treatment (**Fig. 2A**, lanes 2 and 3).
27. RNases may also be inactivated by adding 330 µL 4 M guanidinium thiocyanate, 25 mM Na-citrate, pH 7.0, and 1 M β-mercaptoethanol. Add 20 µg tRNA, precipitate the RNAs with 660 µL of isopropanol, and centrifuge immediately for 15 min at 13,000g (*27*; P. A. Menoud, personal communication).
28. Minor shorter protected fragments are often detected, and they may complicate the interpretation of mapping assays *(28)*. To distinguish between digestion artefacts and rare target RNAs that are only partially complementary to the probe, a synthetic sense transcript fully complementary to the probe may be used as a control target RNA *(29)*. A sense RNA can be included as an external control *(32)*.

Acknowledgments

The author thanks P. Vassalli for his early encouragement to use riboprobes for detecting rare mRNAs. Over the last few years, many colleagues, students, and technicians have contributed to the methods outlined in this chapter, including M. Collart, N. Busso, J.-D. Vassalli, H. Krisch, S. Clarkson, J. Huarte, S. Strickland, P. Sappino, M. Pepper, A. Stutz, G. Moreau, D. Caput, M. Prentki, W. Reith, J. Curran, P. A. Menoud, P. Gubler, F. Silva, V. Monney, D. Gay-Ducrest, and N. Sappino. Research was supported by grants from the Swiss National Science Foundation and by the Canton de Genève. The author is at the University of Geneva Medical School, Department of Pathology, Geneva, Switzerland. E-mail address: dominique.belin@medecine.unige.ch

References

1. Butler, E. T. and Chamberlin, M. J. (1984) Bacteriophage SP6-specific RNA polymerase. *J. Biol. Chem.* **257,** 5772–5788.
2. Melton, D. A., Krieg, P. A., Rebagliati, M. R., Maniatis, T., Zinn, K., and Green, M. R. (1984) Efficient in vitro synthesis of biologically active RNA and RNA hybridization probes from plasmids containing a bacteriophage SP6 promoter. *Nucleic Acids Res.* **12,** 7035–7056.
3. Davanloo, P., Rosenberg, A. H., Dunn, J. J., and Studier, F. W. (1984) Cloning and expression of the gene for bacteriophage T7 RNA polymerase. *Proc. Natl. Acad. Sci. USA* **81,** 2035–2039.
4. Krieg, P. A. and Melton, D. A. (1987) In vitro RNA synthesis with SP6 RNA polymerase. *Methods Enzymol.* **155,** 397–415.
5. Yisraeli, J. K. and Melton, D. A. (1989) Synthesis of long, capped transcripts in vitro by SP6 and T7 RNA polymerases. *Methods Enzymol.* **180,** 42–50.

6. Milligan, J. F. and Uhlenbeck, O. C. (1989) Synthesis of small RNAs using T7 RNA polymerase. *Methods Enzymol.* **180,** 51–62.
7. Breaker, R. B., Banerji, A., and Joyce, G. F. (1994) Continuous in vitro evolution of bacteriophage RNA polymerase promoters. *Biochemistry* **33,** 11,980–11,986.
8. Milligan, J. F., Groebe, D. R., Witherell, G. W., and Uhlenbeck, O. C. (1987) Oligoribonucleotide synthesis using T7 RNA polymerase and synthetic DNA templates. *Nucleic Acids Res.* **15,** 8783–8798.
9. Roitsch, T. and Lehle, L. (1989) Requirements for efficient in vitro transcription and translation: a study using yeast invertase as a probe. *Biochim. Biophys. Acta* **1009,** 19–26.
10. Schenbon, E. T. and Mierendorf, R. C. (1985) A novel transcription property of SP6 and T7 RNA polymerases: dependence on template structure. *Nucleic Acids Res.* **13,** 6223–6234.
11. Nam, S. C. and Kang, C. (1988) Transcription initiation site selection and abortive initiation cycling of phage SP6 RNA polymerase. *J. Biol. Chem.* **263,** 18,123–18,127.
12. Solazzo, M., Spinelli, L., and Cesareni, G. (1987) SP6 RNA polymerase: sequence requirements downstream from the transcription start site. *Focus* **10,** 11,12.
13. Stump, W. T. and Hall, K. B. (1993) SP6 RNA polymerase efficiently synthesizes RNA from short double-stranded DNA templates. *Nucleic Acids Res.* **21,** 5480–5484.
14. Moreau, G. (1991) RNA binding properties of the *Xenopus* LA proteins. Ph. D. dissertation, University of Geneva, Switzerland.
15. Taylor, D. R. and Mathews, M. B. (1993) Transcription by SP6 RNA polymerase exhibits an ATP dependence that is influenced by promoter topology. *Nucleic Acids Res.* **21,** 1927–1933.
16. Sappino, A.-P., Huarte, J., Belin, D., and Vassalli, J.-D. (1989) Plasminogen activators in tissue remodeling and invasion: mRNA localization in mouse ovaries and implanting embryos. *J. Cell Biol.* **109,** 2471–2479.
17. Jostarndt, K., Puntschart, A., Hoppeler, H., and Billeter, R. (1994) The use of [^{33}P]-labeled riboprobes for *in situ* hybridizations: localization of myosin light chain mRNAs in adult human skeletal muscle. *Histochem. J.* **26,** 32–40.
18. Dörries, U., Bartsch, U., Nolte, C., Roth, J., and Schachner, M. (1993) Adaptation of a non-radioactive *in situ* hybridization method to electron microscopy: detection of tenascin mRNA in mouse cerebellum with digoxigenin-labeled probes and gold-labeled antibodies. *Histochemistry* **99,** 251–262.
19. Kriegsmann, J., Keyszer, G., Geiler, T., Gay, R. E., and Gay, S. (1994) A new double labeling technique for combined *in situ* hybridization and immunohistochemical analysis. *Lab. Invest.* **71,** 911–917.
20. Egger, D., Troxler, M., and Bienz, K. (1994) Light and electron microscopic *in situ* hybridization: non-radioactive labeling and detection, double hybridization, and combined hybridization-immunocytochemistry. *J. Histochem. Cytochem.* **42,** 815–822.
21. Pokrovskaya, I. D. and Gurevich, V. V. (1994) *In vitro* transcription: preparative RNA yields in analytical scale reactions. *Anal. Biochem.* **220,** 420–423.
22. Krieg, P. A. (1991) Improved synthesis of full length RNA probe at reduced incubation temperatures. *Nucleic Acids Res.* **18,** 6463.
23. Belin, D., Mudd, E. A., Prentki, P., Yi-Yi, Y., and Krisch, H. M. (1987) Sense and antisense transcription of bacteriophage T4 gene 32. *J. Mol. Biol.* **194,** 231–243.
24. Mead, D. A., Szesna-Skorupa, E., and Kemper, B. (1986) Single-stranded DNA blue T7 promoter plasmids. *Protein Eng.* **1,** 67–74.
25. Macdonald, L. E., Durbin, R. K., and McAllister, W. T. (1994) Characterisation of two types of termination signals for bacteriophage T7 RNA polymerase. *J. Mol. Biol.* **238,** 145–158.

26. Curran, J., Marq, J. B., and Kolakofsky, D. (1992) The Sendai virus nonstructural C proteins specifically inhibit viral mRNA synthesis. *Virology* **189,** 647–656.
27. Hod, Y. (1992) A simplified ribonuclease protection assay. *Biotechniques* **13,** 852–853.
28. Lau, E. T., Kong, R. Y. C., and Cheah, K. S. E. (1993) A critical assessment of the RNase protection assay as a means of determining exon sizes. *Anal. Biochem.* **209,** 360–366.
29. Belin, D., Wohlwend, A., Schleuning, W.-D., Kruithof, E. K. O., and Vassalli, J.-D. (1989) Facultative polypeptide translocation allows a single mRNA to encode the secreted and cytosolic forms of plasminogen activators inhibitor 2. *EMBO J.* **8,** 3287–3294.
30. Vassalli, J.-D., Huarte, J., Bosco, D., Sappino, A.-P., Sappino, N., Velardi, A., Wohlwend, A., Erno, H., Monard, D., and Belin, D. (1993) Protease-nexin I as an androgen-dependent secretory product of the murine seminal vesicle. *EMBO J.* **12,** 1871–1878.
31. Lyakhov, D. L., He, B., Zhang, X., Studier, F. W., Dunn, J. J., and McAllister, W. T. (1997) Mutant bacteriophage T7 RNA polymerases with altered termination properties. *J. Mol. Biol.* **269,** 28–40.
32. Scott, P. A. E., Smith, K., Bichmel, R., and Harris, A. L. (1977) Reliable external control for RNase protection assays. *Nucleic Acids Res.* **95,** 1305–1306.

31

Primer Extension Analysis of mRNA

Maggie Walmsley, Mark Leonard, and Roger Patient

1. Introduction

Primer extension is a relatively quick and convenient means by which gene transcription can be monitored. The technique can be used to determine accurately the site of transcription initiation or to quantify the amount of cap site-specific message produced.

The principle of this technique is shown in **Fig. 1**. In brief, a radiolabeled primer fragment (usually a single-stranded oligonucleotide of approx 20 nt long) is hybridized to its complementary sequence near the mRNA 5' terminus. The primer is then extended by the enzyme reverse transcriptase back to the initiation point (cap site) of the message. The products of the reaction are run out on a denaturing polyacrylamide gel and exposed to autoradiography.

The major advantage of primer extension for RNA analysis is its convenience (compared to S1 mapping or RNase protection). Although RNase protection is usually more sensitive, the convenience of primer extension means that it is often the method of choice for RNA analysis. In the situation where the 5' end of a new gene is being determined, however, it is important to perform both primer extension and either S1 or RNase mapping to identify the cap site unambiguously because all these methods generate artefacts that can result in misassignment of the 5' end. Premature termination of synthesis, perhaps due to RNA secondary structure, leads to less than full-length products in primer extension. Internal cleavage of nucleic acid hybrids, possibly at AT-rich tracts that undergo local denaturation, leads to foreshortened products in S1 and RNase mapping. Because the artefactual bands generated have different causes, it is normally safe to assume that the correct start site has been identified when both methods identify the same cap site.

2. Materials

As with all procedures involving RNA, extreme care must be taken to avoid RNase contamination and degradation of samples *(1)*. In recent years the advent of ribonuclease-inhibiting enzymes such as RNasin have rendered the use of the RNase inhibitor diethylpyrocarbonate (DEPC) redundant, assuming that reasonable care is taken to work cleanly. Work surfaces should be clean and gloves should be worn at all times.

A Schematic Representation Of The Primer Extension Reaction

1) Transfect wild type and marked genes into cells.

   Gene A^{wt} _____   Gene A^{mt} _____..._____

2) Genes transcribed in tissue culture cell nuclei.

   RNA^{wt} _____AAA   RNA^{mt} _____..._____AAA

3) Labelled primer hybridised to mRNAs.

   _____AAA   _____..._____AAA
       ---*   ---*

4) Primer extended by reverse transcriptase.

   _____AAA   _____..._____AAA
   ---------*   ------...---*

5) Precipitate and run on denaturing acrylamide gel.

B An Example Of A Primer Extension Reaction

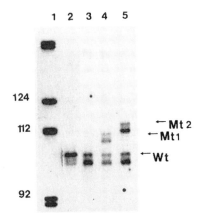

Fig. 1. Primer extension reaction. (**A**) Schematic representation. (**B**) Example.

Stock solutions, glassware, and plastic pipet tips should be autoclaved. Whenever possible, use sterile plasticware in place of glass. Separate stocks of reagents and plasticware should be maintained solely for RNA work.

The most convenient source of single-stranded primer is a chemically synthesized short oligonucleotide, which also yields a labeled probe of high specific activity (*see* **Note 1**). The primer should be 18 nt or longer, lacking a terminal phosphate at the 5' end to facilitate labeling with polynucleotide kinase and should ideally be located within 100–200 bp of the 5' end of the mRNA. Care should be taken that the chosen oligomer does not contain repeats capable of inter- or intraprobe hybridization, as this will reduce the efficiency of annealing to mRNA.

1. 10× Kinase buffer: 500 mM Tris-HCl, pH 7.6, 100 mM MgCl$_2$, 50 mM dithiothreitol (DTT), 1 mM spermidine, 1 mM ethylenediaminetetraacetic acid (EDTA). Alternatively use supplier's buffer.
2. 5× Hybridization buffer: 2 M NaCl, 50 mM piperazine-N,N'-bis[2-ethanesulfonic acid]/(PIPES), pH 6.4 (autoclave before use).
3. 1× Extension buffer: 10 µL of 1 M Tris-HCl, pH 8.3, 10 µL of 200 mM DTT, 10 µL of 120 mM MgCl$_2$, 5 µL of 1 mg/mL actinomycin D (Take care. Actinomycin D is toxic and light sensitive.), 10 µL each of 10 mM dATP, dCTP, dGTP, dTTP, 1 µL (40 U) RNasin and water to 178 µL.
4. Formamide loading dye: 80% deionized formamide, 45 mM Tris-HCl, pH 8.3, 45 mM boric acid, 1.25 mM EDTA, 0.02% xylene cyanole.
5. G25 Sephadex (Amersham Pharmacia Biotech, St. Albans, Herts, UK): Suspend Sephadex in approx 50 vol of TE (10 mM Tris-HCl, pH 8, 1 mM EDTA). Autoclave for 15 min (Sephadex will swell) and allow to cool to room temperature. Store at 4°C in a capped bottle.
6. Carrier (yeast) tRNA (BCL, Lewes, East Sussex, UK): Make up to 25 mg/mL with autoclaved water. Use the maker's bottle to avoid RNase contamination during weighing. Store in sterile plastic tubes at –20°C.
7. 3 M Sodium acetate (NaOAc), brought to pH 5.4 with glacial acetic acid and autoclaved for 15 min.
8. 70% EtOH: 35 mL absolute ethanol, 15 mL autoclaved water made up in a sterile 50 mL plastic Falcon tube.

3. Methods
3.1. Kinasing the Primer

1. In an autoclaved microfuge tube, mix 1–5 pmol single-stranded oligonucleotide (*see* **Note 1**), 1 µL 10× kinase buffer, 35 µCi γ^{32}P adenosine triphosphate (3000 Ci/mmol), 20 U T4 polynucleotide kinase, and autoclaved water to 10 µL.
2. Incubate at 37°C for 30 min.
3. Add 2 µL of 250 mM EDTA.
4. Remove unincorporated labeled nucleotide by centrifugation down a Sephadex G25 column (*see* **Note 2**).
5. Transfer the eluate (which contains the probe) to an autoclaved microfuge tube and heat kill the kinase by incubating at 65°C for 10 min. Extract with an equal volume of phenol:chloroform:isoamyl alcohol (25:24:1) (v/v).
6. To the aqueous phase add sodium acetate and MgCl$_2$ to a final concentration of 0.3 M and 5 mM, respectively, and add 10 µg tRNA.
7. Ethanol precipitate by adding 2.5 vol of 100% ethanol, placing in dry ice for 10 min, and spinning at 13,000 rpm in a microfuge for 30 min at 4°C.

8. Wash the pellet in 70% ethanol, dry under vacuum (approx 2 min), and resuspend at 1–10 fmol/µL in autoclaved water.

3.2. Hybridization Reaction

1. In an autoclaved microfuge tube, mix 1–50 fmol labeled primer (*see* **Note 3**), 4 µL 5× hybridization buffer, 0.1–5 fmol target mRNA (up to 20 µg total RNA, *see* **Note 4**) and autoclaved water to a final volume of 20 µL.
2. Heat the samples to 70°C for 3 min.
3. Hybridize for 1.5–4 h (*see* **Note 4**) in a water bath at approx 54°C or at an empirically determined optimum temperature (*see* **Note 5**).

3.3. Extension Reaction

1. To each sample add 178 µL extension buffer and place the tube on ice.
2. Add 1 µL RNasin and 1 µL reverse transcriptase per sample.
3. Transfer tubes to 42°C and incubate for 1 h (*see* **Note 6**).
4. Add 20 µL 3 *M* NaOAc and precipitate with 2.5 vol EtOH. Transfer to dry ice for 10 min and spin at high speed in a microcentrifuge for 15 min.
5. Wash with 70% EtOH and vacuum dry for 2 min.
6. Take pellet up in 10 µL formamide dyes, being sure to resuspend thoroughly.
7. Heat denature at 90°C for 3 min and chill on ice.
8. Load 8 µL on an appropriate percentage denaturing polyacrylamide gel and electrophorese along with suitable size markers or DNA sequencing ladders.

The labeled extension products are run on a standard denaturing urea-acrylamide gel of suitable percentage (dependent on the distance of the primer from the 5' end of the message) to enable good resolution of cap-site-length transcripts. This is particularly important when the site of transcription initiation is to be determined but is also important for quantitative experiments if multiple start sites exist in the gene of interest (*see* **Notes 7–9**).

4. Notes

1. The sensitivity of the primer extension analysis can be improved by using a uniformly labeled M13 vector-generated probe. However, except for the detection of extremely rare transcripts, this should not be necessary. To generate a probe of maximal specific activity using a kinased oligomer, label up no more than 5 pmol of a single-stranded oligomer (enough for 500 assays) at a time. In our experience, the kinasing reaction is not linear above this point and labeling up larger amounts is counterproductive, leading to a lowering of specific activity.
2. Prepare G25 Sephadex as follows: Plug the nipple of a 2-mL sterile plastic syringe with polymer wool and fill the syringe with a concentrated slurry of Sephadex. Suspend the syringe over a sterile plastic centrifuge tube and spin at 250 g in a swing out rotor for 5 min. Load exactly 100 µL Tris-EDTA (TE) on to the column and spin again for 5 min at 1500 rpm (250*g*). Discard the centrifuge tube and buffer contents, and transfer the syringe to a fresh tube. Load the labeled primer made up to 100 µL with TE and spin for 5 min at 1500 rpm (250*g*). Unincorporated label will be located at the top of the column and labeled primer will be found in the eluate, which should measure close to 100 µL. If the eluate measures significantly less than 100 µL, wash the column with a further 50–100 µL, centrifuge as before, and combine the eluates. Remove 1 µL of the probe, add to 100 µL water and determine the specific activity (SA) by Cerenkov counting. Probes should have an SA of $2–5 \times 10^6$ cpm/pmol.

3. Alternatively, probe and test RNA can be coprecipitated with ethanol. Take the pellet up in autoclaved water and add the remaining components of the hybridization mix.
4. The duration of the hybridization reaction depends on the amount of probe and target present in the reaction mix. In the case of transfected or injected cells where transcripts are abundant, we are dealing with the top end of the range of probe and target concentrations and hybridization over 90 min (or less—the minimum time may be determined empirically) will suffice to drive the reaction to completion. When probing for low-abundance endogenous transcripts, however, less target and probe are being used and hybridization times should be increased to a minimum of 4 h. The hybridization is carried out with the primer in approx 10-fold molar excess over the target RNA. Too great an excess (particularly at suboptimal hybridization stringencies) can result in nonspecific priming. In cases when it is not possible to estimate the abundance of the target mRNA, the following procedure may be followed to ensure that the hybridization is being carried out in primer excess, which is necessary if quantitative results are required. Include three or four extra samples that represent a small titration of an RNA stock that contains the target mRNA (over a range of 1–20-fold, say). If the signals from your test samples fall within the linear range of the titration then the assay has been carried out in primer excess.
5. The optimum temperature for hybridization will depend on the length of primer and its base composition. Formulas for the estimation of RNA-DNA hybrid melting temperatures *(2,3)* are not accurate for short DNA primers. For most probes, the value is in the range of 45–65°C in the buffer given, but if maximum sensitivity is required, pilot experiments should be carried out over this range to determine the optimum temperature for the specific primer-mRNA combination being used.
6. The elongation reaction is carried out at 42°C to reduce the amount of RNA secondary structure that can result in premature termination of the extension reaction.
7. The further the primer is located from the cap site, the more likely the occurrence of premature termination by reverse transcriptase at sites of RNA secondary structure. The presence of discrete "dropoff" bands will result in a decreased cap site signal. Changing the position of the primer to a site 5' of secondary structure barriers will improve the cap site signal.
8. The extension reaction often generates more than one band in the vicinity of the capsite (*see* **Fig. 1B**). These may represent genuine multiple starts of transcription. In addition, methylation of the mRNA at the cap site, or adjacent nucleotide, may cause premature termination in a number of molecules. The ratio of cap site bands can vary between different sources of the mRNA but is constant for a given source.
9. Additional bands may be produced as a result of fold-back cDNA synthesis by reverse transcriptase. The sequence of the 5' end of a new message should be examined closely for the possibility of such fold back structures. Spurious bands may arise by crosshybridization of the primer to endogenous tissue culture cell RNA or carrier tRNA. Negative controls using RNA from nontransfected culture cells and tRNA will identify such bands.

References

1. Blumberg, D. (1987) Creating a ribonuclease-free environment, in *Methods in Enzymology*, vol. 152 (Berger, S. L. and Kimmel, A. R., eds.), Academic, London, New York, pp. 20–24.
2. Thomas, M., White, R. L., and Davis, R. W. (1976) Hybridisation of RNA to double-stranded DNA. *Proc. Natl. Acad. Sci. USA* **73,** 2294–2298.
3. Britten, R. J. and Davidson, E. H. (1985) Hybridisation strategy, in *Nucleic Acid Hybridisation—A Practical Approach* (Hames, B. D. and Higgins, S. J., eds.), IRL, Oxford, Washington, DC, pp. 3–15.

32

S1 Mapping Using Single-Stranded DNA Probes

Stéphane Viville and Roberto Mantovani

1. Introduction

The S1 nuclease is an endonuclease isolated from *Aspergillus oryzae* that digests single- but not double-stranded nucleic acid. In addition, it digests partially mismatched double-stranded molecules with such sensitivity that even a single base-pair mismatch can be cut and hence detected. In practice, a probe of end-labeled double-stranded DNA is denatured and hybridized to complementary RNA molecules. S1 is used to recognize and cut mismatches or unannealed regions and the products are analyzed on a denaturing polyacrylamide gel. A number of different uses of the S1 nuclease have been developed to analyze mRNA taking advantage of this property *(1,2)*. Both qualitative and quantitative information can be obtained in the same experiment *(3)*.

Qualitatively, it is possible to characterize the start site(s) of mRNA, to establish the exact intron/exon map of a given gene (*see* **ref. 4** and **Fig. 1**), and to map the polyadenylation sites. Quantitatively, it can be used to study gene regulation both in vivo and in vitro, for example, in the study of the Eα gene promoter *(5)*.

In this chapter we describe a S1 mapping method based on the preparation and use of a single-stranded DNA probe (*see* **Fig. 2**). This offers many advantages:

1. Oligonucleotides allow the exact choice of fragment for a probe.
2. Oligonucleotide labeling is easy and efficient, resulting in a high specific activity probe.
3. The probe can be prepared from single-stranded DNA (e.g., M13, BlueScript) as well as a double-stranded template.
4. The single-stranded probe avoids problems often encountered setting up hybridization conditions of double-stranded probes *(6)*.
5. The probe is stable for 3–4 wk.

We illustrate the process with examples from the analysis of the Eα promoter.

2. Materials

2.1. Preparation of Single-Stranded DNA Probe from a Single-Stranded DNA Template

1. 10X Kinase buffer: 400 mM Tris-HCl, pH 7.8, 100 mM MgCl$_2$, 100 mM β-mercaptoethanol, 250 µg/mL bovine serum albumin. Store at –20°C.
2. [^{32}P]-γ-ATP: specific activity >3000 Ci/mmol.

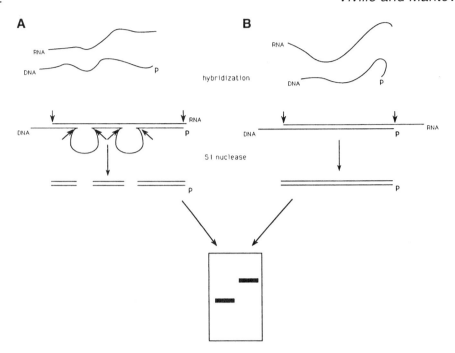

Fig. 1. Illustration of the use of the S1 nuclease. (**A**) Mapping the intron/exon organization of a given gene. The labeled DNA hybridizes to the corresponding mRNA; the introns are cut by the S1 nuclease. (**B**) Mapping of the 5' end of mRNA. The labeled DNA is used as a probe to map the 5' end of a messenger RNA. In both cases the labeled fragments are visualized on a denaturing acrylamide gel.

3. Suitable oligonucleotide (*see* **Note 1**). Prepare a 10 pmol/µL solution in distilled water. Store at –20°C.
4. PNK: Polynucleotide kinase at a stock concentration of 10 U/µL. Store at –20°C.
5. 10X Annealing buffer: 100 mM Tris-HCl, pH 7.5, 100 mM MgCl$_2$, 500 mM NaCl, 100 mM dithiothreitol.
6. DNA template: 1 mg/mL (*see* **Note 2**).
7. 10X dNTP mix: 5 mM dATP, 5 mM dCTP, 5 mM dGTP, and 5 mM dTTP.
8. Klenow: The large fragment of *E. coli* DNA Polymerase I at a stock concentration of 10 U/µL. Store at –20°C.
9. Suitable restriction enzyme (*see* **Note 3**).
10. 6% Polyacrylamide/8 M urea solution in 0.5X TBE and formamide dye.
11. X-ray film: A film of suitable sensitivity, e.g., Kodak XAR.
12. Elution buffer: 50 mM Tris-HCl, pH 7.5, 0.5 mM EDTA.
13. 3 M sodium acetate, pH 7.4.
14. tRNA: Stock solution 10 mg/mL. Store at –20°C.

2.2. Hybridization and S1 Analysis

15. 4X Hybridization buffer: 1.6 M NaCl, 40 mM PIPES, pH 6.4.
16. Deionized formamide.
17. Paraffin oil.
18. S1 buffer: 300 mM NaCl, 30 mM sodium acetate, pH 4.5, 4.5 mM zinc acetate.

S1 Mapping Using Single-Stranded DNA Probes

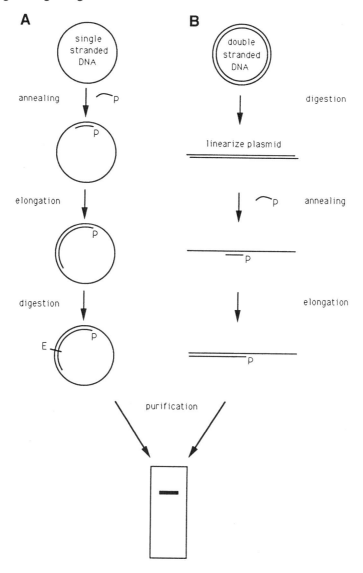

Fig. 2. Schematic illustration of the two methods described in this chapter for the preparation of a single-stranded DNA probe. **(A)** Using a single-stranded plasmid as a template. **(B)** Using a double-stranded DNA plasmid to synthesize the probe.

19. S1 enzyme: A stock concentration of 10 U/μL. Store at –20°C.
20. S1 stop buffer: 2.5 M ammonium acetate, 50 mM EDTA.
21. Isopropanol.

3. Methods

3.1. Preparation of Single-Stranded DNA Probe from a Single-Stranded DNA Template

1. Label the oligonucleotide by incubation of 1 μL of cold oligonucleotide, 10 μL of [^{32}P] ATP, 2 μL of 10X kinase buffer, 6 μL of dH$_2$O, and 1 μL of PNK at 37°C for 45 min. Inactivate the PNK at 95°C for 2 min.

2. Add 2 µL of single-stranded DNA template, 4 µL of 10X annealing buffer, and 14 µL of H$_2$O and incubate for 10 min at 65°C, followed by 10 min at 55°C, then 10 min at 37°C.
3. Add 4 µL of dNTP mix and 1 µL of Klenow and leave at room temperature for 10 min. Inactivate the Klenow by placing the tube at 65°C for 15 min.
4. Correct the mix for the restriction enzyme digest by either adding a suitable amount of sodium chloride or by dilution. Add 20 U of restriction enzyme and incubate at 37°C for 60 min (*see* **Note 3**).
5. Precipitate the sample by adding 2 µL of tRNA stock, 0.1 vol of 3 *M* sodium acetate, and 3 vol of ethanol, and place the sample in dry ice for 10 min. Centrifuge at 12,000g for 15 min in a microfuge.
6. Resuspend the pellet in 20 µL of formamide dye, heat at 95°C for 5 min, and load a 6% polyacrylamide/urea gel. Run the gel until the bromophenol blue reaches the bottom.
7. Locate the single-stranded DNA fragment by exposing the gel for 5 min to a X-ray film. Excise the corresponding band and elute it overnight in 500 µL of elution buffer.
8. Add 50 µL of sodium acetate, 20 µg of tRNA, and 1 mL of ethanol; place at –20°C for 2 h. Centrifuge for 20 min at 12,000g in a microfuge.
9. Count the pellet and resuspend it in formamide (10^5 cpm/µL). Store at –20°C. The probe is now ready for hybridization and S1 analysis.

3.2. Preparation of Single-Stranded DNA Probe from Double-Stranded DNA Template

Follow the same protocol described in **Subheading 3.1.**, but for the following exceptions (*see* **Fig. 2B**).

1. Initially cut 20 µg of double-stranded plasmid with a restriction enzyme of choice. Precipitate and resuspend in H$_2$O at a concentration of 1 mg/mL.
2. The annealing step (**Subheading 3.1.**, **step 2**) must be carried out extremely quickly: After heat inactivation of PNK, add 2–5 µg of restricted plasmid, 4 µL of 10X annealing buffer, and 14 µL of H$_2$O, and incubate for 5 min at 95°C. Immediately place the tube in ice-cold water and then proceed to step 3.
3. Omit the restriction enzyme digestion (**Subheading 3.1.**, **step 4**).

3.3. Hybridization and S1 Analysis (see Fig. 3)

1. In a 0.5 mL Eppendorf tube put 10 µL of probe (10,000 cpm total), 5 µL of 4X hybridization buffer, and 5 µL of RNA (total poly A$^+$ or synthesized in vitro). Add 20 µL of paraffin oil to prevent evaporation and incubate 4–16 h at 37°C (*see* **Note 4**).
2. Add 200 µL of S1 buffer containing 100 U of S1 nuclease, mix well, and place at 37°C for 5–30 min (*see* **Note 5**). Stop the reaction by adding 50 µL of S1 stop buffer, 40 µg of tRNA, and 300 µL of isopropanol. Place in dry ice for 15 min and centrifuge at 12,000g for 15 min.
3. Wash the pellet with 80% ethanol and dry. Resuspend in 3 µL of formamide dye, heat for 5 min at 95°C, and load a 6% polyacrylamide/urea gel. Run until the bromophenol blue reaches the bottom. Expose the gel for 8–48 h (*see* **Fig. 3**).

4. Notes

1. The oligonucleotide should be 20–25 nucleotides long and should be complementary to the DNA strand to be analyzed.
2. Both the single-stranded and double-stranded DNAs are prepared according to standard protocols. The advantage of using a single-stranded DNA template is basically a higher yield of the probe: $1–1.5 \times 10^6$ cpm vs the $4–6 \times 10^5$ cpm expected from a double-stranded plasmid template.

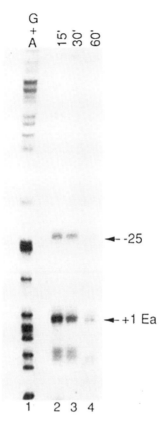

Fig. 3. Time-course of S1 analysis. In vitro synthesized RNA hybridized to Eα probe is incubated at 37°C for the indicated time with 100 U of S1 nuclease.

3. The restriction enzyme should be used to cut 200–600 nucleotides 3' of the position to which the oligonucleotide hybridizes to increase fragment recovery. Longer fragments are harder to recover from acrylamide gels. It does not matter if the restriction enzyme cuts the template DNA more than once as long as it is outside the probe sequence.
4. Hybridization time varies with the type of RNA to be analyzed: Using total cytoplasmic or poly A$^+$ RNA, the samples should be left at 37°C at least 12 h; when hybridizing RNA generated from in vitro transcription, 2–4 h are usually sufficient.
5. An S1 time-course should be performed as shown in Fig. 3 for in vitro synthesized RNA. This will establish the conditions for which all single-stranded nucleic acids, including the free probe, are completely digested. The optimal cutting time for total cytoplasmic and poly A$^+$ RNA is usually longer (30 min) than for in vitro synthesized RNA (15 min).

References

1. Berk, A. J. and Sharp, P. A. (1977) Sizing and mapping of early adenovirus mRNAs by gel electrophoresis of S1 endonuclease digested hybrids. *Cell* **12,** 721–732.
2. Favaloro, J., Treisman, R., and Kamen, R. (1980) Transcription maps of polyoma virus-specific RNA: analysis by two-dimensional nuclease S1 gel mapping. *Meth. Enzymol.* **65,** 718–749.

3. Weaver, R. F. and Weissmann, C. (1979) Mapping of RNA by a modification of Berk-Sharp procedure: the 5' termini of 15 S β-globin mRNA precursor and mature 10 S β-globin mRNA have identical map coordinates. *Nucleic Acids Res.* **7,** 1175–1193.
4. Lopata, M. A., Sollner-Webb, B., and Cleveland, D. W. (1985) Surprising S1-resistant trimolecular hybrids: potential complication in interpretation of S1 mapping analysis. *Mol. Cell. Biol.* **5,** 2842–2846.
5. Viville, S., Jongeneel, V., Koch, W., Mantovani, R., Benoist, C., and Mathis, D. (1991) The Eα promoter: a linker-scanning analysis. *J. Immunol.* **146,** 3211–3217.
6. Dean, M. (1987) Determining the hybridization temperature for S1 nuclease mapping. *Nucleic Acids Res.* **15,** 6754.

33

Measurements of Rate of Transcription in Isolated Nuclei by Nuclear "Run-Off" Assay

Rai Ajit K. Srivastava and Gustav Schonfeld

1. Introduction

Unlike gene expression in prokaryotic cells, which is primarily under transcriptional control, gene expression in eukaryotic cells is subject to both transcriptional and post-transcriptional controls. Since transcription and translation in eukaryotic cells are separated topographically, the regulation of mRNA metabolism can occur at multiple sites, within nuclei and in the cytoplasm. Nevertheless, transcription remains a critical locus of control of eukaryotic gene expression. Transcriptional regulation affects cellular mRNA abundance by affecting rates of transcription. Another important control mechanism that can affect mRNA abundance is the rate of mRNA decay. Thus, steady-state levels of mRNA represents a balance between the rates of transcription and intracellular degradation, e.g., an increase in the abundance of mRNA could result from decreased mRNA degradation, increased mRNA synthesis, or both.

Tissue specific expression and the level of expression of certain genes are routinely determined by measuring the abundance of the corresponding mRNA *(1–3)* and, frequently, mRNA levels are interpreted as reflecting rates of transcription. This is a safe interpretation in some cases *(4–8)*, but not in all cases *(9–15)*. It is, therefore, important to measure the rate of transcription directly, in order to understand the contribution of transcription rates to "setting" the cellular level of any specific mRNA.

Relative rates of transcription are measured by nuclear "run-off" assays using isolated cell nuclei. The assay quantifies the elongations in vitro of nascent mRNA chains already initiated in vivo. Nuclei are isolated from homogenized tissues or cultured cells, and incubated in the presence of four ribonucleotides, one of them radiolabeled. The reaction is stopped by the addition of RNase-free DNase I, and total nuclear RNA that contains newly synthesized labeled RNA is extracted. The extracted RNA is hybridized to a denatured, immobilized cDNA corresponding to the mRNA being measured. After hybridization, the membrane is treated with RNase A and washed to remove the nonspecifically bound RNA. Bound mRNA are hydrolyzed from the filter and counted in a liquid scintillation counter. Nuclear transcription rates of given mRNAs are frequently compared to transcription rates of an "internal standard" such as, β-actin mRNA.

2. Materials

All glassware used for nuclear run-off assay is treated with 0.05% DEPC (diethyl pyrocarbonate) overnight and autoclaved for 1 h. The reagents used are also treated in a similar way and either autoclaved or filter sterilized. The reagents required are spermine, spermidine, creatine phosphate, β-mercapoethanol, HEPES (N-[2-hydroxyethyl]piperazine-N'-[ethanesulfonic acid], recombinant plasmid containing appropriate cDNA fragment, EGTA [ethylene glycol-bis β-amino ethyl] ether), dithiotheritol, nylon cloth (mesh with 40 µM), phenyl methyl sulfonyl fluoride, pancreatic ribonuclease A, heparin sulfate, ribonucleotides (GTP, ATP, CTP), ribonuclease inhibitor (RNasin or human placental ribonuclease inhibitor), α-^{32}P[UTP] (600 µCi/mM, ICN, Biomedicals), DNase I (RNase-free), guanidinium thiocyanate, sodium citrate, sacrcosyl, salt saturated phenol, chloroform/isoamylalcohol (49:1), α-amanitin, isopropanol, deionized formamide, PIPES (piperazine-N,N'-bis[2-ethane sulfonic acid]), sodium dodecyl sulfate, ficoll, polyvinylpyrrolidone, bovine serum albumin, and salmon sperm DNA.

For the isolation of nuclei following reagents are kept ready:

1. Buffer A: 60 mM KCl; 15 mM NaCl; 0.15 mM spermine; 0.5 mM Spermidine; 14 mM β-mercaptoethanol; 0.5 mM EGTA; 2 mM EDTA; and 15 mM HEPES (pH 7.5).
2. Buffer B: Same as buffer A but with 0.1 mM of EGTA and EDTA. (Note: These buffers are made in DEPC-treated water and sterilized by filtration through a sterile millipore filter.)
3. Sucrose solutions: Solution A: 0.3 M sucrose in buffer A; solution B: 1 M sucrose in buffer B; and solution C: 1.5 M sucrose solution in buffer B.
4. Nuclei storage buffer: 20 mM Tris-HCl, pH 7.9; 75 mM NaCl; 0.5 mM EDTA; 0.85 mM DTT; 0.125 mM phenylmethyl sulfonyl fluoride (PMSF); and 50% glycerol.
5. Wash buffer: 20 mM Tris–HCl, pH 7.5; 15 mM NaCl; and 1.1 mM sucrose.
6. Hypotonic buffer: 20 mM Tris-HCl, pH 8.0; 4 mM MgCl$_2$; 6 mM CaCl$_2$; and 0.5 mM DTT.

The above buffers and solutions are made and stored in the refrigerator (0–4°C).

7. Lysis buffer (prepare fresh from stock solutions): 0.6 M sucrose; 0.2% Nonidet P-40; and 0.5 mM DTT.
8. Elongation buffer (2X): Make fresh each time from the stock solutions: 200 mM Tris-HCl, pH 7.9; 100 mM NaCl; 0.8 mM EDTA; 0.2 mM phenylmethyl sulphonyl chloride; 2.4 mM DTT; 2 mg/mL heparin sulphate; 4 mM MnCl$_2$; 8 mM MgCl$_2$; and 20 mM creatine phosphate.
9. Nucleotide mix: Prepare 100 mM solution of GTP, ATP, and CTP and mix in 1:1:1 ratio. Store frozen in aliquots of 500 µL at –20°C.
10. SET buffer (prepare fresh): 5% sodium dodecyl sulfate; 50 mM EDTA; and 100 mM Tris-HCl, pH 7.4.
11. Extraction buffer (store at room temperature): 4 M guanidinium thiocyanate; 25 mM sodium citrate, pH 7.0; 0.5% Sarkosyl; and 0.1 M β-mercaptoethanol.
12. Hybridization buffer (made fresh before use): 20 mM PIPES (piperazine-N,N'-bis-[2-ethanesulfonic acid]), pH 6.7; 50% formamide (deionized with mixed bed resins); 2 mM EDTA; 0.8 M NaCl; 0.2% SDS; 0.02% ficoll; 0.02% polyvinylpyrrolidone; 0.02% bovine serum albumin; and 500 µg/mL denatured salmon sperm DNA.

3. Methods
3.1. Isolation of Nuclei
3.1.1. Preparation of Nuclei from Tissues

1. Homogenize preweighed tissue in 10 vol of ice-cold sucrose solution A (10 mL solution A/1 g tissue). Perform all the steps of nuclei preparation at 0–4°C.
2. Filter the homogenate through either 40-μM nylon cloth or through 4 layers of cheesecloth, and layer over a 10-mL cushion of solution B. Spin for 20 min at 2500g and 4°C in a swinging bucket rotor of Beckman (Fullerton, CA) Centrifuge.
3. Resuspend pelleted crude nuclei in solution B, using 2–3 mL for up to 1 g of tissue sample, layer over 5 mL of solution C, and centrifuge at 45,000 rpm in Beckman SW50 rotor at 4°C for 60 min.
4. Resuspend pellet containing clean nuclei in nuclear storage buffer. This gives nuclei at a concentration of approx $10^6/\mu L$. Isolated clean nuclei are either used immediately after preparation or may be stored frozen in aliquots of about $1–5 \times 10^7$ at –70°C without loss in activity for up to 4 wk.

3.1.2. Preparation of Nuclei from Cultured Cells

1. Wash a confluent culture containing $1–2 \times 10^8$ cells with ice-cold wash solution and collect by centrifugation at 300g.
2. Resuspend the cells in 2.5 mL ice-cold hypotonic buffer and allow to sit on ice for 5 min. Add 2.5 mL of lysis buffer.
3. Break the cells further with the tight-fitting pestle of a homogenizer. Usually 6–10 strokes are enough to break the cells.
4. Pellet nuclei by centrifugation at 1500 rpm for 10 min and resuspend in 2 mL of sucrose solution A.
5. Layer the crude nuclear suspension over 2.5 mL of sucrose solution C and centrifuge for 1 h at 45,000 rpm. Resuspend the nuclear pellet in a nuclear storage buffer and either use immediately for in vitro nuclear run-off assay or store at –70°C for up to 4 wk.

3.2. Elongation of Nascent mRNA Chains

1. To a total volume of 200 µL add 100 µL of 2X elongation buffer, 6 µL of nucleotide mix, 40 U (1 µL) RNase inhibitor, 10^7 nuclei (suspended in nuclei storage buffer), and 100 µCi ^{32}P[UTP].
2. Allow the transcription reaction to proceed at 26°C for 20 min and then stop by the addition of 100 U DNase I (RNase-free). Incubate for an additional 5 min. To determine exclusively the RNA polymerase II dependent transcription, transcription is also performed in presence of α-amanitin (2 µg/mL) and the counts obtained are subtracted from the total counts obtained.
3. Treat the samples with 2 µL proteinase K (10 mg/ mL) and 20 µL SET buffer for 30 min at 37°C. Add 400 µL extraction buffer and 80 µL of sodium acetate (2.0 M, pH 4.0) and vortex the contents for 10 s. Add 700 µL of salt saturated phenol and 150 µL of chloroform:isoamylalcohol, vortex for 10 s and allow to sit on ice for 15 min.
4. Centrifuge at 4°C and 12,000g for 15 min. Transfer the aqueous (top) phase to a clean tube and add 20 µg yeast tRNA and an equal volume of cold isopropanol. Mix tube contents and incubate at –20°C for 20 min. Centrifuge at 12,000g and 4°C for 15 min and wash the pellet with 70% ethanol. Lyophilize and redissolve in 100 µL of hybridization buffer (*see* **Note 1**).

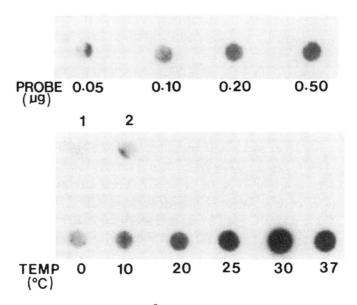

Fig. 1. Nuclear run-off assay, using 10^7 mouse liver nuclei per reaction, was performed as described in the text and probed for the rate of low-density lipoprotein receptor mRNA transcription using 1.3 kb rat liver LDL-receptor cDNA. The top panel shows an increasing hybridization signal with increasing amounts of probe immobilized onto nitrocellulose membrane. Lower panel shows the effect of temperature on the LDL-receptor mRNA transcription. Lane 1, 5 µg plasmid DNA that do not contain LDL-receptor cDNA; Lane 2, transcription performed in the presence of 2 µg/mL α-amanitin.

3.3. Hybridization

The amount of specific mRNA synthesized in the nuclear run-off assay is determined by hybridizing the total synthesized RNA with the specific cDNA probe (*see* **Notes 2** and **3**). This is then compared to transcription of an internal control, e.g., of β-actin mRNA.

1. The double-stranded recombinant plasmid containing the appropriate cDNA fragment (or β-actin cDNA) is denatured with 0.2 M NaOH/2 mM EDTA by incubation for 15 min at 37°C. It is then neutralized by 1 M HEPES (N-[2-Hydroxyethyl]piperazine-N'-[2-ethane sulfonic acid]) buffer (pH 6.5).
2. Five µg of denatured plasmid DNA is applied to nitrocellulose paper using dot blot apparatus and baked for 2 h at 80°C in vacuum. The portions of the filter in which plasmid DNA are spotted are cut out with the help of a sterile cork borer or sharp blade.
3. Prehybridize the membrane for 2 h in 400 µL hybridization buffer. Remove buffer from the tube and replace with total RNA (10^5–10^6 cpm) dissolved in 250 µL of hybridization buffer. Cover hybridization mixture plus membrane with 50–100 µL of mineral oil and hybridize for 24 h at 42°C.
4. After the hybridization, wash the membranes twice with 500 µL of 2X SSC/0.1% SDS for 30 min at room temperature. In order to remove nonspecific binding, treat the membrane with 400 µL of RNase A solution (10 µg/mL in 2X SSC) at 37°C for 30 min. This will remove unhybridized RNA. Wash with 1 mL of 2X SSC twice at room temperature and expose to X-ray film (*see* **Figs. 1** and **2**).

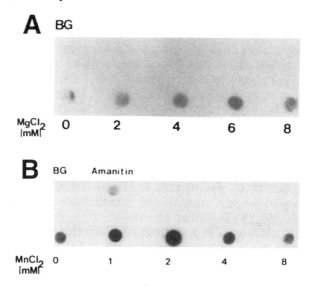

Fig. 2. LDL-receptor mRNA transcription on isolated mouse liver nuclei in the presence of different concentrations of $MgCl_2$ or $MnCl_2$. Each transcription assay was performed with 10^7 nuclei at 28°C for 20 min. The concentration of divalent cations are indicated in the figure.

5. To elute the hybridized RNA, incubate the membranes with 200 µL of 0.3 M NaOH for 15 min at 65°C followed by the addition of 50 µL of glacial acetic acid and 4 mL of scintillation cocktail. Count ^{32}P radioactivity in a liquid scintillation counter. This value provides the relative rates of transcription of a specific mRNA (*see* **Note 4**).

4. Notes

1. The integrity of newly synthesized RNA and the extent of incorporation of label in the nascent RNA chains may be checked prior to proceeding for hybridization:
   a. To determine the extent of label incorporation, an aliquot of synthesized total nuclear RNA is diluted 10–20-fold depending on the label incorporation as judged by Geiger counter. An aliquot (usually 2 concentrations; 5 and 10 µL) of the diluted sample is counted in a liquid scintillation counter.
   b. To determine the quality of the synthesized RNA, an aliquot of newly synthesized RNA (50,000–100,000 cpm) is resolved in 6% denaturing polyacrylamide gel containing 7 M urea. After the electrophoresis, the gel is dried and exposed to X-ray film. If the transcribed RNAs are intact, one can see several distinct RNA bands all along the autoradiogram. If the RNAs are degraded, smaller fragments of RNA appear and there is smudging of the bands.
2. To ensure specificity of the probe, linear concentrations (0.1–1 µg) of recombinant plasmid are immobilized on one set of filters, and identical amounts of a plasmid that does not contain cDNA insert are bound to another set of nitrocellulose membranes. Hybridization is performed with the same amounts of $^{32}P[RNA]$ using both sets of filters. If the cDNA probe is specific, a linear increase in the hybridization signal is obtained with the recombinant plasmid but not with the nonrecombinant plasmid.
3. In another experiment, increasing amounts of $^{32}P[RNA]$ (10^5–10^6 cpm) are hybridized with constant amounts (5 µg) of the recombinant and the nonrecombinant plasmids. Here

again one gets increasing hybridization signals with the recombinant plasmid but not with the nonrecombinant plasmid.
4. For background counts hybridization is carried with the nonrecombinant plasmid and the counts obtained are subtracted from the counts obtained by hybridization with the recombinant plasmid.

References

1. Sorci-Thomas, M., Wilson, M. D., Johnson, F. L., Williams, D. L., and Rudel, L. L. (1989) Studies on the expression of genes encoding apolipoproteins B-100 and B-48 and the low density lipoprotein receptor in non-human primates. Comparison of dietary fat and cholesterol. *J. Biol. Chem.* **264,** 9039–9045.
2. Srivastava, R. A. K., Ito, H., Hess, M., Srivastava, N., and Schonfeld, G. (1995) Regulation of low density lipoprotein gene expression in HepG2 and Caco2 cells by palmitate, oleate, and 25-hydroxycholesterol. *J. Lipid Res.* **36,** 1434–1446.
3. Srivastava, R. A. K., Baumann, D., and Schonfeld, G. (1993) In vivo regulation of low density lipoprotein receptor by estrogen differs at the posttranscriptional level in rat and mouse. *Eur. J. Biochem.* **216,** 527–538.
4. Srivastava, R. A. K. (1996) Regulation of apolipoprotein E by dietary lipids occurs by transcriptional and posttranscriptional mechanisms. *Mol. Cell Biochem.* **155,** 153–162.
5. Brock, M. L. and Shapiro, D. J. (1983) Estrogen regulates the absolute rate of transcription of the *Xenopus laevis* vitellogenin genes. *J. Biol. Chem.* **258,** 5449–5455.
6. McKnight, G. S. and Palmiter, R. D. (1979) Transcriptional regulation of the ovalbumin and conalbumin genes by steroid hormones in chick oviduct. *J. Biol. Chem.* **254,** 9050–9058.
7. Chazenbalk, G. D., Wadsworth, H. L., and Rapoport, B. (1990) Transcriptional regulation of ferritin H messenger RNA levels in FRTL5 rat thyroid cells by thyrotropin. *J. Biol. Chem.* **265,** 666–670.
8. Chinsky, J. M., Maa, M. C., Ramamurthy, V., and Kellems, R. E. (1989) Adenosine deaminase gene expression. Tissue-dependent regulation of transcriptional elongation. *J. Biol. Chem.* **264,** 14,561–14,565.
9. Saini, K., Thomas, P., and Bhandari, B. (1990) Hormonal regulation of stability of glutamine synthetase mRNA in cultured 3T3-L1 adipocytes. *Biochem. J.* **267,** 241–244.
10. Brock, M. L. and Shapiro, D. J. (1983) Estrogen stabilizes vitellogenin mRNA against cytoplasmic degradation. *Cell* **34,** 207–214.
11. Antrast, J., Lasnier, F., and Pairault, J. (1991) Adipsin gene expression in 3T3-F442A adipocytes is post-transcriptionally down-regulated by retinoic acid. *J. Biol. Chem.* **266,** 1157–1161.
12. Jefferson, D. M., Clayton, D. F., Darnell, J. E., Jr., and Reid, L. M. (1986) Post-transcriptional modulation of gene expression in cultured rat hepatocytes. *Mol. Cell Biol.* **4,** 1929–1934.
13. Hod, Y. and Hanson, R. W. (1988) Cyclic AMP stabilizes the mRNA for phosphoenol pyruvate carboxykinase (GTP) against degradation. *J. Biol. Chem.* **263,** 7747–7752.
14. Srivastava, R. A. K., Kitchens, R. T., and Schonfeld, G. (1994) Regulation of the apolipoprotein AIV gene expression by estrogen differs in rat and mouse. *Eur. J. Biochem.* **222,** 507–514.
15. Srivastava, R. A. K., Jiao, S., Tang, J., Pfleger, B., Kitchens, R. T., and Schonfeld, G. (1991) In vivo regulation of low density lipoprotein receptor and apolipoprotein B gene expression by dietary fatty acids and dietary cholesterol in inbred strains of mice. *Biochem. Biophys. Acta* **1086,** 29–43.

34

One-Tube RT-PCR with Sequence-Specific RT Primers

Ulrich Pfeffer and Paola Ferro

1. Introduction

The reverse transcription-polymerase chain reaction (RT-PCR; [1,2]) applies the power of amplification to the study of gene expression: the PCR is carried out on cDNA obtained by reverse transcription of mRNA. The protocol described here has been developed to allow a simple and highly sensitive reaction with a reduced risk of contaminations from preceding PCRs or plasmid preparations (false positives) or RNases (false negatives). The use of specific RT primers increases the specificity of RT-PCR, reverse transcription can be limited to the region of the mRNA that will be amplified avoiding long 3'untranslated regions and RT primers are chosen in order to avoid GC-rich regions that often are copied inefficiently be the enzyme. T_m values of RT primers (nona- to dodecamers) should be between 25°C and 35°C, the 3' nucleotides should be G or C in order to assure stable base pairing at the 3' end and the primers should not show dimerization capabilities. Several primers may be used in the same reverse transcription reaction provided that they are not complementary to each other. In this way, a sample mRNA and an internal standard mRNA, such as β-actin or glyceral-3-phosphate dehydrogenase mRNAs, can be reverse transcribed in the same tube. Reverse transcription is carried out in two steps at 25°C (annealing) and 42°C (extension). The cDNA obtained is either used as a whole or divided for multiple amplifications and/or amplification of the internal standard. PCR primers and *Taq* polymerase are added and cDNA is denatured and then amplified.

When performing semiquantitative analyses, particular care must be taken in order to ensure that none of the components of the system is exhausted during the reaction leading to saturation. The correlation between the amount of starting RNA and the amount of amplification product obtained must be linear for both the sample and the standard RNA.

2. Materials

1. Total RNA may be prepared by any of the common procedures provided that the preparation results are essentially devoid of proteins. In our hands, the method described by Chomczynski and Sacchi *(3)* and the commercial formulations of it or the RNeasy Total RNA kit (Qiagen, Hilden, Germany) work best.

2. RNA should be redissolved in H_2O or in Tris-EDTA, no detergents should be added (*see* **Note 1**).
3. Nucleotide stocks are titrated with NaOH to pH 7.0 or purchased as premade buffered solutions (Pharmacia, Brussels, Belgium).
4. Oligonucleotides should be purified at least by gel filtration and the success of synthesis should be controlled.

    RNase inhibitor and avian myeloblastosis virus (AMV) reverse transcriptase can be obtained from Amersham Pharmacia Biotech (Little Chalfont, UK) and *Taq* polymerase from AGS (Heidelberg, Germany). Enzymes provided by other suppliers were occasionally tested and performed equally well (*see* **Note 2**).
5. PCR buffer (1×): 10 mM Tris-HCl (pH$_{25°C}$ 8.3), 50 mM KCl, 2 mM MgCl$_2$, 0.01% Tween 20 (*see* **Note 3**).
6. Pre-RT PCR and post-RT PCR components and equipment should be physically separated, if possible, in a one-way pursuit. The use of filter pipet tips is recommended (*see* **Note 4**).

3. Methods

3.1. Primer Design

Select PCR sense and antisense primers using gene analysis software (*see* **Note 5**). Make sure that the PCR primers do not show self or cross-homology, partial homology with other regions of the sequence analyzed, or with sequences present in the databank. T_m values should be between 75–85°C, thus allowing annealing temperatures of 65°C. Then select a short RT primer (9–12 nt) closely downstream to the PCR antisense primer. This primer should not be complementary to the PCR primers or to the RT primer of the internal standard that eventually will be analyzed in parallel (*see* **Note 6**). Avoid self-complementary primers; a restriction site present in the RT primer is likely to inactivate it. T_m values of RT primers should be between 25–35°C. Unless specific regions of the messenger are to be analyzed, you can freely choose your primers, but make sure that no GC-rich regions are contained in between the RT primer and the PCR sense primer (complete analyzed region). Choose primers on different exons. If exon/intron borders are not known, choose distant PCR primers and check for amplification from DNA. For some applications, for example RT-PCR from RNA extracted from paraffin-embedded tissue samples, primers must be close to each other in order to allow amplification from partially degraded RNA *(4)*. After synthesis, primers must be purified by gel filtration in order to remove protecting groups and salts. Normally, no further purification occurs, but the actual success of synthesis should be monitored by an analytical chromatography run. If products of precocious chain termination are present, the primers should be purified (*see also* Chapter 74).

3.2. Reverse Transcription (see Notes 7 and 8)

1. Denature 5 µL RNA solution (1 µg total RNA or 0.01 µg polyA$^+$ RNA) at 65°C for 10 min (*see* **Note 9**).
2. In the meantime, prepare the RT mix for the number of samples to be analyzed:
   1X PCR buffer
   1 mM dNTP
   25 pmol RT primer
   20 U RNasin

2 U AMV reverse transcriptase
H₂O to 15 μL (per sample)
20 μL final volume per sample (after addition of RNA)
3. Keep the denatured RNA on ice and add 15 μL RT mix to each sample, vortex, and spin in a microfuge to collect the sample at the bottom of the tube (*see* **Note 10**).
4. Incubate at 25°C, 10 min, and at 42°C, 45 min (*see* **Note 11**). Keep on ice until further processing (*see* **Notes 12** and **13**).

3.3. Polymerase Chain Reaction (see Notes 7 and 8)

1. Prepare the PCR mix for the number of samples to be analyzed:
   1X PCR buffer
   0.2 mM deoxynucleotide triphosphates (dNTP) (*see* **Note 14**)
   25 pmol sense PCR primer
   25 pmol antisense PCR primer
   2.5 U *Taq* polymerase
   H₂O to final volume
   100 μL final volume per sample (after addition of cDNA; *see* **Note 15**)
2. Keep cDNA samples on ice and add the PCR mix to a final volume of 100 μL, vortex, spin to collect the sample at the bottom of the tube, and add two drops of mineral oil.
3. Cycle with the following settings (combined protocol):
   a. Denaturation (*see* **Note 16**)
      94°C, 2 min, 1 cycle
   b. Amplification
      94°C, 30 s
      55–65°C, 30 s (*see* **Note 17**)
      72°C, 30–60 s (*see* **Note 18**)
      25–35 cycles (*see* **Note 19**)
   c. Final extension and storage
      72°C, 7 min
      4°C, ∞
      1 cycle
4. Remove 10 μL of each sample and analyze on a 1–1.5% agarose gel containing 0.4 μg/mL ethidium bromide (Take care: potent carcinogen!; *see* **Note 20**).

4. Notes

1. The analysis may be performed on either total or poly-A⁺-RNA. The quantity necessary to obtain a UV-visible band on the gel depends on the level of expression of the given sample messenger RNA. When starting with 1 μg total RNA or 0.01 μg poly-A⁺-RNA, using half of the cDNA for amplification and performing 35 cycles, as well as low-abundance messengers, should yield a product. Overdrying RNA pellets results in incomplete redissolution. DNA contaminations are to be removed with RNase-free DNAse I if the PCR primers are on the same exon. All solutions for RNA and for reverse transcription should be RNase free.
2. We did not encounter problems with the quality of enzymes purchased from various sources, but note that the enzymes are the most expensive part of the assay.
3. The PCR buffer indicated is one of many possible ones. In most cases, the buffer provided with the *Taq* polymerase is suitable. The MgCl₂ concentration in both reverse transcription and amplification is critical and should be optimized for each new case. In our opinion, 2 mM is a good starting concentration. If under the conditions described no

amplification product can be obtained, the simple variation of the Mg^{2+} concentration will not improve the situation, rather its variation may result in an improvement of already working amplification conditions.

4. If contamination occurs, all aliquots already used should be discarded, the environment should be cleaned, and equipment should be autoclaved where possible (there are autoclavable pipets!). General amplification failure (no product) is normally due to RNase contaminations derived from the hands of the operator or from RNases used in the same lab (e.g., plasmid preps). Run positive controls of RNA preparations from cells that express the gene analyzed. If amplification products of unexpected length occur, consider the possibility of alternatively spliced messenger RNAs. This can be verified by hybridization of the blotted amplification product to a specific probe.

5. We use Primer Detective (Clontech, Palo Alto, CA) and Oligo (Medprobe, Oslo, Norway) software for primer design, but we have not analyzed others that may be equally suited. Not all programs (as Primer Detective) allow the analysis of single primers necessary for the design of the RT primer, and most do not analyze structural features of the region between the primers. In our experience it is sufficient to exclude regions of high G/C content and this can be carried out manually.

6. When performing semiquantitative analyses, an internal standard messenger must be amplified in parallel. The RT reaction can be performed together with the sample, but, in general, coamplification will not be successful. The standard messenger is normally expressed at higher levels than the sample messenger and thus competes for system components and the presence of more PCR primers (4 instead of 2) increases the possibility of nonspecific amplification. However, the major source of error resides in the quantity of RNA starting concentration and in RNA degradation during handling and in pipetting the relatively small amounts of the RT components. These errors are adequately controlled when performing combined RT reactions but separated PCR reactions.

7. All components of the RT mix and of the PCR mix except the enzymes and the RNase inhibitor can be assembled in large master mixes, aliquoted, and stored at –20°C for extended periods. This enhances the sample-to-sample reproducibility and diminishes the pipetting steps to be carried out on the single sample also reducing the risk of contaminations.

8. The major pitfall of RT-PCR methods are contaminations with PCR products of previous analyses and with plasmids containing related sequences. Always perform blank analyses where the RNA solution has been substituted with plain water. Post-PCR processing and plasmid preparations should not be carried out in the same room as the pre-PCR steps; a set of pipets and of other equipment should be reserved for pre-PCR operations.

9. Denaturation of RNA may also be performed for 5 min at 80°C; the use of formamide is not recommended. After denaturation, the RNA is to be kept on ice to avoid renaturation and the further processing is to be carried out soon.

10. Although mixing the reaction components is important, we have observed that vortexing (and subsequent centrifugation) can be omitted. Do not mix by repeated pipetting, this is one of the major sources of pipet contamination!

11. The annealing temperature of the RT primer is 25°C, 42°C is the optimum temperature of AMV reverse transcriptase. Other transcriptases such as MuLV have an optimum temperature of 37°C, but they may also be used at 42°C. The processivity and the capability to read through "difficult" sequences of AMV is somewhat higher.

12. If you wish to interrupt the procedure you can do that at the end of the cDNA synthesis. cDNA is stored at 4°C for prolonged periods, storage at –20°C or lower is not recommended.

13. There is no need to destroy reverse transcriptase at the end of the cDNA synthesis.

14. Take into account the carryover from the reverse transcription reaction when setting up the PCR mix. If you use 10 μL of the cDNA solution, the dNTPs carried over account for half of the amount needed in the PCR reaction.
15. The amount of the cDNA to be used for amplification depends on various considerations: (a) expression level, (b) eventual parallel amplification of an internal standard when the reverse transcription has been carried out in the presence of the appropriate primers, (c) ease of performance: if only the sample messenger RNA is to be analyzed in a large number of samples it may prove easier just to add the PCR mix to the whole cDNA preparation, and (d) eventual storage of a part of the cDNA for further analysis.
16. Denaturation prior to PCR is carried out in the thermal cycler in the presence of the Taq-polymerase. The samples should be introduced into the hot cycler. Two minutes is sufficient for the denaturation of the relatively short cDNAs; any longer results in a loss of enzyme activity. If the thermal cycler does not allow this type of programming, denaturation may be performed separately.
17. Annealing should be performed at 5–10°C below the thermodynamic T_m value of the primer with the lowest T_m. In practice, it may be useful to use long primers (25-mers), which can be used at higher annealing temperatures.
18. The Taq-polymerase should incorporate at least 1500 nt/min so that 30 s of extension are enough in most cases. Longer extension periods often described in the literature lead to premature exhaustion of the enzyme and reduce the number of cycles that can be performed without adding fresh enzyme.
19. In order to enhance the sensitivity of the system, more cycles may be performed. After 40 cycles the enzyme is expected to be exhausted and fresh enzyme should be added.
20. When master mixes are prepared, a large number of samples within the limit of capacity of the thermal cycler can easily be handled in parallel. Results are obtained within a maximum of 5 h with less than 1 h hands-on time.

Acknowledgments

The author thanks Giovanni Levi for continuous support. F. P. is a fellow of the Fondazione Italiana per la Ricerca sul Cancro (FIRC). This work has been supported by a grant from the Ministero della Sanitá to U. P.

References

1. Kawasaki, E. S., Clark, S. S., Coyne, M. Y., Smith, S. D., Champlin, R., Witte, O. N., et al. (1987) Diagnosis of chronic myeloid and acute lymphocytic leukemias by detection of leukemia-specific mRNA sequences amplified *in vitro*. *Proc. Natl. Acad. Sci. USA* **85,** 5698–5702.
2. Rappolee, D. A., Mark, D., Banda, M. J., and Werb, Z. (1988) Wound macrophages express TGF-α and other growth factors *in vivo*: analysis by mRNA phenotyping. *Science* **241,** 708–712.
3. Chomczynski, P. and Sacchi, N. (1987) Single-step method of RNA isolation by acid guanidinium thiocyanate-phenol-chloroform extraction. *Ann. Biochem.* **162,** 156–159.
4. Mies, C. (1994) A simple, rapid method for isolating RNA from paraffin-embedded tissues for reverse transcription-polymerase chain reaction (RT-PCR). *J. Histochem. Cytochem.* **42,** 811–813.

35

Characterization of RNA Using Continuous RT-PCR Coupled with ELOSA

François Mallet

1. Introduction

There is a need for specific and sensitive methods of characterization of RNA adapted to clinical applications. For this purpose, we have designed a procedure consisting of a continuous reverse transcriptase-polymerase chain reaction (RT-PCR) step coupled with a nonradioactive detection on microtiter plates of amplified sequences, which we called enzyme-linked oligosorbent assay (ELOSA).

The continuous RT-PCR protocol *(1)* was based on the use of AMV-RT and *Taq* polymerase; all reaction components were added to a single tube prior to positioning in the thermocycler, what reduces the risk of introducing contaminations during handling. AMV-RT (avian myeloblastosis virus) was chosen for reverse transcription because it can read through stable secondary mRNA structures at relatively high temperature *(2)*, and it can function under the same buffer conditions as *Taq* polymerase *(3)*. Our cycling conditions included the linkage of the primary RT-PCR steps in rapid succession, as achieving a rapid equilibrium between primer and template sequences enhances the efficiency of hybridization *(4)*. Quality of the RNA preparation could be checked by amplification of either a housekeeping gene (for normalization of cellular RNA) or an added positive control (for viral RNA isolated from serum), which informs of the presence of inhibitors of the enzymatic reaction. The continuous RT-PCR procedure was found to be sensitive in detecting at least 10 HIV-1 copies *(1)* or the hepatitis C RNA virus in serum *(5)*.

The ELOSA detection method was based on the sandwich hybridization, in solution and at 37°C, between a single-stranded DNA template and short synthetic oligonucleotides as capture and detection probes *(6)*. The fixation of the capture oligonucleotides was done by a passive adsorption on the surface of the microtiter wells, using phosphate buffer saline containing a high concentration of salts. The detection probe was a horseradish peroxidase-labeled oligonucleotide. ELOSA was shown to be convenient as a routine assay and was applied to HIV-1 detection *(6,7)* and quantitation *(8)*, HLA typing *(9)*, and automated *Mycobacterium tuberculosis* detection *(10)*.

Continuous RT-PCR coupled with ELOSA was shown to be an efficient and simple method for characterization of RNA *(11)*, including competitive-based quantitation *(12)*.

2. Materials

2.1. Amplification

1. 1 µg/5 µL total RNA in water.
2. Diethyl pyrocarbonate (DEPC)-treated-water. Prepare by adding 1 mL of DEPC to 1 L of H_2O. Mix the solution thoroughly and allow to set 30 min overnight followed by autoclaving (*see* **Note 1**).
3. 23.4 U/µL RNAguard (Amersham Pharmacia Biotech, Uppsala, Sweden) (*see* **Note 2**).
4. 10X amplification buffer: 100 mM Tris-HCl, pH 8.3, 500 mM KCl, 0.1% gelatin (*see* **Note 3**).
5. 100 mM $MgCl_2$ in sterile water (Merck) (*see* **Note 3**).
6. 100 mM 4 dNTP mix: (25 mM each dNTP, Pharmacia, Darmstadt, Germany): combination of equal volumes of commercially available 100 mM stock solutions of dATP, dCTP, dGTP, and dTTP.
7. 31 pmol/µL 5' primer in H_2O (*see* **Note 4**).
8. 31 pmol/µL 3' primer in H_2O (*see* **Note 4**).
9. *Taq* DNA polymerase 5 U/µL (Perkin-Elmer/Cetus, Norwalk, CT).
10. AMV reverse transcriptase 25 U/µL (Roche Diagnostics GmBH, Mannheim, Germany) (*see* **Note 5**).
11. Mineral oil (Sigma Chemical Co., St. Louis, MO).

2.2. Detection

1. Microtiter plates (Maxisorb, Nalgene Nunc, Rochester, NY) (*see* **Note 6**).
2. PBS 10X: 1.37 M NaCl, 27 mM KCl, 43 mM Na_2HPO_4, 14 mM KH_2PO_4.
3. Adsorption buffer: Phosphate-buffered saline (PBS) 3X.
4. Washing buffer: PBS 1X, 0.05% Tween 20 (w/v).
5. Hybridization buffer: 0.1 M sodium phosphate, pH 7.0, 0.5 M NaCl, 0.65% Tween 20 (w/v), 0.14 mg/mL salmon sperm DNA (Roche Diagnostics), 2% polyethylene glycol 4000 (*see* **Note 7**).
6. Substrate: OPD (*O*-phenylenediamine, Sigma) 2 mg/mL in 0.05 M citric acid, 0.1 M Na_2HPO_4, pH 4.9, containing H_2O_2 (0.03 volume) (*see* **Note 8**).
7. Denaturation buffer: 2 M NaOH.
8. Neutralization buffer: 2 M Acetic acid.
9. 1 N H_2SO_4 (Merck).
10. Capture probe (150 nM in adsorption buffer) (*see* **Note 9**).
11. Detection probe (15 nM in hybridization buffer) (*see* **Note 9**).

3. Method

3.1. General Considerations

The following guidelines are recommended in order to eliminate the possibility of contamination. Several dedicated areas are required; fresh disposable gloves should be used in each area. The first area is dedicated to the sample treatment; no plasmid preparation and RT-PCR reaction products should enter this area. The second area is dedicated to the preparation of the reaction mixture; no plasmid preparation, no RT-PCR reaction products and no template should enter this area. The use of a laminar flow hood equipped with ultraviolet light is effective as a clean area. This area should have dedicated pipets and pipet tips. Sterile microcentrifuge tubes and pipets tips should be used for all reactions. These items should be sterilized by steam autoclaving, the tubes being sealed. Sterile, cotton-plugged pipet tips are recommended to prevent

Characterization of RNA

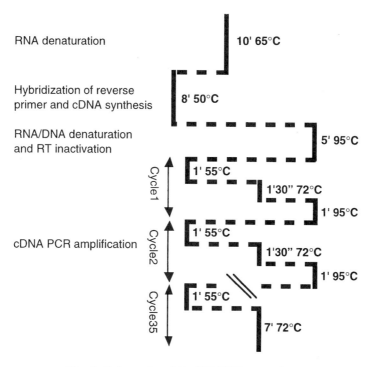

Fig. 1. Schematic of the RT-PCR procedure.

carryover between samples. Pipet tips should not be used more than once. Tubes are sealed between each addition of reagents. Controls with no template should be set up entirely in the clean area and not reopened until ready for analysis. The third area is specifically dedicated to the addition of the template to the reaction mixture; no plasmid preparation and RT-PCR reaction products should enter this area. Care should be taken not to crosscontaminate gloves during template addition. Tubes are handled one by one and sealed immediately after addition of the template. The last dedicated area is the regular laboratory where the detection procedure occurs.

3.2. Continuous RT/PCR Reaction

Reaction conditions have been optimized for the Perkin-Elmer 480A apparatus. The principle of RT-PCR procedure is summarized in **Fig. 1**.

The reaction master mix is prepared in its dedicated area. Master reagent mixtures should be prepared in excess to allow for pipeting losses. As a rule of thumb, at least 10% extra volume is recommended. When adding reagents, pipet up and down several times to mix. For 10 standard RT-PCR tests, prepare a master mix for 11 reactions by adding the following to a sterile 1.5 mL microcentrifuge tube:

1. DEPC-treated water: 864.6 µL.
2. RNAguard: 11 µL.
3. 10X amplification buffer: 110 µL.
4. MgCl$_2$ (*see* **Note 10**): 16.5 µL.
5. 4dNTP mix: 11 µL.
6. 5' primer: 11 µL (*see* **Note 11**).

7. 3' primer: 11 µL (*see* **Note 11**).
8. *Taq* polymerase (*see* **Note 12**): 5.5 µL.

The following steps (**steps 9, 10, 11, 14**) are performed on ice.

9. AMV-RT (*see* **Notes 12** and **13**): 4.4 µL.
10. Gently mix by hand and spin briefly in a microcentrifuge to ensure all contents are in the bottom of the tube (*see* **Note 14**).
11. Distribute 95 µL of the master mix to the bottom of each reaction tube.
12. Cover the RT-PCR master mix with two drops of mineral oil.
13. Move to the area dedicated to template addition.
14. Add 5 µL of total RNA (1 µg) in water (*see* **Note 15**) to the RT-PCR master mix. This may be accomplished by placing the pipet tip through the oil and pipeting the RNA solution into the lower aqueous phase. Pipet carefully up and down several times.

    The final concentrations of components after RNA is added are 1X amplification buffer, 0.234 U/µL RNAguard, 1.5 mM MgCl$_2$, 250 µM each dNTP, 310 nM each primer, 0.025 U/µL of *Taq*-polymerase and 0.1 U/µL of AMV-RT.
15. Gently mix by tapping the tube, and spin briefly in microcentrifuge to obtain a clean oil/aqueous interface and to remove air bubbles.
16. Amplify by RT-PCR using the following cycle profile:
    65°C, 10 min (RNA denaturation) (*see* **Note 16**).
    50°C, 8 min (reverse transcription) (*see* **Note 17**).
    95°C, 4 min (DNA/RNA denaturation and RT inactivation).
    95°C, 1 min—55°C, 1 min—72°C, 1.5 min (×35 cycles).
    72°C, 7 min (final extension).
    8°C (hold).

3.3. Detection of RT/PCR Products Using ELOSA

After amplification, move to the area dedicated to detection. The principle of ELOSA detection procedure is summarized in **Fig. 2**. The use of a multichannel pipet is highly recommended.

Passive adsorption of the capture probes:

1. Drop 100 µL of PBS 3X hybridization buffer containing 150 nM capture probe in each well of the microtiter plate.
2. Incubate 2 h at 37°C or overnight at room temperature.
3. Wash the plate three times with PBS 1X, 0.05% Tween 20 (w/v) (*see* **Note 18**).

Detection of amplified product:

4. Dilute 25 µL out of 100 µL of the amplified double-stranded DNA in 65 µL of hybridization buffer (*see* **Note 19**).
5. Add 10 µL of denaturation buffer (NaOH 0.2 M final concentration) (*see* **Note 19**).
6. Incubate at room temperature for 5 min.
7. Add 10 µL of neutralization buffer (acetic acid 0.2 M final concentration) (*see* **Note 19**).
8. Adjusted the volume to 125 µL using the hybridization buffer (*see* **Note 20**).
9. Drop 50 µL of this solution in a well of a microtiter plate previously coated with a specific capture probe. Test each sample in two separate wells.
10. Immediately thereafter, add 50 µL of the detection probe (15 nM in hybridization buffer) to each well.
11. Incubate 1 h at 37°C.

Characterization of RNA

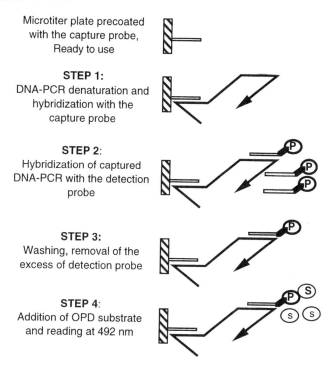

Fig. 2. Schematic of the ELOSA procedure.

12. Wash the microtiter plate three times with washing buffer.
13. Add 100 μL of OPD substrate to each well.
14. After 30 min, stop the reaction by adding 100 μL 1 N H_2SO_4.
15. Read the signal at A492 nm in an automatic microtiter plate reader.

4. Notes

1. Classical precautions concerning RNA handling should be taken, particularly the use of DEPC-treated solutions. Although residual DEPC has been described to inhibit *Taq* polymerase, no effect was observed in RT-PCR using a master mix prepared with or without DEPC-treated reagents.
2. RNAguard was not systematically used, and no differences were observed.
3. Tris-HCL, KCl, and gelatin (Merck) are weighed, dissolved in DEPC-treated water with slight heating to solubilized gelatin, and filtered through a 0.22-μm filter (Millipore, Bedford, MA). $MgCl_2$ stock solution is either autoclaved or filtered using a 0.22-μm filter. The use of commercially available, 10X premade amplification buffers is recommended; generally they are purchased including 15 mM $MgCl_2$ · $MgCl_2$ premade solutions are also commercially available.
4. Oligonucleotides are synthesized on an Applied Biosystems (Foster City, CA) 394 synthesizer by the phosphoramidite method and generally do not require further purification. Primers are chosen empirically by the research worker. Duplex and hairpin formation, false priming sites, and quality of primers are checked using commercial programs (OLIGO4.03, National Bioscience, Plymouth, MN). Pay attention in choosing primers with a Tm difference between generally lower than 3°C. The selection of primers overlapping splice junctions prevents amplification of residual DNA in the RNA preparation;

such overlapping exons primers are designed with the 3' end that covers the inner exon representing 35% of the primer Tm. It is possible to substitute inosine for point mutations when required. For RNA normalization using the aldolase system, prepare 6 pmol/µL primer stock solutions (*see* **Note 11**).

5. The source of AMV-RT appeared not to be important, as the Roche Diagnostics, the Seikagaku (Tokyo, Japan), and Gibco-BRL (Gaithersburg, MD) enzymes worked as well in the procedure. MMuLV-RT, RNase-H plus or minus, did not show the same activity in the continuous RT-PCR assay.
6. The use of irradiated plates is highly recommended to increase adsorption efficiency.
7. The hybridization buffer, passed through a 0.22-µm filter and sterilized by steam autoclaving, is stable for at least 6 mo when stored at 4°C.
8. The substrate solution is prepared extemporaneously.
9. The oligonucleotides used for the detection and capture are synthesized with an amine arm at the 5' end. The addition was performed on a synthesizer with the aminolink II reagent (Applied Biosystems). The horseradish peroxidase labeling of detection oligonucleotides is performed as described by Urdea et al. *(13)*. Preparation of the oligonucleotide-enzyme conjugate is summarized as follows: to conjugate oligonucleotides to horseradish peroxydase (HRP), 400 µg of vacuum-dried 5-aminoalkyl-oligodeoxyribonucleotide were mixed first with 25 µL of 0.1 M sodium borate buffer (pH 9.3) and then with 1 mL of 30 mg/mL 1,4-phenylene diisothiocyanate (DITC, Sigma) prepared in dimethylformamide. The mixture was incubated in the dark for 2 h at room temperature and transferred in a glass tube (corex 30 mL), then 5 mL of H_2O saturated-1-butanol were added and mixed, after which 3 mL of H_2O were added. After three extractions with equal volumes of H_2O saturated-1-butanol, the aqueous phase was vacuum dried and activated oligonucleotides resolubilized in 400 µL of 0.1 M sodium-borate buffer (pH 9.3) containing 9.2 mg of HRP (Roche Diagnostics). The mixture was incubated overnight at room temperature and then was stored several days at 4°C. Oligonucleotide-HRP conjugates were HPLC (high-performance liquid chromatography) purified by ion exchange chromatography, using Protein-Pak DEAE 8HR column (Waters S. A., Saint Quentin sur Yveline, France) and 10–42% buffer B gradient (buffer A: Tris-HCl 20 mM pH 7.6, buffer B: Tris-HCl 20 mM NaCl 2 M, pH 7.6). The oligonucleotide–HRP-containing fraction is immediately dialyzed against H_2O overnight at 4°C, then vacuum dried and mixed with 500 µL Tris 50 mM NH_4Cl 100 mM, pH 7.5. The purity of the detection probe is greater than 95%, and the ratio [oligonucleotide mole/L]/[HRP mole/L] is between 0.9 and 1.3. Detection probes, in Tris 50 mM NH_4Cl 100 mM, pH 7.5, 1 mg/mL salmon sperm DNA, 50% glycerol, are stable several months when stored at –20°C. The HRP-probe concentration must be greater than 150 nM. Probes are chosen and checked like primers (*see* **Note 4**). It is possible to substitute inosine for point mutations when required. Such labeled enzymes are commercially available.
10. Because Mg^{2+} concentration is a critical factor in the stringency of primer annealing, it is the first factor to vary when beginning a study to optimize the specificity and sensitivity of continuous RT-PCR. Mg^{2+} concentration can vary from 1.5–6 mM without altering the activities of both the AMV-RT and *Taq* polymerase enzymes *(1)*.
11. Normalization of 1 µg RNA is achieved by amplifying aldolase mRNA using standard RT-PCR protocol, except that 60 nM each primer are used so that the plateau is not reached. (Primers are described in **Table 1**.)
12. The AMV-RT:*Taq* polymerase ratio affects the sensitivity of the continuous RT-PCR method. It is the second factor to vary when beginning a study to optimize sensitivity. The AMV-RT:*Taq* polymerase ratio can range from 5:2.5–20:2.5, although a significant

Table 1
Primers and Capture and Detection Probes[a]

Target gene	Oligonucleotide function	DNA sequence (5'–3')	Name	Location
Aldolase	PCR	CCCCTTCCGAGGCTAAATCG	U1179	2721–2740
	PCR (reverse)	CTGGTAGTAGCAAGTTC-CTGGCAC	L1178	4177–4161 to 2866–2860
	Capture	CCTTGAATCCACTCGCCAGCC	C4031	2796–2816
	Detection	GCAGAAGGGGTCCTGGTGACG	D4034	2760–2780
HIV-1 MS-mRNA	PCR	TCTAICAAAGCA-ACCCIC	U1082	5580–5591 to 7925–7930
	PCR (reverse)	CCTATCTGTCCCCTCAGCTAC	L972	8254–8234
	Capture	GACCCGACAGGCCCGAAGGAATC	C4033	7947–7969
	Detection	CTACCACCGCTTGAGAGACTTACT	D4035	8071–8094

[a]The *Aldolase* positions are with reference to the HSALDOA sequence (X12447, GenBank). HIV-1 nomenclature is according to HIVHXB2R *(14)*.

Table 2
RT-PCR-ELOSA Normalization of RNA[a]

Sample	Mean OD	Coefficient of variation (%)
1	1768	6.6
2	1631	7.5
3	1664	5.7
4	1227	4.2
5	1166	3.0
6	1430	5.3
All	1481	4.1

[a]Six independent RT-PCR amplifications of 1 µg total RNA were performed for six samples and aldolase amplification products were detected by ELOSA. ODs at 492 nm are expressed as optical density (OD) × 1000. Means and coefficients of variation are indicated. The acceptance criterion for normalization of 1 µg total RNA was fixed at an OD between 700 and 2200, defined by mean ± 2 standard deviations of the 36 RT-PCR-ELOSA (all).

decrease in signal has been observed at a 20:2.5 ratio when using less than 500 copies of template *(1)*.

13. One advantage of this two-enzyme system is that you can omit the RT but perform exactly the same protocol, including cycling. This is particularly useful to control the absence of contaminating DNA when exon-spanning primers are not available.
14. Quantitative competitive RT-PCR is based on the coamplification of different amounts of competitive mutated RNA template along with the RNA target *(11)*. The mimic template differs from the wild-type target by only some mutations. For QC-RT-PCR purpose, the master mix is split into four parts; different amounts of competitive RNA template are then added outside the master mix dedicated area. The standard protocol is applied, except that 45 PCR cycles are performed to reach the plateau.
15. One microgram RNA is used in standard normalized procedure (**Table 2** and **Fig. 3**). However, either a crude RNA preparation or poly (A)$^+$ RNA can be used, prepared using rapid or classical guanidinium isothiocyanate-based protocols. The RT-PCR procedure achieves the same level of amplification from 50 copies of target, whether it is diluted in 10 ng or 1 µg of nonspecific RNA, in buffer conditions that included 1.5 mM Mg^{2+}, and at an AMV-RT to a *Taq* polymerase ratio of 10:2.5.
16. Depending on the required sensitivity and the complexity of the target, the temperature used during the RNA denaturation step can vary from 60–75°C.
17. Depending on the required specificity and the complexity of the target, the reverse transcription step can vary from 5 min at 40°C to 15 min at 65°C. A short step such as 5 min at 37°C can be introduced between the standard denaturation step (10 min 65°C) and the reverse transcription step (8 min 50°C), to allow nonstringent hybridization of reverse primer, or when using degenerated primers. The hybridization temperature in the following PCR should be subsequently modified.
18. Oligonucleotide-coated plates, dried in a vacuum oven and sealed by plastic sheet, are stable for several months at 4°C.
19. This step is a crucial one in ELOSA. Do not try to lower the input of PCR solution under 20 µL and always complete to 90 µL with hybridization buffer. Do not modify the volumes in **steps 5** and **7**. Modifications would decrease denaturation efficiency therefore hybridization efficiency.

Characterization of RNA

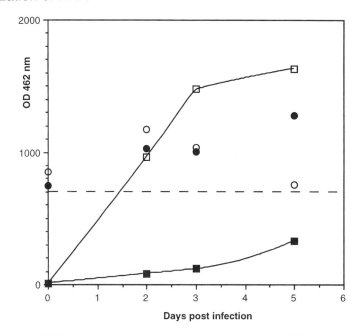

Fig. 3. Kinetic of HIV-1 infection in vitro. A rapid-high virus (RH virus) and a slow-low virus (SL virus) were used. HIV-1 viral stocks with an RT activity of 10^6 cpm were mixed with 2×10^8 PBMC (peripheral blood mononuclear cells) for 2 h, then the infected cells were washed and grown in fresh media. Each day, 3×10^7 cells were collected and total RNA was extracted. RNAs were normalized by amplifying the aldolase RNA (○ and ●) and the multiply spliced HIV-1 mRNA were detected (□ and ■) for the RH virus- (open symbols) and SL virus- (black symbols) infected cells, respectively. ELOSA was performed using 1/75th of the RT-PCR solution (*see* **Note 20**). ODs at 492 nm are expressed as OD × 1000. Acceptance criterion for RNA quality is shown by a dotted line (*see* **Table 2**). Primers and probes are described in **Table 1**.

20. The volume of added hybridization buffer could vary depending of the goal of the research worker. For sensitive detection purpose, a yes/no response is expected, then OD signals can reach the saturation of the microtiter reader. Adjusting the volume to 125 µL is adequate for detection beneath the 10 copies level (the input of 50 µL in a well is equivalent to 10 µL of the initial RT-PCR solution, that is 1/10th). For normalization of RNA using the aldolase system, adjust the volume to 935 µL using the hybridization buffer (input equivalent to 1/75th of the RT-PCR solution) (*see* **Fig. 3**). Correct normalization is achieved when the measured OD range between a minima and a maxima defined as acceptance criteria (*see* **Table 2**; **Fig. 3**). Acceptance criteria should be defined according to the source of the sample, the nature of the sample, and sample treatment *(8)*. For quantitation of mRNA using QC-RT-PCR, a higher dilution is required (input equivalent to 1/200th of the RT-PCR solution) *(11)*; wild-type and mimic templates are discriminated by ELOSA at the capture probe level *(8,11)*.

Acknowledgments

I thank Guy Oriol for excellent technical assistance, Christelle Brun and Nathalie Ferraton for oligonucleotides synthesis and enzyme coupling, C. Guillon for the gift of HIV-1 RNA, and P. Cros and A. Laayoun for helpful discussions.

References

1. Mallet, F., Oriol, G., Mary, C., Verrier, B., and Mandrand, B. (1995) Continuous RT-PCR using AMV-RT and Taq DNA polymerase: characterization and comparison to uncoupled procedures. *BioTechniques* **18,** 678–687.
2. Shimomaye, E. and Salvato, M. (1991) Use of avian myeloblastosis virus reverse transcriptase at high temperature for sequence analysis of highly structured RNA. *Gene Anal. Tech.* **6,** 25–28.
3. Wang, R. F., Cao, W. W., and Johnson, M. G. (1992) A simplified, single buffer system for RNA-PCR. *BioTechniques* **12,** 702–704.
4. Hearst, E. H. (1988) A photochemical investigation of the dynamics of oligonucleotide hybridization. *Ann. Rev. Phys. Chem.* **39,** 291–315.
5. François, M., Dubois, F., Brand, D., Bacq, Y., Guerois, C., Mouchet, C., et al. (1993) Prevalence and significance of hepatitis C (HCV) viremia in HCV antibody-positive subjects from various populations. *J. Clin. Microbiol.* **31,** 1189–1193.
6. Mallet, F., Hebrard, C., Brand, D., Chapuis, E., Cros, P., Allibert, P., et al. (1993) Enzyme-linked oligosorbent assay for detection of polymerase chain reaction-amplified human immunodeficiency virus type 1. *J. Clin. Microbiol.* **31,** 1444–1449.
7. Brossard, Y., Aubin, J. T., Mandelbrot, L., Bignozzi, C., Brand, D., Chaput, A., et al. (1995) Frequency of early *in utero* HIV-1 infection: a blind DNA polymerase chain reaction study on 100 fetal thymuses. *AIDS* **9,** 359–366.
8. Mallet, F., Hebrard, C., Livrozet, J. M., Lees, O., Tron, F., Touraine, J. L., et al. (1995) Quantitation of human immunodeficiency virus type 1 DNA by two PCR procedures coupled with enzyme-linked oligosorbent assay. *J. Clin. Microbiol.* **33,** 3201–3208.
9. Cros, P., Allibert, P., Mandrand, B., Tiercy, J.-M., and Mach, B. (1992) Oligonucleotide genotyping of HLA polymorphism on microtitre plates. *Lancet* **340,** 870–873.
10. Mabilat, C., Desvarenne, S., Panteix, G., Machabert, N., Bernillon, M. H., Guardiola, G., et al. (1994) Routine identification of *Mycobacterium tuberculosis* complex isolates by automated hybridization. *J. Clin. Microbiol.* **32,** 2702–2705.
11. Mallet, F. (1998) Characterisation of RNA using continuous RT-PCR coupled with ELOSA, in *RNA Isolation and Characterisation Protocols* (Rapley, R. and Manning, D. L., eds.), Humana Press, Totowa, NJ.
12. Mallet, F. (1996) Continuous RT-PCR using AMV-RT and *Taq* DNA polymerase, in *PCR: Essential Techniques* (Burke, J., ed.), Wiley, Chichester, pp. 82–85.
13. Urdea, M. S., Warner, B. D., Running, J. A., Stempien, M., Clyne, J., and Horn, T. (1988) A comparison of non-radioisotopic hybridization assay methods using fluorescent, chemiluminescent and enzyme labeled synthetic oligodeoxyribonucleotide probes. *Nucleic Acids Res.* **16,** 4937–4956.
14. Myers, G., Korber, B., Berzofsky, J. A., Smith, R. F., and Pavlakis, G. N. (1991) Human retroviruses and AIDS: a compilation and analysis of nucleic acid and amino acid sequences, Los Alamos National Laboratory, Los Alamos, NM, IA3-IA69.

36

Quantitative Analysis of RNA Species by Polymerase Chain Reaction and Solid-Phase Minisequencing

Anu Suomalainen and Ann-Christine Syvänen

1. Introduction

Methods allowing sensitive and accurate quantitative analysis of defined RNA species are required in a wide variety of gene expression studies. Unlike the traditional hybridization methods, RNase protection or S1 nuclease assays, the methods based on reverse transcription (RT) and polymerase chain reaction (PCR) provide an essentially unlimited sensitivity of detection. A drawback of the PCR-based methods is, however, that they do not allow direct quantification of a sequence present in a sample, because the efficiency of the PCR depends on the amount of the template sequence, and the amplification is exponential only at low concentrations of the template *(1)*. Due to this "plateau effect" of the PCR, the amount of the amplification product does not reflect directly the original amount of the template. Moreover, subtle differences in the reaction conditions may cause significant sample-to-sample variation in the final yield of the PCR. The efficiency of amplification is also affected by the sequence of the PCR primers, as well as the size and, to some extent, the sequence of the PCR product. For these reasons a prerequisite for an accurate quantitative PCR analysis is that an internal standard is coamplified in the same reaction with the target sequence. The standard sequence should be as similar to the target sequence as possible to ensure that the target-to-standard ratio remains constant throughout the amplification. An ideal PCR standard differs from the target sequence only at one nucleotide position, by which the two sequences can be identified and quantified after the amplification. For quantification of RNA the optimal internal standard is RNA, to be able to control not only the efficiency of the PCR, but also that of the cDNA synthesis.

The determination of the relative amounts of the PCR products originating from the target and standard sequences allows calculation of the initial amount of the target sequence *(2–4)*. For many applications it is sufficient to determine the relative amount of the target sequence compared to an endogenous standard sequence originally present in the sample. To be able to determine the absolute amount of a target sequence present, it is necessary to add a known amount of a standard sequence to the sample before amplification. In this case a measure of the amount of the analyzed sample, such as the number of the cells or total amount of RNA is needed. Finally, an accurate method for detecting the PCR products originating from the standard and target sequences is required.

We have developed the solid-phase minisequencing method for the detection and quantification of nucleotide sequence variants differing from each other at one nucleotide position *(5,6)*. In this method a DNA fragment spanning the site of the variable nucleotide is first amplified using one biotinylated and one unbiotinylated PCR primer. The PCR product is carrying biotin at the 5' end of one of its strands is captured on an avidin-coated solid support and denatured. The nucleotides at the variable site in the immobilized DNA strands are identified by two separate primer extension reactions, in which a single labeled deoxynucleotidetriphosphate (dNTP) is incorporated by a DNA polymerase. In our standard format of the assay [^{3}H]dNTPs serve as labels and streptavidin-coated microtiter plates are used as the solid support *(6)*. The results of the assay are numeric counts per minute (cpm) values expressing the amount of the specific [^{3}H]dNTPs incorporated in the minisequencing reactions. When analyzing samples containing a mixture of two sequences, differing from each other by a single nucleotide, the ratio between the two cpm values reflects directly the ratio between the two sequences in the original sample. Because the two sequences are essentially identical, they are amplified with equal efficiency irrespective of the exponential phase of the PCR. The solid-phase format of the assay allows interpretation of the result by direct measurement of the incorporated label without any separation step that could introduce errors into the analysis. These features make the solid-phase minisequencing method close to an ideal tool for PCR-based determination of both relative and absolute amounts of DNA and RNA sequences. **Figure 1** illustrates the principle of the quantitative analysis by the solid-phase minisequencing. The method has been applied for example to accurately compare the transcript levels of mutant and normal alleles of one gene and of two highly homologous genes *(7)*, as well as to accurately determine the absolute amounts of specific transcripts in cell and tissue samples utilizing an internal synthetic RNA *(8)* or DNA standard *(9)*, designed to differ from the target RNA of interest only by one nucleotide. **Table 1** summarizes the alternative strategies by which the solid-phase minisequencing method has been utilized for quantitative PCR analysis of RNA.

2. Materials
2.1. Equipment

1. Programmable heat block and facilities to avoid contamination in PCR.
2. Microtiter plates with streptavidin-coated wells (e.g., Combiplate 8, Labsystems, Helsinki, Finland) (*see* **Note 1**).
3. Multichannel pipet and microtiter plate washer (optional).
4. Shaker at 37°C.
5. Water bath or incubator at 42°C and 50°C
6. Liquid scintillation counter.

2.2. Reagents for RT, PCR, and Minisequencing

All reagents should be of standard molecular biology grade. Use sterile distilled or deionized water.

Quantitative Analysis of RNA Species

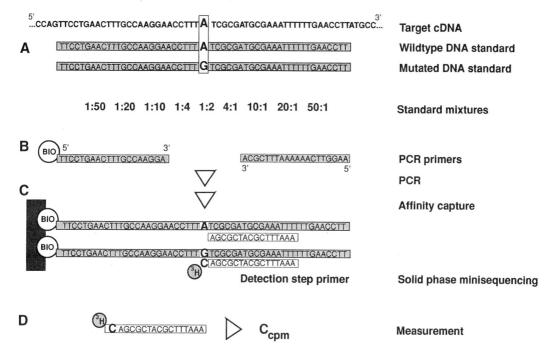

Fig. 1. Quantitative analysis by solid-phase minisequencing utilizing oligonucleotide standards. (**A**) Design of DNA standards (oligonucleotides). The wildtype DNA standard is identical to the target cDNA sequence, and the mutated standard differs from it by one nucleotide. The standards are mixed in known ratios. (**B**) PCR. The 5′ PCR primer carries a biotin residue (BIO). This results in a PCR product carrying biotin in the 5′ end of one of its strands. (**C**) Solid-phase minisequencing analysis. The analysis for one of the nucleotides (G, the mutated nucleotide) is shown. The PCR product is captured in a streptavidin-coated microtiter well and denatured. A detection step primer hybridizes to the single-stranded template, 3′ adjacent to the variant nucleotide. The DNA polymerase extends the primer with the [^{3}H]-labeled dNTP, if it is complementary to the nucleotide present at the variable site. (**D**) Denaturation and measurement of the eluted radioactivity.

1. Avian myeloblastosis virus (AMV) reverse transcriptase (20–25 U/μL, Promega Corp., Madison, WI).
2. Rnasin® ribonuclease inhibitor (20–40 U/μL, Promega Corp.).
3. 10X Concentrated RT buffer: 500 mM Tris-HCl, pH 8.2–8.5, 50 mM MgCl$_2$, 400 mM KCl, and 20 mM dithiothreitol.
4. dNTP mixture: 2 mM dATP, 2 mM dCTP, 2 mM dGTP, and 2 mM dTTP.
5. Diethyl pyrocarbonate-treated water (DEPC H$_2$O).
6. Thermostable DNA polymerase. We use *Thermus aquaticus* (5 U/μL, Promega Corp.) or *Thermus brockianus* (Dynazyme™ II, 2 U/μL, Finnzymes, Espoo, Finland) DNA polymerase (*see* **Note 2**).
7. 10X Concentrated DNA polymerase buffer: 500 mM Tris-HCl, pH 8.8, 150 mM (NH$_4$)$_2$SO$_4$, 15 mM MgCl$_2$, 1% v/v Triton X-100, 0.1% w/v gelatin.
8. dNTP mixture as in **step 4**.
9. PBS/Tween: 20 mM sodium phosphate buffer, pH 7.5, and 0.1% (v/v) Tween 20. Store at 4°C. Fifty milliliters is enough for several full-plate analyses.

Table 1
Summary of Alternative Strategies for Quantitative PCR Analysis of RNA by Solid-Phase Minisequencing

Standard	Reference	Application	Interpretation
Relative RNA quantification			
Two variants of a transcript present as a mixture in a sample: Serve as standards for each other during RT and PCR	6,9	Screening for differences in levels of transcripts of two alleles or two homologous genes	1. Calculate the result from the ratio between the incorporated [^{3}H]dNTPs after correction for differences in specific activities and number of incorporated dNTPs
Genomic DNA: Serves as an external reference for the sequence variant in a natural 1:2 ratio	7		2. Calculate the result by comparing the ratio between the incorporated [^{3}H]dNTPs with that obtained from the genomic DNA
			3. Calculate the result from a complete standard curve constructed with mixtures of the two sequence variants
Absolute RNA quantification			
RNA-fragment			
Differs from the target at one nucleotide.	8	Determination of the absolute amount of target RNA in a defined sample.	As 1 or 3 above.
Prepared by in vitro transcription from a DNA construct (in the future chemical synthesis)		Both the RT and PCR are controlled.	
Mixed with the target RNA before RT.			
DNA-fragment			
Differs from the target at one nucleotide	6	Determination of the absolute amount of the target sequence in a defined sample.	As 1 or 3 above.
Oligonucleotide, cloned DNA or PCR product. Mixed with the template before PCR		Does not control the RT reaction.	

10. TENT (washing solution): 40 mM Tris-HCl, pH 8.8, 1 mM ethylenediaminetetraacetic acid (EDTA), 50 mM NaCl, and 0.1% (v/v) Tween 20. Store at 4°C. Prepare 1–2 L at a time, which is enough for several full-plate analyses.
11. 50 mM NaOH (make fresh every 4 wk), store at room temperature (approx 20°C). Prepare 50 mL.
12. [^{3}H]-labeled deoxynucleotides (dNTPs): dATP to detect a T at the variant site, dCTP to detect a G etc. (Amersham Pharmacia Biotech, Arlington Heights, IL; [^{3}H]dATP, TRK 633; dCTP, TRK 625; dGTP, TRK 627; dTTP, TRK 576), store at −20°C (*see* **Note 3**).
13. Scintillation reagent (e.g., Hi-Safe II, Wallac, Turku, Finland).

2.3. Primer Design

1. PCR primers: Biotinylate one of the PCR primers at its 5' end during the synthesis using a biotin-phosphoramidite reagent (e.g., Amersham Pharmacia Biotech or Perkin Elmer/ABI, Norwalk, CT) (*see* **Note 4**). The 3' primer can be used for the RT, and as the 3'-PCR primer. If oligonucleotides (DNA) are used as quantification standards, the length of oligonucleotides that can be synthesized with acceptable yields sets an upper limit to about 80–100 bp for the PCR product.
2. Detection step primer for the minisequencing analysis is an oligonucleotide complementary to the biotinylated strand, designed to hybridize with its 3' end with the nucleotide adjacent to the variant nucleotide to be analyzed (*see* **Fig. 1**). The detection step primer for our standard protocol is a 20-mer. The primer should be at least five nucleotides nested in relation to the unbiotinylated PCR primer.

2.4. Quantification Standards

The standard should be designed to differ from the target sequence by one nucleotide, which is the one detected in the minisequencing analysis (*see* **Fig. 1**). If a standard curve is to be constructed, a second standard, identical to the target sequence is required (*see* **Subheading 3.4.**) Depending on the application, either DNA or RNA standards may be used (*see* **Table 1**). RNA standards are synthesized by in vitro transcription *(10)* and oligonucleotide standards using a DNA/RNA synthesizer (*see* **Note 5**). PCR products or cloned cDNA fragments can also be used as standards. Measure the molecular concentrations of the standards. The optimal amount of the standard added to a sample depends on the abundance of the target sequence in the original sample. The ratio of the target to the standard sequence should preferably be between 0.1 and 10. If no estimate of the amount of the target sequence is available, it may be necessary to initially titrate the optimal amount of the standard using several amounts of the standard in the analysis (e.g., 10^2, 10^4, 10^6, and 10^8 molecules). For accurate quantification, standards representing both sequence variants should be available, and analysis of mixtures of known amounts of the two standards should be analyzed to construct a standard curve, as demonstrated in **Figs. 1** and **2**.

3. Method

3.1. Reverse Transcription

Any established protocol for RT can be applied. We use the following protocol:

1. Prepare an RT mixture by combining 2 μL of 10 × RT buffer, 5 μL of dNTP mixture, 50 pmol of the 3'-PCR primer, 10 U RNasin, 30 U AMV reverse transcriptase, and DEPC H$_2$O to a final volume of 15 μL. Keep on ice until used.

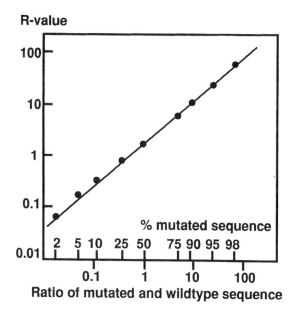

Fig. 2. Standard curve obtained by plotting the R-values (C_{cpm}/T_{cpm}) as a function of the ratio between the standard sequences on a log-log scale.

2. Add the appropriate amount of RNA (usually 50–500 ng) in 5 µL DEPC H_2O to an Eppendorf tube on ice. If an RNA standard is used, add the standard to the sample at this stage, keeping the total volume of the RT reaction at 20 µL. Proceed immediately to the next step.
3. Add 15 µL of the RT mixture to the RNA sample.
4. Incubate for 60 min at 42°C. The samples can be stored at –20°C for 1–3 d.

3.2. PCR for Solid-Phase Minisequencing Analysis

1. The PCR follows the routine protocols, except that the amount of the biotin-labeled primer should be reduced not to exceed the biotin-binding capacity of the microtiter well (*see* **Note 1**). For a 50-µL PCR reaction we use 10 pmol of biotin-labeled primer and 50 pmol of the unbiotinylated primer and as a template, one-quarter of the RT product. The PCR should be optimized (i.e. the annealing temperature and the amount of the template) to be efficient and specific. To be able to use [^{3}H]dNTPs, which have low specific activities, for the minisequencing analysis, one-tenth of the PCR product should produce a single visible band after agarose gel electrophoresis and staining with ethidium bromide.

3.3. Solid-Phase Minisequencing Analysis

1. Affinity capture: Transfer 10 µL aliquots of the PCR product and 40 µL of PBS/Tween to two streptavidin-coated microtiter wells (*see* **Note 6**). Include as negative controls two wells without PCR product. Seal the wells with a sticker and incubate the plate at 37°C for 1.5 h with gentle shaking.
2. Discard the liquid from the wells, and tap the wells dry against a tissue paper.
3. Wash the wells three times at room temperature by adding 200 µL of TENT to each well, discard the washing solution, and empty the wells thoroughly between the washing steps (*see* **Note 7**).

Table 2
Example of the Result of a Solid-Phase Minisequencing Analysis of Mixtures of Two Standard Oligonucleotides

Ratio of wild-type sequence:mutated sequence	T_{cpm} (wt oligo)[a]	C_{cpm} (mut oligo)[b]	R-value C_{cpm}/T_{cpm}
Wild-type	3110	44	.014
50:1	3640	190	.05
20:1	2780	420	.15
10:1	2830	730	.26
4:1	2520	1690	.67
1:2	1650	2810	1.7
1:4	790	3630	4.6
1:10	350	3790	10.8
1:20	210	4760	22.7
1:50	120	4800	40.0
Mutant	43	4580	106.5
H$_2$O	41	23	—

[a]The specific activities of the [^{3}H]dNTPs: dTTP 126 Ci/mmol, dCTP 67 Ci/mmol.
[b]In this case, two [^{3}H]dCTPs were incorporated into the mutant sequence.

4. Denature the captured PCR product by adding 100 µL of 50 mM NaOH to each well, followed by incubation at room temperature for 3 min. Discard the NaOH and wash the wells as in **step 3**.
5. For each DNA fragment to be analyzed prepare two 50 µL mixtures of nucleotide-specific minisequencing solution, one for detection of the wildtype and one for the mutant nucleotide, by mixing 5 µL of 10X DNA polymerase buffer, 10 pmol of detection step primer (e.g., 2 µL of 5 µM primer), 0.2 µCi (usually 0.2 µL) of one [^{3}H]dNTP, 0.1 U of DNA polymerase, and H$_2$O to a total volume of 50 µL. It is obviously convenient to prepare master mixes for the desired number of analyses with each nucleotide (*see* **Note 8**).
6. Add 50 µL of one nucleotide-specific mixture to each well, and incubate the plate at 50°C for 10 min (*see* **Note 9**).
7. Discard the contents of the wells and wash them as in **step 3**.
8. Release the detection step primer from the template by adding 60 µL 50 mM NaOH and incubating for 3 min at room temperature.
9. Transfer the NaOH containing the eluted primer to the scintillation vials, add the scintillation reagent, and measure the radioactivity, i.e., the amount of incorporated label, in a liquid scintillation counter (*see* **Note 10**).
10. The result is obtained as cpm-values. The cpm-value of each reaction expresses the amount of the incorporated [^{3}H]dNTP. Calculate the ratio *(R)* (*see* **Table 2** and **Note 11**) as follows:

$$R = \frac{\text{cpm incorporated in the reaction detecting the wild-type (target) sequence}}{\text{cpm incorporated in the reaction detecting the mutated (standard) sequence}}$$

3.4. Preparation of the Standard Curve

Mix the wild-type and mutated standard sequences in known proportions, e.g., 1:50, 1:20, 1:10, 1:4, 1:2, 4:1, 10:1, 20:1, and 50:1. If an RNA standard is used, synthesize cDNA using the mixtures as the template, amplify the product by PCR and analyze the products by minisequencing. Plot the resulting R values on a log-log scale as a function of the ratio between the sequences present in the original mixture, which should result in a linear standard curve (*see* **Table 2** and **Fig. 2**). This curve can then be utilized for the analysis of the actual samples with an unknown amount of target RNA either to determine the relative or the absolute amount of the target RNA (*see* **Table 1**).

4. Notes

1. The binding capacity of the streptavidin-coated microtiter well that we use is 2–5 pmol of biotinylated oligonucleotide. If a higher binding capacity is desired, avidin-coated polystyrene beads (Fluoricon, 0.99 µM, IDEXX Corp., Portland, ME; biotin-binding capacity over 2 nmol of oligonucleotide/mg beads), or streptavidin-coated magnetic polystyrene beads (Dynabeads M-280, Dynal, Oslo, Norway, streptavidin; biotin-binding capacity 300 pmol/mg) can be used *(11)*. The biotin-binding capacity of a microtiter well allows reliable detection of approx 2% of a sequence-variant present in a sample *(9)*, whereas a detection sensitivity of <0.1% can be obtained with the bead-based format *(12)*.
2. The use of a thermostable DNA polymerase in the single-nucleotide primer extension reaction is advantageous, as a high temperature, also favorable for the simultaneous primer annealing reaction, can be used.
3. Although the [^{3}H]dNTPs are weak β-emitters, their half lives are long (13 yr), and the necessary precautions for working with [^{3}H] should be taken. Also dNTPs or dideoxynucleotides labeled with other isotopes ([^{35}S] or [^{32}P], **ref. 5**) or with fluorophores *(13)* can be used.
4. The efficiency of the 5'-biotinylation of an oligonucleotide on a DNA synthesizer is most often 80–90%. The biotin-labeled oligonucleotides can be purified from the unbiotinylated ones either by high-performance liquid chromatography *(14)*, polyacrylamide gel electrophoresis *(15)* or by disposable ion exchange columns manufactured for this purpose (Perkin-Elmer/ABI). If the biotin-labeled primer is used without purification, the success of the biotinylation can be confirmed after the PCR by affinity capture of an aliquot of the biotinylated PCR product on an avidin matrix with high biotin-binding capacity (*see* **Note 1**). Analyze the supernatant after the capturing reaction by agarose gel electrophoresis. If the biotinylation has been efficient, no product, or a faint product of significantly lower intensity than the unbound PCR product is observed in the supernatant.
5. For use as quantification standards, full-length oligonucleotides should be purified from prematurely terminated ones by high-performance liquid chromatography *(14)* or by size-separation in polyacrylamide gel electrophoresis (*15*; *see* Chapter 15). The molecular concentration of the purified full-length standard DNA or RNA can then be accurately determined. At present it is not feasible to produce RNA fragments chemically by a DNA/RNA synthesizer to serve as RNA standards in a quantitative analysis.
6. Each nucleotide to be detected at the variant site is analyzed in a separate well. Thus at least two wells are needed per PCR product. For quantitative applications we carry out two parallel assays for each nucleotide, i.e., four wells per PCR product.
7. The washing can be performed using an automated microtiter plate washer, or by manually pipetting the washing solution to the wells, discarding the liquid, and tapping the plate against a tissue paper. It is important to thoroughly empty the wells between the washing steps.
8. The minisequencing reaction mixture can be stored at room temperature for 1–2 h. It is convenient to prepare it during the incubation in **step 1**.

9. The conditions for hybridizing the detection step primer are not stringent, and the temperature of 50°C can be applied to analysis of most PCR products irrespectively of the sequence of the detection step primer. If the primer, however, is considerably shorter than a 20-mer or its GC content is low (melting temperature close to 50°C) lower temperatures for the primer annealing may be required.
10. Streptavidin-coated microtiter plates made of scintillating polystyrene are available (ScintiStrips, Wallac). When these plates are used, the final washing, denaturation and transfer of the eluted detection primer to scintillation vials can be omitted, but a scintillation counter for microtiter plates is needed *(16)*.
11. The ratio between the cpm-values for the two nucleotides reflects the ratio between the two sequences in the original sample. The R value is affected by the specific activities of the [^{3}H]dNTPs used, and if either the wildtype or the mutant sequence allows the detection step primer to be extended by more than one [^{3}H]dNTP, this will obviously also affect the R value. Both of these factors can easily be corrected, when calculating the ratio between the two sequences. Another possibility is to construct a standard curve (*see* **Subheading 3.4.**).

References

1. Syvänen, A.-C., Bengtström, M., Tenhunen, J., and Söderlund, H. (1988) Quantification of polymerase chain reaction products by affinity-based hybrid collection. *Nucleic Acid Res.* **16,** 11,327–11,338.
2. Chelly, J., Kaplan, J.-C., Maire, P., Gautron, S., and Kahn, A. (1988) Transcription of the dystrophin gene in human muscle and non-muscle tissues. *Nature* **333,** 858–860.
3. Wang, A. M., Doyle, M. V., and Mark, D. F. (1989) Quantitation of mRNA by the polymerase chain reaction. *Proc. Natl. Acad. Sci. USA* **86,** 9717–9721.
4. Gilliland, G., Perrin, S., Blanchard, K., and Bunn, H. F. (1990) Analysis of cytokine mRNA and DNA: detection and quantitation by competitive polymerase chain reaction. *Proc. Natl. Acad. Sci. USA* **87,** 2725–2729.
5. Syvänen, A.-C., Aalto-Setälä, K., Harju, L., Kontula, K., and Söderlund, H. (1990) A primer-guided nucleotide incorporation assay in the genotyping of apolipoprotein E. *Genomics* **8,** 684–692.
6. Syvänen, A.-C., Sajantila, A., and Lukka, M. (1993) Identification of individuals by analysis of biallelic DNA markers, using PCR and solid-phase minisequencing. *Am. J. Hum. Genet.* **52,** 46–59.
7. Karttunen, L., Lönnqvist, L., Godfrey, M., Peltonen, L., and Syvänen, A.-C. (1996) An accurate method for comparing transcript levels of two alleles or highly homologous genes: application to fibrillin transcripts in Marfan patients fibroblasts. *Genome Res.* **6,** 392–403.
8. Ikonen, E., Manninen, T., Peltonen, L., and Syvänen, A.-C. (1992) Quantitative determination of rare mRNA species by PCR and solid-phase minisequencing. *PCR Methods Appl.* **1,** 234–240.
9. Suomalainen, A., Majander, A., Pihko, H., Peltonen, L., and Syvänen, A.-C. (1993) Quantification of tRNA$_{3243}^{Leu}$ point mutation of mitochondrial DNA in MELAS patients and its effects on mitochondrial transcription. *Hum. Mol. Genet.* **2,** 525–534.
10. Melton, D. A., Krieg, P. A., Rebagliati, M. R., Maniatis, T., Zinn, K., and Green, M. R. (1984) Efficient in vitro synthesis of biologically active RNA and RNA hybridization probes from plasmids containing a bacteriophage SP6 promoter. *Nucleic Acids Res.* **12,** 7035–7056.
11. Syvänen, A.-C. and Söderlund, H. (1993) Quantification of polymerase chain reaction products by affinity-based collection. *Meth. Enzymol.* **218,** 474–490.

12. Syvänen, A.-C., Söderlund, H., Laaksonen, E., Bengtström, M., Turunen, M., and Palotie, A. (1992) N-ras gene mutations in acute myeloid leukemia: accurate detection by solid-phase minisequencing. *Int. J. Cancer* **50,** 713–718.
13. Pastinen, T., Partanen, J., and Syvänen, A.-C. (1996) Multiplex, fluorescent solid-phase minisequencing for efficient screening of DNA sequence variation. *Clin. Chem.* **42,** 1391–1397.
14. Bengtström, M., Jungell-Nortamo, A., and Syvänen, A.-C. (1990) Biotinylation of oligonucleotides using a water soluble biotin ester. *Nucleosides Nucleotides* **9,** 123–127.
15. Wu, R., Wu, N.-H., Hanna, Z., Georges, F., and Narang, S. (1984) In *Oligonucleotide synthesis: a practical approach* (Gait, M. J., ed.), IRL Press, Oxford, p. 135.
16. Ihalainen, J., Siitari, H., Laine, S., Syvänen, A.-C., and Palotie, A. (1994) Towards automatic detection of point mutations: use of scintillating microplates in solid-phase minisequencing. *BioTechniques* **16,** 938–943.

37

Nonradioactive Northern Blotting of RNA

Rainer Löw

1. Introduction

Technical improvements of methods used for the nonradioactive detection of nucleic acids have replaced the ^{32}P-based techniques in many laboratories. The most commonly used labels are digoxigenin, fluorescein, and biotin, which are linked through a spacer to a nucleotide and are incorporated into specific probes by different methods *(1–3)*.

The detection is based on the specific interaction of the labels with appropriate proteins, i.e. specific antidigoxigenin and antifluorescein antibodies or (strept)avidin, conjugated to alkaline phosphatase, or peroxidase. The improvement of chemiluminescent substrates has shifted the sensitivity for detection of a specific target into a range that is comparable to the radioactive methods.

With these tools and the development of improved blotting and detection protocols, it is possible to detect even rare messages in total RNA samples *(3)*. Furthermore, the use of end-labeled oligonucleotides or probes solves one of the most serious problems in nonradioactive detection: the quantitation of transcripts *(4)*.

In this chapter we describe one nonradioactive method for Northern blot analysis using biotinylated probes. The following steps are described in detail: (1) preparation of highly labelled biotinylated-probes generated by polymerase chain reaction (PCR), (2) sample preparation for gel electrophoresis, (3) blotting of the electrophoretically separated RNA, (4) hybridization, and (5) detection of specifically bound probes. As starting material isolated total RNA (for protocols, *see* **refs. 5–7**) is necessary, which should be dissolved in formamide instead of water *(8)*.

2. Materials

All solutions and reagents should be of molecular biology grade. The use of distilled water for the whole blotting and detection procedure is strictly recommended, diethyl pyrocarbonate (DEPC)-treatment is not necessary. All containers and tools (e.g., forceps) should be cleaned with 1% sodium dodecyl sulfate (SDS) and 0.2 *M* NaOH followed by intense watering. Wearing powder-free gloves is recommended during all steps where handling the gel or membrane is necessary (e.g., during setup of the blotting sandwich or transfer of the membrane between the containers).

From: *The Nucleic Acid Protocols Handbook*
Edited by: R. Rapley © Humana Press Inc., Totowa, NJ

2.1. Biotin-Labeling by PCR

1. *Taq*-DNA polymerase (5000 U/mL, Amersham Pharmacia Biotech, Buckinghamshire, UK)
2. 10X *Taq* DNA polymerase reaction-buffer: 100 mM Tris-HCl, pH 9.0 (25°C), 50 mM KCl, 1.5 mM MgCl$_2$, 0.1% Triton X-100 (v/v), 0.2 mg/mL bovine serum albumin (BSA).
3. 10X Biotin 16-dUTP (B-dUTP) labeling mix:

1 mM dATP	10 μL
1 mM dCTP	10 μL
1 mM dGTP	10 μL
1 mM dTTP	5 μL
1 mM B-dUTP	5 μL

   The foregoing labeling mix (250 μM each dNTP) with a 1:1 ratio of biotin 16-dUTP: dTTP may have to be changed when *Taq*-DNA polymerases from other suppliers are used (*see* **Note 1**). Biotin-16dUTP is purchased from Boehringer Mannheim (Mannheim, Germany). The amount of labeling mix is sufficient for eight 50-μL labeling reactions and yields 1–2 μg of labeled probe (depending on the size of the probe). The labeling mix should be stored in aliquots at –20°C to avoid multiple freeze-thaw cycles (stable for 1 yr). Labeled probes can be stored at 4°C for up to 6 mo.
4. Template and oligonucleotides for the labeling reaction: Purified PCR products or cloned (partial) cDNAs can serve as a template (for standard protocols *see* **refs. 6,7**). For efficient amplification, between 0.1 and 1 ng of template are sufficient. For probe amplification, the specific set of primers (vector derived or insert specific) should be used at the highest possible stringency.

2.2. Electrophoretic Separation of Total RNA in Denaturing Agarose Gel

1. 20X MOPS buffer: 0.4 M MOPS, 0.1 M sodium acetate, pH 7.0, 20 mM EDTA
2. DEPC-treated water: 0.1% Diethyl pyrocarbonate in double distilled water. Stir for at least 2–3 h at room temperature and autoclave at 120°C for 45 min. Working in a fume hood is strictly recommended.
3. Formaldehyde: 37% Solution (13 M), (Sigma Chemical Co., St. Louis, MO). Working in a fume hood is strictly recommended.
4. Formamide: Analytical grade (Sigma).
5. 10X sample buffer: 0.05% Bromphenol blue, 50% glycerol. Store at –20°C.
6. Loading mix (per sample): This solution is prepared fresh: 1 μL 20X MOPS, 3.3 μL formaldehyde, 2 μL sample buffer. Sample volume in total is 20 μL.
7. Agarose: Molecular biology grade.
8. Gel chamber: Any appropriate device. Before the first use, the gel chamber, gel plate, and comb are cleaned with 1% sodium dodecyl sodium (SDS), followed by 0.2 M NaOH and intense rinsing with distilled water.

2.3. Alkaline Blotting Procedure

Stock solutions:

1. 20X SSC: 3 M NaCl, 0.3 M NaOAc, pH 7.0.
2. 1 M NaOH.

"Ready-to-use" solution:

1. Transfer buffer: 5X SSC, 10 mM NaOH; pH is not adjusted and should be around 11.6. The volume needed depends on gel size and buffer reservoir. Standard gels (12 × 15 cm) can be blotted with approx 1 L of transfer buffer.
2. Neutralization buffer: 5X SSC, pH 7.0.

Additional material:

1. Sponge: Soft, narrow-pore sponge, about 2 cm thick. A thoroughly cleaning of the sponge before its first use is strongly recommended: soak the sponge 2–3X in 1% SDS followed by intense watering (until it seems that the SDS is removed). After transfer, the sponge is well watered again and dried at room temperature.
2. Whatman paper: Eight pieces of Whatman No. 1 (Clifton, NJ) paper and about 2 cm paper towels cut to the size of the gel.
3. Nylon membrane: Hybond (Amersham Pharmacia Biotech, Buckinghamshire, UK), Tropilon (Tropix, Bedford, MA), or comparable membrane. Both neutral and positively charged membranes are useful (*see* **Note 2**).

2.4. Hybridization

Stock solutions:

1. 20X SSPE: 3 M NaCl, 0.2 M NaH$_2$PO$_4$, 20 mM Na$_2$ EDTA, pH 7.4.
2. 5 M NaCl.
3. 20% SDS (w/v).
4. 5 µg/mL salmon sperm; shared fragment size should be between 300 and 3000 bp *(6)*.
5. Formamide (*see* **Subheading 2.2., step 4**).
6. Polyethylene glycol 6000 (PEG, solid).

"Ready-to-use" solution:

1. From the foregoing stock solutions, the (pre-)hybridization solution is prepared as follows (note that formamide [here 50%] concentration may be varied according to the stringency required during hybridization): 50% formamide, 5% SDS, 6% PEG 6000 (w/v), 1 M NaCl and 250 µg/mL salmon sperm.
   For 20 mL combine: 10 mL formamide, 5 mL SDS, 4 mL NaCl, 1 mL salmon sperm, and 1.2 g PEG 6000.
   The solution is heated to the temperature necessary for prehybridization. The denatured probe is added to this solution after approx 1 h of prehybridization. Prior to addition, it is denatured in 0.5 mL of prehybrization solution at 95°C for 5 min. Probe concentration should be about 0.1 nM.
   Note:
   a. Salmon sperm has to be denatured for 5–10 min at 95°C prior addition to the heated (pre-)hybridization solution (*see* **Note 3**).
   b. Both the prehybridization and hybridization solution can be reused several times. The solutions are stable for 6 mo at 4°C. Before use, the solution should be denatured again at 85°C and cooled down to the required temperature (e.g., 50°C) before it is added to the blot.
2. Wash buffers:
   Low-stringency wash: 2X SSPE, 0.5% SDS (buffer 1).
   High-stringency wash: 0.2–0.4X SSPE, 0.5% SDS (buffer 2).

2.5. Detection

Stock solutions:

1. 10X PBS: 0.4 M Na$_2$HPO$_4$, 0.1 M NaH$_2$PO$_4$, 1 M NaCl, pH 7.4.
2. 20% SDS (w/v).
3. Casein (alkaline phosphatase, RNase and DNase free, Tropix)

4. Substrate (CSPD, CPD-STAR) and avidin (strept)avidin-alkaline phosphatase conjugate are included in the "Southern light"-kit (Tropix).
5. X-ray film: Kodak XOMAT (Kodak, Rochester, NY), Hyperfilm (Amersham Pharmacia Biotech), or any other film appropriate for chemiluminescent detection (contact supplier).

"Ready-to-use" solutions:

1. Blocking buffer: 1X PBS, 0.5% SDS, 0.1–0.2% casein (w/v, alkaline phosphatase free); has to be freshly prepared. Use a microwave oven or heating block to heat the PBS/casein mixture up to 70°C (500 W, 45 s) for dissolving casein while stirring for 10 min. Before use, the hot solution is cooled down to room temperature (*see* **Note 12**).
2. Conjugation buffer: 1:6000 dilution (has to be freshly prepared) of AVIDx-alkaline phosphatase conjugate (Tropix) in blocking buffer. Note: The AP-conjugate should be centrifuged for 1 min and 4°C at 12,000g to remove aggregated proteins from the solution.
3. Wash buffer: 1X PBS, 0.5% SDS (has to be made fresh).
4. Assay buffer: 0.1 M diethanolamine, pH 10, 1 mM $MgCl_2$. The solution is stable at 4°C for about 1 wk when Na-Azid is added (1 mM).
5. Substrate buffer: 1:100 dilution of CSPD (25 mM stock, Tropix) in assay buffer (has to be freshly prepared and is stable for several hours at 4°C).

3. Methods

3.1. Biotin Labeling of Specific Probes by PCR

When biotin-labeled probes are synthesized by PCR, the annealing temperature and extension time will depend on primer sequence(s) and product length, respectively. For illustration, a representative example is given where an insert ligated into a standard vector (pBluescript SK$^+$, Stratagene, La Jolla, CA) is used as template for synthesis of a biotinylated probe of about 500 bp. The vector-based oligonucleotides of T3 and T7 promoter regions prime in sense and antisense orientation to the left and the right from the multiple cloning site in pBluescript vectors. In principle, with this primer, set probes can be amplified for any insert; however, controls should be run to exclude that a chance priming occurs within the insert of interest.

For a 50-µL reaction combine at room temperature 34.5 µL template (1 pg-1 ng) and water, 5 µL 10X reaction buffer, 5 µL 10X bio-dNTP mix, 2.5 µL sense primer (10 pmol/µL), and 2.5 µL antisense primer (10 pmol/µL).

The PCR program used to amplify probes with T3 and T7 primers is as follows: 94°C—5 min (1X), 94°C—30 s, 55°C—60 s, 72°C—30 s (35X), and 72°C—5 min (1X).

3.2. Electrophoretic Separation of Total RNA in Denaturing Agarose Gel

After isolating total RNA according to standard protocols (*5–7*), the washed and dried RNA is resolved in formamide (*8*) to a concentration of 1–2 µg/µL.

1. Preparation of a 1.5% denaturing agarose-formaldehyde gel: To 78.3 mL distilled water (DEPC treated), 1.5 g agarose were weighted into a 200-mL screw-capped bottle. After melting the agarose in a microwave-oven, 5 mL 20X MOPS and 16.6 mL formaldehyde were added. After intense mixing, pour the gel immediately (under a fume hood). After the gel has solidified, evaporate excess formaldehyde for 20 min.
2. Preparation of the sample: Dilute the samples to 13.7 µL in formamide. Add 6.3 µL of loading mix. Denature the samples for 5–10 min at 65°C and cool in ice water for an additional 5 min.
3. Electrophoresis: Electrophorese at 5–7.5 V/cm gel length in 1X MOPS and 1% formaldehyde (v/v). Separation time depends on the resolution required (approx 2–3 h).

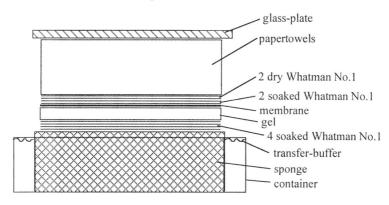

Fig. 1. Assembly for sponge-based capillary transfer of nucleic acids.

3.3. Alkaline Blotting Procedure

After electrophoresis the gel is soaked twice for 10 min in transfer buffer. The assembly for transfer is depicted in **Fig. 1**.

The transfer buffer-soaked sponge (soaking is best done in a beaker) is placed in a clean container filled with transfer buffer. Care should be taken to remove all air bubbles from the sponge surface. Four layers of transfer buffer-soaked Whatman No. 1 blotting paper are placed on top of the sponge, followed by the gel (upside down) and the membrane. On top of the membrane are placed two soaked and two dry Whatman No.1 blotting papers and approx 2 cm of paper towels. The transfer "sandwich" is fixed by a weight (e.g., glass plate) placed on top.

Note: To avoid compression of the gel matrix the weight should not exceed 1–2 g/cm^2 for gels of 1–2%. This increases transfer efficiency significantly. Furthermore, the transfer time is critical and should not exceed 10–15 min/mm of gel thickness (*see* **Note 4**).

After transfer is completed, the membrane is rinsed for 5 min in neutralization buffer and air dried on Whatman paper for 2 min before it is ultraviolet crosslinked. Before starting the prehybridization, the blot has to be dried completely for at least 2–3 h at room temperature (*see* **Note 5**).

3.4. Hybridization

Prehybridization and hybridization are performed in a hybridization oven or a waterbath at the appropriate temperature. Hybridization temperature depends on salt as well as formamide concentration in the solution and homology and GC content between probe and target. Usually hybridization is performed in the range from 37–42°C (which is under standard conditions equal to 60–75% homology between probe and target). Empirical formulas are available to calculate the stringency for specific hybrid formation (*see* **Note 6**).

1. Heat the hybridization solution to the desired temperature.
2. Place the blot into the container or tube and pour the prewarmed hybridization solution over the blot.
3. After 1 h add the denatured probe directly into the prehybridization solution. Hybridization is usually performed overnight at the same temperature as the prehybridization.

4. After overnight hybridization, wash the blot two times at low stringency with buffer 1 for 10 min at room temperature. The final stringency is achieved through a high-stringency wash with buffer 2 at 55–65°C for 30–60 min. The washes must be carried out with gentle shaking at 80–120 rpm (rounds per minute). After this wash the blot is ready for the detection procedure.

3.5. Detection

We present an optimized protocol for the detection of low- to medium-abundant mRNAs in samples of total RNA. For the detection of specific transcripts in poly (A$^+$)-enriched RNA preparations or of abundant transcripts in total RNA, the protocol may be modified (*see* **Note 7**). All incubations should be done on a rotary shaker at low rotation (30–80 rpm, *see* **step 1**) and room temperature.

1. Incubate the blot in blocking buffer twice for 5 min (80 rpm) and once for 30 min (30 rpm). *Note:* excessive blocking reduces the final signal strength, but may improve the signal–noise ratio.
2. Drain the blocking buffer and add the conjugation buffer to the blot. Perform conjugation for 1 h (30 rpm). *Note:* addition of conjugation buffer should be done at one corner of the container and not over the blot surface.
3. After conjugation, transfer the blot to a new container. Allow excess conjugation buffer to drain from its surface. Wash twice with blocking buffer (5 min, 80 rpm) and four times with wash buffer (5 min, 80 rpm).
4. Equilibration to appropriate reaction conditions for alkaline phosphatase is achieved through two washes in assay buffer for 5 min (80 rpm). *Note:* it is essential to remove excess assay buffer before starting the incubation with a substrate buffer to avoid dilution of the latter. This is best done by placing the blot upside-up for 2 min on two layers of assay-buffer-wetted Whatman no. 1 paper.
5. While the blot is lying on the moist Whatman papers, add the substrate buffer to the empty container or a clean glass plate (approx 10–20 µL/cm^2). Subsequently, place the blot upside down on the liquid film of substrate buffer and incubate for 10 min (to avoid partial drying of edges, the blot should be covered by a plastic foil). *Note:* at this step care should be taken that no air bubbles are present at the incubated surface of the blot. Finally excess substrate buffer is removed from the membrane as described in **step 4**.
6. For film exposure place the moist membrane on an assay-buffer-wetted Whatman No. 1 paper between two plastic covers (we use cooking foil, Melitta [Minden, Germany] *see* **Note 8**). Place the covered membrane onto a film exposure cassette and expose X-ray film for 1–2 h (*see* **Note 7**).

Flowsheet for the complete method:
3 h	electrophoretic separation (including casting and running time)
3.5 h	alkaline blotting (including incubation in transfer buffer and transfer and drying of the membrane)
1.5 h	prehybridization (including preparing a fresh solution)
12–18 h	hybridization
4–4.5 h	detection (including the subsequent listed steps)

Detection procedure:
20 min	low-stringency wash (2X 10 min)
30–60 min	high-stringency wash
40 min	blocking (2X 5 min and 1X 30 min incubations)

60 min	conjugation
30 min	washes (2X in blocking buffer and 4X in wash buffer, 5 min each)
10 min	equilibration (2X 5 min in assay buffer)
15 min	substrate incubation (including removing excess buffer before and after incubation)
60–120 min	film exposure

4. Notes

1. Using a different *Taq* DNA-polymerase and labeled nucleotides may require an optimization of label incorporation due to enzyme-specific discrimination. This is done by changing the ratios of the labeled and the replaced nucleotide (e.g., dTTP: Bio-dUTP) while keeping the remaining dNTP mixture constant. Mostly, a 2:1 ratio independent of the nature of label attached to the nucleotide will work. High incorporation of the label does not necessarily yield the most sensitive probe because there may occur a steric hindrance in subsequent detection. Therefore, the sensitivity of the probe has to be determined on dot- or slot-blot before using it for Northern blot. For successful Northern blot detection of low-abundant mRNA 1 pg (or below) of target should be detected.

2. The type of membrane used for transfer does not seem to be critical; however, from our experience the use of neutral nylon membranes yields a better signal-to-noise ratio especially if longer film exposures (e.g., above 2 h) are required.

3. Probe concentration is not critical within the range of 0.5–2.0 n*M*. Note that higher probe concentrations allow to shorten the hybridization time. However, this will also lead to a higher background. Probe concentrations lower than the given (*see* **Subheading 2.4.**) tend to need a longer hybridization time to achieve equal signal strength.

4. For unknown reason(s), transfer time is very critical in the described procedure. For example, after 3 h of blotting, signals that were easily detectable after 1 h of blotting are very weak or not detectable. This is not due to nonspecific RNA hydrolysis (resulting from the alkaline conditions during transfer) because the abundant rRNAs are detectable with methylene blue or ethidium bromide staining on membrane.

    If longer transfer time for a complete transfer is required, e.g., if the distance between the bottom of the slot and the blotting surface of the gel is higher than 0.8–1 mm, more than one membrane should be used. For example, the first membrane can be used as a backup of the blot (usually sufficient for the detection of abundant mRNAs or rRNAs), whereas the second membrane carries most of the transferred RNA and therefore is used for the detection of the desired target. Blotting time for each membrane has to be carried out individually but should not exceed 1.5 h.

5. Blot drying after UV crosslinking increases signal strength and lowers the background significantly. One reason may be that it is necessary to evaporate residual formaldehyde from the membrane. For best results, we dry the blot over night and start hybridization the next day. Alternatively, the blots may be dried at a moderate temperature (e.g., 37°C) for 2–3 h. Recently, we found a second alternative: after UV crosslinking the blot is incubated in a solution of 1% NH_4OH/5X SSC for 20–30 min at room temperature, followed by two washes in 5X SSC for 10 min. After this treatment no background appears on the blots. Furthermore, we observed a slight increase in signal strength compared to an overnight dried blot.

6. Calculation of stringency empirically follows formulas, taking into account the different hybrid stabilities of DNA-DNA and DNA-RNA and RNA-RNA hybrid molecules: DNA-DNA hybrid (*6*):

$$T_m = 81.5 + 16.6 * \log[Na^+] + 0.4 * (\% \, GC) - 0.6 * (\% \, \text{formamide}) - (600/N)$$

DNA-RNA hybrid *(9)*:

$T_m = 79.8 + 18.5 \log [Na^+] + 0.58 (\% GC) + 11.8 (\%GC)^2 - 0.35 (\% \text{formamide}) - (820/L)$

where T_m = melting point of a given hybrid; % *GC* = percent of these nucleotides in DNA molecules; *N/L* = length of the hybrid molecule in bp; $[Na^+]$ = molar concentration of Na^+; and % formamide = percent of formamide in the hybridization solution.

7. Target abundance determines the sensitivity required for detection. For abundant mRNAs hybridization time may be reduced. The detection procedure can also be shortened: the conjugation could be done in 20–30 min, and substrate incubation of 5 min is sufficient. Film exposure for 18S rRNA requires less than 10 min.
   Note: CPD-STAR as an alternative substrate to CSPD requires much shorter exposure times. For example: a 1-h exposure after incubation with CSPD is shortened to 5–10 min with the CPD-STAR.
8. The plastic for exposure has to fulfill two conditions. First, it should not absorb at the emission wavelength of chemiluminescent substrate, and second, it should withstand the substrate buffer. Both should be tested before use.
9. Reduction of background caused by unspecific binding of probe could be achieved with RNase digest after hybridization. Note that the blot is no longer useful for further hybridizations. It is also possible to alter hybridization and washing stringency.
10. In case of spotty background there are different explanations possible:
    a. The AP conjugate was not centrifuged before dilution into blocking buffer.
    b. The AP conjugate is too old.
    c. Blocking was insufficient.
    d. Temperature during conjugation step was too high.
11. In case of a cloudy background, there are different explanations possible:
    a. Casein may be too old.
    b. Temperature during film exposure was above 30°C.
12. The same blotting and detection procedures are routinely used in our lab for Southern blot analysis *(4)*. Stringency conditions required during hybridization and washing are calculated according to the formula given in **step 6**.

References

1. Lanzillo, J. J. (1991) Chemiluminescent nucleic acid detection with digoxigenin-labelled probes: a model system with probes for angiotensin converting enzyme which detect less than one attomole of target DNA. *Anal. Biochem.* **194,** 45–53.
2. Klevan, L. and Gebeyehu, G. (1990) Biotinylated nucleotides for labelling and detection of DNA. *Methods Enzymol.* **184,** 561–577.
3. Löw, R. and Rausch, T. (1994) Sensitive non-radioactive northern blots using alkaline transfer of total RNA and PCR amplified biotinylated probes. *BioTechniques* **17,** 1026–1030.
4. Löw, R. and Rausch, T. (1996) Non-radioactive detection of nucleic acids with biotinylated probes, in *A Laboratory Guide to Biotin-Labelling in Protein and Nucleic Acid Analysis* (Meier, T. and Fahrenholz, F., eds.), Birkhäuser Verlag, Basel, pp. 201–213.
5. Logemann, J., Schell, J., and Willmitzer, L. (1987) Improved method for the isolation of RNA from plant tissues. *Anal. Biochem.* **163,** 16–20.
6. Maniatis, T., Fritsch, E. F., and Sambrook, J. (1989) *Molecular Cloning—A Laboratory Manual*. Cold Spring Harbor Lab. Press, Cold Spring Harbor, NY, Chapters 7.6 and 9.51, pp. 435ff.

7. Ausubel, F., Brent, R., Kingston, R. E., Moore, D. D., Seidman, J. G., Smith, J. A., and Struhl, K. (1987–1995) *Current Protocols in Molecular Biology*, vol. 1, Greene Publishing and Wiley, New York, Appendix A.2.3.
8. Chomczynski, P. (1992) Solubilization in formamide protects RNA from degradation. *Nucleic Acids Res.* **28,** 3791.
9. Casey, J. and Davidson, N. (1977) Rates of formation and thermal stability of RNA:RNA and DNA:DNA duplexes at high concentrations of formamide. *Nucleic Acids Res.* **4,** 1539–1552.

38

Analysis of RNA by Northern Blotting Using Riboprobes

Rai Ajit K. Srivastava

1. Introduction

In studying the expression of a gene in mammalian tissues, it is important to determine the levels of the corresponding mRNA. Several methods have been described for measuring the level of expression of a gene in a tissue. These methods are *in situ* hybridization using cDNA probe or riboprobe *(1)*, slot-blot hybridization on total RNA isolated from tissues *(2)*, and Northern blotting hybridization using either a cDNA *(3)* or a riboprobe *(4)*. Absolute levels of a specific mRNA are also determined by RNase protection assay using cDNA probe *(5)* or a riboprobe *(6)*. *In situ* hybridization provides relative expression of a gene, whereas the slot-blot technique is not sensitive enough for accurate measurements of mRNA. The most widely used technique to measure mRNA levels in mammalian tissues is still Northern blotting analysis, in which total RNA or poly(A$^+$) RNA is separated by electrophoresis in an agarose gel-containing formaldehyde and transferred onto a nitrocellulose or nylon membrane. The transferred RNA on the membrane is then denatured and probed with a labeled cDNA probe or a riboprobe. Although the cDNA probe gives the same information as the riboprobe, the sensitivity of the detection of mRNA is increased severalfold by using a riboprobe *(4,* and **Fig. 1**) because of the increased affinity of riboprobe for the complementary sense strand of mRNA and higher stability of the double-stranded RNA after hybridization. However, the sensitivity of various membranes may differ marginally while a riboprobe is used *(4)*. Thus, the higher sensitivity of the riboprobes in Northern blotting analysis makes them a better choice over cDNA probes especially when detecting a low-abundance message. However, care must be taken while using riboprobes for Northern blotting analysis, as improper use of riboprobe may result in a very high background often by irreversibly binding the riboprobe to the membrane. Therefore, the steps shown in this chapter should be strictly followed in order to avoid background noise.

2. Materials

For RNA analysis, all glassware to be used should be treated properly with 0.05% diethylaminoethyl pyrocarbonate (DEPC) in a hood overnight followed by autoclaving. Deionized water used for preparing solutions or for use in any reaction should be

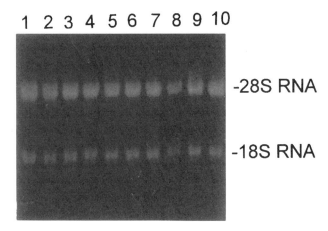

Fig. 1. Ethidium bromide-stained gel of total RNA prepared by using the method described. Fifteen micrograms of 10 different RNA samples were prepared and electrophoresed. The positions of 28S and 18S RNA are marked.

similarly treated. A pipet used for RNase should be avoided while working with RNA samples. In addition to the solutions mentioned in Section 3 following solutions are made and stored appropriately: 4 M guanidinium isothiocyanate, 1 M sodium acetate, pH 7.0, β-mercaptoethanol, 5% sarcosyl, 2 M sodium acetate, pH 4.0, 10% SDS, isopropanol, ethanol, appropriate restriction enzymes, dithiothreitol, ribonucleotides, ribonuclease inhibitor (recombinant RNasin or human placental ribonuclease inhibitor), T7 and SP6 RNA polymerases, α-^{32}P[CTP], RNase-free DNase I, phenol, chloroform, isoamylalcohol, 3 M sodium acetate, pH 4.8, deionized formamide, Sephadex G-50 column, 5 M NaCl, nonfat dry milk, salmon sperm DNA, 1 M MgCl$_2$, 0.5 M ethylenediaminetetraacetic acid (EDTA), 10X MOPS (morpholinopropanesulfonic acid sodium salt), 20X standard sodium citrate (SSC), molecular biology-grade agarose, formaldehyde, and nitrocellulose or nylon membrane.

2.1. Preparation of Cellular RNA

1. Animal tissue samples or cultured cell line: The animal tissues and cultured cells to be used for RNA preparation should be fresh or quick frozen in liquid nitrogen and stored at –70°C.
2. Solution A: 4 M Guanidinium isothiocyanate, 25 mM sodium acetate, pH 7.0, 0.1 M mercaptoethanol. Make this solution fresh from stock solutions every time before use.
3. Solution B: equal volume of 5% sarcosyl and 2 M sodium acetate, pH 4.0. Make fresh every time.
4. Salt-saturated phenol: Prepare 1 M salt-saturated phenol and store at 4°C in a brown bottle to protect from light.
5. Chloroform/isoamylalcohol: Mix chloroform and isoamylalcohol in a 49:1 ratio just before use.
6. Isopropanol: Molecular biology-grade isopropanol.
7. 70% Ethanol: Prepare 70% ethanol using DEPC-treated water and always keep it chilled at –20°C freezer.
8. SDS solution: Prepare 0.2% SDS solution and store at room temperature.

2.2. Preparation of Riboprobe

1. Template plasmid: The cDNA fragment to be used should be subcloned in the polylinker region of a vector. The polylinker region should be flanked by T7 and SP6 RNA polymerase promoters. The recombinant plasmid should be linearized with appropriate restriction enzyme and purified. *See* **Note 1**.
2. 10X Transcription buffer: 400 m*M* Tris HCl (pH 7.5), 60 m*M* MgCl$_2$, 100 m*M* NaCl, and 20 m*M* spermidine. Store in aliquots of 100 µL at –20°C.
3. Nucleotide solution: 10 m*M* solution of each NTP (can be purchased, *see* **Note 2**). Store at –20°C.
4. Nucleotide mix: Prepare nucleotide mix as follows: 1:1:1 ratio mix of adenosine triphosphate (ATP), guanosine triphosphate (GTP), and uridine triphosphate (UTP). Store at –20°C.
5. TCA solution: 10% Trichloroacetic acid, 1.5% sodium pyrophosphate. Store at 4°C.

2.3. Agarose Gel Electrophoresis

1. Agarose: Use molecular biology grade agarose.
2. Deionized formamide: Take 20 mL formamide in an orange cap tube, add 4 g mixed bed resin, and stir for at least 2 h until the color of the resin changes. Aliquot and store at –20°C.
3. 10X MOPS: 0.2 *M* MOPS, 0.05 *M* sodium acetate, 0.01 *M* EDTA. Adjust pH to 7.0 with 1 *M* NaOH. Filter sterilize and store in the refrigerator in the dark.
4. Formaldehyde: Use commercially supplied formaldehyde that is 37% formaldehyde.
5. Ethidium bromide. Prepare 0.5 mg/mL ethidium bromide in DEPC-treated water. Store refrigerated in a brown bottle.
6. Gel loading dye: Bromophenol blue 0.001% and xylene cyanol 0.001% in DEPC-treated water.
7. 20X SSC: 3 *M* NaCl, 0.3 *M* sodium citrate. Adjust pH to 7.0 with HCl. Sterilize by autoclaving.
8. RNA sample buffer: Fifty percent deionized formamide, 6% formaldehyde, 1X MOPS.
9. Membrane: Use a nitrocellulose or nylon membrane.

2.4. Hybridization

Prehybridization and hybridization buffer: 100 m*M* Tris HCl, pH 7.4, 10 m*M* EDTA, 600 m*M* NaCl, 1.5% nonfat dry milk (purchased from the supermarket), 50% formamide. Filter through 0.45 µm a nitrocellulose filter and incubate overnight at 70°C (*see* **Note 3**). Then add 10% SDS to bring final concentration of SDS to 1%.

3. Methods
3.1. Isolation of Cellular RNA

1. Homogenize tissue samples (0.2 g) with 1 mL of solution A. When using cultured cells, wash cells with phosphate buffered saline (PBS) and add 1 mL of solution A for each 3–5 × 10^6 cells. With the help of a 1-mL pipet, mix cells with solution A thoroughly by pipetting in and out several times.
2. To the homogenized tissue samples or the cultured cells add 300 µL of solution B and 1 mL salt-saturated phenol followed by vigorous vortexing for 15 s.
3. Add 250 µL chloroform:isoamylalcohol (49:1) and vortex vigorously for 15 s.
4. Allow the mixture to sit on an ice bath for 5 min, and centrifuge at 12,000*g* for 15 min at 4°C.
5. Remove the upper aqueous layer carefully to another tube, avoiding turbid materials from the interface (*see* **Note 4**).
6. Reextract the supernatants with equal volume of phenol/chloroform, and transfer the upper layer to another tube.

7. Add an equal volume of isopropanol and leave on the ice bath for 30 min or at –20°C for 15 min and pellet RNA by centrifugation at 12,000g for 15 min (see **Note 5**).
8. Wash the RNA precipitate with 500 μL of chilled 70% ethanol.
9. Dry RNA for 5 min in a Speed vac (Savant Instruments, Holbrook, NY).
10. Dissolve dried RNA samples in 0.2% SDS and determine the A_{260}/A_{280} of appropriately diluted RNA sample, which should usually be more than 1.8 (see **Note 6**). After determining the concentration of RNA, store in aliquots at –70°C (see **Note 7**).

3.2. In Vitro Synthesis of a Riboprobe

1. For the in vitro synthesis of a riboprobe, usually 0.5–1 μg of the linearized and purified template recombinant plasmid is used (see **Note 8**). To a properly treated 1.5 mL Eppendorf tube add the following ingredients at room temperature in the following order: 3 μL of 10× transcription buffer, 3 μL of 0.1 M dithiothreitol (DTT), 1 μL of RNasin (ribonuclease inhibitor), 3 μL of nucleotide mix, 5 μL of linearized plasmid, 2.5 μL of ^{32}P[CTP] (cytosine triphosphate) (Amersham Pharmacia Biotech, 20 mCi/mL), sterile water to make the total volume to 30 μL, and 1 μL of T7, SP6, or T3 RNA polymerase (see **Note 9**). Mix the contents by brief centrifugation on a tabletop microcentrifuge. Incubate at 37°C for 1 h.
2. Stop the transcription reaction by adding 1 L of RNase-free DNase I (RQ1 from Promega [Madison, WI] or other RNase-free DNase I works fine). Incubate at 37°C for an additional 15 min.
3. Add 1 L of RNasin and mix by brief vortexing.
4. Remove unincorporated radionucleotide by Sephadex G-50 column as described in here. Both the DNA as well as RNA columns from BMB or from other companies work fine (see **Note 10**). Drain the buffer off from the Sephadex column by gravity in a clean tube supplied with the column.
5. Centrifuge the column at 200 g for 2 min in a swinging bucket rotor to drain off the remaining buffer of the column. Discard the tube in which buffer collected.
6. Place the Sephadex column in another clean tube supplied with the column and carefully apply the transcribed riboprobe to the center of the Sephadex beads (see **Note 11**).
7. Centrifuge again at 200 g for 2 min. Riboprobe collects in the clean tube.
8. Determine the quality of the riboprobe (see **Note 12**).
9. Use the riboprobe fresh or keep it stored in aliquots at –70°C (see **Note 13**).

3.3. Electrophoresis and Transfer of RNA

1. For a 11 × 14-cm gel, prepare agarose in a 250-mL Erlenmeyer flask by melting 1.2 g of agarose in 73 mL of DEPC-treated water in a microwave.
2. Cool the contents to 50–55°C and add 10 mL of prewarmed (37°C) 10X MOPS buffer and 16.2 mL of 37% formaldehyde.
3. Mix gently, but thoroughly, and pour into the gel rig, which has tape on both ends and a comb.
4. Remove the bubbles, if any, with a sterile pasture pipet.
5. Let the gel solidify for at least 20 min.
6. Take 15 μg of total RNA in an Eppendorf tube and add 15 μL of RNA sample buffer. Incubate at 60°C for 15 min.
7. Cool on ice for 5 min, add 2 L of gel loading dye and mix.
8. Centrifuge briefly in a microcentrifuge to collect samples at the bottom of the tube.
9. Prepare 1X MOPS in DEPC-treated water, and prerun the gel for 20 min at 80 V.
10. Load the RNA sample prepared as in **step 6** above in the gel slots, and run at constant voltage of 90 for 3–4 h or till the tracking blue dye reached 1 in. from the bottom of the gel.

11. After the electrophoresis, stain the gel either with ethidium bromide or with acridine orange to visualize 28S and 18S ribosomal RNA (*see* **Note 14**).
12. Cut a piece of GenScreen membrane (HEN® Research Products, Boston, MA) of approximate size of the gel, and soak in 10X SSC buffer.
13. Set up the transfer using the standard method of capillary transfer. Soak a Whatman paper (Whatman, Maidstone, Kent, UK) and put it on the plate so that both ends of the Whatman paper are sitting in the buffer.
14. Flip the gel over so that the loading side faces downward (*see* **Note 15**) and place on the Whatman paper wick. Make a nick on the one corner of the gel to mark lane 1. Place the membrane on top of the gel, and remove all air bubbles with the help of a 10-mL pipet by rolling from one side to another.
15. Put the Whatman paper on the top of the membrane followed by a stack of paper towels (usually 3 in. high).
16. Place a small glass plate on top of the paper towel followed by a 200–300-g weight.
17. Let the transfer proceed overnight.
18. Next morning, take out the membrane and rinse with 2X SSC. Check the gel and the membrane under UV light for complete transfer of the RNA.
19. Take a picture of the membrane and mark 28S and 18S ribosomal RNA on the membrane with a pencil as shown in **Fig. 1**.
20. Bake the membrane at 80°C for 2 h in a vacuum oven, and put in a sealable plastic bag.

3.4. Hybridization

1. For a 11 × 13 cm membrane take 12 mL of hybridization buffer and mix with denatured salmon sperm DNA (150 g/mL) (*see* **Note 16**).
2. Pour the prehybridization buffer in the bag containing the membrane, and seal the bag avoiding any entrapment of bubbles (*see* **Note 17**).
3. Place the blot in a shaking water bath at 60°C for 3 h.
4. After prehybridization is over, heat the riboprobe ($3–5 \times 10^6$ cpm) together with 100 µL of salmon sperm DNA (5 mg/mL) at 90°C for 5 min and cool in an ice bath for 5 min.
5. Add denatured riboprobe in the hybridization bag and seal the bag, avoiding any entrapment of bubbles.
6. Allow hybridization to proceed overnight at 60°C.
7. Next day, cut open the bag and pour the radioactive contents of the bag in a radioactive waste disposal container. Add 20 mL of 1X SSC/0.2% sodium dodecyl sulfate (SDS) in the hybridization bag; after rinsing the membrane discard the solution in the radioactive waste disposal container.
8. Wash with 200 mL of 1X SSC/0.2% SDS at room temperature for 10 min two times.
9. Wash with 200 mL of 0.1X SSC/0.1% SDS for 30 min at 65°C twice.
10. Rinse with 100 mL of 0.1X SSC and let it dry at room temperature. Do not dry completely.
11. Cover the membrane with Saran Wrap and expose to X-ray film.
12. Blot can be reprobed (**Note 18**).

4. Notes

1. Several plasmid vectors are available today that have two different RNA polymerase promoters flanking the polylinker region in which the desired cDNA fragment is subcloned for the purpose of synthesizing RNA transcripts. Some vectors, such as the pGEM3Z series, have T7 and SP6 RNA polymerase promoters, and some vectors, such as BlueScript, have T7 and T3 RNA polymerase promoters. The orientation of the cDNA insert in a vector should be determined by sequencing the double stranded plasmid (*7*) in order to determine where to linearize the recombinant plasmid for riboprobe synthesis (*see* **Fig. 2**).

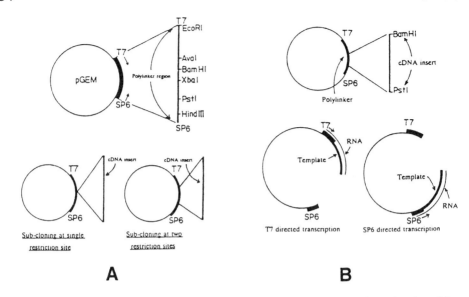

Fig. 2. **(A)** Subcloning of the cDNA fragment into the polylinker region of a plasmid vector. The fragment is shown subcloning either into a single restriction enzyme site or two different sites. **(B)** Linearization of recombinant plasmid for in vitro transcription to synthesize riboprobe.

2. Nucleotide solutions can be purchased from Promega (Madison, WI), Ambion, or Roche Diagnostics (Indianapolis, IN). All these solutions have a pH of 7.5.
3. Alternatively, the prehybridization solution can be treated for 1 h at 90°C.
4. While the upper layer containing RNA is transferred, care must be taken to avoid taking any material from the interface because of the presence of proteins that also contain ribonucleases. For safety, one should take only two-thirds of the upper layer slowly and without disturbing the interface.
5. Precipitating RNA with isopropanol at 4°C for overnight gives a better yield because of the complete precipitation of all sizes of RNA.
6. Instead of 0.2% SDS, the RNA can be dissolved in DEPC-treated deionized water. If the A_{260}/A_{280} ratios of RNA samples are lower than 1.8, it suggests that the RNA is contaminated with proteins or ribonucleoprotein complexes *(8)*. It has been noticed that the method described here does not require proteinase K digestion for separating RNA from the ribonucleoprotein complexes. The quality of the RNA is shown in **Fig. 1**.
7. The quality of the RNA prepared can be checked by performing agarose gel electrophoresis as shown in **Fig. 2**. The ratios of 28S RNA to 18S RNA should be usually 2:1. When degradation of RNA occurs, the ratio decreases.
8. After linearization of the recombinant plasmid by appropriate restriction endonuclease, it should be purified by phenol-chloroform extraction and ethanol precipitation and dissolved in DEPC-treated water.
9. Check the recombinant vector for the appropriate RNA polymerase promoter flanking the polylinker region. RNA polymerases can be purchased from Roche Diagnostics, Promega, Bethesda Research Lab, or other companies.
10. BMB provides Sephadex G-25 and G-50 columns exclusively for RNA, although a DNA column also works fine. Sephadex column from other companies work equally well.
11. After draining the buffer from the Sephadex column by centrifugation, the gel beads become one solid aggregate. The riboprobe reaction mixture should be applied to the cen-

Analysis of RNA by Northern Blotting

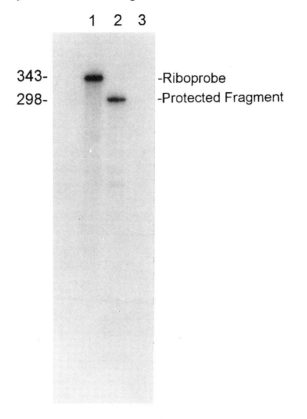

Fig. 3. Determining the quality of the riboprobe. A 298-bp mouse apoE riboprobe was subcloned into the *Pst*I site of the pGEM3Zf(+) vector and linearized with *Hind*III to prepare the riboprobe. Lane 1, the riboprobe; lane 2, the riboprobe hybridized with 5 μg of total mouse liver RNA; lane 3, the riboprobe hybridized with 5 μg of yeast tRNA.

ter of the aggregate by avoiding any contact with the column's sides. This will allow all of the reaction mixture to go through the beads at the time of centrifugation; this will yield a pure riboprobe.

12. The quality of the riboprobe can be determined by running a sequencing gel and observing an appropriate size of one single band of riboprobe, and after hybridization with the total RNA it should protect an appropriate size RNA as shown in **Fig. 3**. Alternatively, the riboprobe can be precipitated with 10% TCA and filtered through a glass fiber filter and the filter counted. There should not be more than 10% difference between the initial riboprobe count and counts obtained after TCA precipitation.
13. After synthesis the riboprobe can be stored frozen at –70°C for up to 2 wk without any loss in the quality.
14. It is always desirable to stain the gel with acridine orange and then destain to visualize 28S and 18S RNA, as acridine orange does not interfere with RNA transfer. If the gel is stained with ethidium bromide, proper destaining avoids any problem of RNA transfer.
15. It has been consistently observed that the transfer of RNA is always complete when flipping over the gel. This is particularly important if the gel has been stained with ethidium bromide.
16. Denature salmon sperm DNA by boiling for 10 min. A stock of salmon sperm DNA can be prepared and stored in aliquots at –20°C.

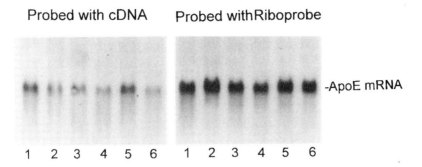

Fig. 4. Northern blot analysis using a riboprobe. Six RNA samples (15 µg each) from mouse liver were electrophoresed on the same gel in two replicates and separated by cutting the membrane from the middle after transfer for hybridization with mouse *apo*AI cDNA probe (left panel) or with mouse apoAI riboprobe (right panel). For hybridization with the riboprobe, 1% nonfat dry milk was used to block the membrane.

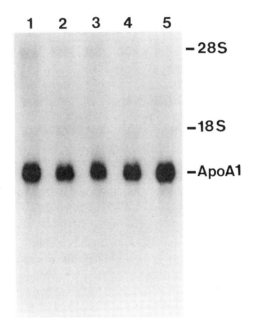

Fig. 5. Northern blot analysis using a riboprobe. Mouse RNA was electrophoresed as described in the legend of **Fig. 1** and probed with mouse apoAI riboprobe using 1.5% nonfat dry milk for blocking the membrane.

17. One percent nonfat dry milk has been used by others for Northern blotting analysis using a riboprobe. However, using 1.5% nonfat dry milk gives less background as compared to 1% nonfat dry milk. The Northern blot analysis shown in **Fig. 4** was performed with 1% nonfat dry milk, and that shown in **Fig. 5** with 1.5 % nonfat dry milk. It is seen that blocking the membrane with 1.5% nonfat dry milk gives a lower background.
18. Unlike the use of the cDNA probe, use of riboprobes in the Northern blotting analysis makes it difficult to reprobe the membrane for the mRNA of similar sizes, as riboprobes are not stripped off completely from the membrane. This is true even after boiling the

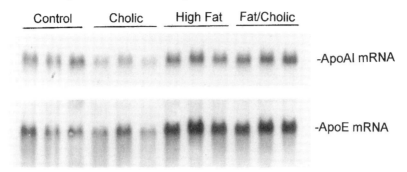

Fig. 6. Reprobing a blot with a riboprobe. The blot shown here was first probed with an apoE riboprobe to detect apoE mRNA (1800 nt), stripped off (stripping does not completely remove apoE riboprobe), and then probed with an apoAI riboprobe to detect apoAI mRNA (1100 nt). Groups of mice were fed the diets indicated in the figure. The hepatic RNA was isolated and analyzed by Northern blotting.

membrane, especially when the concentrations of the target mRNA are very high (4). However, different sizes of mRNA can be reprobed on the same blot after stripping off, as shown in **Fig. 6**.

References

1. Polacek, D., and Becman, M. W., and Schrieber, J. R. (1992) Rat ovarian apolipoprotein E: localization and gonadotropic control of messenger RNA. *Biol. Reprod.* **46,** 65–72.
2. Davis, L. G., Dibner, M. D., and Battey, J. F. (1994) Preparation and analysis of RNA from eukaryotic cells, in *Basic Methods in Molecular Biology,* Elsevier, New York, pp. 319–329.
3. Twoney, T. A. and Krawetz, S. A. (1990) Parameters affecting hybridization of nucleic acids blotted onto nitrocellulose membrane. *BioTechniques* **8,** 478–481.
4. Srivastava, R. A. K. and Schonfeld, G. (1991) Using riboprobes for Northern blotting analysis. *BioTechniques* **11,** 584–587.
5. Williams, D. L., Newman, T. C., Shelnes, G. S., and Gordon, D. A. (1986) Measurements of apolipoprotein E mRNA by DNA-excess solution hybridization with single-stranded probes. *Methods Enzymol.* **128,** 671–689.
6. Srivastava, R. A. K., Pfleger, B., and Schonfeld, G. (1991) Expression of low density lipoprotein receptor, apolipoprotein AI and apolipoprotein AIV mRNA in various mouse organs as determined by a novel RNA-excess solution hybridization assay. *Biochim. Biophys. Acta* **1090,** 95–101.
7. Mierendorf, R. C. and Pfeffer, D. C. (1987) Direct sequencing of denatured plasmid DNA. *Methods Enzymol.* **152,** 556–562.
8. Srivastava, R. A. K., Srivastava, N., and Schonfeld, G. (1992) Expression of low density lipoprotein receptor, apolipoprotein AI, AII, and AIV in various rat organs utilizing an efficient and rapid method for RNA isolation. *Biochem Intern.* **27,** 85–95.

V

GENE LIBRARY CONSTRUCTION AND SCREENING

39

Production of Double-Stranded cDNA for Gene Library Synthesis

Jane Kirk and Steve Mayall

1. Introduction

The synthesis of complementary DNA (cDNA) from an mRNA template by the action of reverse transcriptase is a fundamental technique in molecular cloning. The objective is to make a large number of long cDNA copies. The quality and length of the cDNA product largely depends on the quality of the mRNA used as the starting material. Great care must therefore be taken to avoid mRNA degradation. The choice of primers and RTase used during the synthesis also influences the final product.

A short, double-stranded region is generated by annealing a primer to the mRNA. The DNA strand complementary to the RNA is then generated by the action of avian myeloblastosis virus (AMV) or Moloney murine leukemia virus (MoMuLV) reverse transcriptase. DNA polymerase I used in combination with RNase H replaces the mRNA in the hybrid with small regions of newly synthesized DNA *(1,2)*. The efficiency of cDNA synthesis is measured by the incorporation of radioactive nucleotides into the first and second strands.

The short stretches of double-stranded DNA are ligated into a continuous strand by T4 DNA ligase. At the same time, linkers containing *Eco*RI sites are ligated onto the blunt ends of the cDNA and then digested with *Eco*RI to allow cloning into any similarly cut vector. *Eco*RI sites within the cDNA itself are made resistant to cleavage by prior methylation.

A spermine precipitation in the presence of salt removes a large proportion of the unincorporated linkers *(3)*. Those linkers remaining are removed by a gel purification step, which also permits size selection of the cDNA. The cDNA is then ready to be ligated to a vector of choice.

2. Materials

All solutions should be autoclaved and, where possible, treated overnight at 37°C with 0.01% diethyl pyrocarbonate (DEPC) prior to autoclaving. Do not add DEPC to solutions containing Tris.

All glassware should be DEPC treated and/or baked at 180°C for 3 h. Ideally, sterile, disposable plasticware should be used wherever possible. Single-stranded nucleic acids stick to plastic very readily, and hence tubes should be siliconized before use. Gloves must be worn at all times to preclude the introduction of contaminating RNases. Great care should also be taken to avoid exogeneous DNA contamination. All enzymes and buffers should be stored at –20°C unless indicated otherwise.

2.1. cDNA Synthesis Reactions

1. 10X First-strand synthesis buffer: 500 mM Tris-HCl pH 8.3, 500 mM KCl, 100 mM MgCl$_2$, 10 mM dithiothreitol (DTT). Store at –20°C.
2. Sodium pyrophosphate (NaPPi, 150 mM): Available from Sigma (St. Louis, MO).
3. Human placental RNase inhibitor (100 U/µL): Available from Amersham Life Sciences (Amersham, UK). Store at –20°C.
4. 10X dNTP mix: A solution of dATP, dTTP, dCTP, and dGTP (10 mM each), available from Pharmacia (Uppsala, Sweden). Store at –20°C.
5. Oligo dT$_{12-18}$ primer or random hexanucleotide primers (1 mg/mL): Available from Promega (Madison, WI). Store at –20°C.
6. [α-^{32}P]dCTP (3000Ci/mmole): Available from Amersham. Store at –20°C.
7. Poly (A)$^+$ RNA (200 µg/mL): Stored in DEPC-treated water at –70°C.
8. AMV RTase (20 U/µL): Available from Amersham. Store at –20°C.
9. 2X Second-strand synthesis buffer: 50 mM Tris-HCl pH 8.3, 200 mM KCl, 10 mM MgCl$_2$, 10 mM DTT. Store at –20°C.
10. RNase H (5 U/µL): Available from Amersham. Store at –20°C.
11. DNA polymerase I (5 U/µL): Available from Amersham. Store at –20°C.
12. T4 DNA polymerase (4 U/µL): Available from Amersham. Store at –20°C.
13. Equilibrated phenol: Saturate with 100 mM Tris-HCl pH 8.0 as described *(4)*. Mix with chloroform (1:1 v/v) and store at 4°C. Phenol is available from Rathbone (Walkerburn, Scotland).
14. Saturated chloroform: Saturate with TE buffer (10 mM Tris-HCl pH 8.0, 1 mM ethylenediaminetetraacetic acid [EDTA]) and store at room temperature.
15. DE-81 discs: Available from Whatman (Maidstone, UK).

2.2. Addition of EcoRI Linkers to the cDNA Ends

1. 5X Buffer M: 500 mM Tris-HCl pH 8.0, 500 mM NaCl, 5 mM EDTA. Store at –20°C.
2. 10X Buffer SAM: 800 µM S-adenosyl methionine available from Sigma. Store at –20°C.
3. *Eco*RI methylase (20 U/µL): Available from NEB Labs (Beverly, MA) or Amersham. Store at –20°C.
4. 10X Ligase buffer: 200 mM Tris-HCl pH 7.5, 150 mM MgCl$_2$, 100 mM DTT, 5 mM ATP. Store at –20°C.
5. Phosphorylated *Eco*RI linkers (150 pmol/µL): Available from NEB Labs. Store at –20°C.
6. T4 DNA ligase (400 U/µL): Available from NEB Labs. Store at –20°C.
7. 10X *Eco*RI buffer: 500 mM NaCl, 100 mM Tris-HCl pH 7.5, 100 mM MgCl$_2$, 10 mM DTT. Store at –20°C.
8. *Eco*RI (20 U/µL): Available from NEB Labs or Amersham. Store at –20°C.

2.3. Removal of Unincorporated Linkers and Size Selection of cDNA

1. Spermine wash buffer: 70% Ethanol, 10 mM MgCl$_2$, 0.3 M NaOAc pH 7.0. Store at 4°C.
2. Low-melting-point agarose: FMC SeaPlaque agarose: Available from Flowgen Bioscience (Sittingbourne, UK).
3. 50X TAE buffer: 2 M Tris-acetate, 0.05 M EDTA.

Table 1
First-Strand Synthesis Mix

10X First-strand buffer	2 μL
NaPPi	1 μL
HP RNase inhibitor	1 μL
10X dNTP mix	2 μL
Oligo dT or random primer (see **Note 2**)	1 μL
[α-^{32}P]dCTP	0.5 μL
poly (A)$^+$ RNA (see **Note 3**)	5 μL
H$_2$O	7.5 μL

Table 2
Second-Strand Synthesis Mix

Completed first-strand reaction mix	19 μL
2X Second-strand buffer	50 μL
[α-^{32}P]dCTP	1 μL
RNase H	1 μL
DNA polymerase I	6 μL
H$_2$O	23 μL

4. DNA size marker sample: 0.5 μg DNA ladder in 5% glycerol, 0.04% bromophenol blue: Available from NEB Labs.
5. STE: 100 mM NaCl, 10 mM Tris-HCl pH 8.0, 1 mM EDTA.

3. Methods

3.1. cDNA Synthesis Reactions

1. Mix together the components of first-strand synthesis shown in **Table 1** in the order given (see **Note 1**).
2. Remove 1 μL for the incorporation assay (see **step 8**). Add 1 μL AMV reverse transcriptase (see **Note 4**), mix, and incubate at 42°C for 60 min. Place mixture on ice and remove a further 1 μL for the incorporation assay.
3. Mix together the components of the second-strand synthesis described in **Table 2**. Incubate at 12°C for 60 min, then at 22°C for 60 min, and finally at 70°C for 10 min.
4. Spin the mix for a few seconds in a microcentrifuge. Place the mix on ice and add 1 μL T4 DNA polymerase to blunt the cDNA ends. Mix and incubate at 37°C for 10 min.
5. Add 2 μL 0.5 M EDTA, pH 8.0. Place on ice and remove 1 μL from the mix for the incorporation assay.
6. Extract twice with an equal volume of equilibrated phenol:chloroform, then once with an equal volume of saturated chloroform (see **Note 5**).
7. To remove unincorporated nucleotides, add 100 μL 4 M ammonium acetate and 400 μL ethanol. Vortex and leave on dry ice for 15 min, then allow the sample to warm to 4°C. Spin for 15 min at 4°C in a microcentrifuge at high speed. Remove the supernatant and wash the pellet in ice-cold 70% ethanol. Spin for 5 min at 4°C. Remove the supernatant, air dry, and redissolve in 10 μL TE (see **Note 6**).

8. To measure the incorporation of radioactive nucleotides into the cDNA, spot the retained 1 µL aliquots onto separate DE-81 filters and allow to dry. Wash five times in 200 mL 0.5 M Na_2HPO_4, pH 7.0, by swirling the mixture very gently. Wash the filters two or three times in 200 mL H_2O until clear. Rinse in 100 mL methanol and allow to air dry. Count the filters in a scintillation counter using a toluene-based scintillant and calculate the level of ^{32}P incorporation.

3.2. Addition of EcoRI Linkers to the cDNA Ends

1. To methylate *Eco*RI sites in the cDNA (*see* **Note 7**), add 4 µL 5X buffer M, 2 µL 10X buffer SAM, and 3 µL H_2O to the cDNA. Mix and spin for a few seconds in a microcentrifuge. Add 1 µL *Eco*RI methylase, mix, and incubate at 37°C for 60 min.
2. Inactivate the methylase by heating at 70°C for 10 min and place on ice.
3. To ligate linkers to the cDNA, add 3 µL 10X ligase buffer, 2 µL phosphorylated *Eco*RI linkers, 2 µL T4 DNA ligase, and 3 µL H_2O. Mix and incubate at 12°C for 16 h.
4. Heat at 65°C for 10 min to inactivate the ligase and place on ice.
5. To digest the linkered cDNA, add 9 µL 10X *Eco*RI buffer, 49.5 µL H_2O, and 1.5 µL *Eco*RI. Mix and incubate at 37°C for 5 h, then inactivate at 70°C for 10 min and place on ice.

3.3. Removal of Unincorporated Linkers and Size Selection of cDNA

1. Add 5 µL 2 M KCl and 5 µL 100 mM spermine to the 90 µL digested, linkered cDNA reaction (*see* **Note 8**). Vortex and leave on ice for 30 min.
2. Spin for 15 min in a microcentrifuge at 4°C. Remove the supernatant carefully and add the 1 mL spermine wash buffer. Leave on ice for 30 min. Carefully remove the supernatant again and then repeat the wash.
3. Wash once in 70% ethanol, air dry, and dissolve the pellet in 9 µL TE.
4. Prepare a gel of 1% low-melting-point agarose in 1X TAE containing 0.5 µg/mL ethidium bromide. Leave the gel to set at 4°C for 30 min. Fill the electrophoresis tank with 1X TAE buffer to the height of the gel but do not cover it.
5. Add 1 µL 50% glycerol to the cDNA sample and load onto the gel. The lack of loading dye makes the sample harder to load but easier to visualize small amounts of DNA under UV light. Load an adjacent lane with a DNA-size marker sample containing dye.
6. Electrophorese the gel at constant voltage at 4°C until the bromophenol blue dye has run half the length of the gel. View under long-wave UV light (to minimize UV-induced DNA damage) and cut out a gel slice corresponding to cDNA between 700 and 7000 bp in size.
7. If the slice is large, it may be beneficial to concentrate the cDNA by inserting the slice in a reverse orientation into a second low-melting-point gel and electrophoresing as before until it is concentrated into a tight band.
8. Add 400 µL STE/100 mg of gel slice. Incubate at 65°C for 10 min. Extract once with an equal volume of equilibrated phenol, once with equilibrated phenol:chloroform, and once with saturated chloroform. Avoid the interface, which contains white agarose powder.
9. Precipitate with 0.1 vol of 4 M ammonium acetate and 2.5 vol ethanol. Leave on ice for 30 min and spin for 30 min at 4°C. Wash the pellet with 70% ethanol.
10. Resuspend the pellet in 5 µL TE. Run 0.5 µL on a 1% agarose gel alongside size standards of known concentration to give an indication of cDNA size and concentration. Store the remainder as 1 µL aliquots at –70°C. The stored cDNA should be stable for several months.

4. Notes

1. Components of all reactions should be gently mixed together on ice. Reactions can be scaled up to allow the synthesis of larger quantities of cDNA.
2. Oligo(dT) or random hexanucleotide primers can be used for cDNA synthesis. The oligo(dT) primer, which anneals to the poly (A)$^+$ tail of mRNA, is more commonly used.

This generates a large proportion of full-length cDNAs, although the 5' end of large mRNAs will be underrepresented. Random primers are more useful for certain purposes, such as generating expression libraries, enriching the 5' ends of very long transcripts, or overcoming problems using the oligo(dT) primer because of secondary structure in the mRNA.
3. Sometimes problems with mRNA secondary structure affect cDNA synthesis. These problems may be averted by heating the poly $(A)^+$ RNA to 70°C for 1 min and immediately chilling it on ice prior to the cDNA synthesis reactions.
4. Both AMV and MoMuLV reverse transcriptase can be used for cDNA synthesis. AMV RTase is more commonly used, as fewer units of the enzyme are required. However, it has been reported that under certain conditions, longer transcripts (up to 10 kB long) can be synthesized using MoMuLV reverse transcriptase.
5. The recovery of cDNA from this and all subsequent steps can be monitored using a handheld minimonitor. Expect initial readings of 100–200 cps.
6. Material can be stored at –20°C at this stage. It is advisable to run a small portion (0.5 µL) of the double-stranded cDNA on a 1% agarose gel before proceeding with other steps. This gives an indication of the size range of the cDNA product.
7. The efficiency of the cloning steps should be checked first with a "mock cDNA," rather than risk losing precious cDNA. The mockup should consist of blunt-ended, phosphatased fragments of a similar size to the cDNA.
8. The largest losses of cDNA occur during the spermine precipitation and gel purification steps. These losses are justified because it is imperative to remove excessively small cDNAs and all traces of unincorporated linkers from the reaction. A tiny amount of residual self-ligated linkers in the mix would give a huge molar excess of ends, compared to the linkered cDNA, when ligated to the vector. This would result in a large proportion of the library containing no detectable inserts—essentially an empty library. The problem can usually be remedied by an additional spermine precipitation of the linkered cDNA after gel purification.

References

1. Gubler, U. and Hoffman, B. (1983) A simple and very efficient method for generating cDNA libraries. *Gene* **25,** 263–269.
2. Okayama, H. and Berg, P. (1982) High-efficiency cloning of full-length cDNA. *Mol. Cell. Biol.* **2,** 161–170.
3. Elledge, S. J., Mulligan, J. T., Ramer, S. W., Spottswood, M., and Davis, R. W. (1991) λYES: A multifunctional cDNA expression vector for the isolation of genes by complementation of yeast and *Escherichia coli* mutations. *Proc. Natl. Acad. Sci. USA* **88,** 1731–1735.
4. Sambrook, J., Fritsch, E. F., and Maniatis, T. (eds.) (1989) *Molecular Cloning: A Laboratory Manual*. Cold Spring Harbor Lab. Press, Cold Spring Harbor, NY.

40

Using Rapid Amplification of cDNA Ends (RACE) to Obtain Full-Length cDNAs

Yue Zhang and Michael A. Frohman

1. Introduction

Most attempts to identify and isolate a novel cDNA result in the acquisition of clones that represent only a part of the mRNA's complete sequence (**Fig. 1**). The approach described here to clone the missing sequence (cDNA ends) employs polymerase chain reaction (PCR). Since the initial reports of rapid amplification of cDNA ends (RACE) *(1)* or related techniques *(2,3)*, many labs have developed significant improvements on the basic approach *(4–18)*. The most recent hybrid version of the relatively simple Classic RACE will be described here, as well as a more powerful but technically more challenging "New RACE" protocol, which is adapted from the work of a number of laboratories *(19–26)*. Commercial RACE kits are available from Bethesda Research Laboratories (Gaithersburg, MD) *(11)* and Clontech (Palo Alto, CA) that are convenient but not as powerful as the most recent versions of Classic and New RACE.

1.1. Overview

Why use PCR (RACE) at all instead of screening (additional) cDNA libraries? RACE cloning is advantageous for several reasons. First, it takes weeks to screen cDNA libraries, obtain individual cDNA clones, and analyze the clones to determine if the missing sequence is present; using PCR, such information can be generated within a few days. As a result, it becomes practical to modify RNA preparation and/or reverse transcription conditions until full-length cDNAs are generated and observed. In addition, essentially unlimited numbers of independent clones can be generated using RACE, unlike library screens in which generally a single to a few cDNA clones are recovered. The availability of large numbers of clones provides confirmation of nucleotide sequence and allows the isolation of unusual transcripts that are alternately spliced or that begin at infrequently used promoters.

1.2. Principles

1.2.1. Classic RACE

PCR is used to amplify partial cDNAs representing the region between a single point in a mRNA transcript and its 3' or 5' end (**Fig. 1**). A short internal stretch of

From: *The Nucleic Acid Protocols Handbook*
Edited by: R. Rapley © Humana Press Inc., Totowa, NJ

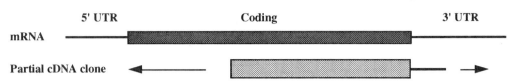

Fig. 1. Schematic representation of the setting in which RACE is useful in cDNA cloning strategies. Depicted is an mRNA for which a cDNA representing only an internal portion of the transcript has been obtained. Such circumstances often arise; e.g, when open reading frame fragments are obtained from expression library, two-hybrid, or Genbank Expressed Sequence Tag searches.

sequence must already be known from the mRNA of interest. From this sequence, gene-specific primers are chosen that are oriented in the direction of the missing sequence. Extension of the partial cDNAs from the unknown end of the message back to the known region is achieved using primers that anneal to the preexisting poly (A) tail (3' end) or an appended homopolymer tail (5' end). Using RACE, enrichments on the order of 10^6 to 10^7-fold can be obtained. As a result, relatively pure cDNA ends are generated that can be easily cloned or rapidly characterized using conventional techniques *(1)*.

To generate 3' end partial cDNA clones, mRNA is reverse transcribed using a "hybrid" primer (Q_T) that consists of 17 nt of oligo(dT) followed by a unique 35-base oligonucleotide sequence (Q_I–Q_O; **Fig. 2A** and **C**), which in many reports is denoted as an "anchor" primer. Amplification is then performed using a primer containing part of this sequence (Q_O) that now binds to each cDNA at its 3' end, and using a primer derived from the gene of interest (*GSP*1). A second set of amplification cycles is then carried out using "nested" primers (Q_I and GSP2) to quench the amplification of non-specific products. To generate 5' end partial cDNA clones, reverse transcription (primer extension) is carried out using a gene-specific primer (*GSP*-RT; **Fig. 2B**) to generate first-strand products. Then, a poly (A) tail is appended using terminal deoxynucleotidyltransferase (TdT) and deoxyadenosine triphosphate (dATP). Amplification is then achieved using (1) the hybrid primer Q_T to form the second strand of cDNA, (2) the Q_O primer, and (3) a gene-specific primer upstream of the one used for reverse transcription. Finally, a second set of PCR cycles is carried out using nested primers (Q_I and *GSP*2) to increase specificity *(5)*.

1.2.2. Classic RACE Variations

In general, as described in **Subheading 1.2.1.**, the gene-specific primer is derived from a short stretch of sequence that is already known from the mRNA of interest. A frequent question is whether degenerate primers, i.e., ones directed against a predicted nucleotide sequence based on a known amino acid sequence, can be used instead. Although such primers increase the quantity of spurious amplification, the approach can work, if other parameters are favorable (i.e., message abundance, GC composition, and cDNA end size; *see* **refs.** *14* and *27*).

At the unknown end of the cDNA, the 5' end can be tailed with C's instead of A's, and then amplified using a hybrid primer with a tail containing G's *(2)*, or a mixture of G's and inosines (I) *(11)*. Although the G:I approach entails synthesizing a primer that

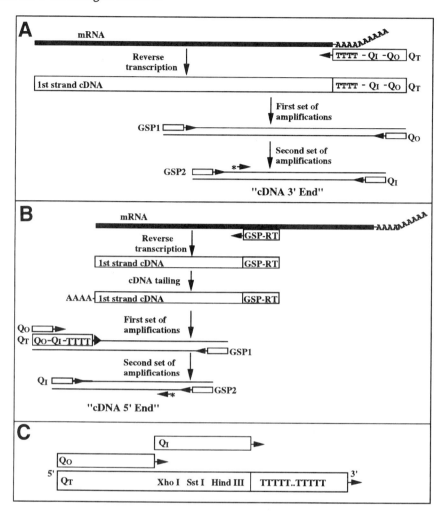

Fig. 2. Schematic representation of Classic RACE. Explanations are given in the text. At each step, the diagram is simplified to illustrate only how the new product formed during the previous step is utilized. GSP1, gene-specific primer 1; GSP2, gene-specific primer 2; GSP-RT, gene-specific primer used for reverse transcription; * →, GSP-Hyb/Seq or gene-specific primer for use in hybridization and sequencing reactions. (**A**) Amplification of 3' partial cDNA ends. (**B**) Amplification of 5' partial cDNA ends. (**C**) Schematic representation of the primers used in Classic RACE. The 52-nt Q_T primer (5' Q_O-Q_I-TTTT 3') contains a 17-nt oligo-(dT) sequence at the 3' end followed by a 35-nt sequence encoding *Hin*dIII, *Sst*I, and *Xho*I recognition sites. The Q_I and Q_O primers overlap by 1 nt; the Q_I primer contains all three of the recognition sites. Primers: Q_T: 5'-ccagtgagcagagtgacgaggactcgagctcaagctttttttttttttttt-3'; Q_O: 5'-ccagtgagcagagtgacg-3'; Q_I: 5'-gaggactcgagctcaagc-3'.

can be used for 5' RACE only (because a T-tailed primer must be used to anneal to the poly (A) tail of the 3' end), there may be sufficient benefits from using a mixed G:I tail to justify the cost, as the G:I region should anneal at temperatures similar to those of other primers normally used in PCR. In contrast, it is believed that homopolymers of either Ts or Gs present problems during PCR because of the very low and very high

$$5'\text{XXXXXXXXX-AAAAAAAAA-}\begin{pmatrix}G\\T\\C\end{pmatrix}\begin{pmatrix}G\\A\\T\\C\end{pmatrix}3'$$

Fig. 3. Lock-docking primer. "X" represents (e.g.) one or more restriction sites at the 5' end of the primer. See text for details.

annealing temperatures, respectively, required for their optimal usage *(1,11)*. On the other hand, the inosine residues function as degenerate nucleotides and will lead to higher spurious amplification, so the magnitude of the benefit of using a mixed G:I primer is unknown.

To minimize the length of homopolymer tail actually amplified, a "lock-docking" primer was developed by Borson et al. *(8)*. In this approach, the final 2 nt on the 3' end of the primer are degenerate. For example, to amplify cDNAs linked to an A tail, the lock-docking primer would appear as shown in **Fig. 3**.

The advantage of this approach is that it forces the primer to anneal to the junction of the natural or appended homopolymer tail and the cDNA sequence. The disadvantage is that it is necessary to synthesize four primers, as most synthesizers can only synthesize primers starting from an unambiguous 3' end.

In another variation, the location of the anchor primer is changed from the end of the unknown region of sequence to random points within the unknown region *(7)*. This is accomplished using a primer containing an anchor region followed by six random nucleotides (5' XXXXXXX-NNNNNN 3'), either for reverse transcription (3' RACE) or for creation of the second strand of cDNA (5' RACE). This approach is valuable when the 3' or 5' ends lie so far away from the region of known sequence that the entire unknown region can not be amplified effectively. Using this approach, cDNA ends of defined sizes are not generated; instead, one obtains a library of randomly sized fragments, all of which initiate at the gene-specific primer. The largest fragments can be cloned and characterized, extending the length of the known sequence, and the process (or standard RACE) repeated until the real unknown end is identified. The development of a "long" PCR may make this approach unnecessary.

1.2.3. New RACE

The most technically challenging step in a classic 5' RACE is to cajole reverse transcriptase to copy the mRNA of interest in its entirety into first-strand cDNA. Because prematurely terminated first-strand cDNAs are tailed by terminal transferase just as effectively as full-length cDNAs, cDNA populations composed largely of prematurely terminated first strands will result primarily in the amplification and recovery of cDNA ends that are not full length either (**Fig. 4A**). This problem is encountered routinely for vertebrate genes, which are often quite GC rich at their 5' ends, and thus frequently contain sequences that hinder reverse transcription. A number of laboratories have developed steps or protocols designed to approach the problem *(19–26)*; the protocol described here and denoted "New RACE" is, for the most part, a composite adapted from the cited reports.

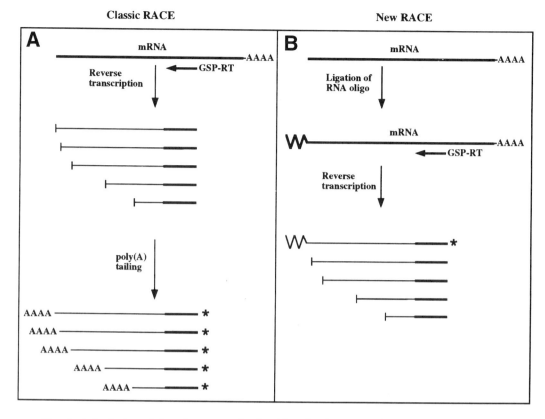

Fig. 4. A depiction of the advantage of using New RACE over Classic RACE. (**A**) In Classic RACE, premature termination in the reverse transcription step results in polyadenylation of less-than-full length first strand cDNAs, all of which can be amplified using PCR to generate less-than-full length cDNA 5' ends. * indicates cDNAs ends created that will be amplified in the subsequent PCR reaction. (**B**) In New RACE, less-than-full length cDNAs are also created but are not terminated by the anchor sequence, and hence cannot be amplified in the subsequent PCR reaction.

New RACE departs from Classic RACE in that the "anchor" primer is attached to the 5' end of the mRNA before the reverse transcription step; hence the anchor sequence becomes incorporated into the first-strand cDNA if and only if the reverse transcription proceeds through the entire length of the mRNA of interest (and through the relatively short anchor sequence) (*see* **Fig. 4B**).

Before beginning New RACE (**Fig. 5A**), the mRNA is subjected to a dephosphorylation step using calf intestinal phosphatase (CIP). This step actually does nothing to full-length mRNAs, which have methyl-G caps at their termini; but it does dephosphorylate degraded mRNAs, which are uncapped at their termini *(21)*. This makes the degraded RNA biologically inert during the ensuing ligation step because the phosphate group is required to drive the reaction. The full-length mRNAs are then decapped using tobacco acid pyrophosphatase (TAP), which leaves them with an active and phosphorylated 5' terminus *(20,24)*. Using T4 RNA Ligase, this mRNA is then ligated to a short, synthetic RNA oligo that has been generated by in vitro transcription of a linear-

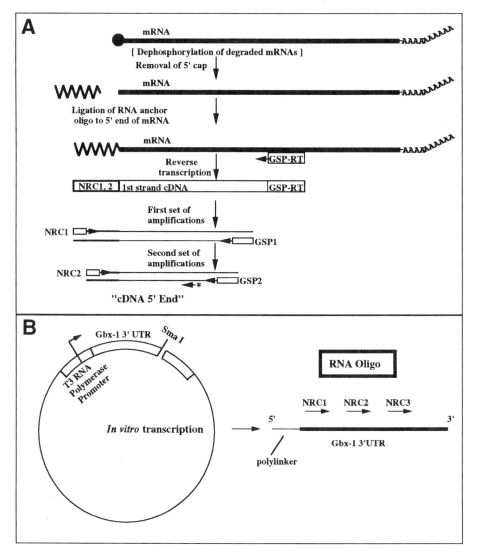

Fig. 5. Schematic representation of New RACE. Explanations are given in the text. At each step, the diagram is simplified to illustrate only how the new product formed during the previous step is utilized. See legend to **Fig. 1** for description of some primers. (**A**) Amplification of 5' partial cDNA ends. (**B**) In vitro synthesis of the RNA oligo used for ligation in New RACE and schematic representation of the corresponding required primers. A 132-nt RNA oligo is produced by in vitro transcription of the plasmid depicted using T3 RNA Polymerase. Primers NRC-1, -2, and -3 are derived from the sequence of the oligo but do not encode restriction sites. To assist in the cloning of cDNA ends, the sequence ATCG is added to the 5' end of NRC-2, as described in the cloning section of the text.

ized plasmid (**Fig. 5B**) *(19)*. The RNA oligo-mRNA hybrids are then reverse transcribed using a gene-specific primer or random primers to create first-strand cDNA. Finally, the 5' cDNA ends are amplified in two nested PCR reactions using additional gene-specific primers and primers derived from the sequence of the RNA oligo.

The New RACE approach can also be used to generate 3' cDNA ends (*21*; *see also* related protocols in **refs. *20* and *22***), and is useful in particular for nonpolyadenylated RNAs. In brief, cytoplasmic RNA is dephosphorylated and ligated to a short, synthetic RNA oligo as described above. Although ligation of the oligo to the 5' end of the RNA was emphasized above, RNA oligos actually ligate to both ends of the cytoplasmic RNAs. For the reverse transcription step, a primer derived from the RNA oligo sequence is used (e.g., the reverse complement of NRC-3, **Fig. 3**). Reverse transcription of the RNA oligos that happen to be ligated to the 3' end of the cytoplasmic RNAs results in the creation of cDNAs that have the RNA oligo sequence appended to their 3' end. Gene-specific primers oriented in the 5' → 3' direction and New RACE primers (e.g., the reverse complements of NRC-2 and NRC-1, **Fig. 5**) can be used in nested PCR reactions to amplify the 3' ends.

2. Materials

2.1. 3' End cDNA Amplification Using Classic RACE

2.1.1. Reverse Transcription to Generate cDNA Templates

1. 5X Reverse transcription buffer: 250 mM Tris-HCl, pH 8.3, 375 mM KCl, 15 mM MgCl$_2$.
2. SuperScript II reverse transcriptase and 5X buffer (Bethesda Research Labs, Gaithersburg, MD; Gibco-BRL, Gaithersburg, MD)
3. RNAse H (Bethesda Research Labs; Gibco-BRL)
4. RNasin (Promega Biotech, Madison, WI)
5. dNTPs (10 mM solutions; PL Biochemicals/Pharmacia, Piscataway, NJ, or Boehringer Mannheim, Mannheim, Germany).
6. TE buffer: 10 mM Tris-HCl, pH 7.5, 1 mM EDTA.

2.1.2. Amplification

1. *Taq* polymerase and 10X buffer (Promega Biotech).
2. (Optional) Instead of using the recommended reaction mixture, a 10X buffer consisting of 670 mM Tris-HCl, pH 9.0, 67 mM MgCl$_2$, 1700 μg/mL bovine serum albumin (BSA), and 166 mM (NH$_4$)$_2$SO$_4$ can be substituted, and reaction conditions are altered as described in **Note 3** below *(1,16)*.
3. PCR cocktail (1X *Taq* polymerase buffer [described in **Subheading 2.1.2., step 2**], each dNTP at 1.5 mM, and 10% dimethylsulfoxide [DMSO]).
4. Oligonucleotide primer sequences are listed in the legend to **Fig. 2**. Primers can be used "crude" except for Q$_T$, which should be purified to ensure that it is uniformly full length.

2.2. 5' End cDNA Amplification Using Classic RACE

2.2.2. Tailing First-Strand cDNA Products

1. Terminal deoxynucleotidyl transferase (TdT) (Gibco-BRL or Boehringer Mannheim)
2. 5X Tailing buffer: 125 mM Tris-HCl, pH 6.6, 1 M KCacodylate, and 1250 μg/mL BSA.

2.3. New RACE 5' End cDNA Amplification

1. CIP and 10X buffer (Boehringer Mannheim).
2. TAP (Epicentre, Madison, WI).
3. RNA transcription kit (Epicentre).
   a. Proteinase K (Boehringer Mannheim).
   b. DNAse (RNAse free) (Epicentre).

4. T4 RNA Ligase (New England Biolabs, Beverly, MA, or Boehringer Mannheim).
5. 10X Buffer: 500 mM Tris, pH 7.9, 100 mM MgCl$_2$, 20 mM DTT/1 mg/mL BSA.

Note that 10X T4 RNA ligase buffers supplied by some manufacturers contain too much adenosine triphosphate (ATP) *(19)*. Check the composition of any commercially supplied 10X buffer and make your own if it contains more than 1 mM ATP (final 1X concentration should be 0.1 mM).

3. Methods
3.1. 3' End cDNA Amplification Using Classic RACE
3.1.1. Reverse Transcription to Generate cDNA Templates
(see **Notes 1** and **9a**)

1. Assemble reverse transcription components on ice: 4 µL 5X reverse transcription buffer, 1 µL dNTP, 2 µL 0.1 M DTT, 0.5 µL Q$_T$ primer (100 ng/µL), and 0.25 µL (10 U) of RNasin.
2. Heat 1 µg of poly (A)$^+$ RNA or 5 µg of total RNA in 13 µL of water at 80°C for 3 min, cool rapidly on ice, and spin for 5 s in a microfuge.
3. Add to reverse transcription components. Add 1 µL (200 U) SuperScript II (Gibco-BRL) reverse transcriptase, and incubate for 5 min at room temperature, 1 h at 42°C, and 10 min at 50°.
4. Incubate at 70°C for 15 min to inactivate reverse transcriptase. Spin for 5 s in a microfuge.
5. Add 0.75 µL (1.5 U) of RNAse H to a tube and incubate at 37°C for 20 min to destroy the RNA template.
6. Dilute the reaction mixture to 1 mL with TE and store at 4°C (3' end cDNA pool).

3.1.2. Amplification (see **Notes 3** and **9c**)

3.1.2.1. First Round

1. Add an aliquot of the cDNA pool (1 µL) and primers (25 pmol each of GSP1 and Q$_O$) to 50 µL of a PCR cocktail in a 0.5-mL microfuge tube.
2. Heat in a DNA thermal cycler for 5 min at 98°C to denature the first-strand products. Cool to 75°C. Add 2.5 U *Taq* polymerase, overlay the mixture with 30 µL of mineral oil (Sigma 400-5, Sigma Chemical Co., St. Louis, MO, preheat it in the thermal cycler to 75°) and incubate at the appropriate annealing temperature (52–60°C) for 2 min. Extend the cDNAs at 72°C for 40 min.
3. Carry out 30 cycles of amplification using a step program (94°, 1 min; 52–60°C, 1 min; 72°C, 3 min), followed by a 15-min final extension at 72°C. Cool to room temperature.

3.1.2.2. Second Round

1. Dilute 1 µL of the amplification products from the first round into 20 µL TE.
2. Amplify 1 µL of the diluted material with primers GSP2 and Q$_I$ using the procedure described in the "First Round" but eliminate the initial 2-min annealing step and the 72°C, 40-min extension step.

3.2. 5' End cDNA Amplification Using Classic RACE
3.2.1. Reverse Transcription to Generate cDNA Templates
(see **Note 2** and **9a**)

1. Assemble reverse transcription components on ice: 4 µL 5X reverse transcription buffer, 1 µL dNTP, 2 µL 0.1 M DTT, and 0.25 µL (10 U) RNasin.

2. Heat 0.5 µL of GSP-RT primer (100 ng/µL) and 1 µg of poly (A)$^+$ RNA or 5 µg of total RNA in 13 µL water at 80°C for 3 min, cool rapidly on ice, and spin for 5 s in a microfuge.
3. Add to reverse transcription components. Add 1 µL (200 U) of SuperScript II reverse transcriptase, and incubate for 1 h at 42°C and 10 min at 50°C.
4. Incubate at 70°C for 15 min to inactivate reverse transcriptase. Spin for 5 s in a microfuge.
5. Add 0.75 µL (1.5 U) of RNAse H to the tube and incubate at 37°C for 20 min to destroy the RNA template.
6. Dilute the reaction mixture to 400 µL with TE and store at 4°C (5' end nontailed cDNA pool).

3.2.2. Appending a Poly (A) Tail to First-Strand cDNA Products (see **Notes 5** and **9b**)

1. Remove excess primer using Microcon-100 spin filters (Amicon Corp., Beverly, MA) or an equivalent product, following the manufacturer's instructions. Wash the material by spin filtration twice more using TE. The final volume recovered should not exceed 10 µL. Adjust volume to 10 µL using water.
2. Add 4 µL 5X tailing buffer, 1.2 µL 25 mM CoCl$_2$, 4 µL 1 mM dATP, and 10 U TdT.
3. Incubate 5 min at 37°C and then 5 min at 65°C.
4. Dilute to 500 µL with TE (5' end-tailed cDNA pool).

3.2.3. Amplification (see **Notes 4** and **11**)

3.2.3.1. FIRST ROUND

1. Add an aliquot of the 5'-end tailed cDNA pool (1 µL) and primers (25 pmol each of GSP1 and Q$_O$ [shown in **Fig. 2B**], and 2 pmol of Q$_T$) to 50 µL of PCR cocktail in a 0.5-mL microfuge tube.
2. Heat in a DNA thermal cycler for 5 min at 98°C to denature the first-strand products. Cool to 75°C. Add 2.5 U *Taq* polymerase, overlay the mixture with 30 µL of mineral oil (Sigma 400-5; preheat it in the thermal cycler to 75°C) and incubate at the appropriate annealing temperature (48–52°C) for 2 min. Extend the cDNAs at 72°C for 40 min.
3. Carry out 30 cycles of amplification using a step program (94°C, 1 min; 52–60°C, 1 min; 72°C, 3 min), followed by a 15-min final extension at 72°C. Cool to room temperature.

3.2.3.2. SECOND ROUND

4. Dilute 1 µL of the amplification products from the first round into 20 µL TE.
5. Amplify 1 µL of the diluted material with primers GSP2 and Q$_I$ using the "First Round" procedure described above, but eliminate the initial 2-min annealing step and the 72°C, 40-min extension step.

3.3. New RACE 5' End cDNA Amplification

The procedure detailed in **Subheading 3.3.1.–3.3.5.** is described using relatively large amounts of RNA and can be scaled down if RNA quantities are limiting. The advantage of starting with large amounts of RNA is that aliquots can be electrophoresed quickly after each step of the procedure to confirm that detectable degradation of the RNA has not occurred, and the dephosphorylated-decapped-ligated RNA can be stored indefinitely for many future experiments.

3.3.1. Dephosphorylation of Degraded RNAs

In general, follow the manufacturer's recommendations for use of the phosphatase.

1. Prepare a reaction mixture containing 50 µg RNA in 41 µL of water, 5 µL 10X buffer, 0.5 µL 100 m*M* DTT, 1.25 µL RNAsin (40 U/µL), and 3.5 µL CIP (1 U/µL).
2. Incubate the reaction at 50°C for 1 h.
3. Add proteinase K to 50 µg/mL and incubate at 37°C for 30 min.
4. Extract the reaction with a mixture of phenol/chloroform, extract again with chloroform, and precipitate the RNA using 1/10th volume of 3 *M* NaOAc and 2.5 vol ethanol. Resuspend the RNA in 43.6 µL of water.
5. Electrophorese 2 µg (1.6 µL) on a 1% TAE agarose gel adjacent to a lane containing 2 µg of the original RNA preparation, stain the gel with ethidium bromide, and confirm visually that the RNA remained intact during the dephosphorylation step.

3.2.2. Decapping Intact RNAs (see **Note 6**)

1. Prepare a reaction mixture containing 38 µg of RNA in 42 µL of water (this is the RNA recovered from **step 1**, **Subheading 3.3.**), 5 µL 10X TAP buffer, 1.25 µL RNAsin (40 U/µL), 1 µL 100 m*M* ATP, and 1 µL TAP (5 U/µL).
2. Incubate the reaction at 37°C for 1 h, and then add 200 µL TE.
3. Extract the reaction with a mixture of phenol/chloroform, extract again with chloroform, and precipitate the RNA using 1/10th vol 3 *M* NaOAc and 2.5 vol ethanol. Resuspend the RNA in 40 µL water.
4. Electrophorese 2 µg on a TAE 1% agarose gel adjacent to a lane containing 2 µg of the original RNA preparation, stain the gel with ethidium bromide, and confirm visually that the RNA remained intact during the decapping step.

3.3.3. Preparation of RNA Oligo

Choose a plasmid that can be linearized at a site approx 100 bp downstream from a T7 or T3 RNA polymerase site (*see* **Fig. 2B**). Ideally, a plasmid containing some insert cloned into the first polylinker site is optimal because primers made from palindromic polylinker DNA do not perform well in PCR. For my experiments, I use the 3'UTR of the mouse gene *Gbx-1* (*28*), which is cloned into the Sst I site of pBS-SK (Stratagene); I linearize with Sma I and transcribe with T3 RNA polymerase to produce a 132 nt RNA oligo, of which all but 17 nt are from *Gbx-1*. Note that adenosines are the best "acceptors" for the 3' end of the RNA oligo to ligate to the 5' end of its target, if an appropriate restriction site can be found. The primers subsequently used for amplification are all derived from the *Gbx-1* 3' UTR sequence. Interested investigators are welcome to the *Gbx-1* NRC primer sequences and plasmid on request.

Carry out a test transcription to make sure that everything is working; then scale up. The oligo can be stored at –80°C indefinitely for many future experiments, and it is important to synthesize enough oligo so that losses due to purification and spot checks along the way will leave plenty of material at the end of the procedure.

1. Linearize 25 µg of the plasmid that is to be transcribed (the plasmid should be reasonably free of RNAses).
2. Treat the digestion reaction with 50 µg/mL proteinase K for 30 min at 37°C, followed by 2X phenol/CHCl$_3$ extractions, one CHCl$_3$ extraction, and an EtOH precipitation.

RACE for Full-Length cDNAs

3. Resuspend the template DNA in 25 μL TE, pH 8.0, for a final concentration of approx 1 μg/μL.
4. Transcription: Mix at room temperature in the following order:

	Test scale	Prep scale
DEPC water	4 μL	80
5X buffer	2	40
0.1 M DTT	1	20
10 mM UTP	0.5	10
10 mM ATP	0.5	10
10 mM CTP	0.5	10
10 mM GTP	0.5	10
Restricted DNA (1 μg/μL)	0.5	10
RNasin (40 U/μL)	0.25	5
RNA polymerase (20 U/μL)	0.25	5

Incubate at 37°C for 1 h

5. DNase template: Add 0.5 μL DNase (RNase free) for every 20 μL of reaction volume and incubate at 37°C for 10 min.
6. Run 5 μL of test or prep reaction on a 1% TAE agarose gel to check. Expect to see a diffuse band at about the right size (or a bit smaller) in addition to some smearing all up and down the gel.
7. Purify the oligo by extracting with phenol/CHCl$_3$ and CHCl$_3$, and then rinse three times using water and a microcon spin filter (prerinsed with water).
8. Run another appropriately sized aliquot on a 1% TAE agarose gel to check integrity and concentration of oligo. Microcon-30 spin filters have a cut off size of 60 nt, and Microcon-100 spin filters have a cut off size of 300 nt. Microcon-10 spin filters are probably most appropriate if the oligo is smaller than 100 nt, and Microcon-30 spin filters for anything larger.

3.3.4. RNA Oligo—Cellular RNA Ligation

1. Set up two tubes—one with TAPped cellular RNA, the other with unTAPped cellular RNA.
   11.25 μL water
   3.00 μL 10X buffer
   0.75 μL RNAsin (40 U/μL)
   2.00 μL 4 μg RNA oligo (3–6 molar excess over target cellular RNA)
   10.00 μL 10 μg TAPped (or unTAPped) RNA
   1.50 μL 2 mM ATP
   1.50 μL T4 RNA Ligase (20 U/μL)

   30.00 μL

2. Incubate for 16 h at 17°C.
3. Purify the ligated oligo-RNA using Microcon-100 spin filtration (3× in water; prerinse filter with RNAse-free water). The volume recovered should not exceed 20 μL.
4. Run one-third of the ligation on a 1% TAE agarose gel to check integrity of the ligated RNA. It should look about as it did before ligation.

3.3.5. Reverse Transcription (see **Note 9a**)

1. Assemble reverse transcription components on ice: 4 μL of 5X reverse transcription buffer, 1 μL of dNTP (stock concentration is 10 mM of each dNTP), 2 μL of 0.1 M DTT, and 0.25 μL (10 U) of RNasin.

2. Heat 1 µL of antisense specific primer (20 ng/µL) or random hexamers (50 ng/µL) and the remaining RNA (approx 6.7 µg) in 13 µL of water at 80°C for 3 min, cool rapidly on ice, and spin for 5 s in a microfuge.
3. Add to reverse transcription components. Add 1 µL (200 U) of SuperScript II reverse transcriptase, and incubate for 1 h at 42°C, and 10 min at 50°C. If using random hexamers, insert a room temperature 10 min incubation period after mixing everything together.
4. Incubate at 70°C for 15 min to inactivate reverse transcriptase. Spin for 5 s in a microfuge.
5. Add 0.75 µL (1.5 U) of RNAse H to the tube and incubate at 37°C for 20 min to destroy the RNA template.
6. Dilute the reaction mixture to 100 µL with TE and store at 4°C (5' end oligo-cDNA pool).

3.3.6. Amplification (see **Note 9c**)

3.3.6.1. First Round

1. Add an aliquot of the 5' end oligo-cDNA pool (1 µL) and primers (25 pmol each of GSP1 and NRC-1) to 50 µL of the PCR cocktail (1X *Taq* polymerase buffer [described in **Subheading 2.1.2.**], each dNTP at 1.5 m*M*, and 10% DMSO) in a 0.5-mL microfuge tube.
2. Heat in a DNA thermal cycler for 5 min at 98°C to denature the first-strand products. Cool to 75°C. Add 2.5 U *Taq* polymerase, overlay the mixture with 30 µL mineral oil (Sigma 400-5; preheat it in the thermal cycler to 75°C) and incubate at the appropriate annealing temperature (52–60°C) for 2 min. Extend the cDNAs at 72°C for 40 min.
3. Carry out 35 cycles of amplification using a step program (94°C, 1 min; 52–60°C, 1 min; 72°C, 3 min), followed by a 15-min final extension at 72°C. Cool to room temperature.

3.3.6.2. Second Round

4. Dilute 1 µL of the amplification products from the first round into 20 µL TE.
5. Amplify 1 µL of the diluted material with primers GSP2 and NRC-2 using the procedure described above under "First Round" but eliminate the initial 2-min annealing step and the 72°C, 40-min extension step.

3.4. Safe and Easy Cloning Protocol

1. Insert preparation: Select a pair of restriction enzymes for which you can synthesize half sites appended to PCR primers that can be chewed back to form the appropriate overhangs, as shown for *Hin*dIII and *Eco*RI, see **Note 7**. For RACE cloning, add "TTA" to the 5' end of Q_I or NRC-2, and add "GCTA" to the 5' end of GSP2. Carry out PCR as usual.
2. After PCR, add proteinase K (10 mg/mL stock) to the PCR reaction to a concentration of 50 µg/mL and incubate at 37°C for 30 min, to remove sticky *Taq* polymerase from the amplified DNA *(29)*.
3. Extract the PCR products with phenol/$CHCl_3$, then $CHCl_3$ (but do not precipitate!) to remove proteins.
4. Filter the PCR products through a Microcon-100–spin column (or Microcon-30 if your product is <150 bp) three times using TE (not water) as the wash buffer to remove unwanted organics, primers, and dNTP.
5. On ice, add the selected dNTP [e.g., dTTP] to a final concentration of 0.2 m*M*, 1/10th vol of 10X T4 DNA polymerase buffer, and 1–2 U T4 DNA polymerase.
6. Incubate at 12°C for 15 min, then 75°C for 10 min to heat inactivate the T4 DNA polymerase. (Optional: Gel isolate DNA fragment of interest, depending on degree of success of PCR amplification.)
7. Vector preparation: Digest vector (e.g., pGem-7ZF [Promega]) using the selected enzymes (e.g., *Hin*dIII and *Eco*RI) under optimal conditions in a total volume of 10 µL.

8. Add a 10-μL mixture containing the selected dNTP [e.g., dATP] at a final concentration of 0.4 mM, 1 μL of the restriction buffer used for digestion, 0.5 μL Klenow, and 0.25 μL Sequenase.
9. Incubate at 37°C for 15 min, then 75°C for 10 min to heat inactivate the polymerases.
10. Gel isolate the linearized vector fragment.
11. For ligation, use equal molar amounts of vector and insert.

4. Notes

1. **Reverse Transcription for 3' RACE.** Poly (A)$^+$ RNA is preferentially used for reverse transcription to decrease background, but it is unnecessary to prepare it if only total RNA is available. An important factor in the generation of full-length 3' end partial cDNAs concerns the stringency of the reverse transcription reaction. Reverse transcription reactions were historically carried out at relatively low temperatures (37–42°C) using a vast excess of primer (approx half the mass of the mRNA, which represents an approx 30:1 molar ratio). Under these low-stringency conditions, a stretch of A residues as short as 6–8 nt will suffice as a binding site for an oligo(dT)-tailed primer. This may result in cDNA synthesis being initiated at sites upstream of the poly (A) tail, leading to truncation of the desired amplification product (*see* **Figs. 4** and **6**). One should be suspicious that this has occurred if a canonical polyadenylation signal sequence is not found near the 3' end of the cDNAs generated. This can be minimized by controlling two parameters: primer concentration and reaction temperature. The primer concentration can be reduced dramatically without decreasing the amount of cDNA synthesized significantly *(30)* and will begin to bind preferentially to the longest A-rich stretches present (i.e., the poly (A) tail). The quantity recommended earlier represents a good starting point; it can be reduced fivefold further if significant truncation is observed.

    In the protocol previously described, the incubation temperature is raised slowly to encourage reverse transcription to proceed through regions of difficult secondary structure. Because the half-life of reverse transcriptase rapidly decreases as the incubation temperature increases, the reaction cannot be carried out at elevated temperatures in its entirety. Alternatively, the problem of difficult secondary structure (and nonspecific reverse transcription) can be approached using heat-stable reverse transcriptases, which are now available from several suppliers (Perkin-Elmer/Cetus, Norwalk, CT; Amersham, Epicentre, Arlington Heights, IL, and others). As in PCR reactions, the stringency of reverse transcription can thus be controlled by adjusting the temperature at which the primer is annealed to the mRNA. The optimal temperature depends on the specific reaction buffer and reverse transcriptase used and should be determined empirically but will usually be found to be in the range of 48–56°C for a primer terminated by a 17-nt oligo(dT) tail.

2. **Reverse Transcription for 5' RACE.** Many of the remarks made earlier are also relevant here and should be noted. There is, however, one major difference. The efficiency of cDNA extension is now critically important, as each specific cDNA, no matter how short, is subsequently tailed and becomes a suitable template for amplification (**Fig. 2A**). Thus, the PCR products eventually generated directly reflect the quality of the reverse transcription reaction. Extension can be maximized by using clean, intact RNA, by selecting the primer for reverse transcription to be near the 5' end of region of known sequence; and in theory by using heat-stable reverse transcriptase at elevated temperatures or a combination of SuperScript II and heat-stable reverse transcriptase at multiple temperatures. Synthesis of cDNAs at elevated temperatures should diminish the amount of secondary structure encountered in GC-rich regions of the mRNA. Random hexamers (50 ng) can be

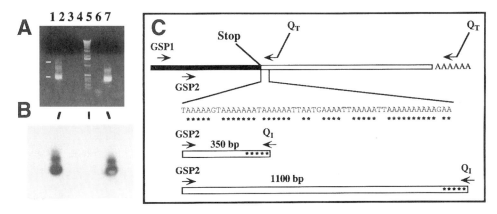

Fig. 6. Amplification of the 3' end of the mouse type II sodium channel gene—an example of cDNA synthesis being initiated at sites upstream of the poly (A) tail. **(A)** Ethidium bromide-stained 1% agarose gel of the PCR product amplified from mouse brain 3' cDNA pool. Lane 1, no template control. Lane 2, PCR product from primer GSP1 to Q_T. Lanes 3 and 4, blank. Lane 5, 1-kb ladder. Lane 6, no template control. Lane 7, nested PCR product from primer GSP2 to Q_I as reamplified from products of lane 2. Products 350 and 1100 bp were generated and are indicated by "-"s. **(B)** Southern blot hybridization with a ^{32}P-labeled oligonucleotide primer located 3' of GSP2 confirming the identity of the products. The hybridizing bands correspond to the 350- and 100-bp fragments observed in panel A. **(C)** Cloning and sequencing of the products revealed that the 350-bp product was a truncation resulting from initiation of reverse transcription in the A-rich region just 3' of the stop codon TAA. The sequence of the A-rich region is shown.

substituted for GSP-RT to create a "universal" 5'-end cDNA pool. A universal pool can be used for amplification of the 5' end of any cDNA created in the reverse transcription. Correspondingly, however, each cDNA is present at a much lower level than if created using the gene-specific reverse transcription described earlier. If you are using random hexamers, insert a room temperature 10 min incubation period after mixing everything together.

3. **3' End Amplification.** It is important to add the *Taq* polymerase *after* heating the mixture to a temperature above the annealing temperature of the primers ("hot start" PCR). Addition of the enzyme prior to this point allows one "cycle" to take place at room temperature, promoting the synthesis of nonspecific background products dependent on low-stringency interactions.

An annealing temperature close to the effective Tm of the primers should be used. The Q_I and Q_O primers work well at 60°C under the PCR conditions recommended here, although the actual optimal temperature may depend on the PCR machine used. Gene-specific primers of similar length and GC content should be chosen. Computer programs to assist in the selection of primers are widely available and should be used. An extension time of 1 min/kb expected product should be allowed during the amplification cycles. If the expected length of product is unknown, try 3–4 min initially.

Very little substrate is required for the PCR reaction. One microgram poly (A)$^+$ RNA typically contains approx 5×10^7 copies of *each* low abundance transcript. The PCR reaction described here works optimally when 10^3–10^5 templates (of the desired cDNA) are present in the starting mixture; thus as little as 0.002% of the reverse transcription mixture suffices for the PCR reaction! Addition of too much starting material to the amplification reaction will lead to production of large amounts of nonspecific product and should be avoided. The RACE technique is particularly sensitive to this problem, as every cDNA in the mixture, desired and undesired, contains a binding site for the Q_O and Q_I primers.

It was found empirically that allowing extra extension time (40 min) during the first amplification round (when the second strand of cDNA is created) sometimes resulted in increased yields of the specific product relative to background amplification, and in particular, increased the yields of long cDNAs vs short cDNAs when specific cDNA ends of multiple lengths were present (1). Prior treatment of cDNA templates with RNA hydrolysis or a combination of RNAse H and RNAse A infrequently improves the efficiency of amplification of specific cDNAs.

Choice of buffer systems: For some applications intended for cloned PCR products, such as expressing cDNAs to generate proteins, it is critically important to minimize the rate at which mutations occur during amplification. In other applications, such as using the cloned DNA as a probe in hybridization experiments, the presence of a few mutations is relatively unimportant and thus it is most convenient to use PCR conditions that maximize the likelihood of generating the desired product the first time a set of primers is used. Unfortunately, PCR conditions that result in a minimum of mutations are finicky and often the desired product can not be generated until the PCR conditions have been optimized, whereas PCR conditions that reliably produce desired products result in a relatively high mutation rate (approx 1% after 30 rounds).

Thus, appropriate conditions must be chosen to generate the PCR products required prior to undertaking cloning steps. PCR conditions that result in a minimum of mutations require the use of nucleotides (dNTP) at low concentrations (0.2 mM). Using the conditions recommended for *Taq* polymerase by Perkin-Elmer/Cetus results in an error rate of approx 0.05% after 30 rounds of amplification. However, the conditions recommended often have to be optimized, meaning that the pH of the buffer and the concentration of magnesium have to be adjusted until the desired product is observed. In addition, inclusion of dimethyl sulfoxide (DMSO) or formamide may be required. For those who do not wish to prepare their own reagents to carry out optimization experiments, such kits are commercially available (e.g., from Invitrogen, San Diego, CA, and Stratagene, La Jolla, CA). PCR conditions that work much more frequently in the absence of optimization steps require the use of DMSO, ammonium sulfate, and relatively high concentrations (1.5 mM) of dNTP, as described previously. It should be noted that the inclusion of DMSO to 10% decreases primer melting temperatures (and thus optimal annealing temperatures) by about 5–6°C.

4. **5' End Amplification.** Many of the remarks made earlier are also relevant here and should be noted. There is, however, one major difference. The annealing temperature in the first step (48–52°C) is lower than that used in successive cycles (52–60°C). This is because cDNA synthesis during the first round depends on the interaction of the appended poly (A) tail and the oligo(dT)-tailed Q_T primer, whereas in all subsequent rounds, amplification can proceed using the Q_O primer, which is composed of approx 60% GC and which can anneal at a much higher temperature to its complementary target.

5. **Using TdT to Tail First-Strand cDNA Products.** To attach a known sequence to the 5' end of the first-strand cDNA, a homopolymeric tail is appended using TdT. We prefer appending poly (A) tails rather than poly (C) tails for several reasons. First, the 3' end strategy is based on the naturally occurring poly (A) tail; thus the same adapter primers can be used for both ends, decreasing variability in the protocol and cost. Second, because A:T binding is weaker than G:C binding, longer stretches of A residues (approx 2X) are required before the oligo(dT)-tailed Q_T primer will bind to an internal site and trun-

cate the amplification product. Third, vertebrate coding sequences and 5' untranslated regions tend to be biased toward G/C residues; thus, use of a poly (A) tail further decreases the likelihood of inappropriate truncation.

Unlike many other situations in which homopolymeric tails are appended, the actual length of the tail added here is unimportant, as long as it exceeds 17 nt. This is because although the oligo(dT)-tailed primers subsequently bind all along the length of the appended poly (A) tail, only the innermost one becomes incorporated into the amplification product, and, consequently, the remainder of the poly (A) tail is lost *(1)*. The truncation appears to happen because *Taq* polymerase is unable to resolve branched structures efficiently. The conditions described in the following procedure result in the addition of 30–400 nt.

6. **Decapping Intact RNAs.** Most protocols call for much more TAP than recommended here. The enzyme is very expensive and it is not necessary!
7. **Analysis of Quality of Amplification Products.** The production of specific partial cDNAs by the RACE protocol is assessed using Southern blot hybridization analysis. After the second set of amplification cycles, the first and second set reaction products are electrophoresed in a 1% agarose gel, stained with ethidium bromide (EtBr), denatured, and transferred to a nylon membrane. After hybridization with a labeled oligomer or gene fragment derived from a region contained within the amplified fragment (e.g., GSP-Hyb/Seq in **Fig. 2A** and **B**), gene-specific partial cDNA ends should be detected easily. Yields of the desired product relative to nonspecific amplified cDNA in the first round products should vary from <1% of the amplified material to nearly 100%, depending largely on the stringency of the amplification reaction, the amplification efficiency of the specific cDNA end, and the relative abundance of the specific transcript within the mRNA source. In the second set of amplification cycles, approx 100% of the cDNA detected by EtBr staining should represent specific product. If specific hybridization is not observed, then trouble shooting steps should be initiated.

    Information gained from this analysis should be used to optimize the RACE procedure. If low yields of specific product are observed because nonspecific products are being amplified efficiently, then annealing temperatures can be raised gradually (approx 2°C at a time) and sequentially in each stage of the procedure until nonspecific products are no longer observed. Alternatively, some investigators have reported success using the "Touchdown PCR" procedure to optimize the annealing temperature without trial and error *(31)*. Optimizing the annealing temperature is also indicated if multiple species of specific products are observed, which could indicate that truncation of specific products is occurring. If multiple species of specific products are observed after the reverse transcription and amplification reactions have been fully optimized, then the possibility should be entertained that alternate splicing or promoter use is occurring.

    a. *Classic RACE only:* If a nearly continuous smear of specific products is observed up to a specific size limit after 5' end amplification, this suggests that polymerase pausing occurred during the reverse transcription step. To obtain nearly full-length cDNA ends, the amplification mixture should be electrophoresed and the longest products recovered by gel isolation. An aliquot of this material can then be reamplified for a limited number of cycles.
    b. *New RACE only:* Expect to see one or two extra nucleotides insert between the RNA oligo 3' end, and the 5' end of the gene of interest—these come from the transcription step using T7, T3, or SP6, which can add an extra nucleotide or two to oligo past the end of the template (template-independent transcription).

        Compare the results you get from unTAPped RNA vs TAPped RNA. Junction sites (where the oligo is connected to the 5' end of your gene) in common arise from ligation of the oligo to degraded RNA; unique junctions in the TAPped RNA population represent candidate transcription start sites. If you have "RNA degradation" sites (e.g., TTT'AAA) in your 5' RNA end, you may have substantial numbers of clones that begin at exactly the same nucleotide but that arise from ligation of the oligo to degraded RNA molecules, not from ligation of the oligo to the true 5' end of the RNA.

Look for TATA, CCAAT, and initiator element (Inr) sites at or around your candidate transcription site in the genomic DNA sequence if it is available—you should usually be able to find either a TATA or an Inr.

8. **Further Analysis and Use of RACE Products**
   a. *Cloning:* RACE products can be cloned like any other PCR products.

   OPTION 1: To clone the cDNA ends directly from the amplification reaction (or after gel purification, which is recommended), ligate an aliquot of the products to plasmid vector encoding a one nucleotide 3' overhang consisting of a T on both strands. Such vector DNA is available commercially (Invitrogen's TA Kit) or can be easily and cheaply prepared *(16,32–35)*.

   OPTION 2: The Classic RACE Q_I primer encodes *Hin*dIII, *Sst*I, and *Xho*I restriction enzyme sites. Products can be efficiently cloned into vectors that have been double cut with one of these enzymes and with a blunt-cutting enzyme such as Sma I (note: remember to "polish" the amplification products with Klenow enzyme or T4 DNA polymerase and separate them from residual *Taq* polymerase and dNTP before carrying out the restriction enzyme digest). If clones are not obtained, determine whether the restriction enzyme chosen is cutting the amplified gene fragment a second time, at some internal location in the new and unknown sequence. A somewhat easier strategy is to append a restriction site (not *Hin*dIII, *Sst*I, or *Xho*I) onto the 5' end of the GSP2 primer to allow for the creation of overhanging strands at both ends of the amplified product.

   OPTION 3: A safer and very effective approach is to modify the ends of the primers to allow the creation of overhanging ends using T4 DNA polymerase to chew back a few nucleotides from the amplified product in a controlled manner and Klenow enzyme (or Sequenase) to partially fill in restriction enzyme digested overhanging ends on the vector, as shown in **Fig. 7** (adapted from **refs. 36** and **37**). For another conceptual variation, see Rashtchian et al. *(10)*.

   The advantages of this approach are that (1) it eliminates the possibility that the restriction enzymes chosen for the cloning step will cleave the cDNA end in the unknown region, (2) vector dephosphorylation is not required since vector self-ligation is no longer possible, (3) which means that insert kinasing (and polishing) is not necessary, and (4) which means that insert multimerization and fusion clones are not observed either. In addition, the procedure is more reliable than TA cloning.

   b. *Sequencing.* RACE products can be sequenced directly on a population level using a variety of protocols, including cycle sequencing, from the end at which the gene-specific primers are located. Note that the Classic RACE products can not be sequenced on a population level using the Q_I primer at the unknown end, as individual cDNAs contain different numbers of A residues in their poly (A) tails and, consequently, the sequencing ladder falls out of register after reading through the tail. The 3' end products can be sequenced from their unknown end using the following set of primers: TTTTTTTTTTTTTTTTA, TTTTTTTTTTTTTTTTG, and TTTTTTTTTTTTTTTTC. The non-T nucleotide at the 3' end of the primer forces the appropriate primer to bind to the inner end of the poly (A) tail *(38)*. The other two primers do not participate in the

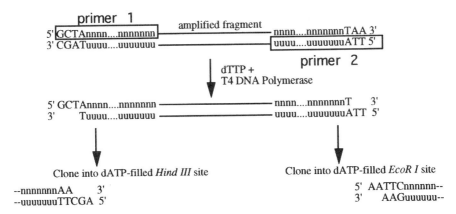

Fig. 7. A safe and easy cloning method. See text for details.

sequencing reaction. Individual cDNA ends, once cloned into a plasmid vector, can be sequenced from either end using gene-specific or vector primers.

c. **Hybridization Probes.** RACE products are generally pure enough that they can be used as probes for RNA and DNA blot analyses. It should be kept in mind that small amounts of contaminating nonspecific cDNAs will always be present. It is also possible to include a T7 RNA polymerase promoter in one or both primer sequences and to use the RACE products in in vitro transcription reactions to produce RNA probes (5). Primers encoding the T7 RNA polymerase promoter sequence do not appear to function as amplification primers as efficiently as the ones listed in the legend to **Fig. 1** (personal observation). Thus, the T7 RNA polymerase promoter sequence should not be incorporated into RACE primers as a general rule.

d. **Construction of Full-Length cDNAs.** It is possible to use the RACE protocol to create overlapping 5' and 3' cDNA ends that can later, through judicious choice of restriction enzyme sites, be joined together through subcloning to form a full-length cDNA. It is also possible to use the sequence information gained from acquisition of the 5' and 3' cDNA ends to make new primers representing the extreme 5' and 3' ends of the cDNA, and to employ them to amplify a *de novo* copy of a full-length cDNA directly from the 3' end cDNA pool (**Figs. 4** and **6**). Despite the added expense of making two more primers, there are several reasons why the second approach is preferred.

First, a relatively high error rate is associated with the PCR conditions for which efficient RACE amplification takes place, and numerous clones may have to be sequenced to identify one without mutations. In contrast, two specific primers from the extreme ends of the cDNA can be used under inefficient but low-error rate conditions (39) for a minimum of cycles to amplify a new cDNA that is likely to be free of mutations. Second, convenient restriction sites are often not available, thus making the subcloning project difficult. Third, by using the second approach, the synthetic poly (A) tail (if present) can be removed from the 5' end of the cDNA. Homopolymer tails appended to the 5' ends of cDNAs have in some cases been reported to inhibit translation. Finally, if alternate promoters, splicing, and polyadenylation signal sequences are being used and result in multiple 5' and 3' ends, it is possible that one might join two cDNA halves that are never actually found together in vivo. Employing primers from the extreme ends of the cDNA as described confirms that the resulting amplified cDNA represents an mRNA actually present in the starting population.

9. Troubleshooting and Controls

a. Problems with Reverse Transcription and Prior Steps

DAMAGED RNA. Electrophorese RNA in 1% formaldehyde minigel and examine integrity of the 18S and 28S ribosomal bands. Discard the RNA preparation if the ribosomal bands are not sharp.

CONTAMINANTS. Ensure that the RNA preparation is free of agents that inhibit reverse transcription, e.g., LiCl and SDS (*see* **ref. 40**) regarding the optimization of reverse transcription reactions).

BAD REAGENTS. To monitor reverse transcription of the RNA, add 20 µCi of ^{32}P-dCTP to the reaction, separate newly created cDNAs using gel electrophoresis, wrap the gel in Saran Wrap and expose it to X-ray film. Accurate estimates of cDNA size can best be determined using alkaline agarose gels, but a simple 1% agarose minigel will suffice to confirm that reverse transcription took place and that cDNAs of reasonable length were generated. Note that adding ^{32}P-dCTP to the reverse transcription reaction results in the detection of cDNAs synthesized both through the specific priming of mRNA and through RNA self-priming. When a gene-specific primer is used to prime transcription (5' end RACE) or when total RNA is used as a template, the majority of the labeled cDNA will actually have been generated from RNA self-priming. To monitor extension of the primer used for reverse transcription, label the primer using T4 DNA kinase and ^{32}P-γATP prior to reverse transcription. Much longer exposure times will be required to detect the labeled primer-extension products than when ^{32}P-dCTP is added to the reaction.

To monitor reverse transcription of the gene of interest, one may attempt to amplify an internal fragment of the gene containing a region derived from two or more exons, if sufficient sequence information is available.

b. Problems with Tailing

BAD REAGENTS. Tail 100 ng of a DNA fragment approx 100–300 bp long for 30 min. In addition, mock tail the same fragment (add everything but the TdT). Run both samples in a 1% agarose minigel. The mock-tailed fragment should run as a tight band. The tailed fragment should have increased in size by 20–200 bp and should appear to run as a diffuse band that trails off into higher molecular weight products. If this not observed, replace the reagents.

MOCK TAIL 25% OF THE CDNA POOL (ADD EVERYTHING BUT THE TDT). Dilute to the same final concentration as the tailed cDNA pool. This serves two purposes. First, although amplification products will be observed using both tailed and untailed cDNA templates, the actual pattern of bands observed should be different. In general, discrete bands are observed using untailed templates after the first set of cycles, and a broad smear of amplified cDNA accompanied by some individual bands is typically observed using tailed templates. If the two samples appear different, this confirms that tailing took place and that the oligo(dT)-tailed Q_T primer is annealing effectively to the tailed cDNA during PCR. Second, observing specific products in the tailed amplification mixture that are not present in the untailed amplification mixture indicates that these products are being synthesized off the end of an A-tailed cDNA template, rather than by annealing of the dT-tailed primer to an A-rich sequence in or near the gene of interest.

c. Problems with Amplification

NO PRODUCT. If no products are observed for the first set of amplifications after 30 cycles, add fresh *Taq* polymerase and carry out an additional 15 rounds of amplification (extra enzyme is not necessary if the entire set of 45 cycles is carried out without interruption at cycle 30). The product is always observed after a total of 45 cycles if efficient amplification is taking place. If no product is observed, carry out a PCR reaction using control templates and primers to ensure the integrity of the reagents.

SMEARED PRODUCT FROM THE BOTTOM OF THE GEL TO THE LOADING WELL. Too many cycles or too much starting material.

NONSPECIFIC AMPLIFICATION, BUT NO SPECIFIC AMPLIFICATION. Check sequence of cDNA and primers. If they all are correct, examine the primers (using a computer program) for secondary structure and self-annealing problems. Consider ordering new primers. Determine whether too much template is being added, or if the choice of annealing temperatures could be improved.

Alternatively, secondary structure in the template may be blocking amplification. Consider adding formamide *(41)* or ^{7}aza-GTP (in a 1:3 ratio with dGTP) to the reaction to assist polymerization. The ^{7}aza-GTP can also be added to the reverse transcription reaction.

THE LAST FEW BASE PAIRS OF THE 5' END SEQUENCE DO NOT MATCH THE CORRESPONDING GENOMIC SEQUENCE. Be aware that reverse transcriptase and T7 and T3 RNA Polymerase can add on a few extra template-independent nucleotides.

INAPPROPRIATE TEMPLATES. To determine whether the amplification products observed are being generated from cDNA or whether they derive from residual genomic DNA or contaminating plasmids, pretreat an aliquot of the RNA with RNAse A.

Acknowledgments

Portions of this chapter have been adapted and reprinted by permission of the publisher from "Cloning PCR Products" by Michael A. Frohman in *The Polymerase Chain Reaction*, pages 14–37. Copyright ©1994 by Birkhauser Boston, and from "On beyond RACE (Rapid amplification of cDNA ends)" by Michael A. Frohman in *PCR Methods and Applications*. Copyright ©1995 by Cold Spring Harbor Press.

References

1. Frohman, M. A., Dush, M. K., and Martin, G. R. (1988) Rapid production of full-length cDNAs from rare transcripts by amplification using a single gene-specific oligonucleotide primer. *Proc. Natl. Acad. Sci. USA* **85,** 8998–9002.
2. Loh, E. L., Elliott, J. F., Cwirla, S., Lanier, L. L., and Davis, M. M. (1989) Polymerase chain reaction with single sided specificity: analysis of T cell receptor delta chain. *Science* **243,** 217–220.
3. Ohara, O., Dorit, R. I., and Gilbert, W. (1989) One-sided PCR: the amplification of cDNA. *Proc. Natl. Acad. Sci. USA* **86,** 5673–5677.
4. Frohman, M. A. (1989) Creating full-length cDNAs from small fragments of genes: amplification of rare transcripts using a single gene-specific oligonucleotide primer, in *PCR Protocols and Applications: A Laboratory Manual* (Innis, M., Gelfand, D., Sninsky, J., and White, T., eds.), pp. 28–38.
5. Frohman, M. A. and Martin, G. R. (1989) Rapid amplification of cDNA ends using nested primers. *Techniques* **1,** 165–173.
6. Dumas, J. B., Edwards, M., Delort, J., and Mallet, J. (1991) Oligodeoxyribonucleotide ligation to single-stranded cDNAs: a new tool for cloning 5' ends of mRNAs and for constructing cDNA libraries by *in vitro* amplification. *Nucleic Acids Res.* **19,** 5227–5233.

7. Fritz, J. D., Greaser, M. L., and Wolff, J. A. (1991) A novel 3' extension technique using random primers in RNA-PCR. *Nucleic Acids Res.* **119,** 3747.
8. Borson, N. D., Salo, W. L., and Drewes, L. R. (1992) A lock-docking oligo(dT) primer for 5' and 3' RACE PCR. *PCR Methods Applic.* **2,** 144–148.
9. Jain, R., Gomer, R. H., and Murtagh, J. J. J. (1992) Increasing specificity from the PCR-RACE technique. *BioTechniques* **12,** 58,59.
10. Rashtchian, A., Buchman, G. W., Schuster, D. M., and Berninger, M. S. (1992) Uracil DNA glycosylase-mediated cloning of PCR-amplified DNA: application to genomic and cDNA cloning. *Anal. Biochem.* **206,** 91–97.
11. Schuster, D. M., Buchman, G. W., and Rastchian, A. (1992) A simple and efficient method for amplification of cDNA ends using 5' RACE. *Focus* **14,** 46–52.
12. Bertling, W. M., Beier, F., and Reichenberger, E. (1993) Determination of 5' ends of specific mRNAs by DNA ligase-dependent amplification. *PCR Methods Applic.* **3,** 95–99.
13. Frohman, M. A. (1993) Rapid amplification of cDNA for generation of full-length cDNA ends: thermal RACE. *Methods Enzymol.* **218,** 340–356.
14. Monstein, H. J., Thorup, J. U., Folkesson, R., Johnsen, A. H., and Rehfeld, J. F. (1993) cDNA deduced procionin—structure and expression in protochordates resemble that of procholecystokinin in mammals. *FEBS Lett.* **331,** 60–64.
15. Templeton, N. S., Urcelay, E., and Safer, B. (1993) Reducing artifact and increasing the yield of specific DNA target fragments during PCR-RACE or anchor PCR. *BioTechniques* **15,** 48–50.
16. Frohman, M. A. (1994) Cloning PCR products: strategies and tactics, in *PCR. The Polymerase Chain Reaction. Methods in Molecular Biology Series* (Mullis, K. B., Ferre, F., and Gibbs, R. A., eds.), pp. 14–37.
17. Datson, N. A., Duyk, G. M., Van Ommen, J. B., and Den Dunnen, J. T. (1994) Specific isolation of 3'-terminal exons of human genes by exon trapping. *Nucleic Acids Res.* **22,** 4148–4153.
18. Ruberti, F., Cattaneo, A., and Bradbury, A. (1994) The use of the RACE method to clone hybridoma cDNA when V region primers fail. *J. Immunol. Methods* **173,** 33–39.
19. Tessier, D. C., Brousseau, R., and Vernet, T. (1986) Ligation of single-stranded oligodeoxyribonucleotides by T4 RNA ligase. *Anal. Biochem.* **158,** 171–178.
20. Mandl, C. W., Heinz, F. X., Puchhammer-Stockl, E., and Kunz, C. (1991) Sequencing the termini of capped viral RNA by 5'-3' ligation and PCR. *BioTechniques* **10,** 484–486.
21. Volloch, V., Schweizer, B., Zhang, X., and Rits, S. (1991) Identification of negative-strand complements to cytochrome oxidase subunit III RNA in Trypanosoma brucei. *Biochemistry* **88,** 10,671–10,675.
22. Brock, K. V., Deng, R., and Riblet, S. M. (1992) Nucleotide sequencing of 5' and 3' termini of bovine viral diarrhea virus by RNA ligation and PCR. *Virol. Methods* **38,** 39–46.
23. Bertrand, E., Fromont-Racine, M., Pictet, R., and Grange, T. (1993) Visualization of the interaction of a regulatory protein with RNA *in vivo*. *Proc. Natl. Acad. Sci. USA* **90,** 3496–3500.
24. Fromont-Racine, M., Bertrand, E., Pictet, R., and Grange, T. (1993) A highly sensitive method for mapping the 5' termini of mRNAs. *Nucleic Acids Res.* **21,** 1683,1684.
25. Liu, X. and Gorovsky, M. A. (1993) Mapping the 5' and 3' ends of tetrahymena-thermophila mRNAs using RNA ligase mediated amplification of cDNA ends (RLM-RACE). *Nucleic Acids Res.* **21,** 4954–4960.
26. Sallie, R. (1993) Characterization of the extreme 5' ends of RNA molecules by RNA ligation-PCR. *PCR Methods Applic.* **3,** 54–56.
27. Skinner, T. L., Kerns, R. T., and Bender, P. K. (1994) Three different calmodulin-encoding cDNAs isolated by a modified 5'-RACE using degenerate oligodeoxyribonucleotides. *Gene* **151,** 247–251.

28. Frohman, M. A., Dickinson, M. E., Hogan, B. L. M., and Martin, G. R. (1993) Localization of two new and related homeobox-containing genes to chromosomes 1 and 5, near the phenotypically similar mutant loci *dominant hemimelia (Dh)* and *hemimelic extra-toes (Hx)*. *Mouse Genome* **91,** 323–325.
29. Crowe, J. S., Cooper, H. J., Smith, M. A., Sims, M. J., Parker, D., and Gewert, D. (1991) Improved cloning efficiency of polymerase chain reaction (PCR) products after proteinase K digestion. *Nucleic Acids Res.* **19,** 184.
30. Coleclough, C. (1987) Use of primer-restriction end adapters in cDNA cloning. *Methods Enzymol.* **154,** 64–83.
31. Don, R. H., Cox, P. T., Wainwright, B. J., Baker, K., and Mattick, J. S. (1991) Touchdown PCR to circumvent spurious priming during gene amplification. *Nucleic Acids Res.* **19,** 4008.
32. Mead, D. A., Pey, N. K., Herrnstadt, C., Marcil, R. A., and Smith, L. A. (1991) A universal method for direct cloning of PCR amplified nucleic acid. *Biotechnology* **9,** 657–663.
33. Marchuk, D., Drumm, M., Saulino, A., and Collins, F. S. (1991) Construction of T-vector, a rapid and general system for direct cloning of unmodified PCR products. *Nucleic Acids Res.* **19,** 1154.
34. Kovalic, D., Kwak, J. H., and Weisblum, B. (1991) General method for direct cloning of DNA fragments generated by the polymerase chain reaction. *Nucleic Acids Res.* **19,** 4650.
35. Holton, T. A. and Graham, M. W. (1991) A simple and efficient method for direct cloning of PCR products using ddT-tailed vectors. *Nucleic Acids Res.* **19,** 1156.
36. Stoker, A. W. (1990) Cloning of PCR products after defined cohesive termini are created with T4 DNA polymerase. *Nucleic Acids Res.* **18,** 4290.
37. Iwahana, H., Mizusawa, N., Ii, S., Yoshimoto, K., and Itakura, M. (1994) An end-trimming method to amplify adjacent cDNA fragments by PCR. *BioTechniques* **16,** 94–98.
38. Thweatt, R., Goldstein, S., and Reis, R. J. S. (1990) A universal primer mixture for sequence determination at the 3' ends of cDNAs. *Anal. Biochem.* **190,** 314.
39. Eckert, K. A. and Kunkel, T. A. (1990) High fidelity DNA synthesis by the *Thermus aquaticus* DNA polymerase. *Nucleic Acids Res.* **18,** 3739–3745.
40. Sambrook, J., Fritsch, E. F., and Maniatis, T. (eds.) (1989) *Molecular Cloning: A Laboratory Manual*, Cold Spring Harbor Lab. Press, Cold Spring Harbor, NY, pp. 82,83.
41. Sarker, G., Kapelner, S., and Sommer, S. S. (1990) Formamide can dramatically improve the specificity of PCR. *Nucleic Acids Res.* **18,** 7465.

41

cDNA Library Construction Using Streptavidin-Paramagnetic Beads and PCR

Kris N. Lambert and Valerie M. Williamson

1. Introduction

cDNA clones of genes expressed in small amounts of material can be hard to obtain because the construction of conventional cDNA libraries requires microgram amounts of poly $(A)^+$ RNA *(1)*. The polymerase chain reaction (PCR), which is commonly used to amplify tiny amounts of DNA *(2,3)*, has been adapted to facilitate the construction of cDNA libraries from small quantities of poly $(A)^+$ RNA *(4–7)*. Most of these cDNA amplification methods require multiple purification or precipitation steps to remove primers and change buffers. These steps result in significant loss of material and compromise the quality of the final library. The method presented here eliminates precipitation and chromatography steps so that all cDNA synthesis and modification reactions are conducted in a single tube.

Briefly, a biotinylated oligonucleotide containing a 3'-dT_{25} tail is bound to streptavidin linked to paramagnetic beads. The poly $(A)^+$ RNA is purified by annealing to the biotinylated oligonucleotide on the beads. The attraction of the beads to a magnet allows for rapid solution changes by placing the sample tube into a magnetic stand to hold the beads against the side of the tube as the solution is pipetted off. The first-strand cDNA is synthesized using the biotinylated oligonucleotide as a primer. The biotinylated oligonucleotide later serves as a 3' end priming site during the PCR amplification. The second-strand cDNA is then synthesized and an adapter is ligated to the 5' end of the cDNA to generate the second priming site. The cDNA is amplified using oligonucleotides complementary to the biotinylated oligonucleotide and the adapter sequences. Using this method, microgram amounts of amplified cDNA can be generated in one to two days from less than one nanogram of poly $(A)^+$ RNA.

2. Materials

2.1. Generation of First-Strand cDNA Linked to Paramagnetic Beads

1. Diethyl pyrocarbonate (DEPC) H_2O: Stir distilled water with 0.1% DEPC for 12 h or longer, then autoclave. Use DEPC-treated H_2O to prepare all solutions for first-strand cDNA synthesis.

Table 1
Primers and Adapter Used in cDNA Library Construction

Name	Sequence
Biotinylated primer[a]	B-5'-GGCGCCCTGGTTCGGCCCACATTGACGTCGACTATCCA(T)$_{25}$-3'
Adapter[b]	5'-CTCGCTCGCCCACGTATATATAGCCTACTATATC-3'
	3'-TGCATATATATCGGATGATATAG-5'-P
Primer A	5'-CTCGCTCGCCCACGTATATATAGCCTACTATATC-3'
Primer B	5'-CTGGTTCGGCCCACATTGACGTCGACTATCCA-3'

P = phosphate, B = biotin.
[a]See **Note 1**.
[b]See **Note 2**.

2. Dynabeads M-280 Streptavidin (product no. 112.05; Dynal, Inc., Great Neck, NY). Store at 4°C. Magnetic microcentrifuge tube stand (product no. 120.01; Dynal).
3. 2X Binding buffer: 20 mM Tris-HCl, pH 7.5, 1M LiCl, 2 mM EDTA. Store at 4°C.
4. Wash buffer: 10 mM Tris-HCl, pH 7.5, 0.15M LiCl, 1 mM EDTA. Store at 4°C.
5. Biotinylated primer (100 pmol/µL), see **Table 1** for sequence (see also **Note 1**). Store at –20°C.
6. Superscript II reverse transcriptase (200 U/µL; Gibco-BRL, Grand Island, NY). Store at –20°C.
7. 5X First-strand buffer: 250 mM Tris-HCl, pH 8.3, 375 mM KCl, 15 mM MgCl$_2$. Store at –20°C.
8. Reverse transcriptase mix: 4 µL 5X first-strand buffer, 10 µL DEPC-treated H$_2$O, 2 µL 0.1 M dithiothreitol (DTT), 1 µL 10 mM dNTP mix (10 mM each dATP, dCTP, dGTP, and dTTP), 1 µL RNasin (10 U/µL; Gibco-BRL); prepare just before use.

2.2. cDNA Amplification Using Oligonucleotide Adapters

1. 5X Second-strand cDNA synthesis buffer: 100 mM Tris-HCl, pH 6.9, 450 mM KCl, 23 mM MgCl$_2$, 0.75 mM β-NAD, 50 mM (NH$_4$)$_2$SO$_4$. Store at –20°C.
2. Second-strand cDNA reaction mix: 23.3 µL distilled H$_2$O, 7.5 µL 5X second-strand cDNA synthesis buffer, 0.75 µL of 10 mM dNTP mix, 0.25 µL *E. coli* RNase H (2 U/µL; Gibco-BRL); 1 µL of *E. coli* DNA polymerase I (10 U/µL; Gibco-BRL), 0.25 µL of *E. coli* DNA ligase (10 U/µL; Gibco-BRL); prepare just before use.
3. T4 DNA polymerase (5 U/µL; Gibco-BRL). Store at –20°C.
4. Adapter (50 pmol/µL). Mix equal amounts of primer A (100 pmol/µL) and primer B (100 pmol/µL). Adapter and primer sequences are shown in **Table 1** (see **Note 2**).
5. Adapter buffer: 66 mM Tris-HCl, pH 7.6, 10 mM MgCl$_2$.
6. 5X Ligation buffer: 330 mM Tris-HCl, pH 7.6, 50 mM MgCl$_2$, 5 mM ATP. Store at –20°C.
7. Adapter ligation mix: 10.2 µL H$_2$O, 1 µL adapter, 4 µL 5X ligation buffer, 2.8 µL 0.1 M DTT, 2 µL T4 DNA ligase (1 U/µL, Gibco-BRL); prepare just before use.
8. 10X PCR reaction buffer: 400 mM Tricine-KOH, pH 9.2, 150 mM KOAc, 35 mM Mg(OAc)$_2$, 750 µg/mL bovine serum albumin (BSA). Store at –20°C.
9. PCR reaction mix: 5 µL 10X PCR reaction buffer, 41 µL H$_2$O, 1 µL of 10 mM dNTP mix, 1 µL (10 pmol) primer A, 1 µL (10 pmol) primer B, 1 µL Klenow*Taq* polymerase mix (Clontech, Palo Alto, CA), prepare just before use.

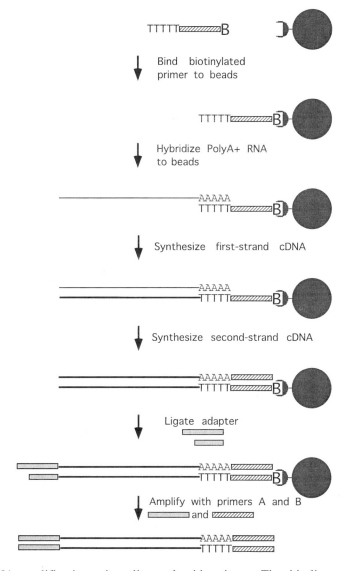

Fig. 1. cDNA amplification using oligonucleotide primers. The thin line represents RNA, the thick line represents cDNA, the cross-hatched bars represent primer B and its complement as well as corresponding sequence of the biotinylated primer. The gray bars represent primer A, its complement and adapter sequences. B = biotin.

3. Methods

3.1. Generation of First-Strand cDNA Linked to Superparamagnetic Beads

The starting material is dried, total nucleic acid prepared by an appropriate method (*see* **Note 3**). A biotinylated oligonucleotide containing a 3' $d(T)_{25}$ tail is prebound to Dynabeads M-280 streptavidin *(8)*. Poly $(A)^+$ RNA is allowed to hybridize to the oligo(dT) tail on the biotinylated primer. The first-strand cDNA is synthesized using the biotinylated oligonucleotide attached to the beads as a primer *(9)* (*see* **Fig. 1**).

1. Binding the biotinylated primer to the Dynabeads: Add 20 µL (200 µg) Dynabeads® M-280 Streptavidin to a 0.2-mL microfuge tube. Place the tube in the magnetic stand. The beads will bind to the side of the tube adjacent to the magnet.
2. Remove the supernatant with a micropipet without disturbing the beads.
3. Remove the tube from the stand and resuspend the beads in 25 µL of 2X binding buffer.
4. Remove the buffer and add a fresh 25 µL of 2X binding buffer to the beads.
5. Add 2 µL (50 pmol/µL) of the biotinylated primer and allow the primer to bind for 15 min at 22°C, keeping beads suspended (see **Note 4**).
6. Wash the beads three times with 25 µL of wash buffer, then resuspend in 25 µL of 2X binding buffer.
7. Poly (A)$^+$ RNA isolation: Resuspend dried total nucleic acid in 25 µL DEPC-treated H$_2$O, heat to 65°C for 2 min, and then cool on ice.
8. Add the total nucleic acid preparation to the beads with the prebound biotinylated oligonucleotide and allow the poly (A)$^+$ RNA to hybridize to the oligo d(T)$_{25}$ on the beads for 15 min at 22°C (see **Note 4**).
9. Remove the binding buffer and unhybridized nucleic acids from the beads.
10. Add 50 µL of 2.5X first-strand buffer to the beads, resuspend, and then completely remove the buffer.
11. First-strand cDNA synthesis: Add 19 µL of reverse transcriptase mix and heat to 37°C for 2 min.
12. Add 1 µL reverse transcriptase and mix.
13. Continue to incubate the reaction at 37°C for 60 min to allow extension from the 3' end of the biotinylated primer (see **Note 4**). The first-strand cDNAs are now linked to the beads.

3.2. cDNA Amplification Using Oligonucleotide Adapters

The second-strand cDNA is synthesized on the beads, and an adapter is ligated to the free end of the cDNA (see **Fig. 1**) *(10,11)*. The cDNA is then amplified by PCR.

1. Second-strand synthesis: Bind the beads and remove 15.5 µL of the reverse transcriptase mix. Add 33 µL second-strand cDNA reaction mix to the remaining 4.5 µL of reverse transcriptase mix, and then incubate the reaction at 16°C for 2 h (see **Note 4**).
2. Blunt-end the cDNA: Add 0.5 µL T4 DNA polymerase to the beads and incubate at 16°C for 5 min.
3. Remove the buffer and wash the beads twice in 75 µL adapter buffer, being sure to completely remove buffer after the last wash (see **Note 5**).
4. Adapter ligation: Add 20 µL adapter ligation mix and incubate at 16°C for 16 h or longer (see **Note 4**).
5. Amplification of cDNA: Bind the beads and remove the ligation mix. Wash the beads once with 75 µL adapter buffer.
6. Remove the adapter buffer and add 50 µL PCR reaction mix. Amplify using the following parameters: 94°C for 30 s; 35 cycles of 94°C for 30 s, 61°C for 30 s, and 68°C for 3 min; 68°C for 30 min.
7. Assessing amplification: Remove 5 µL amplified cDNA and fractionate on a 2% agarose gel. The size range of the PCR products is likely to be slightly smaller than the average size of the starting poly (A)$^+$ RNA reflecting selective amplification of smaller cDNAs. The representation of control genes in the cDNA can be determined by Southern hybridization or PCR amplification of the cDNA with appropriate primers. Control genes that differ in abundance and transcript size are most useful to test for bias for abundantly expressed cDNAs or short transcripts.

3.3. Cloning Amplified cDNA

To generate a small library of thousands of primary transformants, amplified cDNA can be cloned into a plasmid vector using any of several available commercial PCR cloning kits (*see* **Note 6**). If a complex cDNA library is required, ends of cDNA produced here can be filled in with T4 polymerase to produce blunt ends, adapters can be added to the blunt-ended cDNA followed by ligation into a phage vector (*see* **Note 7**).

4. Notes

1. The oligonucleotide primer for first-strand synthesis should be biotinylated when synthesized using a biotin phosphoramidite. The spacer arm between the biotin and the oligonucleotide should be at least six carbons long to prevent steric hindrance during binding to the streptavidin-linked beads.
2. The adapter was made from two oligonucleotides, primer A and its partial complement (*see* **Table 1**), which is phosphorylated at the 5' end. To produce the adapter, combine equal volumes of each oligonucleotide in adapter buffer at 100 pmol/µL in a small beaker and heat the mixture to 80°C, then allow it to slowly cool to room temperature. The adapters are stored at –20°C.
3. The amount of beads can be scaled up proportionally but should not be reduced because there is some nonspecific binding of beads to the pipet tips and microcentrifuge tubes. The nonspecific sticking of the beads can be reduced by using siliconized microcentrifuge tubes and pipet tips. The lower limit of mRNA that can be used in this protocol is not known, but Karrer et al. *(12)* used a similar protocol to construct a cDNA library from the contents of a single plant cell. The method of nucleic acid extraction is dependent on the biological system under study. We found that a stainless steel tissue pulverizer (Fisher Scientific, Pittsburgh, PA) cooled in liquid nitrogen was particularly useful for reducing small amounts of frozen tissue to a fine powder. A number of nucleic acid extraction protocols should be satisfactory. We used a simple phenol/chloroform extraction method *(13)*.
4. In all binding and enzymatic manipulation of nucleic acids on the beads, the beads were kept in suspension by rotating the tubes on a rotating mixer (Labquake shaker, cat. no. 400-110, Labindustries, Berkeley, CA).
5. The beads must be washed thoroughly to remove residual T4 DNA polymerase, which might otherwise digest the cDNA and adapters during the ligation step.
6. We have successfully used the TA cloning kit and the TOPO TA cloning kits from Invitrogen (Carlsbad, CA) to generate small cDNA libraries. Higher numbers of primary transformants can be obtained by using supercompetent *Escherichia coli* instead of the competent cells that come with the kits. While several manufacturers produce supercompetent *E. coli*, cells purchased from Stratagene (La Jolla, CA) have worked well in our hands.
7. For constructing a phage cDNA library we have found it convenient to purchase a cDNA synthesis kit such as the Super Script Choice System from Gibco BRL Products (Gaithersburg, MD). These kits include all the enzymes for first and second strand synthesis as well as the adapters and phage. The kits also can be purchased without phage vectors as a convenient source for all enzymes used in the cDNA library construction protocol presented here.

References

1. Gasser, C. S., Budelier, K. A., Smith, A. G., Shah, D. M., and Fraley, R. T. (1989) Isolation of tissue-specific cDNAs from tomato pistils. *The Plant Cell* **1,** 15–24.
2. Saiki, R. K., Gelfand, D. H., Stoffel, S., Scharf, S. J., Higuchi, R., Horn, G. T., Mullis, K. B., and Erlich, H. A. (1988) Primer-directed enzymatic amplification of DNA with a thermostable DNA polymerase. *Science* **239,** 487–491.
3. Arnheim, N., Li, H., and Cui, X. (1990) PCR analysis of DNA sequences in single cells: single sperm gene mapping and genetic disease diagnosis. *Genomics* **8,** 415–419.
4. Akowitz, A. and Manuelidis, L. (1989) A novel cDNA/PCR strategy for efficient cloning of small amounts of undefined RNA. *Gene* **81,** 295–306.
5. Welsh, J., Liu, J.-P., and Efstratiadis, A. (1990) Cloning of PCR-amplified total cDNA: construction of a mouse oocyte library. *Genet. Anal. Techn. Appl.* **7,** 5–17.
6. Domec, C., Garbay, B., Fournier, M., and Bonnet, J. (1990) cDNA library construction from small amounts of unfractionated RNA: association of cDNA synthesis with polymerase chain reaction amplification. *Anal. Biochem.* **188,** 422–426.
7. Jepson, I., Bray, J., Jenkins, G., Schuch, W., and Edwards, K. (1991) A rapid procedure for the construction of PCR cDNA libraries from small amounts of plant tissue. *Plant Mol. Biol. Rep.* **9,** 131–138.
8. Rosok, O., Odeberg, J., Rode, M., Stokke, T., Funderud, S., Smeland, E., and Lundeberg, J. (1996) Solid-phase method for differential display of genes expressed in hematopoietic stem cells. *Biotechniques* **21,** 114–121.
9. Raineri, I., Moroni, C., and Senn, H. P. (1991) Improved efficiency for single-sided PCR by creating a reusable pool of first-strand cDNA coupled to a solid phase. *Nucleic Acids Res.* **19,** 4010.
10. Gubler, U. and Hoffman, B. J. (1983) A simple and very efficient method for generating cDNA libraries. *Gene* **25,** 263–269.
11. Fernandez, J. M., Mc Atee, C., and Herrnstadt, C. (1990) Advances in cDNA technology. *American Biotechnol. Lab.* **8,** 46–47.
12. Karrer, E. E., Lincoln, J. E., Hogenhout, S., Bennett, A. B., Bostock, R. M., Martineau, B., Lucas, W. J., Gilchrist, D. G., and Alexander, D. (1995) In situ isolation of mRNA from individual plant cells: creation of cell-specific cDNA libraries. *Proc. Natl. Acad. Sci. USA* **92,** 3814–3818.
13. Rochester, D. E., Winer, J. A., and Shah, D. M. (1986) The structure and expression of maize genes encoding the major heat shock protein, hsp70. *EMBO J.* **5,** 451–458.

42

Rapid (Ligase-Free) Subcloning of Polymerase Chain Reaction Products

Alan R. Shuldiner and Keith Tanner

1. Introduction

The polymerase chain reaction (PCR) is a versatile, widely used method for the production of a very large number of copies of a specific DNA molecule *(1,2)*. For some applications, it is advantageous to subclone the PCR product into a plasmid vector for subsequent replication in bacteria *(3–6)*. Subcloning the PCR product into a plasmid vector has several advantages: The amplified fragment can be sequenced with greater reliability, only one allele is sequenced per clone, and the vector containing the PCR product may be used for other molecular biological experiments, e.g., in vitro transcription, transfection, and further amplification in bacteria.

Although conventional strategies, such as blunt-end or sticky-end ligation, can be very successful for subcloning most DNA fragments, the subcloning of DNA generated by PCR is often very difficult *(7–9)*. We describe a rapid and versatile method to subclone PCR products directionally into a specific site of virtually any plasmid vector *(8,9)*. Ligase-free subcloning of PCR products has several advantages over conventional strategies: It does not require DNA ligase, it requires only four primers, two of which are complementary to the plasmid vector, and therefore may be used repeatedly, and it may be accomplished in a single day. Typically, hundreds to thousands of colonies per transformation are obtained (approximate efficiency 5×10^3 to 5×10^4 colonies/microgram of PCR insert). With this method, PCR products of up to 1.7 kb in length have been subcloned successfully.

1.1. Theory of the Method

1.1.1. Plasmid Linearization and Primer Design

The first step in ligase-free subcloning is to linearize the plasmid vector at the desired site with the appropriate restriction endonuclease(s). The second step is to perform PCR on genomic DNA or the cDNA of interest. Like conventional PCR, the primers (primers a and b in **Fig. 1**) must contain sequences at their 3' ends (approx 20–25 nucleotides) that are complementary to opposite strands of the target sequence at a predetermined distance from each other. For ligase-free subcloning, primers a and b must also

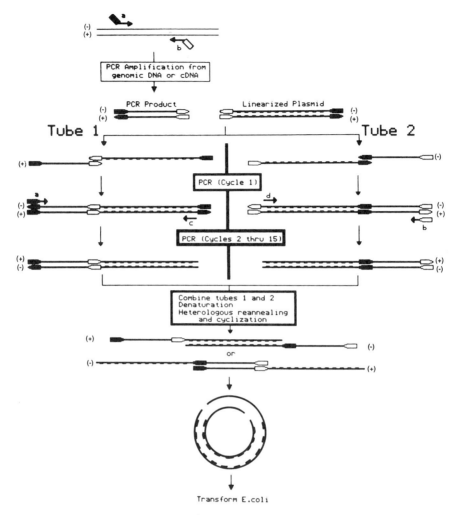

Fig. 1. Schematic of ligase-free subcloning. The 5' addition sequences of primers a and b are designated by closed and open boxes, respectively. DNA sequences corresponding to the PCR-amplified product are shown as straight lines, whereas DNA sequences corresponding to the plasmid vector are shown as hatched lines.

contain sequences at their 5' ends, approx 24 nucleotides in length (designated the 5' addition sequences), that are identical to each of the 3' ends of the linearized plasmid (opened and closed boxes in **Fig. 1**; also, *see* example in **Fig. 2**). Since the two 3' ends of the linearized plasmid are different from each other, the PCR fragment may be subcloned directionally simply by choosing the appropriate 5' addition sequence for each primer.

1.1.2. PCR Amplification and Ligation by Overlap Extension

PCR amplification is accomplished using primers a and b, which results in large amounts of the PCR product containing the 5' addition sequences at each end. Next, the PCR product is freed from excess primers by ultrafiltration and divided into two tubes

Nucleotide sequence of pGEM4Z linearized with SmaI:

```
                                                      primer d                                      primer a
5'-GGGGATCCTCTAGAGTCGACCTGCaggcatgcaagcttgtctctccctatagtgagtcgtattagagc...(  3 Kb)...ttgatttagtgacactataqAATACGAATTCGAGCTCGGTACCC-3'
3'-CCCCTAGGAGATCTCAGCTGGACgtccgtacgttcgaacagagagatcactcagcatatctcg...(  3 Kb)...aacctaaatcactgtgatatcTTATGCTTAAGCTCGAGCCATGGG-5'
      primer b                                                                                  primer c
```

5' addition sequences should be identical to the 3' ends (upper case bold and underlined nucleotides):

```
              Primer a:  5'-AATACGAATTCGAGCTCGGTACCC.....3'
              Primer b:  5'-GCAGGTCGACTCTAGAGGATCCCC.....3'
```

Primers c and d should be complementary to opposite strands of the plasmid internal to the 5' addition sequences (lower case bold and underlined nucleotides):

```
              Primer c:  5'-tatagtgtcacctaaatccaa-3'
              Primer d:  5'-tatagtgagtgctattagagc-3'
```

(Note that primers c and d do not need to be immediately adjacent to primers a and b.)

Fig. 2. Example of primer design for ligase-free subcloning into the *SmaI* site of pGEM4Z (Promega Biotec, Madison, WI).

each containing the linearized plasmid vector. A second PCR reaction is performed. Tube 1 also contains primer a, the same primer that was used during the initial amplification of the PCR product, and primer c, a primer that is complementary to the strand of the plasmid vector that contains the primer a addition sequence (the [+] strand in **Fig. 1**), in a region internal to the primer a addition sequence (**Figs. 1** and **2**). Similarly, tube 2 contains primer b, the same primer that was used for the initial amplification of the PCR product, and primer d, a primer that is complementary to the strand of the plasmid vector that contains the primer b addition sequence (the [–] strand in **Fig. 1**), in a region internal to the primer b addition sequence (**Figs. 1** and **2**).

During the first cycle of the second PCR, denaturation and annealing result in hybridization of the 3' ends of the PCR product (the complements of the 5' addition sequences) to the complementary 3' ends of the linearized plasmid (**Fig. 1**). During the extension step of the first cycle, these overlapping 3' ends act as templates for each other. Extension results in the "ligation" of the PCR fragment to the linearized plasmid (ligation by overlap extension) *(10,11)*.

Amplification during subsequent PCR cycles (cycles 2–20) with the respective primers in each tube results in large amounts of linear double-stranded DNA that contain the plasmid vector ligated to the PCR insert at one end (tube 1 in **Fig. 1**), or to the other end (tube 2 in **Fig. 1**). Although not shown in the figure, hybridization may also occur between the 5' ends of the PCR product and the complementary 5' ends of the linearized plasmid. However, these hybrid products cannot act as templates for *Taq* DNA polymerase.

1.1.3. Denaturation, Heterologous Annealing, and Cyclization

After the second PCR reaction, the reaction mixtures in tubes 1 and 2 are combined and denaturation of double-stranded DNA into single-stranded DNA is accomplished by treating with alkali. After neutralization and dilution of the denatured DNA, the single-stranded DNA may anneal into several different products. Complementary DNA strands may reanneal to form the same linear double-stranded products that were present in tubes 1 and 2 before the denaturation step (not shown in **Fig. 1**). Alternatively, single-stranded DNA products from tube 1 may anneal with single-stranded DNA products from tube 2, resulting in DNAs that are partly double-stranded and contain long single-stranded 5' or 3' overhangs (heterologous annealing in **Fig. 1**). Since these long 5' or 3' overhangs are complementary to each other, annealing at low DNA concentrations results in cyclization.

Although the cyclized product contains two nicks (noncovalently linked ends), it may be used directly to transform competent *E. coli*. Once inside the bacterial cell, the two nicks are covalently joined and the recombinant plasmid is replicated.

2. Materials

1. Plasmid vector.
2. Appropriate restriction endonuclease(s) and buffer(s) for linearization of the plasmid vector.
3. Oligonucleotide primers (*see* **Note 1**). Prepare the four oligonucleotide primers required for ligase-free subcloning by synthesizing chemically with the last dimethoxytrityl (DMT) group off. Cleave the oligonucleotide from the resin by treatment with concentrated ammonia at room temperature, and deprotect by heating to 55°C in concentrated ammonia

overnight. The oligonucleotides may be used for PCR after desalting through a G-25 spin column (Boehringer Mannheim Biochemicals, Indianapolis, IN). Alternatively, the last DMT group may be left on and the oligonucleotides may be purified with NENsorb Prep columns (New England Nuclear, Boston, MA) according to the manufacturer's directions.
4. 10X GeneAmp PCR buffer: 500 mM KCl, 100 mM Tris-HCl, pH 8.3, at 25°C, 15 mM MgCl$_2$, and 0.1 mg/mL of gelatin. GeneAmp is a trademark of Perkin-Elmer/Cetus (Norwalk, CT).
5. dNTPs: 10 mM each.
6. *Taq* DNA polymerase.
7. Paraffin oil.
8. Autoclaved, distilled, deionized water.
9. Phenol equilibrated in TE buffer (10 mM Tris-HCl, pH 8.0, at 25°C, 1 mM EDTA).
10. Chloroform:isoamyl alcohol: 24:1, v/v.
11. tRNA: 10 mg/mL in sterile water.
12. Sterile, 5 M NaCl.
13. Autoclaved sterile Eppendorf tubes.
14. Vacuum dessicator.
15. Ultrafree-MC-100 and Ultrafree-MC-30 ultrafiltration devices (Millipore, Bedford, MA) (*see* **Note 2**).
16. 50 mM Tris-HCl buffer, pH 6.6, at 25°C.
17. DNA size marker: *Hae*III digest of ϕX174.
18. 2 M NaOH.
19. Competent DH5-α *E. coli* (Bethesda Research Laboratories, Gaithersburg, MD; theoretical transformation efficiency >1 × 10^9 transformants/μg PUC19).
21. LB agar plates with the appropriate antibiotic.
22. 10,000X ethidium bromide solution: 5 mg/mL in water. Keep in dark container at 4°C for up to 12 mo. Wear gloves when handling ethidium bromide.
23. High-melting-point agarose.
24. NuSeive GTG agarose (FMC Bioproducts, Rockland, ME).
25. 95% Ethanol.
26. 70% Ethanol.

3. Methods
3.1. Plasmid Linearization

1. Linearize 1–2 μg of the plasmid vector by digestion with the appropriate restriction endonuclease(s). Incubate the plasmid vector, the appropriate buffer, and restriction enzyme(s) in a final volume of 20 μL at the appropriate temperature for 2–3 h in an autoclaved Eppendorf tube.
2. Add 30 μL of sterile water and 1 μL of tRNA (10 mg/mL) to the reaction mixture.
3. Extract by adding 25 μL of phenol that has been pre-equilibrated with TE buffer. Vortex. Add 50 μL of chloroform:isoamyl alcohol. Vortex. Centrifuge in a microfuge at room temperature for 30 s to separate the phases. Remove the upper (aqueous) phase (approx 50 μL) and place in a clean autoclaved Eppendorf tube.
4. Repeat **step 3** once.
5. Precipitate the linearized plasmid DNA by adding 12.5 μL of 5 M NaCl, vortex briefly, then add 125 μL of cold 95% ethanol. Incubate at –20°C for at least 3 h.
6. Recover the linearized plasmid DNA by centrifugation in a microfuge (12,000*g*) at 4°C for 30 min. Pour off the supernatant and wash the pellet with 250 μL of cold 70% ethanol. Carefully decant the supernatant, remove excess liquid from the walls of the Eppendorf tube with a cotton swab, and vacuum-dry the pellet in a vacuum dessicator or Speed-Vac for 10 min.

7. Resuspend the pellet in 20 µL of sterile water. Confirm that the plasmid was linearized and determine the approximate DNA concentration by running 10% (2 µL) of the linearized plasmid DNA on a 1% agarose gel with 100 ng of uncut plasmid and a known amount of a HaeIII-digest of φX174 DNA size marker. DNA should be visualized by staining in ethidium bromide solution for 10–15 min at room temperature and inspection (or photography) during UV transillumination using appropriate precautions (see **Note 3**). The intensity of staining of the sample DNA and the size marker DNA may be compared to each other so that the amount of sample DNA can be roughly quantified.

3.2. PCR Amplification and Ligation by Overlap Extension

1. Assemble the PCR reaction in a final volume of 100 µL containing 1X GeneAmp PCR buffer, 200 µM of each dNTP, 100 nM each of primers a and b, 1.5 U/tube of *Taq* polymerase, and DNA template (see **Note 4**). Overlay with 50 µL of paraffin oil if required.
2. Amplify by PCR using the following cycle profile (see **Note 5**):

Initial denaturation	94°C, 5 min
25–35 main cycles of PCR	55°C, 1 min (annealing)
	72°C, 1 min (extension)
	94°C, 1 min (denaturation)
Final extension	72°C, 10 min

3. To assess the purity of the PCR product, remove 18 µL of the PCR reaction mixture, add 2 µL of 10X loading buffer, and perform electrophoresis on a composite gel consisting of 1% agarose and 2% NuSeive GTG agarose. Estimate the amount of DNA as described in **step 7** in **Subheading 3.1.**
4. Add 320 µL of water to the remaining 82 µL of the PCR reaction mixture and filter through an Ultrafree-MC-100 device by centrifugation at 2000g at room temperature until the retentate volume is about 20–25 µL (approx 3–5 min), being careful not to overfilter (see **Note 6**).
5. After the first ultrafiltration, add 400 µL of water to the retentate and repeat ultrafiltration in the same device to a final retentate volume of approx 20–25 µL, again being careful not to overfilter.
6. Prepare a 100-µL PCR reaction mixture in two separate 0.5-mL Eppendorf tubes as follows (see **Note 7**):

	Tube 1	Tube 2
10X GeneAmp PCR buffer	10.0 µL	10.0 µL
dATP (10 mM)	2.0 µL	2.0 µL
dTTP (10 mM)	2.0 µL	2.0 µL
dCTP (10 mM)	2.0 µL	2.0 µL
dGTP (10 mM)	2.0 µL	2.0 µL
Primer a (5 µM)	2.0 µL	—
Primer c (5 µM)	2.0 µL	—
Primer b (5 µM)	—	2.0 µL
Primer d (5 µM)	—	2.0 µL
Linearized plasmid (50 ng/µL)	2.0 µL	2.0 µL
PCR product (approx 50 ng/µL)	2.0 µL	2.0 µL
Water	73.6 µL	73.6 µL
Taq Polymerase (5 U/µL)	0.4 µL	0.4 µL
Total volume	100.0 µL	100.0 µL

    Overlay with approx 50 µL of paraffin oil if required.
7. Amplify by PCR (20 cycles) as described in **step 2** of **Subheading 3.2.**, except increase the extension time of the main cycles to 1.5 min.

3.3. Denaturation, Heterologous Annealing, and Cyclization

1. After thermocycling is completed, start water boiling in preparation for **steps 3** and **4** *(see below)*.
2. Combine 15 µL of the PCR reaction mixture from tube 1 with 15 µL of the PCR reaction mixture from tube 2. Denature the DNA by adding 3.4 µL of 2 M NaOH. Vortex and incubate for 5 min at room temperature.
3. During the incubation, transfer 400 µL of Tris-HCl buffer (50 mM, pH 6.6, at 25°C) to a 1.5-mL Eppendorf tube and place in the boiling water bath.
4. Add the denatured PCR products from **step 2** (this section) to the heated Tris-HCl buffer. Vortex briefly and place into the boiling water bath for 5 min.
5. Cool to 60°C over approx 15 min (*see* **Note 8**) and incubate at 60°C for 3–24 h to accomplish heterologous annealing and cyclization.
6. Following heterologous annealing and cyclization, allow the reaction mixture from **step 5** to cool to room temperature and place the entire contents in an Ultrafree-MC-30 device. Concentrate to a final volume of approx 10–20 µL by centrifugation at 2000g at room temperature (approx 3–5 min). Check the retentate volume frequently to avoid over-filtration and irreversible loss of DNA (*see* **Note 6**).
7. Transfer the concentrated cyclized product into an autoclaved 0.5-mL Eppendorf tube. The concentrated cyclized product may be stored indefinitely at –20°C for transformation at a future time or may be used immediately. Transform 100 µL of competent DH5-α *E. coli* according to the manufacturer's directions with 1–5 µL of the cyclized product.
8. After transformation, plate the *E. coli* onto LB agar plates containing the appropriate antibiotic and incubate overnight at 37°C. If the plasmid encodes β-galactosidase (LacZ), colonies containing the appropriate recombinant plasmid may be chosen by blue-white selection on agar plates that also contain X-gal (50 µg/plate). Since this approach may be misleading (*see* **Note 9**), we recommend that plasmid mini-preparations be prepared from several white colonies (*12*; *see* Chapter 45), and the presence and size of the DNA insert be confirmed by restriction endonuclease cleavage and gel electrophoresis. Alternatively, colony hybridization with the appropriate radiolabeled probe *(13)* or PCR of plasmid DNA with primers that flank the DNA insert *(14)* may be used to select colonies with plasmids containing the desired DNA insert.

4. Notes

1. Once the plasmid vector has been linearized with the appropriate restriction endonuclease(s), the oligonucleotide primers required for ligase-free subcloning may be designed. Primers a and b should be approx 39–49 bases in length with 20–25 bases at their 3' ends that are complementary to opposite strands of the target DNA at a predetermined distance from each other (**Fig. 1**). In addition, primers a and b must also contain 19–24 nucleotides at their 5' ends (the 5' addition sequences) that are identical to each of the 3' ends of the linearized plasmid vector (**Figs. 1** and **2**). In most of our experiments, 5' addition sequences were arbitrarily chosen to be 24 nucleotides in length. However, 5' addition sequences 19 nucleotides in length may be used if the temperature of the second-stage PCR is decreased to 37°C. 5' addition sequences 14 nucleotides in length or less did not work reliably. Primers c and d should be approx 21 nucleotides in length, and be complementary to opposite strands of the plasmid vector internal to each of the 5' addition sequences (**Figs. 1** and **2**).
2. Centricon-100 and Centricon-30 ultrafiltration devices (Amicon, Danvers, MA) may be substituted for Ultrafree-MC-100 and Ultrafree-MC-30 devices, respectively. When Centricon devices are used, sodium azide, which is used as a preservative, must first be

removed from the membrane by treatment with sodium hydroxide according to the manufacturer's directions. If the sodium azide is not removed, it will kill all of the bacteria during the transformation step.

3. On a 1% agarose gel, the linearized plasmid will run more slowly than the uncut (supercoiled) plasmid. If digestion is complete, it is generally not necessary to gel-purify the linearized plasmid.
4. When genomic DNA is used as starting template, 0.1–0.5 µg/tube will suffice. If plasmid DNA is being used as starting template, 0.1–1 ng/tube is sufficient. If RNA is being used as starting template, reverse transcription must first be performed followed by PCR as outlined. If the cloned PCR product will be used for applications that require precise nucleotide sequence integrity, substitution of *Taq* polymerase with a thermostable DNA polymerase with proofreading ability (i.e., Vent or Pyrococcus DNA polymerase) is preferable to minimize misincorporation errors during the PCR.
5. PCR conditions should be optimized for the specific application.
6. During centrifugation, frequently check the volume of the retentate since overfiltration can result in irreversible binding of DNA to the filter and loss of the PCR product.
7. Although never tested, use of *Taq* polymerase derivatives (or reagents) optimized for long extensions (i.e., TaKaRa LA Taq, Takara Shuzo Co., Ltd.; *Taq*/Pyrococcus GB-D DNA polymerase mixture, Life Technologies [Gaithersburg, MD]; TaqPlus, Strategene) may be advantageous for this step.
8. This is most conveniently done by using a 500-mL beaker filled approximately halfway with water as the boiling water bath and transferring the entire beaker directly into a conventional water bath that has been preset at 60°C. This will result in cooling at the appropriate rate and automatic equilibration at 60°C.
9. In our hands, the percentage of antibiotic-resistant colonies that contain the desired DNA insert varies from 50–100%. If primer-dimers form during the initial PCR and are not removed prior to ligation by overlap extension, they contain the proper 5' addition sequences and are subcloned into the vector with high efficiency resulting in a higher background (i.e., colonies with recombinant plasmids that do not contain the desired DNA insert). If large quantities of primer-dimers are present, we recommend that the conditions of the PCR be modified to minimize these primer artifacts (i.e., decrease the primer concentration and/or decrease the amount of *Taq* polymerase). Alternatively, the PCR product may be gel-purified prior to ligation by overlap extension.

Acknowledgments

We thank Jesse Roth, Charles T. Roberts, Jr., Steven Lasky, and Domenico Accili for their helpful comments on the manuscript. This research was supported in part by grants from the Diabetes Research and Education Foundation, the Juvenile Diabetes Foundation, and the Mallinckrodt Foundation. Dr. Shuldiner is the recipient of a Paul Beeson Physician Faculty Scholar Award from the American Federation of Aging Research.

References

1. Saiki, R. K., Scharf, S., Faloona, F., Mullis, K. B., Horn, G. T., Erlich, H. A., and Arnheim, N. (1985) Enzymatic amplification of beta-globin genomic sequence and restriction site analysis for diagnosis of sickle cell anemia. *Science* **230,** 1350–1354.
2. Saiki, R. K., Gelfand, D. H., Stoffel, S., Scharf, S. J., Higuchi, R., Horn, G. T., Mullis, K. B., and Erlich, H. A. (1988) Primer-directed enzymatic amplification of DNA with a thermostable DNA polymerase. *Science* **239,** 487–491.

3. Scharf, S. J., Horn, G. T., and Erlich, H. A. (1986) Direct cloning and sequence analyses of enzymatically amplified genomic sequences. *Science* **233,** 1076–1078.
4. Lee, C. C., Wu, X., Gibbs, R. A., Cook, R. G., Muzny, D. M., and Caskey, C. T. (1988) Generation of cDNA probes directed by amino acid sequence: cloning of urate oxidase. *Science* **239,** 1288–1290.
5. Higuchi, R. (1989) Using PCR to engineer DNA, in *PCR Technology* (Erlich, H. A., ed.), Stockton, New York, pp. 61–70.
6. Scharf, S. J. (1990) Cloning with PCR, in *PCR Protocols* (Innis, M. A., Gelfand, D. H., Sninsky, J. J., and White, T. J., eds.), Academic, San Diego, pp. 84–91.
7. Kaufman, D. L. and Evans, G. A. (1990) Restriction endonuclease cleavage at the termini of PCR products. *BioTechniques* **9,** 304–306.
8. Shuldiner, A. R., Scott, L. A., and Roth, J. (1990) PCR induced subcloning polymerase chain reaction (PCR) products. *Nucleic Acid Res.* **18,** 1920.
9. Shuldiner, A. R., Tanner, K., Scott, L. A., and Roth, J. (1991) Ligase-free subcloning: a versatile method of subcloning polymerase chain reaction (PCR) products in a single day. *Anal. Biochem.* **194,** 9–15.
10. Higuchi, R., Krummel, B., and Saiki, R. K. (1988) A general method of *in vitro* preparation and specific mutagenesis of DNA fragments: study of protein and DNA interactions. *Nucleic Acid Res.* **16,** 7351–7367.
11. Horton, R. M., Hunt, H. D., Ho, S. N., Pullen, J. K., and Pease, L. R. (1989) Engineering hybrid genes without the use of restriction enzymes: gene splicing by overlap extension. *Gene* **77,** 61–68.
12. Del Sal, G., Manfioletti, G., and Schneider, C. (1989) A one-tube DNA mini-preparation suitable for sequencing. *Nucleic Acids Res.* **16,** 9878.
13. Maas, R. (1983) An improved cloning hybridization method with significantly increased sensitivity for detection of single genes. *Plasmid* **10,** 296–298.
14. Gussow, D. and Clackson, T. (1989) Direct clone characterization from plaques and colonies by the polymerase chain reaction. *Nucleic Acid Res.* **17,** 4000.

43

Subtraction Hybridization cDNA Libraries

Clifford W. Schweinfest, Peter S. Nelson, Michael W. Graber, Rita I. Demopoulos, and Takis S. Papas

1. Introduction

Subtraction-hybridization cDNA libraries (1–4) are libraries enriched for sequences representing mRNAs whose expression in one biological source (e.g., tissues, cell lines) is different from a second source. Single-stranded cDNAs from both sources are allowed to hybridize so that sequences common to the two sources will anneal. The annealed, double-stranded DNAs are "subtracted" from the hybridization solution, leaving a subset of cDNA molecules enriched for sequences preferentially expressed (or repressed) in the biological source of interest. **Figure 1** diagrammatically represents the scheme for subtraction hybridization currently employed in our laboratory.

The subtraction technique is particularly helpful for isolating differentially expressed genes for which there is no *a priori* knowledge (e.g., loss of heterozygosity). It is important that matched sets of tumor and normal tissue be as similar as possible. Nonetheless, tissues from organisms will undoubtedly contain other cell types not necessarily desired (undifferentiated fibroblasts, blood, etc.). A more controlled subtraction can be achieved using cell lines. Here, a subtraction is typically performed on identical cell types, except that one may be cultured under different growth conditions (e.g., growth factor addition) or in the presence of an inducing agent for differentiation or after transfection with a cloned gene. Prior to the subtraction-hybridization technique, differential hybridization was used to identify differentially expressed cDNAs as rare as 0.1% abundance (5,6). In our hands, subtraction hybridization provides a sensitivity sufficient to isolate cDNAs with a 0.01% abundance.

For most subtractions, it is advantageous to start with two cDNA libraries whose inserts are unidirectional and in opposite orientation to each other (*see* **Fig. 1**). In this way, the single-strand phage DNA will contain vectors of the same polarity (hence, nonhybridizing) and inserts of opposite polarity. Therefore, only interlibrary hybridization events will occur. However, an advantage of randomly primed, nondirectional libraries is that they maximize 5' sequence representation within the library. As a general rule, however, we prefer to use directional cDNA libraries since hybridization and subtraction of such libraries maximize enrichment. The protocol that follows is for subtraction with unidirectional libraries.

From: *The Nucleic Acid Protocols Handbook*
Edited by: R. Rapley © Humana Press Inc., Totowa, NJ

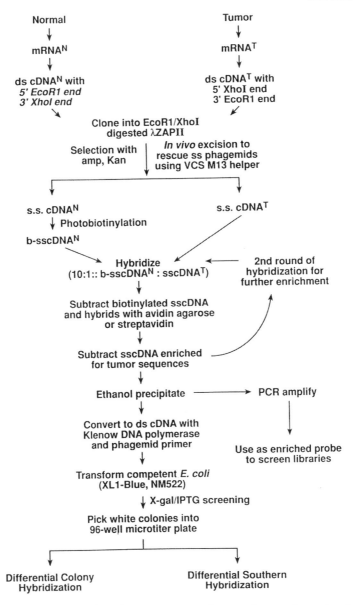

Fig. 1. Flowchart representing the strategy used to perform subtraction hybridization. Although only one direction of subtraction is shown, we typically perform the subtractions in both directions.

2. Materials
2.1. RNA Preparation
1. Guanidine isothiocyanate (GTC, Gibco-BRL, Gaithersburg, MD).
2. CsCl (Gibco-BRL).
3. Lysis solution: 4 M GTC, 100 mM Tris-HCl, pH 7.5, and 0.5% sodium sarkosyl. This GTC solution is made up in RNase-free H$_2$O (H$_2$O treated for 30 min with 0.1% diethyl pyrocarbonate and then autoclaved) then filtered through a 0.45-µm filter, and stored at

4°C. Just before use, β-mercaptoethanol may be added to a concentration of 0.1 M in the aliquot to be used.
4. CsCl solution: 5.7 M CsCl and 0.1 M ethylenediaminetetraacetic acid (EDTA), pH 7.0, are prepared in RNase-free H_2O, and then autoclaved.
5. Mortar and pestle (baked to be RNase free).
6. Dounce homogenizer (Kontes Glass, Vineland, NJ) (baked).
7. 3 M sodium acetate, pH 5.5 (made RNase free).
8. TE: 10 mM Tris-HCl, pH 7.5, and 1 mM EDTA (RNase free).
9. mRNA purification kit (Pharmacia, Piscataway, NJ).
10. Methyl mercury hydroxide (Alfa, Danvers, MA). **Caution:** this is extremely toxic.

2.2. cDNA Synthesis and Subtraction

1. Reverse transcriptase (Gibco-BRL).
2. RNaseH (Gibco-BRL).
3. *E. coli* DNA polymerase I (Boehringer Mannheim, Indianapolis, IN).
4. T4 DNA ligase (Boehringer Mannheim).
5. Polynucleotide kinase (Boehringer Mannheim).
6. Klenow fragment (Boehringer Mannheim).
7. T4 DNA polymerase (New England Biolabs, Beverly, MA).
8. *E. coli* DNA ligase (New England Biolabs).
9. RNasin (Promega, Madison, WI).
10. dNTP: All four deoxynucleotide triphosphates, as well as 5-methyl deoxycytidine triphosphate (^{m5}dCTP) and adenosine triphosphate (Pharmacia), are in solution where possible. The ^{m5}dCTP is made up as a 100-mM solution in RNase-free 10 mM Tris, pH 7.5.
11. 20X First-strand nucleotides: 10 mM dATP, 10 mM dGTP, 10 mM dTTP, and 5 mM ^{m5}dCTP.
12. 50X Second-strand RX nucleotides: 7.5 mM dATP, 7.5 mM dGTP, 7.5 mM dTTP, and 35 mM dCTP.
13. 50X Second-strand XR nucleotides: 7.5 mM dATP, 7.5 mM dGTP, 7.5 mM dTTP, 10 mM ^{m5}dCTP.
14. Linker-primers: Synthesized on an Applied Biosystems (Foster City, CA) 381A DNA synthesizer and purified on an oligonucleotide purification cartridge.
    a. XhoI: 5' GAGAGAGAGAGAACTAGT<u>CTCGAG</u>TTTTTTTTTTTTTTTTT 3'
    b. EcoRI: 5' GAGAGAGAGAGAACTACT<u>GAATTC</u>TTTTTTTTTTTTTTTTTT 3'
    For each, 5 A_{260}/mL = 140 µg = 11 nmol.
15. *Xho · Not* adapter, oligonucleotides:
    a. 5' <u>TCGAG</u>GCGGCCGC 3' 5 A_{260}/mL = 38.2 nmol = 155 µg ("long" oligo)
    b. 3' CCGCCGGCG 5' 5 A_{260}/mL = 60 nmol = 167 µg ("short" oligo)
16. *Eco*RI · *Not* adapter (Pharmacia) is 5'-d[AATTCGCGGCCGCT]-3'.
                                              3'-(GCGCCGGCGA)p-5'
17. 5X Superscript buffer (supplied by Gibco-BRL with Superscript): 250 mM Tris-HCl, pH 8.3, 375 mM KCl, and 15 mM $MgCl_2$.
18. 10X Second-strand buffer: 200 mM Tris-HCl, pH 7.5, 50 mM $MgCl_2$, 1 M KCl, and 100 mM ammonium sulfate.
19. 10X Ligation buffer: 500 mM Tris-HCl, pH 7.5, 100 mM $MgCl_2$, 100 mM dithiothreitol (DTT), 10 mM spermidine, and 500 µg/mL bovine serum albumin (BSA).
20. 10X Kinase buffer: 500 mM Tris-HCl, pH 7.5, 100 mM $MgCl_2$, 50 mM DTT, 1 mM spermidine, and 1 mM EDTA.
21. 10X Annealing buffer: 200 mM Tris-HCl, pH 7.5, and 500 mM NaCl.
22. 10X STE: 100 mM Tris-HCL, pH 7.5, 1.5 M NaCl, and 10 mM EDTA.

23. Sephacryl-200 (Pharmacia, Piscataway, NJ).
24. Isolab (Akron, OH) QS-P columns are used for spin-column chromatography.
25. Phenol:chloroform (1:1).
26. XL1-Blue, PLK-F', SURE, Uni-ZAP XR, helper phage (VCSM13), and Gigapack II Gold (Stratagene, La Jolla, CA).
27. Kanamycin (Gibco-BRL) and ampicillin (Sigma, St. Louis, MO).
28. Phage precipitation solution: 3.5 M ammonium acetate, pH 7.5, and 20% polyethylene glycol (PEG8000).
29. Photoprobe biotin and avidin D agarose resin (Vector Laboratories, Burlingame, CA): Resin is prepared by washing the slurry three to four times in resin buffer (see **step 26**), removing the last wash, and working with the packed resin. Photobiotin and streptavidin from Gibco-BRL can also be used.
30. GE Sunlamp Model RSK with 275 W bulb.
31. HE buffer: 10 mM HEPES, pH 7.5, and 1 mM EDTA.
32. 2-Butanol (Baker, Phillipsburg, NJ).
33. 2X Hybridization mix: 1.5 M NaCl, 50 mM HEPES, pH 7.5, 10 mM EDTA, and 0.2% sodium dodecyl sulfate.
34. Resin buffer: 1 M NaCl and 20 mM HEPES, pH 7.5.
35. 2X YT media: 10 g NaCl, 10 g yeast extract, and 16 g Bacto-tryptone (Difco, Detroit, MI)/L.
36. Superbroth media: 35 g Bacto-tryptone, 20 g yeast extract, and 5 g NaCl, pH 7.5/L.
37. pBluescript primers: Reverse primer, T3 primer, M13 primer, T7 primer, and SK primer (Stratagene).
38. 10X Klenow buffer: 100 mM Tris-HCl, pH 7.5, 70 mM MgCl$_2$, and 10 mM DTT.
39. Sterile 50% glycerol.
40. Gene Amp PCR Kit (Perkin-Elmer/Cetus, Norwalk, CT): This includes a 10X buffer, nucleotides, and *Taq* DNA polymerase.
41. Random Primers Labeling Kit (Gibco-BRL).
42. Quik Hyb hybridization solution (Stratagene).
43. 20X sodium chloride, sodium phosphate, EDTA (SSPE) stock: 3.6 M NaCl, 0.2 M NaH$_2$PO$_4$, and 20 mM EDTA, pH 7.4.
44. TE: 10 mM Tris-HCl, pH 7.5, and 1 mM EDTA.
45. 0.1 M DTT.
46. Superscript reverse transcriptase (Gibco-BRL).
47. 15 mM β nicotinamide adenine dinucleotide (βNAD).
48. α-[^{32}P]dATP.
49. 10 mg/mL BSA.

3. Methods

To obtain successful cDNA subtraction libraries, it is imperative that high-quality cDNA libraries be constructed. This, in turn, requires a high-quality mRNA template. Therefore, we turn now to describing mRNA preparation and cDNA synthesis.

3.1. mRNA Isolation

Tissues to be used for mRNA isolation should be quickly dissected of heterogenous tissue and snap frozen in liquid nitrogen until used. Frozen tissue should be ground to a powder with a mortar and pestle, occasionally adding liquid nitrogen to maintain a frozen "crunchy" state. The powder is then lysed in the GTC reagent. Cell-culture sources should be healthy and well fed before harvesting. Avoid using confluent cultures, if possible. Cells should be harvested quickly, washed one to two times in sterile saline, and lysed immediately in the GTC reagent (*see* **Note 1**). RNA from the GTC-lysed material is purified by centrifugation through a CsCl cushion, and the RNA recovered according to published protocols *(7)*.

1. Grind the frozen tissue to a powder with a mortar and pestle, and then transfer the frozen powder to a Dounce homogenizer.
2. Add the GTC reagent (approx 8 mL/g of starting tissue). As the frozen tissue/powder thaws in the GTC reagent, Dounce homogenize until the sample is uniformly lysed.
3. Layer the lysate on top of a 4–4.5-mL solution of 5.7 M CsCl and 0.1 M EDTA, pH 7.0, in a quick-seal tube for a 50Ti Beckman rotor (Beckman, Palo Alto, CA). Fill the tube to the top with the GTC solution, and seal the tube. Centrifuge at 105,000 g for 15–18 h at 15°C. Alternatively, a swinging bucket rotor, such as an SW41, may be used, but the centrifugation time should be increased to 20 h.
4. After centrifugation, recover the clear, glassy pellet by carefully aspirating away all the liquid in the tube. Resuspend the pellet in 0.5 mL TE.
5. Extract once or twice with an equal volume of phenol: $CHCl_3$ (1:1), then precipitate with 2 vol of ethanol in the presence of 0.3 M sodium acetate for 15 min at –80°C (on dry ice).
6. Collect the precipitate at maximum speed for 15 min at 4°C in a microfuge, and then resuspend the ethanol precipitate in RNase-free H_2O.
7. The mRNA should be purified by at least two rounds of binding and elution from oligo(dT) cellulose. (We find it convenient to use the spin-column kit and method from Amersham Pharmacia Biotech (Piscataway, NJ), especially when multiple samples are being processed.)
8. If the source of tissue or cells is abundant, we typically process 1 mg total RNA and expect yields of about 20 μg mRNA. When the source is nonabundant (e.g., human tissues), a yield of 2% of the input total RNA is assumed (not measured by absorbance) for the purpose of cDNA synthesis (*see* **Note 2**).

3.2. cDNA Library Construction

The synthesis is performed essentially by the method of Gubler and Hoffman *(8)* with some modifications from Stratagene's Uni-Zap Kit (La Jolla, CA) and some of our own.

1. Heat denature 1–2 μg of mRNA in 2 μL RNase-free H_2O at 65°C for 5 min, then chill on ice (*see* **Note 3**).
2. Add 2 μL of 10 mM CH_3HgOH (**caution:** this is very toxic), and incubate for 10 min at room temperature.
3. Add 1 μL of 75 mM β-mercaptoethanol (to sequester the mercury), and incubate for 5 min at room temperature. The denatured mRNA is now in 5 μL and is ready for cDNA synthesis.
4. Prepare a Master Mix no. 1, which contains the following components per each first-strand cDNA synthesis to be performed: 4 μL of 5X Superscript buffer, 2 μL of 0.1 M DTT, 1 μL of RNasin, 0.4 μL of 10 mCi/mL α-[^{32}P] dATP, 3.6 μL of H_2O and 1 μL of 200 U/μL Superscript.
5. Combine the following reagents to perform the first-strand cDNA synthesis: 5 μL denatured mRNA, 1 μL of 20X first-strand nucleotides, 2 μL of 1.4 μg/mL appropriate linker primer (*Xho*I or *Eco*RI linker primer), and 12 μL of Master Mix no. 1.
6. Incubate 1 h at 37°C. You may save 1 μL after first-strand synthesis for later analysis (*see* **Notes 4** and **5**). Dilute it to 10 μL with 10 mM Tris-HCl, pH 7.5, and 1 mM EDTA.

Analyze by trichloroacetic acid (TCA) precipitation and alkaline agarose gel electrophoresis, if desired.

7. Toward the end of first-strand synthesis, prepare a Master Mix no. 2 containing the following components per each second-strand reaction to be performed: 10 µL of 10X second-strand buffer, 3.75 µL of 0.1 M DTT, 0.6 µL of 10 mCi/mL α-[^{32}P] dATP, 1 µL of 15 mM βNAD, 0.5 µL of 10 mg/mL BSA, 5 µL of 5 U/µL *E. coli* DNA polymerase I, 0.5 µL of 2 U/µL RNaseH, 0.25 µL of 4 U/µL *E. coli* DNA ligase, and 56.4 µL of H$_2$O.
8. Immediately after first-strand synthesis, dilute the 20-µL reaction into the Master Mix no. 2 along with appropriate nucleotide mixtures: 20 µL of first-strand reaction, 2 µL of 50X appropriate nucleotides (RX nucleotides for *Xho*I primed, XR nucleotides for *Eco*RI primed) and 78 µL of Master Mix no. 2.
9. Incubate the second-strand reaction 1.5 h at 14°C, then 30 min at room temperature.
10. Add 10 U T4 DNA polymerase, and then incubate 30 min at 37°C. Heat kill the reaction at 65°C for 10 min.
11. Extract once in phenol:CHCl$_3$ (1:1).
12. Purify the samples through a Sephacryl-200 spin column (Pharmacia): A 2-mL (bed volume) Sephacryl-200 column is prepared in an IsoLab QS-P column tube (IsoLab, Inc., Akron, OH). It is equilibrated in 1X STE, allowed to run dry by gravity, and then prespun for 2 min at 400g in a swinging-bucket configuration. The approx 100-µL sample is carefully applied to the top of the column resin (now a cylinder that has somewhat shrunken back from the sides of the column) and spun for 2 min at 400g. Approximately 100 µL are recovered. One to 5 µL may be saved for later analysis (*see* **Notes 4** and **5**).
13. Precipitate the purified cDNA by adding 1/20 vol 3 M sodium acetate, pH 5.5, and 2.5 vol ethanol. Wash the pellet once in 80% ethanol. Lyophilize to dryness.
14. Kinase 10 nmol of the "short" oligo of the *Xho* · *Not* adapter in the following 20-µL reaction mixture: 10 µL of 1 nmol/µL "short" oligo, 2 µL of 10X kinase buffer, 1 µL of 100 mM rATP, 6 µL of H$_2$O, and 1 µL of 10 U/µL polynucleotide kinase.
15. Incubate kinase reaction for 30 min at 37°C.
16. Heat inactivate the polynucleotide kinase by incubating the reaction at 70°C for 30 min.
17. Combine the kinased "short" oligo with the "long" oligo in the following annealing mixture: 20 µL of kinased "short" oligo, 10 µL of nmol/µL "long" oligo, 10 µL of 10X annealing buffer, and 60 µL of H$_2$O.
18. Boil the annealing mixture for 5 min and then allow to slowly cool to <30°C. The *Xho* · *Not* adapter is now ready to ligate to the cDNA. (The *Eco*RI · *Not* adapter is purchased from Pharmacia ready to use.) The annealed adapter is now 100 pmol/µL.
19. Ligate the appropriate adapter to each cDNA (the *Xho*I-primed cDNA receives the *Eco*RI · *Not* adapter and the *Eco*RI-primed cDNA receives the *Xho* · *Not* adapter) by resuspending the lyophilized cDNA (*see* **step 13**) in the following 10 µL reaction: 5 µL of 100 pmol/µL appropriate adapter, 2 µL of H$_2$O, 1 µL of 10X ligase buffer, 1 µL of 10 mM rATP, and 1 µL of 2–5 U/µL T4 DNA ligase.
20. Incubate the ligation reaction overnight at 4°C, and then inactivate the ligase at 68°C for 30 min.
21. Kinase the adapter cDNA in the following 20 µL reaction: 10 µL of adapter-cDNA, 1 µL of 10X kinase buffer, 2 µL of 10 mM rATP, 6 µL of H$_2$O, and 1 µL of 10 U/µL polynucleotide kinase.
22. Incubate the reaction at 37°C for 30 min, and then inactivate the enzyme at 70°C for 30 min.
23. Digest each cDNA at its 3' end (*Xho*I or *Eco*RI) with the appropriate enzyme for 1 h at 37°C in a total volume of 50–60 µL. For the *Xho*I digestion, use 100 U of *Xho*I/µg of

cDNA to be digested (*see* **Note 4** for cDNA quantitation). For the *Eco*RI digestion, divide the cDNA into three equal aliquots and digest in a volume of 20 µL using 40, 80, and 160 U/µg cDNA. The digestions are always performed with the manufacturer's supplied buffers (*see* **Note 6**).

24. Following *Eco*RI digestion, pool the three aliquots and proceed immediately to the next step.
25. Adjust the digested cDNA to 100 µL vol and 1X STE.
26. Extract once with 100 µL phenol:$CHCl_3$ (1:1), and purify through a Sephacryl-200 spin column as in **step 12**. Recovery is approx 100 µL (*see* **Note 7**).
27. Count 1–2 µL of the cDNA by liquid scintillation in order to determine its concentration using the specific activity determined earlier (*see* **Note 4**).
28. Coprecipitate equimolar amounts of the vector (*Eco*RI · *Xho*I digested λZapII) and cDNA with ethanol (*see* **Note 8**). The precipitation mixture is 1 µL of 1 µg/µL vector DNA, an equimolar amount of cDNA (typically <20 µL, *see* **Note 8**), 1X STE up to a volume of 20 µL, 1 µL of 3 *M* sodium acetate, and then 50 µL of 100% ethanol (*see* **Note 9**).
29. Precipitate at –80°C (dry ice powder) for 15 min, and then collect precipitate by centrifugation at maximum speed in a microfuge for 15 min at 4°C.
30. Wash the pellet once in 80% ethanol, and air dry briefly (do not lyophilize).
31. Resuspend the pellet in 5 µL of ligation mixture (0.5 µL of 10X ligation buffer, 0.5 µL of 10 m*M* rATP, 2 U of T4 DNA ligase, and H_2O up to 5 µL final volume).
32. Ligate overnight at 12°C, then allow 2 h at room temperature.
33. Package 1–2 µL of the ligation reaction with Stratagene's Gigapack II Gold exactly according to the manufacturer's protocol. Titer the cDNA libraries, expecting at least 10^5–10^6 PFU/mL for the cDNAs (the *Eco*RI linker-primed library is usually on the low end of this range) and at least 10^6 PFU/mL for the test insert. This library (primary recombinants) must be titered on Stratagene strains PLK-F', XL1-Blue MRF', or SURE, which allow the growth of phage that contain methylated DNA (*see* **Note 10**).

3.3. Mass Rescue of the cDNA Libraries

Rescue is the conversion of the λ library to the single-stranded phagemid library by the process of in vivo excision. During in vivo excision, a helper phage recognizes the initiation site of the origin of replication for the pBluescript phagemid embedded, along with the cloned cDNA, within the λ vector. Replication proceeds, copying the pBluescript phagemid and your cloned cDNA, until the termination site of the origin of replication is reached where the newly synthesized single strand is circularized, packaged as a phagemid, and secreted from the *E. coli* host.

It is important to rescue the once-amplified library in a manner that minimizes possible differential growth of the individual cDNAs while maximizing the yield of recombinant single-stranded phage. It is also helpful, although not imperative, to minimize the amount of helper phage input (and subsequent output) during the rescue in order to generate as pure a yield as possible. The following procedure is our current "state-of-the-art" method for achieving those goals:

1. Combine 3×10^9 XL1-Blue cells in 2X YT medium (10 mL of cells grown to OD_{600} (optical density) of 0.4, then pelleted and resuspended in 5 mL of 10 m*M* $MgSO_4$), 3×10^8 recombinant λZap phage particles from a once-amplified library, and 10^{10} VCS M13 helper phage (Stratagene).
2. Allow 15 min absorption at 37°C.
3. Add 20 mL LB media, then grow, shaking, at 37°C for 2–3 h (do not exceed this time).
4. Heat the sample at 70°C for 20 min.

5. Pellet cells and debris by centrifugation at 6000g for 5–10 min.
6. Decant and save the supernatant containing rescued phage and helper.
7. Combine 1 mL of supernatant and 20 mL of exponentially growing XL1-Blue cells (OD_{600} = 0.4) grown in superbroth.
8. Grow for approx 2 h (until OD = 1.0), and then dilute 50-fold into prewarmed superbroth. After 30–60 min growth at 37°C, add kanamycin and ampicillin to 50 µg/mL each, and grow at 37°C for 4–8 h.
9. Pellet cells and debris. Save the supernatant.
10. Clarify supernatant with a second centrifugation (at a higher speed) to pellet any remaining material.
11. Precipitate the phage from the supernatant by adding 1/4 vol of 3.5 M ammonium acetate, pH 7.5, and 20% polyethylene glycol (PEG 8000).
12. Allow at least 1 h at 4°C for precipitation. Collect the phage by centrifugation for 30 min at 11,000g.
13. For direct isolation of phage DNA, resuspend the phage in 10 mM Tris-HCl, pH 7.5, 10 mM EDTA.
14. Heat 20 min at 70°C.
15. Extract one time each with an equal volume phenol, phenol:$CHCl_3$ (1:1), and then $CHCl_3$.
16. Ethanol precipitate the DNA (*see* **Notes 11–15**).

3.4. Biotinylation

1. Aliquot 100 µL single-stranded (ss) DNA, and adjust volume up to 0.5 mL in HE.
2. Sonicate twice for 60 s.
3. Ethanol precipitate and resuspend in 100 µL HE.
4. Under a safelight, add 100 µL of 1 mg/mL photoprobe biotin to the DNA.
5. Mix and place the open tube, in an ice bath at a distance of 10 cm from a General Electric sunlamp (Model RSK-6) equipped with a 275-W bulb. Irradiate for 15 min.
6. Adjust the solution to 0.1 M Tris-HCl, pH 9.0.
7. Extract twice with an equal volume of 2-butanol.
8. Ethanol precipitate. The pellet should be reddish brown or purple. If not, repeat the photobiotinylation.
9. Resuspend the biotinylated ss DNA (b-ss DNA) in 100 µL HE.

3.5. Subtraction Hybridization

This method is essentially that of Duguid et al. *(2)*.

1. In a total volume of 400 µL or less, combine a 10-fold excess of biotinylated ss DNA with nonbiotinylated ss DNA in the following mixture: 50–100 µg b-ss DNA, 5–10 µg ss DNA (this is the DNA to be enriched), 5 µg poly (A), and 5 µg poly (C).
2. Ethanol precipitate the mixture by adjusting it to 0.3 M sodium acetate and adding 2–2.5 vol of ethanol. Incubate at –80°C for 15 min, then collect the precipitate at 4°C for 15 min at maximum speed in a microfuge, and resuspend in 10 µL H_2O.
3. Add 10 µL of 2X hybridization mix.
4. Seal the mixture into a siliconized 100 µL capillary tube.
5. Boil 1–2 min at 100°C.
6. Allow to hybridize at 68°C for 20 h.
7. After hybridization, carefully shake the contents down to one end of the capillary, break it open, and recover the DNA with a drawn-out capillary or other narrow pipeting device.
8. Dilute the reaction up to 200 µL with HE buffer.

Subtraction Hybridization

9. Adjust to 1 M NaCl and 20 mM HEPES, pH 7.5 (resin buffer = RB).
10. Add 200 μL of packed avidin D agarose resin.
11. Incubate 30 min at room temperature while gently rocking or rotating the mixture.
12. Microfuge 30 s at 3000g. Save the supernatant.
13. Wash the resin three times in 200 μL RB. Save each supernatant.
14. Pool the supernatants and combine with 100 μL fresh-packed resin. Incubate 30 min on a rotator, as in **step 11**.
15. Pellet resin 30 s at 3000g. Save the supernatant.
16. Wash the pellet three times in 100 μL RB. Save the supernatants.
17. Dilute the mixture to 0.5 M NaCl and ethanol precipitate overnight at –20°C.
18. Because the yield of subtracted ss DNA is small, recover the precipitated DNA by centrifugation for 30 min at 68,500g in a SW41 rotor at 4°C.
19. Resuspend the pellet in 20 μL 5 mM Tris-HCl, pH 7.5, 0.1 mM EDTA (*see* **Notes 16–18**).

3.6. Conversion of Subtracted ss cDNA into a Plasmid Library

In order to make permanent subtractive libraries, the ss cDNA is converted to double stranded and transfected into *E. coli*.

3.6.1. Conversion

1. Anneal the subtracted ss cDNA to a primer (reverse primer or T3) in the following 10-μL mixture: 5 μL of subtracted ss cDNA, 1 μL of 10X annealing buffer, 0.5 μL of 10 μM primer, and 3.5 μL of H_2O.
2. Heat to 68°C for 3–5 min, and allow to cool slowly to <30°C.
3. Proceed with synthesis of the second strand: 10 μL of annealed DNA, 5 μL of 10X Klenow buffer, 0.5 μL of 5 mM 4 dNTP, 10 U of Klenow fragment, and H_2O up to a volume of 50 μL.
4. Incubate at 37°C for 2 h.

3.6.2. E. coli Transformation and Library Formation

1. Transform up to 5 μL into competent *E. coli* (e.g., XL1-Blue, NM522) exactly according to the supplier's protocol. It is important to include the X-gal/isopropylthiogalactoside color selection as well.
2. Pick the white colonies into 96-well microtiter plates containing 100 μL of LB + 50 μg/mL ampicillin. Grow overnight on an orbital shaker at 37°C.
3. Add 100 μL sterile 50% glycerol, shake another 15 min, and then freeze at –70°C. Subtraction libraries organized in this way can be replica plated onto a 150-mm Petri dish without thawing the library.

3.7. Screening for Differentially Expressed cDNAs

Depending on the extent to which subtraction removed common sequences and depending on the abundance of a given differentially expressed cDNA, anywhere from a few to a few hundred subtracted library clones may have to be screened. The approach we favor is listed in **Subheading 3.7.1.** For other approaches, *see* **Notes 19** and **20**.

3.7.1. PCR Amplification of Subtracted ss cDNA

1. Utilizing a GeneAmp PCR Kit (Perkin-Elmer) and the subtracted ss cDNA (**Subheading 3.5., step 19**), assemble the following PCR reaction: 5 μL of subtracted ss cDNA, 5 μL 10X PCR buffer, 2.5 μL of 20 μM T3 primer, 2.5 μL of 20 μM M13 primer, 1 μL 10 mM dNTP (all four), 2.5 U of *Taq* polymerase, and H_2O up to 50 μL.

2. Overlay with 50 µL mineral oil unless cycler has heated lid.
3. Amplify using the following regime: 94°C for 7 min, followed by 25 cycles of 94°C for 1 min, 41°C for 1 min, and 72°C for 1 min with a 5-s autoextension/cycle.
4. Recover PCR products by removing as much of the aqueous reaction as possible from underneath the mineral oil.
5. Purify PCR products by spin-column chromatography with Sephacryl-200 (see cDNA synthesis, **Subheading 3.2., step 12**).
6. If necessary, do a second round of PCR on the products from the first round (see **Note 21**).
7. Fifty nanograms of the PCR amplified subtracted cDNA are labeled with ^{32}P-nucleotides exactly according to the instructions provided with the Random Primers Labeling Kit (Gibco-BRL).

3.7.2. Differential Screening of λcDNA Libraries with Subtracted PCR-Amplified Probes (see **Notes 24** and **25**)

Because the subtractions are performed in two directions, the two subtracted cDNAs are separately enriched for sequences preferentially gained or preferentially lost in one library relative to the other. These subtracted DNAs are amplified by PCR as described in **Subheading 3.7.1.** Once amplified and labeled, subtracted DNAs make highly sensitive differential probes to be used on the original libraries.

1. Plate out 50,000–250,000 plaques from the original library at a density of 50,000 PFU/150 mm plate (or 250,000/23 × 23 cm plate), and grow approx 6 h at 37°C.
2. Make duplicate lifts from each plate, and fix the DNA by any preferred method (see **ref. 9**).
3. Prehybridize each filter with 0.033 mL Quik Hyb/cm^2 filter (Stratagene) for 15–30 min at 65°C (see **Note 22**).
4. Remove a small aliquot of prehybridization solution and combine it with the probe (which has been boiled for 5 min, and then chilled on wet ice). Use $2–10 \times 10^6$ cpm/mL of solution.
5. Add the aliquots back to each filter, and hybridize for 2 h at 65°C (see **Note 23**).
6. Wash the filters twice at room temperature in 2X SSPE and 0.2% sodium dodecyl sulfate (SDS), and then twice at 55°C in 0.2X SSPE and 0.2% SDS. Each wash is 30 min.
7. Perform autoradiography.
8. Pick all differential clones, and repeat the hybridizations at progressively lower plaque densities through second and third rounds until pure plaques are obtained. Using this method, we have been able to isolate clones representing mRNAs of 0.008% abundance (**1**).

4. Notes

1. RNazol (Tel-Test, Friendswood, TX) can be used in place of GTC for RNA isolation (**10**). It is a simpler procedure that often results in higher yields. However, in our hands, we have found that RNA isolated from tissues (but not from tissue culture) sometimes will not reverse transcribe. The manufacturer has included "additional steps" that should be taken in order to be able to use the RNA for reverse transcription.
2. Make sure the mRNA is of high quality. The quality of the mRNA can be assessed in several ways, if there is sufficient yield:
   a. A_{260}/A_{280} ratio near 2.0.
   b. Northern blot analysis with probe to any high-molecular-weight mRNA.
   c. Ability to direct the in vitro translation of high-molecular-weight proteins.
   We usually use two of these methods for quality assessment.
3. Directional cDNA libraries are primed from their 3' poly (A) tracts. In order to generate as complete a reverse transcript as possible, it is important that the RNA template be denatured prior to first-strand synthesis.

4. The ratio of α-[^{32}P] deoxyadenosine triphosphate (dATP) to nonradioactive dATP in both first- and second-strand synthesis is designed to be exactly the same. (Note: first-strand nucleotides carry over to the second-strand reaction.) This means that the specific activity of each strand is exactly the same. Consequently, it is very easy to quantitate your yield of cDNA at all steps following ds-synthesis. Although it is not imperative that you use the same amount of radioactivity indicated here, you should maintain identical ratios in first- and second-strand syntheses. Our protocol results in ds cDNA of a specific activity of 6.67×10^5 dpm/μg.

5. Single-stranded cDNA (first-strand synthesis) should be made along with a control RNA, preferably of high molecular weight (a 7.5-kb poly [A]-tailed RNA can be purchased from Gibco-BRL). Aliquots of these first-strand reactions are analyzed by alkaline agarose gel electrophoresis. The control should yield a discrete, largely full-length band, and the tissue RNA should yield a smear ranging from a few hundred nucleotides up to several kilobases. Double-stranded (ds) cDNA can be analyzed by conventional agarose gel electrophoresis. This is always done in order to estimate the size of the cDNA prior to ligation with vector (*see also* **Note 8**).

6. 3'-End cleavage of the linker primer with *Xho*I or *Eco*RI. The *Xho*I linker primer is not methylated and is sensitive to digestion, whereas the *Xho*I sites that may exist within the cDNA are hemimethylated and will not cut with *Xho*I. The *Eco*RI sites, however, behave somewhat differently. Only fully methylated *Eco*RI sites (when the methylated nucleotide is 5-methyl deoxycytidine) are completely resistant to *Eco*RI digestion, whereas hemimethylated sites are partially resistant *(11,12)*. This means that an excess of *Eco*RI must be used to assure cleavage of the linker primer *Eco*RI site—the internal cDNA sites will still be protected. We have calculated that approx 80 U/μg cDNA should be sufficient to digest hemimethylated *Eco*RI sites. In practice, however, we divide this cDNA into three aliquots in order to "bracket" the quantity of enzyme, i.e., 40, 80, and 160 U/μg. After digestion, the aliquots are pooled again. Digest each cDNA with its appropriate enzyme for 1 h at 37°C in a volume of 50–60 μL.

7. Sephacryl-400 may be used here instead of Sephacryl-200. By spinning the cDNA in a 50-μL vol, followed by several (two or three) 50-μL chases with 1X STE, it may be possible to get some size fractionation of the cDNA (largest cDNA elutes first).

8. It is a good idea to analyze an aliquot of ds cDNA by gel electrophoresis in order to determine an average size for the cDNA. In this way, one can estimate more accurately the correct amount of cDNA to ligate to the vector at an equimolar ratio. Typically, the average size is 1–2 kbp. Since the vector (*Eco*RI · *Xho*I digested λZapII) is 40 kbp, this means that 25–50 ng cDNA are ligated with 1 μg of vector. The amount of cDNA coprecipitated with the vector is deduced using the specific activity of the cDNA determined as in **Note 4** and by counting a 1–2 μL aliquot of the cDNA from **step 27**. For example, if one recovers 100 μL of cDNA from the spin column in **step 26**, its radioactive concentration is determined to be 4000 dpm/μL in **step 27**, and its SA is 6.67×10^5 dpm/μg (**Note 4**), then the concentration of cDNA recovered from the spin column is 4000 dpm/μL divided by 6.67×10^5 dpm/μg = 0.006 μg/μL or 6 ng/μL. If the average size of the cDNA is 2 kbp, then 50 ng or 8.3 μL of cDNA are needed to coprecipitate with 1 μg of vector.

9. Also, make sure to coprecipitate the test insert (provided by the manufacturer) and vector as a positive control for ligation and packaging.

10. Primary libraries are not very stable and should be carefully amplified at once using one of the strains described in **step 33**. After amplification (titer >10^9 plaque-forming unit/mL), the library is stable and can be checked for the percentage recombinant by color selection or by direct rescue (in vivo excision) of random clones and gel analysis of their plasmid DNA. Typically, the libraries are greater than 90% recombinant.

11. Helper R408 is not used, as it is not kanamycin resistant. We find that kanamycin selection greatly improves the quality and yield of our phagemid preps.
12. The phagemid DNA should be analyzed by gel electrophoresis. You should expect to see a smear starting at the molecular size of nonrecombinant ss pBluescript (approx 1.6 kb with respect to ds DNA markers) and possibly some helper phage DNA.
13. If the yield seems low (<100 µg/L), it may be necessary to add some more helper at the time of the 50-fold dilution (*see* **step 8**).
14. It is possible to purify the DNA further (e.g., in case there is a lot of helper DNA) by cutting out the ss DNA from a preparative gel and using GeneClean (Bio 101, La Jolla, CA) or GELase (Epicentre Technologies, Madison, WI).
15. Precipitated phage may also be purified by CsCl gradient centrifugation prior to DNA purification.
16. The use of streptavidin (Gibco-BRL) instead of avidin D agarose for subtraction works just as well.
17. Further enrichment may be achieved by rehybridizing the subtracted ss DNA with more biotinylated ss DNA (1–10 µg) followed by another subtraction.
18. In principle, it should also be possible to generate biotinylated nucleic acid for subtraction by using the T3 or T7 promoters in a λZapII (along with the appropriate polymerase) to synthesize uridine triphosphate-biotinylated RNA. This would reduce the need to generate large amounts of driver ss phagemid for the hybridization.
19. Individual colonies from the subtracted library can be grown up, and small-scale plasmid preps performed. These plasmids, individually or in groups up to five, are all digested so as to release their inserts (*Xho*I and *Eco*RI). Equal amounts are electrophoresed on duplicate agarose gels and then transferred to any preferred hybridization membrane. The membranes are probed differentially with probes synthesized from the PCR-amplified subtracted cDNAs (*see* **Subheadings 3.7.1. and 3.7.2.**). We have also used probes synthesized directly as first-strand cDNAs from the original mRNAs, if sufficient material is available. The PCR probes have the advantage of being enriched for sequences that are differentially expressed and are, theoretically, limitless in supply, but they do contain vector sequences. For this reason, it is advisable to excise the vector band (approx 3 kbp) from the gel before transfer and hybridization. First-strand cDNA probes synthesized from the original mRNAs are not enriched (therefore, rare sequences in the mRNA are more difficult to detect by Southern blot hybridization) but are free of vector sequences. Clones that hybridize differentially are analyzed further as good candidates for differentially expressed genes. Differential Southern blot hybridization has the disadvantage of being very tedious when large numbers of clones are analyzed.
20. Differential colony hybridization has the advantage of being able to screen many colonies simultaneously. The drawback is that duplicate colonies are not always quantitatively similar, making subtle differences in autoradiographic intensity somewhat unreliable. Essentially, subtracted colonies from the 96-well microtiter plates are replicated onto duplicate 137-mm filters on 150-mm plates of LB + 50 µg/mL ampicillin and grown overnight at 37°C. The cells are lysed, and the DNA transferred to the filters by any preferred method. Duplicate hybridizations are performed using first-strand cDNA prepared from total mRNA. PCR-generated DNA probes cannot be used because of the presence of large amounts of vector sequences in the colony.
21. Run a 5–10 µL aliquot of the PCR product on a 1% agarose gel, and look for a faint smear of DNA. If none is visible, use 5–10% of the products of the first PCR to initiate a second round of PCR. This time, however, use the SK and T7 primers, which are nested inside the first set of primers. Use the same PCR conditions as the first round of PCR. Purify these

products through Sephacryl-200 spin columns, as before, and analyze by gel electrophoresis once more.
22. **Important:** Prehybridization and hybridization should contain 10–20 µg/mL heat-denatured pBluescript vector in order to compete with the vector sequences that are also present in the PCR-generated probe, which may hybridize to the pBluescript sequences in the λZapII vector in the plaques.
23. We find Stratagene's Quik Hyb solution to be convenient and fast, but you may use any conventional prehybridization and hybridization conditions normally used to screen libraries.
24. It is imperative that all candidate cDNA clones be verified as differential by Northern blot hybridization to RNA, from which the original libraries were made. Growth advantages during phage amplification or rescue or differential amplification during PCR could give a falsely positive (differential) result.
25. It is possible to estimate the efficiency of enrichment during the subtraction hybridization by "spiking" the hybridization with varying known quantities of a previously cloned sequence known not to exist in the two libraries (e.g., bacterial *Kan*® gene from pET30, Noragen, Inc., Madison, WI). In this way, the subtracted material can be assayed for degree of enrichment after transfection into *E. coli*.

Acknowledgments

We thank K. Cannon and Judy Yost for expert typing of the manuscript.

References

1. Schweinfest, C. W., Henderson, K. W., Gu, J.-R., Kottaridis, S. D., Besbeas, S., Panatopoulou, E., and Papas, T. S. (1990) Subtraction hybridization cDNA libraries from colon carcinoma and hepatic cancer. *Genet. Anal. Techn. Appl.* **7,** 64–70.
2. Duguid, J., Rohwer, R. G., and Seed, B. (1988) Isolation of cDNAs of scrapie-modulated RNAs by subtractive hybridization of a cDNA library. *Proc. Natl. Acad. Sci. USA* **85,** 5738–5742.
3. Rubenstein, J. L. R., Brice, A. E. J., Ciaranello, R. D., Denney, D., Porteus, M. H., and Usdin, T. B. (1990) Subtractive hybridization system using single-stranded phagemids with directional inserts. *Nucleic Acids Res.* **18,** 4833–4842.
4. Owens, G. P., Hahn, W. E., and Cohen, J. J. (1991) Identification of mRNAs associated with programmed cell death in immature thymocytes. *Mol. Cell. Biol.* **11,** 4177–4188.
5. Lau, L. F. and Nathans, D. (1985) Identification of a set of genes expressed during the G_0/G_1 transition of cultured mouse cells. *EMBO J.* **4,** 3145–3151.
6. Hirschhorn, R. R., Aller, P., Yuan, Z.-A., Gibson, C. W., and Baserga, R. (1984) Cell-cycle-specific cDNAs from mammalian cells temperature sensitive for growth. *Proc. Natl. Acad. Sci. USA* **81,** 6004–6008.
7. Chirgwin, J. M., Przybyla, A. E., MacDonald, R. J., and Rutter, W. J. (1979) Isolation of biologically active ribonucleic acid from sources enriched in ribonuclease. *Biochemistry* **18,** 5294–5299.
8. Gubler, U. and Hoffman, B. J. (1983) A simple and very efficient method for generating cDNA libraries. *Gene* **25,** 263–269.
9. Sambrook, J., Fritsch, E. F., and Maniatis, T., eds. (1989) *Molecular Cloning: A Laboratory Manual*, 2nd ed. Cold Spring Harbor Lab. Press, Cold Spring Harbor, NY, pp. 2.108–2.117.
10. Chomczynski, P. and Sacchi, N. (1987) Single-step method of RNA isolation by acid guanidinium thiocyanate-phenol-chloroform extraction. *Anal. Biochem.* **162,** 156–159.

11. Nelson, P. S., Papas, T. S., and Schweinfest, C. W. (1993) Restriction endonuclease cleavage of 5-methyl-deoxycytosine hemimethylated DNA at high enzyme-to-substrate ratios. *Nucleic Acids Res.* **21,** 681–686.
12. Brennan, C. A., Van Cleve, M. D., and Gumport, R. I. (1986) The effects of base analogue substitutions on the cleavage by *Eco*RI restriction endonuclease of octadeoxyribonucleotides containing modified *Eco*RI recognition sequences. *J. Biol. Chem.* **261,** 7270–7278.

44

Cloning Polymerase Chain Reaction Products Utilizing the T/A Overhang and a Kit

Melissa Lail-Trecker

1. Introduction

The ever-expanding identification of new gene family members in recent years has depended in large part on the use of the polymerase chain reaction (PCR) technique. Typically PCR products may be cloned into a vector by cohesive- or blunt-end ligation. However, both cohesive- and blunt-end ligation traditionally require additional enzymatic manipulation and/or purification steps before the product may be cloned into the vector. Also, blunt-end ligation is a less efficient process than cohesive-end ligation (1). The Invitrogen (Carlsbad, CA) TA Cloning® Kit combines the efficiency of cohesive-end ligation with the ease of direct cloning of PCR products. A sample may be removed from the completed PCR reaction and transferred directly to the ligation reaction mixture. This procedure is based on the fact that *Taq* polymerase has a nontemplate dependent activity that adds a single deoxyadenosine to the 3' ends of double-stranded DNA (2). The Invitrogen pCR™II vector is supplied as a linearized molecule with 3' T overhangs on either side of the insert site. This allows cohesive-end ligation of the PCR product with the pCRII vector. In this chapter, I describe the steps that I followed in order to clone a 576-bp fragment of the rat *N*-cadherin gene into the pCRII vector. This plasmid was then used to generate RNA probes for use in a ribonuclease protection assay. Although much of the description of the ligation and transformation procedures is based on directions supplied by Invitrogen, some steps differ. Note that variations in protocol and vector have been introduced with more recent kits by Invitrogen. The time involved for this procedure is a total of 4 d: set up PCR reaction in the morning of the first day; after completion, set up ligation that afternoon; ligate overnight. The second day, transform bacteria with ligation reactions; grow transformed cells overnight on plates. The third day, pick colonies off plates, and grow minipreps. Isolate plasmid miniprep DNA on the third or fourth day. The fourth day, restriction cut the miniprep DNA to determine that transformants contain the desired insert.

2. Materials

2.1. Preparation of PCR-Generated Insert

1. Source of insert sequence to be amplified: This may be genomic DNA, cDNA from a reverse transcriptase reaction, or 100–200 ng of plasmid DNA. For this particular experiment, 10 µg of total GH_3 cell cytoplasmic RNA was reverse transcribed to cDNA with reverse transcriptase in a 20-µL reaction and subsequently used as the PCR template.
2. Oligonucleotide primers: The primers used for amplifying the N-cadherin fragment are 18-mers corresponding to amino acids 457–462 and 643–648 in the extracellular domain of mouse N-cadherin. Sequence for the upstream primer is 5' GGA TGT TTG TCC TTA CTG 3'. The downstream primer sequence is 5' TAG TCA CTG GAG ATA AGG 3'. Both primers have T_m of 62.2°C as calculated by the %GC method.
3. 10 X PCR buffer (Perkin-Elmer/Cetus, Norwalk, CT): Alternatively, prepare by this recipe: 100 mM Tris-HCl, pH 8.3; 500 mM KCl; 15 mM $MgCl_2$; 0.01% (w/v) gelatin. Use stock solutions that have been filter sterilized, autoclaved for 15 min on liquid cycle, and designated for PCR use only. For 1 mL of 10 X PCR buffer, combine 100 µL of 1 M Tris-HCl, pH 8.3, 500 µL of 1 M KCl, 15 µL of 1 M $MgCl_2$, and 375 µL of UV-irradiated sterile water. Make up a 1% solution of gelatin in UV-irradiated sterile water. Heat at 60–70°C, mixing occasionally, to dissolve the gelatin. Filter the gelatin solution while it is still warm through a 0.2-µm filter, and add 10 µL of gelatin to each milliliter of 10 X PCR buffer. Store 10 X PCR buffer in 300–500 µL aliquots at –20°C. The buffer may be UV irradiated before use.
4. Sterile UV-irradiated water: Pour 0.2 µm filter-deionized, distilled water into a sterile container. UV irradiate for 2 min in a Stratagene (La Jolla, CA) Stratalinker UV crosslinker (200 mJ/cm^2) *(3)* or at 254 and 300 nm for 5 min *(4)*. Store at room temperature.
5. 10 mM Deoxynucleotide stocks (dATP, dCTP, dGTP, and dTTP): Supplied as 10 mg solids from Pharmacia (Piscataway, NJ). To make 10 mM stocks, resuspend 10 mg of dNTP in 10% less sterile water than is required to give a 10-mM solution. Adjust the pH to 7.0 using sterile NaOH and pH paper. Determine the exact concentration by optical density (OD), using the wavelength and molar extinction coefficient provided by the manufacturer for each dNTP. For example, the A_m (259 nm) for dATP is 15.7×10^3; therefore, a 1:100 dilution of a 10-mM solution of dATP will have an A_{259} of $(0.01 \text{ M} \times 15.7 \times 10^3 \text{ OD U/M}) \times 1/100 = 1.57$. If the actual OD of a 1/100 dilution of the dATP is 1.3, the dATP concentration is $1.3/1.57 \times 10$ mM = 8.3 mM. Store the dNTPs at –20°C in 50–100 µL aliquots. Make a working stock containing 1.25 mM of each dNTP in UV-irradiated sterile water for PCR. Unused working stock may be stored at –20°C for up to 2 wk. **Do not UV-irradiate the dNTP solutions**. Alternatively, dNTP solutions are available from several companies (e.g., Promega, Madison, WI).
6. Light mineral oil.
7. *Taq* DNA polymerase (Perkin-Elmer/Cetus): If *Vent* or *Pfu* polymerase will be used to amplify the insert (*see* **Note 1**).
8. $CHCl_3$.
9. 7.5 M Ammonium acetate: Filter through a 0.2-µm filter and store at room temperature.
10. 95% Ethanol. Store at –20°C.
11. TE, pH 8.0: 10 mM Tris-HCl, pH 8.0, 1 mM ethylenediaminetetraacetic acid (EDTA). Sterilize with a 0.2-µm filter and store at room temperature.

2.2. Ligation and Bacterial Transformation

1. pCRII vector at 25 ng/µL (supplied with the Invitrogen TA Cloning Kit): Prepare insert. Determine approximate concentration of insert (*see* **Note 2**).

Cloning PCR Products

2. T4 DNA ligase and 10X ligation buffer (supplied with Invitrogen TA Cloning Kit): Store at −20°C.
3. Sterile water (supplied with the Invitrogen TA Cloning Kit): Alternatively, pour 0.2 µm filter deionized, distilled water into a sterile container and store at room temperature.
4. One aliquot of TA One Shot™ competent bacterial cells per ligation reaction, plus two aliquots for a "TE only" negative control transformation and a pUC18 positive control transformation (supplied with the Invitrogen TA Cloning Kit). Store at −70°C.
5. 0.5 M β-mercaptoethanol (supplied with the Invitrogen TA Cloning Kit): Store at −20°C.
6. pUC18 test plasmid (supplied with Invitrogen TA Cloning Kit).
7. TE (supplied with the Invitrogen TA Cloning Kit): Alternatively, TE, pH 8.0, may be prepared by the investigator (*see* **Subheading 2.1., step 11**).
8. SOC Media (supplied with the Invitrogen TA Cloning Kit). Store at room temperature.
9. Ampicillin or kanamycin: Prepare a stock solution of 50 mg/mL in water. Sterilize filter. Store in aliquots at −20°C.
10. X-Gal (5-bromo-4-chloro-3-indolyl-β-D-galactoside). Prepare a stock solution of 40 mg/mL in *N,N*-dimethylformamide under a fume hood *(5)*. Store at −20°C.
11. LB agar plates containing ampicillin or kanamycin. Prepare by dissolving 10 g Bacto-tryptone, 5 g yeast extract, 10 g NaCl, and 15 g agar in 1 L of H_2O (5) in a 2.8-L Fernbach flask (Sigma, St. Louis, MO). Heat to near boiling while stirring to dissolve the agar. Autoclave for 20 min on the liquid cycle. After solution has cooled to 50–55°C in a water bath, add ampicillin or kanamycin to 50 µg/mL (1 mL of a 50-mg/mL stock solution). Mix well. Pour molten agar into sterile Petri dishes, using a 20–25–mL/plate. If plates are poured in a sterile hood, they may be allowed to cool partially uncovered until the agar has set. When the agar is solid, cover the plate, invert, and store at 4°C. Antibiotic-containing agar has a shelf life of about 1 mo. If longer storage is required, prepare agar plates without antibiotic and spread with ampicillin or kanamycin (25 µL of a 50-mg/mL stock) at least 1 h before use.

2.3. Plasmid Minipreps for Screening Transformant Colonies

1. 2X TY broth containing 50 µg/mL ampicillin or kanamycin: Prepare 2X TY broth by dissolving 16 g Bacto-tryptone, 10 g of yeast extract, and 5 g NaCl in 1 L of H_2O in an autoclavable flask. Autoclave 20 min on the liquid cycle. Cool and store at 4°C. Aliquot appropriate amount of 2X TY as needed for miniprep procedure and add antibiotic to 50 µg/mL.
2. 80% Glycerol (v/v, in water), 0.2 µm sterile filter: Store at room temperature.
3. GTE buffer: 25 mM Tris-HCl, pH 8.0, 50 mM glucose, 10 mM EDTA. Sterilize with a 0.2-µm filter and store at 4°C.
4. Lysis buffer: 0.2 M NaOH, 1% SDS. Prepare fresh.
5. KOAc buffer: 3 M potassium acetate, 2 M acetic acid. Sterilize with a 0.2-µm filter and store at room temperature.
6. TE, pH 8.0 (*see* **Subheading 2.1., step 11**).
7. Restriction enzyme, *Eco*RI, and the appropriate restriction digest buffer, React 3 (Life Technologies, Gaithersburg, MD). Store at −20°C.

3. Methods

3.1. Preparation of a PCR-Generated Insert

1. Prepare six standard 100 µL PCR reactions, each containing: 200 µM of each deoxynucleotide, 100 pmol of each primer, and 2.5 U of *Taq* polymerase in a 1 X PCR buffer. Two tubes are experimental duplicates, i.e., to amplify the target sequence; a third tube serves as a negative control, the fourth as a positive control, the fifth and sixth as primer

controls. The two experimental tubes (reaction done in duplicate), positive control tube, and primer control tubes will also contain 1/10 of the reverse transcriptase reaction as a DNA template. The negative control tube should contain no cDNA template. The negative control tells you if there is contamination of the PCR solutions with DNA. If primers are available that amplify a known specific sequence in the cDNA, these should be substituted for the experimental primers in the fourth (positive control) tube to check the integrity of the DNA template and the efficiency of the PCR reaction. For example, in this particular experiment primers that amplify a portion of the prolactin gene were used in the positive control tube. The fifth (primer control) tube should contain the upstream but not the downstream primer. The sixth (primer control) tube should contain the downstream but not the upstream primer. *Taq* polymerase should be added last to the reaction mix just prior to the cycling steps. Overlay the final reaction mix with a drop (approx 70 µL) of light mineral oil to prevent evaporation.
2. Amplify by PCR using the following cycle profile (*see* **Note 3**):
    17–25 main cycles   94°C for 1 min
                          55°C for 2 min
                          72°C for 1 min
    Final extension   72°C for 5 min
3. Add 100 µL $CHCl_3$ to each PCR reaction tube to remove mineral oil. Spin for 1 min at approx 10,000g in a microfuge.
4. Remove the top aqueous layer that contains the PCR reaction product and transfer to a fresh microfuge tube. Discard bottom layer as organic waste.
5. At this point, the PCR product may be used directly in a TA ligation reaction (*see* **Subheading 3.2., step 1**), or held on ice until the amount of PCR product available for ligation can be determined (*see* **Note 2** and the following procedure).
6. Precipitate the PCR product by adding 1/2 vol of 7.5 *M* ammonium acetate to the PCR reaction product. Mix briefly. Add 2.5 vol of 95% ethanol. Mix well.
7. Precipitate on dry ice for 1 h, or at –20°C overnight.
8. Spin out precipitate at 11,000–15,000g for 30 min: Discard supernatent while ensuring that DNA pellet remains in the tube. Dry pellet and resuspend in 10–20 µL of either TE or sterile water.
9. At this point, the PCR product may be stored at –20°C or analyzed by gel electrophoresis to determine the size and amount of the amplified product (*see* **Notes 2** and **4**). If multiple-sized products result from the PCR amplification, the band of interest may be purified out of agarose (*see* **Note 5**) and incubated with *Taq* polymerase to add the 3' A overhangs required for TA cloning (*see* **Note 1**).

3.2. Ligation and Bacterial Transformation

1. Combine 2 µL pCRII vector with PCR product insert in a 1:1 or 1:3 ratio. The formula for determining molar amounts of insert to use is as follows:

$$X \text{ ng of PCR product to ligate} = \frac{(Y \text{ bp PCR product}) (50 \text{ ng pCRII vector})}{(\text{size in bp of the pCRII vector})}$$

X ng is the amount of PCR product to be ligated for a 1:1 molar ratio. For the *N*-cadherin insert, $X = 7.3$ ng [$X = (576 \text{ bp} \times 50 \text{ ng}) \div 3932 \text{ bp}$]. Three times X ng would be used for a 1:3 molar ratio. If the PCR product will be used in a TA ligation without first determining the amount needed for a 1:1 and 1:3 molar ratio, set up ligation reactions that use various amounts of the PCR product. Under these circumstances we typically set up two ligation reactions per experimental PCR reaction using 1 or 3 µL of the fresh PCR product in the ligation reaction.

Cloning PCR Products

2. Add the following to the vector and insert to complete ligation reaction setup:
   a. 1 µL 10X ligation buffer
   b. X µL sterile water (to bring reaction volume to 9 µL)
   c. 1 µL T4 DNA ligase
3. A self-ligation reaction may also be performed to test the stability of the pCRII vector. Carry out the reaction as above after substituting sterile water for the insert.
4. Incubate the ligation reactions overnight at 12°C.
5. When ligation reaction is complete, warm LB agar plates to 37°C. Use two plates for each transformation reaction.
6. Spin ligation reaction tubes briefly in a microfuge and place on ice.
7. Thaw 0.5 M β-mercaptoethanol and One Shot competent cells on ice.
8. When cells have thawed, pipet 2 µL of 0.5 M β-mercaptoethanol into each vial of competent cells. Tap gently to mix.
9. Pipet 1 µL of each ligation reaction (or 1 µL of pUC18 test plasmid, or 1 µL TE) into a vial of competent cells. Tap gently to mix. Remaining ligation reaction may be stored at –20°C.
10. Incubate vials on ice for 30 min.
11. Incubate for 30 s at 42°C to heat shock bacterial cells (*see* **Note 6**).
12. Quickly place on ice for 2 min.
13. Add 450 µL of room temperature SOC medium to each vial of transformed cells. Place vials in a microcentrifuge rack and secure with tape.
14. Shake vials at 37°C for 1 h at 225 rpm in a rotary shaking incubator.
15. While vials are shaking, spread 25 µL of X-Gal stock solution (40 mg/mL) on top of LB agar plates with a L-shaped glass spreader. Use sterile technique. Let X-Gal diffuse into the agar for approx 1 h.
16. When incubation is complete, place the transformation vials on ice. Spread 25 µL and 100 µL from each transformation vial on separate, labeled LB agar plates that contain either ampicillin or kanamycin and X-Gal.
17. Invert plates and incubate overnight at 37°C.

3.3. Plasmid Minipreps for Screening Transformant Colonies

1. The plates containing TE-transformed bacteria should be free of colonies. The presence of any colonies could indicate the antibiotic is no longer working or the bacterial cells themselves are antibiotic resistant. Antibiotic resistance should not be a problem with fresh cells obtained from a commercially available kit.
2. The plates containing pUC18 control plasmid-transformed bacteria should have many colonies. This indicates the transformation worked.
3. The plates containing bacteria transformed with the self-ligation reaction should have mainly blue colonies. Any white colonies are due to blunt-end self-ligation of the vector as a result of 3' T-overhang degradation. These colonies should be regarded as false positives. Knowing the percentage of false positives obtained for any given transformation allows an estimation of the number of colonies, which can be expected to be true positives on the experimental plates.
4. The plates containing bacteria transformed with ligated constructs should contain a mix of blue and white colonies. Successful ligation of vector with insert will cause a disruption of the vector's *lacZ* gene and result in white colonies. However, if the insert size is 500 bp or less, the *lacZ* gene may not be disrupted as efficiently. This can result in light blue colonies, which should then be treated as positives. Positive colonies may be screened by direct PCR sequencing, or by preparation of plasmid DNA by the following miniprep procedure.

5. Touch a sterile loop or toothpick to each colony to be screened and inoculate each into a 2-mL aliquot of 2X TY broth containing 50 µg/mL kanamycin or ampicillin.
6. Incubate at 37°C, with shaking (225 rpm) for 4–16 h, until broth is moderately turbid from bacterial growth.
7. Transfer 0.4 mL of each sample into an appropriately labeled 1.5 mL microfuge tube containing 0.1 mL of 80% glycerol. Mix and freeze on dry ice. Store these aliquots at –70°C as the future source of the desired transformants.
8. Transfer the remainder of each sample to another 1.5-mL tube, and centrifuge at 2600g for 2 min in a tabletop microfuge. Aspirate the supernatant without disturbing the pellet. Leave the caps open for the next two steps.
9. Resuspend each pellet in 50 µL of GTE buffer by vortexing. Incubate at room temperature for 5 min.
10. Add 100 µL of lysis buffer to each tube while vortexing. Incubate on ice for 5 min.
11. Add 75 µL of KOAc buffer to each tube, close the caps, vortex to mix well, and incubate on ice for 5 min.
12. Centrifuge at approx 10,000g for 10 min in a tabletop microfuge.
13. Transfer the supernatants to fresh 1.5 mL tubes containing 0.5 mL of 95% ethanol. Mix well and centrifuge again at top speed for 10 min.
14. Decant the supernatants and air dry the DNA pellets for 20–30 min. Resuspend the DNA in 50 µL TE, pH 8.0.
15. To determine which plasmids contain the desired insert, the miniprep DNA may be digested with restriction enzymes that cut within the pCRII vector's polylinker. For example, *Eco*RI restriction sites flank the TA cloning site. *Eco*RI may be used to cut out the insert (*see* **Note 7**). This leaves a minimal amount of vector DNA (6–10 bp) at each end of the insert. For each sample to be analyzed, combine 3–5 µL of miniprep DNA with 1 µL of 10X reaction buffer (React 3, Life Technologies), and 1 µL of *Eco*RI in a total volume of 10 µL. Incubate at 37°C for 30–60 min. Restriction digests were then treated with 1 µL of 10 µg/mL RNase A (Boehringer Mannheim, Indianapolis, IN) for 30 min at 37°C.
16. Analyze the digested DNA samples by agarose gel electrophoresis. Digested plasmids containing the *N*-cadherin insert will produce one approx 576-bp fragment in addition to the 3.9-kb linearized pCRII vector band (*see* **Notes 5** and **8**). Digested plasmids that did not contain the insert will appear as only the 3.9-kb linearized band. If the restriction digest of the plasmid DNA was incomplete, some plasmid may remain intact and supercoiled, and appear as an additional high molecular weight band in the lane.

4. Notes

1. If *Vent* or *Pfu* polymerase is used to amplify the PCR product, then an additional incubation with *Taq* polymerase is required to add 3' A-overhangs to the PCR product. A protocol for this procedure is available from Invitrogen.
2. The approximate concentration of insert may be determined by running one-fifth to one-tenth the PCR reaction product on an agarose gel with three different amounts of λ DNA-*Bst*E II molecular weight markers (New England Biolabs, Beverly, MA). After ethidium bromide staining of the agarose gel, the approximate concentration of PCR product may be estimated by comparing the intensity of the product band to the λ DNA marker bands. Bands of the same intensity will have approximately the same concentration of DNA. Because the percentage of each marker band present in the total amount of λ DNA preparation is known, the concentration of DNA present in a specific marker band is determined by multiplying the relative percentage for a specific marker band times the total amount of λ DNA loaded in that particular lane.

3. Because *Taq* polymerase has no 3' to 5' exonuclease editing function *(5)*, it is important not to overamplify, which may introduce mutations. We used 30 cycles in this particular experiment, but in the interest of minimizing mutations, one should begin with 17–20 cycles.
4. To ensure your PCR reaction, amplify the correct sized product, run a parallel PCR reaction that can be precipitated and analyzed by agarose gel electrophoresis, or remove one-fifth to one-tenth of the PCR reaction product and analyze by agarose gel electrophoresis. In our experience, the PCR reactions may be held at 15°C for up to 12 h without any noticeable decrease in ligation efficiency.
5. If the insert DNA is larger than 200 bp, it may be purified out of the agarose gel after electrophoresis by using the Prep-a-Gene™ DNA purification kit (Bio-Rad, Hercules, CA). Alternatively, the insert may be purified out of agarose by electroelution *(6)*. After this point the purified insert may be labeled by random priming *(7)* for use as a probe for Northern hybridization analysis.
6. Timing and temperature during transformation are critical. If the heat-shock step is too long or the temperature is too high, transformation efficiency may be reduced or cells may die. Also, if the heat-shock temperature is too low, transformation efficiency may be reduced.
7. Choose an enzyme to cut out the insert, which does not cut within the insert.
8. The pCRII vector contains T7 and Sp6 RNA promoters for production of sense and antisense RNA transcripts.

Acknowledgments

This work was performed in the lab of Dr. Bruce A. White and supported by NIH grant DK43064.

References

1. Ausubel, F. M., Brent, R., Kingston, R. E., Moore, D. D., Smith, J. A., Seidman, J. G., and Struhl, K. (eds.) (1987) *Current Protocols in Molecular Biology*, Wiley Interscience, New York, pp. 3.14.1–3.14.4.
2. Clark, J. M. (1988) Novel non-templated nucleotide addition reactions catalyzed by procaryotic and eucaryotic DNA polymerases. *Nucleic Acids Res.* **16,** 9677–9686.
3. Dycaico, M. and Mather, S. (1991) Reduce PCR false positives using the Stratalinker UV crosslinker. *Stratagene Strategies* **4,** 39,40.
4. Sarkar, G. and Sommer, S. S. (1990) Shedding light on PCR contamination. *Nature* **343,** 27.
5. Eckert, K. A. and Kunkel, T. A. (1990) High fidelity DNA synthesis by the Thermus aquaticus DNA polymerase. *Nucleic Acids Res.* **18,** 3739–3744.
6. Zhen, L. and Swank, R. T. (1993) A simple and high yield method for recovering DNA from agarose gels. *BioTechniques* **14,** 894–898.
7. Maniatis, T., Fritsch, E. F., and Sambrook, J. (eds.) (1982) *Molecular Cloning: A Laboratory Manual*. Cold Spring Harbor Lab. Press, Cold Spring Harbor, NY, 1st ed., Chaps. 4, 9.

45

Extraction and Purification of Plasmid DNA

Craig Winstanley and Ralph Rapley

1. Introduction

Plasmids may be isolated by a variety of methods many of which rely on the differential denaturation and reannealing of plasmid DNA compared to chromosomal DNA. Many of these are rapid, small-scale "minipreps" that may be used effectively for plasmid analysis and further manipulation. The methods differ primarily in the means by which cells are lysed.

One commonly used technique developed by Birnboim and Doly involves alkaline lysis *(1)*. Although many slight modifications of the original protocol exist, it essentially relies on bacterial lysis by sodium hydroxide and sodium dodecyl sulfate (SDS), followed by neutralization with a high concentration of low-pH potassium acetate. This gives selective precipitation of the bacterial chromosomal DNA and other high-molecular-weight cellular components. The plasmid DNA remains in suspension and is precipitated with isopropanol.

An alternative to alkaline lysis is the rapid boiling method developed by Holmes and Quigley *(2)*. Here, the cells are lysed partially allowing plasmids to escape, whereas the bacterial chromosomal DNA remains trapped in the cell debris. High temperature is then used to denature the chromosomal DNA, after which reannealing allows the plasmids to reassociate. Centrifugation removes the chromosomal DNA along with the cell debris, leaving the plasmid in suspension, from where it is recovered by isopropanol precipitation.

After the initial characterization, it is possible to purify further some or all of the plasmid DNAs by RNase digestion and extraction with organic solvents. This further purified DNA is suitable for techniques such as DNA sequencing, subcloning or the production of gene probes *(3)*.

2. Materials

2.1. Small-Scale Alkaline Lysis Plasmid DNA Preparations

1. Luria Bertani (LB) broth bacteria culture medium: 1% Tryptone (Difco Laboratories, Detroit, MI), 0.5% yeast extract, 200 mM NaCl. Sterilize by autoclaving in suitable aliquots. To ensure retention of the plasmid, media should be supplemented with the appropriate antibiotic(s).

2. 1.5 mL Microfuge tubes.
3. Sterile tubes: Must have a volume of at least 10 mL to ensure good aeration.
4. Lysis solution: 200 mM NaOH, 1% SDS. Store at room temperature.
5. Resuspension solution: 50 mM glucose, 50 mM Tris-HCl, pH 8.0, 10 mM ethylenediaminetetraacetic acid (EDTA). Keep at 4°C to prevent growth of contaminants.
6. Potassium acetate (neutralizing solution): 3 M potassium/5 M acetate. For 100 mL, take 29.4 g of potassium acetate, add water to 88.5 mL, and 11.5 mL of glacial acetic acid. Store at room temperature.
7. TE: 10 mM Tris-HCl, pH 8.0, 1 mM EDTA.
8. Isopropanol.
9. 70% Ethanol.

2.2. Rapid Boiling Method of Plasmid Extraction

1. LB broth bacteria culture medium: 1% Tryptone, 0.5% yeast extract, 200 mM NaCl. Sterilize by autoclaving in suitable aliquots.
2. STET: 5% (v/v) Triton X-100, 50 mM Tris-HCl, pH 8.0, 50 mM EDTA, pH 8.0, 8% (w/v) sucrose. Store at room temperature.
3. Lysozyme: Dry powder. Store at –20°C.
4. 70% Ethanol.
5. Isopropanol.
6. TE: 10 mM Tris-HCl, pH 8.0, 1 mM EDTA.
7. A boiling water bath: An opened bottom tube rack is required because the tubes must be placed directly in the water to achieve rapid heating.
8. Sterile wooden toothpicks.

2.3. RNA Digestion and Phenol-Chloroform Extraction

1. RNase A: Make up as a solution in water at 10 mg/mL, Heat for 10 min in a boiling water bath or heating block to eliminate any DNase activity. Aliquot and store at –20°C.
2. 0.4 M Ammonium acetate.
3. Chloroform: A 24:1 mix of chloroform and isoamyl alcohol. Store at 4°C.
4. Phenol/chloroform: 25:24:1 mix of TE-equilibrated phenol, chloroform, and isoamyl alcohol. Store at 4°C.
5. 100% Ethanol.
6. Sterile wooden toothpicks.

3. Methods
3.1. Small-Scale Extraction by Alkaline Lysis

1. Take a number of separate sterile tubes and place 2 mL of L-broth into them. Inoculate from individual bacterial 37°C overnight with shaking.
2. Transfer each culture to a labeled 1.5-mL Eppendorf tube, and centrifuge for 30 s at high speed in the microfuge (*see* **Note 1**).
3. Decant the supernatant and place tubes in a rack vertically for 5–10 s. Remove any of the remaining liquid by aspiration with a fine Pasteur pipet.
4. Add 100 µL of resuspension solution into each tube, close the lids, and resuspend the bacteria in each tube by shaking or vortexing to dissociate the bacterial pellet.
5. To each tube add 200 µL of lysis solution and mix by inverting the tube several times (*see* **Note 2**).
6. Allow at least 2–3 min for lysis to take place and leave the tubes to stand for 60 s before opening. This will allow the liquid to return to the bottom of the tube.

Extraction and Purification of Plasmid DNA

7. To each tube add 150 μL of neutralizing solution and invert the tubes several times. At this point bacterial chromosomal DNA is usually seen as a white precipitate.
8. Centrifuge the tubes for 2–5 min at full speed in a microfuge.
9. Place new sterile tubes into a rack, label them, and add 250 μL of isopropanol to each tube.
10. Remove the tubes from the microfuge, being careful not to disturb the precipitate.
11. Remove the supernatant with a 1-mL pipet, avoiding the white precipitate as much as possible (*see* **Note 3**).
12. Transfer the liquid phase into the new set of labeled tubes containing the isopropanol.
13. Vortex the tubes for 5–10 s and centrifuge the tubes in the microfuge for 30 s at high speed. The plasmid DNA precipitates as a white pellet.
14. Decant the supernatant and wash the pellets by adding 750 mL of 70% ethanol, vortex briefly, and centrifuge at high speed for 30 s.
15. Decant the ethanol, and centrifuge again for 10 s to collect the remaining ethanol at the bottom of the tubes. Carefully aspirate the remaining ethanol and leave the tubes to air dry on the bench for 5 min.
16. Dispense 50 μL of TE into each tube, and resuspend the pellet (*see* **Note 4**).
17. Take 10 μL of the resuspended pellet and analyze by agarose gel electrophoresis (*see* Chapter 13).

3.2. Plasmid Isolation by Rapid Boiling Method

1. Set up a culture for each miniprep by inoculating 2–3 mL of L-broth, containing an appropriate antibiotic (e.g., 100 μg/mL ampicillin) with a bacterial colony. Grow overnight at 37°C with vigorous shaking (*see* **Note 5**).
2. Before starting the miniprep, begin boiling the water and make up a fresh solution of 1 mg/mL lysozyme in STET mix.
3. Fill a 1.5-mL labeled microfuge tube with an aliquot from each culture. Pellet the bacteria by centrifugation for 1 min at 12,000g. Carefully aspirate off the supernatant using a drawn-out Pasteur pipet (*see* **Note 6**).
4. Vortex each pellet for a few seconds to break up the pellet. Add 20 μL STET to each tube. The pellet should now easily resuspend by vortexing.
5. Immediately place the tubes in the open-bottom rack, and place in the boiling water for exactly 45 s. Ensure that each tube is at least half submerged.
6. Centrifuge the tubes at 12,000g for 10 min. A large, sticky, loose pellet should form.
7. Remove the pellet from each tube by "fishing" it out with a sterile wooden toothpick. Because the pellet is quite slippery, it is useful to have a paper tissue at the top of the tube to catch the pellet and prevent it from slipping back down into the tube.
8. Add 200 μL isopropanol to each tube, and centrifuge at 12,000g for 5 min.
9. Aspirate the supernatant, and wash the pellet in 500 μL 70% ethanol. Centrifuge the tube for 1 min to compact the pellet, and then aspirate the 70% ethanol.
10. Air dry the pellets for 10 min, and resuspend each one in 100 μL TE buffer. Vortex and shake for 10 min before use to ensure complete dissolution.
11. Use 10 μL (equivalent to 100 ng of plasmid for most vectors) and analyze by gel electrophoresis.

It is possible to scale up procedures for the isolation of plasmid (*see* **Note 7**).

3.3. Phenol/Chloroform Extraction

Once plasmid has been isolated, it is possible to purify the preparation by removing any residual RNA and extracting the solution with phenol-chloroform to remove any contaminating protein.

1. Add 50 μL of 4 M ammonium acetate containing 200 μg/mL RNase A to each miniprep and incubate it at room temperature for 20 min.
2. Add 100 μL of phenol/chloroform to each DNA preparation.
3. Vortex briefly and centrifuge at high speed for 2 min in a microfuge. Remove the top layer containing the DNA and place it in a new sterile tube.
4. Add 100 μL of chloroform to each tube.
5. Vortex briefly and centrifuge at high speed in a microfuge for 2 min. Remove the DNA in the top layer and place it in a second sterile tube (*see* **Note 8**).
6. Add 200 μL of 100% ethanol to each tube.
7. Shake briefly to precipitate the DNA and centrifuge at high speed for 5 min at room temperature.

4. Notes

1. Here, a tight creamy pellet may be seen.
2. The solution should quickly turn transparent and become more viscous indicating bacterial lysis has taken place.
3. Some of the precipitate may float, so it is critical to use a pipet and disposable tips to recover the supernatant rather than pouring it.
4. It is not advisable to vortex, as this may lead to DNA shearing. The sample is best left for 3–5 min with occasional finger flicking of the tube.
5. Where plasmids have a high copy number, the growth time may be reduced to approx 6 h.
6. The short centrifugation time leaves a loose pellet that is easier to resuspend. If the pellet does not readily resuspend, pipet the solution up and down to dislodge it. Do not suck the pellet directly into the pipet tip.
7. Larger-scale versions of the alkaline lysis plasmid isolation method can be achieved from greater quantities of cells by scaling up the amounts of resuspension, lysis, and KAc solution in **steps 4–7**. For example, the bacterial culture from **step 1** (1 mL) can be used to inoculate a fresh broth culture of 500 mL, containing the relevant antibiotics. Following incubation overnight with shaking at 37°C, the cells can be harvested and resuspended in 10 mL resuspension solution containing 5 mg/mL lysozyme (**step 4**). After lysis solution (20 mL, **step 5**) has been added directly to the tube, the contents should be gently mixed by inverting the tube several times and kept on ice for 10 min. Neutralizing solution (15 mL, **step 7**) is added to the tube, which is again mixed by inversion and left on ice for a further 10 min.

   The lysate is then cleared by centrifugation at 30,000g for 30 min at room temperature. The supernatant is removed to a fresh centrifuge tube and subjected to isopropanol precipitation. After the addition of 0.6 vol of isopropanol, the tube contents should be mixed by inversion and left at room temperature for 15 min. Centrifugation at 12,000g for 30 min at room temperature will lead to a pellet that should be subsequently washed in 70% ethanol and air dried by standing the tube inverted for 10–15 min. Pure DNA is obtained by resuspending the DNA pellet in 8 mL TE buffer by gentle shaking, mixing with 100 μL of 10 mg/mL ethidium bromide, adding to approx 8–8.5 g CsCl (to a density of 1.55 g/mL; refractive index approx 1.386) and subjecting to ultracentrifugation in sealed and balanced tubes for 18 h at 20°C. Beckman Quick-Seal (Beckman, Palo Alto, CA) or equivalent tubes, filled with sample and topped up with paraffin oil prior to sealing, should be centrifuged at 45,000–60,000 rpm in a Beckman Type-50 or Type 65 rotor (the equivalent fixed-angled rotors from other manufacturers can also be used). Ethidium bromide staining enables the identification of a plasmid band (covalently closed circular plasmid DNA will be the lower band). The band is removed from the tube by using a syringe.

Ethidium bromide is subsequently removed from the DNA sample by extraction three times with CsCl-saturated isopropanol (the lower layer, containing the DNA, is retained). CsCl is removed by dialyzing the sample against TE at 4°C, with at least three changes of buffer. As an alternative to equilibrium centrifugation using CsCl-ethidium bromide gradients, plasmid DNA can be purified from a lysate by using commercially produced ion-exchange columns such as the QIAGEN-tip 100 columns (Qiagen, Chatsworth, CA).
8. For phenol/chloroform extractions avoid removing material from the interface.

References

1. Birnboim, H. C. and Doly, J. (1979) A rapid alkaline extraction procedure for screening recombinant plasmid DNA. *Nucleic Acids Res.* **7,** 1513–1523.
2. Holmes, D. S. and Quigley, M. (1981) A rapid boiling method for the preparation of bacterial plasmids. *Anal. Biochem.* **114,** 193–197.
3. Maniatis, T., Fritsch, E. F., and Sambrook, J. (eds.) (1989) *Molecular Cloning: A Laboratory Manual*, Cold Spring Harbor Laboratory Press, Cold Spring Harbor, NY, pp. 368,369.

46

Biotinylated Probes in Colony Hybridization

Michael J. Haas

1. Introduction

Colony hybridization is a procedure that allows the detection of cells containing nucleic acid sequences of interest (1). In this method, microbial colonies grown on, or transferred to, a supporting membrane are lysed and their nucleic acids denatured to single strands and fixed in place on the membrane. The membrane is then exposed to a similarly denatured "probe" sequence, which is identical or homologous to all or part of the target sequence, under conditions favoring reannealing. Probe sequences hybridize to complementary sequences on the membrane. Positive hybridization events are then detected by determining the presence and location of probe sequences on the membrane.

The original colony hybridization method described the use of radiolabeled probes and the detection of positive hybridization events by autoradiography (1). However, because of the high waste disposal costs, short half-lives, long autoradiographic exposures, and potential health hazards associated with radioisotopes, there is interest in alternative methods to detect positive hybridizations.

Nonradioactive technology involves the attachment to the nucleic acid probe of a ligand that can subsequently be detected by chemical or enzymatic methods. The vitamin biotin is one such ligand. Biotin can be covalently incorporated into nucleic acids in a manner that does not interfere with their ability to hybridize with homologous sequences. This is accomplished by replacing a nucleoside triphosphate with its biotinylated analog in an in vitro DNA replication or transcription reaction, generating a biotinylated probe sequence (2,3). Hybridization of such a probe to a homologous sequence immobilized on a membrane results in the retention of biotin at that site. Positive hybridization events can then be detected by assaying for biotin.

Enzymatic reaction schemes that generate insoluble colored products at sites where biotin is bound to the filters have been developed for the purpose of biotin detection in these applications. These detection reactions employ either avidin or streptavidin, two functionally identical proteins that bind to biotin with very high affinities and specificities. These proteins are retained at sites where biotinylated probes have hybridized to homologous sequences. Avidin and streptavidin have multiple biotin binding sites per molecule. They therefore retain biotin binding capability even after binding to probe sequences on the membranes. Incubation with a biotinylated form of an enzyme

From: *The Nucleic Acid Protocols Handbook*
Edited by: R. Rapley © Humana Press Inc., Totowa, NJ

(e.g., alkaline phosphatase) for which there exists an assay that generates an insoluble, colored product results in the retention of signal enzyme at sites of positive hybridization. These sites are detected by applying the histochemical assay for the signal enzyme.

To facilitate our work on plasmids with no known phenotype, we have developed a method for the use and detection of biotinylated probes in colony hybridization. It is suitable both for the detection of rare positive hybridization events over a background of nonreactive colonies and for the detection of nonhybridizing colonies in a population containing sequences homologous to the probe. The latter capability could be useful in such applications as the detection of cured (i.e., plasmid-free) cells in a bacterial population containing plasmids.

2. Materials

1. Nitrocellulose filters (82-mm diameter, BA 85) are obtained from Schleicher & Schuell (Keene, NH). (Products from other suppliers may be acceptable.)
2. Formamide: deionized by stirring for 30 min with 10% (w/v) of a mixed-bed ion exchange resin (e.g., Bio-Rad AG 501-X8, 20-50 mesh, Bio-Rad, Hercules, CA), filtering twice through Whatman (Clifton, NJ) no. 1 paper and storing single-use aliquots at $-80°C$.
3. Bovine serum albumin ([BSA], Fraction V, Sigma, St. Louis, MO) is used as obtained. Fatty acid-free albumin gives poor results.
4. Denatured herring sperm DNA is prepared by dissolving in water (10 mg/mL) with stirring at room temperature, shearing by 10 passages through an 18-gauge needle, and immersing in boiling water for 10 min. Aliquots are stored at $-20°C$. Just prior to use, these are incubated for 10 min in a boiling water bath and chilled in ice water.
5. 20X SSC buffer: 3 M sodium chloride, 0.3 M sodium citrate, pH adjusted to 7.0 with sodium hydroxide. Sterilize by autoclaving, store at room temperature.
6. Proteinase K is obtained from Beckman (Somerset, NJ). Other sources may be acceptable. In using alternate sources, the occurrence of blue backgrounds between colonies, and oversize, blurry signals at colony sites after the final color development step indicates insufficient proteolytic activity. Prepare a solution of 200 µg/mL in 1X SSC.
7. 50X Denhardt's solution: 1% (w/v) Ficoll, 1% (w/v) polyvinyl pyrrolidone, 1% (w/v) BSA. Filter-sterilize. Store aliquots at $-20°C$. Do not flame the pipets used to transfer this solution. Denaturation and precipitation of the protein result from the use of hot pipets at this stage.
8. Template DNA for the production of hybridization probes must be pure. Standard methods, such as dye-buoyant density ultracentrifugation, generate acceptable products. Ethidium bromide and cesium chloride are removed prior to use of the DNA *(4)*.
9. Biotin-11-deoxyuridine-5'-triphosphate (BiodUTP) and reagents for its incorporation into DNA by nick translation are obtained commercially. The products from Bethesda Research Laboratories (BRL, Gaithersburg, MD) are acceptable. BRL now provides a prepackaged kit (BioNick) containing necessary supplies and employing biotin-14-dATP as the source of biotin. The concentration of the resulting biotinylated DNA is determined by the histochemical method for biotin *(below)*. Adequate instructions are provided with these kits (*see* **Note 1**).
10. Prehybridization solution: 50% formamide, 5X SSC, 5X Denhardt's solution, 25 mM sodium phosphate, pH 6.5, 300 µg/mL freshly denatured sheared herring sperm DNA. Filter through Whatman no. 1 paper on a Buchner funnel, then through a sterile 0.45-µm filter. Store 10-mL aliquots in glass at $-20°C$. Use only once.

Biotinylated Probes 335

11. Hybridization solution: 45% Formamide, 5X SSC, 5X Denhardt's solution, 20 mM sodium phosphate, pH 6.5, 300 µg/mL freshly denatured, sheared, herring sperm DNA, 200 ng of biotinylated DNA/mL. Before its addition, the biotinylated probe DNA is denatured by incubating for 10 min in a boiling water bath and quick-chilling in an ice bath. Shearing to reduce size is unnecessary, since the products generated by nick translation are sufficiently small. Filter and store the hybridization solution as was done for the prehybridization solution. Hybridization solution can be recovered after use and stored at –20°C. The solution can be reused at least 10 times over a time-span of at least 5 mo, without noticeable reduction in performance. The solution is heat-denatured as described for the herring sperm DNA preparation immediately before each use. (*See* **Note 2**.)
12. Reagents for the detection of filter-bound biotin are obtained from BRL (BlueGene Nonradioactive Nucleic Acid Detection System). Comparable materials are available from Bio-Rad Laboratories.
13. Special equipment required for this protocol are a vacuum oven, a slab gel dryer, a device for the heat sealing of plastic bags (e.g., Seal-A-Meal, Sears Seal-and-Save), thin rubber sheet (such as dental sheet, A. H. Thomas, Philadelphia, PA), and a filtration device designed for the washing of nitrocellulose filters. The latter was originally described by Grunstein and Hogness *(1)* and is available from Schleicher and Schuell as the "Screen-It" colony filter hybridization device.
14. 90% (w/w) Ethanol.
15. Chloroform: Reagent grade.
16. Solutions for the posthybridization washing of filters:
    a. 0.1% (w/v) Sodium dodecylsulfate (SDS) in 2X SSC.
    b. 0.1% (w/v) SDS in 0.2X SSC.
    c. 0.1% (w/v) SDS in 0.16X SSC.
    d. 2X SSC.

3. Methods
3.1. Filter Preparation and Cell Growth

1. Use a soft lead pencil to label nitrocellulose filters with a hash mark and letter or number on one edge to allow subsequent identification and orientation (*see* **Note 3**).
2. Place the labeled filters between sheets of filter paper, wrap in aluminum foil, and autoclave for 10 min.
3. Seal the packets of sterile filters in an air-tight bag and store at 4°C.
4. To inoculate, place a filter on top of solidified media in a Petri dish and spread an appropriately diluted bacterial culture over the surface.
5. Incubate the plates until the cells are approx 1–3 mm in diameter (*see* **Note 4**). Cell densities of approx 800/82 mm diameter filter are compatible with single colony discrimination after hybridization and color assay. If one is attempting to locate positively hybridizing sequences in a generally nonreacting population, and single colony resolution is not required in the first detection step, as many as 10^5 cells can be applied to each filter.
6. Invert the filter and gently lay it onto fresh media just prior to lysis to create a replica of the colony pattern of the filter. (Mark the plate to indicate the orientation of the filter on it.) After an appropriate incubation, this becomes a master plate from which viable analogues of desirable colonies, as identified on the filter after hybridization and processing, can be recovered.

3.2. Cell Lysis

All operations are conducted at room temperature unless otherwise noted. After steps 1–3, gentle suction is applied to the filters (*see* **Note 5**). Steps 2–4 are conducted in glass Petri dishes, one filter per dish. It has not been determined if these steps can be done batchwise. It is difficult to process more than 12 filters at a time.

To achieve lysis, incubate the filters in the following fashion (*see* **Note 6**):

1. Incubate 7 min, colony-side up, on filter paper sheets stacked to a thickness of 4 mm and saturated with fresh 0.5 M NaOH.
2. Incubate 5 min in 1.5 M sodium chloride, 0.5 M Tris-HCl, pH 7.4, 30 mL/filter.
3. Incubate 1 h in prewarmed proteinase K in 1X SSC, 30 mL/filter, 37°C.
4. Incubate 2 × 2 min in 90% (w/w) ethanol, 30 mL/filter (*see* **Note 7**).
5. Air-dry, 20 min.
6. Wash each filter with 100 mL of chloroform using the Screen-It colony hybridization device. A single sheet of filter paper is used as an underfilter.
7. Air-dry (approx 15 min).
8. Sandwich the filters individually between filter paper, wrap loosely in aluminum foil, and bake at 80°C *in vacuo* for 2 h.
9. Store the filters in a vacuum desiccator at room temperature.

3.3. Prehybridization, Hybridization, and Detection of Hybridization

1. For prehybridization, place pairs of filters containing lysed, fixed colonies back to back in sealable plastic bags. Add 20 mL of prehybridization solution, seal the bag, seal it within a second bag, and incubate at 42°C for 2 h. Maintain the proper temperature by submersion in a water bath.
2. After prehybridization, replace the liquid with 20 mL of hybridization solution, exclude air bubbles, reseal the bags, and immerse in the water bath. Brief incubations (1 h) are sufficient for the detection of relatively abundant sequences, such as unamplified plasmid pBR322 in *E. coli*. More extensive incubations (45 h) may be necessary to detect less abundant sequences.
3. Following hybridization, wash the filters sequentially:
   a. Twice in 250 mL of 0.1% (w/v) SDS in 2X SSC, 3 min per wash, room temperature;
   b. Twice in 250 mL of 0.1% (w/v) SDS, 0.2X SSC, 3 min per wash, room temperature;
   c. Twice in 250 mL of 0.1% (w/v) SDS, 0.16X SSC, 15 min per wash, 50°C; and
   d. Briefly in 2X SSC at room temperature.
4. Detection of the sites of hybridization-dependent binding of biotinylated probe to the filters is most readily conducted with commercially available kits. Favorable results have been obtained with the BluGene Nonradioactive Nucleic Acid Detection System from BRL. Follow the manufacturer's instructions when carrying out the following steps. After washing, sequentially expose the filters to streptavidin and biotinylated alkaline phosphatase (or to a conjugate of these two proteins). This causes the immobilization of alkaline phosphatase at sites of positive hybridization.
5. Incubate the filters with 5-bromo-4-chloro-3-indolylphosphate (BCIP) and nitroblue tetrazolium (NBT). Indoxyl generated from BCIP by the action of alkaline phosphatase condenses to form indigo (blue). Indigo then reacts with NBT to form insoluble diformazan (purple).
6. Terminate the reaction when reacting colonies are intensely purple (*see* **Note 8**) by replacing the dye solution with 20 mM Tris-HCl, 5 mM EDTA, pH 7.5. Nonreactive colonies should be light blue on a white background.
7. Store the moist filters in sealed bags. The elapsed time from the end of hybridization to the termination of color development is approx 3 h. **Figure 1** illustrates typical results obtained with this method.

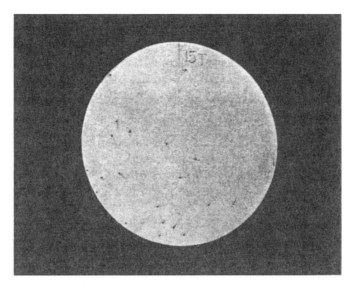

Fig. 1. Specific identification of *E. coli* containing plasmid pBR322. Approximately 225 colonies, consisting of a 10:1 mixture of plasmid-free and plasmid-containing cells, was grown on a nitrocellulose filter. The filter was subjected to the lysis protocol described here, followed by a hybridization with biotinylated pBR322. Sites of positive hybridization were detected by means of streptavidin and alkaline phosphatase. The dark sites correspond to colonies harboring pBR322. Plasmid-free cells give faint signals present at numerous sites on the filter.

4. Notes

1. Nucleic acids can also be biotinylated by nonenzymatic methods with photobiotin, a photoactivatable biotin analog *(5)*, which can be commercially obtained from BRL, Sigma, and other commercial sources. I have not compared the suitability of this method of biotin incorporation with that reported here, but expect that the method would be fully acceptable. FMC (Rockland, ME) markets an alternate nonradioactive sequence detection kit known as Chemiprobe. The basis of this system is a chemical modification of cytosine residues in the probe DNA. After hybridization, the probe is detected by means of a monoclonal antibody that specifically recognizes the sulfonated DNA. Detection of the bound monoclonal antibody is achieved by means of an alkaline phosphatase-conjugated second antibody.
2. The minimal probe concentration necessary for efficient detection of target sequences has not been determined. It has been noted, however, that probe concentrations of 10–20 ng/mL, when coupled with overnight hybridizations, are too low to give strong signals for nonreiterated target DNAs 3 MDa or larger in size. Maas *(6)* reported a simple modification of the Grunstein-Hogness protocol *(1)*, which is reported to increase the sensitivity of the colony hybridization method by 100-fold. This could increase the ability to detect single copy sequences.
3. Cellulose filters give unacceptably diffuse colony patterns after lysis and should not be used. Nylon filters should be acceptable, although I have not examined their suitability.
4. The lysis of colonies larger than specified above is generally acceptable. However, with relatively mucoid strains, such as *Xanthomonas*, the lysis of oversize colonies results in smeared colony patterns. The researcher should investigate the performance of younger cells if such behavior is experienced.

5. The application of gentle suction to the filters following steps 1–3 of the lysis protocol reduces the dispersion of cells from their sites, promoting tighter patterns and stronger signals, and reducing the interference of signals from adjacent colonies with one another. Suction is applied by means of a slab gel dryer and a gentle vacuum source. A single sheet of filter paper serves as an underfilter. On this sheet is placed a template made by cutting into a sheet of flexible rubber holes slightly smaller than the nitrocellulose filters. The filters are placed over these holes and vacuum is applied. A brief suction suffices to remove excess moisture from the filters and to pull lysed colonies down onto them. Six filters can be treated at a time with a standard commercially available gel dryer with an 18×34 cm suction surface.

6. In our initial studies, the filters were swirled in the lysis solutions in an attempt to ensure lysis. After hybridization and application of the color assay, it was found that positively reacting colonies had "tails" extending away from them in a circular pattern across the filters. These tails obscured the signals of adjacent colonies. Tailing was eliminated by omitting the swirling action during lysis. This omission did not noticeably reduce the efficiency or sensitivity of the detection reaction.

7. The ethanol concentration in step 4 of the lysis protocol is a w/w concentration. Ethanol solutions made up v/v, or otherwise in excess of 90% w/w, exceed the ethanol tolerance limits of some batches of nitrocellulose. Filters washed in such solutions may become brittle and be reduced nearly to powder by the end of the hybridization-color assay procedure. The appropriate solution can be made from 100% ethanol.

8. The final color development step must be conducted under dim light (i.e., incubated in a drawer) since the reagents are light sensitive. Examine the filters at frequent intervals (10 min) during this incubation. Stop the reaction when the color of positively reacting colonies is deep purple. Further incubation past this point allows the color of nonreacting colonies to darken to such a degree that they are mistaken for positives. Overdevelopment is the greatest single factor contributing to the appearance of false-positive signals.

References

1. Grunstein, M. and Hogness, D. S. (1975) Colony hybridization: a method for the isolation of cloned DNAs that contain a specific gene. *Proc. Natl. Acad. Sci. USA* **72,** 3961–3965.
2. Langer, P. R., Waldrop, A. A., and Ward, D. C. (1981) Enzymatic synthesis of biotin-labeled polynucleotides: novel nucleic acid affinity probes. *Proc. Natl. Acad. Sci. USA* **78,** 6633–6637.
3. Brigati, D. J., Myerson, D., Leary, J. J., Spalholz, B., Travis, S. Z., Fong, D. K. Y., Hsiung, G. D., and Ward, D. C. (1983) Detection of viral genomes in cultured cells and paraffin-embedded tissue sections using biotin-labeled hybridization. *Virology* **126,** 32–50.
4. Boffey, S. A. (1984) Plasmid DNA isolation by the cleared lysate method, in *Methods in Molecular Biology,* vol. 2: *Nucleic Acids* (Walker, J. M., ed.), Humana, Clifton, NJ, pp. 177–183.
5. Forster, A. C., McInnes, J. L., Skingle, D. C., and Symons, R. H. (1985) Nonradioactive hybridization probes prepared by the chemical labeling of DNA and RNA with a novel reagent, photobiotin. *Nucleic Acids Res.* **13,** 745–761.
6. Maas, R. (1983) An improved colony hybridization method with significantly increased sensitivity for detection of single genes. *Plasmid* **10,** 296–298.

47

Cloning Long Polymerase Chain Reaction Products

Songrong Ren and J. Michael Ruppert

1. Introduction

A variety of methods have been developed and modified to facilitate the cloning of polymerase chain reaction (PCR) products directly into plasmid. However, there are limitations in practice, including the low frequency of correct clones, the expense of commercially available cloning systems, and the need to add sequence or other modifications to primers to enable cloning. The more recent development of long and accurate PCR (LA-PCR) offers the ability to isolate much longer products. Using this technique, the vast majority of full-length eucaryotic mRNA transcripts can be easily amplified by reverse transcription and LA-PCR. In addition, since the coding region for the majority of mammalian genes is contained within approx 20 kilobases (kb) or less of genomic DNA, this technology can be used to rapidly obtain genomic sequence corresponding to known cDNAs.

These important applications of PCR are limited by the ability to efficiently clone amplified products. The low efficiency of cloning seems exacerbated for longer PCR products. In addition, the greater length of unknown sequence between the primers complicates addition of restriction sites to the ends of molecules, as these sites frequently are found within fragments of several kilobases long. In our hands, LA-PCR reactions have failed to generate the desired product after addition of 5'-end sequence to the long (approx 30 base) primers required for this technique. Thus, protocols requiring addition of restriction sites or other sequence to the ends of primers may not be helpful in cloning of longer products.

A procedure adapted from cDNA library construction protocols enables addition of *Eco*RI complementary ends to double-stranded, blunt-ended DNA. The approach requires no prior knowledge of DNA sequence between the primers, requires no restriction enzyme digestion of PCR products, and uses only the enzymes T4 polynucleotide kinase, T4 DNA ligase, and Klenow fragment or DNA Polymerase I. In brief, after rendering PCR product ends blunted and 5' phosphorylated, single adaptors are ligated to each end. The adaptors are then 5' phosphorylated in the same tube, and free adaptors are separated from the *Eco*RI-compatible PCR product using a spin column. Fragments up to 5 kb long can be thus treated, ligated to standard plasmid cloning vectors, and transfected into competent *E. coli*. Consistently, 40–100% of the resulting drug-resistant colonies contain the insert. The method may be useful for cloning products >5.0 kb using bacteriophage lambda or other suitable vectors.

2. Materials
2.1. Equipment

1. Plastic 1 cm³ tuberculin syringes (Becton-Dickinson, Mountain View, CA).
2. 15-mL conical bottom tube (Corning #25319-15, dimensions 17 mm × 118 mm, polypropylene, Corning Glassworks, Corning, NY).
3. Agarose gel electrophoresis apparatus (*see* Chapter 13).
4. Glass wool (Fisher Scientific, Pittsburg, PA, #11-390).
5. Microcon 30 microconcentrators (Amicon, Inc., #42410, Beverly, MA).
6. Vacuum dessicator (Corning #3118-160).

2.2. Reagents and Solutions

1. Agarose (Sigma, St. Louis, MO, #A-6013).
2. Ammonium acetate (10.0 M in water).
3. Adenosine triphosphate (ATP) (10 mM in water, store at –20°C)
4. Dichloro-dimethyl-silane (Sigma, #D-3879) (toxic, use in fume hood).
5. DNA polymerase I, large (Klenow) fragment (New England Biolabs, Beverly, MA, #210S).
6. DNA polymerase I (*E. coli*) (New England Biolabs, #209S).
7. *Eco*RI-*Not*I adaptors (Stratagene, La Jolla, CA, # 901111).
8. Ethanol [100% and 70% (v/v)].
9. GeneClean II® Kit (Bio 101 Inc., La Jolla, CA).
10. IPTG (isopropyl-β-D-thiogalactopyranoside, Amersham Pharmacia Biotech (Piscataway, NJ) #US17886, 0.1 M in water, store at –20°C).
11. Low TE buffer: 3 mM Tris-HCl, 0.2 mM EDTA, pH 7.5.
12. PC-9 (phenol/chloroform/TE9, store at –20°C).
13. Polish and kinase buffer (10X): 500 mM Tris-HCl (pH 7.5), 100 mM MgCl$_2$, 10 mM dithiothreitol, 500 μg/mL bovine serum albumin (BSA), 10 mM ATP, 2 mM each deoxynucleotide triphosphate (dNTP) (store at –20°C).
14. PCR primers (synthesize approx 30 base oligos with 5' phosphates added).
15. Sephacryl S-400-HR (Amersham Pharmacia Biotech, # 17-0609-10).
16. STE buffer (10X): 1 M NaCl, 200 mM Tris-HCl (pH 7.5), 100 mM EDTA (store at 4°C).
17. Sterile deionized water (approx 18 MΩ, e.g., Milli-Q or equivalent, autoclaved).
18. T4 DNA ligase (New England Biolabs, #202S).
19. T4 polynucleotide kinase (New England Biolabs, #201S).
20. TE buffer: 10 mM Tris-HCl, 1 mM EDTA, pH 8.0.
21. TAE electrophoresis buffer (1X): 40 mM Tris, 25 mM sodium acetate, 1 mM EDTA, pH 7.9 (use acetic acid to adjust pH).
22. TE9 (20 mM EDTA, 10 mM NaCl, 0.5 M Tris-HCl, pH 8.9).
23. TM buffer (10X): 500 mM Tris-HCl (pH 7.5), 70 mM MgCl$_2$, 10 mM DTT (store at –20°C).
24. X-Gal (5-bromo-4-chloro-3-indolyl-β-D-galactopyranoside, Fisher, #BP1615-1, 20 mg/mL in *N,N*-dimethyl formamide, store at –20°C).

3. Methods
3.1. Purify the PCR Product with Glass Powder

Purification is carried out using the Geneclean II Kit as described by manufacturer. Briefly:

Cloning Long PCR Products

1. Transfer the PCR product to a microcentrifuge tube, add 3 vol sodium iodide (NaI) stock solution and mix well.
2. Add 15 µL glassmilk (vigorously vortex the silica matrix stock before pipeting). Bind for 30–60 min on ice or at 4°C (invert the tube to mix every 5 min or rotate the tube at 4°C on a tube rotator).
3. To pellet the silica matrix with the bound DNA, microcentrifuge at high speed for 5 s and transfer the NaI supernatant to another tube.
4. Wash pellet three times with new wash buffer: add approx 10–50 vol (200–800 µL) of ice-cold new wash to the pellet, resuspend the pellet in the wash by pipeting up and down, spin for 5 s, and discard the supernatant. Repeat the wash procedure two more times.
5. To elute DNA from glassmilk: resuspend the pellet with 25 µL of Low TE buffer, incubate the tube at 50°C for 2 or 3 min, spin for 30 s to make a solid pellet, and carefully transfer the supernatant containing the eluted DNA to a new tube.
6. Resuspend the pellet with another 25 µL of Low TE and repeat incubation at 50°C for a second elution. Pool the eluates from each elution to one microcentrifuge tube (approx 50 µL).

3.2. Polish and Kinase the Ends of the PCR Product

1. For a 100 µL reaction, add the following components to the microcentrifuge tube with the eluted DNA:

10X polish and kinase buffer	10 µL
DNA polymerase I (10 U/µL)	1 µL
T4 polynucleotide kinase (10 U/µL)	1 µL
water to a final volume of 100 µL	

2. Mix gently and incubate for 1 h at 37°C. Klenow fragment (1 µL of 5 U/µL stock) may be substituted for DNA polymerase I with only a slight loss of cloning efficiency (*see* **Note 2**).
3. Stop the reaction by placing the tube on ice or adding EDTA to a final concentration of 5 m*M*.
4. Prepare PC-9 by mixing 160 mL phenol, 213 mL chloroform, and 107 mL TE9; shake and let phases separate at 4°C for 2 h; repeat this extraction one time; discard aqueous layer (top layer) and store at –20°C. To PC-9 and chloroform extract the DNA sample: thaw PC-9 at room temperature. Dilute the 100-µL sample to 400 µL with TE9 and add 1 vol (400 µL) PC-9. Vortex for 1 min, microcentrifuge for 2 min at maximum speed and transfer the upper aqueous layer to a new microcentrifuge tube. Repeat extraction with PC-9. Extract one final time with chloroform, and transfer the upper aqueous layer to a new microcentrifuge tube.
5. Ethanol precipitate: to the extracted sample, add 1/4 vol (100 µL) of 10 *M* ammonium acetate and mix well; add 2.5 vol (1 mL) of 100% ethanol and incubate on ice for 30 min or overnight at –20°C. Microcentrifuge the sample at high speed for 20 min at 4°C, wash the pellet twice with 70% ethanol, and dry the pellet in a vacuum concentrator.
6. Resuspend the pellet with 5.5 µL of Low TE.

3.3. Ligate to EcoRI-NotI Adaptors

1. Dissolve 20 µg of *Eco*RI-*Not*I adaptors in 40 µL water to make a 500-ng/µL stock solution (store at –20°C). To the microcentrifuge tube with 5.5 µL DNA, add the following components:

*Eco*RI-*Not*I adaptors (500 ng/µL)	1.5 µL
10X TM buffer	1.0 µL
10 m*M* ATP	1.0 µL
T4 DNA ligase (400 U/µL)	1.0 µL

2. Mix and spin down the volume in the microcentrifuge and incubate at least overnight at 8°C (see **Note 3**).
3. Heat inactivate the ligase in a 70°C water bath for 30 min.

3.4. Kinase the EcoRI Ends

1. After heat inactivation, spin down the condensate for 2 s, let the reaction cool at room temperature for 5 min, then kinase the adaptor ends by adding:
10X TM buffer	1 µL
10 m*M* ATP	2 µL
Water	6 µL
T4 polynucleotide kinase (10 U/µL)	1 µL
2. Mix, spin down, and incubate for 30 min at 37°C.
3. Heat inactivate the kinase for 30 min at 70°C, then place on ice.

3.5. Remove Unligated Adaptors

1. Prepare silanized glass wool to serve as a column support: In a fume hood, place a 25-mL glass flask or beaker inside a vacuum dessicator. Using a glass pipet, transfer 6.0 mL silane to the flask (personnel should avoid exposure to silane or vapors).
2. Place a 250-mL glass beaker containing approx 10 g loosely packed glass wool into the dessicator. Place the dessicator lid using vacuum grease and apply a strong vacuum until you see some of the silane vaporize. Silane may damage vacuum pumps, so limit vacuum application time to approx 1 min or less.
3. Close off vacuum and incubate at room temperature overnight. Open the dessicator in fume hood. Transfer the glass wool to a beaker of 100% ethanol, soaking thoroughly. Transfer to a sheet of tin foil, and squeeze excess ethanol out of the glass wool.
4. Dry in a vacuum oven at up to 120°C until dry (approx 2 h) or autoclave on dry cycle until dry. Store at room temperature (stable for months-years).
5. Resuspend the Sephacryl S-400 HR in the stock bottle into a homogeneous slurry by swirling vigorously. Remove approx 1.6 mL slurry to a polystyrene tube for each PCR product.
6. Prepare a 1-mL Sephacryl S-400 spin column: remove the plunger from a 1-mL plastic tuberculin syringe. Insert a small amount of silanized glass wool into the syringe with a flamed forceps. Using the plunger, push the glass wool firmly to the bottom of the syringe (the glass wool plug should be approx 1–2 mm thick). Insert the plunger approx 2 cm into the barrel and draw 1X STE upward past the glass wool. Flick the syringe bottom vigorously with your fingertip to release air bubbles. Set the bottom of the syringe into 1X STE in a tube or vial and remove the plunger completely. Raise the syringe until the fluid level is just above the glass wool.
7. Use a glass Pasteur pipet to fill the syringe to the top with slurry. Avoid air pockets. If multiple columns are being poured, firmly press parafilm onto the syringe tip to serve as a temporary plug and to prevent drying.
8. Place the syringe in a 15-mL, 17 mm × 118 mm conical bottom tube. Spin in a tabletop centrifuge for 2 min at exactly 400*g* and discard the solution in the bottom of the 15-mL tube.
9. Fill the syringe with slurry again to 1/4 in. from the top and spin the column as before. Rinse the column three times by filling to the brim with 1X STE buffer and spinning as in **step 8** (this equilibrates the column and removes any sodium azide preservative). The packed column should be 1.0–1.05 mL and should not pack further with each spin (or you may be spinning too fast).

10. Cut the cap off of a sterile 1.5 mL microcentrifuge tube. Slip the microcentrifuge tube, minus cap, into a clean 15-mL tube to collect the eluted DNA. Place the prepared spin column into the clean 15-mL tube, making sure the tip of the syringe is in the microcentrifuge tube. The spin column is now ready and should be used as soon as possible (within 1–2 min).
11. Dilute the adaptor-ligated and kinased products to 50 µL with low TE, add 5 µL of 10X STE buffer, and mix well. Pipet the sample onto the middle of the prepared spin column, avoiding the edges, as the sample can leach down the sides of the semidry column.
12. Spin for 2 min at 400g. Load 60 µL of 1X STE buffer on the column. Spin the tube as before. Carefully transfer the eluted sample (approx 120 µL) to a clean microcentrifuge tube.
13. Collect a third fraction by loading 60 µL of 1X STE buffer on the column and spinning as before (save this fraction in case the DNA does not elute in the first two fractions, although we have not experienced this difficulty in the past).
14. Purify the eluted DNA by adding TE-9 to 0.4 mL final volume. PC-9 and chloroform extract and ethanol precipitate as before.

3.6. Agarose Gel Purification

1. Purify the sample by agarose gel electrophoresis at 12 V/cm using TAE and 1% agarose (*see* Chapter 13). Stain with ethidium bromide (20 µg/mL) for 5–10 min. Limit ultraviolet light exposure of DNA as much as possible. Cut out the ethidium-stained fragment and purify using Microcon 30 microconcentrators.
2. Purification is carried out as described by the manufacturer. Briefly: assemble the Microcon by inserting micropure insert (which holds the agarose) into the Microcon dialysis chamber and then placing the assembly into the vial provided. Place agarose gel slice (chopped into 1–2 mm^3 pieces) into micropure insert and freeze the entire apparatus for 15 min in −70°C freezer.
3. Warm at 37°C for 3 min, microcentrifuge at high speed for 10 min at room temperature, remove the insert, and discard the insert and the solution at the bottom of the vial.
4. Add 600 µL of Low TE to the Microcon and spin for 5–8 min (make sure that there is no Low TE left above the dialysis membrane). Discard the solution at the bottom of the vial.
5. Place the Microcon dialysis chamber upside down in a new vial, place the assembly into a new 15 mL tube, and recover DNA by adding 40 µL of Low TE and spinning for 8 min at low speed (approx 2800g) in a tabletop centrifuge. Approximately 40 µL of volume should be recovered.
6. Quantitate the recovered DNA on an agarose gel. We typically have about 500 ng at this point, although 5–20 ng should suffice for cloning.

3.7. Clone into EcoRI Site of Plasmid Cloning Vector

1. Digest a standard plasmid cloning vector (e.g., Stratagene Bluescript) with *Eco*RI and treat with calf intestinal alkaline phosphatase.
2. Ligate using vector and 10–80 ng of insert at 1:1 molar ratio.
3. Introduce the ligated plasmid into *E. coli* and select the transformed cells using drug resistance and color selection. We use electroporation-competent XL1-Blue cells (Stratagene). Recombinant clones are identified by the blue/white color assay. Fifteen minutes before plating bacteria, mix 0.12 mL Luria-Bertani medium with 0.04 mL each of X-Gal (20 mg/mL) and IPTG (0.1 M) and spread evenly on a newly poured LB-agar plate. Typically we obtain 200–6000 colonies per ligation reaction, depending on the size of the PCR product. Forty to 100% of these colonies contain the correct insert. Verify the positive (white) clones by plasmid DNA preparation and restriction enzyme digestion.

4. Notes

1. Purify the PCR product by glass powder: For synthesis of long PCR products, see **Subheadings 4.2.–4.6.** and **ref. 1**, and references therein. PCR reactions should be optimized to be as specific as possible. We find it very effective to use EDTA in increments of 0.5 mM to increase the stringency by chelation of free Mg^{2+}. Purification of PCR product using glass powder is to remove dNTP, primers, primer-dimers, oligomers, very small PCR products, and other reaction components that may interfere with subsequent treatment of the PCR product.
2. Polish and kinase the ends of the PCR product: As Klenow contains no 5'-3' exonuclease function, the use of this enzyme during the polishing/kinase reaction is recommended when accurate junctions are required (e.g., for maintaining an open reading frame in the ligated product). When cloning efficiency needs to be maximal, the use of DNA polymerase I holoenzyme may favorably effect the cloning efficiency as discussed in **ref. 1**.
3. Ligate to *Eco*RI-*Not*I adaptors: The recommended temperature of 8°C isn't a typographical error: Stratagene claims this is optimal for ligation of adaptors to cDNA (Stratagene Zap Express™ cDNA Synthesis Kit, #200403). The *Eco*RI-*Not*I adaptors are 5' blunt-end phosphorylated and 5' cohesive-end nonphosphorylated. Thus, only one adaptor can ligate to the end of each PCR product. In addition, the adaptors contain an 8-bp *Not*I site that allows the release of cloned inserts by enzyme digestion. The sequence of the adaptor is:
   5'AATTCGCGGCCGC3'
         GCGCCGGCG5'
4. Kinase the *Eco*RI ends: cDNA library construction protocols successfully rely on this step for 5' end phosphorylation of adaptors (e.g., Stratagene). For our experiments we have ordered primers to be 5'phosphorylated during synthesis, although the kinase step in the protocol is still performed as described. Whether the kinase step can be excluded when PCR primers are synthesized with 5' phosphates has not been tested.
5. Remove unligated adaptors: The Sephacryl spin column produces a high yield and is extremely reproducible. Potential pitfalls include the use of nonsilanized column support that may bind to the PCR product and microbial contamination of the Sephacryl, visible air pockets within the column, and spinning the column faster than recommended. Unused Sephacryl should be stored at 4°C in 0.05% sodium azide (prepare a 10% stock in water), or in 20% ethanol (as shipped) if you aliquot the slurry using a sterile technique. Twenty percent ethanol is not sufficient by itself to inhibit microbes.

   It is very important to phenol-chloroform and chloroform extract the DNA after the spin column to remove the kinase. Kinase often remains active after the heat treatment. In the subsequent ligation reaction, any remaining kinase activity will allow vector to religate. This may cause a high background.
6. Agarose gel purification: Removal of free adaptors after adaptor ligation is crucial for successful cloning. However, cDNA library construction protocols do not include the gel purification step, and instead proceed to ligation after extraction and precipitation of the spun column eluate. Gel purification may therefore be unnecessary, especially if the PCR reaction generates one major product. It seems to us that higher voltage electrophoresis (e.g., 12 V/cm for 2.5 h) is more effective than lower voltage runs at separating adaptors from PCR products. Certainly, gel purification alone is not sufficient to rid the sample of adaptors, and the Sephacryl spin column should not be skipped.

   Elution of the concentrated DNA sample in a microcentrifuge at 13,000g sometimes causes the dialysis membrane to detach from the Microcon apparatus and fall into the eluate tube. This can be avoided by reducing the speed of centrifugation of the inverted apparatus as described. If multiple gel purification experiments are performed, keep the caps closed during processing to prevent the dialysis membrane from drying.

The yield of DNA is high (>70%) at each step except for gel purification. The yield for the gel purification step is approx 20–40%, depending on the size of the fragment. The overall yield for the protocol is 15–20% of starting material, as determined by quantitating a fraction of the starting PCR reaction and the ending gel purified, adapted fragment. We typically obtain 500 ng of adapted fragment.

Reference

1. Ren, S. and Ruppert, J. M. (1997) Ligation of single adaptors allows efficient cloning of long PCR products. *BioTechniques* **22,** 36–40.

48

Cloning DNA Fragments in M13 Vectors

David Walsh

1. Introduction

The genome of the filamentous bacteriophage M13 comprises a single-stranded circular DNA molecule, designated the (+) strand, which is converted into a double-stranded replicative form (RF) on infection of *Escherichia coli*. Cloning vectors based on M13 RF DNA facilitate the production of recombinant (+) strands that carry the sequence of one strand of a foreign DNA insert *(1,2)*. These single-stranded DNA molecules serve as highly efficient templates for nucleotide sequence determination by the dideoxy chain-termination method *(3)*.

Vectors of the M13mp series are designed to allow identification of clones that contain an insert by a simple color test. The vectors contain a multiple cloning site (MCS) located within the *N*-terminal portion of the *E. coli lacZ* gene that expresses the α-portion of β-galactosidase. The α-peptide is capable of complementing a *lac* mutation in the host to produce a functional β-galactosidase. Insertion of a foreign DNA fragment into the MCS normally disrupts the α-peptide sequence, resulting in loss of β-galactosidase activity. Hence, plaques generated by nonrecombinant bacteriophage are blue when plated on medium containing the chromogenic substrate X-gal, whereas recombinant bacteriophage plaques are colorless *(4)*.

When DNA fragments are subcloned from existing bacteriophage or plasmid clones into M13, the cloning strategy will generally be governed by the distribution of restriction sites in the region of interest. Whenever possible, the target fragment should be generated using two restriction enzymes that generate nonidentical overhanging termini to ensure insertion into M13 with a specified end adjacent to the sequencing primer site and at high efficiency. Directional cloning into a pair of M13 vectors that contain the MCS in opposite orientations, such as M13mp18 and M13mp19, allows the nucleotide sequence on opposite strands of the insert to be determined using a single primer. Restriction fragments carrying identical termini can be ligated into the MCS in both orientations and therefore need only be cloned into one of the M13mp vectors. A simple test based on hybridization of bacteriophage (+) strands can be used to identify clones containing the insert in opposite orientations *(5)*. A DNA fragment carrying overhanging ends not compatible with M13 cloning sites can be rendered blunt ended and then cloned into the vector *Sma*I site. This is best achieved using T4 DNA polymerase,

From: *The Nucleic Acid Protocols Handbook*
Edited by: R. Rapley © Humana Press Inc., Totowa, NJ

which can fill 5' overhangs and also has a strong 3'–5' exonuclease activity that will remove the 3' overhangs. This approach can also be used in blunt-end cloning of PCR products amplified using *Taq* DNA polymerase, which frequently carry a single deoxyadenosine nucleotide overhang at the 3' ends *(6)*.

Prior to setting up the ligation, the required DNA insert fragment should be purified from unwanted components present in the reaction mixture, such as other restriction fragments, or, in the case of PCR reactions, residual dNTPs, primers and any artifactual amplification products that would otherwise reduce cloning efficiency. Purification is conveniently achieved by electrophoretic separation through low-gelling-temperature agarose followed by DNA recovery using standard protocols *(7)* or a commercially available DNA purification kit such as Geneclean® (Bio101, Madison, WI).

Following transformation of *E. coli* with ligation products, M13 plaques can be analyzed directly for the presence of inserts by PCR screening using M13 forward and reverse sequencing primers, thus circumventing laborious restriction-digest analysis of RF DNA.

This chapter describes manipulations for the efficient cloning of blunt- and cohesive-ended DNA fragments into M13 vectors and methods that allow quick and reliable identification of the required recombinant clones.

2. Materials

2.1. Recovery of DNA Fragments from Agarose Gels

1. Ultrapure low-gelling-temperature agarose, such as NuSieveR GTGR (FMC Bioproducts, Rockland, ME).
2. 1X TAE buffer: 40 mM Tris-acetate, 1 mm ethylenediaminetetraacetic acid (EDTA), pH 8.0.
3. Geneclean kit, (Bio101).
4. Microcentrifuge.
5. 1X TE buffer: 10 mM Tris-HCl, 1 mM EDTA, pH 8.0.
6. 6X Agarose gel loading buffer: 40% Sucrose, 0.25% bromophenol blue, 0.25% xylene cyanol.

2.2. Conversion of Overhanging 5' and 3' Ends to Blunt Termini

1. T4 DNA polymerase.
2. 10X T4 DNA polymerase buffer: 500 mM NaCl, 100 mM Tris-HCl, pH 7.9, 100 mM MgCl$_2$, 10 mM dithiothreitol (DTT).
3. 1 mg/mL Acetylated bovine serum albumin (BSA).
4. Ultrapure stock of all four deoxynucleotide triphosphates (dNTPs). Make 20 mM stock in sterile water, aliquot, and store at –20°C.
5. Phenol:chloroform:isoamyl alcohol (25:24:1). Mix equal volumes of Tris-buffered phenol, pH > 7.5 (nucleic acid grade), and chloroform:isoamyl alcohol. Store at 4°C in a dark glass bottle.
6. Chloroform:isoamyl alcohol (24:1). Store at 4°C.
7. 3 M Sodium acetate, pH 5.2.
8. 70% and 100% Ethanol.

2.3. Preparation of M13 Vector DNA for Ligation

1. M13 RF DNA purchased from supplier or prepared in-house by CsCl density gradient centrifugation.

Cloning DNA Fragments in M13 Vectors

2. Restriction enzymes at 10 U/μL and 10X reaction buffers.
3. 1 M NaCl.
4. Calf intestinal alkaline phosphatase (CIAP).
5. 10X CIAP buffer: 200 mM Tris-HCl, pH 7.8, 100 mM MgCl$_2$, 10 mM ZnCl$_2$.
6. 0.5 M EDTA, pH 8.0.

2.4. Ligation of DNA Fragments into M13 Vector

1. T4 DNA ligase.
2. 10X T4 DNA ligase buffer: 500 mM Tris-HCl, pH 7.8, 100 mM MgCl$_2$, 100 mM DTT, 250 μg/mL BSA.
3. 10 mM adenosine triphosphate: Dissolve in sterile water and store in aliquots at –20°C. Avoid multiple freeze/thaw cycles.

2.5. Transformation of E. coli

1. 2X YT medium: 16 g/L Bacto-tryptone, 10 g/L Bacto-yeast extract (Difco, Surrey, UK), 10 g/L NaCl, pH 7.0.
2. YT agar plates: 2X YT medium containing 15 g/L Bacto-agar (Difco).
3. YT top agar: 2X YT medium containing 7 g/L Bacto-agar.
4. Competent cells of an appropriate *E. coli* strain, e.g., JM109.
5. Same *E. coli* strain maintained on M9 minimal agar.
6. Isopropyl-β-thiogalactopyranoside (IPTG): 200 mg/mL stock in sterile water. Store at –20°C.
7. 5-Bromo-4-chloro-3-indolyl-β-D-galactopyranoside (X-gal): 20 mg/mL stock in dimethyl formamide. Store at –20°C in dark glass vial.

2.6. PCR Screening of M13 Plaques

1. *Taq* DNA polymerase.
2. *Taq* DNA polymerase buffer 10X: 100 mM Tris-HCl, pH 9.0, 500 mM KCl, 15 mM MgCl$_2$, 1% Triton X–100, 2 mg/mL BSA.
3. 5 mM dNTPs.
4. M13 forward and reverse primers at 50 pmol/μL.
5. Thermal cycler

2.7. Testing Orientation of Inserts

1. 2% Sodium dodecyl sulfate.
2. 0.5 M NaCl.

3. Methods
3.1. Recovery of DNA Fragments from Agarose Gels

1. Dissolve low-gelling-temperature agarose in 1X TAE buffer (1%, w/v) containing 0.5 μg/mL ethidium bromide and pour the gel in a cold room.
2. Add loading buffer to each sample and load onto the gel, leaving an empty lane between samples to avoid cross-contamination.
3. Carry out electrophoresis in 1X TAE buffer, applying no more than 5 V/cm to ensure that the gel does not melt.
4. Visualize the DNA by long-wave UV light. Excise the band of interest using a scalpel and transfer to a preweighed 1.5 mL microcentrifuge tube. Cut the DNA band as close and as quickly as possible. Determine the weight of the agarose gel slice. Gel slices weighing more than 350 mg should be divided between two tubes.

5. Add gel solubilizer to the agarose slice. Use 300 μL for each 100 mg of agarose, up to a maximum of 900 μL. Incubate the tube at 55°C for 5 min to dissolve the agarose, then cool to room temperature.
6. Resuspend Glassmilk (Bio101, Madison, WI) by vortexing. Add 5 μL of the suspension to the solubilized agarose mixture and incubate at room temperature for 5 min, mixing by hand every 1–2 min.
7. Spin the tube for 5 s at 13,000g and remove the supernatant without disturbing the pellet.
8. Resuspend the pellet in 0.5 mL NEW wash buffer by pipeting back and forth. Spin the tube as in **step 5** and discard the supernatant. Repeat this step two more times.
9. After removing supernatant from the third wash, spin the tube again for 5 s and remove any residual liquid with a small-bore pipet tip (**step 7**). Allow the pellet to air dry for 10 min.
10. Resuspend the Glassmilk pellet in 10–20 μL of TE or sterile water and incubate at 55°C for 5 min with periodic agitation. Centrifuge the tube for 30 s at 13,000g. Carefully transfer the supernatant containing eluted DNA into a fresh microcentrifuge tube.

Gel-purified DNA fragments bearing termini compatible with the M13 MCS can now be ligated directly into the vector. Fragments carrying 5'- or 3'-overhanging ends not compatible with vector cloning sites can be converted to blunt-ended fragments ready for insertion into the *Sma*I site, as described in **Subheading 3.2.**

3.2. Conversion of Overhanging 5' and 3' Ends to Blunt Termini

1. Set up a reaction containing: 0.1-1 μg gel-purified DNA fragment, 2 μL 10X T4 DNA polymerase buffer, 100 μM each dNTP, T4 DNA polymerase (5 U/μg DNA), 0.1 mg/mL acetylated BSA, water to 20 μL. Incubate at 37°C for 5 min.
2. Stop the reaction by heating to 75°C for 10 min.
3. Increase the volume to 100 μL with water, and perform one extraction with an equal volume of phenol:chloroform:isoamyl alcohol and one with chloroform:isoamyl alcohol.
4. Remove the aqueous phase to a fresh tube, and add 0.1 vol 3 M sodium acetate, pH 5.2, and 2 vol 100% ethanol. Mix well and store the sample at –20°C for 30 min.
5. Recover the DNA by centrifugation at 13,000g for 10 min at 4°C. If the concentration of DNA is low (< 100 ng/mL), centrifuge for 30 min to enhance recovery.
6. Carefully remove the supernatant and discard. Wash the DNA pellet with 0.5 mL 70% ethanol and spin again at 13,000g for 2 min. Discard the supernatant, vacuum dry the pellet for 2–5 min, and finally resuspend the DNA in 10 μL TE.

DNA fragments are now blunt ended and ready for ligation into the *Sma*I-digested, dephosphorylated M13 vector.

3.3. Preparation of M13 Vector DNA for Ligation

3.3.1. Digestion with Restriction Endonucleases

In directional cloning experiments, where digestion of the vector with two enzymes is required, the ability of each enzyme to cleave toward the end of linear DNA molecules should be considered to determine the preferred order of sequential addition (8). Enzymes that cut less efficiently toward DNA termini should be used first.

1. Set up the following restriction digest: 1 μg M13 RF DNA, 5 U restriction enzyme, 2 μL 10X reaction buffer, water to 20 μL.
2. Incubate for 1 h at the appropriate temperature (25°C for *Sma*I, 37°C for all other MCS enzymes). Remove 1 μL to analyze the extent of digestion by agarose gel electrophoresis

(*see* Chapter 13). If digestion is not complete, add another 5 U of enzyme and continue digestion for a further 1 h.

Linearized M13 vector for use in nondirectional cloning experiments should now be dephosphorylated by treatment with CIAP to reduce recircularization during ligation, as described in **Subheading 3.3.2.** To cut the vector with a second restriction enzyme, proceed as follows:

3. Heat-inactivate the first enzyme at 65°C (phenol-extract heat-stable enzymes).
4. Adjust the ionic strength of the buffer as required by addition of the appropriate volume of 1 M NaCl. If the second enzyme requires a completely different buffer than the first, phenol-extract and sodium acetate/ethanol-precipitate the DNA in between each digest, as described in **Subheading 3.2.**, and resuspend in the appropriate 1X reaction buffer.
5. Add 5 U of the second restriction enzyme and incubate as in **step 2**. At the same time, set up a pilot digest containing circular plasmid DNA and the second restriction enzyme. When the plasmid has been fully linearized, as judged by agarose gel electrophoresis, the linearized M13 DNA should also be digested to completion by the second enzyme.
6. Purify the double-digested M13 DNA by extraction with phenol:chloroform and precipitation with sodium acetate/ethanol as described in **Subheading 3.2.** Resuspend in 10 µL TE.

3.3.2. Dephosphorylation of Vector DNA

1. To the digest containing 1 µg linearized M13 RF DNA add CIAP (0.05 U for 5' overhangs, 0.5 U for 3' overhangs or blunt ends), 5 µL 10X CIAP buffer, and water to 50 µL.
2. Incubate at 37°C for 1 h.
3. Inactivate CIAP by heating the reaction to 75°C for 10 min in the presence of 5 mM EDTA, pH 8.0.
4. Extract the reaction once with phenol:chloroform and recover DNA by sodium acetate/ethanol precipitation as described in **Subheading 3.2.** Resuspend the DNA in 10 µL TE.

3.4. Ligation of DNA Fragments into the M13 Vector

1. Estimate the concentration of DNA in the vector and insert preparations by agarose gel electrophoresis alongside DNA-size markers of known concentration.
2. Set up the ligation reaction, adding components in the following order: 100 ng M13 vector DNA, 1 µL 10 mM ATP, 1 µL 10X T4 DNA ligase buffer, 1–4 µL insert DNA (threefold molar excess for cohesive termini, five- to tenfold molar excess for blunt termini), T4 DNA ligase (0.5 U for cohesive termini, 5 U for blunt termini), and water to 10 µL total volume.
3. Set up a negative control ligation containing the same amount of vector DNA but no insert DNA, and a positive control ligation containing the same amount of vector DNA and an appropriate test fragment that has been successfully cloned into M13 previously.
4. Incubate for 4–16 h at 14°C for cohesive-end ligations or at room temperature for blunt-end ligations.

3.5. Transformation of E. coli

1. Inoculate 5 mL of 2X YT medium with a single colony of JM109 taken from an M9 minimal agar plate and grow overnight at 37°C with constant shaking.
2. The following day, dilute the overnight JM109 culture 1:100 in fresh 2X YT medium. Incubate at 37°C with constant shaking for approx 3–4 h until the exponential culture reaches an OD600 of 0.6.

3. Thaw frozen JM109 competent cells and transfer 100 µL aliquots to prechilled microcentrifuge tubes. Store on ice.
4. Add 5 µL of ligation reactions to competent cells. Set up two controls, one containing 10 pg of uncut M13 RF DNA and the other containing no DNA. Mix by gentle tapping. Incubate on ice for 30 min.
5. Transfer tubes to a 42°C water bath and incubate for 90 s. Return to ice for 10 min.
6. Prepare a set of sterile culture tubes containing 3 mL melted top agar and store at 45°C until required.
7. To each tube containing top agar, add 40 µL of 20 mg/mL X-gal, 4 µL of 200 mg/mL IPTG, 200 µL of the exponential JM109 culture, and 10–100 µL of the transformation mixture. Mix by gentle vortexing, and pour onto a YT agar plate prewarmed to 37°C.
8. Store plates for 10 min on a flat surface at room temperature to allow the top agar to set, then transfer to a 37°C oven and incubate inverted for 8–12 h.
9. Store plates at 4°C for 1–2 h to allow maximum color development in nonrecombinant plaques.

3.6. PCR Screening of Plaques

1. Prepare and label n 0.5 mL microcentrifuge tubes and n sterile culture tubes containing 2 mL 2 × YT medium and 100 µL of an exponential phase JM109 culture.
2. Make up a PCR master mix for (n + 1) reactions containing, per reaction:

Water	41.5 µL
10X *Taq* DNA polymerase buffer	5.0 µL
dNTPs (5 m*M* each)	1.0 µL
M13 forward primer (50 pmol/µL)	1.0 µL
M13 reverse primer (50 pmol/µL)	1.0 µL
Taq DNA polymerase (5 U/µL)	0.5 µL

   Aliquot 50 µL into each microcentrifuge tube.
3. Touch a sterile toothpick to the center of a well-isolated plaque, and swirl the toothpick briefly in the PCR reaction mixture. Then use the toothpick to inoculate the appropriate culture tube. Repeat for each plaque to be tested. Analyze one or more blue (nonrecombinant) plaques as a control.
4. Place culture tubes at 37°C and incubate for 5–6 h with vigorous shaking.
5. Overlay the PCR reaction mixture with mineral oil, if required, and carry out thermal cycling using a program such as the following: 94°C, 3 min (94°C, 1 min/55°C, 1 min/72°C, 1 min) × 25 cycles and 72°C, 5 min.
6. Analyze 10 µL of completed PCR reactions by agarose gel electrophoresis alongside the DNA insert fragment used in ligation.
7. For clones that produce a PCR product of the required size, centrifuge the corresponding bacterial cultures at 13,000*g* for 5 min, and carefully transfer the supernatant to a fresh microcentrifuge tube. Store at 4°C until required.

3.7. Testing Orientation of Inserts

In nondirectional cloning experiments, recombinants containing the insert in opposite orientations need to be identified to facilitate sequencing of both DNA strands. This can be readily accomplished by testing for the ability of the (+) strands of different clones to hybridize to one another.

1. Mix 20 µL of culture supernatants obtained in **Subheading 3.6.** in pairwise combinations in microcentrifuge tubes. Use clones shown by the PCR screen to contain the required

insert. As a control, transfer 40 μL of supernatant from one individual clone into a separate tube.
2. Add 2 μL of 2% SDS to each tube. Vortex briefly and incubate at 65°C for 5 min.
3. Add 4 μL of 0.5 M NaCl to each tube, vortex briefly, and continue incubation at 65°C for 1 h.
4. Spin the tubes briefly to pool evaporated liquid and load samples into wells in a 0.8% agarose gel containing 0.5 μg/mL ethidium bromide.
5. After electrophoresis, visualize the DNA bands by UV light. Pairs of recombinants containing the insert in opposite orientations are identified by the presence of a hybrid structure that migrates more slowly through the gel than does single-stranded phage DNA derived from the control.

Single-stranded M13 DNA for sequencing reactions can now be prepared from appropriate supernatants.

4. Notes

1. Large inserts (> 2.0 kb) tend to be unstable in M13, so it is advisable to construct recombinants containing the smallest possible fragment that carries the sequence of interest, preferably less than 1.5 kb. To clone DNA fragments larger than 1.5 kb, a phagemid vector such as pBluescriptII (Stratagene, Cambridge, UK) should be used instead of M13.
2. In directional cloning experiments, avoid using restriction sites that are directly adjacent in the M13 MCS, e.g., *Bam*HI and *Xba*I, if possible.
3. The small DNA fragment released from the MCS on double digestion of the M13 vector can theoretically religate if the vector is not dephosphorylated. This does not generally seem to occur but, if required, the small fragment can be quickly removed by processing the digest through a device for spin purification of PCR products, e.g., High Pure (Boehringer Mannheim, Mannheim, Germany), or Wizard (Promega, Madison, WI).
4. Ensure that all traces of NEW wash are removed from the Glassmilk pellet since contamination of DNA with gel solubilizer or ethanol may inhibit ligation.
5. Do not exceed the recommended amount of CIAP, as this can result in reduced cloning efficiency.
6. Ligation efficiency can sometimes be improved by either coprecipitating the vector and insert DNAs prior to setting up the reaction, or by including 5% polyethylene glycol (PEG) 8000 in the ligation.
7. Use an *E. coli* recA- (recombination-defective) host strain in order to reduce the occurrence of plasmid multimerization and deletion of insert DNA. Suitable strains include JM109, TG2, and XL1-Blue.
8. The bacterial strain should be maintained on minimal (M9) agar supplemented with thiamine-HCl. This selects for the F' episome, which is required for infection of cells by bacteriophage M13.
9. Occasionally, recombinant clones will be found to contain the DNA insert in only one of two possible orientations. This problem can sometimes be overcome by using a directional cloning strategy or by cloning into a phagemid vector.
10. Recombinant bacteriophage containing DNA insert fragments cloned in-frame with the *lacZ* gene may produce partially active β-galactosidase, giving rise to pale blue plaques that should also be screened.
11. A small proportion of colorless plaques will often be found to contain no insert. These are spontaneous mutants that have lost the ability to accomplish α-complementation.

References

1. Messing, J. (1983) New M13 vectors for cloning. *Methods Enzymol.* **101,** 20–78.
2. Yanisch-Perron, C., Viera, J., and Messing, J. (1985) Improved M13 phage cloning vectors and host strains: nucleotide sequences of the M13mp18 and pUC19 vectors. *Gene* **33,** 103–119.
3. Sanger, F., Nicklen, S., and Coulson, A. R. (1977) DNA sequencing with chain-terminating inhibitors. *Proc. Natl. Acad. Sci. USA* **74,** 5463–5467.
4. Gronenborn, B. and Messing, J. (1978) Methylation of single-stranded DNA in vitro introduces new restriction endonuclease cleavage sites. *Nature* **272,** 375–371.
5. Gardner, R. C., Howarth, A. J., Hahn, P., Brown-Luedi, M., Shepherd, R. J., and Messing, J. (1981) The complete nucleotide sequence of an infectious clone of cauliflower mosaic virus by M13mp7 shotgun sequencing. *Nucleic Acids Res.* **9,** 2871–2888.
6. Clarke, J. M. (1988) Novel non-template nucleotide addition reactions catalysed by prokaryotic and eukaryotic DNA polymerases. *Nucleic Acids Res.* **16,** 9677–9686.
7. Maniatis, T., Fritsch, E. F., and Sambrook, J. (eds.) *Molecular Cloning: A Laboratory Manual.* Cold Spring Harbor Laboratory Press, Cold Spring Harbor, NY.
8. New England Biolabs Catalogue, Reference Appendix.

49

cDNA Library Construction for the Lambda ZAP®-Based Vectors

Marjory A. Snead, Michelle A. Alting-Mees, and Jay M. Short

1. Introduction

Each organism and tissue type has a unique population of messenger RNA (mRNA) molecules. These mRNA populations are difficult to maintain, clone, and amplify; therefore, they must be converted to more stable DNA molecules (cDNA). Successful cDNA synthesis should yield full-length copies of the original population of mRNA molecules. Hence, the quality of the cDNA library can be only as good as the quality of the mRNA. Pure, undegraded mRNA is essential for the construction of large, representative cDNA libraries (*1*). Secondary structure of mRNA molecules can cause the synthesis of truncated cDNA fragments. In this case, treatment of the mRNA with a denaturant, such as methyl-mercuric hydroxide, prior to synthesis may be necessary (*2*). Other potential difficulties include DNA molecules contaminating the mRNA sample. DNA can clone efficiently, and their introns can confuse results. RNase-free DNase treatment of the sample is recommended.

After synthesis, the cDNA is inserted into an *Escherichia coli*-based vector (plasmid or λ), and the library is screened for clones of interest. Since 1980, lambda has been the vector system of choice for cDNA cloning (*3–10*). The fundamental reasons are that in vitro packaging of λ generally has a higher efficiency than plasmid transformation, and λ libraries are easier to handle (amplify, plate, screen, and store) than plasmid libraries. However, most λ vectors have the disadvantage of being poorer templates for DNA sequencing, site-specific mutagenesis, and restriction fragment shuffling, although this trend is reversing to some degree with the continued development of polymerase chain reaction (PCR) techniques.

The development of excisable λ vectors, such as those based on restriction enzyme digestion (*11*), site-specific recombination (*12*), or filamentous phage replication (*13*), has increased the flexibility of DNA cloning. Now it is possible to clone and screen libraries with the efficiency and ease of λ systems, and be able to analyze positive clones with the ease and versatility of a plasmid. The vectors that are compatible with the cDNA synthesis protocol described in this chapter are based on the Lambda ZAP® excision system (Stratagene Cloning Systems, La Jolla, CA) (*13,14*). These vectors use an excision mechanism that is based on filamentous helper phage replication

```
5'    GAGAGAGAGAGAGAGAGAGAGAGACTCGAGTTTTTTTTTTTTTTTTTT    3'
      Protective Sequence        Xho I      Poly(dT)
```

Fig. 1. Forty-eight base pair oligonucleotide hybrid oligo(dT) linker-primer.

(e.g., M13). The choice of vector depends on whether one requires such features as prokaryotic expression, eukaryotic expression, in vitro transcription, in vitro translation, directional cloning, single-strand replication, automated sequencer compatibility, and special antibiotic resistance selection.

Several cloning procedures for constructing cDNA libraries exist *(15–19)*. Here we describe a modification of a directional cDNA cloning protocol *(16)*. This procedure has been successfully used for generating hundreds of directional cDNA libraries representing a vast number of plant and animal species containing poly (A)+ mRNA.

A hybrid oligo(dT) linker-primer containing an *Xho*I site is used to make directional cDNA. This 48-base oligonucleotide was designed with a protective sequence to prevent the *Xho*I restriction enzyme recognition site from being damaged in subsequent steps and an 18-base poly(dT) sequence, which binds to the 3' poly (A) region of the mRNA template (*see* **Fig. 1**).

First-strand synthesis is primed with the linker-primer and is transcribed by reverse transcriptase in the presence of nucleotides and buffer. An RNase H-deficient reverse transcriptase may produce larger yields of longer cDNA transcripts *(20,21)*. The use of 5-methyl dCTP in the nucleotide mix during first-strand synthesis "hemi-methylates" the cDNA, protecting it from digestion during a subsequent restriction endonuclease reaction used to cleave the internal *Xho*I site in the linker-primer.

The cDNA/mRNA hybrid is treated with RNase H in the second-strand synthesis reaction. The mRNA is nicked to produce fragments that serve as primers for DNA polymerase I, synthesizing second-strand cDNA. The second-strand nucleotide mixture is supplemented with dCTP to dilute the 5-methyl dCTP, reducing the probability of methylating the second-strand, since the *Xho*I restriction site in the linker-primer must be susceptible to restriction enzyme digestion for subsequent ligation into the vector.

The uneven termini of the double-stranded cDNA must be polished with cloned *Pfu* DNA polymerase to allow efficient ligation of adapters *(22,23)*. Adapters are complementary oligonucleotides which, when annealed, create a phosphorylated blunt end and a dephosphorylated cohesive end. This double-stranded adapter will ligate to other blunt termini on the cDNA fragments and to other adapters. Since the cohesive end is dephosphorylated, ligation to other cohesive ends is prevented. After the adapter ligation reaction is complete and the ligase has been inactivated, the molecules are phosphorylated to allow ligation to the dephosphorylated vector.

An *Xho*I digestion releases the adapter and protective sequence on the linker-primer from the 3'-end of the cDNA. These fragments are separated from the cDNA on a size fractionation column. The purified cDNA is then precipitated and ligated to the vector. This strategy is illustrated in **Fig. 2**.

cDNA Library Construction

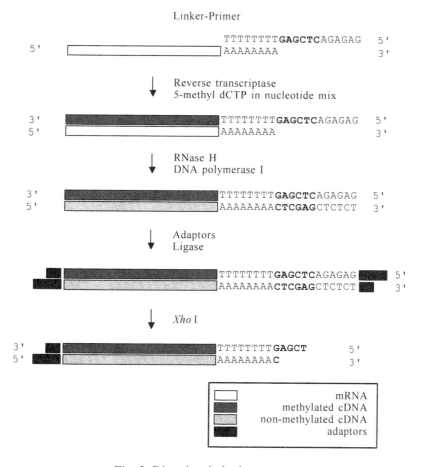

Fig. 2. Directional cloning strategy.

2. Materials

2.1. First-Strand Synthesis

1. 10X First-strand buffer: 500 mM Tris-HCl, pH 7.6, 700 mM KCl, 100 mM MgCl$_2$.
2. First-strand methyl-nucleotide mixture: 10 mM dATP, dGTP, dTTP, and 5 mM 5-methyl dCTP.
3. Linker-primer (3.0 µg at 1.5 µg/µL).
4. Diethylpyrocarbonate (DEPC)-treated water.
5. Ribonuclease inhibitor (40 U).
6. PolyA$^+$ mRNA (5.0 µg in ≤36 µL DEPC-treated water; *see* **Notes 1** and **2**).
7. [α-^{32}P]-Labeled deoxynucleotide (800 Ci/mmol) [α-^{32}P]dATP, [α-^{32}P]dGTP, or [α-^{32}P]dTTP. Do not use [α-^{32}P]dCTP (*see* **Note 3**).
8. Reverse transcriptase (250 U) (RNase H-deficient is recommended *[20,21]*).

2.2. Second-Strand Synthesis

1. 10X Second-strand buffer: 700 mM Tris-HCl, pH 7.4, 100 mM (NH$_4$)$_2$SO$_4$, 50 mM MgCl$_2$.
2. Second-strand dNTP mixture: 10 mM dATP, dGTP, dTTP, and 26 mM dCTP.
3. *E. coli* RNase H (4.0 U).
4. *E. coli* DNA polymerase I (100 U).

2.3. Blunting the cDNA Termini

1. Blunting dNTP mixture (2.5 mM dATP, dGTP, dTTP, and dCTP).
2. Cloned *Pfu* DNA polymerase (5 U).
3. Phenol-chloroform: 1:1 (v/v), pH 7.0–8.0 (*see* **Note 5**).
4. Chloroform.
5. 3 M Sodium acetate.
6. 100% (v/v) Ethanol.

2.4. Ligating the Adapters

1. 70% (v/v) Ethanol.
2. Adapters: 4 µg at 0.4 µg/µL.
3. 5% Nondenaturing acrylamide gel.
4. 10X Ligation buffer: 500 mM Tris-HCl, pH 7.4, 100 mM MgCl$_2$, 10 mM dithiothreitol (DTT).
5. 10 mM rATP.
6. T4 DNA ligase (4 Weiss U).

2.5. Phosphorylating the Adapters

1. 10X Ligation buffer (*see* **Subheading 2.4.**).
2. 10 mM rATP.
3. T4 Polynucleotide kinase (10 U).

2.6. XhoI Digestion

1. *Xho*I reaction buffer: 200 mM NaCl, 15 mM MgCl$_2$.
2. *Xho*I restriction endonuclease (120 U).
3. 10X STE buffer: 1 M NaCl, 100 mM Tris-HCl, pH 8.0, 10 mM EDTA.

2.7. Size Fractionation

1. 1X STE buffer: 100 mM NaCl, 10 mM Tris-HCl, pH 8.0, 1 mM EDTA.
2. Sephacryl® S-500 column filtration medium (Pharmacia).
3. 5% Nondenaturing acrylamide gel.
4. Phenol-chloroform (1:1 [v/v], pH 7.0–8.0) (*see* **Note 5**).
5. Chloroform.
6. 100% (v/v) Ethanol.

2.8. Quantitating the cDNA

1. 70% (v/v) Ethanol.
2. TE buffer: 10 mM Tris-HCl, pH 8.0, 1 mM EDTA.
3. 0.8% Agarose.
4. 10 mg/mL Ethidium bromide.

2.9. Ligating the cDNA to Prepared Vector

1. λ vector (such as Lambda ZAP® and ZAP Express) double-digested and dephosphorylated. Vectors are digested with *Xho*I and a second restriction enzyme that leaves ends compatible with the adapters.
2. 10X Ligation buffer (*see* **Subheading 2.4.**).
3. 10 mM rATP.
4. T4 DNA ligase (4 Weiss U).

cDNA Library Construction

2.10. Packaging and Plating

1. NZY medium, plates and top agarose: 5 g NaCl, 2 g $MgSO_4 \cdot 7H_2O$, 5 g yeast extract, 10 g NZ amine (casein hydrolysate)/L. Add 15 g agar for plates, or add 0.7% (w/v) agarose for top agarose. Adjust the pH to 7.5 with NaOH, and sterilize by autoclaving.
2. Appropriate *E. coli* host strains (such as XL1-Blue MRF' or DH5αMCR) freshly streaked on an LB agar plate containing the appropriate antibiotic (*see* **Note 8**).
3. 10 m*M* $MgSO_4$.
4. Packaging extract (such as Gigapack® packaging extracts [Stratagene] *[23,24]*).
5. SM buffer: 5.8 g NaCl, 2.0 g $MgSO_4 \cdot 7H_2O$, 50.0 mL 1 *M* Tris-HCl, pH 7.5, 5.0 mL 2% (w/v) gelatin/L. Autoclave.
6. Chloroform.
7. LB agar plates: 10 g NaCl, 10 g Bacto-tryptone, 5 g Bacto-yeast extract, 15 g agar/L. Adjust the pH to 7.5 with NaOH, and sterilize by autoclaving.
8. Isopropyl-β-D-thio-galactopyranoside (IPTG), 0.5 *M* in water, and 5-bromo-4-chloro-3-indoyl-β-D-galactopyranoside (X-gal), 250 mg/mL in dimethyl formamide (*see* **Note 10**).

2.11. Amplification of the Primary Library

1. Packaged and titered primary library.
2. Prepared, appropriate *E. coli* host strains.
3. NZY medium, plates and top agarose (*see* **Subheading 2.10.**).
4. SM buffer (*see* **Subheading 2.10.**).
5. Chloroform.
6. Dimethylsulfoxide (DMSO).

3. Methods

3.1. First-Strand Synthesis

The final volume of the first-strand synthesis reaction should be 50 µL. Take this into account when determining the volumes necessary.

1. In an RNase-free microcentrifuge tube, add the reagents in the following order: 5.0 µL 10X first-strand buffer, 3.0 µL methyl-nucleotide mixture, 2.0 µL linker-primer (1.5 µg/µL), X µL DEPC-treated water, 40 U ribonuclease inhibitor.
2. Mix the reagents well. Add X µL of polyA$^+$ mRNA (5 µg), and gently vortex (*see* **Notes 1** and **2**).
3. Allow the mRNA template and linker-primer to anneal for 10 min at room temperature.
4. Add 0.5 µL of [α-^{32}P]-labeled deoxynucleotide (800 Ci/mmol). Do not use [α-^{32}P]dCTP (*see* **Note 3**).
5. Add 250 U of reverse transcriptase. The final volume of the reaction should now be 50 µL.
6. Gently mix the sample, and briefly spin down the contents in a microcentrifuge.
7. Incubate at 37°C for 1 h.
8. After the 1-h incubation, place on ice.

3.2. Second-Strand Synthesis

The final volume of the second-strand synthesis reaction should be 200 µL. Take this into account when determining the necessary volumes.

1. To the first strand reaction (50 µL), add the following components in the following order: 20.0 µL 10X second-strand buffer, 6.0 µL second-strand dNTP mixture, X µL sterile distilled water (DEPC-treated water is not required), 4 U *E. coli* RNase H, 100 U *E. coli* DNA polymerase I.

2. The final volume of the reaction should now be 200 µL. Quickly vortex and spin down the reaction in a microcentrifuge. Incubate for 2.5 h at 16°C.
3. After the 2.5-h incubation, place on ice.

3.3. Blunting the cDNA Termini

1. Add the following reagents to the synthesized cDNA: 23.0 µL blunting dNTP mixture, 2.0 µL cloned *Pfu* DNA polymerase (2.5 U/µL).
2. Mix well, and incubate at 70°C for 30 min. Do not exceed 30 min.
3. Phenol-chloroform/chloroform extract (*see* **Note 5**).
4. Precipitate the cDNA by adding the following: 20 µL 3 *M* sodium acetate, 400 µL 100% (v/v) ethanol.
5. Mix by gently vortexing, and incubate on ice for 10 min or overnight at –20°C.

3.4. Ligating the Adapters

1. Microcentrifuge the precipitated cDNA sample at maximum speed and at 4°C for 1 h.
2. A large white pellet will form at the bottom of the microcentrifuge tube. Carefully remove the radioactive ethanol, and properly discard. Counts left in this supernatant are unincorporated nucleotides.
3. Wash the pellet by gently adding 500 µL of 70% (v/v) ethanol, and microcentrifuge for 2 min.
4. Aspirate the ethanol wash, and lyophilize the pellet until dry.
5. Resuspend the pellet in 9.0 µL of adapters (0.4 µg/µL) by gentle pipeting. Use a Geiger counter to confirm that the cDNA is in solution.
6. Remove 1.0 µL for analysis of cDNA synthesis on a 5% nondenaturing acrylamide gel. This aliquot may be frozen at –20°C (*see* **Notes 1–4**).
7. Add the following components to the tube containing the 8.0 µL of blunted DNA and adapters: 1.0 µL 10X ligation buffer, 1.0 µL 10 m*M* rATP, 1.0 µL T4 DNA ligase (4 U/µL).
8. Mix well and briefly spin in a microcentrifuge. Incubate overnight at 8°C or for 2 d at 4°C.

3.5. Phosphorylating the Adaptors

The final volume of the phosphorylation reaction will be 25 µL. Take this into account when determining the necessary volumes.

1. After ligation, heat-inactivate the ligase by incubating at 70°C for 30 min.
2. Spin down, and allow the reaction to cool at room temperature for 5 min. Add 1.5 µL 10X ligation buffer, 2.0 µL 10 m*M* rATP, X µL sterile distilled water, and 7 U T4 polynucleotide kinase.
3. Incubate at 37°C for 30 min.
4. Heat-inactivate the kinase by incubating at 70°C for 30 min.
5. Spin down, and allow the reaction to cool at room temperature for 5 min.

3.6. XhoI Digestion

The final volume of the digestion reaction will be 60 µL.

1. Add the following components to the phosphorylation reaction (25 µL): 30.0 µL *Xho*I reaction buffer, X µL sterile distilled water, 120 U *Xho*I restriction endonuclease. Be sure the volume of enzyme is ≤10% of the reaction volume.
2. Incubate for 1.5 h at 37°C.
3. Cool the reaction to room temperature, and add 15 µL of 10X STE buffer and 75 µL water.

3.7. Size Fractionation

There are many types of filtration media used to separate DNA molecules. Sephacryl S-500 medium separates efficiently in the 2-kb size range. Drip columns made with Sephacryl S-500 medium separate by size, the larger cDNA molecules eluting from the column first and the small unligated adapters and unincorporated nucleotides eluting later. The cDNA will not have a high number of counts, but will be detectable by a handheld monitor at ≤250 cps.

3.7.1. Drip-Column Preparation

1. Discard the plunger from a 1-mL plastic syringe, and insert a small cotton plug. Push the cotton to the bottom of the syringe.
2. Fill the syringe to the top with Sephacryl S-500 filtration medium.
3. Place the syringe in a rack and allow the column to drip "dry."
4. Fill the syringe up to approx 0.5 cm from the top with medium, and drip through as in **step 3**.
5. Rinse the column with four aliquots of 300 µL of 1X STE buffer (total wash volume of 1200 µL). Drip dry after each addition of buffer.

3.7.2. Collecting Fractions

1. Pipet the cDNA into the washed Sephacryl S-500 drip column, and allow to drip through. This is fraction 1. The recovery volume is approx 150 µL and does NOT contain cDNA (*see* **Note 6**).
2. Load two more aliquots of 150 µL of 1X STE buffer on the column and drip through. These are fractions 2 and 3.
3. Collect fraction 4 in a fresh tube. Load 150 µL of 1X STE buffer and drip as before.
4. Collect fraction 5 as in **step 3**. Two fractions are usually adequate. The size of the cDNA decreases in each additional fraction. Most of the radioactivity will remain in the column owing to unincorporated nucleotides. Discard the radioactive drip column appropriately.
5. Remove 5 µL from each fraction (or up to $^1/_{10}$ of the fraction volume) for analysis of cDNA size on a 5% nondenaturing acrylamide gel. These aliquots can be frozen at –20°C.
6. To remove any residual enzyme from previous reactions, phenol-chloroform/chloroform extract (*see* **Note 5**).
7. Add twice the volume of 100% (v/v) ethanol to precipitate the cDNA.
8. Place on ice for 1 h or at –20°C overnight.

3.8. Quantitating the cDNA

1. Microcentrifuge the fractionated cDNA at maximum speed for 30–60 min at 4°C. Carefully transfer the ethanol to another tube, and monitor with a Geiger counter. Most of the counts should be present in the pellet. Discard the ethanol appropriately.
2. Wash the cDNA pellet with 200 µL of 70% (v/v) ethanol, and microcentrifuge for 2 min.
3. Carefully remove the ethanol wash, and vacuum evaporate until the cDNA pellet is dry.
4. Each fraction can contain 0–250 cps. If the pellet contains 0–10 cps, resuspend the cDNA in 5.0 µL of sterile water. If the pellet contains >10 cps, resuspend the cDNA in 12.0 µL of sterile water.
5. Quantitate the cDNA by UV visualization of samples spotted on ethidium bromide agarose plates (*see* **Note 7**). The cDNA can be stored at –20°C.

3.9. Ligating the cDNA to Prepared Vector

The cloning vector should be double-digested with *Xho*I and an enzyme which leaves ends compatible with the adapters. The vector should also be dephosphorylated to prevent vector-to-vector ligations. The final ligation reaction volume is 5 µL.

1. To a 0.5-mL microcentrifuge tube, add in order: X µL water, 0.5 µL 10X ligation buffer, 0.5 µL 10 mM rATP, 1 µg prepared λ arms, 100 ng cDNA, 0.5 µL T4 DNA ligase (4 Weiss U/µL).
2. Incubate overnight at 4°C.

3.10. Packaging and Plating

The ligation is packaged and transfected into an appropriate *E. coli* host strain.

3.10.1. Preparation of Plating Cells

1. Inoculate 50 mL of NZY medium with a single colony of the appropriate *E. coli* host. Do not add antibiotic.
2. Grow at 30°C with gentle shaking overnight (*see* **Note 9**).
3. Spin the culture at 1000g for 10 min.
4. Gently resuspend the cells in 20 mL sterile 10 mM MgSO$_4$.
5. Determine the concentration of the cells by reading OD$_{600}$ on a spectrophotometer. Store this cell stock at 4°C for no more than 1 wk. To use, dilute cells to OD$_{600}$ = 1.0 in 10 mM MgSO$_4$.

3.10.2. Packaging

Package the ligation reaction following manufacturer's instructions. Stop the reaction by adding 500 µL SM buffer and 20 µL chloroform.

3.10.3. Plating

1. Mix the following components in a Falcon 2059 polypropylene tube: 200 µL appropriate diluted host cells (*see* **Subheading 3.10.1.**), and 1 µL final packaged reaction.
2. Incubate the phage and the bacteria at 39°C for 15 min to allow the phage to attach to the cells.
3. Add 2–3 mL of NZY top agarose (48°C) containing IPTG and X-gal (*see* **Note 10**). Plate onto NZY agar plates, and place the plates upside down in a 39°C incubator.
4. Plaques should be visible after 6–8 h. Background plaques are blue. Recombinant plaques are clear and should be 10- to 100-fold above the background.
5. Count the plaques and calculate the titer. Primary libraries can be unstable. Immediate amplification of at least a portion of the library is recommended to produce a large, stable quantity of a high-titer stock of the library.

3.11. Amplification of the Primary Library

After amplification, the library is suitable for screening by a variety of techniques *(2)*. More than one round of amplification is not recommended, since slower growing clones may be significantly underrepresented.

1. Prepare the host strains (*see* **Subheading 3.10.1.**).
2. Mix aliquots of the packaged library containing approx 50,000 PFU (≤300 µL vol) with 600 µL of host cells in Falcon 2059 polypropylene tubes. Usually, 1 × 10^6 plaques are amplified (20 tubes).

cDNA Library Construction

3. Incubate the tubes containing the phage and host cells for 15 min at 39°C.
4. Mix 8.0 mL of melted NZY top agarose (48°C) with each aliquot of infected bacteria, and spread evenly onto a freshly poured 150-mm NZY plate.
5. Incubate the plates at 39°C for 6–8 h. Do not allow the plaques to grow larger than 1–2 mm.
6. Overlay the plates with 8–10 mL of SM buffer. Store the plates at 4°C overnight with gentle rocking. The phage will diffuse into the SM buffer.
7. Recover the SM buffer containing the bacteriophage from each plate, and pool it in a sterile polypropylene container. Add chloroform to a 5% final concentration and mix well.
8. Incubate for 15 min at room temperature.
9. Remove the cell debris by centrifugation for 10 min at 500g.
10. Recover the supernatant, and transfer it to a sterile polypropylene container. Add chloroform to a 0.3% final concentration, and store at 4°C.
11. Check the titer of the amplified library by making serial dilutions in SM buffer and plating on host cells (*see* **Subheading 3.10.1.**). The average titer is usually 10^9–10^{12} PFU/mL.
12. Frozen stocks can be made by adding DMSO to a final concentration of 7%, mixing well, and freezing at –80°C.

4. Notes

1. The mRNA sample must be highly purified for efficient cDNA synthesis. The mRNA sample may contain inhibitors that can be removed by phenol-chloroform extractions. The presence of DNA or rRNA will give an inaccurate concentration of mRNA leading to an insufficient amount of sample used. Treat the mRNA with RNase-free DNase, or use more mRNA sample.
2. Some populations of mRNA molecules may have tight secondary structures. Methyl-mercuric hydroxide treatment of the RNA sample may be necessary. Perform the following protocol under a fume hood. Resuspend the mRNA in 20 μL of DEPC-treated water, and incubate at 65°C for 5 min. Cool to room temperature, and add 2 μL of 100 mM methyl-mercuric hydroxide. Incubate at room temperature for 1 min, add 4 μL of 700 mM β-mercaptoethanol (dilute stock in DEPC-treated water), and incubate at room temperature for 5 min. The final volume is 26 μL. This denatured mRNA is ready for first-strand synthesis.
3. Do not use [α-^{32}P]dCTP. The 5-methyl dCTP present in the nucleotide mixture will be diluted, and the synthesized cDNA will not be protected from the subsequent restriction digest. Gel analysis may show a false-negative result if the [α-^{32}P]dNTP is degraded, since it may not incorporate into the cDNA even though synthesis is occurring.
4. Gel analysis may show hairpinning of the cDNA, which is caused by a number of factors: an insufficient amount of mRNA was used in the first-strand reaction (*see* **Note 1**), the mRNA population had tight secondary structure (*see* **Note 2**), the second-strand incubation temperature was higher than 16°C (cool the first-strand reaction by placing it on ice before adding the second-strand synthesis reaction components), or an excessive amount of DNA polymerase was used in the second-strand reaction.
5. Phenol-chloroform (1:1 [v/v], pH 7.0–8.0) is recommended. Do not use low-pH phenol routinely used for RNA isolation *(1,2)*. To extract the cDNA sample, add an equal volume of phenol-chloroform (1:1 [v/v], pH 7.0–8.0) and vortex. Microcentrifuge at maximum speed for 2 min. Transfer the upper aqueous layer, which contains the cDNA, to a new sterile tube. Avoid removing any interface. Add an equal volume of chloroform and vortex. Microcentrifuge for 2 min at maximum speed. Save the upper aqueous layer, and transfer it to a new tube.
6. Sephacryl S-500 drip columns can be run "dry." A reservoir at the top of the column is not required. Each 150 μL wash yields approx 150 μL fraction volume. Fractions 1–3 can be

collected in one tube since these fractions do not contain cDNA. The cDNA elutes in fractions 4 (containing fragments ≥1.5 kb) and 5 (containing fragments >500 bp).

7. Ethidium bromide agarose plate quantitation is performed as follows: Using a DNA sample of known concentration (such as a plasmid), make serial dilutions (200, 150, 100, 75, 50, 25, and 10 ng/µL) in TE buffer. Melt 10 mL of 0.8% (w/v) agarose in TE buffer, and cool to 50°C. Under a hood, add 10 µL of 10 mg/mL ethidium bromide, swirl to mix, and pour into a 100-mm Petri dish. Allow the plate to harden. Label the bottom of the Petri dish with a marker to indicate where the sample and standards will be spotted. Carefully spot 0.5 µL of each standard dilution onto the surface of the plate. Do not puncture the agarose. Allow capillary action to pull the small volume from the pipet tip to the surface. Spot 0.5 µL of the cDNA sample onto the plate adjacent to the standards. Allow the spots to absorb into the agarose for 10–15 min at room temperature. Invert the plate and visualize on a UV light box. Compare the spotted sample of unknown concentration with the standards to determine the concentration of the cDNA.
8. Since the cDNA is heavily methylated, introduction into a host with an McrA, McrCB, hsdSMR, Mrr phenotype would be subject to digestion by these restriction systems. Therefore, the choice of packaging extract and an *E. coli* host strain is crucial *(24–29)*.
9. Since λ phage can adhere to dead as well as to viable cells, the lower temperature prevents the bacteria from overgrowing.
10. Most cDNA vectors have color selection by IPTG and X-gal. These components can be added to the top agarose before plating to produce the background blue color. Use 15 µL of 0.5 *M* IPTG (in water) and 50 µL of X-gal at 250 mg/mL in dimethylformamide.

References

1. Chomczynski, P. and Sacchi, N. (1987) Single-step method of RNA isolation by acid guanidinium thiocyanate-phenol-chloroform extraction. *Anal. Biochem.* **162,** 156–159.
2. Sambrook, J., Fritsch, E. F., and Maniatis, T. (eds.) (1989) *Molecular Cloning: A Laboratory Manual*, 2nd ed. Cold Spring Harbor Laboratory Press, Cold Spring Harbor, NY.
3. Han, J. H. and Rutter, W. J. (1987) Lambda gt22, an improved lambda vector for the directional cloning of full-length cDNA. *Nucleic Acids Res.* **15,** 6304.
4. Huynh, T. V., Young, R. A., and Davis, R. W. (1985) *DNA Cloning*, vol. I (Glover, D. A., ed.), IRL, Washington, DC, pp. 49–78.
5. Meissner, P. S., Sisk, W. P., and Berman, M. L. (1987) Bacteriophage lambda cloning system for the construction of directional cDNA libraries. *Proc. Natl. Acad. Sci. USA* **84,** 438–447.
6. Murphy, A. J. M. and Efstratiadis, A. (1987) Cloning vectors for expression of cDNA libraries in mammalian cells. *Proc. Natl. Acad. Sci. USA* **84,** 8277–8281.
7. Palazzolo, M. J. and Meyerowitz, E. M. (1987) A family of lambda phage cDNA cloning vectors, lambda SWAJ, allowing the amplification of RNA sequences. *Gene* **52,** 197–206.
8. Scherer, G., Telford, J., Baldari, C., and Pirrotta, V. (1981) Isolation of cloned genes differentially expressed at early and late stages of Drosophila embryonic development. *Dev. Biol.* **86,** 438–447.
9. Young, R. A. and Davis, R. W. (1983) Efficient isolation of genes by using antibody probes. *Proc. Natl. Acad. Sci. USA* **80,** 1194–1198.
10. Young, R. A. and Davis, R. W. (1983) Yeast RNA polymerase II genes: isolation with antibody probes. *Science* **222,** 778–782.
11. Swaroop, A. and Weissman, S. M. (1988) Charon BS (+) and (–), versatile lambda phage vectors for constructing directional cDNA libraries and their efficient transfer to plasmids. *Nucleic Acids Res.* **16,** 8739.

12. Palazzolo, M. J., Hamilton, B. A., Ding, D. L., Martin, C. H., Mead, D. A., Mierendorf, R. C., Raghavan, K. V., Meyerowitz, E. M., and Lipshitz, H. D. (1990) Phage lambda cDNA cloning vectors for subtractive hybridization, fusion-protein synthesis and Cre-loxP automatic plasmid subcloning. *Gene* **88,** 25–36.
13. Short, J. M., Fernandez, J. M., Sorge, J. A., and Huse, W. D. (1988) Lambda ZAP: a bacteriophage lambda expression vector with in vivo excision properties. *Nucleic Acids Res.* **16,** 7583–7600.
14. Alting-Mees, M., Hoener, P., Ardourel, D., Sorge, J. A., and Short, J. M. (1992) New lambda and phagemid vectors for prokaryotic and eukaryotic expression. *Strategies Mol. Biol.* **5(3),** 58–61.
15. Gubler, U. and Hoffman, B. J. (1983) A simple and very efficient method for generating cDNA libraries. *Gene* **25,** 263–269.
16. Huse, W. D. and Hansen, C. (1988) cDNA cloning redefined: a rapid, efficient, directional method. *Strategies Mol. Biol.* **1(1),** 1–3.
17. Kimmel, A. R. and Berger, S. L. (1989) Preparation of cDNA and the generation of cDNA libraries: overview. *Methods Enzymol.* **152,** 307–316.
18. Krug, M. S. and Berger, S. L. (1989) First strand cDNA synthesis primed with oligo (dT). *Methods Enzymol.* **152,** 316–325.
19. Okayama, H. and Berg, P. (1982) High-efficiency cloning of full-length cDNA. *Mol. Cell. Biol.* **2,** 161–170.
20. Gerard, G. (1989) cDNA synthesis by cloned Moloney Murine Leukemia Virus reverse transcriptase lacking RNaseH activity. *Focus* **11,** 66.
21. Nielson, K., Simcox, T. G., Schoettlin, W., Buchner, R., Scott, B., and Mathur, E. (1993) StratascriptTM RNaseH-reverse transcriptase for larger yields of full-length cDNA transcripts. *Strategies Mol. Biol.* **6(2),** 45.
22. Costa, L., Grafsky, A., and Weiner, M. P. (1994) Cloning and analysis of PCR-generated DNA fragments. *PCR Methods Applic.* **3,** 338–345.
23. Hu, G. (1993) DNA polymerase-catalyzed addition of nontemplated extra nucleotides to the 3' end of a DNA fragment. *DNA Cell Biol.* **12(8),** 763–770.
24. Kretz, P. L., Reid, C. H., Greener, A., and Short, J. M. (1989) Effect of lambda packaging extract mcr restriction activity on DNA cloning. *Nucleic Acids Res.* **17,** 5409.
25. Kretz, P. L. and Short, J. M. (1989) GigapackTMII: restriction free (hsd-, mcrA-, mcrB-, mrr-) lambda packaging extracts. *Strategies Mol. Biol.* **2(2),** 25,26.
26. Bullock, W., Fernandez, J. M., and Short, J. M. (1987) XL1-blue: a high efficiency plasmid transforming recA *Escherichia coli* strain with beta-galactosidase selection. *Biotechniques* **5(4),** 376–379.
27. Kohler, S. W., Provost, G. S., Kretz, P. L., Dycaico, M. J., Sorge, J. A., and Short, J. M. (1990) Development of short-term in vivo mutagenesis assay. The effects of methylation on the recovery of a lambda phage shuttle vector from transgenic mice. *Nucleic Acids Res.* **18,** 3007–3013.
28. Kretz, P. L., Kohler, S. W., and Short, J. M. (1991) Gigapack® III high efficiency lambda packaging extract with single-tube convenience. *Strategies Mol. Biol.* **7(2),** 44,45.
29. Kretz, P. L., Kohler, S. W., and Short, J. M. (1991). Identification and characterization of a gene responsible for inhibiting propagation of methylated DNA sequences in mcrA and mcrB *Escherichia coli* strains. *J. Bacteriol.* **173,** 4707–4716.

50

Expression and Preparation of Fusion Proteins from Recombinant λgt11 Phages

Sheng-He Huang and Ambrose Jong

1. Introduction

The phage λgt11 system has become increasingly popular for expression of cDNAs or genomic DNAs either in phage plaques or in bacteria lysogenized with recombinant phages *(1,2)*. It offers the advantages of high cloning efficiency, high-level expression, the relative stability of β-galactosidase fusion proteins, and simple approaches to purify the fusion proteins. After the desired clone is detected and purified, it is often necessary to obtain preparative amounts of recombinant protein specified by the fusion of the foreign sequence to the carboxyl-terminus of β-galactosidase in λgt11 expression system. The conventional method for preparing fusion proteins from the recombinant λgt11 clones involves production of phage lysogens in *Escherichia coli* strain Y1089 followed by inducing *lacZ*-directed fusion protein expression with isopropyl-β-D-thio-galactopyranoside (IPTG) *(1)*. This method has two limitations: it is time-consuming, and phage lysogeny occurs at a low frequency. We have previously described a method for making fusion proteins from LB agar plates containing *E. coli* Y1090 infected with a high concentration of recombinant λgt11 phages (up to 5×10^6 PFU/150×15-mm plate) *(3)*. A liquid culture method for preparing fusion proteins from *E. coli* Y1090 infected with the λgt11 clones has previously been described *(4)*. More recently, some improvements have been made on the plate method by repeating induction and elution *(5)*. Although the liquid culture method allows the recovery of only 0.2–1% of total proteins *(6,7)*, this method generally yields 5–10% of expressed protein in solution, that is, most lysed cells are trapped in the agar and the expressed proteins are recovered in a small volume of inducing solution, resulting in a higher final concentration of protein. More than 200 µg of fusion protein can be obtained from one plate. Currently, the plate method appears to be a simple and efficient way to express and prepare fusion proteins from recombinant λgt11 phages (**Figs. 1** and **2**). This chapter describes how the β-galactosidase fusion proteins can be made and isolated from the recombinant λgt11 phages with the plate method.

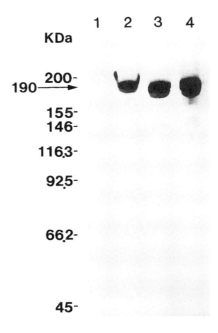

Fig. 1. Induction of the 190-kDa ERP72 fusion protein with various concentrations of IPTG (lane 1, no IPTG; lane 2, 2.5 m*M*; lane 3, 5 m*M*; lane 4, 10 m*M*). Fusion protein was detected by Western blot analysis with anti-ERP72 antibody *(8)*.

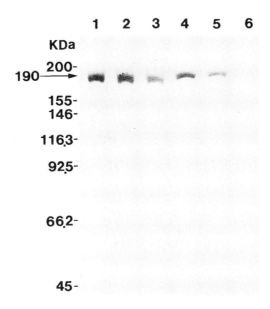

Fig. 2. Repeated induction and elution of the 190-kDa ERP72 fusion protein from agar plates. *E. coli* Y1090 cells were infected and plated by the procedure described in **Subheading 3.** Plates were incubated for 3 h at 42°C, and 5 mL of 5 m*M* IPTG in 0.5X LB containing 10 m*M* MgSO$_4$ added to each plate. Incubation was carried out at 37°C for 3 h (lane 1), and the supernatant saved. Induction and elution were repeated five times with 1-h intervals between inductions (lanes 2–6). Fusion proteins recovered after each induction was analyzed by Western blotting.

2. Materials
2.1. Plating Bacteria

1. Bacterial strain: Y1090 strain is deficient in the lon^- protease. In lon^- cells, β-galactosidase fusion proteins can accumulate at much higher levels than in wild-type cells.
2. Phage: λgt11.
3. LB/ampicillin medium: 10 g bacto-tryptone, 5 g yeast extract, 5 g NaCl/L. After autoclaving and cooling, add 100 mg/L ampicillin from a 100 mg/mL stock solution prior to use.
4. LB plates: 1.5% agar in LB containing 10 mM $MgCl_2$ and 100 µg/mL ampicillin.
5. LB top agar: 0.7% agar in LB containing 10 mM $MgCl_2$ and 100 µg/mL ampicillin.
6. 1 M $MgSO_4$ (autoclave).
7. Phage buffer (SM): 50 mM Tris-HCl, pH 7.5, 100 mM NaCl, and 10 mM $MgCl_2$ or $MgSO_4$.

2.2. Fusion Protein Expression

1. 1 M IPTG: 240 mg in 1 mL H_2O. Store frozen.
2. 0.5X LB containing 10 mM $MgSO_4$.

2.3. Detection of Fusion Proteins

1. 5X Polyacrylamide gel electrophoresis (PAGE) sample buffer: 15% β-mercaptoethanol, 15% sodium dodecyl sulfate (SDS), 1.5% bromophenol blue, and 50% glycerol.
2. 10% Polyacrylamide gels containing 0.2% SDS.

2.4. Preparing Crude Fusion Proteins

1. TEP buffer: 0.1 M Tris-HCl, pH 7.5, 10 mM EDTA, and 1 mM phenylmethylsulfonyl fluoride (PMSF).
2. Saturated or solid $(NH_4)_2SO_4$.

2.5. Immunoaffinity Purification of Fusion Proteins

1. Immunoadsorbent: ProtoSorb *lacZ* Adsorbent (Promega, Madison, WI).
2. Necessary buffers *(6)*.

3. Methods
3.1. Plating Bacteria

1. Grow host cells: Streak the Y1090 on LB agar plates containing 100 µg/mL ampicillin. After growing this plate overnight at 37°C, store it at 4°C. Inoculate the plating cells from a single colony in LB containing 100 µg/mL ampicillin for 8–18 h at 37°C.
2. Infect the cells with phages: Up to 5×10^6 recombinant phages/plate (150 × 15 mm) can be used. Mix 700 µL of the Y1090 culture with up to 200 µL of the phage in SM. Allow the phage to adsorb the cells for 15–30 min at room temperature.
3. Plate cells and phages: Add 7.5 mL of LB top agar (at 45–50°C) to the infected cells, pour onto an LB agar + ampicillin plate (at room temperature), and plate evenly over the surface of the plate. Use slightly dry plates (2 d old) so that better adhesion of the top agar to the bottom agar will be obtained.
4. Grow plaques: Allow the top agar of the plates to solidify for 30 min at room temperature. Incubate the plates at 42°C for 3–3.5 h.

3.2. Fusion Protein Induction

1. Add 5–10 mL of 5 mM IPTG in 0.5X LB containing 10 mM $MgSO_4$ to each plate.
2. Incubate at 37°C for 3 h. Recover the supernatant.

3. Repeat **steps 1** and **2**, except that the incubation time is reduced to 1–2 h. Pool the eluate from each induction.

3.3. Detection of Fusion Proteins

1. Transfer 24 μL of the eluate to a microcentrifuge tube. Spin at 12,000g for 5 min to remove the cellular debris and agar, and then transfer the supernatant to a fresh tube.
2. Add 6 μL of 5X PAGE sample buffer to the tube, and boil for 5 min. Analyze samples on a 10% polyacrylamide gel and visualize proteins with Coomassie blue, or detect the fusion protein on Western blot with specific antibodies.

3.4. Preparing Crude Fusion Proteins

1. Centrifuge the pooled eluates at 15,000g for 10 min to get rid of the debris and save the supernatant.
2. To the supernatant, add either solid ammonium sulfate or 3 vol of saturated ammonium sulfate to 75% saturation, and stir at 4°C for 20 min.
3. Centrifuge at 15,000g for 20 min, discard the supernatant, and redissolve the pellet in cold TEP buffer at approx 20 mg/mL.

3.5. Immunoaffinity Purification of Fusion Proteins (ref. 6)

After diluting the crude fusion proteins with 50 m*M* Tris-HCl buffer, pH 7.3, to about 4 mg/mL total protein, it is convenient to purify the fusion protein by the prepared immunoaffinity column (ProtoSorb *lacZ* Immmunoaffinity Adsorbent) available from Promega.

4. Notes

1. In some cases, allowing expression at 30°C rather than at 37°C could help stabilize expressed fusion proteins.
2. Low amount of IPTG and phages may be used if expression is too high. The expression level of fusion proteins can be estimated by immunoblotting analysis on a plaque lift along with dot blotting a series of known amounts of the antigen on a membrane disk, which is also used to make a plaque lift. The lower detection limits of alkaline phosphatase- and peroxidase-conjugated second antibodies are 20–50 and 200–500 pg of antigens, respectively *(1)*.
3. In certain cases where the fusion protein recovery is extremely low (i.e., <5 μg fusion protein/plate), the yield may be improved by a couple of ways. Some fusion proteins trapped in the top agar can be extracted with TEP buffer and precipitated with ammonium sulfate as above. The recovery may be increased by including a fractionation step (such as gel filtration) prior to immunoaffinity purification.
4. Fusion proteins may be purified by alternative ways. If the immunoaffinity column is not available, the large size of the fusion protein is suitable for preoperative SDS-PAGE and gel-filtration chromatography.

Acknowledgments

This work was supported by the grants GM 39436 and GM 48492 from the National Institutes of Health.

References

1. Mierendorf, R. C., Percy, C., and Young, R. A. (1987) Gene isolation by screening gt11 libraries with antibodies. *Methods Enzymol.* **152,** 458–469.
2. Young, R. A. and Davis, R. W. (1991) Gene isolation with λgt11 system. *Methods Enzymol.* **194,** 230–238.
3. Huang, S. H., Tomich, J., Wu, H. P., Jong, A., and Holcenberg, J. (1989) Human deoxycytidine kinase: sequence of cDNAs and Analysis of Expression in cell lines with and without enzyme activity. *J. Biol. Chem.* **264,** 14,762–14,768.
4. Runge, S. W. (1992) Rapid analysis of λgt11 fusion proteins without subcloning or lysogen induction. *BioTechniques* **12,** 630–631.
5. Huang, S. H. and Jong, A. (1994) Efficient induction and preparation of fusion proteins from recombinant λgt11 clones. *Trends in Genetics* **10,** 183.
6. Promega (1991) *Promega Protocols and Applications Guide*, 2nd ed. Madison, WI.
7. Singh, H., Clerc, R. G., and LeBowitz, J. H. (1989) Molecular Cloning of sequence-specific DNA binding proteins using recognition site probes. *BioTechniques* **7,** 252–261.
8. Huang, S. H., Gomer, C., Sun, G. X., Wong, S., Wu, C., Liu, Y. X., and Holcenberg, J. (1992) Molecular characterization of a 72-kD human stress protein: a homologue to murine ERP72. *FASEB J.* **6,** A1670.

51

Antibody Screening of Bacteriophage λgt11 DNA Expression Libraries

Peter Jones

1. Introduction

Much of our current understanding of the molecular details of the activity and interactions of proteins has stemmed from the ability to isolate cDNA encoding these proteins. Computer-based analysis of deduced amino acid sequence allows predictions to be made about their structure and function that can then be tested experimentally. The starting point for all of this is the ability to isolate the specific cDNA encoding the protein of interest from libraries that contain on the order of 10^5–10^7 recombinants. There are many approaches that can be used to isolate cDNA clones, and the exact strategy applied really depends on the molecular tools available. If a cDNA encoding the particular protein of interest has already been cloned from one species, it can be used to probe cDNA libraries from other species using DNA hybridization techniques. Alternatively, degenerate oligonucleotides can be designed from amino acid sequences from regions of proteins that are conserved across species. The oligonucleotides can be used to screen cDNA libraries by hybridization or used as primers to amplify and clone the cDNA by using the polymerase chain reaction.

Often, when trying to isolate cDNAs encoding novel proteins, the only thing that can be used to any advantage has to be derived from the purified protein itself. The protein can be used for the generation of antibodies and to obtain amino acid sequence. Screening cDNA expression libraries, constructed in vectors like bacteriophage λgt11, with an antibody raised against the protein, represents one of the quickest approaches to isolate its cDNA. The cDNA is prepared from the appropriate source and ligated into the λ vector. It is ligated into a restriction site within the coding region of the β-galactosidase gene that has been engineered into the vector. Large amounts of β-galactosidase are expressed as transcription is driven by a strong promoter. Any coding region of a cDNA that is ligated into the β-galactosidase gene, in the correct orientation and reading frame, will be expressed as a fusion protein with β-galactosidase. Statistically, this will take place at a frequency of one in six ligation events, however, the probability of detecting a specific cDNA is largely determined by its abundance in the mRNA population.

From: *The Nucleic Acid Protocols Handbook*
Edited by: R. Rapley © Humana Press Inc., Totowa, NJ

The basic λgt11 expression system described by Young and Davis *(1,2)* now appears in many commercially available guises, which have been engineered to increase the number of cDNAs that are expressed or facilitate the recovery of cloned cDNAs by phage rescue. Although there have been improvements in the design of λ vectors, the basic protocols in screening the libraries with antibodies are very similar. Briefly, screening involves plating the λ library at a suitable dilution on a lawn of *Escherichia coli*. Expression of the β-galactosidase fusion proteins is then induced by overlaying the growing plaques with nitrocellulose filters that have been impregnated with isopropyl β-D-thiogalactopyranoside (IPTG), an inducer of β-galactosidase gene expression. The proteins expressed by λgt11 clones are transferred onto the nitrocellulose filters, and the filters processed for the detection of antigen in essentially the same way as they are after Western blotting.

The basic methodologies in constructing cDNA libraries are discussed in detail elsewhere *(3,4)*, but unless there is a need to make cDNA libraries from "exotic" sources or to generate a number of cDNAs libraries, it is more usual to purchase a cDNA library from a commercial supplier. The library is kept as a suspension of phage particles in a dilution buffer at a known titer in units of PFU/mL. The λgt11 can then be propagated in a host strain of *E. coli*, generally Y1090. This particular strain is used, since it is deficient in one of the major protease systems of *E. coli, lon,* thus minimizing the possibility of degradation of exogenous proteins.

2. Materials

1. *E. coli* strain Y1090 for library screening.
2. *E. coli* strain C600 for producing large amounts of the fusion protein.
3. Nutrient broth (100 mL): 1 g of Trypticase (BBL), 0.5 g of NaCl in water.
4. Top agar (100 mL): 1 g of Trypticase, 0.5 g of NaCl, 0.65 g of agar.
5. Bottom agar (100 mL): 1 g of Trypticase, 0.5 g NaCl, 1.5 g agar.
6. Ponceau S (Sigma, St. Louis, MO).
7. Chloroform.
8. IPTG (isopropyl-β-D-thiogalactopyranoside).
9. Whatman paper 3MM.
10. Rabbit antiserum or affinity-purified antibody.
11. Swine immunoglobulins raised against rabbit immunoglobulins (Dako, Carpinteria, CA).
12. PAP (soluble complex of horseradish peroxidase and rabbit antihorseradish peroxidase, Dako).
13. PBS: 0.14 M NaCl, 2.7 mM KCl, 1.5 mM KH_2PO_4, 8.1 mM Na_2HPO_4.
14. Tris/saline: 10 mM Tris-HCl, pH 7.5, 0.9% w/v NaCl.
15. Tris/saline/BSA: 10 mM Tris-HCl, pH 7.5, 0.9% w/v NaCl, 0.1% w/v bovine serum albumin (BSA).
16. Tris/saline/Marvel: 10 mM Tris-HCl, pH 7.5, 0.9% w/v NaCl, 3% w/v freeze-dried nonfat milk (e.g., "Marvel," Premier Beverages, Stafford, UK).
17. Tris/saline/Tween-20: 10 mM Tris-HCl, pH 7.5, 0.9% NaCl, 0.1% Tween-20.
18. DAB (3, 4, 3'-4'-tetra-amino biphenyl hydrochloride, BDH); add 20 mg of DAB to 1 mL of water, and mix until dissolved. Add to 99 mL of Tris/saline (*see* **Note 3**).
19. Nitrocellulose filters (Amersham International).
20. SM buffer (1 L): 5.8 g of NaCl, 2.0 g of $MgSO_4·7H_2O$, 50 mL of 1 M Tris-HCl, pH 7.5, 10 mg of gelatine. Autoclave to sterilize.
21. SM phage storage buffer: 10% v/v SM buffer in chloroform.

3. Methods

3.1. Primary Screen of a λgt11 Library

3.1.1. Plating of Library for Screening with Antibodies

1. Inoculate 10 mL of nutrient broth with Y1090. Add 0.1 mL of 20% maltose and 500 mg ampicillin. Grow overnight at 37°C with shaking (*see* **Note 1**).
2. Pellet cells by centrifugation at 4000*g* for 10 min at room temperature, and resuspend in 4 mL of sterile 10 m*M* MgSO$_4$.
3. A suitable number of plaques to screen/85-mm Petri dish is between 20 and 30,000. Dilute a portion of the library in SM buffer to 3×10^5 PFU mL, and add 200 µL of Y1090 to 100 µL of the diluted phage in a plastic universal bottle. Incubate at 37°C for 15 min to allow the phage to adsorb to the cell surface. The plaque size is affected by the amount of Y1090 cells used. If you require larger plaque sizes on subsequent rounds of screening, use 100 µL of Y1090 cells.
4. Add 4 mL of top agar (cooled to 50°C) to the cells, and pour onto the bottom agar plates. The addition of the agar is sufficient to mix the cell suspension, but be careful to avoid formation of air bubbles, which will make the top agar uneven. Prewarming the bottom agar plates at 42°C helps to avoid problems with the top agar setting before it is evenly spread. Allow the top agar to harden (approx 10 min). Number the bottom of the plates and incubate them at 42°C for 3 h to allow the plaques to form.
5. While the plates are incubating at 42°C, soak the nitrocellulose filters in 10 m*M* IPTG for 15 min. This may be carried out in one container, but ensure that all of the filters are properly wetted. Allow the filters to dry at room temperature for 1 h on Whatman 3MM paper. When the filters are dry, number the edges of the filters with a ballpoint pen.
6. The plaques will be very small at this stage and may not be visible to the naked eye. Place the IPTG-treated nitrocellulose filters onto the top agar, taking care not to trap any air bubbles between the top agar and the filter. This is best done by bowing out the center of the filter a little and touching it to the center of the plate first. Then allow the rest of the filter to be gently pulled down onto the plate.
7. Incubate the plates with the filters at 37°C for a further 2–3 h to allow sufficient expression of the cDNA insert.
8. Cool the plates to room temperature, and key the filters to the plate by pushing a syringe needle through the filter and agar of the plate. Mark the positions where the needle has passed through the agar with a marker pen on the underside of the plate. It is easier to align the position of the filters with these ink marks rather than the damaged agar. Remove the filters and perform the immunological screen.
9. Plates can be stored at 4°C for 2–3 wk.

3.1.2. Detection of Clones with Antibodies

1. Peel the filters off the plates, and place them plaque side up in a Petri dish (1/dish). Cover with Tris/saline/Marvel (about 5 mL), and incubate at room temperature for at least 45 min (or 4°C overnight) (*see* **Note 2**).
2. Remove the Tris/saline/Marvel, and wash the filters three times for 10 min in Tris/saline/Tween-20 (3×5 mL) on a rocking platform or similar device. Filters can be washed together in a large container if required.
3. After the final wash, add 5 mL of Tris/saline/Tween-20 containing 0.1% w/v BSA to each of the filters. If the primary antibody is in short supply, a number of filters can be screened in the same dish if care is taken not to trap air between the filters. Add the primary anti-

body to the appropriate dilution (*see* **Note 3** to determine the optimum dilutions of antibody to use in screening the library), and incubate at room temperature for 2 h on a rocking platform or overnight at 4°C.
4. Remove the antibody solution. This can be kept at 4°C for use in further screens. The signal-to-noise ratio often increases with reuse of the antibody. Wash the filters as in **step 2**.
5. To each filter, add 5 mL of Tris/saline/BSA and the appropriate dilution of swine antirabbit antibody (usually about 1:300) to each filter, and incubate for 30 min at room temperature on a rocking platform.
6. Wash the filters as in **step 2**.
7. Add 5 mL of Tris/saline and the appropriate dilution of PAP, and incubate for 30 min at room temperature on a rocking platform.
8. Wash as in **step 2**.
9. Develop color on the filters by the addition of 5 mL of DAB solution to each filter and 2.5 µL of H_2O_2. Color develops in 0.5–5 min (*see* **Note 4**). It may be necessary to adjust the amount of H_2O_2 added to slow down the development times to allow the positive clones to become clearly visible before any background staining appears.
10. Stop color development by washing the filters in water.
11. Once the filters are dry, it is possible to align the filters with the original plates using the key marks that were made previously. Any putative positive plaques are then picked.
12. An agar plug from the region of the plate containing the recombinant phage is picked in the first instance with a 1-mL tip of a micropipet. The end of the pipet tip is cut to a diameter of approx 5 mm before picking the plug. The plug is removed by the suction that can be applied with the pipet. The agar plug containing the recombinant is then ejected into 1 mL of SM buffer containing 100 µL of chloroform (*see* **Note 5**) and the phage allowed to diffuse out of the plug for at least 4 h.
13. It is now necessary to carry out a further round of screening and purification at lower plaque densities until it is possible to pick a single, well-isolated plaque with the end of a Pasteur pipet. This usually requires a further two to three rounds of screening depending on the plaque densities used.

3.1.3. Plaque Purification and Titration

The approximate titer of the phage can be determined by a "spot titer" method:

1. Prepare a lawn of *E. coli* cells by adding 4 mL of molten top agar to 100 µL of cells resuspended in 10 m*M* $MgSO_4$ as described above. Pour the mixture over the bottom agar base, and allow it to set.
2. Make serial dilutions of 10^0, 10^{-3}, 10^{-6}, and 10^{-9} of the phage in SM buffer, and spot 5-µL aliquots of each dilution onto the solidified top agar. Tilt and rotate the plate to allow each spot to spread to a diameter of approx 10 mm, and allow the phage suspension to dry into the top agar. Incubate the plate overnight at 37°C. The approximate titer of the library is then be estimated from the number of plaques at each dilution.
3. Replate the eluted recombinant phage at a density of approx 500–1000 PFU/mL and rescreen with the antibody probe. Repeat the procedure of picking a positive plaque, eluting titrating, and replating at lower plaque densities until it is possible to pick a single well-isolated plaque.
4. Determine the titer of the purified phage, and store in 1 mL of phage storage buffer, in the dark, at 4°C. This can be kept for many years with little loss of phage viability.

3.1.4. Preparation of a High-Titer Stocks of Recombinant Phage

It is necessary to make high-titer stocks of the recombinant phage for the generation of stocks of phage to work with on a day-to-day basis, and also for the preparation of DNA for sequencing or for the preparation of protein lysates from recombinant phage. A relatively quick and easy method for the preparation of high-titer lysates is described below.

1. Infect Y1090 with the purified recombinant phage as described in the screening protocol, and plate at a density of about 50–100 PFU. Pick a single, well-isolated plaque with a Pasteur pipet, and remove the agar plug.
2. Add the agar plug to 50 µL of Y1090 cells resuspended in 10 mM MgSO$_4$ and incubate at room temperature for 5 min. Set up a control at the same time using 50 µL of uninfected Y1090 cells only.
3. Add 2 mL of nutrient broth containing 10 mM MgSO$_4$ to the tubes, and shake them vigorously at 37°C in an orbital shaker for 4–6 h or until lysis occurs. Lysis is complete when the solution becomes clear and contains string-like cell debris. Comparing the appearance of the infected culture with the control makes it easy to determine when this point is reached.
4. After lysis, add 50 µL of chloroform to the tube put on ice for 10 min. It is important to do this in chloroform-resistant containers.
5. Centrifuge the lysate at 4500g for 10 min to pellet the cell debris. Transfer the supernatant to a sterile container, and store in the dark at 4°C. Titers should be 10^{10} mL or better.

3.1.5. λgt11 Infection of E. coli Strain C600 for Western Blot Analysis of β-Galactosidase–cDNA Gene Fusion Products

Once a recombinant phage has been purified, it is expedient to analyze the β-galactosidase–cDNA gene fusion product by Western blotting to confirm the specific crossreaction of the antibody with the fusion protein. Preparative amounts of the fusion protein can be made by creating recombinant λgt11 lysogens, however a quicker and easier method is to take advantage of the amber mutation (S100) in the S gene of λgt11. This mutation renders λgt11 defective in lysing strains of *E. coli* which cannot suppress this mutation (*see* **Note 7**). A method is described below:

1. Add 0.1 mL of an overnight culture of *E. coli* strain C600, grown in nutrient broth containing 0.2% maltose, to 10 mL of fresh broth. Incubate the cells at 37°C, with shaking until they reach an absorbance of between 1.4–1.8 at 450 nm (A_{450} 0.64 = 2 × 10^8 cells/mL). This takes approx 2–3 h.
2. Pellet the cells from 1-mL aliquots of the culture by centrifugation at 12,000g for 2 min in a microfuge, and resuspend the bacterial pellets in 360 µL of nutrient broth containing 10 mM MgSO$_4$.
3. Dilute the recombinant λgt11 in 40 µL of SM buffer, and add the phage to each of the cell suspensions at a multiplicity of infection (M.O.I.) of 10, i.e., ratio of phage particles to bacterial cells is 10:1.
4. Incubate the mixture at 37°C for 3 h with occasional shaking.
5. Pellet the cells by centrifugation at 12,000g for 2 min in a microfuge, and resuspend the pellet in 100 µL of a loading buffer suitable for polyacrylamide gel electrophoresis. Boil the samples for 3 min, and remove any debris by a brief centrifugation. Load aliquots onto gels for analysis by Western blotting (generally, 50 µL contains approx 150 µg of protein).

3.1.6. Purification of Antibody Bound to Fusion Protein

Once a recombinant clone expressing a fusion protein has been purified and the fusion protein clearly shown to crossreact with the antibody by Western blotting, one can be reasonably confident that the cDNA isolated is the one required. Final proof of identity is ultimately dependent on sequencing the cDNA insert. The sequence of the cDNA is either compared with sequences in computer data bases (it is still possible to get a nasty surprise at this stage), or if the protein is novel, it is essential to find overlap be deduced amino acid sequences in the cDNA and peptide sequence from the purified protein. The latter approach is time-consuming and labor-intensive, and since many of the cDNAs initially isolated tend to be relatively small, it can take the sequences of many peptides to find overlaps. To gain further confidence that the time and effort in obtaining peptide data will be well invested, it is a worthwhile demonstrating that the fusion protein can be used to affinity-purify an antibody that will crossreact with the original antigen on a Western blot.

3.1.7. Affinity Purification of Antibody on Nitrocellulose Filters

1. Transfer the fusion protein from cell lysates to nitrocellulose by Western blotting. Load each track of the polyacrylamide gel with as much antigen as can be clearly resolved. The amount used has to be determined for the particular gel system used.
2. After blotting is complete, stain the nitrocellulose filter with Ponceau S for 5 min, and destain in water to detect the position of the fusion protein. Cut out the strip of nitrocellulose filter containing the fusion protein, and wash the filter for 10 min in PBS, 10% BSA, and 0.1% Triton X100. Repeat this washing step another two times.
3. Incubate the antibody (use 5X more concentrated than normal) with the nitrocellulose strip for 2–3 h at room temperature or 4°C overnight.
4. Wash the nitrocellulose filter as described in **step 2**, and then insert it into a precooled tube by curling it around the sides of the tube.
5. Elute the bound antibody by adding 1 mL of precooled 0.2 M glycine-HCl, pH 2.2, and incubating the filter at 4°C for 5 min. Roll the tube to maximize the coverage of the filter.
6. Neutralize the solution by adding 30 µL of 2 M Tris base, and then dialyze against PBS.
7. Dilute the antibody 1:10 for re-probing Western blots. The antibody can be reused several times if stored at 4°C in PBS, containing 5% w/v BSA.

4. Notes

1. Maltose is added to increase the amount of receptor for λgt11 on the cell surface of *E. coli*, but it is not essential.
2. The exact logistics of the number of filters screened and containers used for incubation with antibodies, and for washes can be decided individually. Generally, we do not screen more than 10 filters (300,000 plaques) at a time and find it convenient to keep the filters in separate Petri dishes throughout the procedure.
3. Of the factors that determine the likelihood of success with this approach, assuming that the cDNA library contains a good representation of the mRNA population, by far the most important is the quality of the primary antibody used. In carrying out an antibody screen of a cDNA library, it is preferable to use a polyclonal antibody, since most of the cDNAs will probably be incomplete and may not contain the epitope recognized by a monoclonal antibody. Many of the problems associated with screening expression libraries stem from the crossreactivity of the other antibodies in the serum with *E. coli* proteins. This is quite

a common occurrence, and it is useful to test the preimmune serum from a number of rabbits for crossreactivity with *E. coli* and phage proteins on a phage lift before starting an immunization regime. It is then possible to select rabbits to minimize any problems with high backgrounds or false positives. It is also worthwhile checking the secondary antibodies for crossreactivity. The optimum dilution of the primary antibody to be used in screening the library is one that allows the maximum sensitivity in detecting the antigen of interest without giving rise to high background staining. This optimum dilution can be determined by using segments (usually quarters) of a nitrocellulose filter taken from a plaque lift onto which serial dilutions of the antigen have previously been dried. The individual segments can then be incubated with different dilutions of the primary antibody to determine the optimum dilution. If antisera do produce high backgrounds in screening, it is possible to clean them up to some extent by subtracting antibodies that crossreact with *E. coli* proteins. This is most easily done by preincubating dilutions of antibody with nitrocellulose filters taken from plaque lifts before using them in screening.

4. DAB is a carcinogen, and should be handled and disposed of as dictated by the hazard data. There are a number of less hazardous substrates that can be used instead of the one described here. However, in our hands, DAB has worked best.
5. The chloroform is added to sterilize the phage suspension. Bacteriophage λ is resistant to the effects of chloroform as long as the chloroform is kept stored in the dark to prevent radical formation.
6. It is possible to grid the plate into 12 sections, allowing the titer of three plaque eluates to be determined on one plate. This is convenient, since it avoids the small mountain of plates, which can be generated if titering a number of putative positive plaques if separate plates are used for each dilution.
7. Preparative amounts of the fusion protein can be made from recombinant λgt11 lysogens in the *E. coli* strain Y1089. However, the isolation and purification of a lysogen are time-consuming and often impossible owing to mutations occurring in the λ repressor protein. By infecting an *E. coli* host strain, such as C600, which lacks the *supF* mutation, the phage will replicate many times without causing lysis. This results in high levels of the fusion protein being produced inside the cell, which is easily visible among the proteins of a cell lysate when analyzed by polyacrylamide gel electrophoresis. The very high numbers of λgt11 attained in C600 result in the β-galactosidase fusion protein being expressed constitutively. If constitutive expression of fusion protein, or the fact that C600 is not *lon*⁻, affects the recovery of the protein or the viability of the C600, it is necessary to resort to forming a lysogen in Y1089.

References

1. Young, R. A. and Davis, R. W. (1983) Efficient isolation of genes by using antibody probes. *Proc. Natl. Acad. Sci. USA* **80,** 1194.
2. Young, R. A. and Davis, R. W. (1983) Yeast RNA polymerase II genes: isolation with antibody probes. *Science* **222,** 778.
3. Huynh, T. V., Young, R. A., and Davis, R. W. (1985) Constructing and screening cDNA libraries in λ gt-10 and λ gt-11, in *DNA Cloning,* vol. 1 (Glover, D. M., ed.), IRL, Oxford, pp. 49–87.
4. Sambrook, J., Fritsch, E. F., and Maniatis, T. (1989) *Molecular Cloning, A Laboratory Manual.* Cold Spring Harbor Laboratory, Cold Spring Harbor, NY.

52

Screening cDNA Libraries by Hybridization with Double-Stranded DNA Probes and Oligonucleotides

Caroline A. Austin

1. Introduction

Probably the most commonly used method to screen a cDNA library is hybridization to a labeled DNA probe. This probe may be a single-stranded oligonucleotide or a double-stranded cDNA or polymerase chain reaction (PCR) product. The DNA may be either radioactively or nonradioactively labeled. The sequence of an oligonucleotide probe may be derived from a number of sources, e.g., degenerate probes may be obtained by back translating a peptide sequence of an unknown protein, or they may be a short conserved region of sequence within a cDNA from another member of a multigene family or from a cognate cDNA from another species (*see* **Note 1**). Double-stranded DNA probes may be a partial cDNA obtained by screening another library or a PCR product or a cDNA from another member of a gene family or from another species. This chapter concentrates on the methods for labeling DNA probes and hybridization to filter-bound library DNA.

2. Materials

2.1. Plasmid Library Plating and Colony Lifts

1. Titered plasmid cDNA library.
2. Luria-Bertani (LB) medium: 10 g/L Bacto-tryptone (Difco, Detroit, MI), 5 g/L yeast extract, 10 g/L NaCl. Autoclave to sterilize.
3. LB agar: LB medium with 15 g/L of agar. Autoclave to sterilize. If antibiotics, such as ampicillin, are required, allow agar to cool to 55°C and then add antibiotic to appropriate concentration from stock solution, e.g., add ampicillin to 50–100 µg/mL from a 50-mg/mL stock prior to pouring plates.
4. Plates: Fifteen-centimeter diameter, triple-vent Petri dishes.
5. Nylon filters, such as Hybond N+ (Amersham Pharmacia Biotech, St. Albans, UK) or Protan BA 85 (Schleicher and Schuell, Dassel, Germany).
6. Whatman 3MM paper (Whatman, Maidstone, UK).
7. Three trays or large dishes.
8. Denaturation buffer: 0.5 M NaOH, 1.5 M NaCl.
9. Renaturation buffer: 1 M Tris-HCl, pH 7.0, 1.5 M NaCl.
10. 20X Standard saline citrate (SSC): 3 M NaCl, 0.3 M sodium citrate.

From: *The Nucleic Acid Protocols Handbook*
Edited by: R. Rapley © Humana Press Inc., Totowa, NJ

11. 3X SSC.
12. 3X SSC/0.1% sodium dodecyl sulfate (SDS).

2.2. Screening by Hybridization with Labeled Oligonucleotide

2.2.1. Labeling Oligonucleotide

1. Chosen oligonucleotide (*see* **Note 1**) with free 3'-OH.
2. T4 polynucleotide kinase.
3. 10X T4 kinase buffer: 1 M Tris-HCl, pH 7.6, 100 mM MgCl$_2$, 200 mM 2-mercaptoethanol, store at –20°C.
4. [γ^{32}P]-Adenosine triphosphate (ATP) 3000 Ci/mmol (Amersham Pharmacia Biotech).

2.2.2. Hybridization with a Labeled Oligonucleotide Probe

1. Wash solution: 3X SSC, 0.1% SDS.
2. Wetting solution: 3X SSC.
3. 100X Denhardt's solution: 20 g/L Bovine serum albumin (BSA), 20 g/L Ficoll 400, 20 g/L polyvinylpyrrolidone. Store at –20°C.
4. Prehybridization buffer: 6X SSC, 1X Denhardt's, 0.5% SDS, 100 µg/mL denatured salmon sperm DNA, 0.05% sodium pyrophosphate.
5. Hybridization buffer: 6X SSC, 1X Denhardt's, 0.5% SDS, 100 µg/mL denatured salmon sperm DNA, 20 µg/mL tRNA, 0.05% sodium pyrophosphate.
6. Wash buffer: 6X SSC/0.05% sodium pyrophosphate (SSC/PP).

2.3. Screening by Hybridization with a Double-Stranded DNA Probe

2.3.1. Labeling

Double-stranded probes can be labeled by random priming labeling or by nick translational. Kits for both are available from several companies including Amersham Pharmacia Biotech and Stratagene (La Jolla, CA).

2.3.2. Hybridization of a Double-Stranded DNA Probe

1. 3X SSC.
2. 100X Denhardt's solution: 20 g/L Bovine serum albumin (BSA), 20 g/L Ficoll 400, 20 g/L polyvinylpyrrolidone. Store at –20°C.
3. Prehybridization buffer: 5X SSC, 10X Denhardt's, 7% SDS, 100 µg/mL denatured salmon sperm DNA, 20 mM sodium phosphate.
4. Hybridization buffer: 5X SSC, 10X Denhardt's, 7% SDS, 100 µg/mL denatured salmon sperm DNA, 20 mM sodium phosphate, 8% dextran sulphate.
5. 2X SSC/0.1% SDS.

3. Methods

λ-Phage libraries are by far the most common type of cDNA library and they present a number of advantages, such as ease of constructing large libraries and of plating large numbers of recombinants. However, for some purposes, plasmid libraries are appropriate; in addition, a number of recent specialized screening systems, such as yeast two-hybrid screening, employ plasmid libraries, and it may sometimes be of use to screen such libraries using DNA hybridization techniques. Methods are given for making colony lifts of plasmid libraries in **Subheading 3.1.** The protocols for probing with labeled oligonucleotides or DNA fragments described in **Subheadings 3.2.** and **3.3.**, respectively, are appropriate for screening either plasmid or λ-phage libraries.

In order to ensure that the library used will contain the desired cDNA, it should have been prepared from mRNA extracted from a cell type or tissue in which the message or encoded protein is known to be expressed. Libraries can be made "in house," possibly using one of the commercial kits that are available. Alternatively, libraries may be obtained from other workers or a resource centre, or purchased, as many libraries are now available commercially.

3.1. Plating of cDNA Library and Colony Lifts

3.1.1. Transformation and Plating Plasmid Libraries

Plasmid libraries require very high efficiency transformation either by electroporation *(1)* or using highly competent cells made by the method of Hanahan *(1–3)*. Highly competent and electrocompetent cells can also be obtained commercially from suppliers, such as Stratagene or Promega (Madison, WI). For library screening, the library is plated directly onto filters placed on plates containing the appropriate antibiotic *(1)*; for amplification the library is plated without filters.

1. Plate the library (*see* **Note 2**) by pipeting 0.2 mL of LB containing up to 2×10^5 CFU onto a nylon filter laid on a 15-cm Luria-Bertani (LB) agar plate containing ampicillin at 50–100 g/mL (or other selective antibiotic as appropriate for the plasmid type in which the library is constructed). Spread the liquid evenly over the surface of the filter using a bent glass rod sterilized by dipping in ethanol, flaming with a Bunsen burner, and allowing to cool for 30 s.
2. Leave the plates lid side up on the bench to allow the surplus liquid to soak in or evaporate (about 1 h).
3. Incubate the plates lid side down overnight, by which time distinct colonies should have appeared.

3.1.2. Replica Plating **(3–5)**

1. Remove the master filter from each plate using blunt-ended forceps, and place colony side up on a sheet of dry filter paper such as Whatman 3MM. For each master filter, wet a new nylon filter by placing on a fresh LB plate. Using two blunt-ended forceps, remove the new filter from each plate and place- on top of the master filter. Place a piece of Whatman 3MM paper on top of the pair of filters and then a glass plate. Apply pressure by hand or with a heavy book for a few seconds.
2. Prior to separating the replica and the master, make orientation marks by piercing both the master and replica with a needle. Make the needle hole in an asymmetric pattern to allow later orientation of the replica with the master.
3. Separate the master and replica filters by peeling apart with blunt-ended forceps, and place the replica, colony side up onto a new LB antibiotic plate.
4. Make a second replica from the master in the same way without further growth of the master. Remember to make the orientation holes prior to pulling apart.
5. Place the second replica onto a new LB antibiotic plate.
6. Return the master to its LB plate and store at 4°C.
7. Place both replicas at 37°C until colonies are easily visible (0.5–1 mm diameter). This usually takes 3–5 h.

3.1.3. Lysis of Bacteria on Filters

1. Place the 3MM paper in three trays, the first saturated with 0.5 M NaOH, 1.5 M NaCl, the second with 1 M Tris-HCl, pH 7.0, 1.5 M NaCl, and the third with 3X SSC.
2. Place the filters colony side up on the first tray (0.5 M NaOH, 1.5 M NaCl) and leave for 10–15 min. The NaOH/NaCl soaks through the filter and lyses the bacterial colonies. Remove the filters from tray 1, removing excess liquid from the underside of the filter on the tray edge.
3. Transfer the filters to the second tray containing Whatman 3MM soaked in 1 M Tris-HCl, 1.5 M NaCl, and leave for 2–5 min. Then transfer to the third tray containing 3X SSC for 2–5 min.
4. Transfer to a piece of 3MM paper, and wipe the colony side with a tissue saturated in 3X SSC and 0.1% SDS, to remove bacterial debris (*see* **Note 3**).
5. Remove the filters to a fresh sheet of 3MM paper to air dry.
6. To fix the DNA to the filter, either bake for 1 h at 80°C or UV crosslink in a suitable UV crosslinker, e.g., UV Stratalinker 2400 (Stratagene).

3.2. Screening by Hybridization with Labeled Oligonucleotide

3.2.1. Oligonucleotide Labeling

1. Add 200 ng of oligonucleotide to a microfuge tube containing 2 μL 10X T4 kinase buffer, 100 μCi [γ^{32}P]-ATP, 10 U T4 polynucleotide kinase, and water to 20 μL. Incubate at 37°C for 1 h.
2. Add the whole mixture to the hybridization buffer, or purify labeled oligonucleotide from free ATP (*see* **Note 4**).

3.2.2. Hybridization with a Labeled Oligonucleotide Probe (6)

3.2.2.1. PREWASHING FILTERS

1. Wash filters in 3X SSC, 0.1% SDS at 65°C for 16–20 h. Large volumes and several changes of buffer aid removal of bacterial debris (*see* **Note 3**).
2. After this wash, filters may be prehybridized immediately or stored dry at 4°C.

3.2.2.2. PREHYBRIDIZATION

1. If necessary, wet the filters in 3X SSC.
2. Place the filters in a suitable container (*see* **Note 5**) in prehybridization buffer, allowing 100 μL/cm^2 of filter.
3. Incubate at 37°C for at least 2 h.

3.2.2.3. HYBRIDIZATION

1. Remove prehybridization buffer and add hybridization buffer, allowing 50 μL/cm^2 of filter.
2. Add the labeled oligonucleotide probe to the filters at approx 4 ng/mL.
3. Incubate for 16 h at an appropriate temperature (*see* **Note 6**).

3.2.2.4. WASHING FILTERS

1. Carefully remove the hybridization solution, pour into a thick glass bottle (e.g., Duran), and store at –20°C. This can be reused for second-round screens if these are performed less than a week later.
2. Filters are washed in 6X SSC, 0.05% sodium pyrophosphate (SSC/PP). Place filters in a sandwich box or similar container that is large enough to allow movement of the filters, and wash with large volumes of SSC/PP at appropriate temperatures (*see* **Note 7**).

Screening cDNA Libraries by Hybridization

3. Blot the filters to remove excess liquid, but do not allow to dry. Wrap in Saran Wrap or cling film, and expose to film at –70°C with intensifying screens. Use radioactive or fluorescent markers to aid alignment of the developed autorad with the filters. Strong signals should appear overnight, but exposures of up to a week may be required *(7)* (*see* **Notes 6** and **7**).

3.2.3. Picking Positives and Colony Purification

1. Use the radioactive or fluorescent markers to align the developed autorads with the probed filters, identify each filter on the autorad, and mark the needle holes in the filters on the autorad to enable the film to be aligned with the master filters.
2. Identify positive spots that appear in duplicate, i.e., on both filter lifts. As a guide, genuine positives will be colony/plaque shape and size and will give duplicate positive signals, although one filter may give a stronger signal than the other.
3. For plasmid libraries, pick positive colonies or an area containing a positive. Use a sterile toothpick to transfer the colonies to a series of sterile microfuge tubes containing 200 µL each of LB. Disperse the colonies by vortexing, and replate on 9-cm LB agar plates containing antibiotic, spreading the cell suspension on nylon filters as before. It may save time to replate each putative positive at two or more densities aiming to obtain a plate with 50–200 colonies to increase the chances of being able to pick a well-isolated positive colony at the next stage.
4. Incubate the plates overnight at 37°C, and make replica filters as described in **Subheading 3.1.2.**
5. Continue until well-isolated single colonies can be picked.

3.3. Screening by Hybridization with a Double-Stranded DNA Probe

3.3.1. Labeling

Labeling double-stranded probes can be accomplished by random primer labeling or by nick translation using commercial kits according to the manufacturer's instructions.

3.3.2. Hybridization of a Double-Stranded DNA Probe

3.3.2.1. Prehybridization

1. Wet the filters in 3X SSC.
2. Place the filters in a suitable container (*see* **Note 5**) in prehybridization buffer, allowing 100 µL/cm^2 of filter.
3. Incubate at 65°C for at least 1 h.

3.3.2.2. Hybridization

1. Prepare hybridization buffer allowing 50 µL/cm^2 of filter.
2. Remove prehybridization buffer and add hybridization buffer.
3. Heat the labeled probe in a boiling water bath for 5 min, cool rapidly on ice and then add to the hybridization solution at approx 1 ng/mL.
4. Incubate overnight at 65°C, preferably with shaking or other movement to allow the hybridization solution to move over the filters (*see* **Note 5**).

3.3.2.3. Washing Filters

1. Filters are washed in preheated 2X SSC, 0.1% SDS at 65°C two or three times for 20 min each. For high stringency, wash further in 0.1X SSC, 0.1% SDS at 65°C, for another 20 min.

2. Blot filters dry, and wrap in Saran Wrap or cling film, autoradiograph at −70°C as described in **Subheading 3.2.2.4.**
3. Pick positive colonies or area and rescreen as above until colony-purified positives are obtained.

4. Notes

1. Probably the most common use of oligonucleotides in screening cDNA libraries is to isolate a cDNA for which some partial peptide sequence data are available or to screen for additional members of a gene family where specific amino acid motifs are particularly well conserved. In both cases, the oligonucleotide sequence is derived from a back translated amino acid sequence. Degeneracy can be minimized by selection where possible of less degenerately coded regions of amino acid sequence. In addition, codon usage tables can be consulted to "guess" the correct base at particular points. The paucity of the sequence CG in vertebrate genomes may also allow complexity to be reduced. Oligonucleotides can be synthesized to contain deoxyinosine in place of T or G to reduce probe complexity further.
2. The number of colonies to aim for depends on the complexity of the library and the species of origin. As a rule of thumb, for a vertebrate cDNA library that has not been normalized in any way, and that contains 10^6 or more independent recombinants, around 10^6 colonies should be plated to isolate a moderate to low abundance cDNA. For some purposes, such as screening a library containing cDNA generated with a specific primer, far fewer colonies need to be screened.
3. Bacterial debris on the filters causes background problems, which can lead to false-positive signals. Wiping the filters with very wet tissue in one direction and prewashing of the filters decreases this background.
4. Alternatively, labeled oligonucleotide may be separated from free ATP by such methods as spun-column chromatography. Plug a 1-mL plastic syringe with a small disk cut from a Whatman GF/C filter, and fill with Sephadex G-50M (Amersham Pharmacia Biotech) in TE. The syringe is then placed in a 15-mL Falcon tube and spun at 1000g for 4 min. Add TE (100 µL) to the top of the compressed Sephadex and spin as before. Cut the lid from a 1.5-mL Falcon tube and place at the bottom of the Falcon tube to collect the flowthrough from the column. After adding 80 µL of TE to the probe, mix carefully, pipet it onto the spun column, and spin again as above. Collect the flowthrough, and discard the column as appropriate for radioactive waste. The specific activity of the probe should be in excess of 10^8 cpm/µg.
5. Several types of containers can be used. Screw-cap bottles for hybridization ovens (e.g., Hybaid, Middlesex, UK) are routinely used in many molecular biology laboratories. However, if no such oven is available, a waterbath at the required temperature may be used with the filters either in a plastic sandwich box or sealed in plastic bags.
6. The hybridization temperature for an oligonucleotide probe is crucially important. However, the optimum may have to be determined by trial and error. A rough rule of thumb for oligonucleotide probes is 20°C for a 14 mer, 37°C for a 17 mer, and 42°C for a 23 mer (*6*), although this is affected by the G/C content of the probe. Alternately a temperature 3–5°C below the T_m may be used, where an approximation of the T_m is calculated by $T_m = 2 (A + T) + 4 (G + C)$ (*7*). However, for a degenerate oligonucleotide, it may not be possible to calculate T_m in this way, as the base composition of the hybridizing species is unknown.
7. Washing temperatures also vary depending on the length of the oligonucleotide probe and base content, and may have to be determined empirically: for a 17-mer wash at 37°C for 1 h and at 47°C for a further 10 min, for a 14-mer wash at 30°C for 1 h and at 37°C for a

further 10 min *(6)*. These suggested temperatures assume that there are no such mismatches between the labeled oligonucleotide and the target sequence. As an alternative approach, wash four times for 5 min each at room temperature in 6X SSC, 0.1% SDS, autoradiograph the filters, and then rewash the filters in a series of steps of increasing temperature, autoradiographing after each wash. The later washes should only be for 5 min. With an efficiently labeled probe, an exposure of 4–6 h should be sufficient at –70°C with intensifying screens for the initial exposure.

References

1. Sambrook, J., Fritsch, E. F., and Maniatis, T. (eds.) (1989) *Molecular Cloning: A Laboratory Manual*. Cold Spring Harbor Laboratory Press, Cold Spring Harbor, NY.
2. Hanahan, D. (1983) Studies on transformation of *Escherichia coli* with plasmids. *J. Mol. Biol.* **166,** 557–580.
3. Hanahan, D., Jessee, J., and Bloom, F. R. (1991) Plasmid transformation in *Escherichia coli* and other bacteria. *Methods Enzymol.* **204,** 63–113.
4. Hanahan, D. and Meselson, M. (1980) Plasmid screening at high colony density. *Gene* **10,** 63–67.
5. Grosveld, F. G., Dahl, H. H. M., de Boer, E., and Flavell, R. A. (1981) Isolation of β-globin related genes from a human cosmic library. *Gene* **13,** 227–237.
6. Woods, D. (1984) Oligonucleotide screening of cDNA libraries. *Focus* **6(3),** 1–2.
7. Brown, T. A. (1991) *Molecular Biology LABFAX*. Blackwell Scientific Publications, Oxford, UK.

53

cDNA Library Screening with the Tetramethylammonium Chloride (TMAC) Technique Using Highly Degenerate Oligonucleotide Probes

Bent Honoré and Peder Madsen

1. Introduction

If an unknown protein is purified and available in relatively small amounts, it is possible to determine the sequences of short internal peptides (1). In order to determine the whole sequence of the protein by cDNA cloning, one of the peptides of perhaps five to seven amino acids may be reverse translated into nucleotide sequence resulting in a 15–21-base-long deoxyribonucleotide. Because of codon degeneracy, the number of possible oligonucleotides may be more than several hundred, which must be present in order to insure that the correct sequence is represented. The melting temperature in buffered saline solution of this mixture of oligonucleotides is heterogeneous due to differences in G + C content, as G:C base pairs possessing three hydrogen bonds interact more strongly than A:T base pairs with two hydrogen bonds. Thus, in buffered saline solution one usually chooses a melting temperature that is so low that the oligonucleotide with the lowest G + C content can hybridize. However, in doing so it is possible that oligonucleotides with a higher G + C content may form stable hybrids with mismatches resulting in the cloning of artifact cDNAs. Even though this procedure has been used successfully (2–4), it is more convenient to use a different buffer type that contains tetramethylammonium chloride (TMAC), as it has been reported that this salt selectively binds to and stabilizes A:T base pairs so that their melting temperature becomes similar to that of G:C base pairs (5–7).

The successful cloning of a large number of cDNAs shows that the technique works for 15- to 20-mer oligonucleotides with a degeneracy up to 512 and a G + C content between 27 and 61% in the cloned cDNA (8). Although sequence dependent stability of very short DNA sequences in TMAC have been reported recently (9), such features might be averaged out in the longer oligonucleotides used here. Additionally, we have found that the TMAC technique works with oligonucleotides containing deoxyinosine as a neutral or slightly destabilizing base at highly ambiguous positions.

From: *The Nucleic Acid Protocols Handbook*
Edited by: R. Rapley © Humana Press Inc., Totowa, NJ

2. Materials

2.1. Plating of cDNA Library and Replica Lifts

1. Luria-Bertani (LB) medium: 10 g/L peptone, 5 g/L yeast extract, 5 g/L NaCl and LB-maltose magnesium (MM) medium: LB medium supplemented with 2 g/L maltose and 2 g/L $MgCl_2$ for bacterial growth.
2. LB agar: LB medium with 15 g/L of agar and LB-MM top agar or agarose: LB-MM medium with 7 g/L of agar or agarose for plating on screening plates.
3. Saline magnesium (SM) buffer for diluting λ phages: 100 mM NaCl, 10 mM $MgCl_2$, 50 mM Tris-HCl (pH 7.5).
4. Screening plates, 500 cm^2 (cat. no. 240 835, Nunc, Roskilde, Denmark): 245 × 245 mm.
5. Nylon filters (Hybond N, Amersham Pharmacia Biotech, Amersham UK): 220 × 220 mm and nylon nets to interpose between the filters in hybridization tubes.
6. Denaturation solution: 0.5 M NaOH, 1.5 M NaCl. Store at room temperature.
7. Renaturation solution: 0.5 M Tris-HCl, 1.5 M NaCl, pH 7.5. Store at room temperature.
8. 20X Standard saline citrate (SSC): 3.0 M NaCl, 0.3 M sodium citrate. Store at room temperature.

2.2. Labeling and Purification of Oligonucleotide

1. TE buffer: 10 mM Tris-HCl (pH 8.0), 0.1 mM ethylenediaminetetraacetic acid (EDTA) (pH 8.0). Store at 4°C.
2. T4 Polynucleotide kinase (Amersham Pharmacia Biotech), and 10X kinase buffer: 500 mM Tris-HCl (pH 7.5), 100 mM $MgCl_2$, 50 mM dithiothreitol (DTT), 1 mM spermidine HCl, 1 mM EDTA (pH 8.0). Store aliquots at –20°C.
3. [γ-^{32}P]Adenosine-5'-triphosphate (ATP) from ICN Radiochemicals (Irvine, CA), 7000 Ci/mmol, 167 µCi/µL, 24 pmol ATP/µL or from Amersham Pharmacia Biotech, >5000 Ci/mmol, 150 µCi/µL, 30 pmol ATP/µL.
4. Formamide/dye mix: 10 mM EDTA, 0.5 g/L Bromphenol blue, 0.5 g/L Xylene cyanol in deionized formamide. Store at –20°C.
5. 10X TBE buffer: 1 M Tris-HCl (pH 8.0), 1 M Borate, 0.01 M EDTA (pH 8.0). Store at room temperature.
6. 20% Polyacrylamide/7 M urea: Mix 25 mL of 40% polyacrylamide (380 g/L acrylamide, 20 g/L bis-acrylamide), 21.25 g urea, 5 mL 10X TBE buffer, and add water to 50 mL.
7. Tetramethylethylenediamine (TEMED) (20–25 µL) and 10% ammonium persulfate (APS) (200–250 µL).
8. Kodak X-ray films, Saran Wrap, $NaBH_4$ (10%), plastic bags.

2.3. Hybridization of Oligonucleotide

1. Tetramethylammonium chloride (TMAC) (Fluka [Buchs, Switzerland] or Merck [Darmstadt, Germany]): 5 M in water (see **Note 1**). Store at room temperature. Note that this chemical is hazardous with a strong odor and should only be handled in a fume hood.
2. Yeast RNA (Boehringer-Mannheim, Mannheim, Germany): 100 mg/mL in water (see **Note 2**). Store at –20°C.
3. 100X Denhardt's solution: 20 g/L bovine serum albumin (BSA), 20 g/L Ficoll 400 (Pharmacia, Uppsala, Sweden), 20 g/L polyvinylpyrrolidone. Store at –20°C.
4. Prehybridization buffer: 6X SSC, 5X Denhardt's solution, 20 mM sodium phosphate (pH 7.0) and 1.5 mg/mL yeast RNA. Store at 4°C.
5. Hybridization buffer: 3 M TMAC, 5X Denhardt's solution, 50 mM sodium phosphate (pH 7.0), 1 mM EDTA, 0.6% SDS and 0.25 mg/mL yeast RNA. Store at room temperature.

cDNA Screening with the TMAC Technique

6. Washing buffer 1: 3 M TMAC, 50 mM Tris-HCl (pH 8.0), 0.2% SDS.
7. Washing buffer 2: 2X SSC, 0.1% SDS.
8. 2X YT-medium: 16 g/L peptone, 10 g/L yeast extract, 10 g/L NaCl.
9. Petri dishes (90 mm in diameter, Nunc, Roskilde, Denmark) with LB-agar supplemented, after autoclaving and cooling to 45°C, with 100 µg/mL of ampicillin for growth of bacteria containing pBluescript plasmids with ampicillin resistance (Stratagene, La Jolla, CA).

3. Methods

3.1. Plating of cDNA Library and Replica Lifts

The library used should contain the cDNA encoding the protein to be cloned. The first choice is a cDNA library made with RNA purified from the same cell line or tissue as has been used to sequence the protein, although other cDNA libraries may be used (*see* **Note 3**).

1. Plate the library (e.g., λZap [Stratagene]; *see* Chapter 49) at a density of up to 1.5×10^5 PFU/screening plate (245 mm × 245 mm with 0.25 L of LB agar) using an overnight culture of *Escherichia coli* XL-1-Blue or LE392 and 30 mL of LB-MM top agar or agarose, which is allowed to set for 30 min. The plates are then incubated at 37°C overnight.
2. The next day, place the plates for 1 h at 4°C (*see* **Note 4**). Plaque transfer to filters, denaturation, and renaturation are performed largely as described in the blotting and hybridization protocols for Hybond membranes from Amersham Pharmacia Biotech. In detail, one nylon filter (first replica filter) is marked with a few lines with India ink defining a unique pattern in a band from the edges of the filter and about an inch to the center, and then placed on the plate for 1 min. The pattern of lines are then marked on the reverse side of the plate with India ink. The filter is carefully lifted off the plate, placed with the plaque side up at filter paper, and a second filter (second replica filter) is placed on the plate. Now the line pattern from the plate is marked on the filter with India ink and the filter is left on the plate for 3 min or more. The filters are then placed, still with the plaque side up, on top of filter paper, lightly soaked with denaturation solution and left for 7 min. The nylon filters are then transferred to filter paper lightly soaked with renaturation solution for 3 min. This step is repeated, and finally the filters are washed in 2X SSC, air dried, and crosslinked with UV light from a transilluminator for 1.5 min on each side. For long-term storage of agar plates, *see* **Note 5**.

3.2. Labeling and Purification of Oligonucleotide

The amino acid sequence of the peptide is reverse translated to the fully degenerated nucleotide sequence, and a suitable oligonucleotide is selected. The principles for the selection of the best possible oligonucleotide is described in **Note 6**.

1. Label 20–100 pmol of oligonucleotide in a volume of 10 µL by mixing the oligonucleotide with 1 µL of 10X kinase buffer, 3–5 µL of [γ-^{32}P]ATP (0.5–0.8 mCi, 70–150 pmol), TE buffer to 10 µL, and finally 1 µL of T4 polynucleotide kinase (10 U). The mix is incubated at 37°C for 1 h (*see* **Note 7**).
2. The labeled oligonucleotide may be purified from the unlabeled on a polyacrylamide/7 M urea gel. Twelve microliters of formamide/dye mix is added to the labeling mix, which is then heated to 95°C for 3 min and loaded on a 20% polyacrylamide/7 M urea gel, which has been prerun with 1X TBE buffer at 18 W for 20 min prior to loading.
3. Run the gel at maximum 18 W for a few hours until the blue color disappears. Dismount the gel from the glass plates and cover with Saran Wrap. A few pieces of filter paper are taped to the gel and lightly soaked with 10% NaBH$_4$.

4. Expose an X-ray film for 5–10 s and develop. One strong band that represents the labeled oligonucleotide should be present. This band can then be cut out from the gel with a scalpel by superimposing the X-ray film, which contains marks from the filter paper soaked with NaBH$_4$.
5. Place the gel piece in an Eppendorf tube with 1 mL hybridization buffer and shake. In order to extract the labeled oligonucleotide from the gel, change, save, and pool the buffer several times. The efficiency of the extraction can be checked with a monitor.

3.3. Hybridization of Oligonucleotide

1. Interpose the nylon filters between nylon nets and incubate in prehybridization buffer for 1–2 h. Replace the prehybridization buffer with fresh buffer and incubate for a further 1–2 h (*see* **Note 8**).
2. Wash the filters thoroughly four times for about 15 min in hybridization buffer and finally incubate overnight or longer in hybridization buffer together with the labeled oligonucleotide at the calculated hybridization temperature, which depends on the length of the oligonucleotide (*see* **Note 9**).
3. Pour off the hybridization buffer with labeled probe and save at –20°C (*see* **Note 10**), then wash the filters in washing buffer 1 at the hybridization temperature with four exchanges each for 15 min, followed by 1–2 h of incubation.
4. Wash the filters two times for 2 min in washing buffer 2 at room temperature and seal in plastic bags. The marks on the filters are drawn up with radioactive ink and autoradiography is performed at –70°C with intensifying screens for 24 h to a few weeks.
5. After development of the films they are aligned and stronger spots appearing on both replica films (usually strongest on the first replica) likely represent correct hybridizing clones. Cut about 1 cm^2 or less of the agar/agarose and put in Eppendorf tubes for each positive clone. Add 300 µL of SM buffer, crush the agar/agarose with a glass rod, and shake the tube for 15 min. Dilute 1 µL from this solution with 1500 µL of SM buffer and plate 1, 10 and 100 µL again together with 300 µL of bacteria, and 3 mL of LB-MM top agar on smaller Petri dishes. Put nylon filters on these as previously described. It is usually not necessary to use replica filters, but it may be advantageous for the beginner.
6. Well-isolated positive plaques may be used for in vivo excision of pBluescript plasmids with the cDNA insert from the λZap phages. Cut out the plaque from the plate with a scalpel and place in an Eppendorf tube. Add 300 µL of SM buffer, crush the agar/agarose with a glass rod and shake the tube for 15 min.
7. Perform self-excising of the pBluescript plasmid from the phage DNA as described by Stratagene by mixing 100 µL of the λZap phages, 200 µL of XL-1-Blue bacteria, and 1 µL of the ExAssist helper phage (Stratagene, La Jolla, CA) in a 50-mL tube, incubated at 37°C for 15 min. Add 2X YT medium and grow the mixture in a shaking incubator at 300 rpm for 2–2.5 h. Heat the content to 70°C for 20 min, centrifuge, and save the supernatant containing the pBluescript plasmid, packaged as filamentous phage particles, at 4°C. Mix 1 µL of the phage supernatant with 200 µL of SOLR bacteria (Stratagene) incubated at 37°C for 15 min and finally plate on Petri dishes with agar containing 100 µg/mL of ampicillin. After overnight growth, the colonies may be transferred to Hybond N filters in a similar way to that described for plaque transfer (**Subheading 3.1.**, **step 2**). After the colony transfer, the Petri dishes should be grown again at 37°C for some hours until the colonies are again visible by the naked eye. Plasmid preparations may then be prepared by growing positive colonies in larger scale by standard procedures.

4. Notes

1. The TMAC is hygroscopic, so the water is added directly to the bottles to dissolve the powder. The solution is then filtered to remove undissolved material, and water is finally added to bring the concentration to 5 M by taking the weight as indicated on the bottles. We do not find it necessary to determine the actual concentration by measuring the refractive index of the solution as used in some protocols.
2. The yeast RNA blocking solution is made by dissolving 10 g RNA in 50 mL 0.3 M sodium acetate (pH 6.0). This solution is extracted twice with 0.8 vol phenol, and twice with 0.8 vol of chloroform (note that the lower phase is the RNA containing phase). Precipitation is performed by adding 2 vol of ethanol. The solution is centrifuged, the pellet dried, redissolved in 50 mL 0.3 M sodium acetate (pH 6.0), and 2 vol ethanol are added for reprecipitation. After centrifugation, the pellet is washed several times with ethanol/water (80/20, v/v), dried, and dissolved in water at a concentration of 100 mg/mL as measured by the UV light absorbance at 260 nm.
3. The first choice is to use a cDNA library made with mRNA purified from the tissue or cell line from which the amino acid sequencing results have been obtained. However, the same protein may be expressed in other cell lines or tissues as well. The presence of a protein in a cell line or a tissue may be analyzed by two-dimensional gel electrophoresis, separating the proteins in the first dimension according to pI and in the second dimension according to molecular mass *(10)*. Two proteins from two different sources that migrate to the same position in the gel may indicate that they represent the same protein or are closely related proteins. This can be further confirmed by reaction with a specific antibody against the protein (immunoblotting) or cutting of the Coomassie brilliant blue-stained spot from the dried two-dimensional gel for amino acid microsequencing *(1)*.
4. Bacterial plates, especially those made with top agar, are placed at 4°C in order to solidify the agar further. Without this step, a part of the agar with plaques may stick to the nylon filter with the result that the plaques in that area are lost.
5. The agar plates may be stored for a few days at 4°C. For long-term storage (years) the plates can be put directly at –70°C. The filters can then be stored at –20°C and reused for screening with other oligonucleotides later or immediately if the probe is first stripped off the filters, e.g., by adding water which is boiled, slightly cooled, and brought to 0.1% SDS (blotting and hybridization protocols for Hybond membranes, Amersham Pharmacia Biotech). The filters should never dry out.
6. The oligonucleotide used to screen the library should, in principle, be as long as possible and as little degenerated as possible. We have successfully used oligonucleotides with lengths from 15–20 bases and up to 23 for inosine-containing oligonucleotides *(8)*. In order to lower the degeneracy of the oligonucleotide, it may be advantageous to terminate the oligonucleotide at codon position 2, as the third codon position usually is ambiguous. If possible, avoid oligonucleotides reverse translated from peptides with Leu, Arg, or Ser, as the codons for these amino acids are degenerated at two or all three positions. We have successfully used oligonucleotides with a degeneracy up to 512 *(8)*. If all oligonucleotides are equally represented in such a mixture, the concentration of each different oligonucleotide will be <0.2% of the total concentration. If one or a few nucleotides are unintentionally preferred at the ambiguous positions during the synthesis, the mixture may become 'biased' at some of the ambiguous positions so that the correct oligonucleotide may be underrepresented in the mixture. This means that the concentration of the correct oligonucleotide may be even lower than 0.2% of total. It is important to keep in mind that the quality of the oligonucleotide preparation is crucial for a successful result. We have found that different batches of oligonucleotides that should recognize the same cDNA sequence

may, in fact, hybridize to the plaques with very different strengths and some of them very poorly indeed. Thus, if one batch of oligonucleotides works unsuccessfully, it may be worth trying one or two other batches for screening, perhaps from another supplier. The problem may also be partly overcome by keeping the degeneracy of the oligonucleotide very low, e.g., up to 16 or so by inserting inosines at highly degenerated positions *(11–16)*. The effective length of the oligonucleotide will be shorter, so it requires that the amino acid sequence is sufficiently long (*see also* **Note 9**).

7. The equilibrium constant for the labeling reaction:

$$[\gamma\text{-}^{32}\text{P}]\text{ATP} + 5'\text{-oligonucleotide} \rightleftarrows \text{ADP} + [^{32}\text{P}]\text{oligonucleotide}$$

is about unity *(17)*. The specific activity of the labeled oligonucleotide can be increased by increasing the concentration of $[\gamma\text{-}^{32}\text{P}]\text{ATP}$, whereas the amount of transferred $[^{32}\text{P}]$ to the oligonucleotide can be increased by increasing the oligonucleotide concentration. The unlabeled oligonucleotide will hybridize as strong as the labeled one, and it is therefore necessary to purify the labeled oligonucleotide, e.g., on a denaturing polyacrylamide gel where the labeled oligonucleotide that contains a 5'-phosphate will move faster than the unlabeled oligonucleotide.

8. The prehybridization buffer is used to block the filters with RNA in order to get a low background for the following autoradiography, which may be performed for several days or weeks. The blockage does not work sufficiently in TMAC and is therefore performed in SSC.

9. The hybridization temperature of the oligonucleotide is determined as follows. First the irreversible melting temperature is calculated as described by Jacobs et al. *(7)*:

$$T_i = -682 \times (L^{-1}) + 97°C$$

where L is the number of nucleotides in the oligonucleotide and T_i is the irreversible melting temperature. The hybridization temperature, T_h, is usually about 10°C below the T_i. Thus, for a 15-mer oligonucleotide, the T_i is calculated to 52°C and T_h is accordingly 42°C. If the oligonucleotide used contains inosines, these should be regarded as neutral or slightly destabilizing, e.g., a 20-mer oligonucleotide with three inosines is regarded as a 17-mer or shorter when calculating T_h.

10. The oligonucleotide probe may be reused for the purification of plaques if performed within 1 or 2 wk. The half life of the probe is 14 d.

Acknowledgment

This work was supported by the Danish Medical Research Council.

References

1. Vandekerckhove, J. and Rasmussen, H. H. (1994) Internal amino acid sequencing of proteins recovered from 1D or 2D-gels, in *Cell Biology: A Laboratory Handbook* (Celis, J. E., ed.), Academic Press, San Diego, pp. 359–368.
2. Singer Sam, J., Simmer, R. L., Keith, D. H., Shively, L., Teplitz, M., Itakura, K., Gartler, S. M., and Riggs, A. D. (1983) Isolation of a cDNA clone for human X-linked 3-phosphoglycerate kinase by use of a mixture of synthetic oligodeoxyribonucleotides as a detection probe. *Proc. Natl. Acad. Sci. USA* **80,** 802–806.
3. Lin, F. K., Suggs, S., Lin, C. H., Browne, J. K., Smalling, R., Egrie, J. C., Chen, K. K., Fox, G. M., Martin, F., Stabinsky, Z., Badrawi, S. M., Lai, P.-H., and Goldwasser, E. (1985) Cloning and expression of the human erythropoietin gene. *Proc. Natl. Acad. Sci. USA* **82,** 7580–7584.

4. Honoré, B., Rasmussen, H. H., Celis, A., Leffers, H., Madsen, P., and Celis, J. E. (1994) The molecular chaperones HSP28, GRP78, endoplasmin, and calnexin exhibit strikingly different levels in quiescent keratinocytes as compared to their proliferating normal and transformed counterparts: cDNA cloning and expression of calnexin. *Electrophoresis* **15,** 482–490.
5. Melchior, W. B., Jr. and Von Hippel, P. H. (1973) Alteration of the relative stability of dA-dT and dG-dC base pairs in DNA. *Proc. Natl. Acad. Sci. USA* **70,** 298–302.
6. Wood, W. I., Gitschier, J., Lasky, L. A., and Lawn, R. M. (1985) Base composition-independent hybridization in tetramethylammonium chloride: a method for oligonucleotide screening of highly complex gene libraries. *Proc. Natl. Acad. Sci. USA* **82,** 1585–1588.
7. Jacobs, K. A., Rudersdorf, R., Neill, S. D., Dougherty, J. P., Brown, E. L., and Fritsch, E. F. (1988) The thermal stability of oligonucleotide duplexes is sequence independent in tetraalkylammonium salt solutions: application to identifying recombinant DNA clones. *Nucleic Acids Res.* **16,** 4637–4650.
8. Honoré, B., Madsen, P., and Leffers, H. (1993) The tetramethylammonium chloride method for screening of cDNA libraries using highly degenerate oligonucleotides obtained by backtranslation of amino-acid sequences. *J. Biochem. Biophys. Methods* **27,** 39–48.
9. Riccelli, P. V. and Benight, A. S. (1993) Tetramethylammonium does not universally neutralize sequence dependent DNA stability. *Nucleic Acids Res.* **21,** 3785–3788.
10. O'Farrell, P. H. (1975) High-resolution two dimensional gel electrophoresis of proteins. *J. Biol. Chem.* **250,** 4007–4021.
11. Honoré, B., Madsen, P., Rasmussen, H. H., Vandekerckhove, J., Celis, J. E., and Leffers, H. (1993) Cloning and expression of a cDNA covering the complete coding region of the P32 subunit of human pre-mRNA splicing factor SF2. *Gene* **134,** 283–287.
12. Honoré, B., Leffers, H., Madsen, P., and Celis, J. E. (1993) Interferon-γ up-regulates a unique set of proteins in human keratinocytes. Molecular cloning and expression of the cDNA encoding the RGD-sequence-containing protein IGUP I-5111. *Eur. J. Biochem.* **218,** 421–430.
13. Honoré, B., Rasmussen, H. H., Vorum, H., Dejgaard, K., Liu, X., Gromov, P., Madsen, P., Gesser, B., Tommerup, N., and Celis, J. E. (1995) Heterogeneous nuclear ribonucleoproteins H, H', and F are members of a ubiquitously expressed subfamily of related but distinct proteins encoded by genes mapping to different chromosomes. *J. Biol. Chem.* **270,** 28,780–28,789.
14. Madsen, P., Rasmussen, H. H., Flint, T., Gromov, P., Kruse, T. A., Honoré, B., Vorum, H., and Celis, J. E. (1995) Cloning, expression, and chromosome mapping of human galectin-7. *J. Biol. Chem.* **270,** 5823–5829.
15. Jacobsen, L., Madsen, P., Moestrup, S. K., Lund, A. H., Tommerup, N., Nykjær, A., Sottrup-Jensen, L., Gliemann, J., and Petersen, C. M. (1996) Molecular characterization of a novel human hybrid-type receptor that binds the alpha2-macroglobulin receptor-associated protein. *J. Biol. Chem.* **271,** 31,379–31,383.
16. Petersen, C. M., Nielsen, M. S., Nykjær, A., Jacobsen, L., Tommerup, N., Rasmussen, H. H., Røigaard, H., Gliemann, J., Madsen, P., and Moestrup, S. K. (1997) Molecular identification of a novel candidate sorting receptor purified from human brain by receptor-associated protein affinity chromatography. *J. Biol. Chem.* **272,** 3599–3605.
17. Sambrook, J., Fritsch, E. F., and Maniatis, T. (eds.) (1989) *Molecular Cloning: A Laboratory Manual.* Cold Spring Harbor Laboratory Press, Cold Spring Harbor, New York.

54

Screening Recombinant Libraries by Polymerase Chain Reaction

Michael W. King

1. Introduction

The complexity of the genome of a particular organism or the relative abundance of a particular mRNA, within the cell type from which a cDNA library was constructed, affects the ability with which one can isolate a gene or cDNA clone of interest. With respect to genomic libraries, the number of clones needed to be screened to isolate a single-copy sequence is a function of the complexity of the genome and the average size of the cloned fragments in the library (1). In the case of cDNA libraries, the frequency of a given clone of interest depends on the abundance of the messenger RNA. Highly abundant messages can represent 10% or more of total mRNA, whereas very rare messages can be as low as one in 10^6. In addition, the representation of some sequences in a cDNA library, particularly the 5' ends of large mRNAs, will be less than expected due to the technical difficulties in converting the mRNA into full-length cDNA copies. In some cases, a particular sequence of interest can be depleted or lost at various steps of screening due to its inefficiency to be replicated relative to other clones in the library.

Prior to the advent of the polymerase chain reaction (PCR), the principal technique for screening bacteriophage lambda-based libraries involved screening nitrocellulose filters replicas with radioactively labeled probes. Recently, more highly sensitive methods for screening that utilize the PCR have been described (2,3). This chapter describes a PCR-based sib-selection method for the isolation of clones from recombinant DNA libraries prepared in lambda-based vectors. The technique uses no radioisotopes and can be completed in as few as 7 d. The technique is amenable to the use of highly specific PCR primers as well as degenerate primers designed to isolate families of related clones. In this method, a cDNA or genomic library is initially plated (at a lower density than for filter hybridization) and the phage from each plate is soaked from the plates in SM buffer (4) to generate the starting aliquots for PCR. Positive aliquots from the primary plating are identified and replated at lower and lower densities to generate subaliquots for secondary and tertiary screens until a positive clone is identified by PCR of phage soaked from a single plaque.

2. Materials

1. Recombinant DNA library.
2. Oligonucleotide primers for PCR (see **Subheading 3.1.**).
3. Bacterial culture media, agar plates, and top agar media (see Chapter 45).
4. SM Buffer (see *ref. 4*): 100 mM NaCl, 8 mM MgSO$_4$-7H$_2$O, 50 mM Tris-HCl, pH 7.5.
5. Reagents for PCR (see Chapter 5).
6. Reagents for agarose gel electrophoresis (see Chapter 13).
7. CHCl$_3$.

3. Method

3.1. Design of Primers

1. In designing primers for PCR, one needs to consider several important factors such as making the primers with a near 50% GC content, a high degree of specificity with respect to nucleotide sequences and the absence of primer self complementarity. Primers that are from 18-21 nt long are optimal for this (and most) PCR-based techniques because they allow high annealing temperatures that result in greater specificity in the reaction. In most cases, where possible, the primers should be designed to be maximally useful in a PCR with 2 mM MgCl$_2$ and an annealing temperature above 62°C (see Chapter 73).
2. With the use of degenerate primer pairs, as for the isolation of families of related cDNAs or genes, it is extremely important that the 3'-nucleotide position of each primer set be nondegenerate in order to prevent an increase in nonspecific templating. In most cases, it is also optimal if the annealing temperature used with degenerate primers not be below 55°C, although examples are available where successful isolation of related cDNAs has been carried out using degenerate primer pairs with annealing temperatures of 52°C.
3. For increased specificity, it is possible to add a nested primer that resides within the sequences to be amplified with the primary primer pair. This nested primer can then be used in a second PCR, in combination with one of the primary primers, to test positive PCRs for the presence of correct internal sequences.

3.2. Basics of Phage Growth

1. Each step in this PCR-based screening protocol begins with the overnight culture of the appropriate host *Escherichia coli* (see **Table 1**). The cells should be grown at 37°C with agitation in NZCYM media supplemented with 1 mM MgSO$_4$ and 0.3% maltose *(4)* to ensure optimal phage infection and growth. The minimum volume of the overnight culture of cells depends on the number of plates that will be used. The standard volume of cells is 100 µL/100 mm plate.
2. The next morning, an aliquot of top agar is melted and held at 45°C. The appropriate number of agar plates are prewarmed at 37°C for 15–20 min.
3. The library or plate lysate aliquots are diluted appropriately in SM buffer, then added to an aliquot of the fresh overnight host *E. coli* that corresponds to the total volume needed for plating 100 µL/100 mm plate. Adding the phage to the entire volume of cells instead of to individual aliquots ensures an equal distribution on each plate. The cells and phage are then incubated at 37°C for 10 min to allow the infection cycle to initiate.
4. The infected cells are then separated into individual 100 µL aliquots for plating. The plating is carried out by adding 3 mL of the melted 45°C top agar to each tube, pouring the solution onto individual plates and ensuring even spreading of the top agar prior to it resolidifying.
5. The plates are then inverted and incubated at 37°C for 6–8 h to allow plaques to form. *See* **Subheadings 3.5.–3.8.** for the screening protocol.

Table 1
Outline of Screening Protocol

Day	Activity	Identity
1	Start culture of host *E. coli*	
2	Plate 5–10: 100 mm plates (approx 4000–5000 PFU/plate), grow 6–8 h, soak phage in SM overnight at 4°C	Primary screen
3	Process phage lysates, PCR screen	Primary screen
3	Start culture of host *E. coli*	
4	Plate 5: 100 mm plates, of 1 or all 1°C positives (approx 500 PFU/plate), grow 6–8 h, soak phage in SM overnight at 4°C	Secondary screen
5	Process phage lysates, PCR screen	Secondary screen
5	Start culture of host *E. coli*	
6	Plate 5: 100 mm plates (approx 100 PFU/plate), grow 6–8 h, soak phage in SM overnight at 4°C, is possible to pick single plaques at this stage	Tertiary screen
7	Process phage lysates, PCR screen	Tertiary screen
7	Start culture of host *E. coli*	
8	Plate 1: 100 mm plate (approx 50 PFU/plate), grow 6–8 h, pick single plaques into SM and elute overnight at 4°C	Final screen
9	Screen single-plaque lysates by PCR	Final screen

PFU = plaque-forming units.

3.3. Preparation for Screening

1. It is critically important that, prior to screening any library with this technique, the library be tested for the presence of clones that contain sequences that will amplify with a given primer pair.
2. Test a 1-µL aliquot (undiluted) of the library to be screened using the standard PCR protocol described in **Subheading 3.4.** As described in the Introduction, an amplified library will have approximately 10^7–10^8 phage in a 1-µL aliquot. Therefore, a 1-µL aliquot will contain 100- to 1000-fold more phage than is statistically necessary to screen to find a clone of a given sequence. If the primers are unable to amplify the correct fragment from this amount of phage, then the library is either devoid of clones or the primer pair is not functional as expected.

3.4. Standard PCR

1. All PCRs are performed in a volume of 25 µL containing 1X *Taq* buffer, 2 m*M* $MgCl_2$ (or a concentration appropriate for a given primer pair), 30–35 pmol of each primer (200 ng; assuming primers of 18–21 nt) and 200 µ*M* deoxynucleotide triphosphates (dNTPs).
2. To ensure that the phage particles in the plate lysates are disrupted, the PCRs are "hot started." This is accomplished by an initial denaturation at 95°C for 10 min followed by holding the reaction at 80°C for 30–60 min.
3. The following cycle profiles are used for specific primers (*see* **Subheading 4.** for degenerate primers): 30 cycles: 95°C, 1–2 min (denaturation), n°C, 0.5–2 min (annealing temperature defined by primer sequences), 72°C, n min (extension time depends on product length).

3.5. Primary Screen

1. For the primary screen, it is usually necessary to plate 10 aliquots of the library at a density in the range of 4000–5000 PFU/plate. This density will nearly lyse the entire bacterial lawn. If the titer of the library is unknown, most amplified libraries can be plated using 1 µL of a 10^3 dilution on the 10 plates (i.e., the equivalent of 0.1 µL/plate). In some cases it is possible to screen as few as 5 plates.
2. Incubate the plates upside down at 37°C. Stop the incubation when the plaques begin to merge with one another. This usually takes 6–8 h at 37°C.
3. Overlay the plate with 4 mL of SM buffer and let stand at 4°C overnight. It is possible to incubate the plates with SM buffer for 2 h at 37°C or room temperature for 4–5 h, however, the PCRs are sometimes smeary due to bacterial growth in the SM buffer. Also, the titer of the resultant lysates can be 100- to 1000-fold lower than the lysates prepared by 4°C overnight incubation. The latter fact is important to remember for subsequent screens.
4. Collect the SM buffer as a separate aliquot from each of the plates and remove agar and bacterial debris by centrifugation at 3500 g for 10 min in a JA17 rotor (Beckman high-speed centrifuge, Palo Alto, CA) or an SS34 rotor (Sorval centrifuge [Dupont Instruments, Newtown, CT]).
5. Save 1 mL from each aliquot. Add $CHCl_3$ to 0.3% to prevent bacterial growth in these aliquots and allow for their longer-term storage at 4°C.
6. Use a 1-µL aliquot for the PCR assay (*see* **Subheading 3.4.**).
7. Analyze a 10-µL aliquot of each PCR on an agarose gel to determine which aliquot(s) have amplified the target of interest.
8. In most screens there should be at least 1 plate lysate exhibiting a positive signal by PCR.
9. Provided the library was tested for the presence of DNA that can be amplified with the primer pair being used (*see* **Subheading 3.3.**) it will be possible to find primary lysate that contain positive signals. It may be necessary to continue to screen more primary plates until a positive is found.

3.6. Secondary Screen

1. It is not necessary to titer the primary plate lysates, although this can be done if accurate plaque numbers in the secondary screen are desired. However, based on the fact that the primary plates should have experienced near complete lysis of the bacterial lawn, the secondary screens (using five 100-mm plates) are plated using 1 µL of a 10^3 dilution (i.e., 0.2 µL/plate) of each positive primary lysate. Plating at this density (approx 500 plaques/plate) is to ensure that the secondary lysates are dense enough to ensure enrichment of the clones of interest.
2. Allow plaques to grow as for the primary screen.
3. Prepare the phage lysates as for the primary screen using 3 mL of SM buffer.
4. Process the plate lysates and save a 1 mL aliquot of each with 0.3% $CHCl_3$ for long-term storage.
5. Screen a 1-µL aliquot of each in a 25-µL PCR as for the primary screen.
6. Analyze a 10-µL aliquot of each PCR by agarose gel electrophoresis (*see* Chapter 13).

3.7. Tertiary Screen

1. The tertiary screen is the last plate lysate screen. However, it is possible, in some instances, to proceed directly to the screening of single plaques from the secondary screen.
2. Plate 5 plates (100 mm plates) using 1 µL of a 10^4 dilution of a positive secondary lysate (i.e., 0.2 µL/plate). In some cases it may be necessary to use a 10^5 dilution, depending on the density of plaques in the secondary screen.

3. Allow plaques to grow as for the primary screen.
4. Prepare the phage lysates as for the primary screen using 3 mL of SM buffer.
5. Process the plate lysates and save a 1-mL aliquot of each with 0.3% $CHCl_3$ for long-term storage.
6. Screen a 1-µL aliquot of each in a 25-µL PCR as for the primary screen.
7. Analyze a 10-µL aliquot of each PCR by agarose gel electrophoresis (*see* Chapter 13).

3.8. Single Plaque Screen

1. Plate a single 100-mm plate from a tertiary (or a secondary) positive using 1 µL of 10^3 or 10^4 dilution.
2. The fold dilution used for the single plaque screening plate is less than for a tertiary screen and the amount of the dilution plated (1 µL) is more. This is because the density of plaques in the tertiary screen should have been in the range of less than 250/plate such that the lysates from those plates will have a low titer.
3. Pick individual plaques into 100 µL of SM buffer using sterile glass Pasteur pipets to "scoop" the plaques out of the top agar.
4. Elute the phage particle from the plaque overnight at 4°C.
5. At this step it is best to screen a 3-µL aliquot in the standard 25 µL PCR. Volumes less than 3 µL can be used, but they have a tendency to give variable amplification.

4. Notes

1. The annealing temperature (T_m) used during the PCR, for any given primer, is determined from the base composition of the primer. To calculate the annealing temperature use the following formula: $T_m = 2(A + T) + 4(G + C)$. For example, for a 21-mer primer with 12 G + C and 9 A + T the $T_m = 2(9) + 4(12) = 66°C$. For the majority of PCRs it is optimal to use an annealing temperature that is 2–4°C below the calculated T_m. However, increased specificity is obtained by annealing at the T_m.
2. The elongation time used in the PCR is determined by the length of the resultant product. Given the rate of the majority of DNA polymerases at approx 1000 b/s it would, in theory, be possible to use extremely short elongation times. However, in practice it has been observed that an elongation time of approx 1 min/1000 b is optimal.
3. To analyze PCR products by agarose gel electrophoresis requires different percentages of agarose (dependent on product size) in order to obtain good resolution of products. In general, use 1.5–2% gels for products of less than 500 bp, 1–1.5% for products of 500–1000 bp and 0.7–1% for products greater than 1000 bp.
4. Following the primary screen, each positive plate lysate will most probably represent a different type of cDNA or gene clone unless degenerate primers were utilized. Therefore, to maximize the possibility of obtaining full-length cDNA clones or overlapping genomic clones each primary positive should be carried through to the secondary stage. However, to reduce the screening "load," one single primary positive at a time can be carried through to single-plaque isolation.
5. When specific primers are used, each positive secondary lysate represents the same single type of clone that was present in the primary lysate. Therefore, only one of the positive secondary lysates is carried through to the tertiary screen. Also, only a single tertiary positive is carried through to the single-plaque screen.
6. When degenerate primer pairs are used, the complexity of possible clone types in any given primary positive can be large. For this reason it is necessary to plate at least 10 plates for the secondary screens. Because of this complexity of clones in a primary positive, it is best to carry only one primary positive at a time through to single-plaque isolations.

In addition, since each of the secondary positives will likely represent different types of clones, each of them needs to be screened in the tertiary screen. The latter fact is also likely in many cases at the level of the tertiary (and beyond) screens.

7. The use of degenerate primers in this screening technique will require at least four and possibly as many as six rounds of plate lysate screening prior to the screening of single plaques.

8. The single-plaque lysate is used to prepare phage DNA as well as a permanent stock of the clone by small-scale liquid lysis. Start an overnight culture of the appropriate host *E. coli* in NZCYM plus maltose and MgSO$_4$ (*see* **Subheading 3.2.**). The next morning add 25 µL of the 100-µL of SM (into which the positive plaque was eluted) to 100 µL of overnight cells and 100 µL NZCYM with maltose and MgSO$_4$. Incubate with agitation at 37°C for 10 min. Transfer this culture to 50 mL of NZCYM without maltose or MgSO$_4$. Incubate with agitation at 37°C until the cells in the culture begin to lyse. This takes approx 6 h and is visible as debris in the normally silky appearance of the growing *E. coli*. At this time add CHCl$_3$ to 0.5% and incubate an additional 10 min to accelerate the cell lysis as well as to prevent further growth of the cells.

   Centrifuge the cells and debris at 5800 g for 10 min. Save an aliquot of the supernatant as a stock of the clone, either at 4°C or by adding dimethylsulfoxide (DMSO) to 7% and storing at –80°C. To the remainder of the phage supernatant add RNase A and Dnase I to 1.5 µg/mL and incubate at 37°C for 30 min. Precipitate the phage particles by addition of solid polyethylene glycol (PEG) 6000 to 10% (w/v) and solid NaCl to 0.5 M. Place at 4°C overnight. Collect the precipitate by centrifugation at 11,200 g for 25 min. Resuspend the precipitate in 0.5–1 mL of TE buffer *(4)*. Add proteinase K to 150 µg/mL and incubate at 45°C for 45 min. Extract the released phage DNA with an equal volume of phenol then phenol/CHCl$_3$ and again with CHCl$_3$. Precipitate the DNA by addition of 0.1 vol of 2 M ammonium acetate pH 5 and 2 vol ethanol.

   The DNA should form a stringy precipitate immediately. It is best to remove the precipitating DNA by collecting it on a swirling glass rod. This reduces RNA and protein contamination that may affect restriction enzyme digestion. Centrifuge the precipitate at top speed in a microfuge, remove any supernatant, and resuspend the pellet in 50–100 µL of Tris-EDTA-RNase A buffer *(4)*. Use from 3–10 µL for restriction enzyme digestion.

9. The single-plaque lysate is used for the rescue of phagemid DNA, containing the cDNA clone, from λZap® (Stratagene, Inc., La Jolla, CA) or λZipLox® (Life Technologies, Inc., Gaithersburg, MD) if the cDNA library was constructed with either of these vectors. The protocol for phagemid rescue from λZap is described. Start a culture of XL1-Blue® (Stratagene) overnight in NZCYM plus maltose and MgSO$_2$. Also, start a culture of SOLR® in Luria-Bertani (LB) media *(4)*. The next morning dilute 40 µL of the overnight XL1-Blue cells into 1 mL of NZCYM plus maltose and MgSO$_4$. Incubate with agitation at 37°C for 60 min. Transfer 200 µL of these cells to a new tube, add 25 µL of the 100-µL SM (into which the positive plaque was eluted) and 1 µL of ExAssist® helper phage. Incubate with agitation at 37°C for 15 min. Add 3 mL of LB media *(4)* and incubate with agitation at 37°C for 2.5 h. Centrifuge the solution at 700g (2000 rpm) for 15 min. Transfer the supernatant to a new tube and heat at 70°C for 15 min. Centrifuge at 550 g for 15 min. Save the supernatant in a sterile tube, as this is a stock of the rescued single-stranded phagemid. This solution can be stored at 4°C for up to 2 mo. To obtain colonies with the double-stranded phagemid, add 20–50 µL of the phagemid stock solution to 100 µL of the fresh overnight SOLR cells. Incubate with agitation at 37°C for 15 min. Spread 10–50 µL of the phagemid-inoculated SOLR cells onto a single LB plus ampicillin plate and incubate overnight at 37°C. The double-stranded phagemid DNA can then be isolated from colonies by standard miniprep techniques *(4)*.

References

1. Ausubel, F. M., Brent, R., Kingston, R. E., Moore, D. D., Smith, J. A., Seidman, J. G., and Struhl, K. (eds.) (1987) *Current Protocols in Molecular Biology*. Wiley, New York.
2. Amaravadi, L. and King, M. W. (1994) A rapid and efficient, non-radioactive method for screening recombinant DNA libraries. *BioTechniques* **16,** 98–103.
3. Isola, N. R., Harn, H. J., and Cooper, D. L. (1991) Screening recombinant DNA libraries: a rapid and efficient method for isolating cDNA clones utilizing the PCR. *BioTechniques* **11,** 580–582.
4. Maniatis, T., Fritsch, E. F., and Sambrook, J. (eds.) (1989) *Molecular Cloning: A Laboratory Manual*. Cold Spring Harbor Laboratory Press, Cold Spring Harbor, NY, p. 443.

55

Construction and Screening of Cosmid Libraries

Jens Hanke and Jörg D. Hoheisel

1. Introduction

Cosmid libraries still are an important tool for cloning large genomic regions. Although other *Escherichia coli*-based vector systems like bacteriophage P1, bacterial artificial chromosomes (BACs), and P1-based artificial chromosomes (PACs) allow the cloning of longer inserts, they are more difficult to handle *(1)*. Therefore, even if they are frequently used for the initial generation of libraries from large genomes, cosmid vectors are often still required to generate libraries with smaller, more manageable pieces or libraries from limited DNA amounts, such as bands isolated from pulsed-field gels *(2)*.

In their simplest form cosmid vectors are plasmids with an additional *cos* site, enabling in vitro packaging of DNA in phage particles. This packaging reaction is highly selective with regard to the size of the cloned DNA molecules: only total sizes between 38 and 52 kb fit into the phage. One advantage of cosmids is that the cloned DNA can be brought into the host cells very efficiently, whereas a transformation of naked DNA of the same size results in rather low yields. Once within the host cell, a cosmid behaves like a normal plasmid. Therefore, the usual plasmid sequence features like an origin of replication and an antibiotic resistance gene are required *(3)*. Several modifications and improvements have been made to the vector system like the introduction of a second *cos* sites *(4,5)* or the possibility to propagate cosmids not only in bacteria cells but also in eukaryotic cells (shuttle vectors) *(6)*. Meanwhile, various cosmid vectors exist, frequently adapted to special applications. For physical mapping as a preparation to genomic sequencing, the Lawrist series *(7)* was found to be extremely well suited. Thus, the protocols given below refer to this vector but can be adjusted to others without much problem.

An important, and frequently neglected, issue to efficient cloning is the coverage of a library. With a statistical probability of 99%, every fragment of a given DNA will be present at least once in a library of fivefold coverage *(8)*. Although this is generally satisfactory for the identification and isolation of specific DNA pieces by library screening, it is inadequate for the continuous mapping of large genomic areas *(9,10)*. Furthermore, variations in the cloning efficiency should be considered for which there are several reasons. Partial enzymatic digestion, for example, is prone to produce

From: *The Nucleic Acid Protocols Handbook*
Edited by: R. Rapley © Humana Press Inc., Totowa, NJ

representational variation due to site preferences of the enzymes or the local absence of restriction sites in certain regions *(11)*.

2. Materials

2.1. Vector Preparation

1. The cosmid vectors of the Lawrist series have been described by de Jong et al. *(7)*; for further details contact Pieter J. de Jong, Human Genetics Department, Roswell Park Cancer Institute, Elm & Carlton Streets, Buffalo, NY 14263, U.S.A.; fax no. [+1] (716) 845-8449), email pieter@dejong.med.buffalo.edu. The average insert size was found to be 37 kb with DNA from various sources.
2. 10X TAK buffer: 330 mM Tris-acetate (pH 7.9), 100 mM magnesium-acetate, 660 mM potassium-acetate, 5 mM dithiothreitol (DTT), 33% (v/v) 1 M Tris-acetate (pH 7.9), 32.5% (v/v) 2 M potassium-acetate, 10% (v/v) 1 M magnesium-acetate, 5% (v/v) 100 mM DTT, 19.5% (v/v) H_2O; sterilize by filtration and store in aliquots at –20°C.
3. 0.5 M EDTA (pH 8.0): 16.9% (w/w) EDTA [ethylenediaminetetraacetic acid], 1.9% (w/w) NaOH-pellets, 81.2% (w/w) H_2O.
4. 20% SDS: 20% w/v solution of sodium dodecyl sulphate.
5. 10 mg/mL proteinase K.
6. Phenol: Redistilled phenol equilibrated to a final 0.1 M Tris-HCl (pH 8.0). Initially, equilibrate several times with 0.5 M Tris-HCl to ensure pH 8.0 is reached. Phenol is both caustic and toxic and must be handled with care.
7. Chloroform/isoamyl alcohol (24:1).
8. 3 M sodium acetate (pH 5.2).
9. 70% ethanol.

2.2. Partial Digestion of Genomic DNA

1. 50 mM EDTA (pH 8.0).
2. 5 M NaCl: 24.64% (w/w) NaCl, 75.36% (w/w) H_2O.

2.3. Library Construction (Fig. 1)

1. 10X Ligation buffer: 250 mM Tris-HCl (pH 7.5), 1 M NaCl, 70 mM $MgCl_2$, 10 mM spermidine, 10 mM DTT.
2. 100 mM ATP (pH 7.2): 5.71% (w/w) ATP, 84.86% (w/w) H_2O, 9.43% (w/w) 2 M Tris base. Aliquot and store at –20°C.
3. Packaging mix: Gigapack III Gold Packaging Extract (Stratagene, La Jolla, CA).
4. SM buffer: 5.8 g NaCl, 2 g $MgSO_4 \cdot 7 H_2O$, 50 mL 1 M Tris-HCl (pH 7.5), 5 mL gelatine (2%, w/v), add deionized water to 1 L; autoclave.
5. 2YT growth medium: 1.6% (w/w) Bacto-tryptone (Difco, Surrey, UK), 1% (w/w) Bacto-yeast extract (Difco), 0.5% (w/w) NaCl, 96.9% H_2O; autoclave.
6. 20% Maltose (w/v).
7. 1 M $MgSO_4$.
8. *E. coli*, strain DH5α or an equivalent.
9. 2YT/kanamycin agar plates: 1.6% (w/w) Bacto-tryptone (Difco), 1% (w/w) Bacto-yeast extract (Difco), 0.5% (w/w) NaCl, 1.5% (w/w) agar, 95.4% (w/w) deionized water; autoclave. Let the solution cool down to 65°C and supplement with kanamycin to a final concentration of 30 µg/mL.

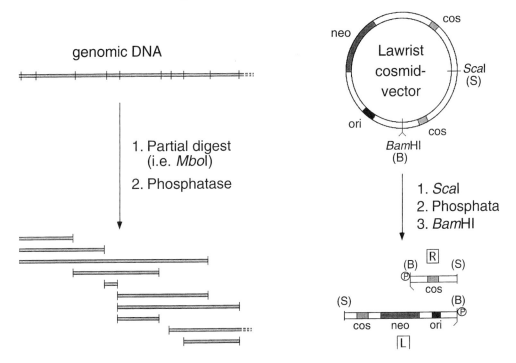

Fig. 1. Cosmid cloning of genomic DNA. Partial digested genomic DNA is ligated between the right and left vector arm. The DNA is subsequently packaged in phage particles and transfected in *E. coli* cells. In the cell, the DNA is circularized via its two *cos* sites. (ori = origin of replication, neo = gene for kanamycin resistance).

10. 10X H.M.F.M. Freezing Medium:
    Solution A: 0.76 g MgSO$_4$ (7 · H$_2$O), 4.50 g Na$_3$-citrate (2 · H$_2$O), 9.00 g (NH$_4$)$_2$SO$_4$, 440.00 g glycerol; add water to 800 mL and autoclave.
    Solution B: 18 g KH$_2$PO$_4$, 47 g K$_2$HPO$_4$; add water to 200 mL and autoclave.
    Mix solutions A and B to make up final solution.
11. 384-Well dishes (four interleaving grids of 96 wells of normal spacing in a standard format dishes; maximal volume per well 70 µL; cat. no.: X5001) can be purchased from Genetix Ltd., UK.

2.4. Filter Production

1. Hybond N$^+$ filter membranes are sold by Amersham Life Sciences (cat. no.: RPN 2222B).
2. 384-Pin replicators (cat. no.: X5050) can be purchased from Genetix Ltd., Christchurch, UK.
3. Filter denaturation buffer (0.5 *M* NaOH, 1.5 *M* NaCl): 1.87% (w/w) NaOH pellets, 8.15% (w/w) NaCl, 89.98% (w/w) H$_2$O.
4. Filter neutralization buffer (1 *M* Tris-HCl [pH 7.6], 1.5 *M* NaCl): 1322.0 g Trizma hydrochloride (Sigma, St. Louis, MO), 194.0 g Trizma base (Sigma), 876.6 g NaCl, 9000 g H$_2$O.
5. Filter processing buffer (50 m*M* Tris-HCl [pH 8.5], 50 m*M* EDTA, 100 m*M* NaCl, 1% (v/v) sodium sarkosyl, 0.25 mg/mL proteinase K or pronase E): 186.1 g EDTA, 58.4 g NaCl, 315 g (300 mL) 30% Na-sarkosyl (Serva, Heidelberg, Germany), 33.3 g Trizma base (Sigma), 35.3 g Trizma hydrochloride (Sigma), 9620 g H$_2$O; add 0.5 g pronase E (Serva) or proteinase K to 1800 mL buffer.

2.5. Library Screening

2.5.1. Radioactive Hybridization

1. Hybridization buffer (0.5 M sodium phosphate [pH 7.2], 7% SDS, 1 mM EDTA): 6.95% (w/w) SDS, 0.20% (w/w) 0.5 M EDTA, 49.75% (w/w) 1 M Na-phosphate (pH 7.2), 43.10% (w/w) H_2O.
2. Church buffer (0.5 M sodium phosphate [pH 7.5], 7% SDS, 1 mM EDTA): 6.95% (w/w) SDS, 0.2% (w/w) EDTA (0.5 M), 49.75% 1 M sodium phosphate (pH 7.5), 43.1% H_2O.
3. Hybridization/prehybridization buffer: Church buffer supplemented with 100 µg/mL tRNA or sonicated salmon sperm DNA.
4. Random primers: 45 A_{260} U/mL pd(N)$_6$ (Pharmacia, Uppsala, Sweden, no. 27-2166-01), 1 mM Tris-HCl (pH 8.0), 1 mM EDTA.
5. TM: 250 mM Tris-HCl, 25 mM $MgCl_2$, 50 mM β-mercaptoethanol.
6. LS: 5 µL random primers, 175 µL TM, 175 µL 1 M HEPES (pH 6.6).
7. 10 mg/mL Bovine serum albumin (BSA).
8. dAGT-Mix: Solution of dATP, dGTP, and dTTP (500 mM each).
9. [$α^{32}P$]-dCTP (10 µCi/µL, i.e., Amersham, Little Chalfont, UK).
10. 10 mg/mL tRNA or sonicated salmon sperm DNA (size should be around 200–500 bp).
11. Isopropanol (2-propanol).
12. 1 µg/µL Sonicated vector DNA (fragments should be around 200–500 bp).
13. 1 M Na_3PO_4 (pH 7.2).
14. Washing buffer: 40 mM Sodium phosphate (pH 7.2), 0.1% (w/v) SDS.
15. Stripping buffer: 5 mM Sodium phosphate (pH 7.2), 0.1% (w/v) SDS.

2.5.2. Nonradioactive Hybridization

1. Dig-11-dUTP nucleotide-mix: 1 mM dATP, 1 mM dCTP, 1 mM dGTP, 0.65 mM dTTP, 0.35 mM DIG-11-dUTP (Roche Diagnostics, Mannheim, Germany).
2. TN: 100 mM Tris-HCl (pH 7.5), 150 mM NaCl.
3. TNB: 100 mM Tris-HCl (pH 7.5), 150 mM NaCl, 1% blocking agent (Boehringer Mannheim).
4. Anti-DIG-AP-solution (alkaline phosphatase coupled with anti-DIG-antibody, Roche Diagnostics, no. 1093274) : 0.15 U/mL anti-DIG-AP in TNB.
5. TNM: 100 mM Tris-HCl (pH 9.0), 100 mM NaCl, 1 mM $MgCl_2$.
6. DM: 100 mM diethanoleamine (pH 9.0), 1 mM $MgCl_2$.
7. AttoPhos-solution: Roche Diagnostics, no. 1681982.
8. 500 mM Na_3PO_4/5% SDS.

3. Methods

3.1. Vector Preparation

1. Dissolve 100 µg of Lawrist-7 DNA in 340 µL of water, add 40 µL of 10X TAK buffer (*see* **Note 1**) and 20 µL of *Sca*I (10 U/µL, New England Biolabs, Beverly, MA). Digest DNA overnight.
2. Check an aliquot on an agarose gel. If the DNA is not completely digested, add more enzyme and continue the reaction.
3. If the digest is complete (*see* **Note 2**) add 2 µL alkaline phosphatase (1 U/µL, Boehringer Mannheim).
4. Incubate for 1 h at 37°C, add another 2 µL of phosphatase and incubate for additional 30 min at 56°C.

5. Transfer reaction to 70°C for 10 min. Add 10 µL 0.5 M EDTA, 10 µL of 20% SDS, and 5 µL proteinase K (10 mg/mL) and incubate for 30 min at 56°C.
6. Extract with 450 µL of phenol. Separate the phases by centrifugation in a microfuge for 5 min.
7. Take off the lower phenol phase and extract the aqueous phase with an equal volume of chloroform/isoamylalcohol (24:1). Separate the phases again by a short spin. Take off the upper aqueous phase.
8. Precipitate the DNA by the addition of 1/10 volume 3 M NaAc (pH 5.2) and 2 volumes of ice-cold ethanol.
9. Dissolve DNA in 160 µL H$_2$O.
10. Add 20 µL of 10X TAK buffer and 20 µL of *Bam*HI (20 U/µL, New England Biolabs).
11. Digest the DNA for 2 h to overnight at 37°C. Check the digest on a gel. Extract completely digested DNA once with phenol and once with chloroform/isoamylalcohol (24:1).
12. Precipitate the DNA with 20 µL 3 M NaAc (pH 5.2) and 450 µL ice-cold ethanol.
13. Spin at maximum speed for 20 min in a microfuge at 4°C. Wash the pellet with 200 µL 70% ethanol.
14. Dry the DNA and dissolve it in 100 µL H$_2$O. Check concentration by fluorescence and the purity by OD$_{260}$/OD$_{280}$ (*see* **Note 3**).

3.2. Partial Digestion of Genomic DNA

1. Take a small aliquot (about 1 µg) of genomic DNA, digest it with *Mbo*I and check the product on a gel to ensure that the DNA is readily cleavable.
2. Dissolve 35 µg of genomic DNA in 179 µL H$_2$O. Add 20 µL of 10X TAK buffer and 1 µL *Mbo*I (5 U/µL, New England Biolabs) (*see* **Note 4**). Incubate at 37°C. Take 40 µL aliquots after 1 min, 2 min, 4 min, 8 min, and 30 min. Add immediately 40 µL of 50 mM EDTA to each aliquot to stop the reaction. Incubate for 10 min at 68°C to inactivate the enzyme (*see* **Note 5**). Extract each of the aliquots once with phenol and once with chloroform/isoamylalcohol (24:1). Precipitate the DNA by adding 4 µL 5 M NaCl and 160 µL ice-cold ethanol and spin for 20 min at 4°C and 13 rpm (10,000*g*).
3. Dissolve the pellets in 44 µL H$_2$O. Add 5 µL of 10X TAK buffer and 1 µL of alkaline phosphatase (1 U/µL).
4. Incubate for 1 h at 37°C and subsequently for another 30 min at 56°C.
5. Transfer the reaction for 10 min to 70°C. Add 2 µL of 500 mM EDTA, 2 µL of 20% SDS and 2 µL proteinase K (10 mg/mL).
6. Incubate for 30 min at 56°C. Extract once with 60 µL phenol and once with 60 µL chloroform/isoamylalcohol (24:1).
7. Precipitate with 1/25 vol 5 M NaCl and 2 vol ethanol. Dissolve the pellets in 14 µL (final concentration 500 ng/µL).
8. Check 2 µL of each digest on a 0.3% agarose gel and choose the appropriate digests (*see* **Note 6**).

3.3. Library Construction

1. Mix 5 µL of partially digested genomic DNA with 2 µg of Lawrist-7 vector arms (*see* **Subheading 3.1., step 4.**), 2.6 µL 10X ligation buffer (*see* **Note 7**) and 0.9 µL 100 mM ATP. Adjust the volume with H$_2$O to 25 µL.
2. Add 1 µL of T4-DNA-Ligase (1 U/µL, Gibco-BRL). Incubate for more than 8 h at 15°C. As a negative control, also prepare one reaction without insert DNA.
3. The ligation of the chosen partial *Mbo*I digests have to be packaged into phage particles. In our lab, Gigapack III Gold Packaging Extract (Stratagene) is routinely used, but we also obtained good results with others (*see* **Note 8**).

4. Thaw one tube of packaging extract between your fingers and add 4 µL of the ligation. Mix gently by stirring with the pipet tip (but do not introduce air bubbles) and incubate for 1.5–2 h at room temperature (do not exceed 2 h).
5. Proceed with other chosen ligations in the same way. If desired, a positive control with wild-type lambda DNA can be carried out.
6. Add 500 µL MS, mixing gently. Add 20 µL chloroform and mix gently by converting the tubes a few times. Spin briefly in a table top centrifuge and transfer the supernatant to a fresh tube. The packaged DNA can be stored at 4°C for several weeks.
7. For the preparation of host bacteria, mix 50 mL 2YT with 500 µL 1 M MgSO$_4$ and 500 µL 20% maltose. Inoculate with the appropriate bacteria strain (i.e., DH5α). Grow them shaking for 4–6 h at 37°C or overnight at 30°C (*see* **Note 9**).
8. Pellet the cells and resuspend them in 10 mL sterile and ice-cold 10 mM MgSO$_4$ (*see* **Note 10**). Before use, determine the OD$_{600}$ and dilute the cells to an OD$_{600}$ of 1 with 10 mM ice-cold MgSO$_4$.
9. Dilute 20 µL of the phage solution (*see* **Subheading 3.3.**, **step 3**) with 180 µL SM. Add 20 µL of this 1:10 dilution to 180 µL SM to obtain a 1:100 dilution. If desired prepare more dilutions. Mix 100 µL of each phage dilution to 100 µL of bacteria solution (*see* **Subheading 3.3.**, **step 4**).
10. Incubate 15 min at 37°C shaking gently to allow the phage to attach to the cells. Add 1 mL of prewarmed 2YT and continue incubating at 37°C for 1 h.
11. Divide each transfection into two aliquots of 600 µL. Plate them on 2YT/kanamycin-agar plates (*see* **Note 11**) and incubate overnight at 37°C. Count the colonies to determine the library's titer.
12. Repeat the transfection with an amount of the ligation reaction calculated to generate the desired numbers of clones.
13. Plate them on a sufficient number of 2YT/kanamycin-agar plates. Plates with a size of 22 × 22 cm should hold 3000–5000 clones.
14. Various protocols exist for the storage of the library *(3)*. In our lab individual clones are picked in 384-well microtiter dishes (Nunc, Roskilde, Denmark) and stored frozen in 2YT/1 × H.M.F.M Freezing Medium at –70°C. Picking can be done either automatically with robotic picking devices (i.e., BioRobotics, Cambridge, UK) or manually.
15. For manual picking of individual clones into microtiter plates a 12-pin picking wheel (described in **ref. 12**) may be extremely useful.

3.4. Filter Production

1. Cut Whatman 3MM paper (Maidstone, UK) to a size of 23 × 23 cm. Soak it with 2YT and remove the access of liquid by rolling a glass pipet over the paper pieces. Put a labeled 22 × 22 cm Highbond N$^+$ filter (Amersham, UK) onto the paper piece. There should be no air bubbles between the filter and the paper.
2. Use a 384-pin tool to transfer clones from the 384-well microtiter dishes to the filter in an ordered array *(13)*. This can either be done by hand or with robotic devices (i.e., BioRobotics). Transfer the filters to a 2YT/kanamycin-agar plate and let colonies grow overnight at 37°C. Alternatively, normal replica filters *(2)* can also be prepared.
3. For the filter processing, cut the Whatman 3MM to the size of your filters. Soak one of the Whatman pieces with filter denaturation buffer and remove access liquid by rolling a glass pipet over the paper pieces. Put the first filter on the Whatman piece for 4 min at room temperature. Repeat the step once over a steaming waterbath.
4. Put the filter on a piece of Whatman soaked with filter neutralization buffer for more than 5 min at room temperature.

Construction and Screening of Cosmid Libraries

5. Transfer the filter very carefully into a 400-mL prewarmed filter processing buffer and incubate at 37°C for 30 min (*see* **Note 12**).
6. Dry the filters on Whatman 3MM for at least one day. Filters should be completely dry (*see* **Note 13**).
7. Filters can be used for more than 50 hybridizations. If you use the filters frequently (once per week) store them between hybridizations at 65°C in Church buffer supplemented with tRNA or salmon sperm DNA. If so, between experiments the prehybridization step (*see* **Subheading 3.5.1., step 1**) can be omitted. For long-term storage, filters should be washed in stripping buffer and dried completely. In this condition, they can be stored at room temperature for years.

3.5. Library Screening
3.5.1. Radioactive Hybridization

1. Prehybridize the filters for more than 2 h at 65°C with 10 mL hybridization buffer (Church buffer supplemented with 100 μg/mL tRNA or sonicated salmon sperm DNA). Prehybridization and hybridization could be performed either in hot sealed plastic bags (check for leaks), boxes, or in rotating glass tubes.
2. Boil 15 μL of DNA (20–500 ng) for 5 min. Chill immediately on ice. Add 18 μL LS, 1.5 μL BSA (10 mg/mL), 1 μL dAGT mix (500 mM each), 1 μL [α^{32}P]-dCTP (10 μCi/μL) and 1.5 μL Klenow (4 U/μL). Incubate from 1 h to overnight at 37°C (*see* **Note 14**).
3. For the removal of unincorporated nucleotides, precipitate labeled DNA by adding 4 μL H$_2$O, 2 μL tRNA (10 mg/mL), 4 μL 0.5 M EDTA (pH 8.0), 10 μL 3 M NaAc (pH 5.2), and 40 μL isopropanol and freeze for 30 min at –70°C.
4. Spin 15 min in a tabletop centrifuge, remove the supernatant, and dissolve the pellet in 165 μL H$_2$O. Solving DNA could be assisted by heating and vortexing and should be checked with a handheld counter (*see* **Note 15**).
5. (*See* **Note 16**): Add 25 μL sonicated vector DNA (1 μg/μL) and 10 μL tRNA (10 μg/μL). Denature double-stranded DNA by boiling for 5 min. Add 25 μL 1 M Na$_3$PO$_4$ (pH 7.2) and incubate for 2 h at 65°C. This allows the sonicated vector DNA to hybridize with sequences of the probe.
6. Remove the prehybridization buffer from the filters. Mix the probe with 10 mL hybridization buffer and add it to the filters. Activity of the hybridization mix should be around 0.5 M cpm/mL. Hybridize more than 10 h at 65°C.
7. Remove the hybridization mix and rinse the filters once with washing buffer at room temperature.
8. Wash them subsequently for 15 to 30 min in washing buffer at 65°C.
9. Put the filters on Whatman 3MM for a few minutes to remove access liquid (*see* **Note 17**).
10. Expose the filters to X-ray film at –70°C with an intensifying screen (*see* **Note 18**).
11. Filters can be regenerated by heating them twice for 30 min in stripping buffer of 90–100°C.

3.5.2. Nonradioactive Hybridization

1. For probe synthesis, prehybridization, hybridization, and filter washing use the same protocol as for radioactive hybridization (*see* **Subheading 3.4., step 1**). The only difference is the nucleotide mix of the labeling reaction, which should contain Dig-11-dUTP in addition to all four normal dideoxy nucleotides.
2. After washing the filters with washing buffer, equilibrate them first in TN and then in TNB (500 mL for three filters).

3. Put filters in 30 mL anti-DIG-AP-solution and incubate for 45 min at room temperature.
4. Wash up to three filters at a time: 5 min in 500 mL TN, twice for 5 min in 500 mL TNM, and 5 min in 500 mL DM.
5. Put the filters on the Whatman 3MM for a few minutes to remove any access of liquid. Put them on a glass plate. Use a sprayer to disperse AttoPhos-solution evenly onto the filters. Immediately cover them airtight in Saran Wrap and incubate for 2–3 h at room temperature.
6. Put filters under a UV lamp of 302–365 nm. Alternatively, filters can also be viewed on a UV plate. Best results are obtained when a 550 nm "cutoff" filter (Kodak, Rochester, NY) is used, which reduces background signals.
7. To regenerate filters, put not more than 5 filters in 1 L of 500 mM Na$_3$PO$_4$/5% SDS overnight at room temperature, rocking gently and subsequently twice in stripping buffer for 30 min at 90°C.

4. Notes

1. TAK is a universal buffer for restriction endonucleases *(14)*. It was chosen because the subsequently used phosphatase also works well in this buffer. Alternatively, other buffers can be used if they work in combination with alkaline phosphatase. Otherwise, a buffer exchange would be necessary.
2. There should be two bands visible (1.4 and 6.8 kb).
3. The concentration and purity of a DNA solution can be determined by measuring OD$_{260}$ and OD$_{280}$ *(3)*. A solution of double-stranded DNA with a concentration of 50 µg/mL has an OD$_{260}$ of 1. Pure DNA has a ratio of OD$_{268}$/OD$_{280}$ of around 2. However, measuring DNA concentration by this method is inaccurate, as frequently the solution is contaminated with DNA or RNA nucleotides. A better method is the quantification of double-stranded DNA by intercalating dye, such as ethidium bromide *(15)*.
4. For limited DNA amounts, a mixture of dam methylase and *Mbo*I should be used as described in **ref.** *16*.
5. Because it is difficult to estimate the efficiency of a partial restriction digest, it is necessary to test several incubation times. You should prepare labeled tubes with 40 µL EDTA in advance to ensure an immediate stop of the reaction.
6. In cosmids, insert fragments from 27–47 kb can be cloned, but it is desirable that the bulk of the fragments used for ligation is bigger. This will greatly reduce the number of chimeric clones, because clones containing more than one inserted fragment will be very likely to be to large to fit in the phage particles. On the other hand, the number of clonable fragments is reduced by this process, but in most experiments the number of clones is not a limiting factor.
7. Ligation buffer should not contain PEG, as it has been shown to inhibit packaging.
8. Packaging components can also be homemade. Protocols can be found in most laboratory manuals (**ref.** *3*).
9. It is important not to grow the bacteria cells past an OD$_{600}$ of 1, as overgrowth increases the number of nonviable cells that would reduce the titer.
10. Cells can be stored at 4°C for up to 48 h.
11. Agar plates should be well dried. Streak out the transfected cells until all the liquid is disappeared.
12. This step should be performed very carefully, as much motion causes the bacteria colonies to be swept away, which will greatly reduce the amount of DNA bound to the filter.
13. Normally the standard procedure is sufficient to attach DNA covalently to the filter. However crosslinking can be improved by UV radiation (120 kJ/cm^2, i.e., Stratalinker, Stratagene, La Jolla, CA: autocrosslink function).

14. The described protocol for DNA random priming (*see* Chapter 20, **ref. *17***) is used in most applications. However, other methods for labeling nucleic acids could be used instead. Nevertheless, it is important to adapt buffer condition and temperature of the hybridization to the length and the GC content of the probe. One advantage of the random priming protocol is its robustness; i.e., instead of 15 µL DNA an excised agarose block with a volumes of up to 100 µL can be used without changing the other reaction components.
15. If competition of vector or repeat sequences is not necessary, precipitation can be omitted. Drying the pellet for too long should be avoided, as this makes dissolving of the DNA more difficult.
16. For the described random priming protocol, DNA of complete clones including the vector sequences can be used. Because this probe portion would cross-hybridize with other clones independent of the insert sequence, it is necessary to block the vector sequence. As mentioned, agarose blocks of excised bands can also be used as templates. If this template DNA does not contain vector sequences, the competition reaction can be omitted.
17. It is important not to dry the filters, because otherwise labeled DNA would be bound irreversibly to the filter and would accumulate with each new hybridization. Nevertheless, any access of washing buffer should be removed, for three reasons: (i) the background is decreased; (ii) because the filters are exposed at −70°C, freezing of the washing buffer will damage the filter material; and (iii) radiation would be quenched by the liquid.
18. Put the filters in X-ray exposition cassettes. To avoid radioactive contaminations of the cassettes, sandwich the filter between two sheets of plastic and put a film on top. Be sure to be able to correlate the orientation of the autoradiography and the filter after film development.

References

1. Monaco, A. P. and Larin, Z. (1994) YACs, BACs, PACs and MACs: artificial chromosomes as research tools. *Trends Biotechnol.* **12(7),** 280–286.
2. Scholler, P., Schwarz, S., and Hoheisel, J. D. (1995) High-resolution cosmid mapping of the left arm of *Saccharomyces cerevisiae* chromosome XII; a first step towards an ordered sequencing approach. *Yeast* **11,** 659–666.
3. Sambrook, J., Fritsch, E. F., and Maniatis, T. (eds.) (1989) *Molecular Cloning: A Laboratory Manual*, 2nd ed. Cold Spring Harbor Laboratory Press, Cold Spring Harbor, NY.
4. Bates, P. F. and Swift, R. A. (1983) Double *cos* site vectors: Simplified cosmid cloning. *Gene* **26,** 137–146.
5. Poustka, A., Rackwitz, H. R., Frischauf, A. M., Hohn, B., and Lehrach, H. (1984) Selective isolation of cosmid clones by homologous recombination in *Escherichia coli*. *Proc. Natl. Acad. Sci. USA* **81,** 4129–4133.
6. Kioussis, D., Wilson, F., Daniels, C., Leveton, C., Taverne, J., and Playfair J. H. L (1987) Expression and rescuing of a cloned human tumor necrosis factor gene using an EBV-based shuttle cosmid vector. *EMBO J.* **6,** 355–361.
7. de Jong, P. J., Chen, C., and Garnes, J. (1989) Application of PCR for the construction of vectors and the isolation of probes, in *Polymerase Chain Reaction* (Erlich, H. A., Gibbs, R. A., and Kazazian, H. H., eds.), Cold Spring Harbor Laboratory Press, Cold Spring Harbor, NY, pp. 205–210.
8. Clarke, L. and Carbon, J. (1976) A colony bank containing synthetic ColE1 hybrid plasmids representative of the entire *E. coli* genome. *Cell* **9,** 91–99.
9. Maier, E., Hoheisel, J. D., McCarthy, L., Mott, R., Grigoriev, A. V., Monaco, A. P., Larin, Z., and Lehrach, H. (1992) Yeast artificial chromosome clones completely spanning the genome of *Schizosaccharomyces pombe*. *Nature Genet.* **1,** 273–277.

10. Hoheisel, J. D. and Lehrach, H. (1993) Use of reference libraries and hybridisation fingerprinting for relational genome analysis. *FEBS Lett.* **325,** 118–122.
11. Hoheisel, J. D., Maier, E., Mott, R., and Lehrach, H. (1995) Integrated genome mapping by hybridisation techniques, in *Analysis of Non-Mammalian Genomes—A Practical Guide* (Birren, B. and Lai, E., eds.), Academic Press, London, pp. 319–346.
12. Larin, Z., Monaco, A. P., Meier-Ewert, S., and Lehrach, H. (1993) Construction and characterisation of yeast artificial chromosome libraries from the mouse genome. *Methods Enzymol.* **255,** 623–637.
13. Lehrach, H., Drmanac, R., Hoheisel, J. D., Larin, Z., Lennon, G., Monaco, A. P., Nizetic, D., Zehetner, G., and Poustka, A. (1990) Hybridisation fingerprinting in genome mapping and sequencing, in *Genome Analysis: Genetic and Physical Mapping* (Davies, K. E. and Tilghman, S., eds.), Cold Spring Harbor Laboratory Press, Cold Spring Harbor, NY, pp. 39–81.
14. O'Farrell, P. H., Kutter, E., and Nakanishi, M. (1980) A restriction map of the bacteriophage-T4 genome. *Mol. Gen. Genet.* **179,** 421–435.
15. Morgan, A. R., Lee, J. S., Pulleyblank, D. E., Murray, N. L., and Evans, D. H. (1979) Review: ethidium fluorescence assays. Part 1. Physicochemical studies. *Nucleic Acids Res.* **7(3),** 547–569.
16. Hoheisel, J. D., Nizetic, D., and Lehrach, H. (1989) Control of partial digestion combining the enzymes Dam methylase and MboI. *Nucleic Acids Res.* **17,** 4571–4582.
17. Feinberg, A. P. and Vogelstein, B. (1983) A technique for radiolabeling DNA restriction endonuclease fragments to high specific activity. *Anal. Biochem.* **132,** 6–13.

56

Generation of Large Insert YAC Libraries

Zoia Larin, Anthony P. Monaco, and Hans Lehrach

1. Introduction

The introduction of yeast artificial chromosomes (YACs) as cloning vectors in 1987 has significantly advanced the analysis of complex genomes *(1)*. The capability of cloning large DNA (100–2000 kb) as YACs has accelerated the construction of physical maps and contig building (a contiguous set of overlapping clones). YAC contigs now cover entire human chromosomes (i.e., Y and 21) *(2,3)* and small genomes (i.e., *Schizosaccharomyces pombe*) *(4)*, and large YAC contigs cover much of the human genome *(5)*. The main advantages of YACs over prokaryotic-based cloning systems are their large insert capacity and ability to maintain sequences that are unstable or not well represented in bacteriophage or cosmid genomic libraries *(6)*. Therefore, YACs complement existing cloning vectors (cosmids, bacteriophage) and new cloning vectors (P1 bacteriophage [P1], bacterial artificial chromosomes [BACs], and P1-derived artificial chromosomes [PACs]; for review, *see* **ref. 7**) in mapping and chromosome walking projects *(6,8)*.

Several laboratories have generated YAC libraries from different eukaryotic genomes including arabidopsis *(9)*, *S. pombe* *(4)*, mouse *(10,11)*, and human DNA *(10,12,13)*. Libraries usually have been constructed in the *Saccharomyces cerevisiae* strain AB1380, but other strains are available with additional genetic markers that may be useful for selection of products following homologous recombination of YACs *(14)*. In addition, recombination deficient yeast strains (*rad1* or *rad52*) have also been used to reduce the problem of chimerism owing to recombination in YACs *(15)*, and these strains stabilize some sequences cloned in YACs *(16)*. Analysis of YACs maintained in *rad52* and *rad1* yeast strains compared to standard strains indicate that the frequency of chimerism is lower *(17)*. Different YAC vectors with centric and acentric arms have been constructed that allow rescue of end fragments in yeast for chromosome walking projects, and a bacteriophage T7 promoter for generation of riboprobes from the rescued end fragments *(14)*. Other YAC vectors incorporate a conditional centromere that allows for amplification of YAC DNA under appropriate conditions *(18)*.

YAC libraries have been constructed by preparing and size fractionating high molecular weight DNA in solution using sucrose gradients *(1,12)*, or in agarose by pulsed field gel electrophoresis (PFGE; *10,13,19*). When DNA is prepared in agarose,

YAC insert sizes are larger on average because shear forces seen with DNA in solution are minimized. However, partial degradation of DNA occurs when melting agarose containing high molecular weight DNA, perhaps due to metal ion contamination or denaturation *(10)*. The presence of polyamines *(10)* or high concentrations of NaCl (100 m*M*) *(20)*, protects DNA in agarose from degradation at the melting step. The authors constructed mouse, human, and *S. pombe* YAC libraries with average insert sizes of 700, 620, and 500 kb, respectively, by incorporating polyamines in the cloning procedure *(10)*.

This chapter describes in detail the protocols the authors used to construct large insert YAC libraries. This includes preparation of pYAC4 vector partial digestion of genomic DNA in agarose blocks, size fractionation by PFGE both before and after ligation to vector, and transformation of the yeast host AB1380.

2. Materials

1. Preparation of vector: All library construction protocols in this chapter are based on the pYAC4 vector *(1)*, available from the American Type Culture Collection. Vector DNA is prepared by large scale plasmid extractions and purification by CsCl gradient centrifugation *(21)*.
2. Restriction enzyme digest buffers: For most restriction digests, buffers recommended by the manufacturer are adequate. The authors recommend T4 polymerase buffer *(21)* when digesting vector DNA because it works with almost all restriction enzymes and calf intestinal alkaline phosphatase (CIP; Boehringer, Mannheim, Germany, 1 U/µL), thus eliminating precipitation of DNA and buffer changes between enzyme reactions. 10X T4 polymerase buffer: 0.33 *M* Tris-acetate, pH 7.9, 0.66 *M* potassium acetate, 0.10 *M* magnesium acetate, 0.005 *M* dithiothreitol (DTT), 1 mg/mL bovine serum albumin (BSA). Store frozen at –20°C in small aliquots.
3. Preparation and lysis of cells in agarose blocks: High molecular weight DNA from fibroblast or lymphoblastoid cell lines, whole blood, or fresh mouse spleen tissue is prepared in low melting point agarose blocks *(22)*, with $2-5 \times 10^6$ cells/block (approx 15–40 µg DNA).
4. *Eco*RI partial digestion reaction buffer: 1 agarose block with DNA 80–100 µL, 50 µL (5 mg/mL) BSA, 50 µL 10X *Eco*RI methylase buffer, 13 µL (0.1 *M*) spermidine, 1 U *Eco*RI, 50–200 U *Eco*RI methylase (NEB), distilled water to 500 µL final volume.
5. 10X *Eco*RI methylase buffer: 800 µ*M* S-adenosyl-methionine (SAM, NEB), 0.02 *M* MgCl$_2$, 1.0 *M* NaCl, 0.5 *M* Tris-HCl, pH 7.5, 0.01 *M* DTT. Store frozen at –20°C in small aliquots.
6. 100X Polyamines: 0.075 *M* spermidine-(HCl)$_3$, 0.030 *M* spermine-(HCl)$_4$. Store frozen at –20°C in small aliquots.
7. 10X Ligase buffer: 0.5 *M* Tris-HCl, pH 7.5, 0.1 *M* MgCl$_2$, 0.03 *M* NaCl, 10X polyamines.
8. YPD medium.
9. Regeneration plates *(23)*: 1.0 *M* sorbitol (Sigma, St. Louis, MO), 2% dextrose, 0.67% yeast nitrogen base without amino acids (Difco, Detroit, MI; add as filter sterile after autoclaving of agar), 1X amino acid supplements (without uracil), 2% agar.
10. 10X Amino acid supplements *(23)*: 200 µg/mL adenine, 200 µg/mL arginine, 200 µg/mL isoleucine, 200 µg/mL histidine, 600 µg/mL leucine, 200 µg/mL lysine, 200 µg/mL methionine, 500 µg/mL phenylalanine, 200 µg/mL tryptophan (light sensitive, filter sterilize and store at 4°C), 1.5 mg/mL valine, 300 µg/mL tyrosine, 200 µg/mL uracil (omit in regeneration and selective plates).
11. SCE: 1.0 *M* sorbitol, 0.1 *M* sodium citrate, pH 5.8, 0.01 *M* EDTA, pH 7.5, 0.03 *M* 2-mercaptoethanol or 0.01 *M* DTT (add fresh).

12. STC: 1.0 M sorbitol, 0.01 M Tris-HCl, pH 7.5, 0.01 M CaCl$_2$.
13. PEG: 20% Polyethylene glycol 6000 (PEG, Serva, Heidelberg, Germany), 0.01 M Tris-HCl, pH 7.5, 0.01 M CaCl$_2$. Make fresh and filter sterilize.
14. SOS: 1.0 M sorbitol, 25% YPD, 0.0065 M CaCl$_2$, 10 µg/mL tryptophan, 1 µg/mL uracil. Make fresh and filter sterilize.
15. YAC selective media and plates: 2% dextrose, 0.67% yeast nitrogen base without amino acids (add filter sterile), 1X amino acid supplements (without uracil and tryptophan), 2% agar for plates.
16. Contour-clamped homogeneous electric field (CHEF) apparatus. The authors recommend the BioRad (Richmond, CA) system.
17. Small horizontal gel electrophoresis apparatus: Use to check restriction enzyme digests of vector and test ligations of vector and genomic DNA.
18. Electrophoresis buffer: For both CHEF and horizontal gels, the authors recommend TBE. 10X TBE: 0.89 M Tris-borate, 0.89 M boric acid, 0.016 M EDTA.
19. Agarose: The authors recommend regular (SeaKem) and low melting point (LMP) (Seaplaque GTG) agarose from FMC. Most gels will be 1% (w/v) (aqueous).
20. Yeast and/or lambda concatamer size markers (BioRad).
21. Agarase (Sigma) dissolved in 50% (v/v) glycerol in water and store at 10 U/µL at –20°C or β-agarase (NEB, Beverly, MA).
22. T4 DNA ligase (NEB) at 400,000 U/mL.
23. T4 polynucleotide kinase (NEB) at 10 U/µL.
24. 1X TE: 0.01 M Tris-HCl, pH 7.5, 0.001 M EDTA, pH 7.5.
25. Proteinase K (Boehringer-Mannheim): Dissolve in water at 10 mg/mL and store in small aliquots at –20°C. Alternatively, use pronase (Boehringer-Mannheim). Add directly at 2 mg/mL.
26. Phenylmethylsulfonylfluoride (PMSF, Sigma): Prepare at 40 mg/mL in ethanol or isopropanol and heat several minutes at 68°C to dissolve. **Caution: Use gloves. It is toxic.**
27. 0.5 M EDTA, pH 8.0.
28. Lyticase (Sigma): Weigh out fresh prior to spheroplast formation (500 U/20 mL of yeast cells in SCE) and dissolve in SCE or water. Lyticase is difficult to get in solution and will need extensive vortexing.
29. 2-Mercaptoethanol (BDH, London, UK): Open in hood and use gloves.
30. For the yeast transformation, a spectrophotometer, a student microscope (10×, 25×, and 40× objectives and phase contrast), and a hemocytometer cell counter are needed.
31. Phenol equilibrated with 0.1 M Tris-HCl, pH 8.0. **Caution: Wear gloves because phenol burns.**
32. Chloroform.
33. 100% Ethanol.
34. Trinitriloacetic acid (BDH): Dissolve in water at 0.15 M and store frozen in small aliquots at –20°C. Used to inactivate CIP.

3. Methods
3.1. Preparation of pYAC 4 Vector

1. Before preparing pYAC4 arms for ligation to genomic DNA, test plasmid preps for deletions of telomere sequences during propagation in *Escherichia coli*. Digest 0.5 µg of the pYAC4 plasmid with *Hin*dIII and check on a 1% agarose gel. Four bands should be visualized: a 3.5, 3.0, 1.9, and 1.4 kb doublet.
2. If there is an additional smaller fragment below the 1.4 kb doublet, then telomere sequences have been deleted from the plasmid and another preparation should be attempted.

3. For preparative vector arms, digest 100–200 µg of pYAC4 with *Eco*RI and *Bam*HI to completion in 500 µL 1X T4 polymerase buffer and check on a 1% agarose gel. Three bands should be visualized: 6.0, 3.7, and 1.7 kb.
4. Heat kill the *Eco*RI and *Bam*HI at 68°C for 10 min.
5. Add directly 0.03–0.06 U/µg vector of CIP and incubate at 37°C for 30 min.
6. Inactivate the CIP with trinitriloacetic acid to 0.015 M at 68°C for 15 min.
7. Extract twice with phenol, once with chloroform, and precipitate with ethanol.
8. Resuspend the vector arms at a concentration of 1 µg/µL in 0.01 M Tris-HCl, pH 7.5, and 0.001 M EDTA (1X TE).
9. Check the efficiency of dephosphorylation of vector ends and the ability of these ends to ligate after phosphorylation. Set up two 20-µL ligation reactions (2 µL 10X ligase buffer without polyamines, 0.5 µg of digested and CIP-treated pYAC4 vector, 1 U T4 DNA ligase), one with and one without 1 U of T4 polynucleotide kinase.
10. Check ligations on a 1% agarose gel:
    a. Without kinase: 3 bands should be visualized as after digestion; and
    b. With kinase.
    The 1.7 kb *Bam*HI fragment can ligate to itself and form several supercoiled forms below 1.7 kb. The upper arms (6.0 and 3.7 kb) should ligate together by their *Eco*RI and *Bam*HI sites and form several larger fragments.

3.2. Partial Digestion of Genomic DNA

1. Partial digestion reactions: Prior to enzyme digestion, wash the blocks containing genomic DNA in 1X TE with 40 µg/mL PMSF at 50°C to inactivate the proteinase K and twice in 1X TE to remove the PMSF. Blocks incubated in pronase instead of proteinase K need only be washed extensively in 1X TE.
2. Perform partial *Eco*RI digestions by incubating blocks with a combination of *Eco*RI and *Eco*RI methylase. To determine the best mixture of the two enzymes, set up analytical reactions of 1 U of *Eco*RI with 0, 20, 40, 80, 160, 320, and 640 U of *Eco*RI methylase.
3. Place individual blocks in *Eco*RI partial digestion buffer (*see* **Subheading 2.**, **step 4**) with the various combinations of *Eco*RI and *Eco*RI methylase and incubate on ice for 1 h.
4. Transfer the reactions to 37°C for 4 h.
5. Add EDTA and proteinase K to 0.02 M and 0.5 mg/mL, respectively, to terminate the reactions, and incubate at 37°C for 30 min.
6. Check partial digests on a 1% agarose gel in a CHEF apparatus with yeast chromosomes as size markers to see which combination of enzymes gives most DNA in the range of 200–2000 kb.
7. Then digest many (6–12) blocks preparatively for the library construction using several of the best enzyme combinations (usually 1 U *Eco*RI and 50–200 U *Eco*RI methylase).

3.3. First Size Fractionation by PFGE

1. Pool blocks containing partially digested DNA in a 50-mL Falcon tube and wash once in 0.01 M Tris-HCl, pH 7.5, and 0.05 M EDTA.
2. Place blocks adjacent to each other in a trough in a 1% LMP agarose gel in 0.5X TBE, and preset for 1 h at 4°C. Place a genomic DNA block in the adjacent gel slot on either side of the trough and place yeast chromosome size markers in the outside gel slots.
3. Overlay the gel slots and trough with 1% LMP agarose. Subject the gel to electrophoresis at 160 V (4.7 V/cm), using a switch time of 30 s (which selects fragments ≥400 kb) for 18 h at 15°C in a CHEF apparatus.

Large Insert YAC Libraries

4. Remove the gel from the CHEF apparatus. Cut away only the outside lanes, including one lane each of partially digested genomic DNA and yeast chromosome size markers, and stain with ethidium bromide (1 µg/mL) for 45 min. Keep the central portion of the preparative gel in 0.5X TBE plus 0.02 M EDTA at 4°C.
5. Under UV light, notch the marker lanes at the edges of the limiting mobility (>400 kb) and take a photograph. Place adjacent to the central portion of the preparative gel, cut out the limiting mobility using the notches in the outside lanes as a guide, and place in a 50-mL Falcon tube. Stain all of the remaining preparative gel with ethidium bromide and take a photograph.

3.4. Ligation to Vector

1. Equilibrate the gel slice (1–2 mL) containing the limiting mobility of size-selected DNA four times (30 min each) in 1X ligase buffer (*see* **Subheading 2.**, **step 7**).
2. Place the gel slice equilibrated in 1X ligase buffer in an Eppendorf tube (<1 mL agarose/tube) and melt at 68°C for 10 min together with digested and CIP-treated pYAC4 vector (*see* **Subheading 3.1.**) in a ratio of 1:1 by weight of genomic DNA.
3. Stir the vector and genomic DNA in molten agarose slowly with a pipet tip and incubate at 37°C for 1–2 h.
4. Add T4 DNA ligase to 4 U/µL, ATP, pH 7.5 and DTT to 0.001 M each in 1X ligase buffer by slow stirring at 37°C. Incubate the reaction at 37°C for an additional 0.5–1 h and then overnight at room temperature. For ligation efficiency controls, *see* **Note 2**.
5. Terminate the reaction by adding EDTA, pH 8.0 to 0.02 M.

3.5. Second Size Fractionation by PFGE

1. Melt the ligation reaction at 68°C for 10 min and cool to 37°C.
2. Carefully pipet the molten agarose with a tip of bore diameter >4 mm into a trough in a 1% LMP agarose gel in 0.5X TBE, and preset for 1 h at 4°C. Place some molten agarose ligation mix in the gel slots adjacent to the trough on each side and place yeast chromosome size markers in the outside gel slots. Overlay the gel slots and trough with 1% LMP agarose.
3. Subject the gel to electrophoresis in a CHEF apparatus using the same conditions as described in **Subheading 3.3.** for the first size fractionation.
4. Excise the limiting mobility as described in **Subheading 3.3.**, **step 5**. If any degradation of DNA is seen at this step, *see* **Note 1**.
5. Equilibrate the gel slice (approx 2–3 mL), containing the limiting mobility from the second size fractionation, four times (30 min each) in 0.01 M Tris-HCl, pH 7.5, 0.03 M NaCl, 0.001 M EDTA, and 1X polyamines.
6. Score the equilibrated gel slice with a sterile scalpel and place less than 1 mL of agarose into individual eppendorf tubes. Melt at 68°C for 10 min, cool to 37°C, and add agarase (Sigma 150–200 U/mL of molten agarose or β-agarase 20 U/mL of molten agarose). Incubate at 37°C for 2–6 h prior to transformation.

3.6. Transformation

Transformation is carried out as described *(24)* using lyticase (Sigma) to spheroplast yeast cells. The yeast strain *S. cerevisiae* AB1380 has largely been used *(1)*, but libraries have been prepared in recombination deficient strains *(15)*.

1. Streak a fresh YPD plate with the appropriate strain from a frozen glycerol stock. Grow at 30°C for 2–3 d. Inoculate a single colony into 10 mL of YPD. Let sit overnight at 30°C.

2. The next evening, inoculate 200 mL of YPD in a 1-L flask with 200 µL of the 10-mL overnight culture. Use a larger inoculum (1/100 or 1/500) if it is a recombination deficient strain, because these cells tend to grow more slowly. Shake at 30°C overnight for 16–18 h.
3. When the $OD_{600\,nm}$ of a 1/10 dilution of the AB1380 culture is between 0.12 and 0.15, split the culture into 50-mL Falcon tubes. Check some of the culture under the microscope for bacterial contamination.
4. Spin the tubes at 400–600g (3000 rpm on tabletop centrifuge) for 5–10 min at 20°C. Decant media and resuspend pellets in 20 mL of distilled, sterile water for each tube.
5. Spin 400–600g for 5–10 min at 20°C. Decant water and resuspend pellets in 20 mL of 1.0 M sorbitol.
6. Spin 400–600g for 5–10 min at 20°C. Decant sorbitol and resuspend pellets in 20 mL SCE.
7. Add 46 µL of 2-mercaptoethanol and take 300 µL from one tube for a prelyticase control. Add 500 U lyticase (Sigma), mix gently, and incubate at 30°C.
8. At 5, 10, 15, and 20 min, test the extent of spheroplast formation of one tube by two independent methods:
    a. Using a spectrophotometer, measure $OD_{600\,nm}$ of a 1/10 dilution in distilled water. When the value is 1/10 of the prelyticase value, spheroplast formation is 90% complete.
    b. Mix 10 µL of cells with 10 µL 2% SDS and check under the microscope using phase contrast.
    When cells are dark ("ghosts") they are spheroplasted.
9. Take the spheroplast formation to 80–90%. This should take 10–20 min. Then spin cells at 200–300g (1100 rpm on tabletop centrifuge) for 5 min at 20°C.
10. Decant SCE and resuspend pellets gently in 20 mL of 1.0 M sorbitol. Spin 200–300g for 5 min at 20°C. Decant sorbitol and resuspend pellets in 20 mL STC.
11. Take a cell count of one tube by making a 1/10 to 1/50 dilution in STC and count on a hemocytometer.
12. Spin cells at 200–300g for 5 min at 20°C and then resuspend in a volume of STC calculated for a final concentration of $4.0–6.0 \times 10^8$ cells/mL when added to genomic DNA.
13. Add approx 0.5–1.0 µg of DNA in digested agarose solution (50–75 µL) to 150 µL of spheroplasts in 15-mL conical polystyrene Falcon tubes. For transformation controls, use:
    a. No DNA;
    b. 10 ng supercoiled YCp50 *(25)*; and
    c. 100 ng restricted and CIP-treated pYAC4.
    Let DNA and spheroplasts sit for 10 min at 20°C.
14. Add 1.5 mL PEG and mix gently by inverting tubes. Let sit for 10 min at 20°C. Spin at 200–300g for 8 min at 20°C.
15. Carefully pipet off PEG solution and do not disturb pellets. Gently resuspend pellets in 225 µL of SOS. Place at 30°C for 30 min.
16. Keep molten top regeneration agar at 48°C. If using small plates, add 5 mL of regeneration top agar (without uracil) to each 225 µL of SOS and cells. If you are using large (22 × 22 cm) plates, pool 10 tubes of 225 µL of SOS and cells to a 50-mL Falcon tube, and add 50 mL of regeneration top agar (without uracil). Mix gently by inverting the tube and pour quickly onto the surface of a prewarmed regeneration plate (without uracil) and let sit. Incubate plates upside down at 30°C for 3–4 d.
17. YAC analysis and replication of transformants. Good transformation efficiencies are between $2–8 \times 10^5$ clones/µg YCp50 and 100–1000 clones/µg genomic DNA. For low transformation efficiencies, *see* **Note 3**. Pick YAC clones individually onto selective plates (without uracil and trypophan, *see* **Subheading 2., step 14**) to test for both vector arms. When using minimal adenine, visualize red color in YAC colonies containing inserts.

Grow YAC clones in selective media and make agarose blocks containing chromosomes to check the size of YAC clones by PFGE. To replicate clones for library screening, pick YAC clones individually into microtiter dishes for screening of pools by polymerase chain reaction (PCR) amplification *(26)* or by colony hybridization after spotting onto filters using manual devices. A multipin transfer device, containing 40,000 closely spaced pins, has been used to efficiently replicate YAC clones from the supportive agar matrix of regeneration plates to the surface of selective plates, for colony hybridization and picking into microtiter dishes *(10)*.

4. Notes

1. Degradation of DNA: If anywhere in the cloning procedure you encounter complete or partial degradation of high molecular weight DNA, use yeast chromosomes in a series of control reactions to pinpoint the problem. Because yeast chromosomes can be visualized as distinct bands on PFGE, degradation can be detected much easier than in partial digests of genomic DNA. Test all buffers and enzymes (*Eco*RI methylase, T4 DNA ligase, proteinase K, agarase) for nuclease activity in mock cloning experiments using yeast chromosomes. Also, melt agarose blocks containing yeast chromosomes in buffers with and without 1X polyamines to test for partial degradation.
2. Ligation controls for vector and genomic DNA: Test the efficiency of ligation of vector arms to partially digested genomic DNA by incubating a small sample of the ligation reaction with and without 1 U T4 polynucleotide kinase. Melt the samples and load them on a small 1% agarose gel to check for no change of vector arms without kinase and disappearance of vector arms to larger sized fragments when incubated with kinase.
3. Transformation efficiency: If your transformation efficiencies are routinely lower than expected, check the following:
   a. Always streak the yeast strain onto a fresh YPD plate before setting up cultures. Cultures grown from old plates (>2 wk) seem to transform less well although they will appear to spheroplast normally.
   b. Try different concentrations of lyticase and percent spheroplast formation for optimum efficiency.
   c. Try various batches of sorbitol and PEG to see if there is any difference in transformation efficiency.
   d. Always use distilled, deionized water to guard against heavy metal ion contamination that can degrade DNA or decrease transformation efficiency.
   e. Check the temperature of room. Transformation is best at 20–22°C and decreases dramatically at temperatures around 30°C.

References

1. Burke, D. T., Carle, G. F., and Olson, M. V. (1987) Cloning of large DNA segments of exogenous DNA into yeast by means of artificial chromosome vectors. *Science* **236,** 806–812.
2. Foote, S., Vollrath, D., Hilton, A., and Page, D. C. (1992) The human Y chromosome: overlapping DNA clones spanning the euchromatic region. *Science* **258,** 60–66.
3. Chumakov, I., Rigault, P., Guillou, S., Ougen, P., Billaut, A., Guasconi, G., et al. (1992) Continuum of overlapping clones spanning the entire human chromosome 21q. *Nature* **359,** 380–387.
4. Maier, E., Howeisel, J., McCarthy, L., Mott, R., Grigoriev, A. P., Monaco, A. P., Larin, Z., and Lehrach, H. (1992) Complete coverage of the *Schizosaccharomyces pombe* genome in yeast artificial chromosomes. *Nature Genet.* **1,** 273–297.

5. Cohen, D., Chumakov, I., and Weissenbach, J. (1993) A first-generation physical map of the human genome. *Nature* **366,** 698–701.
6. Coulson, A., Waterston, R., Kiff, J., Sulston, J., and Kohara, Y. (1988) Genome linking with yeast artificial chromosomes. *Nature* **335,** 184–186.
7. Monaco, A. P. and Larin, Z. (1994) YACs, BACs, PACs and MACs: artificial chromosomes as research tools. *Trends Biotechnol.* **12,** 280–286.
8. Garza, D., Ajioka, J. W., Burke, D. T., and Hartl, D. L. (1989) Mapping the Drosophila genome with yeast artificial chromosomes. *Science* **246,** 641–646.
9. Guzman, P. and Ecker, J. (1988) Development of large DNA methods for plants: molecular cloning of large segments of Arabidopsis and carrot DNA into yeast. *Nucleic Acids Res.* **16,** 11,091–11,105.
10. Larin, Z., Monaco, A. P., and Lehrach, H. (1991) Yeast artificial chromosome libraries containing large inserts from mouse and human DNA. *Proc. Natl. Acad. Sci. USA* **88,** 4123–4127.
11. Burke, D. T., Rossi, J. M., Leung, J., Koos, D. S., and Tilghman, S. M. (1991) A mouse genomic library of yeast artificial chromosome clones. *Mammal. Genome* **1,** 65.
12. Anand, R., Villasante, A., and Tyler-Smith, C. (1989) Construction of yeast artificial chromosome libraries with large inserts using fractionation by pulsed-field gel electrophoresis. *Nucleic Acids Res.* **17,** 3425–3433.
13. Albertsen, H. M., Abderrahim, H., Cann, H. C., Dausset, J., Le Paslier, D., and Cohen, D. (1990) Construction and characterization of a yeast artificial chromosome library containing seven haploid human genome equivalents. *Proc. Natl. Acad. Sci. USA* **87,** 5109–5113.
14. Reeves, R. H., Pavan, W. J., and Hieter, P. (1992) Yeast artificial chromosome modification and manipulation, in *Methods in Enzymology*, vol. 216 (Wu, R., ed.), Humana, Totowa, NJ, pp. 584–603.
15. Chartier, F. L., Keer, J. T., Sutcliffe, M. J., Henriques, D. A., Mileham, P., and Brown, S. D. M. (1992) Construction of a mouse yeast artificial chromosome library in a recombinant-deficient strain. *Nature Genet.* **1,** 132–136.
16. Neil, D. L., Villasante, A., Fisher, R. B., Vetrie, D., Cox, B., and Tyler-Smith, C. (1990) Complete coverage of the *Schizosaccharomyces pombe* genome in yeast artificial chromosomes. *Nucleic Acids Res.* **18,** 421–428.
17. Ling, L. L., Ma, N. S.-F., Smith, D. R., Miller, D. D., and Moir, D. T. (1993) Reduced occurrence of chimeric YACs in recombinant deficient hosts. *Nucleic Acids Res.* **21,** 6045,6046.
18. Smith, D. R., Smyth, A. P., and Moir, D. T. (1992) Copy number amplification of yeast artificial chromosomes, in *Methods in Enzymology*, vol. 216 (Wu, R., ed.), Humana, Totowa, NJ, pp. 603–614.
19. McCormick, M. K., Shero, J. H., Cheung, M. C., Kan, Y. W., Hieter, P. A., and Antonarakis, S. E. (1989) Construction of human chromosome 21-specific yeast artificial chromosomes. *Proc. Natl. Acad. Sci. USA* **86,** 9991–9995.
20. Lee, J. T., Murgia, A., Sosnoski, D. M., Olivos, I. M., and Nussbaum, R. L. (1992) Construction and characterisation of a yeast artificial chromosome library for Xpter-Xq27. 3: a systematic determination of cocloning rate and X-chromosome representation. *Genomics* **12,** 526–533.
21. Maniatis, T., Fritsch, E. F., and Sambrook, J. (1982) *Molecular Cloning: A Laboratory Manual*, Cold Spring Harbor Laboratory Press, Cold Spring Harbor, NY.
22. Herrmann, B. G., Barlow, D. P., and Lehrach, H. (1987) An inverted duplication of more than 650 Kbp in mouse chromosome 17 mediates unequal but homologous recombination between chromosomes heterozygous for a large inversion. *Cell* **48,** 813–825.

23. Rothstein, R. (1985) Cloning in yeast, in *DNA Cloning Volume II* (Glover, D. M., ed.), IRL Press, Oxford, UK, pp. 45–65.
24. Burgers, P. M. J. and Percival, K. J. (1987) Transformation of yeast spheroplasts without cell fusion. *Anal. Biochem.* **163,** 391–397.
25. Hieter, P., Mann, C., Snyder, M., and Davis, R. W. (1985) Mitotic stability of yeast chromosomes: a colony color assay that measures nondisjunction and chromosome loss. *Cell* **40,** 381–392.
26. Green, E. D. and Olson, M. V. (1990) Systematic screening of yeast artificial chromosome libraries by use of the polymerase chain reaction. *Proc. Natl. Acad. Sci. USA* **87,** 1213–1217.

57

YAC Library Storage and Transport

John E. Collins, Sheila Hassock, and Ian Dunham

1. Introduction

The development of the yeast artificial chromosome (YAC) cloning system *(1)* and the construction of YAC libraries with large numbers of genome equivalents *(2–6)* provided a major impetus to mammalian genome mapping. These technical advances enabled the mapping of megabase-sized chromosomal regions *(7,8)*, culminating in the first complete clone maps of single mammalian chromosomes *(9,10)*.

The distribution of YAC libraries, and more recently P1 artificial chromosome (PAC) libraries *(11)*, to multiple laboratories greatly facilitated progress, but managing large clone resources required the development of appropriate tools and protocols. The authors describe a set of protocols to enable the easy manipulation of large numbers of YAC clones.

Traditionally, genomic DNA libraries in lambda or cosmid vectors, were screened by random replica plating and hybridization. The larger size of inserts made possible by the YAC cloning system reduced the number of clones required to give high genome coverage. This allowed a change in approach so that single recombinant clones are stored separately in an ordered array. Libraries are frequently stored with single clones in separate wells of 96- (or 384-) well microtiter plate arrays at –70°C *(see* **Note 1**). Thus, each clone has a unique address within the library, consisting of the microtiter plate number and the well coordinate, A–H, on the vertical axis, and 1–12 on the horizontal axis. This address acts as a permanent reference for the clone, can be stored in a database with associated information, and can be communicated along with the library. It is essential that library reference numbers are used by all workers who use a widespread library so that data is compatible.

YAC libraries may be received by one of the methods described. Initially, the library must be replicated. The first, or archive, copy should remain frozen. The second is a backup copy that is periodically duplicated to remake the working plates. Libraries are replicated using a 96-pin tool (**Fig. 1**). Each microtiter plate is stamped onto YAC selective media and the colonies grown. An inoculum is taken from the agar plate into multiple microtiter plates filled with liquid media.

From: *The Nucleic Acid Protocols Handbook*
Edited by: R. Rapley © Humana Press Inc., Totowa, NJ

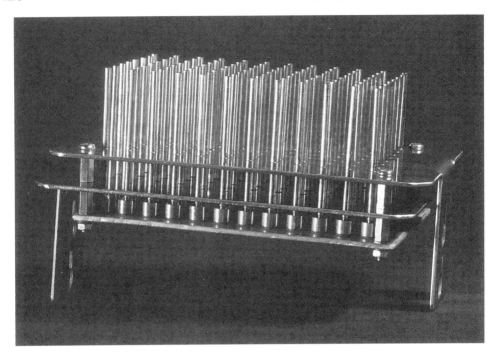

Fig. 1. 96-Pin hand held replicating tool (or hedgehog) (*see* **Subheading 2., item 1**). Note that the two 96-place heads stabilize the pins and minimize horizontal movement and the spacers allow each pin 10 mm of vertical movement. Once sterilized, the tool can be held by the end pieces of the 96-place heads.

2. Materials

1. 96-Pin tool (**Fig. 1**): This is made using 96-inoculation pins (Denley [Billinghurst, Surrey, UK] WR080/02), 2X 96 place heads (Denley WR080/01), a top plate (the authors make their own) all held together with eight spacers, four screws, and four nuts.
2. Wellfill 3 (Denley WF043) or multichannel pipetman for filling microtiter plates.
3. YPD medium: 10 g yeast extract, 20 g peptone, 20 g D-glucose. Add 20 g agar for solid medium.
4. AHC medium: 6.7 g yeast nitrogen base without amino acids, 14 g casein acid hydrolysate, 20 g D-glucose, 20 mg adenine hemisulfate. Add 20 g agar for solid medium. Adjust to pH 5.8 prior to autoclaving.
5. 80% Glycerol, autoclaved.
6. Flat-bottomed microtiter plates (Falcon 3072, Becton Dickinson, Lincoln Park, NJ). Round-bottomed plates are not recommended for YAC libraries as the cells will settle into a small area in the center and it may prove difficult to replicate from these plates.
7. Microtiter plate sealers (Dynatech [Chantilly, VA], cat. no. 001-010-5701).
8. 10-cm Rubber roller, available for all good art materials shops.

3. Methods

All the sterilization procedures and manipulations of YACs should be carried out in a class II microbiological safety cabinet in accordance with local regulations. This is also necessary to minimize contamination problems.

3.1. Initial Sterilization of 96-Pin Tool

1. Invert tool in flowhood (pins up).
2. Spray with absolute ethanol from a wash bottle.
3. Light a Bunsen burner and ignite ethanol covering the 96-pin tool. Ensure that the lit Bunsen is kept well away from the ethanol bottle!
4. Leave 96-pin tool to cool for at least 10 min.

3.2. Cyclic Sterilization Procedure When Using 96-Pin Tool

1. Take 3-microtiter plate size dishes (use either a 15-cm diameter Petri dish or the sterile microtiter plate packaging). Fill one with water (not sterile), one with absolute ethanol, and the third sterile dish with YPD broth ensuring that each tray is filled at least as deep as the depth of media in the microtiter plate.
2. Take the sterile 96-pin tool (*see* **Subheading 3.1.**) and perform the desired YAC manipulation (*see* **Subheading 3.3.**).
3. Rinse the end of the pins in the water to remove any agar or media and stamp them dry on a pad of dry tissues or towels.
4. Place the pins into the absolute ethanol for a few seconds and then invert the tool in the flowhood.
5. Ignite with the ethanol on the tool with the Bunsen flame, briefly waving the flame over each pin head (take care that the dish of ethanol is at the other end of the flowhood). Do not make the tool hot by excessive flaming.
6. Cool the pin heads by placing in the sterile yeast peptone dextrose (YPD). (It is possible to avoid this step if you have two 96-pin tools and use them in rotation allowing them to cool in the air flow after **step 5**.)
7. Shake the tool to remove excess YPD.
8. Perform the next YAC manipulation and repeat steps 3–6 after every YAC manipulation.
9. At the end of each session, sonicate the tool in water in a sonicating water bath for 10 min to remove colony debris.

3.3. YAC Manipulation

3.3.1. Transfer of YACs from YPD Broth to Agar-Filled Microtiter Plate or Agar Plate

1. Place the 96 pins of the sterilized tool into a thawed microtiter plate of YACs.
2. Gently scrape the tool on the bottom of the plate avoiding splashes between wells.
3. Transfer the tool to an AHC agar dish and stamp the YACs onto the plate surface, checking that all the pins are touching the agar. Do not allow the full weight of the tool to rest on the agar as it will sink into the plate, especially if it is still slightly warm.
4. Place the AHC dish at the back of the hood for 5 min until the liquid has dried.
5. Sterilize tool (*see* **Subheading 3.2.**).
6. Incubate agar plate at 30°C for 2 d. If growth is patchy or nonexistent, *see* **Notes 2 and 3**; if contamination is a problem *see* **Notes 4–8**.
7. Repeat manipulation as necessary.

3.3.2. Transfer of YACs from AHC Agar to Single or Multiple Copies of YPD Broth Microtiter Plates

1. Check each AHC agar plate for contaminants such as fungal growth or bacteria. At this stage contaminants may be cut out from the agar with a sterile scalpel, leaving an empty space in the library. The YAC can be added back to the library when it has been recovered usually be streaking separately on selective media (AHC).

2. Fill the required number of microtiter plates with 150 mL YPD broth per well using either a multichannel pipetman or an automatic well filler.
3. Place the 96 pins of the sterile tool over the colonies on the AHC agar dish, checking that all the pins are touching the colonies.
4. Transfer the tool to a prefilled YPD broth microtiter plate. Mix the tool in the broth to remove the cells.
5. If required, return the tool to the same AHC plate to collect more YAC colony and inoculate further copies of the plate as necessary.
6. Sterilize tool (*see* **Subheading 3.2.**).
7. Repeat manipulation as necessary.
8. Incubate microtiter plates at 30°C for 2 d.

3.4. Freezing YACs in Microtiter Plates

1. Look at the microtiter plate from beneath to check that the YACs have grown to cover the bottom of the microtiter plate. Occasionally the YACs grow in clumps that will need to be dispersed using the sterile 96-pin tool before freezing.
2. Mix equal volumes of 80% glycerol and YPD broth to make 0.5 X YPD broth containing 40% glycerol. This dilution makes the glycerol less viscous and easier to manipulate.
3. Add 100 mL of this 40% glycerol mix to each well to give an approximate final glycerol concentration of between 15 and 20% depending on how much of the original YPD broth has evaporated during incubation. Check how much volume is lost through evaporation under your own conditions. If a Wellfill is used, the delivery switch needs to be set at approx 150 mL, which will then add 100 mL to each well because of the viscosity of the glycerol. The amount of glycerol delivered can be measured using an empty microtiter plate and a pipetman before proceeding to add the glycerol to the library plates.
4. Using the sterile tool, gently mix the glycerol with the YACs (*see* **step 1** in **Subheading 3.3.1**). This also disperses any clumps of YACs to form an even coverage of the well bottom.
5. Archive plates should be sealed with plate sealers.
6. Freeze the microtiter trays at –70°C stacked and wrapped in suitable plastic bags to prevent frost. It is prudent to test that the type of bag you are using will survive prolonged freezing at –70°C beforehand.

3.5. Transport of YAC Libraries

YAC libraries can be transported in a number of ways. The best method to suit the exporter and importer may be selected from the following:

1. Each microtiter plate is stamped onto a 15-cm selective media plate. This is the simplest method as the exporter only needs to stamp the YACs from their backup stocks, and the importer has a plate ready to make microtiter plate copies. However, the plates are fragile and bulky, usually requiring careful transport by car.
2. Each microtiter plate is stamped into another microtiter plate filled with 150 mL of YPD agar media. This is the most convenient method for international transport. It is compact and less susceptible to damage during rough handling. However, as the YACs do not tend to grow evenly in such a plate, it is difficult to use this plate for further copies. Thus the importer should recover the YACs in a YPD broth microtiter plate and then make a further selective agar plate that is used to expand the library.
3. YACs may be transported frozen in their microtiter plates. The obvious problem is keeping the large number of plates frozen. The authors have transported several YAC libraries in large expanded polystyrene boxes filled with dry ice. The importer has the advantage of having a ready made archive copy and only two further copies will need to be produced.

4. Notes

1. The authors store bacterial libraries only in 384-well plates. YAC libraries are stored in 96-well plates.
2. If the YACs do not grow, check that the media was made correctly with all the appropriate nutrients (e.g., glucose and adenine). Test a batch of media prior to starting on replication of a library.
3. If the YACs give a patchy growth pattern there are a number of possibilities:
   a. The tool was not cooled enough after sterilization.
   b. The pins are not touching either the agar or the bottom of the microtiter plate. It is possible that the pin heads are sticking to the back plate. This can be solved by cleaning the tool by sonication in water in a deeply filled water bath for 10 min. Tap the tool sharply before taking the colony lifts and visually check that all the pins are level.
   c. The AHC agar plates are poured too thin. A 15-cm diameter Petri dish needs at least 50 mL of agar.
   d. The tool was not scraped enough on the bottom of the microtiter plate (see **Subheading 3.3.1.**).
4. In general, YACs are grown at each stage for 1 or 2 d. This is to minimize the chance of yeast and bacterial contamination competing with the YAC colony. Where fast growing yeast contaminants have been a problem, the authors have added extra adenine to the media, allowing the YACs to grow more rapidly. It is worth noting that the further away from the original YAC library a plate has become, the more likely it is that contamination has occurred with other YACs or another organism. It is therefore important when duplicating and, especially, exporting libraries to try to use the closest possible copy to the original library without unnecessarily disturbing the archive plates.
5. Bacterial contamination may be controlled by adding 50 mg/mL ampicillin, 5 mg/mL tetracycline, or 30 mg/mL kanamycin, or a combination of these to the AHC agar before pouring the plates.
6. Spread of fast-growing yeast contaminants may be limited by adding 100 mg/mL adenine to both broth and agar allowing the YACs to grow more quickly. YACs will grow sufficiently in high adenine media overnight, but will not turn red. However, the best policy is to remove the contaminants completely. This may be achieved by streaking the contents of the well onto AHC plates and after 2 d growth retrieving the red YAC from the contaminant. Unless there is obvious contamination at this point, the plate can be left for another few days at 4°C to allow the red color of the YACs to show fully. The recovered YAC can then be grown up in a microtiter plate and added back to the main library in the correct well. Contaminated wells are cleaned by removing the contents and soaking in 95% ethanol for 10 min. In some cases, two rounds of streaking may be necessary.
7. Most of the widely available YAC libraries do not necessarily contain a single YAC clone in each well, partly because the density of the YACs in the original agar transformation plates from which the YACs were picked was high enough that a picking of a single transformant could not be guaranteed, and partly because manipulation of the microtiter plates invariably leads to some cross contamination between wells. The amount of work required to streak out each clone to a single colony is so great that this has not generally been done. Although this fact confounds screening strategies that rely on single well locations for each YAC, it does mean that the archive copies of the YAC libraries have been grown very little, and, so, any YAC that is prone to deletion has had less chance to be deleted. However, contamination of YACs between the wells of microtiter plates is a continual problem. The only way to obtain a pure clone from a microtiter plate well is to

streak it onto an AHC agar plate and test the single colonies for STS or probe content (*see* Chapter 59).

8. Fungal contamination of agar plates may be removed by cutting out the affected area with a sterile scalpel. YAC colonies lost this way will need to be recovered and added back to the library.

References

1. Burke, D. T., Carle, G. F., and Olson, M. F. (1987) Cloning of large segments of exogenous DNA into yeast by means of artificial chromosome vectors. *Science* **236,** 806–812.
2. Brownstein, B. H., Silverman, G. A., Little, R. D., Burke, D. T., Korsmeyer, S. J., Schlessinger, D., and Olson, M. V. (1989) Isolation of single-copy human genes from a library of yeast artificial chromosome clones. *Science* **244,** 1348–1351.
3. Anand, R. A., Riley, J. H., Butler, R., Smith, J. C., and Markham, A. F. (1990) A 3.5 genome equivalent multiaccess YAC library, construction, characterisation, screening and storage. *Nucleic Acids Res.* **18,** 1951–1956.
4. Albertsen, H. M., Abderrahim, H., Cann, H. M., Dausset, J., Le Paslier, D., and Cohen, D. (1990) Construction and characterisation of a yeast artificial chromosome library containing seven haploid human genome equivalents. *Proc. Natl. Acad. Sci. USA* **87,** 4256–4260.
5. Larin, Z., Monaco, A. P., and Lehrach, M. (1991) Yeast artificial chromosome libraries containing large inserts from mouse and human DNA. *Proc. Natl. Acad. Sci. USA* **88,** 4123–4260.
6. Chartier, F. L., Keer, J. T., Sutcliffe, M. J., Henriques, D. A., Mileham, P., and Brown, S. D. M. (1992) Construction of a mouse yeast artificial chromosome library in a recombination-deficient strain of yeast. *Nature Genet.* **1,** 132–136.
7. Green, E. D. and Olson, M. V. (1990) Chromosomal region of the cystic fibrosis gene in yeast artificial chromosomes: a model for human genome mapping. *Science* **250,** 94–98.
8. Vetrie, D., Vorechovsky, I., Sideras, P., Holland, J., Davies, A., Flinter, F., et al. (1993) The gene involved in X-linked agammaglobulinaemia is a member of the src family of protein-tyrosine kinases. *Nature* **361,** 226–233.
9. Chumakov, I., Riault, P., Guillou, S., Ougen, P., Billaut, A., Guasconi, G., et al. (1992) Continuum of overlapping clones spanning the entire human chromosome 21q. *Nature* (Lond.) **359,** 380–387.
10. Foote, S., Vollrath, D., Hilton, A., and Page, D. C. (1992) The human Y chromosome: overlapping DNA clones spanning the euchromatic regions. *Science* **258,** 60–66.
11. Ioannou, P. A., Amemiya, C. T., Garnes, J., Kroisel, P. M., Shizuya, H., Batzer, M. A., and de Jong, P. J. (1994) A new bacteriophage P1-derived vector for the propagation of large human DNA fragments. *Nat. Gen.* **6,** 84-89. *See also* http://bacpac.med.buffalo.edu/
12. Bentley, D. R., Todd, C., Collins, J., Holland, J., Dunham, I., Hassock, S., et al. (1992) The development and application of automated gridding for efficient screening of yeast and bacterial ordered libraries. *Genomics* **12,** 534–641.

58

YAC Library Screening

Preparation of Hybridization Filters and Polymerase Chain Reaction Pools

Charlotte G. Cole, John E. Collins, and Ian Dunham

1. Introduction

The storage of yeast artificial chromosome (YAC) libraries in ordered microtiter plates required a new approach to screening for clones containing specific DNA sequences. Screening libraries of some 60,000 clones by hybridization to filters prepared from individual 96-well microtiter plates was not a feasible option, prompting development of the polymerase chain reaction (PCR)-based screening approach of Green and Olson *(1)*. Here (and in all subsequently developed PCR-based strategies), YAC libraries are screened by performing the PCR on a series of pools of DNA derived from specific mixtures of yeast clones. Amplification of target DNA sequence from an individual pool indicates the presence of the required YAC within the parent microtiter plates. Further rounds of testing on subsidiary pools are used to reveal the exact location of the YAC. Thus, a library of approx 36,000 clones may be prepared as 24 individual pools of 1536 YACs each for the first round of the PCR, each pool containing yeast DNA from 16 microtiter plates. Screening by the PCR therefore requires the preparation of pools of total yeast DNA derived from several thousand different YAC clones in equal amounts *(2–4)*. Pools containing fewer YACs may also be required for subsequent stages of PCR screening.

Although screening by the PCR has proved successful even for large projects *(4,5)* it is also desirable to have the option of screening whole libraries by hybridization, thus circumventing the need to develop large numbers of suitable primer pairs (sequenced tagged sites [STSs]). Efficient screening of YAC clones by hybridization requires, first, that the DNA derived from many different individual clones is present at high densities on hybridization filters and, second, that these filters can be prepared rapidly and precisely. This has been achieved through the use of robots that automatically grid clones from microtiter plates onto hybridization filters in high density ordered arrays of clones *(2,6–8)*. Thus, the system used in the authors' laboratory grids 1536 YACs onto a single 8 × 12 cm filter *(2,8*; *see* **Fig. 1** in Chapter 59). To enable subsequent identification of individual colonies following autoradiography, it is essential to preserve the ordered

From: *The Nucleic Acid Protocols Handbook*
Edited by: R. Rapley © Humana Press Inc., Totowa, NJ

array precisely. Hence, the clones are gridded and grown on a nylon filter, the resulting colonies spheroplasted and lysed *in situ*, and the DNA denatured and fixed to the filter.

Described herein are methods for the preparation of PCR pools from yeast DNA isolated in agarose plugs (*see* **Note 1**) and of filters for hybridization. The pools described are based on a relatively simple pooling system (*see* **Fig. 2** in Chapter 59) and preparation of pools based on 1/2 filters or rows and columns are not described. However, the method is applicable to any array of YACs grown on nylon filters.

For the preparation of high-density filters for hybridization, the authors strongly advise the use of an automatic gridding system. The method given assumes access to a customized robot of the type described by Bentley et al. *(2)* or McKeown et al. *(8)*. However, nylon filters gridded in different arrays may be treated identically (*see* **Note 2**). Filters generated from YACs spotted onto filters manually using individual "pins" or a hand-held 96-pin replicating "hedgehog" may also be grown and treated in much the same way, with a few caveats, as detailed in **Subheading 3.** (*see also* **Note 3**).

2. Materials
2.1. Hybridization Filters

1. Sterile 80 × 120 mm nylon filters (Hybond N, Amersham Pharmacia Biotech, Arlington Heights, IL; available precut to size cat. no. RPN 119N) (*see* **Note 2**).
2. Sterile rectangular 8 cm × 12 cm Petri dishes with lids (Hybaid "colony picker plates with lids," available from Hybaid [Teddington, Middlesex, UK] on request). For gridding of yeast colonies, these dishes are poured to uniform thickness (50 mL/dish) with YPD agar containing 50 µg/mL ampicillin, 5 µg/mL tetracycline (*see* **Note 4**). Plates can be reused by scraping out the media with a spatula and washing with detergent followed by sterilization with 70% isopropanol and drying in a laminar flowhood.
3. Plastic trays with lids (30 × 40 × 2 cm trays from Jencons [Leighton Buzzard, Bedfordshire, UK], cat. no. 682-008; 390 × 290 mm lid from Marathon [London, UK], cat. no. TT217/132).
4. Yeast spheroplasting solution: 1 M sorbitol, 20 mM EDTA, 10 mM Tris-HCl, pH 7.4, containing freshly added 0.1 mg/mL zymolyase 20T (ICN, High Wycombe, UK) and 14 mM β-mercaptoethanol (*see* **Notes 5** and **6**).
5. Denaturation solution: 0.5 M NaOH, 1.5 M NaCl.
6. Neutralization solution: 0.5 M Tris-HCl, pH 7.4, 1.5 M NaCl.
7. Protease solution: 1/10 dilution of neutralization solution containing 250 µg/mL proteinase K. Sigma (St. Louis, MO) XI-S Protease is of sufficient quality. Store aliquots of stock proteinase K at 25 mg/mL, –20°C, and make protease solution up freshly each time.
8. 50 mM Tris-HCl, pH 7.4.

2.2. PCR Pools

1. Sterile 80 × 120 mm nylon filters Hybond N, Amersham Pharmacia Biotech; available precut to size, cat. no. RPN.119N).
2. *See* **Subheading 2.1.**, **step 2** for high-density pools generated using robot Petri dishes, and for single plate pools, *see* **Subheading 2.1.**, **step 2**, or sterile 15-cm circular Petri dishes poured as noted with or without tetracycline.
3. 50 mM EDTA, pH 8.0.
4. Basic yeast spheroplasting solution: 1 M sorbitol, 20 mM EDTA, 10 mM Tris-HCl, pH 7 4.
5. Yeast spheroplasting solution containing 0.1 mg/mL zymolyase 100T (ICN), 14 mM β-mercaptoethanol.
6. Low-gelling temperature (LGT) agarose (SeaPlaque agarose, FMC, Rockland, ME).

7. 1 mL Disposable (flexible) plastic bulb style pipets, tube diameter approx 2–3 mm.
8. Filter sterilized or autoclaved yeast lysis solution (YLS): 1% lithium dodecyl sulfate, 100 mM EDTA, 10 mM Tris-HCl, pH 8.0 (*see* **Note 7**).
9. $T_{0.1}E$: 10 mM Tris-HCl, pH 8, 0.1 mM EDTA.

3. Methods
3.1. Preparation of Nylon Filters for Hybridization
3.1.1. Growth of YACs on Nylon Filters

1. Thaw working stocks of YACs stored in 15–20% glycerol (*see* **Note 8**).
2. Label the nylon filters in the top left-hand corner corresponding to position A1 of a microtiter plate with a suitable pen (e.g., Edding 1800) and lay onto the YPD agar plates (*see* **Note 4**). Carefully lift the filter and re-lay to remove air bubbles if necessary.
3. Grid the YACs robotically onto the filters following the manufacturers instructions (*see also*, **ref. 2**). Alternatively, stamp the YACs manually onto filters using the 96-pin "hedgehog" as described in Chapter 57, or spot the YACs onto filters using 0.6–2 mm pins (available from Cambridge Repetition Engineers, Cambridge, UK).
4. Grow YACs for approx 27 h at 30°C (or until even growth is observed) (*see* **Note 3**).

3.1.2. Spheroplasting and Lysis of YACs on Filters

1. Soak single layers of 3MM Whatman (Maidstone, UK) paper in spheroplasting solution, using approx 50 mL/780 cm^2 of Whatman paper (26 × 35 cm sheets if using the recommended trays). Pour off excess liquid (approx 5–10 mL) until the paper still "shines" but no pools of liquid remain. Avoid evaporation from Whatman paper prior to **step 2**.
2. Remove the nylon filters from the agar plates taking care not to carry over lumps of agar and carefully lay the filters colony side up onto the freshly soaked Whatman paper. Check carefully to ensure no air bubbles are trapped under the filter (*see* **Note 9**). Place a lid over the tray, seal in a plastic/autoclave bag, and incubate overnight at 37°C.
3. Remove filters from the spheroplasting tray and lay colony side up onto a fresh sheet of 3MM Whatman soaked in denaturation solution. Ensure no air bubbles are trapped under the Whatman paper or the filter. Leave at room temperature for at least 10 min but no more than 20 min. Check constantly for air bubbles. These are observed by the appearance of colonies that fail to lose their red color and may be alleviated by carefully lifting and relaying the filter.
4. Dry the filters for 10 min by laying colony side up onto a fresh piece of 3MM Whatman paper.
5. Carefully submerge each filter colony side up in neutralization solution. Use excess liquid (500 mL in a medium-sized sandwich box). Leave for approx 5 min.
6. Carefully pour off the solution and replace with a 1/10 dilution of neutralization solution for 5 min at room temperature.
7. Incubate the filters colony side up in a sealed box containing protease solution at 37°C for 30–60 min. One hundred milliliters are sufficient for up to 40 filters in a suitable sized (small) sandwich box, but ensure that all filters are covered.
8. Wash the filters by submerging in an excess of 1/10 dilution of neutralization for 5 min, with very gentle shaking. Do not wipe the filters.
9. Wash the filters twice in an excess of 50 mM Tris-HCl, pH 7.4 with very gentle shaking.
10. Following the final rinse, use a pair of tweezers to drag the back of the filter along the edge of the sandwich box to remove excess liquid and any debris stuck on the back of the filter. Lay flat on a fresh piece of 3MM Whatman paper.
11. Air dry for at least 15 min. When nearly dry, place another sheet of 3MM Whatman on top to prevent the filters from rolling up (*see* **Note 10**).

12. Place fully dried filters flat, colony side down, onto a UV transilluminator and irradiate for 2 min at 312 nm (*see* **Notes 11–13**).

3.2. Preparation of Pools of YAC DNA for PCR

3.2.1. Growth of YACs

1. Grid or stamp the YACs onto nylon filters as described in **Subheading 3.1.1.** For pools prepared from YACs gridded at high density, prepare two identical filters for each pool. The authors prepare high-density pools from 16 microtiter plates gridded in a 4×4 array (*see* **Fig. 1** in Chapter 59). For pools prepared from single microtiter plates, stamp each plate onto a single filter using a 96-pin "hedgehog," as described in Chapter 57.
2. Grow the YACs for two nights at 30°C.

3.2.2. Spheroplasting and Lysis of YACs in Agarose Plugs

1. Lift the filters from the agar plates using tweezers, taking care not to remove any agar with the filter. Roll up loosely, colony side inward, and place in a 50-mL Falcon (Becton Dickinson) centrifuge tube containing 25 mL 50 m*M* EDTA.
2. Screw the cap tightly and shake to wash off all the colonies. Remove the washed filter and discard.
3. Pellet the yeast cells by spinning at 3000*g* for 5 min.
4. Discard the supernatant and wash the pellet once more with 25 mL 50 m*M* EDTA, pelleting as in **step 3**.
5. Determine the wet weight of cells (take an average of each pellet assuming approximately even growth and weight of cells) (*see* **Note 14**).
6. Prepare 2% molten LGT agarose in basic yeast spheroplasting solution, cool, and maintain at 45°C. You will need approx 2 mL/YAC filter.
7. Resuspend the cells in 2 vol of prewarmed (37°C) spheroplasting solution containing 0.1 mg/mL 100T zymolyase and 14 m*M* β-mercaptoethanol (e.g., 2 mL/g of cells). Maintain at 37°C.
8. Add 2 mL of molten LGT agarose per gram of cells to the cell suspension, mix well, and draw up the slurry into the barrel of one or more disposable 1-mL plastic pipets (do not suck agarose into the bulb). Stand the pipet(s) upright in the Falcon tube until set (place at 4°C if necessary for speed).
9. Once set, cut the tapered part of pipet away and extrude the agarose "worm" into a new 50-mL Falcon tube.
10. Cover with approx 10 mL of spheroplasting solution containing 0.1 mg/mL zymolyase 100T, 14 m*M* β-mercaptoethanol. Incubate overnight at 37°C with gentle shaking.
11. Replace the solution with 10–15 mL of YLS (*see* **Note 7**). Incubate at 37°C for 30–60 min.
12. Replace with fresh YLS and incubate at 37°C overnight with gentle shaking.
13. Continue replacing with fresh YLS until there is no color left in the agarose (normally once or twice more).
14. The agarose "worms" can now be stored at room temperature in YLS or 0.5 *M* EDTA (*see* **Note 15**).

3.2.3. Rinsing and Dilution of Agarose Plugs

1. Cut off up to 2.5 cm of agarose "worm" and place in a clean 50-mL Falcon tube.
2. Add 25 mL $T_{0.1}E$ and incubate 50°C for 30 min.
3. Pour off $T_{0.1}E$, and repeat step 2 twice more.

4. Pour off $T_{0.1}E$ and add 25 mL fresh $T_{0.1}E$. Wash at room temperature for 30 min with gentle shaking
5. Pour off $T_{0.1}E$ and repeat step 4 twice more.
6. Place the agarose "worm" in a 1.5-mL Eppendorf tube and remove any liquid carried over.
7. Melt the agarose fully at 65°C for approx 15 min.
8. Prewarm 700 µL $T_{0.1}E$ in a 1.5-mL Eppendorf tube to 65°C. Add 100 µL of the molten agarose, vortex briefly to mix, and incubate for an additional 5–10 min at 65°C. Vortex once more.
9. The diluted agarose pool is now ready for use. Store both the diluted PCR pool and the remaining neat melted agarose pool stock at 4°C. For rapid screening of libraries with large numbers of STSs, the pools may be aliquoted into 96-well microtiter plates, thus maximizing the use of multichannel pipets (*see* **Note 16**).

4. Notes

1. Total yeast DNA can be prepared in solution. However, the authors have found that the agarose plug-based method is simpler to perform, gives a more consistent yield of DNA, and the resulting pools give more reliable PCR results when compared to solution DNA preparations.
2. The authors have found that the spheroplasting step was not effective on Hybond N⁺ positively charged membranes.
3. It is essential not to overgrow the yeast colonies, because the centers of larger colonies are not penetrated by the subsequent spheroplasting and lysis solutions. This results in poor or halo-shaped hybridization signals around the edges of colonies only. Filters made from YACs stamped manually using a 96-pin replicating tool are particularly susceptible.
4. Plates can be poured and the filters laid on up to 3 d in advance.
5. Zymolyase does not go into solution easily. Mix the powder vigorously in a small volume of spheroplasting solution for 2–3 min prior to addition to the full volume. Mix again if allowed to stand prior to preparation of the spheroplasting trays.
6. To avoid excessive inhalation of β-mercaptoethenol, all relevant steps should be performed in a fume hood where practically possible.
7. YLS is toxic. Handle with care.
8. For robotically replicated YACs, the authors have observed that a more even growth occurs if gridding is postponed for at least one night following mixing of newly grown YACs with glycerol. This probably results from cells settling out to form an even layer of yeast on the bottom of the dish, rather than from the effects of freeze thawing.
9. Small air bubbles under individual colonies may not be immediately visible. Each filter should be checked again very carefully after several minutes by holding up to eye level. YACs over bubbles will not spheroplast and will be visible as darker colonies the following day.
10. Damp colonies will stick permanently to the second sheet of Whatman paper if laid on too soon.
11. Do not crosslink filters while they are wet/damp.
12. Some filters curl up during drying. Rubbing the surface of a Saran wrap (Dow Chemical Co., Uxbridge, Middlesex, UK)-covered UV-transilluminator hard with a tissue creates static, thus holding the filters down. Cover with Saran wrap and a piece of cardboard to keep flat during crosslinking.
13. Titrate conditions of crosslinking. In the authors' experience, crosslinking YAC filters using a standard UV transilluminator is more suitable and gives superior results compared to more specialized devices.

14. The average wet weight of cells from a single-stamped filter or two high-density gridded filters are approx 0.7 and 1 g, respectively.
15. To prevent precipitation of lithium dodecyl sulfate, all YLS must be rinsed out fully before storage of agarose plugs at 4°C (a minimum of three 30-min washes in $T_{0.1}E$ at 50°C as described in **Subheading 3.2.3., steps 2** and **3**).
16. The use of multichannel pipets vastly increases the rate of PCR screening. Pools are aliquoted into 96-well microtiter plates as described below. Multichannel pipets are used both to set up the PCR reactions directly from the microtiter plate into a 96-well PCR plate, and to load the agarose gels. Round bottomed microtiter plates for pool storage are available from Falcon. PCR plates (96-well) are available from Hybaid or Costar (Cambridge, MA). To increase the efficiency of gel loading, combs are designed to form lanes at twice the frequency of microtiter plate wells, hence, pools are arranged in the microtiter plate with odd numbers in the first column (pool 1 in A1, pool 3 in B1, pool 5 in C1, etc.) and even numbers in the second column (pool 2 in A2, pool 4 in B2, pool 6 in C2, etc.). Aliquot a maximum of 200 µL/well. Repeat this format until the plate is full. To prevent evaporation of the pools during storage, overlay each well with two drops of mineral oil and seal the plates with microtiter plate sealers. Following PCR, use the multichannel pipet to load column 1 into wells 1, 3, 5, 7, 9, 11, 13, and 15. Load PCR reactions in the second column into wells 2, 4, 6, 8, 10, 12, 14, and 16. In this way, the linear order 1, 2, 3, 4, and so on, is re-created on the gel. Pools representing several libraries may be aliquoted into a single microtiter plate. Single plate (secondary) pools may also be aliquoted into microtiter plates.

References

1. Green, E. D. and Olson, M. V (1990) Systematic screening of yeast artificial-chromosome libraries using the polymerase chain reaction. *Proc. Natl. Acad. Sci. USA* **87,** 1213–1217.
2. Bentley, D. R., Todd, C., Collins, J., Holland, J., Dunham, I., Hassock, S., et al. (1992) The development and application of automated gridding for efficient screening of yeast and bacterial ordered libraries. *Genomics* **12,** 534–541.
3. Amemiya, C. T., Alegria-Hartman, M. J., Aslanidis, C., Chen, C., Nikolic, J., Gringrich, J. C., and De Jong, P. J. (1992) A two-dimensional YAC pooling strategy for library screening via STS and Alu-PCR methods. *Nucleic Acids Res.* **20,** 2559–2563.
4. Chumakov, I., Rigault, P., Guillou, S., Ougen, P., Billaut, A., Guasconi, G., et al. (1992) Continuum of overlapping clones spanning the entire human chromosome 21q. *Nature (Lond.)* **359,** 380–387.
5. Foote, S., Vollrath, D., Hilton, A., and Page, D. C. (1992) The human Y chromosome: overlapping DNA clones spanning the euchromatic regions. *Science* **258,** 60–66.
6. Nizetic, D. N., Zehetner, G., Monaco, A. P., Gellen, L., Young, B. D., and Lehrach, H. (1991) Construction, arraying and high density screening of large insert libraries of the human chromosomes X and 21: their potential use as reference libraries. *Proc. Natl. Acad. Sci. USA* **88,** 3233–3237.
7. Copeland, A. and Lennon, G. (1994) Rapid arrayed filter production using the "ORCA" robot. *Nature (Lond.)* **369,** 421,422 (product review).
8. McKeown, G., Watson, A., Karunaratne, K., and Bentley D. (1993) High throughput filter preparation robot. *Genome Science and Technology* (special first issue), Program and Abstracts Genome Sequencing and Analysis Conference V, October 23–27, 1993, p. 56 (abstract C 21).

59

YAC Library Screening

Hybridization and PCR-Based Screening Protocols

Charlotte G. Cole, John E. Collins, and Ian Dunham

1. Introduction

Yeast artificial chromosome (YAC) libraries stored in microtiter plates are available for screening as either complex PCR pools or hybridization filters generated from YACs gridded at high densities (*see* Chapter 57). Different libraries may be available as either PCR pools, hybridization filters, or both. Consequently, screening strategies have been designed that rely solely on either technique, or use a combined approach. Clearly, access to the YAC library microtiter plates and an automatic gridding system allows the user greater flexibility and is an advantage in large mapping projects.

Hybridization filters containing YACs gridded at high density locate each positive YAC to an individual microtiter plate well coordinate in one experiment. The precise arrangement and number of YACs on a filter will vary depending on the robot employed, and different centers will grid the same library in quite different arrays. For the purposes of this chapter, hybridization of nylon filters generated using the customized robot employed in the authors' laboratory are described (**Fig. 1**, **refs.** *1,2*).

A number of different PCR-screening strategies have been described *(1,3–5)*. The more complex PCR pools based on high density row and column pools or combinations of 1/2 plates may be able to locate a positive PCR signal to a single microtiter plate well using relatively few PCR pools. However, these may also give increased background noise. For the purposes of this chapter, use of a relatively simple pooling strategy is described (*see* **Fig. 2**). Here, each primary PCR pool contains total yeast DNA isolated from sixteen 96-well microtiter plates, or 1536 yeast colonies. The secondary PCR pools contain the DNA derived from each of the appropriate 16 plates individually. Once a YAC has been located to a single microtiter plate pools representing the 8 rows and 12 columns of the 96-well microtiter plate can be prepared and screened, thus locating the positive YAC to a single well (**Fig. 3**). Alternatively, a hybridization filter may be prepared and screened using the labeled sequence tagged site (STS) as a probe (*see* **Note 1**).

High Density YAC Filters

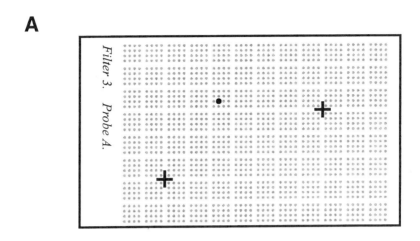

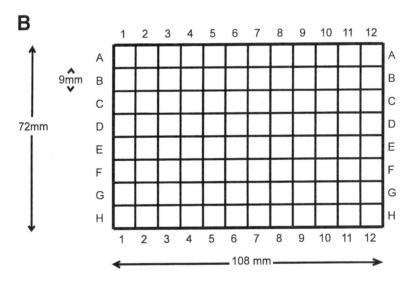

Fig. 1. High-density YAC filters. **(A)** Representation of a hybridization result showing the 4 × 4 array of 1536 clones gridded from 16 microtiter plates onto a single 8 × 12 cm filter. Autoradiographs and filters are labeled in the top left-hand corner, corresponding to position A1 of the microtiter plate. Within each square of 16 YACs the 16 microtiter plates are gridded as follows: row 1, plates 1–4; row 2, plates 5–8; row 3, plates 9–12; row 4, plates 13–16. Filter 3 from a library is shown, therefore this filter contains YACs from microtiter plates 33–48 (*see* **Fig. 2**). The black crosses are drawn on to the autoradiograph to enable accurate interpretation of results with the help of the template shown in (B). **(B)** Scale drawing of the templates used to aid interpretation of high-density YAC filter autoradiographs. Templates are prepared on overhead projector acetate sheets using fine, pale-colored pens. A positive signal is seen in (A) as follows: filter 3, C5, position 10. This signal therefore originates from microtiter plate 42, position C5.

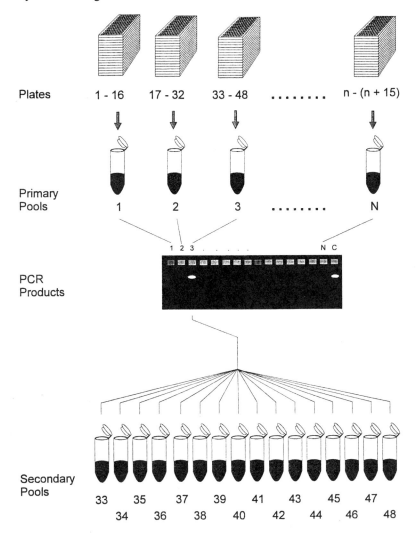

Fig. 2. YAC library pooling scheme. Total yeast DNA from 16 microtiter plates of YAC clones have been combined in each primary PCR pool to give primary pools 1 to N. The PCR is performed on each pool of DNA. Following gel electrophoresis of the PCR products, pool 3 is identified as containing at least one positive YAC with the help of the genomic positive control (lane marked C) and a DNA marker (i.e., 1-kb ladder, not shown). The PCR is performed on the 16 secondary PCR pools (pools 33–48) that correspond to primary pool 3. Gel electrophoresis would be performed to identify a positive secondary pool, thus identifying an individual positive YAC library microtiter plate.

2. Materials

2.1. Hybridization

1. YAC hybridization filters prepared as described in Chapter 58.
2. 20X SSC: 3 M NaCl, 0.3 M Tri-sodium citrate, pH 7.0.
3. 100X Denhardts: 20 mg/mL Ficoll 400-DL, 20 mg/mL polyvinylpyrrolidone 40, 20 mg/mL BSA pentax fraction V.

Rows and Columns Strategy

A Pooling

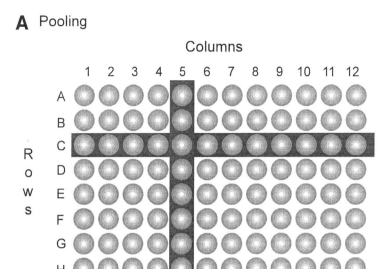

B Rows and columns PCR

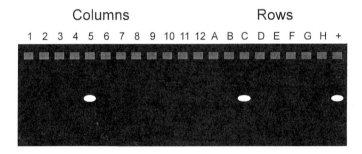

Fig. 3. Row and column strategy. (**A**) Representation of a 96-well microtiter plate containing colony dilutions in $T_{0.1}E$ stamped from a single YAC library plate. An aliquot of YAC colony from each of the 12 columns and 8 rows are combined to generate rows and columns pools. For instance, the eight colony dilutions combined to generate PCR pool "column 5" are highlighted with the vertical block. The 12 colony dilutions combined to generate PCR pool "row C" are highlighted with the horizontal block. (**B**) Representation of results of gel electrophoresis of the 12 column and 8 row PCR pools, showing positive signals from column 5, row C, and a genomic positive control (+). Thus, a positive PCR result detected in the secondary PCR pool representing this plate has been localized to plate position C5.

4. Hybridization buffer: 6X SSC, 1% sarkosyl, 10X Denhardt's solution, 50 mM Tris-HCl, pH 7.4, 10% dextran sulfate.
5. 10–20 ng DNA probe.
6. Total yeast DNA: 40 ng at approx 10 ng/µL (*see* **Note 2**).
7. A-Minus nucleotide mix: 1 mL 1.25 M Tris-HCl/0.125 M MgCl$_2$, 18 µL β-mercaptoethanol, 5 µL 100 mM dCTP, 5 µL 100 mM dTTP, and 5 µL 100 mM dGTP.

8. α-^{35}S-(thio)dATP (600 Ci/mmol).
9. Sheared human placental DNA (10 mg/mL, e.g., Sigma [St. Louis, MO] D-3287) (*see* **Note 3**).
10. Wash solutions:
    a. 2X SSC.
    b. 0.5X SSC, 1% sarkosyl.
    c. 0.2X SSC, 1% sarkosyl.
    d. 0.2X SSC.
11. Autoradiography film (e.g., Dupont [Boston, MA] Cronex 4, cat. no. 6603478) and cassettes with intensifying screens.
12. Luminescent sticker (e.g., glo-gos, Stratagene, La Jolla, CA).

2.2. PCR Screening

1. PCR pools prepared as described in Chapter 58.
2. Genomic DNA (15 ng/µL) derived from the species in question (i.e., human, mouse, etc.) containing the target sequence for use as a positive control.
3. 10X PCR buffer: 670 mM Tris-HCl, pH 8.8, 166 mM enzyme grade $(NH_4)_2SO_4$, 67 mM $MgCl_2$ (*see* **Note 4**).
4. 10X Nucleotide mix: 5 mM each dNTP.
5. 10X Primer mix: 13 µM each primer.
6. 5 mg/mL BSA (Sigma A-4628).
7. 700 mM β-mercaptoethanol: 1 in 20 dilution of 14 M β-mercaptoethanol in sterile water.
8. *Taq* polymerase.
9. PCR grade mineral oil (e.g., Sigma 8042-47-5).
10. $T_{0.1}E$: 10 mM Tris-HCl, 0.1 mM EDTA, pH 8.
11. 2.5% Agarose, 1X TBE minigels containing 0.2 µg/mL ethidium bromide in the gel and the running buffer (*see* **Note 5**).
12. 15-cm Circular YPD-agar plates (*see* **Note 6**).
13. 96-Well microtiter plates.
14. 96-Pin replicating tool (*see* Chapter 57).
15. Access to YAC library microtiter plates.

3. Methods

3.1. Hybridization-Based Screening

3.1.1. Preparation of Labeled Probe

1. Prepare labeled probe using standard techniques (0.25–1 × 10^9 cpm/µg of probe; *see* **Note 7**).
2. Separate labeled probe from unincorporated label if desired (*see* **Note 8**).
3. Add $T_{0.1}E$ to give a final volume of 125 µL (*see* **Note 9**).
4. If competition is required, add 125 µL sheared placental DNA, 125 µL $T_{0.1}E$, and 125 µL 20X SSC. Mix well (*see* **Note 10**).
5. Boil for 5 min, snap chill on water/ice mix. If labeled yeast background is used (*see* **Subheading 3.1.2.**) this can be added to the probe mix before boiling.

3.1.2. Preparation of Labeled Yeast Background (Optional)

1. Prepare ^{35}S-labeled yeast background. Use 40 ng of total yeast DNA in a routine random prime hexamer labeling, but substitute 5 µL [α-^{35}S]dATP in place of [α-^{32}P]dCTP and use A-minus nucleotide mix (*see* **Note 11**).
2. Following labeling, add 100 µL $T_{0.1}E$/25 µL labeled yeast background (do not separate unincorporated label). Store at –20 or –70°C.

3. Allow approx 0.5 µL/10 mL of hybridization buffer. Either add directly to the probe before boiling or boil for 5 min, snap chill on water/ice mix, and add to the hybridization mix separately.

3.1.3. Pretreatment and Hybridization of Nylon Filters

1. Ensure the hybridization filters have been UV crosslinked (*see* Chapter 58, **Subheading 3.1.2., step 12**).
2. Prehybridize for 1 h at 65°C, preferably leave for 2–3 h.
   a. For library screens of up to 25 filters (*see* **Note 12**) the authors use empty rectangular 8 × 12 cm gridding plates (Chapter 58, **Subheading 2.1., item 2**). Use approx 40 mL hybridization buffer for 25 filters, adding the filters individually to the hybridization buffer. Cover with a sheet of plastic (e.g., as used for hybridization bags) and the plate lid. Seal in a sandwich box containing a moist tissue to prevent drying.
   b. For hybridization of one to three filters, the authors use a 15-mL round-bottomed plastic tube. Fill the tube with hybridization buffer, roll up the filter (colony side inward), and use tweezers to slide the filter down to the bottom of the tube where it will unwind. Hybridize upright in an orbital shaker.
3. Add the boiled probe to the filters.
   a. If using filters layered in a box, remove filters from the box draining excess hybridization buffer back into the box. Add the probe to the buffer and mix well. Add filters back into the box one at a time (colony side down), covering each filter with hybridization buffer. Cover with plastic and seal as described in **step 2a**.
   b. If hybridizing filters in 15-mL tubes, add the probe into the center of the tube, re-cap, and invert gently at least 6–10 times to mix.
4. Hybridize with gentle shaking or 18–48 h, 65°C.

3.1.4. Washing and Autoradiography of Filters

1. Pour off hybridization buffer and wash filters in an excess of wash solutions (use at least 500 mL/25 filters).
   a. Wash twice in wash solution 1 for 5 min at room temperature with gentle shaking.
   b. Wash for 30 min in wash solution 2 at 65°C with gentle shaking.
   c. Wash for 30 min in wash solution 3 at 65°C with gentle shaking (*see* **Note 13**).
   d. Wash twice in wash solution 4 for 5 min at room temperature with gentle shaking.
2. Smooth a sheet of Saran Wrap (Dow, Uxbridge, Middlesex, UK) just larger than the cassette onto a flat surface.
3. Briefly touch the back of the filter onto a sheet of 3MM Whatman (Maidstone, UK) to drain excess liquid and lay the filters colony side down onto the smoothed out Saran Wrap (*see* **Note 14**). Do not touch the colony side onto the Whatman paper and avoid lifting the filter from the Saran Wrap since colony debris and associated probe may stick to the Saran Wrap.
4. Cover with a second sheet of Saran Wrap. Do not rub smooth the colony side of the Saran Wrap, particularly after freezing, as the colony debris and associated probe will smear over the filter.
5. Stick a glo-go on to a clear area of Saran Wrap.
6. Preflash the film and autoradiograph for 5 h to 7 d as necessary (*see* **Note 15**).
7. Lay the autoradiography film over the wrapped up filters and mark the position of the corners of the filter and the position of each filter name.
8. Filters may be rewashed if necessary (*see* **Note 14**). Carefully peel the filters off the Saran Wrap (*see* **Note 16**) and wash as required, including a final room temperature wash without sarkosyl. Re-expose as described.

9. Identify positive clones using a reference template if necessary. Prepare the template by using a pale (e.g., not black) fine marker pen to draw a grid on an overhead acetate sheet. If available use an autoradiograph with good yeast background signal as a model (see **Fig. 1**).
10. Mark small crosses on the autoradiograph at the intersections of several 4 × 4 squares of clones (see **Fig. 1**; **Note 17**). Line up the crosses on the film with the intersections on the grid, making sure to keep the orientation of the labeled film and template correct. (The authors label filters in the A1 position.)
11. Each box on the template surrounds the clones gridded in a 4 × 4 array from a specific well position in 16 microtiter plates. If background is not uniformly visible, the position within a square on the template is used to determine the source of the positive signal.

3.1.5. Rescreening of Positive Colonies

1. Thaw the relevant working copies of the library plate.
2. Streak to single colonies on AHC agar plates.
3. Grow for two nights at 30°C.
4. Fill 96-well microtiter plates with 150 µL YPD broth.
5. If purification to single colonies is required pick four to six colonies into individual wells of the microtiter dish (see **Note 18**). Pick a mixture of colonies into a final well. Incubate at 30°C for 2 nights. If single colonies are not desired, pick a mixture of colonies into a single well of the microtiter dish. Grow for 2 nights at 30°C.
6. Duplicate and freeze the microtiter plates as described in Chapter 57, **Subheading 3.4**.
7. Using one of the copies only, stamp or grid the rescreen plates and prepare hybridization filters as described in Chapter 58. Hybridize the filters as described.
8. Use the second plate for archiving the appropriate clones (see **Note 19**).

3.2. PCR-Based Screening

3.2.1. Primary Pool PCR

This method is based on the use of 0.5-mL microcentrifuge tubes. For rapid throughput of STSs, pools are stored in 96-well microtiter plates enabling the use of multichannel pipets to set up the reactions in PCR microtiter plates and to load the agarose gels (see **Note 16**, Chapter 57).

1. Prepare a PCR premix as follows (allow approx one spare tube worth for every 10 yeast primary pools, plus 2 tubes worth for controls): For each reaction tube, mix 1.5 µL 10X PCR buffer, 1.5 µL 10X dNTPs, 1.5 µL 10X primer mix, 0.49 µL 5 mg/mL BSA, 0.21 µL 700 mM β-mercaptoethanol, 0.1 µL *Taq* polymerase (0.5 U), 6.7 µL $T_{0.1}$E. Vortex to mix.
2. Place a 3-µL droplet of each yeast primary pool onto the side of a 0.5-µL Eppendorf tube approx 5 mm below the lip of the tube.
3. Place a 3 µL droplet of $T_{0.1}$E and 3 µL 25 ng/µL genomic DNA onto the edge of the negative and positive control tubes.
4. Add 12 µL of PCR premix into each tube above the DNA (it will mix with the DNA by gravity). You do not need to change the pipet tip between aliquots of premix unless you have touched the DNA.
5. Add a drop of paraffin oil to each tube above the DNA/premix. The DNA/premix should mix and drop to the bottom of the tube below the oil without centrifugation.
6. Perform 35 cycles of PCR (see **Note 20**).
7. Test 5 µL of each PCR product on a 2.5% agarose minigel prepared as described in **Subheading 2**. Include a suitable size marker at the end of each lane (see **Note 21**). Electrophorese until the bromophenol blue (if used in loading buffer) has migrated approx 3–5 cm.

8. Photograph gel under UV.
9. Identify all positive pools (*see* **Notes 20, 22,** and **23**).

3.2.2. Secondary Pool PCR

1. For each positive primary pool, identify the relevant secondary pools. The simple pooling system described in **Fig. 2** identifies 16 secondary pools for each primary pool.
2. Prepare a PCR premix sufficient for each set of 16 secondary pools, plus the positive primary pool and a genomic and $T_{0.1}E$ control. Perform the PCR as described in detail in **Subheading 3.2.1.**
3. Identify the positive secondary pools. Each pool represents the contents of a single microtiter plate in the pooling system just described (*see* **Note 24**).

3.2.3. Preparation of "Rows and Columns" PCR

Preparation of rows and columns PCR pools can be avoided by the use of hybridization to nylon filters (*see* **Note 1**).

1. Thaw YAC library (working) microtiter plate corresponding to each positive secondary pool.
2. Stamp the contents of each microtiter plate onto circular YPD agar plates using the 96-pin hedgehog as described in Chapter 57. Be sure to mark the appropriate position on the plate with "A1," corresponding to position A1 in the microtiter plate.
3. Incubate the plates at 30°C for 2 nights.
4. For each stamped plate, fill each well of two 96-well microtiter dishes with 100 µL PCR-grade $T_{0.1}E$ using a multichannel pipet if available.
5. Using the 96-pin "hedgehog" stamp the YACs from each agar plate into the two filled microtiter dishes. Be sure to preserve the A1 position of the stamped agar plate and the microtiter dish (*see* **Note 25**). Label one plate "rows A–H" and the second plate "columns 1–12."
6. Store the agar plate at 4°C for future use.
7. Take the microtiter plate labeled "rows." Mix using a pipet and transfer 10 µL of colony suspension from each of the 12 wells in row A (A1–A12) to the single well in position A1 of a fresh microtiter dish (**Fig. 3**). You do not need to change pipet tips between each of the 12 wells. If you have access to a multichannel pipeter, **steps 7, 8,** and **9** can be performed simultaneously.
8. Repeat **step 7**, but transfer 10 µL of colony suspension from each of the 12 wells in row B (B1–B12) to position B1 of the new microtiter dish.
9. Repeat **steps 7** and **8** for each of the remaining rows C to H until A1–H1 of the new microtiter dish have been filled with YAC pools representing rows A to H of the YAC library microtiter plate. Using a pipet mix each of the wells to form a uniform colony suspension of all the YACs present.
10. Take the plate labeled "columns." Mix using a pipet and transfer 10 µL of colony suspension from each of the 8 wells in column 1 (A1–H1) to the single well in position A1 of a second fresh microtiter dish. You do not need to change pipet tips between each of the 8 wells. If you have access to a multichannel pipeter, **steps 10, 11,** and **12** can be performed simultaneously.
11. Transfer 10 µL of colony suspension from each of the 8 wells in column 2 (A2–H2) to the single well in position A2 of the new microtiter dish as described in **step 9**.
12. Repeat **steps 10** and **11** for each of the remaining columns 3 to 12 until A1–A12 of the new microtiter dish have been filled with YAC pools representing columns 1 to 12 of the YAC library microtiter plate. Using a pipet mix each of the wells to form a uniform colony suspension of all the YACs present.

13. The microtiter dishes containing the colony suspensions can be stored at 4°C for several months or more. It is advisable to seal the wells to prevent evaporation. The authors use microtiter plate sealers (Dynatech [Chantilly, VA], cat. no. 001-010-5701).

3.2.4. Rows and Columns PCR

1. Set up a PCR premix sufficient for each set of 20 rows and columns PCR reactions (8 rows and 12 columns), plus the positive secondary pool, a genomic and a $T_{0.1}E$ control as described in **Subheading 3.2.1., step 1** (*see* **Note 4**).
2. Aliquot 12 µL of PCR premix into each of, for example, 23 tubes (for a single rows and columns PCR).
3. Mix the colony pools well using a pipet and add 3 µL of colony suspension, 3 µL of genomic DNA, or 3 µL of $T_{0.1}E$ into the PCR premixes as appropriate.
4. Add a drop of mineral oil to each tube.
5. Perform the PCR using appropriate conditions for 35 cycles (*see* **Note 26**).
6. Test 5 µL of PCR product on a 2.5% agarose minigel as described in **Subheading 2.** and **Subheading 3.2.1., step 7**.
7. Photograph under UV and identify the positive pools. You expect one of the 8 rows (A–H) and 1 of the 12 columns (1–12) to be positive. Positive signals of the expected size in, for example, row pool C and column pool 5 determines that the positive YAC is expected to reside in position C5 of the original YAC library plate (**Fig. 3**, *see* **Notes 27–29**).

3.2.5. Confirmation by Colony PCR

The results of a rows and columns pool result must be confirmed by colony PCR on the single "positive" YAC identified (*see* **Notes 4** and **26**).

1. Set up a PCR premix sufficient for each colony to be confirmed, plus a genomic and a $T_{0.1}E$ control as described in **Subheading 3.2.1., step 1** and aliquot 12 µL into each 0.5-mL PCR tube.
2. Use the stamped agar plate representing the relevant YAC library plate (stored at 4°C, **Subheading 3.2.3., step 6**) to add the deduced positive YAC to the PCR premix:
   a. Either: Use a toothpick or yellow tip to suspend a small "blob" of colony in 100 µL of $T_{0.1}E$ and add 3 µL of the resulting colony suspension to the PCR,
   b. Or: Touch a yellow tip just onto the colony and stir directly into the PCR. In this case add 3 µL of $T_{0.1}E$ to the PCR tube to make up the volume to 15 µL.
3. Perform the PCR as appropriate for 30–35 cycles and test on a 2.5% agarose minigel as described in **Subheading 3.2.1., step 7** (*see* **Note 30**).
4. Spreading of YACs between microtiter plate wells during manipulation of the library can result in mixed colonies within a single well. To confirm absolutely the presence of two STSs within a single YAC, streak the YAC to single colonies on AHC agar plates, pick individual colonies into 100 µL $T_{0.1}E$ for use in PCR testing as described. Use the same toothpick to patch the colony onto YPD agar plates for growth and future archiving. Use the colony suspension to test individual colonies with each STS (*see* **Notes 19** and **20**).

4. Notes

1. Generation of numerous row and column pools can be avoided by the preparation of a high density hybridization filter containing all the microtiter plates to be tested with one or more STSs. This is the authors' method of choice when high throughput STS testing is required. Alternatively, a hybridization filter containing a single microtiter plate may be prepared manually. *See* Chapter 57 for preparation of hybridization filters. STSs are labeled by PCR-labeling (*see* **ref. *1***; **Note 7**).

2. **Items 6–8** in **Subheading 2.1.** are optional for labeling yeast background if required.
3. **Item 9** in **Subheading 2.1.** is optional for prereassociation of hybridization probes where required.
4. Many different PCR buffers are available. Some may work better than others for particular STSs. However, the authors have found the buffer/PCR conditions given to be most reliable, particularly when used in a yeast colony-PCR.
5. We use 7.5-cm gels with two 23-well combs placed at 1.5 and 4.5 cm, respectively (Flowgen [Sittingbourne, Kent, UK] minigel apparatus, 23-well combs available from Flowgen on request).
6. Items 12–15, **Subheading 2.2.** are for preparation of single plate rows and columns if required.
7. Labeling may be performed using standard random primed hexamer labeling techniques or by PCR labeling *(1)*. PCR labeling is particularly effective for labeling STS for use in library screens or secondary screens following primary and/or secondary PCR. Use 3–5 µL [α-^{32}P]dCTP (3000 ci/mmol) for a full library screen in up to 40 mL hybridization buffer. Use 0.5–1 µL (1 µL for PCR labeling) of [α-^{32}P]dCTP for hybridization of 1–3 filters in 15 mL buffer.
8. Separation of labeled probe from unincorporated nucleotides is not necessary in order to achieve strong, clean signals and is only required to check labeling efficiency. The authors routinely use labeled probes directly.
9. If PCR labeling is used, add a small amount of a 0.5 % phenol red/dextran blue dye mix to help visualize separation of probe from mineral oil.
10. Contrary to conventional advice, the authors have found it is possible to add the boiled probe/human DNA mix directly to the hybridization filters without prior incubation. This includes labeled *Alu*-PCR probes derived from YACs for use in 2- or 3-filter hybridizations.
11. Labeled yeast DNA background is not always necessary. Its inclusion depends on the type of probe used and the washing stringency. Some probes, e.g., small single copy STSs, may give no background hybridization signals on negative yeast colonies. In contrast, *Alu*-PCR probes and cosmids generally give adequate background when used alone.
12. Before using a newly prepared probe on a large number of filters it is advisable to hybridize and wash a single filter first using an aliquot of labeled probe. Freeze the remaining probe until the filter has been checked for unacceptably high (and potentially irreversible) background hybridization signals.
13. Although these conditions routinely work well, the stringency of washing conditions required will vary depending on the probe employed.
14. Do not dry excessively, since this compromises the effectiveness of rewashing.
15. Unless the filters sound very hot (>20 cpm), the authors routinely test after approx 16 h and take a second exposure as required.
16. It is normal to observe some (radioactive) colony debris stuck to the Saran wrap.
17. If the background is faint these may be difficult to see at first. Look hard, holding the film up to the light at different angles or take a longer exposure.
18. It is advisable to use mixed isolates rather than individual colonies in subsequent tests owing to the possibility of part of the YAC insert in specific colonies deleting during growth. In regions prone to deletions isolation to individual single colonies may never be advisable.
19. Once a colony result has been confirmed, the relevant yeast clone should be grown up for archiving and future testing. The authors pass all clones through selection on an AHC agar plate and pick a mixture of colonies into several microtiter plates. These are grown as

described in Chapter 57 and transferred to sets of microtiter plates containing our entire collection of isolated YACs. The authors recommend preparing a master plate (which is rarely used, and never employed for robotic gridding, etc.), a working master (remade when necessary from the master), and several working plates (remade from the working master).

20. Yeast pools (especially complex primary pools) may give PCR background, whereas human and rodent samples do not. Raising the annealing temperature and lowering the extension time can eliminate this problem. However, persistent yeast-specific fragments may be ignored if they do not interfere with the human-specific band. Yeast-specific fragments cannot be ignored if they overlap in size or they amplify significantly more efficiently and may dominate amplification in a particular PCR tube.
21. To ensure correct sizing of PCR fragments, the authors recommend loading the products of amplification of human DNA at both ends of each comb.
22. Check immediately that the expected size has been amplified.
23. Amplification of significantly more positives than expected from the library complexity indicates a potential problem because these signals are unlikely to result from amplification of a single copy human DNA sequence. It is more likely that they represent either human repeat sequences or intermittent yeast background.
24. The number of positive secondary pools per positive primary pool will depend on the library complexity and the arrangement of pools. The pooling scheme described in Chapter 58 normally yields one (occasionally two) positive secondary pools per primary pool for a <6-hit library.
25. Take care not to transfer agar to the microtiter dish and check yeast colony has transferred to each well.
26. Colony suspensions do not appear to amplify consistently in PCR machines that use heated top plates in place of mineral oil (e.g., Cetus 9600; Eric Green, personal communication). In this case heat the colony suspension (98°C, 10 min), spin, and geneclean (Bio 101) the suspension before use, or add mineral oil to the PCR.
27. If more than one positive in either the 8-row or 12-column PCRs are observed, the position indicated by the strongest row and column PCR is probably correct. Other signals may result from contamination during preparation of the rows and columns pools. Alternatively, one clone may be present twice on the same microtiter plate following spreading during manipulation of the library. However, two strong signals in each row and column set could indicate the presence of different YACs positive for the same STS in a single plate. Test all combinations of rows/columns by colony PCR (*see* **Subheading 3.2.5.**) to identify the correct microtiter plate position(s).
28. Weak PCR signals in hand-prepared rows and columns pools are not uncommon. Test all possible positive colonies by colony PCR (*see* **Subheading 3.2.5.**).
29. If either the row or columns do not give a positive signal, test all 12 or 8 colonies in the relevant positive row or column, respectively, by colony PCR (*see* **Subheading 3.2.5.**).
30. If the expected colony is not positive, the orientation of the row or column pools may have been reversed during their preparation. Test possible alternatives.

References

1. Bentley, D. R.. Todd, C., Collins J., Holland, J., Dunham, I., Hassock, S., et al. (1992) The development and application of automated gridding for efficient screening of yeast and bacterial ordered libraries. *Genomics* **12,** 534–541.
2. McKeown, G., Watson, A., Karunaratne, K., and Bentley D. (1993) High throughput filter preparation robot. *Genome Sci. Technol.* (special first issue), Program and Abstracts Genome Sequencing and Analysis Conference V, October 23–27, 1993, p. 56 (abstract C 21).

3. Green, E. D. and Olson, M. V. (1990) Systematic screening of yeast artificial-chromosome libraries using the polymerase chain reaction. *Proc. Natl. Acad. Sci. USA* **87,** 1213–1217.
4. Amemiya, C. T., Alegria-Hartman, M. J., Aslanidis, C., Chen, C., Nikolic, J., Gringrich, J. C., and De Jong, P. J. (1992) A two-dimensional YAC pooling strategy for library screening via STS and Alu-PCR methods. *Nucleic Acids Res.* **20,** 2559–2563.
5. Chumakov, I., Rigault, P., Guillou, S., Ougen, P., Billaut, A., Guasconi, G., et al. (1992) Continuum of overlapping clones spanning the entire human chromosome 21q. *Nature (Lond.)* **359,** 380–387.

60

Phage-Display Libraries of Murine and Human Antibody Fab Fragments

Jan Engberg, Lene K. Johansen, Michelle Westengaard-Hildinge, Erik S. Riise, and Bjarne Albrechtsen

1. Introduction

This chapter describes efficient procedures for construction, expression, and screening of comprehensive libraries of murine or human antibody Fab fragments displayed on the surface of filamentous phage. Phagemid vectors are used for placing randomly paired light (L) and heavy (H) chain-coding regions under transcriptional control of P_{lac}. The L (or H) chain-coding region is fused in-frame with the truncated phage gene, $\Delta gIII$, coding for a truncated version of the phage surface protein pIII (ΔpIII). After superinfection with helper phage and induction of P_{lac}, Fd (composed of V_H and C_H1 domains), and κ (or λ) L chains assemble into Fab fragments in the periplasm, and the Fab-ΔpIII protein complex is displayed at one end of the phage by displacing one (or more) of the wild-type pIII proteins. Enrichment of Fab phages with affinity for a specific antigen is then done by successive rounds of affinity purification in antigen-coated microtiter wells or immunotubes and reinfection of *Escherichia coli* cells by the eluted bound phages *(1–6)*.

We present protocols for making murine Fab libraries of the IgG isotype and murine and human Fab libraries independently of IgH isotype. In the protocol for making murine Fab libraries of the IgG isotype, Fd and κ coding regions are polymerase chain reaction (PCR) amplified and assembled into a Fab gene cassette by the "jumping PCR" technique before being cloned *(4)*, or these coding regions are cloned separately *(7)*. In the protocol for making murine libraries independently of IgH isotype, V_H gene fragments are amplified using *J*-region primers and joined to $C_H1_{\gamma 1}$ gene fragments. In one variant of the protocol, V_H and $C_H1_{\gamma 1}$ DNA is fused by PCR and linked to PCR-generated κ genes by jumping PCR before cloning *(8)*. In the other variant, κ and V_H DNA is cloned in two steps into a specially developed vector that already contains $C_H1_{\gamma 1}$ fused to $\Delta gIII$ *(7)*. For production of human Fab libraries independently of IgH isotype, V_H and κ (or λ) coding regions are cloned separately into a vector that harbors human $C_H1_{\gamma 1}$ in fusion with $\Delta gIII$ *(9)*.

In addition, protocols for producing and using ultraelectrocompetent cells, for producing Fab phages from libraries, and for selecting antigen binders by panning are presented. This chapter is an updated version of our previous protocols *(10)*.

2. Materials

2.1. Extraction of Total RNA from Mouse Spleen or Human Blood Lymphocytes and Production of cDNA

1. Buffer A: Add 0.36 mL β-mercaptoethanol to 50 mL buffer B just before use.
2. Buffer B: 4 M Guanidinium thiocyanate, 25 mM sodium citrate, pH 7.0, 0.5% sarcosyl (filter sterilized, stable at room temperature for 3 mo).
3. Anticoagulant buffer (ACD from Baxter Health Care Ltd., Thetford, UK).
4. PBS buffer: 8.1 mM Na$_2$HPO$_4$, 1.47 mM KH$_2$PO$_4$, 2.68 mM KCl, 137 mM NaCl, pH 7.4.
5. Lymphoprep from Nycomed Pharma A/S, Oslo, Norway.
6. 2 M Sodium acetate pH 4.1: Mix 41 mL 2 M CH$_3$COOH and 9 mL 2 M NaOAc.
7. Chloroform:isoamyl alcohol (49:1): 49 Vol chloroform and 1 vol isoamyl alcohol.
8. Diethylpyrocarbonate (DEPC) water: double-distilled water containing 0.1% DEPC is incubated at 37°C for 60 min and autoclaved.
9. Water-saturated (DEPC) phenol: phenol is extracted three times with DEPC-treated water.
10. 5 mM dNTP: 5 mM dTTP, 5 mM dATP, 5 mM dGTP, and 5 mM dCTP (100 mM lithium salt solutions, pH 7.0, from Boehringer Mannheim [Mannheim, Germany] are diluted with DEPC-treated water).
11. 5X First-strand buffer (Gibco-BRL, Gaithersburg, MD): 0.25 M Tris-HCl, pH 8.3 at room temperature, 0.375 M KCl and 15 mM MgCl$_2$.
12. Superscript reverse transcriptase (200 U/µL; Gibco-BRL).

2.2. PCR Assembly Method for Constructing Murine Antibody Fab Libraries of the IgG Isotype (Fig. 1)

The aggregate concentration of all PCR primer mixes is 20 pmol/µL (20 µM).

1. Murine H-chain-variable region back primers (*see* **Notes 1** and **2**) (*see* Sequence 1): (One-letter nucleotide symbols are used according to International Union of Biochemistry nomenclature.) The murine variable heavy (MVH) primers consist of 25 individually synthesized oligos representing 88 variants. The nucleotides in bold correspond to the *N*-terminal part of the variable heavy (V$_H$) sequences starting with amino acid codon number 1. Nucleotides 1–13 at the 5' terminus correspond to the C-terminal part of the *pelB* leader. The concentration of each primer in the mixture used for PCR is 0.227 pmol/µL.
2. Murine C$_H$1-chain forward primers (*see* Sequence 2): Nucleotides in bold are complementary to the sequence of the constant heavy (C$_H$1) and hinge junction regions of the γ1, γ2a, and γ2b H-chain isotypes. The stop codon is underlined and the sequence positions 1–26 overlap the 5'-end of the Link fragment, positions 1–26. The concentration of each primer in the MCH1 mix used for PCR is 6.67 µM.
3. Murine κ L-chain back primers (*see* Sequence 3): These primers consist of 25 individually synthesized oligos representing 50 variants. The concentration of each variant in the mixture used for PCR is 0.4 pmol/µL. Nucleotides in bold correspond to the 5'-end of the κ-chain sequences starting with amino acid codon number 1. The *N*-terminal two amino acids of the V region are invariant. Sequence positions 1–24, corresponding to the 3'-end of the *pelB* leader, overlap the Link fragment, positions 94–117 (*see* **step 7**, following).
4. Murine C$_K$-chain forward primer (*see* Sequence 4): The sequence in bold is complementary to the 3' end of the Cκ gene. The *Not*I recognition site is underlined.
5. Extension (linker assembly) primers for Fd chains (*see* **Note 3**): Tag.Back 1 and 2 (*see* Sequence 5): The nucleotides in bold overlap with the MVH primers. The *Sfi*I recognition site is underlined. The concentration of each primer variant in the solution used for PCR is 1.25 µM. Link.For (*see* Sequence 6) is complementary to the 3' end of the Link fragment.

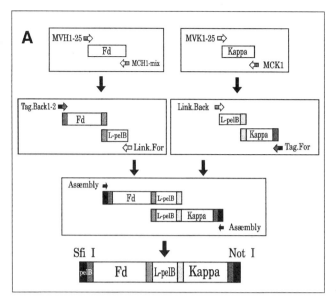
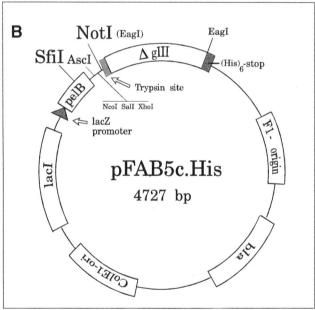

Fig. 1. Overview of primers and steps involved in PCR assembly of murine Fab gene fragments of the IgG isotype and a map of expression vector pFab5c.His. (**A**) Strategy for "Jumping-PCR" assembly. Diagram exemplifying the primary amplification PCR, the linker assembly, and the final assembly, respectively. Boxes show different gene segments included in the construction of the Fab expression cassette: Fd, includes the H chain from the *N*-terminal amino acid to the cysteine residue of the hinge region, which forms the disulphide bridge to the C-terminal cysteine of the kappa L chain; kappa corresponds to the entire variable and constant parts of the light chain; L-pelB = Link, a 117-bp DNA linker fragment that contains a translational stop codon for Fd translation, a ribosome binding site for L-chain expression and the coding region corresponding to the *N*-terminal part of the *pel*B leader. (**B**) pFab5c.His. Phagemid vector used for cloning and expression of murine antibody Fab gene fragments. The Fab antibody DNA cassette is introduced as a *Sfi*I-*Not*I fragment and is expressed from the inducible P_{lac} promoter. The vector is described in **step 16**, **Subheading 2.2.**, and in **ref. 4**.

6. Extension (linker assembly) primers for κ chains: Link.Back (*see* Sequence 7) overlaps the 5' end of the Link fragment. Tag.For (*see* Sequence 8): The sequence in bold corresponds to that of MCK1, positions 1–22. The *Not*I site is underlined.
7. The Link fragment (*see* Sequence 9): The boldfaced sequence at the 5' end overlaps with the MCH1 primers, whereas the sequence in bold at the 3' end is complementary to the MVK primers. The ribosome-binding site and the ATG triplet marking the start of the *pelB* leader sequence is underlined. The Link fragment originates from the λc2 vector (*see* **ref. 4**) and was taken through several PCR cloning steps using different sets of tagged primers in order to introduce the sequence changes necessary for the present assembly system. The Link fragment was cloned into the Bluescript KS+ (Stratagene, La Jolla, CA) vector, giving rise to pLink as described in **ref. 4**.
8. Assembly primer (*see* Sequence 10): This sequence is identical to the 5' end of Tag.For and Tag.Back. The assembly primer solution used for PCR is 5 μ*M*.
9. Ampli*Taq* polymerase and 10X *Taq* buffer I (Perkin-Elmer/Cetus, Norwalk CT).
10. GeneClean (Bio 101 Inc., Vista, CA).
11. Restriction endonucleases: *Sfi*I and *Not*I.
12. Endonuclease buffers: 10X NEB3 (New England Biolabs, Beverly, MA), 10X bovine serum albumin (BSA) (0.1% from Amersham Pharmacia Biotech, Buckinghamshire, UK).
13. Qiagen DNA purification columns (Qiagen GmbH, Hilden, Germany).
14. T_4 DNA ligase (1 U/μL) and 5X T_4 ligase buffer (Gibco-BRL).

```
MVH-1:   G CCG GCC ATG GCC GAG GTR MAG CTT CAG GAG TCA GGA C
MVH-2:   G CCG GCC ATG GCC GAG GTS CAG CTK CAG CAG TCA GGA C
MVH-3:   G CCG GCC ATG GCC CAG GTG CAG CTG AAG SAS TCA GG
MVH-4:   G CCG GCC ATG GCC GAG GTG CAG CTT CAG GAG TCS GGA C
MHV-5:   G CCG GCC ATG GCC GAR GTC CAG CTG CAA CAG TCY GGA C
MVH-6:   G CCG GCC ATG GCC CAG GTC CAG CTK CAG CAA TCT GG
MVH-7:   G CCG GCC ATG GCC CAG STB CAG CTG CAG CAG TCT GG
MVH-8:   G CCG GCC ATG GCC CAG GTY CAG CTG CAG CAG TCT GGR C
MVH-9:   G CCG GCC ATG GCC GAG GTY CAG CTY CAG CAG TCT GG
MVH-10:  G CCG GCC ATG GCC GAG GTC CAR CTG CAA CAA TCT GGA CC
MVH-11:  G CCG GCC ATG GCC CAG GTC CAC GTG AAG CAG TCT GGG
MVH-12:  G CCG GCC ATG GCC GAG GTG AAS STG GTG GAA TCT G
MVH-13:  G CCG GCC ATG GCC GAV GTG AAG YTG GTG GAG TCT G
MVH-14:  G CCG GCC ATG GCC GAG GTG CAG SKG GTG GAG TCT GGG G
MVH-15:  G CCG GCC ATG GCC GAK GTG CAM CTG GTG GAG TCT GGG
MVH-16:  G CCG GCC ATG GCC GAG GTG AAG CTG ATG GAR TCT GG
MVH-17:  G CCG GCC ATG GCC GAG GTG CAR CTT GTT GAG TCT GGT G
MVH-18:  G CCG GCC ATG GCC GAR GTR AAG CTT CTC GAG TCT GGA
MVH-19:  G CCG GCC ATG GCC GAA GTG AAR STT GAG GAG TCT GG
MVH-20:  G CCG GCC ATG GCC GAA GTG ATG CTG GTG GAG TCT GGG
MVH-21:  G CCG GCC ATG GCC CAG GTT ACT CTR AAA GWG TST GGC C
MVH-22:  G CCG GCC ATG GCC CAG GTC CAA CTV CAG CAR CCT GG
MVH-23:  G CCG GCC ATG GCC CAG GTY CAR CTG CAG CAG TCT G
MVH-24:  G CCG GCC ATG GCC GAT GTG AAC TTG GAA GTG TCT GG
MVH-25:  G CCG GCC ATG GCC GAG GTG AAG GTC ATC GAG TCT GG
```

Sequence 1

```
MCH1-G1:  CGACTAGTTTAGAATTCAAGCTGTCGAC TCA ACA ATC CCT GGG CAC AAT TTT
CTT GTC CACC
MCH1-G2A: CGACTAGTTTAGAATTCAAGCTGTCGAC TCA ACA GGG CTT GAT TGT GGG CCC
TCT GGG
MCH1-G2B: CGACTAGTTTAGAATTCAAGCTGTCGAC TCA ACA GGG GTT GAT TGT TGA AAT
GGG CCC G
```

Sequence 2

```
MVK-1:  TTGGCTGCACAACCAGCAATGGCA GAC ATT GTT CTC ACC CAG TCT CC
MVK-2:  TTGGCTGCACAACCAGCAATGGCA GAC ATT GTG CTS ACC CAG TCT CC
MVK-3:  TTGGCTGCACAACCAGCAATGGCA GAC ATT GTG ATG ACT CAG TCT CC
MVK-4:  TTGGCTGCACAACCAGCAATGGCA GAC ATT GTG CTM ACT CAG TCT CC
MVK-5:  TTGGCTGCACAACCAGCAATGGCA GAC ATT GTG YTR ACA CAG TCT CC
MVK-6:  TTGGCTGCACAACCAGCAATGGCA GAC ATT GTR ATG ACA CAG TCT CC
MVK-7:  TTGGCTGCACAACCAGCAATGGCA GAC ATT MAG ATR ACC CAG TCT CC
MVK-8:  TTGGCTGCACAACCAGCAATGGCA GAC ATT CAG ATG AMC CAG TCT CC
MVK-9:  TTGGCTGCACAACCAGCAATGGCA GAC ATT CAG ATG ACD CAG TCT CC
MVK-10: TTGGCTGCACAACCAGCAATGGCA GAC ATT CAG ATG ACA CAG ACT AC
MVK-11: TTGGCTGCACAACCAGCAATGGCA GAC ATT CAG ATG ATT CAG TCT CC
MVK-12: TTGGCTGCACAACCAGCAATGGCA GAC ATT GTT CTC AWC CAG TCT CC
MVK-13: TTGGCTGCACAACCAGCAATGGCA GAC ATT GTT CTC TCC CAG TCT CC
MVK-14: TTGGCTGCACAACCAGCAATGGCA GAC ATT GWG CTS ACC CAA TCT CC
MVK-15: TTGGCTGCACAACCAGCAATGGCA GAC ATT STG ATG ACC CAR TCT C
MVK-16: TTGGCTGCACAACCAGCAATGGCA GAC ATT KTG ATG ACC CAR ACT CC
MVK-17: TTGGCTGCACAACCAGCAATGGCA GAC ATT GTG ATG ACT CAG GCT AC
MVK-18: TTGGCTGCACAACCAGCAATGGCA GAC ATT GTG ATG ACB CAG GCT GC
MVK-19: TTGGCTGCACAACCAGCAATGGCA GAC ATT GTG ATA ACY CAG GAT G
MVK-20: TTGGCTGCACAACCAGCAATGGCA GAC ATT GTG ATG ACC CAG TTT GC
MVK-21: TTGGCTGCACAACCAGCAATGGCA GAC ATT GTG ATG ACA CAA CCT GC
MVK-22: TTGGCTGCACAACCAGCAATGGCA GAC ATT GTG ATG ACC CAG ATT CC
MVK-23: TTGGCTGCACAACCAGCAATGGCA GAC ATT TTG CTG ACT CAG TCT CC
MVK-24: TTGGCTGCACAACCAGCAATGGCA GAC ATT GTA ATG ACC CAA TCT CC
MVK-25: TTGGCTGCACAACCAGCAATGGCA GAC ATT GTG ATG ACC CAC ACT CC
```

Sequence 3

```
MCK1:   TGC GGC CGC ACA CTC ATT CCT GTT GAA GCT CTT GAC
```

Sequence 4

```
Tag.Back1: CA GTC ACA GAT CCT CGC GAA TTG GCC CAG CCG GCC ATG GCC SAN G
Tag.Back2: CA GTC ACA GAT CCT CGC GAA TTG GCC CAG CCG GCC ATG GCC SAN C
```

Sequence 5

```
Link.For: GTC TGC CAT TGC TGG TTG TGC AGC CAA
```

Sequence 6

```
Link.Back: CGA CAG CTT GAA TTC TAA ACT ACT CGA AGG CGC GCC AAG GAG ACA GTC AT
```

Sequence 7

```
Tag.For: CAG TCA CAG ATC CTC GCG AAT TGG TGC GGC CGC ACA CTC ATT CCT G
```

Sequence 8

```
CGACAGCTTGAATTCTAAACTAGTCGAAGGCGCGCCAAGGAGACAGTCATA ATG AAA TAC CTA TTG
CCT ACG GCA GCC GCT GGA TTG TTA TTA TTG GCT GCA CAA CCA GCA ATG GCA
```

Sequence 9

```
Assembly: CA GTC ACA GAT CCT CGC GAA TTG G
```

Sequence 10

15. Glycogen (20 mg/mL from Boehringer Mannheim).
16. The vector pFAB5c.His contains several improvements to the pFAB5c vector previously described by us *(4)*. In the polylinker region downstream of ΔgIII, an adapter was introduced that contained an *Eag*I site followed by six histidine codons and a stop codon (*see* **Fig. 1**). This makes it possible to delete ΔgIII by digestion with *Eag*I (*Eag*I recognizes a subset of the *Not*I recognition sequence present upstream of ΔgIII) and religation, which also regenerates a *Not*I site. Additionally, the His and stop codons are in reading frame with C_K, and the $(His)_6$ tail facilitates easy purification of free Fab molecules on metal affinity columns *(7)*.
17. Luria-Bertani (LB), 2X TY and SOC medium, and TE buffer were prepared as described in **ref. *11***.
18. Antibiotics: Ampicillin is used at a concentration of 100 μg/mL and tetracycline at 10 μL/mL.
19. *E. coli* strain: *TOP10F': mcrA, Δ(mrr-hsdRMS-mcrBC) θ80 lacZΔM15, ΔlacX74, deoR, recA1, araD139, Δ(ara-leu)7697, galU, galK, rpsL, endA1, nupG, F'{lacI^aTn10(Te7^R)}* (British Biotechnology, Oxon, UK).

2.3. PCR Assembly Method for Constructing Murine Antibody Fab Libraries Independent of the IgH Isotype (cf. Fig. 2)

The aggregate concentration of all PCR primer mixes is 20 μ*M*.

1. The primary amplification back primers, MVH1-25, MVK1-25, and the forward primer, MCK1, are described in **steps 1**, **3**, and **4**, **Subheading 2.2.**
2. Murine J region forward primers (*see* Sequence 11): Nucleotides in bold are complementary to the sequence of the J region of all known murine immunoglobulin genes. Italicized nucleotides are complementary to the 5' end of the C_H1 gene fragment. The concentration of each primer in the mixture used for PCR is 5 μ*M*.
3. The linker assembly primers, Tag.Back1-2, Link.For, Link.Back, and Tag.For, are described in **steps 5** and **6**, **Subheading 2.2.**
4. Plasmid pCH1-Link contains the $C_H1_{\gamma1}$ region upstream of the Link fragment in pLink *(8)*.
5. Back primers for PCR production of the C_H1-Link fragment (*see* Sequence 12): CH1-Link-D. Back is complementary to the 5' end of the CH1-Link fragment and is degenerate at the underlined positions in order to match all MJH primers (*see* **step 2**, this section). Italicized nucleotides are complementary to the 5' end of the MJH primers. The concentration of each of the four primers in the mixture used for PCR is 5 μ*M*.
6. The assembly primer, Assembly, is described in **step 8**, **Subheading 2.2.**
7. pFab5c and pFab5c.His are described in **step 16**, **Subheading 2.2.**

```
MJH-1:  GGG GGA GTA GTC TTA GCT GAG GAG ACG GTG ACC GTG G
MJH-2:  GGG GGA GTA GTC TTA GCT GAG GAG ACT GTG AGA GTG G
MJH-3:  GGG GGA GTA GTC TTA GCT GCA GAG ACA GTG ACC AGA G
MWH-4:  GGG GGA GTA GTC TTA GCT GAG GAG ACG GTG ACT GAG G
```

Sequence 11

```
CH1-Link-D-Back: GTC TCY KCA GCT AAG ACT ACT CCC CCG AGC GTC TAT CCA CTG GCC CC
```

Sequence 12

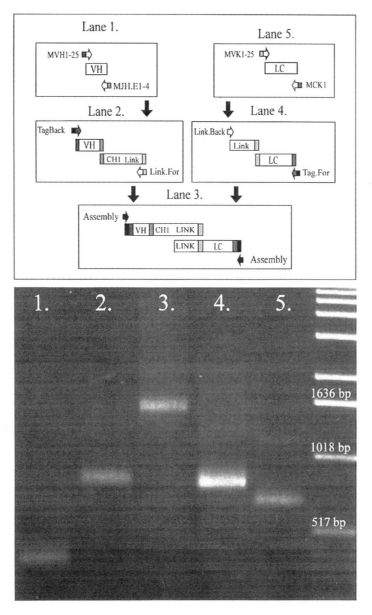

Fig. 2. Overview of primers and steps involved in PCR assembly of murine Fab gene fragments that are independent of IgH isotype, and agarose gel electrophoresis of PCR products. (**A**) *See* legend to **Fig. 1**. CH1 = the first constant domain of the H-chain IgG isotype; CH1-Link = CH1-LinkD, a DNA fragment containing the CH1 region, a translational stop codon for Fd translation, a ribosome binding site for L-chain expression, and the coding region corresponding to the *N*-terminal part of the *pel*B leader. The nucleotide sequence of this part of the *pel*B leader was changed at all third codon positions relative to the *pel*B leader sequence in front of the H-chain to minimize to chance of recombinatorial events between the two *pel*B leader sequences. (**B**) Products from the five PCR reactions in the upper panel were separated on a 1% agarose gel and stained with ethidium bromide. From left to right: V_H, V_H/C_H1-Link, V_H/C_H1-Link/κ, Link/κ, and κ DNA. The size of relevant marker bands are indicated to the right.

2.4. Direct Cloning Method for Making Murine Antibody Fab Libraries Independent of the IgH Isotype (cf. Fig. 3)

The aggregate concentration of all PCR primer mixes is 20 µM.

1. The murine κ L-chain-variable (V_K) back primers KAP1-25, are identical to MVK1-25 described in **step 3, Subheading 2.2.**, except that the KAP primers are four nucleotides shorter at their 5' end.
2. Murine κ L-chain-constant region (C_K) forward primer (*see* Sequence 13): MCK-AscI adds on the recognition sequence for *AscI* (underlined) and tandem stop codons (small capitals) and has a 25 nucleotides (boldfaced) overlap with C_K.
3. Murine L-chain extension primers (*see* Sequence 14): KAP.Ext adds on the *NheI* recognition site (underlined) and has an overlap of 23 nucleotides (italicized) with KAP1-25 (*see* Sequence 15): MCK-AscI.Ext has a 24 nucleotides (italicized) overlap with MCK-AscI.
4. The murine V_H back primers, MVH1-25, are described in **step 1, Subheading 2.2.**
5. Murine J forward primers (*see* Sequence 16): Twenty-six nucleotides (boldfaced) of the MJH.For primers are complementary to the J and C_H1 spanning region. The *SpeI* recognition sequence is underlined. The concentration of each primer variant used for PCR is 5 µM.
6. Murine V_H extension primers, Tag.Back1-2 and MJH.Ext1 (*see* Sequence 17): MJH.Ext1 and the MJH.For primers overlap by 23 nucleotides (italicized). Tag.Back1-2 are described in **step 5, Subheading 2.2.**
7. 10X buffer II (Perkin-Elmer/Cetus).
8. GELase™ (Epicentre Technologies, Madison, WI).
9. Restriction enzymes *AscI*, *HindIII*, *KpnI*, *NheI*, *PmeI*, *SpeI*, *SrfI*, and *StuI*.
10. 10X restriction enzyme buffers: NEB2, 3, and 4 (New England Biolab), and Universal buffer (Stratagene, La Jolla, CA).
11. pFab60. The vector is depicted in **Fig. 3B** and described in detail by us in **ref. 7**.

MCK-AscI: CTTCTCGAGG CGCGCCTCATCA **ACACTCAT TCCTGTTGAA GCTCTTG**

Sequence 13

KAP.Ext: TACTACTACT ACTAGCTAGC *TGCACAACCA GCAATGGCAG AC*

Sequence 14

MCK-AscI.Ext: TACTACTACT *CTTCTCGAGG CGCGCCTCATCAAC*

Sequence 15

MJH.For-1: ATCATCATCC ACTAGTTTTG GCTGAGGAGA CGGTGACCGT GG
MJH.For-2: ATCATCATCC ACTAGTTTTG GCTGAGGAGA CTGTGAGAGT GG
MJH.For-3: ATCATCATCC ACTAGTTTTG GCTGCAGAGA CAGTGACCAG AG
MJH.For-4: ATCATCATCC ACTAGTTTTG GCTGAGGAGA CGGTGACTGA GG

Sequence 16

MJH.Ext: CAGGCGCCAG TGGATAGA*AT CATCATCCAC TAGT*TTTGGC T

Sequence 17

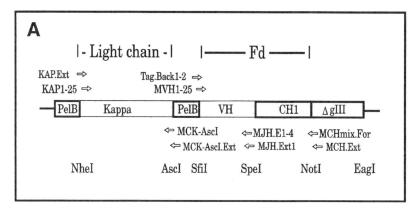

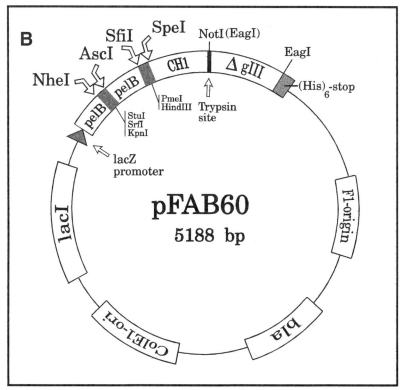

Fig. 3. Overview of primers used to make murine V and Fd-region H-chains and κ L-chain DNA for separate cloning, and a map of expression vector pFab60, which is described in detail in **ref. 7**. **(A)** Oligonucleotide primers for PCR production of Fab gene fragments. Primers above boxes are identical to the DNA top strand, while those below are complementary. Heavy boxes indicate genetic elements present in the vector; light boxes represent elements to be cloned. The entire Fd gene fragment can be cloned between the *Sfi*I and *Not*I sites. Important restriction enzyme sites are indicated. **(B)** Phagemid, pFab60, used for cloning and expression of murine antibody Fab gene fragments. Relevant genetic elements are boxed. Black boxes indicate stuffer DNA between cloning sites: Not all unique restriction sites are indicated.

2.5. Direct Cloning Method for Making Murine Antibody Fab Libraries of the IgG Isotype (cf. Fig. 3)

The aggregate concentration of all PCR primer mixes is 20 µM.

1. The murine κ and extension primers are described in **steps 1** and **3, Subheading 2.4.**
2. The murine V_H and extension back primers are described in **steps 4** and **6, Subheading 2.4.**
3. Murine $C_H1\gamma$ forward primers (*see* Sequence 18): The 24–28 distal nucleotides (boldfaced) of the MCHγ.For primers are identical to the corresponding $C_H1\gamma$ and hinge junction sequence. The *Not*I recognition sequence is underlined. The concentration of each primer in the MCHmix.For solution used for PCR is 6.67 µM.
4. Murine $C_H1\gamma$ forward extension primer (*see* Sequence 19): MCH.Ext and the MCHγ.For primers overlap by 21 nucleotides (italicized). The *Not*I recognition sequence is underlined.

```
MCHγ1.For:  CTCGCGAATT GGTGCGGCCG CACAATCCCT GGGCACAATT TTCTTGTC
MCHγ2a.For: CTCGCGAATT GGTGCGGCCG CGGGCTTGAT TGTGGGCCCT CTGGG
MCHγ2b.For: CTCGCGAATT GGTGCGGCCG CGGGGTTGAT TGTTGAAATG GGCCCG
```
Sequence 18

```
MCH.Ext: CAGTCACAGA TCCTCGCGAA TTGGTGCGGC CGC
```
Sequence 19

2.6. Direct Cloning Method for Making Human Antibody Fab Libraries Independent of the IgH Isotype (cf. Fig. 4)

The aggregate concentration of all PCR primer mixes is 20 µM.

1. Human H-chain, V-region back primers (*see* Sequence 20): These primers consist of 12 individually synthesized oligonucleotides representing 31 variants. The nucleotides in bold correspond to the 5' sequences of the V_H region starting with the codon for amino acid number 1. Primers are invariant for the first five nucleotides from position +1 of the V region. The first 15 nucleotides at the 5' end of the primers correspond to the 3' part of the *pelB* leader. This gives a unique template for HVH.Ext, used in the extension PCR (*see* **step 3**). The concentration of each variant in the solution used for PCR is 0.65 pmol/µL.
2. Human H-chain, J-region forward primers (*see* Sequence 21): The J-region primers consist of three individually synthesized oligonucleotides representing a total of four variants. All nucleotides are complementary to the J region and the first codon at the 5' end corresponds to amino acid residue 114 of C_H1 *(12)*. The concentration of each variant in the mixture used for PCR is 5 pmol/µL.
3. Human H-chain extension primers (*see* Sequence 22): Nucleotides in HVH.Ext corresponding to sequences in HVH1-12 are in bold, and the *Nhe*I site is underlined. Nucleotides in HJH.Ext corresponding to sequences in HJH1-3 are in bold, and the *Apa*I site is underlined. The concentration of each variant in the HJH.Ext mixture used for PCR is 5 pmol/µL.
4. Human κ-chain, V-region back primers (*see* Sequence 23): These primers consist of eight individually synthesized oligonucleotides representing 27 variants. Nucleotides in bold correspond to the 5' sequence of the V_K gene starting with the codon for amino acid number 1. The N-terminal two amino acids of the V region are invariant. The sequences upstream of the V region represent the 3' part of the *pelB* leader gene (located between the

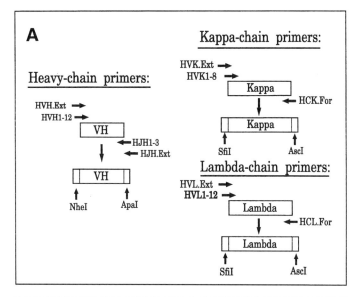
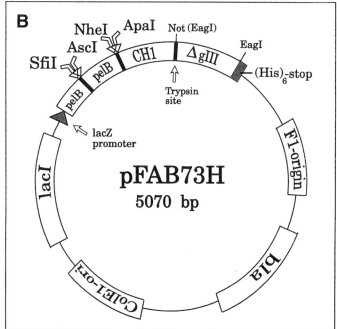

Fig. 4. Overview of primers used to generate human H-chain V-region and κ and λ L-chain DNA for separate cloning and a map of expression vector pFab73H. (**A**) Diagram exemplifying the primary PCR amplifications and extensions of V_H, κ and λ DNA. (**B**) pFab73H. This vector was made as a counterpart of pFab60 for cloning and expression of human antibody Fab fragments (*see* **Fig. 3**; pFab73H does not contain 8-bp recognition sequences for rare-cutting enzymes in the two stuffer sequences). It was derived from pFab4H *(9)*, in several steps, by first introducing the *lacI* gene and then a DNA cassette containing a $(His)_6$ tail and a stop codon in fusion with Δg*III*.

```
HVH-1:   CAG CCA GCA ATG GCA CAG GTN CAG CTG GTR CAG TCT GG
HVH-2:   CAG CCA GCA ATG GCA CAG GTC CAG CTK GTR CAG TCT GGG G
HVH-3:   CAG CCA GCA ATG GCA CAG GTK CAG CTG GTG SAG TCT GGG
HVH-4:   CAG CCA GCA ATG GCA CAG GTC ACC TTG ARG GAG TCT GGT CC
HVH-5:   CAG CCA GCA ATG GCA CAG GTG CAG CTG GTG GAG WCT GG
HVH-6:   CAG CCA GCA ATG GCA CAG GTG CAG CTG GTG SAG TCY GG
HVH-7:   CAG CCA GCA ATG GCA CAG GTG CAG CTG CAG GAG TCG G
HVH-8:   CAG CCA GCA ATG GCA CAG GTG CAG CTG TTG SAG TCT G
HVH-9:   CAG CCA GCA ATG GCA CAG GTG CAG CTG GTG CAA TCT G
HVH-10:  CAG CCA GCA ATG GCA CAG GTG CAG CTG CAG GAG TCC GG
HVH-11:  CAG CCA GCA ATG GCA CAG GTG CAG CTA CAG CAG TGG G
HVH-12:  CAG CCA GCA ATG GCA CAG GTA CAG CTG CAG CAG TCA G
```

Sequence 20

```
HJH1:  GGC TGA GGA GAC RGT GAC CAG GGT
HJH2:  GGC TGA AGA GAC GGT GAC CAT TGT
HJH3:  GGC TGA GGA GAC GGT GAC CGT GGT
```

Sequence 21

```
HVH.Ext:  GCA GCC GCT GGA TTG TTA TTG CTA GCA GCA CAG CCA GCA ATG GCA CAG GT
HJH.Ext:  CAG TCA GAT CTC GCG AAT TGG GCC CTT GGT GGA GGC TGA RGA GAC RGT GAC C
```

Sequence 22

```
HVK-1:  G CCG GCC ATG GCC GAC ATC CAR WTG ACC CAG TCT CC
HVK-2:  G CCG GCC ATG GCC GAC ATC CRG ATG ACC CAG TCT CCW TC
HVK-3:  G CCG GCC ATG GCC GAC ATC GTG MTG ACC CAG TCT CC
HVK-4:  G CCG GCC ATG GCC GAC ATC GTG TTG ACS CAG TCT CCR GG
HVK-5:  G CCG GCC ATG GCC GAC ATC GTG ATG ACY CAG WCT CCA C
HVK-6:  G CCG GCC ATG GCC GAC ATC GTG ATG AYR CAG TCT CCA GC
HVK-7:  G CCG GCC ATG GCC GAC ATC GTG MTG ACW CAG TCT CCA GA
HVK-8:  G CCG GCC ATG GCC GAC ATC GTA ATG ACA CAG TCT CCA CC
```

Sequence 23

```
HCK.For:  GTC TCC TTC TCG AGG CGC GCC TCA CTA ACA CTC TCC CCT GTT GAA GCT
```

Sequence 24

```
HVK.Ext:  CA GTC ACA GAT CCT CGC GAA TTG GCC CAG CCG GCC ATG GCC GAC ATC
```

Sequence 25

AscI and *NheI* sites). This constitutes a unique template for the primers used in the extension PCR (see **step 6**). The concentration of each variant in the solution used for PCR is 0.74 pmol/μL.

5. Human κ-chain, constant-region forward primer (*see* Sequence 24): HCK.For is complementary to codons for the seven carboxy-terminal amino acid residues of the C_K domain, indicated by bold letters. The tandem stop codons and the *AscI* site are underlined.

6. Human κ-chain extension primers, HVK.Ext and HCK.For (*see* **Note 4**) (*see* Sequence 25): The boldfaced part of HVK.Ext overlaps with the 5' end of the primary PCR primers. The *SfiI* site introduced into the extended V_K PCR products is underlined. HCK.For is used for both primary and secondary amplifications.

7. Human λ-chain, V-region back primers (*see* Sequence 26): These primers consist of 12 individually synthesized oligonucleotides representing 24 variants. Nucleotides in bold correspond to the 5' sequences of the V_λ genes starting with the codon for amino acid

```
HVL-1:   G CCG GCC ATG GCC CAG TCT GYC CTG ACT CAG CCT G
HVL-2:   G CCG GCC ATG GCC CAG TCT GCC CTG ACT CAG CCT C
HVL-3:   G CCG GCC ATG GCC CAG TCT GTG CTG ACT CAG CCG TC
HVL-4:   G CCG GCC ATG GCC CAG TCT ATG CTG ACT CAG CCC CAC TC
HVL-5:   G CCG GCC ATG GCC CAG TCT GTG CTG ACT CAG CCA CCC TC
HVL-6:   G CCG GCC ATG GCC CAG TCT GAG CTG ACT CAG GAC CCT GC
HVL-7:   G CCG GCC ATG GCC CAG TCT GAG GTG ACT CAG GAG CC
HVL-8:   G CCG GCC ATG GCC CAG TCT GTG ATG ACY CAG TCT CMA
HVL-9:   G CCG GCC ATG GCC CAG TCT GTG CTG ACT CAG CCA CC
HVL-10:  G CCG GCC ATG GCC CAG TCT GTS BTG ACG CAG CCG CC
HVL-11:  G CCG GCC ATG GCC CAG TCT CAG CTG ACG CAG CCT GC
HVL-12:  G CCG GCC ATG GCC CAG TCT TTA YTG ACT CAA YCG CCC TC
```

Sequence 26

```
HCL.For: GTC TCC TTC TCG AGG CGC GCC TCA CTA TGA ACA TTC YGT AAG GGC MAC
```

Sequence 27

```
HVL.Ext: CA GTC ACA GAT CCT CGC GAA TTG GCC CAG CCG GCC ATG GCC CAG TCT
```

Sequence 28

number 1. The remaining nucleotides are identical to the similarly positioned nucleotides in the κ primers. The concentration of each variant in the solution used for PCR is 0.83 pmol/μL.

8. Human λ-chain, constant-region forward primer (*see* Sequence 27): HCL.For contains four variants and is complementary to codons for the seven carboxy-terminal amino acid residues of the λ constant domain, indicated by bold letters. The tandem stop codons and the *AscI* site are underlined. The concentration of each variant in the solution used for PCR is 5 pmol/μL.
9. Human λ-chain extension primers, HVL.Ext and HCL.For (*see* Sequence 28): The features of HVL.Ext is analogous to those of the κ-chain extension primer, HVK.Ext, described in **step 6**. HCL.For is used for both primary and secondary amplifications.
10. pFab73H is described in the legend to **Fig. 4**.

2.7. Preparation of Ultracompetent Cells

1. *E. coli* Pulser, 0.2- and 0.1-cm electroporation cuvets (Bio-Rad, Hercules, CA).
2. pUC18 DNA (Stratagene).
3. *E. coli* strain: *TG1*: F'{*traD36, lacI,* Δ*(lacZ)M15, proA$^+$B$^+$*}/Δ*(lac-proAB), thi-1, supE,* Δ*(hsdM-mcrB)*5(r_k^-, m_k^+, McrB$^-$).

2.8. Electroporation of Library DNA, Cell Growth, and Storage

1. Glycerol buffer: 10% Glycerol in water.
2. Nunclon Δ Square Dishes (500 cm^2) from Life Technologies (Denmark).

2.9. Production and Titration of Fab Phages from Libraries

1. Phage-precipitation buffer: 20% Polyethylene glycol (PEG$_{6000}$), 2.5 *M* NaCl.
2. Helper phage R408 (Stratagene).
3. IPTG stock solution: 100 m*M* isopropyl-β-D-thiogalactopyranoside in sterile water.

2.10. Selection of Antigen Binders by Panning

1. Washing buffer: 0.5% Tween-20 in phosphate-buffered saline (PBS) buffer.
2. Blocking buffer: 2% Skimmed milk powder (Difco, Detroit, MI) in PBS buffer.
3. Trypsin elution buffer: 1 mg Trypsin (Worthington, Freehold, NJ)/mL PBS buffer.

4. Glycin elution buffer: 0.1 *M* Glycin-HCl, pH 2.2 containing 1 mg/mL BSA.
5. Maxisorp microtiter plates and immunotubes (Life Technologies).
6. Peroxidase-conjugated anti-M13 (Amersham Pharmacia Biotech, Sollentuna, Sweden).

3. Methods

3.1. Extraction of Total RNA from Mouse Spleen or Human Blood Lymphocytes and Production of cDNA

1. Extract total RNA from (a) spleen cells or (b) lymphocytes:
   a. ***Spleen cells:*** Add 10 mL of buffer A to a freshly dissected spleen from an immunized mouse (about 0.5–1 g), transfer to a glass homogenizer, and homogenize with a Teflon™-coated pestle for 1 min in an ice-water bath.
   b. ***Lymphocytes:*** Bleed 100 mL human blood into a 14-mL anticoagulant (ACD) buffer, mix with an equal vol of PBS containing the same concentration of anticoagulant as the diluted blood, and layer on Lymphoprep (Nyeomed, Oslo, Norway) placed in clear centrifuge tubes. Centrifuge at 400*g* for 30 min at room temperature. Collect the lymphocytes at the interphase between plasma and Lymphoprep, centrifuge at 800*g* for 10 min, and resuspend pellet in buffer A (*see* **Note 5**).
2. Transfer spleen-cell homogenate or lymphocytes to a chloroform/phenol-resistant centrifuge tube, immediately add 1 mL 2 *M* NaOAc, pH 4.1, 10 mL water-saturated phenol, and 2 mL chloroform:isoamylalcohol (49:1) and whirlmix vigorously for 30 s.
3. Centrifuge at 10,000*g* for 20 min at 4°C, transfer the upper phase to a fresh centrifuge tube, mix with 10 mL isopropanol, and let stand at –20°C for 30 min or longer.
4. Pellet the RNA by centrifugation at 12,000*g* for 30 min, discard the supernatant, and dissolve the wet pellet in 0.5 mL buffer A.
5. Repeat steps 2–4 using appropriately scaled-down volumes and dissolve pellet in 0.5 mL buffer A, mix with 0.5 mL isopropanol, and keep at –20°C for at least 30 min.
6. Collect the RNA as in **step 4** and dissolve the wet pellet in 0.5 mL DEPC-treated water.
7. Whirlmix for 1 min, make a clearing spin for 5 min at 15,000*g*, and transfer supernatant to a fresh microfuge tube.
8. Add 0.5 mL water-saturated phenol, whirlmix for 1 min, spin for 1 min at 15,000*g*, and transfer the upper phase to a fresh microfuge tube. Add 0.5 mL chloroform and extract in the same manner.
9. Continue the phenol and chloroform extractions until no visible interphase is present.
10. Add 50 µL 3 *M* NaOAc, pH 5.2 (0.1 vol), mix with 1.1 mL 96% ethanol (2 vol), and keep at –20°C until use.
11. Before making cDNA, whirlmix the RNA/alcohol solution, transfer 200 µL to a fresh microfuge tube, and pellet the RNA by centrifugation at 15,000*g* for 15 min.
12. Wash the RNA pellet with 1 mL 70% ethanol and dry it lightly in a Speed Vac (Savant, Farmingdale, NY) centrifuge.
13. Dissolve pellet in 26 µL of DEPC-treated water, remove 1 µL, and determine concentration and purity. The RNA concentration should be about 2 µg/µL with an $A_{260/280}$ ratio of about 2.0.
14. Mix the following in an ice-cold sterile microfuge tube:

DEPC-treated water	2.5 µL
5 m*M* dNTP	5.0 µL
5X First-strand buffer	10.0 µL
0.1 *M* DDT	5.0 µL

15. Add 2.5 µL poly (dT)$_{12-18}$ (500 µg/mL) to 25 µL RNA solution from **step 13** and incubate at 70°C for 10 min, cool on ice, centrifuge at 5000*g* for 5 s, and transfer supernatant to the cDNA mixture of **step 14**, making the total volume 50 µL.

16. Incubate the mixture at 47°C for 2 min, add 3 μL SuperScript reverse transcriptase (Life Technologies), and continue incubation for 1 h at 47°C (*see* **Note 6**).
17. Heat the reaction at 100°C for 3 min, centrifuge at 15,000g for 15 min, transfer supernatant to a fresh microfuge tube, and precipitate as in **step 10**, using scaled-down volumes. Dissolve pellet in 15 μL DEPC-treated water.

3.2. PCR Assembly Method for Constructing Murine Antibody Fab Libraries of the IgG Isotype

The PCR assembly method (*see* **Notes 7** and **8**) for making murine IgG antibody Fab libraries consists of the following:

- Primary amplification of the Fd and κ L-chain genes.
- PCR assembly of each of the primary PCR fragments with Link.
- Final PCR assembly of pairs of Fd/Link and Link/L-chain fragments.
- Cloning of the final PCR product into the expression vector pFAB5c.His.
- Electroporation, growth, and storage of the library are described in **Subheadings 3.7.** and **3.8.**

Figure 1 shows an overview of the PCR primers, the steps of the PCR assembly procedure, and the expression vector pFab5c.His. PCR reactions are hot-started and carried out in PCR tubes. All PCR reaction mixtures (100 μL) are covered with paraffin oil.

1. Primary amplification of Fd gene fragments:
cDNA (*see* **Subheading 3.1.**, **step 17**)	5 μL
10X *Taq* Buffer I	10 μL
10X dNTP (1.25 m*M* each)	8 μL
MVH1-25 (20 μ*M* mix)	1 μL
MCH1-3 (20 μ*M* mix)	1 μL
Sterile water	75 μL

2. Primary amplification of κ L-chain genes:
cDNA (*see* **Subheading 3.1.**, **step 17**)	5 μL
10X *Taq* Buffer I	10 μL
10X dNTP (1.25 m*M* each)	8 μL
MVK1-25 (20 μ*M* mix)	1 μL
MCK1 (20 μ*M*)	1 μL
Sterile water	75 μL

3. Amplification of the Link fragment:
Plasmid pLink DNA (1 ng/μL)	1 μL
10X *Taq* Buffer I	10 μL
10X dNTP (1.25 m*M* each)	8 μL
Link.Back (20 μ*M*)	1 μL
Link.For (20 μ*M*)	1 μL
Sterile water	79 μL

4. PCR program: 94°C 5 min, suspend, add 0.3 μL Ampli*Taq* polymerase (5 U/μL); 94°C 1 min, 55°C 1 min, 72°C 1 min, 30 cycles, 72°C 10 min, refrigerate.
5. Gel purify the primary amplification products as in **steps 6** and **7** (*see* **Note 9**). Fd gene fragments are about 720 bp, κ L-chain genes about 650 bp, and the Link fragment is 117 bp.
6. Both untreated and digested PCR products (and digested vector DNA) are purified on standard agarose gels followed by the GeneClean procedure or on low-melting agarose gels followed by GELase™ treatment and ethanol precipitation according to the manufacturer's recommendations.

7. Vector DNA (approx 5 kb) is purified on 1%, assembled Fd/L-chain DNA on 1.5%, Fd and L-chain DNA on 1.5–2%, and C_H1/Link, V_H and Link DNA on 2% TAE agarose gels.
8. Assembly of Fd and Link DNA (*see* **Note 10**):

1–50 ng purified Fd DNA	5.00 μL
1–20 ng purified Link DNA	2.00 μL
10X *Taq* Buffer I	10.00 μL
10X dNTP (1.25 mM each)	8.00 μL
Tag.Back1-2 (20 μM mix)	1.25 μL
Link.For (20 μM)	1.25 μL
Sterile water	72.50 μL

9. Assembly of L-chain and Link DNA:

About 5 ng purified L-chain DNA	5.00 μL
About 1 ng purified Link DNA	2.00 μL
10X *Taq* Buffer I	10.00 μL
10X dNTP (1.25 mM each)	8.00 μL
Link.Back (20 μM)	1.25 μL
Tag.For (20 μM mix)	1.25 μL
Sterile water	72.50 μL

10. PCR program: 94°C 5 min, suspend, add 0.3 μL Ampli*Taq* polymerase (5 U/μL); 94°C 1 min, 65°C 1 min, 72°C 1 min, 25 cycles, 72°C 10 min, refrigerate.
11. Process the assembled Fd/Link and Link/L-chain fragments, which have increased about 100 bp in size, as described in **steps 6** and **7**.
12. Final assembly of Fd/Link and Link/L-chain DNA. It is important to use a relatively low concentration of assembly primer (10 pmol/100 μL) in this reaction:

1 ng purified Fd-Link DNA	5 μL
1 ng purified L-chain-Link DNA	5 μL
10X *Taq* Buffer I	10 μL
10X dNTP (1.25 mM each)	8 μL
Assembly (5 μM)	2 μL
Sterile water	70 μL

13. PCR program: 94°C 10 min, suspend, add 0.3 μL Ampli*Taq* polymerase (5 U/μL); 94°C 1.5 min, 69°C 1 min, 72°C 2 min, 25 cycles, 72°C 10 min, refrigerate.
14. Treat the assembled 1.6 kb Fd/Link/L-chain fragments as in **steps 6** and **7**.
15. Digest purified PCR fragments with a 50–100 times surplus and vector DNA with about 10 times surplus of the appropriate enzymes for 2–4 h in buffers from New England Biolab at the recommended temperatures.
16. Mix in a microcentrifuge tube and incubate at 50°C for 3 h:

About 0.5 μg purified final-assembly DNA	37.5 μL
10X NEB3	5.0 μL
10X BSA	5.0 μL
*Sfi*I (50 U)	2.5 μL
Sterile water	0.5 μL

17. Add 3 μL *Not*I (25 U) and incubate at 37°C for 3 h more.
18. Gel purify the DNA as in **steps 6** and **7**. The agarose and GeneClean or GELase™ purification step eliminates the small *Not*I and *Sfi*I end fragments. Dissolve the purified fragments in TE buffer at a final concentration of 50 ng/μL.
19. Make pFAB5c.His DNA from 150 mL cells grown overnight in a rich medium, using a Qiagen maxiprep column and the manufacturers guidelines. A portion is digested with

*Sfi*I and *Not*I and processed as described in **steps 15–18**. Digest further with one of the enzymes, whose recognition sequence is present in the stuffer fragment to be excised (*see* **Fig. 1** and **Note 11**).

20. For ligation of insert to vector, mix in a microfuge tube and incubate overnight at 16°C:

Insert DNA from **Subheading 3.2.**, **step 18** (50 ng/µL)	6.0 µL
Vector DNA from **Subheading 3.2.**, **step 19** (50 ng/µL)	6.0 µL
5X T_4 ligase buffer	10.0 µL
Sterile water	25.5 µL
T_4 DNA ligase (2.5 U)	2.5 µL

21. Add 100 µL TE buffer to the ligation mixture and extract once with phenol/chloroform/isoamyl alcohol (25:24:1). Add 0.1 vol 3 *M* NaOAc, pH 5.2, and 1 µL glycogen (*see* **Note 12**), mix, precipitate with 2–2.5 vol ethanol, and, depending on DNA concentration, centrifuge for 10–30 min at 15,000g in a microfuge. Rinse pellet with 70% alcohol three to four times to remove as much salt as possible, dry lightly in a Speed Vac (Savant, Farmingdale, NY) centrifuge, and dissolve the DNA in water at a concentration of 10–20 ng/µL.

22. Proceed with **Subheading 3.8.** to produce the final IgG Fab library.

3.3. PCR Assembly Method for Constructing Murine Antibody Fab Libraries Independent of the IgH Isotype

The PCR assembly method for making murine antibody Fab libraries independent of the IgH isotype consists of the following:

- Primary amplification of V_H and κ L-chain genes.
- PCR assembly of V_H and C_H1-Link fragments and of L-chain and Link fragments.
- Final PCR assembly of pairs of V_H/C_H1-Link and Link/L-chain fragments.
- Cloning of the final PCR products into pFAB5c or pFab5c.His (*see* **Note 13** and **ref.** *8*).
- Electroporation, growth, and storage of the library (*see* **Subheadings 3.7.** and **3.8.**).

Figure 2 shows an overview of the PCR primers, the steps involved in the PCR assembly procedure, and relevant PCR-produced and ethidium bromide (EtBr)-stained DNA fragments.

PCRs are hot started and carried out in PCR tubes. All PCR mixtures (100 µL) are covered with paraffin oil.

1. Primary amplification of V_H gene fragments:

cDNA (*see* **Subheading 3.1.**, **step 17**)	5.0 µL
10X *Taq* Buffer I	10.0 µL
10X dNTP (1.25 m*M* each)	8.0 µL
MVH1-25 (20 µ*M* mix)	2.5 µL
MJH.E1-4 (20 µ*M* mix)	2.5 µL
Sterile water	72.0 µL

2. Primary amplification of κ L-chain genes:

cDNA (*see* **Subheading 3.1.**, **step 17**)	5.0 µL
10X *Taq* Buffer I	10.0 µL
10X dNTP (1.25 m*M* each)	8.0 µL
MVK1-25 (20 µ*M* mix)	2.5 µL
MCK1 (20 µ*M*)	2.5 µL
Sterile water	72.0 µL

3. Amplification of the Link fragment:
Plasmid pLink DNA (1 ng/μL)	1 μL
10X *Taq* Buffer I	10 μL
10X dNTP (1.25 mM each)	8 μL
Link.Back (20 μM)	1 μL
Link.For (20 μM)	1 μL
Sterile water	79 μL

4. Amplification of the C_H1-Link fragment:
Plasmid pCH1-Link DNA (1 ng/μL)	1 μL
10X *Taq* Buffer I	10 μL
10X dNTP (1.25 mM each)	8 μL
CH1-Link-E.Back (20 μM)	1 μL
Link.For (20 μM)	1 μL
Sterile water	79 μL

5. PCR program: 94°C 5 min, suspend, add 0.3 μL Ampli*Taq* polymerase (5 U/μL); 94°C 1 min, 55°C 1 min, 72°C 1 min, 30 cycles, 72°C 10 min, refrigerate. The C_H1-Link fragments are about 460 bp, V_H gene fragments about 390 bp and κ L-chain genes about 650 bp.

6. Gel purify the primary amplification products as described in **steps 6** and **7**, **Subheading 3.2**.

7. Assembly of V_H gene fragments and the C_H1-Link fragment (cf. **step 8**, **Subheading 3.2.** and **Note 10**; *see* **Note 14**):
10 ng purified V_H DNA	5 μL
50 ng purified C_H1-Link DNA	5 μL
10X *Taq* Buffer I	10 μL
10X dNTP (1.25 mM each)	8 μL
Tag.Back1-2 (20 μM mix)	1 μL
Link.For (20 μM)	1 μL
Sterile water	70 μL

8. Assembly of L-chain gene fragments and the Link fragment:
10 ng purified L-chain DNA	5 μL
1 ng purified Link DNA	2 μL
10X *Taq* Buffer I	10 μL
10X dNTP (1.25 mM each)	8 μL
Link.Back (20 μM)	1 μL
Tag.For (20 μM)	1 μL
Sterile water	73 μL

9. PCR program: 94°C 5 min, suspend, add 0.3 μL Ampli*Taq* polymerase (5 U/μL); 94°C 1.5 min, 65°C 1 min, 72°C 1.5 min, 25 cycles, 72°C 10 min, refrigerate.

10. Gel purify the assembled V_H/C_H1-Link and Link/L-chain fragments as in **steps 6** and **7**, **Subheading 3.2.**

11. Final assembly of V_H/C_H1-Link and Link/L-chain DNA. It is important to use a relatively low concentration of assembly primer (10 pmol/100 μL) in this reaction:
5 ng purified V_H/C_H1-Link DNA	5 μL
5 ng purified Link/L-chain DNA	5 μL
10X *Taq* Buffer I	10 μL
10X dNTP (1.25 mM each)	8 μL
Assembly (5 μM)	2 μL
Sterile water	70 μL

12. PCR program: 94°C 10 min, suspend, add 0.3 μL Ampli*Taq* polymerase (5 U/μL), 94°C 2 min, 72°C 2 min, 25 cycles, 72°C 10 min, refrigerate.

13. Gel-purify the assembled V_H/C_H1-Link/L-chain fragments (about 1.6 kb in size) as in **steps 6** and **7**, **Subheading 3.2**.
14. Continue with **steps 15–22**, **Subheading 3.2.**, and **Subheading 3.8.** to produce the IgH-independent Fab library.

3.4. Direct Cloning Method for Making Murine Antibody Fab Libraries Independent of the IgH Isotype

The direct cloning method (*see* **Note 15**) for making murine antibody Fab libraries independent of the IgH isotype consists of the following:

- Primary PCR amplifications of V_H gene fragments and κ genes.
- "Tagging" the primary amplification products with restriction enzyme recognition sequences (extension).
- Cloning κ genes in pFab60 (equal to the pFab60/κ library).
- Cloning V_H gene fragments using pFab60/κ DNA (*see* **Note 16**).
- Electroporation, growth, and storage of the library (*see* **Subheadings 3.7.** and **3.8.**).

Figure 3 shows an overview of the PCR primers and the expression vector, pFab60H. PCRs are hot started and carried out in PCR tubes. All PCR mixtures (50 or 100 μL) are covered with paraffin oil.

1. Primary amplification of κ genes:
c-DNA from 25 μg RNA	5.0 μL
10X buffer II	10.0 μL
MgCl$_2$ (25 mM)	6.0 μL
dNTP (1.25 mM each)	8.0 μL
KAP1-25 (20 μM)	1.0 μL
MCK-AscI (20 μM)	1.0 μL
Sterile water	69.0 μL

2. Primary amplification of V_H gene fragments:
c-DNA from 25 μg RNA	5.0 μL
10X buffer II	10.0 μL
MgCl$_2$ (25 mM)	6.0 μL
dNTP (1.25 mM each)	8.0 μL
MVH1-25 (20 μM)	1.0 μL
MJH.F1-4 (20 μM)	1.0 μL
Sterile water	69.0 μL

3. PCR program: 94°C 10 min, suspend, add 0.3 μL Ampli*Taq* polymerase (5 U/μL); 94°C 1 min, 55°C 1 min, 72°C 1 min, 30 cycles, 72°C 10 min, refrigerate. Primary κ products are about 650 bp, whereas primary V_H products are about 390 bp. Gel purify the primary amplification products as in **steps 6** and **7**, **Subheading 3.2.**

4. κ Extension:
2 ng κ DNA	0.5 μL
10X buffer II	5.0 μL
MgCl$_2$ (25 mM)	5.0 μL
dNTP (1.25 mM each)	8.0 μL
Kap.Ext (20 μM)	0.5 μL
MCK-AscI.Ext (20 μM)	0.5 μL
Sterile water	30.5 μL

5. PCR program: 94°C 10 min, suspend, add 0.3 µL AmpliTaq polymerase (5 U/µL); 94°C 1 min, 60°C 1 min, 72°C 1 min, 10 cycles, 72°C 10 min, refrigerate. Extended κ products are about 710 bp.
6. V_H extension:

0.4 ng V_H DNA	0.5 µL
10X buffer II	5.0 µL
$MgCl_2$ (25 mM)	5.0 µL
dNTP (1.25 mM each)	8.0 µL
Tag.Back1-2 (20 µM)	0.5 µL
MJH.Ext1 (20 µM)	0.5 µL
Sterile water	30.5 µL

7. PCR program: 94°C 10 min, suspend, add 0.3 µL AmpliTaq polymerase (5 U/µL); 94°C 1 min, 55°C 1 min, 72°C 1 min, 15 cycles, 72°C 10 min, refrigerate. Extended V_H products are about 450 bp.
8. Gel purify the secondary amplification products as in **steps 6** and **7**, **Subheading 3.2.**
9. Digest pFab60 and extended κ DNA with *Asc*I and *Nhe*I in a 1:1 mixture of NEB2 and NEB4 + BSA (*see* **step 15, Subheading 3.2.**). Ethanol precipitate the vector DNA and further digest it with *Kpn*I in NEB1 + BSA or *Stu*I in NEB2 (*see* **Notes 11** and **17**). Gel purify DNA as in **steps 6** and **7, Subheading 3.2.**
10. Ligate the digested and purified pFab60 and κ DNA in a molar ratio of about 1:2 and extract and precipitate the ligation products as described in **steps 20** and **21, Subheading 3.2.**
11. Proceed with **Subheading 3.8.** and make pFab60/κ library DNA. Use this DNA as a "vector" to clone the V_H gene fragments.
12. Digest extended V_H DNA and pFab60/κ library DNA first with *Sfi*I at 50°C in NEB2 + 0.01% BSA, then with *Spe*I at 37°C in the same buffer. Digest the pFab60/κ DNA further with *Pme*I in NEB4 + BSA (*see* **Note 17**). Gel purify as in **steps 6** and **7, Subheading 3.2.**
13. Ligate the digested pFab60/κ library and V_H DNA in a molar ratio of about 1:2, and extract and precipitate the ligation products as in **steps 20** and **21, Subheading 3.2.** (The vector harbors the $C_H 1_{\gamma 1}$ gene fragment, *see* **Fig. 3**.)
14. Proceed with **Subheading 3.8.** to produce the IgH-independent Fab library (*see* **Notes 16** and **18**).

3.5. Direct Cloning Method for Making Murine Antibody Fab Libraries of the IgG Isotype

The direct cloning method for making murine antibody Fab libraries of the IgG isotype consists of the following:

- Primary PCR amplifications of Fd gene fragments and κ genes.
- "Tagging" the primary amplification products with restriction enzyme recognition sequences (extension).
- Cloning of κ genes in pFab60 (equal to the pFab60/κ library).
- Cloning of Fd gene fragments using pFab60/κ DNA.
- Electroporation, growth, and storage of the library (*see* **Subheadings 3.7.** and **3.8.**).

Figure 3 shows an overview of the PCR primers and the expression vector, pFab60. PCRs are hot started and carried out in PCR tubes. All PCR mixtures (50 or 100 µL) are covered with paraffin oil.

1. Extended κ gene fragments are produced as described in **steps 1** and **3–5, Subheading 3.4**.
2. Primary amplification of Fd gene fragments:

cDNA from 25 μg RNA	5.0 μL
10X buffer II	10.0 μL
$MgCl_2$ (25 mM)	6.0 μL
dNTP (1.25 mM each)	8.0 μL
MVH1-25 (20 μM)	1.0 μL
MCHmix.For (20 μM)	1.0 μL
Sterile water	69.0 μL

3. PCR program: 94°C 10 min, suspend, add 0.3 μL Ampli*Taq* polymerase (5 U/μL); 94°C 1 min, 55°C 1 min, 72°C 1 min, 30 cycles, 72°C 10 min, refrigerate. Primary Fd products are about 700 bp. Gel purify the primary PCR products as in **steps 6** and **7, Subheading 3.2**.
4. Extension of Fd gene fragments:

1.6 ng Fd DNA	0.5 μL
10X buffer II	5.0 μL
$MgCl_2$ (25 mM)	5.0 μL
dNTP (1.25 mM each)	8.0 μL
Tag.Back1-2 (20 μM)	0.5 μL
MCH.Ext (20 μM)	0.5 μL
Sterile water	30.5 μL

5. PCR program: 94°C 10 min, suspend, add 0.3 μL Ampli*Taq* polymerase (5 U/μL); 94°C 1 min, 55°C 1 min, 72°C 1 min, 15 cycles, 72°C 10 min, refrigerate. Extended Fd products are about 750 bp.
6. Make the pFab60/κ library as described in **steps 8–11, Subheading 3.4**.
7. Digest pFab60/κ and Fd DNA in NEB3 + BSA with *Sfi*I at 50°C, then with *Not*I. Further digest pFab60/κ with *Pme*I at 37°C. Gel purify DNA as in **steps 6** and **7, Subheading 3.2**. (*see* **Note 16**).
8. Ligate the digested pFab60/κ and Fd DNA in a molar ratio of about 1:2 and extract and precipitate the ligation products as described in **steps 20** and **21, Subheading 3.2**.
9. Proceed with **Subheading 3.8.** to produce the IgG Fab library (*see* **Note 18**).

3.6. Direct Cloning Method for Making Human Antibody Fab Libraries Independent of IgH Isotype

The direct cloning method for constructing human antibody Fab libraries independent of IgH isotype consists of the following:

- Primary PCR amplifications of V_H H-chain and κ and λ L chains.
- "Tagging" the primary amplification products with restriction enzyme sites.
- Cloning of V_H gene fragments in pFab73H (equal to the pFab73H/Fd library).
- Cloning of L-chain genes into the pFab73H/Fd library DNA.
- Electroporation, growth, and storage of the library (*see* **Subheadings 3.7.** and **3.8.**).

Figure 4 shows an overview of the PCR primers and the expression vector, pFab73H. PCRs are hot started and carried out in PCR tubes. All PCR mixtures (100 μL) are covered with paraffin oil.

1. Primary amplifications of V_H gene fragments:
cDNA (see **Subheading 3.1., step 17**)	5 µL
10X Buffer I	10 µL
10X dNTP (2 mM each)	10 µL
HVH1-12 (20 µM mix)	1 µL
HJH1-3 (20 µM mix)	1 µL
Sterile water	73 µL

2. Primary amplification of κ genes:
cDNA (see **Subheading 3.1., step 17**)	5 µL
10X Buffer I	10 µL
10X dNTP (2 mM each)	10 µL
HVK1-8 (20 µM mix)	1 µL
HCK.For (20 µM)	1 µL
Sterile water	73 µL

3. Primary amplification of λ genes:
cDNA (see **Subheading 3.1., step 17**)	5 µL
10X Buffer I	10 µL
10X dNTP (2 mM each)	10 µL
HVL1-12 (20 µM mix)	1 µL
HCL.For (20 µM)	1 µL
Sterile water	73 µL

4. PCR program: 94°C 5 min, suspend, add 0.3 µL Ampli*Taq* polymerase (5 U/µL); 94°C 1 min, 55°C 1 min, 72°C 1 min, 30 cycles, 72°C 10 min, refrigerate.

5. Gel purify the primary amplification products as described in **steps 6** and **7, Subheading 3.2.** Primary V_H PCR products are about 390 bp, and primary κ and λ amplification products about 650 bp.

6. For PCR extension of the primary PCR products, proceed as described above using about 1 ng purified template DNA and 1 µL of 20-µM solutions of the following primers:
   For V_H extension: HVH.Ext and HJH.Ext
   For κ extension: HVK.Ext and HCK.For
   For λ extension: HVL.Ext and HCL.For
   PCR program: 94°C 5 min, suspend, add 0.3 µL Ampli*Taq* polymerase (5 U/µL); 94°C 1 min, 55°C 1 min, 72°C 1 min, 15 cycles, 72°C 10 min, refrigerate.

7. Gel purify the secondary amplification products as in **steps 6** and **7, Subheading 3.2.** Extended V_H gene fragments are about 440 bp, and extended κ and λ genes about 700 bp.

8. Digest pFab73H and extended V_H DNA with *Apa*I in NEB4 + BSA at 30°C, then with *Nhe*I in NEB2 at 37°C and gel purify as in **steps 6** and **7, Subheading 3.2.**

9. Ligate, extract, and precipitate the ligation products as described in **steps 20** and **21, Subheading 3.2.**

10. Proceed with **Subheading 3.8.** and make pFab73H/Fd library DNA (the vector harbors the human $C_H1_{\gamma 1}$ gene fragment, see **Fig. 4**). Use this DNA as a "vector" to clone the L-chain genes.

11. Digest the pFab73H/Fd and extended κ and λ DNA with *Asc*I in NEB4, then with *Sfi*I in NEB2 + BSA at 50°C and gel purify as in **steps 6** and **7, Subheading 3.2.**

12. To produce the κ and λ Fab libraries, ligate, extract, and precipitate ligation products in separate reactions as in **steps 20** and **21, Subheading 3.2.**

13. Proceed with **Subheading 3.8.** to produce the human κ and λ IgH-independent Fab libraries (see **Note 18**).

3.7. Preparation of Ultracompetent Cells

High efficiency of electroporation is important for producing large libraries. Using cells prepared by the following protocol, and taking heed of the stated precautions, we routinely obtain efficiencies of $2-6 \times 10^{10}$ CFU/µg supercoiled pUC18 test DNA (Stratagene) with Top10/F' and TG1 cells. These strains give the highest efficiencies using 0.1 cm electroporation cuvets, 25 µL cells at a density of $5-6 \times 10^{10}$/mL, a field strength of 20 kV/cm, and amounts of DNA that give a time constant between 5.0 and 5.5 ms. (For other *E. coli* strains, electroporation conditions should be optimized by varying field strength and cell density. The optimal field strength for most strains is between 16 and 19 kV/cm *[13]*.)

1. Inoculate 1 L rich medium (LB or 2X YT) + tetracycline with newly grown Top10/F' cells (*see* **Note 19**) at an OD_{600} of maximum 0.020. Divide the culture between two baffled 2-L flasks, and shake at 37°C at a minimum of 300 rpm to ensure good aeration (*see* **Note 20**).
2. Transfer cells to four detergent-free 500 mL centrifuge flasks (*see* **Note 20**) at an OD_{600} of 0.8–1.0, and chill by swirling in an ice-water bath for 5 min. Cells, buffer, and *all* equipment are kept ice cold throughout the remaining procedure (*see* **Note 21**).
3. Pellet cells for 6–8 min at 4000–6000 rpm (3000*g*) in an GS-3 rotor (*see* **Note 22**). Decant as much supernatant as possible and resuspend each cell pellet carefully and completely in 5–10 mL 10% glycerol buffer using an ice-cold 10 mL glass pipet. Add 300 mL glycerol buffer to each flask and mix.
4. Repeat **step 3**, but reduce the number of centrifuge flasks to two. Add 300 mL glycerol buffer to each of the two flasks and mix.
5. Repeat **step 4** twice.
6. Resuspend the final pellets from **step 5** in 5–10 mL glycerol buffer, transfer cells to a 50-mL centrifuge tube, add glycerol buffer to a total of 40 mL, mix, and centrifuge cells at 4000–5000 rpm (3000*g*) for 6–8 min (*see* **Note 23**).
7. Resuspend pellet in the remaining supernatant and add buffer to a total of 1.25 mL. Assuming 2×10^8 cells/mL at $OD_{600} = 1$, this corresponds to $6-12 \times 10^{10}$ cells/mL, depending on strain, cell loss, and so forth.
8. Dispense 25–200 µL of cells in chilled microfuge tubes, quick freeze the tubes in ethanol at –80°C (or a dry ice/ethanol bath), and store at this temperature.
9. Test that cells are free of contaminant plasmids, phagemids, or phages.

3.8. Electroporation of Library DNA, Cell Growth, and Storage

This protocol is dimensioned to obtain libraries containing up to 5×10^8 independent CFU.

1. Set the *E. coli* Pulser at 2000 V and thaw ultracompetent Top10/F'cells on ice (*see* **Subheading 3.7.**). Place electroporation cuvets, cuvet holder, cells, and DNA on ice. Dilute cells to $5-6 \times 10^{10}$ cell/mL with ice-cold sterile water.
2. Carefully mix 10–40 ng purified, salt-free DNA from the last step of **Subheadings 3.2.–3.6.** with 25 µL cells.
3. Place the mixture in a 0.1-cm electroporation cuvet taking care not to leave air bobbles. Tap the cuvet gently to get all liquid to cover all of the bottom. Make sure that the cuvet is dry on the outside.
4. Pulse and add instantly 0.5–1.0 mL of SOC medium supplemented with 10 m*M* $MgCl_2$, and mix. The time constant should be in the 5–5.5-ms range (*see* **Note 24**). Transfer the

mixture, or pools of mixtures, to a capped tube and allow phenotypic expression by shaking at 250 rpm for 1 h at 37°C.
5. Withdraw 1 μL, make dilutions, and spread on LB-ampicillin plates to obtain an estimate of the number of CFU for each electroporation or pool of electroporations.
6. For each individual transformation, use 40 mL LB-ampicillin/tetracycline medium supplemented with 1% glucose, which act as catabolite repressor of P_{lac}. We usually do 25–50 electroporations at a time.
7. Propagate cells in liquid culture (a) or on plates (b):
    a. Shake cells at 37°C until an OD_{600} of 1–2 is reached. This takes about 10 h. To increase the effectiveness of the ampicillin selection, cells can be pelleted and resuspended in fresh prewarmed LB-ampicillin/tetracycline medium after 6–8 h of growth.
    b. Alternatively, spread cells from five electroporations on an air-dried Nunclon Δ Square Dish (500 cm^2), and incubate overnight at 37°C. For each Nunclon Δ Square Dish (Life Technologies), scrape off cells using 10 mL LB, transfer to a shake flask, add 100 mL LB-ampicillin/tetracycline medium, shake for 2 h, and combine cells.
8. Pellet two portions of 10^{11} cells (equivalent to 250–500 mL from **step 7a**). Resuspend one portion in LB at 10^{10} cells/mL, and make 10 individual glycerol stocks *(11)*. Prepare Qiagen maxiprep DNA (Qiagen, Chatsworth, CA) from the other.

3.9. Production and Titration of Fab Phages from Libraries

This protocol is dimensioned to libraries that contain about 10^8 independent CFU. If a differently sized library is used, scale up or down correspondingly. Tetracycline is added to ensure that all growing cells have sex pili.

1. Inoculate 400 mL prewarmed LB-ampicillin/tetracycline medium with 2×10^9 cells (200 μL of a glycerol stock from **step 8, Subheading 3.8.**). Grow cells about four generations at 37°C with shaking until OD_{600} reaches 0.5.
2. Add 2×10^{12} R408 helper phages (*see* **Note 25**) to the 400-mL culture. This high multiplicity (about 50) increases the chance that all cells become infected.
3. Shake gently for 20 min at 37°C and add IPTG to a final concentration of 100 μ*M* (*see* **Note 26**).
4. Transfer the culture flask to a shaker at room temperature (*see* **Note 27**), and continue shaking at 300 rpm for a minimum of 10 and a maximum of 16 h (cell lysis occur after 16 h).
5. Pellet the cells by centrifugation at 15,000*g* for 10 min.
6. Transfer supernatant and repeat **step 5**.
7. Transfer supernatant to a fresh centrifuge flask and add 1 vol phage-precipitation buffer to 4 vol phage supernatant. Mix the solution by shaking and incubate at 0 or 4°C for 1 h or longer.
8. Pellet the phage particles by centrifugation at 15,000*g* for 30 min. Resuspend pellet in 8 mL of PBS buffer, transfer to a 10-mL centrifuge tube, and make a clearing spin at 15,000*g* for 10 min.
9. Transfer supernatant to a fresh tube, precipitate phages as in **step 7**, and centrifuge at 15,000*g* for 30 min.
10. Resuspend pellet in 1 mL PBS buffer, transfer to a microfuge tube, make a clearing spin, transfer supernatant to a fresh tube, and precipitate phages as in **step 7**. Resuspend pellet in 1 mL PBS buffer and filtrate through a 0.2- or 0.45-μm filter. Store the phage stock at 4°C or –20°C.
11. Mix 125 μL exponentially growing Top10/F' cells (or other male *coli* cells) at an OD_{600} of 0.8–1.0 with an equal volume of a series of diluted Fab phages. Include controls with

phage dilutions alone and titration cells alone. Place the mixtures at 37°C for 10 min, then chill on ice.

12. Spread 20 and 200 μL from each mixture on LB-ampicillin/tetracycline plates and incubate overnight at 37°C. Also, spread 50 μL cells on an LB plate to make sure that the titration cells are free of wild-type phage.
13. Count colonies, correct for background, and calculate CFU/mL. A titer of about 10^{13} CFU/mL of the concentrated phage stock is typical.

3.10. Selection of Antigen Binders by Panning on Antigen-Coated Plastic (see Note 28)

1. Coat microtiter wells with 100 μL antigen solution (5–50 μg/mL in PBS) overnight at 4°C followed by two to three short rinses with washing buffer and blocked for 2 h at room temperature with blocking buffer using 200 μL/well.
2. Rinse the wells briefly two to three times in washing buffer, and add about 10^{11} Fab phages diluted in 100 μL blocking solution per well.
3. Incubate for 2–3 h with rocking at room temperature. After incubation, the wells are washed and rinsed briefly 10–15 times with washing buffer (see **Note 29**).

    Bound phages are eluted by one of two methods: glycin-buffer treatment (**steps 4** and **5**) or digestion with trypsin (see **step 6**).

4. Add 100 μL 0.1 M glycine elution buffer per well and incubate for 15 min at room temperature.
5. Transfer eluates to fresh microtiter wells containing 8 μL 2 M Tris base to neutralize the acid.
6. Add 100 μL trypsin elution buffer and incubate for 30 min at room temperature.
7. The eluates from **steps 5** or **6** (see **Note 30**) are transferred to vials containing 400 μL exponentially growing Top10/F' cells with an OD_{600} of 0.8–1.0.
8. Incubate at 37°C for 20 min with gentle shaking.
9. Withdraw 5 μL from each culture, make serial dilutions, and spread on ampicillin plates to determine the number of ampicillin-transducing phages in the eluate.
10. Transfer each culture to 40 mL 2X TY medium supplemented with 1% glucose, ampicillin, and tetracycline, and incubate at 37°C with shaking overnight. Alternatively, plate each culture on a Nunclon Δ Square Dish, continue as in **step 7b**, **Subheading 3.8.**, and proceed with **step 11**.
11. Prepare 1-mL glycerol stocks and make plasmid Qiagen miniprep DNA.
12. For preparation of phages for the next round of panning, use 200 μL of the glycerol stock (about 2×10^8 cells) to inoculate 30 mL 2X TY medium containing ampicillin and tetracycline.
13. When the culture reaches an OD_{600} of 0.5–0.8, add 2.5×10^{11} R408 helper phages, and incubate at 37°C with gentle shaking for 20 min.
14. Add IPTG, grow cells, and prepare and titrate Fab phages for the next panning round as described in **steps 3–13**, **Subheading 3.9.**
15. Between each round of panning, determine the number of eluted bound phages relative to the amount of phages added to the wells. Similarly, the ELISA signal generated from a fixed amount of eluted bound phages should be monitored between each round of panning (we use peroxidase-conjugated anti-M13 as a secondary antibody). An increase in these two parameters by a factor of 10 or more should be observed between each round of panning.
16. After three or four rounds of panning, isolate individual clones, and characterize their binding characteristics by ELISA assays.

4. Notes

1. (**Subheading 2.2., step 1**) All primers in the present study were synthesized on an Applied Biosystems oligonucleotide machine, model 394, and checked on sequencing gels. Only high-quality preparations that predominantly contained full-size product were used for further applications.
2. (**Subheading 2.2., step 1**) The rationale behind the design of our rather extensive series of PCR primers for the variable H- and L-chains has been presented previously *(4)*. In short, we believe that the primer sets used should match all available sequence data for these regions and we argued that highly degenerate primers for the variable regions are likely to generate biased libraries and to introduce amino acids not normally found in the variable regions.
3. (**Subheading 2.2., step 5**) To ensure efficient digestion with restriction enzymes, we generally equip PCR products with long flanking regions (20 or more nucleotides between enzyme recognition sequence and ends of fragments).
4. (**Subheading 2.6., step 6**) Because the *Asc*I enzyme is efficient also on sites very near fragment ends *(14)*, HCK.For and HCL.For (**steps 5** and **8**, **Subheading 2.2.**) adds only 12 nucleotides downstream of its recognition sequence (cf. **Note 3**).
5. (**Subheading 3.1., step 1**) We find it important to work with fresh biological material when isolating RNA. If immediate processing of blood samples is impossible, we recommend transporting the sample to the laboratory frozen in anticoagulant, PBS, and buffer A, rather than transporting the sample as frozen material and then adding these ingredients in the laboratory.
6. (**Subheading 3.1., step 16**) From previous experience with reverse dideoxy sequencing of RNA *(15)*, we know that secondary structures in RNA can obstruct the processive action of reverse transcriptase, but that this can be overcome by increasing the reaction temperature. Because the GC content of the IgG mRNAs is relatively high *(12)* and will allow for extensive secondary structure formation, we recommend synthesis of cDNA at 47°C.
7. (**Subheading 3.2.**) The main advantage of the method outlined here and in **Subheading 3.3.** is that the entire cloning procedure requires the use of only two rare-cutting restriction endonucleases (*Sfi*I and *Not*I).
8. (**Subheading 3.2.**) Although the PCR assembly strategy used in **Subheading 3.3.** has been described as somewhat difficult *(16)*, we find it attractive to have available complementary procedures for cloning of Fab gene fragments (compare **Subheadings 3.2.** and **3.3.** with **Subheadings 3.4.** and **3.5.**).
9. (**Subheading 3.2., step 5**) If the yield of the primary PCR reactions are unsatisfactory, check the integrity of the RNA preparation by running a formaldehyde-agarose gel *(10)*. If the bands representing rRNA look distinct and are of the correct molecular size, assume that the mRNAs for H- and L-chain are intact, too. We recommend repeating the phenol/chloroform extractions described in **steps 8** and **9**, **Subheading 3.1.** to improve the success of the cDNA synthesis and subsequent PCR reactions.
10. (**Subheading 3.2., step 8**) This reaction sometimes needs optimization, which is done by varying the relative concentrations of the two templates. We normally obtain the best results by using a 10-times molar excess of the Link fragment to the Fd gene fragment.
11. (**Subheading 3.2., step 19**) As depicted in **Figs. 1** and **3**, the stuffer fragments between pairs of cloning sites in the murine vectors contain 8-bp recognition sequences for restriction enzymes that occur extremely rarely in murine Fab DNA. Self-ligation of the fraction of vector molecules only cut once can be reduced by subsequent digestion with the enzyme, whose recognition sequence is present in the stuffer fragment to be excised. Vectors prepared for ligation with inserts should always be checked for self-ligation as a measure of the extent of cleavage by the relevant restriction enzymes.

Although phage particles without Fab molecules displayed on their surface theoretically should be lost in the panning procedure, some unspecific binding of such phages to the antigen coat of the microtiter plate will occur. This may lead to an outgrowth of the specific binders because of the growth advantage of phages having a smaller genome size and not having to express large quantities of two foreign proteins. This can be prevented by gel purifying the entire library as phagemid DNA before each round of panning (our unpublished results).

12. (**Subheading 3.2.**, **step 21**) Addition of (inert) glycogen to a standard ethanol precipitation ensures quantitative recovery and that the pellet is always visible.

13. (**Subheading 3.3.**) The reasons for choosing the outlined approach rather than attempting to use Fd gene fragments made by MVH and C_H primers of all known isotypes or making single-chain F_V antibody fragments are detailed in **ref. 8**.

14. (**Subheading 3.3.**, **step 6**) Successful assembly of V_H and CH1-Link-D PCR products requires a larger amount of template than does assembly of κ and Link-D products (cf. **step 8**, **Subheading 3.2.** and **Note 10**). The sluggishness of VH/CH1-Link-D assembly may be caused by formation of strong secondary structures within the C_H1 gene fragment, which would also account for the relatively weaker amplification of Fd genes as compared to the V_H gene fragments as observed previously *(4)*.

15. (**Subheading 3.4.**) The main advantage of the direct cloning method is that the risk of cloning PCR artifacts is reduced because considerably fewer rounds of PCR amplification is required (40–45 vs 75 rounds).

16. (**Subheading 3.4.**) Digestion of pFab60/κ library DNA with *Spe*I causes some vector background because 1% of known κ genes contain *Spe*I sites, which produces a subpopulation of linear phagemid DNA with two *Spe*I ends. (Similarly, digestion of pFab60/Fd library DNA with *Nhe*I, gives some background because 0.25% of known Fd gene fragments contain *Nhe*I sites.) The background problem can be avoided by cloning κ and V_H DNA separately in pFab60 and by moving the resulting Fd library DNA as *Sfi*I-*Not*I fragments. Alternatively, the primary PCR product of V_H may be digested with *Spe*I before gel purification and cloning. This will cause a slight reduction in diversity of the library resulting from the loss of V_H fragments containing *Spe*I sites. However, as these fragments will be lost at any rate using the present vector system, it is recommended to perform the *Spe*I digestion and avoid problems resulting from the self-ligation mentioned earlier.

17. (**Subheading 3.4.**, **step 9**) Digest with the rare-cutting *Srf*I enzyme instead of *Kpn*I or *Stu*I, if pFab60/Fd library DNA is used instead of pFab60 DNA. Similarly, use the rare-cutting *Pme*I enzyme instead of *Hin*dIII, if pFab60/κ library DNA is used.

18. (**Subheading 3.4.**, **step 14**) To avoid reduction in the diversity of the first-step library, it is necessary that the number of transformants in the second cloning step is several-fold higher than in the first.

19. (**Subheading 3.7.**, **step 1**) The protocols presented here assume the use of Top10/F' (tet[r]) cells. We generally prefer this strain because it produces Fab phages that give high antigen-binding capacity per unit number of Fab phages *(7)*, electrocompetent cells that give extraordinarily high electroporation efficiencies, and DNA of high quality.

20. (**Subheading 3.7.**, **step 1**) It is critical that all glass and plastic equipment, medium, and buffer are absolutely soap and detergent free. (It is ideal to use a set of dedicated equipment that is washed with water and ethanol only.) Some strains are manyfold more competent if cells are grown at room temperature *(17)*. For Top10/F', we see a twofold increase in competency. For growth at room temperature, it is convenient to start the culture the day (evening) before at an OD_{600} of 0.002–0.005 (strain dependent), and shake vigorously overnight. Using freshly prepared LB that is *not* autoclaved, increases competency for Top10/F' by a factor of 2.

21. (**Subheading 3.7., step 2**) Make sure that all equipment (including centrifuge flasks, pipets, and so forth) and solutions are *ready* and *ice cold* when needed, and work fast but carefully.
22. (**Subheading 3.7., step 3**) Centrifuge cells as slowly as possibly, so as to make it easy to resuspend them *gently and completely*. A recovery of only 50–75% of the start cells (corresponding to 90–95% recovery in each centrifugation step) ensures very gentle treatment of the cells. The optimal centrifugation speed is strain dependent.
23. (**Subheading 3.7., step 6**) It is of *pivotal* importance to remove all salts by repeated washing of the cells, because otherwise arching will occur at the high field strengths necessary for optimal electroporation conditions (*see* introduction to **Subheading 3.7.**).
24. (**Subheading 3.8., step 4**) If arching is a problem then:
    a. Wash cells and/or purify DNA better;
    b. Use a lower field strength;
    c. Dilute cells before electroporation with ice-cold sterile water; or
    d. Use 0.2 cm cuvets and the standard procedure *(18)*.

    If cells are diluted, add correspondingly less SOC medium to avoid overestimating the transformation efficiency or library size, because cells at low densities may start to divide during the 1-h phenotypic expression.
25. (**Subheading 3.9., step 2**) R408 helper phage production: Infect log-phase Top10/F' cells at $OD_{600} = 0.5$ in baffled culture flasks with a 20-fold excess of R408, and shake at 200 rpm overnight at 37°C. Prepare and titrate phages as described in **steps 5–13, Subheading 3.9.**, except that the phage/cell mix of **step 11** is mixed with 3 mL of molten topagar and spread on LB-tetracycline plates. Expect more than 10^{14} PFU/L supernatant.
26. (**Subheading 3.9., step 3**) We have varied the IPTG concentration used for induction of P_{lac} during superinfection between 0 and 2 m*M*, and found no increase in Fab phage production with IPTG concentrations above 100 µ*M*.
27. (**Subheading 3.9., step 4**) It is important to incubate at 22–30°C following superinfection for two reasons:
    a. Folding of active Fab molecules proceeds more successfully at room temperature than at 37°C *(19)*; and
    b. Fab phages will not be lost as a result of re-infection since sex pili are not generated at temperatures below 30°C *(20)*.
28. (**Subheading 3.10.**) Alternatively, use antigen-coated immunotubes, 10^{12} phages in 400 µL blocking buffer, and an end-over-end mixer where relevant. Following incubation, wash and rinse tubes briefly 10–15 times with 4 mL washing buffer. Elute phages with 400 µL trypsin elution buffer and continue with **step 7, Subheading 3.10.**, using 1.5 mL cells. This approach requires a bigger supply of antigen but gives an improved selection *(21)*.
29. (**Subheading 3.10., step 3**) The outlined washing procedure has been used successfully when screening for high-affinity binders (apparent K_d of about $10^{-9}M$) normally present in libraries generated from immunized animals. When dealing with low-affinity binders (apparent K_d of about $10^{-6}M$), the number of washes is reduced to three brief washes in washing buffer.
30. (**Subheading 3.10., step 7**) We have not systematically compared the two elution methods described in **steps 4–6, Subheading 3.10.**, but we like the trypsin elution method, as it is very reliable and removes all bound phages as measured by testing the eluted wells in ELISA assays using antiphage antibodies (peroxidase-conjugated anti-M13). Furthermore, this method is gentle and independent of the strength of the antigen-antibody interactions, and the eluted phages maintain their infectivity even after hours of treatment with the specified amounts of trypsin.

Acknowledgments

The authors thank previous co-workers: Henrik Ørum, Anne Øster, Peter Sejer Andersen, Morten Dziegiel, Leif Kofod Nielsen, Heidi W. Andersen, Svend L. Jørgensen, Bernd Haase, Margit Haahr Hansen, and Mads Bjørnvad for their help during different phases of this work. This work was supported by the Danish Medical and Natural Science Research Councils and the Center of Medical Biotechnology, University of Copenhagen.

References

1. Hoogenboom, R. H., Griffiths, A. D., Johnson, K. S., Chiswell, D. J., Hudson, P., and Winter, G. (1991) Multi-subunit proteins on the surface of filamentous phage: methodologies for displaying antibody (Fab) heavy and light chains. *Nucleic Acids Res.* **19,** 4133–4137.
2. Kang, A. K., Barbas, C. F., Janda, K. D., Benkovic, S. J., and Lerner, R. A. (1991) Linkage of recognition and replication functions by assembling combinatorial antibody Fab libraries along phage surfaces. *Proc. Natl. Acad. Sci. USA* **88,** 4363–4366.
3. Breitling, F., Dübel, S., Seehaus, T., Klewinghaus, I., and Little, M. (1991) A surface expression vector for antibody screening. *Gene* **104,** 147–153.
4. Ørum, H., Andersen, P. S., Riise, E., Øster, A., Johansen, L. K., Bjørnvad, M., Svendsen, I., and Engberg, J. (1993) Efficient method for constructing comprehensive murine Fab antibody libraries displayed on phage. *Nucleic Acids Res.* **21(19),** 4491–4498.
5. Hoogenboom, H. R., Marks, J. D., Griffiths, A. D., and Winter, G. (1992) Building antibodies from their genes. *Immunol. Rev.* **130,** 41–68.
6. Winter, G., Griffiths, A. D., Hawkins, R. E., and Hoogenboom, H. R. (1994) Making antibodies by phage display technology. *Ann. Rev. Immunol.* **12,** 433–455.
7. Johansen, L. K., Albrechtsen, B., Andersen, H. W., and Engberg, J. (1995) pFab60: a new efficient vector for expression of antibody Fab fragments displayed on phage. *Protein Eng.* **8(10),** 1063–1067.
8. Andersen, P. S., Ørum, H., and Engberg, J. (1996) One-step cloning of murine Fab gene fragments independent of IgH isotype for phage display libraries. *Biotechniques* **20,** 340–342.
9. Dziegiel, M., Nielsen, L., Andersen, P. S., Blancher, A., Dickmeiss, E., and Engberg, J. (1995) Phage display used for gene cloning of human recombinant antibody against the erythrocyte surface antigen, rhesus D. *J. Immunol. Methods* **182,** 7–19.
10. Engberg, J., Andersen, P. S., Nielsen, L. K., Dziegiel, M., Johansen, L. K., and Albrechtsen, B. (1996) Phage-display libraries of murine and human antibody fragments. *Mol. Biotechnol.* **6,** 287–310.
11. Sambrook, J., Fritsch, E. F., and Maniatis, T. (eds.) (1989) *Molecular Cloning. A Laboratory Manual*, 2nd ed. Cold Spring Harbor Laboratory Press, Cold Spring Harbor, NY.
12. Kabat, E. A., Wu, T. T., Reid-Miller, M., Perry, H. M., and Gottesman, K. S. (1991) *Sequences of Proteins of Immunological Interest.* U.S. Department of Health and Human Services, U.S. Government Printing Office, Washington, DC, 1356–1460.
13. Zoller, P. (1994) How optimal electroporation efficiency varies for different strains of *E. coli. BIO-RADiations* **90,** 5.
14. Moreira, R. and Noren, C. (1995) Minimum duplex requirements for restriction enzyme cleavage near the termini of linear DNA fragments. *Biotechniques* **19(1),** 56–59.
15. Ørum, H., Nielsen, H., and Engberg, J. (1991) Spliceosomal small nuclear RNAs of Tetrahymena thermophilia and some possible snRNA-snRNA base-pairing interactions. *J. Mol. Biol.* **222,** 219–232.

16. Engelhardt, O., Grabherr, R., Himmler, G., and Rüker, F. (1994) Two step cloning of antibody variable domains in a phage display vector. *Biotechniques* **17,** 44-46.
17. Chuang, S.-E., Chen, A.-L., and Chao, C.-C. (1995) Growth of *E. coli* at low temperature increases the transformation frequency by electroporation. *Nucleic Acids Res.* **23(9),** 1641.
18. Dower, W. J., Miller, J. F., and Ragsdale, C. W. (1988) High efficiency transformation of *E. coli* by high voltage electroporation. *Nucleic Acids Res.* **16(13),** 6127–6145.
19. Plückthun, A. and Skerra, A. (1989) Expression of functional antibody F_v and Fab fragments in *Escherichia coli. Methods Enzymol.* **178,** 497–515.
20. Miller, J. H. (1972) *Experiments in Molecular Genetics.* Cold Spring Harbor Laboratory Press, Cold Spring Harbor, NY, p. 98.
21. Kretzschmar, T., Zimmermann, C., and Geiser, M. (1995) Selection procedures for nonmatured phage antibodies: a quantitative comparison and optimization strategies. *Anal. Biochem.* **224,** 413–419.

VI

DNA Sequencing

61

Preparation and Analysis of DNA Sequencing Gels

Bimal D. M. Theophilus

1. Introduction

DNA sequencing involves a specific application of electrophoresis to resolve the linear single-stranded products of sequencing reactions. A 4–20% polyacrylamide gel is used, normally 0.4 mm thick and at least 40 cm in length.

The sequencing gel is poured into a mold comprising two glass plates separated by spacers running the length of the plates. A variety of methods may be used to seal the sides and bottom edge of the mold. One plate is shorter than the other to form a buffer chamber into which samples can be loaded once the gel has been positioned vertically for electrophoresis (**Fig. 1A**).

Samples may be loaded into spaces made by a standard well-forming comb. Alternatively, a "shark's tooth" comb may be used, which has a straight edge and a jagged edge comprising 24–48 triangular teeth (**Fig. 1B**). The shark's tooth comb is often preferred because there is virtually no separation between adjacent sequence lanes in the final autoradiograph and samples may be loaded with a Pipetman (Anachem, Luton, UK), rather than a 0–10-µL syringe necessary with standard combs. The straight edge of the shark's tooth comb forms a flat, uniform surface across the top of the gel while it is setting. The comb is then reversed to enable sample loading into the "wells" comprising the spaces between the teeth.

The products of the four chain-termination reactions are run on adjacent lanes of the gel. To minimize the formation of secondary structures in the DNA by self-hybridization, the samples are heated to 70°C in the presence of the denaturant formamide before loading, 7 M urea is incorporated into the gel, and the gel is run at around 50°C.

The samples undergo electrophoresis for a time that is determined by the distance from the sequencing primer to the region of interest. The gel is dried and exposed to X-ray film. The sequence of the DNA template is then determined by reading successive bands of increasing size in the resulting sequence "ladder" in the four adjacent tracks of the gel.

2. Materials

1. Sequencing gel apparatus, comprising the following:
   Gel tank.
   Electrical leads (usually incorporated into tank safety covers).

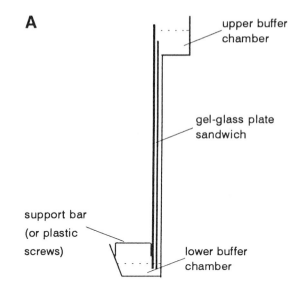

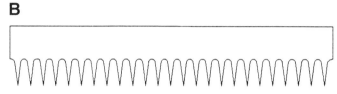

Fig. 1. (**A**) Sequencing gel apparatus. (**B**) Shark's tooth comb.

      Glass plates (usually 21–38 cm wide × 40–100 cm long).
      2X 0.4 mm side spacers.
      1X 0.4 mm bottom spacer (optional).
      0.4 mm standard or shark's tooth comb.
      Clamps or bulldog binder clips.
2. Gel dryer.
3. 10X TBE buffer: 108 g Tris base, 55 g boric acid, 9.3 g Na_2 ethylenediaminetetraacetic acid (EDTA), H_2O, in 1 L deionized H_2O (the pH should be around 8.3, without adjustment).
4. 30% Acrylamide: 28.5 g Acrylamide, 1.5 g *bis*-acrylamide. Make up to 100 mL with deionized H_2O, filter, and store at 4°C (*see* **Note 1**).
5. Freshly made 25% ammonium persulfate: Dissolve 0.25 g in 1 mL distilled H_2O.
6. TEMED (*N,N,N′,N′*-Tetramethylethylenediamine).
7. X-ray film and cassettes.
9. Siliconizing solution (dimethyl dichlorosilane).

3. Methods
3.1. Assembling the Gel Plates
1. Ensure that all parts of the apparatus are thoroughly clean (*see* **Note 2**).
2. Place one glass plate horizontally on a bench, inner side facing upward (*see* **Note 3**). Place the clean, dry spacers along the long edges and along the bottom edge if one is provided for this purpose.
3. Place the other glass plate on top of the spacers so that the two inner sides are facing each other. The bottom edges of the plates and spacers should be aligned.

4. Secure the assembly together along the sides with bulldog clips spaced approx 2 cm apart. If the bottom edge has a spacer, this should be similarly clamped. If not, seal the bottom edge with waterproof tape. The tape may also be used in combination with clips along both the sides and bottom edges for added security, especially near the bottom corners, which are particularly prone to leakage (*see* **Note 4**).

3.2. Pouring the Gel

1. To pour a 6% gel, combine 63 g urea, 15 mL of 10X TBE, and 30 mL of 30% acrylamide stock solution (*see* **Note 5**). Make up the volume to 150 mL with distilled H_2O. This solution can be made up as a stock and stored at 4°C for several weeks (*see* **Note 6**).
2. Add 50 μL each 25% ammonium persulfate and TEMED to 50 mL of the urea/TBE/acrylamide gel mix, which has been allowed to warm to room temperature. Mix by swirling. This volume is sufficient for a 21 cm × 50 cm plate assembly, but an additional aliquot (10–30 mL) may be required for systems that recommend sealing the bottom edge with acrylamide before pouring the main gel (*see* **Note 4**).
3. Without delay, take the gel mix into a 50-mL syringe, attach a needle, and inject the mix between the plates, maintaining a steady flow. During pouring, the plates should be supported by the left hand at a 30° angle and to the side so that the corner into which the mix is injected is uppermost, while the diagonally opposite corner is resting on the bench. Any air bubbles that form should be removed immediately by gently raising the glass plate to lower the level of the liquid, gently tapping the plates with a finger or Pipetman, or using the comb to draw the bubble to the surface.
4. Once the gel solution has reached the top, rest the assembly at about 5° to the horizontal (e.g., on a roll of sticky tape placed near the top of the assembly).
5. Insert the straight edge of the shark's tooth comb about 5 mm into the gel. If a standard well-forming comb is used, the toothed edge should be inserted into the gel. It is also advisable to clamp the glass plates over the comb with two bulldog clips to reduce the risk of leakage across adjacent lanes during sample loading. Check the gel over the next few minutes, and top it up as necessary using the gel mix remaining in the syringe.
6. The gel should set within 1 h, but to maximize resolution, it is recommended to age the gel for at least 3 h before use. If the gel is to be left overnight, place a moistened paper tissue over the comb and cover the upper end of the assembly with cling film to prevent the gel from drying out.

3.3. Running the Gel

1. Remove the bulldog clips and adhesive tape from the long edges of the gel assembly. Specifically designed clamps may be designed to remain in place during electrophoresis. Also remove all components used to seal the bottom edge, e.g., spacer, clips, tape, casting tray, and the like.
2. Remove the comb and secure the plate assembly into the sequencing apparatus using either bulldog clips or the support bar and screws provided.
3. Make up the recommended quantity of 1X TBE buffer (about 1100 mL for a 21 cm × 50 cm gel), and pour it into the upper buffer chamber to about 1 cm from the top. Using a 50-mL syringe and needle, squirt some TBE into the sample wells of the gel to rinse away any unpolymerized acrylamide. If a shark's tooth comb is employed, the straight edge of the gel should be similarly rinsed. The comb should then be washed to remove any acrylamide or urea, and reinserted so that the points of the teeth just pierce the gel by about 1 mm. Once this is done, the comb should not be moved subsequently, as leakage of samples between wells may result. The comb can be secured in place with two miniature bulldog clips if desired.

4. Check that no buffer is leaking from the upper chamber into the lower one (plug any gaps with molten agarose if necessary). Pour the remaining buffer into the lower buffer tank, ensuring that the electrodes are immersed. If a bottom spacer was used during pouring, there may be an air space at the bottom of the gel. This can be removed by squirting TBE into the space with a syringe and attached needle, which has been bent at 45° half way along its length.
5. Load 5 µL of the formamide indicator dye, which is used to stop the sequencing reactions, into a few of the wells and run the gel at an appropriate voltage (e.g., 2000 V for an 21 cm × 50 cm 8% gel; approx 50 W) for 20–60 min until the temperature stabilizes at 55°C. Temperature is best measured by a temperature indicator attached to the outer glass plate. It is important not to let the temperature exceed 65°C, as this may hydrolyze the gel or cause the glass plates to crack. During this time, the level of the buffer in the upper chamber may drop owing to expansion of the apparatus on warming and should be topped up as necessary.
6. Denature the sequencing reaction samples into single strands by heating to 95°C for 3 min. If a microtiter plate has been used for the reactions, incubation should be at 80°C for 10 min, to avoid the risk of melting the plate. Immediately plunge them into ice to prevent reannealing.
7. Turn off the power supply and rinse the loading wells once more with TBE. Load 5 µL of each of the four termination reactions from each template into adjacent wells of the gel. If a shark's tooth comb has been used, samples may be applied with a Pipetman. Otherwise, a 5-µL or 10-µL syringe with a 28- or 30-gage needle may be necessary. If a syringe is used, the needle should be rinsed well between each sample. If sample migration across sample wells is observed, "staggered" loading may be employed in which the gel is run for 2–3 min between each loading of a complete set of four reactions. If there are spare slots in the gel, it is also advisable to load a molecular weight marker at one edge (or two different ones at each edge), both to assist in identifying the position of the sequence and to orientate the final autoradiograph.
8. Reconnect the power supply and run the gel at 50°C until the sample dyes have migrated the required distance. As a guide, bromophenol blue migrates with a DNA fragment of approx 26 nucleotides, and xylene cyanol with a fragment of approx 106 nucleotides, in a 6% gel.

3.4. Gel Drying and Autoradiography

1. At the end of the run, disconnect the power supply and remove the gel. Rinse buffer off the gel, and discard buffer from the gel apparatus into a sink designated for liquid radioactive waste. Remove any clamps or clips that secured the plates together during the run.
2. Remove the siliconized plate by gently prying the plates apart at one end. The gel should adhere to the other plate, but care needs to be taken, as occasionally the gel may adhere to the siliconized plate, or partly to both plates (*see* **Note 7**).
3. Cut a piece of Whatman 3MM paper (Whatman, Kent, UK) to the appropriate size and gently lay it on top of the gel. Rub the back of the Whatman paper with a paper towel and then peel it away from the glass plate. The gel will remain stuck to the Whatman paper.
4. Cover the gel with cling film, smoothing out creases and air bubbles with a paper towel.
5. Dry the gel in a slab gel dryer at 80°C for 30–120 min until the gel is dry.
6. Remove the cling film (this is essential with ^{35}S because it is such a weak beta emitter, but not necessary with ^{32}P), and autoradiograph the gel against a high speed X-ray film in a suitable cassette. If ^{32}P is used, an intensifying screen should be used and the cassette incubated at –70°C. With ^{35}S, incubation can be at room temperature, without a screen. Exposure times vary from about 1–10 d, depending on the type and age of the radioactivity and the quantity of the starting DNA template.

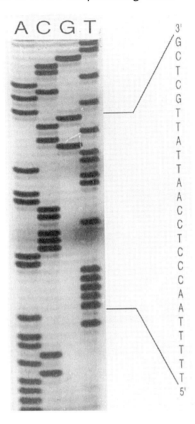

Fig. 2. Autoradiograph of a sequencing gel. The sequence is derived from a single-stranded template isolated from a PCR product amplified from the human factor VIII gene.

3.5. Analysis of Gels

Figure 2 shows an autoradiograph of a sequencing gel. Each vertical position in the gel is occupied by a band in one of the four lanes representing each base. The sequence is read from the bottom of the gel (smaller fragments, closer to primer) to the top (longer fragments, further from primer).

Sometimes the bands on the autoradiograph form a curved pattern across the gel instead of lying in a straight line. This is known as gel "smiling," and occurs when samples near the center of the gel run faster than those at the edges. It is due to the more rapid dissipation of heat near the edges. Features incorporated into the gel apparatus may substantially reduce this problem by the use of a thermostatic plate adjacent to the gel plate, or the design of a buffer chamber that extends over the entire area of the gel enabling the dissipation of heat by convection.

Adjacent bands of DNA may become compressed and appear across all four lanes of the sequencing gel (**Fig. 3**). This is due to intrastrand secondary structure in the DNA arising at regions of dyad symmetry, especially those with a high G + C content. They occur despite protective steps to minimize their formation (*see* **Subheading 1.**). This artifact is commonly observed with double-stranded DNA templates (e.g., plasmids and polymerase chain reaction [PCR] products) and is less of a problem with single-

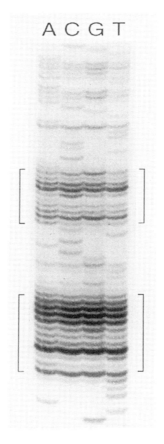

Fig. 3. Areas of compression (brackets) at regions of secondary structure in the DNA template.

stranded templates. DNA sequencing analogs such as dITP (2'-deoxyinosine-5'-triphosphate) and 7-deaza-dGTP (7-deaza-2'-deoxyguanosine-5'-triphosphate), which pair more weakly than the conventional bases may help to resolve compressions *(1,2)*. Alternative methods are to use a single-stranded DNA-binding protein (Amersham 70032, Amersham Pharmacia Biotech, Amersham, UK), or to sequence both strands of the DNA.

The number of bases that can be resolved from a single load can be increased from 200–250 to around 600 by running a buffer gradient gel *(1)*. This involves preparing two gel mixes with the same acrylamide and urea concentrations, one of 0.5X TBE and the other 5X TBE. About a third of the total gel volume is initially prepared by taking up equal volumes of the 0.5X TBE mix, followed by the 5X TBE mix into the same pipet, and then introducing a few air bubbles to mix the solutions at the interphase. This solution is poured into the mold, which is then filled with only 0.5X solution. The success of the gradient formation can be monitored by the addition of 0.05 mg/mL of bromophenol blue to the 5X solution. The increasing ionic concentration in the bottom third results in compression of the lower molecular weight fragments at the bottom of the gel and allows better resolution of high molecular weight fragments near the top. An alternative technique to achieve the same result is to use a wedge-shaped gel, which is poured using spacers that are 0.6–0.75 mm thick at the bottom and taper to 0.25 mm at the top *(3)*. However, because these gels take longer to dry and are more prone to cracking, buffer gradient gels are the preferred choice in most laboratories.

4. Notes

1. Unpolymerized acrylamide is a neurotoxin. Gloves and mask should be worn and care should be taken when handling acrylamide powder or solutions. Depending on the quality of the acrylamide, it may be advisable to stir the acrylamide with a monobed resin (MB-1) to remove contaminating metal ions prior to filtration.
2. Leaks and air bubbles constitute the most problematic aspect of successful gel pouring. The key to avoiding these is thorough cleaning gel plates and spacers and is best performed immediately after use. Clean the plates by scrubbing with a nonabrasive detergent and rinsing with tap water. Wipe the plates with a dry paper towel, followed by a second towel that has been moistened with ethanol. After the initial cleaning with detergent and drying, the siliconized plate should additionally be wiped with a paper towel wetted with siliconizing solution, allowed to dry, and wiped with deionized water. The gel kit and any reusable items used for securing the plates for gel pouring should be washed with a mild detergent and warm water.
3. It is advisable to distinguish the two sides of each glass plate to ensure that the same side is always used for the inner (gel) and outer sides, and that it is always the same side of one of the plates that is siliconized. The design of some manufacturers' apparatuses may ensure this. In other cases, temperature decals, sticky tape, or a permanent marker may be used to identify the outer side.
4. Some gel kits provide alternative sealing methods, e.g., specifically designed clamps for the sides and a casting tray and protocol that uses an acrylamide-saturated paper strip to seal the bottom edge *(4)*. These are usually easier and more reliable than clips and tape.
5. The concentration of acrylamide to be used depends on the distance of the sequence to be resolved from the sequencing primer. A 6% gel is suitable for reading between 25 and 400 nt from the primer. Higher concentrations (12–20%) may be used for sequences within 50 nt, and lower concentrations (4 or 5%) for greater than 400 nt.
6. Some protocols recommend degassing the urea/TBE/acrylamide gel mix immediately prior to use to reduce the chance of air bubbles forming when pouring the gel, but this step is not essential.
7. Many protocols recommend fixing the gel in 10% acetic acid, 10% methanol for 15 min on one glass plate before transfer to the Whatman paper. This procedure removes urea, decreases the time required for drying, and improves resolution. However, it also increases the likelihood of the fragile gel tearing or folding back on itself, and may therefore be omitted.

References

1. Biggin, M. D., Gibson, T. J., and Hong, G. F. (1983) Buffer gradient gels and ^{35}S label as an aid to rapid DNA sequence determination. *Proc. Natl. Acad. Sci. USA* **80,** 3963–3965.
2. Mizusawa, S., Nishimura, S., and Seela, F. (1986) Improvement of the di-deoxy chain termination method of DNA sequencing by use of deoxy-7-deazaguanosine triphosphate in place of dGTP. *Nucleic Acids Res.* **14,** 1319–1324.
3. Reed, A. P., Kost, T. A., and Miller, T. J. (1986) Simple improvements in ^{35}S dideoxy sequencing. *BioTechniques* **4,** 306–308.
4. Wahls, W. P. and Kingzette, M. (1988) No runs, no drips, no errors: a new technique for sealing polyacrylamide gel electrophoresis apparatus. *BioTechniques* **6,** 308–309.

62

DNA Sequencing of Plasmids

George Murphy

1. Introduction

In double-stranded DNA sequencing, the two DNA strands must first be separated to enable the primer to bind to the priming site. This may be done by treating the DNA with alkali. Conventionally, use of the alkali denaturation method involves neutralization of the sample by acid treatment followed by ethanol precipitation and recovery of the DNA by centrifugation. The time-consuming process of ethanol precipitation may be avoided by neutralizing the alkali-treated DNA and recovering the sample in its original volume by passage through a spin-dialysis column. The use of spin-dialysis confers the additional advantage of cleaning up the template by removing traces of low-molecular-weight compounds that may interfere with the sequencing reactions. The method used here to prepare plasmid is a variation on the boiled-lysis method *(1)*, using a rapid approach to avoid phenol extraction.

2. Materials

2.1. Plasmid Preparation

1. Luria-Bertani medium (LB) (Difco, Detroit, MI): 1% (w/v) Bacto-tryptone, 0.5% (w/v) Bacto-yeast extract, and 1% (w/v) NaCl.
2. Suspension buffer: 50 mM Tris-HCl, pH 8.0, 25% (w/v) sucrose.
3. MSTET: 50 mM Tris-HCl, pH 8.0, 50 mM ethylenediaminetetraacetic acid (EDTA), 5% (v/v) Triton-X100, and 5% (w/v) sucrose.
4. Lysozyme solution: 40 mg/mL in 50 mM Tris-HCl, pH 8.0, 50% glycerol (v/v).
5. TE: 10 mM Tris-HCl, pH 8, 1 mM EDTA.
6. Sodium acetate: 3 M sodium acetate, pH 5.0.
7. Isopropanol.
8. Ammonium acetate: 7.5 M ammonium acetate.
9. 96% (v/v) ethanol.

2.2. Denaturation

1. RNase A: Dissolve 10 mg pancreatic RNase A in 10 mU Tris-HCl, pH 8.0, 15 mM sodium chloride. Boil for 15 min and cool slowly. Dispense into 50-pL aliquots and store at −20°C. Dispose of any used solution once thawed.
2. Sodium hydroxide: 1 M Sodium hydroxide, 1 mM EDTA.

3. T0.1E: 10 m*M* Tris-HCl, pH 8.0, 0.1 m*M* EDTA.
4. Sepharose: Equilibrate Sepharose-CL-6B (Pharmacia, Piscataway, NJ) in T0.1E and adjust to a packed gel:supernatant ratio of 2:1. Store at 4°C.
5. Glass beads: 200 µ*M* diameter (Sigma, St. Louis, MO), washed in deionized water, and autoclaved.

2.3. Annealing

1. 5X Buffer: 200 m*M* Tris-HCl, pH 7.5, 100 m*M* MgCl$_2$ and 250 m*M* NaCl.
2. Primer: 10 µg/mL in deionized water.

3. Methods

3.1. Plasmid Preparation

1. Using a toothpick, place a bacterial colony into 10 mL LB in a flat-bottomed screw-top bottle and grow for 15–18 h at 37°C, shaking at 200–250 rpm and using antibiotic selection if required.
2. Centrifuge the cells at 1500*g* for 10 min. Decant the supernatant and leave the inverted bottle to drain for 5 min on absorbent paper.
3. Resuspend the cells by vortexing vigorously in 100 µL suspension buffer, then transfer the suspension to a 1.5-mL microcentrifuge tube.
4. Add 600 µL MSTET solution, then spot 14 pL of lysozyme solution onto the inside of the tube.
5. Cap the tubes, mix the contents thoroughly by shaking, and transfer immediately to a boiling waterbath for 1 min.
6. Place the tubes on ice for 1 min, then centrifuge at 10,000*g* for 30 min at 4°C.
7. Remove the gelatinous pellet with a toothpick. Alternatively, to reduce contamination with chromosomal DNA, use a micropipet to transfer the supernatant to a fresh tube.
8. Add 60 µL sodium acetate and 600 µL isopropanol and leave for 5 min before centrifuging at 10,000*g* for 10 min. Aspirate off the supernatant recentrifuge for 5 s, and remove all traces of liquid. Resuspend in 200 µL TE.
9. Add 100 µL ammonium acetate. Leave on ice for at least 30 min then centrifuge for 15 min.
10. Remove the supernatant to a fresh tube and add 750 µL ethanol. Leave at –20°C for 30 min then centrifuge as in **step 8**. Rinse with 80% ethanol, air dry, and dissolve in 50 µL TE (*see* **Note 1**).

3.2. Denaturation

1. Mix 18 µL of a plasmid miniprep with 2 µL of RNase A and incubate for 15 min at 37°C. Add 5 µL sodium hydroxide and incubate a further 15 min at 37°C. If using cesium gradient purified DNA, use 15 µg in 20 µL TE without RNase treatment.
2. Pierce the base of a 0.5-mL centrifuge tube from the inside with a 21-gage needle so that about two-thirds of the needle bevel emerges. Place the tube inside a 1.5-mL centrifuge tube completely pierced through the bottom with the same needle.
3. Add 25 µL of a slurry of glass beads in water to the bottom of the 0.5-mL tube, followed by 300 µL of the Sepharose slurry, taking care not to disturb the glass beads (*see* **Notes 2** and **3**).
4. Place the assembly onto a suitable receiver tube of at least 5 mL vol and centrifuge in a swing-out rotor at 200*g* for 4 min. Transfer the 0.5-mL tube to an intact 1.5-mL test-tube and use as quickly as possible to prevent drying of the gel matrix (*see* **Note 4**).
5. Add the denatured sample to the top of the gel of the spin-dialysis tube, being careful not to disturb the gel layer. Centrifuge at 200*g* for 4 min and use the dialysate immediately.

3.3. Annealing

1. If sequencing with ^{35}S-dATP, 8 µL of the prepared template is added to 2 µL buffer and 1 µL of primer. For sequencing with ^{32}P-dATP use only 5 µL of template and make the final volume up to 11 µL with water (*see* **Notes 5–7**).
2. Incubate at 37°C for 15 min and centrifuge the tubes briefly to spin down any condensation (*see* **Note 8**).
3. Dispense 2.4 µL to each of four tubes and continue as in Chapter 63, **Subheading 3.3.**, using T7 DNA polymerase as the sequencing enzyme (*see* **Note 9**).

4. Notes

1. The rapid plasmid preparation used here provides template of good sequencing quality. DNA produced by other methods such as alkaline lysis may also be used, but the boiled-lysis methods appear to produce templates that generate fewer artifacts on sequencing. Cesium gradient-purified DNA provides a useful control against which other techniques can be judged, as the DNA is far less contaminated than miniprep DNA. However, some background is often observed when sequencing gradient purified DNA, perhaps because of some nicking of DNA in ethidium bromide solutions when exposed to UV light.
2. Sephacryl S-200 or G50 (Pharmacia) can be substituted for Sepharose C6B. G50 will shrink away from the tube wall on centrifugation, so it is essential to add the sample to the center of the gel surface.
3. Because of possible problems of low recovery of DNA, it is unwise to use spin dialysis with the gel volume described here for samples smaller than a final volume of 20 µL that is using 80% of the volumes given in **Subheading 3.2.**
4. Use the prepared spin-dialysis tubes as quickly as possible after the first centrifugation. If they are to be kept unused for more than a few minutes the tubes should be capped to prevent the gel from drying out.
5. Denatured and spin-dialyzed templates can be stored frozen at –20°C for several days after annealing to the primer. The quality of sequence obtained will not be as good as when the sample is used immediately.
6. The amount of DNA used in the denaturation provides enough material to sequence a short insert from both ends, leaving sufficient DNA for a further reaction if required.
7. The amount of primer indicated is equimolar for 5 µg of a plasmid of 4 kb. Do not use larger amounts of primer because at higher concentrations priming to sequences of lower specificity may occur generating ghost bands on the gel.
8. There is no advantage in annealing at higher temperatures than 37°C and slow cooling to room temperature, as this may cause problems through reannealing of the complimentary strands and premature termination.
9. If following sequencing the DNA bands on the gel are smeary or bands are observed in all four lanes, it is possible that some sodium hydroxide is passing through the column. This can be tested by preparing a spin-dialysis tube as described and adding 25 pL of 10 mg/mL Blue Dextran 2000 and 10 mg/mL Orange G dye in TE to the top of the Sepharose and centrifuging as above. Transfer the spin-column to a fresh tube, add 25 mL TE, and recentrifuge. More than 90% of the Blue Dextran should be present in the first dialysate, and no Orange G should pass through in the second.

Reference

1. Holmes, D. S. and Quigley, M. (1981) A rapid boiling method for the preparation of bacterial plasmids. *Anal. Biochem.* **114**, 193–197.

63

Sequencing DNA Fragments Cloned into M13 and Phagemid Vectors

Neil Brewis

1. Introduction

The dideoxy chain-termination method *(1)* involves enzymatic elongation of a oligonucleotide primer that is annealed to a single-stranded DNA template. Single-stranded DNA of bacteriophage M13 *(2)* or phagemid vectors *(3)* give the most consistently satisfactory sequencing data. However, denatured double-stranded plasmid DNA can also serve as a suitable template and can also yield several hundred nucleotides of sequence information per reaction.

The quality of the template DNA is critical. One of the main causes of noninterpretable sequencing data is the poor quality of the template DNA. Many different methods for preparing DNA have been developed. Here, two quick "miniprep" methods for isolating "sequencing-quality" DNA are presented. One method describes the preparation of single-stranded DNA from M13 phage (or a phagemid vector), whereas the second protocol can be used to prepare double-stranded DNA from a plasmid clone.

Once the DNA template has been prepared, there are three steps in the sequencing protocol. The first involves the annealing of the oligonucleotide primer to the template DNA. If double-stranded DNA is used, the primer-annealing step must be preceded by a denaturation step using sodium hydroxide. In the second step, the primer is extended in the presence of deoxy nucleotides by the action of a DNA polymerase, such as Sequenase 2.0 (Amersham Pharmacia Biotech). In the final step, the reaction mixture is divided into four separate tubes, each containing a different dideoxy nucleotide. The incorporation of a dideoxy nucleotide molecule terminates the growth of the DNA strand. The reactions are stopped by the addition of formamide and ethylenediaminetetraacetic acid (EDTA). Following heat denaturation, the reactions can be loaded onto a sequencing gel.

Sequenase version 2.0 is an excellent polymerase for DNA sequencing. A genetically engineered version of bacteriophage T7 DNA polymerase, Sequenase 2.0 exhibits no 3'–5' exonuclease activity, high processivity and efficiently incorporates nucleotide analogs such as α-thio-dATP, dideoxy nucleotides, deoxy inositol triphosphate, and deaza derivatives of deoxy nucleotides.

From: *The Nucleic Acid Protocols Handbook*
Edited by: R. Rapley © Humana Press Inc., Totowa, NJ

2. Materials

All solutions should be made from reagents of molecular biology grade in purified water.

1. Phenol: Use nucleic acid-grade phenol equilibrated with 10 mM Tris.HCl, pH 8. Store with 0.1% (w/v) hydroxyquinoline at 4°C. Note that phenol is highly corrosive.
2. Chloroform.
3. 3 M Sodium acetate: Adjust to pH 5.2 using acetic acid. Autoclave and store at room temperature.
4. 70% and 100% ethanol.

2.1. Preparation of Single-Stranded DNA

1. 2x YT broth: 1.6% (w/v) Bacto-tryptone (Difco Laboratories, Detroit, MI, 1% (w/v) Bacto-yeast extract, 0.5% (w/v) NaCl. Adjust pH to 7 using 5 M NaOH, make up to 1 L, and sterilize by autoclaving.
2. Stock solutions of ampicillin and kanamycin can both be prepared in water at concentrations of 50 and 10 mg/mL, respectively. Filter sterilize and store at –20°C.
3. Polyethylene glycol (PEG)/NaCl: 20% (w/v) PEG (molecular weight 6000–8000), 2.5 M sodium chloride. Sterilize by filtration and store at 4°C.

2.2. Preparation of Double-Stranded DNA

1. Luria-Bertani (LB) medium: 1% (w/v) Bacto-tryptone, 1% (w/v) NaCl, 0.5% (w/v) Bacto-yeast extract in water. Adjust pH to 7 using 5 M NaOH, make up to 1 L, and sterilize by autoclaving.
2. Ampicillin stock: 50 mg/mL in water. Sterilize by filtration and store at –20°C.
3. 50 mM Glucose, 25 mM Tris-HCl, pH 8, 10 mM EDTA. Autoclave and store at 4°C.
4. 0.2 N NaOH/ 1% (w/v) sodium dodecyl sulfate (SDS): prepare fresh from stock solutions.
5. Potassium acetate pH 5.2: Mix 120 mL 5 M potassium acetate, 23 mL acetic acid, and 57 mL water. Store at 4°C.
6. RNAase: Dissolve pancreatic RNAase at 10 mg/mL in 15 mM NaCl, 10 mM Tris-HCl, pH 7.5. Heat to 100°C for 15 min and allow to cool to room temperature. Store at –20°C.

2.3. Sequencing Reaction

1. Sequenase buffer (5X concentrate): 200 mM Tris-HCl, pH 7.5, 250 mM NaCl, 100 mM MgCl$_2$. Store at –20°C.
2. Oligonucleotide primer dissolved in 20 mM Tris-HCl, pH 8, 0.1 mM EDTA (*see* **Note 1**). Store at –20°C.
3. 1 M NaOH/1 mM EDTA. Store at 4°C.
4. 2 M Ammonium acetate: Adjust pH to 5.4 with acetic acid. Filter sterilize and store at room temperature.
5. Labeling mix (5X concentrate): 7.5 µM dCTP, 7.5 µM dGTP, 7.5 µM dTTP. Store at –20°C.
6. Sequenase Version 2.0 enzyme (13 U/µL) from Amersham Pharmacia Biotech. Store at –20°C.
7. Enzyme dilution buffer: 10 mM Tris-HCl, pH 7.5, 5 mM dithiothreitol (DTT), 0.5 mg/mL bovine serum albumin. Store at –20°C.
8. DTT 0.1 M: Sterilize by filtration and store at –20°C.
9. [^{35}S]dATPαS (10 µM, 10 µCi/µL). Store at –20°C. Radioactive material should be handled and disposed of accordingly.

Sequencing DNA Fragments

10. ddATP Termination mix: 80 µM each dATP, dCTP, dGTP, dTTP, 8 µM ddATP, 50 mM NaCl.
11. ddCTP Termination mix: 80 µM each dATP, dCTP, dGTP, dTTP, 8 µM ddCTP, 50 mM NaCl.
12. ddGTP Termination mix: 80 µM each dATP, dCTP, dGTP, dTTP, 8 µM ddGTP, 50 mM NaCl.
13. ddTTP Termination mix: 80 µM each dATP, dCTP, dGTP, dTTP, 8 µM ddTTP, 50 mM NaCl.
14. Formamide stop solution: 95% (w/v) Analytical grade formamide, 20 mM EDTA pH 8, 0.05% (w/v) Bromophenol Blue, 0.05 % (w/v) Xylene Cyanol FF (both from Sigma Chemicals Co., St. Louis, MO).

Amersham Pharmacia Biotech offers a sequencing kit that contains all buffers, nucleotide mixes, and Sequenase enzyme required for sequencing. Control DNA and primer are also included. Alternatively, nucleotide mixes can be purchased separately.

3. Methods

3.1 Preparation of DNA Template

3.1.1. Preparation of Single-Stranded DNA from M13 Phage (see **Note 2**)

1. Prepare an overnight culture of a suitable host strain such as JM109 (*see* **Note 3**) in 4 mL 2X (TY).
2. Dilute the overnight culture 1 in 100 in 2X YT and aliquot 2 mL into a sterile 15-mL culture tube.
3. Touch the surface of a single plaque with a toothpick and shake the toothpick end in the diluted bacterial culture.
4. Incubate at 37°C for 5 h with vigorous shaking (*see* **Note 4**).
5. Spin a 1.5-mL aliquot in microfuge tube for 5 min.
6. Carefully remove the supernatant to a clean tube and add 200 µL 20% PEG/2.5 M NaCl. Vortex and leave at room temperature for at least 15 min.
7. Spin for 5 min in a microfuge and carefully remove all the supernatant (*see* **Note 5**).
8. Resuspend the pellet in 100 µL Tris-EDTA (TE) by vortexing.
9. Add 50 µL phenol, vortex, let stand for 5 min, and mix again with further vortexing.
10. Spin for 2 min in a microfuge, and in a new tube extract the aqueous phase with 50 µL chloroform.
11. In a new tube add 10 µL of 3 M sodium acetate, pH 5.2, and 250 µL ethanol.
12. Centrifuge for 5 min.
13. Wash the DNA pellet with 400 µL 70% ethanol and spin for 2 min.
14. Dry the pellet with the tube open on the bench.
15. Resuspend in 30 µL TE.

Run a small aliquot on an agarose gel to check the quality and yield of DNA (*see* **Note 6**).

3.1.2. Preparation of Double-Stranded DNA

1. Prepare an overnight culture from a single transformed bacterial colony in 4 mL LB supplemented with 50 µg/mL ampicillin.
2. Decant 1.5 mL of culture into a microfuge tube and spin for 1 min in a microfuge. Aspirate the supernatant and in the same tube pellet a further aliquot of the culture. Completely aspirate the supernatant.
3. Resuspend the bacterial pellet in 150 µL 50 mM glucose, 25 mM Tris-HCl, pH 8, 10 mM EDTA.
4. Add 300 µL of 0.2 M NaOH/1% SDS and invert several times before placing on ice for 5 min.
5. Add 225 µL 3 M potassium acetate, pH 4.8, invert several times, and place on ice for another 5 min.

6. Spin in a microfuge for 5 min.
7. Extract the supernatant with 600 µL of 1:1 phenol:chloroform.
8. In a new tube add 675 µL ethanol, briefly vortex, and pellet the DNA for 5 min using a microfuge.
9. Wash the DNA pellet with 400 µL 70% ethanol and spin for 2 min.
10. Dry the pellet with the tube open on the bench.
11. Resuspend in 30 µL TE containing 20 µg/mL RNAase.

Run a small aliquot on an agarose gel to check the quality and yield of DNA (see **Note 7**).

3.2. Annealing Reaction

3.2.1. Single-Stranded DNA Templates

1. In a microfuge tube mix the following:

Single-stranded DNA template	1 µg (see **Note 8**)
Primer	0.5 pmol (see **Note 9**)
Sequenase buffer (5X)	2 µL
Water	To 10 µL

2. Incubate in a 65°C water bath for 2 min and then allow to cool slowly to 37°C over 30 min (see **Note 10**).

3.2.2. Double-Stranded DNA Templates

1. In a microfuge tube mix the following:

Double-stranded DNA template	3 µg (see **Note 11**)
1 M NaOH/1 mM EDTA	4 µL
Water	To 20 µL

2. Incubate for 5 min at room temperature.
3. Add 2 µL 2 M ammonium acetate, pH 5.4, mix and add 55 µL ethanol. Place the tube on ice for 30 min.
4. Spin in a microfuge for 15 min at 4°C and carefully remove and discard the supernatant.
5. Wash the pellet with 200 µL 70% ethanol, spin in a microfuge for 5 min, and remove and discard the supernatant. Dry the pellet with the tube open on the bench. Use immediately or store at –70°C.
6. Dissolve the DNA in the following:

Primer	1.5 pmol (see **Note 9**)
Sequenase buffer (5X)	2 µL
Water	To 10 µL

7. Incubate at 65°C for 2 min then place at 37°C for 20 min.

3.3. Labeling Reaction

1. Dilute the labeling mix 1 in 5 with water (see **Note 12**).
2. Dilute the Sequenase 1 in 8 with ice-cold enzyme dilution buffer. Store on ice.
3. Add the following to the 10 µL of solution of annealed primer-template:

DTT 0.1 M	1.0 µL
Diluted labeling mix	2.0 µL
[^{35}S]dATPαS (10 µCi/µL)	0.5 µL
Diluted Sequenase enzyme	2.0 µL

4. Mix (see **Note 13**) and incubate at room temperature for 5 min.

3.4. Termination Reactions

1. Label four tubes A, C, G, and T (*see* **Note 14**).
2. Pipet 2.5 µL of the appropriate termination mix into the bottom of each tube (i.e., ddATP termination mix into the tube labeled A, etc.).
3. Prewarm the four tubes to 37°C in a water bath.
4. Add 3.5 µL of the labeling reaction to each of the four tubes and mix.
5. Incubate at 37°C for 5 min.
6. Add 4 µL of formamide stop solution and mix. Store on ice until ready to load the sequencing gel. Samples can be stored at −20°C for one week.
7. Heat the tubes to 80°C for 3 min immediately prior to loading a 2.5-µL aliquot onto the sequencing gel.

4. Notes

1. A variety of so-called "universal primers" and "reverse primers," complementary to sequences in the polylinker region of M13 cloning phages and plasmids, are commercially available. Custom primers can also be made using an automatic DNA chemical synthesizer (*see* **ref. 4**). These should be about 30 nt away from the sequence of interest. Generally, they should be about 17–19 oligomers with a 50% G + C content. Try to avoid palindromic sequences.
2. Single-stranded DNA can also be prepared from phagemid vectors such as pBluescript (Stratagene, La Jolla, CA) using the following procedure. Inoculate a single colony of transformed *Escherichia coli* XL1 blue cells into 4 mL 2x YT supplemented with 75 µg/mL ampicillin and VCSM13 helper phage at about 5×10^6 pfu/mL. Grow with vigorous aeration for 1–2 h at 37°C. Add kanamycin to 70 µg/mL and continue to incubate overnight. Continue as in **step 5**.
3. The host bacteria should be streaked on a minimal agar plate, grown overnight, and a single colony used as the inoculum.
4. Longer than 5 h or temperatures higher than 37°C can lead to reduced yields.
5. After removing most of the supernatant, a short second spin may be advantageous to help remove all the PEG solution.
6. The yield of single-stranded DNA is about 5 µg/mL of culture.
7. The yield of double-stranded DNA is about 4 µg/mL of culture.
8. A small excess of DNA e.g., 2 µg can be used and may help reading sequences close to the primer.
9. 0.5 pmol and 1.5 pmol of a 17 oligonucleotide is equivalent to approx 2.8 and 8.4 ng, respectively.
10. The required rate of cooling can be achieved by incubation in a small beaker of 65°C water that is allowed to stand at room temperature for about 30 min.
11. A small excess of DNA, e.g., 5 µg can be used and may help reading sequences close to the primer.
12. In order to read sequences farther away from the primer, the labeling mix can be diluted less and even used undiluted.
13. Mix by gently pipeting several times up and down.
14. Multiwell microtiter plates with U-shaped wells are ideal for performing the termination reactions. The annealing and labeling reactions are conveniently performed in microfuge tubes. When using microtiter plates, take care to prevent excessive evaporation by replacing the lid.

References

1. Sanger, F., Nicklen, S., and Coulson, A. R. (1977) DNA sequencing with chain-terminating inhibitors. *Proc. Natl. Acad. Sci. USA* **74,** 5463–5467.
2. Messing, J. (1993) M13 cloning vehicles: their contribution to DNA sequencing, in *DNA Sequencing Protocols* (Griffin, H. G. and Griffin, A. M., eds.), Humana Press, Totowa, NJ, pp. 9–22.
3. Zagursky, R. J. and Berman, M. L. (1984) Cloning vectors that yield high levels of single-stranded DNA for rapid DNA sequencing. *Gene* **27,** 183–191.
4. Gerischer, U. and Dürre, P. (1993) Sequencing using custom designed oligonucleotides, in *DNA Sequencing Protocols* (Griffin, H. G. and Griffin, A. M., eds.), Humana Press, Totowa, NJ, pp. 75–82.

64

Direct cDNA Sequencing Using Sequential Linear/Asymmetric Polymerase Chain Reaction

Ivor J. Mason

1. Introduction

Following the isolation of a clonal recombinant phage or plasmid after screening a cDNA library, the first analyses that are routinely performed are the determination of the insert size and the sequence of the 3'- and 5'-ends of the cloned fragment. Prior to the use of polymerase chain reaction (PCR), this was a laborious process, especially when the cDNA library had been constructed using a vector derivative of bacteriophage λ. I recently described a PCR-based protocol that rapidly generated this information from single-phage plaques and, additionally, produced material suitable for subcloning into plasmid vectors (*1*). The procedure, which is outlined in **Fig. 1**, involves an initial PCR reaction using an oligonucleotide primer pair which flank the cDNA insertion site to produce sufficient quantities of double-stranded DNA for sequencing. An aliquot of this reaction is then subjected to further PCR, but with only one of the primers. Any residual second primer is rapidly depleted during this second reaction and large quantities of single-stranded DNA are generated. This material is then used as a template for sequencing using the second oligonucleotide to prime the reaction.

The use of linear PCR alone followed by direct sequencing of the double-stranded product could potentially be employed to analyze phage inserts. However, in my experience, this approach does not consistently generate data of acceptable quality. This is probably because, in part, of the presence of short PCR products, which are the result of incomplete DNA synthesis during the amplification procedure. The second, asymmetric PCR overcomes these problems, as only full-length products of this reaction can anneal the second (sequencing) primer. In addition, the use of single-stranded DNA as a template generally allows each gel to be read further.

2. Materials

All solutions should be made to the standard required for molecular biology using molecular-biology-grade reagents and sterile distilled water.

1. 2X SM buffer: 0.2 M NaCl, 16 mM MgSO$_4$, 40 mM Tris-HCl, pH 7.5, 4% (w/v) gelatin, autoclaved.
2. 10X PCR buffer: 100 mM Tris-HCl, pH 8.4, 0.5 M KCl, 0.01% (w/v) gelatin.

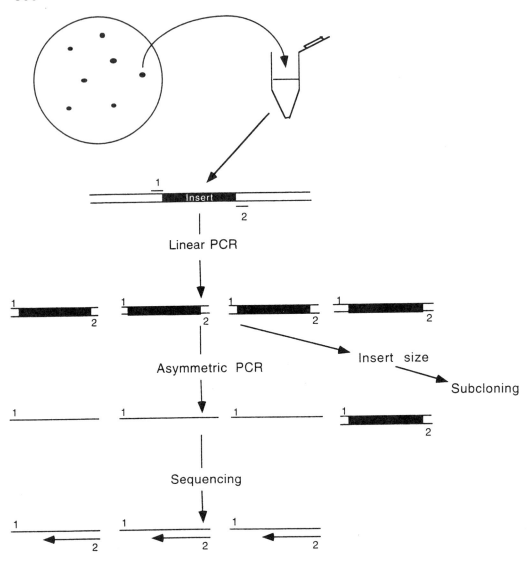

Fig. 1. Schematic diagram of the protocol for sequencing from single-phage plaques.

3. Deoxynucleotide triphosphates: diluted from stock solutions (Boehringer Mannheim, Lewes, UK).
4. *Taq* polymerase: either AmpliTaq™ (Perkin Elmer Cetus, Norwalk, CT) or *Taq* polymerase (Promega, Madison, WI).
5. DNA primers: suitable primers can be purchased from several companies that specialize in molecular biology products.
6. Mineral oil (Merck, Southampton, UK).
7. TE: 10 mM Tris-HCl, pH 8.0, 1 mM ethylenediaminetetraacetic acid (EDTA).
8. DNA sequencing kits: available from several manufacturers. We routinely use Sequenase™ (Amersham Pharmacia Biotech, Amersham, UK) and *Taq*Track (Promega, Madison, WI).
9. SeaPlaque™ agarose (FMC Bioproducts, Rockland, ME).

Direct cDNA Sequencing

3. Methods

3.1. Amplification of the Phage cDNA Insert

The following procedure produces sufficient quantities of double-stranded DNA for several sequencing reactions, allows analysis of the phage insert size by gel electrophoresis, and provides material to facilitate subcloning of the insert into more useful vector systems.

1. Core a single bacteriophage plaque from a plate with a sterile capillary or Pasteur pipet, and transfer it to a microfuge tube containing 400 µL of sterile distilled H_2O. Allow the phage particles to elute from the agar for 1 h (*see* **Note 1**).
2. Amplify the DNA insert from the bacteriophage vector using an appropriate primer combination (**Table 1**). Assemble the reaction as follows in a tube appropriate for the thermal cycler to be used (usually a 0.5-mL microfuge tube):
   a. Distilled water   to 100 µL final volume
   b. 10X PCR buffer   10 µL
   c. 0.25 M $MgCl_2$ 4 µL
   d. 20 mM dNTPs   1 µL
   e. Phage eluate   20 µL
   f. Primer 1   2 µL (17 pmol)
   g. Primer 2   2 µL (17 pmol)
   h. *Taq* polymerase   2 U
3. Overlay the reaction with 50 µL of light mineral oil.
4. Amplify with the following conditions:
   One cycle of 94°C for 2 min.
   30 cycles of 94°C for 30 s; 55°C for 30 s; 72°C for 4 min.
   One cycle of 72°C for 10 min.
5. Examine 10 µL of the reaction product by conventional agarose gel electrophoresis.

3.2. Asymmetric Amplification and Sequencing

An asymmetric amplification is performed on an aliquot of the products of the reaction in **Subheading 3.1.**, and the single-stranded DNA products of this second synthesis are purified and sequenced (*see* **Notes 2–5**).

1. Take 3 µL of the first reaction products (typically 150–750 fmol of double-stranded template DNA; 5–25 ng of a 0.5-kb amplification product) and amplify using 17 pmol of one of the original primers. The second primer is omitted from this reaction, but otherwise amplification conditions are as in **Subheading 3.1., step 4**.
2. Transfer the PCR products of this reaction to a fresh microfuge tube with as little of the mineral oil as possible, add 100 µL of 4 M ammonium acetate, and 400 µL of absolute ethanol.
3. Vortex the mixture, and incubate on ice for 15 min. Centrifuge at 12,000g for 10 min. This process removes most of the unincorporated nucleotides and primer.
4. Wash the pellet briefly with 100 µL 70% (v/v) ethanol, air-dry, and resuspend in 20 µL TE.
5. Sequence this single-stranded DNA using a commercially available DNA sequencing kit (Sequenase™, Amersham Pharmacia Biotech). Sequence half of the resuspended DNA using 840 fmol of the second primer from the original PCR reaction using the manufacturer's protocol for single-stranded DNA template. Analyze the products by standard denaturing polyacrylamide gel electrophoresis (*see* **Notes 6** and **7** *[2]*).

Table 1
Sequences of Some Useful Primer Pairs for Amplifying and Sequencing Using Commonly Encountered Vectors

Vectors	Primer 1	Primer 2
λgt11	5'-GGTGGCGACGACTCCTGGAGCCCG	5'-TTGACACCAGACCAACTGGTAATG
λgt10	5'-CTTTTGAGCAAGTTCAGCCTGGTTAAG	5'-GAGGTGGCTTATGAGTATTTCTTCCAGGG
λZAP and pBluescript	5'-TAATACGACTCACTATAGGG	5'-ATTAACCCTCACTAAAGGGA
pGEM	5'-TAATACGACTCACTATAGGG	5'-GATTTAGGTGACACTATAG
pUC/M13	5'-GTTTTCCCAGTCACGAC	5'-CAGGAAACAGCTATGAC

4. Notes

1. The phage particles can be used for PCR for several months if stored at 4°C. However, the titer of viable phage drops considerably on storage. Therefore, it is recommended to transfer 50 µL of the eluate to an equal volume of 2X SM buffer immediately. Store over 10 µL chloroform at 4°C for a high-titer phage stock.

2. Sequencing large cDNA inserts: The conditions described above have proven suitable for obtaining sequence information from inserts of up to 1.5 kb. However, the conditions for both PCR reactions must be adjusted for larger inserts. This can be reliably achieved by the use of the *Taq* Extender™ PCR additive (Stratagene, La Jolla, CA) in combination with the manufacturer's buffer system. The concentrations of both the templates and primers are not altered, and neither are the times for denaturation (94°C) or annealing of primers (55°C). However, the manufacturer's guidelines for the extension times (72°C) are followed.

3. Sequencing GC-rich regions: Regions of template that are rich in G and C nucleotides frequently cause premature termination of standard sequencing reactions, which appear as bands at the same position in all four lanes on the autoradiograph of the sequencing gel. A simple and reliable way to overcome this problem is to sequence the second aliquot of single-stranded DNA using a thermophilic DNA polymerase. The *Taq*Track sequencing system (Promega) has proven effective in this regard.

4. Sequencing M13 and other phage with a single-stranded DNA genome can be performed on single plaques using the above protocol.

5. Sequencing plasmid DNA: This can be performed using the above procedure, with the exception that a single colony is picked with a sterile tip, touched onto an agar plate to provide a stock of bacteria, and the tip is then agitated in the first PCR reaction mix assembled as below:
   a. Distilled water   to 100 µL final volume
   b. 10X PCR buffer   10 µL
   c. 0.25 M MgCl$_2$ 4 µL
   d. 20 mM dNTPs   1 µL
   e. Primer 1   2 µL (17 pmol)
   f. Primer 2   2 µL (17 pmol)
   g. *Taq* polymerase   2 U

   The rest of the procedure is as described above, and this will generate acceptable data. However, the following protocol provides a more rapid alternative procedure:
   a. A single colony is picked with a sterile tip, touched onto an agar plate, and grown to provide a stock of bacteria. The same tip is then introduced into the PCR reaction mixture as above, and amplification is performed as described in **Subheading 3.1., step 4**.
   b. The reaction products are examined on a 1% (w/v) low-melting-temperature SeaPlaque agarose gel run in TBE buffer as described *(2)*.
   c. Fragments to be sequenced are excised with a razor blade in a minimum volume of agarose, and TE is added to a final volume of 300 µL. The gel is melted at 70°C for 15 min.
   d. The DNA is purified by standard procedures *(2)*. Briefly, the aqueous phase is extracted with an equal volume of water-saturated phenol and transferred to a fresh tube. The phenol is then back-extracted with a further 100 µL of TE, and the two aqueous phases are combined and extracted once with phenol, once with phenol:chloroform (1:1), and twice with chloroform. In each case, extraction is performed with an equal volume of the organic reagent. The aqueous phase is transferred

to a fresh tube, and the DNA is precipitated by the addition of 44 µL 3 M sodium acetate, pH 5.2, and 1 mL ethanol followed by incubation at 4°C for 15 min and centrifugation at 12,000g for 10 min.

   e. The pellet is washed with 70% (v/v) ethanol, dried, and resuspended in 14 µL of distilled water.

   f. Half of the purified product is then incubated with 10 pmol of one of the two PCR primers at 98°C for 5 min, snap-frozen in liquid nitrogen, allowed to thaw, and then incubated at 65°C for 2 min. It is then sequenced using the Sequenase™ kit and the procedure for double-stranded templates.

6. Sequencing close to the primer: To obtain sequence of the region as close as 22 bases from the primer, add 1 µL each of 0.15 M sodium citrate and 0.1 M $MnCl_2$ to the annealing step of the sequencing reaction.

7. Troubleshooting sequencing gels:

   a. No sequence in the body of the gel, but a high-mol-wt band at the top: This is most likely owing to problems with the termination mixes of the sequencing reaction, and they should be replaced.

   b. No signal from the gel at all: Assuming that this is not the result of an error on the part of the experimenter, the most likely causes are:

      i. Failure of the second PCR synthesis: check an aliquot on an agarose gel.

      ii. Inactivation of the DNA polymerase: try another batch.

      iii. Problems with the batch of primer: try a control reaction with the primer/template supplied with the kit, and check the primer sequence if not a commercial preparation.

   c. Faint signals are generally owing to too little template; perform another 30 cycles of PCR on the product of **Subheading 3.2.**

   d. Sequence becomes very faint a short distance from the bottom of the gel: This is generally because of the termination mixes being too concentrated; try diluting them by 50%.

   e. Signals at the same level in all four lanes are generally owing to G–C-rich regions. *See* **Note 3** above.

References

1. Mason, I. J. (1992) Rapid and direct sequencing of DNA from bacteriophage plaques using sequential linear and asymmetric PCR. *BioTechniques* **12,** 60–61.
2. Sambrook, J., Fritsch, E. F., and Maniatis, T. (1989) Molecular Cloning: A Laboratory Manual, 2nd ed. Cold Spring Harbor Laboratory, Cold Spring Harbor, NY.

65

Purification and Enzymatic Sequencing of Polymerase Chain Reaction Products

Frank C. Brosius III, Lawrence B. Holzman, and Xinan Cao

1. Introduction

The advent of direct sequencing of polymerase chain reaction (PCR) products has permitted extremely rapid analysis of DNA mutants and cDNA clones. However, direct PCR sequencing has been problematic for a number of technical reasons, including the presence of impurities and excess oligonucleotide primers used for the PCR amplifications (*1–4*). Therefore, a number of protocols have been devised that address these technical issues, and allow efficient sequencing of either conventional double-stranded PCR products or asymmetrically amplified single-stranded products (e.g., *1–4*). Many of these protocols are described in detail in this volume.

An unrelated, but equally frustrating obstacle to the sequencing of PCR products arises when multiple PCR products are obtained in a single reaction. This can occur because of nonspecific amplification, when using degenerate primers or low stringency annealing conditions, or when the target cDNA is of low abundance. Multiple PCR products can also arise when cDNAs representing several alternatively spliced mRNAs of the sequence of interest are present. In each of these instances, separation and purification of the desired PCR products is necessary before sequencing is possible. The purpose of this chapter is to describe a simple, low-cost method by which multiple PCR products can be purified from agarose gels for direct sequencing (*5*). We have utilized a modification of the procedure of Heery et al. (*6*), which describes a simple, low-speed centrifugation of agarose slices to obtain the PCR products. After phenol-chloroform extraction, the products are ready for DNA sequencing using either of the PCR oligonucleotides as a sequencing primer. A few PCR products cannot be sequenced by this method, but can be sequenced after a further purification step, which is also detailed below.

2. Materials

All chemicals should be of molecular biology grade. All solutions should be made with double-distilled or deionized water. Reagents for sequencing are available as kits.

2.1. Agarose Gel Electrophoresis

1. Ultrapure agarose (Gibco BRL, Life Technologies, Gaithersburg, MD): Use of a low-percentage agarose gel (<1.0%) appears to yield a better-quality sequence (*see* **Note 1**).
2. 1X TAE buffer: 0.04 M Tris-acetate, 0.001 M EDTA: This can be made up in bulk as 20X or 50X stock and stored at room temperature (*see* **Note 2**).
3. Ethidium bromide (EtBr) solution: EtBr is a mutagen. Therefore, adequate safety precautions should be used when handling this reagent (*see* **Note 3**).
4. 10X loading buffer: 0.25% bromophenol blue, 0.25% xylene cyanol FF, 30% glycerol in ddH$_2$O.
5. UV transilluminator.

2.2. Isolation of PCR Products from Individual Gel Slices

1. Silanized glass wool: Glass wool can be silanized by submerging glass wool in a 1:100 dilution of a siliconizing agent, such as Sigmacoat (Sigma, St. Louis, MO) or Prosil 28 (VWR, Chicago, IL) for 15 s, and then rinsing extensively with distilled water, followed by autoclaving for 10 min. Silanized glass wool may be stored at room temperature indefinitely.
2. 1.5-mL Eppendorf centrifuge tubes: Tubes should be capless or have caps and hinges removed.
3. Phenol/chloroform/isoamyl alcohol 25:24:1 solution: Make by mixing 25 vol of buffered phenol with 24 vol of chloroform and 1 vol of isoamyl alcohol. Store under 50 mM Tris-HCl, pH 8.0, at 4°C. Phenol can cause severe burns to skin and mucosal membranes (*see* **Note 3**).
4. 5 M Ammonium acetate.
5. 100% Ethanol.
6. TE: 10 mM Tris-HCl, 1 mM EDTA, pH 8.0.

2.3. PCR Product Denaturation and DNA Sequencing

1. 1 M NaOH.
2. 3 M Sodium acetate, pH 5.2.
3. 5 M Ammonium acetate.
4. 100% Ethanol.
5. 5X Sequenase buffer: 200 mM Tris-HCl, pH 7.5, 100 mM MgCl$_2$, 250 mM NaCl (available in the Sequenase kits from United States Biochemical, Cleveland, OH).
6. 10 pmol 5'- or 3'-oligonucleotide primer used for PCR amplification (*see* **Note 4**).
7. 5X labeling mix (dGTP): 7.5 µM dGTP, 7.5 µM dCTP, 7.5 µM dTTP. There is no dATP in this mix. We often obtained better sequence by using 7.5 µM deaza-dGTP in place of dGTP (*see* **Note 5**).
8. Sequenase version 2.0 modified T7 DNA polymerase (14 U/µL): Dilute eightfold in cold enzyme dilution buffer just before use (*see* **Note 6**).
9. Enzyme dilution buffer: 10 mM Tris-HCl, pH 7.5, 5 mM DTT, 0.5 mg/mL BSA. Also available in Sequenase kits.
10. 0.1 M DTT: Store at –20°C.
11. Dideoxy (dd) G termination mix: 80 µM dGTP, 80 µM dATP, 80 µM dCTP, 80 µM dTTP, 8 µM ddGTP, 50 µM NaCl. We often obtained better sequence by using 7.5 µM deaza-dGTP in place of dGTP in each of the termination mixes (*see* **Note 5**). The termination mixes (either with dGTP or deaza-dGTP) are available as part of the Sequenase kits (United States Biochemical).

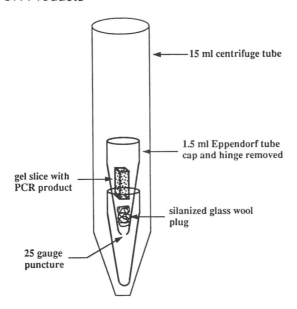

Fig. 1. Schema for isolation of PCR products by low-speed centrifugation of a conventional agarose gel slice.

12. Dideoxy (dd) A termination mix: 80 μM dGTP, 80 μM dATP, 30 μM dCTP, 80 μM dTTP, 8 μM ddATP, 50 mM NaCl.
13. Dideoxy (dd) T termination mix: 80 μM dGTP, 80 μM dATP, 80 μM dCTP, 80 μM dTTP, 8 μM ddTTP, 50 mM NaCl.
14. Dideoxy (dd) C termination mix: 80 μM dGTP, 80 μM dATP, 80 μM dCTP, 80 μM dTTP, 8 μM ddCTP, 50 mM NaCl.
15. Formamide stop solution/gel loading buffer: 95% formamide, 20 mM EDTA, 0.05% bromophenol blue, 0.05% xylene cyanol FF. Also available in Sequenase kits.
16. $[\alpha^{35}S]$-dATP, 1000 Ci/mmol (Amersham Pharmacia Biotech, Arlington Heights, IL)
17. Sequencing gel: 6% denaturing acrylamide gel, 0.4 mm thick.

2.4. Alternate PCR Product Purification Method

1. Acid-phenol reagent: 4 M guanidinium thiocyanate, 25 mM sodium citrate, pH 7.0, 0.5% sarcosyl, 0.2 M sodium acetate, pH 4.0. This is the reagent used by Chomczynski and Sacchi for RNA harvest (7) (*see* **Note 7**).
2. 100% Ethanol.

3. Methods
3.1. Agarose Gel Electrophoresis

1. PCR reactions can be run conventionally. We have often run 100-μL reactions in order to generate approx 0.5 μg of each PCR product (band) for sequencing (*see* **Note 8**).
2. Pour a conventional 0.8% agarose, 1X TAE gel containing 0.5 μg/mL ethidium bromide. Use a comb that will generate wells that can accommodate 111-μL vol (*see* **Note 1**).
3. Add 1/10 vol of agarose gel loading buffer to PCR reaction, and load into wells.
4. Electrophorese at 100 V until adequate separation of PCR bands is achieved.
5. Puncture 1.5-mL Eppendorf centrifuge tubes with a 25-gage needle at the bottom. Plug bottom of tubes with loosely packed silanized glass wool (**Fig. 1**).

6. Place gel directly onto UV transilluminator. Excise bands of interest with a fresh scalpel blade or razor blade, and place into punctured and plugged 1.5-mL Eppendorf tubes (*see* **Fig. 1**). Excise the PCR band in the smallest gel slice possible.

3.2. Isolation of PCR Products from Individual Gel Slices

1. Place 1.5-mL tube with gel slice directly into a second empty 1.5-mL capless tube, and place the entire assemblage inside a 15-mL plastic centrifuge tube.
2. Centrifuge at room temperature for 10 min at 400g in a tabletop centrifuge (swinging bucket); g-force should be calculated using a radius measured to the silanized glass plug (*see* **Note 9**).
3. Extract the centrifugate with an equal volume of phenol/chloroform. Add 2/5 vol of 5 M ammonium acetate to the centrifugate. Ethanol-precipitate by adding 2.5 vol of cold 100% ethanol and placing in –80°C freezer for 15–30 min (*see* **Note 10**).
4. Centrifuge at 13,000g in a microfuge for 30 min.
5. Redissolve pellet in 20-µL of TE. If the PCR product needs to be reamplified, 1 µL of this stock solution can be used for subsequent PCR reactions (*see* **Note 11**).
6. Centrifuge this 20-µL stock solution in a microfuge for 3–5 min to pellet any remaining agarose. Carefully transfer supernatant to another tube for DNA sequencing (*see* **Note 12**).

3.3. PCR Product Denaturation and DNA Sequencing

1. To 14 µL of PCR product stock solution, add 2.8 µL of 1 M NaOH at room temperature for 5 min. Vortex.
2. Place tube in ice H$_2$O bath, and neutralize with 7 µL 5 M ammonium acetate, pH 5.2. Vortex.
3. Add 60 µL 100% ethanol. Invert several times and place in –80°C freezer for 5 min.
4. Centrifuge in a microfuge at 13,000g for 10 min at room temperature. Wash pellet once with 100 µL of cold 70% ethanol. Aspirate and dry pellet. Resuspend in 7 µL ddH$_2$O.
5. Add 2 µL 5X Sequenase buffer and 1 µL (10 pmol) of either the 3'- or 5'-primer used for PCR amplification (*see* **Note 13**).
6. Anneal primer to PCR product at 54°C for 2 min, and then allow to cool to room temperature slowly (*see* **Note 14**).
7. While PCR product–primer mix is cooling, dilute Sequenase enzyme 1:8 in ice-cold enzyme dilution buffer (*see* **Note 6 [8]**).
8. Also dilute 5X labeling mix fivefold with ddH$_2$O.
9. Pipet 2.5 µL of the appropriate termination mix into tubes labeled G, A, T, and C (ddGTP termination mix into tube "G," and so forth).
10. After PCR product–primer mix is cooled, add:
    a. 0.1 M DTT   1.0 µL
    b. Diluted labeling mix   2.0 µL
    c. [α^{35}S] dATP   0.5 µL
    d. Diluted Sequenase   2.0 µL

    Mix thoroughly and incubate for 5–10 min at room temperature (*see* **Note 15**).
11. Heat tubes with termination mixes to 37°C.
12. Add 3.5 µL of the labeling reaction mixture to the tubes with the termination mixes, mix well, and return to 37°C for 20–30 min (*see* **Note 15**).
13. Add 4 µL of formamide stop solution/gel loading buffer to each tube, mix well, and store at 4°C if sequencing gel is to be run the same day or at –20°C if gel is to be run later.
14. Pour a 6% sequencing gel (*see* Chapter 61).
15. Heat mixes to 95°C for at least 5 min before loading on gel.

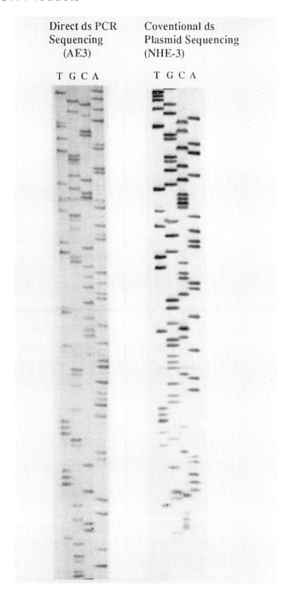

Fig. 2. Typical result of PCR product sequencing using the described protocol vs conventional DNA sequence obtained from a double-stranded plasmid template. The direct PCR sequence gel was exposed to film for 40 h, whereas the conventional plasmid sequence gel was exposed overnight. "AE3" signifies anion exchanger 3, and "NHE-3" signifies sodium proton exchanger 3.

We have successfully employed this method with multiple unrelated primers and PCR products, and generally obtained at least 200 bp of readable sequence. An autoradiograph of a typical sequencing gel using this method is shown in **Fig. 2**. Some of the common problems we have encountered with this method are detailed in **Note 16**.

With a few products, the extent of readable sequence is sometimes limited to approx 150 bp, and occasionally, a primer/PCR product pair produces an unreadable sequence.

In the latter instance, a simple purification of the DNA product has led to generation of readable DNA sequence of at least 100 bp and is presented in the alternate purification protocol below.

3.4. Alternate PCR Product Purification Method

1. Dissolve the PCR product (again at least 0.5 µg) in 100 µL of denaturing solution (7).
2. Add 100 µL of phenol. Vortex and centrifuge in a microfuge at 13,000g for 5 min. Under these conditions, the DNA remains in the interphase and the organic phase (see **Note 7**).
3. Carefully remove and discard the aqueous phase.
4. Add 200 µL of TE. Vortex and centrifuge in a microfuge at 13,000g for 5 min. Remove and retain the aqueous phase. Much of the DNA will now be retained in the aqueous phase (see **Note 7**).
5. Add 22 µL of 3 M sodium acetate to the aqueous phase and 450 µL of 100% ethanol. Let sit at room temperature for 10–15 min. Vortex and centrifuge in a microfuge at 13,000g for 15 min.
6. Resuspend pellet in 14 µL TE, and return to **Subheading 3.3., step 1**.

We do not know which factors are most important for the success of this alternate protocol. It is assumed that further purification of PCR products enhances sequence reaction efficiency. It is also possible that improved denaturation of the PCR products during extraction into the organic phase permits better primer annealing.

4. Notes

1. The use of low-percentage agarose gels (0.8% is our usual concentration) is associated with better PCR sequence results. This may result from decreased agarose contamination of the purified PCR product, but this has not been formally tested. Since most of the PCR products used for sequencing are between 500 and 1000 bp in length, this agarose concentration does not yield ideal separation of the various bands. When separation of PCR products is not optimal or when PCR product yields are low (<0.5 µg), we use 2% Metaphor agarose (FMC Bioproducts, Rockland, ME) to obtain better resolution and increase PCR yield by reamplification. These reamplified products are then purified again on 0.8% conventional agarose gels. We have not yet attempted to sequence PCR products directly after isolation from a Metaphor agarose gel.
2. TAE can either be made up as a 20X or 50X stock, or 1X TAE can be made up in bulk in a carboy and used directly. We have made up 15-L solutions of 1X TAE, which are stable for several months at room temperature.
3. EtBr and phenol-containing solutions should be made up in a fume hood or other confined space. Personnel preparing EtBr solutions should wear adequate protective clothing. Stock solutions of 10 mg/mL EtBr should be stored in a light-protected glass container at 4°C. Such solutions are stable for several years.
4. One of the advantages of this protocol is that the primers used for PCR amplification can also be used for DNA sequencing. We conventionally use 20 mers, which have a G + C content of 50% (T_m approx 60°C). Use of primers with different T_ms should work equally well, but may require appropriate changes in the annealing temperature of primer and the DNA for sequencing (see **Subheading 3.3., step 6**).
5. Deaza-dGTP helps prevent GC compression on sequencing gels and can be used in place of dGTP in all the sequencing reagents. The deaza-dGTP-containing reagents are also available in kit form from US Biochemicals (Cleveland, OH).

6. For routine double-stranded DNA sequencing, we have utilized glycerol enzyme dilution buffer, which can be stored for extended periods at –20°C. The use of this reagent, however, requires the use of glycerol-tolerant DNA sequencing gels, which is described in the Sequenase protocol book. Although we have not used glycerol enzyme dilution buffer and glycerol-tolerant gels for direct sequencing of PCR products, such modifications should not pose difficulties and would help save expensive sequencing enzyme.

7. This acid-phenol reagent is commonly used for RNA harvest, but because DNA is partitioned into the organic phase, it can be utilized to separate and purify plasmid and genomic DNA as well as RNA *(9)*. This extra purification step apparently provides PCR product purification that is superior to that which can be obtained with phenol-chloroform extraction alone and, therefore, improves the sequence from difficult templates. Our method uses a similar method, except for the back-extraction and final precipitation step. Chomczynski's protocol recommends direct precipitation from the organic phase and interphase using ethanol *(9)*. We have not yet tested this direct precipitation step for PCR products that are used for direct sequencing.

8. We routinely estimate PCR product yield by the relative intensity of the PCR band compared to mol-wt ladder bands when UV transilluminated. It is not necessary to measure precisely the PCR product yield. As an alternative to reamplification, multiple separate reactions could be run, and the final PCR products pooled.

9. For example, a *g*-force of 400 is obtained at 1500 rpm in a Beckman GS-6R tabletop centrifuge using a GH-3.8 rotor. It will be necessary to check the specifications of each low-speed centrifuge and rotor to determine the correct rpm.

10. We use ammonium acetate at this step, because some of our less abundant bands need to be reamplified. DNA precipitated with ammonium acetate provides better PCR amplification than does DNA precipitated with sodium acetate *(10)*.

11. Since we use this protocol to help distinguish multiple PCR bands, the lower abundance bands often yield <0.5 µg of DNA. We have found that such small amounts of DNA do not provide adequate template for Sequenase-based sequencing. Therefore, we simply reamplify these products with another 30 cycles of PCR and isolate the product for sequencing again as described in the protocol. We have not tested whether a PCR-based cycle sequencing method would work well with these rare products and, therefore, obviate the need for reamplification.

12. The purpose of this step is to remove any residual agarose or other insoluble contaminants that may inhibit sequencing. It is probably not essential if the extraction and precipitation steps are carefully performed.

13. Either of the primers used to amplify the PCR product of interest can be used for sequencing. The amount of primer used in this step (10 pmol) is 5× that specified in the published Sequenase protocol *(8)*.

14. Many protocols for sequencing of PCR products call for rapid cooling of template and primer after heating to the annealing temperature in order to prevent reannealing of the two PCR strands (e.g., *2,3*). For reasons that are not clear, the slow cooling suggested in this protocol yielded more intense sequence than did more rapid cooling to room temperature. We have not attempted snap-cooling to 4°C or –70°C as recommended by some protocols, and this may increase intensity of the sequence ladder.

15. The labeling reaction duration of 5–10 min is somewhat longer than that recommended in the Sequenase protocol book (2–5 min) *(8)*. Similarly, the termination reaction time of 20–30 min is extended.

16. The major problems encountered with this protocol are high background, relatively low intensity of DNA sequence radioactivity, and frequent "hard stops" as evidenced by bands in each lane at the same mobility. Often the first two problems are overcome by

Fig. 3. Direct PCR sequence gels showing intense bands in all four lanes indicative of "hard stops" owing to premature termination of DNA synthesis. Curiously, bands of the same mobility were seen whether the 5'- or 3'-PCR primer was used as a sequencing primer, suggesting that some factor other than secondary structure or reannealing of the PCR product was responsible. With this particular PCR product preparation, relatively poor sequence ladders were generated, suggesting poor template purification.

repurification using the alternate purification protocol described in **Subheading 3.4.** Also, a 2–3 d exposure to film is sometimes necessary to obtain clearly readable sequence. Also, some of the problems with "signal-to-noise" can be diminished with the use of larger amounts (>1 μg) of starting PCR product. Occasionally, use of a certain primer fails to produce readable sequence despite having a calculated T_m and base composition identical to other primers that work well. In that case, use of an internal primer for sequencing may be necessary. The generation of hard stops remains something of a mystery. As shown in **Fig. 3**, we have obtained intense bands in each sequencing lane while sequencing the same cDNA with either 3'- or 5'-primers. Since the sequence reaction with the 3'-primer proceeds from the opposite end of the PCR product than that with the 5'-primer, it seems unlikely that secondary structure or reannealing of the PCR template is the cause of this phenomenon. Perhaps a contaminating PCR product or other contaminant allows for nonspecific priming and extension up to a region in which the Sequenase prematurely terminates. In some cases, these hard stops may be eliminated by enhanced purification.

Several other protocols for direct sequencing of PCR products in low melt agarose gel slices have been recently published *(11,12)*. These protocols have not been utilized by the authors, but should be considered simple alternatives to the protocol described herein.

Acknowledgment

This work was supported, in part, by a Veterans Administration Merit Review award and a National Institutes of Health program project award (HL18575 Project 2) to F. C. B.

References

1. Dorit, R. L., Ohara, O., and Hwang, C. B.-C. (1991) Direct DNA sequencing of PCR products, in *Current Protocols in Molecular Biology,* vol. II (Ausubel, F. M., Brent, R., Kingston, R. E., Moore, D. D., Seidman, J. G., Smith, J. A., and Struhl, K. S., eds.), John Wiley, New York, pp. 15.2.1–15.2.11.
2. Phear, G. A. and Harwood, J. (1994) Direct sequencing of PCR products. *Methods Mol. Biol.* **31,** 247–256.
3. Cassanova, J.-L., Pannetier, C., Jaulin, C., and Kourilsky, P. (1990) Optimal conditions for directly sequencing double-stranded PCR products with Sequenase. *Nucleic Acids Res.* **18,** 4028.
4. Rao, V. B. (1994) Direct sequencing of polymerase chain reaction-amplified DNA. *Anal. Biochem.* **216,** 1–14.
5. Cao, X., and Brosius, F. C., III. (1993) Direct sequencing of double-stranded PCR products isolated from conventional agarose gels. *BioTechniques* **15,** 384–386.
6. Heery, D. M., Gannon, F., and Powell, R. (1990) A simple method for subcloning DNA fragments from gel slices. *Trends Genet.* **6,** 173.
7. Chomczynski, P. and Sacchi, N. (1987) Single-step method of RNA isolation by acid guanidinium thiocyanate-phenol-chloroform extraction. *Anal. Biochem.* **162,** 156–159.
8. United States Biochemical Corporation (1992) *Step-by-Step Protocols for DNA Sequencing with Sequenase version 2.0 T7 DNA Polymerase,* 6th ed., Cleveland, OH.
9. Chomczynski, P. (1993) A reagent for the single-step simultaneous isolation of RNA, DNA and proteins from cell and tissue samples. *BioTechniques* **15,** 532–534.
10. Coen, D. M. (1992) Quantitation of rare DNAs by PCR., in *Current Protocols in Molecular Biology*, vol. II (Ausubel, F. M., Brent, R., Kingston, R. E., Moore, D. D., Seidman, J. G., Smith, J. A., and Struhl, K. S., eds.), John Wiley, New York, pp. 15.3.1–15.3.6.
11. Khorana, S., Gagel, R. F., and Cote, G. J. (1994) Direct sequencing of PCR products in agarose gel slices. *Nucleic Acids Res.* **22(16),** 3425–3426.
12. Trewick, S. A. and Dearden, P. (1994) A rapid protocol for DNA extraction and primer annealing for PCR sequencing. *BioTechniques* **17,** 842–844.

66

Direct Polymerase Chain Reaction Sequencing with Denaturants

Wei Zhang and Albert B. Deisseroth

1. Introduction

The advance of *Taq*-based polymerase chain reaction (PCR) technology *(1–3)* has had a tremendously positive impact on biomedical research. The combination of PCR and sequencing further revolutionized biological research *(3–11)*. There are two general approaches for sequencing the DNA products amplified by PCR from cDNA or genomic DNA templates.

1. The PCR products can be cloned into a vector, and then the DNA is sequenced by conventional plasmid sequencing. A problem with this approach is that because each clone to be sequenced represents only one molecule in a large population of PCR products; thus, the sequence result of any clone can represent a nonrepresentation sampling of a heterogeneous population of cells composed of both normal and tumor cells. In addition, mutations may arise from incorporation errors that occur during the PCR amplification process *(12,13)*. To overcome the problems arising from sequence amplification errors in single clone, multiple clones are normally sequenced *(1–3)*. This approach is therefore tedious, time consuming, and subjective.
2. The alternative approach is that the PCR product is sequenced directly. The advantage of the latter approach is that the sequencing results represent a population average of the PCR products and any artifacts derived from misincorporation events generated by the *Taq* DNA polymerase are averaged out. This approach is straightforward and less time consuming than the first approach. However, PCR sequencing is associated with a set of unique problems. The first problem is that the two free complementary strands of PCR fragments have a strong reannealing possibility after denaturation that the primer-binding efficiency is very low. The second problem is that PCR fragments with GC-rich regions have even stronger tendency to reanneal and form secondary structures. The third problem is that the primers used for PCR appear to have poorer efficiency in binding to the templates under there conditions. These problems result in high background, weak specific sequencing ladders, ambiguous signals, and premature stops *(14,15)*.

Our analyses have led us to conclude that one way to solve the problems associated with the reannealing of PCR complementary strands was to decrease the probability of formation of hydrogen bonds between the two strands of PCR fragments, especially the GC-rich fragments. Two denaturants, formamide *(16,17)* and dimethyl sulfoxide (DMSO) *(18,19)*, are commonly used for this purpose. The protocol presented here is an easy and effective way to obtain satisfactory sequencing results using formamide in the reaction.

2. Materials

The reagents used should be molecular biology grade, and double-distilled water should be used to make all buffers used in the experiments. The reagents for sequencing are available in the Sequenase kit from U.S. Biochemicals, Amersham Pharmacia Biotech (Arlington Heights, IL).

2.1. Purification of PCR Products from Polyacrylamide Gel

1. Polyacrylamide and *bis*-acrylamide: These reagents are neurotoxic. Masks are recommended when preparing the solution from powder.
2. Tris-boric acid-ethylenediaminetetraacetic acid (EDTA) (TBE) buffer (*see* **Note 2**).
3. A vertical gel apparatus (model V16, Gibco-BRL, Gaithersburg, MD). 1.2 mm-thick spacer.
4. 6X DNA gel loading dye: 0.25% Bromophenol blue, 0.25 xylene cyanole FF, 30% glycerol in water.
5. Spectrum dialysis tubing (*see* **Note 3**) and clamps.
6. A horizontal gel apparatus.
7. A power supply.
8. 3 M Sodium acetate, pH 4.6–6.0.
9. 100% Ethanol.

2.2. Sequencing of the PCR Products

1. 5X Sequenase buffer: 200 mM Tris-HCl, pH 7.5, 100 mM MgCl$_2$, 250 mM NaCl.
2. Formamide. Store at 4°C.
3. 0.1 M dithiothreitol (DTT). Store at –20°C.
4. Labeling mix: A 7.5-μM solution of each deoxynucleotide triphosphate (dNTP). Store at –20°C. Dilute eightfold in water for use in sequencing.
5. ^{35}S-dATP (sequencing grade, Du Pont, Boston, MA). This is a low-energy β-emitter; no screen is needed. However, all waste products should be stored in designated containers and disposed of following institutional policy. The half-life of this isotope is about 3 mo.
6. Sequenase dilution buffer: 10 mM Tris-HCl, pH 7.5, 5 mM DTT, 0.5 mg/mL bovine serum albumin.
7. Sequenase (Version 2.0, U.S. Biochemicals). Should be stored at –20°C at all times. Keep on ice, dilute eightfold with Sequenase dilution buffer shortly before use.
8. Termination mixes: Mixture of 80 μM dNTP plus 8 μM of ddATP for A termination mix, 8 μM ddTTP for T termination mix, 8 μM ddCTP for C termination mix, and 8 μM ddGTP for G termination mix, in 50 mM NaCl.
9. Stop solution: 95% Formamide, 20 mM EDTA, 0.05% bromophenol blue, 0.05% xylene cyanol.
10. Sequencing gel: A 0.4-nm-thick 6% denaturing polyacrylamide gel (*see* **Note 4**).
11. Autoradiographic film: Kodak XAR5 film (Rochester, NY).
12. Cassette. No intensifying screens are needed.

3. Methods

3.1. Purification of PCR Fragments from Acrylamide Gel by Electroelution

We routinely purify the PCR product from a 5% nondenaturing acrylamide gel. PCR primers, free nucleotides, and other contaminants can be eliminated during this proce-

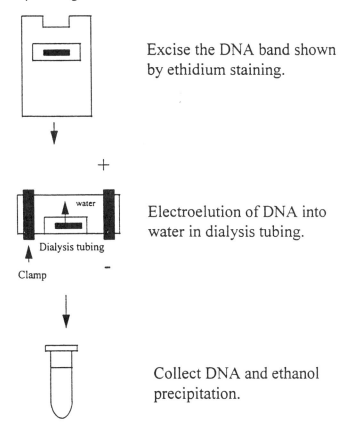

Fig. 1. Purification of PCR products from polyacrylamide gel by electroelution.

dure, and the resulting DNA is very clean for satisfactory sequencing. PCR products can also be purified by commercially available spin columns, especially if the PCR reaction produces a very specified product (one clear band on the gel). However, it has been noticed that some spin columns may have inhibitors to Sequenase; therefore, tests are needed to determine this. PCR products purified from acrylamide gels give consistent results.

1. Prepare a 5% nondenaturing polyacrylamide gel (0.8 mm thick Tris-borate-EDTA [TBE] buffer) in a vertical gel apparatus (same apparatus used for protein gel). Use a comb with wide teeth.
2. Pool the DNA mixtures from several PCR reactions (depending on the yield of PCR reaction). Add 1/5 vol of the 6X DNA gel-loading buffer. Electrophorese at 200 V for several hours, depending on the size of the PCR products.
3. Following conclusion of the electrophoresis, briefly stain the gel with ethidium bromide. Locate the band with long-wave hand-held UV-light box. Cut out the band with a razor blade.
4. Fill a precleaned and soaked dialysis tubing with water and slip the DNA band into the tubing. Remove most of the water, leaving about 500 µL of water in the tube. Squeeze out the air bubbles, and clamp the tube. Move the gel slice to one side of the tube by hand (*see* **Fig. 1**).

5. Place the tubing on a horizontal electrophoresis box (for routine DNA agarose gel electrophoresis) filled with 0.1X TBE. The gel side of the tubing is toward the negative electrode. Electrophorese at 200 V for 30 min. The DNA should be eluted into the buffer in the tubing. The completion of electrophoresis can be easily checked by the hand-held UV-light box.
6. Remove the DNA in water by pipet and transfer into a 1.5-mL Eppendorf tube. Centrifuge at full speed (18,600g) for 10 min to remove gel debris.
7. Transfer the supernatant into a new Eppendorf tube, add 1/10 vol 3 M NaOAc and 2 vol cold ethanol. Mix well and precipitate the DNA for 15 min at –70°C. Then spin down the DNA into a pellet and wash the pellet twice with 70% ethanol. Dry the DNA in a Speed Vac and resuspend the DNA into appropriate volume.
8. Estimate the concentration of DNA by spectrometry or by comparative staining with ethidium bromide.

3.2. Sequencing of PCR Products with the Aid of Formamide

We use the following procedure to sequence PCR fragments having GC-rich regions, or we use the PCR primers for direct sequencing. The majority of reagents used in this procedure are from the U.S. Biological Sequenase kit.

1. Mix 50–200 ng of PCR of fragments and a 5 M excess of primers in a 10-μL 1 × Sequenase buffer containing 10% formamide (*see* **Notes 5–8**). Then boil the mixture for 8 min to denature the PCR fragments.
2. Quickly put the Eppendorf tube on ice or on a dry ice-ethanol bath to fast-cool the reaction (*see* **Note 9**). Leave the reactions on ice or on a dry ice-ethanol bath for 2 min.
3. Spin the tubes briefly at 4°C to collect all the liquid on the bottom of the tubes (*see* **Note 10**). Transfer the tubes onto ice.
4. Add 1 μL of 0.1 M dithiothreitol (DTT), 2 μL of diluted labeling mix, 0.5 μL of ^{35}S-dATP, and 2 μL of freshly diluted Sequenase. Incubate on ice for 10 min (*see* **Note 11**).
5. A set of four tubes should be prepared marked G, A, T, and C. Two and one-half microliters of ddGTP, ddATP, ddTTP, and ddCTP is added to one of the four tubes, respectively. These tubes are prewarmed to 42–45°C for 1 min before the finish of the reaction at **step 4** (*see* **Note 12**).
6. Transfer 3.5 μL of reaction mixture to each of the prewarmed tubes and continue the incubation for 5 min.
7. Add 1 μL of diluted labeling mix to each of the tubes. Incubate for an additional 5 min (*see* **Note 13**).
8. Add 4 μL of stop solution to the reaction. Mix well and leave on ice.
9. Prerun the sequencing gel (Gibco-BRL model 4001) for 30 min with a constant power of 55 W (*see* **Notes 14** and **15**).
10. Heat up the samples at 95°C for 3 min. Meanwhile, stop the prerun and wash the wells four to five times using a syringe (*see* **Note 16**).
11. Load 2.5–3 μL of reaction mixture (*see* **Note 17**). The commonly used loading patterns are CATG and GATC. Continue to run the gel at 55 W.
12. After the gel is run, transfer it onto a piece of 3MM Whatman paper. Cover the gel with a film wrap and dry the gel at 80°C under vacuum.
13. Place a sheet of Kodak X-AR film (Kodak, Rochester, NY) on the top of the gel in a cassette (*see* **Note 18**). Autoradiograph at –70°C overnight. Allow the cassette to warm up and dry the following day. Develop the film in an automatic film developer. Examples of sequencing results are shown in **Figs. 2** and **3**.

Direct PCR Sequencing with Denaturants

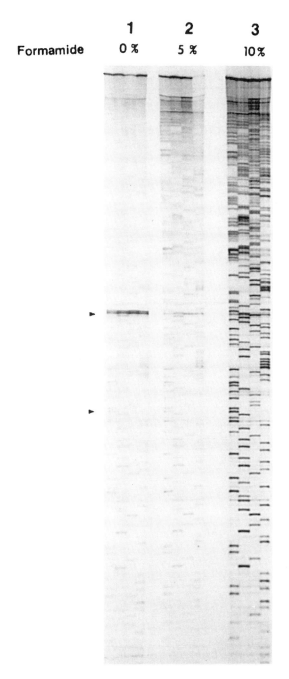

Fig. 2. Two hundred nanogram PCR products of *p53* gene (1.3 kb) were mixed with a 5 *M* excess of sequencing primer in Sequenase buffer with the increasing percentages of formamide indicated in each lane. The annealing mixture was boiled for 10 min and quickly cooled in dry ice-ethanol bath. Sequencing reactions were carried out on ice for 10 min, then aliquoted to prewarmed ddNTP, and reacted for 10 min at 42°C. The reaction was stopped by adding stop solution. The sequencing reactions were heated to 95°C for 3 min before loading. The loading pattern was GATC. The region being sequenced contains 62% GC.

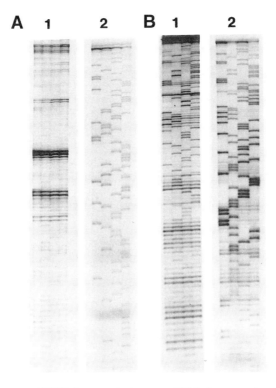

Fig. 3. Fifty nanograms of PCR fragments (approx 120 bp) were mixed with a fivefold molar excess of primers in Sequenase buffer containing no formamide (A1 and B1) or 10% formamide (A2 and B2). **(A)** and **(B)** represent two different PCR fragments with 50% GC content, sequenced using two different PCR primers.

4. Notes

1. Use 200 ng for a 1.2-kb DNA fragment; smaller amounts of DNA can be used for smaller size DNA fragments.
2. Precipitates form in TBE buffers with time; therefore, make new buffer when this happens. Use a new clear glass bottle each time.
3. The molecular cutoff sizes do not appear to matter because DNA is linear and not in globular form. It is, therefore, too big to pass through the pores of the dialysis tubing.
4. We found that gel made fresh from powder gives the best results for the sequencing ladder. Urea, acrylamide, and *bis*-acrylamide should be dissolved into 0.5–1X TBE at 37°C in a water bath. Ammonium persulfate (APS) should then be added and dissolved. This solution should then be vacuum filtered for 5 min to remove undissolved particles and air bubbles. TEMED is then added and mixed gently by pipeting; however, bubbles should be avoided. After the solution is poured, the gel needs to be polymerized for 2 h before use. The recipe for the 6% gel (0.5X TBE, 75 mL) is the following: 3.7 mL 10X TBE, 37.5 g urea, 4.5 g acrylamide, 0.23 g *bis*-acrylamide, 0.06 g APS, 33 mL water, and 20 µL TEMED.
5. We routinely use 200 ng for DNA fragments of about 1 kb and 50 ng for fragments of about 200 bp. The amount of primer needed can be calculated by the following equation:

$$\mu g \text{ of primer} = 5X \text{ } \mu g \text{ of PCR fragment} \times \frac{\text{Size of primer}}{\text{Size of PCR DNA}}$$

Ten-molar excess of primer can also be used in cases in which insufficient annealing is observed.

6. The role of formamide is to weaken the hydrogen bonds between the two strands of the PCR fragment, thus reducing the secondary structure that causes ambiguity in the banding pattern. The percentage of formamide required to generate the improvement of sequencing depends on the percentage of GC nucleotides in the target sequence. It may, therefore, be necessary to empirically determine the amount of formamide needed to resolve secondary structures completely. The following formula can be used to calculate the percentage of formamide needed to reduce the melting temperature of the DNA duplex to that of a fragment with a 50% GC nucleotide: Formamide % = 0.7 (GC%–50). An example of the effect of formamide is shown in **Fig. 2**. It should be pointed out that some nonspecific bands of secondary structure can not be totally abolished, even with 30% formamide, because formamide reduces the efficiency of the findings of the template to primer binding, especially when the GC percentage of the primer is much lower than the GC percentage of the local structure.
7. Ten percent formamide is routinely used when the PCR primer is used to sequence the PCR product. The result is much improved even for non-GC-rich fragments (*see* **Fig. 3**).
8. DMSO can be used in place of formamide. The percentage of dimethyl sulfoxide (DMSO) to use needs to be determined empirically.
9. When the dry ice-ethanol bath is used, do not have flowing ethanol because the ethanol may get inside the tube.
10. This is needed because the solution evaporates during boiling.
11. The reaction is carried out on ice to limit the reannealing of two complementary strands from PCR products.
12. Temperatures of 42–45°C is used instead of 37°C to help overcome the secondary structure.
13. This step helps force the unfinished reaction to completion, thereby reducing premature stops that may cause bands across all four lanes.
14. The TBE buffer used for the running gels should be consistent with the buffer used in the gel. For example, if 0.5X TBE was used to make the gel, 0.5X TBE should be used to run the gel. For an extended long run, 1X TBE is preferred.
15. The gel should be warm (about 50°C) after the prerun. If the plate is not warm, the wrong buffer may have been used.
16. Be careful not to shift the comb by the needle. Also, be careful not to wash the wells too forcefully.
17. If a small leak of sample occurs during loading, start running the electrophoresis after loading of one sample set. Let the DNA run into the gel, then load the next set or sets.
18. Many times gels do not completely dry. If this happens, do not peel off the film wrap used on top of the gel so that the X-AR film will not get stuck to the gel. However, if the gel is completely dry, the film wrap can be peeled off so that stronger signals can be obtained.

References

1. Saiki, R. K., Scharf, S., Faloona, F., Mullis, K. B., Horn, G. T., Erlich, H. A., and Arnheim, N. (1985) Enzymatic amplification of β-globin genomic sequences and restriction site analysis for diagnosis of sickle cell anemia. *Science* **230,** 1350–1354.
2. Mullis, K. B. and Faloona, F. A. (1987) Specific synthesis of DNA in vitro via a polymerase-catalyzed chain reaction. *Methods Enzymol.* **155,** 335–350.
3. Scharf, S. J., Horn, G. T., and Ehrlich, H. A. (1986) Direct cloning and sequence analysis of enzymatically amplified genomic sequences. *Science* **233,** 1076–1078.

4. Innis, M. A., Myambo, K. B., Gelfand, D. H., and Brow, M. A. D. (1988) DNA sequencing with *Thermus aquaticus* polymerase and direct sequencing of polymerase chain reaction-amplified DNA. *Proc. Natl. Acad. Sci. USA* **85,** 9436–9440.
5. Sarkar, F. H., Li, Y. W., and Crissman, J. D. (1993) A method for PCR sequencing of the p53 gene from a single 10-micron frozen or paraffin-embedded tissue section. *Biotechniques* **15,** 36–38.
6. Trumper, L. H., Brady, G., Bagg, A., Gray, D., Lake, S. L., and Griesser, H. (1993) Single-cell analysis of Hodgkin and Reed-Sternberg cells: molecular heterogeneity of gene expression and p53 mutations. *Blood* **81,** 3097–3115.
7. Wong, C., Dowling, C. E., Saiki, R. K., Higuchi, R. G., Ehrlich, H. A., and Kazazian, H. H. (1987) Characterization of β-thalassemia mutations using direct genomic sequencing of amplified genomic DNA. *Nature* **330,** 384–386.
8. Baker, S. J., Fearon, E. R., Nigro, J. M., Hamilton, S. R., Preising, A. C., Jessup, J. M., et al. (1989) Chromosome 17 deletions and p53 gene mutations in colorectal carcinomas. *Science* **244,** 217–221.
9. Hu, G., Zhang, W., and Deisseroth, A. B. (1992) p53 gene mutations in acute myeloid leukemia. *Br. J. Haematol.* **81,** 489–494.
10. Xiong, Y., Hannon, G., Zhang, H., Casso, D., Kobayashi, R., and Beach, D. (1993) p21 is a universal inhibitor of cyclin kinases. *Nature* **366,** 701–704.
11. Scarpetta, M. A. and Uhler, M. D. (1993) Evidence for two isoforms of the endogenous protein kinase inhibitor of cAMP-dependent protein kinase in mouse. *J. Biol. Chem.* **268,** 10,927–10,931.
12. Barnes, W. M. (1992) The fidelity of Taq polymerase catalyzing PCR is improved by an N-terminal deletion. *Gene* **112,** 29–35.
13. Cheng, S., Fockler, C., Barnes, W. M., and Higuchi, R. (1994) Effective amplification of long targets from cloned inserts and human genomic DNA. *Proc. Natl. Acad. Sci. USA* **91,** 5695–5699.
14. Zhang, W., Hu, G., and Deisseroth, A. B. (1991) Improvement of PCR sequencing by formamide. *Nucleic Acids Res.* **19,** 6649.
15. Zhang, W., Reading, C., and Deisseroth, A. B. (1992) Improved PCR sequencing with formamide. *Trends Genet.* **8,** 332.
16. Casey, J. and Davidson, N. (1977) Rates of formation and thermal stabilities of RNA:DNA and DNA:DNA duplexes at high concentrations of formamide. *Nucleic Acids Res.* **4,** 1539–1552.
17. McConaughy, B. L., Laird, C. D., and McCarthy, B. J. (1969) *Biochemistry* **8,** 3289–3295.
18. Chester, N. and Marshak, D. R. (1993) Dimethyl sulfoxide-mediated primer Tm reduction: a method for analyzing the role of renaturation temperature in the polymerase chain reaction. *Anal. Biochem.* **209,** 284–290.
19. Masoud, S. A., Johnson, L. B., and White, F. F. (1992) The sequence within two primers influences the optimum concentration of dimethyl sulfoxide in the PCR. *PCR Methods Appl.* **2,** 89–90.

67

Direct DNA Sequencing of Polymerase Chain Reaction Products Using Magnetic Beads

Joakim Lundeberg, Bertil Pettersson, and Mathias Uhlén

1. Introduction

The use of magnetic particles in many fields of biochemistry, molecular biology, and medicine has been well documented and several magnetic particles are now available for diagnostic and cell separation purposes. The solid-phase approaches has improved robustness by the increased reproducibility with accompanying higher yields. Furthermore, automation has been facilitated, as reaction buffers and additional reagents can be rapidly changed without centrifugation or precipitation steps. The introduction of magnetic particles into DNA sequencing technology has proven to be an attractive alternative compared with traditional means of performing DNA sequencing.

Template preparation for DNA sequencing has relied on cloning of the target sequence into phage or plasmid vectors. After cultivation, sufficient amounts of target DNA can be prepared to enable chain-termination DNA sequencing described by Sanger and coworkers *(1)*. This indirect method—using cloning to prepare DNA for sequencing—has become more efficient by the use of polymerase chain reaction (PCR). By the design of PCR primers containing restriction handles, both a rapid and efficient cloning procedure into suitable sequencing vectors *(2)* is achieved. However, thermostable *Taq* DNA polymerase used in PCR lacks proofreading activity, and therefore, multiple clones may have to be sequenced and analyzed to find the consensus sequence.

Alternatively, an in vitro amplification can be performed on genomic DNA that allows for direct DNA sequencing of the PCR products. Thus, both time-consuming cloning steps as well as control sequencing can be avoided. Therefore, the DNA sequence of a sample can be determined rapidly and will represent the sequence of the sample prior to amplification, as the errors produced by *Taq* polymerase will not significantly contribute to the resulting signal. A clear disadvantage is the need to remove excess PCR primers, enzymes and nucleotides. This can usually be solved by a precipitation step, spin columns, or high-performance liquid chromatography. A more serious disadvantage is the competition between primer annealing and reannealing of the complementary strand, which causes severe DNA sequencing interpretation problems.

From: *The Nucleic Acid Protocols Handbook*
Edited by: R. Rapley © Humana Press Inc., Totowa, NJ

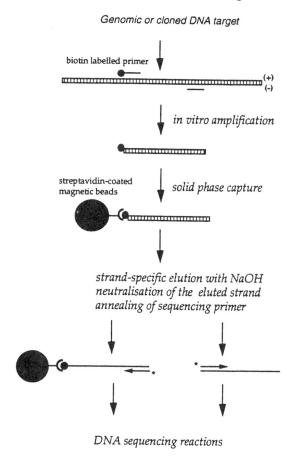

Fig. 1. The solid-phase sequencing concept.

Monodisperse paramagnetic beads coated with streptavidin can be used to capture and purify biotinylated PCR products. Many of the problems in the preparation of sequencing templates are thereby circumvented and a robust system for DNA sequencing (**Fig. 1**) *(3,4)* is achieved. The method allows for complete removal of one strand, nucleotides, DNA polymerase, and PCR primers in a few and simple steps. This procedure results in a pure and single-stranded template, and, in addition, reassociation problems of the complementary strand are minimized. The principle of the solid-phase approach is the use of the strong interaction between biotin and streptavidin, which is extremely strong ($K_d = 10^{-15} M$), temperature stable (up to 80°C), and withstands alkali treatment (0.1 M NaOH). The most preferred manner to introduce a biotin label into a double-stranded DNA fragment is to have one of the PCR primers biotinylated. Biotinylation of primers have been significantly simplified by the introduction of biotin phosphoramidites, which enables a direct coupling onto to the 5' end during primer synthesis (*see* **Note 1**). The capture and immobilization are accomplished by incubating the biotinylated PCR product with the streptavidin-coated magnetic beads for a few minutes. The captured DNA is denatured into two single strands by the addition of 0.1 M NaOH. This results in the elution of the nonbiotinylated strand into the supernatant while the biotinylated strand remains immobilized to the bead surface. The single-stranded template in the supernatant can be recovered by the use of a magnet and subsequent removal of the eluate into a separate tube for neutralization. The benefit of this method is that all reaction components are removed, including the complementary strand, enabling optimal sequencing conditions with no reannealing problems. Note that after neutralization, the eluted nonbiotinylated strand can also be used as a template. Furthermore, the solid-phase approach enables the development of integrated and automated methods for routine sequencing, facilitated by the defined and predictable behavior of the monodisperse magnetic beads *(3,4)*.

All of the available DNA polymerases are suitable for solid-phase DNA sequencing, such as, T7, Klenow, *Taq*, Tth, Bst etc. However, the thermostable enzymes poorly incorporates dideoxynucleotides in a nonconsistent pattern in comparison with native deoxynucleotides *(5,6)*. The variation in relative signal intensities, caused by *Taq* polymerase, make the interpretation by the software algorithm used in automated fluorescent electrophoresis units more difficult *(7)*. Despite the disadvantage of nonuniform signal intensities, the convenience of rapid sequence generation by temperature cycling ("cycle sequencing") has made *Taq* DNA polymerase useful for sequencing of PCR products and cloned target DNA *(8)*. In clinical applications however, the T7 DNA polymerase (Sequenase, Stratagene) is the enzyme of choice, owing to a high processivity, which results in more uniform bands compared to the other polymerases. Peak uniformity is important for accurate base calling, especially for the detection of heterozygosity in genomic material *(6,9)*.

The automation of DNA sequencing in recent years has open new application fields, especially within clinical medicine. The most important developments towards automated routine use is the introduction of PCR and the replacement of isotopic labeling by fluorescent dyes and on-line monitoring of the DNA sequence. The fluorescent sequencing bands are excited by a laser beam and detected in the gel during electrophoresis. There are three main methods to obtain labeled dideoxy DNA fragments: by (1) fluorescent labeled primer, (2) dideoxy chain terminators, and (3) fluorescent labeled dATPs. These alternatives enables many flexible nonradioactive sequencing solutions. The commercially available instruments also enable reliable quantitation of polymorphic and heterozygous positions with the different software packages *(10–12)*. Note, the manual protocols described below have also been used as the basis for developing semi-automated sequencing systems. The template preparation and the sequencing reactions have been implemented on a Beckman Biomek 1000 workstation *(4)* and on an Applied Biosystems Inc. (ABI, Foster City, CA) Catalyst workstation *(13)*.

The following sections describe protocols for:

Amplification of plasmid inserts and genomic DNA targets: General vector primers can be used to amplify inserts in plasmids such as pUC, pBluescript, pEMBL, pGEM etc. A simplified protocol for template preparation of bacterial colonies has therefore been included in combination with a protocol for amplification with universal primer sets containing one biotinylated primer and one nonbiotinylated primer. Amplification of a target gene in genomic DNA can also be performed using a designed primer set with one biotinylated primer (*see* **Note 1**). However, the choice of primers obviously influences the conditions used in the amplification reaction and must be adapted to fit the primer pair chosen for the amplification.

Capture of the PCR product and strand separation: The biotinylated PCR products are directly immobilized to prepared streptavidin-coated paramagnetic beads. Denaturation of the immobilized double-stranded DNA on the beads, followed by elution of the nonbiotinylated strand, yields two single-stranded DNA templates. One template immobilized on the beads and the other in the supernatant.

Solid-phase sequencing protocols: Two alternative sequencing protocols are described as follows. The first is suitable for manual sequencing using radioactive labeled primers or for automated DNA sequencing using the Pharmacia ALF system (Uppsala, Sweden). The second system employs four dye fluorescent labelled primers and requires access to an automated DNA sequencer from Applied Biosystems. The sequencing primers to be used can either be custom designed (complementary to a sequence inside the target DNA being amplified) *(3)* or a universal primer complementary to a sequence introduced by a "handle" sequence in one of the PCR primers *(14)* or one of the primers used in the PCR *(10,16)*. The sequencing primers used for plasmid vectors depends on the choice of PCR primers. PCR primer set A: the immobilized strand can be sequenced with the universal sequencing primer. Alternatively, the immobilized strand can be sequenced with the T3 or the SP6 promotor primers. The eluted (supernatant) strand can be sequenced using the reverse sequencing primer or the T7 promoter primer. PCR primer set B: the immobilized strand is sequenced with the reverse sequencing primer or the T7 promoter primer. The supernatant strand can be sequenced using the universal sequencing primer, T3, or the SP6 promotor primers.

2. Materials

1. Colony lysis buffer 100 mM Tris-HCl pH 8.3 (20°C), 20 mM MgCl$_2$, 500 mM KCl, 1.0% Tween-20
2. 10X PCR Buffer: 100 mM Tris-HCl, pH 8.3 (20°C); 20 mM MgCl$_2$; 500 mM KCl; 1% Tween-20
3. Thermocycler.
4. Ampli*Taq* polymerase (Perkin-Elmer Cetus, Norwalk, CT).
5. Nucleotide solution (2 mM of each dNTP).
6. General PCR vector primers solution containing two primers each having a concentration of 2.5 µM: Primer set A: 5'-biotin-GCTTCCGGCTCGTATGTTGTGTG-3'; 3'-GCGGAA CGTCGTGTAGGGGGAAA-5'; Primer set B: 5'-GCTTCCGGCTCGTATGTTGTGTG-3'; 3'-GCGGAACGTCGTGTAGGGGGAAA-biotin-5'.
7. Dynabeads M-280 streptavidin (10 mg/mL) (Dynal AS, Oslo, Norway).
8. Neodymium-iron-boron magnet (MPC Dynal AS).
9. Washing/binding solution: 10 mM Tris-HCl, pH 7.5 (20°C), 1 mM ethylenediaminetetraacetic acid (EDTA), 2 M NaCl.
10. 1X TE buffer: 10 mM Tris-HCl, pH 7.5 (20°C), 1 mM EDTA.
11. 0.10 M NaOH, freshly made (*see* **Note 2**).
12. 0.333 M HCl, freshly made (*see* **Note 3**).
13. T buffer [0.75 M Tris-HCl, pH 7.5 (20°C)].
14. Water baths or incubators at 0°C, 37°C, and 65°C.
15. A polyacrylamide sequencing gel and electrophoresis equipment or automated sequencers (Pharmacia ALF or ABI Automated Sequencer).
16. Annealing buffer for single-labeled sequencing primer: 280 mM Tris-HCl, pH 7.5 (20°C), 100 mM MgCl$_2$. The label could be either ^{32}P for radioactive sequencing or one dye-fluorescent label such as fluorescein isothiocyanate, FITC (Pharmacia).

17. Annealing buffer for four dye-labeled sequencing primers: 56 mM Tris-HCl, pH 7.5 (20°C), 0.05% Tween-20, 20 mM MgCl$_2$. The general dye-labeled sequencing primers are obtained from ABI Inc.
18. Extension buffer: 300 mM Citric acid, pH 7.0 (20°C), 318 mM dithiothreitol (DTT); 40 mM MnCl$_2$.
19. T7 DNA polymerase with enzyme dilution buffer (Pharmacia).
20. Four nucleotide mixes each containing 40 mM Tris-HCL, pH 7.5 (20°C), 50 mM NaCl, 1.0 mM of each dNTP[a], and 5.0 µM of one specific ddNTP. Thus the A mix contains 5 µM ddATP, the C mix contains 5 µM ddCTP, the G mix contains 5 µM ddGTP, and the T mix contains 5 µM ddTTP.
21. Stop solution (shake 100 mL formamide with 5 g Amberlite (Pharmacia Biotech, Uppsala, Sweden) MB-1 resin and 300 mg dextran blue for 30 min. Filter through 0.45 µm pore-size filter).

3. Methods

3.1. Colony Template Preparation

Dispense 10 µL lysis buffer into each tube. Take a part of one colony from the cultivation plate using a toothpick and place it in the PCR tube. Incubate in the thermocycler at 99°C for 5 min and immediately place the tube on ice.

3.2. Amplification of Plasmid Inserts and Genomic DNA Targets

1. Prepare a polymerase chain reaction master mix in a microcentrifuge tube: 5 µL 10X PCR Buffer; 2 µL (5 pmol) primers (*see* **Notes 1** and **4**); 5 µL dNTP solution; 1 unit Ampli*Taq* and sterile water to 50 µL.
2. Use 1–2 pmol of template DNA (or 5 µL of the colony lysate).
3. Cover with 25 µL light mineral oil, if necessary. For amplification of plasmid inserts, cycle as follows: 96°C 30 s; 72°C 0.5–3.0 min (depending on the length of target sequence) for 30 cycles. A final extension step at 72°C for 10 min is strongly suggested.
4. Analyze the PCR product (2–3 µL) on an agarose gel.

3.3. Preparation of Streptavidin-Coated Magnetic Beads

1. Resuspend the beads by pipeting. Use 20 µL (200 µg) of resuspended beads per PCR template and pipet the suspended beads into a clean 1.5 mL microcentrifuge tube. The beads may be washed in bulk for the total number of PCR templates that need to be purified.
2. Place the tube in the magnetic holder and allow for beads to adhere to the magnet at the side of the wall. Remove the supernatant using a pipet (do not remove the tube from the magnetic holder).
3. Add an equal volume of washing/binding solution and gently pipet to suspend.
4. Repeat, using the magnetic holder, allow the beads to adhere to the side of the tube, and remove the supernatant.
5. Resuspend the beads in washing/binding solution using twice the original volume (i.e., 40 µL). The bead concentration is now 5 µg/µL.

3.4. Capture of the PCR Product and Strand Separation

1. Take 45 µL of the total 50 µL PCR amplification reaction to a fresh 1.5 mL microcentrifuge tube and add 40 µL of the prewashed Dynabeads (*see* **Note 4**).

2. Incubate at room temperature for 15 min. Mix during the immobilization reaction once or twice by gentle pipeting or tapping.
3. Collect the beads by moving the vials to the magnetic holder and remove the supernatant with a pipet.
4. Wash the beads once with 50 µL washing/binding solution (*see* **Note 5**).
5. Wash once with 50 µL 1X TE buffer. Remove the 1X TE buffer carefully, avoiding droplets forming on the walls and the bottom of the tube.
6. Resuspend the beads in *exactly* 10 µL 0.10 M NaOH (*see* **Note 2**).
7. Incubate in room temperature for 5 min.
8. Collect the beads (now with only single-stranded DNA attached) by placing the tube in the magnetic holder and transfer the 10 µL of NaOH supernatant (containing the non-biotinylated strand) to a clean tube. Neutralize the NaOH supernatant with 3 µL 0.333 M HCl (*see* **Note 3**) and mix *immediately*. Add 2 µL of T buffer and save the supernatant strand.
9. Wash the beads once with 50 µL 0.1 M NaOH, once with 50 µL washing/binding solution, and once with 50 µL 1X TE buffer. Remove the 1X TE carefully without leaving any droplets.
10. Resuspend the beads in 13 µL of sterile water (or the appropriate buffer for the sequencing protocol to follow.

3.5. Solid-Phase DNA Sequencing Using T7 DNA Polymerase and One Dye-Labeled Primer

1. Add template DNA (beads with immobilized single-strand DNA or eluted single-strand DNA) to a fresh tube (*see* **Note 6**). Adjust the volume (with sterile water) to 15 µL.
2. Add 2 µL (1 pmol) labeled primer.
3. Add 2 µL of annealing buffer and mix gently with a pipet. Incubate at 65°C for 10 min. Mix gently and leave to cool at room temperature for at least 10 min; mix using two or three times during cooling.
4. Add 1 µL of extension buffer and mix gently.
5. Dilute the T7 DNA polymerase to 1.5 U/µL using *cold* dilution buffer; 2 µL of this diluted stock solution will be required for each template and keep the tube with diluted stock solution (1.5 µ/µL) on ice.
6. Label four new tubes "A", "C", "G", and "T" (*see* **Note 7**). Dispense 2.5 µL of the corresponding dNTP/ddNTP sequencing mixes into the tubes.
7. Warm the dispensed nucleotide sequencing mixes at 37°C for at least 1 min.
8. Add 2 µL of the T7 polymerase diluted stock solution (from **step 5**) to the template mixture (from **step 4**) and mix gently. Immediately add 4.5 µL of this mixture to each of the preincubated nucleotide sequencing mixes.
9. Incubate at 37°C for 5 min.
10. Add 5 µL of stop solution to each reaction and mix gently.
11. Incubate at 95°C for 5 min, then put the tubes on ice.
12. Load the samples onto the polyacrylamide sequencing gel.

3.6. Solid-Phase DNA Sequencing Using T7 DNA Polymerase and Four Dye-Fluorescent Primers for Plasmid Sequencing

1. Dissolve the immobilized single-stranded template in 19 µL of annealing buffer in a microcentrifuge tube (*see* **Note 6**).

2. Dilute the T7 DNA polymerase to 0.7 U/μL using *cold* dilution buffer; 6 μL of this diluted stock solution will be required for each template and keep the tube with diluted stock solution (0.7 μ/μL) on ice.
3. Aliquot the paramagnetic beads with the immobilized template DNA into four tubes labeled "A," "C," "G," and "T." Add ABI's plasmid fluorescent sequencing primer as follows: for A and C take 3 μL beads (in annealing buffer), 1 μL primer, and 0.5 μL extension buffer; for G and T take 6 μL beads (in annealing buffer), 2 μL primer and 1 μL extension buffer. Mix gently and heat to 65°C for 10 min and mix gently and leave to cool at room temperature for at least 10 min, mixing two or three times during cooling.
4. Add an appropriate volume of the relevant dNTP/ddNTP sequencing mixes (A, C, G, or T) to each tube as indicated: for A and C take 1.5 μL nucleotide mixtures; for G and T take 3 μL nucleotide mixtures (see **Note 7**). Keep the samples on ice while doing this.
5. Add the T7 DNA polymerase diluted stock solution, as indicated: for A and C take 1 μL T7 DNA polymerase diluted stock solution; for G and T take 2 μL T7 DNA polymerase-diluted stock solution. Incubate at 37°C for 5 min.
6. Stop the reaction by adding 40 μL *ice cold* 10X TE to each tube and place the tubes on ice.
7. Pool the A, C, G, and T reactions for each sample and collect the beads using the magnet. Discard the supernatant. Wash the beads once with 1X TE buffer and discard the supernatant. Resuspend the beads in 5 μL loading buffer (*see* **Note 8**).
8. Prior to loading the samples onto the ABI automated DNA sequencer, heat the samples to 95°C for 2 min and place on ice.

4. Notes

1. Specific primers for the genomic target gene must be designed and synthesized. A nested primer procedure is strongly recommended to yield a clean product *(14–16)*. Biotinylated oligonucleotides can be obtained from several commercial sources offering oligonucleotide synthesis services. It is of great importance that the biotinylated oligonucleotide is purified from unbound biotin, preferably by reverse phase FPLC or HPLC, as free biotin will occupy binding sites on the beads and reduce the binding capacity of biotinylated PCR products. DNA synthesis can use available biotin phosphoramidites for 5'-end biotinylation of oligonucleotides (Biotin-ON™ Phosphoramidites (Clontech Laboratories, Palo Alto, CA); Biodite (Pharmacia). Aliquots of labeled primers should be stored at –20°C (avoid repeated freeze-thawing).
2. Important: Use a 1.000 ± 0.005 *M* volumetric solution of NaOH and dilute this to 0.1 *M*. Aliquot and store at –20°C.
3. Important: Use a 1.000 ± 0.005 *M* volumetric solution of HCl and dilute this to 0.1 *M*. Aliquot and store at –20°C.
4. The binding capacity of the beads are fragment length dependent. Reduced binding capacity for longer DNA fragments are probably caused by steric hindrance on the bead surface. However, a prolonged binding time increases the degree of capture. For example, a long product can be incubated for 60 min at 43°C alternatively overnight at room temperature. In addition, one may double the amount of beads to decease the effect of steric hindrance on the bead surface. Another option is to change the binding buffer to contain 3 *M* LiCl instead of 1 *M* NaCl, which gives a slight improvement in binding capacity. In addition a threefold excess of nonbiotinylated primer may be used to drive the complete extension of the biotinylated fragment to minimize saturation of free biotin primers onto the streptavidin surface.
5. The immobilized DNA can be stored at 4°C for several weeks.

6. If many samples are to be analyzed, a microtiter plate might be more convenient.
7. 7-Deaza-dGTP (c7dGTP) is preferred instead of dGTP for resolving band compression during electrophoresis.
8. The sequencing reactions may be stored at –20°C if not loaded immediately. Prior to loading, heat the samples to 95°C for 2 min and place on ice.

Acknowledgments

Supported by the Göran Gustafsson Foundation for Research in Natural Sciences and Medicine. The authors thanks Drs. Johan Wahlberg and Thomas Hultman for considerable help with the protocols

References

1. Sanger, F., Nicklen, S., and Coulson, A. R. (1977) DNA sequencing with chain-terminating inhibitors. *Proc. Natl. Acad. Sci. USA* **74(12),** 5463–5467.
2. Scharf, S. J., Horn, G. T., and Erlich, H. A. (1986) Direct cloning and sequence analysis of enzymatically amplified genomic sequences. *Science* **233,** 1076–1078.
3. Hultman, T., Ståhl, S., Hornes, E., and Uhlén, M. (1989) Direct solid phase sequencing of genomic and plasmid DNA using magnetic beads as solid support. *Nucleic Acids Res.* **17,** 4937–4946.
4. Hultman, T., Bergh, S., Moks, T., and Uhlén, M. (1991) Bidirectional solid-phase sequencing of in vitro-amplified plasmid DNA. *BioTechniques* **10(1),** 84–93.
5. Innis, M. A., Myambo, K. B., Gelfand, D. H., and Brow, M. A. D. (1988) DNA sequencing with Thermus aquaticus DNA polymerase and direct sequencing of PCR-amplified DNA. *Proc. Natl. Acad. Sci. USA* **85,** 9436–9440.
6. Leren, T. P., Rødningen, O. K., Røsby, O., Solberg, K., and Berg, K. (1993) Screening for point mutations by semi-automated DNA sequencing using Sequenase and magnetic beads. *BioTechniques* **14(4),** 618–623.
7. Khurshid, F. and Beck, S. (1993) Error analysis in manual and automated DNA sequencing. *Anal. Biochem.* **208,** 138–143.
8. Carothers, A. M., Urlaub, G., Mucha, J., Grunberger, D., and Chasin, L. A. (1989) Point mutation analysis in mammalian gene: rapid preparation of total RNA, PCR amplification of cDNA, and Taq sequencing by a novel method. *BioTechniques* **7,** 494–499.
9. Hedrum, A., Pontén, F., Ren, Z., Lundeberg, J., Pontén, J., and Uhlén, M. (1994) Sequence-based analysis of the human p53 gene based on microdissection of tumor biopsy samples. *Biotechniques* **17,** 1–9.
10. Wahlberg, J., Albert, J., Lundeberg, J., Cox, S., Wahren, B., and Uhlén, M. (1992) Dynamic changes in HIV-1 quasispecies from azidothymidine (AZT) treated patients. *FASEB J.* **6,** 2843–2847.
11. Leitner, T., Halapi, E., Scarletti, G., Rossi, P., Albert, J., Fenyö, E.-M., and Uhlén, M. (1993) Analysis of heterogeneous viral populations by direct DNA sequencing. *BioTechniques* **15(1),** 120–127.
12. Larder, B. A., Kohli, A., Kellam, P., Kemp, S. D., Kronick, M., and Henfrey, R. D. (1993) Quantitative detection of HIV-1 drug resistance mutations by automated DNA sequencing. *Nature* **365,** 671–673.
13. Holmberg, A., Fry, G., and Uhlén, M. (1993) Automated DNA sequencing and analysis techniques, Venter, C., ed., Academic, London, pp. 139–145.
14. Wahlberg, J., Albert, J., Lundeberg, J., Fenyö, E.-M., and Uhlén, M. (1991) Analysis of the V3 loop in neutralization-resistant human immunodeficiency virus type 2 variants by direct solid phase DNA sequencing. *AIDS Res. Hum. Retrov.* **7(12),** 983–990.

15. Wahlberg, J., Lundeberg, J., Hultman, T., and Uhlén, M. (1990) General colorimetric method for DNA diagnostics allowing direct solid phase genomic sequencing of the positive samples. *Proc. Natl. Acad. Sci. USA* **87,** 6569–6573.
16. Pettersson, B., Johansson, K.-E., and Uhlén, M. (1994) Sequence analysis of 16S rRNA from Mycoplasmas by direct solid-phase DNA sequencing. *Appl. Env. Microbiol.* **60(7),** 2456–2461.

68

Polymerase Chain Reaction Cycle Sequencing with Degenerate Primers

Zhiyuan Shen, Jingmei Liu, Robert L. Wells, and Mortimer M. Elkind

1. Introduction

Among the many techniques of cloning new genes, one approach involves degenerate primers *(1–7)*. The approach usually requires three steps:

1. Using degenerate primers to amplify part of the gene of interest by PCR: The degenerate primers' sequences may be designed from known protein sequences or conserved regions of a gene family (e.g., *2,4*). Because deoxyinosine can base pair with all of the four deoxyribonucleotides it has been substituted for specific nucleic acids in degenerate primers to reduce the number of different primer sequences that would otherwise be needed in the reaction *(2,7,8)*.
2. A determination of which amplified PCR product(s) is from the gene of interest: If the target gene and the primers are only partially homologous, a moderate annealing stringency in the PCR reaction is usually necessary to obtain amplification. Moderate stringency may result in multiple PCR products. Although from the size of the PCR products it may be possible to predict which is from the gene of interest, sequencing analysis of the PCR products may be required.
3. The screening of a cDNA library using the correct PCR product as a probe and cloning the gene of interest.

Sequencing the amplified PCR product is one of the most important steps in this approach to gene cloning. To sequence the PCR fragment amplified by degenerate inosine-containing primers, the PCR fragment may be cloned into a sequencing vector, such as M13 bacteriophage. Sequencing is straightforward if primers specific to the vector are used. Theoretically, this method allows any unknown cloned DNA fragment to be sequenced. However, the *Taq* polymerase, which is used to amplify the target fragment, is thought to have relatively high misincorporation rates for dNTPs, approx 10^{-4}. Hence, it is possible that a copy of the product may contain one or more incorrect nucleotides. If such a copy has been cloned into the sequencing vector, the resulting sequencing data would be incorrect for that particular clone. Direct sequencing of PCR products can circumvent this problem because most of the fragments are exact replicas of the target molecule. Thus, the majority of the products used for sequencing would have the right nucleotide at a specified position and result in the correct sequencing ladder. Also, direct sequencing of PCR products avoids the time-consuming cloning of PCR products, and most of the available direct PCR sequencing protocols require relatively small amounts of template.

Many protocols are available for the direct sequencing of PCR products. Most of the protocols require specific sequencing primers. However, this means at least part of the specific base sequence of the template is needed. This requirement may not be met and, in many cases, information about the internal sequence of a gene may be lacking. This shortcoming may apply to the PCR products of degenerate inosine-containing primers of the cDNA of a new gene. Therefore, one may be forced to use the same degenerate inosine-containing primers that were used in the PCR step for direct sequencing. When primers have low degeneracy, they may be treated as sequence-specific primers, and some of the direct-sequencing protocols, such as those described in this volume, may be used with success. When only highly degenerate inosine-containing primers are available, these methods may not succeed.

To sequence a PCR product amplified via the use of a highly degenerate inosine-containing primers, several general factors must be kept in mind.

1. The sequencing primer(s) must anneal specifically to one site on the DNA fragment that is to be sequenced, i.e., a secondary annealing site must be avoided. Therefore, stringent primer annealing temperatures are necessary.
2. A sufficient quantity of the specific primer should anneal to the correct site. Consequently, the primer annealing temperature cannot be too high.
3. Reassociation of the double-stranded DNA template should be minimized. This requirement generally can be met by using optimal PCR protocols for the sequencing reactions.

To carry out the requirements above, a primer-labeling method, in which the primer is labeled at the 5'-end, may be worth considering. Linear PCR is used to generate the labeled dideoxynucleotide-terminated sequences *(9,10)*. The use of this method minimizes problems of template reassociation and/or mismatching of the primer, because the annealing time is relatively short. Also, the annealing temperature is higher than it would be in most protocols that use DNA polymerases other than *Taq*, such as T4 DNA polymerase, but the method requires a 5'-end-labeling step for which ^{35}S is generally not suitable compared to ^{32}P because of its lower specific activity and the lesser efficiency with which some enzymes label 5'-ends with $\alpha\text{-}^{35}S$-ATP vs $\alpha\text{-}^{32}P$-ATP. Only ^{32}P can be used, even though its greater radiation hazard owing to its higher β-particle emission and its shorter half-life make it less convenient. Furthermore, when highly degenerate primers are used, higher primer concentrations in the reaction mixture are needed to insure that sufficient specific priming will occur. The preceding increases the hazard as well as the cost.

To assure that sequencing primer(s) anneal to a DNA template specifically, to eliminate the need for 5'-end labeling, and to avoid reassociation of the double-stranded DNA template, a two-step cycle-sequencing protocol is described to sequence products amplified with degenerate inosine-containing primers. This method uses the same degenerate primers that were used in PCR amplification. The method can be broken down into two steps of linear PCR. The first step is for labeling the primers, and the second is for the random dideoxy-termination. As shown in **Fig. 1**, in the first step, primers were extended and labeled with $\alpha\text{-}^{35}S$-dATP. The extension is limited and

PCR Cycle Sequencing

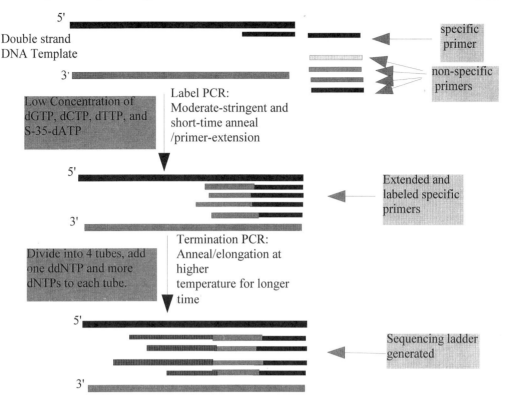

Fig. 1. Procedure of cycle sequencing with degenerate inosine-containing primers. Two linear PCR steps are involved. (1) Label PCR uses low dNTP concentrations, a low temperature, and short times for primer annealing/elongation to produce incomplete extension of specific primers. As a result, specific primers are labeled and extended. The extended and labeled primers have a higher melting temperature than the native printers. (2) Termination PCR using a higher annealing/elongation temperature and is performed with higher dNTP concentrations and in the presence of ddNTPs. Only the extended and labeled primers are involved in the termination reaction.

performed under conditions of high stringency, low dNTP concentration, and a short interval, so that the specific primer in the mixture is favored and a limited length of primer extension is achieved. In the second step, dideoxynucleotide terminations are effected at a more stringent elevated annealing/elongation temperature. The result is that only the extended and labeled primers enter into the termination reactions.

We have used this method to sequence amplified cytochrome p450 cDNA fragments with a highly degenerate inosine-containing primer (1,2). In our case, a set of degenerate primers was used to amplify a presumably novel cytochrome p450 gene(s). The upstream sense primer was a mix of 192, 20 mer, containing three inosines, which theoretically could anneal to 12,288 different sequences. The downstream, antisense primer was a mix of 144, 23 mer, containing five inosines or 147,456 different possible sequences.

In what follows, we will only describe the sequencing reactions. Procedures for sequencing gel electrophoresis can be found in Chapter 61.

2. Materials

1. A thermal cycler: Cetus Perkin-Elmer Model 480 (*see* **Note 1**).
2. 0.5-mL PCR tubes.
3. Mineral oil.
4. Gel-purification kits/reagents, such as: QIAEX Gel Extraction Kit (Qiagen #20020, Chatsworth, CA) or QiaOuick Gel Extraction Kit (Qiagen #28704).
5. All buffers and solutions must be free of DNase.
6. α-^{35}S-dATP (10 μCi/μL 1000 Ci/mmol) (Amersham Pharmacia Biotech, Amersham, UK).
7. Sequencing primers (degenerate primers) dissolved in H_2O, or 0.1X TE buffer.

The sequencing reaction reagents can be homemade. However, we recommend purchasing them from a commercial company to ensure uniform performance. In the following materials, we include the catalog number for US Biochemicals (Cleveland, OH).

8. Reaction buffer (USB #71030): 260 m*M* Tris-HCl, pH 9.5, 65 m*M* $MgCl_2$.
9. Δ*Taq* DNA polymerase (USB #71059) or *Taq* DNA polymerase, (USB# 71057): 32 U/μL.
10. *Taq* DNA polymerase dilution buffer (USB #71051): 10 m*M* Tris-HCl, pH 8.0, 1 m*M* 2-mercaptoethanol, 0.5% Tween-20, and 0.5% Nonidet P-40.
11. Four separate primer label mixes:
    a. dGTP label mix: 3.0 μ*M* (USB #71034).
    b. dATP label mix: 3.0 μ*M* (USB #71036).
    c. dTTP label mix: 3.0 μ*M* (USB #71037).
    d. dCTP label mix: 3.0 μ*M* (USB #71038).
12. Four separate termination mix:
    a. ddG terminator mix: 15 μ*M* each dGTP, dATP, dTTP, dCTP, and 22.5 μ*M* ddGTP (USB #71020).
    b. ddA termination mix: 15 μ*M* each dGTP, dATP, dTTP, dCTP, and 300 μ*M* ddATP (USB #71035).
    c. ddT termination mix: 15 μ*M* each dGTP, dATP, dTTP, dCTP, and 450 μ*M* ddTTP (USB #71040).
    d. ddC terminator mix: 15 μ*M* each dGTP, dATP, dTTP, dCTP, and 75 μ*M* ddCTP (USB #71025).
13. Stop/gel-loading solution (USB #70724): 95% formamide, 20 m*M* EDTA, 0.05% bromophenol blue, and 0.05% xylene cyanol FF.
14. 1X TE buffer: 10 m*M* Tris-HCl, pH 8.0, 1 m*M* EDTA.

3. Methods

3.1. Preparation of DNA as a Sequencing Template (see Note 2)

1. After gel electrophoresis, PCR fragment(s) of interest is cut out from the gel.
2. DNA in the gel is purified with the Qiagen gel-purification kit, and final PCR products are resuspended in a proper amount of 0.1X TE buffer.
3. To estimate the amount of PCR product, run an aliquot of the PCR products on an agarose gel. The amount of DNA may be estimated by a comparison with the amount of DNA that was used in the mol-wt ladder.

In the following steps, always keep tubes on "ice," unless otherwise indicated.

4. Prepare the following labeling PCR mix (*see* **Note 3**):
   H_2O 0–8 μL
   DNA (in 0.1X TE) (need total of 25–100 ng)   1–9 μL
   Reaction buffer   2 μL

Degenerate primers (5–200 μM)	1 μL
dGTP label mix	1 μL
dCTP label mix	1 μL
dTTP label mix	1 μL
α-^{35}S-dATP (10 μCi/μL >1000 Ci/mmol)	0.5 μL
Taq DNA polymerase (4 U/μL) (diluted in *Taq* dilution buffer)	2 μL
Total volume	17.5 μL

Cover the label PCR mix with 10–20 μL of mineral oil.

3.2. Labeling PCR *(see Note 4)*

1. Run the following PCR program: presoak at 94°C for 3–5 min followed by 45 cycles of 95°C for 30 s and 52°C for 30 s.
2. Transfer 15–16 μL of the above labeled mixture to a new tube. Avoid carryover of any mineral oil. This can be done easily by putting the pipeting tip directly below the oil without touching the wall of the tube.
3. Optional *(see* **Note 5**): Load 1–2 μL with 1 μL of gel-loading buffer to a sequencing gel to check the labeling efficiency.
4. Termination PCR mix: For each of the labeled mixes, prepare four tubes labeled as "G," "A," "T," and "C." To each of the tubes, add 4 μL of termination mix "G," "A," "T," or "C" (this can be done toward the end of label-PCR procedure). Add 3.5 μL of the label mix to each of the tubes. Cover the termination PCR mix with 8–10 μL of mineral oil.
5. Termination PCR: Cycle between 95°C for 30 s and 72°C for 90 s *(see* **Note 6**).
6. While the termination PCR is under way, prepare four clean 0.5-mL tubes labeled "G," "A," "T," or "C." To each of them add 4 μL stop/gel-loading solution.
7. Transfer 6–7 μL of termination mix to these tubes with the stop/loading solution. Avoid carryover of mineral oil. Mix and spin down briefly. Store at –20°C (good for up to 1 mo). These samples are ready for the sequencing gel (use 3 μL to load a gel). *See* Chapter 61.
8. Sequencing results: run sequencing gel, perform autoradiography, and read the sequence *(see* **Note 7**).

4. Notes

1. Cetus Perkin-Elmer thermal cycler Model 9600 also may be used. If it is, use 0.1-mL tubes; no mineral oil on the top of the reaction solution is needed. The PCR program should be adjusted accordingly in the procedure.
2. Other methods of DNA preparation are also acceptable as long as "clean" DNA is obtained.
3. Other α-^{35}S-labeled nucleotides may also be used, but the label mix must be changed accordingly. The concentration of the stock of degenerate primers in the reaction is dependent on the degree of degeneracy. In our case, the stock concentration of our >100× degenerate primer was 200 μM. Because radioactive ^{35}S is used for these experiments, always be careful and follow the safety operation procedure for your institute. Check with your radiation safety officer for the authorized amount of radioactivity that you can handle at any one time.
4. Depending on the sequencing primer, the annealing/elongation temperature or time may have to be optimized to give proper primer extension and labeling. The purposes of the label PCR is to have a sufficient amount of specific primer in the primer mix to anneal to a specific site on the DNA template, and to extend the annealed primer for a limited nucleotide length with *Taq* DNA polymerase. The first purpose can be achieved by choice of an optimal annealing temperature and/or time. In our case (200 pmol of the 20 mer with inosine and a degeneracy of more than 100×), we used 52°C, and 30 s. Depending on

circumstances, this temperature and the annealing time may need to be adjusted. The method of generating a limited elongated primer plus labeling of the primer is accomplished by using a shorter annealing/elongation time at a suboptimal temperature (for *Taq* activity), but still a stringent temperature for annealing and a low dNTP concentration. In this way, it is not necessary to know the sequence of the downstream flanking region of the sequencing primer.

5. Loading 1–2 µL of labeled mix to run a gel to check that the length of primer extension and label efficiency is optional. This can be run along with the sequencing sample after all the reactions are finished. Using our p450 degenerate sequencing primers under these labeling PCR conditions, we obtained an average primer extension of 15–25 bp.
6. The temperature used for both the annealing of labeled/extended primers to the DNA template for the elongation/termination reaction was 72°C. Only the prelabeled and pre-extended primers, which are the specific primers in the primer mix, would be allowed during the termination/elongation because of the elevated temperature. If more template DNA is available, fewer cycles may be used.
7. Depending on the length of labeled primers, the readable sequence will vary. For our case of highly degenerate inosine-containing primers of p450 genes (*see* **Subheading 1.** for a description of our p450 primers, a ladder from 25 bp downstream of the primer was readable up to 300 bp.
8. A similar protocol of this method would be to omit one of the four dNTPs in the label step, and use at least one α-^{35}S-labeled dNTP in the labeling mix. This will give an incomplete elongation of the sequencing primer during the labeling step, because the primer extension will stop at the proper position when the omitted nucleotide is not present. The elongated primers may be labeled if the labeled nucleotide is by chance present between the sequence primer and the omitted nucleotide. This method is useful to sequence DNA when some sequence information immediately downstream from the sequencing primer is available. In such a case, one can decide which nucleotide to omit or to label in the label mix.

References

1. Shen, Z., Liu, J., Wells, R. L., and Elkind, M. M. (1993) Cycle sequencing using degenerate primers containing inosines, *BioTechniques* **15(1),** 82–89.
2. Shen, Z., Wells, R. L., Liu, J., and Elkind, M. M. (1993) Identification of a cytochrome p450 gene by reverse transcription-PCR using degenerate primes containing inosines. *Proc. Natl. Acad. Sci. USA* **90,** 11,483–11,487.
3. Shen, Z., Liu, J., Wells, R. L., and Elkind, M. M. (1994) cDNA cloning, sequence analysis, and induction by aryl hydrocarbons of a murine cytochrome p450 gene, Cyplbl. *DNA and Cell Biol.* **13(7),** 763–769.
4. Shen, Z., Denison, K., Lobb, R., Gatewood, J., and Chen, D. J. (1995) The human and mouse homologs of yeast RAD52 genes: cDNA cloning, sequence analysis, assignment to human chromosome 12pl2.2-pl3, and mRNA expression in mouse tissues. *Genomics* **25,** 199–206.
5. Compton, T. (1990) Degenerate primers for DNA amplification, in *PCR Protocol, a Guide to Methods and Applications* (Innis, M. A., Gelfand, D. H., Sninsky, J. J., and White, T. J., eds.), Academic, San Diego, CA, pp. 39–45.
6. Lee, C. C., and Caskey, C. T. (1990) cDNA cloning using degenerate primers, in *PCR Protocol, a Guide to Methods and Applications* (Innis, M. A., Gelfand, D. H., Sninsky, J. J., and White, T. J., eds.), Academic, San Diego, CA, pp. 46–59.
7. Knoth, K. S., Roberds, S., Poteet, C., and Tamkun, M. (1988) Highly degenerate inosine-containing primers specifically amplify rare cDNA using the polymerase chain reaction. *Nucleic Acids Res.* **16,** 10,932.

8. Erlich, H. A., Gelfand, D., and Sninsky, J. J. (1991) Recent advances in the polymerase chain reaction. *Sciences* **252,** 1643–1651.
9. Murray, V. (1989) Improved double strand DNA sequencing using the linear polymerase chain reaction. *Nucleic Acids Res.* **17,** 8889.
10. Smith, D. P., Jonstone, E. M., Little, S. P., and Hsiung, H. M. (1990) Direct DNA sequencing of cDNA inserts from plaques using the linear polymerase chain reaction. *BioTechniques* **9,** 48–52.

69

Direct Automated Cycle Sequencing of Polymerase Chain Reaction Products

Susan E. Daniels

1. Introduction

The polymerase chain reaction (PCR) is well known for being a rapid and versatile method for the amplification of defined-target DNA sequences. This technique can be applied to a variety of research areas, such as the identification and typing of single nucleotide substitutions of DNA sequence polymorphisms, and genetic mapping *(1–4)*.

Since the introduction of PCR *(5)* a variety of methods for sequencing PCR-generated fragments have been described. These are usually based on the Sanger chain terminating dideoxynucleotide sequencing *(6)* rather than the Maxam and Gilbert chemical cleavage method *(7)*. Manual dideoxy sequencing methods are labor intensive, time consuming, involve radioisotopes, and have limitations in sequence ordering. However, a technique combining the PCR and dideoxy terminator chemistry simplifies the process of sequencing and is known as cycle sequencing *(8)*. Automated or fluorescent DNA sequencing is a variation of the traditional Sanger sequencing using the cycle sequencing methodology, where fluorescent labels are covalently attached to the reaction products, and data are collected during the polyacrylamide gel electrophoresis.

The introduction of fluorescently labeled dideoxynucleotides as chain terminators presented the opportunity for the development of reliable cycle sequencing for PCR products. The sequencing reaction with the dye terminators is carried out in a thermal cycler and each of the four dideoxynucleotide triphosphates (ddNTPs) is labeled with a different fluorescent dye. This allows the four chain extension reactions to be carried out within a single tube, sparing considerable labor *(9,10)*. The use of labeled chain terminators allows flexibility of sequencing strategy, as the same primers can be used in the sequencing reaction. This eliminates the time and expense associated with a separate set of modified DNA sequencing primers and is well suited to high throughput sequencing.

Using this method, it is possible to amplify a target DNA sequence, purify the resulting fragment, and obtain sequencing data within 24 h. Also, it has been used to sequence a 500-bp PCR fragment on an automated DNA sequencer with 99.3% accuracy *(10–12)*.

This chapter concentrates on the techniques involved with the direct sequencing reactions rather than on the use of the machine, as each automated DNA sequencer will be provided with an extensive manual for its operation (**Fig. 1**).

2. Materials

All solutions should be made to the standard required for molecular biology. Use molecular-biology-grade reagents and sterile distilled water. Reagents for the cycle sequencing are available commercially.

2.1. Purification of PCR Products before Cycle Sequencing

1. 4 M Ammonium acetate.
2. Isopropanol.
3. 70% (v/v) Ethanol.
4. 10 mM Tris-HCl pH 7.5, 1 mM ethylenediaminetetraacetic acid (EDTA).

2.2. Cycle Sequencing

Prism™ ready reaction DyeDeoxy™ terminator premix (1000 µL) Applied Biosystems (Warrington, UK). 1.58 µM A-dyedeoxy, 94.74 µM T-dyedeoxy, 0.42 µM G-dyedeoxy, 47.37 µM C-dyedeoxy, 78.95 µM dITP, 15.79 µM dATP, 15.79 µM dCTP, 15.79 µM dTTP, 168.42 mM Tris-HCL (pH 9.0) 4.21 mM $(NH_4)_2SO_4$, 42.10 mM $MgCl_2$, 0.42 U/µL Ampli*Taq* DNA polymerase (Stratagene, La Jolla, CA).

2.3. Purification of PCR Products after Cycle Sequencing

1. Chloroform.
2. Phenol:H_2O:chloroform (16:18:14) at room temperature.
3. 2 M Sodium acetate pH 4.5.
4. 100% and 70% (v/v) ethanol at room temperature.

2.4. Six Percent Polyacrylamide Sequencing Gels

1. 10X TBE (890 mM Tris-borate, 890 mM boric acid, 20 mM EDTA pH 8.3).
2. 40 g Urea.
3. 12 mL 40% (w/v) Acrylamide stock solution (19:1 acrylamide/*bis*-acrylamide).
4. 20 mL dH_2O.
5. 1 g Mixed bed ion exchange resin.
6. Tetramethylenediamine (TEMED).
7. 10% (w/v) Ammonium persulphate, freshly made.

2.5. Loading Buffer

1. 50 mM EDTA pH 8.0.
2. Deionized formamide.

3. Methods

3.1. Isopropanol Purification of PCR Products

It is essential to remove excess PCR primers before using DyeDeoxy terminators for cycle sequencing.

1. Aliquot an appropriate amount of the PCR reaction into a 0.6-mL microfuge tube and dilute to a total of 20 µL with distilled water.

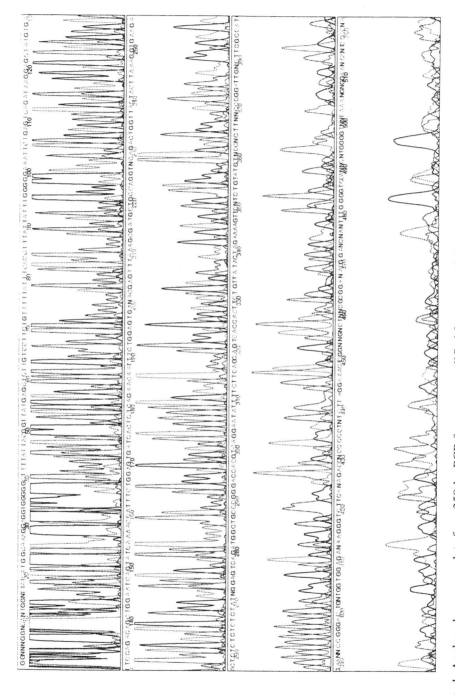

Fig. 1. Analyzed sequence data for a 350-bp PCR fragment amplified from genomic human DNA. The data were obtained using one of the primers used for PCR amplification.

2. Add 20 µL 4 M ammonium acetate into the microfuge tube and mix well.
3. Add 40 µL of isopropanol into the tube, mix well, and leave at room temperature for 10 min then centrifuge the microfuge tube for 10 min at 12,000g.
4. Carefully remove the supernatant and wash the pellet with 70% (v/v) ethanol, then briefly dry the pellet under vacuum.
5. Suspend the pellet with 20 µL of TE.

3.2. Cycle Sequencing of PCR Products

The amount of PCR product should be estimated on an agarose gel before sequencing. Approximately 1 µg of double-stranded DNA template or 0.5 µg of single-stranded DNA template is required for each sequencing reaction.

1. Mix the following reagents in a 0.6-mL microfuge tube, 5 mL DNA template, 1 µL primer (from a 3.2-pmol stock solution) and 4.5 µL sterile dH$_2$O.
2. Add 9.5 µL Prism and ready reaction DyeDeoxy terminator premix.
3. Spin briefly to collect the reaction mix in the bottom of the tube and overlay with approx 50 µL mineral oil.
4. Place the tubes into a thermal cycler (Perkin Elmer Cetus, Norwalk, CT, model 480 or 9600), which has been preheated to 96°C.
5. Immediately begin the cycle sequencing program, which is as follows: rapid thermal ramp to 96°C for 30 s, rapid thermal ramp to 50°C for 15 s, and rapid thermal ramp to 60°C for 4 min for a total of 25 cycles.
6. Try and keep the samples in the dark at 4°C until further processing, as they are light sensitive.
7. Remove the excess DyeDeoxy terminators from the completed sequencing reactions.

3.3. Phenol/Chloroform Extraction of Cycle-Sequencing Products

This step is essential in removing excess primers and unincorporated nucleotides.

1. To each sample add 80 µL of sterile dH$_2$O.
2. Either add 100 µL chloroform to dissolve the oil, or remove the oil by a pipet.
3. Add 100 µL phenol:H$_2$O:chloroform (68:18:14) to the sample and mix well by vortexing.
4. Centrifuge the sample at 12,000g for 1 min. Remove and discard the lower organic phase.
5. Re-extract the aqueous layer (**steps 3** and **4**) except transfer the aqueous upper layer to a clean tube.
6. Add 15 µL of 2 M sodium acetate and 300 µL 100% ethanol to precipitate the extension products.
7. Centrifuge at 12,000g for 15 min at room temperature.
8. Carefully remove the supernatant and wash the pellet with 70% ethanol, then briefly dry the pellet under a vacuum.

3.4. Preparing Samples for Loading

1. Add 4 µL deionized formamide:50 mM EDTA pH 8.0 (5:1) to each sample tube and mix well to dissolve the dry pellet.
2. Centrifuge briefly to collect the liquid at the bottom of the tube.
3. Before loading, heat the samples at 90°C for 2–3 min to denature, then transfer immediately onto ice.
4. Load all of the samples onto the automated DNA sequencer fitted with a 6% polyacrylamide gel using the manufacturer's software.

3.5. Preparation of Six Percent Polyacrylamide-Sequencing Gel

1. Place 40 g urea, 12 mL 40% acrylamide stock, 20 mL dH$_2$O and 1 g mixed bed ion exchange resin into a beaker and stir gently with warming. Continue to stir the solution until all the urea crystals have dissolved.
2. Filter the acrylamide through a 0.2-µM filter, degas for 5 min and transfer to a 100-mL cylinder.
3. Add 8 mL filtered 10X TBE buffer and adjust the volume to 80 mL with dH$_2$O.
4. To polymerize the gel, add 400 µL 10% APS (freshly made) and 45 µL TEMED. Gently swirl to avoid adding air bubbles.
5. Cast the gel according to the instructions provided for the automated DNA sequencer. Analyze data.

4. Notes

1. Both purification steps can also be done by spin columns, such as Centri-Sep™ or Quick Spin™. Although they provide a quicker purification, they are an expensive alternative to those researchers on a tight budget.
2. The dried sequencing pellet can be stored in the dark at 4°C for several days if required. However, once the loading buffer has been added, the samples should be loaded within a few hours.
3. The sequencing gel must polymerize for at least 1 h before use. A good time to prepare the gel is during the PCR of the cycle sequencing reactions.
4. The PCR primers can be used as the DNA sequencing primers and should be at least an 18mer in length. Increasing the length will increase specificity and prevents priming at a secondary site. It will also decrease the chances for nonexact hybridization.
5. The guanine cytosine content of the primer should be between 50 and 60%.
6. After ethanol or isopropanol precipitation, it is very important that the supernatant is carefully aspirated, as the pellets are unstable and might be lost.
7. During the phenol/chloroform extraction, if after the first spin, two separate layers are not seen, then revortex the samples for 1 min and recentrifuge, after which you should get an aqueous and organic phase.
8. Although there are various ready-made acrylamide solutions on the market, it is recommended that you make your own solutions, as better resolution is achieved. However, 40% acrylamide can be purchased ready made and gives good results.
9. It has been shown that the use of formamide in the polyacrylamide gels can resolve compressions *(13)*.
10. If using the Applied Biosystem 373A "Stretch" automated DNA sequencers, it is better to use a 4.75% polyacrylamide gel, and from these machines it is possible to obtain up to 1 kb of sequence with one run.
11. If the thermal cycler in your laboratory is not a Perkin Elmer Cetus, then it will be necessary to optimize your thermal cycler for the sequencing reactions.
12. The addition of blue dextran to the loading buffer will make gel loading easier.

References

1. Orita, M., Suzuki, Y., Sekeiya, T., and Hayashi, K. (1989) Rapid and sensitive detection of point mutations and DNA polymorphisms using the polymerase chain reaction. *Genomics* **5,** 874–879.
2. Kwok, P. Y., Carlson, C., Yager, T. D., Ankener, W., and Nickerson, D. A. (1994) Comparative analysis of human DNA variations by fluorescence based sequencing of PCR products. *Genomics* **23,** 138–144.

3. Martin-Gallardo, A., McCombie, W. R., Gocayne, J. D., Fitzgerald, M. G., Wallace, S., Lee, B. M. B., Lamerdin, J., Trapp, S., Kelly, J. M., Liu, L. I., Dubrick, M., Johnston-Dow, L. A., Kerlavage, A. R., de Jong, P., Carrano, A., Fields, C., and Venter, J. C. (1992) Automated DNA sequencing and analysis of 106 kilobases from human chromosome 19q 13.3. *Nature Genet.* **1,** 34–39.
4. NIH/CEPH Collaborative Mapping Group (1992) A comprehensive genetic linkage map of the human genome. *Science* **258,** 67–86.
5. Mullis, K. B. and Faloona, F. A. (1987) Specific synthesis of DNA in vitro via a polymerase catalysed chain reaction. *Methods Enzymol.* **155,** 335–350.
6. Sanger, F., Nicklen, S., and Coulson, A. R. (1977) DNA sequencing with chain terminating inhibitors. *Proc. Natl. Acad. Sci. USA* **74,** 5463–5467.
7. Maxam, A. M. and Gilbert, W. (1977) A new method for sequencing DNA. *Proc. Natl. Acad. Sci. USA* **74,** 560–564.
8. Carothers, A. M., Urlaub, G., Mucha, J., Grunberger, D., and Chasin, L. A. (1989) Point mutation analysis in a mammalian gene: rapid preparation of total RNA, PCR amplification of cDNA, and Taq sequencing by a novel method. *Biotechniques* **7,** 494–499.
9. McBride, L. J., Koepf, S. M., Gibbs, R. A., Nyugen, P., Salser, W., Mayrand, P. E., Hunkapiller, M. W., and Kronick, M. N. (1989) Automated DNA sequencing methods involving polymerase chain reaction. *Clin. Chem.* **35,** 2196–2201.
10. Tracy, T. E. and Mulcahy, L. S. (1991) A simple method for direct automated sequencing of PCR fragments. *Biotechniques* **11(1),** 68–75.
11. Rosenthal, A. and Charnock Jones, D. S. (1992) New protocols for DNA sequencing with dye terminators. *DNA Seq.* **3,** 61–64.
12 Kelley, J. M. (1994) Automated dye terminator DNA sequencing, in *Automated DNA Sequencing and Analysis* (Adams, M. A., Fields, C., and Venter, J. C., eds.) Academic, London, pp. 175–181.
13. Hawkins, T. L. and Sulston, J. E. (1991) The resolution of compressions in automated fluorescent sequencing. *Nucleic Acids Res.* **19(10),** 2784.

70

Affinity-Capture and Solid-Phase Sequencing of Biotinylated Polymerase Chain Reaction Products

Anu Suomalainen and Ann-Christine Syvänen

1. Introduction

The time and labor required for generating nucleotide sequence information has been significantly reduced since the development of the polymerase chain reaction (PCR). PCR allows the production of sufficient amounts of DNA template in vitro, and consequently the time-consuming cloning procedure to obtain large enough quantities of DNA for sequencing is avoided. The PCR has transformed nucleotide sequencing into a method that can be considered even in routine diagnostics for sequencing large numbers of individual samples.

There are, however, two major problems associated with the sequencing of PCR products. The first one originates from the fact that the PCR and the Sanger dideoxynucleotide sequencing methods are based on the same principle, i.e., a DNA-polymerase catalyzed primer extension reaction with nucleotide triphosphates. The amount of the two primers in a typical PCR reaction is 50- to 100-fold compared to the standard amount of the single primer in a sequencing reaction, and the amount of deoxynucleotide triphosphates (dNTPs) during PCR can be as much as 5000- to 10,000-fold compared to the amount of dNTPs during the labeling reaction or to the amount of the dideoxynucleotide triphosphates (ddNTPs) during the termination step of the sequencing procedure. Complete removal of the primers and the dNTPs from the PCR product is cumbersome, but essential, for successfully performing the sequencing reactions. Another problem arises from the strong tendency of the two strands of a double-stranded PCR product to reanneal, which will compete with the annealing of the sequencing primer. Various approaches to circumvent these problems have been taken *(1–3)*, and many of them are described in this book.

We have solved the problems of direct sequencing of PCR products by employing affinity capture for purifying the sequencing template *(4)*. Our method is based on PCR with one biotinylated and one unbiotinylated primer, after which the PCR product, which will carry biotin in one of its strands, is captured on avidin-coated polystyrene beads taking advantage of the strong interaction between biotin and avidin. Immobilization of the PCR product allows efficient and technically simple removal of the excess primers and dNTPs after PCR by washing, as well as the removal of the unbiotinylated strand by alkaline treatment. The immobilized single-stranded DNA template can then be sequenced by a standard dideoxynucleotide sequencing procedure.

The protocol presented in **Subheading 2.**, which is routinely used in our laboratory, has been simplified from the method for sequencing affinity-captured PCR products originally developed for diagnostic purposes (5). A similar kind of solid-phase sequencing method, in which streptavidin-coated magnetic beads are used as the solid support, was developed by Hultman et al. (6). This method has been automated and has become widely used, especially combined with the use of automated DNA sequencers with fluorescent detection (7). A comb-shaped streptavidin-coated manifold support offers the promise of further simplifying the technical performance of solid-phase sequencing reactions (8).

The major advantage of the avidin-coated polystyrene beads in solid-phase sequencing is their small size (0.8–1.0 µM) and consequently their high biotin-binding capacity. The small size of these beads results in reaction kinetics approaching kinetics of reactions in solution and in an extremely large binding surface (70 cm^2/mg of beads). The biotin-binding capacity of the polystyrene beads exceeds 2 nmol of biotinylated oligonucleotide per milligram of beads, when the corresponding capacities for the streptavidin-coated magnetic beads or standard microtiter plates are 300 pmol/mg beads and 2–5 pmol/well, respectively (9). An additional advantage of the polystyrene beads is their low price compared to that of magnetic ones. The drawback of these beads is the requirement of separation by centrifugation after the washing and denaturation steps.

2. Materials

2.1. Equipment and Reagents

All the reagents should be of standard molecular biology purity grade. Use sterile distilled or deionized water.

1. Facilities and reagents for PCR.
2. One of the PCR primers is biotinylated at its 5' end during the oligonucleotide synthesis (biotin-phosphoramidite reagent, e.g., Amersham Pharmacia Biotech or Perkin Elmer/ABI (Buckinghamshire, UK; Foster City, CA), and the other one is not biotinylated, which results in a PCR product with one biotinylated strand (*see* **Note 1**). In addition, an unbiotinylated sequencing primer is needed (*see* **Note 2**).
3. Avidin-coated polystyrene beads (Fluoricon®, avidin-polystyrene assay particles, 5% w/v, 0.99 µm, IDEXX Corp., Portland, ME). Store at 4°C (*see* **Note 3**).
4. Phosphate-buffered saline (PBS)/Tween buffer: 20 mM sodium phosphate buffer, pH 7.5, 0.1% Tween 20.
5. TENT buffer: 40 mM Tris-HCl, pH 8.8, 0.1% Tween 20, 1 mM ethylenediaminetetraacetic acid (EDTA), 50 mM NaCl. Store at 4°C.
6. NaOH/Tween mixture: 50 mM NaOH, 0.1% Tween 20 (make fresh every 4 wk). Store at room temperature (about 20°C) (*see* **Note 4**).
7. Sequencing reagents. Any standard protocol for sequencing of single-stranded DNA templates, either with a [^{35}S] dNTP as a label or with a [^{32}P]- or fluorescently labeled primer, can be used. We use the following reagents and stock solutions for sequencing with [^{35}S] dATP as a label (store all the reagents at –20°C):
   a. 5X T7 DNA polymerase buffer: 0.2 M Tris-HCl, pH 7.5, 0.1 M $MgCl_2$ and 25 mM NaCl.
   b. 0.1 M dithiotreitol (DTT).

Affinity-Capture and Solid-Phase Sequencing

    c. Labeling mixture: 7.5 μM dGTP, dTTP, dCTP.

    d. [^{35}S] dATP (Amersham Pharmacia Biotech, Amersham, UK).
T7 DNA polymerase (Amersham Pharmacia Biotech, Piscataway, NJ).

    e. Enzyme dilution buffer: 20 mM Tris-HCl, pH 7.5, 5 mM DTT, 100 μg/mL bovine serum albumin and 5% glycerol.

    f. Four ddNTP termination mixtures: 80 μM dATP, dTTP, dCTP, dGTP, containing 8 μM of one ddNTP, in 50 mM NaCl.

    g. Formamide dye: 95% formamide (v/v), 20 mM EDTA, 0.05% bromphenol blue (w/v), 0.05% xylene cyanol (w/v).

8. Facilities for polyacrylamide gel electrophoresis (PAGE).

3. Method

3.1. PCR for Solid-Phase Sequencing

PCR is done according to standard protocols, with one biotinylated primer and the other one not biotinylated. We use 100 μL reaction volumes to obtain a sufficient amount of the PCR product. We have successfully sequenced products shorter than 2 kb (*see* **Note 5**). The annealing temperature and the amount of the template added should be optimized to result in a specific amplification product, i.e., one clearly visible band when 1/10 of the PCR product is analyzed by agarose gel electrophoresis stained with ethidium bromide. If the PCR is specific and efficient, no further purification of the PCR product is needed (*see* **Note 6**).

3.2. Affinity Capture and Direct Sequencing of the PCR Product

1. Mix the following components:
   10 μL of washed avidin-coated polystyrene beads
   25–60 μL PCR product (*see* **Note 7**)
   TENT buffer to a total volume of 100 μL
2. Capture the biotinylated PCR product on the avidin-coated beads for 30 min at 37°C.
3. Centrifuge down the beads carrying the captured PCR product at 13,000g in a microfuge for 2 min, and remove the supernatant.
4. Denature the captured PCR product by adding 100 μL of the NaOH/Tween mixture, mix thoroughly, and incubate at room temperature for 5 min. Centrifuge as in **step 3** and remove the supernatant. At this stage, the biotinylated strand of the PCR product remains bound to the beads.
5. Wash the beads twice by adding 200 μL of the TENT buffer, mix well, centrifuge as in **step 3**, and remove carefully the supernatant (*see* **Note 8**).

The template is now ready for the sequencing reactions. Any protocol for sequencing single-stranded DNA templates can be applied (*see* **Note 9**). We use the following protocol:

6. Suspend the bead pellet carrying the captured template DNA with 7.5 μL of dH$_2$O, and add 0.5 μL (10 pmol) of the sequencing primer, and 2 μL of 5X T7 DNA polymerase buffer. Mix and let the primer anneal at 37–42°C for 30 min (*see* **Note 10**).
7. It is convenient to prepare the dilutions for the following steps during the annealing **step 6**. Dilute the labeling mixture 1:4 with dH$_2$O. Dilute the T7 DNA polymerase with the enzyme dilution buffer to a concentration of 1.25 U/μL. Keep the dilutions on ice until used. Label Eppendorf tubes for each termination reaction (C, A, T, G), and pipet 2.5 μL of the relevant ddNTP termination mixture to each tube.

8. Mix 1 µL of 0.1 M DTT, 2 µL of diluted labeling mixture, and 0.5 µL of [^{35}S] dATP; add this mixture to the tube containing the template and the annealed primer. Add 2 µL of diluted T7 DNA polymerase and incubate the mixture for 4 min at room temperature (*see* **Note 11**).
9. Preheat the termination mixtures to 37°C for 1 min.
10. Transfer 3.5 µL of the sequencing reaction mixture to each preheated termination mixture and incubate at 37°C for 3–4 min.
11. Stop the sequencing reactions by adding 4 µL of the formamide dye into each sample at room temperature, and store the samples at –20°C until use. The samples can be stored at least 1–2 wk.
12. Denature the samples at 80°C for 3 min and analyze 4 µL of each sample by denaturing 5–6% PAGE (*see* **Note 12**).

4. Notes

1. The efficiency of the 5'-biotinylation of an oligonucleotide on a DNA synthesizer is most often 80–90%. The biotin-labeled oligonucleotides can be purified from the unbiotinylated ones either by high-performance liquid chromatography *(10)*, PAGE *(11)*, or by ion exchange columns manufactured for this purpose (Perkin-Elmer/ABI). If the biotin-labeled primer is not purified, the biotinylation should be confirmed after the PCR by affinity capture of the biotinylated PCR product, followed by detection of possible unbound products by agarose gel electrophoresis as follows: capture 10 µL of the PCR product (or 1/10 of the reaction volume) on the avidin-coated beads in a total volume of 30 µL, as described in this chapter, **Subheading 3.2.** Electrophorese the supernatant, containing the possible unbiotinylated PCR product, adjacent to 1/10 of the original PCR product, in an agarose gel containing ethidium bromide. If the biotinylation has been efficient, no product, or a faint product of significantly lower intensity than the unbound PCR product, is observed in the supernatant.
2. The sequencing primer can be the unbiotinylated PCR primer or a nested primer. The use of a nested sequencing primer ensures that the possible unspecific PCR products do not serve as templates in the sequencing reaction.
3. Wash the avidin-coated beads before use to eliminate the possible detached avidin as follows: mix the required amount of beads with 1 mL of the PBS/Tween buffer. Vortex well and centrifuge at 13,000*g* for 2 min. Discard the supernatant and resuspend the pellet to the original volume with PBS/Tween buffer.
4. The avidin-coated beads are easy to handle when the solutions contain a detergent (e.g., 0.1% Tween 20) and the concentration of NaCl is not above 150 m*M*.
5. If the PCR product is long, the 3'-end of the captured single-stranded template may hybridize with homologous sequences within the template and serve as a sequencing primer. In our hands this has not happened with PCR products shorter than 2 kb.
6. If the PCR product is to be purified from an agarose gel, purification protocols including steps of phenol extraction are not suitable for biotin-labeled PCR products. We have successfully used commercial glass-absorption purification kits for PCR products (e.g., Wizard PCR Preps, Promega Corp., Madison, WI).
7. The accurate molar amount of template DNA is not crucial in this protocol, but about 1 pmol of the template DNA is optimal.
8. The removal of all the supernatant is crucial to avoid an increase of the reaction volume in the following sequencing procedure.
9. When sequencing protocols other than the one presented here are applied, it should be noted that the volume of the template DNA captured on the beads is zero, as the particles are nonporous.

10. The protocol presented here can equally well be used with a [^{32}P]-labeled sequencing primer, or with fluorescent labeling. Either a sequencing primer that has been fluorescently labeled at its 5' end during the oligonucleotide synthesis or fluorescent dATP (Amersham Pharmacia Biotech) can be used, when the products of the sequencing reactions are analyzed using an automated DNA sequencer.
11. Sequence close to the primer (15–20 bp) is obtained by adding 1 µL of 0.1 M MnCl$_2$, in 0.15 M sodium isocitrate, into the mixture. The labeling mixture can also be diluted more, e.g., 1:8, and the labeling time can be reduced to 1–2 min to obtain sequence close to the primer.
12. Note that the beads can be loaded on the sequencing gel along with the sample.

References

1. Casanova, J.-L., Pannetier, C., Jaulin, C., and Kourilsky, P. (1990) Optimal conditions for directly sequencing double-stranded PCR products with Sequenase. *Nucleic Acids Res.* **18,** 4028.
2. Gyllensten, U. B. and Erlich, H. (1988) Generation of single stranded DNA by the polymerase chain reaction and its application to direct sequencing of the HLA-DQ alpha-locus. *Proc. Natl. Acad. Sci. USA* **85,** 7652–7656.
3. Stoflet, E. S., Koeberl, D. D., Sarkar, G., and Sommer, S. S. (1988) Genomic amplification with transcript sequencing. *Science* **239,** 491–494.
4. Syvänen, A.-C., Bengtström, M., Tenhunen, J., and Söderlund, H. (1988) Quantification of polymerase chain reaction products by affinity-based hybrid collection. *Nucleic Acids Res.* **16,** 11,327–11,338.
5. Syvänen, A.-C., Aalto-Setälä, K., Kontula, K., and Söderlund, H. (1989) Direct sequencing of affinity-captured amplified human DNA: application to the detection of apolipoprotein E polymorphism. *FEBS Lett.* **258,** 71–74.
6. Hultman, T., Ståhl, S. Hornes, E., and Uhlén, M. (1989) Direct solid phase sequencing of genomic and plasmid DNA using magnetic beads as solid support. *Nucleic Acids Res.* **17,** 4937–4946.
7. Hultman, T., Bergh, S., Moks, T., and Uhlén, M. (1991) Bidirectional solid-phase sequencing of in vitro amplified plasmid DNA. *BioTechniques* **10,** 84–93.
8. Lagerkvist, A., Stewart, J., Lagerström-Fermér, M., and Landegren, U. (1994). Manifold sequencing: efficient processing of large sets of sequencing reactions. *Proc. Natl. Acad. Sci. USA* **91,** 2245–2249.
9. Syvänen, A.-C. and Söderlund, H. (1993) Quantification of polymerase chain reaction products by affinity-based collection. *Methods Enzymol.* **218,** 474–490.
10. Bengtström M., Jungell-Nortamo, A., and Syvänen, A.-C. (1990) Biotinylation of oligonucleotides using a water soluble biotin ester. *Nucleosides Nucleotides* **9,** 123–127.
11. Wu, R., Wu, N.-H., Hanna, Z., Georges, F., and Narang, S. (1984) In *Oligonucleotide Synthesis: A Practical Approach* (Gait, M. J., ed.), IRL Press, Oxford, UK, p. 135.

71

DNA Sequencing by the Chemical Method

Eran Pichersky

1. Introduction

The chemical method of sequencing DNA *(1)* has some advantages and some disadvantages compared with the enzymatic method *(2)*. The major disadvantage is that it takes more time to produce the same amount of sequence. This is so for two main reasons. First, the DNA has to be end-labeled and then reisolated prior to the actual chemical sequencing reactions, a process that usually requires an additional day. Also, because more DNA is used in the reaction, and because the lower specific activity of the sequenced DNA requires the use of an intensifying screen in the autoradiography, bands are not as sharp as in the enzymatic method, and therefore it is difficult to obtain reliable sequence past about nucleotide 250 (unless very long gels are run).

Nevertheless, the chemical method is often useful for several reasons. It enables one to begin sequencing anywhere in the clone where a restriction site that can be labeled occurs without any further subcloning. The sequence thus obtained can then be used to synthesize oligonucleotide primers for enzymatic sequencing. In addition, in cases of regions that give poor results in the enzymatic reactions (because of secondary structures that inhibit the polymerase enzyme), the chemical method almost always resolves the problem and yields the correct sequence.

The chemical sequencing reaction has acquired a reputation of being difficult. We believe this reputation in undeserved. In our hands, the chemical method is consistently successful in producing results as reliable as those obtained by the enzymatic method. It has been our experience that many protocols in molecular biology include unnecessary steps. The likely explanation is probably that when researchers encountered difficulties, they added these steps as a solution to the problem, often on the assumption that the additional steps would not hamper the process, even if they did not help. This is clearly not the case here. In developing the method presented here from preexisting protocols, we have eliminated many steps. In general, we have found that the quality of the sequence has improved with the progressive elimination of these steps. It is still possible that some steps included here are not necessary; certainly no additional steps need to be added. And, of course, the end result has been that the protocol as presented here is very short and the entire process of sequencing (starting with end-labeled DNA) and gel electrophoresis can be accomplished in one (long) day.

2. Materials

1. G Buffer: 50 mM Sodium cacodylate, pH 8.0.
2. CT/C Stop: 0.3 M Sodium acetate, 1 mM ethylenediaminetetraacetic acid (EDTA), pH 7. 0.
3. GA Stop: 0.3 M Sodium acetate, pH 7.0.
4. G Stop: 1.5 M Sodium acetate, 1 M 2-mercaptoethanol, pH 7.0.
5. 10% Formic acid.
6. Dimethyl sulfate (DMS) (*see* **Notes 1** and **4**).
7. 100% Ethanol.
8. Hydrazine (95% anhydrous) (*see* **Notes 1** and **4**).
9. 5 M NaCl.
10. ddH$_2$O (dd = double distilled).
11. Carrier DNA: 1 mg/mL in ddH$_2$O (any DNA will do; we use plasmid DNA).
12. 10 mg/mL tRNA in ddH$_2$O (any tRNA).
13. The DNA fragment to be sequenced, end-labeled at one end only, in ddH$_2$O.
14. 10% Piperidine (dilution prepared on the day of the experiment).
15. Loading buffer: 100% Formamide, 0.1% (w/v) bromophenol blue, 0.1% (w/v) xylene cyanol FF.

3. Method

1. For each DNA fragment to be sequenced, mark four 1.5-mL Eppendorf reaction tubes and add the following solutions:
   G tube: 1 µL carrier DNA, 200 µL G buffer, 5 µL labeled DNA.
   GA tube: 1 µL carrier DNA, 10 µL ddH$_2$O, 10 µL labeled DNA.
   CT tube: 1 µL carrier DNA, 10 µL ddH$_2$O, 10 µL labeled DNA.
   C tube: 1 µL carrier DNA, 15 µL 5 M NaCl, 5 µL labeled DNA.
2. Mark up four 1.5 mL Eppendorf stop tubes and add the following solutions:
   G stop tube: 2 µL tRNA, 50 µL G stop solution, 1 µL ethanol.
   AG stop tube: 2 µL tRNA, 200 µL AG stop solution, 1 µL ethanol.
   CT stop tube: 2 µL tRNA, 200 µL CT/C stop solution, 1 µL ethanol.
   C stop tube: 2 µL tRNA, 200 µL CT/C stop solution, 1 µL ethanol.
3. To start the reactions (*see* **Note 2**), add the following:
   G tube: 1 µL DMS, mix, and let the reaction proceed for 5 min at room temperature (*see* **Notes 1** and **4**).
   AG tube: 3 µL 10% formic acid and mix (15 min at 37°C) (*see* **Note 4**).
   CT tube: 30 µL hydrazine and mix (9 min at room temperature) (*see* **Notes 1** and **4**).
   C tube: 30 µL hydrazine and mix (11 min at room temperature) (*see* **Notes 1** and **4**).
4. Stop each reaction by pipeting the contents of the corresponding stop tube into the reaction tube (use the same Pasteur pipet; a slight cross contamination of stop solutions has no effect, but do not touch the contents of the reaction solutions with the pipet). Cap the reaction tubes, shake briefly but vigorously, and place in a dry ice-ethanol bath (–80°C) for 3–10 min (3 min are enough, but the tubes can be left there for up to 10 min if other reactions are not done yet; do not leave for longer than 10 min) (*see* **Note 3**).
5. Centrifuge at 4°C for 7 min, discard the supernatant, aspirate the rest of the liquid with a drawn Pasteur pipet, and then add 1 mL of 100% ethanol to the tube, invert twice, and centrifuge for 2 min at room temperature. Aspirate as before, and dry in a vacuum for 10 min.
6. To each reaction tube, add 100 µL of the 10% piperidine solution (do not shake the tubes as there is no need to resuspend the DNA) and place the uncapped tubes in a 90°C heat block. After 15–30 s, cap the tubes and let stand for 30 min.

7. Remove the tubes from the heat block, let stand at room temperature for 2–5 min; centrifuge briefly to get the condensation to the bottom. Puncture one hole in the cap with a syringe, then place in dry ice-ethanol bath for 5 min.
8. Place the tubes in a vacuum centrifuge and lyophilize for 2 h. The vacuum should be below 100 mT.
9. Prior to gel electrophoresis, add 10 µL of loading buffer to each tube, resuspend the sample by shaking and then a brief centrifugation, and place the tubes in the 90°C heat block for 10 min. Load 1–2 µL per sample (*see* **Note 5**).

4. Notes

1. Quality of chemicals: In general, standard laboratory-grade chemicals should be used. Some chemicals, however, could be the cause of problems when not sufficiently pure or when too old (presumably degradation products are the culprits). We have only had problems with two chemicals (*see* **step 3**): dimethyl sulfate and hydrazine. Note also that these two chemicals, together with piperidine, are hazardous chemicals. In addition to observing the rules pertaining to the handling of radioactive chemicals, all reactions involving these three chemicals should be carried out in a fume hood (**Subheading 3.**).
2. Reaction times: We typically do all chemical reactions together, timing them so that they end at the same time. If one is sequencing five different fragments, all 20 tubes can be spun together in a single run (we have a microcentrifuge with 20 slots). When stopping all 20 reactions at about the same time, some reactions are invariably going to run longer than the allotted time. This is usually not a problem because the reaction times indicated are general, and they can be extended by up to 30% without much noticeable effect.
3. Precipitation: We always use 100% ethanol. There is no need to use any other concentration of ethanol at any step of the process, and 100% ethanol has the advantage because it evaporates fast. The goal is to get the DNA to precipitate with as little salt coprecipitation as possible. This is accomplished by aspirating all the liquid after the ethanol precipitation step, and again after the ethanol wash step. The pellet forms nicely on the side of the tube, and it is easy to put the end of the stretched Pasteur pipet all the way to the bottom of the tube and aspirate all the liquid. Repeated cycles of resuspension and precipitations are inadvisable. Prolonged incubation in the dry ice-ethanol bath is also strongly discouraged. DNA precipitates well at room temperature, but one gets more salt precipitation at low temperature, thus making things worse, not better. We almost always precipitate DNA at room temperature; the only reason **step 4** calls for incubation at –80°C is to inhibit further reaction with the reactive reagents that at this stage have not yet been removed.
4. Troubleshooting:
   a. G reaction: This reaction is usually very clean, but it is the reaction most sensitive to prolonged incubation and to the quality of the reactive reagent, DMS. If reaction proceeds longer than the allotted time or if old or bad-quality dimethyl sulfate is used, excessive and nonspecific degradation of DNA will occur. Also, when several Gs occur in a row, the 3'-most Gs (lower bands if the 3' end was labeled with the Klenow enzyme) may appear weaker.
   b. AG reaction: This is usually a trouble-free reaction.
   c. CT and C reactions: The quality of the hydrazine should be good (it does not have to be exceptional), otherwise excessive and nonspecific degradation will occur. Sometimes faint bands will be seen in the C and CT lanes when the base is G (a strong band is then observed in the G lane). The likely explanation is that the pH in the reaction tubes is too low (there is no buffer in the C and CT reaction tubes, but carryover with the DNA sample might cause this to happen). However, these faint bands are not nearly as strong

as the signal in the G lane or as bona fide bands of C and T bases. Also, the T bands in the CT lane are often not as strong as the C bands. This is probably caused by inhibition of the reaction by residual salt (in the C reaction, salt is added specifically to obtain complete inhibition).

5. Gel electrophoresis: We use a 60 × 40 cm (0.3-mm thick) gel of 6% acrylamide (20:1 acrylamide:bisacrylamide, 50% urea, 50 mM Tris, 50 mM borate, 1 mM EDTA *[3]*). We run the gel at constant power (65 W), with an aluminium plate to disperse the heat. The samples are loaded twice (a "long run" and a "short run"): the second loading is done when the xylene cyanol dye of the first sample is approx two-thirds of the way down the gel, then electrophoresis is halted when the bromophenol blue of the second sample reaches the bottom of the gel. The complete run takes 5–6 h, and it allows us to read, in the short run, the sequence from about nucleotide 25 to nucleotide 120–150, and in the long run the sequence from nucleotide 100 to about 220–250. Additional sequences may be obtained by running longer gels, by loading the sample a third time, or by a variety of other methods if so desired.

References

1. Maxam, A. M. and Gilbert, W. (1980) Sequencing end-labeled DNA with base specific chemical cleavages. *Methods Enzymol.* **65,** 499–560.
2. Sanger, F., Nicklen, S., and Coulson, A. R. (1977) DNA sequencing with chain terminating inhibitors. *Proc. Natl. Acad. Sci. USA* **74,** 5463–5467.
3. Maniatis, T., Fritsch, E. F., and Sambrook, J. (eds.) (1982) *Molecular Cloning, A Laboratory Manual.* Cold Spring Harbor Laboratory Press, Cold Spring Harbor, NY.

72

One-Step One-Lane Chemical Sequencing of DNA

Giovanna Costanzo, Ernesto Di Mauro, and Rodolfo Negri

1. Introduction

There are several chemical methods for sequencing DNA. The classical method, introduced in 1977 by Maxam and Gilbert *(1)*, uses four different chemical reactions to attack, modify, and open the heterocyclic bases at guanine (G), adenine and guanine (A + G), thymine and cytosine (T + C), and cytosine (C) in a radioactive end-labeled DNA fragment. The reactions result also in breakage of the glycosidic bond. A second reaction with hot piperidine is required to cleave the DNA backbone at these positions. The reaction products are thus analyzed by electrophoresis in four different lanes of a polyacrylamide gel. Several alternatives to the basic protocol were developed, as reviewed in *(2)*.

In another approach, a single or a combination of two chemical reactions are used to cleave the end-labeled DNA fragments simultaneously at all four bases followed by analysis of the broken fragments in a single lane of the gel (reviewed in *3*). Under certain conditions, different cleavage rates for all four bases can be obtained, the nucleotide sequence being determined by measuring the band intensity in one single electrophoretic lane. Among these methods, those based on the reaction with formamide and its derivative *N*-methylformamide *(4–10)* have several advantages on the others:

1. A single reaction is required (in a template in which guanines are replaced by inosines) to produce four different base-specific signal intensities (i.e., I > A > C > T) *(6)*.
2. The reaction not only opens the bases and cleaves the glycosidic bonds but also cleaves the phosphodiester bonds so that no additional reactions are required.
3. The products of the reaction can be loaded directly in the gel without further handling.
4. No highly toxic or carcinogenic compounds are involved.
5. Most of the commercially available fluorochromes are stable in the reaction conditions, as expected from the fact that formamide is the chemical agent generally used for denaturing the samples.

We present here several possible applications of this sequencing reaction in genomic sequencing, genomic mutation screening, plasmid minipreps analysis, partial sequencing for generating marker lanes, and rapid checking of single-stranded oligonucleotides.

From: *The Nucleic Acid Protocols Handbook*
Edited by: R. Rapley © Humana Press Inc., Totowa, NJ

2. Materials

2.1. Equipment

1. ABI 373A DNA sequencer (Perkin Elmer/Applied Biosystem Division, Foster City, CA).
2. Dry Block (Techne Ltd., Cambridge, UK): with three aluminium blocks for 1.5 mL snap-cap tubes.
3. 2400 Termal Cycler (Perkin Elmer).
4. TFX-35LC UV Transilluminator, equipped with 6X 15 W–365 nm and 6X 15 W–254 nm lamps (Vilber-Lourmat, Marne La Vallée, France).

2.2. Reagents

1. Genomic DNA from 20 mL of human blood or 5 mL of late-log yeast culture prepared by genomic-tips (Qiagen, Santa Clarita, CA).
2. MicroAmp reaction tubes (0.2 mL) (Perkin Elmer).
3. 0.45 Micropure™ separators (Amicon, Beverly, MA).
4. *Taq* DNA polymerase (native) at a concentration of 5 U/µL (Perkin Elmer /Cetus, Norwalk, CT, USA).
5. Deoxynucleotides mix: 2.5 mM each of dATP, dGTP, dCTP, and dTTP.
6. Inosine-guanine deoxynucleotides mix: 2.5 mM each of dATP, dCTP, dTTP, 2 mM dITP, and 250 µM dGTP.
7. *Taq* DNA polymerase storage buffer: 100 mM KCl, 20 mM Tris-HCl, pH 8, 0.1 mM ethylenediaminetetraacetic acid (EDTA), 1 mM DDT, 0.5% Tween-20, 05% Nonidet P40, 50% (v/v) glycerol.
8. *Taq* 10X buffer: 500 mM KCl, 100 mM Tris-HCl, pH 9, 1% Triton X-100.
9. 6-FAM- and TAMRA-labeled oligonucleotides custom made in solid phase according to *(11)* (M-Medical, Florence, Italy), using 6-FAM amidite and TAMRA amidite (Perkin Elmer) stored in aliquots (100 pmol/µL) at –20°C in the dark.
10. 3 M Sodium acetate, pH 5.2.
11. Ethanol: Absolute and 70% (v/v).
12. 10X TBE: 108 g Tris base, 55 g boric acid, and 9.3 g Na$_2$EDTA dissolved and made up to 1 L in deionized water.
13. Formamide buffer: 80% Ultrapure, deionized formamide (Fluka, Buchs, Switzerland), 10 mM EDTA, 0.05 % bromophenol blue, and 0.05 % xylene cyanol.
14. Loading buffer: 80% Ultrapure, deionized formamide (Fluka), 10 mM EDTA.
15. Preparative denaturing gel: 20 × 30 cm × 1 mm thick, containing 7 M urea, 1X TBE, and 6% polyacrylamide at a 19:1 ratio of acrylamide to *bis*-acrylamide.
16. 6% Sequencing gel for automatic sequencer prepared as recommended in *(12)*.
17. Gel-drying film (V713A) (Promega, Madison, WI).
18. Thin-layer chromatography (TLC) plates silica gel 60 precoated with F$_{254}$ (Merck, Darmstadt, Germany).
19. DNA standard curve: A 500-bp restriction fragment quantified by OD$_{260}$ and appropriately diluted to produce 10 ng, 30 ng, 100 ng, 300 ng, and 1 µg aliquots ready to be loaded on agarose minigels.
20. 6-FAM standard curve: 0.3, 1, 3, 10, and 30 pmol aliquots of 6-FAM-labeled oligonucleotide in formamide buffer (30 µL).
21. TAMRA standard curve: 3, 10, and 30 pmol aliquots of TAMRA-labeled oligonucleotide in formamide buffer (30 µL).
22. Acrylamide elution buffer: 0.3 M sodium acetate pH 7.5, 1 mM EDTA.

23. *N*-Methylformamide solution: Made fresh before use by adding 3 µL of 1 *M* $MnCl_2$ to 1 mL of *N*-methylformamide (Fluka).
24. DNA restriction fragment labeled at one 3' end by fill-in with Klenow DNA polymerase and (α^{32}P) dNTP.
25. Single-stranded oligonucleotide labeled at the 5' end by polynucleotide kinase and (γ^{32}P) ATP purified on a preparative denaturing gel.

3. Methods

3.1. One-Step One-Lane Genomic Sequencing

3.1.1. Preparation of PCR Templates from Genomic DNA

The following protocol has been used to isolate and purify PCR products in the 0.3–1 kb range from human or yeast genomic DNA prior to one-lane chemical sequencing.

1. Amplify 20–50 µg of genomic DNA from human blood or 5–10 µg of yeast genomic DNA by 30 polymerase chain reaction (PCR) cycles with 1 nmol of a pair of oligonucleotides primers sitting at a distance of 0.3–1 kb from one another in a final volume of 20–50 µL, containing 1/10 vol of deoxynucleotide mix, 1/10 vol of 10X *Taq* buffer, and an appropriate concentration of $MgCl_2$. The annealing temperature should be optimized for the specific oligonucleotides pair.
2. Nonspecific DNA fragments, primers, and nucleotides are removed from genomic PCR products on a 1% agarose minigel in 1X TBE and ethidium bromide (also load in the gel: a DNA standard curve for an approximate evaluation of the quantity of the DNA produced, size markers).
3. Excise the appropriate PCR DNA band from the gel with a scalpel using a UV transilluminator at 254 nm, minimizing the time of UV exposure as much as possible.
4. Place the band-containing slice inside a 0.45-micropure separator at –20°C for 15 min.
5. Spin at 13,000*g* for 15 min in an Eppendorf centrifuge. Remove the separator and discard it, recover eluted DNA in the filtered vial.
6. Add 1/10 vol of 3 *M* sodium acetate pH 5.2 and 2 vol of absolute ethanol. Invert several times to mix and precipitate the DNA 15 min at –20°C.
7. Pellet the DNA by centrifugation at 13,000*g* for 15 min. Discard the supernatant and rinse the pellet with 70% ethanol.
8. Dry the pellet briefly under vacuum and dissolve the DNA with deionized water at the approximate concentration of 50–100 ng/µL, assuming a 60% final DNA recovery.

3.1.2. Fluorescent End-Labeling and Amplification of DNA by PCR

This procedure is necessary for two reasons: end-labeling of the products and partial replacement of guanines with inosines in the DNA substrate to be sequenced. *N*-Methylformamide attacks inosine more efficiently than adenine and guanine (*6*).

1. Amplify 50–100 ng of PCR template from genomic DNA (15 cycles) with 100 pmol of a pair of oligonucleotide primers, one of which is labeled with a TAMRA or a 6-FAM fluorochrome (for the choice of the oligonucleotide positions, *see* **Note 1**). The amplification is performed in 10 µL containing 1 µL of 10X *Taq* buffer and 1 µL of inosine-guanine deoxynucleotides mix (*see* **Note 2**) and $MgCl_2$ as appropriate. Annealing temperature should be optimized for the specific oligonucleotide primers pair. From this step on, sample exposure to direct light should be limited as much as possible.
2. Add 20 µL of formamide buffer, heat at 92°C for 2 min, and load on a preparative denaturing gel together with 6-FAM and/or TAMRA standard curves (*see* **Note 4**).

3. Run the gel until the bromophenol blue is 3/4 of the length.
4. Transfer the gel on a gel-drying film and place it on an UV transilluminator.
5. Locate full-length fluorescent product under 365 nm illumination. Evaluate approximate quantity by comparison with standard curves (in our hands 0.3 pmol of 6-FAM-labeled product and 3 pmol of TAMRA-labeled product are clearly visible).
6. Excise the band with a scalpel and place it in a 2-mL snap-cap Eppendorf tube. Add 500 µL of acrylamide elution buffer and elute for 4 h at 37°C.
7. Recover the liquid and add 2 vol of absolute ethanol. Mix and precipitate DNA 15 min at −20°C.
8. Spin at 13,000g for 15 min in an Eppendorf centrifuge, discard the supernatant, rinse the pellet with 70% ethanol, and dry briefly under vacuum.
9. Resuspend the eluted DNA in deionized water at an approximate concentration of 50–100 ng/µL, assuming a 60% final DNA recovery for a fragment length lower than 500 nucleotides and 30% for longer fragments. Typical yields are 5–30 pmol.

3.1.3. One-Step One-Lane Sequencing of Amplified End-Labeled Products

Products obtained according to **Subheading 3.1.2.** are reacted with *N*-methylformamide and loaded in an automatic sequencer for one-lane sequence analysis.

1. Take 0.5 pmol (1 µL) of 6-FAM-labeled product and add deionized water to 5 µL and 45 µL of *N*-methylformamide solution. Place the sample in a 0.2-mL microamp tube and heat at 99°C for 90 min in a thermal cycler with heated cover. Cool to 4°C.
2. Spin the tube for 10 s in an Eppendorf centrifuge. Add 1/10 vol (5 µL) of sodium acetate (pH 5.2) and 2 vol (100 µL) of absolute ethanol, mix, and precipitate the DNA at room temperature for 15 min. Pellet the DNA at 13,000g in an Eppendorf centrifuge. Discard supernatant, rinse the pellet with 70% ethanol, and dry briefly under vacuum (*see* **Note 5**).
3. Resuspend in 5 µL of loading buffer, heat at 92°C for 2 min and load in a 6% sequencing gel of a 373A automated sequencer. Raw data are plotted in one channel: select **C** for 6-FAM-labeled samples and **G** for TAMRA-labeled sample as **bases** in the sample sheet. With the present software, automatic base calling is not possible and the sequence must be manually determined by comparison the peak's heights (*see* **Figs. 1–3**). Typically, 200–300 nucleotides can be read per gel lane with an average error rate close to 1%.

3.1.4. Comparison of Samples and Screening for Genomic Mutations

One advantage of this sequencing system is that it allows the analysis of more than one sample in the same lane, provided that each sample is labeled with a different fluorochrome (**Fig. 3**). Owing to carryover effects among the fluorochromes used in the 373A sequencer, we have so far worked out only a 2-fluorochromes protocol with two samples per lane, labeled with 6-FAM or TAMRA. The emissions of these two fluorochromes do not mutually interfere enough to perturb the analysis.

1. Perform two separate sequencing reactions for two samples labeled respectively with 6-FAM and with TAMRA. Follow the protocol of **Subheading 3.1.3.** to step 2 with the following modification: start with 1 pmol of the 6-FAM-labeled product and 3 pmol of the TAMRA-labeled product.
2. Mix 2.5 µL each of the two reacted samples resuspended in 5 µL of loading buffer, heat at 92°C for 2 min, and load in a single well. Raw data are plotted in two channels selecting **G** and **C** as **bases** in the sample sheet.

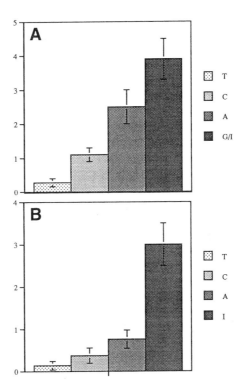

Fig. 1. Average peak intensity (arbitrary units) for the bases after reaction with N-methylformamide. Panel A: Average peak intensity for a template amplified with the inosine-guanine deoxynucleotides mix (see **Subheading 2.**). Panel B: Average peak intensity for a template amplified with a deoxynucleotides mix containing dITP but no dGTP. Data were obtained on 500 sequenced bases, standard deviation is indicated.

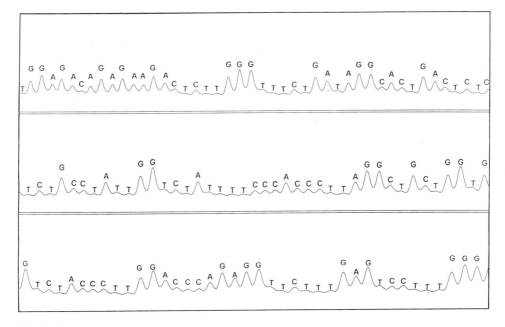

Fig. 2. One-step one-lane genomic sequencing. Portion of the raw data obtained for genomic sequencing of the human β-globin gene (see **Subheading 3.1.3.**).

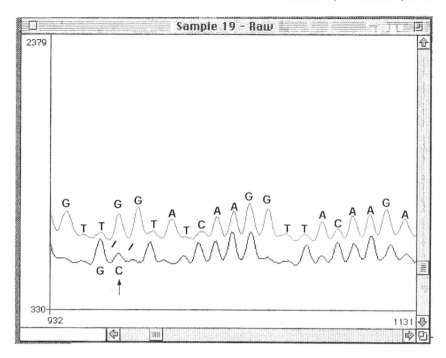

Fig. 3. Screening for mutations. Portion of the raw data obtained from single-lane analysis of two different samples (*see* **Subheading 3.1.4.**). The arrow points at one genomic homozygous G-C mutation. The different migration due to the different fluorochromes used is evident.

3.1.5. Reading Both Strands of a Sample in the Same Lane

1. Follow the protocol in **Subheading 3.1.2.** with the following modification: at step 1 use 100 pmol of a pair of oligonucleotide primers one of which is labeled with 6-FAM and the other with TAMRA.
2. Follow the protocol in **Subheading 3.1.3.** until step 4. Modify step 2 by having raw data plotted in two channels (selecting **G** and **C** as **bases** in the sample sheet).

3.2. Screening of Plasmid Minipreps by One-Lane Chemical Sequencing

This protocol can be used to select positive clones in a random mutagenesis assay *(5,13)*. The DNA template is prepared using the boiling method plasmid miniprep as described in *(14)*. Starting from a single colony, follow this protocol to **step 10** *(14)*.

1. Resuspend the pellet in 10 µL deionized water.
2. Take 2 µL of DNA and use it as a template for fluorescent end-labeling and amplification of DNA by PCR (*see* **Subheading 3.1.2.**) using an appropriate (*see* **Note 6**) pair of primers, one of which is labeled with a TAMRA or a 6-FAM fluorochrome.
3. Follow the protocols in **Subheadings 3.1.2.** and **3.1.4.** consecutively, loading two sequencing products in each lane (48 samples can be analyzed on the same gel with a 24-tooth comb).

3.3. Partial Sequencing of Radioactively End-Labeled DNA to Generate A + G > C Markers or for Fast Checking of DNA Fragments Sequence

This protocol allows simple and fast production of a partial sequence ladder from terminally labeled DNA. Because no amplification step is involved, it is not possible to unambiguously distinguish A_s from G_s because G_s has not been replaced by I_s.

3.3.1. Partial Sequencing (A + G > C) of 3' End-Labeled Fragments (**Fig. 4**)

1. 100,000 cpm of a 3' end-labeled DNA restriction fragment (1–50 ng of DNA) are reacted in 10 µL of formamide buffer for 10 min (fragments between 200 and 1000 bp long) or for 20 min (40–200 bp) at 110°C in a dry heating block.
2. Directly load 5 µL on a 6% sequencing gel.

3.3.2. Partial Sequencing (A + G > C) of 5' End-Labeled Fragments

1. 100,000 cpm of a 5' end-labeled DNA restriction fragment or 300,000 cpm of a 5' end-labeled and gel-purified, single-stranded oligonucleotide less than 40 bases long (1–50 ng of DNA) are reacted in a 0.2-mL microamp tube with 10 µL of 80% *N*-methylformamide solution for 90 min at 99.9°C; cool to 4°C.
2. Add 1/10 vol sodium acetate (pH 5.2) and 2 vol absolute ethanol, mix, and precipitate the DNA at –20°C for 15 min (*see* **Note 7**).
3. Spin at 13,000*g* for 15 min, discard the supernatant, rinse with 70% ethanol, and dry briefly under vacuum.
4. Resuspend in 5 µL formamide buffer, heat for 2 min at 92°C, and load on a sequencing gel.

4. Notes

1. The primers in this amplification can be the same used for the first genomic amplification or can be more internal. Care should be taken in choosing primers with a minimum content in C_s because the C-I pairing would lower the annealing temperature. With primers containing more than 2 C_s in 20 bases, touch-down PCR should be performed.
2. The inosine-guanine deoxynucleotides mix contains a deoxyribonucleotides triphosphate mixture of a 1:8 G to I ratio, therefore lowering the peak intensity corresponding to G positions relative to that obtained when I alone is incorporated (**Fig. 1**). Peaks of high intensity can introduce errors in sequence interpretation due to an overevaluation of adjacent pyrimidine peaks caused by shoulder effects.
3. Gel purification is required at this step in order to eliminate short molecules produced by premature termination or primer mispairing. As a less time consuming, but more expensive, alternative the terminal transferase treatment described in *(7)* can be used.
4. Although the full-length product is often the most represented band, a size marker at a concentration of at least 3 µg/band should be included whenever possible. This amount of single-stranded DNA can be easily identified as a darker area under short UV illumination (254 nm), placing the gel on top of a TLC silica gel plate coated with fluoresceine.
5. This precipitation step is included because *N*-methylformamide is not dense enough for a good sample layering in the sequencing gel. As an alternative, if a large amount of labeled sample is available, react 5 pmol 6-FAM- or 15 pmol TAMRA-labeled products with 10 µL of *N*-methylformamide solution, take 1 µL, dilute with 4 µL of formamide buffer, heat 2 min at 92°C, and load in the sequencing gel.
6. The labeled oligonucleotide primers should be selected at a distance of 20–200 bp from the putative mutation site.

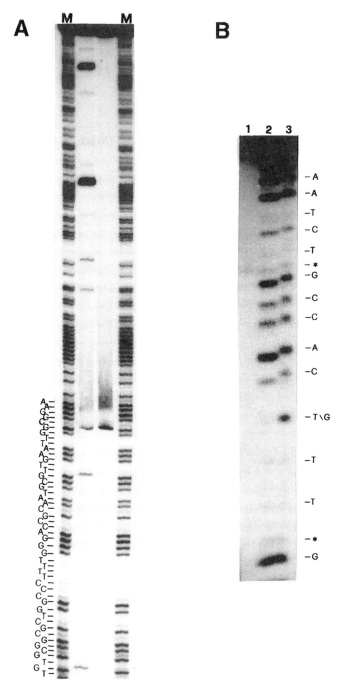

Fig. 4. Two examples of partial sequencing (A + G > C). Panel A: Partial sequencing of a 3' end-labeled fragment used as a marker lane, lane M (see **Subheading 3.3.1.**). Panel B: Partial sequencing (see **Subheading 3.3.2.**) of two 5' end-labeled 16-bases long oligonucleotides differing in one single indicated (T/G) position. Lane 1: untreated; lanes 2 and 3: *N*-methyl-formamide treated oligonucleotides. The asterisks point to bands with anomalous migration due to incomplete cleavage at G residues *(5)* occurring in a small fraction of the molecules and visible only in concentrated polyacrylamide gels.

7. For single-stranded oligonucleotides less than 40 bases long, add (instead of **steps 2–4**) 20 µL of formamide buffer, vortex, take 5 µL, heat at 92°C for 2 min, and load on an 18% sequencing gel.

References

1. Maxam, A. M. and Gilbert, W. (1977) A new method for sequencing DNA. *Proc. Natl. Acad. Sci. USA* **74,** 560–564.
2. Ambrose, B. J. B. and Pless, R. C. (1987) DNA sequencing: chemical methods. *Methods Enzymol.* **152,** 522–538.
3. Rosenthal, A. (1993) DNA sequencing by chemical degradation using one, two, and four different fluorophores. *Methods Mol. Biol.* **23,** 261–280.
4. Costanzo, G., Negri, R., Di Nicola-Negri, E., De Leo, R., and Di Mauro, E. (1997) Detection of human genomic mutations by chemical single-reaction DNA sequencing. *Trends Genet. Tech. Tips on Line* **T40030** (http://www. elsevier. nl/ locate.tto).
5. Negri, R., Costanzo, G., Saladino, R., and Di Mauro, E. (1996) One-step, one-lane chemical sequencing by N-methylformamide in the presence of metal ions. *BioTechniques* **21,** 910–917.
6. Saladino, R., Mincione, E., Crestini, C., Negri, R., Di Mauro, E., and Costanzo, G. (1996) Mechanism of degradation of purine nucleosides by formamide. Implications for chemical DNA sequencing. *J. Am. Chem. Soc.* **118,** 5615–5618.
7. Di Mauro, E., Costanzo, G., and Negri, R. (1994) One-lane chemical sequencing of PCR amplified DNA: the use of terminal transferase and of the base analogue inosine. *Nucleic Acids Res.* **22,** 3811–3812.
8. Negri, R., Ferraboli, S., Barlati, S., and Di Mauro, E. (1994) Chemical method for DNA sequence determination from the 5' extremity on PCR amplified fragments. *Nucleic Acids Res.* **22,** 111,112.
9. Ferraboli, S., Negri, R., Di Mauro, E., and Barlati S. (1993) One lane chemical sequencing of 3' -fluorescent labeled DNA. *Anal. Biochem.* **214,** 566–570.
10. Negri, R., Costanzo, G., and Di Mauro, E. (1990) A single reaction method for DNA sequence determination. *Anal. Biochem.* **197,** 389–395.
11. Hakura, K., Rossi, J. J., and Wallace, R. B. (1984) Synthesis and use of synthetic oligonucleotides. *Annu. Rev. Biochem.* **53,** 323–356.
12. Polyacrylamide gels and DNA sequencing (1991) *A. B. I. User Bull.* **16,** 2–12.
13. Benedetti, P., Fiorani, P., Capuani, L., and Wang, J. C. (1993) Camptotechin resistance from a single mutation changing glycine 363 of human DNA topoisomerase I to cystein. *Cancer Res.* **53,** 4343–4348.
14. Sambrook, J., Fritsch, E. F., and Maniatis, T. (eds.) (1989) *Molecular Cloning A Laboratory Manual*, Cold Spring Harbor Laboratory Press (2nd ed.), Cold Spring Harbor, NY.

VII

BASIC POLYMERASE CHAIN REACTION METHODS

73

Polymerase Chain Reaction

Basic Principles and Routine Practice

Lori A. Kolmodin and J. Fenton Williams

1. Introduction
1.1. PCR Definition

The polymerase chain reaction (PCR) is a primer-mediated enzymatic amplification of specifically cloned or genomic DNA sequences *(1)*. This PCR process, invented by Kary Mullis over 10 years ago, has been automated for routine use in laboratories worldwide. The template DNA contains the target sequence, which may be tens or tens of thousands of nucleotides in length. A thermostable DNA polymerase, *Taq* DNA polymerase, catalyzes the buffered reaction in which an excess of an oligonucleotide primer pair and four deoxynucleoside triphosphates (dNTPs) are used to make millions of copies of the target sequence. Although the purpose of the PCR process is to amplify template DNA, a reverse transcription step allows the starting point to be RNA (2–5).

1.2. Scope of PCR Applications

PCR is widely used in molecular biology and genetic disease research to identify new genes. Viral targets, such as HIV-1 and HCV can be identified and quantitated by PCR. Active gene products can be accurately quantitated using RNA-PCR. In such fields as anthropology and evolution, sequences of degraded ancient DNAs can be tracked after PCR amplification. With its exquisite sensitivity and high selectivity, PCR has been used for wartime human identification and validated in crime labs for mixed-sample forensic casework. In the realm of plant and animal breeding, PCR techniques are used to screen for traits and to evaluate living four-cell embryos. Environmental and food pathogens can be quickly identified and quantitated at high sensitivity in complex matrices with simple sample preparation techniques.

1.3. PCR Process

The PCR process requires a repetitive series of the three fundamental steps that defines one PCR cycle: double-stranded DNA template denaturation, annealing of two oligonucleotide primers to the single-stranded template, and enzymatic extension of the primers to produce copies that can serve as templates in subsequent cycles. The target copies are double-stranded and bounded by annealing sites of the incorporated primers. The 3' end of the primer should complement the target exactly, but the 5' end can actually be a noncomplementary tail with restriction enzyme and promoter sites that will also be incorporated. As the cycles proceed, both the original template and the amplified targets serve as substrates for the denaturation, primer annealing, and primer extension processes. Since every cycle theoretically doubles the amount of target copies, a geometric amplification occurs. Given an efficiency factor for each cycle, the amount of amplified target, Y, produced from an input copy number, X, after n cycles is

$$Y = X(1 + \text{efficiency})^n \qquad (1)$$

With this amplification power, 25 cycles could produce 33 million copies. Every extra 10 cycles produces 1024 more copies. Unfortunately, the process becomes self-limiting and amplification factors are generally between 10^5- and 10^9-fold. Excess primers and dNTPs help drive the reaction that commonly occurs in 10 mM Tris-HCl buffer (pH 8.3 at room temperature). In addition, 50 mM KCl is present to provide proper ionic strength and magnesium ion is required as an enzyme cofactor (*6*).

The denaturation step occurs rapidly at 94–96°C. Primer annealing depends on the T_m, or melting temperature, of the primer:template hybrids. Generally, one uses a predictive software program to compute the T_ms based on the primer's sequence, their matched concentrations, and the overall salt concentration. The best annealing temperature is determined by optimization. Extension occurs at 72°C for most templates. PCR can also easily occur with a two temperature cycle consisting of denaturation and annealing/extension.

1.4. Carryover Prevention

PCR has the potential sensitivity to amplify single molecules, so PCR products that can serve as templates for subsequent reactions must be kept isolated after amplification. Even tiny aerosols can contain thousands of copies of carried-over target molecules that can convert a true negative into a false positive. In general, dedicated pipeters, plugged pipet tips, and separate work areas should be designated for pre and post-PCR work. As with any high sensitivity technique, the judicious and frequent use of positive and negative controls is required (*7–9*). Through the use of dUTP instead of dTTP for all PCR samples, it is possible to design an internal biochemical mechanism to attack the PCR carryover problem. PCR products will then be dU-containing and can be cloned, sequenced, and analyzed as usual. Pretreatment of each PCR reaction with uracil N-glycosylase (UNG) will destroy any PCR product carried over from previous reactions, leaving the native T-containing sample ready for amplification (*10*).

1.5. Hot Start

PCR is conceptualized as a process that begins when thermal cycling ensues. The annealing temperature sets the specificity of the reaction, assuring that the primary primer binding events are the ones specific for the target in question. In preparing a PCR reaction on ice or at room temperature, however, the reactants are all present for nonspecific primer annealing to any single-stranded DNA present. Since *Taq* DNA polymerase has some residual activity even at lower temperatures, it is possible to extend these misprimed hybrids and begin the PCR process at the wrong sites! By withholding a key reaction component, such as *Taq* DNA polymerase, until an elevated temperature can be reached, the possibility of mispriming is avoided. This can be accomplished by a manual addition of enzyme above 65–70°C during the first heating ramp to denaturation at 94°C. Alternatively, an inactive form of the enzyme, AmpliTaq Gold®, can be added to all reactions to prevent misprimed extensions. Adding a pre-PCR heat step at 92–95°C for 9–12 min synchronously reactivates the inactive enzyme and achieves an "invisible" hot start. In both cases, the lowest temperature experienced by the reaction components is the stringent primer annealing temperature, assuring best specificity *(11,12)*.

1.6. PCR Achievements

PCR has been used to speed the gene discovery process and for early detection of viral diseases. Single sperm cells to measure crossover frequencies can be analyzed and four-cell cow embryos can be typed. Trace forensic evidence of even mixed samples can be analyzed. Single copy amplification requires some care, but is feasible for both DNA and RNA targets. True needles in haystacks can be found simply by amplifying the needles. PCR facilitates cloning of DNA sequencing and forms a natural basis for cycle sequencing by the Sanger method *(13)*.

2. PCR Enzymes
2.1. AmpliTaq® DNA Polymerase

AmpliTaq DNA Polymerase (PE Biosystems, Foster City, CA) is a highly characterized recombinant enzyme for PCR. It is produced in *E. coli* from the *Taq* DNA polymerase gene, thereby assuring high purity. It is commonly supplied and used as a 5 U/µL solution in buffered 50% glycerol *(14)*.

2.2. Biophysical Properties

The enzyme is a 94-kDa protein with a 5'-3' polymerization activity that is most efficient in the 70–80°C range. This enzyme is very thermostable, with a half-life at 95°C of 35–40 min. In terms of thermal cycling, the half-life is approx 100 cycles. PCR products amplified using AmpliTaq DNA polymerase will often have single base overhangs on the 3' ends of each polymerized strand, and this artifact can be successfully exploited for use with T/A cloning vectors (*see* Chapter 44).

2.3. Biochemical Reactions

AmpliTaq DNA polymerase requires magnesium ion as a cofactor and catalyzes the extension reaction of a primed template at 72°C. The four dNTPs (consisting of dATP, dCTP, dGTP, and dTTP or dUTP) are used according to the base-pairing rule to extend the primer and thereby to copy the target sequence. Modified nucleotides (ddNTPs, biotin-11-dNTP, dUTP, deaza-dGTP, and fluorescently labeled dNTPs) can be incorporated into PCR products.

2.4. Associated Activities

AmpliTaq DNA polymerase has a fork-like structure-dependent, polymerization-enhanced, 5'-3' nuclease activity. This activity allows the polymerase to degrade downstream primers and indicates that circular targets should be linearized before amplification. In addition, this nuclease activity has been employed in a fluorescent signal-generating technique for PCR quantitation *(15)*. AmpliTaq DNA polymerase does not have an inherent 3'-5' exonuclease or proofreading activity, but produces amplicons of sufficiently high fidelity for most applications.

3. PCR Primers

3.1. Design Criteria

PCR primers are short oligodeoxyribonucleotides, or oligomers, that are designed to complement the end sequences of the PCR target amplicon. These synthetic DNAs are usually 15–25 nucleotides long and have ~50% G + C content. Since each of the two PCR primers is complementary to a different individual strand of the target sequence duplex, the primer sequences are not related to each other. In fact, special care must be taken to assure that the primer sequences do not form duplex structures with each other or hairpin loops within themselves. The 3' end of the primer must match the target in order for polymerization to be efficient, and allele-specific PCR strategies take advantage of this fact. To screen for mutants, a primer complementary to the mutant sequence is used and results in PCR positives, whereas the same primer will be a mismatch for the wild type and not amplify. The 5' end of the primer may have sequences that are not complementary to the target and that may contain restriction sites or promotor sites that are also incorporated into the PCR product. Primers with degenerate nucleotide positions every third base may be synthesized in order to allow for amplification of targets where only the amino acid sequence is known. In this case, early PCR cycles are performed with low, less stringent annealing temperatures, followed by later cycles with high, more stringent annealing temperatures. (*See* Chapter 74.)

A PCR primer can also be a homopolymer, such as oligo $(dT)_{16}$, which is often used to prime the RNA PCR process. In a technique called RAPDs (Randomly Amplified Polymorphic DNAs), single primers as short as decamers with random sequences are used to prime on both strands, producing a diverse array of PCR products that form a fingerprint of a genome *(16)*. Often, logically designed primers are less successful in PCR than expected, and it is usually advisable to try optimization techniques for a practical period of time before trying new primers frequently designed near the original sites.

3.2. T_m Predictions

DNA duplexes, such as primer-template complexes, have a stability that depends on the sequence of the duplex, the concentrations of the two components, and the salt concentration of the buffer. Heat can be used to disrupt this duplex. The temperature at which half the molecules are single-stranded and half are double-stranded is called the T_m of the complex. Because of the greater number of intermolecular hydrogen bonds, higher G + C content DNA has a higher T_m than lower G + C content DNA. Often, G + C content alone is used to predict the T_m of the DNA duplex; however, DNA duplexes with the same G + C content may have different T_m values. Computer programs are available to perform more accurate T_m predictions using sequence information (nearest neighbor analysis) and to assure optimal primer design.

Since the specificity of the PCR process depends on successful primer binding events at each amplicon end, the annealing temperature is selected based on the consensus of melting temperatures (within ~2–4°C) of the two primers. Usually the annealing temperature is chosen to be a few degrees below the consensus annealing temperatures of the primers. Different strategies are possible, but lower annealing temperatures should be tried first to assess the success of amplification of the target and then higher annealing temperatures can be investigated to find the stringency required for best product specificity.

4. PCR Samples
4.1. Types

The PCR sample type may be single- or double-stranded DNA of any origin—animal, bacterial, plant, or viral. RNA molecules, including total RNA, poly (A$^+$) RNA, viral RNA, tRNA, or rRNA, can also serve as templates for amplification after conversion to so-called complementary DNA (cDNA) by the enzyme reverse transcriptase (either MuLV or recombinant *Thermus thermophilus,* rTth DNA polymerase) *(17,18)*.

4.2. Amount

The amount of starting material required for PCR can be as little as a single molecule, compared to the millions of molecules needed for standard cloning or molecular biological analysis. As a basis, up to nanogram amounts of DNA cloned template, up to microgram amounts of genomic DNA, or up to 10^5 DNA target molecules are best for initial PCR testing.

4.3. Purity

Overall, the purity of the DNA sample to be subjected to PCR amplification need not be high. A single cell, a crude cell lysate, or even a small sample of degraded DNA template is usually adequate for successful amplification. The fundamental requirements of sample purity must be that the target contain at least one intact DNA strand encompassing the amplified region and that the impurities associated with the target be adequately dilute so as to not inhibit enzyme activity. However, for some applications, such as long PCR, it may be necessary to consider the quality and quantity of the DNA sample *(19,20)*. For example,

1. When more template molecules are available, there is less occurrence of false positives caused by either crosscontamination between samples or "carryover" contamination from previous PCR amplifications;
2. When the PCR amplification lacks specificity or efficiency, or when the target sequences are limited, there is a greater chance of inadequate product yield; and
3. When the fraction of starting DNA available to PCR is uncertain, it is increasingly difficult to determine the target DNA content *(21)*.

5. Thermal Cycling Considerations
5.1. PCR Vessels

PCR must be performed in vessels that are compatible with low amounts of enzyme and nucleic acids and that have good thermal transfer characteristics. Typically, polypropylene is used for PCR vessels and conventional, thick-walled microcentrifuge tubes are chosen for many thermal cycler systems. PCR is most often performed at a 10–100 µL reaction scale and requires the prevention of the evaporation/condensation processes in the closed reaction tube during thermal cycling. A mineral oil overlay or wax layer serves this purpose. More recently, 0.2 mL thin-walled vessels have been optimized for the PCR process and oil-free thermal cyclers have been designed that use a heated cover over the tubes held within the sample block.

5.2. Temperature and Time Optimization

It is essential that the reaction mixtures reach the denaturation, annealing, and extension temperatures in each thermal cycle. If insufficient hold time is specified at any temperature, the temperature of the sample will not be equilibrated with that of the sample block. Some thermal cycler designs time the hold interval based on block temperature, whereas others base the hold time on predicted sample temperature.

In a conventional thick-walled tube used in a cycler controlled by block temperature, a 60-s hold time is sufficient for equilibration. Extra time may be recommended at the 72°C extension step for longer PCR products. Using a thin-walled 0.2-mL tube in a cycler controlled by predicted sample temperature, only 15 s is required. To use existing protocols or to develop protocols for use at multiple labs, it is very important to choose hold times according to the cycler design and tube wall thickness.

6. Conditions for Successful PCR
6.1. Metal Ion Cofactors

Magnesium chloride is an essential cofactor for the DNA polymerase used in PCR, and its concentration must be optimized for every primer/template pair. Many components of the reaction bind magnesium ion, including primers, template, PCR products and dNTPs. The main 1:1 binding agent for magnesium ion is the high concentration of dNTPs in the reaction. Since it is necessary for free magnesium ion to serve as an enzyme cofactor in PCR, the total magnesium ion concentration must exceed the total dNTP concentration. Typically, to start the optimization process, 1.5 m*M* magnesium chloride is added to PCR in the presence of 0.8 m*M* total dNTPs. This leaves about 0.7 m*M* free magnesium ion for the DNA polymerase. In general, magnesium ion should be varied in a concentration series from 1.5–4 m*M* in 0.5-m*M* steps *(1,21)*.

6.2. Substrates and Substrate Analogs

Taq DNA polymerase incorporates dNTPs very efficiently, but can also incorporate modified substrates when they are used as supplemental components in PCR. Digoxigenin-dUTP, biotin-11-dUTP, dUTP, c^7deaza-dGTP, and fluorescently labeled dNTPs all serve as substrates for *Taq* DNA polymerase. For conventional PCR, the concentration of dNTPs remains balanced in equimolar ratios, but for mutagenesis, unequal concentrations should be used.

6.3. Buffers/Salts

The PCR buffer for *Taq* DNA polymerase consists of 50 m*M* KCl and 10 m*M* Tris-HCl, pH 8.3, at room temperature. This buffer provides the ionic strength and buffering

PCR: Basic Principles and Routine Practice

capacity needed during the reaction. It is important to note that the salt concentration affects the T_m of the primer/template duplex, and hence the annealing temperature. Cosolvents, such as DMSO and glycerol, have been successfully used in PCR buffers when the targets have very high denaturation temperatures *(22)*.

6.4. Cycles

The number of cycles of PCR should be optimized with respect to the number of input target copies. In a typical PCR, 10^{12} copies represents the plateau in the maximum amount of amplification possible. From a single copy, the most efficient PCR would reach plateau in 40 cycles ($10^{12} \approx 2^{40}$). PCR may be 80–95% efficient, so the amplification factors are nearer $(1.9)^n$, where n is the number of cycles. It is usually advisable to run the minimum number of cycles needed to see the desired specific product, since unwanted nonspecific products will interfere if the number of cycles is excessive *(1,21)*.

6.5. Enzyme/Target

In a standard aliquot of *Taq* DNA polymerase used for a 100-μL reaction, there are about 10^{10} molecules. Each PCR sample should be evaluated for the number of target copies it contains or may contain. For example 1 ng of lambda DNA contains 1.8×10^7 copies. For low input copy number PCR, the enzyme is in great excess in early cycles. As the amplicon accumulates in later cycles, the enzyme becomes limiting and it may be necessary to give the extension process incrementally more time. Thermal cyclers can reliably perform this automatic segment extension procedure in order to maximize PCR yield *(1,21)*.

7. PCR Protocols

7.1. Reagents and Supplies

The protocol described below illustrates the basic principles and techniques of PCR and can be modified to suit other particular applications. The example chosen uses the PE Biosystems HIV Primer pair, SK145 and SK431, in conjunction with the PE Biosystems Gene Amp® PCR Reagent Kit and PCR Carry-Over Prevention Kit (PE Biosystems), to amplify a 142-bp DNA fragment from the conserved *gag* region of HIV-1 using the AmpliWax® PCR Gem-facilitated hot start process *(11,19)*.

Reagents	Stock concentrations
10X PCR Buffer II	500 mM KCl, 100 mM Tris-HCl, pH 8.3
MgCl$_2$ solution	25 mM
dNTPs	10 mM stocks of each of dATP, dCTP, dGTP; 20 mM stock of dUTP; all neutralized to pH 7.0 with NaOH
Primer 1: SK145	25 μM in 10 mM Tris-HCl, pH 8.3 5'-AGTGGGGGGACATCAAGCAGCCATGCAAAT-3'
Primer 2: SK431	25 μM in 10 mM Tris-HCl. 5'-TGCTATGTCAGTTCCCCTTGGTTCTCT-3'
AmpErase™ UNG	Uracil N-glycosylase, 1.0 U/μL pH 8.3 in 150 mM NaCl, 30 mM Tris-HCl, pH 7.5, 10 mM EDTA, 1.0 mM DTT, 0.05% Tween-20, 5% (v/v) glycerol.

Table 1
Lower Reagent Mix

Reagent	Volume, 1X mix, μL	Final concentration, per 100 μL volume
Sterile water	13.5	N/A
10X PCR buffer II	4.0	1X
25 mM MgCl$_2$	10.0	2.5 mM
10 mM dATP	2.0	200 μM
10 mM dCTP	2.0	200 μM
10 mM dGTP	2.0	200 μM
20 mM dUTP	2.0	400 μM
25 μM primer 1 (SK145)	2.0	0.5 μM
25 μM primer 2 (SK431)	2.0	0.5 μM
1 U/μL AmpErase UNG	0.5	1 U/reaction
Total volume	40.0	

Reagents	Stock concentrations
AmpliTaq DNA polymerase	5 U/μL in 100 mM KCl, 20 mM Tris-HCl, pH 8.0, 0.1 mM EDTA, 1 mM DTT, 0.5% Tween 20, 0.5% Nonidet P40, 50% (v/v) Glycerol
HIV-1 positive control DNA	10^3 copies/μL in 10 μg/mL human placental DNA, 1 mM EDTA, 10 mM NaCl, 10 mM Tris-HCl, pH 8.0
AmpliWax PCR gems	Gem 100s for 50–100 μL reactions or Gem 50s for 20–50 μL reactions

Use 0.5 mL PE Biosystems GeneAmp® PCR microcentrifuge tubes and a PE Biosystems GeneAmp® PCR instrument system.

7.2. Methods

In the AmpliWax PCR gem-facilitated hot start process, a solid wax layer is formed over a subset of PCR reactants, called the lower reagent mix, that encompasses 30–50% of the total reaction mix volume. The remaining reactants, called the upper reagent mix, comprise the remaining 50–70% of the total reaction mix volume and are added above the wax layer. In the first thermal cycle, the wax layer melts during the temperature ramp to the denaturation temperature and is displaced by the more dense upper reagent mix. Thermal convection adequately mixes the combined lower and upper reagent mixes, whereas the melted wax layer acts as a vapor barrier during each PCR cycle.

1. Assemble the lower reagent mix shown in **Table 1**.
2. For 100 μL reactions, add 40 μL of the lower reagent mix (which can be made up as a batch mix) into the bottom of each GeneAmp PCR reaction tube. Avoid splashing liquid onto the tube. If any liquid is present on the tube walls, spin the tube briefly in a microcentrifuge.
3. Carefully add one AmpliWax PCR Gem 100 to each tube containing the lower reagent mix. Melt the wax gem by incubating each reaction tube at 75–80°C for 3–5 min. Solidify the wax at room temperature (25°C) for 3–5 min.

Table 2
Upper Reagent Mix

Reagent	Volume, 1X mix, μL	Final concentration, per 100 μL volume
Sterile water	43.0–52.9	N/A
10X PCR Buffer II	6.0	1X
5 U/μL AmpliTaq DNA polymerase	0.5	2.5 U/reaction
1 U/μL AmpErase UNG	0.5	1 U/reaction
10^3 Copies/μL positive control DNA	0.1–10.0	10^2–10^4 copies
Total volume	60.0	

4. Assemble upper reagent mix as in **Table 2**.
5. For 100 μL reactions, carefully aliquot 60 μL of the upper reagent mix to each GeneAmp PCR reaction tube above the wax layer. Avoid splashing liquid onto the tube wall. If any liquid is present on the tube wall, tap the tube gently to collect all droplets into the upper reagent layer. Do not spin the tube in a microcentrifuge, because this may dislodge the wax layer.
6. Amplify the PCR reactions within a programmable thermal cycler. For the Perkin Elmer DNA Thermal Cycler 480, program and run the following linked files:
   a. Step cycle file: 95°C for 1 min, 60°C for 2 min for 2 cycles; link to file (2).
   b. Step cycle file: 94°C for 1 min, 60°C for 1 min for 38 cycles; link to file (3).
   c. Time delay file: 60°C for 10 min for 1 cycle; link to file (4).
   d. Soak file: 10°C for "forever" (an infinite hold)
7. Very gently insert a pipet tip through the center of the solid wax layer to form a small hole. To withdraw the reaction sample, use a fresh tip.

8. Analysis of PCR Products

8.1. Ethidium Bromide/Agarose Gel Electrophoresis

PCR products can be easily and quickly analyzed using a 3% NuSieve GTG agarose (FMC Bioproducts, Rockland, ME) and 1% Seakem GTG agarose (FMC Bioproducts) gel run in either TBE (89 mM Tris-borate, 2 mM EDTA) or in TAE (40 mM Tris-acetate, 2 mM EDTA, pH ~8.5) stained with ~0.5 μg/mL ethidium bromide. Use a 123-base pair (bp) or 1-kilobase pair (kbp) ladder as a convenient marker for size estimates of the products *(24)* (*see* Chapter 13).

8.2. Other Analytical Methods

A variety of other detection methods are available for PCR product analysis, such as ethidium bromide-stained 8–10% polyacrylamide gels run in TBE buffer, Southern gels or dot/slot blots, subcloning and direct sequencing, HPLC analysis, and the use of 96-well microplates, to name a few. The reverse dot-blot method combines PCR amplification with nonradioactive detection *(25)*. A fluorescence assay has also been optimized to detect amplification using the 5'–3' nuclease activity of *Taq* DNA polymerase *(15)*.

9. PCR Troubleshooting

9.1. Most Obvious Pitfalls

Even though the PCR process has greatly enhanced scientific studies, a variety of problems with the process, easily revealed by ethidium bromide-stained agarose gel electrophoresis, can and may need to be considered when encountered. For example, unexpected molecular weight size bands (nonspecific banding) or smears can be produced. These unexpected products accumulate from enzymatic extension of primers that annealed to nonspecific target sites. Second, primer-dimer (approx 40–60 bp in length, the sum of the two primers) can be produced. Primer-dimer can arise during PCR amplification when: the DNA template is left out of the reaction, too many amplification cycles are used, or the primers are designed with partial complementary at the 3' ends. An increase in primer-dimer formation will decrease the production of desired product. Third, *Taq* DNA polymerase, which lacks the 3' to 5' exonuclease "proofreading" activity, will occasionally incorporate the wrong base during PCR extension. The consequences of *Taq* misincorporations usually have little effect, but should be considered during each PCR amplification.

9.2. How to Investigate "Failures"

PCR amplification for user-selected templates and primers are considered "failures" when: no product bands are observed, the PCR product band is multibanded, or the PCR product band is smeared. These "failures" can be investigated and turned into successful PCR by manipulation of a number of variables, such as enzyme and salt concentrations, denaturation and anneal/extend times and temperatures, primer design, and hot start procedures *(26)*.

When no desired PCR product band is observed, initially verify the enzyme addition and/or concentration by titrating the enzyme concentration. Second, the magnesium ion concentration is also critical, so care should be taken not to lower the magnesium ion molarity on addition of reagents (i.e., buffers containing EDTA will chelate out the magnesium ion). The denaturation and anneal/extend times and temperatures may be too high or too low, causing failures, and can be varied to increase reaction sensitivity. Finally, the chemical integrity of the primers should be considered. In cases where the PCR product band is multibanded, consider raising the anneal temperature in increments of 2° and/or review the primer design and composition.

If a smear of the PCR product band is seen on an ethidium-bromide-stained agarose gel, consider the following options initially, individually or in combinations: decreasing the enzyme concentration, lowering the magnesium ion concentration, lengthening and/or raising the denaturation time and temperature, shortening the extension time, reducing the overall cycle number, and decreasing the possiblity of carryover contamination. Finally, in PCR amplifications where the PCR product band is initially observed, and on later trials a partial or complete loss of the product band is observed, consider testing new aliquots of reagents and decreasing the possibility of carryover contamination.

References

1. Innis, M. A., Gelfand, D. H., Sninsky, J. J., and White, T. J., eds. (1990) *PCR Protocols. A Guide to Methods and Applications.* Academic Press, San Diego, CA.
2. Mullis, K. B. and Faloona, F. A. (1987) Specific synthesis of DNA *in vitro* via a polymerase chain reaction. *Methods Enzymol.* **155,** 335–350.
3. Saiki, R. K., Gelfand, D. H., Stoffel, S., Scharf, S. J., Higuchi, R., Horn, G. T., Mullis, K. B., and Erlich, H. A. (1988) Primer-directed enzymatic amplification of DNA with a thermostable DNA polymerase. *Science* **239,** 487–491.
4. Saiki, R. K., Scharf, S. J., Faloona, F., Mullis, K. B., Horn, G. T., Erlich, H. A., and Arnheim, N. (1985) Enzymatic amplification of β-globin genomic sequences and restriction site analysis for diagnosis of sickle cell anemia. *Science* **230,** 1350–1354.
5. Scharf, S. J., Horn, G. T., and Erlich, H. A. (1986) Direct cloning and sequence analysis of enzymatically amplified genomic sequences. *Science* **233,** 1076–1087.
6. Wang, A. M., Doyle, M. V., and Mark, D. F. (1989) Quantitation of mRNA by the polymerase chain reaction. *Proc. Natl. Acad. Sci. USA* **86,** 9717–9721.
7. Kwok, S. and Higuchi, R. (1989) Avoiding false positives with PCR. *Nature* **339,** 237,238.
8. Orrego, C. (1990) Organizing a laboratory for PCR work. *PCR Protocols. A Guide to Methods and Applications* (Innis, M. A., Gelfand, D. H., Sninsky, J. J., and White, T. J., eds.), Academic Press, San Diego, CA, pp. 447–454.
9. Kitchin, P. A., Szotyori, Z., Fromholc, C., and Almond, N. (1990) Avoiding false positives. *Nature* **344,** 201.
10. Longo, N., Berninger, N. S., and Hartley, J. L. (1990) Use of uracil DNA glycosylase to control carry-over contamination in polymerase chain reactions. *Gene* **93,** 125–128.
11. Chou, Q., Russell, M., Birch, D. E., Raymond, J., and Bloch, W. (1992) Prevention of pre-PCR mis-priming and primer dimerization improves low-copy-number amplifications. *Nucleic Acids Res.* **20,** 1717–1723.
12. Birch, D. E., Kolmodin, L., Laird, W. J., McKinney, N., Wong, J., Young, K. K. Y., Zangenberg, G. A., and Zoccoli, M. A. (1996) Simplified hot start PCR. *Nature* **381,** 445–446.
13. Innis, M. A., Myambo, K. B., Gelfand, D. H., and Brow, M. A. D. (1988) DNA sequencing with *Thermus aquaticus* DNA polymerases and direct sequencing of polymerase chain reaction-amplified DNA. *Proc. Natl. Acad. Sci. USA* **85,** 9436–9440.
14. Abramson, R. D. (1995) Thermostable DNA polymerases, in *PCR Strategies* (Innes, M. A., Gelfand, D. H., and Sninsky, J. J., eds.), Academic Press, San Diego, CA, pp. 39–57.
15. Holland, P. M., Abramson, R. D., Watson, R., and Gelfand, D. H. (1991) Detection of specific polymerase chain reaction product by utilizing the 5'-3' exonuclease activity of *Thermus aquaticus* DNA polymerase. *Proc. Natl. Acad. Sci. USA* **88,** 7276–7280.
16. Sobral, B. W. S. and Honeycutt, R. J. (1993) High output genetic mapping of polyploids using PCR generated markers. *Theor. Appl. Genetics* **86,** 105–112.
17. Myers, T. W. and Gelfand, D. H. (1991) Reverse transcription and DNA amplification by a *Thermus thermophilus* DNA polymerase. *Biochemistry* **30,** 7661–7666.
18. Myers, T. W. and Sigua, C. L. (1995) Amplification of RNA, in *PCR Strategies* (Innes, M. A., Gelfand, D. H., and Sninsky, J. J., eds.), Academic Press, San Diego, CA, pp. 58–68.
19. Cheng, S., Fockler, C., Barnes, W. M., and Higuchi, R. (1994) Effective amplification of long targets from cloned inserts and human genomic DNA. *Proc. Natl. Acad. Sci. USA* **91,** 5695–5699.

20. Cheng, S., Chen, Y., Monforte, J. A., Higuchi, R., and Van Houten, B. (1995) Template integrity is essential for PCR amplification of 20- to 30-kb sequences from genomic DNA. *PCR Meth. Appl.* **4,** 294–298.
21. Erlich, H. A., ed. (1989) *PCR Technology, Principles and Applications for DNA Amplification.* Stockton, New York.
22. Landre, P. A., Gelfand, D. H., and Watson, R. H. (1995) The use of cosolvents to enhance amplification by the polymerase chain reaction, in *PCR Strategies* (Innis, M. A., Gelfand, D. H., and Sninsky, J. J., eds.), Academic, San Diego, CA, pp. 3–16.
23. AmpliWax PCR Gem 100 and PCR Gem 50. Package Insert. **BIO-66,** 55631-10/93. Perkin Elmer Corporation, Foster City, CA.
24. Sambrook, J., Fritsch, E. F., and Maniatis, T. (1989) *Molecular Cloning: A Laboratory Manual,* 2nd ed. Cold Spring Harbor Laboratory, Cold Spring Harbor, NY, pp. 6.20, 6.21, B.23, B.24.
25. Saiki, R. K., Walsh, P. S., Levenson, C. H., and Erlich, H. A. (1989) Genetic analysis of amplified DNA with immobilized sequence-specific oligonucleotide probes. *Proc. Natl. Acad. Sci. USA* **86,** 6230–6234.
26. Kolmodin, L., Cheng, S., and Akers, J. (1995) *Gene*Amp XL PCR kit. *Amplifications: A Forum for PCR Users* (The Perkin-Elmer Corporation) **13,** 1–5.

74

Primer Selection and Design for Polymerase Chain Reaction

Wojciech Rychlik

1. Introduction

One of the most important factors affecting the quality of polymerase chain reaction (PCR) is the choice of primers. Several rules should be observed when designing primers and, in general, the more DNA sequence information available, the better the chance of finding an "ideal" primer pair. Fortunately, not all primer selection criteria need be met in order to synthesize a clean, specific product, as the adjustment of PCR conditions (such as composition of the reaction mixture, temperature, and duration of PCR steps) may considerably improve the reaction specificity. Amplification of 200–400 bp DNA is the most efficient and, in these cases, one may design efficient primers simply by following a few simple rules described in this chapter. It is more difficult to choose primers for efficient amplification of longer DNA fragments, and use of an appropriate primer analysis software is worthwhile.

The important parameters to be considered when selecting PCR primers are the ability of the primer to form a stable duplex with the specific site on the target DNA and no duplex formation with another primer molecule or no hybridization at any other target site. The primer stability can be measured in the length (base pairs) of a DNA duplex, the GC/AT ratio, kcal/mol (duplex formation free energy), or in degrees Celsius (melting temperature). The most accurate methods for computing helix stability are based on nearest-neighbor thermodynamic parameters (1). Calculation of T according to the nearest-neighbor method is complicated, and therefore not practical to use without computer software. Similar duplex stability accuracy, however, may be achieved by calculating the free energy of duplex formation (ΔG). This calculation is simple and can be performed manually.

Subheading 2. describes the following: an example of ΔG calculation, needed for accurate determination of duplex stability; general rules for PCR primer selection; primer design based on a peptide sequence; and primer design for subcloning PCR products.

Table 1
Free Energy Values of a Nearest-Neighbor Nucleotide[a]

First (5') nucleotide	Second nucleotide			
	dA	dC	dG	dT
	ΔG (kcal/mol)			
dA	−1.9	−1.3	−1.6	−1.5
dC	−1.9	−3.1	−3.6	−1.6
dG	−1.6	−3.1	−3.1	−1.3
dT	−1.0	−1.6	−1.9	−1.9

[a]Calculated according to **Eq. (1)** in 25°C.

2. Methods
2.1. Calculations of DNA Duplex Stability

The method of predicting free energy of duplex formation (ΔG) for DNA oligomers, described in the following equation, is a simplified method of Breslauer et al. *(1)*. It is based on **Eq. 1**:

$$\Delta G = \Delta H - T\Delta S \quad (1)$$

where ΔH and ΔS are the enthalpy and entropy of duplex formation, respectively, and T is the temperature in degrees Kelvin.

Table 1 lists the ΔG values of nucleotide pairs.

For simplicity, all calculations are made with T set to 298.15 K (25°C). The relative stability of base pairing in a duplex is dependent on the neighboring bases *(1)*. Thus, for example, to calculate the ΔG of the d(ACGG/CCGT) duplex formation, add the ΔG values of the three nucleotide pairs as follows:

$$\Delta G \text{ (ACGG)} = \Delta G \text{ (AC)} + \Delta G \text{ (CG)} + \Delta G \text{ (GG)} \quad (2)$$

$$\Delta G \text{ (ACGG)} = -(1.3 + 3.6 + 3.1) = -8.0 \text{ (kcal/mol)} \quad (3)$$

This method is especially useful for determining primer compatibility owing to formation of 3'-terminal duplexes, discussed in **Subheading 2.2**. Use the same approach when calculating the ΔG of a hairpin loop structure, except that the ΔG increment for the loop must be added. For loop sizes 3–8 nt, I use the following values (averaged from **refs. 2** and **3**): 3 nt, 5.2 kcal/mol; 4 nt, 4.5; 5 nt, 4.4; 6 nt, 4.3; 7 and 8 nt, 4.1 kcal/ mol. More data can be found in **ref. 2**.

2.2. Selection of PCR Primers
2.2.1. General Rules
2.2.1.1. DIMER FORMATION

PCR primers should be free of significant complementarity at their 3'termini, as this promotes the formation of primer-dimer artifacts that reduce product yield. Formation of primer-dimer artifacts may also cause more serious problems, such as nonspecific DNA synthesis owing to an unbalanced primer ratio (asymmetric PCRs fail more fre-

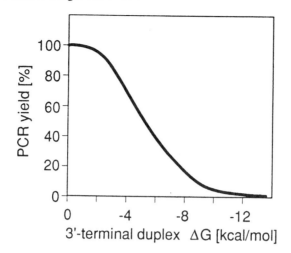

Fig. 1. Dependence of PCR yield on the ΔG of 3'-terminal primer duplexes. The ΔG values were calculated as described in **Subheading 2.1.**

quently than "standard" reactions). **Figure 1** illustrates the PCR yield dependence on the ΔG of 3'-terminal duplexes.

These values are approximate, as the yield also depends on the annealing temperature, the specificity of primers, and other parameters not considered here. The high dependence of yield on dimer formation tendency is the result of the very high processivity of *Taq* polymerase. Duplexes need not be stable to prime DNA synthesis. Very little time is required for the enzyme to recognize a 3'-terminal duplex and start polymerization.

2.2.1.2. SELF-COMPLEMENTARITY

In general, oligonucleotides forming intramolecular duplexes with negative ΔG should be avoided. Although self-complementary PCR primers with hairpin loop ΔG approaching –3 kcal/mol (at 25°C) are suitable in certain cases, a hairpin loop-forming primer is troublesome when its 3' end is "tied up," since this can cause internal primer extension, thus eliminating a given primer from the reaction. Hairpins near the 5' end, however, do not significantly affect the PCR.

2.2.1.3. MELTING TEMPERATURE: STABILITY

There is a widely held assumption that PCR primers should have about a 50% GC/AT ratio. This is not correct. An 81% AT-rich primer (with a second primer of a similar composition and human genomic DNA as substrates) produced a single, specific, 250-bp PCR product (70% AT rich). Without getting into the complex calculations of product and primer Tmls, PCR primers should have a GC/AT ratio similar to that of the amplified template.

A more important factor is the Tm difference between the template and the stable primer. PCR is efficient if this difference is minimized. Note that the Tm of DNA also depends on its length. This is why researchers typically design primers that are too long and unnecessarily too stable. Longer oligos, however, are less likely to be suitable in

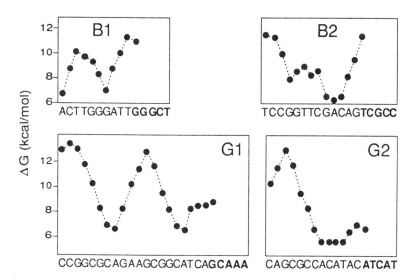

Fig. 2. Internal stability of two poorly functioning (B1, B2) and two efficient (G1, G2) sequencing primers. Primer G1 and G2 performed above average (with almost any other compatible primer) in PCR. The ΔG values were calculated for all pentamers in each primer. The last symbol in each inset represents the ΔG value of the subsequence written in bold (the 3'-terminal pentamer).

terms of dimer formation and self-complementarity and, therefore, generally scarce in a given sequence. If the expected PCR product is 500 bp or less, select 18–21 nt primers. For the synthesis of a 5-kb fragment, choose 24–26 nt oligonucleotides. In the latter case, however, it is difficult to choose a compatible primer pair without the aid of primer selection software to check dimer formation, self-complementarity, and the specificity of primers. When a long DNA fragment is amplified, there is a good chance that an oligonucleotide selected "by eye" will prime from other than the intended target site, yielding nonspecific product(s). The likelihood of false priming can be significantly reduced by observing the internal stability rule, described in **Subheading 2.2.1.4.**

2.2.1.4. INTERNAL STABILITY

Primers that are stable at their 5' termini but somewhat unstable on their 3' ends perform best in sequencing and PCR as well. This primer structure effectively eliminates false priming. These findings, based on primer internal stability, are supported by the experimental data presented in **Fig. 2**. A primer with low stability on its 3' end will function well in PCR because the base pairings near and at the 3' end with nontarget sites are not sufficiently stable to initiate false priming. Therefore, the 5' and the central part of the primer must also form a duplex with the target DNA site in order to prime efficiently. Conversely, oligonucleotides with stable, GC-rich, 3' termini need not to anneal with the target along their entire length in order to efficiently prime, often resulting in nonspecific product synthesis. Examples of efficient PCR (and sequencing) primers are presented in **Fig. 2** (primers G1 and G2).

Notice the high 3' end stability of nonspecific primers (B1 and B2) and low stability of specific primers. The optimal annealing temperature range is unusually broad when primers exhibiting moderately low 3' terminal stability are used. This improves the chances of running the PCR at optimal conditions without prior optimization. It is worth noting that the quality of the PCR product depends on the template (secondary structure, substrate complexity, product length, and Tm), as well as on the annealing time and temperature *(4)*. In certain conditions, primers with high 3' terminal stability perform satisfactory in PCR. Nevertheless, oligonucleotides with 3' terminal pentamers less stable than –9 kcal/mol (*see* **Subheading 2.1.** for calculations) are more likely to be specific PCR (and especially sequencing) primers. Primers with a GC clamp, located anywhere except at the 3' end, perform well in PCR.

2.2.1.5. UNIQUE PRIMERS

In order to amplify a single, specific DNA fragment, the primer's sequence should not repeat in the template *(5)*. Although it is highly unlikely that the entire primer matches perfectly at more than one site on the template, primers with 6–7 nt-long nonunique 3' termini are not uncommon. This may create problems when a "false" priming site is located inside the amplified region. In these cases, a nonspecific product formation is observed (especially in later cycles), because the PCR of shorter DNA fragments is usually more efficient. Note that the more unstable the primer's 3' end, the lower the likelihood of false priming (*see* **Subheading 2.2.1.4.**). When working with mammalian genomic sequences, it is helpful to check the primer of interest for complementarity with Alu sequences or with other common repetitive elements. For a similar reason, homooligomers (like –AAAAAA–) and dinucleotide repeats (like –CACACA–) should be avoided, especially on the 3' ends of primers.

2.2.2. Specific Applications

2.2.2.1. PRIMER DESIGN BASED ON PEPTIDE SEQUENCES

When designing primers from peptide sequences, the use of degenerate primers rather than "guessmers" is preferred. Although it has been reported that up to 1024 degenerate primers have been used successfully *(6)*, regions of high degeneracy should be avoided. There are many (unreported) cases in which less degenerate primers have not worked. It is generally assumed that PCR is acceptably efficient when using primers with 15–20% base-pair mismatches with the template. Mismatches at a primer's 3' end, however, cause more serious problems than the same mismatch ratio at the 5' end. The PCR yield using a primer with two mismatches within the last four bases is drastically reduced. Studies of Kwok et al. *(7)* indicate, however, that primers with 3' terminal "T"-mismatches can be efficiently utilized by *Taq* polymerase when the nucleotide concentration is high. At 0.8 mM, most 3' end mismatches are acceptable *(7)*, although nonspecific product formation is high, and the fidelity DNA synthesis is reduced *(8)*. There is a low level of priming from mismatched bases even at low nucleotide concentrations *(9)*, and therefore, increasing the annealing time to 3–5 min in the initial PCR cycles may yield a desired product of a better quantity than when using standard annealing times and high dNTP concentrations. A total nucleotide concentration of 0.2 mM, or below, is recommended when unique primers are used, as high concentrations increase the misincorporation rate *(8,10)*. When degenerate oligonucleotides are used, PCRs should be run at higher primer concentrations (1–3 µM instead of 0.2 µM) because most oligos in the mixture will not prime specifically and only contribute to a high background.

Table 2
Cleavable Efficiencies of Short DNA Fragments

Enzyme	Excess bp[a]	%Cleavage after 2 h	%Cleavage after 20 h	Enzyme	Excess bp[a]	%Cleavage after 2 h	%Cleavage after 20 h
Acc I	3	0	0	Not I	8	25	90
Afl III	1	0	0		10	25	>90
	2	>90	>90	Nsi I	3	10	>90
Asc I	1	>90	>90	Pst I	1	0	0
BamH I	1	10	25		4	10	10
	2	>90	>90	Pvu I	1	0	0
Bgl II	1	0	0		2	10	25
	2	75	>90		3	0	10
	3	25	>90	Sac I	1	10	10
BssH II	2	0	0	Sac II	1	0	0
	3	50	>90		3	50	90
BstE I	1	0	10	Sca I	1	10	25
Cla I	1	0	0		3	75	75
	2	>90	>90	Sma I	0	0	10
	3	50	50		1	0	10
EcoR I	1	>90	>90		2	10	50
Hae III	1	>90	>90		3	>90	>90
Hind III	2	0	0	Spe I	1	10	>90
	3	10	75		2	10	>90
Kpn I	1	0	0		3	0	50
	2	>90	>90		4	0	50
Mlu I	1	0	0	Sph I	1	0	0
	2	25	50		3	0	25
Nco I	1	0	0		4	10	50
	4	50	75	Stu I	1	>90	>90
Nde I	1	0	0	Xba I	1	0	0
Nhe I	1	0	0		2	>90	>90
	2	10	25		3	75	>90
	3	10	50		4	75	>90
Not I	2	0	0	Xho I	1	0	0
	4	10	10		2	10	25
	6	10	10		3	10	75

[a]Number of base pairs added on each side of the recognition sequences.

2.2.2.2. Primer Design for Subcloning

The addition of a (mismatched) restriction site at the 5' terminus is the most useful method. Add a few "dummy" 5'-terminal bases beyond the recognition site so that the restriction endonucleases can cut the DNA. Try not to extend a potential dimer structure (inherent to restriction sites) beyond the recognition site. There are no general rules as to how many nucleotides to add. A list of cleavage efficiencies of short oligonucleotides has been published *(11)*; the summary is listed in **Table 2**.

An alternative to incorporating a full-restriction enzyme recognition site is to use oligonucleotide primers with only half a palindromic recognition site at the 5' termini of each phosphorylated primer. After amplification, the PCR product should be concatamerized with ligase and then digested with the appropriate enzyme *(12)*. This is an efficient method, actually forcing a researcher to use high-fidelity synthesis conditions *(8,13)*, i.e., low nucleotide concentration, low number of cycles, short extension times, and no "final extension." In these conditions, the formation of 3' overhangs, preventing efficient ligation, is minimal.

If the amplified product is to be subcloned, and the restriction site not needed, use unphosphorylated primers for the reaction and then ligate the product with an SmaI-digested vector in the presence of low concentrations of SmaI (a blunt-end cutter compatible with the ligation conditions). Again, high-fidelity PCR conditions should be used, as mentioned earlier, to minimize formation of 3' overhangs.

When high-fidelity synthesis is less essential, one may utilize the template-independent activity of *Taq* polymerase to create 3'-"A" overhangs in the PCR product and use a vector with 3'-"T"-overhangs *(14,15)*. This method is very efficient when high concentration of nucleotides and long extension times are used, followed by prolonged incubation at the extension temperature after the last cycle.

References

1. Breslauer, K. J., Frank, R., Blocker, H., and Markey, L. A. (1986) Predicting DNA duplex stability from the base sequence. *Proc. Natl. Acad. Sci. USA* **83,** 3746–3750.
2. Freier, S. M., Kierzek, R., Jaeger, J. A., Sugimoto, N., Caruthers, M. H., Neilson, T., and Turner, D. H. (1986) Improved free-energy parameters for predictions of RNA duplex stability. *Proc. Natl. Acad Sci. USA* **83,** 9373–9377.
3. Groebe, D. R. and Uhlenbeck, O. C. (1988) Characterization of RNA hairpin loop stability. *Nucleic Acids Res.* **16,** 11,725–11,735.
4. Rychlik, W., Spencer, W. J., and Rhoads, R. E. (1990) Optimization of the annealing temperature for DNA amplification in vitro. *Nucleic Acids Res.* **18,** 6409–6412.
5. Rychlik, W. and Rhoads, R. E. (1989) A computer program for choosing optimal oligonucleotides for filter hybridization, sequencing and in vitro amplification of DNA. *Nucleic Acids Res.* **17,** 8543–8551.
6. Lee, C. C. and Caskey, C. T. (1990) CDNA cloning using degenerate primers, in *PCR Protocols* (Innis, M. A., Gelfand, D. H., Sninsky, J. J., and White, T. J., eds.), Academic, New York, pp. 46–53.
7. Kwok, S., Kellogg, D. E., McKinney, N., Spasic, D., Goda, L., Levenson, C., and Sninsky, J. J. (1990) Effects of primer-template mismatches on the polymerase chain reaction: human immunodeficiency virus type 1 model studies. *Nucleic Acids Res.* **18,** 999–1005.
8. Eckert, K. A. and Kunkel, T. A. (1990) High fidelity DNA synthesis by the Thermus aquaticus DNA polymerase. *Nucleic Acids Res.* **18,** 3739–3744.
9. Petruska, J., Goodman, M. F., Boosalis, M. S., Sowers, L. C., Cheong, C., and Tinoco, I., Jr. (1988) Comparison between DNA melting thermodynamics and DNA polymerase fidelity. *Proc. Natl. Acad Sci. USA* **85,** 6252–6256.
10. Kawasaki, E. (1990) Amplification of RNA, in *PCR Protocols* (Innis, M. A., Gelfand, D. H., Sninsky, J. J., and White, T. J., eds.), Academic, New York, pp. 21–27.
11. New England BioLabs, 1990–1991 Catalog, "Cleavage close to the end of DNA fragments," p. 132.

12. Jung, V., Pestka, S. B., and Pestka, S. (1990) Efficient cloning of PCR generated DNA containing terminal restriction endonuclease recognition sites. *Nucleic Acids Res.* **18,** 6156.
13. Eckert, K. A. and Kunkel, T. A. (1991) The fidelity of DNA polymerase used in PCR, in *Polymerase Chain Reaction: A Practical Approach* (McPherson, M. J., Quirke, P., and Taylor, G. R., eds.), IRL, Oxford, UK, pp. 227–246.
14. Marchuk, D., Drumm, M., Saulino, A., and Collins, F. S. (1991) Construction of T-vectors, a rapid and general system for direct cloning of unmodified PCR products. *Nucleic Acids Res.* **19,** 1154.
15. Holton, T. A. and Graham, M. W. (1991) A simple and efficient method for direct cloning of PCR products using ddt-tailed vectors. *Nucleic Acids Res.* **19,** 1156.

75

One-Step Optimization Using Touchdown and Stepdown Polymerase Chain Reaction

Kenneth H. Roux and Karl H. Hecker

1. Introduction

Polymerase chain reaction (PCR) optimization and troubleshooting can consume considerable energy and resources because of the finicky and often unpredictable nature of the reactions. Small variations in any of the many variables in a given reaction can have a pronounced effect on the resultant amplicon profile. Reactions that are too stringent yield negligible product and reactions that are not stringent enough yield artifactual amplicons. Variables include concentrations of Mg^{2+}, H^+, dNTPs, primers, and template, as well as cycling parameters. With regard to the latter, the value selected for the annealing temperatures is most critical. Unfortunately, even with the most sophisticated algorithms (e.g., OLIGO) it is often difficult to predict the amplification optima *a priori* leaving no other choice but to employ empirical determination.

Touchdown (TD) PCR *(1,2)* and its sister technique, stepdown (SD) PCR *(2)*, represent a markedly different approach that, in a single amplification regimen, inherently compensates for suboptimal reagent concentrations and less than perfect cycling parameters. Instead of guessing (or using imprecise calculations) to arrive at an appropriate temperature for the primer extension segment of the cycle, one can cast a wider net by using progressively lower annealing temperatures over consecutive cycles. The goal is to select a broad range of annealing temperatures that begins above the estimated T_m and ends below it (*see* **Note 1**). Typically, one runs a TD PCR program at 2 cycles/°C declining over a 10–20°C range at 1°C intervals. In this way, the first primer-template hybridizations and primer extensions will be those with the highest specificity, i.e., presumably, the combination that gives the desired amplicon. Although the annealing temperature continues to drop in subsequent cycles to levels that normally would promote spurious amplification, the desired product, having already experienced several cycles of amplification, will be in a position to out-compete most lower T_m (spurious) amplicons. If, for example, there is only a 3°C difference between the T_m of the target amplicon and the T_m of the first-primed spurious amplicon, the desired product will have undergone up to a 64-fold (2^6) amplification.

From: *The Nucleic Acid Protocols Handbook*
Edited by: R. Rapley © Humana Press Inc., Totowa, NJ

Our experience has been that TD PCR is applicable to a wide range of PCR situations *(2,3)*. At one extreme, TD PCR generally yields a single strong amplicon from genomic DNA even when the primer-template combinations are grossly mismatched (*see* **Note 2**). Mismatching might occur when attempting to amplify specific members from a complex multigene family, using nucleotide sequence information deduced from an amino acid sequence, or amplifying across species lines *(4–6)*. Primer-template base pair mismatches are permissible and can even be near (but probably not at) the 3' end of the primer *(3)*. TD PCR can also compensate for suboptimal buffer composition (e.g., Mg^{2+} concentration) *(2)*. On the other hand, reactions that are already optimal, as assessed by conventional PCR, will usually yield equally strong amplicons even when using a broad temperature range TD PCR protocol in which the annealing temperature dips well below the T_m. Stated another way, TD PCR appears to greatly aid marginal reactions while not imposing significant penalties on already robust reactions. Hence, TD PCR can be used routinely in lieu of conventional PCR and need not be viewed solely as an optimization procedure *(7)*.

One potential drawback to TD PCR stems from the complexity of the programming (*see* **Note 3**). Because of the numerous (10–20) annealing temperatures used, a large segment of the programming capacity of conventional thermal cyclers can be encumbered. Also, attempts to adjust the annealing temperature range can involve considerable reprogramming (*see* **Note 4**). Some newer thermal cyclers circumvent these problems by permitting the programming of automatic incremental temperature changes in progressive cycles. We have recently tested modified versions of TD PCR, which we term SD PCR, that utilize simplified programming *(2)*. For SD PCR, one uses fewer but larger annealing temperature steps with proportionately more cycles per step. For example, a program might consist of three or four steps, at three to four cycles per step, with 3–5°C temperature differences between steps. SD PCR is not quite as universally applicable as TD PCR, but is adequate for many applications. The ease in programming may frequently be worth the trade-off.

2. Materials

2.1. Touchdown PCR With Mismatched Primer-Template Pairs

1. Template DNA (rabbit genomic liver 125 ng/μL).
2. Primers. Stock solutions are at 200 μg/mL in H_2O. The following primer pairs yield a 445-bp amplicon. The sites of mismatches are in capital letters. The sequence of the corresponding genomic homologous strand (shown in brackets) are for comparative purposes (*see* **Note 5** for comments on degenerate primer design).
    a. Primers: 5'cttgccaGtaatatAcgccctgcTaaCTtg3'; 5'ggatcttctgttgatgtctgactGttGgAg 3'
    b. Homologous genomic sequences: [5'cttgccaAtaatatCcgccctgcCaaTCtg 3']; [5'ggatcttct-gttgatgtctgactAttTgTg 3'];
3. 10X PCR buffer: 500 m*M* KCl, 100 m*M* Tris-HCl, pH 9.0, 1% Triton X-100, 2 m*M* of each dNTP (A, T, C, and G), and 15 m*M* $MgCl_2$ (*see* **Note 6**).
4. Ampli*Taq* DNA Polymerase (Perkin Elmer, Norwalk, CT).
5. Sterile mineral oil.
6. Standard wall 0.6-mL capped conical tubes.
7. Equipment and reagents for 1.5% agarose gel electrophoresis.

2.2. Stepdown PCR with a Mismatched Degenerate Primer

1. Template DNA (rabbit genomic liver 125 ng/µL).
2. Primers: Prepare stock solutions at 200 µg/mL. The following primer pair yields a 703-bp amplicon. The sites of mismatches are in capital letters. Degeneracies are separated by a slash and are in parenthesis. The sequence of the genomic homologous strand corresponding to the primer with degeneracies is presented (shown in brackets) for comparative purposes.
    a. Primers: 5'agggatcgggtgaaaggggtctcagc3'; 5'ttAtgagcattcat(a/G)aacttctggagg 3'.
    b. Homologous genomic sequence: 5'agggatcgggtgaaaggggtctcagc3'; [5'ttGtgagcattcatAaacttctggagg 3'].
3. 10X PCR buffer. 500 mM KCl, 100 mM Tris-HCl, pH 9.0, 1% Triton X-100, 2 mM of each dNTP (A, T, C, and G), and 15 mM MgCl$_2$.
4. Ampli*Taq* DNA Polymerase.
5. Equipment and reagents for 1.5% agarose gel electrophoresis.

3. Methods

Programming of the thermal cycler for TD and SD PCR are described first, followed by specific PCR conditions used for both reactions.

3.1. TD PCR Programming

1. Set thermal cycler to denature for 1 min at 94°C, anneal for 2 min, and primer extend for 3 min at 74°C.
2. Follow the cycling program with a 7-min primer extension step and a 4°C soak step (*see* **Note 7**).
3. Set the annealing stage for 2 cycles/°C beginning at 55°C and decreasing at 1°C increments to 41°C (i.e., 30 total cycles in 15 steps) to be followed by ten additional cycles at 40°C.

3.2. SD PCR Programming

1. Set thermal cycler to denature for 1 min at 94°C, anneal for 2 min, and primer extend for 3 min at 74°C as detailed below. Follow the cycling program by a 7-min primer extension step and a 4°C soak step (*see* **Note 7**). Program the annealing stage for six cycles per temperature step beginning at 70°C and decreasing at 3°C increments to 58°C (i.e., 30 total cycles in five steps) to be followed by ten additional cycles at 55°C (*see* **Note 8** for programming considerations).
2. Analyze and reamplify as described below.

3.3. PCR Setup

1. Set up master mix for 50 µL reactions as indicated in **Table 1**.
2. Dispense master mix to 0.6-mL standard wall PCR tubes. Add 50 µL mineral oil to each tube and place in thermal cycler (Perkin Elmer DNA Thermal Cycler). Begin initial cycle and add 2.5 µL (1.25 U) of a 1:10 dilution of polymerase to each tube only after the temperature exceeds 80°C in the thermal cycler (i.e., the hot start protocol, *see* **Note 9**). Cap tubes and continue cycling.
3. Following amplification, monitor results by running 3 µL on a 1.5% agarose/ethidium bromide gel and view by UV illumination. If a product is not evident or the desired amplicon is of insufficient amount, amplify for an additional 5–10 cycles (with TD PCR, you are not sure exactly how many nonproductive and suboptimal cycles preceded the

Table 1
PCR Master Mix

Components	Stock concentration	Amount per 50 μL reaction, μL	Final concentration
dNTP mix	2 mM, ea.	5	0.2 mM
PCR buffer	10X	5	1X
Primer 1	200 ng/μL	1	4.0 ng/μL
Primer 2	200 ng/μL	1	4.0 ng/μL
Template	125 ng/μL	3	7.5 ng/μL
MgCl$_2$	25 mM	3	1.5 mM
H$_2$O	—	29.5	—
		Subtotal: 47.5	

start of efficient amplification) at lowest annealing temperature or consider repeating the amplification with the TD PCR annealing temperature range shifted downward by 5°C.

4. If the desired product is still not evident, consider conventional or TD nested PCR on a 1:100 to 1:1000 dilution of the initial TD PCR reaction.

4. Notes

1. The T_m for the primer-template combination can be roughly estimated using the formula:

$$T_m = 2(A + T) + 4(G + C) \tag{1}$$

   For primer-template combinations with known or suspected mismatches, 5–20°C should be subtracted from the normal annealing temperature. Of course, more sophisticated programs, such as OLIGO Primer Analysis Software (National Biosciences, Inc., Plymouth, MN) *(10,11)*, may also be used to calculate the T_m.

2. An estimation of the T_m is particularly difficult when using primers and templates containing mismatched base pairs. We have successfully amplified, to a single intensely staining band, primer-template pairs containing 3–5 mismatches with the template *(3)*. In all cases, we used a TD PCR program in which the annealing temperatures dropped from 55–41°C at 2 cycles/°C.

3. If using a thermal cycler that has a programmable automatic temperature variation feature, set the annealing stages to decline by 0.5°C/cycle. For a standard thermal cycler, program 2 cycles/°C drop (for example, cycles 1 and 2, 65°C; cycles 3 and 4, 64°C, and so forth). In both instances, the TD portion of the program should be followed by ten cycles at a fixed annealing temperature about 10°C below the estimated T_m. One should bear in mind that for situations in which the template is not fully complementary to the primers, once amplification commences, the amplicon will be fully complementary to the primers and thus will have a greater T_m than initially estimated. On the other hand, we have noted that final stage amplification at 10°C or more below the estimated T_m (rather than the 4–5°C usually recommended for standard PCR) can significantly increase the yield of otherwise marginal reactions *(2)*. Avoid the temptation of adding too many cycles to this terminal fixed annealing temperature stage of the program. Excessive cycling can degrade the product and lead to spurious banding and high-mol-wt smearing *(14)*.

4. A convenient way to adjust the TD temperature range segment of a conventional thermal cycler (i.e., having linked or sequential file programming) is to program files covering a

wide range of annealing temperatures (20–25°C). The specific subset of files to be used in any given amplification protocol can be bypassed by simply linking the initial 5 min denaturation file to the file having the highest annealing temperature to be used and linking the file containing the lowest annealing temperature in the TD range to the terminal primer extension file. Be sure to keep note of these changes because they must be undone before the next alteration in the range. If this approach will tie up too many files, you may wish to reprogram those files containing the segments to be deleted from the range (say 5°C from the bottom) and using the freed file capacity to add new files to the top (beginning) of the program. Again, the initial denaturation stage and terminal primer extension steps must be linked to the beginning and end of the new range, respectively. Thermal cyclers in which individual files cannot be linked but which rely on a single long multistep program are even less versatile and may require complete reprogramming. Fortunately, TD and SD PCR are very forgiving and a single temperature range can be applied to a wide variety of situations.

5. Design the best primer set based on the information available. When designing primers to amplify genes of uncertain complementarity, try to cluster the sequence of greatest certainty near the 3' ends of the primers. Degeneracy derived from multiple nucleotides or inosine residues at positions of uncertainty are permissible *(4,5,8)* but not necessary. Note that some polymerases other than *Taq* cannot prime from inosine-containing primers *(9)*.
6. Because of the minimal effort involved, it is generally advantageous to vary one of the buffer components (usually Mg^{2+}) during the initial optimization.
7. Most time and temperature characteristics of the cycling program (denaturation, primer extension) will be the same as conventional PCR for your system if using primer-template combinations other than the examples described.
8. The temperature range of SD PCR may be divided into 3–5 more or less equal increments (steps) and be programmed accordingly. If you can afford to tie up the programming capacity of the thermal cycler, more steps are better than fewer. Proportion the total number of cycles to be used in the SD segment of the program equally among the steps. Add ten cycles at a fixed temperature well below the T_m as described above. When background problems are expected to be minimal, a simple two step SD PCR protocol can still be advantageous. Here, the initial stage has perhaps six to ten cycles at a fixed temperature 5°C above that which would normally be used in standard PCR (i.e., slightly above the calculated T_m). Even though a full 30 cycles at this elevated temperature would not be expected to yield a detectable amplicon, we have found *(2)* that sufficient, highly specific amplification is occurring to allow the desired amplicon to dominate the amplification throughout the remainder of the standard temperature cycles, thus reducing the possibility that unwanted amplicons will be generated.
9. Because TD PCR is based on the use of high temperature to prevent spurious priming, it is imperative that hot start procedures be followed *(12,13)*. If multiple samples are to be run, add an extended denaturation step to the beginning of the program or use the hold option.

References

1. Don, R. H., Cox, P. T., Wainwright, B. J., Baker, K., and Mattick, J. S. (1991) 'Touchdown' PCR to circumvent spurious priming during gene amplification. *Nucleic Acids Res.* **19,** 4008.
2. Hecker, K. H. and Roux, K. H. (1996) High and low annealing temperatures increase both specificity and yield in touchdown and stepdown PCR. *BioTechniques* **20,** 478–485.
3. Roux, K. H. (1994) Using mismatched primer-template pairs in touchdown PCR. *BioTechniques* **16,** 812–814.

4. Knoth, K., Roberds, S. Poteet, C., and Tamkun, M. (1988) Highly degenerate inosine-containing primers specifically amplify rare cDNA using the polymerase chain reaction. *Nucleic Acids Res.* **16,** 10932.
5. Patil, R. V. and Dekker, E. E. (1990) PCR amplification of an *Escherichia coli* gene using mixed primers containing deoxyinosine at ambiguous positions in degenerate amino acid codons. *Nucleic Acids Res.* **18,** 3080.
6. Batzer, M. A., Carlton J. E., and Deininger, P. L. (1991) Enhanced evolutionary PCR using oligonucleotides with inosine at the 3'-terminus. *Nucleic Acids Res.* **19,** 5081.
7. Roux, K. H. (1995) Optimization and troubleshooting in PCR. *PCR Meth. Applic.* **4,** S185–S194.
8. Peterson, M. G., Inostroza, J., Maxon, M. E., Flores, O., Adomon, A., Reinberg, D., and Tjian, R. (1991) Structure and functional properties of human general transcription factor IIE. *Nature* **354,** 369–373.
9. Knittel, T. and Picard, D. (1993) PCR with degenerate primers containing deoxyinosine fails with *Pfu* DNA polymerase. *PCR Meth. Applic.* **2,** 346,347.
10. Rychlik, W. and Spencer, W. J. (1989) A computer program for choosing optimal oligonucleotides for filter hybridization, sequencing and in vitro amplification of DNA. *Nucleic Acids Res.* **17,** 8543–8551.
11. Rychlik, W. (1994) New algorithm for determining primer efficiency in PCR and sequencing. *J. NIH Res.* **6,** 78.
12. D'Aquila, R. T., Bechtel, L. J., Viteler, J. A., Eron, J. J., Gorczyca, P., and Kaplin, J. C. (1991) Maximizing sensitivity and specificity of PCR by preamplification heating. *Nucleic Acids Res.* **19,** 3749.
13. Erlich, H. A., Gelfand, D., and Sninsky, J. J. (1991) Recent advances in the polymerase chain reaction. *Science* **252,** 1643–1651.
14. Bell, D. A. and DeMarini, D. (1991) Excessive cycling converts PCR products to random-length higher molecular weight fragments. *Nucleic Acids Res.* **19,** 5079.

76

Cloning Gene Family Members Using Polymerase Chain Reaction with Degenerate Oligonucleotide Primers

Gregory M. Preston

1. Introduction
1.1. What Are Gene Families?

As more and more genes are cloned and sequenced, it is apparent that nearly all genes are related to other genes. Similar genes are grouped into families, such as the collagen and globin gene families. There are also gene superfamilies. Gene superfamilies are composed of genes that have areas of high homology and areas of high divergence. Examples of gene superfamilies include the oncogenes, homeotic genes, and myosin genes. In most cases the different members of a gene family carry out related functions. A detailed protocol for the cloning by degenerate oligonucleotide polymerase chain reaction (PCR) of members of the *Aquaporin* family of membrane water channels *(1,2)* will be discussed here.

1.2. Advantages of PCR Cloning of Gene Family Members

There are several considerations that must be taken into account when determining the advantages of using PCR to identify members of a gene family over conventional cloning methods of screening a library with a related cDNA, a degenerate primer, or an antibody. It is recommended that after a clone is obtained by PCR one uses this template to isolate the corresponding cDNA from a library, because mutations can often be introduced in PCR cloning. Alternatively, sequencing two or more PCR clones from independent reactions will also meet this objective. The following is a list of some of the advantages of cloning gene family members by PCR.

1. Either one or two degenerate primers can be used in PCR cloning. When only one of the primers is degenerate, the other primer must be homologous to sequences in the phage or bacteriophage cloning vector *(3,4)* or to a synthetic linker sequence, as with RACE PCR (*see* Chapter 78). The advantage to using only one degenerate primer is that the resulting clones contain all of the genetic sequence downstream from the primer (be it 5' or 3' sequence). The disadvantage to this anchor PCR approach is that one of the primers is recognized by every gene in the starting material, resulting in single-strand amplification

of all sequences. This is particularly notable when attempting to clone genes that are not abundant in the starting material. This disadvantage can often be ameliorated in part by using a nested amplification approach with two degenerate primers to preferentially amplify desired sequences.

2. It is possible to carry out a PCR reaction on first-strand cDNAs made from a small amount of RNA, and in theory, from a single cell. Several single-stranded "minilibraries" can be rapidly prepared and analyzed by PCR from a number of tissues at different stages of development, or cell cultures under different hormonal conditions. Therefore, PCR cloning can potentially provide information about the timing of expression of an extremely rare gene family member, or messenger RNA splicing variants, that may not be present in a recombinant library.

3. Finally, the time and expense required to clone a gene should be considered. Relative to conventional cloning methods, PCR cloning can be more rapid, less expensive, and in some cases, the only feasible cloning strategy. It takes at least 4 d to screen 300,000 plaques from a λgt10 library. With PCR, an entire library containing 10^8 independent recombinants (~5.4 ng DNA) can be screened in one reaction. Again, to ensure authenticity of your PCR clones, you should either use the initial PCR clone to isolate a cDNA clone from a library, or sequence at least two clones from independent PCR reactions.

1.3. Degenerate Oligonucleotide Theory and Codon Usage

Because the genetic code is degenerate, primers targeted to particular amino acid sequences must also be degenerate to encode the possible permutations in that sequence. Thus, a primer to a six-amino-acid sequence that has 64 possible permutations can potentially recognize 64 different nucleotide sequences, one of which is to the target gene. If two such primers are used in a PCR reaction, then there are 64 × 64 or 4096 possible permutations. The target DNA will be recognized by a small fraction (1/64) of both primers, and the amplification product from that gene will increase exponentially. However, some of the other 4095 possible permutations may recognize other gene products. This disadvantage can be ameliorated by performing nested amplifications and by using "guessmer" primers. A guessmer primer is made by considering the preferential codon usage exhibited by many species and tissues (*see* **Subheading 3.1.**). For instance, the four codons for alanine begin with GC. In the third position of this codon, G is rarely used in humans (~10.3% of the time) or rats (~8.0%), but often used in *Escherichia coli* (~35%) *(5)*. This characteristic of codon usage may be advantageously used when designing degenerate oligonucleotide primers.

1.4. Strategy for Cloning Aquaporin Gene Family Members

In a related methods chapter *(3)* I described the cloning by degenerate primer PCR of *Aquaporin*-1 (formerly CHIP28) from a human fetal liver λgt11 cDNA library starting with the first 35 amino acids from the *N*-terminus of the purified protein. A full-length cDNA was subsequently isolated from an adult human bone marrow cDNA library *(4)*, and following expression in *Xenopus* shown to encode a water selective channel *(6)*. We now know that the *Aquaporin* family of molecular water channels includes genes expressed in diverse species, including bacteria, yeast, plants, insects, amphibians, and mammals *(1,2,7)*. We have recently used degenerate oligonucleotide primers designed to highly conserved amino acids between the different members of

Table 1
The Degenerate Nucleotide Alphabet

Letter	Specification
A	Adenosine
C	Cytidine
G	Guanosine
T	Thymidine
R	puRine (A or G)
Y	pYrimidine (C or T)
K	Keto (G or T)
M	aMino (A or C)
S	Strong (G or C)
W	Weak (A or T)
B	Not A (G, C, or T)
D	Not C (A, G, or T)
H	Not G (A, C, or T)
V	Not T (A, C, or G)
N	aNy (A, G, C, or T)
I	Inosine[a]

[a]Although inosine is not a true nucleotide, it is included in this degenerate nucleotide list since many researchers have employed inosine-containing oligonucleotide primers in cloning gene family members.

the *Aquaporin* family to clone novel *Aquaporin* gene family cDNAs from rat brain (*AQP*4) and salivary gland (*AQP*5) libraries *(8,9)*. In **Subheading 3.**, I will describe the creation of a new set of degenerate primers that we are currently using to clone, by degenerate primer PCR, *Aquaporin* homologs from a number of different tissues and species. **Subheading 3.** has been broken up into three parts.

1. **Subheading 3.1.** describes the designing of the degenerate primers.
2. **Subheading 3.2.** describes the PCR-amplification with degenerate primers.
3. **Subheading 3.3.** describes the subcloning and DNA sequencing of the specific PCR-amplified products.

2. Materials

2.1. Design of Degenerate Oligonucleotide Primers

No special materials are required here, except the amino acid sequence to which the degenerate primers are going to be designed and a codon usage table *(5)*. If the degenerate primers are going to be designed according to a family of related amino acid sequences, these sequences should be aligned using a multiple sequence alignment program. A degenerate nucleotide alphabet (**Table 1**) provides a single letter designation for any combination of nucleotides. Some investigators have successfully employed mixed primers containing inosine where degeneracy was maximal, assuming inosine is neutral with respect to base pairing, to amplify rare cDNAs by PCR *(10,11)*.

2.2. PCR Amplification with Degenerate Primers

For all buffers and reagents, distilled deionized water should be used. All buffers and reagents for PCR should be made up in distilled deionized 0.2-µ filtered water that has been autoclaved (PCR-water) using sterile tubes and aerosol blocking pipet tips to prevent DNA contamination (*see* **Note 1**). All plastic supplies (microfuge tubes, pipet tips, and so on) should be sterilized by autoclaving or purchased sterile.

1. 10X PCR reaction buffer: 100 mM Tris-HCl, pH 8.3, at 25°C, 500 mM KCl, 15 mM MgCl$_2$, 0.1% w/v gelatin. Incubate at 50°C to melt the gelatin, filter sterilize, and store at –20°C (*see* **Note 2**).
2. dNTP stock solution (1.25 mM dATP, dGTP, dCTP, dTTP) made by diluting commercially available deoxynucleotides with PCR-water.
3. Thermostable DNA polymerase, such as Amplitaq DNA Polymerase (Perkin Elmer Cetus, Norwalk, CT) supplied at 5 U/µL.
4. Mineral oil.
5. A programmable thermal cycler machine, available from a number of manufacturers, including Perkin Elmer Cetus, MJ Research, and Stratagene.
6. Degenerate oligonucleotide primers should be purified by reverse-phase high performance liquid chromatography (HPLC) or elution from acrylamide gels, dried down, resuspended at 20 pmol/µL in PCR-water, and stored at –20°C, preferably in aliquots.
7. The DNA template can be almost any DNA sample, including a single-stranded cDNA from a reverse transcription reaction, DNA from a phage library, and genomic DNA. The DNA is heat denatured at 99°C for 10 min and stored at 4 or –20°C.
8. Chloroform (*see* **Note 3**).
9. Tris-saturated phenol (*see* **Note 3**), prepared using ultra pure redistilled crystalline phenol as recommended by the supplier (Gibco-BRL [product #5509], Gaithersburg, MD). Use polypropylene or glass tubes for preparation and storage.
10. PC9 (*see* **Note 3**): Mix equal volumes of buffer-saturated phenol, pH >7.2, and chloroform, extract twice with an equal volume of 100 mM Tris-HCl, pH 9.0, separate phases by centrifugation at room temperature for 5 min at 2000g, and store at 4 to –20°C for up to 1 mo.
11. 7.5 M AmAc for precipitation of DNA. Ammonium acetate is preferred over sodium acetate because nucleotides and primers generally do not precipitate with it. Dissolve in water, filter through 0.2-µm membrane, and store at room temperature.
12. 100% ethanol, stored at –20°C.
13. 70% ethanol, stored at –20°C.
14. *TE*: 10 mM Tris, 0.2 mM EDTA, pH 8.0. Dissolve in water, filter through 0.2-µm membrane, and store at room temperature.
15. 50X TAE: 242 g Tris-HCl base, 57.1 mL acetic acid, 18.6 g Na$_2$(H$_2$O)$_2$EDTA. Dissolve in water, adjust volume to 1 L, and filter through 0.2-µm membrane. Store at room temperature.
16. *Hae*III digested φX174 DNA markers. Other DNA molecular weight markers can be used depending on availability and the size of the expected PCR-amplified products.
17. 6X gel loading buffer (GLOB): 0.25% bromophenol blue, 0.25% xylene cyanol FF, 1 mM EDTA, 30% glycerol in water. Store up to 4 mo at 4°C.
18. Agarose gel electrophoresis apparatus and electrophoresis grade agarose. For the optimal resolution of DNA products <500 bp in length, NuSieve GTG agarose (FMC BioProducts) is recommended.
19. Ethidium bromide (*see* **Note 3**). 10 mg/mL stock of ethidium bromide (EtBr) prepared in water and stored at 4°C in a brown or foil wrapped bottle. Use at 0.5–2.0 µg/mL in water for staining nucleic acids in agarose or acrylamide gels.

20. For the elution of specific PCR-amplified DNA products from agarose gels, several methods are available, including electroelution and electrophoresis onto DEAE-cellulose membranes *(12,13)*. Several commercially available kits will also accomplish this task. I have had some success with GeneClean II (Bio 101, La Jolla, CA) for PCR products >500 bp in length, and with QIAEX (Qiagen, Chatsworth, CA) for products from 50–5000 bp. If you do not know the approximate size of the PCR-amplified products and wish to clone all of them, the QIAquick-spin PCR purification kit is recommended (Qiagen) since this will remove all nucleotides and primers before attempting to clone. This kit is also recommended for purification of PCR products for secondary PCR-amplification reactions.

2.3. Cloning and DNA Sequencing of PCR-Amplified Products

1. From **Subheading 2.2.**, **items 8–14** and **20**.
2. pBluescript II phagemid vector (Stratagene). A number of comparable bacterial expression vectors are available from several companies.
3. Restriction enzymes: *Eco*RV (for blunt-end ligation).
4. Calf intestinal alkaline phosphatase (CIP) (New England Biolabs, Beverly, MA).
5. Klenow fragment of *E. coli* DNA polymerase I (sequencing grade preferred) and 10 mM dNTP solution (dilute PCR or sequencing grade dNTPs).
6. T4 DNA ligase (1 or 5 U/µL) and 5X T4 DNA ligase buffer (Gibco-BRL).
7. Competent DH5α bacteria. Can be prepared *(12,13)* or purchased. Other bacterial strains can be substituted.
8. Ampicillin: 50 mg/mL stock in water, 0.2-µ filtered, stored in aliquots at –20°C (*see* **Note 4**).
9. LB media: 10 g bacto-tryptone, 5 g bacto-yeast extract, and 10 g NaCl dissolved in 1 L water. Adjust pH to 7.0. Sterilize by autoclaving for 20 min on liquid cycle.
10. LB-Amp plates: Add 15 g bacto-agar to 1000 mL LB media prior to autoclaving for 20 min on the liquid cycle. Gently swirl the media on removing it from the autoclave to distribute the melted agar. **Be careful:** The fluid may be superheated and may boil over when swirled. Place the media in a 50°C water bath to cool. Add 1 mL of ampicillin, swirl to distribute, and pour 25–35 mL/90-mm plate. Carefully flame the surface of the media with a Bunsen burner to remove air bubbles before the agar hardens. Store inverted overnight at room temperature, then wrapped at 4°C for up to 6 mo.
11. IPTG: Dissolve 1 g isopropylthiogalactoside in 4 mL water, filter through 0.2-µm membrane, and store in aliquots at –20°C.
12. X-Gal: Dissolve 100 mg 5-bromo-4-chloro-3-indolyl-β-D-galactopyranoside in 5 mL dimethylformamide and stored at –20°C in a foil wrapped tube (light sensitive).
13. Plasmid DNA isolation equipment and supplies *(12,13)* or plasmid DNA isolation kits, available from many manufacturers.
14. Double-stranded DNA sequencing equipment and supplies *(12,13)*, or access to a DNA sequencing core facility.

3. Methods
3.1. Design of Degenerate Oligonucleotide Primers

1. The first step in designing a degenerate primer is to select a conserved amino acid sequence, then determine the potential nucleotide sequence (or the complement of this sequence for a downstream primer), considering all possible permutations. If the amino acid sequence is relatively long, you can potentially design two or more degenerate primers. If only one is made, make it to sequences with a high (50–65%) GC content, because these primers can be annealed under more stringent conditions (e.g., higher temperatures). **Figure 1** shows an alignment of the amino acid sequences for several members of the

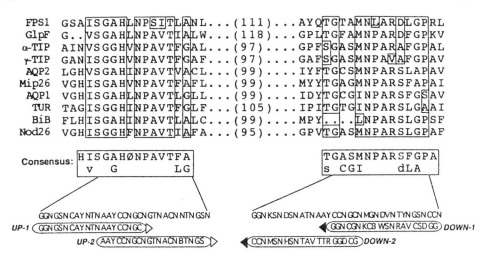

Fig. 1. Design of degenerate primers to amplify *Aquaporin* gene family members. (**Top**) The amino acid sequences of ten MIP family proteins, including the *S. cerevisiae* FPS1 *(23)*, *E. coli* GlpF *(24)*, α- and γ-tonoplast intrinsic proteins (TIP) of *Arabidopsis thaliana* *(25)*, the vasopressin-responsive water channel of rat renal collecting tubules (AQP2) *(26)*, the major intrinsic protein (MIP) of bovine lens fiber membranes *(27)*, human *Aquaporin*-1 *(4)*, turgor responsive gene (TUR) 7a from *Pisum stivum* *(28)*, the *Drosophila* neurogenic *big brain* protein *(29)*, and the *Rhyzodium* root Nodulin-26 peribacteroid membrane protein *(30)* were aligned by the PILEUP program of progressive alignments *(31)* using a gap weight of 3.0 and a gap length of 0.1 running on a VAX computer system. The two most highly conserved regions are shown, separated by the number of intervening amino acids. The most highly conserved amino acids are enclosed. (**Middle**) Below the aligned sequences, the consensus amino acid sequences are shown. (**Bottom**) From part of the consensus amino acid sequences, the degenerate nucleotide sequences were determined (using the degenerate nucleotide alphabet from **Table 1**) followed by the sequences for the degenerate oligonucleotide primers.

Aquaporin gene family in the two most highly conserved regions. Also shown is the consensus amino acid sequence, the degenerate nucleotide sequence, and the sequence of the primers we are currently using to isolate *Aquaporin* gene family members. Interestingly, not only are these two regions highly conserved among the different members of this gene family, but they are also highly related to each other, with the conserved motif being (T/S)GxxxNPAxx(F/L)G, that has been speculated to have resulted from an ancient internal duplication in a primordial bacterial organism, since this repeat has persisted in *Aquaporin* homologs from bacteria through plants and mammals *(1,6,14)*. These two regions are functionally related, both contributing to the formation of the water pore in *Aquaporin*-1 *(15)*.

2. The next step is to determine the number of permutations in the nucleotide sequence. There are 192 permutations ($[2 \times 4] \times 3 \times 4 \times 2$) in the sequence 5'-YTN-ATH-GGN-GAR-3' which encodes the hypothetical amino acid sequence *Leu-Ile-Gly-Glu*. We can reduce the degeneracy by making educated guesses in the nucleotide sequence, i.e., by making a guessmer. The 3'-end of a primer should contain all possible permutations in the amino acid sequence, since *Taq* DNA polymerase will not extend a prime with a mismatch at the extending (3') end. If the above primer was to a human gene, a potential guessmer would be 5'-CTB-ATY-GGN-GAR-3', which only contains 64 permutations.

This guessmer is proposed by taking into account the preferential codon usage for leucine and isoleucine in humans *(5)*.

3. The degeneracy of a primer can be reduced further by incorporating inosine residues in the place of *N*. The advantages of using inosine-containing primers is that they have a reduced number of permutations, and the inosine reportedly base pairs equally well with all four nucleotides, creating a single bond in all cases *(10)*. The disadvantage is that inosines reduce the annealing temperature of the primer. I have not employed inosine-containing primers in these studies.
4. It is often convenient to incorporate restriction endonuclease sites at the 5'-ends of a primer to facilitate cloning into plasmid vectors *(4,8,9)*. Different restriction sites can be added to the 5'-ends of different primers so the products can be cloned directionally. However, not all restriction enzymes can recognize cognate sites at the ends of a double-stranded DNA molecule equally well. This difficulty can often be reduced by adding a two to four nucleotide 5'-overhang before the beginning of the restriction enzyme site (*see* **Note 5**). Some of the best restriction enzymes sites to use are *Eco*RI, *Bam*HI, and *Xba*I. Catalogs from New England Biolabs have a list of the ability of different restriction enzymes to recognize short base-pair sequences. A potential pitfall of this approach would be the occurrence of the same restriction site within the amplified product as used on the end of one of the primers. Therefore, only part of the amplified product would be cloned.
5. The final consideration you should make is the identity of the 3' most nucleotide. The nucleotide on the 3'-end of a primer should preferably be G or C, and not be N, I, or T. The reason for this is that thymidine (and supposedly inosine) can nonspecifically prime on any sequence. Guanosines and cytidine are preferred since they form three H-bonds at the end of the primer, a degree stronger than an A:T base pair.

3.2. PCR Amplification and DNA Purification

The template for these reactions can be the DNA in a phage library or the first-strand cDNA from a reverse transcription reaction on RNA. A phage library with a titer of 5×10^9 pfu/mL would contain, in a 5-µL aliquot, 2.5×10^7 pfu (~1.5 ng of DNA). Prior to PCR-amplification, the DNA is heat denatured at 99°C for 10 min.

3.2.1. PCR Reaction (see **Notes 1** and **6**)

In all cases, the DNA template should also be PCR-amplified with the individual degenerate primers to determine if any of the bands amplified are derived from one of the degenerate primer pools. A DNA-free control is required to assess if there is contaminating DNA in any of the other reagents.

1. Pipet into 0.5-mL microcentrifuge tubes in the following order: 58.5 µL PCR-water that has been autoclaved; 10 µL 10X PCR reaction buffer (*see* **Note 2**); 16 µL 1.25 m*M* dNTP stock solution; 5.0 µL primer up-1; 5.0 µL primer down-1; and 5.0 µL heat-denatured library or cDNA (1–100 ng). If several reactions are being set up concurrently, a master reaction mix can be made up consisting of all the reagents used in all of the reactions, such as the PCR-water, reaction buffer, and dNTPs.
2. Briefly vortex each sample and spin for 10 s in a microfuge. Overlay each sample with 2–3 drops of mineral oil.
3. Amplify by hot-start PCR using the following cycle parameters. Pause the thermocycler in **step 4**-cycle 1, and add 0.5 µL Ampli*taq* DNA polymerase to each tube. 95°C, 5 min (initial denaturation); 94°C, 60 s (denaturation); 50°C, 90 s (annealing; *see* **Note 7**); 72°C, 60 s (extension); cycle 29 times to **step 2**; 72°C, 4 min; and 10°C hold.

3.2.2. DNA Isolation and Gel Electrophoresis Analysis

1. Remove the reaction tubes from the thermal cycler and add 200 µL chloroform. Spin for 10 s in a microfuge to separate the oil-chloroform layer from the aqueous layer. Carefully transfer the aqueous layer to a clean microfuge tube.
2. Remove the Ampli*Taq* DNA polymerase by extracting the aqueous phase twice with 100 µL PC9 (*see* **Note 3**). Spin for 2 min in a microfuge to separate the lower organic layer from the upper aqueous layer and transfer the aqueous layer to clean microfuge tube. This step is essential before digesting the DNA with restriction enzymes for directional cloning (*see* **Subheading 3.3.**) since the polymerase can precipitate, and in the presence of nucleotides, fill in recessed 3' termini on DNA.
3. AmAc-EtOH precipitation: To a 100 µL DNA sample add 50 µL 7.5 M AmAc (50% vol). Vortex briefly to mix. Precipitate the DNA with 350 µL 100% ethanol (2–2.5 vol). Vortex the samples for 15 s and ice for 15 min. Spin down the DNA at 12,000g for 15 min at 4°C in a microfuge. Decant the aqueous waste. Add 250 µL 70% ethanol. Vortex briefly and spin another 5 min at 4°C. Decant the ethanol and allow the pellets to dry inverted at room temperature, or dry in a Speed-Vac for 2–10 min.
4. Resuspend in 20 µL PCR-water.
5. The next step is to resolve an aliquot (2–10-µL) of the PCR fragments by gel electrophoresis. Small DNA products (<300 bp) can be resolved at high resolution on 5–10% polyacrylamide gels *(12,13)*. Moderate-sized PCR products (150–1000 bp) should be resolved on 2–4% NuSieve agarose gels (in 1X TAE buffer). Larger PCR products (>500 bp) can be resolved on 0.8–2% agarose gels (1X TAE buffer).
6. After the bromophenol blue dye has reached the end of the gel, soak the gel for 5–30 min in about 10 vol of water containing 1 µg/mL EtBr (*see* **Note 3**). Then view and photograph the gel under UV light. As shown in **Fig. 1**, there is little variability in the distance between the NPA motifs with the known members of the *Aquaporin* gene family. PCR-amplification of the known *Aquaporin*s cDNAs using the internal degenerate primers would generate products from 345–415 bp. A typical result is shown in **Fig. 2**.

3.2.3. Secondary PCR Amplifications and DNA Purification

Based on the results from gel electrophoresis of the PCR-amplified DNA products, a decision must be made on what to do next. The options are the following.

1. Amplify by PCR from the initial DNA sample under different conditions.
2. Amplify by PCR from a different DNA sample under the same conditions. (Different MgCl$_2$ concentration, annealing temperature, or primers, *see* **Notes 2, 6,** and **7**.)
3. Gel purify a band(s) of DNA from the gel for cloning or to reamplify by PCR.
4. Purify all PCR-amplified DNA fragments for cloning or to reamplify by PCR.
5. Reamplify by PCR with the same or an internal pair of degenerate primers.

Options 1 and 2 are self explanatory. If you want to gel purify a particular band or group of bands from an agarose gel, a number of procedures and kits are available (*see* **Subheading 2.2.**). If you plan on immediately cloning a PCR band(s), you may want to run the rest of the initial PCR reaction on another gel to increase the recovery of DNA. It is also possible to recover specific DNA fragments from an acrylamide gel *(3,12,13)*. To purify all PCR-amplified DNA fragments from the remaining sample, a number of methods are available, including the QIAquick-spin PCR purification kit, which can be used instead of **steps 1–3** in **Subheading 3.2.2.** (Qiagen). Finally, aliquots

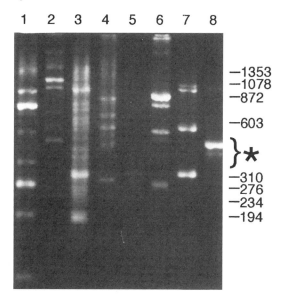

Fig. 2. Gel electrophoresis analysis of PCR-amplified DNA. DNA isolated from a human kidney cDNA library in bacteriophage λgt10 was amplified with degenerate primers up-1 (lanes 1, 5, and 6), up-2 (lanes 2, 7, and 8), down-1 (lanes 3, 5, and 7), and down-2 (lanes 4, 6, and 8). Reactions containing 5×10^6 *pfu* of heat-denatured phage DNA, 100 pmol of degenerate primers, and 1.5 mM MgCl$_2$ in a 100 µL volume were subject to 40 cycles of PCR-amplification under the following parameters: 94°C for 60 s, 48°C for 90 s, and 72°C for 60 s. Following chloroform extraction and ethanol precipitation, the DNA was resuspended in 20 µL of water, and 5 µL was electrophoresed into a 4% NuSieve agarose gel in 1X TAE. The gel was stained with ethidium bromide and photographed. The relative mobility of *Hae*III digested φX174 DNA markers is shown on the right. The bracket shows the size range of known members of this gene family from the primers employed.

of DNA purified from a gel or from the initial PCR reaction (1–10%) can be reamplified by PCR with either the same or an internal pair of degenerate oligonucleotide primers (*see* **Note 1**).

When attempting to identify a gene family homolog from a tissue that is known to express a homolog(s), a number of tricks can be tried to enrich the final PCR sample for new homologs. Since the degenerate oligonucleotide primers are designed from the sequence of the known gene family members, these primers will likely be biased for those homologs. *Aquaporin*-1 is abundant in the capillaries around the salivary glands and throughout the body, but absent in the salivary gland *(16)*. To identify a salivary homology of the *Aquaporin* gene family, we used a rat salivary gland cDNA library, that also contained *Aquaporin*-1 cDNAs, presumably from the surrounding capillaries. We first amplified the cDNA library with an external set of degenerate primers, digested the PCR-amplified DNAs with the restriction enzyme *Pst*I (which cuts between the NPA motifs of rat *AQP*1), and reamplified with an internal pair of primers. We again digested with *Pst*I to digest the rat *AQP*1 DNAs, then cloned and sequenced the DNA fragments between 350 and 450 bp *(9)*. This strategy would not work if the resulting cDNA (*AQP*5) also contained a *Pst*I site. By trying different restriction enzymes that cut DNA infrequently (6–8 bp-recognition sites), a number of new homologs will preferentially be identified. Alternatively, after cloning the DNA products into bacterial expression vectors, bacterial colony lift hybridization can be used to identify colonies containing inserts for known gene family members *(3,12,13)*.

3.3. Cloning and DNA Sequencing of PCR-Amplified Products

3.3.1. Preparation of Vector for Ligation

1. For blunt-end ligations, digest 1 µg pBluescript II KS phagemid DNA (Stratagene) with 10 U *Eco*RV in a 50 µL vol. Incubate at 37°C for 2 h. For cohesive-end ligations, similarly digest the vector with the appropriate restriction enzyme(s).
2. For both blunt-end ligations and cohesive-end ligations where the vector has been digested with only one restriction enzyme, it is necessary to remove the 5'-phosphate from the vector to inhibit the vector from self ligating. This is accomplished by treating the vector with CIP according to the manufacturer's recommendations. Note that 1 µg of a 3 kbp linear DNA molecule contains 1 pmol of 5'-overhangs (*Bam*HI), blunt-ends (*Eco*RV), or 3'-overhangs (*Pst*I), depending on the enzyme that digested it. Afterward add EDTA to 5 m*M* and heat-kill the enzyme at 65°C for 1 h. Adjust the volume to 50–100 µL with *TE* and extract once with Tris-saturated phenol, twice with PC9, and twice with chloroform. Back extract each organic layer with 50 µL *TE* and pool with the final sample. AmAc-EtOH precipitate (*see* **Subheading 3.2.2.**) and resuspend in 10 µL water.
3. If the insert is going to be directionally cloned into the vector, just extract once with 50 µL PC9, AmAc-EtOH precipitate (*see* **Subheading 3.2.2.**), and resuspend in 10 µL water.

3.3.2. Preparation of Inserts for Ligation

Ampli*Taq* and other thermostable DNA polymerases often fail to completely fill in the ends of the double-stranded DNA products, thus leaving recessed 3' termini that can be filled in with the Klenow fragment of *E. coli* DNA polymerase I. This should be done whether or not the DNA is going to be digested with restriction enzymes added to the ends of the primers for directional cloning (*see* **Subheading 3.1.**).

1. AmAc-EtOH precipitate the DNA (*see* **Subheading 3.2.2.**) and resuspend in 15 µL water.
2. Add 2 µL 10X restriction enzyme reaction buffer. Klenow DNA polymerase works well in most restriction enzyme digestion buffers (10X REact 2 or 3 from Gibco-BRL). If the DNA is going to be subsequently digested with a restriction enzyme(s), use the buffer for that enzyme.
3. Add 2 µL 10 m*M* dNTP solution. Then add Klenow DNA polymerase (1 U/µg DNA) and incubate at room temperature for 15 min.
4. Heat-inactivate the enzyme at 75°C for 10 min. If the DNA is going to be directly used in ligation reactions it is not necessary to purify the DNA from the unincorporated dNTPs, since they will not inhibit T4 DNA ligase. To concentrate the DNA sample, proceed with **step 6**.
5. PCR products containing restriction sites on their ends should now be digested with the restriction enzymes. Incubate in the appropriate buffer, using 20 U of enzyme/µg of DNA and incubating for 2–4 h at the proper temperature.
6. Extract the DNA once or twice with PC9 and precipitate with AmAc-EtOH as described above (*see* **Subheading 3.2.2.**). Resuspend the final pellet in 5–10 µL water.

3.3.3. DNA Ligation and Bacterial Transformation

1. At this point it is often advantageous to run a small aliquot of the different DNA fragments on a gel to assess their approximate concentrations and purity. Ideally you want at least a 2:1 molar ratio of insert to vector in the ligation reactions. If necessary return to the above procedures to isolate more DNA for the ligation reaction.
2. Set up the ligation reactions with the vector and insert similar to the following:
    a. Reaction 1: 1 µL vector (10 ng; vector control);
    b. Reaction 2: 1 µL vector + 1 µL insert (~10 ng insert);
    c. Reaction 3: 1 µL vector + 4 µL insert.

    Then add 2 µL of 5X T4 DNA ligase buffer (Gibco-BRL) and water to 9.5 µL. If the buffer is more than 4-mo-old, the ATP may be depleted. Therefore, add fresh ATP to a final concentration of 1 mM.
3. For cohesive-end ligations add 0.5 µL of T4 DNA ligase (1 U/µL), gently mix, spin 5 s in a microfuge, and incubate at 15°C for 10–20 h. For blunt-end ligations add 1 µL of T4 DNA ligase (5 U/µL), gently mix, spin 5 s in a microfuge, and incubate at 25°C (or room temperature) for 1–12 h. Stop the reaction by heating at 75°C for 10 min and store the samples at –20°C.
4. Set up a bacterial transformation with competent DH5α bacteria or a comparable strain of bacteria. Be sure to include a positive control (10 ng undigested vector DNA) and a negative control (water). To 1.5-mL microfuge tubes, add half of the ligation mix (5 µL) or 5 µL of control DNA or water and 50 µL of competent bacteria (thawed slowly on ice); incubate on ice for 30 min. Heat-shock at 42°C for 2 min. Return to ice for 1 min. Add 200 µL of LB media containing 10% glycerol. Mix gently and allow bacteria to recover and express the ampicillin resistance gene by incubating at 37°C for 1 h.
5. Prewarm LB-Amp plates at 37°C for 45 min. About 30 min before plating the bacteria on the plates, add 40 µL of X-Gal and 4 µL IPTG and quickly spread over the entire surface of the plate using a sterile glass spreader. Spread 20–200 µL of the transformation reactions on these plates. Allow the inoculum to absorb into the agar and incubate the plates inverted at 37°C for 12–24 h (*see* **Note 4**). Afterward, placing the plates at 4°C for 2–4 h will help enhance the blue color development.

3.3.4. Plasmid DNA Minipreps and DNA Sequencing

1. Colonies that contain active β-galactosidase will appear blue, whereas those containing a disrupted *lac*Z gene will be white. Set up minicultures by inoculating individual white colonies into 2 mL of LB media containing ampicillin. After growing at 37°C overnight, isolate the plasmid DNA. Resuspend the DNA in 20–50 µL water or *TE*.
2. Digest 5–20 µL of the DNA with the appropriate restriction enzymes and analyze by agarose gel electrophoresis (*see* **Subheading 3.2.3.**).
3. Perform double-stranded DNA sequencing on recombinants containing inserts in the expected size range. (*See* Chapter 61.)

4. Notes

1. All PCR reactions should be set up in sterile laminar flow hoods using pipet tips containing filters (aerosol-resistant tips) to prevent the contamination of samples, primers, nucleotides, and reaction buffers by DNA. If the PCR reaction is going to be reamplified by PCR, all possible intervening steps should also be performed in a sterile hood with the same precautions to prevent DNA contamination. These precautions should also be extended to all extractions and reactions on the nucleic acid (RNA or DNA) through the last PCR reaction. Likewise, all primers, nucleotides, and reaction buffers for PCR should

be made up and aliquoted using similar precautions. All buffers for PCR should be made with great care using sterile disposable plastic or baked glass, and restricted for use with aerosol-resistant pipet tips.

2. Standard PCR reaction buffers contain 15 mM MgCl$_2$ (1.5 mM final concentration). In many cases, changes in the MgCl$_2$ concentration will have significant consequences on the amplification of specific bands. In PCR-amplifying the four exons of the *AQP*1 gene, MgCl$_2$ concentrations between 0.7 and 1.0 mM gave the best results *(17,18)*, however MgCl$_2$ concentrations between 0.5 and 5.0 mM have been reported.

3. Organic solvents and ethidium bromide are hazardous materials. Always handle with tremendous caution, wearing gloves and eye protection. Contact your hazardous waste department for proper disposal procedures in your area.

4. Ampicillin-resistant bacteria secrete β-lactamase into the media, which rapidly inactivates the antibiotic in regions surrounding the growing bacterial colony. Thus, when bacteria are growing at a high density or for long periods (>16 h), ampicillin-sensitive satellite colonies will appear around the primary colonies (which are white in blue-white selections). This problem can be ameliorated (but not eliminated) by substitution of carbenicillin for ampicillin on agar plates.

5. When designing primers with restriction enzyme sites and 5'-overhangs, note that the 5'-overhang should not contain sequences complementary to the sequence just 3' of the restriction site, because this would facilitate the production of primer-dimers. Consider the primer 5'-ggg.agatct.CCCAGCTAGCTAGCT-3', which has a *Xba*I site proceeded by a 5'-ggg and followed by a CCC-3'. These 12 nucleotides on the 5'-end are palindromic, and can therefore easily dimerize with another like primer. A better 5'-overhang would be 5'-cac.

6. When cloning a gene from a recombinant library by PCR, remember that not all genes are created equally. Genes with high G:C contents have proven more difficult to clone than most. Several researchers have made contributions in a search for factors to enhance the specificity of PCR reactions. Nonionic detergents, such a Nonident P-40, can be incorporated in rapid sample preparations for PCR analysis without significantly affecting *Taq* polymerase activity *(19)*. In some cases, such detergents are absolutely required in order to reproducibly detect a specific product *(20)* presumably because of inter- and intrastrand secondary structure. More recently, tetramethylammonium chloride has been shown to enhance the specificity of PCR reactions by reducing nonspecific priming events *(21)*. Commercially available PCR enhancers are also available.

7. A critical parameter when attempting to clone by PCR is the selection of a primer annealing temperature. This is especially true when using degenerate primers. The primer melting temperature (T_m) is calculated by adding 2° for A:T base pairs, 3° for G:C base pairs; 2° for N:N base pairs, and 1° for I:N base pairs. Most PCR chapters suggest you calculate the T_m and set the primer annealing temperature to 5–10°C below the lowest T_m. Distantly related gene superfamily members have been cloned using this rationale *(22)*. However, I have found that higher annealing temperatures are helpful in reducing nonspecific priming, which can significantly affect reactions containing degenerate primers.

Acknowledgments

I thank my colleagues, especially Peter Agre and William B. Guggino, for their support and helpful discussions. This work was supported in part by NIH grants HL33991 and HL48268 to Peter Agre.

References

1. Reizer, J., Reizer, A., and Saier, M. H., Jr. (1993) The MIP family of integral membrane channel proteins: sequence comparisons, evolutionary relationships, reconstructed pathways of evolution, and proposed functional differentiation of the two repeated halves of the proteins. *Crit. Rev. Biochem. Mol. Biol.* **28,** 235–257.
2. Knepper, M. A. (1994) The *aquaporin* family of molecular water channels. *Proc. Natl. Acad. Sci. USA* **91,** 6255–6258.
3. Preston, G. M. (1993) Use of degenerate oligonucleotide primers and the polymerase chain reaction to clone gene family members, in *Methods in Molecular Biology,* vol. 15: *PCR Protocols: Current Methods and Applications* (White, B. A., ed.), Humana, Totowa, NJ, pp. 317–337.
4. Preston, G. M. and Agre, P. (1991) Isolation of the cDNA for erythrocyte integral membrane protein of 28 kilodaltons: member of an ancient channel family. *Proc. Natl. Acad. Sci. USA* **88,** 11,110–11,114.
5. Wada, K.-N., Aota, S.-I., Tsuchiya, R., Ishibashi, F., Gojobori, T., and Ikemura, T. (1990) Codon usage tabulated from the GenBank genetic sequence data. *Nucleic Acids Res.* **18,** 2367–2411.
6. Preston, G. M., Carroll, T. P., Guggino, W. B., and Agre, P. (1992) Appearance of water channels in *Xenopus* oocytes expressing red cell CHIP28 protein. *Science* **256,** 385–387.
7. Chrispeels, M. J. and Agre, P. (1994) *Aquaporin*s: water channel proteins of plant and animal cells. *TIBS* **19,** 421–425.
8. Jung, J. S., Bhat, B. V., Preston, G. M., Guggino, W. B., Baraban, J. M., and Agre, P. (1994) Molecular characterization of an *aquaporin* cDNA from brain: candidate osmoreceptor and regulator of water balance. *Proc. Natl. Acad. Sci. USA* **91,** 13,052–13,056.
9. Raina, S., Preston, G. M., Guggino, W. B., and Agre, P. (1995) Molecular cloning and characterization of an *aquaporin* cDNA from salivary, lacrimal and respiratory tissues. *J. Biol. Chem.* **270,** 1908–1912.
10. Knoth, K., Roberds, S., Poteet, C., and Tamkun, M. (1988) Highly degenerate, inosine-containing primers specifically amplify rare cDNA using the polymerase chain reaction. *Nucleic Acids Res.* **16,** 10,932.
11. Chérif-Zahar, B., Bloy, C., Kim, C. L. V., Blanchard, D., Bailly, P., Hermand, P., Salmon, C., Cartron, J.-P., and Colin, Y. (1990) Molecular cloning and protein structure of a human blood group Rh polypeptide. *Proc. Natl. Acad. Sci. USA* **87,** 6243–6247.
12. Sambrook, J., Fritsch, E. F., and Maniatis, T., eds. (1989) *Molecular Cloning: A Laboratory Manual.* Cold Spring Harbor Laboratory, Cold Spring Harbor, NY.
13. Ausubel, F. M., Brent, R., Kingston, R. E., Moore, D. D., Seidman, J. G., Smith, J. A., and Struhl, K., eds. (1994) *Current Protocols in Molecular Biology.* Greene Publishing/Wiley-Interscience, New York.
14. Wistow, G. J., Pisano, M. M., and Chepelinsky, A. B. (1991) Tandem sequence repeats in transmembrane channel proteins. *TIBS* **16,** 170,171.
15. Jung, J. S., Preston, G. M., Smith, B. L., Guggino, W. B., and Agre, P. (1994) Molecular structure of the water channel through *aquaporin* CHIP: the hourglass model. *J. Biol. Chem.* **269,** 14,648–14,654.
16. Nielsen, S., Smith, B. L., Christensen, E. I., and Agre, P. (1993) Distribution of the *aquaporin* CHIP in secretory and resorptive epithelia and capillary endothelia. *Proc. Natl. Acad. Sci. USA* **90,** 7275–7279.
17. Smith, B. L., Preston, G. M., Spring, F. A., Anstee, D. J., and Agre, P. (1994) Human red cell *Aquaporin* CHIP, I. molecular characterization of ABH and Colton blood group antigens. *J. Clin. Invest.* **94,** 1043–1049.

18. Preston, G. M., Smith, B. L., Zeidel, M. L., Moulds, J. J., and Agre, P. (1994) Mutations in *aquaporin*-1 in phenotypically normal humans without functional CHIP water channels. *Science* **265,** 1585–1587.
19. Weyant, R. S., Edmonds, P., and Swaminathan, B. (1990) Effects of ionic and nonionic detergents on the *Taq* polymerase. *BioTechnology* **9,** 308,309.
20. Bookstein, R., Lai, C-C., To, H., and Lee, W-H. (1990) PCR-based detection of a polymorphic *Bam*HI site in intron 1 of the human retinoblastoma (RB) gene. *Nucleic Acids Res.* **18,** 1666.
21. Hung, T., Mak, K., and Fong, K. (1990) A specificity enhancer for polymerase chain reaction. *Nucleic Acids Res.* **18,** 4953.
22. Zhao, Z.-Y. and Joho, R. H. (1990) Isolation of distantly related members in a multigene family using the polymerase chain reaction technique. *Biochem. Biophys. Res. Comm.* **167,** 174–182.
23. Aelst, L. V., Hohmann, S., Zimmermann, F. K., Jans, A. W. H., and Thevelein, J. M. (1991) A yeast homologue of the bovine lens fiber MIP gene family complements the growth defect of a *Saccharomyces cerevisiae* mutant on fermentable sugars but not its defect in glucose-induced RAS-mediated cAMP signalling. *EMBO J.* **10,** 2095–2104.
24. Muramatsu, S. and Mizuno, T. (1989) Nucleotide sequence of the region encompassing the *glpKF* operon and its upstream region containing a bent DNA sequence of *Escherichia coli*. *Nucleic Acids Res.* **17,** 4378.
25. Höfte, H., Hubbard, L., Reizer, J., Ludevid, D., Herman, E. M., and Chrispeels, M. J. (1992) Vegetative and seed-specific forms of Tonoplast Intrinsic Protein in the vacuolar membrane of *Arabidopsis thaliana*. *Plant Physiol.* **99,** 561–570.
26. Fushimi, K., Uchida, S., Hara, Y., Hirata, Y., Marumo, F., and Sasaki, S. (1993) Cloning and expression of apical membrane water channel of rat kidney collecting tubule. *Nature* **361,** 549–552.
27. Gorin, M. B., Yancey, S. B., Cline, J., Revel, J.-R., and Horwitz, J. (1984) The major intrinsic protein (MIP) of the bovine lens fiber membrane: characterization and structure based on cDNA cloning. *Cell* **39,** 49–59.
28. Guerrero, F. D., Jones, J. T., and Mullet, J. E. (1990) Turgor-responsive gene transcription and RNA levels increase rapidly when pea shoots are wilted: sequence and expression of three induced genes. *Plant Mol. Biol.* **15,** 11–26.
29. Rao, Y., Jan, L. Y., and Jan, Y. N. (1990) Similarity of the product of the *Drosophila* neurogenic gene *big brain* to transmembrane channel proteins. *Nature* **345,** 163–167.
30. Fortin, M. G., Morrison, N. A., and Verma, D. P. S. (1987) Nodulin-26, a peribacteroid membrane nodulin is expressed independently of the development of the peribacteroid compartment. *Nucleic Acids Res.* **15,** 813–824.
31. Feng, D.-F. and Doolittle, R. F. (1990) Progressive alignment and phylogenetic tree construction of protein sequences, in *Methods in Enzymology, vol. 183*: *Molecular Evolution: Computer Analysis of Protein and Nucleic Acid Sequences* (Doolittle, R. F., ed.), Academic, New York, pp. 375–387.

77

Construction of Synthetic Genes by Polymerase Chain Reaction

Patrick J. Dillon and Craig A. Rosen

1. Introduction

Although the polymerase chain reaction (PCR) *(1,2)* is invaluable for the cloning and manipulation of existing DNA sequences, PCR also makes it possible to create new DNA fragments consisting of a nucleic acid sequence that is specified entirely by the investigator. In this chapter we describe a simple two-step PCR method for the rapid construction of synthetic genes *(3)*. This method is based on early observations by Mullis et al. *(4)* in which multiple overlapping oligonucleotides could be used to generate synthetic DNA through several sequential rounds of Klenow based PCR amplification. The method described in this chapter utilizes the thermostable *Taq* polymerase and allows for the generation of synthetic genes in as little as 1 d. This method has proven useful in studies in which synthetic genes were constructed for the HIV-2 Rev protein *(3,5)* and the Wilms' tumor locus zinc finger protein *(6)*. Furthermore, this method has been successfully employed in extensive mutagenesis of the HIV-1 rev response element *(7)*.

Examples for the use of designing synthetic genes include (1) the generation of unique chimeric constructs to study structure-function relationships and domain-swapping effects for a variety of related and unrelated proteins; (2) large-scale alterations or mutational analysis of motifs presenting either proteins or transcriptional elements (e.g., promoters, terminators, and so forth); (3) the creation of unique or novel promoters or proteins; and (4) saturation mutagenesis of genes through the use of random nucleotide incorporation or the use of deoxyinosine in the design of the gene sequence.

The principles of this two-step PCR method for the construction of synthetic genes are outlined in **Fig. 1**. In this method, two sequential PCR reactions are used, the first PCR reaction generates a template DNA corresponding to the synthetic gene, which is then amplified in a second PCR reaction. Before starting this procedure, the investigator must design the construct and determine the nucleic acid sequence of the desired synthetic gene. Once this has been accomplished, oligonucleotides that span the length of the gene must be designed and synthesized. In general, an even number of oligonucleotides should be synthesized and should contain overlaps that are between 15 and 30 nt long. The orientation of the oligonucleotides should be similar to that in panel A

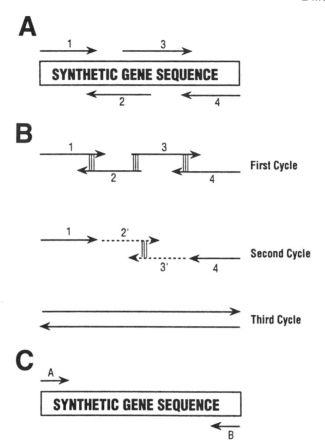

Fig. 1. Description of two-step PCR method for construction of synthetic genes. (**A**) Schematic of design and orientation of overlapping oligonucleotides for first PCR reaction. (**B**) Diagram of oligonucleotide extensions during initial cycle of first PCR. (**C**) Schematic of design and orientation of flanking primers used in the second PCR reaction.

of **Fig. 1**. It is imperative that the outermost oligonucleotides correspond to opposite strands and be positioned so that they will extend inward toward each other over the gene.

The number and length of individual oligonucleotides will vary according to the size of the synthetic DNA to be generated. Typically, oligonucleotides should be between 60 and 125 nt long. For example, four oligonucleotides can be used to synthesize a 325-bp DNA, whereas eight oligonucleotides can be used to generate a 765-bp construct. For this method, crude oligonucleotide preparations are used and it is not necessary for any additional purification of the oligonucleotides.

Once the oligonucleotides have been obtained, the first step of the method is to mix the overlapping oligonucleotides in a standard PCR reaction. Panel B in **Fig. 1** shows how four overlapping oligonucleotides would be extended through the first few cycles of PCR. The first PCR should be carried out with enough cycles to generate a double-stranded PCR product that spans the full length of the synthetic gene. The second step in the method is to take a small aliquot of the first PCR reaction and amplify the synthetic gene in a second-strand PCR reaction that contains short flanking primers (A,B) as illustrated in panel C. The sequence of the flanking primers should contain restriction sites to facilitate cloning. This procedure provides ample amounts of DNA for subsequent cloning into appropriate vectors. This method is an extremely powerful tool for the manipulation of nucleic acid and construction of synthetic genes.

2. Materials

1. 10X PCR buffer: 500 m*M* KCl, 100 m*M* Tris-HCl, pH 8.0, 15 m*M* MgCl$_2$
2. 10X dNTP solution: 2 m*M* each dATP, dCTP, dGTP, and dTTP.
3. *Taq* DNA polymerase.
4. Sterile water.
5. Sterile mineral oil.
6. Overlapping oligonucleotides that span the length of the DNA segment to be synthesized (*see* **Note 1**). An even number of oligonucleotides should be used and contain overlaps of at least 15 nt (*see* **Note 2**).
7. Flanking oligonucleotide primers that contain suitable restriction sites for cloning.
8. Agarose gel for analysis of PCR products (*see* Chapter 13).

3. Methods

1. Set up the first PCR reaction as follows:
    a. 10 µL of 10X PCR buffer
    b. 10 µL of 10X dNTP solution
    c. 0.5 µg each of overlapping oligonucleotides
    d. 2.5 U of *Taq* DNA polymerase;
    e. Sterile water to a final volume of 100 µL
    f. Overlay sample with 50 µL of sterile mineral oil
2. Amplify by PCR using the following cycle profile (*see* **Notes 3**, **4**, and **5**):
    a. Initial denaturation 94°C, 5 min
    b. 10 main cycles 94°C, 1 min (denaturation)
       55°C, 1 min (annealing)
       72°C, 1 min (extension)
    c. Final extension 72°C, 5 min
3. Set up second PCR reaction as follows:
    a. 10 of 10X PCR buffer
    b. 10 µL of 10X dNTP solution
    c. 1 µL of first PCR reaction as template
    d. 1 µg of each flanking primer
    e. 2.5 U *Taq* polymerase
    f. Sterile water to final volume of 100 µL
    g. Overlay sample with 50 µL of sterile mineral oil
4. Run the second PCR using the following cycle profile:
    a. Initial denaturation 94°C, 5 min
    b. 25 main cycles 55°C, 1 min
       72°C, 1 min
       94°C, 1 min
    c. Final extension 72°C, 5 min
5. Analyze 10 µL of the first and second PCR reactions by agarose gel electrophoresis (*see* Chapter 13). A faint smear should be present in the first PCR reaction, and a band corresponding to the size of the desired product should be present in the second PCR reaction (*see* **Note 6**).

6. Digest the product from the second PCR reaction and clone into a suitable vector (*see* **Note 7**).

4. Notes

1. It is not necessary to purify oligonucleotides when using this method. In addition, there is no need to phosphorylate the oligonucleotides, as no ligation steps are used in this protocol.
2. Although it is suggested that an even number of overlapping oligonucleotides be used, an odd number may be used as long as the outermost oligonucleotides are on opposite strands and will extend inward toward each other.
3. The number of cycles needed for the first PCR reaction can be varied depending on the number of oligonucleotides used. In theory, only three cycles should be necessary for full-length template synthesis using four oligonucleotides, whereas four cycles would be necessary if eight oligonucleotides were used.
4. The flanking primers should not be included in the first PCR reaction, as their addition results in the generation of many different-sized products that do not amplify well in the second PCR reaction.
5. It should be noted that the nucleic acid sequences of the overlaps may influence the annealing temperatures used during the first PCR reaction.
6. This method has been successful for the generation of synthetic constructs over 750 bp long.
7. When using this protocol for generating synthetic constructs, it is advisable to sequence the final product to assure that the sequence is correct. The error rate for this method should approximate that observed for other PCR protocols using *Taq* polymerase.

References

1. Saiki, R. K., Gelfand, D. H., Stofel, S., Scharf, S. J., Higuchi, R., Horn, G. T., Mullis, K. B., and Ehrlich, H. A. (1988) Primer-directed enzymatic amplification of DNA with a thermostable DNA polymerase. *Science* **239,** 487–491.
2. Saiki, R. K., Scharf, S., Faloona, F., Mullis, K. B., Horn, G. T., Ehlrich, H. A., and Amheim, N. (1985) Enzymatic amplification of P-globin genomic sequences and restriction site analysis for diagnosis of sickle cell anemia. *Science* **230,** 1350–1354.
3. Dillon, P. J. and Rosen, C. A. (1990) A rapid method for the construction of synthetic genes using the polymerase chain reaction. *BioTechniques* **9,** 298–299.
4. Mullis, K., Faloona, F., Scharf, S., Saiki, R., Hom, G., and Erlich, H. (1986) Specific enzymatic amplification of DNA in vitro: the polymerase chain reaction. *Cold Spring Harbor Symp. Quant. Biol.* **51,** 263–273.
5. Dillon, P. J., Nelbock, P., Perkins, A., and Rosen, C. A. (1990) Function of the human immunodeficiency virus types I and 2 Rev proteins is dependent upon their ability to interact with a structural region present in the env gene mRNA. *J. Virol.* **64,** 4428–4437.
6. Rauscher III, F. J., Morris, J. F., Joumay, O. E., Cook, D. M., and Cuffan, T. (1990) Binding of the Wilms' tumor locus zinc finger protein to the EGR-1 consensus sequence. *Science* **250,** 1259–1262.
7. Olsen, H. S., Beidas, S., Dillon, P. J., Rosen, C. A., and Cochrane, A. W. (1991) Mutational analysis of the HIV-1 Rev protein and its target sequence, the rev response element. *J. Acquired Immun. Defic. Syndrome* **4,** 558–567.

78

Rapid Amplification of cDNA Ends

David Bertioli

1. Introduction

Rapid amplification of cDNA ends (RACE) amplifies either terminal of a cDNA, even where they have unknown sequence *(1,2)*. RACE only requires knowledge of a short sequence within the mRNA of interest. It is often used for cloning the remainder of incomplete cDNAs.

In PCR, RACE uses a "specific primer" designed using the known mRNA sequence together with a "general primer" complementary either to the mRNA poly (A) tail (for 3'-RACE) or to a homopolymer added to the 3'-end of the cDNA (for 5'-RACE) (**Fig. 1**). Because homopolymers do not make good PCR primers, and to facilitate cloning of RACE products, the general primers contain a sequence with a restriction endonuclease site at their 5'-end. The cDNA template for the PCR may be produced either using an oligo-dT primer (for 3'- or 5'-RACE) or using a primer complementary to the known sequence within the mRNA (for 5'-RACE only). Where RACE gives a mixture of products, an aliquot of this mixture may be used as the template for another PCR using a second specific primer (nested within the first) and the original general primer (nested RACE).

Several modifications of RACE have been investigated *(3–6)*. The method discussed here has been published *(7)* and was designed to be simple and cheap. Examples of 5'-RACE products made using the method are shown in **Fig. 2**. This RACE method differs from that of Frohman et al. *(1)* in several ways including the use of:

1. Using dG instead of dA for the homopolymer tail. This enables the use of a higher annealing temperature in 5'-RACE and avoids nonspecific amplification of total cDNA, which occurs when the nonspecific dT-based primer is complementary to both ends of the cDNAs.
2. Hexadecyltrimethylammonium bromide (CTAB) precipitation to simultaneously purify, size select, and protect nucleic acids during the procedures.
3. Total RNA as the usual substrate, although mRNA may also be used. When total RNA is used, the structural RNAs act as carriers for the much less abundant mRNAs and cDNAs. Although the structural RNAs are substrates for terminal transferase, their presence does not seem to have adverse effects.

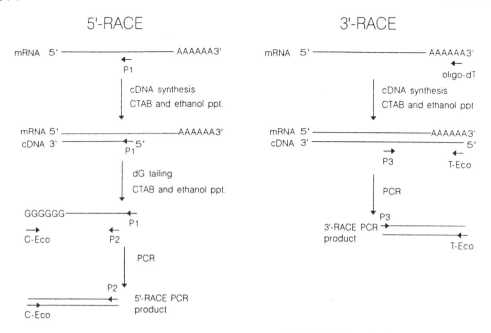

Fig. 1. A diagrammatic summary of the RACE method.

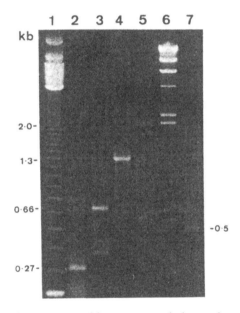

Fig. 2. Some RACE products separated by agarose gel electrophoresis (from **ref. 7**). Lane 1, 123-bp ladder; lane 6, lambda DNA digested with *Hin*dIII. Other samples are RACE reactions designed to amplify the cDNA end corresponding to the 5'-end of the mRNA of lane 2, a "purple pod"-specific mRNA from *P. sativum*; lane 3, β-subunit of pyrophosphate-dependent phosphofructokinase from the potato; lane 4, RNA-2 of arabis mosaic virus; lane 5, acetylcholinesterase from the house fly; lane 7 as lane 4 but cDNA synthesized by AMV RT. Note the smaller RACE product in lane 7 derived from cDNA synthesized by AMV RT compared with the equivalent product in lane 4 from cDNA synthesized by superscript (RNAase H$^-$). Bars indicate amplified cDNA. Numbers next to bars indicate approximate size of RACE products in kilobytes.

2. Materials

1. RNA from the tissue of interest: Many extraction methods could be used, but it is important that the RNA used be intact and essentially free of DNA (*see* **Note 1**). Our usual method is as described by Logemann et al. *(8)*.
2. Reagents for reverse transcription: Moloney murine leukemia virus RNase H⁻ reverse transcriptase (Superscript™; Life Technologies Inc., Gaithersburg, MD). The use of this enzyme is recommended (*see* **Note 2**).
3. 5X Reverse transcription buffer: 250 mM Tris-HCl (pH 8.3), 375 mM KCl, 15 mM MgCl$_2$ and 0.1 M dithiothreitol (DTT), as supplied by Life Technologies.
4. 20 mM dNTPs: An aqueous solution of 20 mM of each dNTP (dGTP, dATP, dTTP, and dCTP).
5. Oligonucleotide primer for cDNA synthesis: This may be dT12–18 (for 5'- or 3'-RACE) or a 15–25mer primer designed to be complementary to the target mRNA (for 5'-RACE only). Use of the latter is more likely to give a single RACE product.
6. Reagents for homopolymer tailing of first strand cDNA.
7. Terminal transferase (Life Technologies Inc.).
8. 5X Tailing buffer (0.5 M potassium cacodylate, pH 7.2, 10 mM CoCl$_2$ 1 mM DTT) as supplied by Life Technologies Inc.
9. An aqueous solution of 100 mM dGTP.
10. Reagents for PCR (*see* Chapter 72).
11. Oligonucleotide primers for PCR: The sequences of the general primers contain a homopolymer at the 3'-end and a restriction endonuclease site at the 5'-end, for instance:
    For 5'-RACE, CAAGGAATT(dC)14 (C-*Eco*RI)
    For 3'-RACE, GCGCCAGGAATTCGC(dT)17 (T-*Eco*RI)
    The specific primers are designed using the known sequence from the target mRNA. For 5'-RACE the PCR primer must be nested to the 3' side of the primer used for cDNA synthesis.
    The PCR primer should be homologous to about 20 bases of the target mRNA/cDNA. To facilitate cloning, a nonhomologous region with a restriction site may be included 5' to these 20 homologous bases (*see* **Note 3**).
12. 10% (w/v) CTAB (it may be necessary to warm this to keep the CTAB in solution [*see* **Note 4**]).
13. 3.5 M and 1.2 M NaCl.
14. Ethanol.
15. 75% (v/v) Ethanol.
16. Reagents for agarose gel electrophoresis (*see* Chapter 13).

3. Method

3.1. First-Strand cDNA Synthesis

Use a 20-mL reaction volume. The quantity of RNA used is not critical and we have used anything from approx 40 mg to approx 100 ng total RNA (although, if at all possible, avoid using such small quantities). It is preferable to do two or three reverse transcriptions with 10-fold different quantities of total RNA. Add the following to a microcentrifuge tube:

1. Add cDNA synthesis primer: approx 500 ng oligo dT12–18 or 15 pmol of a primer complementary to the mRNA (approx 200 ng of a 20mer), total RNA- and RNase-free water to 12.5 mL.

2. Heat to 70°C for 5 min and place on ice.
3. Centrifuge to collect condensation.
4. Add the following:
   a. 4 mL 5X reverse transcriptase buffer
   b. 2 mL 0.1 M DTT
   c. 0.5 mL 20 mM dNTPs
   d. 1 mL (200 U) Superscript reverse transcriptase.
5. Mix well and incubate at 37°C for 1 h.
6. Purify and size select nucleic acids by CTAB and ethanol precipitations.
7. To the reverse transcription reaction add 2.5 mL 3.5 M NaCl.
8. Mix and add 2.5 mL 10% CTAB.
9. Recover CTAB:nucleic acid complexes by centrifugation in a microfuge at high speed at room temperature for 30 min.
10. Discard supernatant and redissolve pellet in 100 mL 1.2 M NaCl.
11. Precipitate nucleic acids by adding 2.7 vol ethanol.
12. Recover nucleic acids by centrifugation in microfuge, wash with 70% ethanol, dry, and resuspend in 15.25 mL water. This may be used directly for 3'-RACE PCR (**step 18**). For 5'-RACE continue to **step 13**.
13. To the microfuge tube add:
    5 mL 5X tailing buffer
    3.75 mL 100 mM dGTP
    1 mL terminal transferase 15 U/mL
14. Mix and incubate for 1 h.
15. Purify nucleic acids by CTAB and ethanol precipitations. To the tailing reaction add 3.125 mL 3.5 M NaCl.
16. Mix and add 3.125 mL 10% CTAB.
17. Recover CTAB: Nucleic acid complex and ethanol precipitate as described in **step 2**.
18. Use the cDNA mix (1 mL) as the template in PCR. To a microcentrifuge tube add the following:
    5 mL of 10X Reaction buffer (100 mM Tris-HCl (pH 8.8) at 25°C, 500 mM KCl, 15 mM MgCl$_2$, 1% Triton X-100)
    0.5 mL 20 mM solution of each dNTP
    0.5 mL *Taq* polymerase (5 U/mL)
    1.5 mL of general oligonucleotide primer (50 ng/mL)
    1.5 mL of specific primer (50 ng/mL)
    Water to 50 mL
19. The following conditions are used for PCR before modifying them if necessary.
    Initial denaturation 94°C for 3 min
    30 cycles of 96°C for 1 min, 55°C for 2 min, 72°C for 3 min, final extension 72°C for 10 min.
20. Analyze a 10-mL aliquot of the PCR products by agarose gel electrophoresis (*see* Chapter 13), and if possible by Southern blotting (*see* Chapter 16) before attempting to clone the products (*see* **Note 5**).

    If the PCR gives many nonspecific products then it may be necessary to use nested RACE. For this use aliquots (three or four 10-fold dilutions from 1 mL downward) of the first PCR as template for a further PCR, which uses as primers the same general primer and a second specific primer nested to the 3' of the first specific primer.

4. Notes

1. To check the integrity of RNA it is unnecessary to use denaturing agarose gel electrophoresis. Instead, use formamide sequencing buffer *(9)* as 1X gel loading buffer and heat the sample with buffer to 85°C before analysis by electrophoresis in 1% agarose buffered in 0.5X TBE with ethidium bromide. Because the quantity of RNA used for this RACE protocol is not critical, it is unnecessary to quantitate the RNA accurately.
2. Using the reverse transcriptase Superscript™, which lacks RNAase H activity, gave larger RACE products than avian myoblastosis virus reverse transcriptase (AMV RT).
3. For 5'-RACE it is important not to use the same primer for cDNA synthesis as for the specific primer in PCR. This is because the annealing of the primer for cDNA synthesis is at 37°C, and all the products of reverse transcription will have this primer at their 5'-end. Therefore, undesired products are very likely to be produced during PCR.
4. CTAB precipitation provides a convenient method for the simultaneous purification and size selection of nucleic acids. CTAB binds nucleic acids to form an insoluble complex, the formation of which is influenced by salt concentration. When the concentration of NaCl is above 1 M no complex formation occurs, when below 0.2 M all nucleic acids are included in the complex and, when between 0.3 M and 0.4 M, the incorporation of small, single-stranded nucleic acids, such as oligonucleotide primers, is very inefficient *(10)*. Using CTAB has the added benefit of protecting nucleic acids (particularly the structural RNAs, which act as carrier) from degradation *(11)*.
5. For cloning RACE products it is preferable to restriction digest the ends and then purifying by gel elution before ligation.

References

1. Frohman, M. A., Dush, M. K., and Martin, G. R. (1988) Rapid production of full-length cDNAs from rare transcripts: amplification using a single gene-specific oligonucleotide primer. *Proc. Natl. Acad. Sci. USA* **85,** 8998–9002.
2. Ohara, O., Dorit, R. L., and Gilbert, W. (1989) One-sided polymerase chain reaction: The amplification of cDNA. *PNAS-USA* **86,** 5673–5677.
3. Harvey, R. J. and Darlison, M. G. (1991) Random-primed cDNA synthesis facilitates the isolation of multiple 5'-cDNA ends by RACE. *Nucleic Acids Res.* **19,** 4002.
4. Borson, N. D., Salo, W. L., and Drewes, L. R. (1992) A lock-docking oligo(dT) primer for 5' and 3' RACE PCR. *PCR Methods Appl.* **2,** 144–148.
5. Jain, R., Gomer, R. H., and Murtagh, J. J. (1992) Increasing specificity from the PCR-RACE technique. *Biotechniques* **12,** 58–59.
6. Troutt, A. B., McHeyzer-Williams, M. G., Pulendran, B., and Nossal, G. J. V. (1992) Ligation-anchored PCR: a simple amplification technique with single-sided specificity. *PNAS-USA* **89,** 9823–9825.
7. Bertioli, D. J. and Burrows, P. R. (1995) A simple method for the amplification of cDNA ends from small amounts of tissue. *Methods Mol. Cell. Biol.*, in press.
8. Logemann, J., Schell, J., and Lothar, W. (1987) Improved method for the isolation of RNA from plant tissues. *Anal. Biochem.* **163,** 16–20.
9. Sambrook, J., Fritsch, E. F., and Maniatis, T. (eds.) (1989) *Molecular Cloning: A Laboratory Manual*, 2nd ed. Cold Spring Harbor Laboratory Press, Cold Spring Harbor, NY.
10. Belyavsky, A., Vinogradova, T., and Rajewsky, K. (1989) PCR-based cDNA library construction: general cDNA libraries at the level of a few cells. *Nucleic Acids Res.* **17,** 2919–2923.
11. Macfarlane, D. E. and Dahle, C. E. (1993). Isolating RNA from whole blood—the dawn of RNA-based diagnosis? *Nature* **362,** 186–188.

79

Multiplex Polymerase Chain Reaction

Jerald Radich

1. Introduction

Often, studies require the polymerase chain reaction (PCR) amplification of several nucleic acid targets. Examples include the amplification of several exons of the same gene for mutation detection (e.g., p53 or dystrophin) *(1–4)*, the amplification of several different genomes for infectious pathogen identification *(5–9)*, identification of the simultaneous expression of transcripts of interest *(10)*, quantification of message or gene copy number *(11,12)*, and the amplification of a target and control gene to test for the nucleic acid integrity *(10)*. When these assays must be performed repetitively on multiple samples, the multiplex PCR, which allows for simultaneous amplification of 2–50 different targets *(13–15)*, is potentially very useful from a perspective of time, and money. However, because multiplex PCR calls for the simultaneous amplification of several different targets, PCR primers optimized for different targets can potentially interact both with other primers (forming the dreaded "primer-dimers"), or with PCR products ("amplicons") produced by the successful and intended amplification of another target. This can potentially decrease the amplification efficiency of some, or all, targets. Thus, multiplex PCR often necessitates the investment of considerable time and effort. It probably should only be used in situations where the advantage of streamlining multiple assays clearly outweighs the investment of time and resources.

Two strategies can be employed in creating a multiplex PCR reaction. The first is completely empirical and can be used when PCR protocols already exist for separate nucleic acid targets and one wishes to try to put these reactions together into a single multiplex. No special insight is needed at the outset using this approach (however, luck helps). However, if the multiplex PCR is to be designed "from scratch," some thought should be given to target size, amplification kinetics, and primer design.

2. Materials
2.1. PCR Reagents

The basic ingredients are those used in a PCR reaction. Thus, the following reagents are starting points; the actual reaction mix is based on the optimization of each individual PCR reaction (*see* Chapter 73 for PCR optimization strategies).

1. 500 ng–1 µg of DNA/RNA.
2. 25 pmol oligonucleotide PCR primers.
3. PCR buffer: 50 mM KCL, 10 mM Tris-HCL (pH 9.0), 0.1% Triton X-100, 1.5 mM MgCl$_2$.
4. 200 µM Dioxynucleotides (dNTPs).
5. 1.25 U *Taq* polymerase.
6. For simultaneous RT-PCR reaction, add 10 U RNAsin, 7.5 U avian myeloblastosis virus reverse transcriptase (AMV RT) (Boehringer Mannheim, Indianapolis, IN) (*see* **Note 4, step 1**).
7. Diethyl pyrocarbonate (DEPC) H$_2$O to 50 µL total volume.

2.2. Thermocycler for PCR

The optimal PCR conditions of a given target can vary greatly from thermocycler to thermocycler. It is probably easiest to develop the multiplex reaction on the same thermocycler as was used to develop the individual PCR reactions. If this is impossible, it is best to find a thermocycler that can be dedicated to the multiplex assay so that this variable is eliminated.

2.3. Detection of PCR Product

The different PCR amplicons produced by the multiplex PCR reaction must be discriminated from one another by some means, and the easiest is by size. The method used is largely determined by availability, considerations of sensitivity, and cost. Each method has a different threshold of size discrimination, and this must be considered in determining the size of PCR amplicons of the various multiplex targets (*see* **Subheading 3.2.**).

3. Methods
3.1. PCR Reactions

To a 50-µL Eppendorf tube add the following:

1. 5 µL (10×) PCR buffer, 25 pmol of each oligonucleotide PCR primer; 200 µM dioxynucleotides (dNTPs), 500 ng–1 µg of each individual template (DNA/RNA); 1.25 U *Taq* polymerase, DEPC H$_2$O to 50 µL total volume.
2. For simultaneous RT-PCR reaction, add 10 U RNAsin, 7.5 U AMV RT (Boehringer Mannheim) (*see* **Note 4, step 1**).
3. Add 15 µL of light mineral oil, unless using a thermocycler with a heated lid, in which case this may be omitted.
4. Undertake thermal cycling for desired time and temperatures (*see* **Subheading 3.2.**).
5. Analyze multiplex PCR using agarose or polyacrylamide gel electrophoresis (Chapters 13 and 15).

3.2. Optimization of Multiplex PCR
3.2.1. Choose Nucleic Acid Targets (see **Note 4, Step 2**)

In general it is easier to keep all targets the same nucleic acid (mRNA or DNA). Consideration must be made concerning the lengths of the PCR amplicons corresponding to each of the targets. In general the targets should be similar in size so that a much smaller target does not out compete the larger targets for amplification reagents. However the PCR targets should be designed so that the various amplicons are easily discernible by whatever the preferred method of detection (i.e., agarose v. polyacrylamide gel, etc.)

3.2.2. Design PCR Primers

If the PCR reactions are already optimized for each target, it is reasonable to take a chance and simply use the existing primers. If, however, the multiplex PCR reaction is being designed from scratch, it is best to design primers with similar annealing temperatures, with sequences not complementary to sequences on the other targets (*see* Chapter 74). This can be done manually but is greatly facilitated by any of a number of available sequence analysis software.

In addition, adding a "random" nucleotide sequence to the 5' tail of the primers will tend to equilibrate differences in the optimal annealing temperature of the primers, and facilitate optimization of the multiplex PCR *(14)*. This has been described with the addition of the 20-mer "universal primer" to the 5' end the PCR primers. This has the secondary advantage of allowing the universal primer to function as a sequencing primer if direct nucleotide sequencing is desired.

3.2.3. Develop PCR Conditions for Each Target

The strategy for optimizing a single PCR is outlined in Chapter 73. It is essential to optimize each reaction to get a single, clean PCR product for each target before attempting a multiplex. In addition, after optimization it is quite useful to perform dilution experiments to determine the lowest amount of target sequence needed to generate the appropriate PCR product (*see* **Note 4, step 3**).

3.2.4. Add Primers from Separate Targets Together

Introduce a new set of primers for targets T_1, T_2, T_3, etc., one set at a time. A reasonable starting point is the start with the reaction conditions for a given PCR, and add the optimal amount of primers for the next target to the mix without adding addition dNTPs or *Taq* polymerase, or modifying buffer conditions.

3.2.5. Optimize Multiplex

The best approach here depends on the results of the first experiment.

3.3. Troubleshooting Guide for Multiplex PCR

1. All intended targets amplify correctly and produce only the intended amplicon products. Perform a dilution experiment of each of the targets to compare the sensitivities from the single PCR with the multiplex. If the sensitivity of the target(s) has decreased to a unacceptable level, *see* **Note 4, step 3**.
2. Too many bands. If each target has been optimized, this outcome is likely from interactions of the primers to each other, and/or priming and amplification of one of the amplicons produced by other primers. Efforts must be made to increase the specificity of the reaction, perhaps at the expense of sensitivity. Potential methods to apply are the same as would be used to optimize a single target PCR (*see* Chapter 75), and include using a "hot start," increasing the temperature of annealing and lowering the concentration of one set of primers.
3. One target is amplified greater than the other(s). This results from differences in reaction efficiency. An easy first approach is to decrease the ratio of the primers of the most effi-

ciently amplified target compared to the least efficient target. In general if the reaction has produced only the anticipated products, simply increasing the amount of the least efficient target primers may work; if there is already evidence of nonspecific amplification, a good start is to decrease the primer concentration of the most efficient amplified target.

4. All targets amplify poorly. All the targets may amplify with equal efficiency but are competing for resources, and thus the total amplification product for all targets is low. Try adding "fuel" to the PCR reaction: increase the amounts of dNTPs, primers, and polymerase. However, if there is evidence of primer-dimers, then methods to eliminate these, such as using a "hot start," increasing the annealing temperature, or actually decreasing PCR primer concentrations may decrease dimer formation and actually promote target amplification. Once the multiplex is optimized for a given set of targets, add another set of primers for the next desired target, and repeat the process.

4. Notes

1. RT-PCR: The traditional method to perform RT-PCR is to perform the RT reaction separately, then take an aliquot to another PCR reaction using the cDNA as template (*see* Chapter 34). We have found that in most situations the RT and PCR can be performed in the same reaction using PCR buffer conditions and the 3' PCR primer to make cDNA off the RNA *(10)*. The sensitivities of the two-step and one-step RT-PCR are often similar, as in the latter reaction the loss of cDNA production caused by the suboptimal RT conditions is offset by using all of the cDNA made for the subsequent PCR reaction.

2. Target choices: If possible, target sizes from 100–400 bp are best, as these sizes are easily amplified by PCR, are large enough to discriminate from primer-dimers, and are appropriate for many applications, such as direct sequencing and single-strand conformation polymorphism. Size differences between targets should be such that amplicons can clearly be resolved by agarose or polyacrylamide (PA) gel. If you wish to perform quantification by comparing the intensity of amplicon signals on the gel, we have found that a PA gel is more reliable and gives better resolution compared to agarose.

3. Sensitivity: For many applications the sensitivity of detecting the targets of concern is a major issue. In the multiplex the sensitivity to detect a given target may decrease because another target amplifies more efficiently and thus "strips" reagents away. This is why it is important to know the optimal sensitivity of each individual PCR reaction prior to the multiplex so this can be compared to the sensitivity of that target in the multiplex assay.

References

1. Chamberlain, J. S., Gibbs, R. A., Ranier, J. E., Nguyen, P. N., and Caskey, C. T. (1988) Deletion screening of the Duchenne muscular dystrophy locus via multiplex DNA amplification. *Nucleic Acids Res.* **16,** 11,141–11,156.
2. Cremonesi, L., Belloni, E., Magnani, C., Seia, M., and Ferrari, M. (1992) Multiplex PCR for rapid detection of three mutations in the cystic fibrosis gene. *PCR Methods Appl.* **1,** 297–298.
3. Ozcelik, H. and Andrulis, I. L. (1995) Multiplex PCR-SSCP for simultaneous screening for mutations in several exons of p53. *BioTechniques* **18,** 742–744.
4. Lohmann, D., Horsthemke, B., Gillessen-Kaesbach, G., Stefani, F. H., and Hofler, H. (1992) Detection of small RB1 gene deletions in retinoblastoma by multiplex PCR and high-resolution gel electrophoresis. *Hum. Genet.* **89,** 49–53.
5. Zazzi, M., Romano, L., Brasini, A., and Valensin, P. E. (1993) Simultaneous amplification of multiple HIV-1 DNA sequences from clinical specimens by using nested-primer polymerase chain reaction. *AIDS Res. Hum. Retroviruses* **9,** 315–320.

6. Repp, R., Rhiel, S., Heermann, K. H., Schaefer, S., Keller, C., and Ndumbe, P. (1993) Genotyping by multiplex polymerase chain reaction for detection of endemic hepatitis B virus transmission. *J. Clin. Microbiol.* **31,** 1095–1102.
7. Covone, A. E., Caroli, F., and Romero, G. (1992). Screening Duchenne and Becker muscular dystrophy patients for deletions in 30 exons of the dystrophin gene by three-multiplex PCR. *Am. J. Hum. Genet.* **51,** 675–677.
8. Herrera, E. A., Perez, O., and Segovia, M. (1996) Differentiation between Mycobacterium tuberculosis and Mycobacterium bovis by a multiplex-polymerase chain reaction. *J. Appl. Bacteriol.* **80,** 596–604.
9. Wilton, S. and Cousins, D. (1992) Detection and identification of multiple mycobacterial pathogens by DNA amplification in a single tube. *PCR Methods Appl.* **1,** 269–273.
10. Lee, A., Kirk, J., Edmands, S., and Radich, J. (1995) Multiplex PCR of bcr-abl fusion transcripts in Philadelphia positive acute lymphoblastic leukemia. *PCR Methods Appl.* **4,** 283–287.
11. Pertl, B., Weitgasser, U., Kopp, S., Kroisel, P. M., Sherlock, J., and Adinolfi, M. (1996) Rapid detection of trisomies 21 and 18 and sexing by quantitative fluorescent multiplex PCR. *Hum. Genet.* **98,** 55–59.
12. Liu, E., Thor, A., He, M., Barcos, M., Ljung, B. M., and Benz, C. (1992) The HER2 (c-erbB-2) oncogene is frequently amplified in *in situ* carcinomas of the breast. *Oncogene* **7,** 1027–1032.
13. Edwards, M. C. and Gibbs, R. A. (1994) Multiplex PCR: advantages, development, and applications. *PCR Methods Appl.* **3,** S65–S75.
14. Shuber, A. P., Grondin, V. J., and Klinger, K. W. (1995) A simplified procedure for developing multiplex PCRs. *Genome Res.* **5,** 488–493.
15. Wang, D., Fan, J., Siao, C.-J., et al. (1997) Toward a third generation genetic map of the human genome based on bi-allelic polymorphisms. The 10th Annual Cold Spring Harbor Meeting on Genome Mapping and Sequencing, Cold Spring Harbor, NY, May 14–17, p. 17.

80

Inverse Polymerase Chain Reaction

Sheng-He Huang

1. Introduction

Since the first report on cDNA cloning in 1972 *(1)*, this technology has been developed into a powerful and universal tool in the isolation, characterization, and analysis of both eukaryotic and prokaryotic genes. But the conventional methods of cDNA cloning require much effort to generate a library that is packaged in phage or plasmid and then survey a large number of recombinant phages or plasmids. There are three major limitations in those methods. First, substantial amount (at least 1 µg) of purified mRNA is needed as starting material to generate libraries of sufficient diversity *(2)*. Second, the intrinsic difficulty of multiple sequential enzymatic reactions required for cDNA cloning often leads to low yields and truncated clones *(3)*. Finally, screening of a library with hybridization technique is time consuming.

Polymerase chain reaction (PCR) technology can simplify and improve cDNA cloning. Using PCR with two gene-specific primers, a piece of known sequence cDNA can be specifically and efficiently amplified and isolated from very small numbers ($<10^4$) of cells *(4)*. However, it is often difficult to isolate full-length cDNA copies of mRNA on the basis of very limited sequence information. The unknown sequence flanking a small stretch of the known sequence of DNA cannot be amplified by the conventional PCR. Recently, anchored PCR *(5–7)* and inverse PCR (IPCR) *(8–10)* have been developed to resolve this problem. Anchored PCR techniques have a common point: DNA cloning goes from a small stretch of known DNA sequence to the flanking unknown sequence region with the aid of a gene-specific primer at one end and a universal primer at other end. Because of only one gene-specific primer in the anchored PCR, it is easier to get a high level of nonspecific amplification by PCR than with two gene-specific primers *(10,11)*. The major advantage of IPCR is to amplify the flanking unknown sequence by using two gene-specific primers.

At first, IPCR was successfully used in the amplification of genomic DNA segments that lie outside the boundaries of known sequence *(8,9)*. There is a new procedure that extends this technique to the cloning of unknown cDNA sequence from total RNA *(10)*. Double-stranded cDNA is synthesized from RNA and ligated end to end (**Fig. 1**). Circularized cDNA is nicked by selected restriction enzyme or denatured by NaOH treatment *(12,13)*. The reopened or denatured circular cDNA is then amplified by two

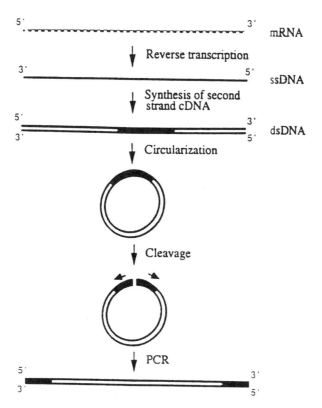

Fig. 1. Diagram of IPCR for cDNA cloning. The procedure consists of five steps: reverse transcription, synthesis of second-stranded cDNA, circularization of double-stranded cDNA, reopen the circle DNA, and amplification of reverse DNA fragment. The black and open bars represent the known and unknown sequence regions of double-stranded cDNA, respectively.

gene-specific primers. Recently, this technique has been efficiently used in cloning full-length cDNAs *(14–16)*. The following protocol was used to amplify cDNA ends for the human stress-related protein ERp72 *(10)* (**Fig. 2**).

2. Materials

2.1. First-Strand cDNA Synthesis

1. Total RNA prepared from human CCRF/CEM leukemic lymphoblast cells *(17,18)*.
2. dNTP Mix (10 mM of each dNTP).
3. Random primers (Boehringer Mannheim, Indianapolis, IN). Prepare in sterile water at 1 μg/μL. Store at –20°C.
4. RNasin (Promega, Madison, WI).
5. Actinomycin D (1 mg/mL): Actinomycin D is light sensitive and toxic. It should be stored in a foil-wrapped tube at –20°C.
6. Moloney Murine Leukemia Virus (MMLV) reverse transcriptase.
7. 5X First-strand buffer: 250 mM Tris-HCl, pH 8.3, 375 mM KCl, 50 mM MgCl$_2$, 50 mM dithiothreitol (DTT), and 2.5 mM spermidine. The solution is stable at –20°C for more than 6 mo.

Inverse PCR

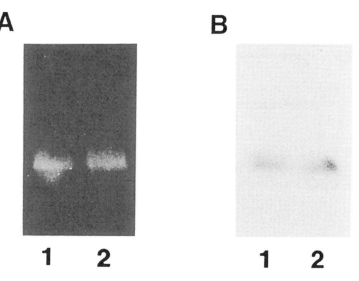

Fig. 2. Application of IPCR to amplifying the joining region (280 bp) from 5' (160 bp) and 3' (120 bp) sequences of human ERp72 cDNA. Amplified DNAs from CCRF/CEM cells sensitive (lane 1) and resistant (lane 2) to cytosine arabinoside stained by ethidium bromide (**A**) or hybridized with ^{32}P-labeled ERp72 cDNA (**B**). *See* text for the sequences of the primers and the parameters of IPCR.

2.2. Second-Strand Synthesis

1. 10X Second-strand buffer: 400 mM Tris-HCl, pH 7.6, 750 mM KCl, 30 mM MgCl$_2$, 100 mM (NH$_4$)$_2$SO$_4$, 30 mM DTT, and 0.5 mg/mL of bovine serum albumin (BSA). The solution is stable at –20°C for at least 6 mo.
2. 1 mM NAD.
3. 2 U/μL RNase H.
4. 5 U/μL *Escherichia coli* DNA polymerase I.
5. 1 U/μL *E. coli* DNA ligase.
6. Nuclease-free water.
7. T4 DNA polymerase.
8. 200 mM EDTA, pH 8.0.
9. GeneClean (Bio 101, La Jolla, CA).
10. TE buffer: 10 mM Tris-HCl, pH 7.6, 1 mM EDTA. Sterile filter.
11. DNA standards: Prepare 1-mL aliquots of a purified DNA sample at 1, 2.5, 5, 10, and 20 μg/mL in TE buffer. Store at –20°C for up to 6 mo.
12. TE/ethidium bromide: 2 μg/mL of ethidium bromide in TE buffer. Store at 4°C for up to 6 mo in a dark container.

2.3. Circularization and Cleavage or Denaturation

1. 5X Ligation buffer (supplied with T4 DNA ligase).
2. 1 U/μL T4 DNA ligase.
3. 4 μg/μL T4 RNA ligase.
4. 15 μM Hexaminecobalt chloride.
5. Phenol:CHCl$_3$:isoamyl alcohol (25:24:1).
6. 3 M sodium acetate, pH 7.0.

7. Absolute ethanol.
8. 70% Ethanol.

2.4. IPCR

1. 10X PCR buffer: 100 mM Tris-HCl, pH 8.3, 500 mM KCl, 15 mM MgCl$_2$, 0.01% (w/v) gelatin.
2. 15 mM MgCl$_2$.
3. Deoxyoligonucleotides were synthesized on an Applied Biosystems (Foster City, CA) 380B DNA synthesizer and purified by OPEC column from the same company. The primer pairs were selected from the 5' and 3' sequence of the cDNA coding for human ERp72 stress-related protein (5'-primer: 5'-TTC CTCCTCCTCCTCCTCTT-3'; 3'-primer: 5'-ATCTAAATGTCTAGT-3') *(10)*.
4. Light mineral oil.
5. *Taq* DNA polymerase.

3. Methods

3.1. First-Strand cDNA Synthesis *(19)*

Perform reverse transcription in a 25-µL reaction mixture, adding the following components: 5.0 µL 5X first-strand buffer, 2.5 µL dNTP mix, 2.5 µL random primers, 1.0 U RNasin, 1.25 µL actinomycin D, 250 U MMLV reverse transcriptase, 15–25 µg of total RNA (heat denature RNA at 65°C for 3 min prior to adding to reaction), and nuclease-free water to 25 µL final volume.

3.2. Second-Strand Synthesis *(20)*

1. Add components to the first-strand tube on ice in the following order: 12.5 µL 10X second strand buffer, 12.5 µL NAD (1 mM), 0.5 µL RNase H (2 µ/µL), 5.75 µL *E. coli* DNA polymerase I (5 µ/µL), 1.25 µL *E. coli* ligase (1 µ/µL), and 67.5 µL nuclease-free water.
2. Incubate at 14°C for 2 h.
3. Heat the reaction mix to 70°C for 10 min, spin for a few seconds, and then put in ice.
4. Add 4 U of T4 DNA polymerase and incubate at 37°C for 10 min to blunt the ends of double-stranded cDNA.
5. Stop the reaction with 12.5 µL of 0.2 M EDTA and 200 µL sterile water.
6. Concentrate and purify the sample with GeneClean. Resuspend the DNA in 100–200 µL of sterile water.
7. Estimate the DNA concentration by comparing the ethidium bromide fluorescent intensity of the sample with that of a series of DNA standards on a sheet of plastic wrap *(21)*. Dot 1–5 µL of sample onto the plastic wrap on a UV transilluminator. Also dot with 5 µL of DNA standards. Add an equal volume of TE buffer containing 2 µg/mL of ethidium bromide, and mix by repipetting up and down. Use proper UV shielding for exposed skin and eyes.

3.3. Circularization and Cleavage (see Notes 1–4)

1. Set up the circularization reaction mix containing the following components: 100 µL (100 ng DNA) of the purified sample, 25 µL of 5X ligation buffer, and 6 µL of T4 DNA ligase. Finally, add 2 µL of T4 RNA ligase or 15 µL of 15 µM hexaminecobalt chloride (*see* **Note 5**).
2. Incubate at 18°C for 16 h.

3. Boil the ligated circular DNA for 2–3 min in distilled water or digest with an appropriate restriction enzyme to reopen circularized DNA.
4. Purify the DNA sample with GeneClean as described in **step 6** in **Subheading 3.2.** or extract with water-saturated phenol/CHCl$_3$ and then precipitate with ethanol *(20)*.

3.4. IPCR (see Note 6)

1. Add $^1/_{10}$ of the purified cDNA to 100 µL of amplification mix (22): 10 µL 10X PCR buffer, 10 µL 15 m*M* MgCl$_2$, 10 µL dNTP mix (2.5 m*M* of each), 10 µL 5'-primer (10 pmol/µL), 10 µL 3'-primer (10 pmol/µL), 10 µL cDNA, 39.5 µL nuclease-free water, and 0.5 µL *Taq* DNA polymerase (2.5 µ/µL).
2. Cap and vortex the tubes to mix. Spin briefly in a microfuge. Cover each reaction with a few drops of light mineral oil to prevent evaporation.
3. Put a drop of mineral oil into each well of the thermal cycler block that will hold a tube. Load the reaction tubes.
4. Amplify by PCR using the following cycle profile: 25 cycles: 94°C for 1 min (denaturation), 65°C for 2 min (annealing), and 72°C for 4 min (elongation).

4. Notes

1. For maximum efficiency of intramolecular ligation, a low concentration of cDNA should be used in the ligation mix. High density of cDNA may enhance the level of heterogeneous ligation, which creates nonspecific amplification.
2. Cleavage or denaturation of circularized double-stranded cDNA is important since circular double-stranded DNA tends to form supercoil and is poor template for PCR *(23)*. Circularized double-stranded DNA is only good for amplification of a short DNA fragment.
3. The following three ways can be considered to introduce nicks in circularized DNA. Boiling is a simple and common way. However, because of the unusual secondary structure of some circular double-stranded DNA, sometimes this method is not sufficient in nicking and denaturing circular double-stranded DNA. A second method is selected restriction enzyme digestion. The ideal restriction site is located in the known sequence region of cDNA. In most cases, it is difficult to make the right choice of a restriction enzyme because the restriction pattern in unidentified region of cDNA is unknown. If an appropriate enzyme is not available, EDTA-oligonucleotide-directed specific cleavage may be tried *(24,25)*. Oligonucleotide linked to EDTA-Fe at T can bind specifically to double-stranded DNA by triple-helix formation and produce double-stranded cleavage at the binding site.
4. Alkali denaturation has been successfully used to prepare plasmid DNA templates for PCR and DNA sequencing *(12,13,26)*. This method should be feasible in denaturing circularized double-stranded cDNA.
5. Inclusion of T4 RNA ligase or hexaminecobalt chloride can enhance the efficiency of blunt-end ligation of double-stranded DNA catalyzed by T4 DNA ligase *(27)*.
6. IPCR can be used to efficiently and rapidly amplify regions of unknown sequence flanking any identified segment of cDNA or genomic DNA. This technique does not need construction and screening of DNA libraries to obtain additional unidentified DNA sequence information. Some recombinant phage or plasmid may be unstable in bacteria and amplified libraries tend to lose them *(23)*. IPCR eliminates this problem.

Acknowledgments

This chapter is an updated version of a chapter previously published in the *Methods in Molecular Biology* series. I acknowledge the contributions of John Holcenberg, Chun-Ha Wu, and Bing Cai from the earlier version. I acknowledge John Holcenberg for his invaluable comments and generous support. I especially thank Chun-Hua Wu and Bing Cai for their technical assistance and Kai-Jin Wu for her art work.

References

1. Verma, I. M., Temple, G. F., Fan, H., and Baltimore, D. (1972) In vitro synthesis of double-stranded DNA complimentary to rabbit reticulocyte 10S RNA. *Nature* **235,** 163–169.
2. Akowitz, A. and Mamuelidis, L. (1989) A novel cDNA/PCR strategy for efficient cloning of small amounts of undefined RNA. *Gene* **81,** 295–306.
3. Okayama, H., Kawaichi, M., Brownstein, M., Lee, F., Yokota, T., and Arai, K. (1987) High-efficiency cloning of full-length cDNA: construction and screening of cDNA expression libraries for mammalian cells. *Methods Enzymol.* **154,** 3–28.
4. Brenner, C. A., Tam, A. W., Nelson, P. A., Engleman, E. G., Suzuki, N., Fry, K. E., and Larrick, J. W. (1989) Message amplification phenotyping (MAPPing): a technique to simultaneously measure multiple mRNAs from small numbers of cells. *BioTechniques* **7,** 1096–1103.
5. Frohman, M. A. (1990) RACE: rapid amplification of cDNA ends, in *PCR Protocols: A Guide to Methods and Applications* (Innis, M. A., Gelfand, D. H., Sninsky, J. J., and White, T. J., eds.), Academic, San Diego, CA, pp. 28–38.
6. Shyamala, V. and Ames, G. F.-L. (1989) Genome walking by single-specific-primer polymerase chain reaction: SSP-PCR. *Gene* **84,** 1–8.
7. Huang, S.-H., Jong, A. Y., Yang, W., and Holcenberg, J. (1993) Amplification of gene ends from gene libraries by PCR with single-sided specificity. *Methods Mol. Biol.* **15,** 357–363.
8. Ochman, H., Gerber, A. S., and Hartl, D. L. (1988) Genetic applications of an inverse polymerase chain reaction. *Genetics* **120,** 621–625.
9. Triglia, T., Peterson, M. G., and Kemp, D. J. (1988) A procedure for in vitro amplification of DNA segments that lie outside the boundaries of known sequences. *Nucleic Acids Res.* **16,** 8186.
10. Huang, S.-H., Hu, Y. Y., Wu, C.-H., and Holcenberg, J. (1990) A simple method for direct cloning cDNA sequence that flanks a region of known sequence from total RNA by applying the inverse polymerase chain reaction. *Nucleic Acids Res.* **18,** 1922.
11. Delort, J., Dumas, J. B., Darmon, M. C., and Mallet, J. (1989) An efficient strategy for cloning 5' extremities of rare transcripts permits isolation of multiple 5'-untranslated regions of rat tryptophan hydroxylase mRNA. *Nucleic Acids Res.* **17,** 6439–6448.
12. Cusi, M. G., Cioé, L., and Rovera, G. (1992) PCR amplification of GC-rich templates containing palindromic sequences using initial alkali denaturation. *BioTechniques* **12,** 502–504.
13. Lau, E. C., Li, Z.-Q., and Slavkin, S. C. (1993) Preparation of denatured plasmid templates for PCR amplification. *BioTechniques* **14,** 378.
14. Green, I. R. and Sargan, D. R. (1991) Sequence of the cDNA encoding bovine tumor necrosis factor-α: problems with cloning by inverse PCR. *Gene* **109,** 203–210.
15. Zilberberg, N. and Gurevitz, M. (1993) Rapid Isolation of full length cDNA clones by "inverse PCR": purification of a scorpion cDNA family encoding α-neurotoxins. *Anal. Biochem.* **209,** 203–205.
16. Austin, C. A., Sng, J.-H., Patel, S., and Fisher, L. M. (1993) Novel HeLa topoisomerase II is the IIβ isoform: complete coding sequence and homology with other type II topoisomerases. *Biochim. Biophys. Acta* **1172,** 283–291.
17. Delidow, B. C., Lynch, J. P., Peluso, J. J., and White, B. A. (1993) Polymerase chain reaction: basic protocols. *Methods Mol. Biol.* **15,** 1–29.

18. Davis, L. G., Dibner, M. D., and Battey, J. F. (1986) *Basic Methods in Molecular Biology*, Elsevier, New York.
19. Krug, M. S. and Berger, S. L. (1987) First strand cDNA synthesis primed by oligo(dT). *Methods Enzymol.* **152,** 316–325.
20. Promega (1991) *Protocols and Applications*, 2nd ed., Madison, WI, pp. 199–238.
21. Sambrook, J., Fritch, E. F., and Maniatis, T. (1989) *Molecular Cloning*, 2nd ed., Cold Spring Harbor Laboratory, Cold Spring Harbor, NY.
22. Saiki, R. K., Gelfand, D. H., Stoffel, S., Scharf, S. J., Higuchi, R., Horn, G. T., Mullis, K. B., and Erlich, H. A. (1988) Primer-directed enzymatic amplification of DNA with a thermostable DNA polymerase. *Science* **239,** 487–491.
23. Moon, I. S. and Krause, M. O. (1991) Common RNA polymerase I, II, and III upstream elements in mouse 7SK gene locus revealed by the inverse polymerase chain reaction. *DNA Cell Biol.* **10,** 23–32.
24. Strobel, S. A. and Dervan, P. B. (1990) Site-specific cleavage of a yeast chromosome by oligonucleotide-directed triple-helix formation. *Science* **249,** 73–75.
25. Dreyer, G. B. and Dervan, P. B. (1985) Sequence-specific cleavage of single-stranded DNA: oligodeoxynucleotide-EDTA.Fe(II). *Proc. Natl. Acad. Sci. USA* **82,** 968–972.
26. Zhang, H., Scholl, R., Browse, J., and Somerville, C. (1988) Double strand DNA sequencing as a choice for DNA sequencing. *Nucleic Acids Res.* **16,** 1220.
27. Sugino, A., Goodman, H. M., Heynecker, H. L., Shine, J., Boyer, H. W., and Cozzarelli, N. R. (1977) Interaction of bacteriophage T4 RNA and DNA ligases in joining of duplex DNA at base-paired ends. *J. Biol. Chem.* **252,** 3987.

81

Long-Range Polymerase Chain Reaction

William Waggott

1. Introduction
1.1. Background to Long-Range PCR Amplification

Conventional polymerase chain reaction (PCR) enables reliable amplification of 3–4 kb of DNA *(1)* while attempts at optimization has enabled 15.6 kb of λ DNA to be amplified *(2)*. The maximum amplifiable length of PCR is limited by the low fidelity of the *Thermus aquaticus* (*Taq*) DNA polymerase *(3)*, the most commonly used thermostable polymerase. It is believed that inadvertent nucleotide misincorporations during the PCR extension steps cause chain terminations *(3)*. The *Taq* polymerase lacks proofreading properties *(4)* and thus is unable to correct such misincorporations. The higher extension K_M value for a misincorporated nucleotide is thought to cause detachment of the *Taq* polymerase from template DNA.

The turning point in the development of long and accurate PCR amplifications has been the discovery that proofreading enzymes with exonuclease activity enables a tenfold increase in the PCR product length. Utilizing λ bacteriophage DNA as a template, Barnes first reported the high fidelity amplification of 35 kb of DNA *(5)*. A proofreading enzyme with 3'–5' exonuclease activity, *Pyrococcus furiosus (Pfu)* DNA polymerase, was used in conjunction with *Taq*. This record was later exceeded by the amplification of 44 kb of lambda DNA and a 22 kb fragment from human genomic DNA *(6)* using similar methodology. It is believed that the *Pfu* polymerase circumvents the low fidelity of the *Taq* polymerase by correcting the nucleotide misincorporations, paving the way for the *Taq* polymerase to complete the polymerization reaction.

1.2. Utility of Long-Range PCR

Long accurate amplification has enhanced the utility of PCR in genetic and clinical studies. A selection of research reports utilizing long-range PCR are listed in **Table 1**.

In gene isolation, long-range PCR would minimize laborious microbiological and vector manipulation procedures by enabling direct isolation of cDNA inserts. A 50-kb insert may be amplified in two halves starting from the flanking vector sequences. Studies into gene structure would be assisted by enabling amplifications across intron/exon boundaries. PCR-based chromosome walking would also be greatly enhanced.

From: *The Nucleic Acid Protocols Handbook*
Edited by: R. Rapley © Humana Press Inc., Totowa, NJ

Table 1
A Selection of Reports Utilizing Long-Range PCR

Ref.	Target	Cloning and characterization
18	Hepatitis B genome	Structural characterization of heterogeneous populations of complete virion-encapsidated full length HBV DNAs from sera of transplant patients
19	Mouse L-isoaspartyl/D-aspartyl methyl-transferase gene	Determined the exon positions of methyl-transferase gene by restriction mapping of long PCR fragments
20	Long inserts cloned on lambda EMBL3 phage vector	Rapid mapping of the restriction sites of long inserts cloned in lambda EMBL3 phage vector
21	Retinoblastoma gene (RB1)	Coamplification of 24 RB gene exons prior to multiplex mutational scanning by two-dimensional denaturing gradient gel electrophoresis (DGGE) Clinical diagnosis
14,15	Nucleophosmin-Anaplastic lymphoma kinase gene fusion	Gene fusion at the t(2;5)(p23;q35) breakpoint, a diagnostic gene rearrangement for a type of non-Hodgkin's Lymphoma
7	Human polyomavirus JC (JCV)	PCR from clinical specimens of progressive multifocal leukoencephalopathy patients prior to RFLP analysis and sequencing for viral typing and characterization
16	Various gene fusions found at translocation breakpoints	Amplification of junctional sequences in chromosome translocations of B cell neoplasms
11	Trinucleotide repeat expansion	CTG trinucleotide repeat expansions associated with myotonic dystrophy was PCR amplified to facilitate molecular diagnosis and studies of the disease

Whole viral (7) and mitochondrial (8,9) genomes have been amplified to search for gene deletions and mutations. Amplification of FRDA gene intron in Friedreich's ataxia (10) has shown that the intronic GAA triplet expansion is a potential pathogenic pathway. Presymptomatic disease diagnosis may thus be possible in this and other diseases sharing similar modes of pathogenesis such as Huntington's disease and myotonic dystrophy (11).

Recent development of a long range reverse transcriptase PCR (RT-PCR) system has extended the scope for assaying for rare transcripts and tissue specific gene expression. Amplifications of full length transcripts of hepatitis A virus (12) and transcripts as large as 13.5 kb has recently been demonstrated. Long-range PCR has also impacted on the differential display technique, allowing larger transcripts to be studied (13).

Detection of gene rearrangements found at chromosome translocation breakpoints in hematological malignancies may be greatly facilitated by long-range PCR and may be helpful in clinical diagnosis. Molecular diagnosis of such abnormality usually relies on RT-PCR or Northern blotting for the detection of novel chimeric transcripts resulting from gene fusion. At the DNA level, such gene rearrangements span distances

Table 2
Representative Long-Range PCR Kits and Notable Features

Product name	Manufacturer	Proofreading enzyme	Second polymerase enzyme	Ref.
Expand™ Long Template PCR System	Boehringer Mannheim	*Pwo* 3'–5' exonuclease	*Taq*	*18,24*
TaKaRa LA PCR™ Kit Version 2	TaKaRa	3'–5' exonuclease	LA*Taq*	*16,25*
Gene Amp XL PCR Kit	Perkin-Elmer	Vent 3'–5' exonuclease	r*Tth*	*10,26*
Taq Plus™	Stratagene	*Pfu* 3'–5' exonuclease	*Taq*	*27*
Elongase	Life Technologies	GD-B *Pyrococcus* 3'–5' exonuclease	*Taq*	not available

beyond the amplification limit of conventional PCR and thus can only be analyzed by Southern blotting. Thus, in designing a genomic DNA-based PCR detection system for a recently characterized nucleophosmin-anaplastic lymphoma kinase (NPM-ALK) gene fusion, of unknown intron structure, we *(14)* and others *(15)* have adopted the long-range PCR methodology. Long-range PCR has also been utilized in the detection of other gene rearrangements *(16)*.

1.3. Model Systems: Detection of NPM-ALK Gene Fusion of t(2;5)(p23;q35) Chromosomal Abnormality in Anaplastic Large Cell Lymphoma (ALCL) and Amplification of 23 kb of Genomic DNA

In this chapter, application of long-range PCR to the detection of NPM-ALK gene fusion from genomic DNA is outlined *(14)* utilizing an Expand™ Long Template PCR System (Boehringer Mannheim) (*see* **Table 2** and **Note 1**). The NPM-ALK gene fusion was recently identified at the genomic breakpoint of the t(2;5)(p23;q35) chromosome translocation, found in anaplastic large cell lymphoma *(17)*. In order to further characterize the intronic region of the translocation breakpoint in seven cases of ALCL, a long-range PCR strategy was adopted. In a second example, a 23-kb fragment of the human β-globin gene region is amplified from genomic DNA using a 20 kb Plus System (Boehringer Mannheim, Mannheim, Germany).

2. Materials
2.1. Template DNA Preparation

1. Cultured cell lines SU-DHL-1 and Karpas 299.
2. Human peripheral blood.
3. Nucleon 1 DNA extraction kit (Scotlab, Strathclyde, Scotland, UK).
4. QIAGEN Genomic-tip 20G/100G (Qiagen, Surrey, UK).

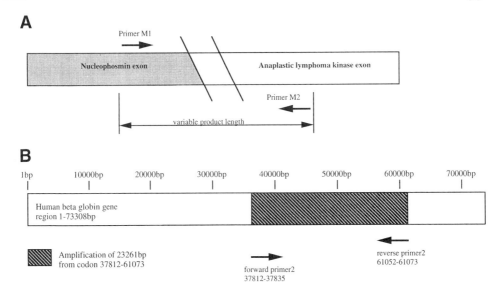

Fig. 1. **(A)**. Schematic diagram NPM-ALK gene fusion at t(2;5)(p23;q35). **(B)** Human β-globin gene region on chromosome 11 (DNA sequence source: National Center of Biotechnology Information, NCBI ID:U01317), indicating PCR primer positions and the PCR amplified regions.

2.2. Thermocycling Machine

Thermal cyclers capable of extending the extension time for successive cycles are recommended. Perkin-Elmer Gene Amp 9600/Biometra Trio Thermoblock Thermal Cycler (Norwalk, CT).

2.3. PCR Primers (see Note 2)

1. Amplification of NPM-ALK gene fusion.
   a. Forward primer 5' NPM (100 ng/µL stock), M1: 5'-TCC CTT GGG GGC TTT GAA ATA ACA CC-3'.
   b. Reverse primer 3' ALK (100 ng/µL stock), M2: 5'-CGA GGT GCG GAG CTT GCT CAG C-3'.
   These PCR primers were designed from the respective exons of the NPM and ALK genes involved in the translocation breakpoint. The position of the PCR primers with respect to the known genomic sequences are illustrated in **Fig. 1A**.
2. Amplification of human β-globin gene.
   a. Forward primer 2 (158 ng/µL stock), 5'-CAC AAG GGC TAC TGG TTG CCG ATT-3'.
   b. Reverse primer 2 (145 ng/µL stock), 5'-AGC TTC CCA ACG TGA TCG CCT T-3'.
   The locations of these primers are indicated in **Fig. 1B**.

2.4. PCR Reaction Tubes

Thin-walled 0.2 or 0.5-mL reaction tubes may be used. Thin-walled tubes ensure efficient heat exchange between the thermoblock and the reactions.

2.5. Long-Range PCR Systems

1. Expand Long Template PCR System (Boehringer Mannheim).
2. Expand 20 kb Plus System (Boehringer Mannheim).

Table 3
Composition of Reaction Mixes

Reaction components (stock concentrations)	Volume (µL) NPM-ALK Expand™ Long Template	Volume (µL) β-Globin 20 kb Plus System
Master mix 1		
dNTPs; 100 mM nucleotide mix	1.0	1.0
Forward primers	1.0	1.0
5' NPM;100 ng/µL stock: primer M1		
5' β-globin;158 ng/µL stock: forward primer 2		
Reverse primers	1.0	1.0
3' ALK;100 ng/µL stock: primer M2		
3' β-globin;145 ng/µL stock: reverse primer 2		
0.2 mg/µL human genomic DNA	1.0	1.0
Sterile redistilled water	23.0	23.0
Final volume	25.0	25.0
Master mix 2		
PCR buffer mix	5.0[a]	10.0
Enzyme mix	0.75	0.75
Sterile double-distilled water	19.25	14.25
Final volume	25.0	25.0

[a]See **Note 4**.

3. Methods

3.1. Template DNA Preparation

1. Extract DNA for cultured cell lines SU-DHL-1 and Karpas 299 cell lines with t(2;5) (p23;q35) chromosome translocation using a Nucleon 1 DNA extraction kit.
2. Use normal human peripheral blood as a control.

3.2. Long-Range PCR Methodology

1. Thaw the reaction components and place on ice.
2. Prepare two separate reaction mixes as outlined in **Table 3**, separating the primer, template, and nucleotides from the enzyme and buffer mix. This is important for long-range PCR using DNA polymerase with a 3'–5' exonuclease activity, e.g., *Pwo*, which would otherwise digest the free primers.
3. Gently agitate the reaction mixes to thoroughly mix the components before combining the two halves. Mix thoroughly again and centrifuge briefly to collect the solution.
4. Overlay 30 µL of paraffin oil.
5. Place tubes on PCR machine and start thermocycling immediately.

3.3. Thermocycling Profiles and Other Variables

1. Thermocycling profiles are adapted to suit the size of the target DNA template. For amplification of 23 kb of genomic DNA, an extension time of 18 min is required.
2. The concentration of dNTP, MgCl$_2$, template, and enzyme concentrations are modified according to the template size and whether the template is genomic DNA or λ DNA. These parameters should be adjusted as outlined in the manufacturers' instructions.

3.4. Thermocycling Profile

1. Amplification of the NPM-ALK gene fusion product from genomic DNA samples of ALCL cases.
   PCR conditions were adapted for the amplification of up to 5 kb of genomic DNA.
   a. Initial denaturation at 92°C for 2.0 min.
   b. Denature 92°C for 10 s; anneal 65°C for 30 s; elongation 68°C for 45 s for a total of 10 cycles.
   c. Denature 92°C for 10 s; anneal 65°C for 30 s; elongation 68°C for 45 s for a total of 10 cycles.
   d. Denature 92°C for 10 s; anneal 65°C for 30 s; elongation 68°C for 45 s, increase elongation time by 20 s increments in each successive cycle for a total of 10 cycles.
   e. Final extension time at 68°C for 7 min.
   Store at 4°C until analysis by agarose gel electrophoresis.
2. Amplification of β-globin gene from human genomic DNA.
   PCR conditions were adapted for the amplification of up to 20–35 kb of genomic DNA.
   a. Initial denaturation at 92°C for 2.0 min.
   b. Denature 92°C for 10 s; anneal 65°C for 30 s; elongation 68°C for 45 s for a total of 10 cycles.
   c. Denature 92°C for 10 s; anneal 62°C for 30 s; elongation 68°C for 18 min, increase elongation time by 20 s increments in each successive cycle for 19 cycles for a total of 20 cycles.
   d. Final extension time at 68°C for 7 min.
   Store at 4°C until analysis by agarose gel electrophoresis.

3.5. PCR Product Analysis

1. Mix 15 µL of each PCR reaction products with 3 µL of loading dye.
2. Load onto a 0.8% agarose 1X TBE minigel gel. The long β-globin PCR products can be run at 60 V for 3 h (**Fig. 2A**). For smaller products up to 3.0 kb long an 1% agarose 1X TBE minigel has achieved satisfactory resolution run at 100 V for 60 min (**Fig. 2B**).

4. Notes

1. Methodologies in long-range PCR are based in the majority on commercially available kits. A large number of such kits utilize various proofreading enzymes in conjunction with thermostable DNA polymerases. Representative commercial products and their notable features are listed in **Table 2**. Long amplifications can also be achieved using proofreading enzymes alone. Although polymerases with proofreading activity generally have inferior processivity to *Taq* (thus limiting their application to products of up to 5 kb), they have better fidelity to *Taq*/proofreading enzyme combinations *(22)*. These single enzyme systems are thus recommended for applications in which fidelity is of great importance.
2. PCR primers should ideally have balanced T$_m$ (between 63°C and 68°C) not differing by more than 1°C or 2°C and be between 21–34 nucleotides in length. Care must be taken to avoid opportunities for dimer and secondary structure formation.

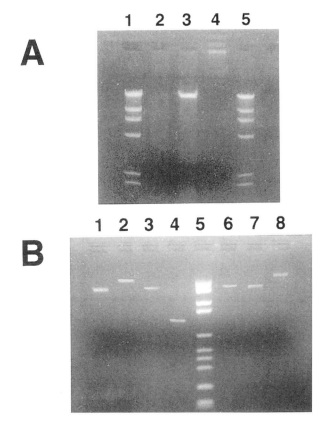

Fig. 2. (**A**) Amplification of 23261 bp of human β-globin gene from genomic DNA. Lanes 1 and 5, DNA size marker lanes, λ DNA digested with *Hin*dIII, 23,130, 9416, 6557, 4361, 2322, and 2027 bp. Lane 2, β-globin amplification from human genomic DNA extracted using conventional methodology from human peripheral blood. Lane 3, β-globin amplification from human genomic DNA extracted using Qiagen genomic tip for preparation of high molecular weight DNA from human peripheral blood. Lane 4, Negative control. (**B**) Amplification of NPM-ALK gene fusion from genomic DNA. Lanes 1–4 and 6–8: NPM-ALK amplification from genomic DNA from tumor tissue of seven ALCL non-Hodgkin's lymphoma patients. Lane 5, DNA size marker lane, 2176, 1766, 1230, 1033, 653, 517, 453, 394, 298, and 234 bp.

3. Extremely long PCR amplifications from genomic DNA (20 kb or longer) requires DNA of high integrity *(23)*. The integrity of DNA may be checked by running the DNA on a 0.3% agarose gel. Single-stranded nicks or double-stranded breaks caused by manipulation procedures in conventional DNA extraction protocols reduce the copy number of long target sequences and are not suitable for long amplification (**Fig. 2A**).
4. The Expand PCR System offers a choice of three buffer systems to suit the size of the DNA template to be amplified.
    a. Buffer 1: 0.5–12 kb of λ DNA and 0.5–25 kb of genomic DNA.
    b. Buffer 2: 12–15 kb of λ DNA and 25–30 kb of genomic DNA.
    c. Buffer 3: >15 kb of λ DNA and >30 kb of genomic DNA.
    The 20 kb Plus System has a single buffer.

References

1. Erlich, H. A., Gelfand, D., and Sninsky, J. J. (1991) Recent advances in the polymerase chain reaction. *Science* **252,** 1643–1651.
2. Kainz, P., Schmiedlechner, A., and Strack, H. B. (1992) In vitro amplification of DNA fragments greater than 10 kb. *Anal. Biochem.* **202,** 46–49.
3. Innis, M. A., Myambo, K. B., Gelfand, D. H., and Brow, M. (1988) DNA sequencing with *Thermus aquaticus* DNA polymerase and direct sequencing of polymerase chain reaction amplified DNA. *Proc. Natl. Acad. Sci. USA* **85,** 9436–9440.
4. Tindall, K. R. and Kunkel, T. A. (1988) Fidelity of DNA synthesis by the *Thermus aquaticus* DNA polymerase. *Biochemistry* **27,** 6008–6013.
5. Barnes, W. M. (1994) PCR amplification of up to 35 kb DNA with high fidelity and high yield from lambda bacteriophage templates. *Proc. Natl. Acad. Sci. USA* **91,** 2216–2220.
6. Cheng, S., Fockler, C., Barnes, W. M., and Higuchi, R. (1994) Effective amplification of long targets from cloned inserts and human genomic DNA. *Proc. Natl. Acad. Sci. USA* **91,** 5695–5699.
7. Agostini, H. T. and Stoner, G. L. (1995) Amplification of the complete polyomavirus JC genome from brain, cerebrospinal fluid and urine using pre-PCR restriction enzyme digestion. *J. Neurovirol.* **1,** 316–320.
8. Li, Y. Y., Hengstenberg, C., and Maisch, B. (1995) Whole mitochondrial genome amplification reveals basal level multiple deletions in mtDNA of patients with dilated cardiomyopathy. *Biochem. Biophys. Res. Comm.* **210,** 211–218.
9. Reynier, P., Pellissier, J. F., Harle, J. R., and Malthiery, Y. (1994) Multiple deletions of the mitochondrial DNA in polymyalgia rheumatica. *Biochem. Biophys. Res. Comm.* **205,** 375–380.
10. Campuzano, V., Montermini, L., Molto, M. D., Pianese, L., Cossee, M., Cavalcanti, F., Monros, E., Rodius, F., Duclos, F., Monticelli, A., Zara, F., Canizares, J., Koutnikova, H., Bidichandani, S. I., Gellera, C., Brice, A., Trouillas, P., Demichele, G., Filla, A., Defrutos, R., Palau, F., Patel, P. I., Didonato, S., Mandel, J. L., Cocozza, S., Koenig, M., and Pandolfo, M. (1996) Friedreichs ataxia—an autosomal recessive disease caused by an intronic GAA triplet repeat expansion. *Science* **271,** 1423–1427.
11. Cheng, S., Barcelo, J. M., and Korneluk, R. G. (1996) Characterization of large CTG repeat expansions in myotonic dystrophy alleles using PCR. *Hum. Mutat.* **7,** 304–310.
12. Tellier, R., Bukh, J., Emerson, S. U., and Purcell, R. H. (1996) Amplification of the full-length hepatitis A virus genome by long reverse transcription PCR and transcription of infectious RNA directly from the amplicon. *Proc. Natl. Acad. Sci. USA* **93,** 4370–4373.
13. Diachenko, L. B., Ledesma, J., Chenchik, A. A., and Siebert, P. D. (1996) Combining the technique of RNA fingerprinting and differential display to obtain differentially expressed messenger RNA. *Biochem. Biophys. Res. Comm.* **219,** 824–828.
14. Waggott, W., Lo, Y. M. D., Bastard, C., Gatter, K. C., Leroux, D., Mason, D. Y., Boultwood, J., and Wainscoat, J. S. (1995) Detection of NPM-ALK DNA rearrangement in CD30 positive anaplastic large-cell lymphoma. *Br. J. Haematol.* **89,** 905–907.
15. Sarris, A. H., Luthra, R., Papadimitracopoulou, V., Waasdorp, M., Dimopoulos, M. A., McBride, J. A., Cabanillas, F., Duvic, M., Deisseroth, A., Morris, S. W., and Pugh W. C. (1996) Amplification of genomic DNA demonstrates the presence of the t(2-5)(p23-q35) in anaplastic large-cell lymphoma, but not in other non-Hodgkins lymphomas, Hodgkins disease, or lymphomatoid papulosis. *Blood* **88,** 1771–1779.
16. Akasaka, T., Muramatsu, M., Ohno, H., Miura, I., Tatsumi, E., Fukuhara, S., Mori, T., and Okuma, M. (1996) Application of long distance polymerase chain reaction to detection of junctional sequences created by chromosomal translocation in mature B-cell neoplasms. *Blood* **88,** 985–994.

17. Morris, S. W., Kirstein, M. N., Valentine, M. B., Dittmer, K. G., Shapiro, D. N., Saltman, D. L., and Look, A. T. (1994) Fusion of a kinase gene, ALK, to a nucleolar protein gene, NPM, in non-Hodgkins lymphoma. *Science* **263**, 1281–1284.
18. Gunther, S., Li, B. C., Miska, S., Kruger, D. H., Meisel, H., and Will, H. (1995) A novel method for efficient amplification of whole hepatitis B virus genomes permits rapid functional analysis and reveals deletion mutants in immunosuppressed patients. *J. Virol.* **69**, 5437–5444.
19. Maclaren, D. C. and Clarke, S. (1996) Rapid mapping of genomic p1 clones—the mouse L-isoaspartyl/D-aspartyl methyltransferase gene. *Genomics* **35**, 299–307.
20. Machida, M., Manabe, M., Yasukawa, M., and Jigami, Y. (1996) Application of long-distance PCR to restriction site mapping of a cloned DNA fragment on the lambda-EMBL3 phage vector. *Biosci. Biotechnol. Biochem.* **60**, 1011–1013.
21. Li, D. Z. and Vijg, J. (1996) Multiplex coamplification of 24 retinoblastoma gene exons after pre-amplification by long-distance PCR. *Nucleic Acids Res.* **24**, 538,539.
22. Cline, J., Braman, J. C., and Hogrefe, H. H. (1996) PCR fidelity of *Pfu* DNA polymerase and other thermostable DNA polymerases. *Nucleic Acids Res.* **24**, 3546–3551.
23. Cheng, S., Chen, Y. M., Monforte, J. A., Higuchi, R., and Vanhouten, B. (1995) Template integrity is essential for PCR amplification of 20 kb to 30 kb sequences from genomic DNA. *PCR Methods Appl.* **4**, 294–298.
24. Melov, S., Lithgow, G. J., Fischer, D. R., Tedesco, P. M., and Johnson, T. E. (1995) Increased frequency of deletions in the mitochondrial genome with age of *Caenorhabditis elegans*. *Nucleic Acids Res.* **23**, 1419–1425.
25. Ermak, G., Jennings, T., Robinson, L., Ross, J. S., and Figge, J. (1996) Restricted patterns of CD44 variant exon expression in human papillary thyroid carcinoma. *Cancer Res.* **56**, 1037–1042.
26. Stewart, A., Gravitt, P. E., Cheng, S., and Wheeler, C. M. (1995) Generation of entire human papillomavirus genomes by long PCR- frequency of errors produced during amplification. *PCR Methods Appl.* **5**, 79–88.
27. Ling, M. F. and Robinson, B. H. (1995) A one-step polymerase chain reaction site-directed mutagenesis method for large gene cassettes with high efficiency, yield, and fidelity. *Anal. Biochem.* **230**, 167–172.

VIII

ANALYZING GENES, MUTATIONS, AND PROTEIN INTERACTIONS

82

Nonradioactive Differential Display of Messenger RNA

Thomas C. G. Bosch and Jan U. Lohmann

1. Introduction

Changes in cell behavior are driven by changes in gene expression. Thus, in order to understand the mechanisms regulating cell behavior, one has to identify and characterize differentially expressed genes. Standard methods currently used to isolate differentially expressed genes include subtractive hybridization *(1,2)*, differential hybridization *(3)*, and single-cell polymerase chain reaction (PCR) *(4)*. Differential display of mRNA by PCR (DD-PCR) is a new and powerful procedure for quantitative detection of differentially expressed genes *(5,6)*. Advantages over alternative approaches include: quantitative identification of differences in gene expression between different cell fractions; simultaneous detection of both upregulation and downregulation of genes; requirement of only small amounts of messenger RNA; and drastically reduced time of analysis. Since on average there are about 15,000 individual mRNA species present in any individual cell, some steps of selection have to be taken before the transcript population can be displayed simultaneously. Therefore, in the DD-PCR procedure mRNAs from different cell fractions are reverse transcribed (RT) using an oligo-dT-NN anchor primer. Since this 3' primer will hybridize only with transcripts carrying the corresponding two NN-bases in front of the poly(A) tail, only $1/12$ (i.e., about 1200) of all transcripts are reverse transcribed into cDNA. Accordingly, for reverse transcription of all messages 12 different 3' primer have to be used independently. PCR of the resulting cDNA is then carried out using radiolabeled dNTPs, the oligo-dT-NN anchor primer as 3' primer and a short arbitrary oligonucleotide as 5' primer. By using 10 mer as 5' primers, about 150–300 products will be obtained in a PCR using 1200 different cDNAs as templates. Thus, for quantitative screening of the transcript pool of a particular cell, 20 arbitrary 5' primers have to be used in 240 independent PCR reactions. After electrophoretic separation of the resulting fragments on a polyacrylamide gel, differential gene expression is visualized by autoradiography. Differentially expressed cDNA species can be recovered from the gel using the autoradiogram for band localization. DD-PCR has previously been employed to identify differentially expressed genes in preimplantation mouse embryo *(7)*, in human endothelial cells treated with fibroblast growth factor *(8)*, in normal and tumor cells *(9)*, and in *Salmonella* treated with peroxide *(10)*. A recent review of the DD-PCR method can be found in McClelland et al. *(6)*.

Although straightforward, the original procedure suffers from a number of drawbacks—poor reproducibility and false positives due to the technical difficulty in recovering a unique radiolabeled DNA species from the polyacrylamide gel *(8,11,12)*. Since success rates in differential display are likely to be increased considerably by directly staining the cDNA populations in the polyacrylamide gel, here we describe a simple and nonradioactive method of differential display of cDNA that allows both rapid screening of a large number of samples for differentially expressed genes and efficient isolation of unique cDNA species (REN-Display, **ref. *13***).

mRNA is isolated from two or more cell populations, RT and PCR amplified using one 3' anchored oligo-dT-NN primer and one 5' 10 mer arbitrary primer. The PCR products are resolved in adjacent lanes on horizontal polyacrylamide gels under denaturing conditions. cDNA bands are visualized by silver staining. Differentially expressed mRNA species are identified by comparing the pattern of silver-stained cDNAs in adjacent lanes. To detect false positives, 2–3 identical PCR reactions are carried out in parallel. Differentially expressed mRNA species can then be directly isolated from the gel and further characterized. Due to the visibility of the cDNA in the original gel, the probability of recovering a mRNA species of interest is increased considerably compared to the standard radioactive procedure. However, even single silver-stained bands contain usually more than one transcript species. Therefore, cloning of the PCR products must be followed by analyzing several (up to 10) independent transformants.

The outline of the nonradioactive differential display method is shown in **Fig. 1**. The REN technique is particularly valuable when applied to complex biological systems and developmental and physiological states, since it requires little amounts of biological material, allows rapid (72 h) investigations of a large number of samples, and can be carried out even in routine diagnostic laboratories.

2. Materials

2.1. mRNA Preparation and Reverse Transcription

All solutions are prepared with water that has been treated with 0.1% diethyl pyrocarbonate.

1. QuickPrep mRNA Purification Kit (Amersham Pharmacia Biotech, Freiburg, Germany).
2. First-strand cDNA Synthesis Kit (Amersham Pharmacia Biotech).
3. 3' tailing primers (e.g., 5'-$T_{(12)}$AG-3').

2.2. PCR

1. Thermocycler.
2. *Taq* polymerase.
3. Nucleotide triphosphates.
4. Arbitrary 10-mer oligonucleotides (Operon, Alameda, CA).
5. Sequencing stop solution: 95% formamide, 20 mM Na$_2$EDTA, 0.05% (w/v) bromphenol blue, 0.05% (w/v) xylene cyanol.

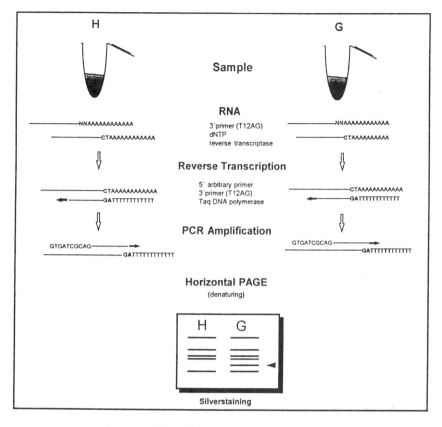

Fig. 1. Outline of the REN display procedure.

2.3. Horizontal Gel Electrophoresis

1. CleanGel 12, 5% 48S (ETC Electrophorese Technik, Kirchentellinsfurt, Germany).
2. DNA Disc Buffer system (ETC Electrophorese Technik).
3. Multiphor II Electrophoresis Unit (Amersham Pharmacia Biotech).
4. Gelpool/Paperpool (Amersham Pharmacia Biotech) or other appropriate electrophoresis unit.
5. Silver staining solutions *(14)*, prepared in a final volume of 250 mL:
   a. Fixing solution: 10% acetic acid (v/v).
   b. Staining solution: 0.1% $AgNO_3$ (w/v) + 250 µL 37% formaldehyde.
   c. Developing solution: ice-cold 2.5% Na_2CO_3 + 250 µL 37% formaldehyde + 250 µL 2% Na-thiosulfate solution.
   d. Stop solution: 2% glycine + 0.5% EDTA-disodium.
   e. Impregnation: 5% glycerol (v/v).

3. Methods

3.1. mRNA Preparation and Reverse Transcription (see Notes 1 and 2)

1. Isolate mRNA or total RNA from about 5×10^5 cells using, for example, the Pharmacia QuickPep mRNA Purification Kit. Final volume: 20 µL.
2. Calculate concentration of RNA by removing an aliquot (2 µL) of RNA suspension and determining the A_{260} in 100 µL water. Yield should be approx 1 µg mRNA. The samples can be stored at –20°C at this point.

3. For reverse transcription use 250 ng of mRNA or 3 µg of total RNA and First-Strand cDNA Synthesis Kit from Amersham Pharmacia Biotech.
4. Mix 5 µL bulk reaction mix, 1 µL DDT, 1.25 µL of 25 µM stock of tailing primer (e.g., 5'-$T_{(12)}$AG-3 ').
5. Heat 7.75 µL of RNA solution (250 ng of mRNA or 3 µg of total RNA) at 65°C for 10 min. Thereafter place on ice.
6. Add 7.75 µL of mRNA solution to reaction mix. Final volume: 15 µL.
7. Reaction time is 1 h at 37°C.
8. Stop reaction by heating for 5 min at 90°C.

3.2. PCR

1. Dilute the resulting cDNA mixture 1:50 with water. Use 3 µL of a dilution as template for a 10 µL PCR reaction (concentrations are of stock solutions) (*see* **Notes 3** and **4**).
2. Mix 3 µL cDNA (1:50), 1 µL 10X PCR buffer, 1 µL random 10-mer primer (5 µM), 1 µL tailing primer (25 µM), 2 µL dNTP (100 µM), 0.1 µL *Taq* (0.5 U), 1.9 µL H_2O.
3. Cover reaction with 30 µL mineral oil.
4. PCR parameters: initial denaturing step for 5 min at 94°C, followed by 43 cycles with cycle times of 30 s at 94°C, 60 s at 42°C, and 30 s at 72°C.
5. Stop reaction by adding 7 µL sequencing stop solution to 10 µL reaction mix. The samples can be stored at –20°C at this point.

3.3. Horizontal Polyacrylamide Gel Electrophoresis

1. Hydrate CleanGels (ETC) in 40 mL gel buffer containing 7 *M* urea for at least 1 h on a shaker using Gelpool.
2. Cut two pieces of electrode wicks lengthwise in half and soak them in 40 mL electrode buffer using Paperpool.
3. Apply 2 mL kerosine oil to center of Multiphor II cooling plate.
4. Remove excess buffer from the gel using filter paper.
5. Lay hydrated gel on the plate.
6. Lay electrode wicks on hydrated gel.
7. Boil sample for 4 min. Chill samples on ice before loading.
8. Load 8 µL of sample solution in wells. Store remaining solution at –20°C.
9. Running conditions: Start at 400 V_{max} and 11 mA_{max} for 30 min at room temperature (22°C). Thereafter run at 18 mA_{max} for 2 h.
10. Stop electrophoresis when Xylene cyanol band reaches the anode wicks.

3.4. Visualization of PCR Products by Silver Staining (see Notes 5 and 6)

1. Fix for 30 min in 250 mL fixation solution.
2. Wash 3×2 min in 250 mL H_2O.
3. Stain for 30 min in 250 mL staining solution.
4. Rinse gel for 30 s in H_2O.
5. Develop for 5 to 15 min in 250 mL ice-cold developing solution.
6. Stop reaction by incubating for 10 min in 250 mL stop solution.
7. Impregnate by incubating for 10 min in 250 mL impregnation solution.

A typical result showing the differential pattern of silver-stained cDNAs from head and gastric tissue of the freshwater polyp *Hydra* is shown in **Fig. 2**. Poly(A)$^+$ RNA was isolated and reverse transcribed using the 3' anchor primer $T_{(12)}$AG. PCR amplification of the reverse transcription reaction was carried out in duplicates using the 3' anchor

Identification of Genes by Display of mRNA

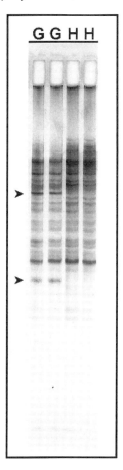

Fig. 2. Differential display of hydra genes from head and gastric cells. Lanes marked G contain PCR products derived from mRNA of gastric tissue. Lanes marked H contain PCR products derived from head specific cells. Arrowheads identify two genes specifically expressed in gastric tissue and absent in head specific cells.

primer and an arbitrary 10 mer 5' primer (5'-GGGTAACGCC-3'). Two differentially expressed transcripts can be detected by this primer (arrowheads in **Fig. 2**).

3.5. Elution of Candidate Bands and Reamplification (see Note 7)

1. Transfer silver-stained cDNA directly from the gel into a PCR tube using two sterile pipet tips.
2. Add 6 μL water to the isolated polyacrylamide slice.
3. Smash the gel slice and vortex briefly.
4. Centrifuge at high speed in a microfuge for 30 s and boil for 2 min to support elution.
5. Vortex and centrifuge again.
6. Prepare four sterile PCR tubes containing 0 μL, 2 μL, 2.7 μL, and 3 μL of water.
7. Add elution mix to the tubes to reach a final volume of 3 μL (in the last 3 μL tube; just dip the used pipet tip into the water).
8. From the elution mixtures, use the total of 3 μL of each dilution as template for a 10 μL PCR reaction. Mix 3 μL eluted PCR product, 1 μL 10 X PCR buffer, 1 μL random

10 mer primer (5 µM), 1 µL tailing primer (2 µM), 0.2 µL dNTP (10 mM), 0.1 µL *Taq* (0.5 U), 3.7 µL H$_2$O.
9. Cover reaction with 30 µL mineral oil.
10. PCR conditions: initial denaturing step 5 min at 94°C, 43 cycles of PCR with cycle times of 30 s at 94°C, 60 s at 42°C, and 30 s at 72°C.

3.6. Confirmation of Reamplification and High-Yield Second Reamplification (see Note 8)

1. For control of successful reamplification of the desired fragment, transfer 6 µL of the PCR reamplification mixture (not all!) into a new tube.
2. Freeze the remaining 4 µL of reamplification products at –20°C.
3. Add 4 µL sequencing stop solution to the 6 µL aliquot.
4. Follow **Subheadings 3.3.** and **3.4.** for loading reamplified fragments next to original PCR products (stored in –20°C).
5. If the PCR fragment of interest was successfully amplified, use an aliquot of the frozen PCR product (*see* **step 2**) for second reamplification and further characterization (e.g., cloning and sequencing) (*see* **Note 9**).
6. To obtain large amounts of PCR product, e.g., for cloning, it is useful to perform a second reamplification step. Use the same concentrations as for first reamplification in a total volume of 100 µL.
7. As template use 2 µL of first reamplification product.
8. Purify PCR product by preparative agarose gel electrophoresis.

4. Notes

1. Critical for success in REN display experiments is to avoid both ribonuclease and nucleic acid contamination. Therefore, adopt semi-sterile techniques including wearing gloves and using water treated with 0.1% diethylpyrocarbonate.
2. Work quickly and keep all reagents on ice.
3. Concentration of cDNA-mixture is also essential for successful PCR.
4. Use dilutions of at least 1:25 to avoid interference with dNTPs and primers from the first strand cDNA synthesis mixture.
5. For silver-staining of cDNA use clean glassware for all incubations and ice-cold developer.
6. Note: 2% sodium thiosulfate stock solution is only stable for about one week at 4°C.
7. For elution of candidate bands the size of the gel slice used for elution should be as small as possible since it may contain toxic substances that inhibit the reamplification procedure.
8. After reamplification and cloning of the desired PCR fragment, screen multiple clones.
9. Any band recovered from the gel can contain more than one cDNA fragment. Thus, in order to detect the clone encoding the differentially expressed transcript, multiple cloned fragments have to be screened and analyzed by Northern blotting (*see* Chapter 37).

References

1. Hedrick, S. M., Cohen, D. I., Nielsen E. A., and Davis, M. M. (1984) Isolation of cDNA clones encoding T-cell specific membrane-associated proteins. *Nature* **308,** 149–153.
2. Travis, G. H. and Sutcliffe, J. G. (1988) Phenol emulsion-enhanced DNA-driven subtractive cDNA cloning: isolation of low-abundance monkey cortex-specific mRNAs. *Proc. Natl. Acad. Sci. USA* **85,** 1696–1700.
3. Dworkin, M. B. and Dawid, I. B. (1980) Construction of a cloned library of expressed embryonic gene sequences from *Xenopus laevis*. *Dev. Biol.* **76,** 435–448.

4. Lambolez, B., Audinat, E., Bochet, P., Crepel, F., and Rossier, J. (1992) AMPA receptor subunits expressed by single Purkinje cells. *Neuron* **9,** 247–258.
5. Liang, P. and Pardee, A. B. (1992) Differential display of eukaryotic messenger RNA by means of the polymerase chain reaction. *Science* **257,** 967–971.
6. McClelland, M., Mathieu-Daulde, F., and Welsh, J. (1995) RNA fingerprinting and differential display using arbitrarily primed PCR. *Trends Genet* **11,** 242–246.
7. Zimmermann, J. W. and Schultz, R. M. (1994) Analysis of gene expression in the pre-implantation mouse embryo: use of mRNA differential display. *Proc. Natl. Acad. Sci. USA* **91,** 5456–5460.
8. Li, F., Barnathan, E. S., and Karikó, K. (1994) Rapid method for screening and cloning cDNAs generated in differential mRNA display: application of Northern blot for affinity capturing of cDNAs. *Nucl. Acids Res.* **22,** 1764–1765.
9. Watson, M. A. and Fleming, T. P. (1994) Isolation of differentially expressed sequence tags from human breast cancer. *Cancer Res.* **54,** 4598–4602.
10. Wong, K. K. and McClelland, M. (1994) Stress-inducible gene of Salmonella typhimurium identified by arbitrarily primed PCR of RNA. *Proc. Natl. Acad. Sci. USA* **91,** 639–643.
11. Liang, P., Averboukh L., Keyomarsi K., Sager R., and Pardee A. B. (1992) Differential display and cloning of messenger RNAs from human breast cancer versus mammary epithelial cells. *Cancer Res.* **52,** 6966–6968.
12. Bauer, D., Muller, H., Reich, J., Riedel, H., Ahrenkiel, V., Warthoe, P., and Strauss, M. (1993) Identification of differentially expressed mRNA species by an improved display technique (DDRT-PCR). *Nucl. Acid Res.* **21,** 4272–4280.
13. Lohmann, J., Schickle, H. P., and Bosch, T. C. G. (1995) REN, a rapid and efficient method for nonradioactive differential display and isolation of rnRNA. *BioTechniques* **18(2),** 200–202.
14. Bassam, B. J., Caetano-Annolés, G., and Gresshoff, P. M. (1991) Fast and sensitive silver staining of DNA in polyacrylamide gels. *Anal. Biochem.* **196,** 80–83.

83

Gene Isolation by Exon Trapping

David B. Krizman

1. Introduction

The technology of exon trapping, sometimes called exon amplification, strives to exploit the phenomenon of mRNA splicing to discover genes directly from genomic DNA. There are three distinct exon trapping methodologies that differ simply in the genomic target of interest. The original experimental design was to capture isolated 3'-splice sites residing within fragments of genomic DNA *(1)*, whereas later approaches focused on either complete internal exons *(2–5)* or entire 3'-terminal exons *(6)*. The requirement of complete, intact exons as targets has proven absolutely essential, and only the trapping of complete internal or 3'-terminal exons is practical.

Plasmids, phage, cosmids, P1s, BACs, YACs, or pooled clones of any type containing mammalian genomic DNA can be used as substrate for either exon trapping approach. Both internal and 3'-terminal exon trapping protocols have successfully trapped exons from all of the aforementioned substrates *(6–9)*. Small numbers of exons are trapped from individual clones, such as plasmid, phage, or cosmid clones, whereas larger numbers can be trapped if using collections of clones, such as phage or cosmid pools. Trapping from YACs is desirable when wishing to discover exons from larger genomic regions that may be difficult to cover with the small-insert cloning vectors.

The choice of trapping system to use is an important consideration. The end result from internal exon trapping is usually a single exon that can be quite small (20–200 bases) in length. An advantage to this system is that nearly all of the exonic sequence will contain valuable protein coding information that can be very useful when performing database searches. These searches usually render some homology with known genes and can possibly lend insight into the function of the gene from which the exon was trapped. A disadvantage is that a probe generated from the trapped exon may be very small and difficult to use in cDNA hybridization screening experiments designed to retrieve the full-length cDNA.

In contrast, the use of 3'-terminal exon trapping generates sequence derived from the last exon of a gene and these sequences can range anywhere from 200–5000 bases in length. Most of the sequence from such trapping experiments is 3'-untranslated region and thus little, if any, useful protein coding sequence. This is a disadvantage when performing database searches, because few strong homologies will be found.

However, this may be a distinct advantage if using this clone in hybridization based cDNA screening experiments owing to the unique nature of these sequences. Also, these sequences can be very useful for obtaining full-length cDNA clones by the technology of 5'-RACE *(10)*. One of the hallmarks of database search results is the existence of multiple stop codons within 3'-untranslated regions of genes. Thus, most trapped 3'-terminal exons will contain many such codons.

Internal exon trapping is accomplished by the subcloning of restricted fragments of genomic DNA into one of many plasmid-based trapping vectors, each engineered to produce vector-derived mRNA molecules when transfected into specific mammalian cells that support vector transcription. Internal exons present in subcloned genomic fragments are incorporated within the transcription unit of the trapping vector, and subsequently included in a chimeric mRNA that results from splicing and processing of vector-derived nascent transcripts. An exon trapped in such a manner is amplified by reverse transcription of RNA purified from transfected mammalian cells followed by the polymerase chain reaction (RT/PCR). Primers used for the RT/PCR reaction are designed to be specific to mRNA species derived only from the trapping vector. This functions to amplify products specifically from the vector and not endogenous mRNA species from the transfected mammalian cell. Once the PCR product is generated, it can be subcloned and sequenced for further analysis.

A typical internal exon is small (average 127 bp in size), and flanked on the 5'-end by a 3'-splice (splice acceptor) and on the 3'-end by a 5'-splice (splice donor). There is usually an open reading frame all the way through the sequence, and most of them will show some degree of homology to known protein coding regions when used to search available sequence databases.

To date, there are four internal exon trapping vectors available, and each vector shares many common features. Most are plasmid-based, and contain the ampicillin resistance gene, bacterial origin of replication, and a trapping cassette consisting of a eukaryotic enhancer/promoter driving transcription of a two-exon transcription unit with a multiple cloning site between the two exons. The first of these exons functions as a 5'-terminal exon, whereas the second exon functions as a 3'-terminal exon capable of directing polyadenylation. Foreign fragments of genomic DNA are inserted between the two exons and any internal exons present are subsequently trapped between the two vector exons by the cellular splicing mechanism.

In addition to the basic features discussed that are common to all internal exon trapping vectors, a variety of custom modifications have been made to individual vectors. The following internal exon trapping vectors are available:

1. pSPL3: This vector is a modification of the original internal exon trapping vector pSPL1 and is the most commonly used trapping vector *(2,7)*. This plasmid consists of the SV40 early region, rabbit β-globin exons, human HIV/Tat 5'- and 3'-splice sites, HIV/Tat intron, and the SV40 poly (A) signal. A multiple cloning site (MCS) exists within the intron between the two chimeric exons.
2. pL53In: The transcription of this plasmid vector is driven by an RSV-LTR. The first exon is derived from the human phosphatase gene, whereas the second exon and the intervening intron are derivatives of the rat preproinsulin gene. A single unique *Kpn*I site acts as the cloning site within the intron *(3)*.

Exon Trapping

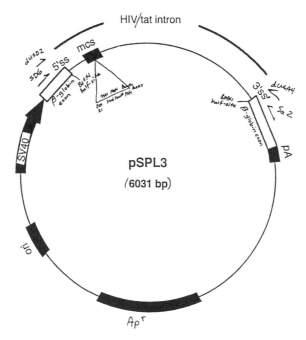

Fig. 1. Diagrammatic representation of internal exon trapping vector pSPL3.

3. pMHC2: This plasmid vector initiates transcription from the SV40 early region. The trapping cassette consists of part of exon 10, intron 10, and part of exon 11 from the human p53 gene together with the SV40 polyadenylation signal. Foreign DNA is inserted into a unique *Bgl*II cloning site within the intron *(4)*.
4. LambdaGET: This is a phage-based vector designed to clone and analyze larger fragments of target genomic DNA than is possible with the plasmid-based trapping vectors *(5)*. This vector is derived from pL53In, and uses the same trapping cassette. This vector offers the added advantage of high-efficiency cloning of foreign target DNA owing to the phage approach of subcloning inserts.

The most commonly used internal exon trapping vector is pSPL3 and is illustrated in **Fig. 1**. It is 6031 bases in length, and contains the Amp^R gene and a bacterial origin of replication for propagation in *Escherichia coli*. The trapping cassette consists of the SV40 early region to direct transcription and replication in the African green monkey cell line Cos7. This cell line harbors a replication-defective mutant SV40 virus that expresses the large T-antigen, which functions to initiate both replication and transcription from the SV40 early region. Exon 1 of pSPL3 is a chimeric exon constructed from rabbit β-globin exon sequences and the human HIV/Tat 5'-splice site. The intron is from the HIV/Tat intron, whereas the last exon begins with HIV/Tat 3'-splice site followed by rabbit β-globin exon sequence and ending with the SV40 poly (A) signal sequence to direct correct cleavage and polyadenylation. *Bst*XI half-sites are present on either side of the intron that functions to allow removal of molecules not containing an internal exon when digested with this enzyme. An MCS is present within the intron, and the following sites are available for subcloning fragments of genomic DNA: *Eco*RI, *Sst*I, *Xho*I, *Not*I, *Xma*III, *Pst*I, *Bam*HI, and *Eco*RV.

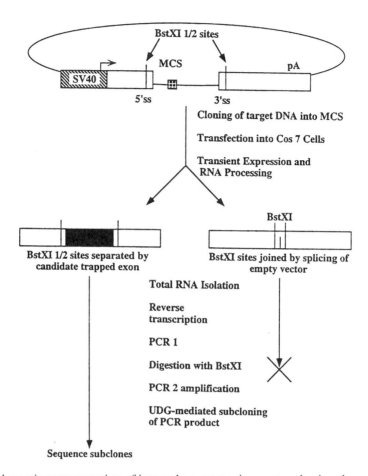

Fig. 2. Schematic representation of internal exon trapping protocol using the vector pSPL3.

The experimental use of this vector will be described here in detail; however, the basic protocol for all internal exon trapping vectors is the same. The general approach to the use of pSPL3 is diagrammed in **Fig. 2**. Genomic fragments of DNA from any source, including individual cosmids, pooled cosmids, P1s, or YACs, are shotgun cloned into one or more of the restriction sites within the MCS and recombinants selected in *E. coli*. All recombinants are picked, pooled, and transiently transfected into Cos7 cells. Total RNA is collected 24 h later and reverse-transcribed using a vector-specific oligonucleotide to yield first-strand cDNA. The primary round of a nested PCR approach to amplify trapped exons is performed using vector-specific primers for six cycles followed by digestion with *Bst*XI. This digestion functions to remove vector-only splicing events that will lead to false-positive scoring. A secondary PCR is performed using nested vector-specific primers to amplify trapped exons. The secondary PCR primers contain an additional 12 bases at their 5'-ends to specifically enable cloning of the PCR product by UDG-mediated high-efficiency cloning. PCR results are analyzed by agarose gel electrophoresis.

The genomic target of 3'-terminal exon trapping is the last exon of a gene and is accomplished by ligation of restricted genomic fragments of DNA with the trapping

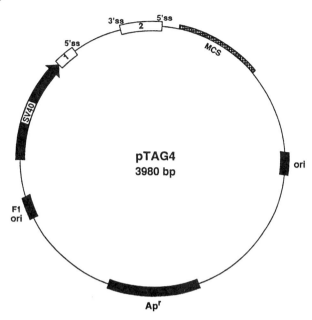

Fig. 3. Diagrammatic representation of 3'-terminal exon trapping vector pTAG4.

vector pTAG4. Ligation products are subsequently transfected directly to Cos7 cells that are able to support transcription from this vector. This vector was engineered to contain a trapping cassette that is an incomplete transcription unit that lacks only a last exon. The foreign DNA fragments are expected to donate a 3'-terminal exon to the vector to complete the transcription unit and generate a stable vector-derived mRNA molecule on transfection into Cos7 cells. Exons trapped in this manner can be amplified by a modification of the RT/PCR reaction termed 3'-rapid amplification of cDNA ends (3'-RACE) *(10)* that was designed to amplify mRNA species specifically from the 3'-end of the molecule using the poly(A) tail as an anchor. Primers used for the PCR reaction are specific for vector exons, thus imparting the specificity needed when using mRNA preparations from transfected mammalian cells. PCR product is subcloned and sequenced for further analysis.

A typical 3'-terminal exon is larger, averaging 627 bp, than an internal exon and consists of mostly a 3'-untranslated region flanked on the 5'-end by a 3'-splice site (splice acceptor), and on the 3'-end by a cleavage site and the consensus poly (A) signal AATAAA or ATTAAA. On a sequence database search, a 3'-terminal exon will contain little protein coding sequence. Thus, very little sequence homologies based on these regions will be seen. In fact, the existence of stop codons in all frames is suggestive of a 3'-terminal exonic sequence.

There is only one 3'-terminal exon trapping vector, pTAG4, and it is illustrated in **Fig. 3**. The vector is 3980 bases in length with an Amp^R gene and bacterial origin of replication for propagation in *E. coli*. The trapping cassette consists of the SV40 early region to drive transcription when transfected into Cos7 cells. Exons 1 and 2, as well as intron 1, of pTAG4 are naturally occurring leader exon/intron sequences from the human adenovirus 2 genome. pTAG4 lacks a last exon containing a poly (A) signal.

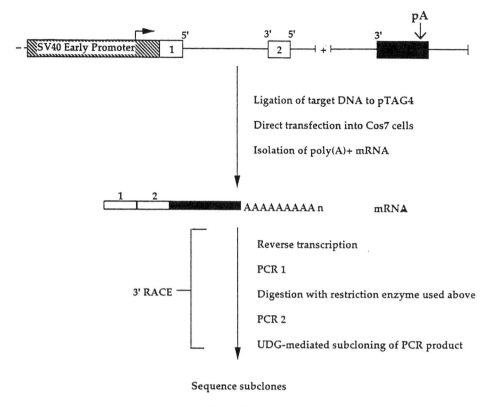

Fig. 4. Schematic representation of 3'-terminal exon trapping using the vector pTAG4.

Thus, no mature polyadenylated mRNA is produced from the vector when transfected into Cos7 cells. An MCS resides downstream of the 5'-splice site of exon 2 and contains the following unique restriction enzyme sites: *Eco*RI, *Bam*HI, *Bgl*II, *Bss*HII, *Sph*I, *Nhe*I, *Eag*I, *Not*I, *Pst*I, *Nar*I, *Mlu*I, and *Spl*I.

The experimental use of this vector is described in detail, and this process is diagrammed in **Fig. 4**. The target DNA is digested to completion with one of the restriction enzymes within the MCS to be used for the ligation reaction. pTAG4 is double-digested with *Ava*II, and the same restriction enzyme used to digest the target DNA. The target DNA is ligated to pTAG4, and the ligation reaction is directly transfected into Cos7 cells. After transient expression, mRNA is harvested and reverse-transcribed using an oligo(dT)-based adapter primer for the 3'-RACE technique. Exons are amplified by a nested PCR approach. The primary PCR reaction uses one primer specific to the tail sequence on the adapter RT primer, whereas the other is specific to vector exon sequence. Primary PCR product is digested with the same restriction enzyme used for preparative digestion of the target DNA and pTAG4. This step functions to remove any false positives resulting from reverse transcription of unspliced precursor RNA or residual contaminating DNA present in the mRNA preparation. The secondary PCR uses nested primers containing 12-bp for UDG cloning and to give a greater degree of specificity to the reaction. PCR results are analyzed by agarose gel electrophoresis.

The complexity of the secondary PCR product from both exon trapping protocols will be proportional to the amount and complexity of target DNA used for the original ligation reactions. For example, a small number *(1–3)* of ethidium bromide-stained bands will be seen when secondary PCR product is analyzed by agarose gel electrophoresis in a trapping experiment with single cosmid or phage clones. If trapping from a YAC clone or a pool of smaller clones, the numbers of bands seen on a gel will be proportionally greater. Thus, subcloning of the products may involve excision of single bands from the gel (single clones) to a shotgun approach using total PCR product (YAC clone and pools of clones).

Product from the secondary PCR reactions should be subcloned for further analysis by either UDG-mediated cloning or TA cloning. Vectors designed to perform both types of subcloning are commercially available, and the protocols should be followed according to manufacturer's recommendations. The primers used for the secondary PCR reactions in both exon trapping protocols are designed to impart UDG cloning capabilities on PCR product, and this approach is generally more efficient and yields greater numbers of subclones containing inserts. For these reasons, UDG cloning is recommended; however, TA cloning can be used and should yield sufficient numbers of subclones for sequencing.

Subclones containing candidate exons should first be sequenced to aid in determining exon validity. Interpretation of sequencing results differs between the two exon trapping approaches. Internal exon trapping yields small products of usually 25–200 bp in length. Each unique sequence from a trapping experiment should be used in a computer search of existing nucleotide databases. Computer search results from an individual valid internal exon will yield evidence of splice events on both sides of the novel sequence as determined by removal of vector intron sequences and will show a complete open reading frame all the way through the sequence in at least one frame. Primers can then be designed to the unique sequence for use in a number of subsequent experiments including:

1. Mapping the candidate exon to its genomic region of origin;
2. RT/PCR expression studies; and
3. Generating a specific probe for cDNA screening and Southern/Northern blot mapping.

Alternatively, standard restriction enzyme digestion and gel purification of an insert representing a candidate exon from a single subclone can be used for hybridization experiments. However, owing to the small size of many trapped exons, it may be more informative to perform PCR experiments using DNA and RNA from interesting sources as templates.

3'-Terminal exon trapping yields candidate exons that are much larger in size (200–2500 bp) than those obtained from internal exon trapping. The sequence of an individual subclone will show evidence of a splice event at the 5'-end of the sequence and a poly(A) tail at the 3'-end of the sequence. In approx 90% of the cases, a consensus poly(A) signal consisting of one of the hexanucleotides AATAAA or ATTAAA will be present 12–30 bases upstream of the poly(A) tail. On nucleotide database search, multiple termination codons will be present in most frames. The great majority of 3'-terminal exon sequence is a 3'-untranslated region. Thus, very little coding sequence exists. As discussed for internal exons, both PCR-based and hybridization-based experiments to map the exon and perform gene expressions studies can be performed. Full-length cDNA clones can also be obtained in hybridization screens. Hybridization studies may prove more fruitful using 3'-terminal exons as probes vs internal exon simply owing to their greater length.

2. Materials

1. One or more of the following restriction endonucleases: *Eco*RI, *Sst*I, *Xho*I, *Not*I, *Xma*III, *Pst*I, *Bam*HI, *Eco*RV, *Bgl*II, *Bss*HII, *Sph*I, *Nhe*I, *Eag*I, *Nar*I, *Mlu*I, *Spl*I, and *Bst*XI.
2. Internal exon trapping vector pSPL3 or 3'-terminal exon trapping vector pTAG4.
3. 3 M Sodium acetate.
4. 70% and 100% Ethanol.
5. TE buffer: 10 mM Tris-HCl, 1 mM EDTA, pH 7.4.
6. Calf intestinal phosphatase (CIP).
7. Phenol/chloroform/isoamyl alcohol, 25:24:1.
8. 5X T4 ligase buffer: 250 mM Tris-HCl, pH 7.9, 50 mM MgCl$_2$, 5 mM dATP, 5 mM dithiothreitol (DTT), 25% (w/v) polyethylene glycol (PEG) 8000.
9. T4 DNA ligase.
10. LB/amp medium (80 µg/mL amp).
11. LB/amp plates (80 µg/mL amp, 15% agar).
12. Cos7 cells (CRL 1651, American Type Culture Collection, Rockville, MD).
13. Cationic lipid medium for transient transfection.
14. Internal exon trapping primers:
    SA2: 5' ATCTCAGTGGTATTTGTGAGC 3'
    SD6: 5' TCTGAGTCACCTGGACAACC 3'
    dUSA4: 5' CUACUACUACUACACCTGAGGAGTGAATTGGTCG 3'
    dUSD2: 5' CUACUACUACUAGTGAACTGCACTGTGACAAGCTGC 3'
15. 5X First-strand buffer: 250 mM Tris-HCl, pH 8.3, 375 mM KCl, 15 mM MgCl$_2$.
16. RT enzyme.
17. 10X *Taq* polymerase buffer: 500 mM KCl, 100 mM Tris-HCl, pH 8.3, 10 mM 4dNTP: 10 mM each dNTP in H$_2$O, 50 mM MgCl$_2$.
18. *Taq* DNA polymerase.
19. 1.2% Agarose gel in 1X TBE or 1X TAE with 400 µg/mL concentration ethidium bromide.
20. 0.8% Agarose gel in 1X TBE or 1X TAE with 400 µg/mL concentration ethidium bromide.
21. 3'-Terminal exon trapping primers:
    AP: 5' AAGGATCCGTCGACATCGATAATACGAC(T)$_{17}$ 3'
    SV40P: 5' AGCTATTCCAGAAGTAGTGA 3'
    UAP: 5' CUACUACUACUAGTCGACATCGATAATACGAC 3'
    Ad2: 5' CAUCAUCAUCAUCAGTACTCTTGGATCGGA 3'
22. 0.1 M DTT.

3. Methods

3.1. Target DNA Preparation

The starting target DNA for both exon trapping protocols can reside in any cloning vector (phage, cosmid, P1, BAC, or YAC vector) used for propagating genomic DNA. The common characteristic of each is that the target DNA must be sufficiently purified from host DNA (*E. coli* and yeast) to keep nonspecific background to a minimum. It is not necessary to purify the DNA insert from the cloning vector, since the vectors do not contain exons and introns and only rarely will lead to false-positive scoring. Many protocols exist that are designed to purify cloned DNA within the various cloning vectors, and they have been described elsewhere. Thus, treatment of target DNA in this chapter consists of describing the preparation of DNA that has already been isolated by an existing technique.

1. Digest a minimum of 1 μg of target DNA with one of the available restriction endonucleases, depending on which exon methodology has been chosen. For internal exon trapping, the following enzymes can be used: *Eco*RI, *Sst*I, *Xho*I, *Not*I, *Xma*III, *Pst*I, *Bam*HI, and *Eco*RV. For 3'-terminal exon trapping, the following enzymes can be used: *Eco*RI, *Bam*HI, *Bgl*II, *Bss*HII, *Sph*I, *Nhe*I, *Eag*I, *Not*I, *Pst*I, *Nar*I, *Mlu*I, or *Spl*I (*see* **Note 1**).
2. Phenol-extract and precipitate target DNA using 0.1 vol 3 *M* sodium acetate and 2.5 vol 100% ethanol. Wash the pellet with 70% (v/v) ethanol.
3. Resuspend DNA in TE buffer at a final concentration of 250 ng/μL (*see* **Note 2**).
4. Run 250 ng of the digested target DNA on a 0.8% agarose gel to assess the success of DNA cutting. Complete digestion of the target is desirable; however, a small amount of undigested will not interfere with the protocol.

3.2. Internal Exon Trapping

1. Digest the trapping vector pSPL3 with the same restriction endonuclease used to digest the target DNA (*see* **Fig. 1**).
2. Dephosphorylate the vector with CIP according to manufacturer's recommendations (*see* **Note 3**).
3. Phenol/chloroform-extract and ethanol-precipitate linearized, CIP-treated pSPL3 with 0.1 vol 3 *M* sodium acetate and 2.5 vol 100% ethanol. Wash pellet in 70% ethanol, and resuspend the pellet in TE buffer at 250 ng/μL concentration.
4. Set up the following shotgun cloning reaction: 250 ng pSPL3, 250 ng to 1 μg target DNA, 2 μL 5X T4 DNA ligase buffer, 1 μL T4 DNA ligase (1 U/μL), H$_2$O to final volume 10 μL.
5. Mix gently, and incubate at room temperature for 1 h. Alternatively, the ligation reaction can proceed overnight at 15°C (*see* **Note 4**).
6. Transform *E. coli* with the ligation reactions (vector-only control and experimental) by either chemical or electrocompetent cells. The final transformation volume should be 1 mL.
7. Inoculate 5 mL of LB/amp medium with 0.5 mL of the experimental transformation, and incubate overnight at 37°C.
8. With the remaining 0.5 mL of transformation, plate 10 and 100 μL on two separate LB/amp plates. In parallel plate 10 and 100 μL of the vector-only control transformation, and incubate all plates overnight at 37°C. By comparing the numbers obtained from both transformations this will assess the degree of vector-only ligation (*see* **Note 5**).
9. Prepare DNA from the 5-mL overnight culture of experimental recombinants by an alkaline lysis miniprep procedure.
10. Sixteen hours before transient transfection, plate Cos7 cells on 3.5-cm 6-well tissue-culture plates at a density of roughly 300,000 cells/well (*see* **Note 6**). Use 1 well/transfection.
11. Use 1 μg of the purified recombinant DNA for each transfection. Cationic lipid-mediated transfection can be used for introduction of DNA into the cells and should be used according to manufacturer's recommendations. Many companies now market lipid reagents for transient transfection, and all are acceptable for this purpose.
12. After transfection, incubate cells at 37°C for 16–24 h.
13. Prepare total RNA from transfected cells by acidic-phenol/guanidinium thiocyanate extraction, and resuspend in 10 μL of TE buffer (*see* **Note 7**).

14. Prepare to synthesize first-strand cDNA. Add the following to a 0.5-mL microcentrifuge tube: 1 µL 20 µ*M* oligonucleotide SA2, 1–3 µg total RNA, DEPC-treated H$_2$O to 12-µL final volume.
15. Incubate mixture for 5 min at 70°C and then place on ice. Microcentrifuge briefly at high speed, and add the following components at room temperature: 4 µL 5X first-strand buffer, 2 µL 0.1 *M* DTT, 1 µL 10 m*M* dNTP.
16. Mix gently, microcentrifuge briefly at high speed, and incubate for 5 min at 42°C. Add 1 µL (200 U) RT, mix gently, and incubate an additional 30 min at 42°C.
17. Incubate RT reaction at 55°C for 5 min. Then add 1 µL of RNaseH, mix gently, and incubate an additional 10 min at 55°C (*see* **Note 8**).
18. Microfuge briefly and place on ice. The cDNA pool can be stored indefinitely at –20°C. Use 5 µL for primary PCR amplification.
19. Set up the primary PCR reaction according to the following: 5 µL cDNA, 5 µL 10X *Taq* polymerase buffer (MgCl$_2$-free), 1.5 µL 50 m*M* MgCl$_2$, 1 µL 10 m*M* 4dNTP, 2.5 µL 20 µ*M* oligonucleotide SA2, 2.5 µL 20 µ*M* oligonucleotide SD6, sterile H$_2$O to 47-µL final volume. Mix gently and overlay with 50 µL mineral oil.
20. Preheat thermal cycler to 94°C. Place the reaction tube in the cycler, and incubate for 5 min at 94°C.
21. Reduce cycler temperature to 80°C, and add 2.5 U *Taq* DNA polymerase that has been diluted into H$_2$O to a final volume of 3 µL.
22. Perform PCR amplification according to the following parameters: 94°C for 1 min, 72°C for 5 min, and 60°C for 1 min, 6 cycles; and 72°C for 10 min, 1 cycle; 4°C, hold.
23. Add 25 U *Bst*XI and incubate overnight at 55°C.
24. In the morning, spike the digestion with 5 more units of *Bst*XI and incubate for 2 h at 55°C (*see* **Note 9**).
25. Perform the secondary PCR reaction according to the following: 5 µL *Bst*XI-digested primary PCR product, 5 µL 10X *Taq* polymerase buffer (MgCl$_2$-free), 1.5 µL 50 m*M* MgCl$_2$, 1 µL 10 m*M* 4dNTP, 1 µL 20 µ*M* oligonucleotide dUSA4, 1 µL 20 µ*M* oligonucleotide dUSD2, sterile H$_2$O to 47 µL final volume. Mix gently and overlay with 50 µL mineral oil.
26. Preheat thermal cycler to 94°C. Place the reaction tube in the cycler, and incubate for 5 min at 94°C.
27. Reduce cycler temperature to 80°C, and add 2.5 U *Taq* DNA polymerase that have been diluted into H$_2$O to a final volume of 3 µL.
28. Perform PCR amplification according to the following parameters: 94°C for 30 s, 72°C for 2 min, and 60°C for 30 s, 30 cycles; and 72°C for 10 min, 1 cycle; 4°C, hold.
29. Electrophorese 10 µL of the secondary PCR reaction on a 1.2% agarose gel, and identify reactions that contain PCR product by ethidium bromide staining (*see* **Note 10**). These PCR products should be used for subcloning by the UDG-mediated PCR cloning approach.

3.3. 3'-Terminal Exon Trapping

The use of the trapping vector pTAG4 in the direct ligation/transfection procedure described here is illustrated in **Fig. 5**. Digestion of pTAG4 with *Ava*II and one of the restriction sites within the MCS functions to leave a blunt end upstream of the SV40 enhancer/promoter and a sticky end downstream of the second exon. The *Ava*II site is very difficult to religate, whereas the sticky end is capable of readily ligating to target DNA with the same sticky end. In the example outlined in **Fig. 5**, pTAG4 is double-digested with *Ava*II/*Eco*RI and ligated directly to *Eco*RI-digested target DNA to form linear concatamers consisting of target DNA restriction fragments flanked on either

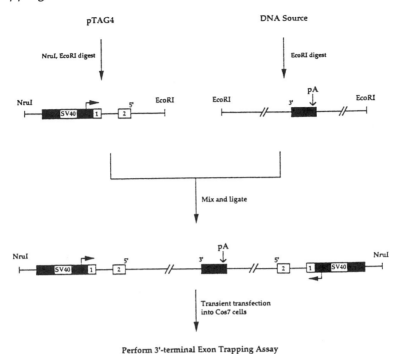

Fig. 5. Diagrammatic representation of the direct ligation/transfection procedure used in the 3'-terminal exon trapping protocol.

side by pTAG4. The ligation reaction is directly transfected into Cos7 cells to induce transcription from the linear concatamers for amplification of trapped exons.

1. Prepare the trapping vector pTAG4 (**Fig. 3**) by double restriction digest with the enzyme *Ava*II in conjunction with the same enzyme used to digest the target DNA. Gel-purify the digested vector in a 0.8% agarose gel to remove the small DNA fragment liberated from the MCS after double digestion. A number of protocols can be used to purify the vector from the agarose gel slice including glass bead purification and electrophoresis.
2. Resuspend pTAG4 in TE at a final concentration of 500 ng/µL.
3. Set up the following ligation reaction (*see* **Note 11**): 500 ng pTAG4, 500 ng target DNA, 1 µL 5X T4 DNA ligase buffer, 1 µL T4 DNA ligase (1 U/µL), H$_2$O to a final volume of 5 µL.
4. Mix gently and incubate overnight at 16°C.
5. Sixteen hours before transient transfection, plate Cos7 cells on 3.5-cm 6-well tissue-culture plates at a density of roughly 300,000 cells/well. Use 1 well/transfection.
6. Use the entire 5-µL ligation reaction volume to transfect into 1 well of the Cos7 cells. The same lipid-mediated transfection protocol as used for internal exon trapping is used for this procedure.
7. After transfection, incubate cells at 37°C for 16–24 h.
8. Prepare poly (A)+ RNA from each transfection (*see* **Note 12**).
9. Approximately 1 µg of poly (A)$^+$ mRNA will be obtained from a single transfection of which 500 ng will be used to synthesize first-strand cDNA. Set up the following reverse transcription reaction: 1 µL 500 ng/µL oligonucleotide AP, 500 ng poly (A)$^+$ mRNA, 1 µL 20 m*M* EDTA, DEPC-treated H$_2$O to 20 µL final volume (*see* **Note 13**).

10. Incubate mixture for 5 min at 70°C, and then place at 42°C for 5 min. Add the following mixture (preheated to 42°C) to the RNA reaction tube: 10.5 µL DEPC-treated H$_2$O, 10 µL 5X first-strand buffer, 5 µL 0.1 M DTT, 2.5 µL 10 mM dNTP, 2 µL RT (200 U/µL).
11. Incubate the reverse transcription reaction at 42°C for 30 min.
12. Incubate RT reaction at 55°C for 5 min, then add 1 µL of RNaseH, mix gently, and incubate an additional 10 min at 55°C.
13. Microfuge briefly and place on ice. The cDNA pool can be stored indefinitely at –20°C. Use 5 µL for primary PCR amplification.
14. Set up the primary PCR reaction according to the following: 5 µL cDNA pool, 72.5 µL sterile H$_2$O, 10 µL 10X *Taq* buffer (MgCl$_2$-free), 5 µL 25 mM MgCl$_2$, 2.5 µL 10 mM 4dNTP, 1 µL 50 µM oligonucleotide SV40P, 1 µL 50 µM oligonucleotide UAP. Mix and overlay with 100 µL mineral oil (*see* **Note 14**).
15. Preheat thermal cycler to 94°C. Place the reaction tube in the cycler, and incubate for 5 min at 94°C.
16. Reduce cycler temperature to 80°C, and add 2.5 U of *Taq* DNA polymerase that have been diluted into H$_2$O to a final volume of 3 µL.
17. Perform PCR amplification according to the following parameters: 94°C for 30 s, 72°C for 2 min, and 55°C for 30 s, 20 cycles; and 72°C for 5 min, 1 cycle; 4°C, hold.
18. Digest the product from the primary PCR reaction with the same restriction enzyme used to digest vector and target DNA. Set up a restriction enzyme reaction according to the following: 17 µL primary PCR product, 2 µL 10X buffer, 1 µL restriction enzyme, for a total of 20 µL. Mix gently, and incubate for 1 h at 37°C (*see* **Note 15**).
19. Dilute the restriction digestion by addition of 150 µL sterile H$_2$O.
20. Set up the following secondary PCR amplification using 1 µL of the above digestion as template: 1 µL digested primary PCR product, 76.5 µL sterile H$_2$O, 10 µL 10X *Taq* buffer (MgCl$_2$-free), 5 µL 25 mM MgCl$_2$, 2.5 µL 10 mM 4dNTP, 1 µL 50 µM oligonucleotide Ad2, 1 µL 50 µM oligonucleotide UAP. Mix and overlay with 100 µL mineral oil.
21. Preheat thermal cycler to 94°C. Place the reaction tube in the cycler, and incubate for 5 min at 94°C.
22. Reduce cycler temperature to 80°C, and add 2.5 U of *Taq* DNA polymerase that have been diluted into H$_2$O to a final volume of 3 µL.
23. Perform PCR amplification according to the following parameters: 94°C for 30 s, 72°C for 2 min, and 55°C for 30 s, 30 cycles; and 72°C for 5 min, 1 cycle; 4°C, hold.
24. Electrophorese 10 µL of the secondary PCR reaction on a 1.2% agarose gel, and identify reactions that contain PCR product by ethidium bromide staining (*see* **Note 16**). These PCR products should be used for subcloning and sequencing.

4. Notes

1. The amount of starting target DNA will depend on the type of vector used to carry the genomic DNA of interest. It is difficult and time consuming to gel-purify YAC DNA; thus, as little as 1 µg can be digested if more cannot be obtained. In contrast, it is relatively easy to purify microgram quantities of cosmid DNA, and more of this substrate can be used for the initial restriction digest.
2. This final concentration of digested target DNA will be suitable for either trapping protocol.
3. Alternatively, two different restriction endonucleases can be used to prepare vector and target DNA, and thus, CIP treatment of pSPL3 would not be necessary. If two different enzymes are used, it will be necessary to gel-purify the linearized vector from the small fragment generated from the MCS on a 0.8% agarose gel.
4. A control ligation reaction should also be included containing all of the above components without the target DNA. This functions to assess the amount of vector-only ligation.

5. Expect 10–100 colonies from the 10 µL experimental plate. If >10% of the resulting colonies do not contain target DNA insert, as determined by comparing number of colonies in the control ligation vs the experimental ligation, the vector should be prepared again.
6. Cos7 cells can be obtained from ATCC. It is recommended that a single T75 flask be maintained and passaged in order to carry the line for use in transfections. The cells grow in DMEM supplemented with 10% fetal calf serum.
7. Many companies market a product for total RNA isolation that is fast, efficient, and yields RNA of good quality. Appropriate RNase-free precautions should be taken when working with RNA, such as using only DEPC-treated water for making solutions (except those containing Tris), the use of gloves, and the use of sterile plasticware instead of glass.
8. RNaseH treatment digests the RNA strand of a DNA:RNA hybrid. The presence of RNA in a PCR reaction has been shown to inhibit the reaction. Digestion is also carried out at 55°C to eliminate possible snap-back structures and second-strand products that can result from residual reverse transcriptase activity. RT enzyme is inactive at this temperature.
9. The double-stranded PCR product is digested with *Bst*XI to remove two classes of background products. The first results from pSPL3-derived RNA molecules containing only vector sequences. The second results from amplification of RNA molecules that have used a cryptic 5'-splice site within the vector. Subcloning of the target DNA into either *Eco*RI or *Eco*RV sites will inactivate the overlapping *Bst*XI sites and make it necessary to digest putative exon-containing clones with *Nde*I to determine if the vector cryptic 5'-splice site has been used. This step is used to increase efficiency and reduce background inherent within the system.
10. Positive results from internal exon trapping are indicated by any PCR product of a size greater than the 147 bp generated from the vector only. This vector-only PCR product will be present even after *Bst*XI digestion, and thus any incorporation of a trapped exon will yield PCR products larger than 147 bp.
11. The concentration of DNA in this ligation reaction should be at 200 µg/mL to induce concatamerization.
12. Many companies market kits for mRNA selection that are both fast and extremely efficient. Appropriate RNase-free precautions should be taken when working with RNA.
13. The oligonucleotide AP is 45 bases in length, and consists of 17 T residues at the 3'-end that binds to and primes reverse transcription at poly A tails of mRNA species. The remainder of the AP primer is an engineered sequence that does not base pair with any endogenous sequence from Cos7 cells. The resulting cDNA primed with AP can be amplified with 5'-primers specific for vector sequences and 3'-primers specific for the engineered tail sequence of the AP primer. This is the basis for the 3'-RACE technology.
14. The oligonucleotide SV40P is specific for the portion of the cDNA that is derived from the SV40 promoter, whereas the UAP is specific for the tail region of the AP primer.
15. This digestion step functions to remove specific background products that result from unspliced precursor RNA or residual DNA that can reverse transcribe at A/T-rich regions with the AP primer.
16. The resulting PCR product from 3'-terminal exon trapping will not contain any vector-only derived products, and thus, all PCR product should be considered trapped sequences.

References

1. Duyk, G. M., Kim, S., Myers, R. M., and Cox, D. R. (1990) Exon trapping: A genetic screen to identify candidate transcribed sequences in cloned mammalian genomic DNA. *Proc. Natl. Acad. Sci. USA* **87,** 8995–8999.
2. Buckler, A. J., Chang, D. D., Graw, S. L., Brook, J. D., Haber, D. A., Sharp, P. A., and Housman, D. E. (1991) Exon amplification: a strategy to isolate mammalian genes based on RNA splicing. *Proc. Natl. Acad. Sci. USA* **88,** 4005–4009.

3. Auch, D. and Reth, M. (1991) Exon trap cloning: using PCR to rapidly detect and clone exons from genomic DNA fragments. *Nucleic Acids Res.* **18,** 6743–6744.
4. Hamaguchi, M., Sakamoto, H., Tsuruta, H., Sasaki, H., Muto, T., Sugimura, T., and Terada, M. (1992) Establishment of a highly sensitive and specific exon-trapping system. *Proc. Natl. Acad. Sci. USA* **89,** 9779–9783.
5. Nehls, M., Pfeifer, D., and Boehm, T. (1994) Exon amplification from complete libraries of genomic DNA using a novel phage vector with automatic plasmid excision facility: application to mouse neurofibromatosis-1 locus. *Oncogene* **9,** 2169–2175.
6. Krizman, D. B. and Berget, S. M. (1993) Efficient selection of 3'-terminal exons from vertebrate DNA. *Nucleic Acids Res.* **21,** 5198–5202.
7. Church, D. M., Stotler, C. J., Rutter, J. L., Murrell, J. R., Trofatter, J. A., and Buckler, A. J. (1994) Isolation of genes from complex sources of mammalian genomic DNA using exon amplification. *Nature Genet.* **6,** 98–105.
8. Huntington Disease Collaborative Research Group (1993) A novel gene containing a trinucleotide repeat that is expanded and unstable on Huntington's Disease chromosomes. *Cell* **72,** 971–983.
9. Krizman, D. B., Hofmann, T. A., DeSilva, U., Green, E. D., Meltzer, P. S., and Trent, J. M. (1995) Identification of 3' terminal exons from yeast artificial chromosomes. *PCR Methods Appl.* **4,** 322–326.
10. Frohman, M. A., Dush, M. A., and Martin, G. R. (1988) Rapid production of full-length cDNAs from rare transcripts: amplification using a single gene-specific oligonucleotide primer. *Proc. Natl. Acad. Sci. USA* **85,** 8998–9002.

84

DNA Rescue by the Vectorette Method

Marcia A. McAleer, Alison Coffey, and Ian Dunham

1. Introduction

A major advance in physical mapping of the human genome was the development of yeast artificial chromosome (YAC) vectors *(1)*. This has enabled the cloning of pieces of DNA several hundred kilobases in length *(2)*. The availability of such large cloned genomic DNA fragments means that by ordering a series of overlapping YAC clones, a contiguous stretch of DNA, several megabases in length, can be isolated around a genomic region of interest (e.g., the region of a chromosome linked to a particular disease gene). The successful isolation of terminal sequences of a given YAC can be very useful in assembling an ordered "contig" of YAC clones. Such terminal clones may be used directly as hybridization probes or sequenced and used to generate sequence tagged sites (STSs) to identify overlaps between, and isolate other, members of the contig. Several methods have been used to this end, including PCR with vector-specific primers in combination with primers designed either for repetitive elements, such as *Alu* sequences *(3)*, or in combination with random nonspecific primers *(4)*. However, these techniques rely on a suitable repetitive element or random primer sequence occurring close enough to the end of the YAC so as to be amplified by PCR. Furthermore, probes isolated in this manner may well contain highly repetitive sequences that, if unsuccessfully blocked, will increase nonspecific signal in any subsequent hybridization procedures *(5)*.

The vectorette method was originally described by Riley et al. *(6)*. YAC DNA is digested with a restriction enzyme, and the resulting fragments are ligated to a linker molecule to create a vectorette "library," i.e., a complex mixture of restriction fragments with linker ligated to each end. Within this library are fragments that contain the YAC vector/genomic DNA junction, which includes the terminal sequences of the YAC (**Fig. 1**). The linker molecule consists of two long (>50 nucleotides) preannealed oligonucleotides incorporating a suitable 5'-overhang corresponding to the restriction enzyme used in the initial YAC digest. Blunt-ended linkers may also be used. Although the oligonucleotides comprising the linker are complementary at the 5'- and 3'-ends, there is a region of noncomplementarity in the middle where the two strands are unable to pair and a vectorette "bubble" is formed. The PCR is then performed on this mixture using one of two vector-specific primers (designed either for the centric or acentric

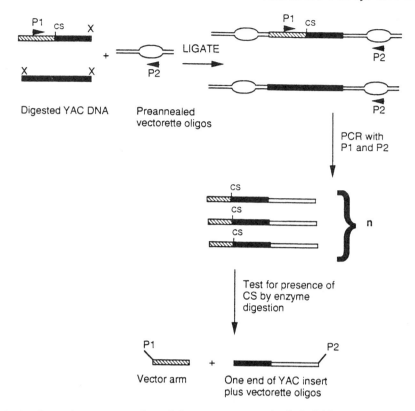

Fig. 1. A schematic representation of the vectorette method. Solid boxes represent genomic DNA, and the hatched boxes represents YAC vector sequence. YAC DNA is digested with a restriction enzyme, X. Following ligation to annealed vectorette oligos, products are amplified with a vectorette-specific primer (P2) and a primer specific for one or other of the YAC vector arms (P1). Only fragments containing vector/insert junction are amplified. Confirmation of the presence of the cloning site (CS) within the amplified fragment can be obtained by digestion of the hybrid fragment with the enzyme that cuts at the cloning site, releasing a fragment diagnostic of the vector arm (**Table 2**) together with one or more fragments corresponding to the genomic DNA insert. CS = cloning site.

vector arms) in combination with a linker-specific primer. The linker-specific primer corresponds to the sequence of the linker ligated to the 5'-end of each DNA strand and has no complement on the other strand of the "bubble." It is therefore unable to anneal to template until the complementary sequence has been generated by priming off the vector-specific sequence. Thus, only those fragments containing binding sites for the vector-specific primer (i.e., DNA including and immediately adjacent to the cloning site of the YAC vector) will be successfully amplified by the PCR. The amplification products may then be used as DNA probes, for DNA sequencing, or may be cloned into a suitable vector.

A recent adaptation of the vectorette method has been used to isolate possible gene fragments from selected regions of the genome without prior knowledge of gene sequence *(7)*. This method is termed Island Rescue PCR (IRP), and relies on the fact that nearly all housekeeping genes and over 40% of tissue-specific genes have a CpG island in or near the 5'-end of the gene *(8)*. Such CpG islands have a significantly increased C + G content compared to the bulk of genomic DNA. These CpG islands can be detected in native human genomic DNA, by rare-cutting restriction endonucleases that recognize unmethylated CpG-containing sequences. The principles of the vectorette method described above are used, except the YAC DNA in this instance is digested with restriction endonucleases that specifically recognize CpG-containing sequences, e.g., *Sac*II, *Eag*I. Therefore YAC DNA will be cut at CpG-rich sites, which may be associated with a gene. The mixture is then ligated to the preannealed vectorette oligos, and PCR in this instance is driven by an *Alu*-specific primer together with the vectorette oligo described above. Northern blot analysis may then be used to test that amplified sequences are associated with expressed mRNAs. There are two main drawbacks to this method. First, since DNA in yeast is not differentially methylated, all CpG-containing restriction sites will be cut whether or not they are associated with an unmethylated island in native genomic DNA. Therefore, a portion of the amplified fragments may not be associated with an expressed mRNA. Second, as with all *Alu*-PCR based methods, there is a requirement for an *Alu* sequence close enough to the restriction site to allow amplification by the *Taq* polymerase. However, in terms of transcript mapping, where the previously described methods (direct selection/cDNA enrichment *[9]*, exon trapping *[10]*, probing cDNA libraries directly with radiolabeled YAC DNA *[11]*) all have limitations, IRP may prove to be a rapid and useful technique for the identification of transcriptional units within complex sources of DNA.

Although the vectorette method was originally developed for rescuing the vector-insert junctions of YACs, it may be used to isolate sequences adjacent to any known sequence (e.g., the identification of intron/exon boundaries in a specified gene *[12]*). This chapter describes in detail the application of the vectorette method to isolating terminal sequences from YACs.

2. Materials

All solutions should be made to the standard required for molecular biology, i.e., using sterile distilled water and molecular-biology-grade reagents.

1. T4 DNA ligase, 1 U/μL and 5X T4 DNA ligase buffer (0.25 *M* Tris-HCl, pH 7.6, 50 m*M* MgCl$_2$, 5 m*M* ATP, 5 m*M* DTT, 25% [w/v] polyethylene glycol-8000) (Gibco BRL, Paisley, Scotland).
2. The sequences of the oligonucleotides used in this chapter are given in **Table 1** and are taken from **ref.** *(6)*. The vector-specific primers are designed against pYAC4 (these can be replaced with appropriate vector primers or *Alu*-specific primers if performing IRP). The vectorette oligonucleotides described are suitable for blunt-ended ligations. If desired, a suitable overhang at the 5'-end of the "top" strand oligonucleotide may be incorporated to facilitate "sticky ended" ligations.

    Oligonucleotides were synthesized by phosphoramidite chemistry on an Applied Biosystems 392 DNA/RNA synthesizer. After deprotection (7 h at 55°C), oligonucleotides are dried in a centrifugal evaporator (alternatively, the standard ethanol precipitation procedure may be used). Oligonucleotides used in PCR are resuspended in H$_2$O to a concentration of 20 μ*M*. Vectorette oligonucleotides are purified by HPLC (12% polyacrylamide gel electrophoresis may also be used). Prior to use, equimolar quantities of the "top" and "bottom" oligonucleotides are preannealed in 25 m*M* NaCl by heating at 65°C

Table 1
Oligonucleotide Sequences for Vectorette PCR

Vectorette oligonucleotides (for blunt-ended ligations)
  "Top" strand
    CAAGGAGAGGACGCTGTCTGTCGAAGGTAAGGAACGGACGAGAGA
    AGGGAGAG
  "Bottom" strand
    CTCTCCCTTCTCGAATCGTAACCGTTCGTACGAGAATCGCTGTCCTC
    TCCTTG
  Universal vectorette primer 224
    CGAATCGTAACCGTTCGTACGAGAATCGCT
pYAC4-specific primers
  Centric ("left") arm

1089	CACCCGTTCTCGGAGCACTGTCCGACCGC
Sup4-2	GTTGGTTTAAGGCGCAAGAC
pYACL	AATTTATCACTACGGAATTC

  Acentric ("right") arm

1091	ATATAGGCGCCAGCAACCGCACCTGTGGCG
Sup4-3	GTCGAACGCCCGATCTCAAG
pYACR	CCGATCTCAAGATTACGGAATTC

All oligonucleotide sequences are written in the 5'→3' direction.

for 5 min and left to cool to room temperature. A working concentration of 1 µM is used in ligations. All oligonucleotides are stored at –20°C.

3. PCR is performed using a GeneAmp PCR reagent kit (Perkin Elmer, Warrington, UK) in 10 mM Tris-HCl (pH 8.3), 50 mM KCl, 1.5 mM MgCl$_2$, 0.01% (w/v) gelatin containing 200 µM of each dNTP and 1.0 µM of each primer. Amplitaq is added to a concentration of 1.25 U/50 µL reaction and Perfect Match (Stratagene, Cambridge, UK) to a concentration of 5 U/50 µL reaction, and overlaid with mineral oil (Sigma, Poole, UK). DNA amplification is performed in an Omnigene thermocycler (Hybaid, Teddington, UK).

3. Methods

1. Take half an agarose plug (approx 50–100 µL containing 1–2 µg DNA) of miniprep YAC DNA (DNA in solution may also be used; *see* **Note 1**) and wash as follows: 3 × 20 min in 10 mM Tris-HCl, 0.1 mM EDTA, pH 7.4 (1 mL/plug) at 50°C. 1 × 20 min in 10 mM Tris-HCl, 0.1 mM EDTA, pH 7.4 (1 mL/plug) at room temperature.
2. Preincubate plugs for 30 min at 37°C in 100 µL of the appropriate enzyme buffer (*see* manufacturer's recommendation).
3. Remove buffer, and replace with 100 µL of fresh enzyme buffer containing 20–30 U of restriction enzyme (*see* **Note 2**), and incubate overnight at the recommended temperature (usually 37°C). After digestion, the plug may be cut into three, and one portion electrophoresed through a 1.0% agarose mini gel alongside a similar amount of untreated YAC DNA to test for complete digestion. One slice may be stored dry at 4°C and redigested if incomplete digestion has occurred.
4. Incubate 1/3 of the agarose plug from **step 3** in 1 mL of 1X ligation buffer for 1 h on ice.
5. Replace with 100 µL fresh 1X ligation buffer. To this add 10 µL of preannealed blunt-ended vectorette linker (at 1 µM: *see* **Subheading 2.**, **step 2**) i.e., 10 pmol of linker.

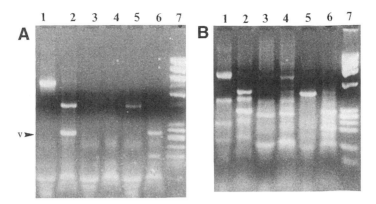

Fig. 2. Three vectorette "libraries" were created using the blunt-ended restriction enzymes: *Pvu*II (lanes 1 and 2), *Stu*I (lanes 3 and 4), and *Rsa*I (lanes 5 and 6). PCR was performed using oligos specific for the centric arm of the pYAC4, 1089, and the universal vectorette oligo, 224. In A, 5 µL of Perfect Match have been added to each PCR, whereas in B, this has been omitted. Ten microliters of untreated product were loaded on a 2.5% agarose minigel in lanes 1, 3, and 5, whereas samples in lanes 2, 4, and 6 were first digested with *Eco*RI to release the vector arm from the genomic fragment. Lane 7 contains *Hae*III fragments of ΦX RF DNA (Gibco BRL, Paisley, Scotland).

*Stu*I-digested YAC has failed to produce a PCR product (A, lanes 3 and 4), probably through the lack of an enzyme site close to the vector/insert junction. *Pvu*II- and *Rsa*I-digested YAC yields products of approx 800 and 500 bp (lanes 1 and 5), respectively, which on digestion with *Eco*RI release vector fragments (V) of the predicted size 287 bp together with the terminal *Pvu*II and *Rsa*I fragments of the YAC insert (500 and 200 bp).

6. Heat to 65°C for 15 min to melt the agarose plug, and then equilibrate at 37°C (approx 5 min).
7. When the reaction mix is equilibrated, add 1 µL of T4 DNA ligase (1 U/µL) and incubate at 37°C. After 1 h add 400 µL 10 m*M* Tris-HCl, 0.1 m*M* EDTA, pH 8.0, and mix thoroughly. The vectorette library may now be stored in aliquots at –20°C.
8. Two sets of PCR mixes need to be prepared for each vectorette library constructed. The first contains a primer, 1091, specific for the "right" arm of the YAC vector (i.e., the acentric arm encoding the URA3 gene) together with the vectorette-specific oligo (224), whereas the second contains a primer, 1089, specific for the "left" arm of the YAC vector (i.e., the centric arm, which contains the CEN4 gene) together with 224 (*see* **Table 1**). Each reaction is carried out in 50 µL buffer described in **Subheading 2., step 3**, including 5 µL of Perfect Match (*see* **Note 3** and **Fig. 2**) using the following cycling conditions: 94°C for 1 min, 1 cycle, followed by 93°C for 1 min, 65°C for 1 min, and 72°C for 3 min, 38 cycles, and followed by 72°C for 5 min, 1 cycle. For IRP, *see* **Note 4** for suggested primer sequences.
9. Confirmation that PCR products originate from the terminal sequences of YAC clones can be obtained by demonstrating the presence of the YAC vector cloning site in the hybrid fragment. This is done by digesting the PCR product with a restriction enzyme that cleaves within the cloning site. To 9 µL of PCR product add 1 µL 10X restriction enzyme buffer (*see* manufacturer's recommendation), 10 U of enzyme, and incubate for 1 h at 37°C. When the vector is pYAC4, 10 U of *Eco*RI may be added directly to 9 µL of PCR product, without addition of enzyme buffer. Restriction fragments can be visualized on a

Table 2
Positions of Primer Sequences Described in Table 1 with Respect to the *Eco*RI Sequence in the Cloning Site of pYAC4

Centric arm		Acentric arm	
1089→*Eco*RI	287 bp	1091→*Eco*RI	172 bp
Sup4-2→*Eco*RI	40 bp	Sup4-3→*Eco*RI	29 bp
pYACL→*Eco*RI	17 bp	pYACR→*Eco*RI	20 bp

2.5% agarose minigel containing ethidium bromide (0.5 µg/mL) (ethidium bromide is a powerful mutagen and gloves should be worn at all times) (**Fig. 2**). The distances from the primer sequences described in **Subheading 2., step 3** to the *Eco*RI cloning site of pYAC4 are given in **Table 2**.

10. A second PCR may be performed to reduce the amount of vector DNA contained in the amplified product. A nested vector-specific primer that anneals closer to the cloning site (**Tables 1** and **2**) is used in combination with the vectorette-specific oligo. Either use 1 µL of the primary PCR or toothpick the fragment found to cut with *Eco*RI in **step 9** (**not** the restriction digestion product) directly from the agarose gel into a PCR containing: for "left" arm products: Sup4-2 + 224 or pYACL + 224, and for "right" arm products: Sup4-3 + 224 or pYACR + 224. The same cycling conditions as those described in step 8 are used, but the annealing temperature is reduced to 59°C and only 20 cycles are performed. Ten microliters may be visualized on a 2.5% agarose minigel.

11. PCR products may now either be sequenced directly, radiolabeled and used as a hybridization probe (*see* **Note 5**), or subcloned using a suitable cloning system, such as pCR-Script™ SK(+) (Stratagene) or TA-cloning™ system (Invitrogen, Leek, The Netherlands).

4. Notes

1. Use approx 1 µg of solution DNA for each restriction enzyme digest. Reactions should be performed in the buffers recommended by the manufacturers for 4 h at the specified temperature. Before ligation (**Subheading 3., step 5**), enzymes should be heat-inactivated (65°C for 15 min is usually sufficient), extracted with phenol:chloroform (equal volume of ratio 1:1), ethanol-precipitated by standard methods (2 vol 95% ethanol with 1/10 vol 3 *M* sodium acetate, pH 5.6) and resuspended to a concentration of 250 ng/µL. Ligations can be performed in a volume of 10 µL with 1 µL preannealed vectorette oligos.

2. It is important to check that there are no recognition sites for a given restriction endonuclease between the sequences corresponding to the vector-specific primers and the cloning site. If such a site were present, it would be cleaved in the initial digest and a vector-only fragment would be amplified. Suitable enzymes for pYAC4 are *Rsa*I, *Pvu*II, and *Stu*I.

3. The addition of Perfect Match to the PCR reduces the number of nonspecific bands generated (compare **Fig. 2A** with **B**), although some laboratories have found little difference on its omission.

4. IRP is a variant of *Alu*-vectorette PCR that can be used to generate probes from YACs as an alternative to *Alu*-PCR. For IRP, the universal vectorette primer 224 is used together with primer sequences that recognize a human *Alu* repeat. For example: 5'-GGATTACAGGCGTGAGCCAC-3' and 5'-GATCGCGCCACTGCAC TCC-3' (both sequences taken from **ref. 7**). The thermocycling conditions described in **Subheading 3., step 8** may also be used for these two sets of primers.

5. Probes generated by this method may contain highly repetitive sequences. Therefore, it is advisable to pre-reassociate the labeled probe with total human genomic DNA prior to any hybridization procedure. Make probe up to 250 μL with H$_2$O. Add 125 μL 10 mg/mL sonicated total human DNA (Sigma) and boil for 5 min. Snap-chill on ice for 5 min, and then add probe to hybridization as normal.

References

1. Burke, D. T., Carle, G. F., and Olson, M. V. (1987) Cloning of large segments of exogenous DNA into yeast by means of artificial chromosome vectors. *Science* **236**, 806–812.
2. Cohen, D., Chumakov, I., and Weissenbach, J. (1993) A first-generation physical map of the human genome. *Nature* **366**, 698–701.
3. Nelson, D. L., Ledbetter, S. A., Corbo, L., Victoria, M. F., Ramirez-Soli, R., Webster, T. D., Ledbetter, D. H., and Caskey, C. T. (1989) *Alu* polymerase chain reaction: A method for rapid isolation of human-specific sequences from complex DNA sources. *Proc. Natl. Acad. Sci. USA* **86**, 6686–6690.
4. Wesley, C. S., Myers, M. P., and Young, M. W. (1994) Rapid sequential walking from termini of cosmid, P1 and YAC inserts. *Nucleic Acids Res.* **22**, 538–539.
5. Cole, C. G., Patel, K, Shipley, J., Sheer, D., Bobrow, M., Bentley, D. R., and Dunham, I. (1992) Identification of region-specific yeast artificial chromosomes using pools of *Alu* element-mediated polymerase chain reaction probes labelled via linear amplification. *Genomics* **14**, 931–938.
6. Riley, J., Ogilvie, D., Finniear, R., Jenner, D., Powell, S., Anand, R., Smith, J. C., and Markham, A. F. (1990) A novel, rapid method for the isolation of terminal sequences from yeast artificial chromosome (YAC) clones. *Nucleic Acids Res.* **18**, 2887–2890.
7. Valdes, J. M., Tagle, D. A., and Collins, F. S. (1994) Island rescue PCR: A rapid and efficient method for isolating transcribed sequences from yeast artificial chromosomes and cosmids. *Proc. Natl. Acad. Sci. USA* **91**, 5377–5381.
8. Larsen, F., Gundersen, G., Lopez, R., and Prydz, H. (1992) CpG islands as gene markers in the human genome. *Genomics* **13**, 1095–1107.
9. Lovett, M., Kere, J., and Hinton, L. (1991) Direct selection: A method for isolation of cDNAs encoded by large genomic regions. *Proc. Natl. Acad. Sci. USA* **88**, 9628–9632.
10. Buckler, A. J., Chang, D. D., Graw, S. L., Brook, J. D., Haber, D. A., Sharp, P. A., and Housman, D. E. (1991) Exon amplification: A strategy to isolate mammalian genes based on RNA splicing. *Proc. Natl. Acad. Sci. USA* **88**, 4005–4009.
11. Elvin, P., Slynn, G., Black, D., Graham, A., Butler, R., Riley, J., Anand, R., and Markham, A. F. (1990) Isolation of cDNA clones using yeast artificial chromosome probes. *Nucleic Acids Res.* **18**, 3913–3917.
12. Roberts, R. G., Coffey, A. J., Bobrow, M., and Bentley, D. R. (1993) Exon structure of the human dystrophin gene. *Genomics* **16**, 536–538.

85

Random Amplified Polymorphic DNA (RAPDs)

Scott Tingey

1. Introduction

Because polymorphisms are noted by the presence or absence of amplification products from a single allele, the random amplified polymorphic DNA (RAPD) technique tends to provide only dominant markers. Individuals containing two copies of one allele are not distinguished by amplification *(1,2)* from those with only one copy. Dominant markers provide little linkage information for markers linked in repulsion. Therefore, when mapping, it is advisable to only work with markers linked in coupling, e.g., in a backcross or recombinant-inbred population, haploid, or gametophytic tissue or, alternately, in an F2 population with markers amplified from only one parent.

RAPD markers are especially practical when used in techniques that are designed to reveal polymorphism in targeted regions of a genome. Near isogenic lines *(3,4)* can be screened to identify regions of the genome introgressed from the donor parent. Bulked segregant analysis (BSA) *(5)* has been developed as an assay for quickly identifying markers linked to any specific region of any genome. In theory, two bulked DNA samples are gathered from a segregating population, where each bulk is composed of individuals identical for a specific trait or region, but randomized in all unlinked regions. Since many segregating individuals are used to generate the bulks, there is only a minimal chance that regions unlinked to the target region will differ between the bulked samples. A similar approach uses genetic mapping information to create bulks based on genotype. In this way Giovannoni et al. *(6)* and Reiter et al. *(7)* were able to target RAPD markers to a defined region of the tomato and Arabidopsis genome, respectively. Combined with the RAPD assay, screening near isogenic lines or using BSA can quickly add markers to a specific region of the genome.

Using an arbitrary primer as short as 5 nucleotides Gresshoff et al. *(8)* were able to produce a detailed and relatively complex DNA profile. This approach, termed DNA amplification fingerprinting (DAF), which uses silver staining of DNA and acrylamide gels, promises to obtain fingerprints with greater ease than is possible by conventional techniques. The RAPD protocol is relatively quick and easy, fluorescence is used in lieu of radioactivity, and the assay saves time, effort, and the myriad risks and precautions associated with Southern blot technology.

2. Materials

1. Approximately 20 ng of genomic DNA per assay (*see* **Note 1**).
2. A single 10-base oligodeoxynucleotide primer of random sequence, containing 50–70% G + C. Twenty picograms of primer are required for each assay (*see* **Note 1**).
3. 10X *Taq* buffer: 100 mM Tris-HCl, pH 8.3, 500 mM KCl, 19 mM MgCl$_2$, 1% gelatin.
4. 20 mM MgCl$_2$.
5. 2 mM Solution of all four deoxynucleotide triphosphates (dNTPs): dATP, dCTP, dGTP, and dTTP.
6. *Taq* Polymerase: 5 U/µL (Perkin-Elmer/Cetus, Norwalk, CT).
7. Reaction premix: Make up enough mixture for all the samples to be assayed. For a single 25-µL reaction mix 15.8 µL of sterile distilled water, 2.5 µL of 10X *Taq* buffer, 1.25 µL of 2 mM dNTPs, 0.25 µL of 20 mM MgCl$_2$ (*see* **Note 2**) and 0.2 µL of 5 U/µL *Taq* polymerase. Mix thoroughly.
8. Mineral oil.

3. Methods

1. Pipet 20 µL of reaction premix into each tube, add approx 20 ng of template DNA, and 20 pg of primer (*see* **Notes 1**, **3**, and **4**) to bring the reaction volume to 25 µL. (The final reaction conditions are 10 mM Tris-HCl, pH 8.3, 50 mM KCl, 2.1 mM MgCl$_2$ (*see* **Note 2**), 0.01% gelatin, 100 µM each dNTP, 0.2 µM primer, and 1 U of *Taq* polymerase/25 µL reaction.)
2. Overlay each reaction with 50 µL mineral oil.
3. This reaction is placed in a Perkin-Elmer/Cetus thermocycler or equivalent for 45 cycles. Each cycle is composed of 1 min at 94°C (denaturation step), 1 min at 35°C (annealing step), and 2 min at 72°C (extension step). The final cycle is culminated by 5 min at 72°C and then held at 4°C until assayed (*see* **Note 5**).
4. The sizes of the amplification products are determined by gel electrophoresis in a 1.4% agarose gel and visualized by staining with ethidium bromide (*see* **Fig. 1**, **Note 6**, and Chapter 13).

4. Notes

1. The concentrations of DNA and primers are such that the combined volume of 20 ng of DNA and 20 pg of primer is 5 µL.
2. Fluctuation in the final concentration of MgCl$_2$ above or below 2.1 mM can often cause a marked decrease in amplification products. The concentration of MgCl$_2$ that will produce the most consistent results will usually depend on the quality of the template DNA.
3. The ratio of primer to template DNA in the RAPD reaction is critical. A shift in concentration of either the primer or template DNA can result in the inconsistent amplification of discrete loci, increased background, or failure of amplification. Template DNA concentration should be carefully titered against a standard concentration of primer. This will reveal the concentration of DNA that gives the most reproducible amplified products. A titration should be performed when using DNA from different organisms, or when using DNA of varying quality, including DNA isolated with different extraction procedures. It is not necessary to retiter each DNA sample as long as the concentration of DNA and the quality of DNA remains constant.
4. Primer quality is important. If the amplification reaction is initiated with a primer containing significant levels of premature termination products, the RAPD reaction will usually not work. If care is taken during the synthesis and storage of the oligodeoxynucleotide, purification is usually not required beyond deprotection and desalting of the primer.

5. In order to obtain reproducible results, one should maintain consistent thermocycling conditions. Fluctuation of cycle time or number of cycles, or the use of different brands of thermocyler will often result in different amplification products.
6. Occasionally one notices a hazy smear partially or completely obscuring the amplified bands on the agarose gel. This is usually caused by a failure to saturate the DNA template with primer and can be corrected by adjusting the ratio of primer to template DNA.

References

1. Williams, J. G. K., Kubelik, A. R., Livak, K. J., Rafalski, J. A., and Tingey, S. V. (1990) DNA polymorphisms amplified by arbitrary primers are useful as genetic markers. *Nucleic Acids Res.* **18,** 6531–6535.
2. Welsh, J. and McClelland, M. (1990) Fingerprinting genomes using PCR with arbitrary primers. *Nucleic Acids Res.* **18,** 7213–7218.
3. Paran, I., Kesseli, R., and Michelmore, R. W. (1991) Identification of restriction fragment length polymorphisms and random amplified polymorphic DNA markers linked to downy mildew resistance genes in lettuce, using near-isogenic lines. *Genome* **34,** 1021–1027.
4. Martin, G. B., Williams, J. G. K., and Tanksley, S. D. (1991) Rapid identification of markers linked to a pseudomonas resistance gene in tomato by using random primers and near isogenic lines. *Proc. Natl. Acad. Sci. USA* **88,** 2336–2340.
5. Michelmore, R. W., Paran, I., and Kesseli, R. V. (1991) Identification of markers linked to disease resistance by bulked segregant analysis: a rapid method to detect markers in specific genomic regions using segregating populations. *Proc. Natl. Acad. Sci. USA* **88,** 9828–9832.
6. Giovannoni, J. J., Wing, R. A., Ganal, M. W., and Tanksley, S. D. (1991) Isolation of molecular markers from specific chromosomal intervals using DNA pools from existing mapping populations. *Nucleic Acids Res.* **19,** 6553–6558.
7. Reiter, R. S., Feldman, K. A., Williams, J. G. K., Rafalski, J. A., Tingey, S. V., and Scolnik, P. A. (1992) Genetic linkage of the arabidopsis genome: methods for mapping with recombinant inbreds and random amplified polymorphic DNAS. *Proc. Natl. Acad. Sci. USA* **89,** 1477–1481.
8. Caetano-Anolles, G., Bassani, B. J., and Gresshoff, P. M. (1991) High resolution DNA amplification fingerprinting using very short arbitrary oligonucleotide primers. *Biotechnology* **9,** 553–557.

86

Restriction Fragment Length Polymorphism

Mohammad S. Enayat

1. Introduction

DNA sequence changes within a gene results either in polymorphism or mutation causing different diseases. Some of these polymorphisms that occur with a high frequency within the population can be a useful tool for gene tracking for a given disease. Such investigations have initially been done by Southern blot techniques, but where possible have now been replaced by polymerase chain reaction (PCR)-based methodology. The nucleotide substitutions can be identified in two ways:

1. By use of restriction enzyme analysis or restriction fragment length polymorphisms (RFLP).
2. Allele specific oligonucleotide hybridization (ASO-H) or similar techniques.

Another type of polymorphism, a polymorphic tandem dinucleotide repeat sequence or variable number tandem repeat (VNTR) can also be used for gene tracing in a familial disease. In these cases a segment containing the repeats is amplified and the fragment size differences are detected by gel electrophoresis.

Hemophilia A or Factor VIII deficiency is the most common inherited bleeding disorder in humans. This X chromosome-linked disorder affects approximately 1 in every 10,000 males and within the families of these patients, the females are at risk of being carriers of this disorder. Factor VIII is a component of the intrinsic coagulation pathway and the *FVIII* gene is a large gene, encompasses 186 kb at Xq28. It has 26 exons encoding a mRNA of 9 kb *(1)*. Both RFLP and VNTR analysis have been extensively used in carrier detection and antenatal diagnosis in families with classical or familial hemophilia A. So far 10 useful polymorphisms have been identified within (intragenic) or flanking (extragenic) the *FVIII* gene (**Table 1**). Seven of these polymorphisms are diallelic RFLP and one, within the intron 7, is a nucleotide substitution (G/A) usually detected by ASO-H *(2)*.

The most useful intragenic polymorphic sites with high heterozygosity in different ethnic populations are in intron 18 and 22, recognized with the BclI and XbaI restriction enzymes, respectively. A closely linked polymorphism recognized by the BglII restriction is also highly informative but with 5% theoretical chance of recombination. All of these three RFLPs have originally been identified and analyzed by Southern blotting. However, this method is time consuming and may need the radioactive method

Table 1
DNA Polymorphisms Within or Flanking the Factor VIII Gene from Peake et al. (6)

Restriction enzyme	Site	Detection PCR	Detection Probe	Heterozygosity in Caucasian
*Bcl*I	Intron 18	+	+	0.43
*Xba*I	Intron 22	+	+	0.49
*Hin*dIII	Intron 19	+	+	0.38
*Msp*I	Intron 22	–	+	0.01
*Taq*I	5'	–	+	0.40
*Bgl*I	3'	–	+	0.25
*Msp*I	3'	–	+	0.43
(CA repeat)	Intron 13	+	–	(10 alleles approx 0.80)
(CA repeat)	Intron 22	+	–	(6 alleles approx 0.55)
(G/A)	Intron 7	+	–	0.33

for DNA band visualization. To this end, the BclI intragenic RFLP method has now been replaced by a fast and nonradioactive PCR analysis *(3,4)*.

2. Use of Southern Blotting in BclI RFLP Analysis of Hemophilia A

This method involves a series of techniques some of which are dealt in detail elsewhere in this book. These techniques include extraction of DNA from blood samples; digested with appropriate restriction enzyme and electrophoresis; preparation, extraction, isolation, purification and radio-labeling of the DNA probe; Southern blotting; hybridization; and finally autoradiography for DNA band visualization.

2.1. Materials Used in Restriction of DNA with the BclI Enzyme

1. Assay buffer (X10): The composition of this buffer varies from one manufacturer to another. For example, the composition of Amersham Pharmacia Biotech (Amersham Pharmacia Biotech, Buckinghamshire, UK) reaction buffer called One-Phor-All buffer PLUS (OPA) is: 10 m*M* Tris-HCl (pH 8.0).
2. 200 μg/mL bovine serum albumin (BSA).
3. Loading buffer (LB): 30% Xylene in 30% glycerol.
4. See **Notes 1–4** for restriction enzyme.

2.2. Southern Blotting

1. 0.25 *M* HCl
2. Standard saline citrate (SSC) (20X): NaCl 175.3 g/L and trisodium citrate 88.2 g/L. Adjust to pH 7.0.
3. Denaturation buffer: NaCl 87.66 g/L and NaOH 20 g/L.
4. Neutralization buffer: Tris 60.55 g/L and NaCl 87.66 g/L. Adjust to pH 7.4 with concentrated HCl.
5. 0.4 *M* NaOH.
6. Amersham Hybond N$^+$ (nylon) (Amersham Pharmacia Biotech, Buckinghamshire, UK) as transfer membrane
7. Whatman 3MM chromatography paper (Whatman International, Ltd., Maidstone, Kent, UK).

3. Methods
3.1. Restriction of DNA

1. Isolate and purify DNA (*see* Chapter 13).
2. Pipet into a small Eppendorf tube a desired (about 25 µg) amount of DNA and dilute to 20 µL with distilled water.
3. Add 3 µL of the appropriate 10X assay buffer, 5 µL of 200 µg/mL BSA and appropriate number of units (usually 10 U) of the restriction enzyme diluted in dilution buffer (usually supplied with the enzymes) as desired, in a volume of 2 µL (*see* **Notes 2** and **3**).
4. Mix by pipeting and incubate at desired temperature for at least 60 min, preferably 3–4 h.
5. Centrifuge contents in benchtop microcentrifuge at full speed to recover the full content of the tubes.
6. Add 6 µL of loading buffer mixture and mix thoroughly.
7. Load the samples into a suitable size submarine gel well without touching the sides (*see* **Note 4**).
8. Electrophorese the gel at 30 V for 8 h, but usually overnight.
9. Observe the gels on a UV transilluminator and make a permanent record by taking a photograph of the gel. If the DNA has been digested properly, a smearing from well to the bottom of the gel should be present.

3.2. Southern Blotting

This method is used for determination of the molecular sizes of the DNA fragments after digestion with restricted enzymes and gel electrophoresis. DNA fragments are transferred to a nylon membrane for reaction with a labeled probe for band visualization and molecular weight sizing of each of the fragments.

All the procedures are done at room temperature and the buffers used do not have to be sterile.

1. After electrophoresis trim away unwanted areas in the gel. Mark on the corner for gel orientation and identification.
2. Soak the gel in 0.25 M HCl for 15 min with gentle agitation on an orbital shaker (*see* **Note 5**).
3. Wash the gel twice with denaturation buffer for 30 min.
4. Neutralize the gel by replacing fluid with neutralization buffer and soak as in **step 2** for 30 min and repeat (*see* **Note 6**).
5. While the gel is in final soak, construct a bridge for blotting. Cut a piece of the 3MM paper to the same width as the base glass plate but long enough to form a wick into the buffer compartment over the edges of the bridge.
6. After final soak in neutralization buffer, pour off the excess fluid and take up the gel onto a spare piece of 3MM paper.
7. Place the gel onto the bridge with the DNA side up.
8. Smooth out gel gently with a gloved finger to remove any air bubbles between the bridge and the gel.
9. Cut piece of Hybond N$^+$ membrane to approximate size of the gel and place on the gel. Trim to the exact size of the gel, again ensuring that no air bubbles are trapped underneath the membrane.
10. Cut two or three pieces of paper to the size of the gel and then presoak briefly in 2X SSC. Layer on top of membrane.
11. Surround the bridge/gel with Saran Wrap to prevent buffer bypass and evaporation.
12. Cut a stack of paper towels to size and place on top of the presoaked papers. Finally, compress with a glass plate and a 1–1.5-kg weight.

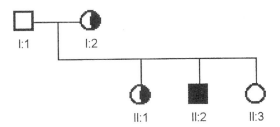
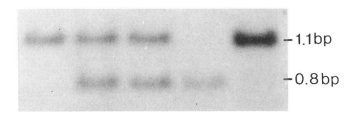

Fig. 1. A family with history of hemophilia A was investigated for BclI polymorphism and found to be fully informative. Patient (II:2) has inherited the hemophilic haplotype (0.8 bp) from her mother, who is the daughter of a hemophiliac and an obligate carrier. The patient's sister (II:1) is also a carrier and has the hemophilic haplotype, while his other sister (II:3) is unaffected. She has inherited from her mother the unaffected 1.1-bp haplotype.

13. Add transfer buffer (approx 400 mL) 0.4 M NaOH and allow the DNA to transfer overnight.
14. After blotting, carefully remove the membrane and soak in 2X SSC to remove any adherent agarose.
15. Blot dry briefly the membrane, which is now ready for either storage at 4°C or immediate hybridization.

3.3. Hybridization and Autoradiography

There are many different methods of hybridization that are dealt with elsewhere in this book (*see* Chapter 27). However, after hybridization the filter is probed with a ^{32}P-labeled DNA fragment from the *FVIII* gene. This genomic probe called p114.12, is a 647-bp Stu I/Sac I restricted *FVIII* *(4)*. The probed filter is exposed to an X-ray film (Hyperfilm MP, Amersham) for 4–7 d at –70°C in a cassette fitted with intensifying screen. In this polymorphism, a restriction fragment of variable length of 879 bp and/or 1165 bp can be detected in Southern blots of genomic DNA. About 42% of females are heterozygous at this locus. **Figure 1** shows the Southern blot using restricted DNA samples with BclI and probed with a ^{32}P-labeled p114.12 probe.

3.4. Use of PCR and BclI RFLP Analysis in Hemophilia A

The same RFLP, identified by BclI Southern blotting, has now been demonstrated by PCR followed by digestion with the restriction enzyme *(5)*. The PCR product of this highly polymorphic allele gives a 142-bp (– allele) and 99+43 bp (+ allele) after restriction with the enzyme (*see* **Fig. 2**).

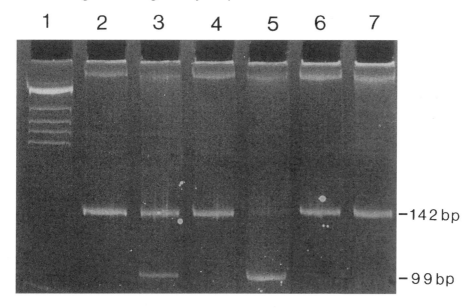

Fig. 2. BclI-restricted (lanes 3, 5, and 7) and -unrestricted (lanes 2, 4, and 6) PCR products, showing 142 bp (–) and 99 bp (+) fragments. Lane 1 is the molecular weight marker.

4. Notes

1. Keep the restriction enzyme cold at all times—if removed from the freezer it should be immediately kept on ice. In the majority of cases it can be used straight from the freezer.
2. Reaction volume here is fixed at 30 µL, as it is manageable. Generally the smaller the volume the better. Note that if the reaction volume is changed:
   a. Change the volume 10X assay buffer.
   b. Ensure that the enzyme added is <10% reaction volume.
   c. LB mixture added is 1/5th reaction volume.
3. Less enzyme can be used if the incubation period is lengthened.
4. To ensure adequate digestion before full-size gel electrophoresis, a minigel should be run. During digestion remove 2–3 µL of reaction mix and add 6 µL LB mixture. Load into 1% minigel and run at 50–60 mA for 1 h. View under UV illumination to check digestion.
5. HCl acid denaturation allows large >13-kb fragments to be transferred more efficiently by breaking the DNA into smaller fragments. Do not leave in HCl for more than 30 min or smaller DNA fragments will also be broken up into <300 bp, significantly reducing the ability to DNA to bind covalently to the membrane.
6. Gels may be left in neutralization buffer for longer than 1 h with no adverse effects if kept at 4°C so as to limit diffusion. Maximum time in neutralization buffer is 4 h.

References

1. White, G. C. and Shoemaker, C. B. (1989) Factor VIII and haemophilia A. *Blood* **73,** 1–12.
2. Peake, I. (1995) Molecular genetics and counselling in haemophilia. *Thromb. Haemost.* **74,** 40–44.
3. Gitschier, J., Drayna, D., Tuddenham, E. G. D., White, R. L., and Lawn, R. M. (1985) Genetic mapping and diagnosis of haemophilia A achieved through a Bcl I polymorphism in the factor VIII gene. *Nature* **314,** 738–740.

4. Gitschier, J., Lawn, R. M., Rotblat, F., and Goldman, E. (1985) Antenatal diagnosis and carrier detection of haemophilia A using factor VIII gene probe. *Lancet* **1,** 1093–1094.
5. Kogan, S. C., Doherty, N., and Gitschier, J. (1987) An improved method for prenatal diagnosis of genetic diseases by analysis of amplified DNA sequences: application to haemophilia A. *New Engl. J. Med.* **317,** 980–990.
6. Peake, I. R., Lillicrap, D. P., Boulyjenkov, V., Briet, E., Chan. Ginter, E. M., Kraus, E. M., Ljung, R., Mannucci, P. M., Nicolaides, K., and Tuddenham, E. G. D. (1993) Report of a joint WHO/WFH meeting on the control of haemophilia: carrier detection and prenatal diagnosis. Blood Coag. Fibrinol. **4,** 313–344.

87

Detection of Mutations in DNA and RNA by Chemical Cleavage

Richard G. H. Cotton

1. Introduction

Chemical cleavage of mismatch (CCM) is one of the mismatch cleavage methods, all of which allows scanning of 1–2 kb lengths in one test (reviewed in **ref. 1**). Heteroduplexes are formed between wild-type and mutant DNA (or RNA) and reacted with hydroxylamine, which modifies matched or unmatched C bases (*see* **Fig. 1**) and separately with osmium tetroxide, which modifies mismatched or unmatched T bases. The DNA is then cleaved at the modified base with piperidine *(2)*. Fragments are sized on sequencing gels and often on sequencing machines *(3)* to indicate the site of the mutation. Recent protocols dictate that all four strands be labeled (e.g., *see* **ref. 3**) to give two chances of detecting the mutation. This is because the only mismatches shown to be negative in this protocol are one-third of TG mismatches and labeled mutant strands allow these mismatches to be detected by cleavage at the complementary CA mismatch. Thus, this method almost guarantees detection of all mutations by cleavage at mismatched or unmatched (for deletions or insertions) bases.

1.1. Strategy

Because all classes of C and T mismatches (C.C, C.T, C.A, T.T, T.G, and T.C) are cleaved *(2)*, complete screening of a double-stranded target for point mutations can be achieved using probes of both senses (**Figs. 1** and **2a**). Deletions will be detected by cleavage of unmatched C and T bases (**Fig. 2c**) or indirectly because of reactive bases nearby (**Fig. 2d**). Insertions will be detected indirectly by increased reactivity of nearby matched C and T bases in the probe next to or near the loop of the unlabeled DNA or RNA in the heteroduplex (**Fig. 2f**). In single-stranded targets, such as messenger RNA, increased reactivity of matched C and T bases near or next to the mismatched (**Fig. 2b**) or unmatched (**Figs. 2d** and **2f**) bases become more important for complete screening. However, to be certain of detecting all classes of mutations directly, cDNA needs to be made for heteroduplex formation with probes of both senses. For two chances of detecting each mutation and avoiding problems of unreactive TG mismatches, wild-type and mutant are usually labeled and heteroduplexes formed using equivalent concentrations of both.

From: *The Nucleic Acid Protocols Handbook*
Edited by: R. Rapley © Humana Press Inc., Totowa, NJ

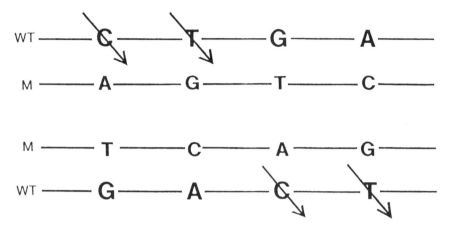

Fig. 1. All mutations can be detected by use of probes of both senses in making the heteroduplexes (top and bottom strands).

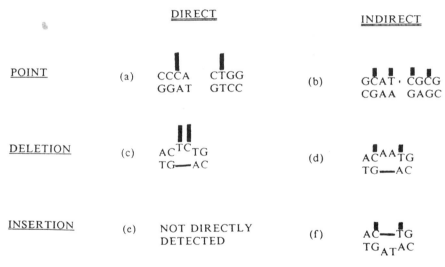

Fig. 2. Modes of detection of mutations. Heteroduplexes are shown with the probe strand in the upper position and test DNA or RNA in the lower position. Vertical lines indicate relative reactivity of T and C bases, with osmium tetroxide and hydroxylamine. Reactivity of matched nearby bases is less than unmatched or mismatched bases.

The technique has two modes of use with either (1) uniformly labeled probe or (2) end-labeled probe. Either mode can be used when the variation expected is minimal, such as one mutation in the region covered by a probe. However, if multiple differences are expected, e.g., 1 base in 10 is likely to vary, an end-labeled probe will generate a single and unique band for each reactive C or T in the probe, thus generating a pattern of difference between the pieces of nucleic acid. When using an end-labeled probe, the low reactivity of all matched C and T bases with hydroxylamine and osmium tetroxide, respectively, offers a background C and T track to help locate the mismatch.

The method has been applied to DNA amplified either via plasmids *(4)* or the polymerase chain reaction (PCR) technique *(5)*, and directly to RNA and viral RNA isolated from mammalian cells, but not as yet directly to unamplified genomic DNA.

2. Materials

1. 2X DNA/DNA annealing buffer: 1.2 M NaCl, 12 mM Tris-HCl, pH 7.5, and 14 mM MgCl$_2$. Store at room temperature.
2. 5 M Hydroxylamine solution: 1.39 g Solid hydroxylamine hydrochloride or hydroxylammonium chloride is dissolved in 1.6 mL of warmed distilled water in a glass test tube; 1.75 mL diethylamine is added dropwise to bring the pH to 6. Store at 4°C for up to 1 wk.
3. 4% Osmium tetroxide solution: (N.B. Osmium tetroxide irritates mucous membranes and should be used in a fume hood.) Break a 0.5-g ampule (Johnson Matthey, Materials Technology, Royston, Herts, UK) and place in 12.5-mL distilled water in a glass bottle with a lid and a good seal. Stand for 2–3 d at 40°C to dissolve, and store. A fresh 1 in 5 dilution is made on the day of use. Alternatively, 4% solution from Aldrich can be diluted appropriately.
4. 10X Osmium tetroxide buffer: 100 mM Tris-HCl, pH 7.7, 10 mM ethylenediaminetetraacetic acid (EDTA), 15% pyridine. Store at –20°C.
5. HOT (hydroxaline osmium tetroxide) stop buffer: 0.3 M Na acetate, pH 5.2, 0.1 mM EDTA, 25 μg/mL tRNA. Store at –20°C.
6. DNA/RNA annealing buffer: 80 Parts filtered deionized formamide is added to 20 parts of the following, immediately before use: 200 mM PIPES, pH 6.5, 5 mM EDTA, and 2 M NaCl. Filter for storage at room temperature.
7. Piperidine: 1 M Piperidine diluted immediately before use, from 10 M stock. (N.B.: Piperidine is toxic and should be used in a fume hood.) Pyrrolidone is a noncontrolled alternative *(6)*.
8. Tris EDTA (TE): 20 mM Tris-HCl, pH 7.4, 1 mM EDTA.

3. Methods

3.1. DNA and RNA Preparation

1. Prepare unlabeled DNA (about 10 μg) by standard methods, such as plasmid amplification *(4)* or PCR amplification *(5)*, and cut with appropriate enzymes.
2. Prepare unlabeled RNA (about 5 μg total cellular) by standard methods *(4)*. When RNA is abundant, e.g., collagen mRNA and viral RNA, total cellular RNA can be used as RNA source.
3. Prepare ^{32}P-labeled probe DNA by:
   a. Appropriating a restriction enzyme digest of about 1 μg, so that 5' overhangs can be filled with Klenow fragment *(4)*. Prepare probes of both senses for the same region by filling 5' overhangs at both ends. Prepare single-sense probes by making one cut with an enzyme not producing a 5' overhang that can be labeled, or if it does, cut off the label with an appropriate second enzyme. This provides a single end-labeled probe.
   b. Filling in an M13 clone using the universal primer and radiolabeled base, and subsequently cut out the insert *(2)*.
   c. Kinase labeling a PCR amplified fragment *(4)*.
   d. Adding radiolabeled bases during the PCR reaction *(7)*.
   e. Fluorescently label by using labeled PCR primers (e.g., **ref. 3**).

3.2. Heteroduplex Formation

3.2.1. DNA/DNA

1. Take labeled control and unlabeled test DNA (0.1–1 µg) in 20–100 µL TE and mix, so that there is at least a 12X excess of unlabeled DNA. To this add an equal volume of 2X annealing buffer. For two chances of detecting any mutation, equivalent labeled mutant and wild-type DNA are used.
2. Place the mixture in a boiling water bath for 5 min and then transfer to 42°C for 60 min.
3. Precipitate the heteroduplex once with ethanol, wash once with 70% ethanol, and dry. Resuspend the DNA in distilled water, so that 1 µL contains 1000 dpm.
4. Prepare a homoduplex-labeled control plus unlabeled control identically for the control reaction.

3.2.2. DNA/RNA

1. Mix DNA probe and test RNA (0.1–1 µg) so that there is at least a 12X excess of RNA, precipitate the mixture, wash with 70% ethanol, and dry the pellet.
2. Add 40 µL of DNA/RNA annealing buffer. Incubate at 90°C for 5 min, then at 55°C for 2 h, and then ethanol precipitate. Wash the pellet with 70% ethanol and dry. Resuspend the pellet in distilled water so that 1 µL contains 1000 dpm.

3.3. Modification Reactions

Osmium tetroxide and hydroxylamine tubes should be handled in a fume hood until the ethanol precipitates have been washed.

1. Distribute 6000 dpm in 6 µL to each tube. Total DNA or DNA/RNA in each reaction should be 0.2–2.6 µg.
2. Hydroxylamine reaction (two tubes): Add 20 µL of hydroxylamine solution, mix, and incubate at 37°C for 10 and 30 min for partial cleavage and 60 min for complete cleavage.
3. Osmium tetroxide reaction (two tubes). Add 2.5 µL of 10X osmium tetroxide buffer and 15 µL of diluted osmium tetroxide solution. Mix (do not centrifuge, as a yellow precipitate appears), and incubate at 37°C for 1 and 5 min for partial cleavage, and 20 and 60 min for complete cleavage.
4. Stopping reactions: Stop reactions with 200 µL of HOT stop buffer and 750 µL of ethanol and precipitate the DNA. Wash the pellet with 70% ethanol and dry.

3.4. Piperidine Cleavage (8)

Add 50 µL of 1 M piperidine to each tube. Vortex for 10 s, heat at 90°C for 30 min, and cool on ice. Add 50 µL of 0.6 M sodium acetate buffer, pH 5.2, and 300 µL of ethanol to precipitate the DNA. Wash with 70% ethanol and dry. Take up the samples with formamide dyes and analyze by electrophoresis on sequencing gels (*see* Chapter 60 and **ref. 4**).

4. Notes

1. The probe length is limited only by the analytical technique and the fidelity of the heteroduplex formation. Probes up to 1.7 kb have been analyzed on sequencing gels.
2. Probe overlap: When screening kilobase lengths, probes need to be overlapped by 20–30 bases to avoid missing mismatches at the overlap. This is because of considerable breathing of the duplex at each end.

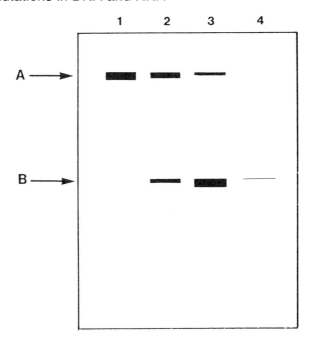

Fig. 3. Schematic representation of detection of a mutation in probe A hybridized to mutant DNA. Cleavage of the end-labeled probe by one of the HOT chemicals plus piperidine at a mismatch results in the fragment of reduced size (B) in the sequencing gel. Lanes 1–4 represent increasing chemical reaction time. The diagram illustrates that if too little incubation (lane 1) or too much incubation (lane 4) is given, band B (which signals the mutation) may not be seen.

3. Time of incubation: Pilot work needs to be done on the time needed for analysis of particular quantities of unlabeled DNA/DNA probe. If most of the probe remains uncleaved, there has been too little reaction, and if it is all cleaved, there may be too much reaction. This is because matched bases are modified and cleaved at a rate of about 1/100 of that of mismatched bases, allowing nonspecific probe destruction if the incubation time is too long. A time course is necessary between these limits to be sure of a complete assessment (*see* **Fig. 3**). Usually the concentrations of chemicals given is correct for approximately 50 ng total weight of DNA.
4. Heteroduplex formation: Sometimes when changing operators or laboratories, it has been found that heteroduplexes have not been formed. This has been thought to be caused by the boiling water bath coming off the boil before melting, or use of too large a volume of solution for heteroduplex formation (>1 mL). Formation of a heteroduplex with a known mismatch should be included as a control that will be cleaved if, in fact, that heteroduplex has been formed.
5. Osmium tetroxide: Potency of this reagent between two laboratories has varied. It is not clear whether this is owing to different reagent batches or aging of an initial solution. A range of concentrations should be tried when setting up the method. Solutions should be freshly made each 3 mo before the solution takes on a green hue, although it is still active at this time. Eppendorf tubes darken as a result of the reaction. Osmium tetroxide from one manufacturer has been ineffective, as it either does not cleave the T mismatches or the background is too high. The purity of the chemical that was successful from two different manufacturers was 99.8 and 99.9% (Aldrich, Milwaukee, WI, and BDH,

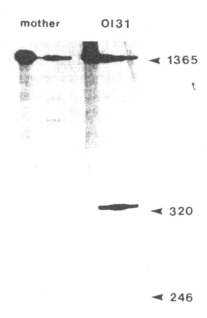

Fig. 4. Analysis of patient OI31 *(10)*. Fibroblast RNA was annealed to a 1365-bp end-labeled *Nco*I-*Eco*RI fragment from a human collagen α-l(I) cDNA clone Hf404. Treatment with hydroxylamine was as described in **Subheading 3**. RNA from the mother and patient OI31 was analyzed as indicated.

London, UK, respectively). In one case, the reagent was inactivated when the stock solution was prepared in a plastic bottle.

6. False positives and negatives: There have been no false positives identified so far. With the initial protocol *(2)* there were some false negatives as one-third of TG mismatches were unreactive with osmium tetroxide. This has led to the need for the labeling of mutant and wild-type DNA to give each mutation two chances of detection (*see* **ref. 3**, e.g.).
7. Restriction enzyme sites have been added to PCR primers to facilitate probe labeling *(9)*.
8. Use with PCR and sequencing: The CCM method has been found to be especially useful with PCR-amplified products. This is because (a) in the case of a heterozygote cDNA at a particular locus, often one allele has been found to be more frequent in the product than the other, making interpretation of mutant sequence difficult after direct sequencing and (b) M13 cloning of PCR products from heterozygotes has, on occasion, shown many PCR errors, and there is a need to distinguish these from the actual mutation. Application of the CCM method samples the average of the product and indicates where to expect the mutations and which base is changed, i.e., distinguishes PCR artifacts from the real mutation.
9. Detection of homozygosity or heterozygosity: It is logical that cleavage kinetics should be different when mismatches are present in one or both alleles. Thus, in the case of homozygotes, the probe band should disappear to nothing at a uniform rate, but in the case of a heterozygous situation, the probe band should rapidly diminish to half intensity and then diminish more slowly and in parallel with the band signaling the mismatch. This situation is observed in the heterozygous mutation 0131 in **Fig. 4**. However, a test has recently been developed *(11)* for testing allele status. This involves making the probe from the test DNA and hybridizing it with unlabeled test DNA. If the wild-type allele is present in the test sample, i.e., it is a heterozygote, cleavage will occur, but not if it is homozygous for the mutation in question.

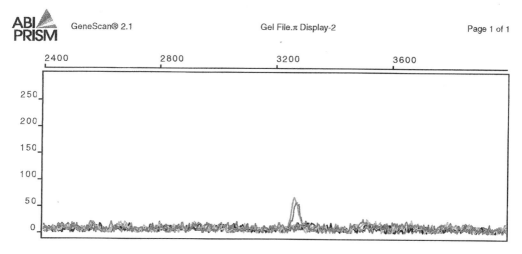

Fig. 5. A computer-generated composite trace from the output of the ABI 377 sequencer exon 13 of six patients. The trace shows the peaks from two patients with the same polymorphism. (Figure contributed by Dr. T. Ellis and used with permission.)

10. Practical example: Several points are illustrated in **Fig. 4**. The major band at 320 bp represents direct detection of a mismatched C found in the heteroduplex as a result of a G→C mutation. It is notable that this mutation is heterozygous, as osteogenesis imperfecta is dominant. As mentioned in **Note 9**, the kinetics of band change is typical of a heterozygote. The minor band at 246 bp represents indirect detection of a polymorphism by reaction of a nearby matched C. Thus, such detection is easily possible when the mutation is only on one allele.
11. Criteria for CCM screening: Before a stretch of DNA can be deemed to be negative for the presence of mutations, it must appear identical with an identically prepared control homoduplex using the following criteria: (a) a range of reaction times are performed as per **Fig. 3**, (b) highest and lowest bands are visualized on long and short runs; respectively; and (c) heteroduplexes containing probes of both senses are treated with hydroxylamine and osmium tetroxide (d) mutant and wild-type DNA is labeled.
12. It has been shown that by treating the heteroduplexes first with hydroxylamine, washing, and then treating with osmium tetroxide in the same tube and then with piperidine the result is the same *(12)*, but the chemical causing the cleavage cannot be ascertained without further work.
13. A recent analysis with a fluorescent protocol similar to that in **ref. 3** can be seen in **Fig. 5**. This shows an exon of the *BRCA1* gene labeled at each end with the same fluorophore. The computer-generated figure shows the traces of six patients superimposed with the detection of a polymorphism in two of them.
14. Applications of the method. Recent applications include (a) C1-inhibitor *(13)*, (b) the *VDR* gene *(14)*, (c) amyloid precursor protein *(15)*, (d) Factor VIII *(16,17)*, and (e) HPRT *(18)*, (f) Factor IX *(19)*, (g) DAPA *(20)*, and (h) Dengue Virus *(21)*.

References

1. Cotton, R. G. H. (1997) *Mutation Detection*, Oxford University Press, Oxford, UK.
2. Cotton, R. G. H., Rodrigues, N. R., and Campbell, R. D. (1988) Reactivity of cytosine and thymine in single base-pair mismatches with hydroxylamine and osmium tetroxide and its application to the study of mutations. *Proc. Natl. Acad. Sci. USA* **85**, 4397–4401.

3. Cotton, R. G. H. and Campbell, R. D. (1989) Chemical reactivity of matched cytosine and thymine bases near mismatched and unmatched bases in a heteroduplex between DNA strands with multiple differences. *Nucleic Acids Res.* **17,** 4223–4232.
4. Maniatis, T., Fritsch, E. F., and Sambrook, J. (eds.) (1989) *Molecular Cloning. A Laboratory Manual,* 2nd ed., Cold Spring Harbor Laboratory Press, Cold Spring Harbor, NY.
5. Saiki, R. I. C., Scharf, S., Faloona, F., Mullis, K. G., Horn, C. T. Ehrlich, H. A., and Arnheim, N. (1985) Enzymatic amplification of β-globin genomic sequences and restriction site analysis for diagnosis of sickle cell anemia. *Science* **230,** 1350–1354.
6. Shi, Y. and Tyler, B. M. (1989) Pyrrolidone, a noncontrolled substance, can replace piperidine for the chemical sequencing of DNA. *Nucleic Acids Res.* **17,** 3317.
7. Schowaller, D. B. and Sommer, S. S. (1989) The generation of radiolabeled DNA probes with polymerase chain reaction. *Anal. Biochem.* **177,** 90–94.
8. Maxam, A. M. and Gilbert, W. (1977) A new method for sequencing DNA. *Proc. Natl. Acad. Sci. USA* **74,** 560–564.
9. Scharf, S. J., Horn, G. T., and Erlich, H. A. (1986) Direct cloning and sequence analysis of enzymatically amplified genomic sequences. *Science* **233,** 1076, 1077.
10. Dahl, H.-H. M., Lamande, S. R., Cotton, R. G. H., Cole, W. G., and Bateman, J. F. (1990) A rapid chemical cleavage method for the detection and localization of base changes in RNA. *Proceeding of the UCLA symposia on molecular and cellular biology; Biotechnology and human genetic predisposition to disease,* vol. 126. Cantor, C. R., Caskey, C. T., Hood, L. E., Kamely, D., and Ommen, G. S., eds. Alan R. Liss, NY, pp. 209–216.
11. Dianzani, I., Forrest, S. F., Camaschelia, C., Gottardi, E., and Cotton, R. G. H. (1991) Heterozygote identification by chemical cleavage of mismatch. *Am. J. Hum. Genet.* **49,** 175–183.
12. Grompe, M., Muzny, D. N. F., and Caskey, C. T. (1989) Scanning detection of mutations in human ornithine transcarbamylase (OTC) by chemical mismatch cleavage. *Proc. Natl Acad. Sci. USA* **86,** 5888–5892.
13. Verpy, E., Biasotto, M., Brai, M., Misiano, G., Meo, T., and Tosi, M. (1996) Exhaustive mutation scanning by fluorescence assisted mismatch analysis discloses new genotype-phenotype correlations in angiodema. *Am. J. Hum. Genet.* **59,** 308–319.
14. Zerwekh, J. E., Hughes, M. R., Reed, B. Y., Breslau, N. A., Heller, H. J., Lemke, M., Nasonkin, I., and Pak, C. Y. C. (1995) Evidence for normal vitamin D receptor messenger ribonucleic acid and genotype in absorptive hypercalciuria. *J. Endocrinol. Metab.* **80,** 2960–2965.
15. Lidell, M. B., Bayer, A. J., and Owen, M. J. (1995) No evidence that common allelic variation in the amyloid precursor protein (APP) gene confers susceptibility to Alzheimer's disease. *Hum. Mol. Genet.* **4,** 853–858.
16. Bidchandrani, S. I., Lanyon, W. G., Shiach, C. R., Lowe, G. D. O., and Connor, J. M. (1995). Detection of mutations in ectopic factor VIII transcripts from nine haemophilia A patients and the correlation with phenotype. *Hum. Genet.* **95,** 531–538.
17. Rudzki, Z., Duncan, E. M., Casey, G. J., Neumann, M., Favaloro, E. J., and Lloyd, J. V. (1996) Mutations in a subgroup of patients with mild haemophilia A and a familial discrepancy between the one-stage and two-stage factor VIII: C methods. *Br. J. Haem.* **94,** 400–406.
18. Tsuboi K., Nose T., Okinaka R. T., and Chen D. J. (1995) Non radioactive mismatch analysis to detect small mutations in human hypoxanthine guanine phosphoribosyl transferase cDNA. *Jpn. J. Med. Sci. Biol.* **48,** 163–175.
19. Montandon, A. J., Green, P. M., Gianneli, R., and Bentley, D. R. (1989) Direct detection of point mutations by mismatch analysis: application to haemophilia B. *Nucleic Acids Res.* **17,** 3347–3358.

20. Howells, D. W., Forrest, S. M., Dahl, H.-H. M., and Cotton, R. G. H. (1990) Insertion of an extra codon for threonine is a cause of dihydropteridine reductase deficiency. *Am. J. Hum. Genet.* **47,** 279–285.
21. Cotton, R. G. H. and Wright, P. J. (1989) Rapid chemical mapping of dengue virus variability using RNA isolated directly from cells. *J. Virol. Methods* **26,** 67–76.

88

Mutation Screening Using PCR-SSCP

Silver Staining and Isotopic Protocols

Philip J. Saker

1. Introduction

Screening for mutations prior to sequencing can reduce the time and costs of identifying mutations. When the DNA sequence is known, the technique of detecting mutations as single-stranded conformational polymorphisms (SSCP) is a convenient method of screening for possible mutations. SSCP was originally developed by Orita et al. *(1)*. It has the ability of detecting a single base change, and has been applied to a number of genes, including the insulin receptor *(2)*, GLUT 4 *(3)*, glucokinase *(4)*, and the mitochondrial genome *(5)*.

The principle of SSCP analysis is relatively simple. In nondenaturing conditions, single-stranded DNA (ssDNA) has a folded conformation that is determined by intramolecular interactions and therefore its base sequence. When electrophoresed on a nondenaturing polyacrylamide gel, the ssDNA will have a specific mobility depending on this base sequence. Any difference in the base sequence of an ssDNA sample, due to a mutation or polymorphism, will be detected as a mobility shift, and will produce a different band pattern when compared to the normal "wild-type" (**Fig. 1**).

SSCP is a convenient and a relatively rapid method to identify those subjects likely to possess a mutation. Subsequent sequencing is required to determine whether a particular "abnormal" band pattern is due to a mutation or polymorphism. When optimized, SSCP is reported to have a sensitivity of between 85–95% *(1,6)*. The conformation of the ssDNA in the gel may be altered by a number of conditions that have to be optimized to detect the mutations. These variables include:

1. Properties and pore-size of the gel; these are determined by the percentage of polyacrylamide, ratio of acrylamide/*bis* (cross-linker), and the temperature at which polymerization is carried out, including the temperature of the reagents.
2. Presence and percentage of glycerol; this has a weak denaturing action that partially opens the folded ssDNA structure so that more surface area is available for the gel to "sense" conformational changes.
3. Temperature at which electrophoresis is performed.
4. pH of the buffer used to make the gel.
5. pH and ionic strength of the buffer system.

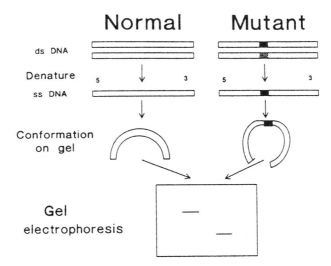

Fig. 1. Principles of SSCP. Heat denatured dsDNA to form ssDNA. The conformation of ssDNA is determined by its base sequence. A mutation or polymorphism will alter this, and affect its electrophoretic mobility. This will be seen as a difference in the SSCP band pattern compared to that of a normal ("wild-type") ssDNA sequence.

The SSCP band patterns may be detected by using radioactivity, silver-staining or fluorescence. This protocol will concentrate on silver-staining, because of its ease of use and safety within a laboratory not designated for using radioactive isotopes.

2. Materials

All solutions are made using deionized-distilled water. Purity of 16–18 $M\Omega$ per cm is recommended. Ultrapure or molecular biology grade chemicals and reagents are used.

2.1. Sample Amplification

1. 100 ng of DNA.
2. 50 pmol of each primer.
3. 0.2 mM of each dNTP.
4. Buffer: 10 mM Tris HCl (pH 8.3 at 25°C), 50 mM KCl, 1.5 mM MgCl$_2$, 1% TritonX 100.
5. 1 U of *Taq* polymerase.
6. For detection of SSCP bands using radioactivity Hot PCR is performed. Precautions need to be employed when handling radioisotopes, particularly the prevention of aerosols and avoidance of contamination. Local safety regulations need to be followed, along with the manufacturer's guidelines.
   The foregoing reaction is used with the addition of 1 µCi of [α^{32}P] dCTP (3000 Ci/mmol, 10 mCi/mL) *(4)* or 0.2 mM of each dCTP, dGTP and dTTP, 2 nmol cold dATP; 0.4 µl [^{35}S] dATP (1000 Ci/nmol, 10 µCi/mL) *(6)*.

2.2. Gel Preparation

1. 10X TBE: 1 L; 108 g Trizma-base, 55 g boric acid, 9.3 g ethylenediaminetetraacetic acid (EDTA), pH 8.0. Store at room temperature (RT).
2. Ammonium persulfate: Store tightly sealed at room temperature (RT) for up to 1 yr. Use 25%$^w/_v$ solution, make fresh daily. Store solution at 4°C.

3. TEMED: N,N,N',N'-tetramethylethylenediamine. Store at 4°C. After 10–12 mo there is a significant reduction in activity.
4. Glycerol: store at RT.
5. Acrylamide/*bis* solution: 40% 49:1 acrylamide/*bis* solution. Store at 4°C. WARNING: Acrylamide monomer is a neurotoxin (polyacrylamide is not toxic) absorbable through the skin; always wear gloves, avoid creating aerosols and dusts. If in contact with skin, wash with soap and rinse thoroughly with water.
6. Deionized-distilled water.

2.3. Sample Preparation

1. Gel loading dye: 0.1% bromophenol blue, 0.1% xylene cyanol, 90% formamide and 20 mM EDTA in deionized-distilled water. Store at 4°C.
2. Deionized-distilled water.

2.4. Silver Staining

1. Fixative 1: 40% methanol/10% acetic acid in deionized-distilled water. Store at RT (*see* **Note 20**).
2. Fixative 2: 10% ethanol/5% acetic acid in deionized-distilled water. Store at RT (*see* **Note 20**).
3. Oxidizer: 0.0032 M potassium dichromate and 0.0032 M nitric acid. Store at 4°C. WARNING: Strong oxidizer, avoid contact with reducing agents; irritant, avoid contact with skin and eyes.
4. Silver reagent: 0.012 M silver nitrate. Store at 4°C. WARNING: Poisonous, caustic to eyes, skin, and mucous membranes.
5. Developer: 0.28 M sodium carbonate and 0.5 mL of formalin per liter. Store at 4°C, although it may be stored at RT for 1 month. WARNING: Poisonous. Irritant, vapor and dust irritates eyes, mucous membranes, and skin.
6. Whatman filter paper (Whatman, Maidstone, Kent, UK).

2.5. Autoradiography

1. Fixative 1: 10% methanol/5% acetic acid in deionized-distilled water.
2. Whatman filter paper.
3. Kodak X OMAT film (Kodak, Rochester, NY).
4. X-ray cassette, intensifying screens are not required.
5. –70°C Freezer.
6. Darkroom.

3. Methods

3.1. Sample Amplification

1. Amplify region of interest, usually the exons and approximately 30 bases flanking it, using the relevant primers and PCR. A fragment of 150–200 bp is optimal for SSCP (*see* **Note 1**). A typical PCR is carried out in a 50 µL volume using a thermal profile of 94°C 5 min; then 35 cycles of 94°C 30 s, 60°C 30 s, 72°C 30 s; and finally 72°C 10 min (*see* **Note 3**).
2. Confirm amplification on a 1.5% agarose gel with ethidium bromide staining, followed by illumination with UV light (*see* **Note 10**; *see* Chapter 13).

3.2. Preparation of Polyacrylamide Gels (see Note 18)

For SSCP, denatured samples are run on polyacrylamide gels at two different conditions. This is generally 10% polyacrylamide gel with 5% glycerol with electrophoresis at 25°C, and 10% polyacrylamide gel without glycerol with electrophoresis at 4°C.

Table 1
Typical Volumes Required to Prepare a Polyacrylamide Gel

Electrophoresis temperature	4°C, mL	25°C, mL
10X TBE	3.0 mL	3.0 mL
Deionized-distilled water	19.5 mL	18.0 mL
Glycerol	—	1.5 mL
40% of 19:1 acrylamide/*bis* solution	7.5 mL	7.5 mL
TEMED	24.0 µL	24.0 µL
25% Ammonium persulfate solution	190.0 µL	190.0 µL

1. Ensure deionized-distilled water is used for SSCP as chloride ions interfere with silver staining.
2. Allow acrylamide/*bis* solution and TEMED to warm to room temperature.
3. Make fresh 10X TBE and 25% ammonium persulphate solution (*see* **Note 14**).
4. Set up glass plates and spacers prior to casting gel.
5. A 30-mL gel mix is sufficient for a 20 × 20 cm gel with 0.4-mm thick spacers (*see* **Note 6**). For a 10% polyacrylamide gel run at 4°C and 25°C, the volumes listed in **Table 1** are required.
6. The foregoing reagents are put into a Pyrex beaker. Ensure that the glycerol is thoroughly mixed in solution prior to addition of acrylamide/*bis*, TEMED and ammonium persulphate. Gently, but thoroughly, mix the gel solution.
7. The gels are poured at room temperature (*see* **Note 4**) and the toothed comb inserted into the top of the gel. The gels are left to polymerize for at least 2 h at RT prior to use (*see* **Note 5**).
8. After polymerization, carefully remove comb and flush out wells with 1X TBE before placing gel in electrophoresis equipment (*see* **Note 7**).

3.3. Sample Preparation and Electrophoresis

Gel-loading dye containing formamide prevents the renaturation of the ssDNA after denaturation. Double-stranded DNA (dsDNA) is run with the samples to act as a marker and to distinguish its band from those created by ssDNA (*see* **Note 11**). If available, a positive control (PC) sample and normal "wild-type" sample are also loaded onto each gel (*see* **Notes 8** and **9**). The loading positions of the dsDNA and PC samples aid identification and orientation of the gel.

1. Samples are prepared as follows in labeled tubes:
   Sample tubes: 16 µL dH$_2$O, 2 µL SSCP dye, 2 µL PCR product
   PC tubes: 16 µL dH$_2$O, 2 µL SSCP dye, 2 µL PCR product
2. The tubes are capped and centrifuged for a few seconds at 4°C.
3. The samples are denatured at 95°C for 6 min. This may be done in a heated block, but it is more convenient to use a PCR thermal cycler.
4. To prevent renaturation, the tubes are immediately placed on ice for 10 min.
5. Centrifuge the tubes again for 1 min at 4°C, and then immediately put back on ice.
6. Prepare the dsDNA sample in a labeled tube: 17 µL dH$_2$O, 2 µL SSCP dye, 1 µL PCR product.
7. The samples are loaded onto the polyacrylamide gel using a duck-billed pipet tip (*see* **Note 12**).

PCR-SSCP Mutation Screening

8. The gel is typically run at 25 W for approx 18 h in 1X TBE (*see* **Notes 13**, **15**, and **16**). A 200-bp ssDNA sample will run approximately three-quarters the way down the gel. The power may be altered to vary the run-time and separation (*see* **Note 17**).

3.4. Visualization of SSCP Bands

3.4.1. Silver Staining of Gels

SSCP bands may be visualized by silver staining (*see* **Note 27**). Clarity may depend on the amount of sample loaded onto the gel. The gel may have a dark background or be stained totally black if chloride ions are present in the water used to make the solutions. The following protocol is based on a method by Merril et al. *(7)*, and is commercially available in a modified kit (silver stain kit, Bio-Rad Laboratories, Hemel Hempstead, Hertfordshire, UK). The process is temperature-dependent, particularly the developer stage, and should be carried out at room temperature (*see* **Note 22** and **Table 2**).

The following steps have to be performed with care so as not to damage the gel. Gloves should be washed in deionized-distilled water to remove powder and other contaminants which may discolor the gel. If possible, lightly hold the gel at the top only; excessive handling and pressure will affect the staining.

1. Remove gel from glass plates and place face down in a 21 × 21 × 5 cm Pyrex baking dish.
2. Add 400 mL of fixative 1; remove after 30 min (*see* **Notes 19**, **21**, **23–25** and **26**).
3. Add 400 mL of fixative 2; remove after 15 min.
4. Repeat **step 3**.
5. Add 200 mL of oxidizer; remove after 5 min.
6. Add 400 mL of deionized water, gently agitate; remove after 5 minutes.
7. Repeat **step 6** twice.
8. Add 200 mL of silver reagent; remove after 20 min.
9. Add 400 mL of deionized water; remove after 1 min.
10. Add 200 mL of developer, gently agitate; remove after approx 30 s when the solution turns yellow, or brown precipitate appears.
11. Add 200 mL of developer, may be gently agitated; remove after 5 min.
12. Repeat **step 11** if bands require further development.
13. Add 400 mL of Stop solution, leave for at least 5 min, then remove.
14. Gently place piece of Whatman filter paper, of sufficient size, onto the back of the moist gel. Starting from the top, carefully lift from baking dish. Gently place clingfilm onto front of gel and dry on a gel dryer.
15. The gel may be stored or photographed.

Figure 2 shows a silver-stained SSCP gel.

3.4.2. Autoradiography

1. Remove gel from glass plates and place face up in a 21 × 21 × 5 cm Pyrex baking dish.
2. Add 400 mL of fixative 1; remove fixative after 30 min.
3. Gently place Whatman filter paper, of sufficient size, onto the front of the moist gel. Starting from the top, carefully lift from the baking dish. Gently place clingfilm onto the gel and dry on a gel dryer.
4. Place dried gel into an X-ray cassette and, in a dark room, load Kodak X OMAT film so that it is exposed to the gel surface. Leave at 70°C for 1–5 h.
5. Develop film (*see* **Note 28**).

Table 2
Troubleshooting Guide for Silver Staining

Problem	Solution
Gray or brown precipitate appearing as smudges or swirling on gel surface. May become mirrorlike. Bands may be faint or absent.	1. Nonspecific deposition of silver due to carryover of oxidizer or silver reagent. Increase number and duration of wash steps. 2. Temperature too low. Ensure all reagents are at least 23°C. 3. Mirroring can be due to developer precipitate sticking to gel surface. First volume of development solution must be decanted as soon as precipitate appears.
Dark uniform background, usually yellow.	Incomplete removal of oxidizer. Increase wash steps to remove all traces of yellow before addition of silver reagent.
Mottled background.	1. Contaminants in water. Check purity. 2. Incomplete removal of gel buffer components, increase timing of fixatives 1 and 2.
Slow or no development.	1. Development rate is temperature-dependent. Developer solution may be heated to 50°C to speed up development. 2. Developer solution is too old. 3. Ensure developer solution is mixed thoroughly.
Gel continues developing, or becomes darker when drying on Whatman filter paper.	Increase timing of **step 13** of **Subheading 3.4.1.** Repeat two to three times. As a safeguard, photograph gel prior to drying.
Incomplete staining.	Complete silver staining, then recycle gel. Soak in deionized-distilled water for 30 min and then repeat **steps 8–13** of **Subheading 3.4.1.**
Large discolored spots on gel.	Pressure on the gel will cause the gel to stain darker at the contact point. Avoid crushing the gel with fingers, etc.

4. Notes

1. Studies have shown that the optimum PCR length for SSCP is 150–200 bp, and that there is a reduction in sensitivity as the fragment increases (*6*). There may also be a minimum fragment length.
2. It is possible to amplify a longer PCR product and then cut this with a restriction enzyme into appropriate fragments. This is particularly useful when exons are separated by only a short length of intron, or if an exon is large. The two fragments can then be run together on the same SSCP gel. Twice the amount of digested PCR sample needs to be loaded.
3. If possible, optimize the PCR so that nonspecific primer binding does not occur as this could interfere with SSCP and cause confusion when interpreting the SSCP bands.
4. Use of a 50-mL plastic syringe eases pouring of gels.

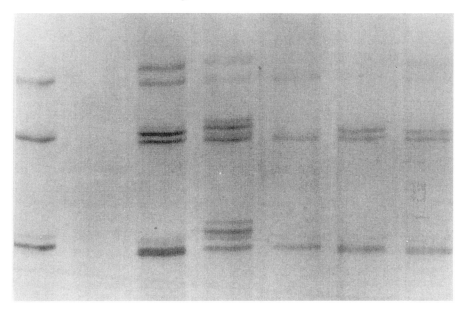

Fig. 2. SSCP gel using silver stain to detect bands. SSCP analysis of exon 8 of the glucokinase gene. Lane 1 is a negative control sample; lane 2 is blank; lane 3 is a positive control sample (Gly^{299}→Arg) from a member of pedigree BX **(12)**; lane 4 is a positive control sample from a French pedigree; lane 5 is a patient with a normal SSCP band pattern; lanes 6 and 7 are two patients with similar abnormal SSCP band patterns to the member of pedigree BX (lane 3). Sequencing confirmed that these two patients (lanes 6 and 7) possessed the Gly^{299}→Arg mutation **(13)**.

5. Polymerized gels may be stored at 4°C prior to use; paper towels soaked in 1X TBE should be placed across the top of the combs, and then the gels wrapped in cling film. If run at 25°C, allow them to warm-up prior to loading.
6. Thinness or low percentage acrylamide may make the gels too weak to handle, which will be a particular problem when silver-staining.
7. Prior to loading of samples, the gels may be pre-run for approx 30 min.
8. Amplified DNA sample from a subject known to possess a mutation in the fragment being screened and a normal "wild-type" sample should be run to act as positive and negative controls, respectively. This will ensure that each gel is capable of resolving different conformers, and as indication that no problems have occurred with the gel, parameters or equipment during electrophoresis.
9. If appropriate positive controls are unavailable for the particular exon or gene being screened, it is preferable to run other samples of a similar length which possess a mutation, with the understanding that the conditions for resolving these conformers may not be optimal for the region being screened.
10. If the PCR product is of weak intensity on the agarose gel, it may be possible to load more product onto the SSCP gel rather than having to repeat the PCR.
11. The dsDNA may aid identification of the gel if more than one gel is run at once; load dsDNA in well 1 of gel 1, in well 2 of gel 2, and so on.
12. The same pipet tip may be used for loading each sample if it is thoroughly flushed out with deionized-distilled water between samples.

13. To maintain the temperature at which electrophoresis is performed, it is recommended one use a gel tank with a water-cooled chamber and a water-circulator with temperature control (e.g., LTD6G from Grant Instruments, Cambridge, Royston, Hertfordshire, UK).
14. Fresh 25% ammonium persulphate solution is recommended, but a stock may be made and store at 4°C if preferred.
15. Varying the temperature at which electrophoresis is performed, the percentage of polyacrylamide and content of glycerol in the gel can affect the presence or absence of SSCP bands.
16. 6% Polyacrylamide gels are often used as an alternative to 10% gels, and may be run for only 6 h at 25 W. The disadvantage is that they are more delicate and liable to tear during silver staining.
17. In our experience, the optimized conditions may vary between laboratories and even technicians. Other research groups have reported similar experiences (8).
18. Some researchers have found that gels other than polyacrylamide, e.g., MDE gel (Hydrolink, FMC BioProducts, Rockland, ME) are better at resolving SSCP variants (9).
19. After placing the gels in a glass baking dish, they may be stored for at least 2 d in fixative 1 and the container sealed with cling film.
20. It is not essential to add 5% acetic acid in fixative 1 or 2.
21. Adjust duration of immersion for thicker or thinner gels accordingly.
22. Trouble-shooting (modified from the Bio-Rad protocol handbook with permission) (*see* **Table 2**).
23. To pour off solutions from gel, one can gently place the thumb and index finger of both hands at the top corners of the gel to hold it in the dish and prevent its movement. Angle baking dish so that the solution pours out from the end nearest the bottom of the gel. The gel will loosely adhere to the bottom of the dish.
24. Gel may roll up when immersed in solutions, this will affect the staining process. To unroll, place the thumb and index finger of both hands at the top and bottom of the gel. Slowly move hands out from the center, to the sides of the gel.
25. To prevent roll up of gels, slowly pour solutions down the sides of the Pyrex baking dish.
26. Bubbles under gel will effect staining. Gently agitate gel to remove large bubbles from underneath gel.
27. Automated systems for electrophoresis and silver-staining have been developed and used for SSCP (10,11).
28. If bands are faint, expose the film for a longer time. Alternatively, there may be a problem with the activity of the radioisotope used. Addition of a higher concentration of labeled nucleotide in the PCR may compensate for the reduction in activity.

Acknowledgments

I would like to thank Andrew Hattersley for his drawing shown in **Fig. 1**, and guidance during my own development of PCR-SSCP.

References

1. Orita, M., Suzuki, Y., Sekiya, T., and Hayashi, K. (1989) Rapid and sensitive detection of point mutations and DNA polymorphisms using the polymerase chain reaction. *Genomics* **5,** 874–879.
2. O'Rahilly, S., Choi, W. H., Patel, P., Turner, R. C., Flier, J. S., and Moller, D. E. (1991) Detection of mutations in insulin-receptor gene in NIDDM patients by analysis of single-stranded conformation polymorphisms. *Diabetes* **40,** 777–782.

3. Choi, W. H., O'Rahilly, S., Buse, J. B., Rees, A., Morgan, R., Flier, J. S., and Moller, D. E. (1991) Molecular scanning of insulin-responsive glucose transporter (Glut4) gene in NIDDM subjects. *Diabetes* **40,** 1712–1718.
4. Vionnet, N., Stoffel, M., Takeda, J., Yasuda, K., Bell, G. I., Zouali, H., Lesage, S., Velho, G., Iris, F., Passa, Ph., Froguel, P., and Cohen, D. (1992) Nonsense mutation of the glucokinase gene causes early-onset non-insulin-dependent diabetes mellitus. *Nature* **356,** 721,722.
5. Thomas A. W., Morgan, R., Rees, A. E., and Alcolado, J. C. (1994) Rapid and reliable detection of mtDNA mutations in patients with maternally inherited diabetes. *Diabetic Medicine (Supplement 1)* **A18,** S7.
6. Sheffield, S. C., Beck, J. S., Kwitek, A. E., Sandstrom, D. W., and Stone, E. M. (1993) The sensitivity of single-strand conformation polymorphism analysis for the detection of single base substitutions. *Genomics* **16,** 325–332.
7. Merrill, C. R., Goldman, D., Sedman, S. A., and Ebert, M. H. (1981) Ultrasensitive stain for proteins in polyacrylamide gels shows regional variation in cerebrospinal fluid proteins. *Science* **211,** 1437,1438.
8. Elbein, S. C., Sorensen, L. K., and Schumacher, C. (1993) Substitution in exon 17 of the insulin receptor gene in a pedigree with familial NIDDM. *Diabetes* **42,** 429–434.
9. Chiu, K. C., Tanizawa, Y., and Permutt, M. A. (1993) Glucokinase gene variants in the common form of NIDDM. *Diabetes* **40,** 579–582.
10. Hager, J., Blanche, H., Sun, F., Vionnet, N., Vaxillaire, M., Poller, W., Cohen, D., Czernichow, P., Velho, G., Robert, J.-J., Cohen, N., and Froguel, P. (1994) Six mutations in the glucokinase gene identified in MODY by using a non-radioactive sensitive screening technique. *Diabetes* **43,** 730–733.
11. Thomas, A. W., Morgan, R., Majid, A., Rees, A., and Alcolado, J. C. (1995) Detection of mitochondrial DNA mutations in patients with diabetes mellitus. *Diabetologia* **38,** 376–379.
12. Stoffel, M., Patel, P., Lo, Y. M. D., Hattersley, A. T., Lucassen, A. M., Page, R., Bell, J. I., Bell, G. I., Turner, R. C., and Wainscoat, J. S. (1992) Characterisation of a missense glucokinase mutation in maturity-onset diabetes of the young (MODY) and mutation screening in late-onset diabetes. *Nature Genetics* **2,** 153–156.
13. Saker, P. J., Hattersley, A. T., Barrow, B., Hammersley, M., et al. (1996) High prevalence of a missense mutation of the glucokinase gene in gestational diabetic patients due to a founder effect in a local population. *Diabetologia* **39,** 1325–1328.

89

Detecting Point Mutations by Denaturing-Gradient Gel Electrophoresis

Stephen R. Dlouhy, Patricia Wheeler, James A. Trofatter, Peter J. Stambrook, and Jay A. Tischfield

1. Introduction

Denaturing-gradient gel electrophoresis (DGGE) detects DNA sequence differences. Thus, it can be used to screen for point mutations or other types of mutation prior to DNA sequence analysis. The technique, first described by Fischer and Lerman *(1)*, entails electrophoresis of DNA fragments at high temperature (approx. 60°C) in an acrylamide gel that contains a gradient of denaturant (formamide and urea). As a DNA fragment migrates in the gel, it encounters increasing concentrations of denaturant and at some point, it will become partially or totally single-stranded (melted, denatured). The position at which the DNA melts is determined by its nucleotide sequence and composition. Partial or complete denaturation causes a marked decrease in the electrophoretic mobility of the DNA and results in bands that are usually quite sharp. Two DNA fragments of the same size but of different sequence frequently will denature at different points within the gradient, and will therefore be separable by DGGE. In general, A-T-rich sequences denature at lower denaturant concentrations than G-C rich sequences. The method is applicable for fragments that are about 50–1000 bp long *(2,3)*.

DGGE is useful as a rapid screen for mutations in a gene, although it does not establish the precise position or type of mutation(s) involved. Rather, it provides a comparative analysis between samples. The technique is versatile because sequence changes can be detected even if they are not at restriction endonuclease recognition sites. However, because DNA fragments denature in distinct segments or "domains," only those sequence differences that reside within the early melting domains of a particular fragment can be detected by DGGE *(2–4)*. In addition, some base substitutions in native duplex DNA do not result in differences in denaturation even if they reside within an early melting domain *(5–7)*. Thus, although the separation of two fragments by DGGE can be indicative of a sequence difference between them, failure to observe separation does not exclude such differences. To overcome this difficulty, the efficiency of detection of mutations/sequence differences by DGGE can be increased by (1) the production of heteroduplexes, (2) digestion of the DNA with different restriction enzymes to alter melting domains, or (3) the attachment of terminal GC-rich regions known as GC

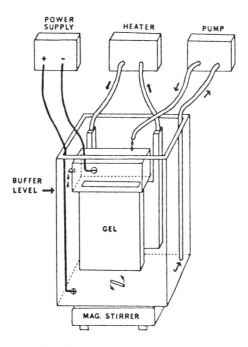

Fig. 1. Schematic representation of the major components of an apparatus used for denaturing gradient gel electrophoresis *(see text for details)*. Although only a single gel is shown, with some electrophoresis systems it is possible to run multiple gels simultaneously. For simplicity, some aspects of the apparatus are not shown, such as the cover, insulating Styrofoam (optional), and supports for the heat exchanger and gel assembly. The buffer level for the lower (outer) chamber is indicated.

clamps (*see* **Subheading 3.**). DGGE has been applied to the detection of mutations in cloned and genomic DNAs, including DNA amplified by the polymerase chain reaction (PCR) *(8–14)*.

2. Materials
2.1. Apparatus

1. Overview of the apparatus: It is important that the gel be uniformly heated. This is accomplished by submerging it in heated, well-mixed buffer. Furthermore, the electrophoresis buffer must be recirculated in order to maintain constant pH. Several commercially available electrophoresis systems are potentially suitable for DGGE, including units from Hoefer (SE 600, SE 620, or SE 660), Bio-Rad (Hercules, CA, Protean II), or Green Mountain Lab Supply (Waltham, MA). Alternatively, one can construct a suitable apparatus *(2)*. For brevity and simplicity, only the use of a commercial system will be described.

    A schematic of a DGGE apparatus, following the basic principles of a Hoefer SE 660 unit, is shown in **Fig. 1**. The major components of the system are:
    a. An electrophoresis chamber, including the gel and a heat exchanger.
    b. A pump for recirculating the electrophoresis buffer.
    c. An external, circulating, heated water bath.
    d. An electrophoresis power supply capable of generating up to 250 V and up to 150 mA. To reduce heat loss from the apparatus, the chamber can be insulated with Styrofoam sheets (1–2 cm thick) on the outside of the buffer chamber.

2. Electrophoresis chamber: Lower and upper buffer chambers, heat exchanger, and gel preparation materials (glass plates, spacers, combs, and pouring frame), electrodes, and so on, are part of the commercial electrophoresis systems, and are assembled according to the supplier's instructions.
3. Buffer recirculation pump: It is best to use a peristaltic or equivalent pump with the capability of varying the flow rate. Recirculation of buffer at 200 mL/min works well for the Hoefer SE 660.
4. Circulating, heated water bath (Heater): When using an external, circulating, heated water bath, heat loss occurs between the water bath and gel chamber. Use of a Neslab RT3-210D circulating water bath (flow rate ca. 3 L/min) in conjunction with an insulated Hoefer SE 660 results in about a 3–4°C drop in temperature between the heated bath and the interior of the electrophoresis chamber. Any comparable apparatus should suffice. Alternatively, for a homemade apparatus, an aquarium style heater can be used directly in the chamber.
5. Miscellaneous materials: Tubing, clamps, and connections for buffer recirculation and for external circulating, heated water bath; gradient maker (chamber vol ca. 10–30 mL); magnetic stir plate with a surface area large enough to stably support the electrophoresis chamber; UV light box; thermometer for inside the electrophoresis chamber; microsyringe (e.g., Hamilton); micropipet (e.g., Pipetman); blotting membrane (e.g., Nytran); and an electroblot chamber, such as a Bio-Rad Trans-Blot. Tubing for connecting the circulating heater should be heat- resistant (e.g., autoclavable) since it will expand with prolonged use.

2.2. Solutions

1. 40% Acrylamide stock solution (37.5:1 acrylamide:bisacrylamide): Dissolve 194.8 g electrophoresis grade acrylamide and 5.2 g bisacrylamide in distilled water and bring vol to 500 mL. Keep in a brown glass bottle at 4°C. CAUTION: Acrylamide is a neurotoxin, use gloves and surgical mask when handling powder, and gloves when handling solutions.
2. 50× stock TAE gel buffer: 2 M Tris, 1 M sodium acetate, 50 mM EDTA, pH 7.4. The pH is adjusted with glacial acetic acid.
3. 100% Denaturant stock solution: 6.5% Acrylamide, 40% formamide, and 7 M urea in 1× TAE buffer. Mix 81.25 mL of 40% acrylamide stock, 10 mL of 50× TAE buffer, 210 g urea, and 200 mL deionized formamide. Fluka formamide (47670) can be used directly without additional deionization. Adjust the final vol to 500 mL.
4. Denaturant-free stock solution: 6.5% Acrylamide in TAE buffer. Mix 81.25 mL of 40% acrylamide stock, 10 mL of 50× TAE buffer, and 468.75 mL distilled water.
5. Ammonium persulfate (APER, $(NH_4)_2S_2O_8$) (10%): Bring 1 g ammonium persulfate to 10 mL with distilled water. Make fresh. CAUTION: APER is a strong oxidant.
6. TEMED (N,N,N',N'-tetramethylethylenediamine). CAUTION: TEMED is flammable and causes irritation.
7. Sample loading solution (5X): 80% glycerol, 10 mM Tris-HCl, pH 7.4, 25 mM EDTA, 0.25% bromphenol blue.

3. Methods
3.1. Preliminary Considerations

There are variations in the design and preparation of denaturing gradient gels, and in the treatment of samples prior to electrophoresis. Each has its advantages and shortcomings relevant to specific applications.

1. Orientation of denaturing gradient: A DGGE gel can be oriented so that the denaturant gradient is either parallel with or perpendicular to the direction of electrophoresis.

The parallel method is more useful for side by side comparisons of multiple samples, and will be described in detail. Perpendicular gradient gels are discussed in **Subheading 4.1.**

2. Denaturation point and gradient concentration range: In order to make gradient gels that are optimal for detection of mutation in a particular fragment, it is necessary to determine the approximate denaturant concentration required to cause partial melting/denaturing of the fragment. This concentration can be established empirically with a parallel gradient gel(s) *(see below)* or with a perpendicular gradient gel (*see* **Subheading 4.1.**). Alternatively, one can use a computer algorithm to predict melting behavior *(3,4)*.

   To estimate the denaturation point with a parallel gradient gel, use moderate voltage (60–90 V); a broad concentration range of denaturant (e.g., 10–90%); electrophoresis of long duration (18–24 h); and multiple, temporally staggered loads of the same sample (e.g., load at $t = 0$, 5, and 10 h). Alternatively, the same sample can be loaded onto two different gradients (e.g., 10–60% and 50–90%) and electrophoresis conducted at higher voltage (130–160 V) for shorter periods (6–12 h). The appearance and position of bands at the end of the electrophoresis provides data concerning the denaturation point of the fragment. Fragments will be significantly retarded at the concentration of denaturant that is sufficient to cause melting of early melting domains. Also, when denatured, the DNA bands do not trail at the sides of the channel (i.e., "smile") and are usually quite sharply focused (e.g., **Fig. 2A**). The denaturing concentration for any fragment is estimated by determining the position in the gradient at which the fragment denatured. For example, if a fragment denatures approximately halfway down a 20–60% gel, it is denaturing at a concentration of about 40% denaturant.

   Once the approximate denaturing concentration has been established for a particular fragment(s), the extremes of the gradient should be narrowed. This will increase separation between fragments (cf **Figs. 2A** and **2B**). For best results, use a gradient in which the fragment(s) of interest will denature approximately halfway down the gel. For example, if broad-range parallel or perpendicular gel analysis indicates that a particular fragment denatures at approx 60% denaturant, for subsequent analysis of the fragment, change the gradient to 50–70%.

3. Heteroduplex formation: Some DNA fragments that differ from each other by a single base pair cannot be distinguished by DGGE, even if the sequence difference resides within an early melting domain (cf lanes 5 and 6 in **Fig. 2B**). Such single base-pair differences, however, can frequently be resolved following formation of heteroduplexes.

   When a mixture of two DNA fragments that differ at one or more base pairs is denatured and allowed to anneal, the products will include each of the parental homoduplexes and two heteroduplexes that contain a region(s) of mismatch. Such mismatches destabilize the helix and result in a reduced melting temperature compared to that of each of the homoduplexes. As a consequence, the heteroduplexes melt at a lower concentration of denaturant than the homoduplexes (**Fig. 2C**). Detection of heteroduplexes during DGGE establishes that one or more sequence differences probably exist between the parental homoduplexes. Thus, analysis of heteroduplexes can reveal sequence differences that are cryptic in the analysis of homoduplexes (e.g., *see* **Fig. 2**, lanes 5, 6, and 9). As a model, we have analyzed all permutations of base substitutions at the terminal 2 nucleotides of the splice acceptor site in the third intron of the mouse *adenine phosphoribosyltransferase (aprt)* gene.

   All possible transitions and transversions in DNA at that site can be detected by DGGE analysis of heteroduplexes *(7)*.

   Heteroduplexes are produced by mixing the DNA fragments of interest in high salt buffer (300 mM NaCl, 1 mM EDTA, 30 mM Tris-HCl, pH 8), heating to 98°C for 10 min, and allowing annealing to occur at 65°C (for 3–4 h). After annealing, 2.5 vol of ethanol

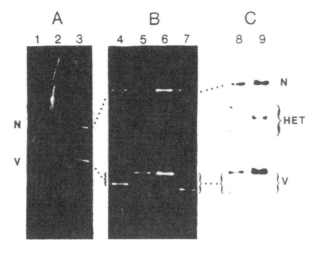

Fig. 2. Parallel DGGE analysis of cloned mouse *adenine phosphoribosyltransferase (aprt)* fragments (parts modified from *[7]*). Variant fragments (V) (560 bp) and 428 bp nonvariant fragments (N) were obtained by *Bam*H1 plus *Xho*1 double digestion of cloned *aprt* DNA as previously described *(7)*. A: Portion of an ethidium bromide stained, 40–90% gradient gel. Lanes 1, 2, and 3 illustrate V fragments that differ from one another at the same site. Lane 1: wild-type (contains a G at the variable site); lane 2: mutation to A; lane 3: mutation to T. Note that the V fragments in lanes 2 and 3 are indistinguishable. B: Portion of an ethidium bromide stained 50–70% gradient gel demonstrating the increased band separation that can be achieved by narrowing the gradient range. All possible base substitutions at one site are shown for the V fragment. Lanes 4, 5, and 6 contain the same samples as 1, 2, and 3, respectively. Lane 7 has a G to C point mutation at the site. Note that the V fragments in lanes 5 and 6 are still indistinguishable, even though the gradient range has been changed. C: Probed electroblot demonstrating the heteroduplexes (HET) that were produced following denaturation and annealing of a mixture of DNAs that differ in sequence. Lane 8: Mixture of DNA samples from lanes 6 and 7, denatured, and allowed to anneal prior to DGGE. Lane 9: Mixture of DNA samples from lanes 5 and 6, denatured, and allowed to anneal prior to DGGE. In any mixture of two different DNA samples, two different heteroduplexes will be produced. The two heteroduplexes produced in lane 8 are evident. Two heteroduplexes also are produced in lane 9, but they both denature at the same position in the gradient, producing only a single visible band *(7)*.

are added directly to the mixture (no additional salt is required), the DNA is precipitated at –20°C or –70°C, and resuspended in appropriate buffer for further restriction enzyme digestion or DGGE.
4. GC clamping: This approach can be used to increase the efficiency of mutation detection by DGGE *(5,6,13–15)*. A GC-rich region termed a "GC clamp" (100% intermixed G + C) is added to one end of the DNA fragment. Because the clamp is very thermostable, it serves as the last melting domain. This enables mutation detection in the segment of a DNA fragment that was previously a late melting domain. For example, Myers et al. *(6,14)* have shown that attachment of a GC clamp to a fragment of the beta globin promoter enables DGGE detection of mutations in two melting domains of the fragment whereas, without the clamp, DGGE can only detect mutation in the first melting domain. For analysis of cloned DNA, GC clamps can be directly ligated to the fragment(s) *(14)*. For analysis of PCR-amplified DNA, a convenient way to produce GC-clamped fragments is by inclu-

sion of a 40–45-bp G + C rich sequence at the 5' end of one or both of the two primers *(16)*. GC clamps on both primers can be used if a PCR fragment is to be cut by restriction enzyme digestion prior to DGGE.

3.2. Sample Preparation

Fragments should be between 50 and 1000 bp, so that only one or a few melting domains are present. This is most easily accomplished by restriction endonuclease digestion of the DNA. If the sequence or a restriction map of the DNA is available, one or a combination of enzymes that will yield fragments within this size range is chosen. If no such information is available, use restriction enzymes that cut often, such as *Hae* III, *Alu* I, *Msp* I, *Taq* I, and so on, in order to generate small fragments. Use of different enzymes, in separate experiments, is advantageous, since sequence context and melting domains will be altered. Thus, one may increase the efficiency of detection of sequence changes *(4,7)*.

The optimum amount of DNA loaded on the gel will vary, depending on the experiment. For cloned DNA, 200 ng of a single fragment is readily visible in ethidium bromide stained denaturing gels. Much less DNA (cloned) can be used if the DNA is transferred (Southern blotted) to a solid matrix, such as Nytran, and hybridized with a radiolabeled probe, or if the fragments themselves are radioactively labeled.

In some studies, denatured genomic DNA fragments have been analyzed following hydridization to cloned, radiolabeled single-strand DNA with subsequent DGGE and autoradiography *(8,9)*. Alternatively, genomic DNA fragments have been blotted onto a membrane after DGGE fractionation, and analyzed by hybridization with radiolabeled probes (Mark Gray, personal communication). A more convenient method, however, entails the analysis of genomic fragments that have been amplified by PCR.

3.3. Amplification of Samples by PCR

Either cloned or genomic DNA amplified by PCR is suitable for DGGE (**Fig. 3**). This approach is particularly advantageous for analysis of mutations in genomic DNA because PCR amplified fragments can be visualized in ethidium bromide-stained denaturing gels, thus obviating the need for blotting and hybridization with radioactive probes. Standard methods for performing PCR should produce usable fragments. If PCR amplification is used, one should select conditions that result in the production of a single fragment (i.e., use 20–30-mer primers and high annealing and extension temperatures). This will help avoid confounding multiple bands in the DGGE gel. Even when a single fragment has been amplified, minor bands or smears sometimes are seen with DGGE analysis of PCR-amplified DNA (*see* **Fig. 3A**, lane 3). This is owing, in part, to errors introduced during amplification. In fact, DGGE has been used as a method to assess the error rate of different polymerases that may be used for PCR *(15)*.

To analyze a large gene or DNA region by PCR amplification and DGGE, a set of appropriate small fragments must be produced. This can be accomplished with multiple sets of primers to divide the region into a series of smaller fragments for amplification, or by the use of restriction enzymes to cleave large PCR-amplified fragments. Thus, although it is not always necessary to use restriction endonucleases to digest amplified DNA prior to DGGE, such digestion is useful (i) if long sequences are amplified, (ii) if one is interested in the behavior of a particular subfragment,

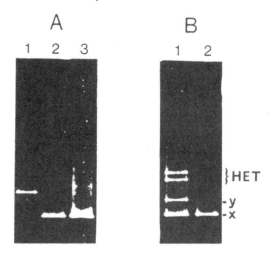

Fig. 3. Parallel DGGE analysis of the mouse *aprt* V fragment obtained from PCR-amplified DNA. In these analyses, the V fragment (560 bp) is homologous to that shown in **Fig. 2** because it was obtained by restriction endonuclease digestion (*Bam*HI plus *Xho*I) of a larger fragment (773 bp) that was amplified. A: DGGE analysis of DNA that has been PCR-amplified from dilute plasmid DNA preparations (ca. 5 ng). Lane 1: Wild-type V fragment; lanes 2 and 3: V fragment with point mutation (A to G) at the nucleotide adjacent to that contained in the DNA shown in Fig. 2. B: DGGE analysis of DNA that has been PCR amplified from genomic DNA. Genomic DNA was isolated from a human cell line (HTD114 derivative) that contains two copies of a mouse *aprt* transgene. Lane 1: DNA from a clone derived after ethylmethanesulfonate (EMS) treatment of the cell line. Lane 2: DNA from the cell line prior to EMS treatment. Samples for both lanes were denatured and allowed to anneal prior to DGGE. Lane 1 demonstrates DGGE detection of an EMS-induced mutation (actually a reverse mutation) in the V fragment of one of the two mutant *aprt* transgenes that are present in this cell line. HET: heteroduplexes; X homoduplex of mutant V fragment; Y: homoduplex of wild-type (reverted) V fragment.

or (iii) if it is necessary to alter general sequence context and melting domains (*see* **Subheading 3.2.**).

Figure 3B illustrates the use of DGGE for the detection of a single-base-change in a fragment produced by amplification of genomic DNA. In this example, the target of amplification is a mouse *aprt* transgene, two identical copies of which have been stably transfected into a human cell line. Thus, only a single band is seen on a DGGE gel (lane 2). However, a more complex pattern (lane 1) with multiple bands is shown for a cell clone derived after treatment with the mutagen ethylmethanesulfonate (EMS). In this clone, there are now two different copies of the mouse *aprt* gene (fragments x and y), and heteroduplexes (HET) are seen. Although transgenes are analyzed in this example, lane 2 illustrates the type of pattern that is also seen for individuals who are homozygous for a particular region or "allele" of DNA (no evidence of mutation/sequence differences). Complex, multiple-band patterns (e.g., lane 1) are seen for individuals who are heterozygous for a particular DNA region. As discussed previously, the formation of heteroduplexes is frequently advantageous for the detection of mutation. However, heteroduplexes are sometimes, but not always, spontaneously formed during the amplification procedure. Therefore, to ensure heteroduplex formation, we denature and anneal in high-salt buffer.

3.4. Making a Parallel Gradient Gel

Preparation of a 20–80% gel is described as an example.

1. Thoroughly clean the glass plates, first with soapy water and then with ethanol. Assemble the apparatus according to the manufacturer's protocol. With the Hoefer system, only side spacers (0.75 or 1.5 mm thick) are used. Sealing with grease to avoid leaks is usually not needed except at the bottom corners where there is contact between the plates and the gel-stand gasket. Gel vol is determined empirically. If leakage is a problem, the outside edges of the plates can be sealed with agarose prior to pouring the gel.
2. Bottom plug gel: A narrow bottom plug gel is poured (**Fig. 4**). The plug can be acrylamide or agarose. To form an acrylamide plug (1.5–2.5 cm high), prepare an acrylamide solution (e.g., 10% v/v acrylamide stock, 90% v/v denaturant stock) and add appropriate volumes of catalysts (*see* **step 9**, this section). Apply from the top of the clamped plates along one edge using a syringe and allowed to flow to the bottom and polymerize. No overlay solution is needed. Alternatively, melted agarose (1.5% in 1× TAE) can be allowed to rise between the plates from below by capillary action while the plates sit in a trough.
3. Denaturing gradient gel: The setup for making gradients is shown in **Fig. 4**. Each chamber of the gradient maker should contain approx one-half of the vol needed to fill the gel plate assembly. The upstream chamber contains the solution with the low denaturant concentration and the downstream chamber contains the solution with the high denaturant concentration. The denaturant concentration in each chamber is chosen to produce the desired gradient (*see* discussion in **Subheading 3.1., step 2**).
4. Check polymerization of the plug gel.
5. Prepare an 80% denaturant solution: Mix 9.6 mL of 100% denaturing stock solution and 2.4 mL of nondenaturing stock solution in a small glass beaker on ice.
6. Prepare a 20% denaturant solution: Mix 2.4 mL of 100% denaturing stock solution and 9.6 mL of nondenaturing stock solution in a small glass beaker on ice.
7. Attach tubing from the gradient maker to the gel plate assembly and close both valves/clamps.
8. Set up a magnetic stirrer or vibrating mixer for the downstream side of the gradient maker.
9. Add catalyst (60 µL of 10% APER and 5 µL of TEMED) to the high concentration (80%) denaturant solution in the beaker on ice, mix well by swirling, and pour into the downstream side of the gradient maker.
10. To remove air bubbles, open the valve that separates the two chambers of the gradient maker and allow a small amount of the high concentration denaturant solution to flow into the upstream chamber; then, close the valve. If excess high concentration denaturant gets into the upstream chamber, return it to the downstream chamber with a pipet.
11. Be certain that the mixing device is operating in the downstream chamber.
12. Add catalyst (as in **step 9**) to the low concentration denaturant solution, mix well, and pour into the upstream chamber.
13. Open the valve between the chambers.
14. Open the valve or clamp between the downstream chamber and the gel plate assembly. The solution can be delivered to the gel by gravity or peristaltic pump. Fill slowly to the top of the plates. Filling times between 10 and 30 min work well.
15. Insert a slot former (10–20 slot comb) immediately after pouring the gel. Polymerization takes between 15 and 45 min. An alternative to placing the comb directly into the gradient

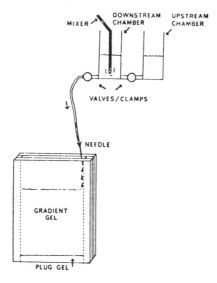

Fig. 4. Schematic of arrangement for pouring gradient gels *(see text for details)* (not drawn to scale).

gel is to pour the gradient (as in **step 14**), but leave about a 2–2.5-cm space above and overlay the gel solution with a small amount of butanol. After the gradient has polymerized, an upper gel (0% denaturant) can be poured and the comb inserted.

16. The gel can be run immediately after polymerization or it can be left covered with plastic wrap overnight at room temperature.

3.5. Electrophoresis

1. Temperature equilibration of chamber: At least 1 h prior to use, electrophoresis buffer is put into the lower buffer chamber, a stir bar placed in the chamber, the magnetic stirrer turned on, and a thermometer placed in the chamber. The circulating heating bath is turned on and set at a temperature sufficient to heat the electrophoresis chamber to 60°C (determined empirically). For example, a circulating bath temperature of 63.5°C is required for a Hoefer SE660 (insulated) Neslab RT3-210D combination.
2. Loading the gel and starting electrophoresis: After the comb has been removed, the wells should be rinsed (once or twice) with electrophoresis buffer, and the samples mixed with loading buffer (1× final concentration). The gel can be loaded while in the gel-pouring stand by using an adjustable micropipet or a microsyringe. If a microsyringe is used, it should be rinsed three times between samples. Once loaded, the gel and upper buffer chamber assembly are placed into the lower chamber and buffer is added to the upper chamber. If a microsyringe is used to load samples, the gel can be loaded after it has been put into the electrophoresis chamber. This is useful for temporally staggered loading of samples once electrophoresis has begun. Once the loaded gel is in place, a cover is placed over the apparatus to reduce evaporation and heat loss and the buffer recirculation pump started. If the gel is loaded while outside the chamber, it will be cool and the temperature of the chamber will drop slightly when the gel is placed inside. Chamber temperature usually recovers quickly to 60°C. For many analyses, electrophoreses can be started before the chamber temperature equilibrates to 60°C.

Checklist:
- a. _____Gel loaded and in place.
- b. _____Buffer recirculating pump on.
- c. _____Heat source (circulating water bath) on.
- d. _____Magnetic stirrer in chamber working.
- e. _____Thermometer in chamber

3. Gel electrophoresis: Parameters for electrophoresis depend on the fragment(s) being analyzed and the gradient used (see **Subheading 3.1.2.**). Voltages can range between 60 and 160, and run times between 6 and 24 h (see **Note 3**). These should be determined empirically (see **Note 5**). Because fragment mobility is significantly retarded following domain denaturation, moderately extended run times generally are acceptable because the denatured fragments remain in relatively fixed positions in the gel and separations between fragments are maintained.

4. Postelectrophoresis: The gel is stained with 0.5 µg/mL ethidium bromide *(17)* and the DNA bands visualized on a UV transilluminator (see **Notes 2** and **4**). DGGE gels can also be electroblotted onto nylon membranes, such as Nytran, and hybridized with radiolabeled probe.

4. Notes

1. Perpendicular gel electrophoresis: In this configuration, the denaturant concentration gradient is perpendicular to the direction of electrophoresis *(2,6,14)*. This enables one to more easily determine the position in the gradient at which denaturation occurs. A single sample, containing one or a mixture of DNA fragments, is applied across the top of the gel. After electrophoresis, the DNA forms an S-shaped curve in which the inflection of the curve indicates the point at which denaturation is occurring (**Fig. 5**). Two fragments that are identical in size but differ in sequence in an early melting domain will comigrate in low concentration denaturant because the DNA migrates only according to molecular size. However, as the fragments encounter higher denaturant concentrations, they separate (note the split in the curve in **Fig. 5**). Because fragments migrate according to molecular size on the low denaturant side of the gel, a mixture of different sized fragments can be analyzed simultaneously in a single perpendicular gel and the denaturation points determined for each fragment.

   In preparing a perpendicular gradient, the glass plates should be assembled so that the gel can be poured with the plates on their side. With the Hoefer SE 660, this can be accomplished by using the same plate assembly apparatus as for a parallel gel, except that an additional spacer, slightly shorter than the width of the gel, is inserted between the plates. Assemble the glass plates and spacers as for a parallel gel, but place the additional spacer (lightly greased) between the top of the plates and abutting the right spacer such that the upper right corner will be leakproof. This will leave a small gap in the upper left corner of the frame. Clamp the top edge with a binder clip. Turn the plates clockwise 90°. Pour the gradient through the gap as described previously. No comb is inserted. Once the gel is polymerized, the gel is rotated back to the vertical position and will have a gap along one edge. Prior to loading the sample, the gap should be filled with an agarose or acrylamide gel, or otherwise sealed. The top spacer is removed, the DNA sample (100–200 µL) is loaded along the entire top edge (see **Fig. 5**), the apparatus is assembled, and electrophoresis is initiated.

2. Samples in lanes at the sides of DGGE gels usually have curved bands after denaturation.
3. The bromophenol blue marker dye usually runs off of the gel before the run is finished.

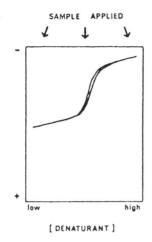

Fig. 5. Diagrammatic representation of the results of a perpendicular gradient analysis of a mixture of two DNA fragments. The splitting of the band at the inflection of the curve indicates a sequence difference between the two samples. The position of the inflection indicates the denaturant concentration where melting occurs.

4. Some DNA fragments do not appear as discrete bands, but rather as streaks or smears because of complex denaturation behavior.
5. Inclusion of ethidium bromide in the gel is not useful for evaluating the status of a run.
6. Methylation can alter DNA melting behavior (6). This is important if one tries to compare PCR-amplified DNA with cloned DNA that has been derived from a source (e.g., bacterial strain) capable of methylation (our unpublished results).

Acknowledgments

This work was supported by National Institutes of Health grants DK38185 and CA36897 to J. A. Tischfield and P. J. Stambrook.

References

1. Fischer, S. G. and Lerman, L. S. (1979) Length-independent separation of DNA restriction fragments in two-dimensional gel electrophoresis. *Cell* **16,** 191–200.
2. Myers, R. M., Maniatis, T., and Lerman, L. S. (1987) Detection and localization of single base changes by denaturing gradient gel electrophoresis. *Methods Enzymol.* **155,** 501–527.
3. Lerman, L. S. and Silverstein, K. (1987) Computational simulation of DNA melting and its application to denaturing gradient gel electrophoresis. *Methods Enzymol.* **155,** 482–501.
4. Lerman, L. S., Silverstein, K., and Grinfeld, E. (1986) Searching for gene defects by denaturing gradient gel electrophoresis. *Cold Spring Harbor Symp. Quant. Biol.* **51,** 285–297.
5. Fischer, S. G. and Lerman, L. S. (1983) DNA fragments differing by single-base pair substitutions are separated in denaturing gradient gels: Correspondence with melting theory. *Proc. Natl. Acad. Sci. USA* **80,** 1579–1583.
6. Myers, R. M., Fischer, S. G., Lerman, L. S., and Maniatis, T. (1985) Nearly all single base substitutions in DNA fragments joined to a GC-clamp can be detected by denaturing gradient gel electrophoresis. *Nucleic Acids Res.* **13,** 3131–3146.
7. Dlouhy, S. R., Schaff, D. A., Trofatter, J. A., Liu, H. S., Stambrook, P. J., and Tischfield, J. A. (1989) Denaturing gradient gel analysis of single-base substitutions at a mouse adenine phosphoribosyltransferase splice acceptor site. *Mol. Carcinogenesis* **2,** 217–225.

8. Myers, R. M., Lumelsky, N., Lerman, L. S., and Maniatis, T. (1985) Detection of single base substitutions in total genomic DNA. *Nature* **313,** 495–498.
9. Noll, W. W. and Collins, M. (1987) Detection of human DNA polymorphisms with a simplified denaturing gradient gel electrophoresis technique. *Proc. Natl. Acad. Sci. USA* **84,** 3339–3343.
10. Cariello, N. F., Scott, J. K., Kat, A. G., Thilly, W. G., and Keohavong, P. (1988) Resolution of a missense mutant in human genomic DNA by denaturing gradient gel electrophoresis and direct sequencing using in vitro DNA amplification: $HPRT_{Munich}$. *Am. J. Hum. Genet.* **42,** 726–734.
11. Amselem, S., Duquesnoy, B. S., Attree, O., Novelli, G., Bousnina, S., Postel-Vinay, M-C., and Goossens, M. (1989) Laron dwarfism and mutations of the growth hormone-receptor gene. *N. Engl. J. Med.* **321,** 989–995.
12. Theophilus, B. D. M., Latha, T., Grabowsky, G. A., and Smith, F. I. (1989) Comparison of RNase A, a chemical cleavage and GC-clamped denaturing gradient gel electrophoresis for the detection of mutations in exon 9 of the human acid β-glucosidase gene. *Nucleic Acids Res.* **17,** 7707–7722.
13. Sheffield, V. C., Cox, D. R., Lerman, L. S., and Myers, R. M. (1989) Attachment of a 40-base-pair G + C rich sequence (GC clamp) to genomic DNA fragments by the polymerase chain reaction results in improved detection of single-base changes. *Proc. Natl. Acad. Sci. USA* **86,** 232–236.
14. Myers, R. M., Fischer, S. G., Maniatis, T., and Lerman, L. S. (1985) Modification of the melting properties of duplex DNA by attachment of a GC-rich DNA sequence as determined by denaturing gradient gel electrophoresis. *Nucleic Acids Res.* **13,** 3111–3129.
15. Keohavong, P. and Thilly, W. (1989) Fidelity of DNA polymerases in DNA amplification. *Proc. Natl. Acad. Sci. USA* **86,** 9253–9275.
16. Myers, R. M., Sheffield, V. C., and Cox, D. R. (1989) Mutation detection by PCR, GC-clamps, and denaturing gradient gel electrophoresis, in *PCR Technology* (Erlich, H. A., ed.) Stockton, NY, pp. 71–88.
17. Maniatis, T., Fritsch, E. F., and Sambrook, J. (eds.) (1982) *Molecular Cloning: A Laboratory Manual,* Cold Spring Harbor Laboratory Press, Cold Spring Harbor, NY, pp. 173–177.

90

Analysis of Nucleotide Sequence Variations by Solid-Phase Minisequencing

Anu Suomalainen and Ann-Christine Syvänen

1. Introduction

The Sanger dideoxy-nucleotide sequencing method has been simplified by a number of methodological improvements, such as the use of the polymerase chain reaction (PCR) technique for generating DNA templates in sufficient quantities, followed by affinity-capture techniques for convenient and efficient purification of the PCR fragments for sequencing. Furthermore, the cyclic Sanger sequencing reactions are easy to automate with laboratory robots, and instruments have been developed for automatic online analysis of fluorescent products of the sequencing reactions. Despite these technical improvements, the requirement for gel electrophoretic separation remains an obstacle when sequence analysis of large numbers of samples is needed, as in DNA diagnosis or in the analysis of sequence variation for genetic, evolutionary or epidemiological studies.

We have developed a method for analysis of DNA fragments differing from each other in one or a few nucleotide positions *(1)*, denoted solid-phase minisequencing, in which gel electrophoretic separation is avoided. Analogous to the methods for solid-phase sequencing of PCR products, the solid-phase minisequencing method is based on PCR amplification using one biotinylated and one unbiotinylated primer, followed by affinity-capture of the biotinylated PCR product on an avidin- or streptavidin-coated solid support. The nucleotide at the variable site is detected in the immobilized DNA fragment by a primer extension reaction: a detection step primer that anneals immediately adjacent to the nucleotide to be analyzed is extended by a DNA polymerase with a single labeled nucleotide complementary to the nucleotide at the variable site (**Fig. 1**). The amount of the incorporated label is measured, and it serves as a specific indicator of the nucleotide present at the variable site.

We have used the solid-phase minisequencing method for detecting numerous mutations causing human genetic disorders, for analyzing allelic variation in genetic linkage studies and for identification of individuals *(2–4)*. The protocol described below is generally applicable for detecting any variable nucleotide. The method is well suited for analyzing large numbers of samples because it comprises simple manipulations in a microtiter plate or test tube format, and the result of the assay is obtained as an objective

From: *The Nucleic Acid Protocols Handbook*
Edited by: R. Rapley © Humana Press Inc., Totowa, NJ

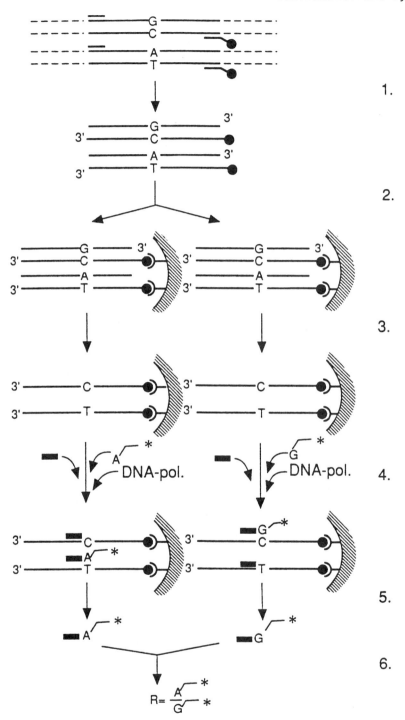

Fig. 1. Steps of the solid-phase minisequencing method. (1) PCR with one biotinylated (black ball) and one unbiotinylated primer. (2) Affinity-capture of the biotinylated PCR product in streptavidin-coated microtiter wells. (3) Washing and denaturation. (4) The minisequencing primer extension reaction. (5) Measurement of the incorporated label. (6) Calculation of the result.

numeric value, which is easy to interpret. Furthermore, the solid-phase minisequencing method allows quantitative detection of a sequence variant present as a minority of less than 1% in a sample *(2,3,5,6)*. We have utilized the sensitive quantitative analysis for detecting point mutations in malignant cells present as a minority in a cell population *(5)* and for analyzing heteroplasmic mutations of mitochondrial DNA *(3,6)*. The high sensitivity is an advantage of the minisequencing method, compared to dideoxy-nucleotide sequencing, in which a sequence variant must be present as 10–20% of a mixed sample to be detectable. A limitation of the solid-phase minisequencing method is that it is restricted to analyzing variable nucleotides only at positions predefined by the detection step primers used. The method is based on the use of equipment and reagents that are available from common suppliers of molecular biological products, facilitating easy setup. In the future, high-throughput analysis of nucleotide sequence variation will be performed by rapid, automatic methods based on homogeneous detection principles, or alternatively using methods in microarray or chip formats. The minisequencing reaction principle is applicable for both types of assay formats *(7,8)*.

2. Materials
2.1. Equipment

1. Programmable heat block, and facilities to avoid contamination in PCR.
2. Microtiter plates with streptavidin-coated wells (e.g., Combiplate 8, Labsystems, Helsinki, Finland) (*see* **Note 1**).
3. Multichannel pipet and microtiter plate washer (optional).
4. Shaker at 37°C.
5. Water bath or incubator at 50°C.
6. Liquid scintillation counter.

2.2. Reagents

All the reagents should be of standard molecular biology grade. Use sterile distilled or deionized water.

1. Thermostable DNA polymerase. We use *Thermus aquaticus* (5 U/µL, Promega or Perkin-Elmer-ABI, Norwalk, CT) or Thermus brockianus (Dynazyme™ II, 2 U/µL, Finnzymes, Espoo, Finland) DNA polymerase. Store at –20°C. (*See* **Note 2**.)
2. 10X Concentrated DNA polymerase buffer: 500 mM Tris-HCl, pH 8.8, 150 mM $(NH_4)_2SO_4$, 15 mM $MgCl_2$, 0.1% (v/v) Triton X-100, 0.01% (w/v) gelatin or 10X concentrated buffer supplied with the DNA polymerase enzyme. Store at –20°C.
3. dNTP mixture: 2 mM dATP, 2 mM dCTP, 2 mM dGTP, and 2 mM dTTP. Store at –20°C.
4. PBS/Tween: 20 mM sodium phosphate buffer, pH 7.5, and 0.1% (v/v) Tween-20. Store at 4°C. 50 mL is enough for several full plate analyses.
5. TENT (washing solution): 40 mM Tris-HCl, pH 8.8, 1 mM EDTA, 50 mM NaCl, and 0.1% (v/v) Tween-20. Store at 4°C. Prepare 1–2 L at a time, which is enough for several full-plate analyses.
6. 50 mM NaOH (make fresh every 4 wk), store in plastic vial at room temperature (approx 20°C).
7. [^{3}H]-labeled deoxynucleotides (dNTPs): dATP to detect a T at the variant site, dCTP to detect a G etc. (Amersham Pharmacia Biotech; [^{3}H]dATP, TRK 633; dCTP, TRK 625; dGTP, TRK 627; dTTP, TRK 576), store at –20°C (*see* **Note 3**).
8. Scintillation fluid (e.g., Hi-Safe II, Wallac, Turku, Finland), store at room temperature (approx 20°C).

2.3. Primer Design

1. PCR primers: One PCR primer of each pair is biotinylated at its 5' end during the synthesis using a biotin-phosphoramidite reagent (e.g., Amersham Pharmacia Biotech or Perkin Elmer-ABI) (*see* **Note 4**).
2. The detection step primer for the minisequencing analysis is a 20-mer oligonucleotide complementary to the biotinylated strand of the PCR product, designed to hybridize with the 3' end immediately adjacent to the variant nucleotide to be detected (*see* **Fig. 1**). The minisequencing primer should be at least five nucleotides nested in relation to the unbiotinylated PCR primer.

3. Method

3.1. PCR for Solid-Phase Minisequencing Analysis

The PCR is performed according to routine protocols, except that the amount of the biotin-labeled primer used is reduced not to exceed the biotin-binding capacity of the microtiter well (*see* **Note 1**). For a 50 µL PCR reaction we use 10 pmol of biotin-labeled primer and 50 pmol of the unbiotinylated primer. The PCR should be optimized (i.e., the annealing temperature and template amount) to be efficient and specific. To be able to use [^{3}H] dNTPs, which are low-energy β-emitters, for the minisequencing analysis, one-tenth of the PCR product should produce a single visible band after agarose gel electrophoresis, stained with ethidium bromide. There is no need for purification of the PCR product prior to the minisequencing analysis.

3.2. Solid-Phase Minisequencing Analysis

1. Affinity capture: Transfer 10 µL aliquots of the PCR product and 40 µL of the PBS/Tween solution to two streptavidin-coated microtiter wells (*see* **Note 5**). Include a control reaction, i.e., a well with no PCR product. Seal the wells with a sticker and incubate the plate at 37°C for 1.5 h with gentle shaking.
2. Discard the liquid from the wells and tap the wells dry against a tissue paper.
3. Wash the wells three times at room temperature as follows: pipet 200 µL of TENT solution to each well, discard the washing solution and empty the wells thoroughly between the washings (*see* **Note 6**).
4. Denature the captured PCR product by adding 100 µL of 50 m*M* NaOH to each well and incubate at room temperature for 3 min. Discard the NaOH and wash the wells as in **step 3**.
5. Prepare for each DNA fragment to be analyzed two 50 µL mixtures of nucleotide-specific minisequencing solution, one for detection of the normal and one for the mutant nucleotide (*see* **Note 7**). Mix 5 µL of 10X *Taq* DNA polymerase buffer, 10 pmol of detection step primer, 0.2 µCi (usually equals to 0.2 µL) of one [^{3}H] dNTP, 0.1 U of *Taq* DNA polymerase, and dH$_2$O to a total volume of 50 µL. It is obviously convenient to prepare master mixes for the desired number of analyses with each nucleotide.
6. Pipet 50 µL of one nucleotide-specific mixture per well, incubate the plate at 50°C for 10 min in a water bath or 20 min in an oven (*see* **Note 8**).
7. Discard the contents of the wells and wash them as in **step 3**.
8. Release the detection step primer from the template by adding 60 µL 50 m*M* NaOH and incubating for 3 min at room temperature.
9. Transfer the eluted primer to the scintillation vials, add scintillation reagent, and measure the radioactivity, i.e., the amount of incorporated label, in a liquid scintillation counter (*see* **Note 9**).

10. The result is obtained as counts per minute (cpm) values. The cpm value of each reaction expresses the amount of the incorporated [^{3}H] dNTP. Calculate the ratio (R) between the mutant and normal nucleotide cpms. In a sample of a subject homozygous for the mutant nucleotide the R will be > 10, in a homozygote for the normal nucleotide $R < 0.1$, and in the case of a heterozygote R varies between 0.5 and 2.0, depending on the specific activities of the [^{3}H] dNTPs (see **Note 10**).

4. Notes

1. The binding capacity of the streptavidin-coated microtiter well (Labsystems) is 2–5 pmol of biotinylated oligonucleotide. Other solid supports with higher biotin binding capacity, such as avidin-coated polystyrene beads (Fluoricon Assay Particles, IDEXX Corp., Portland ME; biotin-binding capacity over 2 nmol of oligonucleotide/mg beads) *(1)*, or streptavidin-coated magnetic polystyrene beads (Dynabeads M-280, Dynal, Norway; biotin-binding capacity 300 pmol/mg) *(5)* or avidin-coated manifolds *(9)*, can also be used. The biotin-binding capacity of a microtiter well allows reliable detection of up to 2% of a sequence-variant present in the sample *(6)*, whereas a detection sensitivity of less than 0.1% can be obtained with the bead-based format *(5)*.
2. It is advantageous to use a thermostable DNA polymerase for the single-nucleotide primer extension reaction, because a high temperature, favorable for the simultaneous primer annealing reaction, can be applied. Thermostable DNA polymerases engineered for efficient incorporation of ddNTPs, such as the ThermoSequenase™ (Amersham Pharmacia Biotech) enzyme, should be preferably used for incorporating labeled ddNTPs.
3. Although the specific activities of the [^{3}H]dNTPs are low and their half lives are long (13 yr), the necessary precautions for working with [^{3}H] should be taken. Also dNTPs or dideoxy-nucleotides (ddNTPs) labeled with other isotopes ([^{35}S] or [^{32}P]) *(1)*, with haptens *(1,10)* or with fluorescent groups can be used *(9)*.
4. If the biotin-labeled oligonucleotides are used without purification from unbiotinylated ones by high performance liquid chromatography, polyacrylamide gel electrophoresis, or by ion exchange columns manufactured for this purpose (Perkin-Elmer-ABI), the level of biotinylation should be confirmed after the PCR. This can be performed by affinity capture of the biotinylated PCR product on an avidin matrix with high biotin-binding capacity, followed by detection of possible unbound products by agarose gel electrophoresis.
5. Each nucleotide to be detected at the variant site is analyzed in a separate well. Thus at least two wells are needed per PCR product.
6. It is important that the PCR reagents are removed efficiently after the capturing reaction, to avoid unspecific nucleotide incorporation in the minisequencing reaction. The washings can be performed in an automated microtiter plate washer, or by manually pipeting the washing solution to the wells, discarding the liquid and tapping the plate against a tissue paper.
7. The minisequencing reaction mixture can be stored at room temperature for 1–2 h. It is convenient to prepare it during the affinity capture (**step 1**).
8. The conditions for hybridizing the detection step primer are not stringent, and the temperature of 50°C can be applied to analysis of most PCR products irrespectively of the sequence of the detection step primer. If the primer, however, is considerably shorter than a 20-mer or its GC content is low (melting temperature close to 50°C) lower temperatures for the primer annealing may be required.
9. Streptavidin-coated microtiter plates made of scintillating polystyrene are available (Wallac). In this case the final washing, denaturation and transfer of the eluted detection primer can be omitted, but a scintillation counter for microtiter plates is needed *(11)*.

10. The ratio between the cpm-values for the two nucleotides reflects the ratio between the two sequences in the original sample. Therefore the solid-phase minisequencing method can be used for quantitative PCR analyses *(2–6)*. The *R*-value is affected by the specific activities of the [^{3}H] dNTPs used, and if either the mutant or the normal sequence allows the detection step primer to be extended by more than one [^{3}H] dNTP, this will obviously also affect the *R*-value. Both of these factors can easily be corrected for, when calculating the ratio between the two sequences. Alternatively, a standard curve is constructed by mixing the two sequences in known ratios, and plotting the obtained *R*-values as a function of the ratios to obtain a linear standard curve *(3,5,6)*. The test results can then be interpreted from the standard curve without the need of taking the specific activities of the number of [^{3}H] dNTPs incorporated into account.

References

1. Syvänen, A.-C., Aalto-Setälä, K., Harju, L., Kontula, K., and Söderlund, H. (1990) A primer-guided nucleotide incorporation assay in the genotyping of apolipoprotein E. *Genomics* **8**, 684–692.
2. Syvänen, A.-C., Sajantila, A., and Lukka, M. (1993) Identification of individuals by analysis of biallelic DNA markers, using PCR and solid-phase minisequencing. *Am. J. Hum. Genet.* **52**, 46–59.
3. Suomalainen, A., Kollmann, P., Octave, J.-N., Söderlund, H., and Syvänen, A.-C. (1993) Quantification of mitochondrial DNA carrying the tRNA$_{8344}$^{Lys} point mutation in myoclonus epilepsy and ragged-red-fiber disease. *Eur. J. Hum. Genet.* **1**, 88–95.
4. Syvänen, A.-C. (1999) From gels to chips: "Minisequencing" primer extension for analysis of point mutations and single nucleotide polymorphisms. *Human Mutation* **13**, 1–10.
5. Syvänen, A.-C., Söderlund, H., Laaksonen, E., Bengtström, M., Turunen, M., and Palotie, A. (1992) N-ras gene mutations in acute myeloid leukemia: accurate detection by solid-phase minisequencing. *Int. J. Cancer* **50**, 713–718.
6. Suomalainen, A., Majander, A., Pihko, H., Peltonen, L., and Syvänen, A.-C. (1993) Quantification of tRNA$_{3243}$^{Leu} point mutation of mitochondrial DNA in MELAS patients and its effects on mitochondrial transcription. *Hum. Mol. Genet.* **2**, 525–534.
7. Chen, X. and Kwok, P. Y. (1997) Template-directed dye-terminator incorporation (TDI) assay: a homogeneous DNA diagnostic method based on fluorescence resonance energy transfer. *Nucleic Acids Res.* **15**, 347–353.
8. Pastinen, T., Perola, M., Niini, P., Terwilliger, J., Salomaa, V., Vartiainen, E., Peltonen, L., and Syvänen, A.-C. (1998) Array-based multiplex analysis of candidate genes reveals two independent and additive risk factors for myocardial infarction in the Finnish population. *Hum. Mol. Gen.* **7**, 1453–1462.
9. Pastinen, T., Partanen, J., and Syvänen, A.-C. (1996) Multiplex, fluorescent solid-phase minisequencing for efficient screening of DNA sequence variation. *Clin. Chem.* **42**, 1391–1397.
10. Sitbon, G., Hurtig, M., Palotie, A., Lönngren, J., and Syvänen, A.-C. (1997) A colorimetric minisequencing assay for the mutation in codon 506 of the coagulation factor V gene. *Thrombosis and Haemostasis* **77**, 701–703.
11. Ihalainen, J., Siitari, H., Laine, S., Syvänen, A.-C., and Palotie, A. (1994) Towards automatic detection of point mutations: use of scintillating microplates in solid-phase minisequencing. *BioTechniques* **16**, 938–943.

91

The Amplification Refractory Mutation System

John M. Old

1. Introduction

The amplification refractory mutation system (ARMS) is a simple and rapid method of detecting point mutations, restriction fragment length polymorphisms, and small deletions or insertions of a DNA sequence. The method was first described by Newton et al. *(1)* for the analysis of single nucleotide differences in DNA from patients with α – 1 antitrypsin deficiency and has since been applied for the carrier detection and prenatal diagnosis of many other genetics disorders including cystic fibrosis *(2)*, β thalassaemia *(3)*, and sickle cell disease *(4)*.

The technique is based on the principle of allele-specific priming of the polymerase chain reaction (PCR). An oligonucleotide primer can be designed so that it only functions as a primer for the polymerase chain reaction when it anneals to its specific target DNA sequence. The presence of the specific target sequence in a DNA sample is revealed quickly and simply by the visualization of the product after agarose gel electrophoresis and ethidium bromide staining. For example, to detect the single nucleotide change (G → C), the mutation-specific primer is complementary to the target DNA sequence with its 3' terminal nucleotide (G) complementary to the point mutation (C). The 3' nucleotide of the primer will form a mismatch (G-G) with DNA containing the normal target sequence and prevent the synthesis of amplification product. Not all 3' terminal mismatches completely prevent the extension of the oligonucleotide primer, and therefore a deliberate mismatch with the target sequence is generated by changing the primer sequence at the second, third, or even fourth nucleotide from the 3' end in order to enhance the specificity of the primer (*see* **Note 1** for details about designing ARMS primers).

The ARMS technique involves two polymerase chain reactions in the one reaction mixture. The ARMS primer requires a second primer (called the common primer) to generate the allele-specific product and in addition, two control primers must be included in the reaction mixture in order to generate an unrelated product that indicates the reaction is working correctly (see **Note 2**). **Figure 1** illustrates the use of the ARMS technique to detect a point mutation (G → C) in the β-globin gene.

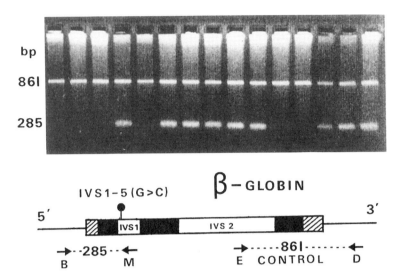

Fig. 1. Diagram showing the β-globin gene and the position of the mutation IVSI-5 (G → C) in the intervening sequence (IVS) 1, together with the positions of the mutant ARMS primer (M) to detect IVSI-5, primer B, and the control pair of primers D and E. The result of screening 16 DNA samples from individuals with β-thalassemia trait with primers β, M, D, and E are shown herein. The 285-bp product seen in nine tracks indicates the presence of IVSI-5 (G → C).

2. Materials

1. The PCR buffer (Cetus, Norwalk, CT): 50 mM KCl, 10 mM Tris-HCl (pH 8.3 at room temperature), 1.5 mM MgCl$_2$, and 100 μg/mL gelatin. A 10X stock buffer can be prepared by adding together 0.5 mL of 1 M Tris-HCl, pH 8.3, 1.25 mL 2 M KCl, 75 μL 1 M MgCl$_2$, 5 mg of gelatin, and 3.275 mL of distilled water. The stock buffer is heated at 37°C until the gelatin dissolves and is then frozen in aliquots.
2. Stock deoxynucleotide mixture: 1.25 mM Each dNTP. Add together 50 μL of a 100-mM solution of each dNTP and 3.8 mL of distilled water. The 1.25-mM dNTP stock should be stored at –20°C in 0.8-mL aliquots.

    Ready-made 100 mM dNTP solutions are now easily obtainable commercially (e.g., from Boehringer Mannheim, Mannheim, Germany), although purists may still wish to make up their own stocks from dNTP salts (which are less expensive) as follows: Dissolve to approx 200 mM, neutralize carefully with 0.05 M Tris base to a pH of 7.0 by checking droplets on pH paper, read the optical density of a diluted aliquot at the correct wavelength (259 nm-A, 253 nm-G, 271 nm-C, 260 nm-T), calculate the concentration from the extinction coefficient (for a 1-cm path length: 1.54×10^4-A, 1.37×10^4-G, 9.1×10^3-C, 7.4×10^3-T), and finally adjust to 100 mM with distilled water.
3. PCR reaction mixture stock solution (4 mL). Add together 0.5 mL 10X Cetus buffer, 0.8 mL 1.25 mM dNTP stock solution, and 2.7 mL distilled water.
4. Dilute aliquots of primer stock solutions to make working solutions at a concentration of 1 OD U/mL. Primers may be synthesized commercially by OSWEL DNA service (Southampton, UK) and are purified by high-performance liquid chromatography. Store at –20°C.
5. *Taq* DNA polymerase: AmpliTaq (Applied Biosystems, Warrington, UK) is the gold standard but cheaper thermostable DNA polymerases may be used successfully (e.g., Advanced Biotechnologies, Leatherhead, UK).

3. Methods (see Notes 3 and 4)

The standard PCR conditions recommended by Roche for their Ampli*Taq* enzyme can be used as follows. These are: 10 m*M* Tris-HCl, pH 8.3, 50 m*M* KCl, 1.5 m*M* MgCl$_2$, 0.01% gelatin, 0.2 µ*M* each primer, and 200 µ*M* each of dATP, dCTP, dGTP, and dTTP. The reaction volume can be scaled down from the recommended 100 µL to just 25 µL, which then requires only 0.5 U Ampli*Taq* enzyme and 0.1–0.5 µg of genomic DNA to complete the reaction.

Set up the reaction mixture as follows using a dedicated set of automatic pipets that are never used for handling the amplified product:

1. Pipet 20 µL of the PCR reaction mixture stock solution into a 0.5-mL Eppendorf tube.
2. Add 1 µL of each control primer, 1 µL of the common primer, and 1 µL of the appropriate ARMS primers.
3. Add 0.5 U Ampli*Taq* enzyme. This is supplied at a concentration of 5 U/µL and therefore only 0.1 µL is required. This is difficult to estimate, and therefore it is best to make a dilute solution of enzyme at a concentration of 0.5 U/µL using distilled water.

    If the same four primers are being used in every tube (e.g., when screening a number of DNA samples for one mutation), then all four primers and enzyme can be mixed together (1 µL each primer and 0.1 µL U enzyme per tube) and then 4.1 µL of primer/enzyme mixture can be added to each tube (*see* **Notes 5** and **6**).
4. Add 1 µL of DNA solution (at approx 0.1–0.5 mg/mL).
5. Add 25 µL of light paraffin oil.
6. Place the tubes in a PCR machine and program for 25 cycles at:
    a. 94°C for 1 min
    b. 65°C for 1 min
    c. 72°C for 1.5 min
    with a final extension period of 3 min at 72°C after the last cycle (*see* **Note 7**).
    The annealing temperature of 65°C is appropriate for primers of 30 nucleotides in length. No initial denaturation step in necessary before the program of 25 cycles unless the target DNA is GC rich.
7. Remove the tubes from the PCR machine and pipet 5 µL of Ficoll/bromophenol blue dye (15%/0.05%) into each tube.
8. Vortex the mixture to incorporate the blue dye and spin for 5 s in a microfuge.
9. Remove a 20-µL aliquot of the blue aqueous mixture underneath the paraffin oil layer and load into the well of submerged agarose minigel. For most purposes a 3% agarose gel is used made up of 1.5% agarose and 1.5% Nusieve agarose (FMC BioProducts, Kent, UK, marketed through Flowgen, Shenstone, UK) (*see* **Note 8**).
10. After electrophoresis at 100 V for approx 45 min in standard Tris-acetate buffer, the gel is stained in ethidium bromide solution (0.5 µg/mL and photographed on a UV transilluminator (312 nm) using an electronic camera system or a Polaroid CU-5 instant camera fitted with an orange filter (e.g., Kodak Wratten 22A, Kodak, Rochester, NY).

4. Notes

1. ARMS primers should be 30 or more oligonucleotides long in order to use a high annealing temperature of 65°C for the prevention of nonspecific amplified products. The mutation-specific primer has its 3' terminal base complementary to the mutation, the normal-specific primer should be complementary to the corresponding normal sequence. If the 3' terminal mismatch is a strong mismatch then a weak additional mismatch is engineered at either the second, third, or fourth nucleotide from the 3' end. If the 3' terminal

mismatch is a weak one, then a strong mismatch should be added to increase the specificity of the ARMS reaction. Strong mismatches are C-C, G-A, and A-A; weak mismatches are T-T, T-C, T-G, G-G, and A-C *(5)*.

2. The common and control primers should also be 30 bases long and selected to have a G + C content of approx 50%. It is important to ensure that there are no known mutations or polymorphic DNA sequence located in the DNA sequences chosen for any of the primers. Any such occurrence is liable to reduce the effectiveness of the primer and may lead to an erroneous result. The size of the control product should be such that it runs well away from the expected ARMS products. ARMS primers can be multiplexed to detect several different mutations using the same common primer provided that the ARMS products can be distinguished by size *(6)*. This principle forms the basis of the method known as the mutagenically separated polymerase chain reaction (MS-PCR) where two ARMS primers of different lengths are used in competition to detect both mutant and normal sequence in one reaction tube *(7)*.

3. Both positive and negative controls for each mutation being screened for must always be amplified and run on the gel alongside the test samples. Primers designed to detect the normal DNA sequence at the site of a mutation must be tested using control DNA from an individual homozygous for the mutation in question.

4. The relationship of fragment intensities should be constant for all DNA samples of the same genotype. Any deviation to the expected pattern of band intensities should be treated as suspect (e.g., a trace of maternal DNA contamination in a fetal DNA sample or a false-positive ARMS-PCR result) and the sample retested.

5. Sometimes an ARMS primer may produce a positive signal that is less intense relative to the control fragment than expected. If the same faint band is observed in the negative control sample, the signal is probably a false positive result occurring when either there has been a subtle change of reaction conditions or if the ARMS primer has started to lose its specificity. The latter has occurred occasionally with the stock primer solutions and possibly results from some degradation of the 3' end of the oligonucleotide, in which case the primer has had to be resynthesized.

6. No amplification product from the control primers may result from the DNA samples being too dilute (reprecipitate in a smaller volume), the DNA being too concentrated (try a 1:10 dilution), or the DNA containing impurities. The latter may be overcome by cleaning up the DNA with a repeat phenol extraction, but if the DNA sample is too small and precious for such treatment, the addition of spermidine to the PCR reaction mixture at a final concentration of 1 mM usually allows amplification to proceed.

7. Always use a limited number of cycles to avoid false-positive results with ARMS primers and also the possibility of amplifying any contaminating maternal DNA in a fetal DNA sample. If, as a last resort, a larger number of cycles is required, amplify diluted DNA controls so that band intensities between samples remain comparable.

8. To separate small DNA fragments very similar in size, increase the percentage agarose gel to 4% (3% NuSieve/1% agarose) or even 4.5% (3.5% NuSieve/1% agarose).

References

1. Newton, C. R., Graham, A., and Heptinstall, L. E. (1989a) Analysis of any point mutation in DNA. The amplification refractory mutation system (ARMS). *Nucl. Acids Res.* **17**, 2503–2516.
2. Newton, C. R., Heptinstall, L. E., Summers, C., Super, M., Schwartz, M., Anwar, R., Graham, A., Smith, J. C., and Markham, A. F. (1989b) Amplification refractory mutation system for prenatal diagnosis and carrier assessment in cystic fibrosis. *Lancet* **ii**, 1481–1483.

3. Old, J. M., Varawalla, N. Y., and Weatherall, D. J. (1990) The rapid detection and prenatal diagnosis of β thalassaemia in the Asian Indian and Cypriot populations in the UK. *Lancet* **336,** 834–837.
4. Old, J. M. (1996) Haemoglobinopathies. Community clues to mutation detection, in *Methods in Molecular Medicine, Molecular Diagnosis of Genetic Diseases* (Elles, R., ed.), Humana Press Inc., Totowa, NJ, pp. 169–183.
5. Kwok, S., Kellogg, D. E., McKinney, N., Spasic, D., Goda, L., Levenson, C., and Sninsky, J. J. (1990) Effects of primer-template mismatches on the polymerase chain reaction: human immunodeficiency virus type I model studies. *Nucl. Acids Res.* **18,** 999–1005.
6. Tan, J. A., Tay, J. S., Lin, L. I., Kham, S. K., Chia, J. N., Chin, T. M., Aziz, N. B., and Wong, H. B. (1994) The amplification refractory mutation system (ARMS): a rapid and direct prenatal diagnostic techniques for β-thalassaemia in Singapore. *Prenat. Diagn.* **14,** 1077.
7. Chang, J. G., Lu, J. M., Huang, J. M., Chen, J. T., Liu, H. J., and Chang, C. P. (1995) Rapid diagnosis of β-thalassaemia by mutagenically separated polymerase chain reaction (MS-PCR) and its application to prenatal diagnosis. *Brit. J. Haematol.* **91,** 602–607.

92

DNase I Footprinting

Benoît Leblanc and Tom Moss

1. Introduction

DNase I footprinting was developed by Galas and Schmitz in 1978 as a method to study the sequence-specific binding of proteins to DNA *(1)*. In this technique a suitable uniquely end-labeled DNA fragment is allowed to interact with a given DNA-binding protein and then the complex is partially digested with DNase I. The bound protein protects the region of the DNA with which it interacts from attack by the DNase. Subsequent molecular weight analysis of the degraded DNA by electrophoresis and autoradiography identifies the region of protection as a gap in the otherwise continuous background of digestion products (for examples, *see* **Fig. 1**). The technique can be used to determine the site of interaction of most sequence-specific DNA-binding proteins but has been most extensively applied to the study of transcription factors. Since the DNase I molecule is relatively large compared to other footprinting agents, its attack on the DNA is more readily prevented by steric hindrance. Thus DNase I footprinting is the most likely of all the footprinting techniques to detect a specific DNA–protein interaction. This is clearly demonstrated by our studies on the transcription factor xUBF (*see* **Fig. 1B**). The xUBF interaction with the *Xenopus* ribosomal DNA enhancer can be easily detected by DNase I footprinting but has not yet been detected by other footprinting techniques.

DNase I footprinting can not only be used to study the DNA interactions of purified proteins but also as an assay to identify proteins of interest within a crude cellular or nuclear extract *(2)*. Thus it can serve much the same function as a gel-shift analysis in following a specific DNA-binding activity through a series of purification steps. Since DNase I footprinting can often be used for proteins that do not "gel-shift," it has more general applicability. However, because of the need for a protein excess and the visualization of the footprint by a partial DNA digestion ladder, the technique requires considerably more material than would a gel-shift.

DNase I (EC 3.1.4.5) is a protein of roughly 40 Å diameter. It binds in the minor groove of the DNA and cuts the phosphodiester backbone of both strands independently *(3)*. Its bulk helps to prevent it from cutting the DNA under and around a bound protein. However, a bound protein will usually have other effects on the normal cleavage by DNase I, resulting in some sites becoming hypersensitive to DNase I

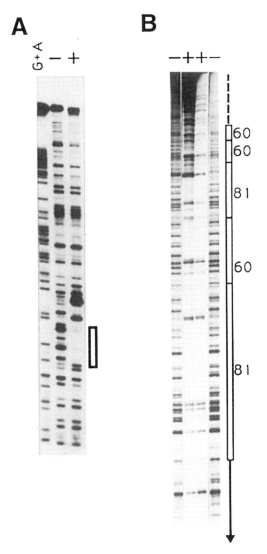

Fig. 1. Examples of DNase I footprints. (**A**) Footprint (open box) of a chicken erythrocyte DNA binding factor on the promoter of the H5 gene *(2)* (figure kindly donated by A. Ruiz-Carrillo). (**B**) Interaction of the RNA polymerase I transcription factor xUBF with the tandemly repeated 60 and 81 bp. *Xenopus* ribosomal gene enhancers. Both A and B used 5' end-labeled fragments. (–) and (+) refer to naked and complexed DNA fragments and (G + A) to the chemical sequence ladder.

(*see* **Figs. 1** and **2**). It is also not uncommon to observe a change in the pattern of DNase cleavage without any obvious extended protection (**Fig. 2**).

Unfortunately, DNase I does not cleave the DNA indiscriminately, some sequences being very rapidly attacked whereas others remain unscathed even after extensive digestion *(4)*. This results in a rather uneven "ladder" of digestion products after electrophoresis, something that limits the resolution of the technique (*see* naked DNA tracks in **Figs. 1** and **2**). However, when the protein-protected and naked DNA ladders are run

DNase I Footprinting

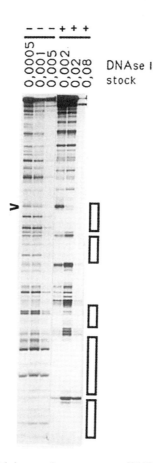

Fig. 2. Course of digestion with increasing amounts of DNase I. Here xUBF was footprinted on the *Xenopus* ribosomal promoter using a 5' end-labeled fragment. The numbers above the tracks refer to the DNase I dilution, in U/µL employed, and (–) and (+) refer to the naked and complexed DNAs respectively. The predominant footprints are indicated by open boxes.

alongside each other, the footprints are normally quite apparent. To localize the position of the footprints, G + A and/or C + T chemical sequencing ladders of the same end-labeled DNA probe *(5)* should accompany the naked and protected tracks (*see* **Note 9**). Since a single end-labeled fragment allows one to visualize interactions on one strand only of the DNA, it is usual to repeat the experiment with the same fragment labeled on the other strand. DNA fragments can be conveniently 5' labeled with T_4 kinase and 3' labeled using Klenow, T_4 polymerase (fill out), or terminal transferase *(6)*. A combination of the 5' and 3' end-labeling allows both DNA strands to be analyzed side by side from the same end of the DNA duplex.

DNase I footprinting requires an excess of DNA-binding protein over the DNA fragment used. The higher the percent occupancy of a site on the DNA, the clearer a footprint will be observed. It is therefore important not to titrate the available proteins with too much DNA. This limitation can in part be overcome when a protein also generates a gel-shift. It is then feasible to fractionate the partially DNase digested protein–DNA complex by nondenaturing gel electrophoresis and to excise the shifted band (which is then a homogeneous protein-DNA complex) before analyzing the DNA by denaturing gel electrophoresis as in the standard footprint analysis.

Footprinting crude or impure protein fractions usually requires that an excess of a nonspecific competitor DNA be added. The competitor binds nonspecific DNA-binding proteins as effectively as the specific labeled target DNA fragment and hence, when present in sufficient excess, leaves the main part of the labeled DNA available for the sequence-specific protein. Homogeneous and highly enriched protein fractions usually do not require the presence of a nonspecific competitor during footprinting. When planning a footprinting experiment, it is a prerequisite to start by determining the optimal concentration of DNase I to be used. This will be a linear function of the amount of nonspecific DNA competitor but more importantly and less reproducibly, this will be a function of the amount and purity of the protein fraction added. As a general rule, more DNase is required if more protein is present in the binding reaction, whether or not this protein binds specifically. Thus, very different DNase concentrations may be required to produce the required degree of digestion on naked and protein-bound DNA. A careful titration of the DNase concentration is therefore essential to optimize the detection of a footprint and can even make the difference between the detection or lack of detection of a given interaction.

The following protocol was developed to study the footprinting of the *Xenopus* ribosomal transcription factor xUBF, which is a rather weak DNA-binding protein, with a rather broad sequence specificity. The protocol is not original, being derived from several articles *(1,7)*. It does, however, represent a very practical approach that can be broadly applied. We recommend that the reader also refers to the available literature for more information on the quantitative analysis of protein-DNA interactions by footprinting *(8)*.

2. Materials

1. 2X Binding buffer: 20% glycerol, 0.2 mM EDTA, 1 mM DTT, 20 mM HEPES, pH 7.9, and 4% polyvinyl alcohol (*see* **Note 1**).
2. Poly d(AT): 1 mg/mL in TE (10 mM Tris-HCl, pH 8.0, 1 mM EDTA). Keep at –20°C (*see* **Note 2**).
3. End-labeled DNA fragment of high-specific activity (*see* **Note 3**).
4. Cofactor solution: 10 mM MgCl$_2$, 5 mM CaCl$_2$.
5. DNase I stock solution: A standardized vial of DNase I (D–4263, Sigma, St. Louis, MO) is dissolved in 50% glycerol, 135 mM NaCl, 15 mM sodium acetate, pH 6.5, at 10 Kunitz U/µL. This stock solution can be kept at –20°C for many months (*see* **Note 4**).
6. 1 M KCl.
7. Reaction stop buffer: 1% SDS, 200 mM NaCl, 20 mM EDTA, pH 8.0, 40 µg/mL tRNA (*see* **Note 5**).
8. 10X TBE buffer: 900 mM Tris-borate, pH 8.3, 20 mM EDTA.
9. Loading buffer: 7 M urea, 0.1X TBE, 0.05% of xylene cyanol, and bromophenol blue.
10. Sequencing gel: 6% acrylamide, 7 M urea, 1X TBE.
11. Phenol-chloroform (1:1) saturated with 0.3 M TNE (10 mM Tris-HCl, pH 8.3, 1 mM EDTA, 0.3 M NaCl).
12. Ethanol 99% and ethanol 80%. Keep at –20°C.
13. 1 M pyridine formate, pH 2.0. Keep at 4°C.
14. 10 M piperidine.

DNase I Footprinting

3. Methods

The footprinting reaction is done in three stages: binding of the protein to the DNA, partial digestion of the protein–DNA complex with DNase I, and separation of the digestion fragments on a DNA sequencing gel.

1. The binding reaction is performed in a total volume of 50 µL containing 25 µL of 2X binding buffer, 0.5 µL of 1 mg/mL poly d(AT), 2–3 ng of end-labeled DNA fragment (~15,000 cpm) (*see* **Note 6**), the protein fraction and 1 M KCl to bring the final KCl concentration to 60 mM. The maximum volume of the protein fraction that can be used will depend on the salt concentration of this solution. The reaction is performed in a 1.5-mL Eppendorf tube.
2. Incubate on ice for 20 min.
3. During the binding reaction, dilute the DNase I stock solution in distilled water at 0°C. We suggest working concentrations of about 0.0005–0.1 Kunitz U/µL, depending on the level of protein present (*see* **Note 7** and **step 5**). A good range is the following: 0.0005; 0.001; 0.005; 0.002; 0.02; 0.08 Kunitz U/µL.
4. After the incubation, transfer the reaction tubes in batches of eight to a rack at room temperature and add 50 µL of the cofactor solution to each.
5. Add 5 µL of the appropriate DNase I dilution to a tube every 15 s (from the 0.0005–0.005 Kunitz U/µL stocks for naked DNA; from the 0.002–0.08 ones for DNA + proteins).
6. After 2 min digestion, each reaction is stopped by the addition of 100 µL of the stop solution (*see* **Note 8**).
7. After all the reactions have been processed, extract each reaction once with phenol-chloroform as follows: add 1 vol phenol-chloroform (1:1) saturated with 0.3 M TNE, vortex briefly, and centrifuge in a desktop microcentrifuge for about 10 min. Recover the top phase and transfer to a new microcentrifuge tube.
8. Add 2 vol (400 µL) of ethanol 99% (–20°C) and allow nucleic acids to precipitate at –80°C for 20 min.
9. Microcentrifuge for 15 min, ~10,000g, and remove the supernatant with a Pasteur pipet. Check the presence of a radioactive pellet with a Geiger counter before discarding the ethanol.
10. Add 200 µL of 80% ethanol (–20°C) to the pellet and microcentrifuge again for 5 min. After removing the supernatant, dry the pellets in a vacuum dessicator.
11. Resuspend each pellet in 4.5 µL loading buffer, vortex, and centrifuge briefly.
12. A G + A ladder and a molecular weight marker should be run in parallel with the samples on a sequencing gel (*see* **Note 9**). The G + A ladder can be prepared as follows (*5*): ~200,000 cpm of end-labeled DNA are diluted into 30 µL H$_2$O (no EDTA). 2 µL of 1 M pyridine formate, pH 2.0, are added and the solution incubated at 37°C for 15 min. One hundred fifty microliters of 1 M piperidine are added directly and the solution incubated at 90°C for 30 min in a well-sealed tube (we use a 500-µL microcentrifuge tube in a thermal cycler). Add 20 µL of 3 M sodium acetate and 500 µL of ethanol and precipitate at –80°C for 10–20 min. Microcentrifuge (10,000g, 10 min) and repeat the precipitation. Finally, redissolve the pellet in 200 µL of H$_2$O and lyophilize. Resuspend in loading buffer and apply about 5,000 cpm/track.
13. Prerun a standard 6% acrylamide, sequencing gel (43 × 38 cm, 0.4 mm thick, 85 W) for 30 min before loading each of the aliquots from the DNase I digestion, plus the markers. Running buffer is 1X TBE. Wash the wells thoroughly with a syringe, denature the DNA for 2 min at 90°C, and load with thin-ended micropipet tips. Run the gel hot to keep the DNA denatured (*see* **Note 10**). After the run, cover the gel in plastic wrap and expose it

overnight at –70°C with an intensifying screen. We use either a Cronex Lightning Plus (Dupont, Wilmington, DE) or Kyokko Special (Fuji, Japan) screens, the latter being about 30% less sensitive but also less expensive. Several different exposures will probably be required to obtain suitable band densities.

4. Notes

1. This binding buffer has been shown to work well for the transcription factor NF-1 *(6)*, and in our laboratory for both the hUBF and xUBF factors and thus should work for many factors. Glycerol and PVA (an agent used to reduce the available water volume and hence concentrate the binding activity) are not mandatory. The original footprinting conditions of Galas and Schmitz *(1)* for the binding of the *lac* repressor on the *lac* operator were 10 mM cacodylate buffer, pH 8.0, 10 mM MgCl$_2$, 5 mM CaCl$_2$, and 0.1 mM DTT. Particular conditions of pH, cofactors, and ionic strength may need to be determined for an optimal binding of different factors to DNA.
2. Since poly d(IC), another nonspecific general competitor, has been shown to compete quite efficiently with G-C rich DNA sequences, poly d(AT) is preferred here. The choice of an appropriate nonspecific competitor (whether it is synthetic, as in this case, or natural, e.g., pBR322 or calf thymus DNA) may have to be determined empirically for the protein studied. When working with a pure or highly enriched protein, no competitor is usually needed. The DNase I concentration must then be reduced accordingly (to about naked DNA values).
3. Single-stranded breaks in the end-labeled DNA fragment must be avoided because they give false signals indistinguishable from genuine DNase I cleavage and hence can mask an otherwise good footprint. It is therefore advisable to check the fragment on a denaturing gel before use. Always use a freshly labeled fragment (3–4 d at the most) because radiochemical nicking will degrade it.
4. These standardized vials allow for very reproducible results. Glycerol will keep the enzyme from freezing, as repeated freeze–thaw cycles will greatly reduce its activity.
5. Do not be tempted to use too much RNA since it causes a very annoying fuzziness of the gel bands that prevents resolution of the individual bands.
6. The use of 5'end-labeling with kinase in the presence of crude protein extracts can sometimes lead to a severe loss of signal because of the presence of phosphatases. In these cases 3'end-labeling by "fill out" with Klenow or T$_4$ polymerase is to be prefered.
7. For naked DNA and very low amounts of protein, working stocks diluted to 0.0005–0.005 Kunitz U/µL give a good range of digestion.
8. It is convenient to work with groups of eight samples during the DNase I digestion. Cofactor solution is added to eight samples at a time and then the DNase I digestions begun at 15 s intervals: 15 s after adding DNase to the eighth sample, stop solution is added to sample 1 and then to the other samples at 15 s intervals.
9. In comparing a chemical sequencing ladder with the products of DNase I digestion, one must bear in mind that each band in the sequencing ladder corresponds to a fragment ending in the base *preceding* the one read because chemical modification and cleavage destroys the target base. For example, if a DNase I gel band corresponds in mobility to the sequence ladder band read as G in the sequence ACGT, then the DNase I cleavage occurred between the bases C and G. DNase I cleaves the phosphodiester bond, leaving a 3'-OH, whereas the G + A and C + T sequencing reactions leave a 3'-PO$_4$, causing a mobility shift between the two types of cleavage ladders. This is a further potential source of error. However, in our experience the shift is less than half a base and hence cannot lead to an error in the deduced cleavage site.

10. Sequencing gels are not denaturing unless run hot (7 M urea produces only a small reduction in the T_m of the DNA). A double-stranded form of the DNA fragment is therefore often seen on the autoradiogram, especially at low levels of DNase I digestion (*see* **Fig. 2**) and can sometimes be misinterpreted as a hypersensitive cleavage. By running a small quantity of undigested DNA fragment in parallel with the footprint this error can be avoided.

Acknowledgments

The authors wish to thank A. Ruiz-Carrillo for providing the autoradiogram in **Fig. 1A**. The work was supported by the Medical Research Council of Canada (MRC). T. Moss is presently an F.R.S.Q. "Chercheur-boursier" and B. Leblanc was until recently supported by a grant from the F.C.A.R. of Québec.

References

1. Schmitz, A. and Galas, D. J. (1978) DNase I footprinting: a simple method for the detection of protein-DNA binding specificity. *Nucleic Acids Res.* **5,** 3157–3170.
2. Rousseau, S., Renaud, J., and Ruiz-Carrillo, A. (1989) Basal expression of the histone H5 gene is controlled by positive and negative *cis*-acting sequences. *Nucleic Acids Res.* **17,** 7495–7511.
3. Suck, D., Lahm, A., and Oefner, C. (1988) Structure refined to 2 Å of a nicked DNA octanucleotide complex with DNase I. *Nature* **332,** 464–468.
4. Drew, H. R. (1984) Structural specificities of five commonly used DNA nucleases. *J. Mol. Biol.* **176,** 535–557.
5. Maxam, A. M. and Gilbert, W. (1980) Sequencing end-labeled DNA with base-specific chemical cleavages, in *Methods in Enzymology,* vol. 65 (Grossman, L. and Moldave, K., eds.), Academic, New York, pp. 499–560.
6. *Current Protocols in Molecular Biology*, Chapter 3 (1991) (Ausubel, F. M., Brent, R., Kingston, R. E., Moore, D. E., Smith, S. A., and Struhl, K., eds.), Greene and Wiley-Interscience, New York.
7. Walker, P. and Reeder, R. H. (1988) The *Xenopus laevis* ribosomal gene promoter contains a binding site for nuclear factor-1. *Nucleic Acids Res.* **16,** 10,657–10,668.
8. Brenowitz, M., Senear, D. F., and Kingston, R. E. (1991) DNase footprint analysis of protein-DNA binding, in *Current Protocols in Molecular Biology* (Ausubel, F. M., Brent, R., Kingston, R. E., Moore, D. E., Smith, S. A., and Struhl, K., eds.), Greene and Wiley-Interscience, New York, pp. 12.4.1–12.4.11.

93

Identification of Protein–DNA Contacts with Dimethyl Sulfate

Methylation Protection and Methylation Interference

Peter E. Shaw and A. Francis Stewart

1. Introduction

Dimethyl sulfate (DMS) is an effective and widely used probe for sequence-specific protein–DNA interactions. It is the only probe routinely used both for in vitro (methylation protection, methylation interference) and in vivo (DMS genomic footprinting) applications since it rapidly reacts with DNA at room temperature and readily penetrates intact cells *(1)*. DMS predominantly methylates the 7-nitrogen of guanine and 3-nitrogen of adenine. Thus, reactivity with G residues occurs in the major groove and with A residues in the minor groove. In standard Maxam and Gilbert protocols *(2)*, the methylated bases are subsequently converted to strand breaks and displayed on sequencing gels.

Methylation protection and interference are essentially combinations of the gel retardation assay *(3,4)* with the DMS reaction of the Maxam and Gilbert sequencing procedure. Protein–DNA interactions are reflected either as changes in DMS reactivities caused by bound protein (methylation protection) or as selective protein binding dictated by methylation (methylation interference).

In methylation protection, protein is first bound to DNA that is uniquely end-labeled, and the complex is reacted with DMS. DMS reactivities of specific residues are altered by bound protein either by exclusion, resulting in reduced methylation, increased local hydrophobicity, resulting in enhanced methylation, or by local DNA conformational changes, such as unwinding, resulting in altered reactivity profiles *(5–7)*. After the DMS reaction, free DNA is separated from protein-bound DNA by gel retardation, and both DNA fractions are recovered from the gel. Methylated residues are converted into strand scissions, and the free and bound DNA fractions are compared on a sequencing gel. A complete analysis requires examination of both strands. This is accomplished by preparing two DNA probes, each uniquely labeled at one end, and carrying both probes through the protocols. A binding site characterized by methylation protection will therefore appear as a cluster of altered DMS reactivities.

From: *The Nucleic Acid Protocols Handbook*
Edited by: R. Rapley © Humana Press Inc., Totowa, NJ

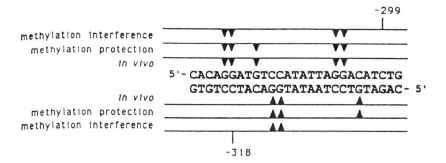

Fig. 1. Comparison of methylation interference and protection patterns formed by factors binding at the c-fos serum response element (SRE) in vitro and in vivo. G residues identified by methylation interference *(9)*, methylation protection, and in vivo genomic footprinting *(7)* are indicated. An additional G on both the upper and lower strands is implicated in the protein–DNA interaction by methylation protection.

In methylation interference, DNA is first reacted with DMS, purified, and then presented to protein *(8,9)*. Under the reaction conditions used, methylation is partial, yielding approximately one methylated base per DNA molecule. Thus, the protein is presented with a mixture of DNA molecules that differ with respect to the positions of methyl groups. Some methyl groups will interfere with protein binding since they lie in or near the binding site. Gel retardation separates the mixture into two fractions: free DNA, which, as long as DNA is in excess over binding activity, represents the total profile of methylation reactivity, and bound DNA, which will not contain any molecules with methyl groups incompatible with binding. Both free and bound DNA fractions are recovered, methylated residues are converted to strand scissions, and the fractions are compared on a sequencing gel. The binding site is observed as the absence of bands in the bound sample corresponding to the positions where methylation interferes with binding.

These two uses of DMS may not deliver identical results. For example, **Fig. 1** presents a comparison obtained from experiments with the serum response element binding factors p67^{SRF} p62^{TCF} and their binding site in the human c-*fos* promoter (SRE). Since the use of DMS in vivo for genomic footprinting is limited to the equivalent of methylation protection, a direct comparison between in vivo and in vitro footprints requires the use of methylation protection rather than the more widely used methylation interference assay.

These two techniques are, however, very similar in practical terms and, thus, are presented together. Both techniques rely on preestablished conditions that permit a protein–DNA complex to be resolved in a gel retardation assay (*see* Chapter 94) and on routinely used DNA sequencing methodologies, for which the reader will find a detailed treatment *(2)*.

2. Materials

1. Dimethyl sulfate (Merck, Rahway, NJ), analytical grade.
2. Piperidine (Sigma, St. Louis, MO), analytical grade: Use freshly made 1:10 dilution in double-distilled water.

Methylation Protection and Interference

3. Phenol/chloroform: 50% v/v, buffered with 50 mM Tris-HCl, pH 8.0.
4. NA45 paper (Schleicher and Schuell, Dassel, Germany).
5. Whatman 3MM paper or GB 002 paper (Schleicher and Schuell).
6. Gel retardation equipment.
7. Electroblotting apparatus for Western transfer (e.g., Trans-Blot, Bio-Rad, Hercules, CA).
8. Standard DNA sequencing gel electrophoresis equipment.
9. Vacuum gel dryer (optional).
10. TBE buffer: 108 g Tris base, 55 g boric acid, 40 mL 0.5 M EDTA, pH 8.3/L for 10X stock; working solution is 89 mM Tris base, 89 mM boric acid, 2 mM EDTA.
11. NA45 elution buffer: 10 mM Tris-HCl, pH 8.0, 1 mM EDTA, 1 M NaCl.
12. Carrier DNA: Salmon testis DNA or calf thymus DNA, sheared (Sigma; dissolved at 3 mg/mL in 10 mM Tris-HCl, pH 8.0, 1 mM EDTA).
13. Sequencing gel loading buffer: 90% formamide, 10 mM EDTA, 0.1% w/v bromophenol blue, 0.1% w/v xylene cyanol blue.
14. Gel retardation loading buffer: 20% Ficoll, 20 mM EDTA, 0.1% w/v bromophenol blue.
15. 2X DMS buffer: 120 mM NaCl, 20 mM Tris-HCl, pH 8.0, 20 mM MgCl$_2$, 2 mM EDTA.
16. DMS stop buffer: 1.5 M sodium acetate, pH 7.0, 1 M 2-mercaptoethanol; store frozen.
17. X-ray film (e.g., Kodak X-Omat, Rochester, NY).
18. 0.25 M Dithiothreitol (DTT).
19. Sequencing gel stock A: 1X TBE, 14.25% w/v acrylamide (Bio-Rad), 0.75% w/v *bis*-acrylamide, 8 M urea. Sequencing gel stock B: As A except omit acrylamide and *bis*-acrylamide. Store in brown glass at 4°C.
20. 10% w/v ammonium persulfate made weekly.
21. *N,N,N',N'*-Tetramethylethylenediamine (TEMED; Bio-Rad).
22. 5% v/v dimethyldichlorosilane (Sigma) in chloroform.

3. Methods
3.1. Methylation Protection

1. Incubate 300,000 cpm of uniquely end-labeled DNA probe and a corresponding amount of protein together, under conditions previously optimized by gel retardation analysis (*see* Chapter 94 and **Note 1**), in a total vol of 100 µL.
2. Add 1 µL of DMS and incubate at room temperature (the incubation time depends on the length of the DNA probe and is empirical: As a guide 200-bp fragment $t = 1.5$ min; for 50-bp oligo duplex $t = 3.0$ min).
3. Add $^1/_{10}$ vol of 250 mM DTT, mix gently, add 1/10 vol of gel retardation loading buffer, mix gently, load onto a 2-mm thick retardation gel in 1X TBE (or alternative buffer), and run as optimized for analytical retardation gels. However, the load should be spread over up to ten times more well area (*see* **Note 2**).
4. After electrophoresis, separate the glass plates carefully so that the gel adheres to one plate and cover the gel with clingwrap. Expose to X-ray film long enough to reveal complexes clearly (i.e., a few hours to overnight). The alignment of the film to the gel must be reliably marked.
5. Put the developed film on a lightbox. Remove the clingwrap from the gel and align the gel on the X-ray film. Cut pieces of NA45 paper sufficiently large to cover individual complexes in the gel, yet small enough to fit into 1.5-mL tubes when rolled up. Wet the paper pieces in retardation gel running buffer and, with the help of tweezers, position one over each complex of interest in the gel, as visualized from the underlying film. Also position a similar sized piece of paper over some of the uncomplexed DNA. NA45 paper can be labeled with pencil before wetting.

6. Carefully cover the gel and paper pieces with two sheets of 3MM paper wetted in 1X TBE (or alternative gel running buffer from **step 3**). Lay a scotchbrite pad from the electroblotting apparatus on top of the paper and turn the gel over. Carefully remove the second glass plate, cover the other side of the gel with 3MM paper, scotchbrite as before, and insert the package into an electro-transfer apparatus as described in the manufacturer's instructions with the NA45 paper toward the anode. Transfer in 1X TBE (or alternative buffer) at 80 V for 1.5 h (*see* **Note 3**).
7. Stop the transfer, unpack the gel carefully with the NA45 paper on top, and transfer each piece to a labeled 1.5-mL tube containing 600 μL of elution buffer (check that the radioactivity has transferred to the paper using a hand monitor). Incubate at 70°C for 1 h.
8. Remove NA45 paper from each tube, check that at least half the radioactivity has eluted into the buffer (do not expect quantitative elution but at least 50% should come off; *see* **Note 4**), add 20 μg of carrier DNA, extract with 600 μL phenol/chloroform, and precipitate with 1 vol of isopropanol (*see* **Note 5**). Wash the precipitate once in 70% ethanol and dry under vacuum.
9. Dilute piperidine 1:10 in water and add 50 μL to each pellet (*see* **Note 6**). Vortex briefly, incubate at 90°C for 30 min (tubes must be clamped or weighted down to prevent the lids from opening), and then dry under vacuum. Take up the samples in 100 μL of water and repeat the drying process.
10. Measure the Cerenkov counts in each tube; then redissolve the samples in water (e.g., 10 cpm/μL) and transfer equivalent counts (1000 cpm is optimal; *see* **Note 7**) from each into fresh tubes. Dry down and redissolve the cleaved DNA in 5 μL sequencing gel loading buffer (*see* **Note 8**).
11. Prepare a standard sequencing gel. The percentage of the gel will depend on the DNA probe length, that is, for 30 mers or less, 10–15% gels should be used. For longer probes, lower percentage gels should be used (*see* **Note 9**). A standard sequencing gel is poured between two 40 × 20-cm glass plates separated by 0.4-mm spacers. One of the glass plates is cut to produce a 2-cm deep, 16-cm long U-shaped recess that leaves 2-cm wide ears on either side. This edge will form the top of the gel assembly. Clean the glass plates thoroughly, including a final wipe with ethanol, and then wipe one of the plates with 5% dimethylchlorosilane. This step must be performed in a fume hood and secure gloving must be worn. When this plate is dry, polish it with tissues (*see* **Note 10**). Seal the assembly of plates and spacers with waterproof tape along the two sides and bottom, and then clamp with bulldog clips. Make sure that the slot former fits in snugly. Mix sequencing gel stocks A and B to produce 60 mL of the desired acrylamide concentration and warm up to room temperature. Add 60 μL TEMED, mix, then add 600 μL 10% ammonium persulfate, and pour immediately. Pouring is aided by tilting the gel assembly so that the solution is poured into the corner of the U-shaped recess and from this reservoir flows down one edge. Make sure that no air bubbles are trapped during pouring. If air bubbles are present, stand the assembly vertically and gently tap the glass next to the bubble. When the assembly is full, lay it down and insert the slot former, again ensuring that no air bubbles are trapped. Place clamps over the top end of the assembly so that the glass plates are squeezed onto the slot former. Polymerization of the remaining unpoured gel mix should occur within 5 min. Allow the gel at least 1 h to polymerize thoroughly.
12. After removing the sealing tape, clamp the gel assembly into a gel tank. Remove the slot former, top up the reservoirs with the running buffer (1X TBE), and clean the wells. Preelectrophorese at approx 50 W or 40 V/cm (*see* **Note 11**). Denature probes at 95°C for 5 min, snap cool in ice, and load onto the gel. Run the gel until optimal separation of sequence is achieved.

13. Stop electrophoresis, remove the gel from the tank, and lift off the glass plate that was silanized. Place the other glass plate with gel into 20% ethanol, 10% acetic acid for 10 min. Drain briefly; then overlay the gel with two sheets of 3MM paper and carefully peel it off the glass plate. Cover the gel surface with cling wrap and dry on a vacuum gel dryer (*see* **Note 12**). Expose the dry gel to X-ray film with intensifying screens as necessary (*see* **Note 13**).

3.2. Methylation Interference

1. Mix 300,000 cpm of end-labeled probe, 100 µL of 2X DMS buffer, and water to 200 µL. Add 2 µL of DMS and incubate at room temperature (the same guidelines as given in **Subheading 3.1.2.** apply for the reaction time). Stop the reaction by the addition of 50 µL cold DMS stop mix and precipitate with 850 µL cold ethanol. Redissolve the probe in 200 µL cold 0.3 M sodium acetate, pH 7.0; add 700 µL cold ethanol and reprecipitate. Wash the precipitate twice in 80% ethanol, dry, and redissolve the probe in water or binding buffer to give about 20,000 cpm/µL.
2. Incubate the probe with protein for gel retardation as previously optimized for gel retardation analyses of the complexes in question in a total vol of 100 µL.
3. Add $^1/_{10}$ vol of gel retardation loading buffer, mix gently, load onto a 2-mm thick retardation gel in 1X TBE (or alternative buffer), and run as optimized for analytical gels. However, the load should be spread over up to ten times more well area (*see* **Note 2**).
4. Continue with **step 4** and all subsequent steps as described for methylation protection (*see* **Subheading 3.1.**).

4. Notes

1. To have sufficient counts to complete the procedure, proceed with at least 10 times the amount of material required for a simple gel retardation analysis (i.e., at least 300,000 cpm).
2. A common difficulty with these methods is the persistence of contaminants that accompany DNA after the preparative retardation gel. These contaminants interfere with the migration of DNA on the sequencing gel, producing blurred and distorted patterns. In order to minimize this problem, it is worth ascertaining the load limit of the retardation gel so that the protein–DNA complex will not smear, but will be well resolved and therefore concentrated in the gel before elution.
3. It is also possible to use a semidry electro-transfer apparatus (e.g., Bio-Rad Trans-blot SD) to transfer the DNA from the gel retardation gel onto NA45 paper. In this case, the transfer time and potential are both reduced.
4. In some instances, it may prove difficult to elute the DNA from the NA45 paper, in which case raising the salt concentration or the temperature may improve elution. (Extending the incubation time does not seem to help.) If not, the batch of NA45 may be to blame, or it is even conceivable that the DNA–protein complex in question is adsorbed too tightly onto the paper. It is not possible to phenol extract the NA45 paper in order to remove bound protein–DNA.
5. Retain the isopropanol supernatants until you are sure the samples have precipitated quantitatively. Add more carrier DNA if required.
6. The strand scission protocol described here should not convert methylated A residues into strand breaks. It is often observed, however, that breakages at As occur with reasonable efficiency. The following modification will produce efficient cleavage at both G and A residues. After the preparative retardation gel, resuspend the dried, purified DNA in 30 µL 10 mM sodium phosphate, pH 6.8, 1 mM EDTA. Incubate for 15 min at 92°C. Then add 3 µL 1 M NaOH and incubate for 30 min at 92°C, followed by 320 µL 500 mM NaCl,

50 µg/mL carrier DNA, and 900 µL ethanol. Chill and centrifuge to pellet the radioactivity. Wash the pellet in 70% ethanol and dry. Proceed as above from **step 10, Subheading 3.1.**

7. As discussed in **Note 2** above, it is similarly advisable to load as little material onto the sequencing gel as practicable. The practical lower limit for the sequencing gel is approx 1000 cpm/lane, but desperate individuals know no bounds.
8. Bromophenol blue and, to a lesser extent, xylene cyanol blue have been observed to cause exclusion distortions in the sequencing gels, and either one can be excluded from the sequencing loading buffer if this occurs in a critical part of the gel. When dye(s) are omitted from the loading buffer, load the dye-containing loading buffer in a spare lane to act as a tracker.
9. The choice of percentage of the sequencing gel depends on the distance between the radioactive label and the site of the protein–DNA interaction. If the end-labeled DNA fragment is relatively long and multiple binding sites are to be resolved, a gradient or wedge sequencing gel can be used in **step 11** of **Subheading 3.1.**
10. The polished, silanized glass plate can be baked in a 180°C oven. This fixes the silanization through repeated use so that the plate need not be silanized every time. It is better to silanize the plate with the U-shaped cut.
11. The exact settings for electrophoresis of sequencing gels vary greatly and should be determined empirically by assessing the temperature of the glass plates after running for about 1 h. Sequencing gels should be run so that their temperature remains stable between 45 and 55°C.
12. It is not essential to dry down the sequencing gel because, after one glass plate has been removed, it can be covered with cling wrap and exposed to X-ray film at –70°C with one screen. This alternative should only be considered if the signal is sufficiently strong or if a gel dryer is not available.
13. An appropriate complement for the final result is to perform the Maxam and Gilbert G + A reactions *(2)* on the end-labeled probe. On the sequencing gel, these reactions should provide unambiguous sequence information and, in case difficulties are encountered, clues as to the steps that are problematic.

References

1. Church, G. M. and Gilbert, W. (1984) Genomic sequencing. *Proc. Natl. Acad. Sci. USA* **81,** 1991–1995.
2. Maxam, A. and Gilbert, W. (1980) Sequencing end-labeled DNA with base-specific chemical cleavages. *Meth. Enzymol.* **65,** 499–560.
3. Fried, A. and Crothers, D. M. (1981) Equilibria and kinetics of lac repressor-operator interactions by polyacrylamide gel electrophoresis. *Nucleic Acids Res.* **9,** 6505–6525.
4. Garner, M. M. and Revzin, A. (1981) A gel electrophoresis method for quantifying the binding of protein to specific DNA regions: application to components of the *E. coli* lactose operon regulatory system. *Nucleic Acids Res.* **9,** 3047–3059.
5. Gilbert, W., Maxam, A., and Mirzabekov, A. (1976) Contacts between the LAC repressor and DNA revealed by methylation, in *Control of Ribosome Biosynthesis, Alfred Benzon Symposium IX* (Kjelgaard, N. O. and Maaloe, O., eds.), Academic, New York, pp. 139–148.
6. Johnsrud, L. (1978) Contacts between *Escherichia coli* RNA polymerase and a lac operon promoter. *Proc. Natl. Acad. Sci. USA* **75,** 5314–5318.
7. Herrera, R. E., Shaw, P. E., and Nordheim, A. (1989) Occupation of the c-fos serum response element in vivo by a multi-protein complex is unaltered by growth factor induction. *Nature* **340,** 68–70.

8. Siebenlist, U. and Gilbert, W. (1980) Contacts between *E. coli* RNA polymerase and an early promoter of phage T7. *Proc. Natl. Acad. Sci. USA* **77,** 122–126.
9. Shaw, P. E., Schröter, H., and Nordheim, A. (1989) The ability of a ternary complex to form over the serum response element correlates with serum inducibility of the human c-fos promoter. *Cell* **56,** 563–572.

94

The Gel Shift Assay for the Analysis of DNA–Protein Interactions

John D. Taylor, Alison J. Ackroyd, and Stephen E. Halford

1. Introduction

The gel shift assay is one of the most powerful methods for the analysis of DNA–protein interactions *(1,2)*. The assay itself is simple. DNA and protein are mixed together, the solution subjected to electrophoresis through polyacrylamide, and the gel is then analyzed for DNA, usually by autoradiography of radiolabeled DNA *(3,4)*. Binding of the protein to the DNA can result in a complex that has a different electrophoretic mobility from the free DNA. In general, the mobility of the complex is retarded relative to the unbound DNA and thus the assay is often called gel retardation. However, with circular DNA substrates (typically, minicircles of 200–400 bp), the DNA–protein complex can migrate faster than the free DNA *(5,6)*. The separation of the complex from the free DNA, and therefore the detection of the complex, is dependent on a variety of factors. These must be determined experimentally for each system. However, the ease with which the assay can be performed means that the optimal conditions can be discovered quickly. Factors that influence the electrophoretic mobility of DNA–protein complexes include the molecular weight of the protein and the DNA *(7,8)*, the ionic strength and the pH of the electrophoresis buffer *(9)*, the concentration of the gel matrix, and the temperature. Particularly useful accounts of how modifications to the assay can affect the mobility of DNA–protein complexes have been published *(2,10)*.

The principle of the gel shift assay is that the entry of the mixture of free DNA and DNA–protein complex into the gel matrix results in the physical separation of the two species. In the subsequent electrophoresis, the protein can make no difference to the mobility of the free DNA and, provided that the bound DNA remains associated with the protein, it too will have a characteristic mobility. The gel matrix may stabilize the complex by hindering the diffusion of the protein away from the DNA *(4,7,8)*, although it has been shown that complexes reversibly dissociate and reassociate within the gel *(11)*. However, even if the bound DNA dissociates from the protein during electrophoresis, it can never "catch up" with the DNA that was free at the start of the run. Thus, the method has the potential of "freezing" the equilibrium between bound and free DNA at the moment of entry into the gel. The concentration of each species can

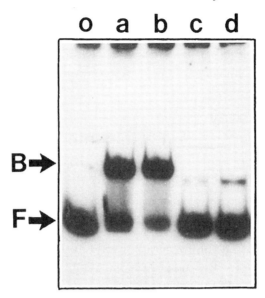

Fig. 1. Autoradiograph from a gel shift assay, showing binding to one specific site. The reactions contained both a 203-bp DNA fragment ([^{32}P]-labeled, 5 nM) with one binding site for the resolvase from Tn3 and unlabeled calf thymus DNA. These were mixed with either no additional protein (lane 0), the resolvase from Tn3 (in the lanes marked 3 above the gel; 100 nM in a, 200 nM in b), or helix-swap resolvase (in the lanes marked H; 100 nM in a, 200 nM in b). The samples were then analyzed by electrophoresis through polyacrylamide. F and B, on the left of the autoradiograph, mark the positions of the 203 bp DNA, either free or bound to resolvase. Resolvase possesses a helix-turn-helix motif and helix-swap resolvase was constructed from Tn21 resolvase by replacing all of the amino acids in its recognition helix with the equivalent sequence from Tn3 resolvase. Native Tn21 resolvase has virtually no affinity for this segment of DNA from Tn3 but the replacement of its recognition helix allows for some binding to Tn3 DNA, although the yield of the complex is much lower than with the cognate Tn3 resolvase. (Data from **ref. 13**: Reprinted with permission from *J. Mol. Biol.* Copyright [1990] Academic).

then be determined. Assays of this type can yield the equilibrium constant for the binding of the protein to its DNA ligand and also the kinetics of the interaction, the latter by analyzing samples at different times after mixing the DNA with the protein *(3,6,7,12)*. An important parameter in kinetic experiments is the "electrophoretic dead time," the time taken for the complex to migrate from the solution loaded in the well to the gel itself. This needs to be made as short as possible *(8)*.

If the DNA substrate has one specific site that binds the protein much more tightly than any other site, a gel shift experiment should reveal just one DNA–protein complex (**Fig. 1**). However, the appearance of a single retarded band does not necessarily mean that a specific complex has been formed. One test for specificity is to add unlabeled competitor DNA to a binding reaction containing [^{32}P]-labeled substrate DNA *(3,7,13)*. On addition of competitor DNA, all nonspecific complexes with the labeled DNA should be titrated away. Any remaining complexes should represent the specific DNA–protein association, and this can confirmed by observing their disappearance on further

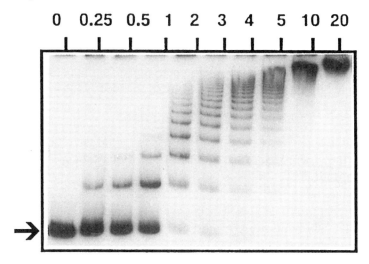

Fig. 2. Autoradiograph from a gel shift assay, showing binding to multiple nonspecific sites. The reactions contained a 381-bp DNA fragment ([^{32}P]-labeled, 0.1 nM), with either no added protein (lane 0; the arrow on the left of the gel marks the free DNA), or with EcoRV restriction endonuclease at the concentrations (nM) indicated above each lane: The binding buffer contained EDTA. The DNA contains one copy of the EcoRV recognition sequence but, instead of observing a single DNA-protein complex, the EcoRV restriction enzyme produces a series of complexes. The first retarded band is the DNA bound to 1 molecule of enzyme, the second with 2 molecules of enzyme, and so forth. In the absence of Mg^{2+} ions, to prevent cleavage of the DNA, this restriction enzyme binds all DNA sequences, including its recognition site, with the same equilibrium constant. (Data from **ref. 17**: Reprinted with permission from *Biochemistry*. Copyright [1991] American Chemical Society).

titrations with unlabeled substrate DNA. Indeed, the interaction of a protein with a specific DNA sequence is best analyzed in the presence of nonspecific competitor DNA as shown in **Fig. 1**. Moreover, by using unlabeled competitor with labeled substrate, the gel shift assay can be applied to unfractionated nuclear extracts, to detect proteins that bind specifically to the substrate DNA *(14)*. Which proteins in the crude extract are bound to the substrate can then be determined immunologically *(15)*.

In addition to separating bound from free DNA, the gel shift assay can also resolve DNA–protein complexes with different stoichiometries; that is, the DNA with 1, 2, 3, or more molecules of protein bound to it *(4,16,17)*. The gel shift experiment then reveals a series of bands of DNA with progressively reduced electrophoretic mobilities (*see* **Fig. 2**). The stoichiometry of each complex can be determined: The change in mobility from one complex to the next is usually proportional to the change in molecular weight caused by adding one more molecule of protein *(2,16,17)*. A more rigorous, although technically demanding, approach to determining stoichiometries involves labeling the protein with ^{14}C and the DNA with ^{32}P so that the molar ratios, in each band on the gel, can be measured directly *(9,18)*. The gel can also be stained for protein, with Coomaasie blue or silver, and another approach is quantitative Western blotting *(19)*. The ability of gel shifts to resolve DNA species multiply bound by one protein (as in **Fig. 2**), or DNA bound by two or more different proteins *(3,7)*, is one of the major advantages of this technique over many other methods for DNA–protein interactions. For example, in filter-binding (*see* Chapter 93), retention can be established by just one molecule of protein binding to the DNA. Moreover, gel shift assays can also resolve complexes having the same stoichiometry but different conformations, in which the DNA in the complex is either looped *(5–7,20)* or bent *(21,22)*.

1.1. Choice and Preparation of DNA Fragment

If specific binding is to be investigated and the potential binding sequence is known, the first requirement is a DNA fragment of suitable length that contains the site. It is not essential to radiolabel the DNA. After a gel shift experiment, the DNA can be visualized by staining the gel with ethidium bromide *(4)*. However, it is preferable to use low DNA concentrations, lower than can be detected with ethidium. This necessitates radiolabeled DNA followed by autoradiography. There is no minimum length of DNA below which it cannot be used for gel shifts, provided that it contains the complete binding site, although there is effectively a maximum length. If the DNA is >500 bp, only a small shift in mobility is caused by binding one molecule of protein, unless the protein is very large (such as DNA gyrase or RNA polymerase). However, some DNA–protein interactions involve several proteins associating with an array of binding sites and, in these cases, longer DNA fragments can be used in gel shift experiments. For the electrophoresis of large nucleoprotein assemblies, agarose, rather than polyacrylamide, is the preferred matrix *(23)*. Mixtures of polyacrylamide and agarose have also been used as the matrix *(24)*.

The minimum length of DNA will need to be determined experimentally, but usually it will be the target sequence and a few bp on either side. Short DNA molecules can be readily made in an oligonucleotide synthesizer and the synthetic oligonucleotides should be purified by electrophoresis or HPLC. Alternatively, if the target sequence is not known, it can be identified by gel shift experiments on oligonucleotides of random sequence *(25)*. Larger DNA fragments, 100–500 bp, are generally prepared by either restriction digests or PCR reactions on plasmid vectors. They are then isolated from the vector by electrophoresis through polyacrylamide *(26)*. If a restriction fragment is to be used, it is best prepared from a plasmid that has been purified on CsCl/ethidium gradients and it is advantageous to use restriction enzymes that leave 5'-extensions. These ends can be filled in by the Klenow fragment of DNA polymerase I in the presence of the appropriate α-[^{32}P]dNTP, with the other dNTPs as required *(26)*. A less attractive alternative is to dephosphorylate the DNA with alkaline phosphatase and then label it with polynucleotide kinase and γ-[^{32}P]ATP. The labeled DNA *must* be separated from the unincorporated label by gel filtration or by repeated precipitations of the DNA with ethanol *(26)*.

Virtually all gel shift experiments require not only the substrate DNA but also other DNA samples. As noted above, the specificity of a DNA–protein interaction can be examined by adding unlabeled competitor DNA to the radiolabeled substrate *(3,4,14)*. The competitor should not contain any copies of the specific binding site for the protein. A nonspecific DNA of known sequence is one possibility: Phage λ DNA and many plasmids may meet this requirement. Sheared *E. coli* or calf thymus DNA have also been used, as have polynucleotides and heparin *(12,27)*. Unnatural polynucleotides with simple repeat sequences, such as poly d(I-C), are particularly useful *(27)*. The amount of competitor DNA that needs to be added to remove nonspecific bands will depend on the specificity of the interaction. On the addition of a large excess of cold DNA, even specific complexes will dissociate because the sum of the affinities for all of the nonspecific binding sites will then be greater than the affinity for the specific site *(28)*. Another test for specificity is to use an isogenic DNA fragment that differs from the original fragment by 1 or 2 bp *(9,17)*.

Certain experiments demand further variations in the DNA fragments. For example, in the analysis of DNA bending by gel shift, a permuted series of DNA fragments will be needed, all of the same length but with the binding site for the protein at different locations relative to the center of the DNA molecule. Similarly, in the study of DNA looping interactions, where one protein binds simultaneously to two separate sites on the DNA, a series of DNA fragments may be needed with different lengths of DNA between the two sites *(5,20)*.

2. Materials

1. Standard apparatus for the electrophoresis of polyacrylamide gels and an electrophoresis dc power supply capable of 200 V. If possible, do not use apparatus that has been used previously for SDS-PAGE on proteins (*see* **Note 1**).
2. Electrophoresis running buffer (TBE): 0.089 M Tris base, 0.089 M boric acid, and 2 mM EDTA. Make this up at 5 times the required strength from 54 g of Tris base, 27.5 g of boric acid, 20 mL of 0.5 M EDTA, pH 8.0, and water to a final volume of 1 L: Dilute fivefold with water before use. (All water throughout this procedure is double-distilled; *see* **Note 2**.)
3. Acrylamide (29/1 acrylamide/N',N'-methylene *bis*-acrylamide). Make up a stock solution at 30% (w/v) from 290 g of acrylamide and 10 g of *bis*-acrylamide with water to bring the final volume to 1 L, then stir until dissolved. Add a couple of spoonfuls of a mixed-bed resin (such as Amberlite MBI, BDH), leave stirring for 30 min, filter, and store at 4°C (*see* **Note 3**). Also needed for the polyacrylamide gel are TEMED (N,N,N',N'-tetramethylene diamine), and a freshly prepared solution of ammonium persulfate (10% w/v in water).
4. ^{32}P-radiolabeled substrate DNA, and additional DNA samples as required for the type of gel shift experiment being carried out (*see* **Subheading 1.1.**). Depending on the nature of the substrate, the method chosen for its preparation and for the incorporation of the label, a wide range of the standard reagents and equipment for molecular biology will be needed at this stage (either the vector DNA with the fragment of interest and restriction endonucleases, or synthetic oligonucleotides, Klenow polymerase, both unlabeled and [^{32}P]-labeled nucleotide triphosphates, and so on). Standard procedures *(26)* can be followed in the preparation of both the substrate and control DNA samples.
5. The DNA-binding protein. (*See* **Note 4**.)
6. Binding buffer: For example, 50 mM Tris-HCl, pH 7.5, 100 mM NaCl, 10 mM β-mercaptoethanol, 0.1 mM EDTA, and 500 µg/mL of BSA (bovine serum albumin). To prepare this, make up a solution of Tris-HCl, NaCl, and EDTA, all at twice the required concentrations, and sterilize it by autoclaving. Once cool, add the β-mercaptoethanol to again give twice the required concentration: This solution is described below as 2X binding buffer. Prepare separately a stock solution of nuclease-free BSA at 5 mg/mL by first dissolving the protein (from any supplier) in 10 mM Tris-HCl, 0.1 mM EDTA, pH 7.5; readjust the pH to 7.5 and heat the solution to 67°C for at least 2 h; store at 4°C (*see* **Note 5**).

7. Loading buffer: The buffer added to samples prior to loading the gel is the same as binding buffer but supplemented with 40% (w/v) sucrose and 100 µg/mL bromophenol blue (*see* **Note 6**).
8. Gel fixer solution (10% v/v glacial acetic acid, 10% v/v methanol, 80% water), 3MM chromatography paper (Whatman), and a gel drier.
9. Autoradiography cassettes, preflashed X-ray film (Hyperfilm MP, Amersham International, Arlington Heights, IL), and developer for the autoradiograph (*see* **Note 7**).
10. If the autoradiographs are to be assessed quantitatively, a scanning densitometer may be required.

3. Methods
3.1. The Gel Shift Assay

1. Make the gel well in advance of performing the actual assay. First, both the gel tank and its component parts (glass plates, spacers, and comb) must be rigorously cleaned and dried, especially if the equipment has been used previously for SDS-PAGE. The gel plates should be cleaned with 95% ethanol and one plate should be "silanized" by coating it with a solution of 1% v/v dichlorodimethylsilane in $CHCl_3$ and allowing this to evaporate in the fume chamber.
2. Prepare a polyacrylamide gel in the appropriate buffer (*see* **Note 2**) by the standard procedure *(26)* (*see* **Note 8**). Use a comb that creates wells that are substantially larger than the volume of the sample so that the sample, once loaded in the well, occupies the minimal height above the gel matrix, that is, wide wells are better than narrow ones. The reason for this is that it facilitates the transfer of the DNA-protein complexes from the sample solution into the gel matrix, and it also minimizes streaking from the edges of the bands.
3. Allow the polyacrylamide gel to set for at least 2 h. However, it is advisable to remove the comb as soon as the gel has set, and thoroughly wash the wells with buffer. Then mount the gel in the gel tank and fill the chambers with electrophoresis buffer.
4. Prerun the gel at constant voltage, at the same voltage as will be used in the subsequent assay. The minimum length of time for prerunning is until the current becomes invariant with time. The gels are prerun first to remove all traces of the ammonium persulfate that was used in the polymerization of the gel; second, to distribute evenly throughout the gel any cofactors that may have been added to the electrophoresis buffer (especially if these were reagents that could not be added to the buffer used to make up the gel); and third, and most important, to ensure that the gel is at a constant temperature. The latter is absolutely essential for gels run in the coldroom: There is no point in working in the coldroom unless both the gel and the buffers are at thermal equilibrium before the samples are loaded. Just before loading the samples (*see below*, **step 8**), layer a solution of binding buffer across the wells in order to minimize the change in conditions that the sample undergoes when loaded on the gel.
5. Use a series of sterile 0.5-mL Eppendorf tubes for the binding reactions. Each reaction will have a final volume of 20 µL. In each tube, place 10 µL of 2X binding buffer, 2 µL of nuclease-free BSA (stock at 5 mg/mL), the desired amount of the [^{32}P]-labeled DNA substrate, and, if needed, the competitor DNA. In general, the amount of the radiolabeled DNA should be the minimum that gives a band that can be observed readily on the autoradiograph (this can be as little as 0.1 ng/tube). Before use, the stocks of both substrate and competitor DNA should be diluted with buffer so that the addition of the DNA causes the minimal change to the composition of the binding reaction: 10 mM Tris-HCl, 0.1 mM EDTA, pH 7.5, can be used for this dilution. Add sterile water so that the volume in each tube, after the subsequent addition of the protein, is 20 µL. Incubate the tubes in a water bath at the required temperature for a few minutes.

6. Add the DNA-binding protein to all but one of the tubes. Ideally, this should be the same small volume to each tube (typically, <10% of the assay volume), but at different dilutions of the protein (*see* **Note 9**). To the tube to which no protein was added, add the same volume of the buffer used for diluting the protein. The protein should obviously be diluted in a buffer in which it remains stable. However, less obvious is the fact that some DNA-binding proteins seem to be inactivated by "dilution shock" on the addition of a small volume of protein to a large volume of buffer. In these cases, use serial dilutions with small steps between each stage.
7. Thoroughly mix the solution of DNA and protein. This can be done by vortexing, but sometimes DNA-protein complexes that are observed by gel shift after gentle mixing cannot be detected after vortex mixing *(4)*. An alternative is to flush the sample repeatedly in a micropipet tip. The binding reaction is then allowed to equilibrate at the required temperature (*see* **Note 10**).
8. Add 8 μL of loading buffer to each sample and immediately load an aliquot (15-μL) from each on the polyacrylamide gel. The additional loading buffer, the loading itself, and starting the electrophoresis all need to be done as quickly as possible. One procedure is to start by gently placing the loading buffer on top of all of the separate samples. The buffer is viscous and will maintain its position above each sample for a short time. The individual tubes are then tapped to mix their contents immediately before loading. The loading is best done with plastic tips on a micropipet, using special (flexible) gel-loading tips if necessary to ensure an even layer across the bottom of each well. To minimize the time that the complexes take to enter the gel, some workers load the gel while it is running at a very low voltage. However, you can be electrocuted if you follow their procedure (*see* **Note 11**).
9. Turn up the electrophoresis power supply as quickly as possible after loading the samples on the gel. Run the gel, typically at 10 V/cm, until the free DNA is close to the bottom of the gel (*see* **Note 12**). The best voltage for a particular application, and the time of the run, are determined by trial and error. However, running the gel at too high a voltage will heat the gel excessively and denature the protein. Recirculation of the electrophoresis buffer can be advantageous, especially with buffers of low capacity, and it can be essential if cofactors are included in the buffer (*see* **Note 2**). The buffer should be recirculated at a flow rate of 50% of the total buffer vol/h.
10. After the gel is run, dismantle the apparatus. Remove the gel from the plates, place it in a bath of gel fixer for 30 min, then layer it onto Whatman 3MM paper and dry it on the gel dryer. The gel is finally autoradiographed with preflashed X-ray film.
11. If quantitative data are required, scan the autoradiograph in a densitometer. Determine the proportion of the free DNA and the bound complexes, relative to the total amount of DNA in each sample, by measuring the area of each peak in the densitometric scans. *See* **Note 13**.

3.2. Analysis of Results

If the gel shift assay reveals a single DNA–protein complex as a result of the binding of the protein to one specific site on the DNA (as in **Fig. 1**), the binding reaction between DNA (D) and protein (P) to form the complex (DP) can be described by:

$$D + P \leftrightarrow DP$$

The concentrations of bound and free DNA that are measured directly by the gel shift assay, $[D_b]$ and $[D_f]$ respectively, can then be related to an equilibrium constant:

$$K = [D_b]/\{[D_f] \times [P_o - D_b]\}$$

where $[P_o]$ is the total concentration of protein in the assay. Provided that the protein concentration is much higher than the concentration of the DNA substrate, so that $[P_o - D_b]$ is effectively equal to $[P_o]$, the increase in the amount of bound DNA with increasing levels of protein should follow a rectangular hyperbola. Hence, K can be evaluated from the variation in $[D_b]$ with $[P_o]$ either graphically (by an Eadie plot; $[D_b]$ against $[D_b]/[P_o]$) or, better, by using a computer program that fits data directly to rectangular hyperbolas (for example, ENZFITTER from Biosoft, Cambridge, UK). However, if any dissociation of the DNA-protein complex occurs during electrophoresis, the value for $[D_b]$ that is measured from the amount of DNA in the retarded complex (i.e., band B in **Fig. 1**) will be less than the true value for the complex. However, the value for $[D_f]$ measured from the amount of free DNA (i.e., band F in **Fig. 1**), will still be the true value: DNA that dissociates from the protein during the run cannot "catch up" with the DNA that was free at the start of the run. In this situation, it is better to evaluate K from the decrease in $[D_f]$ rather than the increase in $[D_b]$.

If the gel shift assay reveals multiple DNA-protein complexes (as in **Fig. 2**), it may be a result of nonspecific binding of the protein anywhere along the DNA. One test for this type of binding is to repeat the gel shift assay with DNA fragments of different lengths: The number of complexes resolved by gel shift should increase as the length of the DNA is increased. For example, the *Eco*RV restriction enzyme produces three complexes with a 55 bp DNA (the DNA bound to 1, 2, or 3 molecules of protein), six complexes with a 100 bp DNA, and >12 complexes with a 235 bp DNA *(17)*. The binding equilibria are then given by the equation:

$$D + P \leftrightarrow DP_1 \leftrightarrow DP_2 \leftrightarrow DP_3 \ldots DP_n$$

where n is the maximum number of protein molecules that can fit on the DNA. However, even though the concentrations of D_f, DP_1, DP_2, through to DP_n can all be measured by gel shift assays, the intrinsic equilibrium constant for the binding of the protein to an individual site on the DNA cannot be determined by the procedures used above for analyzing specific binding.

The analysis of nonspecific binding must take account of the following *(28)*: First, the DNA will contain a large number of binding sites for the protein (each bp is in effect the start of another binding site); second, since each molecule of protein will cover a certain number of bp, the sites overlap one another and only a fraction of them can be filled at any one time; third, the number of sites that are left after the binding of one or more molecules of protein depends on the location of the protein on the DNA (i.e., is the distance between the protein and the end of the DNA, or the distance between two protein molecules, long enough to accommodate another molecule of protein?); and fourth, if there are cooperative interactions between protein molecules on the DNA, the protein may bind preferentially to a site immediately adjacent to one already occupied by protein. One procedure for evaluating equilibrium constants from the concentrations of all of the complexes separated by gel shift has recently been published *(17)*.

4. Notes

1. Almost any apparatus for polyacrylamide gel electrophoresis, either homemade or commercial, can be used for gel shift assays. However, it is best to have one where the gel plates are virtually immersed in running buffer (for example, the Mini-Protean II system

from Bio-Rad, Hercules, CA), to prevent localized overheating in the gel. Additional features that are required for some, but not all, gel shift experiments are buffer recirculation and thermostatic control of the gel surface. However, the latter is not essential even in the analysis of complexes that are stable only at low temperatures. Running the gel in a cold room is often an adequate substitute.

2. The electrophoresis running buffer is a key factor in successful gel shift assays. Ideally, one wants a buffer in which the DNA–protein complex is so stable that none of it dissociates during electrophoresis. In many cases, the transfer of the DNA–protein complex from a binding buffer containing NaCl to a gel matrix in TBE achieves this goal. Other systems need electrophoresis buffers with lower ionic strengths, for example, 0.04 M Tris-acetate, 1 mM EDTA, pH 8.0. Moreover, the separation of the DNA–protein complex from the free DNA can depend on the pH of the electrophoresis buffer *(9)*. In addition, the buffer may require additional components to maintain the stability of the complex. For instance, a stable complex with a restriction enzyme demands the absence of Mg^{2+} ions, which can be ensured by having EDTA in the buffer *(17)*, whereas complexes with DNA gyrase are seen by gel shift only if the buffer contains Mg^{2+} *(29)*. Many proteins bind a cofactor before they bind specifically to DNA (for example, tryptophan with the *trp* repressor *[9]*), and these require their cofactor in the electrophoresis buffer. Buffer recirculation is then needed to maintain a constant concentration of cofactor throughout the gel.

3. *Acrylamide is a potent neurotoxin.* Always wear gloves when working with acrylamide and, when handling the dry powder, a face mask is also essential.

4. For certain types of gel shift experiments, it is essential that the protein is purified to homogeneity. One example is the analysis of the DNA sequence-specificity of a DNA-binding protein by using gel shifts to measure its binding to a series of DNA fragments that differ from each other by only 1 or 2 bp within the target sequence. Major artifacts can be generated by using this approach with partially purified preparations of the protein (*see* **ref. 30** for the elimination of one such artifact). But for other experiments, such as the identification of a protein that binds to a given DNA sequence, it is unnecessary to purify the protein. Indeed, as noted above (**Subheadings 1. and 1.1.**), the latter experiments can be carried out with unfractionated nuclear extracts *(14,27)*, provided that the target DNA is mixed with a suitable competitor, such as poly d(I-C).

5. The buffer for the binding reaction will obviously vary from one DNA-binding protein to another. Nearly all DNA–protein interactions are highly dependent on the salt concentration *(31)*; they can also vary with pH, and they often require some cofactor that must be added to the binding buffer as well as the electrophoresis buffer (*see above*, **Note 2**). The stability of the protein in different buffers will also influence the choice of binding buffer; in some cases, glycerol or spermidine must be added as well as BSA, in order to stabilize the protein.

6. The purpose of adding sucrose to binding buffer, to create loading buffer, is simply that this is the minimal perturbation to increase the density of the solution. This is to facilitate loading into the wells of the polyacrylamide gel and to prevent the binding reaction from mixing with electrophoresis buffer before the sample has entered the gel. The sucrose can be replaced by either glycerol or Ficoll400 (Pharmacia-LKB, Piscataway, NJ), and it is worthwhile using these to check that the binding equilibrium is not perturbed by sucrose. Moreover, if the binding buffer contains glycerol (*see* **Note 5**), it may be possible to dispense with loading buffer and load the samples directly on to the gel. Bromophenol blue is just a marker for electrophoresis and is not really necessary: In gel shift assays, electrophoresis should be continued until the free DNA is close to the end of the gel, and this must determined experimentally. The size of the DNA that comigrates with bromophenol blue will vary with the percentage of polyacrylamide in the gel *(26)*.

7. If a storage phosphor system *(32)* is available, use it in place of film autoradiography. It will provide quantitative data directly, by measuring the amount of radioactivity in each band on the gel, and it avoids the problems associated with nonlinear film response (*see* **Note 13**).
8. The gel strength (% acrylamide) for optimal resolution of complexes can only be found by trial and error. It will be a function of the sizes of both the DNA and the protein. For DNA fragments of >200 bp, initial experiments should be done with 5 or 6% gels. With short synthetic oligonucleotides, gels at 10 or 15% may be better, provided that the protein is not so large that the complex then cannot enter the gel. Avoid situations in which the complex sticks in the well—it should migrate a measurable distance into the gel. Another factor that can affect the mobility of DNA–protein complexes is the ratio of acrylamide to *bis*-acrylamide, and it may be necessary to alter the 29/1 ratio given here (*see* **Subheading 2.**, **step 3**).
9. If the binding assay is carried out as described here, by titrating a fixed amount of DNA with varied amounts of protein, then the range of protein concentrations should extend from the situation in which very little of the DNA is complexed with protein to that in which essentially all of the DNA is complexed (e.g., **Fig. 2**). As a first experiment, it is best to assay over a wide range of protein concentrations. Subsequent assays, once shifts have been observed, can then be carried out over a narrower range.
10. For kinetic experiments, the samples are subjected to electrophoresis as soon as the relevant time has elapsed after mixing the DNA with the protein. Generally, the association rates of proteins to specific sites on DNA are extremely rapid *(31)*, and, for equilibrium experiments, an incubation time of 15 min is usually sufficient. However, it is important to test that the system has reached equilibrium by carrying out assays with different incubation times. In some cases, it may be helpful to first incubate the sample at a higher temperature (e.g., 37°C) before reequilibrating it at a lower temperature.
11. If you chose to load the gel while it is running at low voltage, *take every possible precaution to avoid electrocution*. If it is to be attempted at all, wear dry plastic gloves and use only plastic tips. The voltage across the gel *must* be <1V/cm. *Do not, under any circumstances, attempt to load a gel at too high a voltage.*
12. The sample to which no protein has been added gives one marker for the free DNA. But in preliminary experiments, it is also advisable to run in a separate lane some DNA size markers to act as standards. Observed shifts can then be compared between different experiments.
13. If the autoradiographs are to be analyzed by densitometry, it is essential to work in the range where the film gives a linear response to the amount of radioactivity. This range can be established by running on a gel different amounts of a known sample of [^{32}P]-labeled DNA. On preflashed film, the areas under the peaks in the densitometric scans increase linearly with increasing amounts of DNA, but only up to a certain amount of DNA. Beyond that amount, successive increments of DNA produce proportionally smaller increments in peak areas. If the film is not preflashed, the scale will also be nonlinear with very low amounts of DNA. Only those band intensities that lie within the linear range should be used to obtain quantitative data. One way to ensure that the data are within the linear range is to autoradiograph the same gel for a number of different exposure times. Strong black bands from a long exposure allow for easy visualization, but densitometry generally requires faint bands from a short exposure.

References

1. Revzin, A. (1987) Gel electrophoresis assays for DNA-protein interactions. *BioTechniques* **7**, 346–355.
2. Fried, M. (1989) Measurement of protein-DNA interaction parameters by electrophoresis mobility shift assay. *Electrophoresis* **10**, 366–376.
3. Garner, M. and Revzin, A. (1981) A gel electrophoresis method for quantifying the binding of proteins to specific DNA regions: application to components of the *Escherichia coli* lactose operon regulatory system. *Nucleic Acids Res.* **9**, 3047–3060.
4. Fried, M. and Crothers, D. M. (1981) Equilibria and kinetics of lac repressor-operator interactions by polyacrylamide gel electrophoresis. *Nucleic Acids Res.* **9**, 6505–6525.
5. Krämer, H., Amouyal, M., Nordheim, A., and Müller-Hill, B. (1988) DNA supercoiling changes the spacing requirement of two *lac* operators for DNA loop formation with lac repressor. *EMBO J.* **7**, 547–556.
6. Lobell, R. B. and Schleif, R. F. (1990) DNA looping and unlooping by *ara*C protein. *Science (Washington, DC)* **250**, 528–532.
7. Fried, M. and Crothers, D. M. (1983) CAP and RNA polymerase interactions with the *lac* promoter: binding stoichiometry and long range effects. *Nucleic Acids Res.* **11**, 141–148.
8. Fried, M. and Crothers, D. M. (1984) Kinetics and mechanism in the reaction of gene regulatory proteins with DNA. *J. Mol. Biol.* **172**, 263–282.
9. Carey, J. (1988) Gel retardation at low pH resolves *trp* repressor-DNA complexes for quantitative study. *Proc. Natl. Acad. Sci. USA* **85**, 957–979.
10. Ceglarek, J. A. and Revzin, A. (1989) Studies of DNA-protein interactions by gel electrophoresis. *Electrophoresis* **10**, 360–365.
11. Revzin, A., Ceglarek, J., and Garner, M. (1986) Comparison of nucleic-acid interactions in solution and in polyacrylamide gels. *Anal. Biochem.* **153**, 172–177.
12. Shanblatt, S. and Revzin, A. (1984) Kinetics of RNA polymerase-promoter complex formation: effects of nonspecific DNA-protein interactions. *Nucleic Acids Res.* **12**, 5287–5306.
13. Avila, P., Ackroyd, A. J., and Halford, S. E. (1990) DNA binding by mutants of Tn*21* resolvase with DNA recognition functions from Tn*3* resolvase. *J. Mol. Biol.* **216**, 645–655.
14. Strauss, F. and Varshavsky, A. (1984) A protein binds to a satellite DNA repeat at three specific sites that would be brought into mutual proximity by DNA folding in the nucleosome. *Cell* **37**, 889–901.
15. Kristie, T. M. and Roizman, B. (1986) α4, the major regulatory protein of herpes simplex virus type 1, is stably and specifically associated with promoter-regulatory domains of α genes and of selected other viral genes. *Proc. Natl. Acad. Sci. USA* **83**, 3218–3222.
16. Hudson, J. M., Crowe, L., and Fried, M. (1990) A new DNA binding mode for CAP. *J. Biol. Chem.* **265**, 3219–3225.
17. Taylor, J. D., Badcoe, I. G., Clarke, A. R., and Halford, S. E. (1991) *Eco*RV restriction endonuclease binds all DNA sequences with equal affinity. *Biochemistry* **30**, 8743–8753.
18. Hendrickson, W. and Schleif, R. (1985). A dimer of *ara*C protein contacts three adjacent major groove regions of the *araI* DNA site. *Proc. Natl. Acad. Sci. USA* **82**, 3129–3133.
19. Hoess, R., Abremski, K., Irwin, S., Kendall, M., and Mack, A. (1990). DNA specificity of the *cre* recombinase resides in the 25 kDa carboxyl domain of the protein. *J. Mol. Biol.* **216**, 873–882.
20. Krämer, H., Niemöller, M., Amouyal, M., Revet, B., von Wilcken-Bergmann, B., and Müller-Hill, B (1987) *Lac* repressor forms loops with linear DNA carrying two suitably spaced *lac* operators. *EMBO J.* **6**, 1481–1491.

21. Wu, H-M. and Crothers, D. M. (1984) The locus of sequence directed and protein-induced DNA bending. *Nature* **308,** 509–513.
22. Liu-Johnson, H-N., Gartenberg, M. R., and Crothers, D. M. (1986) The DNA binding domain and bending angle of *E. coli* CAP protein. *Cell* **47,** 995–1005.
23. Berman, J., Eisenberg, S., and Tye, B. K. (1987) An agarose gel electrophoresis assay for the detection of DNA-binding activities in yeast cell extracts. *Meth. Enzymol.* **155,** 528–537.
24. Topol, J., Ruben, D. M., and Parker, C. S. (1985) Sequences required for in vitro transcriptional activation of a *Drosophila* hsp 70 gene. *Cell* **42,** 527–537.
25. Pollock, R. and Treisman, R. (1990) A sensitive method for the determination of protein-DNA binding specificity. *Nucleic Acids Res.* **18,** 6197–6204.
26. Sambrook, J., Fritsch, E. F., and Maniatis, T. (1989) *Molecular Cloning, A Laboratory Manual*, 2nd Ed. Cold Spring Harbor Laboratory, Cold Spring Harbor, NY.
27. Varshavsky, A. (1987) Electrophoretic assay for DNA-binding proteins. *Meth. Enzymol.* **151,** 551–565.
28. McGhee, J. D. and von Hippel, P. H. (1972) Theoretical aspects of DNA-protein interactions: cooperative and noncooperative binding of large ligands to a one-dimensional homogeneous lattice. *J. Mol. Biol.* **86,** 469–489.
29. Maxwell, A. and Gellert, M. (1984) The DNA dependence of the ATPase activity of DNA gyrase. *J. Biol. Chem.* **259,** 14,472–14,480.
30. Carey, J., Lewis, D. E. A., Lavoie, T. A., and Yang, J. (1991) How does trp repressor bind to its operator? *J. Biol. Chem.* **266,** 24,509–24,513.
31. Lohman, T. M. (1986) Kinetics of protein-nucleic acid interactions: use of salt effects to probe mechanisms of interaction. *CRC Crit. Rev. Biochem.* **19,** 191–245.
32. Johnston, R. F., Pickett, S. C., and Barker, D. L. (1990) Autoradiography using storage phosphor technology. *Electrophoresis* **11,** 355–360.

95

Yeast Two-Hybrid Library Screening

Ian G. Cowell

1. Introduction

The two-hybrid system was originally devised by Fields and Song as a protein interaction detection system in yeast *(1)*. Subsequently, it has been employed in many laboratories as a means of screening cDNA and genomic fusion libraries for protein interaction partners *(2–8)*. The method relies on the fact that transcription factors such as the yeast GAL4 factor consist of separable DNA-binding and transcriptional regulatory domains, the former being required to direct the latter to appropriate promoters where transcriptional activation is effected, usually by direct or indirect interaction of the activation domain with the general transcription machinery. The essence of the two-hybrid system is the in vivo reconstitution of a functional transcriptional activator from two interacting polypeptides, one fused to a sequence-specific DNA binding domain and the other to a potent transcriptional activation domain that is detected by the activation of a reporter gene or genes (*see* **Fig. 1**). For the purposes of identifying new protein partners, DNA encoding the polypeptide for which partners are sought (bait polypeptide) is ligated into a yeast shuttle vector to create a fusion protein with the DNA binding domain of GAL4 (*see* **Fig. 2A**) or sometimes the bacterial LexA DNA binding protein *(6–10)*. Library cDNA is ligated into a second shuttle vector to create an activation domain-tagged cDNA library (*see* **Fig. 2B**). In the original method as proposed by Chien et al. *(2)*, the bait (GAL4 DNA binding domain) construct and activation domain fusion library were cotransformed into a yeast strain containing an integrated *lacZ* gene driven by a GAL4-responsive promoter *(2)*. Yeast-containing plasmids encoding interacting polypeptides were detected by the presence of β-galactosidase activity (blue colonies in the presence of the chromogenic substrate X-gal). The method was subsequently developed to facilitate the screening of large mammalian cDNA libraries *(11)*. This innovation employs a *his3* test yeast strain (that is auxotrophic for histidine) that contains in addition to the *lacZ* reporter, an integrated *HIS3* gene that is transcribed at a significant level only in the presence of the reconstituted activator (*see* **Fig. 1**), thus allowing selection of positives on minimal plates lacking histidine.

From: *The Nucleic Acid Protocols Handbook*
Edited by: R. Rapley © Humana Press Inc., Totowa, NJ

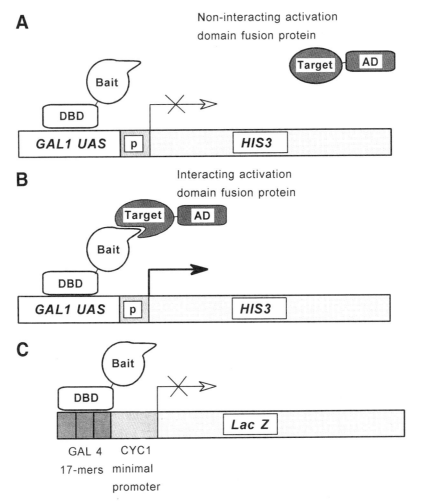

Fig. 1. The principle of the yeast two-hybrid system. The two-hybrid system is a method of detecting specific protein-protein interactions between two polypeptides in vivo. The "bait" and "target" polypeptides are fused to the DNA binding and transcriptional activation domains of the transcription factor GAL4, respectively (other DNA binding and activation domains have been used). In vivo interaction of bait and target polypeptides results in formation of a functional activator and activation of GAL4 responsive reporter genes. Yeast strains used for two-hybrid analysis such as YPB2 and HF7c contain two integrated GAL4-responsive reporter genes *HIS3* and *lacZ*. Activation of the *HIS3* reporter gene confers the ability to grow on media lacking histidine. (**A**) In the situation where bait and target proteins do not interact, the bait polypeptide is targeted to the promoter of the *HIS3* reporter gene through specific protein-DNA interactions between the GAL4 DNA-binding domain (DBD) and the GAL1 UAS, but the activation domain is not recruited to the promoter, resulting in no transcription of the *HIS3* gene. (**B**) Where there is a functional interaction between bait and target polypeptides, a functional activator is formed on the GAL1 UAS. *HIS3* expression ensues resulting in histidine prototrophy. (**C**) The *lacZ* reporter gene in strains such as YPB2 or HF7c is driven by an artificial GAL4-dependent promoter. Interaction of bait and target polypeptides results in expression of β-galactosidase activity that can easily be assayed.

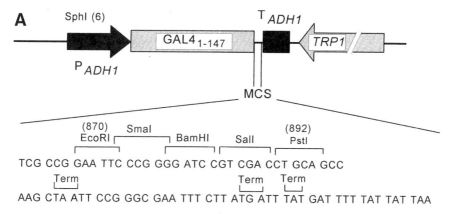

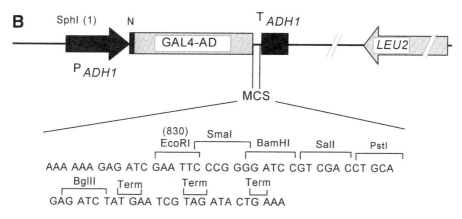

Fig. 2. Cloning sites in the DNA-binding domain fusion vector pGBT9 and activation domain vector pGAD424. (**A**) The region of pGBT9 around the segment encoding the DNA binding domain of GAL4 (residues 1–147). P$_{ADH1}$, *ADH1* promoter; MCS, multicloning site; T$_{ADH1}$, *ADH1* transcriptional terminator. The *TRP1* auxotrophic marker gene is also indicated. The first TCG codon in the sequence in the lower part of the figure corresponds to residue 147 of GAL4. Restriction sites are indicated as are termination codons (Term). (**B**) The region of pGAD424 around the segment encoding the GAL4 activation domain (residues 768–881). The first GAG codon in the sequence corresponds to codon 881 of GAL4. The segment designated N indicates the SV40 large T nuclear localization sequence engineered into the N-terminus of GAL4 *(2)*.

2. Materials

2.1. Maintenance of Yeast Strains

1. A solution of 40% glucose (dextrose) in water. Autoclave to sterilize.
2. YPD (yeast, peptone, dextrose) medium: Per liter: 20 g Peptone, 10 g yeast extract. Adjust pH if necessary to pH 5.8–pH 6.5. Add 15 g agar for plates. Autoclave and add 50 mL of glucose (dextrose) from a sterile 40% stock.
3. YPDA. Prepare YPD as in **step 2** and after autoclaving add adenine sulphate to 40 mg/L from a filter sterilized stock.
4. Yeast nitrogen base (Difco, Detroit, MI, cat. no. 0919-15-3).

Table 1
10X Dropout Solution

Component	Milligrams/liter
L-Isoleucine	300
L-Valine	1500
Adenine hemisulphate	200
L-Arginine-HCl	200
L-Histidine-HCl	200
L-Leucine	1000
L-Lysine-HCl	300
L-Methionine	200
L-Phenylalanine	500
L-Threonine	2000
L-tryptophan	200
L-Tyrosine	300
Uracil	200

5. Amino acids: L-tryptophan, L-histidine-HCl, L-arginine-HCl, L-methionine, L-tyrosine, L-leucine, L-isoleucine, L-lysine-HCl, L-phenylalanine, L-glutamic acid, L-aspartic acid, L-valine, L-threonine, L-serine (all obtainable from Sigma Chemical Co., St. Louis, MO).
6. Adenine sulphate (Sigma, cat. no. A2196).
7. Uracil.
8. 10X Dropout solution: Add the components in **steps 5–7** in the proportions given in **Table 1**. Omitting the appropriate ingredient (*see* **Subheading 3.1.**) Dissolve in MilliQ water (Millipore, Bedford, MA) and autoclave to sterilize. Store for up to 1 yr at 4°C.
9. Synthetic (SD) medium and SD agar (per liter of medium): dissolve 6.7 g Difco yeast nitrogen base in 850 mL MilliQ deionized water and add 20 g agar (omit the agar for liquid media). Autoclave at 120°C, at 15 lb, for no more that 15 min. Add 100 mL of the appropriate 10X dropout solution (*see* **Table 1**) and 50 mL 40% glucose. For agar medium, pour plates on the thick side and leave on the bench the right way up with lids on for 2 to 3 d to dry. Plates may then be wrapped and stored for up to several months at 4°C. If no short autoclave cycle is available, for SD agar make up the nitrogen base and agar in 425 mL of water each and autoclave separately, combine while still hot, and then add 100 mL 10X dropout solution and 50 mL 40% glucose.

2.2. Yeast Phenotype Verification

1. Two-hybrid yeast strains (*see* **Note 1**).
2. SD plates lacking trp, leu, his, or ura (*see* **Subheadings 2.1. and 3.1.**).
3. YPDA and YPD plates.
4. Disinfectant-free toothpicks.

2.3. Yeast Transformation and Library Screening

2.3.1. Transformation with Bait Plasmid

1. Fresh plate of (verified phenotype) yeast such as HF7c (*see* **Note 1**).
2. YPD (*see* **Subheading 2.1.**).
3. Sterile water.
4. 50% Polyethylene glycol (PEG) 4000: Make 200–500 mL in water, autoclave.

Yeast Two-Hybrid Library Screening

5. 10X TE: 100 mM Tris, pH 7.8, 10 mM ethylenediaminetetraacetic acid (EDTA). Half fill a number of glass universal bottles and autoclave.
6. Lithium acetate: 1 M Prepared in water. Adjust to pH 7.5 with dilute acetic acid and autoclave.
7. 1X TE/LiAc: Mix 1 mL of 10X TE and 1 mL of 1 M LiAc in a 30-mL sterile universal bottle and add 8 mL sterile water.
8. PEG/LiAc: Mix 1 mL of 10X TE, 1 mL of 1 M LiAc, and 8 mL 50% PEG in a 30-mL sterile universal bottle.
9. Dimethylsulfoxide (DMSO).
10. 10 mg/mL Sonicated salmon sperm DNA.
11. 9 cm SD plates without tryptophan (–T plates), see **Subheading 2.1.**
12. DNA Binding domain (bait) fusion construct (see **Note 2**).

2.3.2. Yeast Transformation with Activation Domain Library

1. Amplified activation domain library (see **Note 3**).
2. YPD Medium, 2.5 L (see **Subheading 2.1.**).
3. –TL Medium, 1.5 L (see **Subheadings 2.1.** and **3.1.**).
4. –TLH Medium, 250 mL (see **Subheadings 2.1.** and **3.1.**).
5. 1X TE/LiAc, 20 mL (see **Subheading 2.3.1.**).
6. PEG/LiAc, 140 mL: Prepare fresh by combining in a sterile container: 14 mL 10X TE, 14 mL 1 M LiAc, 112 mL 50% PEG.
7. Sterile water.
8. Several 9 cm –TL plates.
9. 10 mg/mL Salmon sperm DNA.

2.3.3. Library Plating

1. Several 9 cm –TL plates.
2. At least 50 15 cm –THL plates (see **Note 4**).
3. 1 M 3-Aminotriazole made up in water. Prepare 10–20 mL and filter sterilize (see **Note 4**).

2.4. Further Analysis and Elimination of False Positives

2.4.1. Gridding Out HIS$^+$ Transformants

1. –THL plates (9 cm).
2. 7 cm Filter paper circles (Whatman #1, Whatman, Maidstone, Kent, UK). Imprinted with a grid of 1 cm squares. Number in pencil, wrap in an aluminium envelope, and autoclave to sterilize.
3. Sterile toothpicks.

2.4.2. β-Galactosidase Filter Assays

1. X-Gal (5-bromo-4-chloro-3-indolyl-β-D-galactopyranoside): dissolved in N,N-dimethylformamide (DMF) at 20 mg/mL. Store in 1 mL aliquots in 1.5 mL microfuge tubes at –20°C.
2. Z Buffer: 16.1 g/L Disodium hydrogen phosphate, 7H$_2$O, 5.5 g/L sodium dihydrogen phosphate, 1H$_2$O, 0.75 g/L potassium chloride, 0.25 g/L magnesium sulphate, 7H$_2$O. Make 1 L. Autoclave to sterilize.
3. Working X-gal solution: 10 mL Z Buffer, 167 µL X-gal stock, 27 µL β-mercaptoethanol.

2.4.3. Curing Transformant of the Bait Plasmid

1. –L SD Medium, 500 mL.
2. –L Plates.
3. –LT Plates.

3. Methods

It is advisable for workers unfamiliar with yeast manipulations to read Chapter 1 of volume 194 of Methods in Enzymology "Getting started with yeast" *(12)*.

3.1. Maintenance of Yeast Strains and Transformants

To select for and maintain yeast plasmid transformants, yeast strains with characteristic auxotrophic markers are grown on synthetic media lacking the cognate nutrient. The required enzymatic activity is then provided by a plasmid-borne marker gene. Common transformation markers include *trp1*, *leu2*, *his3*, and *ura3* affecting tryptophan, leucine, histidine, and uracil biosynthesis, respectively. Synthetic medium (SD) consists of a minimal medium or nitrogen base plus glucose combined with a "dropout solution," which contains amino acid and nucleotide supplements (*see* **Table 1**). One or more of the dropout solution components are omitted to make a selective medium. In this chapter, SD lacking leucine will be referred to as –L medium, that lacking tryptophan as –T medium, that lacking both leucine and tryptophan –TL medium and so on.

3.2. Yeast Phenotype Verification

Before embarking on transformation and library screening it is worthwhile checking the phenotype of the yeast strain to be used. We have used the strain HF7c obtained commercially from Clontech, Palo Alto, CA (*see* **Note 1**). A number of other strains have been used by other workers (*see* **Note 1**). HF7c is characterized by *trp1*, *leu2*, and *his3* auxotrophic markers and contains a second *HIS3* gene driven by the GAL4 responsive *GAL1* promoter. This strain also contains an inserted *lacZ* reporter gene driven by an artificial GAL4-responsive promoter.

1. Using a sterile inoculating loop, scrape some cells from the top of a frozen glycerol stock (*see* **Note 5**) and streak them onto a YPDA plate. Incubate at 30°C for 2–3 d until colonies about 1–2 mm across have formed.
2. Take four or five well-isolated colonies and, using sterile toothpicks, streak some of the cells from each colony onto a sector of a second YPD or YPDA plate, and in parallel onto –Ura, –Trp, –Leu, and –His synthetic-medium plates and on –His plates containing 2 mM 3-AT.
3. Incubate at 30°C for 3–4 d. Cells will grow on the YPD plate and on –Ura plates (if the strain being used is *URA+*) but should not grow on either –Trp or –Leu plate. Some slow growth may occur on –His plates owing to leaky expression of the *HIS3* reporter gene. However, no growth should occur on the plates containing 3-AT.
4. Colonies displaying the expected phenotype should be used as the stock for the library screening described in **Subheading 3.3.**

3.3. Yeast Transformation and Library Screening

Construction of the bait plasmid encoding the DNA-binding domain fusion is mentioned here only in brief. DNA binding domain fusion vectors are discussed in **Note 2** and **Table 2**. The following are written assuming the use of a bait and activation library plasmids carrying *trp1* and *leu2* selection markers, respectively (*see* **Tables 2** and **3**), and a *trp1*, *leu2* reporter strain such as HF7c (*see* **Table 4**) carrying a *HIS3* reporter gene, allowing preliminary screening for positive two-hybrid interaction by selection for histidine prototrophy.

Table 2
DNA-Binding Domain Fusion Vectors Used in the Two-Hybrid System

Plasmid	Size	DNA-binding domain	Selection	Useful cloning sites	Other characteristics	Refs.
pMA424	12.0 kb	GAL4	HIS3	EcoRI, BamHI, SalI		(9)
pGBT9[a]	5.4 kb	GAL4	TRP1	EcoR1, SmaI, BamHI, SalI, PstI		(11)
pAS2[b]	8.5 kb	GAL4	TRP1	NdeI, NcoI, SmaI, BamHI, SalI, PstI	Contains CYH^S2 gene conferring cycloheximide sensitivity on transformed cells	(4,13)
pBTM116	5.4 kb	LexA	TRP1	EcoR1, SmaI, BamHI, SalI, PstI		(11)

[a]Available commercially from Clontech and licensed from the Research Foundation of the State University of New York, Albany.
[b]Available commercially from Clontech and licensed from Baylor College, Houston, TX.

Table 3
Activation Domain Fusion Vectors Used in the Two-Hybrid System

Plasmid	Size	Activation domain	Selection	Useful cloning sites	Other characteristics	Refs.
pGAD2F	13.00 kb	GAL4[a]	LEU2	BamHI	pGAD1F, 2F, and 3F contain the BamHI site in three different frames with respect to the activation domain	(2)
pGAD10	6.60 kb	GAL4[a]	LEU2	BglII, XhoI, BamHI, EcoRI		(11)
pGAD424[b]	6.60 kb	GAL4[a]	LEU2	EcoRI, SmaI, BamHI, SalI, PstI, BglII		(9,11)
pJG4-5		B42	TRP1	EcRI, XhoI, HindIII	Fusion protein expression is inducible with galactose, reducing potential toxicity problems	(6)
pVP16	8.10 kb	VP16	LEU2	BamHI, NotI		(9)
pACT	7.65 kb	GAL4[a]	LEU2	BglII, EcoRI, BamHI	Automatic excision from the λ vector λ-ACT	(4)

[a]A nuclear localization signal was engineered at the N terminus of the GAL4 activation domain in the GAD series (2) and λ-ACT (4).
[b]Available commercially from Clontech and licensed from the Research Foundation of the State University of New York, Albany.

Table 4
Yeast Reporter Strains Used for Two-Hybrid Library Screening[a]

Strain	Mating type	Transformation markers	Reporter genes	Refs.
YPB2	MATa	trp1, leu2	$GAL1_{UAS}$-$LEU2_{TATA}$-HIS3, $(GAL\ 17\text{-}mers)_3$-$CYC1_{TATA}$-lacZ	(11)
CTY1	MATα	trp1, leu2	GAL1-lacZ, GAL1-HIS3	(11)
Y185	MATa	trp1, leu2	GAL1-lacZ, GAL1-HIS3	(4)
Y190	MATa	trp1, leu2	GAL1-lacZ, GAL1-HIS3	(13)
HF7c[b]	MATa	trp1, leu2	$GAL1_{UAS\text{-}TATA}$-HIS3, $(GAL\ 17\text{-}mers)_3$-$CYC1_{TATA}$-lacZ	(14)
CG-1945[b]	MATa	trp1, leu2	$GAL1_{UAS\text{-}TATA}$-HIS3, $(GAL\ 17\text{-}mers)_3$-$CYC1_{TATA}$-lacZ	
L40	MATa	trp1, leu2	$(LexAop)_4$-HIS3, $(LexAop)_8$-lacZ	(7,9)
Y187[c]	MATα	trp1, leu2, his3	GAL1-lacZ	(13)

[a]Only strains with a two-hybrid activated reporter gene conferring a means of selection for positive interaction are listed.
[b]Available commercially from Clontech and are licensed from the Research Foundation of the State University of New York.
[c]Available commercially from Clontech, licensed from Baylor University. Used in yeast mating assay, see **Note 12**.

3.3.1. Transformation with Bait Plasmid (see **Note 2**)

This protocol is derived from the published methods of Schiestl and Gietz *(15)* and Gietz et al. *(16)*.

1. Use a single colony of phenotype-verified reporter strain such as HF7c (*see* **Table 4**) to inoculate 10 mL of YPD. Shake overnight at 30°C.
2. Measure the OD_{600} of the culture, which should be about 1.5–2 and dilute the overnight culture into 50 mL YPD at 30°C to give a final OD_{600} of 0.2.
3. Incubate for a further 3 h at 30°C.
4. Pellet the cells by centrifuging at $1000g$ for 3 min using a benchtop centrifuge.
5. Resuspend the cells in 25 mL sterile water and pellet as in **step 4**.
6. Gently resuspend the cells in 0.6 mL TE/LiAc.
7. For each transformation add to a sterile 1.5 mL microfuge tube 0.5 µg plasmid DNA, and 100 µg sonicated salmon sperm DNA, Mix and add 100 µL yeast cell suspension from **step 6**.
8. Mix gently but thoroughly and add 600 µL PEG/LiAc solution. Vortex to mix and place the tubes in a 30°C incubator for 30 min. Mix gently several times during this incubation.
9. Add 70 µL DMSO to each tube, mix gently, and place the tubes in a 42°C water bath for 15 min.
10. Briefly chill on ice and pellet cells by spinning for 5 s in a microfuge at high speed.
11. Aspirate the supernatant and resuspend the cells in 500 µL sterile TE and spread the cells onto appropriate selective plates (Trp minus, –T for most bait plasmid/reporter strain combinations). Spread 100–200 µL/9 cm plate or 200–400 µL/15 cm plate.
12. Incubate for 2–4 d for colonies to appear.
13. Keep stocks of the transformed strain on selective medium plates. Use one of the *TRP+* colonies to prepare competent cells for large-scale transformation as described in **Subheading 3.3.2.** It is also advisable to make a glycerol stock of the transformed strain (*see* **Note 4**).

Before going any further it is advisable to carry out some preliminary tests with the bait construct as described in **Note 6**.

3.3.2. Yeast Transformation with Activation Domain Library

This protocol is derived from the published methods of Schiestl and Gietz *(15)* and Gietz et al. *(16)*.

1. Inoculate one of the Trp+ colonies from **Subheading 3.3.1.** into 5 mL selective –T medium. Incubate at 30°C overnight with shaking.
2. Inoculate the whole 5 mL culture into 300 mL prewarmed –T medium in a 1-L flask and shake for 24 h at 30°C (*see* **Note 7**).
3. Measure the OD_{600}, which should be approximately 1.2–1.5, and add sufficient culture to each of two 2.5-L flasks, each containing 500 mL of YPD, to attain an OD_{600} of 0.3.
4. Shake at 30°C for 4 h.
5. Pellet the cells by centrifugation at $2000g$ for 5 min using a Beckman JA-10 or similar rotor.
6. Decant the supernatant and resuspend the cell pellet in 500 mL water.
7. Pellet the cells as in **step 5**, decant the supernatant and resuspend the washed cells in 20 mL TE/LiAc solution and transfer to a sterile 250-mL flask.
8. Add premixed 1 mL denatured salmon sperm DNA (10 mg/mL) and up to 500 µg of activation library plasmid DNA and mix (*see* **Note 3**).

9. Add 140 mL PEG/LiAc solution and mix again.
10. Incubate at 30°C for 30 min with shaking.
11. Transfer to a sterile 2-L flask and add 17.6 mL DMSO. Swirl to mix while adding the DMSO.
12. Stand the flask in a 42°C water bath for 6 min to heat shock the cells, swirling occasionally.
13. Transfer the flask to a water bath at 20°C to cool.
14. Pellet the cells at 2000g. Decant the supernatant and gently resuspend the cells in 200 mL YPD.
15. Pellet the cells as in **step 14** and resuspend in 1 L of the same medium prewarmed to 30°C and shake at that temperature for 1 h.
16. Remove 1 mL of the cell suspension from **step 15**, wash the cells once in –TL medium, and resuspend in 1 mL of the same medium. Plate 1, 10, and 100 µL of the cell suspension on –TL plates to determine the primary transformation efficiency (*see* **Note 8**).
17. Pellet the remainder of the cells as in **step 5** and wash with 500 mL –TL medium.
18. Resuspend the cell pellet in 1 L warm –TL medium and incubate with gentle shaking for 4 h.
19. Pellet the cells as in **step 5** and wash twice in 100 mL –TLH medium
20. Resuspend the final pellet in 10 mL of the same medium.

3.3.3. Library Plating

1. To estimate the final transformation and plating efficiency, dilute 10 µL of the cell suspension (from **Subheading 3.3.2.**, **step 20**) into 1 mL of TE and spread 1, 10, and 100 µL onto single 9-cm or 15-cm –TL plates (*see* **Note 8**).
2. Spread 100 µL of the cell suspension from **step 20** in **Subheading 3.3.2.** on each of a series of –THL plates. This requires at least 50 plates to screen enough colonies to cover all of the primary transformants (*see* **Note 9**).
3. Leave the plates right-side up on the bench with lids on for 1–2 h for any surplus liquid to dry. Transfer plates, lid-side down to a 30°C incubator.
4. HIS^+ colonies should appear after 2–4 d, although up to 10 d may be required for weak positives to appear.

3.4. Further Analysis and Elimination of False Positives

False positives are a big problem in two-hybrid screening. Assuming that the bait construct has already been tested to check that it lacks activation potential alone (*see* **Note 6**), false positives may arise through fortuitous binding of the activation domain fusion to the test promoter, nonspecific binding to the bait, or through one of a number of less well-understood routes. The first step after HIS^+ colonies appear is to grid them onto fresh –THL plates before testing for β-galactosidase activity (*see* **Subheadings 3.4.1.** and **3.4.2.**). The *lacZ* reporter in strain HF7c for example is driven by a different GAL4 responsive promoter than the *GAL1-HIS3* reporter. His^+ and $LacZ^+$ colonies are then carried forward for further analysis. Because it is possible for more than one activation domain fusion plasmid to be present in each β-gal positive colony, such colonies should be streaked onto –TL plates to allow segregation and then be retested for β-galactosidase activity.

The next stage is to check that reporter gene activity requires both activation domain and DNA binding domain constructs. This is achieved by curing the tentative positives of the DNA binding domain plasmid and retesting the resulting *trp* cells for β-galactosidase activity and/or His prototrophy. Only transformants that are *trp*, *LEU*, and β-galactosidase⁻ are candidates for genuine positives. A rapid method for curing the bait plasmid has been devised *(13)* using the bait vector pAS2 (*see* **Table 2**), which contains a gene conferring cycloheximide sensitivity on resistant reporter strains such as Y190 *(13)* or CG-1945. Plating on -L plates containing cycloheximide then selects for bait plasmid loss.

Further testing then involves retransformation of the *trp*, *LEU*, and β-galactosidase⁻ activation domain fusion-containing strains with the original bait plasmid along with a series of controls including the original bait vector and one or more different fusion constructs, which would not be expected to interact with the same target. An ideal control would be a mutant bait domain, known to be inactive in some way. The retransformed strains would then be tested again for β-galactosidase activity or His prototrophy. Those positives that are β-gal positive when retransformed with the bait plasmid but not with the controls are candidates for real positives.

The final step is to isolate the activation domain plasmid DNA and to confirm the protein-protein interaction by independent means.

3.4.1. Gridding Positive Colonies

1. Lay a series of 7 cm sterile gridded filter papers (*see* **Subheading 2.4.1.**) directly onto a set of –THL plates.
2. Using sterile toothpicks, grid His⁺ colonies from **Subheading 3.3.3.** onto the plates from **step 1** and onto a second set of –THL plates, using the same pattern for both sets. The second set of plates will be the master set of tentative positives.
3. Incubate at 30°C for 2–4 d until colonies or small blobs of cells have appeared on the filter papers.

3.4.2. β-Galactosidase Filter Assay

1. Using flat-ended forceps, carefully lift the filters from **Subheading 3.4.1.** away from the agar plates one at a time and float, colony-side up, in a pool of liquid nitrogen. After a few seconds submerge the filter for 10 s and then remove from the liquid nitrogen and place the filter colony-side up on a tissue paper to thaw (*see* **Note 10**).
2. Once the filter is thawed, carefully place it on a second filter paper soaked in freshly prepared Z buffer containing X-gal (*see* **Subheading 2.4.2.**).
3. Lift the filter sandwich colony-side up onto a 9-cm petri dish, replace the lid, and incubate at 30°C for 30 min to several days (*see* **Note 11**).
4. β-Galactosidase-positive colonies can be matched with the colonies on the master plates from their grid positions and can be carried forward for further analysis.

3.4.3. Curing Bait Plasmid (see also [13])

Unless using a strain/vector combination that allows selection for loss of the bait construct (*see* **Subheading 3.4.** and **ref. 13**), transformants may be cured of the bait plasmid as described in **steps 1–4** below.

1. For each β-gal positive transformant, set up an overnight culture in 2 mL –L medium. Use one drop of the overnight culture to inoculate a further 2 mL of –L medium and incubate overnight again.
2. Take 100 µL of the second overnight culture and dilute to 1 mL with sterile water and plate out 50–100 µL of the culture onto a series of –L plates.
3. Using sterile toothpicks grid out 20–30 colonies from each plate onto –L and –TL plates.
4. Transformants that grow on –L but fail to grow on –TL plates are cured of the bait plasmid (which carries a *TRP* marker gene). These transformants should be retested for β-galactosidase activity and those that are now negative in this assay should be saved for further verification (*see* **Subheadings 3.4.2.** and **3.4.3.**).

3.4.4. Recovery of Plasmid DNA from Transformed Yeast

Further analysis and identification of cDNA inserts requires isolation of activation domain plasmid DNA. Only plasmid DNA heavily contaminated with genomic DNA and other impurities can easily be isolated from yeast, therefore plasmid DNA is shuttled into a suitable *Escherichia coli* strain for preparation. *E. coli* strains such as HB101 are useful leucine auxotrophs owing to the *leuB* mutation, which can be complemented by the yeast *LEU2* gene and so activation domain plasmids with LEU2 as their transformation marker can be selected for by plating transformed *E. coli* cells on minimal plates lacking leucine. A method for the recovery from yeast of plasmid DNA suitable for transformation of *E. coli* can be found in **ref. *(17)***. Electroporation or high-efficiency chemical means should be used for *E. coli* transformation with this material.

4. Notes

1. Two of the most commonly used reporter strains for two-hybrid library screening are *YPB2(11)* (*MATa, ura3-52, his3-200, ade2-101, lys2-801, trp1-109, leu2-3, 112, gal4-542, gal80-358, LYS2::GAL1$_{UAS}$-LEU2$_{TATA}$-HIS3, URA3::(GAL4 17-mers)$_3$-CY1C$_{TATA}$-LacZ* and HF7c(14) (*MATa, ura3-52, his3-200, ade2-101, lys2-801, trp1-109, leu2-3, 112, gal4-542, gal80-358, LYS2::GAL1$_{UAS,TATA}$-HIS3, URA3::(GAL4 17-mers)$_3$-CYC1$_{TATA}$-LacZ*) (*see* **Table 1**), both constructed for use with GAL4 DNA binding domain bait plasmids. The *HIS3* reporter gene in HF7c reportedly has a lower level of background expression (*see also* **Note 4**). For bait constructs using the LexA DNA-binding domain, the strain L40 has been described *(7,9)* (partial genotype: *MATa, trp1, leu2, his3, LYS::lexA-HIS3, URA3::lexA-lacZ*). Some of the other yeast strains that have been used for two-hybrid library screening are listed in **Table 4** and **ref. *11***.

2. **Table 2** lists some published DNA-binding domain fusion plasmids (bait plasmids). The important features of each of theses plasmids are as follows. (a) A yeast transformation marker (*TRP1* in most cases), (b) a DNA segment encoding sequence-specific DNA binding domain (the DNA-binding domain from the yeast transcription factor GAL4 in most cases) driven by a constitutive promoter such as the ADH1 promoter and placed immediately upstream of a series of restriction sites for cloning of insert cDNAs, (c) the origin of replication from the yeast 2 μ plasmid, (d) The *bla* gene for ampicillin resistance in *E. coli* and an *E. coli* origin. The plasmid pAS2 *(13)* possesses a number of additional features (*see* **Table 2**). The sequence of the multiple cloning sites in pGBT9 *(11)* are shown in **Fig. 2A**; the multicloning sites of pGBT9 and pAS2 *(13)* are identical between the *Sma*I and *Pst*II sites, but the pAS2 MCS also contains *Nde*I and *Nco*I sites (containing in-frame ATG codons) just upstream of the *Sma*I site (*see* **ref. 13**).

3. The quality of the activation domain-tagged cDNA library is obviously of central importance for successful two-hybrid screening. It is outside the scope of this chapter to describe in detail the construction of such libraries beyond the points made in this paragraph. First, two approaches can be made in the construction of activation domain-tagged plasmid libraries. The first is to generate a relatively large amount of cDNA from the target tissue or cell line. Sufficiently large libraries may then be generated by direct ligation into the activation domain vector (*see* Hannon et al. *[3]*, for example). Following plasmid transformation by electroporation, transformed *E. coli* are plated onto a large number of 15 cm Luria-Bertani-amp plates (approx 50–100) ideally at about 1X 10^5 colonies per plate. After overnight incubation, cells are collected by scraping them from the plates and plasmid DNA prepared by standard alkaline lysis *(18)*. In the second method *(4)*, a library is constructed in the λ-phage vector λ-ACT, which contains an embedded copy of the two-

hybrid activation domain-tagged plasmid pACT. In vivo excision allows the direct generation of the pACT activation domain library that may be amplified as described in the previous part of this paragraph. Whichever method is used for activation domain-tagged library construction, it is argued that random primed first-strand cDNA synthesis is preferable, as this eliminates biases toward C-terminal sequences. Rather than generating libraries themselves, many workers will obtain libraries from other workers or from commercial sources (Clontech has in their catalog a number of activation domain libraries constructed largely in pGAD10). Plasmid libraries obtained commercially or from other workers will usually require amplification prior to yeast transformation (for which 500 µg of plasmid DNA is required per transformation). Caution should be exercised here, and *E. coli* transformation should aim to generate at least as many colonies as the estimated original complexity of the library. A high-efficiency transformation procedure such as electroporation should be used, plating the results of several independent transformations containing 10–50 ng of library DNA each on a large number of 15 cm plates.

4. The *HIS3* reporter gene in strains such as YPB2 (*see* **Table 4**) is leaky resulting in slow growth on –TLH plates in the absence of GAL4 activity. This can produce a significant background when the transformed two-hybrid library is plated. Background growth can be eliminated by the addition of 5 m*M* to 20 m*M* 3-aminotriazole (3-AT) to plates. The reporter strains HF7c and CG1945 *(14)* contain a slightly different *HIS3* reporter (*see* **Table 4**) and suffer less this way. Nevertheless, use of 3-AT may still be necessary to fully eliminate troublesome background growth of colonies (*see also* **Note 6**).

5. Glycerol stocks of yeast strains are prepared by adding to 0.5 mL of a midlog culture 0.5 mL of YPDA or appropriate SD medium containing 20% glycerol. The glycerol-containing culture can then be stored in cryogenic storage vials at –80°C. Glycerol stocks remain viable for in excess of 1 yr at this temperature.

6. Prior to library screening, the bait construct should be tested for fortuitous transcriptional activation properties in the absence of interacting activation domain-tagged plasmid. Perform a series of small scale transformations as in **Subheading 3.3.1.** using (a) the bait construct alone, (b) the bait construct and the activation domain vector from which the library to be screened was constructed, and (c) the empty bait vector alone. Plate out on appropriate selective plates and test for β-galactosidase activity as in **Subheadings 3.4.1.** and **3.4.2**. In addition, grid the transformants on plates lacking histidine to check for fortuitous activation of the *HIS3* reporter gene. None of the transformants should give significant reporter activity although some slow growth on the minus histidine plates may occur after several days in which case the use of 3-AT is advisable. To determine the concentration of 3-AT to use, transformants from (b) should be gridded on –TLH plates containing 0, 2, 5, 10, or 20 m*M* 3-AT and the lowest concentration that inhibits background growth should be used for library screening. Use of higher 3-AT concentrations may prevent the detection of weak positives.

7. The cells should be at approx midlog phase at this point. If the OD_{600} is below 1.2, continue incubation until the optical density falls into this range.

8. The primary transformation efficiency using this method should be in excess of 1X 10^4 colonies per microgram of plasmid DNA. The protocol given here should therefore give greater than 5 million transformants. The protocol includes a 4-h recovery period. This time can be optimized for different strains. Because some growth will occur during this period, a second series of aliquots should be plated after this recovery period to determine the plating efficiency.

9. This is a large number of plates, but do not be tempted to plate more cells per plate as growth of HIS^+ colonies is inhibited by too dense plating. Even at 100 µL per plate, each plate will receive 50,000 transformants.

10. Freeze thawing permeablizes the cells.
11. Strong signals may appear in 1 h, whereas weak positives will take at least overnight for β-galactosidase activity to be apparent.

References

1. Fields, S. and Song, O. (1989) A novel genetic system to detect protein-protein interactions. *Nature* **340**, 245–246.
2. Chien, C. T., Bartel, P. L., Sternglanz, R., and Fields, S. (1991) The two-hybrid system: a method to identify and clone genes for proteins that interact with a protein of interest. *Proc. Natl. Acad. Sci. USA* **88**, 9578–9582.
3. Hannon, G. J., Demetrick, D., and Beach, D. (1993) Isolation of the Rb-related p130 through its interaction with Cdk2 and cyclins. *Genes Dev.* **7**, 2378–2391.
4. Durfee, T., Becherer, K., Chen, P. L., Yeh, S. H., Yang, Y., Kilburn, A. E., Lee, W. H., and Elledge, S. J. (1993) The retinoblastoma protein associates with the protein phosphatase type 1 catalytic subunit. *Genes Dev.* **7**, 555–569.
5. Hardy, C. F. J., Sussel, L., and Shore, D. (1992) A RAP1-interacting protein involved in transcriptional silencing and telomere length regulation. *Genes Dev.* **6**, 801–814.
6. Gyuris, J., Golemis, E., Cherkov, H., and Brent, R. (1993) Cdi1, a human G1 and S phase protein phosphatase that associates with Cdk2. *Cell* **75**, 791–803.
7. Vojtek, A. B., Hollenberg, S. M., and Cooper, J. A. (1993) Mammalian ras interacts directly with the serine/threonine kinase Raf. *Cell* **74**, 205–214.
8. Dunaief, J. L., Strober, B. E., Khavari, P. E., Ålin, K., Luban, J., Begemann, M., Crabtree, G. R., and Goff, S. P. (1994) The retinoblastoma protein and BRG1 form a complex and cooperate to induce cell cycle arrest. *Cell* **79**, 119–130.
9. Hollenberg, S. M., Sternglanz, R., Gheng, P. F., and Weintraub, H. (1995) Identification of a new family of tissue-specific basic helix-loop-helix proteins with a two-hybrid system. *Mol. Cell Biol.* **15**, 3813–3822.
10. Dalton, S. and Triesman, R. (1992) Characterisation of SAP-1, a protein recruited by serum response factor to the c-fos serum response element. *Cell* **68**, 597–612.
11. Bartel, P. L., Chien, C. T., Sternglanz, R., and Fields, S. (1993) Using the two hybrid system to detect protein-protein interactions, in *Cellular Interactions in Development: A Practical Approach.* (Hartley, D. A., ed.) Oxford University Press, Oxford, UK, pp. 153–179.
12. Sherman, F. (1991) Getting started with yeast. *Methods Enzymol.* **194**, 3–21.
13. Harper, J. W., Adami, G. R., Wei, N., Keyomarsi, K., and Elledge, S. (1993) The p21 Cdk-interacting protein Cip is a potent inhibitor of G1 cyclin-dependent kinases. *Cell* **75**, 805–816.
14. Feilotter, H. E., Hannon, G. J., Ruddel, C. J., and Beach, D. (1994) Construction of an improved host strain for two hybrid-screening. *Nucleic Acids Res.* **22**, 1502–1503.
15. Schiestl, R. H. and Gietz, R. D. (1989) High efficiency transformation of intact cells using single-stranded nucleic acids as a carrier. *Curr. Genet.* **16**, 339–346.
16. Gietz, D., St. Jean, Woods, R. A., and Schiestl, R. H. (1992) Improved method for high efficiency transformation of intact yeast cells. *Nucleic Acids Res.* **20**, 1425
17. Hoffman, C. S. and Winston, F. (1987) A ten-minute DNA preparation from yeast efficiently releases autonomous plasmids for transformation of Escherichia coli. *Gene* **57**, 267–272.
18. Sambrook, J., Frisch, E. F., and Maniatis, T. (1989) *Molecular Cloning: A Laboratory Manual*, Cold Spring Harbor Laboratory Press, Cold Spring Harbor, NY.

96

The Southwestern Assay

Jacques Philippe

1. Introduction

Determination of cellular phenotypes results from the expression of a limited number of genes whose products interact to establish a unique environment. The mechanisms by which individual cells can selectively express only a few of all the genes in a specific cell have been the focus of intense research during the last 10 yr. It has become apparent that developmental, tissue-specific, and hormone-regulated gene expression is, for the most part, controlled at the level of transcriptional initiation. This involves the interaction of specific DNA binding proteins with control elements present in the gene promoters. To better understand the process of gene transcription, characterization of these DNA binding proteins is a mandatory step.

Detection of sequence-specific DNA binding proteins has been achieved by a variety of techniques, including nitrocellulose filter binding *(1)*, DNase I footprinting *(2)* (*see* Chapter 92), and the electrophoretic mobility shift assay (gel retardation; **ref. 3**) (*see* Chapter 94). These techniques, however, are limited since they do not inform on the proteins themselves. To this end alternative methodologies have been developed to directly characterize DNA-binding proteins. Ultraviolet light can be used to crosslink binding proteins to labeled DNA elements *(4)*. The molecular weight of the protein can then be estimated by SDS polyacrylamide gel electrophoresis (SDS-PAGE). This chapter describes a second approach that avoids the need to crosslink the protein–DNA complex, and hence any inaccuracies of size determination resulting from this process. This procedure was first developed by Bowden et al. *(5)* and since it closely follows the principles of the Western assay, it has been referred to as the Southwestern assay. In this technique, crude protein extracts are first separated by SDS-PAGE and blotted onto a nitrocellulose membrane. Specific DNA binding proteins are detected by incubating the membrane with a labeled DNA probe that is derived from the protein recognition site. This technique has now been refined into a simple, reliable, and versatile assay, and is commonly used not only to characterize nuclear proteins from cellular extracts, but also as a preliminary step to set up conditions for *in situ* detection of DNA-binding proteins expressed by recombinant bacteriophages *(6)* or to select DNA sequences recognized by a specific DNA-binding protein. By analogy, other procedures derived from the same principles, but based on protein–protein or protein–RNA interactions, have been established *(5,8; see* **Notes**).

2. Materials

All solutions should be made with distilled, deionized water.

1. Running gel solution: 33.5% acrylamide/0.3% *bis*-acrylamide (*see* **Note 1**). Bring 33.5 g of acrylamide and 0.3 g of *N,N'* methylene*bis*-acrylamide to 100 mL. Filter and store at 4°C. Protect from light. The solution can be kept refrigerated for at least 3 mo. Wear a mask and gloves to weigh and handle acrylamide and *bis*-acrylamide since they are potent neurotoxic agents that may be absorbed through the skin.
2. Running gel buffer: 1 M Tris-HCl, pH 9.1. Dissolve 12.1 g of Tris-base in 80 mL. Adjust pH to 9.1 with concentrated HCl and make the volume up to 100 mL. Filter and refrigerate.
3. 10% SDS: Wear a mask to handle SDS powder.
4. 10% ammonium persulfate: 0.5 g of ammonium persulfate to 5 mL. Store at 4°C for up to 3 wk; for longer storage aliquot and freeze at –20°C.
5. TEMED.
6. Stacking gel solution: 30% acrylamide/0.44% *bis*-acrylamide. Bring 3 g of acrylamide and 0.44 g of *N,N'* methylene*bis*-acrylamide to 10 mL. Filter and store at 4°C. Protect from light.
7. Stacking gel buffer: 0.5 M Tris-HCl, pH 6.2. Dissolve 1.5 g of Tris-base, pH 6.8, with concentrated HCl.
8. Tank buffer: 25 mM Tris-base, 250 mM glycine, 0.1% SDS. Dissolve 12 g of Tris-base, 57.6 g of glycine, and 4 g of SDS in 4 L. It is not necessary to pH this solution.
9. 2X sample buffer: 125 mM Tris-HCl, pH 6.8, 4% SDS, 20% glycerol, 1.44 M β-mercaptoethanol, 0.1% bromophenol blue. For 10 mL, use 2.5 mL of 0.5 M Tris-HCl, pH 6.8, 4 mL of 10% SDS, 2 mL of glycerol, 1 mL of β-mercaptoethanol, and a few crystals of bromophenol blue; make up to volume with water. Store at –20°C.
10. Suitable electrophoresis apparatus: A vertical electrophoresis apparatus. Gels are poured between two clamped plates. These are available from many manufacturers, e.g., Hoefer (San Francisco, CA) and Bio-Rad (Hercules, CA) (*see* **Note 2**).

2.2. Transfer of Proteins

11. Nitrocellulose membrane: Good quality nitrocellulose, e.g., BA 85, Schleicher and Schuell (*see* **Note 3**).
12. Transfer tank buffer: 25 mM Tris-base, 192 mM glycine, 20% v/v methanol. Dissolve 12.1 g of Tris-base, 57.6 g of glycine in 3 L of water. Add 800 mL of methanol and bring to 4 L with water.
13. Transfer apparatus: A rectangular tank with a large network of platinum electrodes on each side to generate an even field. The gel/nitrocellulose sandwich is held vertically in place between two sponge pads by tight-fitting cassettes (*see* **Fig. 1**). These are available from many manufacturers, e.g., Hoefer and Bio-Rad.

2.3. Probe Labeling

14. Probe DNA: Two complementary synthetic oligonucleotides that anneal to provide the recognition site for the protein under investigation (*see* **Note 4**). Dilute to 2 ng/μL and store at –20°C.
15. γ-[^{32}P]ATP: specific activity >3000 Ci/mmol.
16. 10X T4 kinase buffer *(8)*: 200 mM Tris-HCl, pH 7.6, 50 mM MgCl$_2$, 50 mM DTT, and 500 μg/mL BSA (DNase-free). Store in small aliquots at –20°C.
17. T4 polynucleotide kinase: 10 U/μL. Store at –20°C.
18. 10X Ligase buffer: 500 mM Tris-HCl, pH 7.6, 100 mM MgCl$_2$, 100 mM DTT, 10 mM ATP, and 500 μg/mL BSA (DNase-free). Store in small aliquots at –20°C.

The Southwestern Assay

19. T4 DNA ligase: 1 U/mL. Store at –20°C.
20. G50 Sephadex spun column: (*see* **Note 5**).

2.4. Probe Binding

21. 10X Binding buffer: 250 m*M* HEPES, pH 7.9, 30 m*M* MgCl$_2$, 500 m*M* KCl, 1 m*M* DTT. To make a 10X stock solution, use 25 mL of 1 *M* HEPES (titrated to pH 7.9 with 1 *M* NaOH), 3 mL of 1 *M* MgCl$_2$, and 50 mL of 1 *M* KCl; adjust to 100 mL with water. Store at 4°C. Add DTT to the diluted solution just before use.
22. Denaturation buffer: 6 *M* guanidine hydrochloride in 1X binding buffer. For a 100-mL solution, dissolve 57.3 g in 1X binding buffer, but add DTT just before use.
23. Blocking buffer: 5% nonfat dry milk in 1X binding buffer.
24. Poly (dI · dC): A competitor used to block nonspecific DNA binding. Store at –20°C as a 1 mg/mL stock (*see* **Note 6**).
25. Specific and nonspecific oligonucleotide competitors: Synthetic oligonucleotides made as for the probe (*see* **Subheading 3.3.**) but with unlabeled ATP and at 50-fold the concentration, i.e., 100 ng/µL (*see* **Note 7**).
26. Probe solution: 0.25% nonfat milk and 10 µg/mL of poly(dI · dC) in 1X binding buffer.

3. Methods

Proteins from many sources can be analyzed by the Southwestern assay (*see* **Note 8**). Loadings should be adjusted according to the source. I routinely load 75 µg of crude nuclear extract per electrophoretic lane, but this may vary depending on the source of the extract, the abundance of the binding protein, and the affinity of the protein for the DNA sequence. If very high concentrations of extract are used, preliminary gels should be run to ensure that overloading of the gel does not occur.

To confirm the specificity of the proteins detected using this assay, it is necessary to demonstrate that binding is abolished by specific competitors but unaffected by those that are known to be nonspecific. For each probe it is therefore necessary to run at least three lanes of protein extract. The nitrocellulose membrane is cut into strips for each lane; the first lane is incubated with probe alone, the second with probe and specific competitor, and the third with probe and a nonspecific competitor.

3.1. SDS Gel Electrophoresis

1. Mix 12 mL of running gel solution, 15.2 mL of running gel buffer, 11.2 mL of water, 0.4 mL of 10% SDS, and 0.33 mL of 10% ammonium persulfate. Just before pouring, add 20 µL of TEMED and swirl the components. Pour between the glass plates of the gel apparatus, but leave the last 4 cm for the stacking gel. Overlay the mix with water to ensure even polymerization. After approx 30 min the gel should have polymerized. Pour off the overlaying water and dry with an absorbent paper towel.
2. Mix 1.3 mL of stacking gel solution, 2.5 mL of stacking gel buffer, 6.1 mL of water, 0.1 mL of 10% SDS, and 33 µL of 10% ammonium persulfate. Just before pouring, add 10 µL of TEMED. Fill the remaining space above the running gel and insert the comb. Once the gel has polymerized, remove the comb, wash out the wells with water, and assemble the gel apparatus. Ensure that no bubbles are trapped at the bottom of the gel when the tank buffer is added. Do not prerun the gel before loading.
3. Typically each sample should contain 75 µg of protein. Dilute each nuclear extract to 10 µL with the appropriate extraction buffer and mix with 10 µL of 2X sample buffer. These samples can be scaled up for the number of gel loadings, e.g., for a triple loading 60 µL may be prepared. Prepare two samples containing marker proteins of known molecular weights.

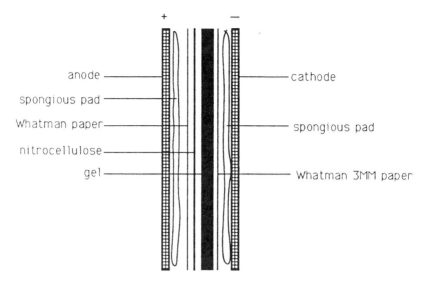

Fig. 1. Arrangement for electrophoretic transfer: + and – designate the polarity of the plastic supports containing the gel/nitrocellulose sandwich. These supports are oriented toward the anode and cathode of the electrophoretic transfer tank, respectively.

4. Boil the samples for 3 min to dissociate proteins and inactivate proteases (*see* **Note 9**). Insoluble material should be removed by centrifugation at 12,000*g* for 5 min to avoid streaking during electrophoresis.
5. Load 20 µL for each sample onto the gel (*see* **Note 10**). Apply a voltage of 8 V/cm for the run through the stacking gel and increase to 15 V/cm for the running gel. Run at room temperature. The run should take 4 h and is complete when the marker dyes reach the bottom.
6. Remove the glass plates and mark the orientation of the gel by cutting off one corner.

3.2. Transfer of Proteins

Wear gloves through the rest of the procedure.

1. Equilibrate the gel in transfer buffer for 30 min. The size of the gel may change slightly owing to the methanol in the buffer.
2. While the gel is equilibrating, cut the nitrocellulose membrane and two pieces of Whatman 3MM paper to the size of the gel. Wet these and the sponge pads of the transfer apparatus in transfer buffer.
3. Assemble the transfer apparatus as follows (*see* **Fig. 1**). Lay the cathode flat and onto it pace one of the sponge pads. On top place a sheet of Whatman 3MM paper, followed by the gel, the nitrocellulose membrane, the second sheet of Whatman 3MM paper, the second sponge pad, and finally the anode. Using a glass pipet as a roller, squeeze out any trapped air bubbles. Immerse the sandwich into the transfer chamber filled with buffer. Ensure the nitrocellulose membrane is between the gel and the anode.
4. Although voltage and current readings depend on many variables, for a standard size gel, transfer with a constant voltage of 30 V (initial current of approx 180 mA) for 14 h at 4°C. Cool and stir constantly (*see* **Note 11**).
5. After transfer rinse the filter in water and dry at room temperature before DNA binding (*see* **Note 12**).

3.3. DNA Labeling

The probe should ideally contain multiple copies of the control element to enhance sensitivity. The multiple sites allow a single molecule to interact with more than one protein and alleviate problems of rapid probe dissociation. The probe is generated by ligating labeled oligonucleotides to create a concatemer (*see* **Note 13**).

Since the probe is labeled with ^{32}P, necessary precautions should be taken when handling it.

1. Add 5 µL (10 ng) of each complementary oligonucleotide to 3 µL of 10X kinase buffer, 6 µL of γ-[^{32}P]ATP, and 10 µL of water. Add 1 µL of T4 polynucleotide kinase and incubate at 37°C for 1 h.
2. Place the reaction in a beaker containing 500 mL of water at 90°C and then place the beaker in ice. Let it stand for 2–3 h until the water temperature reaches 20°C.
3. Add 3.3 µL of 10X ligation buffer and 1 µL of T4 DNA ligase. Incubate overnight at 15°C.
4. Purify the catenated DNA from unincorporated label using a G50 Sephadex spun column (*see* **Note 5**).

This should produce a concatenated probe suitable for high affinity binding to the immobilized DNA binding protein (*see* **Note 14**).

3.4. DNA Binding

The proteins blotted onto nitrocellulose are denatured and then renatured. This is thought to help appropriate folding or multimerization of the denatured proteins to occur and reduce nonspecific DNA interactions (*see* **Fig. 2A** and **Notes 15–17**).

1. Cut the membrane into strips corresponding to each electrophoretic lane and immerse in denaturation solution. Choose a small glass or plastic dish to spare solutions; 20–40 mL is adequate. Agitate gently for 10 min at 4°C.
2. Dilute with the same volume of 1X binding buffer and shake for 5 min at 4°C. This dilutes the solution to 3 M guanadine HCl.
3. Repeat step 2 four times to give 1.5 M, 0.75 M, 0.38 M, and 0.185 M guanadine HCl sequential dilutions. Agitate each time for 5 min at 4°C. Finally wash in 1X binding buffer.
4. Place the membrane in 40 mL of blocking buffer and agitate gently for 1 h at 4°C.
5. Add the labeled probe to sufficient volume of probe solution for binding. This should be carried out in the smallest volume possible (0.1–0.2 mL/cm^2). By using heat sealable bags for each strip of membrane, the volume of solution can be minimized. A 50-fold excess of specific or nonspecific competitors should be included in the solutions of the relevant strips. Agitate gently for 2 h at 4°C.
6. Wash membrane strips for 5 min in 100 mL of 1X binding buffer at 4°C. Repeat twice. Dry strips on Whatman paper for 1–2 min and autoradiograph for at least 12 h.

Figure 2 shows a typical result using this technique. To maximize the sensitivity of the assay it is important to carry out each step efficiently. Problems with probe synthesis, protein transfer, or renaturation of the protein will certainly reduce the sensitivity of the assay. Some problems, however, may be inherent to the protein under investigation (*see* **Notes 9–11**). A major problem encountered is if the protein is a heterodimer and therefore unable to reform the DNA binding domain after separation by SDS-PAGE (*see* **Note 18**). There are a number of other applications of the technique discussed in **Note 19**.

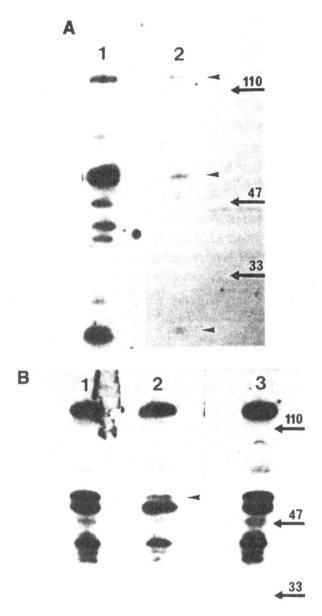

Fig. 2. Southwestern assay. Denaturation/renaturation: 75 µg nuclear extracts were separated by SDS-PAGE, transferred, and probed with a concatenated 42-bp element from the rat glucagon gene promoter. The mol wt of the protein markers are indicated (A) Lane 1; the membrane was not subject to denaturation/renaturation. Lane 2; the membrane underwent denaturation/renaturation with guanidine hydrochloride, as described in **Subheading 3.4.** The denaturation/renaturation treatment reduced the ten bands detected in lane 1 to the three specific bands in lane 2 (marked by arrow point). (B) Competition: To illustrate the effectiveness of competition, the membranes were not subject to denaturation/renaturation treatment but probed with the same probe as in (A). A 50-fold excess of multimerized specific competitor (unlabeled wildtype DNA) was added to probe for lane 2 and 50-fold excess of multimerized nonspecific competitor (a mutant element) was added to probe for lane 3. A single band is systematically and reproducibly competed (arrowhead).

4. Notes

1. Both the percentage of polyacrylamide and the ratio of acrylamide to *bis*-acrylamide need to be determined to optimize separation and transfer of proteins, particularly for high molecular weight proteins. The lower the total monomer the higher the porosity of the gel. In this protocol I use a 10% gel with an acrylamide:*bis*-acrylamide ratio of 110:1; this resolves proteins of molecular weights from 10–200 kDa *(10)*.
2. Polyacrylamide gels for protein electrophoresis can be run on 1.5-mm thick standard (16 × 18 cm) or mini (8 × 10 cm) gels. The recipe is given for a standard size gel; the volume should be scaled down according to the gel size.
3. Although nylon membranes have a higher protein binding capacity than nitrocellulose, they give unacceptably high background even with increasing the nonfat milk concentration of the blocking buffer to 5%. If the background is unacceptable with nonfat milk (at 0.25–5% concentrations), it may be necessary to optimize the signal-to-noise ratio by using lipid-free BSA as a blocking agent *(1)*.
4. The sequence of the oligonucleotides used for the probe must first have been established by other means, e.g., deletion analysis and footprinting studies. They should be synthesized so that after annealing there is a complementary sequence overhang (or "sticky end"). This aids the concatemerization process described in **Subheading 3.3.**
5. G50 Sephadex spun columns are used to purify the radiolabeled DNA. This can be prepared in a 1-mL syringe or bought ready prepared (e.g., Boehringer Mannheim, Indianapolis, IN). The elution limit for double-stranded DNA is less than 70-bp and recovery is greater than 90%. Alternatives are to use ion exchange columns (e.g., Elutip, Schleicher and Schuell) or to precipitate the DNA using ethanol.
6. DNA from many sources, e.g., Salmon sperm DNA, may also be used as a nonspecific block. Again, use at a final concentration of 10 µg/mL.
7. To confirm the identity of the signals detected in the assay, it is necessary to determine their specificity for the probe. Both specific and nonspecific competition must be carried out (*see* **Fig. 2B**). At its simplest the specific competitor can be unlabeled probe. Additional specific competitors, derived from related recognition sites, should be used if available. Nonspecific competitors at their best should correspond to mutants of the recognition site that are known not to bind the protein. If not available, oligonucleotides unrelated to the recognition site should be used.

    The competitors should be synthesized in exactly the same manner as the probe. In this protocol, competitors are used at a 50-fold excess. It is possible that this may vary for different binding proteins.
8. One of the major advantages of the Southwestern assay resides in its power to discriminate a specific DNA-binding protein among many thousands. Whole cell, nuclear and, in the case of the glucocorticoid receptor, cytosolic extracts have been used as the starting source of the proteins. Other sources, such as fusion proteins that contain DNA binding domains expressed by recombinant bacteriophage or truncated proteins synthesized from cloned cDNA in vitro or in bacteria, are also perfectly suitable.
9. During sample preparation, the boiling step might cause some proteins to precipitate and be lost from the assay. This step therefore may be omitted if the Southwestern assay is unsuccessful.
10. Some nuclear nonhistone proteins have a tendency to form aggregates and remain at the top of the gel. Addition of 8 M urea before loading may sometimes allow them to enter the gel.
11. Efficient protein transfer is critical for good results. Different factors can be modified to optimize transfer efficiency.

a. Gel composition (see **Note 1**).
b. Transfer time and current. These variables may have to be adapted for complete transfer. Overnight applications with high voltages should only be attempted with caution and heating should be minimized. High currents can be used satisfactorily for short transfer times (e.g., 0.5 A for 2–3 h). Reducing the buffer strength may allow the use of higher voltages.
c. Buffer type. The standard buffer can adversely affect transfer of basic proteins. These may require pH 9.5–10 to elute from the gel. In this case, methanol may have to be omitted, but this can diminish binding to the nitrocellulose membrane. Similarly, addition of 0.1% SDS to the standard buffer may help elution of proteins from the gel but may reduce binding efficiency to the membrane and also increase the relative current and cause heating.

12. The transfer efficiency may be checked by staining the gel and a strip on the membrane containing one of the marker lanes with Coomassie blue. Do this by fixing the gel for 30 min in 100 mL of gel fix (50% methanol, 10% acetic acid), stain for 30 min in 100 mL of stain solution (0.2% Coomassie blue, 50% methanol, 10% acetic acid), and destain in 100 mL of 10% methanol, 10% acetic acid for three changes of 10 min each. The gel should be dried and the filter rinsed in distilled water.

13. In this protocol the concatenated DNA is synthesized by simple ligation of monomers. It also may be made by subcloning the monomers into a vector. Probes generated in this second way must be labeled by nick translation, but should be of a higher specific activity.

14. To check the efficiency of the ligation reaction, run an aliquot of the labeling reaction on a nondenaturing polyacrylamide gel. Dry the gel down and autoradiograph. This will indicate the degree of concatemerization. Multimers of between 3 and 5 oligonucleotides should be sufficient to give a good probe.

15. The renaturation procedure described in the protocol is based on conditions established for *in situ* detection of sequence specific DNA binding fusion proteins from λgt11 expression libraries *(6)*. Alternative protocols to renature proteins have been described in the literature, but it is unclear to what extent these procedures allow renaturation and whether they may result in better protein DNA interactions.

16. Protein renaturation can also be carried out directly in the gel before transfer. After electrophoresis, the gel is incubated in 200 mL of renaturation buffer (50 mM NaCl, 10 mM Tris-HCl, pH 7.2, 0.2 mM EDTA, 0.1 mM DTT, and 4 M urea) with gentle agitation for 2 h at room temperature. This incubation will remove SDS from the proteins, but the extent of renaturation is unknown *(12)*. The gel is then equilibrated in transfer buffer for 30 min. The proteins are electrophoretically transferred onto a nitrocellulose membrane and directly placed in blocking buffer. After urea treatment, it has been observed that there is a decrease in transfer efficiency (personal observation; *13*). This may be owing to the reduced solubility of the protein samples when SDS is replaced by urea resulting in proteins becoming trapped in the gel. Compared to the guanidine-HCl treatment, more nonspecific protein–DNA binding is observed after urea incubation.

17. For some binding proteins, a specific renaturation step may not be necessary since removal of SDS from proteins during transfer may be sufficient to allow the proteins to renature *(14)*.

18. Failure to detect a specific DNA binding protein may indicate that the protein is a heterodimer. Dissociation of the subunits in the process of SDS-PAGE may destroy the binding activity. This possibility may be tested by addition of nuclear or cellular extracts during the binding step. If a subunit or cofactor is required for high affinity binding the addition of extract may restore the binding *(15)*.

19. Variants to the Southwestern assay:
    a. Two-dimensional Southwestern assay *(10)*: Characterization of DNA-binding proteins can also be attempted after two-dimensional electrophoresis. Distinct isoelectric point variants can thus be occasionally revealed. Two-dimensional electrophoresis separates proteins by charge in the first dimension and by size in the second, allowing discrimination of proteins of similar size but varying charge. Protein transfer, renaturation, and DNA binding are conducted as in the standard protocol.
    b. Double replica Southwestern: Although double replica from a single gel is not recommended in initial attempts to detect a nuclear protein, once conditions are established for a specific protein it may be possible to transfer proteins onto two different filters. One way to do this is to sandwich the gel between two nitrocellulose membranes and blot the proteins at 250 mA in Tris/glycine buffer but with frequent polarity changes *(15)*. A simple alternative is to place two membranes on top of the gel before transfer. Sufficient protein may travel through the first membrane and be retained on the second.
    c. Variants on the Western blot: Multiple variants of the Western blot based on the same principles but utilizing different probes have been developed. High affinity interactions may be detected between proteins and RNA (Northwestern) and other proteins (Farwestern). (*See* **refs. 5** and **6**).

References

1. Lin, S. Y. and Riggs, A. D. (1975) The general affinity of lac repressor for *E. Coli* DNA: implication for gene regulation in prokaryotes and eukaryotes. *Cell* **4,** 107–111.
2. Galas, D. and Schmidt, A. (1978) DNase footprinting, a simple method for the detection of protein-DNA binding specificity. *Nucleic Acids Res.* **5,** 3157–3170.
3. Garner, M. M. and Revzin, A. (1981) A gel electrophoresis method for quantifying binding of proteins to specific DNA regions: applications to components of the *E. coli* lactose operon regulatory system. *Nucleic Acids Res.* **9,** 3047–3059.
4. Chodosh, L. A., Carthew, R. W., and Sharp, P. A. (1986) A single polypeptide possesses the binding and transcription activities of the adenovirus major late transcription factor. *Mol. Cell. Biol.* **6,** 4723–4733.
5. Bowen, B., Steinberg, J., Laemmli, U. K., and Weintraub, H. (1980) The detection of DNA-binding proteins by protein blotting. *Nucleic Acids Res.* **8,** 1–20.
6. Vinson, C. R., LaMarco, K. L., Johnson, P. F., Landschulz, W. H., and McKnight, S. L. (1988) In situ detection of sequence-specific DNA binding activity specified by a recombinant bacteriophage. *Genes Dev.* **2,** 801–806.
7. Keller, A. D. and Maniatis, T. (1991) Selection of sequences recognized by a DNA binding protein using a preparative Southwestern blot. *Nucleic Acids Res.* **19,** 4675–4680.
8. Hoeffler, J. P., Lustbader, J. W., and Chen, C. Y. (1991) Identification of multiple nuclear factors that interact with cAMP-response element-binding protein and activating transcription factor-2 by protein-protein interactions. *Mol. Endo.* **5,** 256–266.
9. Sambrook, J., Fritsch, E. F., and Maniatis, T. (1989) *Molecular Cloning. A Laboratory Manual.* Cold Spring Harbor Laboratory, Cold Spring Harbor, NY.
10. Dreyfuss, G., Adam, S. A., and Choi, Y. D. (1984) Physical change in cytoplasmic messenger ribonucleoproteins in cells treated with inhibitors of mRNA transcription. *Mol. Cell. Biol.* **4,** 415–423.
11. Papavassiliou, A. G. and Bohmann, D. (1992) Optimization of the signal-to-noise ratio in Southwestern assays by using lipid-free BSA as blocking reagent. *Nucleic Acids Res.* **20,** 4365–4366.

12. Silva, C. M., Tully, D. B., Petch, L. A., Jewell, C. M., and Cidlowski, J. A. (1987) Application of a protein blotting procedure to the study of human glucocorticoid receptor interactions with DNA. *Proc. Natl. Acad. Sci. USA* **84,** 1744–1748.
13. Jack, R. S., Gehring, W. J., and Brack, C. (1981) Protein component from Drosophila larval nuclei showing sequence specificity for a short region near a major heat shock protein gene. *Cell* **24,** 321–331.
14. Miskimins, W. K., Roberts, M. P., McClelland, A., and Ruddle, H. (1985) Use of a protein blotting procedure and a specific DNA probe to identify nuclear proteins that recognizes the region of the transferrin receptor gene. *Proc. Natl. Acad. Sci. USA* **82,** 6741–6744.
15. Matsuno, K., Suzuki, T., Takiya, S., and Suzuki, Y. (1989) Complex formation with the fibroin gene enhancer through a protein–protein interaction analyzed by a modified DNA-binding assay. *J. Biol. Chem.* **264,** 4599–4604.
16. Hübscher, U. (1987) Double replica Southwestern. *Nucleic Acids Res.* **15,** 5486.

97

Nonradioactive Methods for the Detection of RNA-Protein Interaction

Asier Echarri, María Eugenia González, Iván Ventoso, and Luis Carrasco

1. Introduction

RNA in biological systems is associated with proteins. Recent work in eukaryotes has identified common motifs present in families of RNA-binding proteins. Usually, RNA-binding proteins recognize both sequence and structure at their target sites. Therefore, identification of proteins that interact with a specific RNA sequence contributes to the understanding of biological processes. Thus, the genome of human immunodeficiency virus (HIV) encodes proteins, Tat, Rev and NC, that bind to specific viral RNA motives *(1,2)*. These interactions mediate different steps of virus replication i.e., transactivation of transcription, nuclear export of viral transcripts, or packaging of two RNA genomes into the mature virion. In addition, we recently reported that Nef, an accessory protein encoded by HIV-1, belongs to the family of RNA-binding proteins *(3)*. Using different-size variant proteins and point-mutated proteins, it was found that the amino terminal Arg-rich domain of Nef is involved in the RNA-binding activity. Nef proteins from HIV-2 and SIV (simian immunodeficiency virus) also showed RNA-binding capacity *(4)*.

In this chapter we describe methodologies for the direct detection of RNA-protein interactions using biotinylated riboprobes. The northwestern assay involves the interaction of labeled RNA with proteins immobilized on a membrane support. This assay allows the identification of the specific protein present in a crude extract that interacts with a given RNA molecule. The UV crosslinking and gel retardation assays mimic the in vivo binding reaction in solution conditions using purified proteins and RNA molecules. Gel retardation assay separates RNA-protein complexes from free RNA in a gel matrix. This technique is very useful for dissociation kinetics and ionic strength analysis of the binding reaction. A UV crosslinking assay is particularly appropriate for detecting labile protein-nucleic acid complexes, because the UV irradiation fixes the complex in amounts sufficient for further analysis by sodium dodecyl sulfate/polyacrylamide gel electrophoresis (SDS-PAGE). All these techniques use biotinylated probes to detect RNA-protein complexes on blotted membranes. The advantages of these nonradioactive methods when compared to their radioactive counterparts are their safety, low cost, easy handling, and long-term storage of labeled riboprobes *(5,6)*.

From: *The Nucleic Acid Protocols Handbook*
Edited by: R. Rapley © Humana Press Inc., Totowa, NJ

2. Materials

RNases are ubiquitous and heat stable; care should be taken to avoid nuclease contamination of either reagents and equipment. All solutions should be treated with diethylpyrocarbonate (DEPC), an inhibitor of RNases, or dissolved in DEPC-treated dH$_2$O (*see* **Note 1**). Gloves should be worn.

2.1. In Vitro Synthesis of Biotin-Labeled Riboprobes

1. 5 µg Linearized template DNA in RNase-free dH$_2$O.
2. 2.5 m*M* ATP, CTP, and GTP, 10 m*M* UTP (Amersham Pharmacia Biotech) and 10 m*M* biotin-21-UTP (Clontech, Palo Alto, CA).
3. 40 U T7 or SP6 RNA Polymerase, 50 U RNasin ribonuclease inhibitor and RQ1 RNase-free DNase (Promega, Madison, WI).
4. 5X Transcription buffer: 200 m*M* Tris-HCl, pH 7.5, 30 m*M* MgCl$_2$, 10 m*M* spermidine, 50 m*M* NaCl.
5. 100 m*M* DL-dithiothreitol (DTT).
6. Sephadex G-50 column (Pharmacia).
7. DEPC-treated dH$_2$O (*see* **Note 1**).

2.2. Northwestern Assay

1. 10X Ponceau S: 2% (w/v) ponceau S, 30% (v/v) trichloroacetic acid, 30% (v/v) sulfosalicylic acid.
2. Binding buffer: 10 m*M* Tris-HCl, pH 7.5, 1 m*M* EDTA, 50 m*M* NaCl, 0.1% (v/v) Triton-X-100, 1X Denhardt's solution (50X: 1% Ficoll (Pharmacia, Piscataway, NJ), 1% polyvinylpyrrolidone, and 1% bovine serum albumin [BSA]).
3. Transfer buffer: 50 m*M* Tris-HCl, pH 8.3, 380 m*M* glycine, and 20% (v/v) methanol.
4. Streptavidin-peroxidase (POD) conjugate at 500 U/mL (Boehringer Mannheim, Mannheim, Germany).
5. Solution A: 100 m*M* Tris-HCl, pH 8, 2 m*M* H$_2$O$_2$.
6. Solution B: 2.5 m*M* luminol, 75 µ*M* D(-)luciferin (Boehringer Mannheim). Prepare a solution of 2.5 m*M* luminol dissolved in 100 m*M* Tris-HCl, pH 8, (store at 4°C in dark bottle), and a stock solution of 35 m*M* D(-)luciferin dissolved in 100 m*M* Tris-HCl, pH 8, (store in 25 µL aliquots at –70°C protected from light). To prepare solution B, mix 10 mL 2.5 m*M* luminol and 22 µL 35 m*M* D(-)luciferin (*see* **Note 4**).
7. Nitrocellulose membrane (Bio-Rad, Hercules, CA).
8. SDS 15% (w/v) polyacrylamide gel *(7)*.
9. DEPC-treated dH$_2$O (*see* **Note 1**).

2.3. Gel Retardation Assay

1. 100 m*M* sodium phosphate, pH 7: mix 57.7 mL of 0.5 *M* Na$_2$HPO$_4$ with 42.3 mL 0.5 *M* NaH$_2$PO$_4$, add DEPC-treated dH$_2$O to 500 mL.
2. 99% Glycerol.
3. 250 m*M* HEPES, pH 7.5.
4. 100 m*M* KCl.
5. 20 U RNasin ribonuclease inhibitor.
6. 20X SSC: 3 *M* NaCl, 300 m*M* Na$_2$Citrate·2H$_2$O, Adjust pH to 7 with HCl.
7. Nitrocellulose membrane.
8. Agarose (low melting point) dissolved in 10 m*M* sodium phosphate, pH 7, at 1% (w/v). To prepare 250 mL of agarose use a 1-L bottle and dissolve by heating in a microwave until the solution is completely clear.

9. Paper towels.
10. Whatman 3MM paper (Whatman, Maidstone, Kent, UK).
11. BSA.
12. Phosphate-buffered saline (PBS): 137 mM NaCl, 2.7 mM KCl, 4.3 mM Na$_2$HPO$_4$·7H$_2$O, 1.4 mM KH$_2$PO$_4$.
13. Buffer P: PBS plus 0.05% (v/v) Tween 20.
14. Streptavidin-POD conjugate at 500 U/mL.
15. Solution A and B (*see* **Subheading 2.2.**, **steps 5** and **6**).
16. Vacuum oven.
17. Loading buffer (0.25% bromophenol blue, 0.25% xylene cyanol FF, 30% glycerol).
18. DEPC-treated dH$_2$O (*see* **Note 1**).

2.4. UV Crosslinking

1. UV lamp with 254 nm wavelength.
2. 96 Microwell plate.
3. 250 mM HEPES, pH 7.5.
4. 100 mM KCl.
5. Streptavidin-POD conjugate at 500 U/mL.
6. Solution A and B (*see* **Subheading 2.2.**, **steps 5** and **6**).
7. RNase A at 1.5 µg/µL.
8. 5X Buffer sample: 29.2 mL 1 M Tris-HCl, pH 6.8, 22 mL 99% glycerol, 8.3 g sodium dodecyl sulfate (SDS), 7.75 g DTT, 50 mg bromophenol blue, to 100 mL with dH$_2$O.
9. SDS 15% (w/v) polyacrylamide gel *(7)*.
10. Buffer P: PBS plus 0.05% (v/v) Tween 20.
11. Transfer buffer (*see* **Subheading 2.2.**, **step 3**).
12. DEPC-treated dH$_2$O (*see* **Note 1**).

3. Methods

3.1. In Vitro Synthesis of Biotin-Labeled Riboprobes

Time required: *see* **Note 3**.

1. Add the following components at room temperature in the order listed:

5X Transcription buffer	20 µL
100 mM DTT	10 µL
RNasin ribonuclease inhibitor	50 U
2.5 mM GTP, ATP, and CTP	20 µL
10 mM UTP	2.5 µL
10 mM biotin-21-UTP	2.5 µL
5 µg Linearized template DNA	
T7 or SP6 RNA polymerase	20 U
DEPC-treated dH$_2$O	to 100 µL

2. Mix and incubate for 45 min at 37°C.
3. Add again 20 U of T7 or SP6 RNA polymerase and incubate for a further 45 min at 37°C.
4. Add 1 U RQ1 RNase-free DNase per µg of template DNA. Incubate 30 min at 37°C.
5. The unincorporated nucleotides are eliminated from the reaction mixture by chromatography on a Sephadex G-50 column (*see* **Notes 2** and **11**).
6. Determine riboprobe concentration by measuring optical density at a wavelength of 260 nm. Concentration should be between 100–500 ng/µL (*see* **Note 9**).

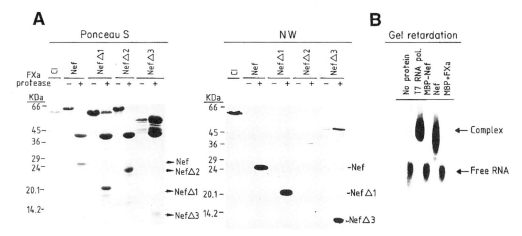

Fig. 1. A. Northwestern assay of HIV-1 nef proteins with biotinylated riboprobe. The left panel shows ponceau S staining of different proteins separated by SDS-PAGE and transferred to nitrocellulose membrane. After renaturation of the proteins in binding buffer, the membrane was incubated with 600 ng of biotinylated riboprobe (7830-8932 nucleotides of BH10 HIV-1 RNA), with streptavidin-POD peroxidase and then with detection solution. An autoradiography (2 min exposure) is shown in the right panel (NW). Potyvirus CI protein was used as a positive control. All proteins (except CI) were purified as fusion constructs with maltose binding protein (MBP, 41 KDa). MBP was separated from different nef proteins by Factor Xa protease digestion (+) or not (−). Nef (from HIV-1 bru) is the wild-type protein (27 KDa); nef Δ1 lacks 33 amino acids from the carboxy terminus (20 KDa), nef Δ2 lacks 22 amino acids from the amino terminus (23 KDa), and nef Δ3 lacks 104 amino acids from the carboxy terminus (13 KDa). B. Gel retardation assay with biotinylated riboprobe. Fifty nanograms of biotinylated riboprobe (7830–8932 nucleotides of BH10 HIV-1 RNA) were incubated with different proteins, loaded on the agarose gel, transferred by capillary action to nitrocellulose membrane, and incubated with detection solution. Autoradiography (1 min exposure) shows the biotinylated-free RNA or RNA-protein complexes. MBP-nef, nef, T7 RNA polymerase (positive control) and MBP + Factor Xa protease were tested for their ability to bind RNA; only nef and T7 RNA polymerase form complexes with RNA.

3.2. Northwestern Assay

Time required: *see* **Note 3**.

1. Protein samples (0.5–5 µg) are diluted 1:1 in 2X buffer sample (*see* **Note 7**) and run in a standard SDS-15% PAGE at 30 mA (*see* **Note 8**). Do not boil protein samples.
2. The gel is transferred to a nitrocellulose filter in transfer buffer at 200 mA for 16 h (*see* **Note 10**) *(8)*.
3. Stain the filter with ponceau S (*see* **Fig. 1A**, left panel) and mark the position of molecular weight marker with a glasswriter (*see* **Note 12**).
4. For renaturation of the proteins incubate the nitrocellulose filter in a RNase-free tray (*see* **Note 1.2**) containing binding buffer at room temperature with agitation on a platform shaker for 2 h (change binding buffer every 30 min).
5. Incubate the renatured proteins with the riboprobe at 20–50 ng riboprobe/ml binding buffer (usually 1–5 µL of transcription reaction in 10 mL of binding buffer) at room temperature for 1 h.

6. Wash the filter three times (2 min each) with binding buffer to remove the unbound riboprobe.
7. Incubate with streptavidin-POD conjugate, diluted 1:20,000 in binding buffer, for 30 min. Cover the dish with aluminum paper until **step 10**.
8. Wash three times (2 min each) with binding buffer.
9. Incubate for 1 min in detection solution (mix 10 mL of freshly made solution A with 10 mL of freshly made solution B; *see* **Subheading 2.2.**, **steps 5** and **6**).
10. Air dry the filter, cover with a plastic sheet, and expose to X-ray films (**Fig. 1A**, right panel), usually for 30 s to 10 min (*see* **Note 5**).

3.3. Gel Retardation Assay

Time required: *see* **Note 3**.

1. Decontaminate all plasticware with H_2O_2 (*see* **Note 1.b.**).
2. Prepare 1% (w/v) agarose gel of appropriate length (approx 15 cm) and depth (approx 8 mm) and let solidify at 4°C. Use 10 m*M* sodium phosphate, pH 7, as electrophoresis buffer and prerun the agarose gel for 15 min at 4°C at 20 V *(9)*.
3. Set up binding reaction in a microcentrifuge tube:

250 m*M* HEPES, pH 7.5	1.6 µL
100 m*M* KCl	2 µL
Biotinylated RNA	20–200 ng
RNasin ribonuclease inhibitor	20 U
Purified protein	0.5–2 µg (<5 µL)
DEPC treated dH_2O	to 20 µL

4. Mix and incubate at room temperature for 15 min.
5. Add 2.2 µL 99% glycerol to the reaction mixtures and load the samples. Load in an adjacent lane 5 µL of loading buffer. Run at 20 V for 16 h or until the bromophenol blue has migrated 80% of the gel length, at 4°C.
6. Incubate the agarose gel in 20X saline sodium citrate (SSC) for 1 h with agitation at room temperature.
7. Meanwhile, cut a piece of nitrocellulose filter to the gel size. Wet with dH_2O and soak in 20X SSC for 30 min.
8. Transfer the gel to the filter by capillary action *(10)* as follows. Place a piece of Whatman 3MM paper wetted with 20X SSC on a piece of glass to form a support that is longer and wider than the agarose gel. Place the support inside a large baking dish. Fill the dish with 20X SSC. Immerse the Whatman 3MM paper edges into the 20X SSC. Place the gel on the support, avoiding air bubbles. Surround the gel with Saran Wrap. Place the wet nitrocellulose membrane on top of the gel, avoiding air bubbles (mark position of the wells on the nitrocellulose membrane with a pencil or glasswriter). Place on top of nitrocellulose, in the order listed, four pieces of Whatman 3MM paper, 10 cm of paper towels, a glass plate, and a 500-g weight. Allow overnight capillary transfer.
9. Dry the nitrocellulose by baking for 2 h at 80°C in a vacuum oven.
10. Incubate the nitrocellulose filter in buffer P plus 3% (w/v) BSA for 30 min.
11. Wash once with buffer P for 1 min.
12. Incubate with streptavidin-POD conjugate dissolved 1:20,000 in buffer P for 30 min. Cover the dish with aluminum foil until **step 15**.
13. Wash with buffer P for 10 min. Repeat this step twice.
14. Incubate for 1 min in detection solution (mix 10 mL of freshly made solution A with 10 mL of freshly made solution B; *see* **Subheading 2.2.**, **steps 5** and **6**).
15. Air dry in the dark, cover with plastic sheet, and expose the filter to X-ray films for 30–300 s (*see* **Fig. 1B** and **Note 5**).

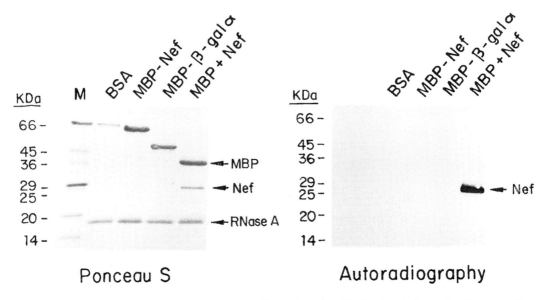

Fig. 2. UV crosslinking assay of HIV-1 nef protein using biotinylated riboprobe. Ponceau S staining (left panel) of the nitrocellulose membrane and autoradiography (right panel) of the membrane after treatment with detection solution. Eight hundred nanograms of biotinylated RNA (1-829 nucleotides of BH10 HIV-1) were used in each reaction. Autoradiography (3 min exposure) is shown. Only free nef protein interacts with RNA.

3.4. UV Crosslinking Assay

Time required: *see* **Note 3**.

1. Set up binding reaction in a microcentrifuge tube (*see* **Note 7**):
250 mM HEPES, pH 7.5	1.6 µL
100 mM KCl	2 µL
Biotinylated RNA	300–1500 ng
Purified protein	0.5–2 µg (<5 µL)
DEPC-treated dH$_2$O	to 20 µL
2. Mix and incubate at room temperature for 15 min.
3. Transfer the reaction to a 96-microwell plate and place the plate on ice during exposure to UV. Irradiate the reaction mixture for 10 min at 4°C using a 254-nm UV lamp placed 4 cm above the samples. Protect from irradiation outside the sample by covering the lamp with aluminum foil during this step (*see* **Note 6**).
4. Add to the reaction 1 µL of 1.5 µg/µL RNase A and incubate for 30 min at 37°C.
5. Add 5 µL 5X buffer sample and separate proteins by SDS-15%PAGE at 30 mA.
6. Transfer the gel to a nitrocellulose membrane at 200 mA in transfer buffer overnight.
7. Stain the proteins with 1X ponceau S (**Fig. 2**, left panel). Mark molecular weight standards on the nitrocellulose filter with a glasswriter (*see* **Note 12**).
8. Incubate the filter in buffer P plus 3% (w/v) BSA for 30 min.
9. Wash once with buffer P for 1 min.
10. Incubate with streptavidin-POD conjugate dissolved 1:40,000 in buffer P for 30 min. Cover the dish with aluminum foil until the last step.
11. Follow **Subheading 3.3.**, **steps 13–15** (**Fig. 2**, right panel).

4. Notes

1. DEPC reacts rapidly with amines and cannot be used to treat solutions containing buffers such as Tris. DEPC can modify purine residues in RNA by carboxymethylation. DEPC is suspected to be a carcinogen.
   a. DEPC treatment: Add the DEPC to dH$_2$O or to the indicated solution to a concentration of 0.1% (v/v). Mix well and allow to stand overnight in a fumehood. Autoclave the liquids to eliminate the DEPC.
   b. Preparation of RNase-free plasticware: Wash the electrophoresis tanks, combs, and casting trays with 3% (v/v) H$_2$O$_2$ for 1 h and rinse with DEPC-treated dH$_2$O.
2. Do not use phenol extraction, as biotinylated probes are soluble in the phenol layer.
3. Duration of procedures:
   a. In vitro synthesis of biotin labeled riboprobe: 1 d
      Transcription   1.5 h
      DNA digestion   30 min
      Chromatography   15 min
   b. Northwestern assay: 2 d
      SDS-PAGE (10 × 10 cm)   3 h
      Transfer   overnight
      Renaturation and detection of RNA   4 h
   c. Gel retardation assay: 3 d
      Gel preparation: 4 h
      Running the samples: overnight
      Transfer   1.5 h + overnight
      Detection of RNA   4 h
   d. UV crosslinking assay: 2 d
      Reaction, UV light and RNA digestion   1.5 h
      Running the gel   3 h
      Transfer   overnight
      Detection of RNA   2 h
4. D(-)Luciferin solution sometimes does not become completely clear but works well nonetheless.
5. Usually 1 min exposure is enough to detect biotinylated RNA. Exposition can be extended for 3 h. However if the signal is strong, long exposures could decrease the band intensity.
6. Longer periods of irradiation will not improve the signal intensity. Protein degradation and/or oligomerization may occur if longer periods of irradiation are used.
7. Streptavidin-POD conjugate can interact with some proteins such as RNase A. Moreover, some proteins, such as propionyl-CoA carboxylase and pyruvate carboxylase, can interact with biotin. In the case of Northwestern and UV crosslinking assays, a negative control without biotinylated riboprobe is recommended.
8. Electrical currents greater than 30 mA can cause irreversible denaturation of the protein. A 10 × 10 cm SDS-PAGE at 30 mA requires 3 h electrophoresis.
9. Biotinylated transcripts can be quantified by serial dilution of 1 µL transcription reaction in 20X SSC spotted to nitrocellulose with a dot-blot apparatus (Bio-Rad). Filters are dried at 80°C for 2 h *in vacuo* and the biotinylated riboprobes are detected as in **Subheading 3.3., steps 10–15** is described.
10. Protein transfer from the gel to the nitrocellulose filter is performed wet, because semidry transfer causes heating of the samples and may inhibit subsequent renaturation of the transferred proteins *(11)*.
11. Alternatively, ethanol precipitation can be performed.

12. Ponceau S staining fades after renaturation and treatment of the nitrocellulose membrane with detection solution.
13. Potential problems:
    a. In vitro synthesis of biotin labeled riboprobes:
        i. Low amounts of riboprobe: Make sure that reaction components are at room temperature. Add larger amount of linearized plasmid dissolved in RNase free dH_2O to the reaction mixture and make sure that no NaCl is present in the template solution (NaCl concentration higher than 30 mM could inhibit RNA polymerase activity). Make sure that all solutions are RNase free.
        ii. Presence of transcripts longer than expected: DNA template should be completely digested with an enzyme that generates a protruding 5' terminus. If it is impossible to use such a restriction enzyme, the linear DNA template should be "blunt ended."
    b. Northwestern assay: If the riboprobe is correct, the problems could be:
        i. The positive control does not bind RNA: Solutions and/or labware contaminated with RNase is a possible cause. Treat all the solutions with DEPC and the plasticware with 3% (v/v) H_2O_2. Check that luminol, luciferin, and H_2O_2 are at a correct concentration and stored in the dark.
        ii. A nonspecific background signal is obtained: Reduce the riboprobe concentration or increase the duration of the washes.
    c. Gel retardation assay:
        i. Biotinylated RNA cannot be visualized: Incorrect transfer. Use a heavier weight for the transfer. The paper towels perhaps may not have been in contact with the transfer buffer.
        ii. Spots in autoradiography: Air bubbles between nitrocellulose and the agarose gel were present.
    d. UV crosslinking assay:
        i. The proteins disappear: Check the transfer buffer and current during the transfer procedure. There is excess UV light irradiation.
        ii. Band smearing: Incomplete RNase A digestion.
        iii. Low band intensity: Concentrate the riboprobe by ethanol precipitation to 1 μg/μL.

Acknowledgments

The expert technical assistance of Mr. M. A. Sanz is acknowledged. The authors thank J. A. García and A. Fernández for kindly giving us CI protein. DGICYT project number PB94-0148 and the institutional grant to the CBM of Fundación Ramón Areces are acknowledged for their financial support. A. E. is a holder of a Gobierno Vasco fellowship and I. V. is a holder of a Comunidad Autónoma de Madrid fellowship.

References

1. Luban, J. and Goff, S. P. (1991) Binding of human immunodeficiency virus type 1 (HIV-1) RNA to recombinant HIV-1 gag polyprotein. *J. Virol.* **65,** 3203–3212.
2. Mattaj, I. W. (1990) A selective review of RNA-protein interactions in eukaryotes. *Mol. Biol. Rep.* **14,** 151–155.
3. Echarri, A., González, M. E., and Carrasco, L. (1996) Human immunodeficiency virus (HIV) nef is an RNA binding protein in cell-free systems. *J. Mol. Biol.* **262,** 640–651.
4. Echarri, A., González, M. E., and Carrasco, L. (1997) The N-Arg-rich region of human immunodeficiency virus types 1 and 2 and simian immunodeficiency virus Nef is involved in RNA binding. *Eur. J. Biochem.* **246,** 38–44.

5. Rodríguez, P. L. and Carrasco, L. (1996) Biotin labeled riboprobes to study RNA-binding proteins, in *A Laboratory Guide to Biotin-Labeling in Biomolecule Analysis* (Meier, T. and Fahrenholz, F., eds.) Birkhäuser Verlag, Basel, Switzerland, 215–225.
6. Rodríguez, P. L. and Carrasco, L. (1994) Nonradioactive northwestern analysis using biotinylated riboprobes. *Biotechniques* **17,** 702–705.
7. Sambrook, J., Fritsch, E. R., and Maniatis, T., eds. (1989) *Molecular Cloning: A Laboratory Manual*, 2nd ed. Cold Spring Harbor Laboratory Press, Cold Spring Harbor, NY.
8. Harlow, E. and Lane, D. (1988) *Antibodies. A Laboratory Manual*, Cold Spring Harbor Laboratory Press, Cold Spring Harbor, NY.
9. Lin, J. H., Chang, M. F., Baker, S. C., Govindarajan, S., and Lai, M. M. C. (1990) Characterization of hepatitis delta antigen: Specific binding to hepatitis delta virus RNA. *J. Virol.* **64,** 4051–4058.
10. Southern, E. M. (1975) Detection of specific sequences among DNA fragments separated by gel electrophoresis. *J. Mol. Biol.* **98,** 503–517.
11. Chen, X., Sadlock, J., and Schon, E. A. (1993) RNA-binding patterns in total human tissue proteins: Analysis by Northwestern blotting. *Biochem. Biophys. Res. Commun.* **191,** 18–25.

98

Transcriptional Activation Analysis by the Chloramphenicol Acetyl Transferase (CAT) Enzyme Assay

David R. Hodge, Delores M. Thompson, Alexandra Panayiotakis, and Arun Seth

1. Introduction

Gene expression is controlled by cis-regulatory elements. Generally, the most important elements that are required for transcription are contained in the promoter sequences located upstream of a gene (1). Eukaryotic RNA polymerase requires several accessory factors, such as TFIID, TFIIA, TFIIB, and TFIIF, as well as transcription factors that augment or regulate developmental expression of various genes (2,3). Tissue-specific factors have been shown to bind DNA in a sequence-specific manner and interact with other transcription factors in order to regulate gene expression (3,4). The sequences that many transcription factors bind to are located in both promoter and enhancer elements (5,6). To study regulation of cloned promoter sequences, one needs to introduce these promoter sequences into specific cells, and by linking them to an appropriate reporter gene, one can estimate promoter activity by an increase in reporter gene activity either by enzymatic assay or mRNA expression. Various reporter genes can be used to measure transcription activity, such as chloramphenicol acetyl transferase (CAT), luciferase, β-galactosidase, and human growth hormone (7). To examine the transcription enhancement activity of a putative promoter/enhancer sequence, the cloned DNA fragment being evaluated is inserted upstream of a particular reporter gene-containing vector. The recombinant promoter-reporter gene construct is then introduced into an appropriate cell type, either by $CaPO_4$ transfection or electroporation (8–10). Transcription activity can then be measured directly by estimating enzymatic activity or mRNA expression (11,12). Generally, a control plasmid is also introduced and used to normalize transfection efficiency (13). After making corrections for this transfection efficiency and averaging replicate experiments, the expression of the reporter genes appear to be directly proportional to the transcriptional activity of the promoter sequences employed for such studies.

A "reporter" vector contains several essential components that are required for its function. A map of a typical reporter vector, pHDCAT, is presented in **Fig. 1**. To propagate the plasmid DNA in bacteria, the vector contains the *E. coli* origin (ori) of replication.

From: *The Nucleic Acid Protocols Handbook*
Edited by: R. Rapley © Humana Press Inc., Totowa, NJ

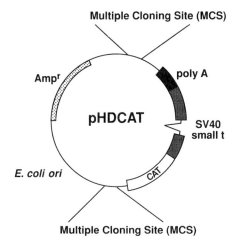

Fig. 1. Map of a typical reporter vector. The pHDCAT vector contains various elements (MCS, poly A signals, *E. coli*-ori, Amp^r) to allow for cloning and propagation in bacteria, and for proper transcription in mammalian cells.

In addition, an antibiotic-resistance gene (usually β-lactamase, which provides ampicillin resistance) is included in the vector for drug selection. Most vectors also contain a multiple cloning site (MCS) that is located either upstream from the reporter gene or following the polyA signals, thus facilitating the cloning of both promoter and enhancer sequence element(s) *(14)*.

Since addition of an unknown promoter sequence may perturb the stability of a "reporter" gene mRNA, it is necessary in some cases to include elements (polyA signals and intron sequences) that allow correct splicing and also increase the stability of the transcript. Most commonly utilized polyA signals and intron sequences are derived from the SV40 small t-antigen (**Fig. 1**). Although the exact functional role of the intronic sequences has not been fully understood, their presence seems to increase the efficiency and stability of the transcription product *(15)*. By conducting parallel transfections of "reporter" vectors lacking the element under study, it is possible to obtain a background level of "reporter" gene stimulation. Although a slight amount of background reporter activation is unavoidable, the negative control level is often negligible and easily compensated for by standardizing signal-to-noise ratios. Moreover, a "reporter" vector (pUMSVO-CAT) with virtually no background signal is also available *(16)*. This plasmid contains the murine c-*mos* proto-oncogene transcription terminal signal (UMS) located upstream from the promoter sequences under study *(17)*. This "reporter" vector (pUMSVO-CAT) is useful for studying weak promoters that would otherwise be difficult to assay because of nonspecific background activity displayed by other vectors.

Transfection of a vector containing putative regulatory elements linked to a "reporter" gene allows for the determination of the reporter gene expression in both a quantitative and a temporal fashion. The methodology utilized is dependent on the particular requirements of the investigator, e.g., the type of sequences being analyzed, the required sensitivity of the assay, and the avoidance of naturally occurring cellular products that would interfere with the assay. For example, alkaline phosphatase would not be practi-

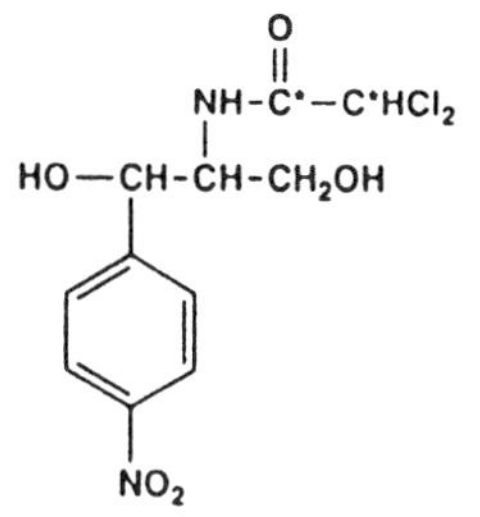

$$\text{HO-CH-CH-CH}_2\text{OH} \quad \begin{array}{c} \text{O} \\ \| \\ \text{NH-C*-C*HCl}_2 \end{array}$$

(attached to p-nitrophenyl group)

Chloramphenicol, D-threo-(dichloroacetyl 1, 2-C^{14})

$$\text{Cam} \xrightarrow[\text{CAT}]{\text{AcCoA} \quad \text{CoA}} \text{Ac-Cam} \xrightarrow[\text{CAT}]{\text{AcCoA} \quad \text{CoA}} \text{Ac}_2\text{-Cam}$$

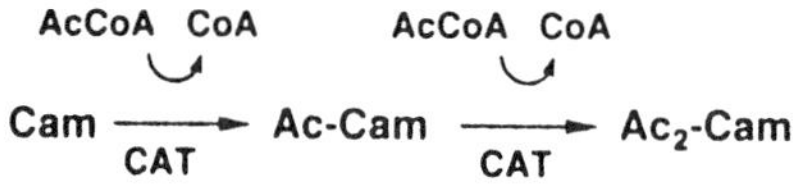

Fig. 2. The chemical structure of Cam, D-threo-(dichloroacetyl 1,2-C^{14}). Cam is acetylated by CAT enzyme using acetyl coenzyme A as an acyl group donor.

cal as a reporter gene because of its natural ubiquitous pattern of expression. A discussion of the CAT "reporter" gene, followed by a brief description of the luciferase (luc), the β-galactosidase (β-gal), and the human growth hormone (hGH) reporter genes, will help to illustrate the type of vector that may be best suited to an individual's need.

The most widely used "reporter" gene assay employs a bacterial gene that encodes the CAT enzyme. Originally derived from a transposon, the CAT enzyme functions by utilizing acetyl coenzyme-A as an acetyl donor for the covalent addition of acetyl groups to ^{14}C-labeled chloramphenicol (**Fig. 2**). The chloramphenicol (Cam) molecule has two potential acetylation sites; usually the CAT-mediated enzymatic reaction results in both 1-acetyl and 3-acetyl-Cam. In addition, sometimes diacetylated Cam derivative (1,3-acetyl-Cam) (**Fig. 3**) can also appear under conditions of strong CAT enzyme expression, which could result from efficient promoter activation, excess reactants, or simply excess plasmid transfected into the cells. The acetylated and nonacetylated forms of Cam are usually separated by thin-layer chromatography. A typical pattern of thin-layer autoradiochromatogram is shown in **Fig. 3**. The extent of activation can be directly correlated with the intensities of the different spots of the acetylated Cam derivatives, which can then be easily quantified by using an AMBIS radioanalytic imaging system. A practical example of an application for the CAT assay, in order to assess transcriptional activation of the *Endo*A enhancer with Ets proteins, is described in **Note 6**.

The activity of a promoter may also be assessed by use of the CAT-enzyme linked immunosorbent assay (CAT-ELISA) *(18)*. The CAT-ELISA does not require the use of radioactive Cam substrate. Instead, it takes advantage of the immunological detection of the actual amount of CAT protein in cell extracts, and not merely enzyme activity, which could be affected by some of the harsh protein extraction procedures, like the freeze-thaw step. This ELISA method is faster, and sensitivity is equivalent to the

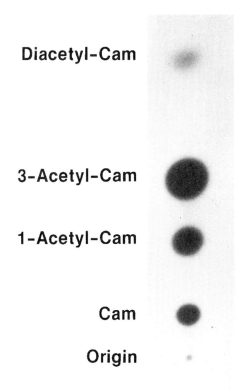

Fig. 3. A typical thin-layer chromatogram of CAT assay. Reaction products of a CAT assay are analyzed by thin-layer chromatography. TLC plate after development is exposed to X-ray film. The acetylated and nonacetylated forms of Cam are indicated.

use of isotopically labeled substrates; moreover, this assay measures all the CAT product, not just the active enzyme.

The CAT liquid assay is a very simple procedure for detecting promoter activity that exploits the differences in solubility between Cam and its acetylated derivatives *(11)*. Cam is relatively insoluble in organic solvents, whereas the acetylated derivatives are soluble in the organic solvents. Briefly, [^{3}H]-acetyl-coA and Cam are added to the cell extract; the entire mixture is then overlaid with an immiscible organic-based scintillation fluid containing fluor. Since the acetylation of Cam occurs via CAT enzyme activity, the [^{3}H]-acetyl-Cam then diffuses into the organic phase; this procedure allows for repetitive activity counts and the generation of a linear plot of activity. Since this liquid system uses less expensive tritiated acetyl-CoA, rather than ^{14}C-labeled products, and requires only a scintillation counter for detection, it is quick, reproducible, and economical.

Briefly, several other "reporter" genes are also utilized in addition to CAT. The use of firefly luciferase enzyme (*luc*) has gained popularity since its first use as a "reporter" gene in 1986 *(19)*. For studying weak promoters, a vector similar to the pUMSVO-CAT is available that utilizes the luciferase "reporter" gene (20). Also available is a vector that contains the luciferase "reporter" gene and the *lacZ* ΔM15 α-complementation peptide to permit the use of blue/white colony screening for rapid subcloning of putative promoter and enhancer sequences, and the point-mutation analysis of inserts *(21)*.

Another "reporter" system employs the bacterial β-galactosidase gene. β-gal is also used as a normalization "reporter"; however, since this enzyme is sensitive to heat, care must be taken during cell-extract preparation *(13)*. When conducting transfection experiments, it is helpful to determine the efficiency of transfection by including the β-gal *(lac-Z)* expression vector, such as pCH110 (Pharmacia-LKB, Piscataway, NJ), along with the other plasmids, but at a much lower concentration. The β-galactosidase enzyme catalyzes the colorimetric change in o-nitrophenyl-β-D-galactopyranoside (ONPG), allowing for the quantitation of promoter activity based on optical absorbance at 420 nm *(22)*. The use of human growth hormone (hGH) as a "reporter" gene enables the temporal monitoring of a promoter activity, without the requirement for preparation of the cell extracts, since the hGH product is secreted into the cell-culture medium. The analysis of the hGH product is usually performed by radioimmune assay (RIA) *(23,24)*; however, this assay may be simplified by utilizing a rapid ELISA detection system *(25,26)*.

2. Materials
2.1. Chemicals and Solutions

1. Chloroform.
2. Methanol.
3. Bakerflex TLC (Scientific Products cat. #5560).
4. Protein dye (Bio-Rad [Hercules, CA] cat. #40074): Dilute (1:5) with PBS, and pass through a 0.45-μm filter before use.
5. Bio-Rad protein standard (Bio-Rad cat. #500-0005): Prepare as a 1 mg/mL stock solution in PBS, and store at –20°C.
6. 10X HB salts solution: 1.37 M NaCl, 0.05 M KCl, 0.007 M Na$_2$HPO$_4$, and 0.06 M dextrose in water. Sterilize by passing through a 0.2-μ filter.
7. Solution H: Prepare by adding 10 mL of 10X HB salts and 2.0 mL of 1M HEPES buffer to 38 mL of H$_2$O. Adjust to pH 7.0, and pass through a 0.2-μ filter before use.
8. Glycerol shock solution: Add 5 mL of solution H to 3.5 mL of H$_2$O and 1.5 mL of glycerol; pass through a 0.2-μm filter.
9. *Ortho*-nitrophenol-β-D galactopyranoside (ONPG) stock solution: Prepare by dissolving at 4 mg/mL in a solution of 60 mM Na$_2$HPO$_4$ and 40 mM NaH$_2$PO$_4$. Store the ONPG stock solution at 4°C under light-proof conditions.
10. β-gal Reaction buffer: 60 mM Na$_2$HPO$_4$, 40 mM NaH$_2$PO$_4$, 10 mM KCl, 1 mM MgCl$_2$, and 50 mM β-mercaptoethanol in double-distilled H$_2$O.
11. The COS-7 and HeLa cell lines were obtained from the American Type Culture Collection (ATCC) (Rockville, MD). The cells are grown in Dulbecco's Modified Eagle's Medium (DMEM), supplemented with 10% calf serum, penicillin (50 μM), and streptomycin (50 μM); passage each week. All cell-culture reagents, PBS, and 1 M HEPES may be obtained from Gibco-BRL (Gaithersburg, MD); 60-mm Petri dishes (#3002) and tubes (#2063) may be purchased from Falcon (Los Angeles, CA). The 96-well titer plates (#3596) may be purchased from Costar (Cambridge, MA).
12. ^{14}C-chloramphenicol.
13. Ethyl acetate.
14. Acetyl CoA (Pharmacia): 10 mg/500 μL H$_2$O.
15. Extraction buffer: 0.04 M Tris-HCl, pH 7.4, 1 mM EDTA, and 0.15 M NaCl.
16. 0.25 M Tris-HCl, pH 7.8.
17. PBS: 137 mM NaCl, 2.7 mM KCl, 10 mM Na$_2$HPO$_4$, and 1.8 mM KH$_2$PO$_4$, pH 7.4.

3. Methods

3.1. Transfection

1. Plate out cells to be used in the transfection protocol 1 d in advance to allow for approx three-fourths confluency (*see* **Note 1**).
2. Two hours prior to transfection, feed the cells with 2 mL of DMEM containing 0.01 M HEPES, and incubate at 37°C with CO_2.
3. Prepare the $CaPO_4$ DNA precipitate by adding 1 µg of the reporter plasmid DNA (1 µg/µL), 10 µg of ETS plasmid DNA (pSGETS1 or pSGETS2) (1 µg/µL), and 1 µg/µL of pCH110 (β-gal) to 138 µL H_2O, with 150 µL of 0.5 M $CaCl_2$ in a Falcon tube (#2063); these solutions are then mixed by gently tapping the tube.
4. To this tube, add dropwise 300 µL of Solution H, and incubate at room temperature for 30 min to 1 h until the precipitate has formed.
5. Aspirate HEPES-DMEM media from the plates containing the cells to be transfected, and add 1.5 mL of fresh HEPES-DMEM medium to each dish.
6. The transfection precipitate (600 µL) is then added slowly and allowed to mix with the HEPES-DMEM media already present by gently rocking the entire plate. Once the precipitate and medium are thoroughly mixed, the cells are incubated at 37°C with CO_2 for 4–5 h (*see* **Note 2**).
7. Aspirate the transfection solution completely from the dishes, and add 1 mL of glycerol shock solution at room temperature, allowing the plates to incubate at room temperature for 3 min (*see* **Note 3**).
8. Aspirate the glycerol shock solution quickly, and wait for 1 min. Then add 2 mL of DMEM to rinse the dish.
9. Finally, add 4 mL of complete DMEM (supplemented with serum and antibiotics), and incubate the dishes at 37°C with CO_2.

3.2. Protein Extraction

1. At 48–72 h post-transfection, aspirate the DMEM, and rinse cells with 2 mL of PBS.
2. Aspirate PBS, and add 1.5 mL of extraction buffer to the dish. Allow to sit at room temperature for 5 min.
3. Collect the cells by scraping and transferring to a microcentrifuge tube followed by spinning at 13,000g at 4°C for 5 min.
4. Discard supernatant fluid. To the cell pellet add 160 µL of 0.25 M Tris-HCl, pH 7.8, and vortex.
5. In a dry-ice ethanol bath, freeze-thaw the tubes three times, 5 min each cycle (thawing is carried out in a 37°C water bath).
6. Spin the tubes in a microcentrifuge for 5 min at 4°C.
7. Transfer the supernatant (containing proteins) to a new microfuge tube, and keep protein solution at 4°C or store at –20°C until use.

3.3. Protein Determination

1. Using a 96-well titer plate (Costar), label lanes A-H (vertically) and 1-12 (horizontally).
2. Add 100 µL of PBS to wells A-1 and A-2.
3. Add 150 µL of PBS and 50 µL of Bio-Rad protein standard (1 mg/mL stock) to wells H-1 and H-2.
4. To row H-3-H-12, add 187.5 µL of PBS and 12.5 µL of protein solution derived from transfected cells (i.e., containing CAT protein).
5. To rows A-3-A-12 and rows B-G, add 100 µL of PBS.

Table 1
Quantitation of 1- and 3-Acetyl Cam Spots on AMBIS Radioisotopic Imaging System

Sample	Counts	CPM	SD CPM	% of Spots	2 SD %
Control	2171	447.1	9.6	0.3	4.3
pBLCAT ENDO A + ETS1	18,719	3855.1	28.2	2.8	1.5
pBLCAT ENDO A + ETS2	29,066	5986.0	35.1	3.8	1.2

6. Beginning with row H, dilute serially using 100 µL with a 12-tip pipetor excluding blanking wells A-1 and A-2 (each well now contains 100 µL).
7. Add 100 µL of diluted Bio-Rad dye (dilute dye 1:5 with PBS and 0.45-µm filter) to all wells, except A-1 and A-2. Read plates on a microtiter plate reader.

3.4. β-Galactosidase Expression "Normalization" Assay

1. Add 100 µL ONPG dye stock (4 mg/mL) and 50 µL protein extract to 500 µL ONPG reaction buffer, and incubate at 37°C for 30–90 min, or until a light yellow color is visible in tubes.
2. Stop the reaction by adding 500 µL of 1 M Na_2CO_3 to each tube.
3. Determine the optical densities spectrophotometrically for each tube at 420 nm.
4. This value is used to correlate the protein concentration with the activity of the β-galactosidase enzyme in the cell extract (*see* **Note 10**).

3.5. CAT Assay

1. Add 100 µg of cell extract to a microcentrifuge tube.
2. Adjust the total volume in the tube to 155 µL with 0.25 M Tris-HCl, pH 7.8.
3. Add 5 µL of C^{14}-chloramphenicol and 20 µL of acetyl CoA, and mix well.
4. Incubate the tube overnight (16 h) at 37°C.
5. Extract the reaction mixture at room temperature with 1 mL of ethyl acetate (J. T. Baker, Photrex) by vortexing for 1 min.
6. Spin tubes in a microcentrifuge for 5 min. Transfer the supernatant to a new microfuge tube, and dry the sample in a SpeedVac for 1 h.
7. To the pellet add 25 µL of ethyl acetate, and vortex to dissolve the pellet completely.
8. Spot the 25 µL of ethyl acetate on a Bakerflex TLC plate 1 in. from the bottom using a 10-µL glass capillary pipet (*see* **Note 5**).
9. Run ascending TLC in 125 mL (95:5) of chloroform:methanol in a covered glass tank lined with a 3M paper saturated in chloroform:methanol for 70 min.
10. Remove TLC from the tank with forceps, and air-dry for 15 min.
11. Expose the TLC plates to high-speed X-ray film for several days, or quantitate the amount of acetylated Cam on an AMBIS® radioisotopic imaging system. TLC plates "scanned" by AMBIS allow for the generation of data in a tabular form, indicating total number of counts in every spot and a statistical analysis of the assay (**Table 1**).

4. Notes

1. The best transfection efficiencies are obtained when the monolayers are not completely confluent (4×10^5 cells/60-mm dish/3 mL DMEM), and cover approx three-fourths of the plates.
2. The transfection precipitate should have a fine, granular appearance on the dish when observed through an inverted microscope.

3. It is important to time the glycerol shock step accurately. If the glycerol shock solution step exceeds 3 min, the cells will lose viability, with a concomitant loss of transfection efficiency.
4. The use of high-quality, ultrapure reagents ensures that the results are consistent between experiments.
5. When spotting the desiccated reaction products onto the TLC plate, care must be exercised not to disturb the silica coating of the plate. Usually, the best results are obtained by loading discrete portions of the sample, and then gently blowing the spot dry with filtered air or nitrogen gas prior to loading the next aliquot. An air-tight chromatography "tank" fitted with a tight lid is required to prevent loss of solvent and subsequent aberrant migration of reaction products.
6. All the transfection buffers and solutions are sterilized by passing them through 0.22-µm Nalgene filters, including the TE buffers used for diluting plasmid DNAs.
7. In all cases, the appropriate amount of plasmid DNAs must be empirically determined for each individual experimental condition. This type of assay provides the most reliable results, when reaction times and reagents are kept to a minimum. Under these conditions, the formation of acetylated forms of chloramphenicol are, typically, directly proportional to the transcriptional strength of the promoter under study.
8. In order to determine the transfection efficiency, a small amount of β-gal expression plasmid, pCH110 (Pharmacia), is included in the transfection solution. The cell extract containing the CAT and the β-gal is extracted, and a portion of the protein is assayed for β-gal enzymatic activity using the "Normalization" protocol described in **Subheading 3.4.**
9. The 420-nm optical density value obtained for each tube is used to correlate the protein concentration with the activity of the β-galactosidase enzyme in the cell extract. Also, the transformation efficiency may vary between dishes. By assuming that the activity of the β-galactosidase enzyme in the cell extract is roughly equal to the amount of CAT activity and utilizing the differences in optical densities as a correction factor, one can accurately adjust the amount of cell extract used in each CAT reaction. This is important because the amount of the total protein present in the cell lysis supernatant may not always be indicative of the active protein in the solution or of the transfection efficiency.
10. Ets is a family of transcriptional factors that activate transcription from various cellular and viral promoter/enhancer sequences *(27,28)*. *EndoA* is a type II keratin that forms intermediate filaments in conjunction with *endoB*, a type I keratin. The 3' end of the *EndoA* gene contains an enhancer that has 12 *Ets*-binding sites, which are arranged in six tandem repeats *(29)*. A CAT reporter plasmid containing the *EndoA* enhancer sequences was constructed by cloning a PCR-generated fragment from mouse genomic DNA into the pBLCAT2 vector *(30)*. Several constructs were sequenced to verify the presence of all six Ets tandem repeats, and two clones that contain enhancer in forward (pBLCAT-ENDOA-10) and backward (pBLCAT-ENDOA-14) orientation were selected for further study. The DNA-binding data showed that the ETS1 and ETS2 proteins bind to the ENDOA enhancer in a sequence-specific manner *(30)*.

The *EndoA* reporter plasmids were then cotransfected with increasing amounts of pSGETS1 or pSGETS2 expression vectors in NIH3T3 or COS cells. As seen in **Fig. 4**, the cells transfected with pBLCAT2-ENDOA-10 and pSGETS1 or pSGETS2 containing the ETS cDNAs, resulted in activation of CAT expression, whereas the pSG5 vector containing no ETS1 or ETS2 insert did not show such CAT activation. Thus, the DNA binding (not shown) and this CAT transactivation data indicate that the ETS proteins (ETS1 and ETS2) are able to regulate expression of the ENDOA enhancer.

CAT Enzyme Analysis

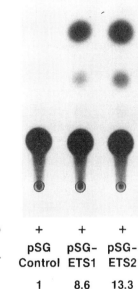

Reporter (pBLCATENDO A)	+	+	+
Expression Vector	pSG Control	pSG-ETS1	pSG-ETS2
Fold Activation	1	8.6	13.3

Fig. 4. Transactivation of ENDOA enhancer by ETS proteins. CAT assays were performed as described in **Subheading 3.** Lane 1 contains pSG5 (control) vector and pBLCAT ENDOA; lane 2 contains pSGETS1 and pBLCAT ENDOA; lane 3 contains pSGETS2 and pBLCAT ENDO A. Fold activation was determined by scanning the plate on AMBIS radioanalytical imaging system and is indicated below every lane and in **Table 1**.

References

1. Seth, A. and Papas, T. S. (1993) Principles of molecular cell biology of cancer: general aspects of gene regulation, in *Cancer: Principles and Practice of Oncology* (DeVita, V. T., Jr., Hellman, S., and Rosenberg, S. A., eds.), J. B. Lippincott, Philadelphia, pp. 23–34.
2. Klausner, R. D. and Hartford, J. B. (1989) *Cis-trans* models for post-transcriptional gene regulation. *Science* **246,** 870–872.
3. Shenk, T. (1981) Transcriptional control regions: nucleotide sequences requirements for initiation by RNA polymerase I and II. *Curr. Etop. Microbiol. Immunol.* **93,** 25–40.
4. Zwartkruis, F., Hoeijmakers, T., Deschamps, J., and Meijlink, F. (1991) Characterization of the murine Hox-2.3 promoter: involvement of the transcription factor USF (MLTF). *Genes Dev.* **33(3),** 179–190.
5. Haberstroh, L., Galindo, J., and Firtel, R. A. (1991) Developmental and spatial regulation of a Dictyostelium prespore gene: cis-acting elements and a cAMP-induced, developmentally regulated DNA binding activity. *Development* **113,** 947–958.
6. Serfling, E., Jasin, M., and Schaffner, W. (1985) Enhancers and eukaryotic gene transcription. *Trends Genet.* **1,** 224–230.
7. Alam, J. and Cook, J. L. (1990) Reporter genes: application to the study of mammalian gene transcription. *Anal. Biochem.* **188,** 245.
8. Pahl, H. L., Burn, T. C., and Tenen, D. G. (1991) Optimization of transient transfection into human myeloid cell lines using a luciferase reporter gene. *Exp. Hematol.* **10,** 1038–1041.
9. Ray, J. and Gaga, F. H. (1992) Gene transfer into established and primary fibroblast cell lines: comparison of transfection methods and promoters. *Biotechniques* **13,** 598–603.

10. Nickoloff, J. A. and Reynolds, R. J. (1992) Electroporation-mediated gene transfer efficiency is reduced by linear plasmid carrier DNAs. *Anal. Biochem.* **205(2),** 237–243.
11. Martin, J. D. (1990) Application of the two-phase assay for chloramphenicol acetyl transferase (CAT) to transfections with simian virus 40-CAT plasmids. *Anal. Biochem.* **191,** 242–246.
12. Sankaran, L. A. (1992) Simple quantitative assay for chloramphenicol acetyltransferase by direct extraction of the labeled product into scintillation cocktail. *Anal. Biochem.* **200,** 180–186.
13. Miller, J. H. (1972) in *Experiments in Molecular Genetics*, Cold Spring Harbor Laboratory, Cold Spring Harbor, NY, p. 352.
14. Sambrook, J., Fritsch, E. F., and Maniatis, T. (1989) *Molecular Cloning*, 2nd ed., Cold Spring Harbor Laboratory, Cold Spring Harbor, NY.
15. Rosenthal, N. (1987) Identification of regulatory elements of cloned genes with functional assays. *Methods Enzymol.* **152,** 704–720.
16. Salier, J.-P. and Kurachi, K. (1989) A CAT expression vector with virtually no background: pUMSVO-CAT. *Biotechniques* **7,** 30,31.
17. McGeady, M. L., Wood, T. G., Maizel, J. V., and Vande Woude, G. F. (1986) Sequences upstream from the mouse *c-mos* oncogene may function as a transcriptional terminal signal. *DNA* **5,** 289–298.
18. Gendloff, E. H., Bowen, B., and Buchholz, W. G. (1990) Quantitation of chloramphenicol acetyl transferase in transgenic tobacco plants by ELISA and correlation with gene copy number. *Plant Mol. Biol.* **14,** 575–583.
19. Ow, D. W., Wood, K. V., DeLuca, M., Dewet, J. R., Helinski, D. R., and Howell, S. H. (1986) Transient and stable expression of the firefly luciferase gene in plant cells and transgenic plants. *Science* **234,** 856–859.
20. deMartin, R., Strasswimmer, J., and Philipson, L. (1993) A new luciferase promoter insertion vector for the analysis of weak transcriptional activities. *Gene* **124,** 137,138.
21. Caricasole, A. and Ward, A. (1993) A luciferase-reporter vector with blue-white selection for rapid subcloning and mutational analysis of eukaryotic promoters. *Gene* **124,** 139,140.
22. Hall, C. V., Jacob, P. E., Ringold, G. M., and Lee, F. (1983) Expression and regulation of *Escherichia coli* lac Z gene fusions in mammalian cells. *J. Mol. Appl. Gen.* **2,** 101.
23. Ribela, M. T., Murata, Y., Morganti, L., Toniolo, D., and Bartolini, P. (1993) The use of recombinant human growth hormone for radioiodination and standard preparation in radioimmunoassay. *J. Immunol. Methods* **159,** 269–274.
24. Wan Nazaimoon, W. M., Satgunasingam, N., and Khalid, B. (1990) Development of an in-house radioimmunoassay for human growth hormone. *Malays. J. Pathol.* **12,** 13–20.
25. Dinesen, B. (1991) Immunochemical aspects of growth hormone assays. *Horm. Res.* **36,** 11–16.
26. Gervasi, G., Samy, M., and Scholler, R. (1990) Comparison of four human growth hormone (hGH) immunoassay kits and analysis of recognition of circulating forms. *Pathol. Biol.* **38,** 912–919.
27. Seth, A., Robinson, L., Thompson, D. M., Panayiotakis, A., Smyth, F. E., Watson, D. K., and Papas, T. S. (1993) Transactivation of *GATA*-1 promoter with ETS1, ETS2 and ERGB/Hu-FLI-1 proteins: stabilization of the ETS1 protein binding on *GATA*-1 promoter sequences by monoclonal antibody. *Oncogene* **8,** 1783–1790.
28. Seth, A., Hodge, D. R., Thompson, D. M., Robinson, L., Panayiotakis, A., Watson, D. K., and Papas, T. S. (1993) ETS family proteins activate transcription from HIV-1 LTR. *AIDS Res. Hum. Retroviruses* **9,** 1017–1023.

29. Seth, A., Ascione, R., Fisher, R. J., Mavrothalassitis, G. J., Bhat, N. K., and Papas, T. S. (1992) The *ets* gene family. *Cell Growth Differ.* **3,** 327–334.
30. Seth, A., Robinson, L., Panayiotakis, A., Thompson, D. M., Hodge, D., Zhang, X. K., Watson, D. K., Ozato, K., and Papas, T. S. (1994) The EndoA enhancer contains multiple ETS binding site repeats and is regulated by ETS proteins. *Oncogene* **9,** 469–477.

IX

Mutagenesis, Transcription, and Translation In Vitro

ized by R. Rapley © Humana Press Inc., Totowa, NJ

99

Generating Nested Deletions with Exonuclease III

George Murphy

1. Introduction

Exonuclease III (Exo III) will digest double-stranded DNA in a 3' to 5' direction if the DNA is blunt ended or possesses a 5' overhang. It will not digest if there is a 3' overhang of three or more bases, or if the 3' end has had thiophosphate-containing bases incorporated into it. In order to generate a set of insert deletions using Exo III, it is necessary to cut the polylinker twice with different restriction enzymes so that the cut end nearest the primer site possesses a 3' overhang, or has thiophosphate residues at the 3' end, and the end of the polylinker attached to the insert possesses a 5' overhang or blunt end. If this can be achieved, digestion with Exo III will result in progressive deletion of the 3' end of the insert, leaving a single-stranded 5' overhang that can be removed by treatment with mung bean nuclease *(1)* or exonuclease VII (Exo VII). The blunt ends thus formed are ligated, a suitable host is transformed, and colonies are picked at random and screened for insert size *(2)*, following which a suitable range of inserts is then sequenced.

One of the advantages of using directed deletions is the accuracy with which insert sizes can be measured by restriction enzyme digestion. Size selection is an essential part of this technique, as there is a very broad distribution in the size of the insert obtained at each timepoint, even if a clearly defined band is observed when the rate of deletion is being measured. A rapid screening procedure employing the isolation of crude plasmid from colony streaks, or more easily from small cultures grown up in microliter wells, is outlined here. The number of clones that can be screened in a working day using this method is limited only by the running of the gels, as the isolation of crude plasmid takes about 30 min. Selected transformants can then be miniprepped and a more accurate estimation made of their insert size by restriction enzyme digest.

2. Materials

2.1. Exonuclease Deletions

2.1.1. Linearizing the Insert

1. 10X TA buffer: 330 m*M* Tris-acetate (pH 7.9), 660 m*M* potassium acetate, 100 m*M* magnesium acetate, 40 m*M* spermidine, and 5 m*M* dithiothreitol (DTT). NaCl can be added to the TA buffer for those enzymes requiring a higher salt concentration.

From: *The Nucleic Acid Protocols Handbook*
Edited by: R. Rapley © Humana Press Inc., Totowa, NJ

2. Phenol/chloroform: Equal volumes of phenol (equilibrated with TE) and chloroform: isoamyl alcohol (25:1 v/v).
3. Chloroform.
4. Sepharose-CL-6B: Equilibrate the Sepharose as below (Pharmacia, Uppsala, Sweden) in T0.1E and adjust to a packed gel:buffer ratio of 2:1.
5. 10X Exo III buffer: 500 mM Tris-HCl, pH 8.0, 50 mM MgCl$_2$, and 10 mM DTT.
6. TE: 10 mM Tris-HCl (pH 8.0), 1 mM ethylenediaminetetraacetic acid (EDTA).
7. TO.LE: 10 mM Tris-HCl (pH 8.0), 0.1 mM EDTA.

2.1.2. Protecting the Primer Site with Thiophosphates

1. 10X TM: 100 mM Tris-HCl, pH 8.0, 50 mM MgCl$_2$.
2. Thio-dNTPs: 20 mM in each thiophosphate dATP, dCTP, dGTP, or dTTP pH 8.0.
3. Klenow fragment of DNA polymerase (Gibco-BRL, Richmond, LA).
4. Phenol/chloroform.
5. Chloroform.
6. Sepharose-CL-6B (Amersham Pharmacia Biotech).

2.1.3. Digestion with Exo III

1. Exo III (Gibco-BRL).
2. Stop buffer: 10 mM Tris-HCl, pH 8.0, 10 mM EDTA.
3. 10X Exo VII buffer: 500 mM potassium phosphate, pH 7.9, 90 mM EDTA, and 10 mM DTT.

2.1.4. Blunting the DNA

2.1.4.1. BLUNTING WITH MUNG BEAN NUCLEASE

1. Mung bean nuclease buffer: 30 mM Sodium acetate, 50 mM NaCl, 1 mM ZnCl$_2$, 1 mM cysteine, 0.001% (v/v) Triton, and 5% (v/v) glycerol.
2. Mung bean nuclease (Gibco-BRL).
3. Phenol/chloroform.
4. Chloroform.
5. Sodium acetate: 3 M Sodium acetate, pH 5.0.
6. 96% Ethanol.
7. TE: 10 mM Tris-HCl (pH 8.0), 1 mM EDTA

2.1.4.2. BLUNTING WITH EXO VII EXONUCLEASE

1. Exo VII buffer.
2. Exo VII (Gibco-BRL).
3. Sodium acetate.
4. Ethanol.
5. Sepharose-CL-6B (Amersham Pharmacia Biotech).
6. TE: 10 mM Tris-HCl (pH 8.0), 1 mM EDTA
7. 10X TA buffer.
8. T4 DNA polymerase (Gibco-BRL).
9. dNTPs: Solution 2.5 mM in each of dATP, dCTP, dGTP, and dTTP.

2.1.5. Ligation

1. 10X Ligation buffer: 500 mM Tris-HCl, pH 7.5, 100 mM MgCl$_2$, 100 mM DTT, 10 mM spermidine, 10 mM ATP, and 1 mg/mL bovine serum albumin.
2. T4 DNA ligase (Gibco-BRL).

2.2. Template Screening

1. LB: 1% Bacto-tryptone (Difco, Detroit, MI), 0.5% Bacto-yeast extract, and 1% NaCl (all w/v).
2. Protoplasting buffer: 20 mM Tris-HCl, pH 7.5, 5 mM EDTA, 50 mM NaCl, 20% w/v sucrose, 100 pg/mL RNase A, and 100 µg/mL lysozyme.
3. Lysis buffer: 2.5 mM EDTA, 2% SDS, 10% w/v sucrose, and 0.04% w/v bromophenol blue.
4. Gel buffer: 1X TBE, 0.05% SDS (w/v).
5. Gel stain: 0.5 µg/mL ethidium bromide in water.

3. Methods

3.1. Exonuclease Deletions

To prepare the DNA for Exo III treatment the sample is first digested with two restriction enzymes, one leaving a blunt end or a 5' overhang at the end of the insert to be deleted, the other protecting the vector primer site from deletion by leaving a 3' overhang of at least three bases. As an alternative, the digestion is performed in two steps. In the first step, the site to be used to protect the priming site is 3' endfilled with thiophosphates. In the second step, the insert is prepared for deletion by digesting off its 3' thiophosphate protection with an enzyme leaving a blunt end or a 5' overhang (*see* **Notes 1** and **2**).

3.1.1. Linearizing the Insert

1. Mix 15 µg DNA, water, 88 µL 10X TA buffer, and 40 U of both restriction enzymes in a total volume of 80 µL. Digest for 2 h at 37°C (*see* **Notes 3** and **4**).
2. Add 50 µL phenol/chloroform and vortex for 1 min before centrifugation at 10,000g for 4 min. Repeat the extraction with 50 µL chloroform.
3. Spin dialyze the sample on a column prepared from 500 µL of Sepharose slurry (*see* **Note 5**). Add 14 µL 10X Exo III buffer, adjust the volume to 140 µL, and use 70 µL for digestion with Exo III, store the remainder of the sample at –20°C.

3.1.2. Protecting the Primer Site with Thiophosphate

1. Perform **steps 1–3** from **Subheading 3.1.1.** as far as the spin dialysis, but using in **step 1** 40 U of the restriction enzyme distal to the site to be used to expose the insert (i.e., the restriction site closest to the primer site of the vector).
2. Add 10 µL 10X TM and 2 µL of thiophosphate-dNTPs, followed by 15 U Klenow fragment of DNA polymerase I in a total volume of 100 µL. Incubate 30 min at room temperature.
3. Heat at 65°C for 10 min then cool to room temperature.
4. Add 10 µL of 10X TA, 40 U of an enzyme that will produce a blunt end or a 5' overhang at the end of the insert, releasing the thiophosphate-labeled site, and water to 120 µL. Incubate for 1 h at 37°C.
5. Extract with phenol/chloroform and chloroform as above then spin dialyze the sample on a column prepared from 600 µL Sepharose slurry (Amersham Pharmacia Biotech). Continue as in **Subheading 3.1.1.**, **step 2.**

3.1.3. Digestion with Exo III

The prepared sample is then treated with Exo III to progressively delete through the insert and samples removed at appropriate times, including a zero-time sample.

1. Preincubate the sample for 2 min at 37°C, then add 125 U Exo III. Mix rapidly and immediately remove 10 µL to (1) 10 µL stop buffer if using mung bean nuclease to generate blunt ends, or (2) to 1.5 µL 10X Exo VII buffer and 4 µL water if using Exo VII. Place on ice.
2. Remove a 10-µL sample as above every minute for 6 min. Heat the samples for 10 min at 65°C then cool to room temperature (*see* **Notes 6–8**).

3.1.4. Blunting the DNA

Two methods may be used to generate blunt ends for ligation. The use of mung bean nuclease is more rapid than the method with Exo VII but has the disadvantage of causing occasional loss of the primer site due to "nibbling back" or destruction of DNA caused by digestion from nicks or through an excessive ratio of enzyme to substrate. Use of Exo VII has the additional advantage that up to five times more colonies are obtained per microgram of DNA than in the nuclease SI method.

3.1.4.1. BLUNTING WITH THE MUNG BEAN NUCLEASE

1. Add 100 µL mung bean buffer containing 100 U/mL mung bean nuclease to each sample. Incubate at 37°C for 30 min.
2. Extract with 100 µL phenol/chloroform, followed by 50 µL chloroform, centrifuging each time at 10,000g for 4 min.
3. Add 12 µL sodium acetate and 400 µL ethanol to each sample and leave at –70°C for 1 h before centrifugation at 10,000g for 10 min. Rinse with cold ethanol, dry the pellet, and dissolve in 20 µL TE.

3.1.4.2. BLUNTING WITH EXO VII NUCLEASE AND T4 DNA POLYMERASE

1. Add 1.5 U Exo VII to each tube and incubate for 1 h at 37°C, then heat at 65°C for 15 min.
2. Add 10 µL water and spin dialyze the sample on a spin column made from 250 µL Sepharose slurry (Pharmacia) to remove the phosphate that will otherwise interfere with ligation, add 3 µL sodium acetate and 100 µL ethanol, and leave at –70°C for 1 h before centrifugation at 10,000g for 10 min. Rinse with cold ethanol, dry the pellet, and dissolve in 10 µL TE.
3. Add 2 µL 10X TA buffer, 1 µL dNTPs, 5 U of T4 DNA polymerase, and water to 20 µL. Incubate at 25°C for 30 min. Heat at 65°C for 10 min then cool to room temperature.

3.1.5. Measurement of the Rate of Deletion and DNA Ligation

In order to ascertain the rate of deletion, and to determine which DNA samples should be ligated and used for transformation, part of the sample is run out on an agarose minigel. Large inserts in small vectors may be separated directly, but it is often more accurate to cut out the remaining insert with a suitable restriction enzyme with a site on the polylinker distal to the end subjected to deletion.

1. Take 8 µL of the blunted sample and add 1 µL 10X TA and 10 U of an polylinker enzyme cleaving at the undeleted end of the insert or Pvu II, which cuts in the flanking *lac* regions of many vectors. Digest for 1 h at 37°C and then separate on a suitable agarose gel. Ligate suitable timepoints (*see* **Note 9**).
2. Add 3 µL of the remaining sample to 3 µL 10X ligation buffer, 3 U T4 DNA ligase and water to 30 µL.
3. Incubate at 15°C overnight. Use 3 µL of the ligation mixture to transform 200 µL of competent cells, and plate out the entire sample.

3.2. Template Screening

Following deletion and transformation, colonies are picked at random from each timepoint and screened for insert size. A rough approximation of size is obtained through sizing intact plasmid using protoplast lysis. This is followed by making minipreps and screening them by restriction digest.

3.2.1. Screening by Streaking Colonies

1. Pick 12 colonies from each timepoint using sterile toothpicks. Streak each colony onto a fresh Luria Bertani (LB) plate prepared with a suitable antibiotic and marked on the bottom into 12 sectors. Grow overnight at 37°C.
2. Scrape off about 0.5–1 cm of each streak with a pipet tip, picking up as little agar as possible, and transfer to a 0.5-mL microcentrifuge tube containing 10 µL of protoplasting buffer, vortexing vigorously to disperse the sample. Leave the tip in the sample for loading onto the gel. Incubate for 15–30 min at room temperature (*see* **Note 10**).
3. At the same time, prepare insert size markers, combining streaks of cells transformed either with the vector or with vector containing the full-length insert.
4. Load the samples onto a 0.7% agarose gel, into wells preloaded with 5 µL of lysis buffer. Use a supercoiled DNA ladder (Gibco-BRL) as a size marker (*see* **Notes 11** and **12**).
5. Leave the gel for 5 min after loading to ensure lysis of the cells, then run the gel at 50 V for 15 min before increasing to 100 V and running until the dye band runs off the bottom of the gel.
6. Stain the gel and photograph the DNA bands.

3.2.2. Screening Using Microtiter Plates

1. Add 100 µL LB plus ampicillin to each required microliter well and inoculate with a single colony. Incubate at 37°C overnight, shaking at about 150 rpm. To prevent evaporation of the medium it is best to place autoclave tape over the wells.
2. Remove 10 µL of the medium to a second microliter plate containing 5 µL of protoplasting buffer in each well. Incubate and run on an agarose gel as above.

4. Notes

1. A potential problem with this strategy, particularly with long inserts, is the need to find two restriction enzyme sites in the region between the insert and the primer site that will satisfy the foregoing conditions and are also not present in the insert. The more sites there are in the polylinker, the easier this process becomes. For this reason vectors such as Bluescript (Stratagene, La Jolla, CA) are ideal for this approach. If inserts can be cloned into the central sites of the Bluescript polylinker then it may be possible to delete the insert from either end, with the additional advantage that three priming sites exist at each end of the polylinker.
2. By judicious selection of enzyme sites close to the insert and using thiophosphates to block exonuclease activity, it may be possible to delete through an insert in several stages if enough sites are left on the primer side of the polylinker to provide a second linearization satisfying the criteria for Exo III digestion.
3. It is essential that the initial digestions with restriction enzyme to open up the insert for deletion go to completion. The presence of nonlinearized plasmid will eventually result in the recovery of large numbers of full-length inserts during template screening.
4. The amount of DNA to be digested depends on the length of insert that it is desired to remove. The values suggested in the methods are sufficient to provide about 1 µg DNA

per timepoint over 6 min of treatment, with a deletion rate of around 200 bp/min, that should provide enough deletion to sequence an insert of around 1.5 kb. The suggested amounts and volumes should be increased proportionately for longer inserts. There is a broadening of the size range of the insert in samples taken at later timepoints.

5. It is unnecessary to use cesium gradient-purified DNA for this technique, although use of such DNA has the advantage that enough material is available for several trials to determine the rate of deletion. In this laboratory, deletions are mainly performed on miniprep DNA to avoid the time-consuming process of caesium gradient banding. The use of spin columns to clean up linearized DNA provides excellent substrates for digestion. However, the digestion rates of miniprep plasmid DNA tend to be more variable than when using more purified material. If using miniprep DNA the plasmid from two overnight 10-mL cultures are pooled in a volume of 100 µL. This provides enough material for six trials of the digestion rate if 30 µL is used per linearization.

6. The activity of Exo III falls off with length of incubation, and so if a longer insert is being treated later, samples should be taken at longer intervals than the initial samples. For example, if digesting an insert of 5 kb the amount of DNA, exonuclease, and the reaction volumes should be tripled, and samples 7–12 removed at 2-min intervals, with samples 13–18 being removed every 5 min.

7. Substantial variation in the rate of digestion is often observed because of factors such as quality of the Exo III and purity of the DNA substrate. If insufficient or excessive digestion rates are observed, the digestion may be repeated using the retained portion of the linearized plasmid and altering the amount of enzyme used.

8. Because Exo III is moderately temperature sensitive, the rate can also be controlled by altering the digestion temperature, a particularly useful parameter if only a few nucleotides need to be removed (e.g., in studies on promoter location). If, under the conditions used, 200 bp are removed every minute at 37°C, then about 120 bp will be removed at 30°C and 50 bp at 20°C.

9. When the rate of Exo III deletion is measured by cutting out the insert before gel separation, any change in mobility of the vector band indicates substantial digestion through the primer site. If this is observed, the samples should not be ligated, as the primer sites will have been lost.

10. If the initial screening for insert size is done by streaking out the transformant, some practice is required in judging the amount of cells to take. After vortexing, the solution should be turbid rather than dense. Better results are obtained when too little, rather than too many, cells are used. The minimum sample that can still be seen on the gel would be about half of the cells from a large colony on the original plate.

11. When screening the transformants by protoplast lysis, it is important not to use a submerged gel. Buffer should be added to the gel tank until it is a few millimeters below the gel surface. Failure to do this will result in the DNA being sucked out of the well by changes in surface tension as the sodium dodecyl sulfate in the lysis buffer mixes with the gel buffer.

12. After a first round of screening by protoplast lysis, it is preferable to screen again by making miniprep DNA and measuring the insert size more accurately by restriction enzyme digest. If a restriction enzyme site on the end of the polylinker has been used to protect the primer site, it will clearly be removed during the linearization process. The Pvu II sites flanking the polylinker in pUC and Bluescript plasmids (Stratagene) provide useful sites for digestion in such circumstances, or other rare sites in the vector close to the insert may be used.

References

1. Henikoff, S. (1984) Unidirectional digestion with exonuclease III creates targeted breakpoints for DNA sequencing. *Gene* **28,** 351–359.
2. Wen-Qin, X. and Potts, M. (1989) Quick screening of plasmid deletion clones carrying inserts of desired sizes for DNA sequencing. *Gene Anal. Tech.* **6,** 17–20.

100

Primer-Directed Site-Specific Mutagenesis

Michael J. O'Donohue and G. Geoff Kneale

1. Introduction

Site-directed mutagenesis provides a powerful means for probing protein structure and function. Using this approach, one can introduce specific amino acid changes at any given position in the protein sequence and test the functional consequences of these mutations in vitro or in vivo. If an in vivo test is available, strategies that introduce random base changes at a given codon (or indeed in a larger region of the protein) are particularly valuable.

Since the use of primer mutagenesis was first reported *(1)* the technique has rapidly developed. The principle of primer-directed mutagenesis is relatively simple and can be summarized schematically (*see* **Fig. 1**). The more recent alterations to the basic technique have largely been concerned with achieving greater mutagenic efficiency (*see* **refs.** *2–4* for the most common methods). All of these techniques select against the wild-type DNA. The last of these methods *(4)* involves the incorporation of thionucleotides into the mutant DNA strand (which protects against cleavage by certain restriction enzymes in vitro). The protocol that will be described here is based on this method (*see* **Note 2**). The major steps of this procedure are shown in **Fig. 2**. Alternative procedures allowing the introduction of site specific mutations include cassette mutagenesis *(6,7)* and mutagenesis by the polymerase chain reaction (PCR) *(8–10)*.

Any of the mutagenesis techniques cited here can be extended to carry out more complex mutagenic studies, such as site saturation mutagenesis. In this strategy, a codon defining a particular amino acid in a protein is substituted by a degenerate triplet that encodes either all of the other nineteen amino acids or a subset thereof *(11–14)*. We describe a method based on primer mutagenesis that is rendered more efficient by the use of a strand-selection procedure. This method is presented in **Subheading 2.** as a variant of the standard site-directed mutagenesis experiment since, with the exception of the oligonucleotide design, the steps are identical.

2. Materials

1. Agarose gel: 0.8% Electrophoresis grade agarose (w/v) in 0.5X Tris-boric-acid-ethylenediaminetetraacetic acid (TBE) buffer. The agarose is dissolved in 100 mL of TBE buffer by heating the solution for 3 min, with intermittent swirling, in a microwave oven

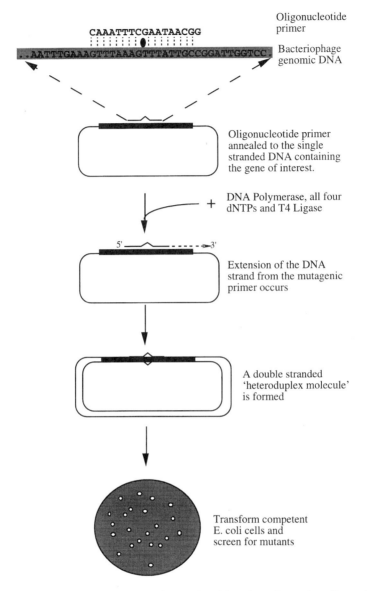

Fig. 1. Site-directed mutagenesis. An oligonucleotide primer is used to direct the mismatching of bases when annealed to the single-stranded DNA template. Extension of the primer using DNA polymerase I (or Klenow fragment) converts the molecule into a double-stranded heteroduplex, which can be used to directly transform a suitable *E. coli* host.

at a power setting of 500 W. The agarose solution is allowed to cool to 60°C before pouring it into a gel-casting tray. The comb is put in place and the gel is allowed to set.
2. Buffer A: 700 mM NaCl, 700 mM Tris-HCl, pH 8.0.
3. Buffer B: 100 mM dithiothreitol (DTT), 60 mM MgCl$_2$, 300 mM NaCl.
4. Buffer C: 100 mM DTT, 60 mM MgCl$_2$, 800 mM NaCl, 400 mM Tris-HCl, pH 8.0.
5. Centrex™ filter unit-centrifugal filter (Schleicher and Schuell, Dassel, Germany) with 0.45 μm pore size nitrocellulose filters.
6. Centrifugal evaporator: Savant Speedvac (Hicksville, NY), or similar.

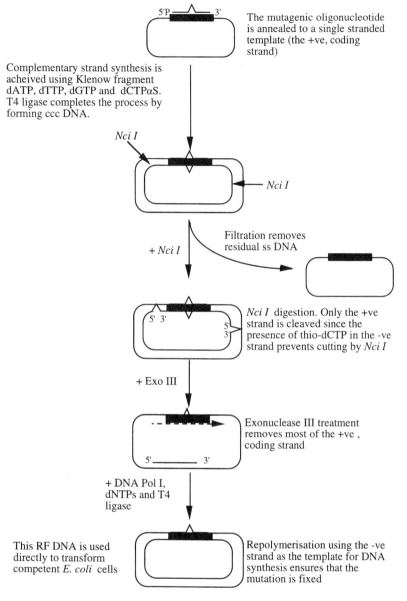

Fig. 2. Oligonucleotide-directed mutagenesis using phosphorothioate DNA. Polymerization of the complementary strand is carried out in the presence of dCTPαS. Following the removal of residual single-stranded template molecules, the double-stranded heteroduplex is incubated with NciI. Because the restriction enzyme cannot cleave the phosphorothioate DNA, the heteroduplex is nicked in the template strand. The template strand is removed by incubation with exonuclease III and subsequent repolymerization fixes the mutation.

7. Competent *Escherichia coli* cells: *E. coli* DH5αF' can be made competent for transformation by DNA using any of the recommended techniques (*15,16*, **Note 3**). A protocol is described in **Subheading 3.5**. Other *E. coli* strains displaying an F+ genotype, such as TG1, JM109, and MV1193, may be used in place of DH5αF' if this strain is not available.

8. Dimethylsulfoxide (DMSO): Spectroscopy-grade DMSO should be frozen in small aliquots in polypropylene tubes at –80°C immediately after the bottle has been opened in order to avoid oxidation. Once an aliquot is thawed for use, the remaining DMSO should be discarded.
9. Double-stranded (RF) DNA marker: An aliquot of an RF DNA solution (100 ng/µL) is combined with an equal aliquot of 5X loading buffer. The volume of the mixture is then adjusted to give a final concentration of 1X loading buffer. A 10-µL aliquot of this solution can be loaded on to an agarose gel to act as a marker.
10. Enzymes: Polynucleotide kinase, Klenow fragment, DNA polymerase I, and Exonuclease III are routinely purchased from Pharmacia (Uppsala, Sweden). The preferred source of the restriction enzyme *Nci*I is New England Biolabs (Beverly, MA).
11. Ethanol: Analytical reagent-grade absolute ethanol is used, which is stored, in small quantities, at –20°C.
12. 80% Ethanol solution: Prepare using distilled H_2O and absolute ethanol (Analytical reagent grade) and store at –20°C.
13. Ethidium bromide solution: Prepare a concentrated stock solution of ethidium bromide (10 mg/mL solution) using ultrapure water. **WARNING:** Ethidium bromide is a carcinogen and should therefore be handled with extreme care.
14. Frozen storage buffer (FSB): Prepare a of solution of potassium acetate (1 M) in ultrapure water using a solution of KOH to obtain a final pH of 7.5. Sterilize by filtration through a sterile, 0.2 µm, detergent-free filtration unit, then use this stock solution to prepare FSB. For FSB, make a potassium acetate (10 mM)/glycerol (10% w/v) solution containing KCl (7.4 g/L), $MnCl_2 \cdot 4H_2O$ (8.9 g/L), $CaCl_2 \cdot 2H_2O$ (1.5 g/L) and hexamminecobalt trichloride (0.8 g/L) and adjust the pH to 6.4 with HCl (1 N). Sterilize by filtration through a sterile, 0.2 µm, detergent-free filtration unit then store at 4°C.
15. H_2O: We routinely use ultrapure water produced by a Millipore purification system (Bedford, MA).
16. 10X Kinase buffer: 1 M Tris-HCl pH 8.0, 100 mM $MgCl_2$, 70 mM DTT, 10 mM ATP. Prepare the buffer using H_2O and sterilize by filtration.
17. 5X Loading buffer: 50% Glycerol (v/v), 100 mM EDTA, pH 8.0, 0.125% bromophenol blue (w/v), 0.125% xylene cyanol (w/v).
18. Microcentrifuge: A centrifuge capable of reaching a maximum speed of approx 11,000g will be required that can accommodate 1.5 mL microcentrifuge tubes.
19. Mg^{2+} solution: Prepared by dissolving 203 g of $MgCl_2 \cdot 6H_2O$ and 247 g of $MgSO_4 \cdot 7H_2O$ in ultrapure water to a final volume of 1 L. The solution is sterilized by filtration through a sterile, 0.2-µm, detergent-free filtration unit. Ten milliliters of this solution is added to 1 L of SOB-Mg to make SOB. SOB agar (Bacto-agar at 15 g/L [Difco, Detroit, MI]) and SOB soft agar (Bacto-agar at 7.5 g/L) are made by adding 10 mL of the Mg^{2+} solution to 1 L of molten medium.
20. 100 mM $MgCl_2$: Prepare using ultrapure H_2O. Prior to use, the solution is sterilized by autoclaving.
21. 5 M NaCl: Prepare using ultrapure H_2O. Before use, the solution is sterilized by autoclaving.
22. 500 mM NaCl: Dilute the 5 M sodium chloride solution 10 times with ultrapure water.
23. Nucleotide mix A: 3.1 mM each of dATP, dGTP, dTTP, and dCTPαS, 12.5 mM rATP.
24. Nucleotide mix B: 200 mM each of dATP, dGTP, dTTP, and dCTPαS, 8.3 mM rATP.
25. rATP solution: We routinely use fresh, molecular biology grade rATP (or equivalent).
26. Single-stranded template DNA: A 1 µg/µL solution of single-stranded template DNA containing the gene of interest is prepared using a standard protocol for the preparation of bacteriophage or phagemid DNA *(15)*. The single-stranded DNA is dissolved in H_2O

before use and the concentration of the solution is determined by measuring the absorbance at 260 nm. One can assume that 1 OD_{260nm} unit represents a concentration of approx 33 μg/mL.

27. Single-stranded DNA marker: This marker is prepared using the same method as that employed to prepare the double-stranded DNA marker. In this case, a 100-ng/μL single-stranded DNA solution is used instead of the RF DNA solution.
28. SOB-Mg: This culture medium is prepared by combining tryptone (20 g/L), yeast extract (5 g/L), NaCl (0.58 g/L), and KCl (0.19 g/L) in water of the highest quality possible. This solution is then sterilized by autoclaving. Before use, it is recommended to sterilize by filtration using a sterile, 0.2 μm, detergent-free filtration unit. For solid SOB-Mg, Bacto-agar (Difco) (15 g/L) is added to the solution before autoclaving.
29. Sodium acetate solution: 3 M CH_3COONa, pH 6.5
30. Staining solution: A few drops of the ethidium bromide solution in 100 mL of ultrapure water contained in a sealable plastic tray (a sandwich-type box is ideal). This solution can be reused several times before disposal. Postelectrophoretic staining of agarose gels limits ethidium bromide contamination in the laboratory.
31. TBE buffer: 90 mM Tris base, 90 mM boric acid, 2.5 mM EDTA, pH 8.3.
32. The mutagenic oligonucleotide: Our oligonucleotides are routinely purchased from Eurogentec (Abingdon, Oxon, UK) although a large number of companies offer a similar range of synthesis services. For site-saturation mutagenesis, we have used a Cruachem P250A automated DNA synthesizer (Cruachem, Glasgow, Scotland), which is capable of synthesizing randomized oligonucleotides. *See* **Notes 4–7** for the design of a suitable oligonucleotide sequence.
33. Tris buffer: 500 mM Tris-HCl, pH 8.0.
34. Water baths: At least three water baths will be required. One should be set at 16°C in the cold room and the other two should be set at 37°C and 70°C, respectively.

3. Methods
3.1. Oligonucleotide Synthesis

An oligonucleotide is synthesized using an automated DNA synthesizer. If the oligonucleotide is to be used to generate a single-base mutation, the oligonucleotide will be fully complementary to a region in the gene of interest, apart from a single-base mismatch that will define the mutation. Oligonucleotides used for this purpose should have a minimum length of 17 bases but may be longer depending on circumstances (*see* **Note 4**). For site-saturation mutagenesis, the oligonucleotide will be complementary to a region in the gene of interest but will contain a degenerate sequence of three bases in the center of the oligonucleotide (if the coding sequence is on the plus strand, these three bases correspond to the anticodon of the amino acid that one wants to mutate; *see* **Note 7**).

3.2. Oligonucleotide Purification and Phosphorylation

1. For primer-directed mutagenesis we find that postsynthesis purification of the mutagenic primer is unnecessary. However, no problems should be encountered if a purification step is included as part of a commercial synthesis service. If the oligonucleotide purification is to be carried in the laboratory, we have found that NEN-Sorb purification columns (NEN-Dupont, Boston, MA) produce good results. In order to employ these columns, however, the last trityl group to be added during the oligonucleotide synthesis must be left on.

2. For site-saturation mutagenesis, purification maybe disadvantageous. Because the mutagenic oligonucleotide is degenerate in three positions, the synthesis product will represent a mixture of oligonucleotide species, which may be differentially purified.
3. 5' Phosphorylation of an oligonucleotide can be performed during the chemical synthesis. However, if this is not possible, or if the enzymatic method is preferred, 0.125 OD_{260nm} U of the mutagenic oligonucleotide is combined with 3 µL of 10X kinase buffer in a microcentrifuge tube. Add 25 µL of ultrapure water and 2 U of T4 polynucleotide kinase. Mix by repeated pipetting and incubate at 37°C for 15 min (*see* **Note 8**). Terminate the reaction by incubating the reaction mixture at 70°C for 10 min. Store the sample at –20°C until required.

3.3. Heteroduplex Formation

1. Combine 5 µg of single-stranded template DNA (M13 or phagemid DNA, e.g.) with 4 pmol of phosphorylated mutagenic oligonucleotide and 3.4 µL of buffer A in a screw-top microcentrifuge tube. Adjust the total volume of the mixture to 17 µL with H_2O.
2. Place the tube in a 100°C water bath. Remove the water bath from the heat source and cool by incubation at 4°C until the temperature of the water has fallen below 35°C. Finally, place the tube on ice. This procedure ensures optimal annealing of the oligonucleotide to the template.
3. To the mixture add a 6.25 µL aliquot of 100 m*M* $MgCl_2$, 5 µL of 500 m*M* Tris buffer, and 5 µL of nucleotide mix A. Add 6 U each of Klenow fragment and T4 DNA ligase and adjust the total volume to 50 µL with H_2O. Mix briefly by gentle repeated pipetting and incubate the reaction at 16°C overnight (at least 16 h). This is sufficient to ensure that the primer has been extended all the way round and ligated to form covalently closed, circular, double-stranded DNA.
4. Combine 1 µL of the reaction mixture with 2 µL of 5X loading buffer and adjust the sample volume to 10 µL with water. Load the sample into a well of a 0.8% agarose gel and electrophorese, along with the double- and single-stranded DNA markers, for 30 min at a constant voltage of 100 V (the exact conditions will vary depending on the apparatus used), using TBE buffer as the electrophoresis buffer. Once the migration is complete, the gel is transferred to the tray, containing the staining solution, which is placed on a slow-moving agitator for 10 min. After, observe the gel by UV illumination. If one finds that the conversion of the single-stranded template DNA to heteroduplex DNA has been successful, the mutagenesis procedure can continue (*see* **Note 8**).
5. Add 170 µL of water and 30 µL of 5 *M* NaCl to the heteroduplex DNA mixture. Apply this solution to the top of a Centrex™ filter unit. With the lower half of the filter unit in place, centrifuge at 15,000*g* (1500 rpm) for 10 min at room temperature in a Sorvall RC-5B employing an HB-4 swing-out rotor. Wash the filter by adding 100 µL of 500 m*M* NaCl to the upper part of the filter unit. Centrifuge again using the same conditions as before. Remove the lower half of the filter unit containing the sample. This procedure removes excess single-stranded template DNA (*see* **Note 9**).
6. Add 700 µL ice-cold ethanol and 28 µL sodium acetate solution and incubate at –20°C for 30 min. Recover the precipitated DNA by centrifugation in a microcentrifuge at 15,000*g* (13,000 rpm) for 15 min at 4°C (*see* **Note 10**). Wash the DNA pellet with 1 mL ice-cold 80% ethanol and centrifuge as before. Carefully discard the supernatant and dry the pellet in a centrifugal evaporator.

3.4. Strand Selection

1. Resuspend the DNA in 25 µL of 90 m*M* Tris-HCl, pH 8.0. At this stage, 15 µL of the DNA solution is stored at –20°C and the mutagenesis is continued using only a 10-µL aliquot of the DNA sample.

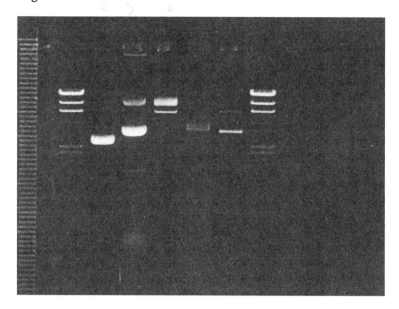

Fig. 3. Analysis of the samples of agarose gel electrophoresis. The samples from the mutagenesis procedure were electrophoresed on a 1% agarose gel containing ethidium bromide at a constant voltage of 100 V. The DNA bands were visualized by illuminating the gel on a UV transilluminator (254 nm). Lanes (1, 7) λ HindIII marker DNA; (2): single-stranded template DNA marker DNA (20 ng/µL); (3) double-stranded (replicative form) template DNA marker (20 ng/µL); (4) DNA sample after filtering and nicking with NciI; (5) DNA sample after digestion with exonuclease III; (6) DNA sample after repolymerization and ligation.

2. Add a 7.5 µL aliquot of buffer B to the 10 µL sample and adjust the total volume to 75 µL with H_2O. Add 5 U of NciI and incubate at 37°C for 90 min. After incubation, remove a 10-µL sample of this reaction mixture for subsequent analysis by agarose gel electrophoresis.
3. Combine the remainder of the solution (65 µL) with 9 µL of buffer C and adjust the total volume to 88 µL with H_2O. Add 50 U of exonuclease III (25 U/µL). After mixing by repeated pipettings, incubate the reaction mixture at 37°C. The time of this incubation is determined by the position of the NciI site(s) with respect to the mutagenic mismatch. If M13 DNA is used, a digestion time of 30 min should suffice. However, exonuclease III will digest about 100 bases per minute so it is possible to estimate a suitable digestion time (*see* **Note 11**). Following digestion, it is necessary to heat inactivate the Exonuclease III by incubation at 70°C for 15 min. Following a 30-s centrifugation at 15,000g (13,000 rpm) at room temperature in a microcentrifuge, remove a 15-µL sample for subsequent analysis by agarose gel electrophoresis.
4. Add 9.5 µL of nucleotide mix B, 6 µL of 100 mM $MgCl_2$ solution, 3 U of DNA polymerase I and 2 U of T4 DNA ligase to the reaction mixture. Adjust the total volume to 95 µL with H_2O and incubate the sample at 16°C for at least 3 h. Remove a final 15-µL sample for analysis by agarose gel electrophoresis.
5. At this stage, to each of the samples taken from **steps 2–4** should be added 5X loading buffer (giving a final concentration of 1X loading buffer). Electrophorese the samples on a 0.8% agarose gel using the same conditions as those described in **step 4**. Again, the double- and single-stranded DNA markers should be also loaded on to the gel (*see* **Note 12** and **Fig. 3**).

6. Use aliquots of the remainder of the reaction mixture (e.g., 1 and 10 µL) to transform competent *E. coli* cells in separate experiments using any suitable transformation procedure.
7. In order to screen for mutants, it is prudent to sequence 10 isolates by Sanger dideoxy sequencing. Of course, if a phenotypic selection procedure is available then the task of finding mutant derivatives may be simpler (*see* **Note 13**).

3.5. Transformation with the Heteroduplex

The following protocol may be used to make competent *E. coli* for transformation:

1. Streak out a F+ *E. coli* strain on a SOB-Mg agar plate and incubate for 16 h at 37°C in order to obtain isolated colonies.
2. The next day, pick several colonies and disperse them in 1 mL of SOB by vigorous agitation on a vortex mixer. Use this inoculum to inoculate SOB medium (10 mL/colony picked) in an Erlenmeyer flask where the culture volume to flask volume ratio is between 1:10 and 1:30. Incubate at 30°C at 200 rpm until the cell density reaches $6–9 \times 10^7$ cells/mL (for DH5αF', this corresponds to an OD_{550nm} of 0.5).
3. Transfer the cells to prechilled tubes suitable for centrifugation and incubate on ice for 15 min. Remove an aliquot to determine the viable cell density by serial dilution and plating on SOB agar plates.
4. Pellet the cells at 750–1000 g at 4°C for 12 min. Discard the supernatant and drain the pellet thoroughly using a micropipet to remove the last drops. Resuspend the pellet in one-third of the culture volume of FSB by moderate agitation. Incubate on ice for 15 min.
5. A second centrifugation is performed as in **step 4**, and the pellet is once again thoroughly drained. The pellet is resuspended in FSB (1/12.5 of the original culture volume) and DMSO is added to a final concentration of 3.5% (v/v). The DMSO is squirted into the middle of the tube and the cell suspension is immediately swirled and then incubated on ice for 5 min. This procedure is repeated with a second, equal aliquot of DMSO and the cell suspension is incubated on ice for a further 15 min.
6. The cell suspension is divided into 210 µL aliquots, flash frozen in liquid nitrogen or in a dry ice/ethanol bath, and transferred to –80°C for storage.
7. For transformation, thaw a tube of competent cells on ice and transfer the cells to a prechilled 10-mL tube (preferably polypropylene). Add the DNA in of volume which does not exceed 20 µL and swirl gently to mix. Incubate the tube on ice for 10–30 min.
8. Heat shock the cells for 90 s at 42°C and then chill immediately on ice for 5 min.
9. Add 200 µL of a fresh *E. coli* culture and 2–3 mL of molten soft SOB agar, which is held at 45°C until use. Mix by inversion of the tube and pour immediately onto SOB agar plates. Once the soft agar is solid, incubate the plates inverted, overnight at 37°C.

4. Notes

1. A more recent publication by Jung et al. (*20*) describes a method of mutagenesis which employs a hybrid procedure, where elements of the methods described by Kunkel et al. (*2*) and Taylor et al. (*4*) are brought together to allow mutagenesis of double-stranded plasmid DNA. In principle (although we have not attempted this ourselves), the oligonucleotide annealing protocol described in this new method, as well as the subsequent separation of the denatured plasmid DNA from the nondenatured molecules, could be incorporated into the procedure that we describe here. In this way, one could extend the method to the mutation of sequences contained in double stranded plasmids.
2. A similar protocol to that described here has essentially been adopted in a site-directed mutagenesis kit (Sculptor, Amersham). The buffers and nucleotide mixes as well as all of the enzymes required to perform the mutagenesis reaction are supplied in this kit.

3. As for any other procedures involving a transformation step, the described mutagenesis procedure requires a reasonably good efficiency of transformation in order to obtain a sufficient number of colonies at the end. **Reference 16** is an excellent review article in which most of the currently used techniques for transforming *E. coli* are described.
4. For a single-base substitution, an oligonucleotide composed of as few as 17 bases can be used. For a more complex procedure, such as site-saturation mutagenesis, a larger oligonucleotide of at least 27 bases should be used. When designing any mutagenic oligonucleotide, the bases that will direct the mutation should be placed at the center of the oligonucleotide. The primer sequence should be checked to ensure it does not contain an *Nci*I site. An oligonucleotide primer containing an *Nci*I site will allow the heteroduplex to be cut, as the primer does not contain phosphorothioate nucleotides. If an *Nci*I site is found in the primer sequence, one should either redesign the primer (if this is possible) or employ an alternative restriction enzyme in the place of *Nci*I. Eckstein and co-workers *(17–19)* have investigated several restriction enzymes that display an inability to cleave DNA containing phosphorothioate nucleotides. However, if an alternative restriction enzyme is employed, several alterations to the basic protocol may have to be made.
5. Once the mutagenic oligonucleotide has been designed it is advisable to check that the percentage of A-T content in each of the primer arms is not too high and that there are no other competing primer sites in the template DNA.

   High A-T content may lead to instability of the oligonucleotide when it anneals to the template, thus, as a rule of thumb, the melting temperature (T_m) of each "arm" of the mutagenic oligonucleotide (the sequences either side of the mismatch) should not be lower than 37°C. The T_m can be estimated for short oligonucleotides (16–40 mers) using the Wallace rule: $T_m = 4\ (\%\ \text{G-C}) + 2\ (\%\ \text{A-T})$.

   In order to check for competing primer sites in the template DNA, a computer DNA analysis program will be necessary. One should verify that there are no other contiguous sequences that have more than 70% of bases in common with the mutagenic oligonucleotide. If potential competing sites are identified, then the sequence of the mutagenic oligonucleotide should be altered. This can be achieved either by increasing the length of the oligonucleotide or by shifting the sequence of the oligonucleotide with respect to the target site. Competing sites on the template for the mutagenic oligonucleotide can also be detected by Sanger dideoxy sequencing. The template is employed in a sequencing reaction in which the mutagenic oligonucleotide is used as the sequencing primer. If there is more than one target site for the oligonucleotide, multiple DNA sequences will be observed on the resulting autoradiograph. Normally, one can expect a sequence as good as that which could be obtained using a normal, fully complementary primer. However, as long as the correct sequence is predominant, with other sequences being only faintly represented, the mutagenic oligonucleotide may still be useful.
6. It has been suggested that greater success with site-directed mutagenesis could be obtained if a more rational approach to oligonucleotide design were adopted *(21)*. For this, a more sophisticated analysis must be carried out using a computer analysis package such as Oligo (NBI, Plymouth MN). This particular computer software allows one to analyze the internal stability of the oligonucleotide, its T_m and composition, and its potential to form hairpins and/or dimers that may interfere with the overall efficiency of the oligonucleotide for mutagenesis procedures.
7. When designing an oligonucleotide for site-saturation mutagenesis, careful consideration should be given to the following points:
   a. If the oligonucleotide employs the degenerate sequence NNN (where N is G, A, T or C) to specify the codon that is to be mutated, on mutagenesis all 64 codons will be made. Therefore, theoretically 1 in 64 mutant isolates will display a wild-type geno-

type. Furthermore, if the mutagenesis is not 100% efficient (which is probable), some isolates that display the wild-type genotype will represent unmutated templates which will be indistinguishable from the mutated wild-type isolates in their DNA sequence. If it is necessary to calculate the efficiency of mutagenesis, this may represent a problem (but see part b).

b. It may be desirable to limit the number of codons that can be created if one wants to observe the relative frequency of occurrence of any particular amino acid substitution. This may be especially important in experiments where one is employing a positive screening strategy (i.e., screening for protein activity) in order to identify those amino acid substitutions that do not abolish the functional viability of the protein. In this case, restricting the number of possible mutant codons, usually at the third base position in the codon (i.e., the first base of the "anticodon" in the mutagenic primer in it's usual orientation), will facilitate normalization of the results, so that the wild-type codon is not represented in the primer population. Any wild-type sequences that are found therefore must represent the background of unmutated template DNA. An alternative strategy is to introduce a silent "marker" base change into the oligonucleotide primer outside of the codon of interest (22).

On many automated DNA synthesizers it is possible to specify nucleotide mixtures other than those specified by N (25% of each of the four bases). For example, the random sequence NNS (where S is G or C) would specify only 32 codons but would still direct the substitution of an amino acid by all 20 amino acids in a mutagenesis experiment. To take a concrete example, if the wild-type codon were TAT (tyrosine), then this particular codon would be eliminated by allowing only G or C (A could also be included if required) at the third base (i.e., C or G at the first base of the anticodon). All 20 amino acids can be coded by this combination although their frequencies will be different. Tyrosine would also be allowed, but from the codon TAC; any transformants sequenced as TAT must therefore be due to incomplete destruction of the template, and these can be eliminated from any statistical analysis.

8. Successful 5' phosphorylation of the oligonucleotide is essential if one is to obtain closed circular double-stranded DNA after the initial primer extension reaction. Therefore, on analyzing sample 1 by agarose gel electrophoresis, one should take particular note of the amount of nicked circle DNA that is present. The nicked circle DNA will not be suitable for the mutagenesis reaction. In principle, this problem should not be encountered if the oligonucleotide is chemically phosphorylated (see **Subheading 3.2.**). However, in the case of enzymatic phosphorylation, the efficiency of the kinasing reaction can be maximized by using fresh, good quality enzyme and fresh rATP. If only 40–50% of the double-stranded DNA from the primer extension reaction is in a closed circular form, one may still proceed with the mutagenesis and expect to obtain reasonable mutant yields.

9. We have found the centrex filter unit does not fit perfectly into the centrifuge bucket. However, it appears that as long as the unit can remain in an approximately vertical position in the bucket when the centrifuge is at rest no other action is necessary. Routinely, we use a HB 4 (Dupont) swing out rotor with the buckets modified using plastic inserts that would normally accommodate a 14 mL polypropylene tube.

   WARNING: Take care to ensure that the filter unit cannot fall out of the buckets during the centrifugation process.

10. At this stage a DNA pellet may be visible. However, it is good practice to orientate the microcentrifuge tube in the same way whenever a centrifugation step is required. This ensures that even when the pellet is invisible the approximate position of the DNA is known therefore reducing the possibility of an accidental loss of the pellet.

11. If necessary, a series of exonuclease III digestions can be performed using small samples of the nicked heteroduplex DNA that was stored (*see* **Subheading 3.4., step 2**). The digestions can be terminated after different incubation times and the results can be analyzed on a 1% agarose gel in order to determine the optimum digestion time (*see* **Note 12**). The exonuclease III digestion is unlikely to completely destroy the nonmutant strand, and therefore small "primers" for the repolymerization reaction should always remain. However, if the subsequent repolymerization reaction fails as a result of the complete destruction of the nonmutant strand by exonuclease action then one can reprime the single-stranded template. This can be achieved using the M13 universal primer or any other suitable primer that may be available.

12. The analysis of each stage of the mutagenesis procedure by agarose gel electrophoresis is extremely important and should therefore be carried out routinely. **Figure 3** shows an example of the results one should obtain if the mutagenesis procedure is successful. After filtration and nicking there should be only one DNA species corresponding to the nicked circle DNA that is present in the double-stranded marker. If single-stranded template DNA, or double-stranded closed circular DNA, remains at this stage one can expect that the efficiency of mutagenesis will be considerably decreased. In this case, it may be necessary to screen more isolates in order to identify mutants if a phenotypic screen is unavailable. In the example shown there is also a small amount of linear DNA present; this does not affect the efficiency of mutagenesis.

    Similarly, the incidence of wild-type isolates will be increased if the exonuclease III treatment is inefficient. Following incubation with exonuclease III, the nicked circle DNA band seen in the previous sample should disappear, replaced by a mixed population of smaller DNA molecules similar to those in lane 5 of the example. These degraded double-stranded DNA molecules will migrate further on the agarose gel than the double-stranded nicked circle DNA.

    Finally, the repolymerization and ligation step should regenerate double-stranded closed circular DNA. This reaction does not always produce only double-stranded closed circular DNA. Often after this step, a significant proportion of the DNA may be in a nicked circle form. This is not important, however, as the DNA will still be suitable for transformation of competent *E. coli* cells. Equally, even if there is only a small amount of DNA visible one can still obtain a large quantity of isolates, as 1 ng of double-stranded closed circular DNA should produce more than 100 transformants.

13. If it is necessary to screen large quantities of isolates by Sanger dideoxy sequencing, it may be better initially to employ a single-track protocol. Using this strategy, only one dideoxynucleotide is used and mutants are identified by the loss or gain of one or more bands on the autoradiograph. The advantages of using this protocol are that more isolates can be screened using one quarter of the sequencing reagents that one would normally employ. This strategy is applicable whenever one is sure of the nature of the mutation that is being created (i.e., normal primer directed mutagenesis). In the case of site saturation, whether one uses a single-track strategy for screening will depend on the design of the mutagenic oligonucleotide. Clearly, if the sequence directing the mutation is 5' NNN 3' then the mutagenesis could result in the loss or gain of a band in either of the four tracks, and therefore only a four-track sequencing protocol will suffice for the screening procedure.

References

1. Hutchison, C. A., Phillips, S., Edgell, M., Gillam, S., Jahnke, P., and Smith, M. (1978) Mutagenesis at a specific position in a DNA sequence. *J. Biol. Chem.* **253(18)**, 6551–6560.
2. Kunkel, T. A. (1985) Rapid and efficient site-specific mutagenesis without phenotypic selection. *Proc. Natl. Acad. Sci. USA* **82**, 488–492.

3. Deng, W. P. and Nickoloff, J. A. (1992) Site-directed mutagenesis of virtually any plasmid by eliminating a unique site. *Anal. Biochem.* **200,** 81–88.
4. Taylor, J. W., Ott, J., and Eckstein, F. (1985) The rapid generation of oligonucleotide-directed mutations at high frequency using phosphorothioate-modified DNA. *Nucleic Acids Res.* **13,** 8765–8785.
5. Sayers, J. R. and Eckstein, F. (1989) Site-directed mutagenesis, based on the phosphorothioate approach, in *Protein Function: A Practical Approach* (Creighton, T. E., ed.), IRL, Oxford, UK, pp. 279–295.
6. Worrall, A. F. (1994) Site-directed mutagenesis by the cassette method, in *Methods in Molecular Biology 30* (Kneale G. G., ed.), Humana Press Inc., Totowa, NJ, pp. 199–210.
7. Kegler-Ebo, D. M., Docktor, C. M., and DiMaio, D. (1994) Codon cassette mutagenesis: a general method to insert or replace individual codons by using universal mutagenic cassettes. *Nucleic Acids Res.* **22(9),** 1593–1599.
8. Barettino, D., Feigenbutz, M., Valcarcel, R., and Stunnenberg, H. G. (1994) Improved method for PCR-mediated site-directed mutagenesis. *Nucleic Acids Res.* **22(3),** 541–542.
9. Mikaelian, I. and Sergeant, A. (1992) A general and fast method to generate multiple site-directed mutations. *Nucleic Acids Res.* **20(2),** 376.
10. Picard, V., Ersdal-Badju, E., Lu, A., and Clark Bock, S. (1994) A rapid and efficient one-tube PCR-based muatgenesis technique using *Pfu* DNA polymerase. *Nucleic Acids Res.* **22(13),** 2587–2591.
11. Goff, S. A., Short-Russell, S. R., and Dice, J. F. (1987) Efficient saturation mutagenesis of a pentapeptide coding sequence using mixed oligonucleotides. *DNA* **6(4),** 381–388.
12. Alber, T, Bell, J. A., Dao-Pin, S., Nicholson, H., Wozniak, J. A., Cook, S., and Matthews, B. W. (1988) Replacements of Pro[86] in phage T4 lysosyme extend an α-helix but do not alter protein stability. *Science* **239,** 631–635.
13. Zabin, H. B. and Terwilliger, T. C. (1991) Isolation and characterization of temperature-sensitive mutants of the bacteriophage f1 gene V protein. *J. Mol. Biol.* **279,** 257–275.
14. O'Donohue, M. J., Scarlett, G. P., and Kneale, G. G. (1993) Tyr 26 and Phe 73 are essential for full biological activity of the Fd gene V protein. *FEMS Microbiol. Lett.* **109,** 219–224.
15. Sambrook, J., Fritsch, E. F., and Maniatis, T. (eds.) (1989) Molecular Cloning—A laboratory Manual, Cold Spring Harbor Laboratory Press, Cold Spring Harbor, NY.
16. Hanahan, D., Jessee, J., and Bloom, F. R. (1991) Plasmid transformation of *Escherichia coli* and other bacteria. *Methods Enzymol.* **104,** 63–113.
17. Taylor, J. W., Schmidt, W., Costick, R., Okrusek, A., and Eckstein, F. (1985) The use of phosphorothioate-modified DNA in restriction enzyme reactions to prepare nicked DNA. *Nucleic Acids Res.* **13,** 8749–8764.
18. Nakayame, K. L. and Eckstein, F. (1986) Inhibition of restriction endonuclease *Nci* I cleavage by phosphorothioate groups and its application to oligonucleotide-directed mutagenesis. *Nucleic Acids Res.* **14,** 9679–9698.
19. Sayers, J. R., Schmidt, W., and Eckstein, F. (1988) 5'-3' Exonucleases in phosphorothioate-based oligonucleotide-directed mutagenesis. *Nucleic Acids Res.* **16,** 9027–9039.
20. Jung, R., Scott, M. P., Oliviera, L. O., and Nielsen, N. C. (1992) A simple and efficient method for the oligodeoxyribonucleotide-directed mutagenesis of double-stranded plasmid DNA. *Gene* **121,** 17–24.
21. Piechocki, M. P. and Hines, R. N. (1994) Oligonucleotide design and optimized protocol for site-directed mutagenesis. *Biotechniques* **16(4),** 702–707.
22. Hermes, J. D., Parekh, S. M., Blacklow, S. C., Koster, H., and Knowles, J. R. (1989) A reliable method for random muatgenesis: the generation of mutant libraries using spiked oligodeoxyribonucleotide primers. *Gene* **84,** 143–151.

101

Site-Directed Mutagenesis Using a Uracil-Containing Phagemid Template

Christian Hagemeier

1. Introduction

The ability to introduce specific changes into almost any given DNA sequence has revolutionized the analysis of cloned genes. This technique has enabled researchers to identify regions necessary for the regulation of gene expression. Also, it was and still is instrumental to learn about the importance of functional domains or even single amino acids of proteins. Of the many useful methods available for site-directed mutagenesis, in this chapter I describe a protocol that is based on the "Kunkel Method" *(1,2)*. This method allows the generation of point mutations, deletions, and insertions for a given DNA sequence with high efficiency. This procedure has been used successfully many times and most usefully to map protein interaction sites within the activation domain of the transcription factor E2F *(3)*.

1.1. Overview and Background Information

The method relies on synthetic oligonucleotides as carriers of the specific alteration one wishes to introduce into the target DNA sequence. The oligonucleotide is annealed to its cognate region of single-stranded DNA and elongated by DNA polymerase. This generates a mutant second-strand containing the specific alteration introduced through the synthetic oligonucleotide. This part of the method is very much similar to virtually all other procedures based on site-directed in vitro mutagenesis. However, it is the principle of how to select against the nonmutant strand that distinguishes this method from all other protocols.

The selection is based on the use of template DNA containing a small percentage of uracil residues instead of thymine. This is achieved by passaging the vector carrying the target DNA through an *Escherichia coli* strain lacking the enzyme dUTPase (dut^-) (**Fig. 1A,B**). The *dut*-deletion results in bacteria with high concentrations of dUTP and as a consequence, dUTP competes with thymine for incorporation into DNA. The *E. coli* dut^- strain cannot repair the faulty incorporation since it also lacks the enzyme uracil *N*-glycosylase that normally leads to a removal of uracil residues from DNA (ung^-). As a result, the dut^- ung^- *E. coli* produce double-stranded vector DNA containing a proportion of uracil residues.

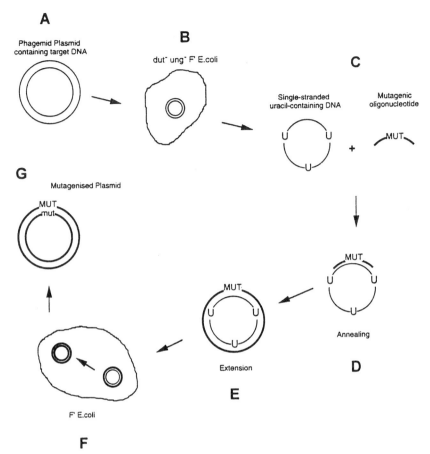

Fig. 1. Schematic presentation of the for site-directed mutagenesis procedure using a uracil-containing phagemid template. *See* **Subheading 1.1.** for detailed information.

The vector is designed such that single-stranded template DNA can be rescued from the bacteria with a helper phage (*see the following*, **Fig. 1C**). The single-stranded template is then used in the aforementioned standard site-directed mutagenesis reaction (**Fig. 1D,E**). It results in double-stranded vector DNA in which the nonmutant template strand, but not the in vitro synthesized strand with the desired mutation, contains uracil residues (**Fig. 1E**). By simply transforming an aliquot of the in vitro reaction into an *E. coli ung*$^+$ strain, the uracil-containing template strand will be degraded (**Fig. 1F**). Therefore, most of the progeny should contain the desired mutation (**Fig. 1G**).

1.2. General Considerations

As outlined herein, the choice of the right bacterial strain is absolutely essential for this mutagenesis procedure to succeed. But also the type of vector used in the reaction should be carefully considered. Usually, single-stranded template is obtained by subcloning the DNA fragment of interest into single-stranded phage vectors (e.g., M13). However, in order to save the time-consuming subcloning steps involved it would be most desirable to perform the mutagenesis reaction in the same vector that will then be used in an assay for the loss/gain of function the mutation will produce. Therefore, prior to the mutagenesis procedure it is worth cloning the DNA of interest into a vector harboring the intergenic region of a filamentous phage (phagemid). Like in the M13-system, this region then allows packaging of single-stranded vector DNA under two conditions: the use of a bacterial strain harboring an F1 episome (F'), and the use of helper phage for packaging the single-stranded vector DNA.

Thus, with the right choice of bacterial strain (dut^- ung^- F') and vector DNA (phagemid) the production of mutant clones can be achieved in a minimum of time. Since this mutagenesis reaction has a very high efficiency (usually greater than 50%) screening for mutants can be undertaken directly by sequencing. Together, the high efficiency of this method and no need for subcloning steps make this mutagenesis procedure a valuable and less expensive alternative to PCR-based methods.

2. Materials
2.1. Preparation of Uracil-Containing Single-Stranded DNA
2.1.1. Preparation of Helper Phage Stock

1. *E. coli*: Any strain harboring an F' episome should be sufficient. However, I use JM101, which usually gives a reliably high titer phage preparation.
2. 2X YT medium: For 1 L medium, dissolve 16 g Bacto-tryptone, 10 g yeast extract, and 10 g of NaCl in deionized H_2O. Aliquot into volumes of 100 mL, autoclave, and store at room temperature.
3. Bacteriophage: R408 (Stratagene, La Jolla, CA).
4. 0.45-µm Filter and 10-mL syringe.

2.1.2. Transformation of Competent E. coli (dut^- ung^- F')

1. Competent bacteria of an *E. coli* dut^- ung^- F' strain (e.g., RZ1032 or CJ236).
2. 2X YT medium: *see* **Subheading 2.1.1.**
3. 2X YT/Amp plates: Add 15 g of agar to 1 L of 2X YT medium prior to autoclaving. When cooled to approx 50°C add 100 µg/mL ampicillin. Pour the solution into sterile 90-mm Petri dishes such that the bottom of the plates is well covered.

2.1.3. Preparation of Single-Stranded DNA

1. 2X YT/Amp medium: Prepare medium as described in **Subheading 2.1.1.** Prior to use add 100 µg/mL ampicillin.
2. Bacteriophage R408 as described in **Subheading 2.1.1.**
3. PEG/ammonium acetate solution: 3.5 M ammonium acetate, pH 7.5, 20% (w/v) polyethylene glycol (PEG) 8000. Store at room temperature.
4. TE buffer: 10 mM Tris-HCl, pH 8.0, 1 mM EDTA. Store at room temperature.
5. Phenol: Phenol equilibrated to pH 7.0–7.5. Aliquot into volumes of 100 mL and store in dark bottles at 4°C.
6. Chloroform/isoamyl alcohol: Mixture of 24 parts (v/v) of chloroform and 1 part (v/v) of isoamyl alcohol. Store in a dark bottle at 4°C.
7. 3 M sodium acetate equilibrated to pH 5.2. Autoclave and store at 4°C.
8. Ice-cold ethanol (100%).

2.2. In Vitro Mutagenesis Procedure

2.2.1. Phosphorylation of the Synthetic Oligonucleotide

1. Synthetic mutagenic oligonucleotide: This oligonucleotide is complementary to the target DNA of the single-stranded vector and carries the desired mutation. Its concentration should be adjusted to 10 pmol/µL (*see* **Note 1**).
2. 10X Kinase buffer: 500 mM Tris-HCl, pH 7.4, 100 mM MgCl$_2$, 50 mM dithiothreitol (DTT). Store at –20°C.
3. ATP: 10 mM ATP in TE buffer. Store in small aliquots at –20°C.
4. T4 Polynucleotide kinase (10 U/µL).

2.2.2. Mutagenesis Reaction

1. Single-stranded target DNA (*see* **Subheading 2.1.4.**).
2. Phosphorylated mutagenic oligonucleotide (*see* **Subheading 2.2.1.**).
3. Annealing buffer: 500 mM NaCl, 100 mM Tris-HCl, pH 8.0, 100 mM MgCl$_2$, 50 mM DTT. Store at –20°C.
4. dNTP: mix 2.5 mM each of dATP, dTTP, dGTP, and dCTP. Store in small aliquots at –20°C (*see* **Note 2**).
5. ATP: 10 mM ATP in TE buffer. Store in small aliquots at –20°C.
6. 10X Ligase buffer: 660 mM Tris-HCl, pH 7.5, 50 mM MgCl$_2$, 50 mM DTT. Store at –20°C.
7. T4 DNA polymerase (1 U/µL).
8. T4 DNA ligase (1 U/µL).

2.3. Selection Against the Nonmutant Strand

1. Frozen competent bacteria of an *E. coli ung*$^+$ strain (e.g., JM101).
2. 2X YT medium: *See* **Subheading 2.1.1.**
3. 2X YT/Amp plates: *See* **Subheading 2.1.2.**
4. Double-stranded vector DNA from the mutagenesis reaction.

3. Methods

3.1. Preparation of Uracil-Containing Single-Stranded DNA

3.1.1. Preparation of Helper Phage Stock

A single preparation of a helper phage stock usually will give an overall yield sufficient for hundreds of mutagenesis reactions. At the same time the phage stock is extremely stable and can be kept at 4°C for many months or years. Therefore, once this step has been accomplished further mutagenesis procedures will normally start with the transformation of phagemid DNA into competent *E. coli* (*see* **Subheading 3.1.2.**).

1. Grow an overnight culture of *E. coli* (F'; e.g., JM101) in 2X YT medium at 37°C.
2. On the next morning dilute 100 µL of the overnight culture into 10 mL of fresh 2X YT medium in a small sterile conical flask. Continue to grow the culture at 37°C until the OD$_{600}$ reaches about 0.2 (mid-log-phase).
3. Infect the culture with 1 µL of R408 stock (Stratagene) and continue incubation for 6 h with vigorous shaking at 37°C (*see* **Note 3**).
4. Pellet the cells at 5000g for 10 min.
5. Pass the supernatant through a 0.45-µm filter into a sterile tube with screwcap using a 10-mL syringe.

Uracil-Containing Phagemid Template

6. Incubate the tube at 60°C for 10 min to kill residual bacteria (*see* **Note 4**).
7. Aliquot the phage preparation into 500-μL volumes in sterile reaction vials and store tubes at 4°C. The phage stock will be stable for months or years (*see* **Note 5**).

3.1.2. Transformation of Competent E. coli (dut⁻ ung⁻ F')

For the preparation of uracil-containing single-stranded phagemid DNA, the vector DNA has to be transformed into a *dut⁻ ung⁻* F' strain of *E. coli* (e.g., RZ1032 or CJ236).

1. Mix 50 μL of competent *E. coli* (*dut⁻ ung⁻* F') cells with approx 1–10 ng of phagemid vector containing the target DNA and incubate on ice for 30 min.
2. Transfer the tube to a 42°C water bath and heat shock for 1 min.
3. Add 150 μL of 2X YT and incubate for 30 min at 37°C with shaking.
4. Plate the contents of the tube onto one 2X YT/Amp plate and incubate overnight at 37°C. The yield should be approx 50–300 colonies.

3.1.3. Preparation of Single-Stranded Phagemid DNA

1. Pick two colonies from the plate and inoculate one small conical flask with 5 mL of 2X YT/Amp. Incubate at 37°C until OD_{600} is about 0.2 (mid-log-phase after 2.5–3 h).
2. Infect the culture with 10 μL of phage stock and continue the incubation with vigorous shaking for 5–6 h at 37°C (*see* **Note 6**).
3. Transfer the culture into 3×1.5 mL reaction vials and pellet the bacteria in a microfuge for 5 min. Meanwhile, add 250 μL of 20% PEG in 3.5 M ammonium acetate to three more tubes.
4. Transfer 1 mL of supernatant from each tube into the PEG-containing reaction vials. Take care not to disturb the pellet! Mix the contents by briefly vortexing the tubes and precipitate the phage for 20 min at room temperature.
5. Pellet the phage in a microcentrifuge for 10 min at 4°C.
6. Carefully remove the supernatant and briefly respin the tubes to collect any remaining PEG solution. Make sure to aspirate all the supernatant using a drawn out Pasteur pipet attached to a vacuum line.
7. Resuspend each bacteriophage pellet in 100 μL of TE and pool contents of tubes in one reaction vial.
8. Extract once with 150 μL of phenol and once with an equal volume of chloroform:isoamyl alcohol.
9. Transfer the aqueous phase from the final extraction to a fresh tube. Take great care not to disturb the interphase! A clean preparation is more important than a maximum yield. Add 0.1 vol of 3 M sodium acetate and 2 vol of ice-cold ethanol. Mix well and allow DNA to precipitate for 30 min at –20°C.
10. Recover the DNA by centrifugation in a microfuge for 10 min at 4°C. Wash the pellet with 80% ethanol, respin the tube, and remove the supernatant with a drawn out Pasteur pipet attached to a vacuum line (*see* **Note 7**). Dry the pellet in a desiccator or speedvac for 3 min and dissolve the pellet in 50 μL TE.
11. Use an aliquot of 4 μL (should equal about 0.5 μg) for analyzing the single-stranded DNA preparation on a 1% agarose gel. Two bands of about equal intensity should be visible. The upper band refers to single-stranded plasmid DNA, the lower to single-stranded phage DNA.

3.2. In Vitro Mutagenesis Procedure

3.2.1. Phosphorylation of the Synthetic Oligonucleotide

To ligate the oligonucleotide to the 3' terminus of the newly synthesized strand in the mutagenesis reaction, it needs to be phosphorylated on its own 5' terminus.

1. Adjust concentration of the oligonucleotide to 10 pmol/μL.
2. Use 5 μL in the phosphorylation reaction and further add 1 mL of 10X kinase buffer, 1 μL of 10 mM ATP, and 2.5 μL of T4 polynucleotide kinase (10 U/μL).
3. Incubate the reaction for 30 min at 37°C.
4. Heat inactivate the enzyme at 65°C for 10 min. Spin the tube briefly to collect all droplets. The oligonucleotide can now be used directly in the mutagenesis reaction (*see* **Note 8**).

3.2.2. Mutagenesis Reaction

This reaction involves three steps. First, the phosphorylated oligonucleotide is annealed to the uracil-containing single-stranded DNA. Second, the oligonucleotide is extended along the template DNA from its 3' end. Third, the extended strand is ligated to the 5' end of the oligonucleotide.

1. Anneal 10 pmol of the phosphorylated oligonucleotide to 1 μg of single-stranded phagemid DNA by mixing 2 μL of the phosphorylation reaction (*see* **Subheading 3.2.1.**), 7 μL of the single-stranded DNA preparation (*see* **Subheading 3.1.3.**), and 1 μL of annealing buffer (*see* **Note 9**).
2. Incubate the reaction for 5 min at 65°C and then allow the reaction to slowly cool down to room temperature for about 30 min (*see* **Note 10**).
3. Add to the annealing reaction on ice: 4 μL of a 2.5 mM dNTP mix, 1 μL 10 mM ATP, 1 μL 10X ligase buffer, 2.5 U of T4 DNA polymerase, 2 U of T4 DNA ligase, and deionized H$_2$O to a total volume of 20 μL (*see* **Note 11**).
4. Incubate the reaction mix at room temperature for 10 min to allow efficient initiation of the polymerase reaction. Then incubate for a further 2 h at 37°C.

3.3. Selection Against the Nonmutant Strand

This step finally selects against the nonmutant strand. The plasmids are simply transformed into *ung$^+$ E. coli,* which will specifically degrade the uracil-containing (nonmutant) strand. The bacteria then synthesize a new strand along the template. The new double-stranded plasmid should contain the desired mutation in both strands.

1. Mix 50 μL of competent *ung$^+$ E. coli* (e.g., JM101) with 3 μL of the mutagenesis reaction and incubate on ice for 30 min.
2. Transfer the tube to a 42°C water bath and heat shock for 1 min.
3. Add 150 μL of 2X YT and incubate for 30 min at 37°C with shaking.
4. Plate the contents of the tube onto one 2X YT/Amp plate and incubate overnight at 37°C.

Owing to the high efficiency of the procedure on average one to two of three colonies should contain plasmid with the desired mutation. Therefore, pick three colonies per mutagenesis reaction, perform a standard DNA "miniprep," and sequence the DNA directly, without prior screening for mutants. If, for some reason, double-stranded sequencing does not give a satisfying result, single-stranded DNA easily can be rescued for sequencing according to the method described in **Subheading 3.1.3.**

4. Notes

1. This method allows the introduction of point mutations, insertions (loop in), and deletions (loop out) into the target DNA. In general the complementary region of the mutagenic oligonucleotide should be proportional to the number of mismatches. For instance, for point mutations involving up to five adjacent bases, the annealing of 12 bases either side of the mutated region is sufficient. An insertion of 12 bases should have 15 matches either

side and a larger deletion should have 20 bases on either side annealing to the template DNA. (I have successfully deleted 250 bp under these conditions.) Having cytosine and guanine bases at the termini of the mutagenic oligonucleotide ensures a tighter bond of the ends of the oligonucleotide to its template DNA. This is beneficial for the initiation of the polymerase reaction (at the 3' terminus) and the ligation of the synthesized strand to the 5' terminus of the oligonucleotide.
2. To avoid the risk of contaminating uracil, only high quality dNTPs should be used in the extension reaction.
3. Sufficient aeration of the culture is very important for bacteriophage growth. Therefore, even small volumes should be incubated in a conical flask and shaken at high speed (300 cycles/min).
4. Any bacteria passing this step would produce thymine-containing phage during the preparation of single-stranded DNA (*see* **Subheading 3.1.1.**).
5. Generally the supernatant will contain phage at a titer of about 5×10^{10} plaque forming units (PFU)/mL and phage titering is not essential. However, if preparations of single-stranded phagemid DNA give a low yield, phage titers should be checked.
6. The phage stock usually contains about 5×10^{10} PFU/mL or $5 \times 10^{8}/10$ μL. Ten milliliters of a bacterial culture with an OD_{600} of about 0.2 will contain approx 5×10^{9} cells. Therefore, this configuration results in an MOI (multiplicity of infection) of 0.1. Consequently, after 5–6 h of further bacterial growth virtually all phage progeny will have evolved from the dut^- ung^- strain of *E. coli* harboring uracil-containing DNA.
7. At this stage the pellet may not be visible. Try to use a flat bottom tube. In that case, after centrifugation the pellet will be in the top quarter of the bottom (with respect to its orientation in the microfuge) and all the supernatant can be removed easily with the tip of the Pasteur pipet being placed in the bottom quarter. If you use a flat bottom tube try to position the tube in the same orientation in the microfuge during the wash and respin steps. This avoids loosening of the pellet from the wall of the tube during centrifugation.
8. The successful incorporation of a phosphate group can be monitored by trace-labeling the oligonucleotide with [γ^{32}P]ATP in the same reaction. The product can then be analyzed on a polyacrylamide gel.
9. The concentration of plasmid DNA is not too important. A ratio of 4:1 of phage and plasmid DNA will still work well in the mutagenesis reaction. One microgram of plasmid DNA is equivalent to about 0.5 pmol.
10. This can be achieved easily by using a beaker (with the tube placed in a floater) in a water bath. After 5 min at 65°C place the beaker on the bench and allow the reaction to cool to room temperature that will take about 30 min.
11. Traditionally, the Klenow fragment of *E. coli* DNA polymerase I has been used for the extension reaction. However, T4 (or T7) DNA polymerase offers the advantage that it can not displace the mutagenic oligonucleotide from the template DNA. Since the T4 DNA polymerase has a strong 3' exonucleotide activity the concentration of dNTPs should not be below 500 μM. A high dNTP concentration counteracts the exonuclease activity and leads to a more efficient extension reaction by the enzyme.

References

1. Kunkel, T. A. (1985) Rapid and efficient site-specific mutagenesis without phenotypic selection. *Proc. Natl. Acad. Sci. USA* **82,** 488–492.
2. Kunkel, T. A., Roberts, J. D., and Zakour, R. A. (1987) Rapid and efficient site-specific mutagenesis without phenotypic selection. *Methods Enzymol.* **154,** 367–382.
3. Hagemeier, C., Cook, A., and Kouzarides, T. (1993) The retinoblastoma protein binds E2F residues required for activation *in vivo* and TBP binding *in vitro*. *Nucleic Acids Res.* **21,** 4998–5004.

102

Site-Directed Mutagenesis Using Double-Stranded Plasmid DNA Templates

Jeffrey Braman, Carol Papworth, and Alan Greener

1. Introduction

In vitro site-directed mutagenesis is an invaluable technique for studying protein structure-function relationships, gene expression and vector modification. Several methods for performing this technique have appeared in the literature *(1–5)*. These procedures generally require multiple enzymatic steps, specialized vectors, and convenient restriction sites or subcloning of the sequence of DNA to be mutated into a bacteriophage vector like M13 to produce and recover single-stranded DNA. These manipulations present major limitations to the routine use of these methods because they are time consuming and tedious.

A site-directed mutagenesis procedure was developed by Deng and Nickoloff *(6;* **Fig. 1**) that eliminated the need for subcloning and generating single-stranded DNA templates by employing double stranded plasmid DNA. In this chapter, we describe several modifications that have been made to improve the reliability of this procedure and to increase the mutation efficiency *(7)*. The most significant change that has been made is the construction of a new mismatch repair-deficient *Escherichia coli* host strain that is EndA⁻. The strain is referred to as XL*mutS*. The endA mutation eliminates an endonuclease that degrades miniprep plasmid DNA, prepared by either the boiling or alkaline lysis procedures. Removal of this endonuclease greatly improves the quantity and quality of recovered plasmid DNA, thus improving the efficiency and reproducibility of mutated plasmid generation. Additional changes that have been made include the use of T7 DNA polymerase instead of T4 DNA polymerase *(8)* and optimization of the selection and mutagenic oligonucleotide primer-to-plasmid template ratio.

2. Materials

1. pWhitescript (pWS) control plasmid DNA was produced at Stratagene (La Jolla, CA). The plasmid contains a stop codon (TAA) at the position where a glutamine codon (CAA) would normally appear in the α-complementing portion of the β-galactosidase gene of pBluescript II SK(–) corresponding to amino acid 9 of the protein. XL1-Blue *E. coli* cells transformed with this plasmid appear white on LB agar plates containing ampicillin and supplemented with IPTG and X-gal because active β-galactosidase has not been synthesized.

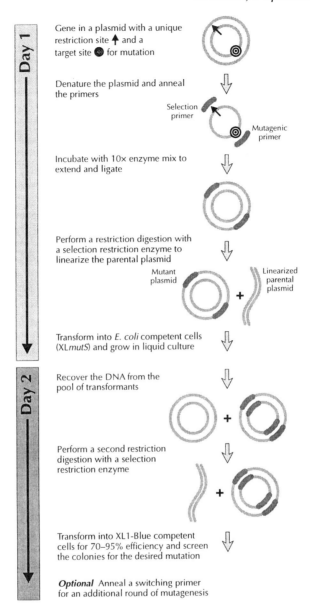

Fig. 1. Overview of the double-stranded, site mutagenesis protocol.

Annealing of a "blue" mutagenic oligonucleotide primer to denatured pWS followed by polymerization and ligation results in a point mutation that converts the stop codon of pWS (TAA) back to the glutamine-encoding codon (CAA). Transformed XL1-Blue cells containing the mutated pWS plasmid appear blue on LB agar plates containing ampicillin supplemented with IPTG and X-gal because active β-galactosidase has been synthesized.
2. pUC18 Control plasmid (0.1 ng/μL) was also obtained from Stratagene.
3. The "blue" mutagenic oligonucleotide primer was produced at Stratagene to test the efficiency of the procedure. The mutagenic oligonucleotide primer changes the stop codon (TAA) in the β-galactosidase gene of pBluescript II SK(–) to a glutamine-encoding codon

(CAA) at amino acid 9 of the protein. This results in the conversion of white to blue bacterial colonies when pWS is converted to pBluescript and active β-galactosidase is synthesized; "blue" primer: GTG AGG GTT AAT TGC GCG CTT GGC GTA ATC ATG G.
4. High-transformation-efficiency *E. coli* strains were produced at Stratagene and their genotypes are as follows:
   a. XL*mutS* strain: *Δ(mcrA)183 Δ(mcrCB-hsdSMR-mrr)173 endA1 supE44 thi-1 gyrA96 relA1 lac mutS::Tn10* (Tet^r) (*F' proAB lacI^{q}ZΔM15* Tn5 [Kan^r]).
   b. XL1-Blue strain: *recA1 endA1 gyrA96 thi-1 hsdR17 supE44 relA1 lac* (*F' proAB lacI^{q}ZΔM15* Tn*10* [Tet^r]).
5. 10X Mutagenesis buffer: 100 m*M* Tris-acetate, pH 7.5, 100 m*M* magnesium-acetate, 500 m*M* potassium acetate.
6. Enzyme dilution buffer: 20 m*M* Tris-HCl, pH 7.5, 10 m*M* KCl, 10 m*M* β-mercaptoethanol, 1 m*M* dithiothreitol (DTT), 0.1 m*M* EDTA, 50% (v/v) glycerol.
7. Deoxynucleotide (dNTP) mix: 2.86 m*M* dATP, 2.86 m*M* dCTP, 2.86 m*M* dGTP, 2.86 m*M* TTP, 4.34 m*M* rATP, 1.43X mutagenesis buffer.
8. 10X Enzyme mix: 0.25 U/μL of native T7 DNA polymerase, 1.0 U/μL of T4 DNA ligase, 0.6 μg/μL of single-stranded binding protein, in enzyme dilution buffer.
9. 1.4 *M* β-mercaptoethanol.
10. SOB medium (per L): 20 g of tryptone, 5 g of yeast extract, 0.5 g of NaCl. Autoclave. Add 10 mL of 1 *M* MgCl$_2$ and 10 mL of 1 *M* MgSO$_4$/L of SOB prior to use. Filter sterilize.
11. SOC Medium (per 100 mL): Add 1 mL of 2 *M* filter-sterilized glucose solution or 2 mL of 20% (w/v) glucose to 100 mL of SOB medium prior to use.
12. 2X YT Broth (per L): 10 g of NaCl, 10 g of yeast extract, 16 g of tryptone. Adjust to pH 7.5 with NaOH.
13. LB Agar (per L): 10 g of NaCl, 10 g of tryptone, 5 g of yeast extract, 20 g of agar. Adjust to pH 7.0 with 5 *N* NaOH. Add deionized H$_2$O to a final volume of 1 L. Autoclave. Pour into Petri dishes (25 mL/100-mm plate). Store plates at 4°C.
14. LB-Ampicillin Agar (per L): One liter of LB agar. Autoclave. Cool to 55°C. Add 1 mL of 50 mg/mL filter-sterilized ampicillin. Pour into Petri dishes (25 mL/100-mm plate).
15. LB-Ampicillin-methicillin agar (use for reduced satellite colony formation [per L]): One liter of LB agar. Autoclave. Cool to 55°C. Add 0.4 mL of 50 mg/mL filter-sterilized ampicillin. Add 1.6 mL of 50 mg/mL filter-sterilized methicillin. Pour into Petri dishes (25 mL/100-mm plate).
16. Falcon 2059 polypropylene tubes (15 mL).
17. 5-Bromo-4-chloro-3-indoyl-β-D-galactopyranoside (X-gal).
18. Isopropyl-β-D-thio-galactopyranoside (IPTG).

3. Methods
3.1. Mutagenic and Selection Oligonucleotide Primer Design

Mutagenic primers introduce specific experimental mutations, and mutagenic oligonucleotide primers for use in this protocol must be designed individually according to the desired mutation required. The following considerations should be made for designing mutagenic and selection oligonucleotide primers:

1. Both the mutagenic and the selection oligonucleotide primers must anneal and alter the same strand of the plasmid that is to be altered.
2. Mutagenic and selection oligonucleotide primers should be between 25 and 45 bases in length.
3. The mismatched portions of mutagenic oligonucleotide primers should be in the middle of the primer with approx 10–15 bases of correct sequence on either side.

4. Mutagenic and selection oligonucleotide primers should have a minimum GC content of 40% and should terminate in one or more C or G bases.
5. Mutagenic and selection oligonucleotide primers must be 5' phosphorylated and should be purified by either FPLC or polyacrylamide gel electrophoresis prior to use.
6. Selection against sites having GATC, such as *Bgl*II, *Bam*HI, or *Pvu*I, has been found to be inefficient and primers using these sites are not recommended.
7. The distance between the selection site and the mutation target site oligonucleotide primers should be as far apart as possible for best results.

3.2. Selection and Switch Oligonucleotide Primer Sequences

Examples are given of selection and switch oligonucleotide primers and their sequences. The selection oligonucleotide primers change a unique, nonessential restriction site to the switch primer site and the switch oligonucleotide primer changes its restriction site back to the unique, nonessential restriction site.

1. *ALw*NI Selection oligonucleotide primer and *Nru*I switch primer: The *ALw*NI restriction site is located at basepair number 1569 in the *Col*EI origin of replication in the pBluescript II SK(–) phagemid. *ALw*NI to *Nru*I: CGC CAC TGG CAG CAG TCG CGA GTA ACA GGA TTA GCA GAG and *Nru*I to *ALw*NI: CGC CAC TGG CAG CAG CCA CTG GTA ACA GGA TTA GCA G.
2. *Kpn*I selection oligonucleotide primer and *Srf*I switch primer: The *Kpn*I restriction site is located at base pair number 657 in the polylinker of the pBluescript II SK(–) phagemid. *Kpn*I to *Srf*I: CTA TAG GGC GAA TTG GGT GCC CGG GCC CCC CTC GAG GTC G and *Srf*I to *Kpn*I: CTA TAG GGC GAA TTG GGT ACC GGG CCC CTC GAG GTC G
3. *Sca*I Selection oligonucleotide primer and *Mlu*I switch primer: The *Sca*I restriction site is located at basepair number 2526 in the ampicillin-resistance gene of the pBluescript II SK(–) phagemid. *Sca*I to *Mlu*I: CTG TGA CTG GTG ACG CGT CAA CCA AGT C and *Mlu*I to *Sca*I: GCT TTT CTG TGA CTG GTG AGT ACT CAA CCA AGT C.

3.3. Protocol for Mutagenesis

3.3.1. Step I: Annealing the Primers to the DNA

Simultaneously anneal the selection and mutagenic oligonucleotide primers to the double-stranded target plasmid DNA by preparing the following control and experimental reactions in 1.5-mL microcentrifuge tubes. For the purpose of demonstration, a control reaction is described that includes the use of the pWS plasmid DNA (*see* **Subheading 2.**) to be mutated and the *Kpn*I selection oligonucleotide primer (*see* **Subheading 3.2.**).

1. Control reaction: 1 µL (490 ng or 0.25 pmol) of pWS plasmid DNA (*see* **Note 1**), 5 µL (330 ng or 25 pmol) of *Kpn*I selection primer (*see* **Note 2**), 5 µL (281 ng or 25 pmol) of "blue" mutagenic primer (*see* **Note 2**), 2 µL of 10X mutagenesis buffer, 7 µL of double-distilled water (ddH$_2$O) to a final volume of 20 µL.
2. Experimental reaction: 0.25 pmol of the plasmid of interest (*see* **Note 1**), 25 pmol of the selection primer (*see* **Note 2**), 25 pmol of the mutagenic primer (*see* **Note 2**), 2 µL of 10X mutagenesis buffer, ddH$_2$O to a final volume of 20 µL.
3. Place the microcentrifuge tubes in a boiling water bath for 5 min and then immediately place the tubes on ice for 5 min. Centrifuge briefly in a tabletop centrifuge to collect the condensate.
4. Incubate the microcentrifuge tubes at room temperature (23–25°C) for 30 min.

3.3.2. Step II: Extending the Primers and Ligating the New Strands

For both the control and experimental reactions, extend the primers with T7 DNA polymerase and ligate the new strands with T4 DNA ligase as indicated:

1. Prepare a fresh 1:10 dilution of the 10X enzyme mix by diluting 1 µL of the 10X enzyme mix in 9 µL of enzyme dilution buffer (*see* **Note 3**).
2. To each microcentrifuge tube, add 7 µL of the deoxynucleotide mix. The deoxynucleotide mix must be added to the reaction first or the 3'- to 5'-exonuclease activity of native T7 DNA polymerase will degrade the primers. Stir the 1:10 enzyme dilution with a pipet tip and add 3 µL of this fresh 1:10 enzyme dilution to the reaction. Stir the reaction mixture again with a pipet tip and centrifuge briefly.
3. Incubate the microcentrifuge tubes at 37°C for 1 h.
4. Inactivate the T7 DNA polymerase and T4 DNA ligase by incubating the microcentrifuge tubes at 70–80°C for 10–15 min to prevent religation of the digested strands during the subsequent digestion in **Subheading 3.3.3.** Cool the reactions to room temperature.

3.3.3. Step III: Digesting with a Restriction Enzyme

After the reactions cool, digest both the control and experimental reactions with a restriction enzyme to eliminate those plasmids that did not anneal with the selection oligonucleotide primer.

1. Control reaction: Add the following components to the control reaction: 20 U of *Kpn*I restriction enzyme, ddH$_2$O to a total reaction vol of 60 µL. (Because *Kpn*I is active in 0.5 mutagenesis buffer, no buffer is added and the control reaction is diluted to twice its volume, thus diluting the buffer concentration from 1–0.5X. For the buffer requirements of other restriction enzymes, please consult a restriction enzyme buffer chart available from the respective manufacturer of the restriction enzyme.)
2. Incubate the control reaction at 37°C for 1 h (*see* **Note 4**).
3. Experimental reaction: Digest the DNA at the selection restriction site by adding 20 U of the restriction enzyme. In order to prevent incomplete digestion owing to the glycerol concentration, the amount of restriction enzyme must not exceed 10% of the total digestion reaction volume, which includes the 3-µL volume of the fresh 1:10 enzyme dilution from **step 2** in **Subheading 3.3.2.**, and the required volume containing 20 U of the restriction enzyme. For example, if 2.5 µL is the required volume of restriction enzyme needed to provide 20 U, then the total enzyme volume is 5.5 µL (i.e., 3 + 2.5 µL). Because the total volume of enzyme (i.e., 5.5 µL) must not exceed 10% of the total digestion reaction volume, the final reaction volume must be 55 µL or more. The buffer must also be adjusted to the appropriate salt concentration for the restriction enzyme corresponding to the selection primer in use. Be sure to take into account the fact that the original 30 µL of the reaction is already 1X in mutagenesis buffer. For an example, see the control reaction description in **step 1**.
4. Incubate the experimental reaction at 37°C for 1 h.

3.3.4. Step IV: Transforming into XLmutS Competent Cells

For both the control and experimental reactions, follow the transformation procedure outlined as follows (*see* **Notes 5–10**).

1. Thaw the XL*mutS* competent cells on ice.
2. Gently mix the cells by hand. Aliquot 90 µL of the XL*mutS* competent cells into two prechilled 15-µL Falcon 2059 polypropylene tubes.

3. Add 1.5 µL of the β-mercaptoethanol to the 90-µL aliquots of XL*mutS* competent cells to yield a final concentration of 25 m*M*.
4. Swirl the contents of the Falcon 2059 polypropylene tubes gently. Incubate the cells on ice for 10 min, swirling gently every 2 min.
5. Add 1/10 of the volume of the control and experimental reactions that have been digested with a restriction enzyme (*see* **Subheading 3.3.3.**) to each Falcon 2059 polypropylene tube and swirl gently. Use 6 µL for the control reaction.
6. Incubate the Falcon 2059 polypropylene tubes on ice for 30 min (*see* **Note 11**).
7. Heat pulse the Falcon 2059 polypropylene tubes in a 42°C water bath for 45 s. The length of time of the heat pulse is critical for obtaining the highest efficiencies.
8. Incubate the transformation mixture on ice for 2 min.
9. Add 0.9 mL of preheated SOC medium and incubate the Falcon 2059 polypropylene tubes at 37°C for 1 h with shaking at 225–250 rpm.
10. Plate 100 µL of the control reaction transformation on LB-ampicillin (100 µg/mL) agar plates, containing X-gal and IPTG, to verify the *lac* phenotype. For color selection, spread 20 µL of 0.2 *M* IPTG and 20 µL of 10% (w/v) X-gal on LB ampicillin agar plates 30 min before plating the transformants (*see* **Note 12**).
11. Plate 1, 5, 25, and 200 µL of the transformation mixture, using a sterile spreader, onto the agar plates containing the appropriate antibiotic (*see* **Note 13**). If the transformants are ampicillin resistant, the transformation mixture may also be plated on LB-ampicillin (20 µg/mL)-methicillin (80 µg/mL) agar plates, if satellite colonies are observed.
12. Incubate the agar plates overnight at 37°C (*see* **Note 14**).

3.3.5. Step V: Enriching for Mutant Plasmids

1. Enrich for mutated plasmids by adding the remaining transformation mixture that was not plated into 3 mL of 2X YT broth, supplemented with an appropriate antibiotic for the experimental plasmid.
2. Grow the culture overnight at 37°C with shaking.
3. For both the control and experimental reactions, perform a miniprep plasmid DNA isolation from 1.5 mL of the overnight culture in **step 2** using a standard protocol *(9)*.
4. Perform restriction enzyme digestion of the miniprep plasmid DNA for both the control and experimental reactions as follows: Digest 10 µL (approx 500 ng) of the resulting miniprep plasmid DNA with the same selection restriction enzyme and appropriate buffer as described in **Subheading 3.3.3.** The reaction volume should be 10X the volume of added enzyme. Use at least 20 U of the restriction enzyme (*see* the following control reaction for an example).
5. Control reaction: 10 µL (approx 500 ng) of miniprep DNA, 1 µL of 10X mutagenesis buffer (0.5X final concentration), 20 U of *Kpn*I restriction enzyme, ddH$_2$O to a final volume of 20 µL. Incubate the digestion reaction for 1–2 h at 37°C.

3.3.6. Step VI: Final Transformation into XL1-Blue Competent Cells

For both the control and experimental reactions, transform the digested DNA into the XL1-Blue competent cells or into a cell line of choice as outlined in the following.

1. Control reaction: Transform 40 µL of XL1-Blue competent cells with 1/10 of the volume (2 µL) of the digested DNA described in step 4 of **Subheading 3.3.5.**
2. Experimental reaction: Transform 1/10 of the volume of the digested DNA (but do not exceed 4 µL) into 40 µL of XL1-Blue competent cells or any desired competent cell line by following the transformation protocol given here and by referring to **Notes 5–10**:

a. Thaw the competent cells on ice.
b. Gently mix the competent cells by hand. Aliquot 40 µL of the competent cells into a prechilled 15-mL Falcon 2059 polypropylene tube.
c. Add 0.68 µL of the β-mercaptoethanol to the 40 µL of competent cells to yield a final concentration of 25 mM.
d. Swirl the contents of the Falcon 2059 polypropylene tube gently. Incubate the cells on ice for 10 min, swirling gently every 2 min.
e. Add 1/10 of the volume of the experimental reaction (but do not exceed 4 µL) to the Falcon 2059 polypropylene tube and swirl gently. As an additional control, add 1 µL of the pUC18 control plasmid to a 40-µL aliquot of the XL1-Blue competent cells and swirl gently.
f. Incubate the Falcon 2059 polypropylene tube on ice for 30 min (*see* **Note 11**).
g. Heat pulse the Falcon 2059 polypropylene tubes in a 42°C water bath for 45 s. The length of time of the heat pulse is critical for obtaining the highest efficiencies.
h. Incubate the transformation mixture on ice for 2 min.
i. Add 0.45 mL of preheated SOC medium and incubate the Falcon 2059 polypropylene tubes at 37°C for 1 h with shaking at 225–250 rpm.
j. The control reaction transformation should be plated on LB-ampicillin agar plates, containing X-gal and IPTG, to verify the *lac* phenotype. For color selection, spread 20 µL of 0.2 M IPTG and 20 µL of 10% (w/v) X-gal on LB-ampicillin (100 µg/mL) agar plates 30 min before plating the transformants (*see* **Note 12**). If plating 5 µL, expect to see 150–300 colonies.
k. Plate 1, 5, 25, and 200 µL of the experimental reaction transformation mixture, using a sterile spreader, onto the agar plates containing the appropriate antibiotic. If the transformants are intended to be ampicillin resistant, the transformation mixture may also be plated on LB-ampicillin (20 µg/mL)–methicillin (80 µg/mL) agar plates, containing IPTG and X-gal, if satellite colonies are observed (*see* **Note 15**).
l. Incubate the plates overnight at 37°C (*see* **Note 16**).
m. Observations that have been made during the course of developing this protocol and solutions to various problems associated with these observations are listed in **Notes 17–20**.

4. Notes

1. To determine the mass of template required to give 0.25 pmol (1 pmol = 1×10^{-12} mol), use the formula: mass of template required = [(0.25 × 10^{-12} mol) (660 g/mol/bp) (number of basepairs in the plasmid)]. For example, if the plasmid is 2960 bp, then mass of template required = [(0.25 × 10^{-12} mol) (660 g/mol/bp) (2960 bp)] = (4.88 × 10^{-7} g) = 488 ng.
2. The amount of selection and mutagenic oligonucleotide primer used should be 100X the picomole amount of template. To determine the mass of primer required to give 25 pmol, use the formula: mass of oligonucleotide primer = [(25 × 10^{-12} mol) (330 g/mol/base) (number of bases in the primer)]. For example, if the primer is 40 bases long, then mass of primer = [(25 × 10^{-12} mol) (330 g/mol/base) (40 bases)] = (3.30 × 10^{-7} g = 330 ng). To determine the concentration of oligonucleotide primer solutions, assume that a primer solution measuring 1 OD_{260} U is at a concentration of 33 µg/mL.
3. It is crucial to use the fresh 1:10 enzyme dilution within 6–8 h of preparation.
4. In order to keep the reaction volume as low as possible to obtain maximum digestion and transformation efficiency, use the highest concentration of high-quality restriction enzyme available.
5. Storage conditions: The competent cells are very sensitive to even small variations in temperature and must be stored at the bottom of a –80°C freezer. Transferring tubes from

one freezer to another may result in a loss of efficiency. The XLmutS and XL1-Blue competent cells from Stratagene should be placed at –80°C directly from the dry ice shipping container. Cells stored in this manner should retain their guaranteed efficiency for 6 mo.

6. Aliquoting cells: When aliquoting, keep the competent cells on ice at all times. It is essential that the Falcon 2059 polypropylene tubes are placed on ice before the competent cells are thawed and that the cells are aliquoted directly into the prechilled tubes. When transforming the competent cells, it is important to use at least 90 µL of XL1mutS competent cells for each transformation and 40 µL of XL1-Blue competent cells for each transformation. Using smaller volumes will result in lower efficiencies.
7. Use of Falcon 2059 polypropylene tubes: It is important that Falcon 2059 polypropylene tubes are used for the transformation protocol, since other tubes may be degraded by the β-mercaptoethanol. In addition, the incubation period during the heat-pulse step is critical and has been calculated for the thickness and shape of the Falcon 2059 polypropylene tubes.
8. Use of β-mercaptoethanol: β-mercaptoethanol has been shown to increase transformation efficiencies two- to threefold.
9. Quantity of DNA added: Highest transformation efficiencies are observed by adding 1 µL of 0.1 ng/mL of supercoiled DNA to 100 µL of competent cells. An increasing number of colonies will be obtained when plating up to 50 ng, although the overall efficiency may be lower.
10. Length of the heat pulse: There is a defined "window" of highest transformation efficiency resulting from the heat pulse in **Subheading 3.3.4., step 7** and **Subheading 3.3.6., step 2**. Optimal efficiencies are observed when cells are heat pulsed for 45–50 s. Heat pulsing for at least 45 s is recommended to allow for slight variations in the length of incubation. Efficiencies decrease sharply when incubating for <45 s or for >50 s.
11. Preheat the SOC medium to 42°C at this step.
12. Do not mix IPTG and X-gal, since these chemicals will precipitate. X-gal should be prepared in dimethylformamide (DMF). Prepare IPTG in sterile dH$_2$O.
13. When spreading bacteria onto the plate, tilt and tap the spreader to remove the last drop of the cells. For the XLmutS competent cells, if plating is >100 µL, the cells can be spread directly onto the plates. If plating is <100 µL of the transformation mixture, increase the volume of the transformation mixture to be plated to a total volume of 200 µL using SOC medium.
14. Approximately 45–50% of the colonies from the control reaction transformation should display the blue phenotype, and the expected colony numbers for the experimental transformations should be between 200 and 600 colonies per transformation plate.
15. When spreading bacteria onto the plate, tilt and tap the spreader to remove the last drop of the cells. For the competent cells, if plating is >100 µL, the cells can be spread directly onto the plates. If plating is <100 µL of the transformation mixture, increase the volume of the transformation mixture to be plated to a total volume of 200 µL using SOC medium.
16. Approximately 70–95% of the control reaction transformation colonies should display the blue phenotype.
17. *Observation*: There are no colonies produced after the first transformation.
    Suggestion: Verify that the correct antibiotic selection was used. The control plasmid uses ampicillin, but the experimental plasmid may require another antibiotic. Check the quantity of the target plasmid DNA, which should be 0.25 pmol.
18. *Observation*: There are >1000 colonies produced after the first transformation.
    Suggestion: The expected colony numbers for the experimental reaction transformations should be between 200 and 600 colonies per transformation plate. If the expected colony numbers exceed 1000 following the first transformation, then incomplete digestion with

the restriction enzyme may have occurred. Repeat the digestion step as outlined in **Subheading 3.3.3.**

19. *Observation*: The total number of colonies produced after the second transformation is low.
    Suggestion: The plasmid DNA yield from the miniprep may be low. Check the yield by electrophoresing a 10 µL sample of the digested DNA on a 1% (w/v) agarose gel and comparing the ethidium bromide staining intensity with a known standard. Check the transformation efficiency of the competent cells with the pUC18 control plasmid DNA.

20. *Observation*: There are a large number of transformant colonies, but the mutation efficiency is low.
    Suggestion: Linearization of the parental plasmid may not have been complete. Ensure that the correct restriction enzyme is used for the selection oligonucleotide primer and that the restriction buffer is optimal for this enzyme. The miniprep plasmid DNA may contain impurities that could interfere with the restriction enzyme digestion. An additional purification step may be required to remove any RNA or protein remaining in the preparation. The extent of the restriction enzyme digestion can be monitored by visualizing a small aliquot of the digested DNA on a 1% (w/v) agarose gel. If necessary, add more restriction enzyme and digest the mixed plasmid DNA longer. Alternatively, there could be a problem with the incorporation of the selection and/or mutagenic primer into the newly synthesized DNA strand. If neither the experimental nor control reaction resulted in a reasonable number of mutants, there may be a problem with the T7 DNA polymerase and/or the T4 DNA ligase. Prepare a fresh 1:10 dilution of the 10X enzyme mix (*see* **Subheading 2.**) and use 3 µL of the fresh dilution in the reactions. Heat the reaction adequately to destroy the T4 DNA ligase after the extension and ligation step (*see* **Subheading 3.3.2.**). If the control reaction gave a reasonable number of mutants whereas the experimental reaction did not, there may be a problem with the experimental primers. Check to make sure the mutant primer is designed to hybridize to the same strand as the selection primer. Ensure that both of the primers have been phosphorylated and that both primers are of high purity and accurate concentration (*see* **Subheading 3.1.**). Occasionally, the mutation efficiency will be lower if the mutagenic primer is introducing a large insert or a deletion. This may also be the case when the target plasmid is very large (>8 kb). Try isolating a pool of plasmid DNA from the cells of the second transformation and repeat the restriction enzyme digestion step (*see* **Subheading 3.3.3.**) and then the transformation step (*see* **Subheading 3.3.4.**) a third time. This will further enrich for the mutant plasmid.

References

1. Kunkel, T. A. (1985) Rapid and efficient site-specific mutagenesis without phenotypic selection. *Proc. Natl. Acad. Sci. USA* **82,** 488–492.
2. Sayers, J. R., Schmidt, W., Wendler, A., and Eckstein, F. (1988) Strand specific cleavage of phosphorothioate containing DNA by reaction with restriction endonucleases in the presence of ethidium bromide. *Nucleic Acids Res.* **16,** 803–814.
3. Vandeyar, M. A., Weiner, M. P., Hutton, C. J., and Batt, C. A. (1988) A simple and rapid method for the selection of oligonucleotide-directed mutants. *Gene* **65,** 129–133.
4. Stanssens, P., Opsomer, C., McKeown, Y. M., Kramer, W., Zabeau, M., and Fritz, H.-J. (1989) Efficient oligonucleotide-directed construction of mutations in expression vectors by the gapped duplex DNA method using alternating selectable markers. *Nucleic Acids Res.* **17,** 4441–4454.
5. Lewis, M. K. and Thompson, D. V. (1990) Efficient site directed *in vitro* mutagenesis using ampicillin selection. *Nucleic Acids Res.* **18,** 3439–3443.

6. Deng, W. P. and Nickoloff, J. A. (1992) Site-directed mutagenesis of virtually any plasmid by eliminating a unique site. *Anal. Biochem.* **200,** 81–88.
7. Papworth, C., Greener, A., and Braman, J. (1994) Highly efficient double-stranded, site-directed mutagenesis with the Chameleon™ kit. *Strategies* **7,** 38–40.
8. Bebenek, K. and Kunkel, T. A. (1989) The use of native T7 DNA polymerase for site-directed mutagenesis. *Nucleic Acids Res.* **17,** 5408.
9. Sambrook, J., Fritsch, E. F., and Maniatis, T. (eds.) (1989) *Molecular Cloning: A Laboratory Manual*, 2nd ed., Cold Spring Harbor Laboratory Press, Cold Spring Harbor, NY.

103

Site-Directed Mutagenesis with LA-PCR™ Technology

Atsushi Shimada and Osamu Takeda

1. Introduction

Protocols for site-directed mutagenesis are widely used in molecular biology and include many polymerase chain reaction (PCR)-based methods that have been developed in order to achieve efficient mutagenesis of a target DNA sequence *(1–8)*. This chapter describes an efficient and economic PCR-based site-directed mutagenesis method, which is designed to introduce a series of mutations into long DNA cloned in pUC vectors (pUC 18, 19, 118, 119), pBluescript® vectors (Stratagene, Cambridge, UK) (pBluescript® II SK(±), KS (±)) and pGEM® vectors (Promega, Madison, WI) (pGEM®-3Zf,4Z) by adding the advantage of LA-PCR technology™ (TaKaRa Shuzo Co., Kyoto, Japan). The protocol uses a combination of a primer designed for introducing a mutation at the target sequence, with primers that may be reused for each mutagenesis reaction (**Fig. 1**). By using this method, a series of site-directed mutations may be undertaken that only require a single primer for each desired change, and furthermore, no reiterative transformation steps are necessary *(9)*. As based on the Long and Accurate (LA)-PCR-technology with the improved enzyme, *TaKaRa LA Taq*, this method provides the high fidelity and introduction of site-directed mutation into longer DNA can be achieved.

2. Materials

1. *Taq* DNA polymerase, 5 U/µL (TaKaRa Shuzo).
2. 10X PCR buffer: 100 m*M* Tris-HCl, pH 8.3, 500 m*M* KCl, 15 m*M* MgCl$_2$.
3. *TaKaRa LA Taq* DNA polymerase, 5 U/µL (TaKaRa Shuzo).
4. 10X LA-PCR buffer II (25 m*M* MgCl$_2$ containing) (TaKaRa Shuzo).
5. 10X *EX Taq* buffer (20 m*M* MgCl$_2$ containing) (TaKaRa Shuzo).
6. dNTP Stock: 2.5 m*M* each of dATP, dCTP, dGTP, and dTTP (TaKaRa Shuzo).
7. Oligonucleotides: MUT 1-6 (*see* **Fig. 2**), MUT B1-6 (*see* **Fig. 3**), MUT G1-6 (*see* **Fig. 4**) (2.5 pmol/µL), M13 M4 (2.5 and 10 pmol/µL), M13 RV (2.5 and 10 pmol/µL), and R1 (2.5 pmol/µL) (*see* **Fig. 2**).

   MUT primers for pUC vectors

MUT 1	5' CATGATTACGAGTTCTAGCT 3'
MUT 2	5' GATCATCTATAGTGGACCTG 3'

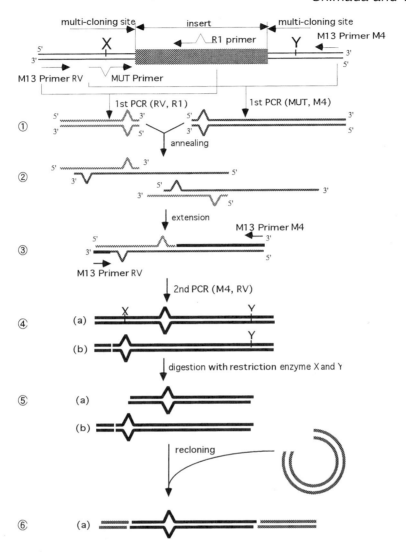

Fig. 1. Principle of PCR in vitro mutagenesis. (1) First-round PCR of target DNA following cloning of sequence into multicloning site of one of the pUC, pBluescript, pGEM series of vectors. One of the MUT primers is chosen to destroy a restriction site, based on both the direction of R1 primer (primer for introducing a mutation) and the restriction site used for cloning the target DNA sequence. The first round PCR is carried out by the combination of R1 primer and M13 primer RV (or M13 primer M4), MUT primer, and M13 primer M4 (or M13 primer RV) separately in two tubes. (2) After DNA purification to remove excess primers, the amplified products are mixed, heat denatured, and annealed. (3) *LA Taq* or *Taq* DNA polymerase is added to complete the DNA heteroduplex. (4) The second-round PCR is performed using M13 primer M4 and M13 primer RV flanking oligonucleotides, which will result in two types of amplified products, (a) and (b). (5) The amplified products are digested with two restriction enzymes, one of which shall recognize the site (X) that had been destroyed by the MUT primer and the other which recognizes the appropriate site (Y) within the multicloning site. (6) Reclone the digested fragment into the vector digested with the same two restriction enzymes. Only the fragment (a) that contains the mutation introduced by R1 primer sequence will be recloned.

Mutagenesis with LA-PCR™ Technology

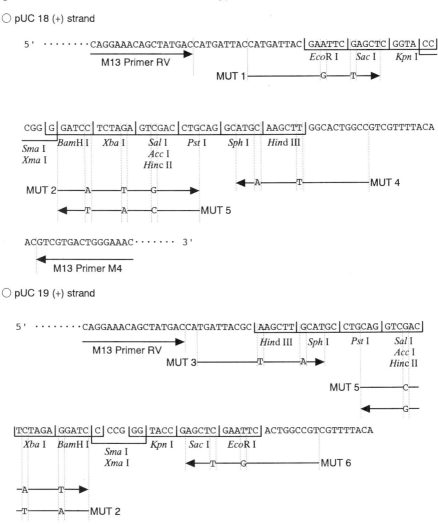

Fig. 2. Position of MUT primers, M13 primer M4, and RV in pUC vectors. The MUT primers are designed to anneal within the multicloning site of pUC vectors; each has a single base mismatch in the recognition sequence of the restriction enzymes with which it is associated. M13-M4 and M13-RV primers also bind to sequences within the multicloning site, but flank the MUT series of oligonucleotides.

MUT 3	5' TGATTACGCCTAGCTTACAT 3'
MUT 4	5' GGCCAGTGCCTAGCTTACAT 3'
MUT 5	5' CAGGTCCACTATAGATGATC 3'
MUT 6	5' ACGGCCAGTGAGTTCTAGCT 3'
M13 M4	5' GTTTTCCCAGTCACGAC 3'
M13 RV	5' CAGGAAACAGCTATGAC 3'

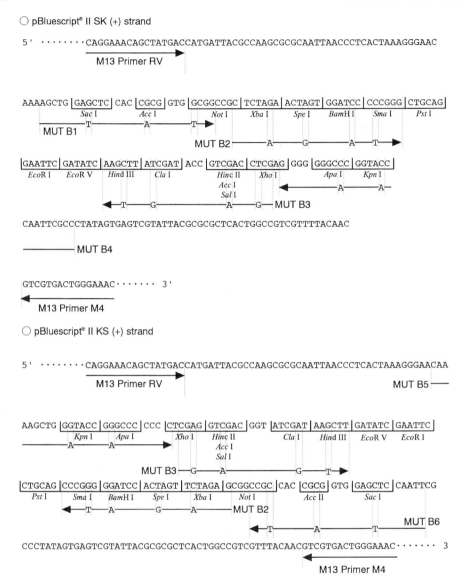

Fig. 3. Position of MUT B primers, M13 primer M4, and RV in pBluescript® vectors. The MUT B primers are designed to anneal within the multicloning site of pBluescript vectors; each has a single base mismatch in the recognition sequence of the restriction enzymes with which it is associated. M13-M4 and M13-RV primers also bind to sequences within the multicloning site, but flank the MUT B series of oligonucleotides.

MUT B primers for pBluescript® vectors

MUT B1	5' AGCTGGATCTCCACCACGGTGTCGG 3'
MUT B2	5' CTCTAAAACTGGTGGATACCTCGGG 3'
MUT B3	5' CGGGGTAGACGGTATCGGTAATCTT 3'
MUT B4	5' GGCGAATTGGATACCGAGCCCCCCC 3'
MUT B5	5' CAAAAGCTGGATACCGAGCCCCCCC 3'
MUT B6	5' AATTGGATCTCCACCACGGTGTCGG 3'

Mutagenesis with LA-PCR™ Technology

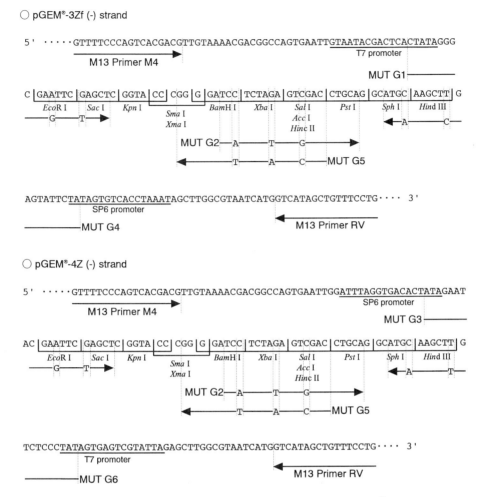

Fig. 4. Position of MUT G primers, M13 primer M4, and RV in pGEM® vectors. The MUT G primers are designed to anneal within the multicloning site of pGEM vectors; each has a single base mismatch in the recognition sequence of the restriction enzymes with which it is associated. M13-M4 and M13-RV primers also bind to sequences within the multicloning site, but flank the MUT G series of oligonucleotides.

MUT G primers for pGEM® vectors

MUT G1	5' CTATAGGGCGAGTTCTAGCT 3'
MUT G2	5' TCATCTAT AGTGGACCTGCA 3'
MUT G3	5' ATAGAATACGAGTTCTAGCT 3'
MUT G4	5' TAGAATACTCCAGCTTACAT 3'
MUT G5	5' GTCCACTATAGATGATCCCC 3'
MUT G6	5' ATAGGGAGACTAGCTTACAT 3'

For the design of Primer R1, see **Note 1**.

8. Plasmid DNA, into which the target sequence is cloned, prepared using standard protocols (*10*).
9. Phenol saturated in Tris-ethylenediaminetetraacetic acid (EDTA) (TE) buffer.
10. Chloroform.

11. 3 M Sodium acetate, pH 4.8 to 5.2.
12. Appropriate restriction enzymes.
13. SUPREC-01: Filter cartridge for rapid recovery of DNA from agarose gels (TaKaRa Shuzo).
14. SUPREC-02 or Microcon-100. Filter cartridge for rapid purification and/or concentration of DNA samples (TaKaRa Shuzo).
15. T4 DNA ligase (350 U/μL) with 10X reaction buffer (660 mM Tris-HCl, pH 7.6, 66 mM MgCl$_2$, 100 mM dithiothreitol [DTT], and 1 mM adenosine triphosphate [ATP]) (TaKaRa Shuzo).
16. *Escherichia coli* JM109 competent cells (TaKaRa Shuzo).
17. TE buffer: 10 mM Tris-HCl, pH 8.0, 1 mM EDTA.
18. PCR Thermal Cycler MP (mineral oil free, TaKaRa Shuzo).

3. Methods

The procedure requires that the target DNA sequence is cloned into a pUC vector (pUC 18, 19, 118, 119) or pBluescript vectors (pBluescript II SK(±), KS(±)) or pGEM vectors (pGEM-3Zf, 4Z).

3.1. Preparation of First-Round PCR Products

1. Design and prepare the mutagenic primer (R1) to introduce the desired change into the target DNA sequence (*see* **Note 1**).
2. Prepare the first round of PCR reactions using the primer combinations described in (a) and (b) (*see* **Figs. 5** and **6** and **Note 2**):
   a. Combination of R1 primer and M13 primer M4* (or M13 primer RV): 10 pg-1 ng plasmid template DNA, 1 μL R1 primer, 1 μL M13 primer (2.5 pmol/μL), 5 μL 10X PCR buffer or 10X LA-PCR buffer II or 10X *EX Taq* PCR buffer, 4–8 μL dNTP stock, 0.25 μL *Taq* DNA polymerase or 0.5 μL *TaKaRa LA Taq* DNA polymerase, make up to 50 μL with distilled sterilized water (*see* **Note 3**).
   b. Combination of a MUT primer (MUT 1–6 or MUTB 1–6 or MUT G1–6) and M13 primer RV* (or M13 primer M4): 10 pg-1 ng plasmid template DNA, 1 μL MUT primer, 1 μL M13 primer (2.5 pmol/μL), 5 μL 10X PCR buffer or 10X LA-PCR buffer II or 10X *EX Taq* PCR buffer, 4–8 μL dNTP stock, 0.25 μL *Taq* DNA polymerase or 0.5 μL *TaKaRa LA Taq* DNA polymerase, make up to 50 μL with distilled sterilized water (*see* **Notes 3** and **4**).
   *Different M13 primers must be used in (a) and (b).
3. Overlay each tube with mineral oil (approx 50 μL. When using an oil-free type thermal cycler, this is not necessary.) and perform first-stage PCR using the following program for 25 cycles: denaturation, 94°C for 30 s; primer annealing, 55°C for 2 min; primer extension, 72°C for x min (*see* **Notes 5** and **6**).
4. Remove a small aliquot (5 μL) from **steps 2a** and **2b** for analysis by agarose gel electrophoresis (1–3% gel) to determine if the PCR reactions were successful (*see* **Note 7**).
5. Purify the PCR products using standard phenol-chloroform extraction/ethanol precipitation. Alternatively, products available commercially such as SUPREC-02 or Microcon-100 filter cartridges may be used. Each of the purified PCR products should be resuspended in 50 μL TE.

3.2. Heteroduplex Formation and Second-Round PCR

1. Prepare the following reaction mixture and then overlay with 50 μL of mineral oil (when using oil-free type thermal cycler, this is not necessary): First-round PCR product (a) 0.5 μL; first-round PCR product, (b) 0.5 μL; 10X PCR buffer or 10X LA-PCR buffer II or 10X *EX Taq*

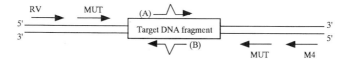

Insertion site of target DNA	Direction of R1 primer	Applicable MUT primers
pUC18 (pUC118)		
EcoR I, Sac I	(A)	MUT4 or MUT5
Kpn I, Sma I	(A)	MUT4 or MUT5
	(B)	MUT1
Acc I, BamH I Hinc II, Sal I Xba I	(A)	MUT4
	(B)	MUT1
Pst I	(A)	MUT4
	(B)	MUT1 or MUT2
Sph I, Hind III	(B)	MUT1 or MUT2
pUC19 (pUC119)		
Hind III, Sph I	(A)	MUT6 or MUT2
Pst I	(A)	MUT6 or MUT2
	(B)	MUT3
Acc I, BamHI Hinc II, Sal I Xba I	(A)	MUT6
	(B)	MUT3
Kpn I, Sma I	(A)	MUT6
	(B)	MUT3 or MUT5
EcoR I, Sac I	(B)	MUT3 or MUT5

Insertion site of target DNA	Direction of R1 primer	Applicable MUT primers
pBluescript ®II SK		
Sac I, Acc II, Not I	(A)	MUT B3 or MUT B4
Xba I, Spe I BamH I Sma I	(A)	MUT B3 or MUT B4
	(B)	MUT B1
Pst I, EcoR I EcoR V	(A)	MUT B3 or MUT B4
	(B)	MUT B1 or MUT B2
Hind III, Cla I Hinc II, Acc I Sal I, Xho I	(A)	MUT B4
	(B)	MUT B1 or MUT B2
Apa I, Kpn I	(B)	MUT B1 or MUT B2
pBluescript® II KS		
Sac I, Acc II, Not I	(B)	MUT B3 or MUT B5
Xba I, Spe I BamH I Sma I	(A)	MUT B6
	(B)	MUT B3 or MUT B5
Pst I, EcoR I EcoR V	(A)	MUT B2 or MUT B6
	(B)	MUT B3 or MUT B6
Hind III, Cla I Hinc II, Acc I Sal I, Xho I	(A)	MUT B2 or MUT B6
	(B)	MUT B5
Apa I, Kpn I	(A)	MUT B2 or MUT B6

Fig. 5. Selection of MUT and MUT B primers.

PCR buffer (same as used in first-round PCR), 5 µL; dNTP stock, 4–8 µL; make up to 47.5 µL (case of using *LA-Taq*) or 47.75 µL (case of using *Taq*) with distilled sterilized water (*see* **Notes 3** and **4**).

2. Perform the following annealing/extension step with the mix in **step 1** using a thermal cycler with the program: denaturation, 94°C for 10 min; cool to 37°C over a period of 60 min; maintain at 37°C over 15 min.
3. Add 0.25 µL *Taq* or 0.5 µL *LA-Taq* and mix gently. Spin down any droplets on the tube wall and then incubate at 72°C for *x* min (*see* **Note 5**).
4. To this reaction mix add: 1 µL M13 primer M4 (10 pmol/µL), 1 µL M13 primer RV (10 pmol/µL).
5. Carry out PCR using the following program for 5–10 cycles: denaturation, 94°C for 30 s; primer annealing, 55°C for 2 min; primer extension, 72°C for *x* min (*see* **Note 5**).
6. Remove an aliquot (5 µL) following the amplification reaction and confirm that a specific DNA product of the correct size has been generated.
7. Purify the remaining amplified DNA by phenol/chloroform extraction followed by ethanol precipitation. Resuspend in 10 µL distilled sterilized water.

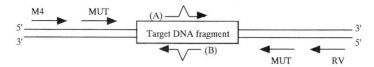

Insertion site of target DNA	Direction of R1 primer	Applicable MUT primers
<pGEM® vectors>		
pGEM®-3Zf		
EcoR I, Sac I	(A)	MUT G4 or MUT G5
Kpn I, Sma I	(A)	MUT G4 or MUT G5
	(B)	MUT G1
BamH I, Xba I Sal I, Acc I, Hinc II	(A)	MUT G4
	(B)	MUT G1
Pst I	(A)	MUT G4
	(B)	MUT G1 or MUT G2
Sph I, Hind III	(B)	MUT G1 or MUT G2
pGEM®-4Z		
EcoR I, Sac I	(A)	MUT G5 or MUT G6
Kpn I, Sma I	(A)	MUT G5 or MUT G6
	(B)	MUT G3
BamH I, Xba I Sal I, Acc I, Hinc II	(A)	MUT G6
	(B)	MUT G3
Pst I	(A)	MUT G6
	(B)	MUT G2 or MUT G3
Sph I, Hind III	(B)	MUT G2 or MUT G3

Fig. 6. Selection of MUT G primers.

3.3. Recloning of the Mutated Target DNA Sequence

The second-round PCR product is then digested with two restriction enzymes (X and Y in **Fig. 1**), which are dictated by the MUT primer/M13 primer combination used in the first-round PCR reaction (*see* **Note 2**).

1. To 10 µL of the purified second-round PCR product, add 2 µL of appropriate 10X restriction buffer, 6 µL of distilled sterilized water, mix, and then add 1 µL of each specified restriction enzyme. Mix gently and then incubate at 37°C for at least 1 h.
2. Restrict the DNA of an appropriate cloning vector with identical enzymes. Take 1 µg vector DNA, add 5 µL of appropriate 10X restriction buffer, 1 µL of each restriction enzyme, and make up to an appropriate volume with distilled water. Mix gently and incubate as before.

3. Purify the restricted PCR product and cloning vector by phenol/chloroform extraction and ethanol precipitation. Resuspend each purified DNA in 2 µL TE buffer or distilled sterilized water.
4. Separate the restricted cloning vector by agarose gel electrophoresis. Excise the desired band from the gel and recover the DNA using a SUPREC-01 filter cartridge.
5. Purify the eluted DNA by phenol/chloroform extraction and ethanol precipitation. Resuspend the DNA pellet in 20 µL TE buffer or distilled sterilized water.
6. To 1 µL of restricted cloning vector add 2 µL PCR product (*see* **Note 8**), 1 U (cohesive end) or 100 U (blunt end) T4 DNA ligase, 2 µL 10X ligation buffer, and then make up to 20 µL with sterile distilled water. Mix gently and incubate at 16°C for 16 h.
7. Transform 20 µL (maximum) of the ligation reaction into 100 µL JM109 competent *E. coli* cells using standard protocols. Following transformation, at least three clones should be submitted for DNA sequencing to confirm that the desired mutation has been introduced.

4. Notes

1. For the design of the mutagenic primer R1, the following should be considered to ensure that the oligonucleotide will anneal efficiently to the specified target. If a single- or double-point mutation is to be introduced, then 10 flanking bases should be incorporated into the primer sequence on either side of the mutation site. If the flanking regions are A + T rich, or substantial alterations are involved, this can be extended to 15 bases or more. Ensure that the designed R1 primer does not have significant complementarily to itself or to its possible partners for PCR, particularly at their 3' ends. This will avoid "primer dimer" formation in which two primers hybridize to each other forming a very effective substrate for PCR. If possible, keep the G + C composition of the primer to about 50–60% and avoid long stretches of the same base. Usually using LA-PCR, the length of primer should be 25 mers or more.
2. The choice of the MUT primer used when designing the reaction will be dependent on: the cloning site used for insertion of the target DNA sequence (*see* **Figs. 5** and **6**); the direction of the R1 primer; the mutated fragment must be recloned following mutagenesis, which should be undertaken with unique restriction sites (enzymes X and Y in **Fig. 1**), i.e., that neither of these enzymes should have sites within the cloned target DNA sequence.
   a. For enzyme X (*see* **Fig. 1**), any of the restriction endonuclease recognition sequences mutated by the particular MUT primer used in the first round PCR reaction (*see* **Figs. 2–4**) may be chosen. The choice of enzymes that may be selected for by a particular MUT primers are:

   MUT 1   *Eco*RI, *Sac*I
   MUT 2   *Bam*HI, *Xba*I, *Sal*I, *Acc*I, *Hinc*II
   MUT 3   *Sph*I, *Hin*dIII
   MUT 4   *Sph*I, *Hin*dIII
   MUT 5   *Bam*HI, *Xba*I, *Sal*I, *Acc*I, *Hinc*II
   MUT 6   *Eco*RI, *Sac*I

   MUT B1 *Sac*I, *Acc*I, *Not*I
   MUT B2 *Xba*I, *Spe*I, *Bam*HI, *Sma*I
   MUT B3 *Hin*dIII, *Cla*I, *Hinc*II, *Acc*I, *Sal*I, *Xho*I
   MUT B4 *Apa*I, *Kpn*I
   MUT B5 *Apa*I, *Kpn*I
   MUT B6 *Sac*I, *Acc*I, *Not*I

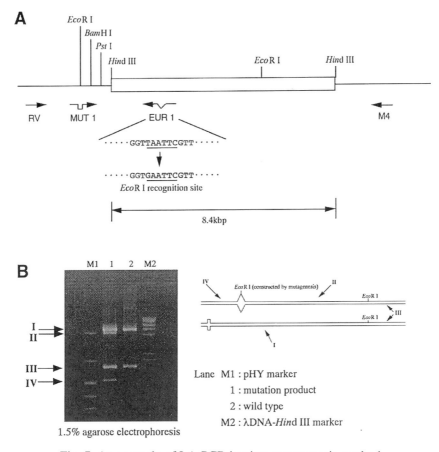

Fig. 7. An example of LA-PCR in vitro mutagenesis method.

MUT G1 *Eco*RI, *Sac*I
MUT G2 *Bam*HI, *Xba*I, *Sal*I, *Acc*I, *Hin*cII
MUT G3 *Eco*RI, *Sac*I
MUT G4 *Sph*I, *Hin*dIII
MUT G5 *Bam*HI, *Xba*I, *Sal*I, *Acc*I, *Hin*cII
MUT G6 *Sph*I, *Hin*dIII

   b. For enzyme Y (*see* **Fig. 1**), any of the restriction enzymes with sites in the multicloning site on the side flanking both the inserted DNA fragment and the recognition sequence for enzyme X may be utilized.
3. Choose the buffer and DNA polymerase according to the length of DNA fragment to be amplified:
    10X LA-PCR buffer II (25 m*M* MgCl$_2$ containing) and *TaKaRa LA Taq* DNA polymerase for amplification of ≥2 kbp
    10X *EX Taq* PCR buffer (20 m*M* MgCl$_2$ containing) and *TaKaRa LA Taq* DNA polymerase for amplification of <2 kbp
    10X PCR buffer and *Taq* DNA polymerase for amplification of <500 bp.
4. Choose using volume of dNTP stock/50 µL reaction scale:
    When using 10X LA-PCR buffer II: 8 µL (final 400 µ*M*)
    When using 10X *EX-Taq* buffer: 5 µL (final 250 µ*M*)
    When using 10X PCR buffer: 4 µL (final 200 µ*M*)

5. The heat time at 72°C depends on the length of amplified DNA fragment. Approximately 1 min at 72°C is recommended as the standard for amplification of 1 kbp.
6. MUT 2, MUT 5, MUT G2, and MUT G5 primers are sometimes difficult to anneal to the template at 55°C. In such cases, lower the annealing temperature to 45°C.
7. If any nonspecific bands appear on the agarose gel following electrophoresis of the first- and second-round PCR products, the desired band in each case should be excised from the gel and purified. This may be achieved by electroelution or by using commercial products such as SUPREC-01 filter cartridges.
8. Recommended amounts of DNA are vector:insert = 0.03 pmol: 0.1–0.3 pmol (0.03 pmol of pUC 18 DNA corresponds to about 50 ng).
9. An example of the utility of this method can be introducing one-point mutation into 8.4 kbp DNA fragment in pUC 18 vector to make the *Eco*RI restriction enzyme site (*see* **Fig. 7**).

Acknowledgment

U.S. patent 5,436,149 for LA-PCR Technology is owned by TaKaRa Shuzo Co., Ltd, Kyoto, Japan.

References

1. Ito, W., Ishiguro, H., and Kurosawa, Y. (1991) A general method for introducing a series of mutations into cloned DNA using the polymerase chain reaction. *Gene* **102,** 67–70.
2. Hemsley, A., Arnheim, N., Toney, M. D., Cortopassi, G., and Galas, D. J. (1989) A simple method for site-directed mutagenesis using the polymerase chain reaction. *Nucleic Acids Res.* **17,** 6545–6551.
3. Higuchi, R., Krummel, B., and Saiki, R. (1988) A general method of in vitro preparation and specific mutagenesis of DNA fragments: study of protein and DNA interactions. *Nucleic Acids Res.* **16,** 7351–7367.
4. Ho, S. N., Hunt, H. D., Horton, R. M., Pullen, J. K., and Pease, L. R. (1989) Site-directed mutagenesis by overlap extension using the polymerase chain reaction. *Gene* **77,** 51–59.
5. Weiner, M. P., Costa, G. L., Schoettlin, W., Cline, J., Mathur, E., and Bauer, J. C. (1994) Site-directed mutagenesis of double-stranded DNA by the polymerase chain reaction. *Gene* **151,** 119–123.
6. Ling, M. and Robinson, B. H. (1995) A one-step polymerase chain reaction site-directed mutagenesis method for large-cassettes with high efficiency, yield, and fidelity. *Anal. Biochem.* **230,** 167–172.
7. Datta, A. K. (1995) Efficient amplification using "megaprimer" by asymmetric polymerase chain reaction. *Nucleic Acids Res.* **23,** 4530–4531.
8. Seraphin, B. and Kandels-Lewis, S. (1996) An efficient PCR mutagenesis strategy without gel purification step that is amenable to automation. *Nucleic Acids Res.* **24,** 3276–3277.
9. Shimada, A. (1996) PCR-based site directed mutagenesis. *Methods Mol. Biol.* **57,** 157–165.
10. Sambrook, J., Fritsch, E. F., and Maniatis, T. (eds.) (1989) *Molecular Cloning. A Laboratory Manual*, 2nd ed., Cold Spring Harbor Press, Cold Spring Harbor, NY.

104

Recombination and Mutagenesis by Overlap Extension PCR

Robert J. Pogulis, Abbe N. Vallejo, and Larry R. Pease

1. Introduction

The polymerase chain reaction (PCR) *(1,2)* is now a fundamental tool of molecular biology. Although PCR provides the basis for a variety of sensitive analytical techniques, it can also be used in a synthetic capacity to generate large quantities of specific DNA fragments. The alteration of amplified DNA sequences is also possible, since synthetic oligonucleotide primers become incorporated into the final PCR product. Although the 3' ends of these primers must match the target DNA sequence, the 5' ends may contain modifications. Sequence modifications in the primers will therefore be present in the ends of the amplified DNA fragment, offering a straightforward, although limited, ability to introduce site-directed mutations during PCR.

The "overlap extension" technique *(3,4)* provides a means of introducing mutations into the center of a PCR fragment. In this case, primers are designed to provide two different PCR products with a region of common sequence. The two overlapping fragments are then fused in a subsequent PCR amplification. Depending on the choice of primers and templates, mutant or chimeric DNA molecules can be generated. Overlap extension represents a simple and versatile approach to genetic engineering. Subcloning of the wild-type sequence into a single-stranded vector is unnecessary and mutagenic efficiency approaches 100%. In addition, "splicing by overlap extension" (SOE) allows segments from different genes to be recombined without relying on restriction sites.

1.1. Mechanism of Overlap Extension

The general mechanism of overlap extension, as applied to site-directed mutagenesis, is illustrated in **Fig. 1**. In separate reactions, two overlapping subfragments of the target sequence are amplified. Each reaction uses one flanking primer (a or d) and one internal primer containing the desired mutation (b or c). Since the two internal primers overlap, the overlapping fragments generated in the first PCRs can be fused in a second round of PCR. When fragments AB and CD are mixed, denatured, and reannealed, the 3' end of the upper strand of fragment AB and the 5' end of fragment CD's lower strand can hybridize. In the presence of a DNA polymerase, each of the overlapping strands acts as a primer on the other, and extension of the overlap results in a full-length mutant

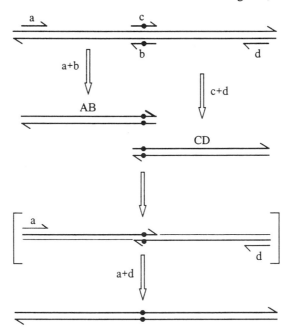

Fig. 1. Mutant product AD. Site-directed mutagenesis by overlap extension. DNA strands and synthetic oligonucleotide primers are represented by lines, with arrows indicating the 5'–3' orientations. The mutation is depicted by a solid circle. Primers are labeled with lowercase letters. PCR products are named by pairs of uppercase letters corresponding to the primers used to generate them. The intermediate species enclosed in brackets is formed by annealing at the overlap region. The overlap is extended from the 3'-end of each strand (gray dotted line) by *Taq* polymerase to generate a fusion product that can be amplified with primers a and d.

product (AD). Even if overlap extension occurs at relatively low frequency, only the fusion product will be amplified with primers a and d.

The flexibility of the overlap extension technique is illustrated in **Figs. 2** and **3**. Insertional mutations can be generated by simply adding the sequence to be inserted to the 5'-ends of primers b and c (**Fig. 2A**). Deletions are created by designing b and c primers that contain sequence on either side of the deletion (**Fig. 2B**). Last, **Fig. 3** depicts the mechanism of SOEing. In this case, sequence from gene II is added to the 5'-end of primer b, and the first two reactions utilize different templates. Thus, the intermediate products overlap and can be fused by overlap extension to generate the

Fig. 2. *(opposite page)* Insertion product AD. Use of overlap extension PCR to generate insertion and deletion mutations. (**A**) Insertional mutation. The 5' ends of primers b and c contain the inserted sequence, which also serves as the region of overlap. (**B**) Deletion mutation. The sequence to be deleted is represented by heavy lines. The sequences on either side of the deletion are represented by gray and solid lines. Primers b and c contain sequence from each side of the region to be deleted.

Fig. 3. *(opposite page)* Fusion product AD. Splicing by Overlap Extension (SOEing). PCR products AB and CD are derived from different genes. The overlap region is provided by the addition of CD sequence to primer b.

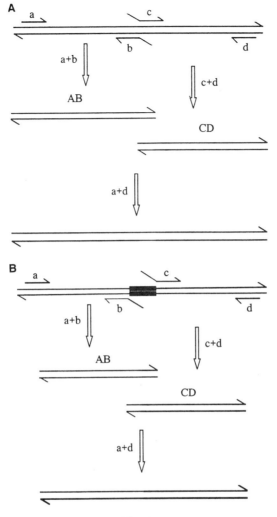

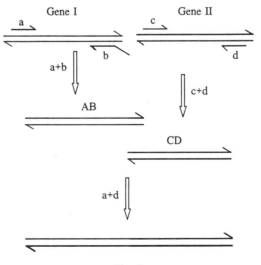

Fig. 2

Fig. 3

recombinant product AD. It is also possible to perform recombination and mutagenesis simultaneously, by combining the strategies shown in **Figs. 1** and **2C** *(5)*.

2. Materials

1. A thermocycler ("PCR machine").
2. *Taq* DNA polymerase (*see* **Notes 1** and **2**).
3. 10X PCR buffer: 100 m*M* Tris-HCl, pH 8.3, 500 m*M* KCl, and 15 m*M* MgCl$_2$ (*see* **Note 3**).
4. 10X dNTP solution: 2 m*M* each, dATP, dCTP, dGTP, and dTTP, pH 7.5 (prepared from commercially available aqueous stocks).
5. Oligonucleotide primers (*see* **Subheading 3.1.** and **Notes 4** and **5**).
6. DNA template(s) (*see* **Subheading 3.2.**).
7. Light mineral oil (depending on thermocycler).
8. Microfuge tubes that fit precisely into the heating block of the thermocycler.
9. Agarose gel electrophoresis supplies and apparatus.
10. A method for purifying amplified fragments from agarose gels—the authors use GeneClean and/or Mermaid kits (Bio 101, La Jolla, CA), depending on the size of the fragment(s) to be isolated.

3. Methods

3.1. Primer Design (see Notes 4 and 5)

1. All of the standard considerations in PCR primer design *(6)* apply to SOE primers as well. For example, they must not be complementary to one another (except for primers b and c, which are complimentary by design), or capable of any obvious secondary structure formation.
2. The internal primers used in an overlap extension protocol (primers b and c in **Figs. 1–3**) serve two functions. They must contain sequences complementary to the template in order to serve as PCR primers ("priming regions"), as well as sequences that provide a region of overlap between the two intermediate PCR products ("overlap regions"). In the case of site-directed mutagenesis (**Fig. 1**), the priming and overlap regions may completely coincide. The mismatched bases should, however, be positioned near the middle of each internal primer to ensure proper priming. In the case of gene SOEing (**Fig. 3**), the 3' end of an internal primer serves as the priming region and its 5' end contains the overlap region.
3. We routinely design our internal primers so that the region of overlap in the fusion reaction has an estimated melting temperature (T_m) of 50°C, according to the following formula *(7)*:

$$T_m(°C) = [4(G + C) + 2(A + T)]$$

   The priming region of an internal primer should also have a T_m of at least 50°C.
4. The flanking primers (a and d) do not contribute to the region of overlap. Since they serve only to amplify the intermediate products and the recombinant product, no special considerations apply to their design. In some cases, restriction sites may be added to the 5' ends of these primers to facilitate cloning of the final product (*see* **Note 6**). The flanking primers should be capable of annealing at 50°C.

3.2. Source and Preparation of Template(s)

Most often, the starting template is a cloned version of the gene(s) to be modified, although other types of PCR templates, such as first-strand cDNA (reverse-transcribed from RNA) can also be used *(8)*. However, if the initial template is a complex mixture of DNA molecules, the risk of obtaining the wrong product and/or introducing mutations in the first amplification is increased. The advantage of using cloned DNA is that high concentrations of a defined, specific template can be used. This allows for generation of the expected final product in fewer cycles of amplification, which decreases the chance of misincorporation by *Taq* polymerase *(3)*. Since the reaction plateaus after the production of some maximum amount of product, it is not necessary to minimize the actual number of heating/cooling cycles to obtain the benefit derived from high template concentrations. The authors have successfully used plasmids prepared by cesium chloride gradient centrifugation or "miniprep" methods based on the alkaline lysis procedure.

3.3. Production of Intermediate Products AB and CD

1. Set up two separate amplification reactions to generate the intermediate products AB and CD, as illustrated in **Fig. 1**:

Component	PCR #1	PCR #2
10X PCR buffer	10.0 µL	10.0 µL
10X dNTP solution	10.0 µL	10.0 µL
100 pmol 5' primer	a	c
100 pmol 3' primer	b	d
Template	0.5 µg	0.5 µg
Taq polymerase	2.5 U	2.5 U
H$_2$O	to 100 µL	to 100 µL

    For information regarding Mg^{2+} concentration in PCR buffer, *see* **Note 3**. For mutagenesis, the same template is used in both reactions; for SOEing, two different templates are used (*see* **Figs. 1** and **2**).
2. Overlay the reaction mixtures with two to three drops of light mineral oil, and amplify by PCR using the following cycle profile (*see* **Note 7**): 20–25 cycles: 94°C, 1 min (denaturation); 50°C, 2 min (annealing); 72°C, 2 min (extension).

3.4. Purification of Intermediate Products AB and CD

Gel purification of the intermediate products leads to a cleaner overlap extension reaction and increased product yield. Most important, gel purification removes the starting templates, which could be amplified to generate wild-type product if carried over into the overlap extension PCR.

1. Run each reaction mixture (100 µL) on an agarose gel. Usually, there is only one major product, so 1% agarose works well in most situations. If greater resolution is required, NuSieve agarose (FMC Bioproducts, Rockland, ME) can be used at concentrations up to 4%. It is often helpful to analyze a 5-µL sample of each intermediate PCR product by gel electrophoresis first—this can aid in estimating the approximate yield and choosing the best conditions for the preparative gel.
2. Excise the bands of interest (products AB and CD), and recover the DNA. This can be accomplished by a variety of methods, including "freeze-squeeze" *(9)* and electroelution *(10)*. We typically employ either the GeneClean or Mermaid kits (Bio 101), depending on the size of the fragment to be isolated: Mermaid is specifically designed for the purification of fragments ≤300 bp, whereas GeneClean works well for fragments larger than this.

3.5. Production of Fusion Product, AD

1. Set up the overlap extension PCR as follows. Component: 10.0 µL 10X PCR buffer; 10.0 µL 10X dNTP solution; 100 pmol 5' primer, a (*see* **Note 3**); 100 pmol 3' primer, d (*see* **Note 8**). Templates: intermediate AB, 1/4 of product from **Subheading 3.4.**; intermediate CD, 1/4 of product from **Subheading 3.4.**; *Taq* polymerase, 2.5 U; H_2O to 100 µL.
2. Overlay the reaction mixtures with two to three drops of light mineral oil, and amplify as before: 20–25 cycles: 94°C, 1 min (denaturation); 50°C, 2 min (annealing); 72°C, 2 min (extension).
3. Analyze a 5-µL sample of the final PCR product by agarose gel electrophoresis. A single fragment, representing the mutant product AD, of the anticipated size should be present.
4. Clone the fusion product AD for further analysis.

It is usually necessary to sequence the cloned mutant gene, since even a single unanticipated nucleotide change could affect the structure and function of the encoded protein. A "cassette" approach can reduce the amount of sequencing required: PCR-mediated mutagenesis is performed on a 300–500-bp gene segment, which is subsequently ligated back into a vector containing the remainder of the gene. As DNA sequencing technology advances, the need to sequence the products of overlap extension PCR is becoming less of a limitation. We sequence "miniprep" plasmid DNA (Wizard Minipreps, Promega, Madison, WI) directly using the Sequenase kit (US Biochemical, Cleveland, OH) or via an automated sequencing system (Applied Biosystems [Foster City, CA] Sequencer Model 373A).

Although the concept of overlap extension PCR is straightforward, the power of this approach to genetic engineering is significant. It enables gene sequences to be altered quickly and easily with nearly 100% efficiency. Complex constructs involving the fusion of multiple gene segments have been generated by SOEing *(4,5)*. In another extension of SOE, mixed populations of internal (b and c) primers have been used to effect random mutagenesis in the overlap region *(11,12)*.

4. Notes

1. The *Taq* polymerase error frequency associated with the overlap extension protocol outlined here is approx 1/4000 nucleotides *(3)*. A slightly higher error frequency (approx 1/2000) may be observed during complex constructions involving multiple SOE reactions *(4)*. Although these error frequencies are sufficiently low to make overlap extension practical for routine use, several precautions are in order. As indicated in **Subheading 3.2.** and **Note 3**, template and Mg^{2+} concentrations are important considerations.
2. It has been reported that other thermostable DNA polymerases (*Vent* polymerase, New England Biolabs, Beverly, MA; *Pfu* DNA polymerase, Stratagene, La Jolla, CA) with higher fidelity than *Taq* can be used for overlap extension *(13–16)*.
3. Mg^{2+} concentration is a critical parameter in PCR, and overlap extension PCR is no exception. "Standard" reaction conditions include 1.5 mM (final concentration) $MgCl_2$. In cases where standard conditions either do not yield significant amounts of product, or result in a high background of nonspecific products, we titrate $MgCl_2$ concentration over a range of 0.5–2.5 mM, in 0.5 mM increments. Also, since *Taq* polymerase error rates increase with increasing Mg^{2+} concentration, it is desirable to use the lowest $MgCl_2$ concentration possible. We find that many amplifications from cloned templates will tolerate a drop to 1.0 mM $MgCl_2$.

4. Oligonucleotides used in our laboratory are synthesized with an Applied BioSystems DNA Synthesizer. The final step in the synthesis involves treatment with concentrated NH_4OH, which is subsequently evaporated away under vacuum. The dried DNA pellet is resuspended in water, and desalted over a Sephadex G-25 column (NAP-10 columns, Pharmacia, Uppsala, Sweden).

5. On a few occasions we have found several clones derived from the same overlap extension reaction that contain a deoxythymidine (T) residue at the nucleotide (same strand) immediately 5' of a mutagenic primer. This may be explained by the fact that *Taq* polymerase can add untemplated deoxyadenosine (A) residues to the 3' ends of amplified DNA fragments *(17)*. An additional A residue at the 3' end of one of the overlapping DNA strands (**Fig. 1**: upper strand of AB, lower strand of CD) would result in a T on the opposite strand, immediately 5' to the mutagenic primer. In our experience, this is a rare event (observed in less than 1 in 10 constructions), probably because a mismatch at the 3' terminal nucleotide significantly inhibits priming efficiency. However, others have reported that unwanted mutations owing to untemplated 3' nucleotide addition pose a significant problem *(18,19)*. If this type of mutation does become a problem, design of primers b and c to begin immediately 3' to T residues in their respective strands ensures that an untemplated 3' A residue would simply become part of the overlap region.

6. Cloning of SOE-generated DNA fragments is facilitated by the use of flanking primers (a and d) designed to include one or more restriction sites at their 5' ends. The restriction site sequences should be three to nine bases from the 5' termini for optimal digestion of the overlap extension product. We have successfully used *Bam*HI, *Bgl*II, *Cla*I, *Eco*RI, *Hin*dIII, *Sal*I, and *Xho*I (20 bp from 5' end).

7. The exact PCR cycle profile used does not seem to be critical. In addition to the profile suggested in **Subheadings 3.3.** and **3.5.**, profiles using a shorter (30-s) denaturation step or an extension step ranging from 1–3 min also work.

8. If the yield of fragments AB and CD from the first PCRs is good, but the fusion product AD is being amplified inefficiently, withholding the primers (a and d) for two to three cycles sometimes results in more efficient formation of fragment AD by overlap extension. Addition of PCR primers after this reaction has been allowed to begin may prevent undesirable side reactions that compete for the available reagents.

References

1. Mullis, K., Faloona, F., Scharf, S., Saiki, R., Horn, G., and Erlich, H. (1986) Specific enzymatic amplification of DNA in vitro: the polymerase chain reaction. *Cold Spring Harbor Symp. Quant. Biol.* **51**, 263–273.
2. Saiki, R. K., Gelfand, D. H., Stoffel, S., Scharf, S. J., Higuchi, R., Horn, G. T., Mullis, K. B., and Erlich, H. A. (1988) Primer-directed enzymatic amplification of DNA with a thermostable DNA polymerase. *Science* **239**, 487–491.
3. Ho, S. N., Hunt, H. D., Horton, R. M., Pullen, J. K., and Pease, L. R. (1989) Site-directed mutagenesis by overlap extension using the polymerase chain reaction. *Gene* **77**, 51–59.
4. Horton, R. M., Hunt, H. D., Ho, S. N., Pullen, J. K., and Pease, L. R. (1989) Engineering hybrid genes without the use of restriction enzymes: gene splicing by overlap extension. *Gene* **77**, 61–68.
5. Horton, R. M., Cai, Z., Ho, S. N., and Pease, L. R. (1990) Gene splicing by overlap extension: tailor-made genes using the polymerase chain reaction. *Biotechniques* **8**, 528–535.
6. Rychlik, W. (1993) Selection of primers for polymerase chain reaction, in *PCR Protocols: Current Methods and Applications* (White, B. A., ed), Humana, Totowa, NJ, pp. 31–40.

7. Suggs, S. V., Hirose, T., Miyake, T., Kawashima, E. H., Johnson, M. J., Itakura, K., and Wallace, R. B. (1981) Use of synthetic oligo-deoxyribonucleotides for the isolation of cloned DNA sequences, in *Developmental Biology Using Purified Genes* (Brown, D. D. and Fow, C. F., eds.), Academic, New York, pp. 683–693.
8. Davis, G. T., Bedzyk, W. D., Voss, E. W., and Jacobs, T. W. (1991) Single chain antibody (SCA) encoding genes: one-step construction and expression in eukaryotic cells. *Biotechnology* **9,** 165–179.
9. Tautz, D. and Renz, M. (1983) An optimized freeze-squeeze method for the recovery of DNA fragments from agarose gels. *Anal. Biochem.* **132,** 14–19.
10. Sambrook, J., Fritsch, E. F., and Maniatis, T. (eds.) (1989) *Molecular Cloning: A Laboratory Manual,* 2nd ed., Cold Spring Harbor Laboratory, Cold Spring Harbor, NY.
11. Morrison, H. G. and Desrosiers, R. C. (1993) A PCR-based strategy for extensive mutagenesis of a target DNA sequence. *Biotechniques* **14,** 454–457.
12. Kirchhoff, F. and Desrosiers, R. C. (1993) A PCR-derived library of random point mutations within the V3 region of simian immunodeficiency virus. *PCR Methods Appl.* **2,** 301–304.
13. Hanes, S. D. and Brent, R. (1991) A genetic model for interaction of the homeodomain recognition helix with DNA. *Science* **251,** 426–430.
14. Cease, K. B., Potcova, C. A., Lohoff, C. J., and Zeigler, M. E. (1994) Optimized PCR using Vent polymerase. *PCR Methods Appl.* **3,** 298–300.
15. Juncosa-Ginestra, M., Pons, J., Planas, A., and Querol, E. (1994) Improved efficiency in site-directed mutagenesis by PCR using a *Pyrococcus* sp. GB-D polymerase. *Biotechniques* **16,** 820–823.
16. Picard, V., Ersdal-Badju, E., Lu, A., and Bock, S.C. (1994) A rapid and efficient one-tube PCR-based mutagenesis technique using *Pfu* DNA polymerase. *Nucleic Acids Res.* **22,** 2587–2591.
17. Clark, J. M. (1988) Novel non-templated nucleotide addition reactions catalyzed by prokaryotic and eukaryotic DNA polymerase. *Nucleic Acids Res.* **17,** 3319.
18. Landt, O., Grunert, H. P., and Hahn, U. (1990) A general method for rapid site-directed mutagenesis using the polymerase chain reaction. *Gene* **96,** 125–128.
19. Kuipers, O. P., Boot, H. J., and de Vos, W. M. (1991) Improved site-directed mutagenesis method using PCR. *Nucleic Acids Res.* **19,** 4558.

105

Site-Directed Mutagenesis and Gene Fusion by Megaprimer PCR

Sailen Barik

1. Introduction

In the last few years, a variety of polymerase chain reaction (PCR)-based mutagenesis procedures have been developed *(1–16)*. Among these, the three-primer two-PCR methods *(1–3)* represented by the original "megaprimer" technique *(1)*, appear to be the simplest and most versatile ones available. A number of recent improvements and modifications of the megaprimer method have contributed to an increase in its yield, creation of a larger variety of mutations, and reduction of unwanted mutational errors *(17–23)*. The rationale of the basic method is shown schematically in **Fig. 1A**, where A and B represent the "flanking" primers that can map either within the cloned gene or outside the gene (i.e., within the vector sequence) and M represents the internal "mutant" primer containing the desired base change. The first PCR is performed using the mutant primer (e.g., M1 in **Fig. 1**) and one of the flanking primers (e.g., A). The double-stranded product is purified and used as one of the primers (hence the name "megaprimer") in the second PCR along with the other flanking primer (B). The wild-type cloned gene is used as a template in both PCR reactions. The final PCR product containing the mutation can be used in a variety of standard applications, such as cloning in expression vectors and sequencing, or in more specialized applications, such as production of the gene message in vitro if primer A (or the template sequence downstream of primer A) also contains a transcriptional promoter (e.g., that of SP6 or T7 phage). Both primers A and B are usually designed to contain convenient restriction sites so that the final, mutant PCR product can be restricted and cloned. This basic procedure can be adopted to create site-specific insertions (**Fig. 1B**), deletions (**Fig. 1C**), or gene fusions (**Fig. 1D**) by designing appropriate "mutant" primers (M) for the first PCR.

Because the megaprimer can be quite large (it may approach the size of the whole gene) and is incorporated internal to the gene, successful and error-free PCR in the second round often requires special considerations described in this chapter. The reader is therefore strongly urged to go through the whole chapter, including the **Notes** section before proceeding with the actual experiment.

From: *The Nucleic Acid Protocols Handbook*
Edited by: R. Rapley © Humana Press Inc., Totowa, NJ

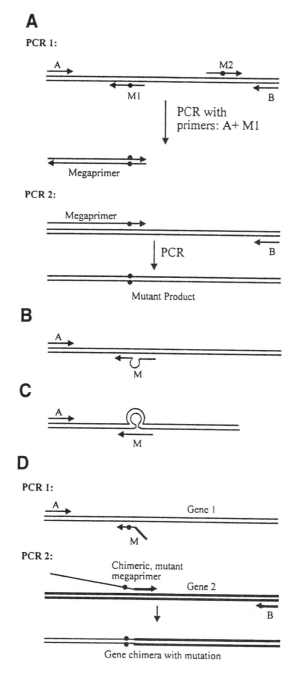

Fig. 1. The megaprimer method of site-directed mutagenesis. The general procedure is shown in A. For insertion (B) and deletion (C) mutagenesis, only PCR 1 is shown; the corresponding PCR 2 will use the megaprimer produced in PCR 1 and primer B. Primers A, B, M1, and M2 (as well as the priming strand of the megaprimer, AM1) are indicated by single lines with arrowhead, whereas the double lines represent the template. For gene fusion (D), the two different genes are shown as narrow and thick double lines. The dots shown on some M primers indicate the desired mutations (base changes) to be introduced into the product via the megaprimer. The M primer in panel B (insertion) and the template in panel C (deletion) are shown as looped out, solely to indicate the region to be inserted or deleted; such structures may not necessarily, and probably do not, form in actual PCR.

2. Materials

1. DNA template containing the cloned gene (e.g., in pUC or pGEM vector) to be mutated, about 1 μg.
2. Oligonucleotide primers A, B, and M1: The "upstream" primer A in the message sense, and the "downstream" primer B in the antimessage sense. Include restriction sites, preferably unique, in these primers so that the final product can be restricted and cloned. Also *see* **Subheading 3.1.** and **Notes 1**, **2**, **5**, and **6**.
3. Standard PCR reagents.
4. Standard reagents and apparatus for agarose gel electrophoresis.
5. A system for purifying DNA from agarose gels. We routinely use a gel-extraction kit (Qiaex II Gel Extraction kit, Qiagen Inc., Chatsworth, CA) (*see* **Note 7** for other methods).
6. SpeedVac to concentrate PCR samples (Savant Instruments, Farmingdale, NY).

3. Methods
3.1. Primer Design

1. For technical reasons described herein, avoid making megaprimers that approach the size of the final, full-length product (gene) AB (**Fig. 1**); in brief, if M1 is too close to B, it will make separation of AB and AM1 (leftover megaprimer) difficult after the second round of PCR. Ideally, the megaprimer should be shorter than the full-length gene by more than 200–500 bp, depending on the exact length of the gene. [Example: If the gene (AB) is 2 kb, megaprimer (AM1) can be up to approx 1.5 kb long, as 2 kb and 1.5 kb can be separated reasonably well in agarose gels. However, if the gene is 8 kb, the megaprimer should not be bigger than, say, 7 kb, as 8- and 7-kb fragments would migrate so close to each other.]
2. When the mutation is to be created near B (**Fig. 1**), one should make an M primer of the *opposite* polarity, such as M2 (**Fig. 1**), and synthesize the BM2 megaprimer (rather than AM1).
3. When the mutation is very near one terminus of the gene (say, within 1–50 nucleotides [nt]), there is no need to use the megaprimer method. One can simply incorporate the mutation in either A or B primer and do a straightforward PCR. In borderline situations, such as when the mutation is, say, 120 nt away from the 5' end of the gene, incorporation of the mutation in primer A may make the primer too big to synthesize, or else it will make the megaprimer AM1 too short to handle conveniently. In such a case, simply back up primer A a few hundred bases further into the vector sequence in order to make AM1 megaprimer longer. In general, remember that primers A and B can be located anywhere on either side of the mutant region and try to utilize that flexibility as an advantage when designing these primers.
4. In addition to the standard rules of primer design (such as, GC content matching the template, sequence specificity, extra "clamp" sequence for restriction, absence of self-complementarity, etc.) *(24)*, attention should be paid to the following aspects. As stated earlier, primers A and B should contain unique restriction sites for ease of cloning. The minimum length of an extra sequence needed for efficient restriction varies from one restriction enzyme to another; a list for some selected enzymes has recently been published *(25)*. As regards the M primer, two additional considerations are important. First, the mutational mismatch should not be too close to the either end of the primer. Mismatch at the very 3' nucleotide of any primer virtually abolishes amplification by *Taq* polymerase *(26,27)*. Because, in the megaprimer procedure, both ends of the M primer eventually get extended (one in PCR 1, the other in PCR 2), the mismatch should be at least 6 (six) bases away from either end of the primer. Second, as described in detail in **Note 2**, the 5' end of the M primer should preferably be located such that there is at least one

(two or more is better) T residue in the template strand of the same sense just upstream of this end of the primer *(18,19,21)*. If a T is not available, try to have the "wobble" base of a codon just upstream of the 5' end of the M primer so that substitution of this base with A will not alter the amino acid *(18,19,21*; *see* **Note 2**).

5. Essentially the same considerations apply in designing primers for insertion, deletion, and fusion of two gene sequences (i.e., M in **Fig. 1B–D**). Using variations of the megaprimer method as depicted in **Fig. 1**, we have inserted, deleted, and fused hundreds of base-pairs. Thus, for all practical purposes, there appears to be no upper limit for gene lengths that can be manipulated, as long as approx 15–20 nt at each end of the M primer anneal to the template, thus allowing enough stringency in both PCRs 1 and 2 *(26–28*; S. Barik and R. Honkanen, unpublished data). More importantly, one can combine gene fusion with mutagenesis of either or both genes by incorporating the desired mutation(s) in the M primer (**Fig. 1D**).

3.2. PCR 1: Synthesis and Purification of the Megaprimer

1. Assemble a 100-µL standard PCR as follows:

H_2O	70 µL
10X PCR buffer (Mg^{+2} free)	10 µL
25 mM $MgCl_2$	10 µL
4 dNTP mix (2 mM each)	10 µL
DNA template	10–50 ng
Primer A	50 pmole (about 0.3 µg)
Primer M1	50 pmole (about 0.3 µg)
Taq polymerase	(2.5 U) 0.5 µL

   For increased specificity, we routinely "hot-start" the PCR as follows. Add all the components *except* the *Taq* polymerase, vortex the mixture, and then spin briefly in a microfuge (about 10 s). At this time, proceed to do the PCR if your thermal cycler has a heated lid (bonnet); otherwise, first overlay your reaction with mineral oil (80–100 µL).

2. Amplify using the following PCR parameters: 94°C, 3 min (initial denaturation); 94°C, 2 min (denaturation); t°C, 2 min (annealing); 72°C, n min (elongation); 35 cycles; and 72°C, 1.5 × n min (final extension). *See* **Notes 3** and **4** for more exact parameters. When the PCR arrives at the first annealing step (in cycle 1), open the tube briefly (do not remove tube from the heat block), and add the *Taq* polymerase and mix well by pipetting back and forth. Reclose the cap and let the PCR continue.

3. To increase the yield of the megaprimer, perform two such 100 µL reactions.

4. Following PCR, remove oil overlay (if applicable) as follows. Freeze the PCR tubes, then thaw just enough so that the oil overlay melts, but the aqueous reaction stays frozen. Remove as much oil as possible. Then use a drawn-out round tip to transfer approx 80 µL of the lower, aqueous layer carefully to a clean Eppendorf tube, wiping the outside of the tip to remove any adhering oil. It is important to remove the oil completely, otherwise, the sample will float up when loaded in horizontal agarose gels in **Subheading 3.2.**, **step 4**.

5. Reduce volume to approx 20 µL in a SpeedVac concentrator or by some other means of evaporation or lyophilization, so that the PCR fits in a single lane in an agarose gel described in **step 6**.

6. Gel-purify the megaprimer. Any method that can separate the megaprimer from the small primers (A and M1) will work (*see* **Note 7**). Gel electrophoresis is routinely used in our laboratory and is described here. Electrophorese PCR 1 in standard agarose gels made in Tris-boric acid EDTA (TBE) buffer in the presence of ethidium bromide. Use an appropriate concentration of agarose (0.7–1.2%) depending on the length of the megaprimer

product to be purified. Perform electrophoresis until a good separation of the megaprimer and the small primer has been achieved. Locate the megaprimer band by UV light and cut out the gel slice. The DNA can be extracted from the gel slice by using any established method in the laboratory. A variety of commercial kits are available for this purpose that employs either chemical extraction of the gel (such as the Qiaex II Gel Extraction Kit, Qiagen) or electroelution (such as using the electroeluter by Kodak-IBI, New Haven, CT). Follow the manufacturer's directions.

3.3. PCR 2

1. Assemble a second PCR (100 µL) as follows:

H_2O	70 µL
10X PCR buffer (Mg^{+2} free)	10 µL
25 mM $MgCl_2$	10 µL
4 dNTP mix (2 mM each)	10 µL
DNA template	0.1–0.2 µg
All of the megaprimer purified above	20–50 µL

   Make up volume to 100 µL with H_2O.

   Note the higher-than-usual amount of plasmid template used in PCR 2 (the same template that was used in PCR 1 earlier) *(17)*. Often, this tends to improve the yield. Generally, use all the megaprimer (recovered in **step 4**, **Subheading 3.2.**) in this PCR in order to achieve a good yield of the product *(19)*. Mix well, and start the reaction as described for PCR 1, using 2.5 U of *Taq* polymerase and the hot-start technique. Following five cycles, when the reaction is again at an annealing step, quickly add 50 pmole of primer B, much the same way as *Taq* polymerase was added *(22)*. Let the PCR continue for another 30 cycles. Use the maximum allowable temperature for annealing in the thermal cycle, as dictated by the smaller primer B. Ignore the megaprimer for annealing considerations, as its T_m will be too high for the smaller primer.
2. Do a standard PCR in a parallel well using the same parameters, the same wild-type template, and primers A and B (but no megaprimer). Use 5 µL of this PCR as a size marker when analyzing PCR 2 by gel electrophoresis in order to recognize the correct product in case there are multiple ones due to some mispriming that occasionally occurs in PCR 2.
3. Purify the mutant PCR product as described in **Subheading 3.2.**, **steps 4–6**. It is now ready for restriction, ligation etc. by the use of standard procedures. If the yield is poor, reamplify a portion of the gel-purified mutant product using primers A and B in a standard PCR.

4. Notes

1. Note that the double-stranded megaprimer is directly used in the second round of PCR; in other words, prior separation of the two strands is unnecessary. Melting the megaprimer is essentially achieved in the denaturation steps of the second PCR. Although both strands of the megaprimer have the potential to anneal to the respective, complementary strands of the template, the basic rules of PCR amplification automatically ensures that only the correct strand (one that extends to the other primer, B, in **Fig. 1**) will be amplified into the double-stranded product.
2. The problem of nontemplated insertion and its solution: This issue has been discussed previously *(19,21)*; some of the solutions have been subsequently confirmed in specific applications. In brief, *Taq* polymerase has a natural tendency to incorporate nontemplated residues, particularly A, at the 3' end of the daughter polynucleotide strand at a certain frequency *(19,29)*. In the subsequent cycles of the PCR, these residues are copied and

amplified into the final double-stranded product. This is generally not a problem in standard PCR where the termini of the product are usually cleaved off by restriction enzymes for cloning purposes. However, in the megaprimer method, the 3' end of the megaprimer is directly incorporated into the final product. Therefore, nontemplated A residues in the megaprimer will eventually show up in a certain percentage of the final product and cause a mutation that may be undesirable. The frequency of such an "error" is usually low and megaprimers with a mismatch at the 3' end will not prime well; however, the frequency may be appreciable in some cases *(19)*. There are two kinds of solutions to this problem; one kind, exemplified by (a), does not introduce the nontemplated base or removes it; the other kind, described in (b), does not remove the nontemplated base but tolerates the alteration.

a. Tolerate the alteration: *This is the recommended method.* It relies on clever primer design and does not require any extra step *(19,21)*. There are two ways of achieving this. As an example, suppose the relevant region of the wild-type sequence is (the amino acids are shown at the bottom in single-letter codes):

   5' — AAA CTG CCA ACT CCG TCA TAT CTG CAG — 3'
   3' — TTT GAC GGT TGA GGC AGT ATA GAC GTC — 5'
         K   L   P   T   P   S   Y   L   Q

   and the Ser (TCA) is to be mutated to Ala (GCA). A mutant primer in the message sense (like M2 in **Fig. 1**) may have the sequence 5' CA ACT CCG **G**CA TAT CTG CAG 3' (the boldface G being the mutant base). However, when this M2 primer and primer B are used in PCR, the nontemplated A incorporated at the M2 end of the product (megaprimer) will result in the sequence:

   5' TCA ACT CCG GCA TAT CTG CAG —
   3' AGT TGA GGC CGT ATA GAC GTC —

   (the nontemplated A/T is underlined). When incorporated into the final product, this megaprimer will produce the following mutant (the underlined amino acids are altered from the wild-type sequence):

   5' — AAA CTG TCA ACT CCG GCA TAT CTG CAG — 3'
         K   L   S   T   P   A   Y   L   Q

   resulting in an undesired Ala → Ser change (boldface). To avoid this, make the following M2 primer: 5' G CCA ACT CCG GCA TAT CTG CAG 3' so that there is a T residue upstream of the 5' end of M2 on the template; any extratemplated T in this strand of the megaprimer will therefore match with the T residue in the wild-type sequence and will not cause any mutation.

   When no T residues are available, use the wobble base of a codon. This is possible when the primary purpose of the clone is to produce a protein product; thus, substitution of a codon with another, synonymous codon is permissible (make sure that the resultant change in the nucleotide sequence is acceptable in terms of introduction or loss of restriction sites, etc.). Now, make the following M2 primer: 5' ACT CCG GCA TAT CTG CAG 3', so that the codon upstream of it is CCA. The nontemplated T will change this codon to CCT; however, since they both code for proline, the protein will remain unaltered.

b. If the sequence of the template is such that proper M primer design as suggested is not possible, use a thermostable DNA polymerase that has a 3'-exonuclease activity, such as *Pfu* (Stratagene Cloning Systems, La Jolla, CA) or Vent™ polymerases (New England Biolabs, Beverly, MA). We have used *Pfu* polymerase in standard PCR conditions described in **Subheading 3.** with success. For Vent, follow the manufacturer's instructions; a typical 100 μL PCR will contain 20 m*M* Tris-HCl, pH 8.8 (at 25°C),

10 mM KCl, 10 mM (NH$_4$)$_2$SO$_4$, 2 mM MgCl$_2$, 0.1% Triton X-100, 100 µg/mL acetylated BSA, 200 µM of each dNTPs, 2 U of Vent polymerase, and standard amounts of template and primers. For either polymerase, it is all the more important that the mismatch in the M primer should be far (at least 8 nt) from either end *(16)*.

3. As a rule, elongation time (at 72°C) in a PCR cycle should be proportional to the length of the product. An approximate guideline is 1 min of elongation per kilobyte, i.e., 100 nt = 10 s; 500 nt = 40 s; 1 kb = 1 min 10 s; 2 kb = 2 min 20 s, etc.

4. Annealing temperature (e.g., *t*°C in PCR 1) is primarily governed by the base composition of the primers. A golden rule is to calculate the T_m of the primer as follows: add 2°C for each A or T, and 4°C for each G or C, then deduct 4°. Example: for a 22-nt primer with 10 G + C and 12 A + T, the T_m is (10 × 4 + 12 × 2) = 64°C; therefore, anneal at 60°C. However, the upper limit of the annealing temperature in PCR is in the range of 72–75°C, since it is the elongation temperature of the *Taq* polymerase. More detailed thermodynamic criteria of primer selection can be found elsewhere *(24)*.

5. When a gene is cloned within two unique sites (a multicloning site, e.g.) of a vector, primers A and B can be made at these sites. The mutant product can then be restricted and ligated at the same two sites in the same vector, and the final clone will retain the original flanking sequences of the gene. This is especially important when sequences upstream of A (e.g., a Shine-Dalgarno sequence or an upstream activating sequence [UAS]) or downstream of B (e.g., a polyadenylation site or a RNase processing site) are essential for gene expression or regulation and must remain unaltered.

6. Studies of structure-function relationship of a protein require the generation of a battery of mutant proteins altered at specific amino acid residues. In such cases, primers A and B can be kept constant and a variety of mutant primers (M1, M2, etc.) can be used to produce the various mutants.

7. Although we have routinely use both gel extraction and electroelution procedures for purification of PCR products, one may prefer other methods for more specialized and routine applications, a few of which are as follows:
   a. Centricon-100 spin filtration column (Amicon, Beverly, MA) may be used directly with the PCR reaction to separate the megaprimer product from the small PCR primers, provided the megaprimer is sufficiently bigger than the smaller primers. Megaprimer purified by this method will also contain the template; however, this is of no concern since the same template is going to be used in the second round of PCR. Follow the recommended procedure *(30)*.
   b. Freeze-squeeze method *(1,31)*: The method works best for fragments smaller than 500 bp. The agarose gel slice containing the DNA fragment is taken in an Eppendorf tube, frozen in a dry ice-ethanol bath (10 min) or in a –70°C freezer (20 min) and then spun at room temperature in a microcentrifuge for 15 min. The recovered (30–70%) megaprimer DNA can be used directly in PCR.
   c. GeneClean method (Bio 101, La Jolla, CA), which is somewhat similar to the Qiagen method, works best for DNA segments larger than 500 bp. Follow the detailed instructions that come with the kit. A somewhat modified version has also been published *(1)*.

8. In a recent method *(31)*, DNA fragments in low melting point agarose slices have been directly used in PCR reaction apparently without any problem. Since a good quantity of the megaprimer is important for successful PCR 2, this is worth trying. This will also bypass the need to recover it from the gel.

9. As in any cloning procedure, the final mutants obtained by the megaprimer method must be confirmed by DNA sequencing. This can be done either by directly sequencing the PCR product *(33,34)* or after cloning the mutant product in plasmid vectors *(35)*. When using the dideoxy method, PCR primers A or B can be used as sequencing primers as well.

10. For gene fusion only, one can combine PCR 1 and 2 in a single PCR containing both templates, and primers A, M (the chimeric primer), and B *(20)*, provided that primers A and B do not anneal to their noncognate templates. Clearly, combining PCR 1 and 2 is not feasible for substitution, deletion, or insertion mutagenesis (i.e., those procedures that use the same template for both PCR 1 and 2), since primers A and B will simply amplify the wild type template, and primer M will be essentially ignored.

Acknowledgments

Research in the author's laboratory was supported by National Institutes of Health Grant AI37938.

References

1. Sarkar, G. and Sommer, S. S. (1990) The "megaprimer" method of site-directed mutagenesis. *Biotechniques* **8,** 404–407.
2. Landt, O., Grunart, H.-P., and Hahn, U. (1990) A general method for rapid mutagenesis using the polymerase chain reaction. *Gene* **96,** 125–128.
3. Perrin, S. and Gilliland, G. (1990) Site-specific mutagenesis using asymmetric polymerase chain reaction and a single mutant primer. *Nucleic Acids Res.* **18,** 7433–7438.
4. Hemsley, A., Arnheim, N., Toney, M. D., Cortopassi, G., and Galas, D. J. (1989) A simple method for site-directed mutagenesis using the polymerase chain reaction. *Nucleic Acids Res.* **17,** 6545–6551.
5. Higuchi, R., Krummel, B., and Saiki, R. K. (1988) A general method of in vitro preparation and specific mutagenesis of DNA fragments: study of protein and DNA interactions. *Nucleic Acids Res.* **16,** 7351–7367.
6. Horton, R. M., Ho, S. N., Pullen, J. K., Hunt, H. D., Cai, Z., and Pease, L. R. (1993) Gene splicing by overlap extension. *Methods Enzymol.* **217,** 270–279.
7. Jones, D. H. and Howard, B. H. (1991) A rapid method for recombination and site-specific mutagenesis by placing homologous ends on DNA using polymerase chain reaction. *Biotechniques* **10,** 62–66.
8. Tomic, M., Sunjevaric, I., Savtchenko, E. S., and Blumenberg, M. (1990) A rapid and simple method for introducing specific mutations into any position of DNA leaving all other positions unaltered. *Nucleic Acids Res.* **18,** 1656.
9. Vallette, F., Mege, E., Reiss, A., and Adesnik, M. (1989) Construction of mutant and chimeric genes using the polymerase chain reaction. *Nucleic Acids Res.* **17,** 723–733.
10. Stemmer, W. P. and Morris, S. K. (1992) Enzymatic inverse PCR: a restriction site independent, single-fragment method for high-efficiency, site-directed mutagenesis. *Biotechniques* **13,** 214–220.
11. Diaz, J. J., Rhoads, D. D., and Roufa, D. J. (1991) PCR-mediated chemical mutagenesis of cloned duplex DNAs. *Biotechniques* **11,** 204–206.
12. Costa, G. L., Bauer, J. C., McGowan, B., Angert, M., and Weiner, M. P. (1996) Site-directed mutagenesis using a rapid PCR-based method. *Methods Mol. Biol.* **57,** 239–248.
13. Watkins, B. A. and Reitz, M. S. (1996) Using PCR for rapid site-directed mutagenesis in large plasmids. *Methods Mol. Biol.* **57,** 217–228.
14. Shimada, A. (1996) PCR-based site-directed mutagenesis. *Methods Mol. Biol.* **57,** 157–165.
15. Merino, E., Osuna, J., Bolivat, F., and Soberon, X. (1992) A general PCR-based method for single or combinatorial oligonucleotide-directed mutagenesis on pUC/M13 vectors. *Biotechniques* **12,** 508–510.
16. Picard, V. and Bock. S. C. (1997) Rapid and efficient one-tube PCR-based mutagenesis method. *Methods Mol. Biol.* **67,** 183–188.

17. Barik, S. and Galinski, M. (1991) "Megaprimer" method of PCR: increased template concentration improves yield. *Biotechniques* **10,** 489–490.
18. Kuipers, O. P., Boot, H. J., and de Vos, W. M. (1991) Improved site-directed mutagenesis method using PCR. *Nucleic Acids Res.* **19,** 4558.
19. Barik, S. (1993) Site-directed mutagenesis by double polymerase chain reaction: megaprimer method. *Methods Mol. Biol.* **15,** 277–286.
20. Pont-Kingdon, G. (1994) Construction of chimeric molecules by a two-step recombinant PCR method. *Biotechniques* **16,** 1010–1011.
21. Barik, S. (1995) Site-directed mutagenesis by PCR: substitution, insertion, deletion, and gene fusion. *Methods Neurosci.* **26,** 309–323.
22. Datta, A. K. (1995) Efficient amplification using "megaprimer" by asymmetric polymerase chain reaction. *Nucleic Acids Res.* **23,** 4530–4531.
23. Aiyar, A., Xiang, Y., and Leis, J. (1996) Site-directed mutagenesis using overlap extension PCR. *Methods Mol. Biol.* **57,** 177–191.
24. Rychlik, W. (1993) Selection of primers for polymerase chain reaction. *Methods Mol. Biol.* **15,** 31–40.
25. New England BioLabs, Inc. (1998/1999 Catalog). Cleavage close to the end of DNA fragments.
26. Sarkar, G., Cassady, J., Bottema, C. D. K., and Sommer, S. S. 1990. Characterization of polymerase chain reaction amplification of specific alleles. *Anal. Biochem.* **186,** 64–68.
27. Kwok, S., Kellog, D. E., McKinney, N., Spassic, D., Goda, L., Levenson, C., and Sninsky, J. J. (1990) Effects of primer-template mismatches on the polymerase chain reaction: human immunodeficiency virus type I model studies. *Nucleic Acids Res.* **18,** 999–1005.
28. Barik, S. (1993) Expression and biochemical properties of a protein serine/threonine phosphatase encoded by bacteriophage λ. *Proc. Natl. Acad. Sci. USA* **15,** 10,633–10,637.
29. Clark, J. M. (1988) Novel non-templated nucleotide addition reactions catalyzed by procaryotic and eucaryotic DNA polymerases. *Nucleic Acids Res.* **16,** 9677–9686.
30. Krowczynska, A. M. and Henderson, M. B. (1992) Efficient purification of PCR products using ultrafiltration. *Biotechniques* **13,** 286–289.
31. Stoflet, E. S., Koeberl, D. D., Sarkar, G., and Sommer, S. S. (1988) Genomic amplification with transcript sequencing. *Science* **239,** 491–494.
32. Zintz, C. B. and Beebe, D. C. (1991) Rapid re-amplification of PCR products purified in low melting point agarose gels. *Biotechniques* **11,** 158–162.
33. Dicker, A. P., Volkenandt, M., and Bertino, J. R. (1993) Manual and automated direct sequencing of product generated by polymerase chain reaction. *Methods Mol. Biol.* **15,** 143–152.
34. Sarkar, G. and Bolander, M. E. (1995) Semi-exponential cycle sequencing. *Nucleic Acids Res.* **23,** 1269–1270.
35. Maniatis, T., Fritsch, E. F., and Sambrook, J. (eds.) (1982) *Molecular Cloning: A Laboratory Manual.* Cold Spring Harbor Laboratory Press, Cold Spring Harbor, NY, pp. 113–119.

106

Transcription In Vitro Using Bacteriophage RNA Polymerases

Elaine T. Schenborn

1. Introduction

Synthesis of specific RNA sequences in vitro is simplified because of the availability of bacteriophage RNA polymerases and specially designed DNA vectors. RNA polymerases encoded by SP6, T7, or T3 bacteriophage genomes recognize particular phage promoter sequences of their respective viral genes with a high degree of specificity *(1–3)*. These RNA polymerases also transcribe DNA templates containing their cognate promoters under defined conditions in vitro *(4,5)*. Standard reaction conditions for transcription in vitro can be adjusted for synthesis of large amounts of RNA or for smaller amounts of labeled RNA probes.

Larger-scale in vitro synthesis produces RNA that mimics biologically active RNA in many applications. The following examples represent some of the different uses for RNA synthesized in vitro. RNA transcripts are particularly well suited for the study of RNA virus gene regulation. For example, the in vitro transcribed RNA genomes of poliovirus *(6)* and cowpea mosaic virus *(7)* produce infectious particles in transfected cells. For other types of studies, messenger RNA-like transcripts are used as substrates to study RNA processing activities, such as splicing *(8)* and 3'-end maturation *(9,10)*. RNA transcripts synthesized in vitro are also widely used as templates for protein synthesis in cell-free extracts designed for in vitro translation *(11)* (*see* Chapters 107–109). Transfer RNA-like transcripts have been used as substrates to study RNase P cleavage specificities *(12)*, and other mechanisms of RNA cleavage have been investigated using RNA substrates and ribozymes synthesized in vitro *(13)*. Gene regulation studies using antisense RNA also have taken advantage of the ease of in vitro RNA synthesis. In vitro translation of a targeted message has been shown to be inhibited in the presence of antisense RNA in vitro *(14)*, and in vivo translation has been blocked in *Xenopus* oocytes by antisense RNA *(15)*. The ability to synthesize discrete RNA templates in vitro also facilitates studies of RNA and protein interactions *(16,17)*.

The generation of radioactively labeled RNA hybridization probes is a widely used application for RNA synthesized in vitro. RNA probes are synthesized predominantly by incorporation of a radiolabeled ribonucleotide, ^{32}P-, ^{3}H-, or ^{35}S-rNTP, into the transcript. Nonisotopic probes can be synthesized by incorporation of biotinylated *(18)* or digoxigenin *(19)* modified bases. For Northern blots, single-stranded RNA probes are

From: *The Nucleic Acid Protocols Handbook*
Edited by: R. Rapley © Humana Press Inc., Totowa, NJ

generally more sensitive than the corresponding DNA probe because of the higher thermal stability of RNA:RNA hybrids compared to RNA:DNA hybrids and the absence of self-complementary sequences in the probe preparation *(4)*.

RNA probes also are more sensitive than DNA probes for the detection of DNA sequences transferred to membranes from Southern blots, plaque lifts, and colony lifts *(20)*. The lower background and increased signal sensitivity of RNA probes are possible because of higher stability of RNA:DNA hybrids compared to DNA:DNA hybrids. This increased stability allows more stringent conditions to be used for the hybridization and washing procedures *(21)*. Another advantage of RNA probes is that RNase A can be added after the hybridization reaction to eliminate nonspecific binding of the probe to the membrane. High sensitivity also has been achieved with RNA probes used for *in situ* hybridization *(22)* and localization of genes in chromosome spreads *(23)*. RNase mapping is another application that takes advantage of the superior properties of RNA probes for hybridization to complementary sequences. In this application, a radiolabeled RNA probe is hybridized in solution to cellular RNA, then the nonhybridized, single-stranded regions of the probe are later digested with RNase A and RNase T1, and the protected, hybridized regions are identified by gel analysis. This type of mapping is used to quantitate low-abundance species of RNA, and to map exons, transcription start sites, and point mutations *(4,24)*.

The DNA templates used for in vitro transcription contain the cloned sequence of interest immediately "downstream" of an SP6, T7, or T3 promoter sequence. Plasmid vectors are commercially available with the phage promoter sequence adjacent to a cloning region. One example is the pGEM® series of vectors (Promega, Madison, WI) designed with multiple cloning sites flanked by opposed SP6 and T7 promoters, allowing the synthesis of either sense or antisense RNA from a single recombinant plasmid. Discrete RNAs, corresponding to the cloned sequence of interest, are synthesized as "run-off" transcripts from a linear DNA template. To prepare the linear template, the recombinant plasmid DNA is cut with a restriction enzyme cleaving within, or shortly downstream of, the cloned insert. The linear DNA is then added to the reaction mixture for in vitro synthesis of RNA (*see* **Fig. 1**).

2. Materials

1. Transcription buffer (5X): 200 mM Tris-HCl, pH 7.9, 30 mM MgCl$_2$, 10 mM spermidine, and 50 mM NaCl. Store at –20°C.
2. ATP, GTP, CTP, UTP: 10 mM stocks prepared in sterile, nuclease-free water and adjusted to pH 7.0. Store at –20°C.
3. 100 mM DTT: Store at –20°C.
4. RNasin® Ribonuclease Inhibitor: (Promega): Store at –20°C.
5. Nuclease-free water: Prepare by adding 0.1% diethyl pyrocarbonate (DEPC) to the water. Autoclave to remove the DEPC. **Caution:** DEPC is a suspected carcinogen.
6. TE buffer: 10 mM Tris-HCl, pH 8.0, and 1 mM EDTA. Prepare with stock solutions that are nuclease-free.
7. TE-saturated phenol/chloroform: Mix equal parts of TE buffer and phenol, and allow phases to separate. Mix 1 part of the lower, phenol phase with 1 part of chloroform:isoamyl alcohol (24:1).
8. Chloroform:isoamyl alcohol (24:1): Mix 24 parts of chloroform with 1 part isoamyl alcohol.
9. Ammonium acetate: 7.5 and 2.5 M.
10. 3 M sodium acetate, pH 5.2.

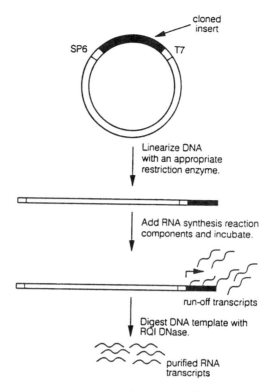

Fig. 1. Synthesis of RNA by transcription in vitro from a linear DNA template.

11. Ethanol: Absolute (100%) and 70%.
12. Enzymes: SP6, T3, or T7 RNA polymerase at 15–20 U/μL.
13. RNase-free DNase: RQ1 (Promega).
14. Restriction enzyme and appropriate buffer to linearize plasmid DNA template.
15. DE81 filters: 2.4 cm diameter (Whatman).
16. 0.5 M Na_2HPO_4, pH 7.0.
17. $m^7G(5')ppp(5')G$: 5 mM (Promega).

Microcentrifuge tubes, pipet tips, glassware: To provide a nuclease-free environment, use sterile, disposable microcentrifuge tubes and pipet tips whenever possible for the preparation and storage of reagents. Larger volumes of reagents can be stored in bottles that have been baked at 250°C for four or more hours to inactivate RNases.

3. Methods

Throughout these procedures, precautions should be taken to protect against ribonuclease contamination. These precautions include the use of sterile, nuclease-free reagents and materials, and the use of disposable gloves to prevent accidental contamination of samples with ribonucleases present on the skin.

Three steps are required for synthesis of RNA in vitro:

1. Preparation of the DNA template.
2. Transcription reaction.
3. Enrichment of the RNA product.

3.1. Preparation of the DNA Template

The sequence of interest is cloned by established methods into an appropriate vector, downstream of a promoter sequence for SP6, T7, or T3 RNA polymerase. The recombinant plasmid DNA is purified, and either added directly to the in vitro transcription reaction or linearized prior to the run-off transcription reaction. Transcription of supercoiled plasmid DNA results in the synthesis of high-mol-wt RNA, which contains vector sequences. Discrete RNA sequences of interest, without vector sequence, are generated by run-off transcription from linear templates prepared in the following manner:

1. Determine the restriction site downstream of, or within, the cloned insert, which will generate the desired run-off transcript. Whenever possible, select a restriction enzyme that produces 5' overhanging or blunt ends. If an enzyme that generates a 3' overhang is selected, *see* **Note 1**. Set up the restriction digest according to the enzyme supplier's directions.
2. Check for completeness of digestion by agarose gel electrophoresis. During this analysis, keep the DNA sample on ice. If digestion is complete, proceed with **step 3**. Otherwise, add additional restriction enzyme to the DNA, incubate an additional 30 min, and repeat the agarose gel analysis. Ensure that restriction enzyme does not exceed 10% of final volume. Restriction enzymes are supplied in glycerol, which at high concentration can inhibit enzyme activity.
3. Extract the DNA by adding an equal volume of TE-saturated phenol/chloroform, vortex for 1 min, and centrifuge at 12,000g for 2 min. Transfer the upper phase to a fresh tube, and add 1 vol of chloroform:isoamyl alcohol (24:1). Vortex for 1 min, and centrifuge at 12,000g for 2 min.
4. Precipitate the DNA by transferring the upper, aqueous phase to a fresh tube, and adding 0.1 vol of 3 M sodium acetate, pH 5.2, and 2 vol of absolute ethanol. Cool 30 min at –70°C, and centrifuge at 12,000g for 5 min.
5. Carefully pour off the supernatant, wash the pellet briefly with 1 mL of 70% ethanol, spin at 12,000g for 2 min, and remove the supernatant. Dry briefly in a vacuum desiccator. Resuspend the pellet in nuclease-free water or TE buffer to a final DNA concentration of approx 1 mg/mL.

3.2. Synthesis of Radiolabeled RNA Probes (see Notes 2–5)

RNA probes at a specific activity of $2–9 \times 10^8$ cpm/µg can be generated by transcribing DNA in the presence of a limiting concentration (12–24 µM) of one radiolabeled ribonucleotide and saturating concentrations (0.5 mM) of the other three rNTPs (*see* **Notes 2** and **3**). The following example uses 50 µCi of α-[^{32}P]CTP at a specific activity of 400 Ci/mmol/20 µL reaction, providing a final concentration of 6 µM of α-[^{32}P]CTP. An additional 12 µM of unlabeled CTP is added to bring the total concentration to 18 µM CTP. Expect approx 1 mol of RNA/mol of DNA template to be synthesized under these conditions.

1. To a sterile microcentrifuge tube, add the following components at room temperature in the order listed. This order of addition prevents precipitation of the DNA by spermidine: 4 µL of 5X transcription buffer, 2 µL of 100 mM DTT, 20 U RNasin® Ribonuclease Inhibitor, 4 µL of ATP, GTP, and UTP (2.5 mM each; prepare by mixing 1 vol of each individual 10 mM stock of ATP, GTP, and UTP, and 1 vol of water), 2.4 µL of 100 µM CTP (dilute 10 mM stock 1:100 with water), 1 µL of DNA template (up to 2 µg; 1–2 mg/mL in

nuclease-free water or TE), 5 µL of α-[^{32}P]CTP (400 Ci/mmol; 10 mCi/mL). Bring to a final vol of 19 µL with nuclease-free water.
2. Initiate the reaction by adding 1 µL of SP6, T7, or T3 RNA polymerase (at 15–20 U/µL).
3. Incubate for 60 min at 37–40°C.
4. Remove 1 µL from the reaction at this point to determine the percent incorporation and specific activity of the probe. The remainder of the sample can be digested by RQ1 RNase-free DNase (**Subheading 3.6.**).

3.3. Determination of Percent Incorporation and Probe Specific Activity

1. Remove 1 µL of the labeled probe, and dilute into 19 µL of nuclease-free water. Spot 3 µL of this 1:20 dilution onto 4 DE81 filters. Dry the filters at room temperature or under a heat lamp.
2. Place two filters directly into separate scintillation vials, add scintillation fluid, and count. Calculate the average cpm per filter, and determine the total cpm per microliter of original reaction as follows:

$$\text{Total cpm/µL of original reaction} = \text{average cpm per filter} \times (20\text{-fold dilution}/3 \text{ µL}) \tag{1}$$

3. Wash the unincorporated nucleotides from the remaining two filters by placing the filters in a small beaker containing 50–100 mL of 0.5 M Na$_2$HPO$_4$ (pH 7.0). Swirl the filters occasionally for 5 min, then decant, and replace with fresh buffer. Repeat the wash procedure two more times. Dip the filters briefly into 70% ethanol, and dry at room temperature or under a heat lamp.
4. Place each filter into a scintillation vial, add scintillation fluid, and count. Calculate the amount of labeled nucleotide incorporated into RNA (incorporated cpm) per microliter of original reaction as follows:

$$\text{Incorporated cpm/µL of original reaction} = \text{average cpm per filter} \times (20\text{-fold dilution}/3 \text{ µL}) \tag{2}$$

This value will also be used in estimating the probe specific activity in **step 6**.
5. Calculate the percent incorporation from the values determined above in **steps 2** and **4**.

$$\% \text{ Incorporation} = (\text{incorporated cpm/total cpm}) \times 100 \tag{3}$$

The percentage of incorporation under the conditions described generally ranges from 70 to nearly 100%. A low incorporation of radiolabeled nucleotide (for example, below 50%) reflects a low yield of RNA product (*see* **Note 5**).
6. Calculate the specific activity of the probe as cpm/µg RNA synthesized. To do this, first calculate the total incorporated cpm in the reaction:

$$\text{Total incorporated cpm} = (\text{incorporated cpm/µL of reaction}) \times 20 \text{ µL reaction vol} \tag{4}$$

Next we need to calculate the total nmoles of nucleotide in the reaction to determine how many micrograms of RNA were synthesized; 50 µCi of α-[^{32}P]CTP at 400 µCi/nmol corresponds to 0.12 nmol of ^{32}P-CTP/reaction. Adding in the 12 µM of unlabeled CTP (0.24 nmol) gives a total of 0.36 nmol of CTP. If a maximum 100% incorporation was achieved and CTP represents one-fourth of all the nucleotides in the probe, then the total amount of nucleotides incorporated into the probe would be (0.36 nmol × 4) or 1.44 nmol. Assuming an average FW/nucleotide of 330, the amount of RNA synthesized in this example would be 1.44 nmol × (330 ng/nmol) = 475 ng of RNA synthesized. If the percentage of incorporation calculated from **step 5** was 80%, for example, then the actual amount of RNA synthesized in the reaction would be 475 ng × 0.80 = 380 ng RNA.

$$SA = \text{total incorporated cpm/µg RNA} \tag{5}$$

In this example, the total incorporated CPM would be divided by 0.380 µg RNA.

3.4. Synthesis of Large Quantities of RNA (see Notes 2–6)

Using the following reaction conditions in which all four rNTPs are at a saturating concentration, yields of 5–10 µg of RNA/µg of DNA template can be obtained (*see* **Note 6**). This represents up to 20 mol of RNA/mol of DNA template. Incubation with additional polymerase after the initial 60-min reaction can increase the yield of RNA up to two-fold. The following reaction can be scaled up or down as desired.

1. To a sterile microcentrifuge tube, add the following components at room temperature in the order listed. This order of addition prevents precipitation of the DNA by spermidine: 20 µL of 5X transcription buffer, 10 µL of 100 m*M* DTT, 100 U RNasin Ribonuclease Inhibitor, 20 µL of ATP, GTP, UTP, and CTP (2.5 m*M* each; prepare by mixing 1 vol of each individual 10 m*M* stock of ATP, GTP, UTP, and CTP), 2–5 µL of DNA template (5–10 µg total; 1–2 mg/mL in nuclease-free water or TE). Add nuclease-free water to a final vol of 98 µL.
2. Initiate the reaction by adding 2 µL of SP6, T7, or T3 RNA polymerase (at 15–20 U/µL).
3. Incubate for 60 min at 37–40°C.
4. Add an additional 2 µL of SP6, T7, or T3 RNA polymerase. Incubate for 60 min at 37–40°C.

The DNA template can now be digested by RQ1 RNase-free DNase (**Subheading 3.6.**).

3.5. Synthesis of 5' Capped Transcripts

Some RNA transcripts require a $m^7G(5')ppp(5')G$ cap at the 5' end for higher translation efficiency, either in cell-free extracts or in *Xenopus* oocytes *(25)*. Methylated capped transcripts also have been reported to function more efficiently for in vitro splicing reactions *(8)* and are more resistant to ribonucleases in nuclear extracts. The following reaction can be scaled up or down as desired.

1. To a sterile microcentrifuge tube, add the following components at room temperature in the order listed. This order of addition prevents precipitation of the DNA by spermidine: 4 µL of 5X transcription buffer, 2 µL of 100 m*M* DTT, 20 U RNasin Ribonuclease Inhibitor, 4 µL of ATP, UTP, and CTP (2.5 m*M* each; prepare by mixing 1 vol of each individual 10 m*M* stock of ATP, UTP, and CTP, and 1 vol of water), 2 µL of GTP (0.5 m*M*; dilute 10 m*M* stock 1:20 with water), 2 µL of the cap analog $m^7G(5')ppp(5')G$ (5 m*M*), and 1 µL of DNA template: 1–2 µg (1–2 mg/mL in nuclease-free water or TE). Add nuclease-free water, if necessary, to a final vol of 19 µL.
2. Initiate the reaction by adding 1 µL of SP6, T7, or T3 RNA polymerase (at 15–20 U/µL).
3. Incubate for 60 min at 37–40°C.

The DNA template can now be digested by RQ1 RNase-free DNase (**Subheading 3.6.**).

3.6. Digestion of the DNA Template Post-Transcription

To achieve maximal sensitivities with RNA probes, the DNA template must be eliminated after the transcription reaction. Elimination of the DNA template also may be required for the preparation of biologically active RNAs. DNase can be used to digest the DNA template, but during this enzymatic step, it is critical to maintain the integrity of the RNA. RQ1 DNase (Promega) is certified to be RNase-free and is recommended for the following protocol.

1. After the in vitro transcription reaction, add RQ1 RNase-free DNase to a concentration of 1 U/µg of template DNA.
2. Incubate for 15 min at 37°C.
3. Extract with 1 vol of TE-saturated phenol/chloroform. Vortex for 1 min, and centrifuge at 12,000g for 2 min.
4. Transfer the upper, aqueous phase to a fresh tube. Add 1 vol of chloroform:isoamyl alcohol (24:1). Vortex for 1 min and centrifuge as in **step 3**.
5. Transfer the upper, aqueous phase to a fresh tube. At this point, a small aliquot can be taken for electrophoretic analysis on a denaturing gel, and the remainder of the sample can be precipitated (**Subheading 3.7.**).

3.7. Precipitation of RNA

1. Add 0.5 vol of 7.5 M ammonium acetate to the aqueous RNA sample prepared in **Subheading 3.6.** If the RNA sample was not digested with RQ1 DNase, extract the RNA after the transcription reaction with TE-saturated phenol/chloroform followed by a chloroform extraction, as described in **Subheading 3.6., steps 3–5**.
2. Add 2.5 vol of ethanol, mix, and place at –70°C for 30 min.
3. Centrifuge at 12,000g for 5 min. Carefully remove the supernatant.
4. Resuspend the RNA pellet in 100 µL of 2.5 M ammonium acetate and mix.
5. Repeat the ethanol precipitation as described in **steps 2** and **3** above.
6. Dry the pellet briefly under vacuum, and resuspend in 20 µL or other suitable volume of sterile TE or nuclease-free water. Store the RNA at –70°C.

4. Notes

1. Extraneous transcripts complementary to the opposite strand and vector sequences are generated from DNA templates with 3' overhanging ends (26). The ends of these templates can be made blunt in the following manner using the 3'–5' exonuclease activity of the Klenow fragment of DNA polymerase I. Set up the transcription reaction, but without nucleotides and RNA polymerase. Add 5 U of Klenow fragment/µg DNA, and incubate for 15 min at 22°C. Then initiate the transcription reaction by adding nucleotides and RNA polymerase, and incubate for 60 min at 37–40°C.
2. Incomplete transcripts are more likely to be generated under the conditions used for probe synthesis, in which the concentration of a radiolabeled nucleotide becomes limiting. Of the four nucleotides, rGTP yields the highest percentage of full-length transcripts when present in limiting concentrations (4). However, for best results, radiolabeled rGTP should be used within 1 wk of the reference date. rATP yields the lowest percentage of full-length transcripts and lowest incorporation when present at a limiting concentration (5). In some cases, the amount of full-length transcripts increases when the incubation temperature is lowered to 30°C. Another possible cause for incomplete transcripts can be the presence of a sequence within the DNA template that acts as a terminator for that particular polymerase. In this case, one can subclone the sequence of interest behind a different RNA polymerase promoter.
3. The specific activity of a probe can be increased by using more than one radiolabeled nucleotide per reaction at a limiting concentration. Also, more than 5 µL of the

Table 1
SA and Concentration of rNTPs Used for Transcription In Vitro

Nucleotide	Specific activity	μCi/reaction	Final conc.
α-[^{32}P] rNTP	400 Ci/mmol	50 μCi	6 μM
α-[^{35}S] rNTP	1300 Ci/mmol	300 μCi	12 μM
5,6[^{3}H] rNTP	40 Ci/mmol	25 μCi	31 μM

radionucleotide can be used per 20 μL reaction if the nucleotide is first aliquoted into the reaction tube and dried down under vacuum. **Table 1** lists the specific activity (SA) and final concentration (final conc.) of radionucleotides commonly used in RNA probe synthesis, in a 20-μL reaction volume. Thiol-substituted rNTPs are incorporated less efficiently by the RNA polymerases than the corresponding ^{32}P or ^{3}H rNTPs *(5)*.

4. Biotinylated rNTP can be added during the transcription reaction, but the yield of RNA may be lowered. Alternatively, RNA can be modified after transcription using photo-activatable biotin *(27)*.
5. A low yield of RNA product can be caused by several conditions, including precipitation of DNA by spermidine in the transcription buffer, RNase contamination, carryover of residual contaminants or salts in the DNA preparation, or inactive RNA polymerase.
6. High yields of RNA synthesized by SP6 or T7 RNA polymerase have been reported using a transcription buffer containing 80 mM HEPES-KOH, pH 7.5, 2 mM spermidine, 10–40 mM DTT, 3 mM each rNTP, 12–16 mM MgCl$_2$, and 1200–1800 U/mL RNA polymerase. Under these conditions, yields up to 80 μg of RNA/μg DNA were reported *(28)*.

References

1. Butler, E. T. and Chamberlin, M. J. (1982) Bacteriophage SP6-specific RNA polymerase. *J. Biol. Chem.* **257,** 5772–5778.
2. Davanloo, P., Rosenberg, A. H., Dunn, J. J., and Studier, F. W. (1984) Cloning and expression of the gene for bacteriophage T7 RNA polymerase. *Proc. Natl. Acad. Sci. USA* **81,** 2035–2039.
3. Jorgensen, E. D., Joho, K., Risman, S., Moorefield, M. B., and McAllister, W. T. (1989) Promoter recognition by bacteriophage T3 and T7 RNA polymerases, in *DNA–Protein Interaction in Transcription* (Gralla, J. D., ed.), Liss, New York, pp. 79–88.
4. Melton, D. A., Krieg, P. A., Rebagliati, M. R., Maniatis, T., Zinn, K., and Green, M. R. (1984) Efficient *in vitro* synthesis of biologically active RNA and RNA hybridization probes from plasmids containing a bacteriophage SP6 promoter. *Nucleic Acids Res.* **12,** 7035–7056.
5. Krieg, P. A. and Melton, D. A. (1987) *In vitro* RNA synthesis with SP6 RNA polymerase. *Methods Enzymol.* **155,** 397–415.
6. Kaplan, G., Lubinski, J., Dasgupta, A., and Racaniello, V. R. (1985) *In vitro* synthesis of infectious poliovirus RNA. *Proc. Natl. Acad. Sci. USA* **82,** 8424–8248.
7. Eggen, R., Verver, J., Wellink, J., DeJong, A., Goldbach, R., and van Kammen, A. (1989) Improvements of the infectivity of in vitro transcripts from cloned cowpea mosaic virus cDNA: impact of terminal nucleotide sequences. *Virology* **173,** 447–455.
8. Krainer, A. R., Maniatis, T., Ruskin, B., and Green, M. R. (1984) Normal and mutant human β-globin pre-mRNAs are faithfully and efficiently spliced *in vitro*. *Cell* **36,** 993–1005.

9. Krieg, P. A. and Melton, D. A. (1984) Formation of the 3' end of histone mRNA by post-transcriptional processing. *Nature* **308,** 203–206.
10. Georgiev, O., Mous, J., and Birnstiel, M. (1984) Processing and nucleo-cytoplasmic transport of histone gene transcripts. *Nucleic Acids Res.* **12,** 8539–8551.
11. Krieg, P. A. and Melton, D. A. (1984) Functional messenger RNAs are produced by SP6 *in vitro* transcription of cloned cDNAs. *Nucleic Acids Res.* **12,** 7057–7070.
12. Burgin, A. B. and Pace, N. R. (1990) Mapping the active site of ribonuclease P RNA using a substrate containing a photoaffinity agent. *EMBO J.* **9,** 4111–4118.
13. Heus, H. A., Uhlenbeck, O. C., and Pardi, A. (1990) Sequence-dependent structural variations of hammerhead RNA enzymes. *Nucleic Acids Res.* **18,** 1103–1108.
14. Nicole, L. M. and Tanguay, R. M. (1987) On the specificity of antisense RNA to arrest *in vitro* translation of mRNA coding for Drosophila hsp 23. *Biosci. Rep.* **7,** 239–246.
15. Melton, D. A. (1985) Injected antisense RNAs specifically block messenger RNA translation *in vivo*. *Proc. Natl. Acad. Sci. USA* **82,** 144–148.
16. Witherell, G. W., Wu, H.-N., and Uhlenbeck, O. C. (1990) Cooperative binding of R17 coat protein to RNA. *Biochemistry* **29,** 11,051–11,057.
17. Turek, C. and Gold, L. (1990) Systematic evolution of ligands by exponential enrichment: RNA ligands to bacteriophage T4 DNA polymerase. *Science* **249,** 505–510.
18. Langer, P. R., Waldrop, A. A., and Ward, D. C. (1982) Enzymatic synthesis of biotin-labeled polynucleotides: novel nucleic acid affinity probes. *Proc. Natl. Acad. Sci. USA* **78,** 6633–6637.
19. Aigner, S. and Pette, D. (1990) *In situ* hybridization of slow myosin heavy chain mRNA in normal and transforming rabbit muscles with the use of a nonradioactively labeled cRNA. *Histochemistry* **95,** 11–18.
20. Sambrook, J., Fritsch, E. F., and Maniatis, T. (eds.) (1989) *Molecular Cloning, A Laboratory Manual*, 2nd ed., Cold Spring Harbor Laboratory, Cold Spring Harbor, NY.
21. Casey, J. and Davidson, N. (1977) Rates of formation and thermal stabilities of RNA:DNA and DNA:DNA duplexes at high concentrations of formamide. *Nucleic Acids Res.* **4,** 1539–1552.
22. Uhlig, H., Saeger, W., Fehr, S., and Ludecke, D. K. (1991) Detection of growth hormone, prolactin and human beta-chorionic gonadotropin messenger RNA in growth-hormone-secreting pituitary adenomas by *in situ* hybridization. *Virchows Arch. Pathol. Anat. Histopathol.* **418,** 539–546.
23. Matthaei, K. I. and Reed, K. C. (1986) Chromosome assignment in somatic hybrids by *in situ* hybridization with tritium labeled Riboprobe® RNA probes. *Promega Notes* **5,** 5–6.
24. Zinn, K., DiMaio, D., and Maniatis, T. (1983) Identification of two distinct regulatory regions adjacent to the human β-interferon gene. *Cell* **34,** 865–879.
25. Contreras, R., Cheroutre, H., Degrave, W., and Fiers, W. (1982) Simple, efficient *in vitro* synthesis of capped RNA useful for direct expression of cloned eukaryotic genes. *Nucleic Acids Res.* **10,** 6353–6362.
26. Schenborn, E. T. and Mierendorf, R. C. (1985) A novel transcription property of SP6 and T7 RNA polymerases: dependence on template structure. *Nucleic Acids Res.* **13,** 6223–6236.
27. Forster, A. C., McInnes, J. L., Skingle, D. C., and Symons, R. H. (1985) Non-radioactive hybridization probes prepared by the chemical labelling of DNA and RNA with a novel reagent, photobiotin. *Nucleic Acids Res.* **13,** 745–761.
28. Gurevich, V. V., Pokrovskaya, I. D., Obukhova, T. A., and Zozulya, S. A. (1991) Preparative *in vitro* mRNA synthesis using SP6 and T7 RNA polymerases. *Analyt. Biochem.* **195,** 207–213.

107

In Vitro Translation of mRNA in a Rabbit Reticulocyte Lysate Cell-Free System

Louise Olliver and Charles D. Boyd

1. Introduction

The identification of specific messenger RNA molecules and the characterization of the proteins encoded by them has been greatly assisted by the development of in vitro translation systems. These cell-free extracts comprise the cellular components necessary for protein synthesis, i.e., ribosomes, tRNA, rRNA, amino acids, initiation, elongation and termination factors, and the energy-generating system (1). Heterologous mRNAs are faithfully and efficiently translated in extracts of HeLa cells (2), Krebs II ascites tumor cells (2), mouse L cells (2), rat and mouse liver cells (3), Chinese hamster ovary (CHO) cells (2), and rabbit reticulocyte lysates (2,4), in addition to those of rye embryo (5) and wheat germ (6). Translation in cell-free systems is simpler and more rapid (60 min vs 24 h) than the in vivo translation system using *Xenopus* oocytes.

The synthesis of mRNA translation products is detected by their incorporation of radioactively labeled amino acids, chosen specifically to be those occurring in abundance in the proteins of interest. Analysis of translation products usually involves specific immunoprecipitation (7), followed by polyacrylamide gel electrophoresis (8) and fluorography (9) (see **Fig. 1**).

In vitro translation systems have played important roles in the identification of mRNA species and the characterization of their products, the investigation of transcriptional and translational control, and the cotranslational processing of secreted proteins by microsomal membranes added to the translation reaction (10,11) (see Chapters 106, 108, and 109). This chapter describes the rabbit reticulocyte lysate system for in vitro translation of mRNA.

Although the endogenous level of mRNA is lost in reticulocyte lysates, it may be further reduced in order to maximize the dependence of translation on the addition of exogenous mRNA. This reduction is achieved by treatment with a calcium-activated nuclease that is thereafter inactivated by the addition of EGTA (4). The system is thus somewhat disrupted with respect to the in vivo situation and is particularly sensitive to the presence of calcium ions. The resulting lysate, however, is the most efficient in vitro translation system with respect to the exogenous mRNA-stimulated incorporation of radioactive amino acids into translation products. It is therefore particularly appropriate for the study of translation products. The system is sensitive, however, to regulation

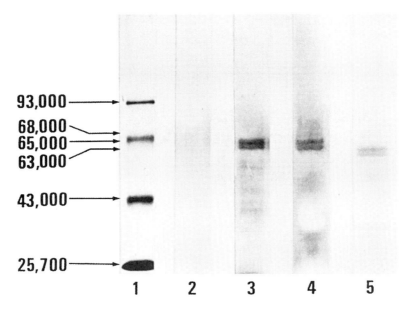

Fig. 1. SDS polyacrylamide gel electrophoretic analysis of in vitro translation products. In vitro translation products were derived from exogenous mRNA in an mRNA-dependent reticulocyte lysate cell-free system. Following electrophoresis on 8% SDS polyacrylamide gels, radioactive protein products were analyzed by flourography. Lane 1: [^{14}C]-labeled proteins of known molecular weights, i.e., phosphorylase a (93K), bovine serum albumin (68K), ovalbumin (43K), α-chymotrypsinogen (25.7K). Lanes 2–5 represent [^{3}H]-proline-labeled translation products of the following mRNAs: Lane 2: endogenous reticulocyte lysate mRNA, Lane 3: 0.3 μg of calf nuchal ligament polyadenylated RNA. Lane 4: 0.3 μg of calf nuchal ligament polyadenylated RNA, and immunoprecipitated with 5 μL of sheep antiserum raised to human tropoelastin, Lane 5: 0.3 μg of calf nuchal ligament polyadenylated RNA and cotranslationally processed by 0.3 A_{260}nm microsomal membranes.

by a number of factors, including hemin, double-stranded RNA, and depletion of certain metabolites. The effects of these factors on regulation of translation of various mRNAs may therefore be investigated. Despite the efficiency of reticulocyte lysates, the competition for initiation of translation by various mRNA species may differ from the in vivo situation. Products therefore may not be synthesized at in vivo proportions; the wheat germ extract cell-free system reflects the in vivo situation more faithfully. Nuclease-treated rabbit reticulocyte lysate cell-free systems are available as kits from a number of commercial suppliers.

2. Materials

All in vitro translation components are stored at –70°C. Lysates, microsomal membranes, and [^{35}S]-labeled amino acids are particularly temperature-labile and therefore should be stored in convenient aliquots at –70°C; freezing and thawing cycles must be minimized. Solutions are quick-frozen on dry ice or in liquid nitrogen prior to storage.

1. Folic acid: 1 mg/mL folic acid, 0.1 mg/mL vitamin B_{12}, 0.9% (w/v) NaCl, pH 7.0; filtered through a 0.45-μm filter and stored in aliquots at –20°C.

2. 2.5% (w/v) phenylhydrazine, 0.9% (w/v) sodium bicarbonate, pH 7.0 (with NaOH). Stored no longer than 1 wk at –20°C in single dose aliquots. Thawed unused solution must be discarded. Hydrazine degrades to darken the straw color.
3. Physiological saline: 0.14 M NaCl, 1.5 mM MgCl$_2$, 5 mM KCl. Stored at 4°C.
4. 1 mM hemin.
5. 0.1 M CaCl.
6. 7500 U/mL micrococcal nuclease in sterile distilled water. Stored at –20°C.
7. Rabbit reticulocyte lysate: This is prepared essentially as described by Pelham and Jackson *(4)*. Rabbits are made anemic by intramuscular injection of 1 mL folic acid solution on d 1, followed by six daily injections of 0.25 mL/kg body weight of 2.5% phenylhydrazine solution (*see* **Note 1**). At a reticulocyte count of at least 80%, blood is collected on d 7 or 8 by cardiac puncture into a 200-mL centrifuge tube containing approx 3000 U of heparin, and mixed well. Preparation should continue at 24°C.
    a. Blood is centrifuged at 120g, 12 min, 2°C, and plasma removed by aspiration.
    b. Cells are resuspended in 150 mL *ice cold* saline and washed at 650g for 5 min. Washing is repeated three times.
    c. The final pellets are rotated gently in the bottle, then transferred to Corex tubes (which are only half-filled). An equal volume of saline is added, the cells gently suspended, then pelleted at 1020g for 15 min at 2°C. The leukocytes (buffy coat) are then removed by aspiration with a vacuum pump.
    d. In an ice bath, an equal volume of ice-cold sterile deionized distilled water is added and the cells lysed by vigorous vortexing for 30 s (*see* **Note 2**). The suspension is then immediately centrifuged at 16,000g for 18 min at 2°C.
    e. At 4°C the supernatant is carefully removed from the pellet of membranes and cell debris. This lysate is then quick frozen in liquid nitrogen in aliquots of approx 0.5 mL.

    The optimum hemin concentration is determined by varying its concentration from 0–1000 µM during the micrococcal nuclease digestion. Lysate (477.5 µL), 5 µL of 0.1 M CaCl$_2$, and 5 µL of nuclease (75 U/mL final concentration) is mixed. A 97.5 µL volume of this is incubated with 2.5 µL of the relevant hemin concentration at 20°C for 20 min. A 4 µL 0.05 M solution of EGTA is added to stop the digestion (*see* **Note 3**). The optimum hemin concentration is that allowing the greatest translational activity (incorporation of radioactive amino acids) in a standard cell-free incubation (*see* **Subheading 3.**). A quantity of 25 µM is generally used to ensure efficient chain initiation.

    Lysates are extremely sensitive to ethanol, detergents, metals, and salts, particularly calcium. Stored at –70°C, reticulocyte lysates remain active for more than 6 mo.
8. L-[^{3}H]- or L-[^{35}S]-amino acids. A radioactive amino acid, labeled to a high specific activity (140 Ci/mmol tritiated, or approx 1 Ci/mmol [^{35}S]-labeled amino acids), is added to the translation incubation to enable detection of the translation products. An amino acid known to be abundant in the protein of interest is chosen. Radioactive solutions should preferably be aqueous; those of low pH should be neutralized with NaOH; ethanol should be removed by lyophilization, and the effect of solvents on lysate activity should be tested. [^{35}S] degrades rapidly to sulfoxide and should be aliquoted and stored at –70°C to prevent interference by sulfoxides.
9. Messenger RNA. Total RNA may be extracted from various tissues by a number of methods (*see* Chapter 2). mRNA stored in sterile dH$_2$O at –70°C is stable for more than a year. Contamination by ions, metals, and detergents should be avoided.

    Phenol may be removed by chloroform:butanol (4:1) extractions; salts are removed by precipitation of RNA in 0.4 M potassium acetate, pH 6.0, in ethanol. Ethanol should be removed by lyophilization. Convenient stock concentrations for translation are 1.5 mg/mL total RNA or 150 µg/mL polyA$^+$ RNA.

10. Translation cocktail: 250 mM HEPES, pH 7.2, 400 mM KCl, 19 amino acids at 500 mM each (excluding the radioactive amino acid), 100 mM creatine phosphate.
11. 20 mM magnesium acetate, pH 7.2.
12. 2.0 M potassium acetate, pH 7.2.
13. Sterile distilled H$_2$O.

Sterile techniques are used; RNase contamination is avoided by heat treatment of glassware (250°C, 12 h) or by treatment of heat-sensitive materials with diethylpyrocarbonate, followed by rinsing in distilled water. Sterile gloves are worn throughout the procedure.

3. Method

In vitro translation procedures are best carried out in autoclaved plastic microfuge tubes (1.5 mL); a dry incubator is preferable to waterbaths for provision of a constant temperature. All preparations are performed on ice.

1. Prepare (on ice) the following reaction mix (per inculcation): 0.7 µL of dH$_2$O, 1.3 µL of 2.0 M potassium acetate (*see* **Note 4**), 5 µL (10–50 µCi) of radioactive amino acid (*see* **Note 5**), and 3 µL of translation cocktail. Components are added in the above order, vortexed, and 10 mL is aliquoted per incubation tube on ice (*see* **Note 6**).
2. Add 10 µL (300 mg) of total mRNA (*see* **Notes 7** and **8**). A control incubation without the addition of exogenous mRNA detects translation products of residual endogenous reticulocyte mRNA.
3. A 10 µL volume of lysate is added last to initiate translation. If required, 0.5 A$_{260}$ nm U of microsomal membranes are also added at this point for cotranslational processing of translation products (*see* **Notes 9** and **10**).
4. The mixture is vortexed gently prior to incubation at 37°C for 60 min. The reaction is stopped by placing the tubes on ice (*see* **Note 6**).
5. Detection of mRNA-directed incorporation of radioactive amino acids into translation products is performed by determination of acid-precipitable counts.

    At the initiation and termination of the incubation, 5-µL aliquots are spotted onto glass fiber filters that are then air-dried. Filters are then placed into 10 mL/filter of the following solutions:
    a. 10% (v/v) cold trichloroacetic acid (TCA) for 10 min on ice.
    b. 5% (v/v) boiling TCA for 15 min, to degrade primed tRNAs.
    c. 5% (v/v) cold TCA for 10 min on ice.

    The filters are then washed in 95% (v/v) ethanol, then in 50% (v/v) ethanol-50% (v/v) acetone, and finally in 100% (v/v) acetone. The filters are dried at 80°C for 30 min. TCA-precipitated radioactivity is determined by immersing the filters in 5 mL of toluene-based scintillation fluid and counting in a scintillation counter.

    Exogenous mRNA-stimulated translation can be expected to result in a five- to 30-fold increase over background of incorporation of [^{3}H]- or [^{35}S]-labeled amino acids, respectively, into translation products.
6. An equal volume of 2% (w/v) SDS, 20% (w/v) glycerol, 0.02% (w/v) bromophenol blue, 1 M urea is added to the remaining 20 µL of translation mixture. This is made 0.1 M with respect to dithiothreitol, heated at 95°C for 6 min, and slowly cooled to room temperature prior to loading onto a polyacrylamide gel of appropriate concentration (between 6 and 17%). After electrophoresis, radioactive areas of the gel are visualized by fluorography (**Fig. 1**).

4. Notes

1. Maximum anemia may be achieved by reducing the dose of phenylhydrazine on d 3, then increasing it on following days. The reticulocyte count is determined as follows:
   a. 100 µL of blood is collected in 20 µL of 0.1% heparin in saline.
   b. 50 µL of blood heparin is incubated at 37°C for 20 min with 50 µL of 1% (w/v) brilliant cresyl blue, 0.6% (w/v) sodium citrate, 0.7% (w/v) sodium chloride.
   c. Reticulocytes appear under the microscope as large, round, and with blue granules. Erythrocytes are small, oval, and agranular.
2. The volume of water (in mL) required to lyse the reticulocyte preparation is equal to the weight of the pellet in the tube.
3. Endogenous mRNAs of lysates are degraded by a calcium-activated nuclease that is inactivated by EGTA. Lysates are therefore sensitive to calcium ions, the addition of which must be avoided to prevent degradation of added mRNAs by this activated nuclease.
4. Optimum potassium concentrations may vary from 30–100 mM depending on mRNAs used and should be determined prior to definitive translations. Similarly, specific mRNAs may require altered magnesium concentrations, although a concentration of 0.6–1.0 mM is generally used.
5. Specific activities greater than those described in **Subheading 2.** may result in depletion of the amino acid concerned, with subsequent inhibition of translation.
6. Vigorous vortexing decreases efficiency of translation, therefore do so gently when preparing the reaction mix.
7. The optimum mRNA concentration should be determined prior to definitive experiments by varying the mRNA concentrations while keeping other variables constant. Care should be taken to avoid excess mRNA; polyadenylated RNA in excess of 1 µg has been noted to inhibit translation.
8. Heating of mRNA at 70–80°C for 1 min followed by quick cooling in an ice bath, prior to addition to the incubation mixture, has been shown to increase the efficiency of translation of GC-rich mRNA; for example, heating elastin mRNA at 70–80°C prior to translation resulted in a 100% increase, compared with unheated mRNA, of incorporation of radioactivity into translation products *(12)*.
9. Cotranslational processing of translation products may be detected by the addition of dog pancreas microsomal membranes to the translation incubation. These may be prepared as described by Jackson and Blobel *(11)* or may be ordered with a commercial translation kit. Microsomal membranes should be stored in aliquots of approx 5 A_{260} nm U in 20 mM HEPES, pH 7.5, at –70°C. Repeated freezing and thawing must be avoided.
10. The addition of spermidine at approx 0.4 mM has been noted to increase translation efficiency in certain cases *(12)*, possibly by stabilizing relevant nucleic acids. However, this effect may also be lysate-dependent and should be optimized if necessary for individual lysate preparations.

References

1. Lodish, H. F. (1976) Translational control of protein synthesis. *Annu. Rev. Biochem.* **45**, 39–72.
2. McDowell, M. J., Joklik, W. K., Villa-Komaroff, L., and Lodish, H. F. (1972) Translation of reovirus messenger RNAs synthesized in vitro into reovirus polypeptides be several mammalian cell-free extracts. *Proc. Natl. Acad. Sci. USA* **69**, 2649–2653.
3. Sampson, J., Mathews, M. B., Osborn, M., and Borghetti, A. F. (1972) Hemoglobin messenger ribonucleic acid translation in cell-free systems from rat and mouse liver and Landschutz ascites cells. *Biochemistry* **11**, 3636–3640.

4. Pelham, H. R. B. and Jackson, R. J. (1976) An efficient mRNA-dependent translation system from reticulocyte lysates. *Eur. J. Biochem.* **67,** 247–256.
5. Carlier, A. R. and Peumans, W. J. (1976) The rye embryo system as an alternative to the wheat-system for protein synthesis in vitro. *Biochem. Biophys. Acta* **447,** 436–444.
6. Roberts, B. E. and Paterson, B. M. (1973) Efficient translation of tobacco mosaic virus RNA and rabbit globin 9S RNA in a cell-free system from commercial wheat germ. *Proc. Natl. Acad. Sci. USA* **70,** 2330–2334.
7. Kessler, S. W. (1981) Use of protein A-bearing staphylococci for the immunoprecipitation and isolation of antigens from cells, in *Methods in Enzymology* (Langone, J. J. and Van Vunakis, H., eds.), Academic, New York, pp. 441–459.
8. Laemmli, U. K. (1970) Cleavage of structural proteins during the assembly of the head of bacteriophage T4. *Nature* **227,** 680–685.
9. Banner, W. M. and Laskey, R. A. (1974) A film detection method for tritium-labeled proteins and nucleic acids in polyacrylamide gels. *Eur. J. Biochem.* **46,** 83–88.
10. Shields, D. and Blobel, G. (1978) Efficient cleavage and segregation of nascent presecretory proteins in a reticulocyte lysate supplemented with microsomal membranes. *J. Biol. Chem.* **253,** 3753–3706.
11. Jackson, R. C. and Blobel, G. (1977) Post-translational cleavage of presecretory proteins with an extract of rough microsomes, from dog pancreas, with signal peptidase activity. *Proc. Natl. Acad. Sci. USA* **74,** 5598–5602.
12. Karr, S. R., Rich, C. B., Foster, J. A., and Przybyla, A. (1981) Optimum conditions for cell-free synthesis of elastin. *Coll. Res.* **1,** 73–81.

108

In Vitro Translation of mRNA in a Wheat Germ Extract Cell-Free System

Louise Olliver, Anne Grobler-Rabie, and Charles D. Boyd

1. Introduction

The wheat germ extract in vitro translation system has been used widely for faithful and efficient translation of viral and eukaryotic messenger RNAs in a heterologous cell-free system *(1–9)*. With respect to the yield of translation products, the wheat germ extract is less efficient than most reticulocyte lysate cell-free systems (*see* Chapters 106, 107, and 109). There are advantages, however, of using wheat germ extracts:

1. The in vivo competition of mRNAs for translation is more accurately represented, making the wheat germ system preferable for studying regulation of translation *(1)*.
2. Particularly low levels of endogenous mRNA and the endogenous nuclease activity *(10)* obviate the requirement for treatment with a calcium-activated nuclease. There is, therefore, less disruption of the in vivo situation and contamination with calcium ions is less harmful. The identification of all sizes of exogenous mRNA-directed translation products is facilitated because of the low levels of endogenous mRNA present.
3. There is no posttranslational modification of translation products; primary products are therefore investigated, although processing may be achieved by the addition of microsomal membranes to the translation reaction.
4. The ionic conditions of the reaction may be altered to optimize the translation of large or small RNAs *(2)* (*see* **Note 1**).

Translational activity is optimized by the incorporation of an energy-generating system of ATP, GTP, creatine phosphate, and creatine kinase *(3)*. Wheat germ is inexpensive and commercially available (*see* **Note 2**); preparation of the extract is rapid and simple, resulting in high yields. Wheat germ extract cell-free system kits are also commercially available.

2. Materials

Components of the wheat germ in vitro translation system are heat-labile and must be stored in aliquots of convenient volumes at −70°C. Freeze-thaw cycles must be minimized. Sterile techniques are used throughout. RNase contamination is prevented by heat-sterilization (250°C, 8 h) of glassware and tips, and so on, or by diethyl pyrocarbonate treatment of glassware, followed by thorough rinsing of equipment in sterile distilled water.

1. Wheat germ extract: This is prepared essentially as described by Roberts and Paterson *(4)*. The procedure must be carried out at 4°C, preferably in plastic containers since initiation factors stick to glass. Fresh wheat germ (approx 5 g) (*see* **Note 2**) is ground with an equal weight of sand and 28 mL of 20 m*M* HEPES, pH 7.6, 100 m*M* KCl, 1 m*M* magnesium acetate, 2 m*M* CaCl$_2$, and 6 m*M* 2-mercaptoethanol, added gradually. This mixture is then centrifuged at 28,000g for 10 min at 2°C, pH 6.5. This pH prevents the release of endogenous mRNA from polysomes and therefore removes the requirement for a preincubation to allow polysome formation *(4,5)*. The supernatant (S-28) is then separated from endogenous amino acids and plant pigments that are inhibitory to translation, by chromatography through Sephadex G-25 (coarse) in 20 m*M* HEPES, pH 7.6, 120 m*M* KCl, 5 m*M* magnesium acetate, and 6 m*M* 2-mercaptoethanol. Reverse chromatography will prevent the loss of amino acids. Fractions of more than 20 A$_{260}$ nm/mL are pooled before being stored in aliquots at a concentration of approx 100 A$_{260}$ nm/mL, at –70°C. The extract remains translationally active for a year or more.
2. L-[^{3}H]- or L-[^{35}S]-amino acids: 10–50 µCi of an appropriate amino acid (abundant in the protein[s] of interest) is added to the reaction to allow detection of translation products. Convenient specific activities are 140 Ci/mmol tritiated, or 1 Ci/mmol [^{35}S]-amino acids, respectively (*see* **Note 3**). Aqueous solutions should be used since ethanol, salts, detergents, and various solvents interfere with translation. Ethanol should be removed by lyophilization and the effects on translation of other solutions should be determined prior to their use. [^{35}S]-labeled amino acids must be stored in small aliquots at –70°C where they remain stable for up to 6 mo, after which time sulfoxide products of degradation inhibit translation.
3. Messenger RNA: The extraction of both total and polyadenylated RNA has been described by a number of authors *(10–12)*. Total RNA (1.5 mg/mL) or 150 µg/mL polyadenylated RNA (in sterile distilled water) are convenient stock concentrations. RNA is stable for more than a year at –70°C. Contamination with potassium (*see* **Note 1**), phenol, and ethanol must be prevented by 70% (v/v) ethanol washes, chloroform:butanol (4:1) extractions, and lyophilization respectively.
4. 10X energy mix: 10 m*M* ATP, 200 µ*M* GTP, 80 m*M* creatine phosphate. Potassium salts of the nucleotide triphosphates should be used and the final pH adjusted (if necessary) to 7.4–7.6 with sodium hydroxide.
5. 0.5–1.0 *M* potassium acetate (*see* **Note 1**), 25 m*M* magnesium acetate.
6. 20 m*M* dithiothreitol.
7. 0.6–1.2 m*M* spermine or 4.0–8.0 m*M* spermidine (*see* **Note 4**).
8. 0.2 *M* HEPES, pH 7.4–7.6 (*see* **Note 5**).
9. 200–500 µg/mL creatine kinase (*see* **Note 6**).

3. Method

All preparations are carried out on ice. After use, components are quick-frozen on dry ice. Reactions are carried out in sterile plastic microfuge tubes.

1. Mix the following solutions (all components are v/50 µL): 5 µL of energy mix, 5 µL of potassium and magnesium acetate, 5 µL of dithiothreitol, 5 µL of HEPES, 5 µL of spermine, 10 µL of 0.3–8.0 µg mRNA in dH$_2$O, (*see* **Note 7**), 10 µL of wheat germ extract, 10 µL of creatine kinase (0.8–1.0 A$_{260}$ U), and 5 µL of creatine kinase. If a number of incubations are to be made, a master mix of the first five solutions may be prepared and 25 µL aliquoted into each reaction tube. Creatine kinase is added last to ensure that no energy is wasted. The solutions are mixed by tapping the tube or by gentle vortexing. Microsomal membranes (0.5 A$_{260}$ U) may be added before the creatine kinase to detect cotranslational modification of translation products (*see* **Note 8**).

2. Incubate at 28°C for 1 h (*see* **Note 9**). The reaction is terminated by placing the tubes at 4°C.
3. Incorporation of radioactive amino acids into mRNA-derived translation products is detected by TCA precipitation of an aliquot of the reaction (*see* Chapter 107 and **Note 10**). Incorporation of radioactivity into translation products is generally not as well-stimulated by mRNA added to wheat germ extracts as it is in described reticulocyte lysates.
4. The remaining in vitro translation products may be analyzed further by standard techniques, including tryptic mapping and ion-exchange chromatography, but specific products may be analyzed by immunoprecipitation followed by SDS-polyacrylamide gel electrophoresis.

4. Notes

1. Wheat germ extract translational activity is particularly sensitive to variation in the concentration of potassium ions. At concentrations lower than 70 mM, small mRNAs are preferentially translated, whereas larger mRNAs are translated at potassium acetate concentrations of 70 mM or greater *(2,5)*. Polypeptides of up to 200 kDa are synthesized under correct ionic conditions *(9)*. Furthermore, chloride ions appear to inhibit translation such that potassium acetate should preferably be used *(5)*. In this context, residual potassium should be removed from RNA preparations, by 70% (v/v) ethanol washes.
2. Inherent translational activity varies with the batch of wheat germ. Israeli mills (for example, "Bar-Rav" Mill, Tel Aviv) supply wheat germ, the extracts of which are usually active.
3. Most of the endogenous amino acids are removed by chromatography through Sephadex G-25 (coarse) (*see* Chapter 31). Depending on the batch of wheat germ extract, addition of amino acids (to 25 μM) and/or tRNA (to 58 μg/mL) may be necessary to optimize translational activity. Wheat germ extract is particularly sensitive to amino acid starvation; use of radioactive amino acids at specific activities greater than those suggested may result in inhibition of translation because of amino acid starvation.
4. The use of either spermine or spermidine generally stimulates translation, and is essential for the synthesis of larger polypeptides *(5)*, probably by stabilizing longer mRNAs. Omission of either compound will increase the optimum magnesium acetate concentration to 4.0–4.3 mM.
5. HEPES has been shown to buffer the wheat germ extract in vitro translation system more effectively than Tris-acetate *(4)*. Use of the latter will alter the optimum potassium and magnesium concentration.
6. Commercial preparations of creatine kinase differ with respect to the levels of nuclease contamination. This must be considered when larger amounts of the enzyme are to be used.
7. Heating of large mRNAs at 70°C for 1 min followed by rapid cooling on ice increases the efficiency of their translation in wheat germ extract in vitro translation systems.
8. Cotranslational processing of translation products may be detected by the addition of dog pancreas microsomal membranes to the translation incubation. They may be prepared as described by Jackson and Blobel *(12)* or may be ordered with a commercial translation kit. Microsomal membranes should be stored in aliquots of approx 5 A_{260} nm U in 20 mM HEPES, pH 7.5, at –70°C. Repeated freezing and thawing must be avoided.
9. mRNA-stimulated incorporation of radioactive amino acids into translation products is linear, after a 5 min lag, for 50 min and is complete after 90 min. The system is labile at temperatures >30°C; optimum activity is achieved at 25–30°C depending on the batch of wheat germ extract. An incubation temperature of 28°C is generally used.
10. In order to obtain maximum translational activity, it is necessary to determine the optima for each preparation of wheat germ extract; mRNA concentration, potassium and magnesium concentrations, and incubation temperature. Take into account the concentration of salts in the wheat germ extract column eluate.

References

1. Steward, A. G., Lloyd, M., and Arnstein, H. R. V. (1977) Maintenance of the ratio of α and β globin synthesis in rabbit reticulocytes. *Eur. J. Biochem.* **80,** 453–459.
2. Benveniste, K., Wilczek, J., Ruggieri, A., and Stern, R. (1976) Translation of collagen messenger RNA in a system derived from wheat germ. *Biochemistry* **15,** 830–835.
3. Huntner, A. R., Farrell, P. J., Jackson, R. J., and Hunt, T. (1977) The role of polyamines in cell-free protein in the wheat germ system. *Eur. J. Biochem.* **75,** 149–157.
4. Roberts, B. E. and Paterson, B. M. (1973) Efficient translation of tobacco mosaic virus RNA and rabbit globin 9S RNA in a cell-free system from commercial wheat germ. *Proc. Natl. Acad. Sci. USA* **70,** 2330–2334.
5. Davies, J. W., Aalbers, A. M. J., Stuik, E. J., and van Kammen, A. (1977) Translation of cowpea mosaic RNA in cell-free extract from wheat germ. *FEBS Lett.* **77,** 265–269.
6. Boedtker, H., Frischauf, A. M., and Lehrach, H. (1976) Isolation and translation of calvaria procollagen messenger ribonucleic acids. *Biochemistry* **15,** 4765–4770.
7. Patrinou-Georgoulas, M. and John, H. A. (1977) The genes and mRNA coding for the theory chains of chick embryonic skeletal myosin. *Cell* **12,** 491–499.
8. Larkins, B. A., Jones, R. A., and Tsai, C. Y. (1976) Isolation and in vitro translation of zein messenger ribonucleic acid. *Biochemistry* **15,** 5506–5511.
9. Schroder, J., Betz, B., and Hahlbrock, K. (1976) Light-induced enzyme synthesis in cell suspension cultures of *petroselinum*. *Eur. J. Biochem.* **67,** 527–541.
10. Pelham, H. R. B. and Jackson, R. J. (1976) An efficient mRNA-dependent translation system from reticulocyte lysates. *Eur. J. Biochem.* **67,** 247–256.
11. Darnbrough, C. H., Legon, S., Hunt, T., and Jackson, R. J. (1973) Initiation of protein synthesis: evidence for messenger RNA-independent binding of methionyl-transfer RNA to the 40S ribosomal subunit. *J. Mol. Biol.* **76,** 379–403.
12. Jackson, R. C. and Blobel, G. (1977) Post-translational cleavage of presecretory proteins with an extract of rough microsomes, from dog pancreas, with signal peptidase activity. *Proc. Natl. Acad. Sci. USA* **74,** 5598–5602.

109

The *Xenopus* Egg Extract Translation System

Glenn M. Matthews and Alan Colman

1. Introduction

A full analysis of the post-translational modifications that a given protein undergoes during transit through the secretory pathway may, in some cases, only be performed by analysis of the natural protein, expressed in its normal tissue.

Often this is not possible since the quantity produced is too small to give strong signals after, for example, radiolabeling and immunoprecipitation. Expression of cDNA clones in *Xenopus* oocytes or cultured cells has been widely used to determine not only the nature of covalent modifications, but also the fate, that is, whether membrane bound, secreted, or resident in the secretory pathway, of a wide range of proteins.

A major barrier to analyses performed by expression in living cells is that some method is required to visualize the protein under investigation, normally by means of an antibody capable of being used for immunoprecipitation. Often a useful antibody may be difficult to produce or, during the early stages of analysis of a newly cloned sequence, it may be desirable to verify the secretory phenotype and characterize the primary posttranslational modifications that occur to the protein before raising antisera. It is at this stage where cell-free systems can be most useful.

The "traditional" translocating in vitro systems, where reticulocyte lysate (*see* Chapter 107) or wheat-germ extracts (*see* Chapter 108) are combined with canine pancreatic membranes, solve the problem of background translation, and have been enormously useful in the development of current knowledge of the mechanisms of translocation and early processing events. For routine analysis, however, they can often be difficult to use, since the translocation capacity can be limited and the membranes are often quite fragile, making protease protection and fractionation experiments difficult.

The *Xenopus* egg extract *(1)* has a high capacity for translocation, signal sequence cleavage, and *N*-glycosylation, and so gives consistent processing patterns across a wide range of added mRNA concentrations, whereas the stability of the membranes present allows the use of sucrose gradient fractionation and protease protection to verify the location of translation products. The ability of the extract to support assembly of multimeric proteins and perform, to a limited extent, *O*-glycosylation and mannose-6-phosphorylation further extend its utility as an analytical tool.

Preparation of the extract, which is based on the method described by Murray *(2)* for the preparation of extracts for cell-cycle studies, involves centrifugal lysis of *Xenopus* eggs. In this crude form, with only ^{35}S-methionine added, a translation reaction produces approx 60 μg/mL of protein from the endogenous mRNA. This figure is increased to 140 μg/mL on addition of creatine phosphate to 7 mM and 10% by volume of an S-100 fraction of rabbit reticulocyte lysate. After ribonuclease treatment to remove endogenous mRNAs, approximately half of this activity can be restored by the addition of poly A$^+$ mRNA. When a single synthetic mRNA, encoding a secretory protein, is used to program translation, yields in excess of 10 μg/mL of translocated and processed product can be obtained. The extract can be frozen for storage, allowing many independent experiments to be performed on a single batch. After freezing, however, the addition of creatine phosphate is necessary for activity, whereas reticulocyte lysate S-100 and 1 mM spermidine stimulate translation. This combination restores 25–60% of original activity. When performed in the presence of 800 μCi/mL of ^{35}S-methionine, this range of activities allows most translation products to be easily detected by overnight fluorography of SDS gels.

2. Materials

1. Frogs: *Xenopus* stock.
2. High-Salt Modified Barth's X (MBS) It is supplemented by addition of 1.28 g of NaCl/L to give a final concentration of 110 mM.
3. Folligon (serum gonadotrophin) and Chorulon (chorionic gonadotrophin) are obtained from Intervet (Cambridge, UK) and dissolved in the solvent provided.
4. Dejellying solution: 2% cysteine HCl titrated to pH 7.7 with NaOH.
5. Extraction buffer: 100 mM KCl, 0.1 mM CaCl$_2$, 1 mM MgCl$_2$, 50 mM sucrose, and 10 mM HEPES-KOH, pH 7.7 (titrated at 10 mM).
6. Versilube VF50 (General Electric): This oil has a density between that of the eggs and of the extraction buffer.
7. Cytochalasin B (Sigma, Poole, UK): Stock at 10 mg/mL in DMSO; store at 4°C
8. Aprotinin (Boehringer, Lewes, UK): Stock at 10 mg/mL in water; store at –20°C.
9. RNase A (Boehringer): Stock at 1 mg/mL in water; store in aliquots at –20°C.
10. Ribonuclease inhibitor (Boehringer): This is normally supplied at 50 U/μL.
11. Dithiothreitol, Boehringer (DTT): Stock at 1 M; stored in aliquots at –20°C. This is freshly diluted to 100 mM before addition to the extract.
12. tRNA (calf liver, Boehringer): Stock at 5 mg/mL in water; store at –20°C in aliquots.
13. Creatine phosphate (Boehringer): Stock at 350 mM in water; store at –20°C in small aliquots.
14. [^{35}S]-Methionine (SJ 204, Amersham Pharmacia Biotech, Little Chalfont, UK): Divide into 100-μCi aliquots on first thawing, and store at –70°C.
15. Spermidine (Sigma): Stock at 120 mM; store at –20°C.
16. Rabbit reticulocyte lysate/S-100 extract: It is only necessary to prepare the S-100 fraction if it is desirable to exclude exogenous ribosomes from the reaction. Whole (nuclease-treated) reticulocyte lysate can be added in place of the S-100 fraction if this is not important. We normally use the reticulocyte lysate supplied by Bethesda Research Laboratories (Paisley, UK), but nuclease-treated material from any source would probably be as effective. To prepare an S-100 fraction, 100-μL portions of reticulocyte lysate are centrifuged at 100,000g in a TLA-100 rotor for 2 h, 80 μL of the supernatant is recovered, taking care to avoid the ribosomal pellet, and flash-frozen in liquid nitrogen as 10-μL aliquots, before storage at –70°C.
17. Triton X-100 (Surfact-Amps X-100, Pierce, Chester, UK): This is supplied as a 10% solution in 10-mL vials. After opening, unused material can be stored in the dark at 4°C.

Xenopus *Egg Extract*

18. Phenylmethylsulfonyl fluoride (PMSF, Sigma): Stock at 100 mM in propan-2-ol. Store at 4°C. Aqueous solutions of PMSF lose activity very rapidly, and should be made up immediately before use.
19. 1% Triton X-100, 1 mM PMSF: Freshly made from the stock solutions above.
20. 2X T buffer: 100 mM KCl, 10 mM Mg acetate, 200 mM NaCl, and 40 mM Tris-HCl, pH 7.6. Sterilize by filtration and store at –20°C. This is blended with sucrose from a 40% stock solution to generate 1X T + 10% and 1X T + 20% sucrose solutions. Unused portions of T + 10 and T + 20 can be stored at –20°C.
21. Proteinase K (Boehringer): Stock solutions at 25 mg/mL in autoclaved 50% glycerol appear to be stable on storage at –20°C. Before use in protease protection experiments, the stock is diluted to 1 mg/mL in 10% sucrose and allowed to stand at room temperature for 15 min to digest any contaminating enzymes, such as lipases.
22. Sodium carbonate: Prepare a fresh 1 M stock for each alkaline sucrose gradient experiment. Check that the pH of a 100-mM solution is 11. This can then be diluted (1 in 5 with water) to 200 mM for treatment of membrane fractions and can be blended with 40% sucrose (1 vol 1 M sodium carbonate, 4 vol water, and 5 vol 40% sucrose) to form the alkaline 20% sucrose cushion.
23. 1 M HCl.
24. Acetyl-Asn-Tyr-Thr-amide (custom synthesized by Alta Bioscience, University of Birmingham): This is relatively insoluble in water, so a 100-mM stock solution must be prepared in DMSO. Dilutions from this can then be made in water.
25. Centrifuges: We routinely use a Beckman TL-100 for the crushing spin, which has the advantage of efficient refrigeration, an appropriate tube/rotor format, and rapid acceleration to the set speed. Conditions specified below, therefore, all relate to this model. Many other machines could probably be used instead, but the only alternative we have tried is an Eppendorf model 5414 microfuge in a 4°C cold room. This gave good results, but the yield was compromised because of the use of a fixed-angle rotor. For the low-speed spin before the crushing step, almost any refrigerated bench-top model with a swing-out rotor should be adequate.
26. Cold room: Extracts prepared in a 4°C cold room are invariably more active than those exposed to room temperature during preparation.

3. Methods
3.1. Preparation of the Extract (see Notes 1–7)
3.1.1. Preparation of the Basic Extract

A flow diagram indicating the stages of the preparation of the extract is shown in **Fig. 1**. Before starting to make an extract, ensure that all buffers, tubes, and rotors are cooled to 4°C and that all necessary materials are at hand, since the faster the whole procedure can be performed, the better the final extract. Unless a temperature is otherwise stated, all the procedures, after the dejellying step, should be performed on ice and, ideally, in a cold room.

1. Large adult female *Xenopus laevis* are primed by injection with 50–100 U of Folligon, on d 1. Three to five days later, they are induced to lay by injection of 500–750 U of Chorulon, late in the evening of the day before the extract is to be made. Frogs are left overnight to shed eggs into high-salt MBS, which prevents activation.
2. Transfer approx 30 mL of loosely packed eggs to a 250-mL glass beaker, rinse a few times with high-salt MBS to remove debris, and aspirate off any obviously dead eggs with a

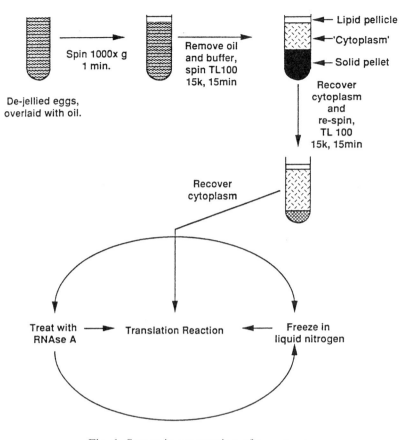

Fig. 1. Stages in preparation of an extract.

Pasteur pipet. Pour away as much supernatant buffer as possible, and add 100 mL of dejellying solution, repeat this two or three times. Agitate the suspension of eggs occasionally by swirling for 5–10 min. The dissolution of the jelly coats can be easily seen as a marked reduction in the volume occupied by the eggs. When this has occurred, rinse once more in dejellying solution and then transfer, by multiple washes, to ice cold extraction buffer.

3. Fill four 2-mL polyallomer TLS 55 centrifuge tubes by gently pipeting the eggs using a wide-bore Pasteur pipet, transferring as little buffer as possible. Allow the eggs to settle for a minute or so, then carefully remove the supernatant buffer, top the tubes up with Versilube VF50, and centrifuge at $500g$ for 1 min at 4°C in a swing-out rotor. Depending on the rotor format, it may be useful to place the tubes inside another container, e.g., pre-cooled 7-mL bijoux tubes, for easier handling, thermal insulation during transfer, and to contain any spillage. The eggs should pack tightly, but not lyse during this step, and the buffer should be separated from the eggs by a layer of oil.

4. Remove the supernatant buffer and then the oil by careful aspiration with a 200-µL micropipet, load the tubes into a TLS 55 rotor, and crush the eggs by centrifugation at $20,000g$ for 15 min at 4°C. This produces a multilayered lysate of which the desired product, the viscous amber middle layer normally referred to as cytoplasm, constitutes around 40% by volume. Recover this layer by inserting a Pasteur pipet through the lipid pellicle.

5. Pool the product from all tubes, estimate the volume, and add 5 µL/mL of 10 mg/mL cytochalasin B (mL). Mix by gentle pipeting, transfer to fresh polycarbonate 1.5 mL TLS

55 centrifuge tubes, and centrifuge again at 20,000g for 15 min at 4°C. The cytoplasm should now occupy most of the volume, but this must be removed very slowly to avoid disturbing the pellet.
6. Add 1 µL of 10 mg/mL aprotinin/mL of extract, and mix gently but thoroughly.

The extract should now be ready for translation reactions, frozen immediately, or mRNA depleted as described in **Subheading 3.1.2.**

3.1.2. Depletion of Endogenous mRNAs

As with the preparation of the basic extract, this is best performed in a cold room, where the activity of the ribonuclease is more readily controlled.

1. Dilute stock RNase A to 100× the final desired concentration.
2. Add 1 µL of diluted RNase to a series of screw-capped 1.5-mL microfuge tubes, and then add 100 µL of extract, mixing thoroughly by pipeting gently. Incubate at 10°C for 15 min.
3. Transfer to ice, and add 1 µL of 100 mM DTT to each tube, followed by 50 U of ribonuclease inhibitor. Mix well, and incubate at 10°C for a further 10 min. Then add 2 µL of 5 mg/mL calf liver tRNA. This product should now be used for translation reactions or frozen in liquid nitrogen for storage as soon as possible.

3.2. The Translation Reaction (see Notes 8–14)

1. Thaw frozen extracts at room temperature until just liquid, and then place on ice.
2. Meanwhile, distribute mRNAs to be translated into 0.5- or 1.5-mL microfuge tubes on ice.
3. To each 100 µL aliquot of extract add: 10 µL reticulocyte lysate S-100, 1 µL 120 mM spermidine, 2.5 µL 350 mM creatine phosphate, and 100 µCi ^{35}S-methionine. Add aliquots (10–50 µL) of this mixture to the tubes containing mRNAs, mixing well, and then incubate at 21°C for 1 h.
4. If highly radioactive synthetic mRNA was used to program the reactions, treat, at the end of the reaction, with 10 µg/mL RNase A for 15 min at 21°C to remove the radioactive background, which the mRNA contributes to further analyses.
5. For storage, the reaction can be stopped by freezing at this stage.
6. For analysis by gel electrophoresis or TCA precipitation (1), reaction products should be diluted in 4 vol of 1% Triton X-100, 1 mM PMSF before either adding an equal volume of twofold concentrated SDS-PAGE sample buffer or spotting to filters.

3.3. Analysis of Translation Products (see Notes 13–15)

Simple gel electrophoresis of translation products can often give an indication of a secretory phenotype. Signal sequence cleavage reduces molecular weight by 2–3 kDa (see **Fig. 2A**), whereas N-glycosylation normally reduces mobility relative to unmodified protein produced in wheat-germ extract or reticulocyte lysate (see **Fig. 3A**). A simple mobility shift could, however, be caused by a range of other factors, and it is normally necessary to confirm that the protein has been translocated into membranes by performing either protease protection or sucrose fractionation experiments. Further fractionation, on alkaline sucrose gradients, can be used to determine whether the protein is free within the lumen of the endoplasmic reticulum or integral to the membrane, whereas specific inhibition of N-glycosylation can demonstrate that this is responsible for any apparent increase in molecular weight. Examples of protease protection and neutral sucrose gradient fractionation experiments are shown in **Fig. 2B** and **C**, respectively, and inhibition of N-glycosylation is shown in **Fig. 3B**.

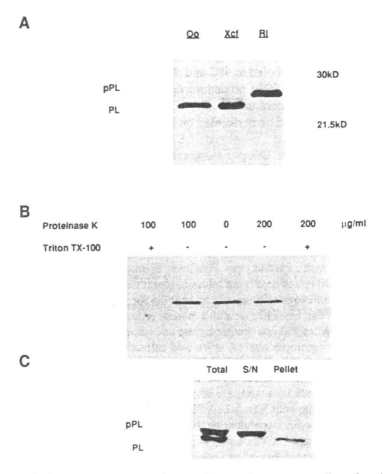

Fig. 2. Translation, protease protection, and neutral sucrose gradient fractionation of bovine prolactin. (**A**) Translation products obtained from synthetic prolactin mRNA translated in *Xenopus* oocytes (Oo), the *Xenopus* egg cell-free extract (Xcf), and rabbit reticulocyte lysate (Rl). Prolactin is not *N*-glycosylated, so signal sequence cleavage of preprolactin (pPL) to prolactin can be seen as a reduction in molecular weight in the *Xenopus* systems relative to the reticulocyte lysate product. The *Xenopus* oocyte sample was immunoprecipitated before electrophoresis, whereas the in vitro products were loaded directly onto the gel. (**B**) Result of a protease protection experiment performed on a *Xenopus* egg extract translation mixture, again programmed with bovine prolactin mRNA. (**C**) Result of a neutral sucrose gradient fractionation of a *Xenopus* egg extract translation programmed with excess prolactin mRNA to saturate the translocation apparatus and provide a marker for unsegregated protein.

3.3.1. Protease Protection

1. Remove 3×10 µL aliquots from the translation reactions to be assayed, and place on ice. If it is necessary to use less than this, for example, because the products are to be analyzed by a variety of other methods, dilute the reaction by addition of up to 4 vol of 1X T buffer + 10% sucrose.
2. Add 1 µL of 10% Triton X-100 to one of each set of tubes in order to disrupt the membranes present and thus provide a positive control for proteolysis.

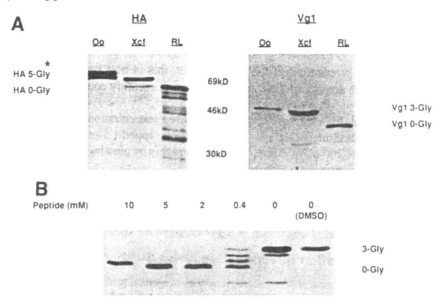

Fig. 3. Translation of *N*-glycosylated proteins and the partial inhibition of glycosylation. **(A)** Products obtained from the translation of *Xenopus* Vg1 and influenza virus hemagglutinin (HA) synthetic mRNAs in the *Xenopus* oocyte (Oo), *Xenopus* egg cell-free extract (Xcf), and rabbit reticulocyte lysate (Rl). The occupation of five glycosylation sites on HA (HA 5-Gly) and three on Vg1 (Vg1 3-Gly) can be clearly seen to have retarded these proteins relative to the unglycosylated reticulocyte lysate products (0-Gly). **(B)** Result of *Xenopus* cell-free extract translation reactions, programmed with synthetic mRNA encoding Vg1, performed in the presence of the indicated concentration series of the tripeptide Asn-Tyr-Thr. Total inhibition of *N*-glycosylation can be seen above 2 m*M* peptide, whereas the partial reaction, at 0.4 m*M* tripeptide, allows the number of *N*-glycosylation sites occupied to be directly determined.

3. Add 1 µL of 1 mg/mL proteinase K solution in 10% sucrose to the tube with Triton added and one of the other two (the third aliquot, with no additions, serves as a control for stability during the procedure), and incubate on ice for 1 h.
4. Freshly dilute stock 100 m*M* PMSF with 3 vol of 10% sucrose: Add 1 µL of this to each reaction to give a final concentration of 2–2.5 m*M*, and continue the incubation on ice for a further 15 min.
5. Add 100 µL of 1X SDS-PAGE sample buffer, including 1% Triton X-100, and heat in a boiling water bath for 5 min. If the sample was diluted before protease treatment, add a smaller volume of 2X SDS-PAGE buffer to ensure that the final mixture contains 10% of extract by volume.
6. Analyze by SDS-PAGE, including an untreated sample of the same reaction at an equivalent dilution, as a marker.

3.3.2. Neutral Sucrose Gradient Fractionation

Sucrose gradient fractionation can be useful not only as an analytical technique, but can also provide a significant degree of purification before, for example, performing an activity assay on the translation products. An example of the results of this method of fractionation is shown in **Fig. 2C**.

If this procedure is to be followed by the alkaline sucrose fractionation described below, it is advisable to start with a translation reaction of at least 50 µL since the volumes will be easier to handle.

1. On ice, dilute the translation reaction (by gentle pipeting) into 10 vol of 1X T buffer plus 10% sucrose. Retain a small portion as a marker for SDS-PAGE analysis, and carefully layer the remainder onto a 1-mL step of 1X T buffer plus 20% sucrose in a 1.5-mL polycarbonate centrifuge tube.
2. Centrifuge at 30,000 rpm (40,000g) in a TLS 55 rotor for 30 min at 4°C.
3. Recover the top (10% sucrose) layer, containing the cytosolic proteins, by aspiration: It is rarely necessary to recover all of this, and it is easier to avoid the dilution caused by mixing with the 20% step if only about half to three-quarters of this fraction is recovered. Remove and discard the rest of the sucrose buffer, taking care not to disturb the membrane pellet, which can be seen as a brown spot on the bottom of the tube.
4. If membrane stability is required in later analysis, such as protease protection or alkaline sucrose fractionation, gently resuspend the membrane pellet in T buffer plus 10% sucrose in a volume equivalent to half that originally loaded onto the gradient. Alternatively, dissolve the membranes in 1% Triton X-100, 1 mM PMSF.
5. Analyze equivalent portions of the total reaction and each fraction by SDS-PAGE.

3.3.3. Alkaline Sucrose Gradient Fractionation

Alkaline treatment disrupts the membrane vesicles, liberating lumenal proteins, without dissolving the lipid bilayer. After fractionation on a sucrose step gradient, the lumenal proteins remain in the supernatant, whereas membrane-bound components are pelleted.

1. To 100 µL of membranes from **Subheading 3.3.2., step 4**, add an equal volume of 200 mM Na$_2$CO$_3$. Incubate on ice for 30 min.
2. Layer onto a 250-µL step of 0.1 M Na$_2$CO$_3$ in 20% sucrose in a 1.5-mL polycarbonate centrifuge tube, and centrifuge for 1 h at 100,000g at 4°C in a TLS 55 rotor.
3. Recover the supernatant (again, it is best to accept a loss here and only remove 100–150 µL), discard the 20% sucrose step, and redissolve the pellet in 1% Triton X-100, 1 mM PMSF.
4. Before analysis by SDS-PAGE, neutralize the supernatant by addition of HCl. Approximately 10% by volume of 1 M HCl is required, and this should be added slowly, with mixing, to prevent local precipitation of proteins. The progress of this titration can be checked by spotting a small volume onto suitable pH indicator paper.

3.3.4. Inhibition of N-Glycosylation

Traditionally, the number of N-glycosylation sites occupied on a polypeptide chain has been measured by partial endoglycosidase H digestion. The use of the competitor tripeptide (acetyl)-Asn-Tyr-Thr-(amide) to inhibit N-glycosylation partially in the *Xenopus* extract, however, provides an alternative approach to this question.

1. Prepare a twofold dilution series of tripeptide in water, covering the range from 50–1.5 mM (6 points).
2. Assemble, on ice, a translation reaction of at least 80 µL final vol, including the mRNA for the protein under investigation.
3. Add 9 µL of the translation reaction to 1 µL of each dilution of tripeptide and to 1 µL of 50% DMSO (as a control for the effect of DMSO on the reaction). Incubate these mix-

tures, together with the remainder, which serves as a positive control, at 21°C for 1 h, and then analyze by SDS-PAGE, including a reticulocyte lysate translation product as a marker for unglycosylated preprotein. Partial inhibition of glycosylation will be seen at the low end of the concentration series.

4. Notes

1. As with any translation system, ribonuclease contamination should be avoided. All tubes, tips, and where possible, reagents, must be autoclaved or treated by some other means to inactivate ribonuclease. The extract is a complex mixture that, during a translation reaction, is supporting a wide range of processes. It is therefore important to ensure that no inactivating contaminants are present in the materials added, by using the highest quality reagents available and for stock solutions to be made up in double-glass distilled or reverse-osmosis-treated water.
2. Egg quality varies from one female to another and, sometimes, with the time of year. Even the poorest eggs will give an extract that incorporates ^{35}S-methionine into proteins before ribonuclease treatment. Some batches are, however, compromised beyond this stage, particularly after freezing. A good guide to the quality of a batch of eggs is their ability to undergo fertilization. Extracts produced from eggs that do not fertilize well may benefit from the addition of sucrose to 200 mM from a 2 M stock, before freezing.
3. Very occasionally, a batch of eggs will fail to lyse well. When this happens, a significant proportion (30% or more) of the extract volume will be gray, instead of amber, after the first spin. These extracts never perform well and should be abandoned at this stage.
4. There are many steps involved in generating an extract, so the chances of a batch being compromised by a faulty reagent or an error in handling are significant. Unfortunately, it is not possible to assay an extract during preparation, but performing reactions on samples at each stage of the process (i.e., fresh and frozen material both before and after RNase treatment) helps, although retrospectively, to locate any problem areas. For comparison, it is best to add spermidine, creatine phosphate, and reticulocyte lysate S-100 to all samples assayed. Extracts that have not been frozen should incorporate ^{35}S-methionine without these being present.
5. To freeze, divide the extract into aliquots of 100 µL or less in ice-cold microfuge tubes (this should already be the case if the extract has been ribonuclease-treated), and plunge into liquid nitrogen for a min. After freezing, do not allow the extract to thaw until it is to be used. Activity, relative to that of an aliquot thawed and tested immediately after freezing, is unaffected by storage at –70°C for a number of months, but storage in liquid nitrogen is preferable, if this is available.
6. To assay for recovery from freezing, an aliquot from a batch of frozen extract should simply be thawed and used to set up a translation reaction. It is tempting to economize on reagents by setting up a translation reaction with all reagents present, then splitting this into two, and freezing and thawing one aliquot. This should be avoided, however, since the dilution owing to the additions will compromise activity, probably by destabilizing a proportion of the membranes present.
7. RNase A purity varies considerably according to source and batch, so the conditions quoted here should be regarded as a starting point for titration. To titrate a batch of enzyme, prepare an extract, and treat aliquots of extract with a range of final ribonuclease concentrations from 0.1–5 µg/mL. Neutralize these with ribonuclease inhibitor as described, and then perform translation reactions on each in the presence and absence of a well-characterized synthetic mRNA. SDS-PAGE analysis should show a declining background with rising RNase concentration, whereas, at high concentrations, the signal resulting from

the added mRNA will be lost. Choose the "crossover" point where the signal-to-noise ratio is highest for future use. This concentration should be appropriate for subsequent extracts. Ensure that as much as possible of the batch of ribonuclease is then stored in aliquots, so that this procedure does not have to be repeated too often.

8. Dilution of the reaction mixture beyond a final volume of about 130% of the volume of extract present causes a reduction in activity, so in general, any additions to the extract, including mRNAs, should be made in the smallest volume possible. If a significant degree of dilution is a necessary part of the experiment, then wherever possible, reagents to be added should be dissolved in 10% sucrose to ensure membrane stability.

9. Generally, the performance of mRNAs from different sources parallels that seen in the *Xenopus* oocyte. Most mRNAs derived from higher eukaryotes translate well, whereas those from prokaryotes are not as effective. As with any other translation system, natural mRNAs are more efficiently translated. We find that, almost invariably, synthetic mRNAs are translated more efficiently if they are transcribed from the vector pSP64T *(3)*. A reliable protocol for transcription of synthetic mRNA is described in **ref. *1***.

10. The amount of mRNA to be added to a translation reaction depends largely on its origin and activity. In general, a final concentration of around 50 µg/mL of synthetic mRNA produced from a cDNA cloned into pSP64T or 100–200 µg/mL of poly A$^+$ mRNA gives maximal signal without saturating the capacity of the extract to modify the translation products posttranslationally. Beyond these levels, the efficiency of segregation of secretory proteins begins to decline, and *N*-glycosylation becomes less efficient. Translocation without signal sequence cleavage has, however, not been observed, even at very high mRNA levels.

11. The methionine pool of the extract is approx 35 µM (± 10%). The quantity of protein produced in a reaction can therefore be estimated from the percentage incorporation of ^{35}S-methionine into TCA-precipitable material. When a reaction is programmed with a single cloned mRNA, the methionine content of the product will be known. In the case of a complex mixture, such as poly A$^+$ mRNA, being used, a reasonable estimate of average methionine content is 2%. Incorporation of all the methionine present in the extract would therefore indicate that 240 µg/mL of protein had been synthesized.

If the aim of a reaction is to produce the largest possible quantity of protein, the yield can be increased by around 50% by the addition of excess amino acids. This is achieved by adding 5% by volume of a solution containing 700 µM methionine and 2 mM of all other amino acids to the extract, since increasing the methionine concentration more than twofold has no further effect and only serves to reduce further the specific activity of the radiolabel.

12. ^{35}S-methionine is used routinely to follow translation reactions, because it is readily available and generates good signals owing to the relatively low level of methionine in the extract. Some proteins, however, contain little or no methionine, and so require an alternative strategy. Since the pool size for methionine in the extract is comparable with that measured for the *Xenopus* oocyte, it is probably reasonable to expect the same to apply to other amino acids. Amino acids with small pool sizes in the oocyte include cysteine, leucine, histidine, and proline, whereas there are very large pools of lysine, aspartic acid, threonine, serine, glutamic acid, glycine, and alanine. *See* **ref. *4*** for a more detailed list.

The specific activity of any amino acid can be increased by simply adding more radioactive material, but the effect on the final dilution of the extract in the reaction should be considered and, if necessary, the radiolabel should be concentrated before use. Tritiated amino acids are often supplied as dilute solutions and invariably require concentration. Generally, it is not advisable to use crude extracts of mixed radioactive amino acids, such as Translabel, in cell-free translation systems.

13. If fractionation or protease protection experiments are planned, these should, ideally, be performed immediately after the translation reaction, but we have produced acceptable results from reactions that have been stored at –70°C after freezing in liquid nitrogen.
14. We normally use the Bio-Rad Mini Protean II apparatus to analyze translation products by SDS-PAGE, using the reagents recommended by the manufacturer. The high protein content of the extract limits the amount of material that can be loaded to the equivalent of 1 µL of whole extract/5-mm wide slot on a 0.75-mm thick gel. Sucrose gradient pellets contain about a fifth of the protein present in the whole extract, and the proportion loaded can be increased accordingly, if this is not restricted by the need to load comparable amounts of unfractionated material. We routinely treat gels containing ^{35}S-labeled proteins with En3Hance (DuPont) before drying and exposure to Kodak XAR 5 film at –70°C.
15. In protease protection experiments, some transmembrane proteins with cytoplasmic tails of significant length will show a reduction in size owing to trimming by the protease.

References

1. Matthews, G. M. and Colman, A. (1991) A highly efficient, cell-free translation/translocation system prepared from *Xenopus* eggs. *Nucleic Acids Res.* **19,** 6405–6412.
2. Murray, A. W. (1991) Cell cycle extracts, in *Methods in Cell Biology*, vol. 36 (Kay, B. K. and Peng, H. B., eds.), Academic, San Diego, pp. 581–605.
3. Kreig, P. A. and Melton, D. A. (1984) Functional messenger RNAs are produced by SP6 in vitro transcription of cloned cDNAs. *Nucleic Acids Res.* **12,** 7057–7070.
4. Colman, A. (1984) Translation of eukaryotic messenger RNA in *Xenopus* oocytes, in *Transcription and Translation—A Practical Approach* (Hames, B. D. and Higgins, S. J., eds.), IRL, Oxford, pp. 271–302.

110

Manipulation of Baculovirus Vectors

Claire L. Merrington, Mark J. Bailey, and Robert D. Possee

1. Introduction

The prototype baculovirus is the *Autographa californica* nuclear polyhedrosis virus (AcNPV), which has been the focus for most studies developing expression vector systems. In contrast to other reviews on this topic *(1–3)* this chapter aims to describe the various practical aspects of the system and to highlight particular difficulties that may be encountered by workers new to the field. Comprehensive reviews are also available that describe the biology of baculoviruses *(4–7)*.

Baculoviruses produce two structurally distinct forms. Both have a nucleocapsid core containing the double-stranded, covalently closed, circular DNA genome (AcNPV; approx 134 kbp *[8]*). This is surrounded by a lipoprotein envelope that delineates the virus particle. Budded virus particles (BV) form as nucleocapsids, which bud through the plasma membrane of infected cells. During this process the 64-kd glycoprotein (gp64) surface antigen, (probably the BV cell attachment protein) is acquired *(9)*. In the second process, nucleocapsids are enveloped without gp64 within the nucleus to form occlusion-derived virus (ODV). The virus particles are further packaged into polyhedra or occlusion bodies comprising a single virus-encoded polypeptide (polyhedrin) of approx 30 kDa. The BV enters insect cells in culture and after virus DNA has reached the nucleus the genome is expressed in early and late phases. The late phase is subdivided into the expression of late and subsequently very late genes. Early genes are transcribed by host RNA polymerase II in the absence of viral gene products, whereas late and very late genes are transcribed by an α-amanitin-insensitive RNA polymerase. Late-gene expression occurs concomitant with or subsequent to viral DNA replication and requires the presence of virus-encoded gene products. These factors, designated LEFS, probably act to direct viral DNA replication and may also constitute the novel RNA polymerase *(10,11)*. BV production begins in the late phase with occlusion body formation taking place in the very late phase. During the very late phase, two virus-encoded proteins are produced in large quantities. These are the polyhedrin protein, described earlier, and the p10 gene product, which is a nonstructural polypeptide of 10 kDa thought to play some role in the formation of the polyhedra. Both proteins are nonessential for the replication and formation of virus particles *(12–14)* and therefore have been targeted for use in expression vectors. The strategy for using baculoviruses

From: *The Nucleic Acid Protocols Handbook*
Edited by: R. Rapley © Humana Press Inc., Totowa, NJ

as vectors involves manipulation of a portion of the virus genome in a bacterial plasmid. This is designated the transfer vector, which serves as a vehicle for inserting the foreign gene into the virus genome via homologous recombination after cotransfection of insect cells with infectious genomic DNA. Examples of transfer vectors are provided in **Table 1**.

2. Materials

2.1. Cell Culture General

1. TC-100 medium, 10% fetal calf serum (FCS), penicillin, and streptomycin.
2. *Spodoptera frugiperda* cells (Sf 21/Sf 9).
3. Phosphate-buffered saline (PBS) (Dulbecco's A).
4. Trypan blue 0.05% in PBS.
5. Dimethyl sulfoxide (DMSO) (Sigma, St. Louis, MO) and liquid nitrogen for cell storage.
6. Inverted microscope; must be of high quality.
7. Tissue culture hood.
8. 28°C Incubator.
9. Tissue culture plasticware.
10. Hemocytometer.

2.2. Titration, Propagation, and Purification of Viral DNA

1. 2% Low-gelling temperature agarose (Sea Plaque, FMC Bioproducts, Rockland, ME) in distilled water.
2. Neutral red 0.01% in PBS.
3. TE: 10 mM Tris-HCl pH 8.0, 1 mM EDTA.
4. 10% and 50% sucrose in TE.
5. Virus lysis buffer: 2% sodium N-lauryl sarcosine, 1 mM EDTA.
6. 50% w/w CsCl solution in TE containing ethidium bromide (25 µg/mL). CAUTION: Ethidium bromide is mutagenic.
7. Cell lysis buffer: 50 mM Tris-HCl pH 7.8, 5% 2-mercaptoethanol; 10 mM EDTA; 0.4% sodium dodecyl sulfate.
8. Ribonuclease.
9. Proteinase K.
10. Phenol:chloroform (50:50), equilibrated with 100 mM Tris-HCl, pH 8.0. CAUTION: Phenol is caustic.
11. 3 M Sodium acetate, pH 5.2.
12. 100% and 70% Ethanol.

2.3. Preparation of Linear Viral DNA

1. BacPAK6, AcRP6-SC or AcNPV.*lacZ* DNA of known concentration.
2. 10X Restriction enzyme buffer (usually supplied by the manufacturer).
3. *Bsu*36I.
4. Sterile water.

2.4. Transfection and Generation of Recombinant Virus

2.4.1. Cotransfection Using Lipofectin

1. TC100 medium without added serum.
2. TC100 plus 5% FCS (TC100/5% FCS).
3. Transfer vector.

4. Linearized viral DNA.
5. Lipofectin (Gibco-BRL).

2.4.2. Cotransfection by Calcium Phosphate Precipitation

1. Linearized viral DNA.
2. Transfer vector.
3. 2 M CaCl$_2$, filter sterilized.
4. 0.1 M D-glucose, filter sterilized.
5. Transfection mix: HEPES buffer; 40 mM HEPES, pH 7.05, 2 mM sodium phosphate, 10 mM potassium chloride, 280 mM sodium chloride.

3. Methods

Lepidopteran insect cells are routinely grown in TC100 medium. This is a complex medium requiring skill and experience to prepare; Gibco-BRL (Gaithersburg, MD) and Flow Laboratories (McLean, VA) now conveniently sell this product in liquid form or as a powder that can be prepared, filter sterilized, and stored at 4°C for some months. It is supplemented before use with 5 or 10% (v/v) FCS; 50 U penicillin mL^{-1} and 50 μg streptomycin mL^{-1} may be added if desired. The complete medium is stable for at least 3 mo at 4°C. The FCS should be tested before purchasing in bulk; it may have to be heat inactivated (56°C for 30 min) prior to use.

3.1. Maintenance of Insect Cells

Spodoptera frugiperda IPLB-Sf-21AE cells (Sf21) *(15)*, were originally derived from insect pupal ovaries and have been used for all of our work. A cloned version of this cell line (Sf9) was derived in Professor Max Summer's laboratory and is available from the American Tissue Culture Collection (Manassas, VA). Sf9 cells have been adapted to grow in serum-free medium; Sf900II medium is available commercially (Gibco). *Trichoplusia ni* TN-368 and BTI-TN-5B1-4 (also known as High 5) also support the replication of AcNPV and produce recombinant proteins to higher levels and with more authentic posttranslational modification than Sf9 cells *(16,17)*. However, these cell lines are not easily adaptable to growth in suspension culture and tend to form large aggregates *(16)*. These cells must, therefore, be grown in monolayer culture to achieve the optimum level of protein production. We commonly use Sf21 cells between passage 150–180 for our work. The higher level is not a limiting factor to cell growth and, with care, cells can be maintained indefinitely. However, we have observed a decrease in virus production when cells of high passage number are used, and for reproducibility we prefer to use a defined passage range. Cells may be grown either as monolayer or suspension culture at 28°C or 21°C in atmospheric air. The doubling time corresponds to 18–24 h depending on culture conditions and temperature. Cells may be counted in an improved Neubauer chamber by diluting in medium and trypan blue (final concentration 0.05% w/v). It is essential to keep cells in a healthy condition prior to use in transfection experiments and plaque assays (*see* **Subheading 3.1.1.**).

3.1.1. Monolayer Culture

In general it is advisable to seed cells at a density of 2×10^4 cm^{-2} and allow them to grow to become confluent before subculturing. However, the cells can be seeded at

Table 1
Vectors of Use

Vector	Size	Features	Reference
Polyhedrin promoter based			
pAcYM1	9.2 kbp	Retains complete 5' leader sequence and first nucleotide of polyhedrin ATG. BamHI cloning site.	21
pAcCL29	7.8 kbp	As pAcYM1, less sequence flanking the polyhedrin gene. Able to produce single-stranded DNA in *Escherichia coli* for site-directed mutagenesis.	22
pBacPAK8/9	5.5 kbp	Polyhedrin coding sequences replaced with MCS. Identical except for order of cloning sites.	23
p10 promoter based			
pAcUW1	4.5 kbp	*p10* Promoter at *p10* locus; *p10* coding sequences deleted. Recombinant virus polyhedra positive. BglII cloning site.	24
pAcUW21	9.3 kbp	Complete *p10* promoter upstream of complete polyhedrin gene and promoter at polyhedrin locus. BglII cloning site. Single-stranded DNA capability.	24
Vectors employing late promoters			
pAcMP1	10.1 kbp	Polyhedrin gene plus promoter replaced by basic protein promoter. BamHI cloning site.	25
pAcATM3	7.9 kbp	*gp64* Promoter in place of polyhedrin gene and promoter. BamHI cloning site. Single-stranded DNA capability.	R. D. Possee, unpublished
Multiple promoter expression vectors			
pAcUW2B	10.7 kbp	Dual promoter vector. *p10* Promoter upstream of polyhedrin gene and promoter in opposite orientation. Recombinant viruses polyhedra positive. BglII cloning site.	24
pAcUW31	8.5 kbp	Dual promoter vector. Both polyhedrin and *p10* promoters at polyhedrin locus. BamHI and BglII cloning sites, respectively. Single-stranded DNA capability.	PharMingen
pAcAB3	10.1 kbp	Triple promoter vector. Contains two copies of *p10* promoter, one of polyhedrin promoter at polyhedrin locus. Useful for coexpression/assembly studies. Cloning sites SmaI, XbaI, BglII.	26

Vector	Size	Description	Source/Ref
pAcAB4	10.2 kbp	Quadruple promoter expression vector. Two copies each of p10 and polyhedrin promoters. Useful for coexpression/assembly studies. Cloning sites BamHI, SmaI, XbaI, BglII.	26
Vectors containing E. coli lacZ			
pBlueBac4	4.8 kbp	Polyhedrin promoter-based, etl promoter drives lacZ expression. MCS insertion site.	Invitrogen
Fusion/tagged protein production			
pAcG1 pAcG2T pAcG3X	8.5 kbp	Based on pAcCL29. Contain sj26 gene, encoding glutathione-S-transferase downstream of polyhedrin promoter. Permit production of GST fusion proteins. BamHI, SmaI, EcoRI cloning sites.	27,28
pBlueBacHis2	4.9 kbp	Similar to pBlueBac4, but contains His6 sequence downstream of the polyhedrin promoter. Facilitates purification of recombinant protein. MCS insertion site.	Invitrogen
pBacPAKHis	5.4 kbp	Contains six His residues downstream of polyhedrin promoter. Facilitates purification of recombinant protein.	Clontech
Eukaryotic ligand display			
pAcSurf2	8.5 kbp	Contains amino terminus of gp64 gene downstream of polyhedrin promoter. Permits display of ligands on the surface of the virion. MCS insertion site.	29

Based on plasmids containing a selectable marker, β-lactamase which replicate to high copy numbers in E. coli. All these vectors may be used in conjunction with linearized BacPAK6 DNA, apart from pAcUW1, which is used in conjunction with linearized AcUW1.lacZ (lacZ at p10 locus; normal polyhedrin locus [53]) and pBlueBac4, which may be used with linearised BacPAK5 DNA (similar to BacPAK6, except it contains polyhedrin gene in place of lacZ [36]). Various baculovirus manuals are available to supplement this list (54,55).

lower densities to permit the maintenance of cells at room temperature for up to 10 d. The cells are harvested by gently scraping them from the flask with a sterile rubber policeman into fresh medium. It is imperative that the cells not be subjected to trypsin, as this will severely reduce viability.

3.1.2. Suspension Culture

This system is particularly useful for growing larger stocks of virus or recombinant protein. Cells are seeded at 1×10^5 mL^{-1} in TC100/5% (v/v) FCS in round-bottomed flasks and stirred with a magnetic bar at 100 rpm. It is essential to maintain the supply of oxygen to the cells and so the surface area exposed to the atmosphere should be maximized. Therefore, it is advisable that only 50% of the available volume in a flask be used (e.g., 500 mL in a 1000-mL flask). Cultures of up to 2000 mL can be grown in this way but with some reduction in total cell number due to oxygen limitation, unless air is bubbled through the medium. The maximum cell number obtained is usually between $2-4 \times 10^6$.

3.1.3. Storage of Cells in Liquid Nitrogen

Cells should be harvested when they are in their exponential phase of growth and diluted into TC100 medium containing 10% (v/v) FCS and 10% (v/v) DMSO to a final concentration of $1-4 \times 10^6$ mL^{-1}. Cryotubes are filled with 1–2 mL volumes and chilled on ice before placing in the gaseous phase of liquid nitrogen for approx 2 h. The cells are subsequently immersed in the liquid phase of the nitrogen for long-term storage. In this state they remain viable indefinitely. The cells are revived by thawing rapidly at 37°C, diluting one-fifth in fresh medium and dispensing into a 25-cm^2 flask. After incubation at 28°C for 24 h, the medium should be changed and the cells incubated further until ready for subculture.

3.2. Titration of Virus in a Plaque Assay

This is possibly the most critical technique to master when using the baculovirus expression system for the first time. The production of clear, well-defined plaques is vital to permit the easy identification of recombinant virus. Furthermore, accurate titers of virus stocks are required for the calculation of the multiplicity of infection (MOI; infectious virions per cell) to be used when optimizing expression and comparing yields from different viruses. In our experience workers new to the field have more difficulty with this method than any other. The stages involved are described as follows.

1. Actively growing (subconfluent) cells should be harvested and accurately counted as previously described; viability must exceed 98%.
2. Dilute the cells to a concentration of 0.75×10^6 mL^{-1} and dispense 2 mL into each 35-mm tissue culture dish (NUNC). Incubate at 28°C for 1–3 h on a level, vibration-free surface before use. Alternatively, the cells may be seeded at a concentration of 10^6 per dish and used the following day (*see* **Note 2**).
3. Remove all media and add, dropwise, 100 µL appropriately diluted virus to the center of the dish. Gently rock the plate to ensure even distribution of the virus, then incubate at room temperature for 1 h.
4. The virus inoculum must then be removed with a Pasteur pipet after tilting the dish to one side to facilitate drainage.

5. Prepare overlay medium by mixing sterile 2% (w/v) low-gelling temperature agarose, cooled to 37°C with an equal volume of TC100/5% FCS, prewarmed to the same temperature. Add 1.5 mL overlay medium to the side of each dish, allow to set for 20 min, and then supplement with 1.0 mL TC100/5% FCS as a liquid feeder layer (plaques do not form well unless this extra medium is added). Incubate dishes in a humid environment at 28°C for 72–96 h.
6. Plaques may be observed without staining by directing a light source to the side of each dish and viewing against a black background. Alternatively, the cells may be stained by adding 1.0 mL neutral red (0.01% [w/v] in PBS) directly to the dishes and incubating for 2 h at 28°C. Thereafter, all liquid is removed by inverting the dish over a sterile beaker and then blotting residual stain with a paper towel. Allow to destain, by standing, for at least 5 h, or preferably overnight at 21°C. Uninfected cells retain the red stain, whereas infected areas (plaques) appear clear. By viewing the stained monolayer over a lightbox, plaques can be counted and the titer of the virus stock easily calculated. Techniques for the identification of recombinant virus will be discussed in **Subheading 3.6.**

3.3. Virus Propagation

The propagation of AcNPV in insect cells will be described from the isolation of a single plaque to a large (1000 mL) suspension culture.

1. Remove the agarose from a clear, well-isolated plaque with a sterile Pasteur pipet and disperse into 0.5 mL TC100/5% FCS containing antibiotics. The recovery of virus varies but should be about 10^4–10^5 PFU/mL.
2. This stock is amplified by inoculating 100 µL of the virus onto 5×10^5 cells in 35-mm culture dishes, after removal of the medium. After 1 h at room temperature TC100/5% FCS is added and the cells are incubated at 28°C for 3–4 d until the cells display a cytopathic effect. The harvested medium is clarified by low-speed centrifugation (1000g) and the virus stock titrated as described in **Subheading 3.2.**, the titer should be 10^6–10^7 PFU/mL.
3. To amplify the stock further, prepare 80 cm² flasks with 5×10^6 cells and inoculate with virus at a MOI of 0.1–0.5 PFU/cell; incubate as before until all cells are infected. This virus stock should be 5×10^7–10^8 PFU/mL.
4. The highest titer virus stocks (>10^8 PFU/mL) may be produced by inoculating 50–1000 mL suspension cultures (5×10^5 cells/mL) with 0.1–0.5 PFU/cell and incubating at 28°C for 3–4 d.

3.4. Purification of Virus DNA

The quality and purity of this DNA is essential if successful transfection of insect cells is to be achieved. For the majority of procedures, wild-type BacPAK6 DNA will be required.

1. It is usually necessary to have a total of between 0.5–1.0×10^9 infected cells to permit purification of useful quantities of virus. This quantity of cells should be prepared using a suspension culture as described in **Subheading 3.3.**
2. The cells are pelleted at 1000g for 10 min and then the virus particles are pelleted at 75,000g for 1 h at 4°C. The pellets are conveniently resuspended in 1 mL TE after overnight soaking at 4°C and residual cell debris is removed using brief (5 min) low-speed centrifugation (1000g).
3. The virus is then applied to a 20-mL 10–50% (w/v) discontinuous sucrose gradient and sedimented to the interface (25,000g Beckman SW41) for 1 h at 4°C.

4. The thick, white virus band is harvested by downward displacement, diluted with TE, and pelleted at 75,000g for 1 h at 4°C.
5. The pellet is resuspended in approx 1.6 mL TE. The virus is disrupted by the addition of 0.4 mL of virus lysis buffer (20% w/v sodium-N-lauryl sarcosine, 10 mM EDTA) and incubation at 60°C for 30 min.
6. The lysate is added to the top of a 50% (w/w) CsCl gradient in TE containing ethidium bromide (25 µg/mL). Spin 200,000g (45,000 rpm), 20°C for 24 h (Beckman Ti70). The virus DNA may also be purified using a swing-out titanium rotor (e.g., Beckman SW41); in this case the lysed virus is applied to the top of a 5-mL 50% (w/w) CsCl cushion and the DNA is sedimented at 200,000g for 16 h at 20°C.
7. Harvest the lower (supercoiled) and upper (nicked circular and linear) bands and pool prior to extraction with butanol to remove the ethidium bromide.
8. Dialyze extensively against TE at 4°C before measuring the concentration and then store at 4°C; do not freeze the DNA, as this reduces infectivity. The yield should be about 100–200 µg. Throughout the purification it should be remembered that the genome is large and easily damaged by pipetting and other harsh mechanical treatment.

3.5. Insertion of Foreign Genes into the Transfer Vector

The use of a transfer vector such as pAcYM1 requires that the foreign gene be excisable with a restriction enzyme that produces *Bam*HI compatible sticky ends (**Fig. 1**). Alternatively, the insert and digested vector must be treated with a suitable enzyme to produce flush ends for blunt-end ligation. For some vectors, the gene must also have its own translation initiation and termination codons. If only part of a coding sequence is to be expressed, synthetic signals must be added prior to insertion; vectors retaining the polyhedrin ATG may also be used. In our experience the amount of 3' noncoding sequence retained after the translation stop codon is not important. However, it is advisable to remove the 5' noncoding region before the ATG of the foreign gene, particularly if it is GC rich; PCR, Bal31, or exonuclease III digestion are very effective for this purpose. Such manipulations are adequately described elsewhere *(18,19)*. Once the foreign gene has been inserted into the transfer vector, it is advisable to sequence across both ends of the gene by double-stranded dideoxy plasmid sequencing using primers generated to complement sequences at each end of the inserted gene.

It is also advisable to use very pure DNA for transfection. Therefore, plasmids should be prepared according to Sambrook et al. *(18)* and purified by CsCl gradient centrifugation.

3.6. Transfection of Insect Cells and the Generation of Recombinant Virus

3.6.1. Cotransfection Using Lipofectin

1. For each cotransfection, seed 1×10^6 Sf21 cells in 2 mL TC100/5% FCS into a 35-mm dish. Incubate at 28°C for 3 h.
2. Mix together the viral DNA and transfer vector in a sterile polystyrene container. Add 0.5 mL TC100 lacking serum.
   If you are using linearized viral DNA, use 100–200 ng viral DNA and 500–1000 ng transfer vector.
3. Add 5 µL lipofectin to 0.5 mL TC100 lacking serum in a separate polystyrene container. Add to the DNA solution. Incubate at room temperature for 15 min to allow the DNA:liposome complexes to form.

Manipulation of Baculovirus Vectors

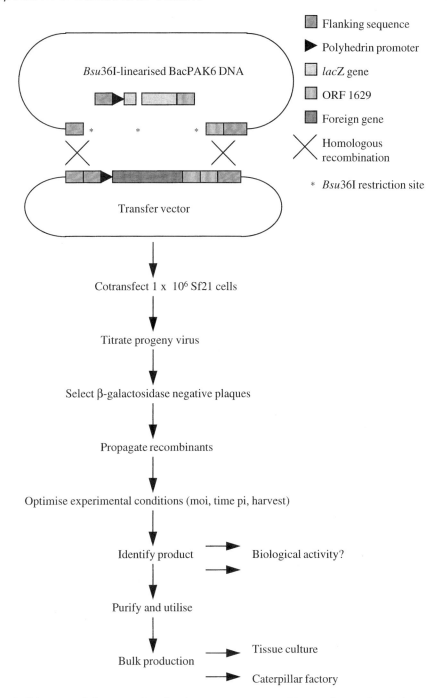

Fig. 1. Diagram of the steps involved to generate recombinant virus using linearized viral DNA. *Bsu*36I-digested BacPAK6 DNA is mixed with the recombinant plasmid vector. After transfection of the insect cells, homologous recombination occurs between the virus genome and the region flanking the insert in the vector DNA, producing an intact ORF1629 and, therefore, infectious viral DNA. Progeny virus replicate and are selected as colorless plaques against a very low background of blue parental virus plaques. Recombinant virus express the insert via the polyhedrin transcriptional and translational machinery.

4. Remove the medium overlaying the cell monolayer. Wash the cells twice by gently adding 1–2 mL TC100 lacking serum, rocking the dish once, then removing the medium and discarding.
5. Add 1 mL TC100 lacking serum to the cells. When ready, add the DNA:lipofectin solution (1 mL) and swirl gently. Incubate the cells at 28°C for at least 5 h or overnight.
6. Remove the medium overlaying the cells and add 2 mL TC100/5% FCS. Incubate at 28°C for 48–72 h.
7. Harvest the medium into a sterile container and store at 4°C. Use in plaque assay for identification and selection of recombinant viruses.

3.6.2. Cotransfection by Calcium Phosphate Precipitation

The best results are obtained by titrating the recombinant plasmid DNA against the viral genomic DNA; routinely we use 2, 10, and 25 µg of input plasmid DNA with 1 µg virus DNA for reliable recombination. The quality of the cells is also very important, they must be in the exponential phase of growth and have a viability greater then 98% (*see* **Note 4**).

1. Seed 1.5×10^6 Sf21 cells in 2 mL TC100/5% FCS into a 35-mm dish, incubate at 28°C for 3 h; alternatively, seed the dishes with 10^6 cells and incubate at the same temperature overnight before using.
2. Make up transfection mix as follows: mix 500 µL HEPES buffer, 100 µL 0.1 M D-glucose, AcNPV DNA (1 µg), add plasmid DNA (1–10 µg), and mix. Then add H_2O to make a volume of 935 µL.
3. While vortexing, add 65 µL 2 M $CaCl_2$ dropwise (this should take no longer than 10 s to avoid shearing the virus DNA). Incubate at room temperature for 30 min to produce a fine precipitate.
4. Remove all medium from the cells and add the transfection mix. Incubate at room temperature for 30 min. Add 1 mL TC100/5% FCS and incubate at 28°C for 2–4 h. Remove the transfection mix/medium and add a further 2 mL TC100/5% FCS; incubate at 28°C for 48–72 h.

3.7. Recombinant Virus Selection
3.7.1. Visual Identification of Plaques

The selection of a recombinant in the progeny virus from a transfection mix against the background of the polyhedrin-positive plaques requires patience and careful observation. The stained or unstained plaques may be viewed by directing light sideways on the dish and holding it against a dark background. Polyhedrin-negative plaques appear clear. On occasion certain biological properties may facilitate the selection process. For instance, the fusogenic properties of the glycoprotein of the vesicular stomatitis virus resulted in the production of giant cell syncytia when recombinant virus expressed the protein *(20)*. Ring with a fine felt-tip pen any putative polyhedrin negative plaques and view under a good high-resolution inverted microscope. Isolate the plaques as described previously and retitrate until the stock is clear of contaminating wild-type virus. Pick at least five clear plaques, as sometimes AcNPV mutates to a polyhedrin-negative phenotype.

3.7.2. Distinguishing Between lacZ-Positive and lacZ-Negative Plaques

This is accomplished by staining with the chromogenic substrate for β-galactosidase, X-gal. If β-galactosidase-positive plaques are required, stain the plaque assays with X-gal only. Alternatively, if β-galactosidase-negative plaques are to be picked from a background of blue, β-galactosidase-positive plaques, dual stain with both X-gal and neutral red.

1. Perform the plaque assay (*see* **Subheading 3.2.**) up to the staining step.
2. Replace the 1-mL liquid overlay with 1 mL TC100/5% FCS containing 15 µL/mL 2% X-gal in dimethylformamide. Incubate at 28°C for 5 h or overnight.
3. To dual stain the plaque assay with both X-gal and neutral red, add 1 mL neutral red stain (prepared as in **Subheading 3.2.**) after 3 h of the incubation. Incubate for a further 2 h at 28°C, then drain the dishes prior to viewing the plaques.
4. Pick the desired plaques in the same manner as for wild-type virus (*see* **Subheading 3.3.**).

3.8. Extraction of Recombinant Virus DNA from Infected Cells for Southern Blot Analysis

A useful and very rapid method for analyzing recombinant virus DNA is to extract nucleic acid from infected cell cultures.

1. Tissue culture dishes (35 mm diameter) are seeded with 1×10^6 cells as described previously and incubated for 3 h at 28°C. The medium is removed and virus is added to the center of each dish at an MOI of 10 PFU/cell. After 1 h at room temperature the inoculum is removed and replaced with 2 mL TC100/5% FCS.
2. The cultures are incubated at 28°C for 24 h, harvested and the cells washed with PBS in 1.5 mL microfuge tubes using a low-speed spin in the centrifuge.
3. The cells are then resuspended in 250 µL TE (10 mM Tris-HCl, pH 7.8, 0.1 mM EDTA) and 250 µL of lysis buffer (50 mM Tris-HCl, pH 7.8, 5% 2-mercaptoethanol, 10 mM EDTA and 0.4% SDS) added. Ribonuclease A is added to a final concentration of 40 µg/mL and the lysate incubated at 37°C for 30 min. Proteinase K is then added to a final concentration of 100 µg/mL and the incubation continued for a further 30 min at the same temperature.
4. The lysate is extracted twice with phenol/chloroform (50:50; equilibrated with 100 mM Tris-HCl, pH 7.8); vigorous shaking must be avoided to prevent shearing of the virus DNA. The DNA is precipitated from the final aqueous phase by the addition of sodium acetate to 0.3 M and 2 vol of ethanol. The nucleic acid is pelleted using a microfuge and washed twice with 75% ethanol, air dried, and resuspended in 100 µL TE. The DNA may be more readily resuspended after storage at 4°C overnight.
5. A small quantity of this material (5–10 µL; equivalent to about 100 ng) is sufficient for digestion with restriction enzymes and subsequent analysis in agarose gels and Southern blot hybridizations. After staining the gel with ethidium bromide the virus-specific bands are readily apparent.

4. Notes

1. Cells must at all times be kept in a healthy ($1–20 \times 10^5$ mL^{-1}) and viable state. They are the key to success.
2. Plaque assays need to be performed on level, vibration-free surfaces; do not add hot agarose (above 40°C) because it kills cells. Do not heat or cold shock cells with nonequilibrated media.
3. Ensure that all materials are detergent free especially in cell culture.
4. DNA must be pure and uncontaminated for transfections.
5. Secreted proteins may be, and often are, present in small quantities in the culture supernatant, therefore, it may be necessary to concentrate by ammonium sulphate cutting. Growth in serum-free medium for 24–72 h postinfection may facilitate purification by removing, or reducing, contaminating serum albumin.

References

1. Maeda, S. (1989) Expression of foreign genes in insects using baculovirus vectors. *Annu. Rev. Entomol.* **34,** 351–372.
2. Bishop, D. H. L. (1992) Baculovirus expression vectors. *Semin. Virol.* **3,** 253–264.
3. Kidd, M. and Emery, V. C. (1993) The use of baculoviruses as expression vectors. *Appl. Biochem. Biotech.* **42,** 137–159.
4. Blissard, G. W. and Rohrmann, G. F. (1990) Baculovirus diversity and molecular biology. *Annu. Rev. Entomol.* **35,** 127–155.
5. Volkman, L. E. and Keddie, B. A. (1990) Nuclear polyhedrosis virus pathogenesis. *Semin. Virol.* **1,** 249–256.
6. Adams, J. R and McClintock, J. T. (1991) Baculoviridae. Nuclear polyhedrosis viruses. Part 1. Nuclear polyhedrosis viruses of insects, in *Atlas of Invertebrate Viruses* (Adams, J. R. and Bonami, J. R., eds.), CRC, Boca Raton, FL.
7. Cory, J. S. (1993) Biology and ecology of baculoviruses, in *Opportunities for Molecular Biology in Crop Protection* (Beadle, D. J., Bishop, D. H. L., Copping, L. G., and Holloman, D. W., eds.), British Crop Protection Council, Monograph 55, Farnham, UK.
8. Ayres, M. D., Howard, S. C., Kuzio, J., Lopez-Ferber, M., and Possee, R. D. (1994) The complete DNA sequence of *Autographa californica* nuclear polyhedrosis virus. *Virology* **202,** 586–605.
9. Volkman, L. E., Goldsmith, P. A., Hess, R. T. D., and Faulkner, P. (1984) Neutralization of budded *Autographa californica* NPV by a monoclonal antibody: identification of the target antigen. *Virology* **133,** 354–362.
10. Lu, A. and Miller, L. K. (1995) The roles of eighteen late expression factor genes in transcription and DNA replication. *J. Virol.* **69,** 975–982.
11. Passarelli, A. L., Todd, J. W., and Miller, L. K. (1994) A baculovirus gene involved in late gene expression predicts a large polypeptide with a conserved motif of RNA polymerases. *J. Virol.* **68,** 4673–4678.
12. Smith, G. E., Vlak, J. M., and Summers, M. D. (1983) Physical analysis of *Autographa californica* nuclear polyhedrosis virus transcripts for polyhedrin and 10,000-molecular weight protein. *J. Virol.* **45,** 215–225.
13. Vlak, J. M., Klinkenberg, F., Zaal, K. J. M., Usmany, M., Klinge-Roode, E. C., Geervliet, J. B. F., Roosien, J., and Van Lent. J. W. M. (1988) Functional studies on the p10 gene of *Autographa californica* nuclear polyhedrosis virus using a recombinant expressing a p100 Beta-galactosidase fusion gene. *J. Gen. Virol.* **69,** 765–776.
14. Williams, G. V., Rohel, D. Z., Kuzio, J., and Faulkner, P. (1989) A cytopathological investigation of *Autographa californica* nuclear polyhedrosis virus p10 gene function using insertion/deletion mutants. *J. Gen. Virol.* **70,** 187–202.
15. Vaughn, J. L., Goodwin, R. H. Tompkins, G. J., and McCawley, P. (1977) The establishment of two cell lines from the insect *Spodoptera* frugipera (Lepidoptera: Noctuidae). *In Vitro* **13,** 213–217.
16. Wickham, T. J. and Nemerow, G. R. (1993) Optimization of growth methods and recombinant protein production in BTI-Tn5B1-4 insect cells using the baculovirus expression vector. *Biotechnol. Prog.* **9,** 25–30.
17. Davis, T. R., Shuler, M. L., Granados, R. R., and Wood, H. A. (1993) Comparison of oligosaccharide processing among various insect cell lines expressing a secreted glycoprotein. *In Vitro Cell Dev. Biol. Anim.* **29A,** 842–846.
18. Sambrook, J., Fritsch, E. F., and Maniatis, T. (eds.) (1989) *Molecular Cloning: A Laboratory Manual*. Cold Spring Harbor Laboratory Press, Cold Spring Harbor, NY.
19. Heinikoff, S. (1984) Unidirectional digestion with ExoIII creates targeted break points for DNA sequencing. *Gene* **28,** 351–359.

20. Bailey, M. J., Mcleod, D. A., Kang, C. Y., and Bishop, D. H. L. (1989) Glycosylation is not required for the fusion activity of the G Protein of VSV in insect cells. *Virology* **169,** 323–331.
21. Matsuura, Y., Possee, R., Overton, H. A., and Bishop, D. H. L. (1987) Baculovirus expression vectors: the requirement for high level expression of protein, including glycoproteins. *J. Gen. Virol.* **68,** 1233–1250.
22. Livingstone, C. and Jones, I. (1989) Baculovirus expression vectors with single strand capability. *Nucleic Acids Res.* **17,** 2366.
23. Kitts, P. A. and Possee, R. D. (1993) A method for producing recombinant baculovirus expression vectors at high frequency. *Biotechniques* **14,** 810–817.
24. Weyer, U., Knight, S., and Possee, R. D. (1990) Analysis of very late gene expression by *Autographa californica* nuclear polyhedrosis virus and the further development of multiple expression vectors. *J. Gen. Virol.* **71,** 1525–1534.
25. Hill-Perkins, M. S. and Possee, R. D. (1990) A baculovirus expression vector derived from the basic protein promoter of *Autographa californica* nuclear polyhedrosis virus. *J. Gen. Virol.* **71,** 971–976.
26. Belyaev, A. S. and Roy, P. (1993) Development of baculovirus triple and quadruple expression vectors. Co-expression of three or four bluetongue virus proteins and the synthesis of bluetongue virus-like particles in insect cells. *Nucleic Acids Res.* **21,** 1219–1223.
27. Davies, A. H., Jowett, J. B. M., and Jones, I. M. (1993) Recombinant baculovirus vectors expressing glutathione-S-transferase fusion proteins. *Biotechnology* **11,** 933–936.
28. Wang, Y. H., Davies, A. H., and Jones, I. M. (1995) Expression and purification of Glutathione-S-transferase tagged HIV-1 gp120: no evidence of an interaction with CD26. *Virology* **208,** 142–146.
29. Boublik, Y., Di Bonito, P., and Jones, I. M. (1995) Eukaryotic virus display: engineering the major surface glycoprotein of the *Autographa californica* nuclear polyhedrosis virus (AcNPV) for the presentation of foreign proteins on the virus surface. *Biotechnology* **13,** 1079–1084.

111

Procedures for the Analysis and Purification of His-Tagged Proteins

Richard E. Kneusel, Joanne Crowe, Melanie Wulbeck, and Joachim Ribbe

1. Introduction

The heterologous expression of recombinant proteins is a valuable tool in the study of gene expression, and has resulted in the development of many systems to express and purify hybrid proteins. Most of these systems are based on the fusion of the protein of interest with a naturally occurring protein (glutathione *S*-transferase, maltose binding protein, or protein A) and using their natural affinity to substrates (glutathione, amylose, or immunoglobulins) coupled to columns in the purification step. Among the main drawbacks with these systems are that the affinity tag may affect protein structure and function, and that it is not possible to purify insoluble proteins.

One very effective method for purifying recombinant fusion proteins utilizes immobilized metal ion-affinity chromatography (IMAC), which takes advantage of the adsorption of polyhistidine (6xHis) containing proteins and peptides to a nickel-chelate-nitrilotriacetic acid (Ni-NTA) column. This highly selective interaction allows purification of tagged proteins or protein complexes from <1% to >95% homogeneity in just one step *(1,2)*. The tight association between the tag and the Ni-NTA resin allows contaminants to be easily washed away under stringent conditions, yet the bound proteins can be gently eluted by competition with imidazole or by a slight reduction in pH. Because the interaction is independent of the tertiary structure of the tag, 6xHis-tagged proteins can be purified even under the strongly denaturing conditions required to solubilize many membrane proteins and inclusion bodies.

The affinity tag for purification on the Ni-NTA resin consists of just six consecutive histidine residues. It is very poorly immunogenic or nonimmunogenic in all species, except some monkeys and, because it is uncharged at physiological pH, rarely affects tertiary structure. The 6xHis tag has seldom interfered with the structure or function of the purified protein. Comparative X-ray crystallographic investigations indicate that 6xHis-tagged proteins have the same structure as the nontagged protein *(3)*.

The advantages of 6xHis/Ni-NTA purification have been combined with a high level bacterial expression system to create an elegant, yet simple strategy, allowing protein purification whether the expressed protein is at low or high levels, denatured, or associated

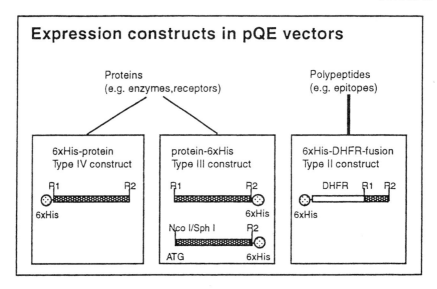

Fig. 1. Expression constructs in pQE vectors.

with other proteins, DNA, or RNA. It is currently used in a wide variety of applications, ranging from the large scale purification of proteins for antibody production, to the purification of antibodies, subunits, and substrates through their interactions with the tagged proteins.

1.1. Vectors

pQE expression vectors allow high-level expression of proteins or peptides with a 6xHis affinity tag. The tag may be placed at the N-terminus of the protein to create a Type IV construct, at the C-terminus of the protein to create a Type III construct, or at the C-terminus of a protein utilizing its original ATG start codon to create a Type ATG construct (pQE-60) (**Fig. 1**). If small peptides are to be expressed, they can be fused to mouse dehydrofolate reductase (DHFR) to create a Type II construct. The poorly immunogenic DHFR stabilizes the peptide during expression and enhances its antigenicity.

The pQE plasmids were derived from plasmids pDS56/RBSII and pDS781/RBSII-DHFRS (1). They contain the following elements as shown for two typical vectors pQE-30 (Type IV) and pQE-40 (Type II) (**Fig. 2**):

1. An optimized, regulatable promoter/operator element N250PSN250P29, consisting of the *Escherichia coli* phage T5 promoter, recognized by the *E. coli* RNA polymerase containing two *lac* operator sequences for tight regulation.
2. A synthetic ribosome binding site, RBSII, designed for optimal recognition and binding.
3. Optimized 6xHis affinity tag coding sequence.
4. The mouse DHFR coding sequence (in some vectors only).
5. A multicloning site (available in all reading frames).
6. Translation stop codons in all reading frames.
7. The transcriptional terminator "t0" from phage λ.
8. The nontranslated open reading frame for chloramphenicol acetyltransferase.
9. The transcriptional terminator t1 of the *E. coli* rrnB operon.
10. The replication region and the gene for b-lactamase of plasmid pBR322.

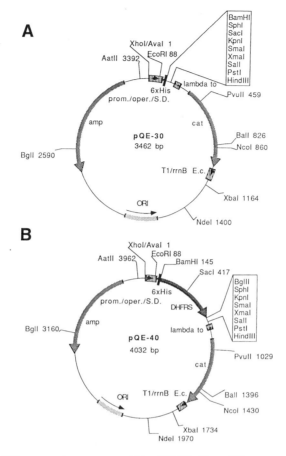

Fig. 2. Typical pQE expression vectors. (**A**) pQE-30 (Type IV construct). The polycloning region is directly 3' to the 6xHis tag sequence. (**B**) pQE-40 (Type II construct) contains a DHFRS sequence between the 6xHis tag and the polycloning region. DHFRS stabilizes short protein sequences.

1.2. Host Strains

The *E. coli* host cells M15[pREP4] and SG13009[pREP4] contain multiple copies of the plasmid pREP4. This carries the gene for neomycin phosphotransferase (*neo*) conferring kanamycin resistance, and the *lac*I gene encoding the *lac* repressor. The multiple copies of pREP4 present in the host cells ensure high levels of *lac* repressor and tight regulation of protein expression. The plasmid is maintained in *E. coli* cells in the presence of kanamycin at a concentration of 25 µg/mL *(1)*. Expression from pQE vectors is rapidly induced by the addition of IPTG, which inactivates the repressor. The level of IPTG used for induction can be varied to control the level of expression.

The small size of the 6xHis tag makes it ideally suited for inclusion in a variety of other expression systems. It works well in prokaryotic, mammalian, yeast, baculovirus, and other eukaryotic systems. The six histidine residues can be easily inserted into the expression construct, at the C- or N-terminus of the protein, by PCR, mutagenesis, or ligation of a small synthetic fragment.

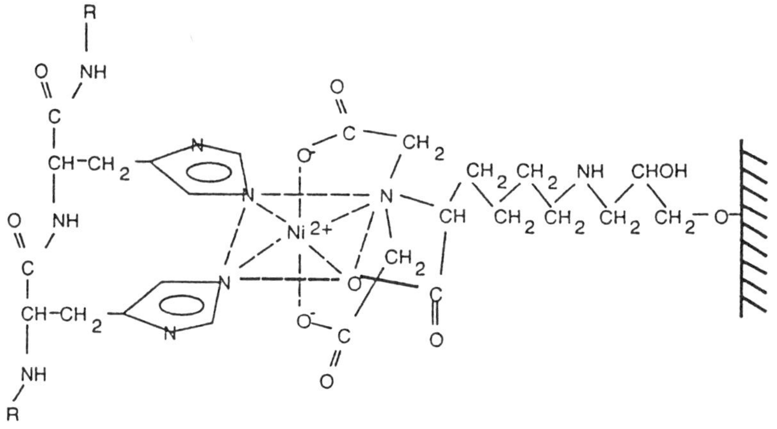

Fig. 3. Model for the binding of neighboring 6xHis residues to Ni-NTA resin.

1.3. Resin

Immobilized metal chelate affinity chromatography was first used to purify proteins in 1975 *(4)* and has become a widely used technique owing to its efficiency and ease of use. The chelating ligand iminodiacetic acid (IDA) can be charged with metal ions such as Zn^{2+} and Ni^{2+}, and then used to purify a variety of different proteins and peptides that could bind to the immobilized ions *(5)*. IDA, however, has only three chelating sites and does not effectively bind metal ions with six coordination sites. The ions are washed out of the resin on loading with chelating proteins and peptides or during the washing steps, which results in low binding capacity, low yields, and impure products.

NTA is a novel chelating adsorbent developed that overcomes these problems (**Fig. 3**). NTA binds the metal ions more stably by occupying four of the ligand binding sites in the coordination sphere of the Ni^{2+} ion and leaving two sites free to interact with the 6xHis tag *(6)*. As a result, Ni-NTA resin binds proteins 100–1000 times more tightly than Ni-IDA, allowing the purification of proteins constituting <1% of total cellular protein to >95% homogeneity in just one step *(2)*.

This chapter describes a protein miniprep procedure to enable the investigator to confirm correct protein expression, and protocols for bulk purification of proteins from *E. coli* under both nondenaturing and denaturing conditions. Although each procedure works very well for most proteins, some modifications may be necessary if host systems other than *E. coli* are used. The purification power of the 6xHis Ni-NTA system will be enhanced if the conditions are optimized for each individual protein.

2. Materials

1. The pQE-vectors, *E. coli* host strains, Ni-NTA resin, and Ni-NTA spin columns are available exclusively from Qiagen GmbH (Hilden, Germany); Qiagen Inc. (Chatsworth CA); Qiagen Ltd. (Dorking, UK); Qiagen AG (Basel, Switzerland); and their distributors.
2. Culture media: Use LB-medium and its modifications, 2X YT or Super Broth, containing 100 µg/mL ampicillin and 25 µg/mL kanamycin for growth of M15 cells containing pQE expression and pREP4 repressor plasmids (*see* **Note 1**): LB medium: 10 g/L bacto-tryptone, 5 g/L bacto-yeast extract, and 5 g/L NaCl. 2X TY medium: 16 g/L bacto-tryptone,

Procedures for His-Tagged Proteins

10 g/L bacto-yeast extract, and 5 g/L NaCl. Super medium: 25 g/L bacto-tryptone, 15 g/L bacto-yeast extract, and 5 g/L of NaCl.
3. IPTG: Prepare a stock solution with a concentration of 1 M IPTG.
4. Buffer A: 6 M guanidinium-HCl, 0.1 M NaH$_2$PO$_4$, 0.01 M Tris-HCl. Adjust the pH to 8.0 with NaOH.
5. Buffer B: 8 M urea, 0.1 M NaH$_2$PO$_4$, 0.01 M Tris-HCl, pH adjusted to 8.0 with NaOH. Owing to the dissociation of urea, the pH must be adjusted immediately before use.
6. Buffer C: Same composition as buffer B, but pH adjusted to 6.3 with HCl. Owing to the dissociation of urea, the pH must be adjusted immediately before use.
7. Buffer D: Same composition as buffer B, but pH adjusted to 5.9 with HCl. Owing to the dissociation of urea, the pH must be adjusted immediately before use.
8. Buffer E: Same composition as buffer B, but pH adjusted to 4.5 with HCl. Owing to the dissociation of urea, the pH must be adjusted immediately before use.
9. 5X SDS-PAGE sample buffer: 15% β-mercaptoethanol, 15% SDS, 1.5% bromophenol blue, and 50% glycerol.
10. 12.5% Polyacrylamide gels containing 0.2% SDS *(7)*.
11. Sonication buffer 1: 50 mM NaH$_2$PO$_4$, 300 mM NaCl, 1 mM PMSF, pH 8.0.
12. Sonication buffer 2: 50 mM NaH$_2$PO$_4$, 300 mM NaCl, 20 mM imidazole, 1 mM PMSF, pH 8.0.
13. Lysozyme: stock solution 10 mg/mL.
14. RNase: stock solution 200 mg/mL.
15. DNase: stock solution 60 mg/mL.
16. Wash buffer: 50 mM NaH$_2$PO$_4$, 300 mM NaCl, 20 mM imidazole, 1 mM PMSF, pH 8.0.
17. Elution buffer: 50 mM NaH$_2$PO$_4$, 300 mM NaCl, 250 mM imidazole, pH 8.0.

3. Methods

3.1. Rapid Screening of Miniexpression Cultures

The following is a basic protocol for screening small expression cultures by purifying 6xHis-tagged proteins on Ni-NTA spin columns. Purification is performed under denaturing conditions, allowing the isolation of any tagged protein, independent of its solubility within the cell. Denaturing the 6xHis protein completely exposes the 6xHis tag, leading to optimized binding and higher yields compared to the purification under native conditions.

Lysing cells in buffer B allows usually the solubilization of most proteins and inclusion bodies, and the lysate to be analyzed directly by SDS-PAGE. For solubilization of very hydrophobic receptor or membrane proteins, buffer A containing GuHCl may have to be used, sometimes also in combination with detergents. *See* **Note 2** for the treatment of samples in buffer A prior to SDS-PAGE.

Some proteins may be subject to degradation during cell harvest, lysis, or even during growth after induction. Addition of PMSF (0.1–1 mM) or other protease inhibitors is recommended. PMSF treatment during cell growth may result, however, in reduced expression levels.

1. Inoculate 10 mL of LB broth containing 100 µg/mL ampicillin and 25 µg/mL kanamycin with a fresh colony of M15[pREP4] containing the pQE expression plasmid. Grow at 37°C overnight.
2. Dilute the noninduced overnight culture 1:50 with fresh LB broth containing 100 µg/mL ampicillin and 25 µg/mL kanamycin. Grow at 37°C with vigorous shaking until the A$_{600}$ reaches 0.7–0.9 (*see* **Note 3**).

3. Add IPTG to a final concentration of 1 m*M*, and grow the culture at 37°C for four more hours (*see* **Note 4**).
4. Harvest the cells by centrifugation for 10 min at 4000*g*, and discard supernatants.
5. Resuspend cells in 1/10 vol of buffer B. Lyse cells by gently vortexing, taking care to avoid frothing, or stir cells for 1 h at room temperature (*see* **Note 5**).
6. Centrifuge the lysate at 10,000*g* for 10 min at room temperature to remove the cellular debris, and transfer the supernatant to a fresh tube. Save 20 µL of lysate for SDS-PAGE analysis (*see* **Note 6**).
7. Pre-equilibrate an Ni-NTA spin column with 600 µL of buffer B. Centrifuge for 2 min at 2000 rpm (*see* **Note 7**).
8. Load up to 600 µL of the cleared lysate supernatant containing the 6xHis-tagged protein onto the pre-equilibrated Ni-NTA spin column. Centrifuge for 2 min at 2000 rpm. Save flow through for SDS-PAGE analysis (*see* **Note 7**).
9. Wash the Ni-NTA spin column 2X with 600 µL buffer C. Centrifuge for 2 min at 700*g*. Save the flowthrough for SDS-PAGE analysis (*see* **Note 8**).
10. Elute the protein with 2X 200 µL buffer E. Centrifuge for 2 min at 700*g* and collect the eluates in separate tubes (*see* **Notes 9** and **10**).
11. Take 20 µL samples of all fractions and add 5 µL of 5X PAGE sample buffer. Heat for 7 min at 95°C. Analyze samples on a 12.5% polyacrylamide gel and visualize proteins by staining with Coomassie blue (*see* **Note 2**).

3.2. Native Purification Protocol: Ni-NTA Agarose With and Without 20 mM Imidazole in the Cell Lysis Buffer

This protocol is for use with Ni-NTA agarose to purify up to 2.5 mg of soluble 6xHis-tagged protein. Before purifying proteins under nondenaturing conditions, it is important to determine how much of the protein is soluble in the cytoplasm, and how much is in insoluble precipitates or inclusion bodies. Therefore, parallel purification under denaturing conditions is recommended.

The amount of purified protein will depend on the expression level. This protocol is designed for an expression level from approx 10 mg/L. The flow rate during loading, washing, and elution should not exceed 1–2 mL/min.

1. Grow and induce a 60-mL culture (**NOTE**: 125 mL culture is equivalent to 600 µg 6xHis-tagged protein).
2. Harvest the cells by centrifugation at 4000*g* for 10 min. Resuspend the pellet in 2 mL sonication buffer 1 (no imidazole) or 2 (including 20 m*M* imidazole). Freeze sample in dry ice/ethanol (or overnight at –20°C), and thaw in cold water. Alternatively, add lysozyme to 1 mg/mL, and incubate on ice for 30 min. This treatment leads to the most efficient cell lysis.
3. Sonicate on ice (1 min bursts, 1 min cooling, 200–300 W). Monitor lysis by measuring the amount of released nucleic acids (A_{260}). Lysis is complete when A_{260} remains constant.
4. If the lysate is very viscous, add RNase A to 10 µg/mL and DNase I to 5 µg/mL, and incubate on ice for 10–15 min. Alternatively, draw the lysate through a narrow-gage syringe needle several times. Centrifuge at >10,000*g* for 20 min at 4°C, and collect the supernatant. Save 20 µL for SDS-PAGE.
5. Add 1 mL of a 50% slurry of Ni-NTA agarose, previously equilibrated in sonication buffer 1 or 2, and stir at 4°C for 60 min (*see* **Note 11**).

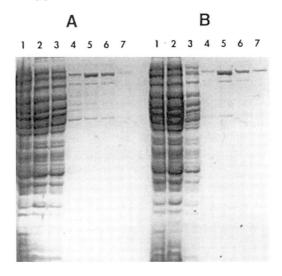

Fig. 4. Purification of 6xHis-tagged 85-kDa protein. The protein was purified using the Ni-NTA spin procedure (3.2) under native conditions with (A) and without (B) imidazole in the binding buffer, 5 μL cell lysate (lane 1), flowthrough (lane 2), first wash (lane 3), second wash (lane 4), and first-third eluates (lanes 5–7) were separated on a 10% denaturing gel. The gel was stained with Coomassie.

6. Load the lysate and Ni-NTA agarose into a column and collect the column flowthrough for SDS-PAGE. Work at 4°C if possible (*see* **Note 12**).
7. Wash with 5 mL sonication buffer, or until the A_{280} of the flowthrough is below 0.01 (*see* **Note 13**). Collect wash fractions for SDS-PAGE.
8. Wash with 2X 5 mL wash buffer, or until the flow through A_{280} is below 0.01 (*see* **Note 13**).
9. Elute the protein with 3 x 500 μL elution buffer (*see* **Note 14**). Collect 500-μL fractions, and analyze 5-μL samples on SDS-PAGE (**Fig. 4**).

3.3. Denaturing Purification of Insoluble Proteins

This protocol is for use with Ni-NTA agarose to purify up to 5–10 mg of 6xHis-tagged protein. Purification under denaturing conditions is often more efficient than purification under native conditions and is essential when proteins cannot be solubilized without denaturation. The amount of purified protein will depend on the expression level. This protocol is designed for an expression level ca. 10 mg/L. For optimal results, the protocol should be scaled up or down according to the expression level of the 6xHis-tagged protein. The flow rate during loading washing and elution should not exceed 1 mL/min.

1. Grow and induce a 500-mL culture as described in **Subheading 3.1.** Harvest the cells by centrifugation at 4000*g* for 10 min. Store at –70°C if desired.
2. Thaw cells for 15 min and resuspend in buffer B at 5 mL/g wet wt (*see* **Note 15**). Stir cells for 1 h at room temperature. Centrifuge lysate at 10,000*g* for 15 min at room temperature. Collect supernatant, and save 20 μL for SDS-PAGE.
3. Add 2 mL of a 50% slurry of Ni-NTA agarose, previously equilibrated in buffer B. Stir at room temperature for 45 min, and then load resin carefully into a column. Collect flowthrough for SDS-PAGE (*see* **Note 6**).

4. Wash with 10 mL buffer B. If necessary, wash further until the flowthrough A_{280} is below 0.01.
5. Wash with 6 mL buffer C, or until the flowthrough A_{280} is below 0.01.
6. Elute the protein with 10 mL of buffer E (*see* **Note 10**). Collect 500-µL fractions, and analyze by SDS-PAGE (*see* **Notes 2, 13**, and **16**).

3.4. Protein Refolding

Heterologous expression of proteins in *E. coli* often leads to the formation of insoluble aggregates, i.e., inclusion bodies, that can only be solubilized by treatment with high concentrations of a denaturant (6 *M* guanidine-HCl or 8 *M* urea). Solubilization is often the only course for initial analysis, but protein function is often disrupted making activity tests impossible. The sole viable alternative for subsequent analysis is renaturing or folding the protein to recover its unique three-dimensional conformation.

Proteins fold solely as a result of thermodynamic drive, which leads to the formation of specific ionic and hydrophobic interactions characteristic of the native state, i.e., its native state is more stable than the unfolded state.

There are certain factors that must be taken into account in order to renature proteins after solubilization: the native functional state should be stabilized, barriers that block the folding pathway should be minimized, and intermolecular aggregation should be limited.

3.4.1. Recommendations for Renaturation

Since individual proteins behave differently on renaturation, refolding conditions must be determined empirically. The following are a few recommendations that may be helpful in designing refolding experiments. Detailed examples of refolding procedures are available in Wingfield et al. *(8)*.

1. Maintain low protein concentration (0.5–1 mg/mL).
2. Include thiol reagents: Disulfide bonds contribute to the stability of the native conformation of many proteins. Secondary structure formation may be so favorable that the correct cysteine residues spontaneously form the pair, i.e., little oxidizing pressure is required. If, however, this is not the case, stronger oxidizing pressure may be required. A redox pair of reduced glutathione (GSH) and oxidized glutathione (GSSH) creates the necessary oxidizing potential to make and break disulfide bonds in folding intermediates, thereby allowing the optimal, native conformation to be reached. A GSH:GSSG ratio of 10:1 (at a concentration of 2–5 m*M* GSH) has been shown to be effective for a number of proteins.
3. Remove denaturants slowly by diluting or dialysis: Glycine often has remarkable solubilization properties (50 m*M*, pH 9.0, 5 m*M* EDTA). If guanidine-HCl must be used as a denaturant, urea should be included in the renaturation buffer (2 *M*) to stabilize the protein on refolding. Very low concentrations of detergents may also be included (0.1–0.5% NP-40 or 0.005% [v/v] Tween-20).
4. Include cosolvents: Many cosolvents, such as glycerol and ethylene glycol, as well as glucose and sucrose can stabilize proteins. Certain anions (e.g., phosphate and sulfate) and cations (e.g., MES, HEPES) also have positive effects. They exhibit little or no effect on the folding rate constant, but decrease the folding rate constant. They act by stabilizing hydrophobic interactions, but care should be taken because they can also stabilize aggregate formation.

- 5–20% Glycerol;
- 10% Sucrose;
- 100 mM KCl or 150–500 mM NaCl, 2 mM MgCl$_2$; and
- 10-50 mM HEPES-KOH (pH 7.5).

5. Suppress proteolytic degradation with protease inhibitors: 0.5 mM PMSF, 5–100 μg/mL aprotinin (0.1% aprotinin), 2 μg/mL pepstatin, or 2–5 μg/mL leupeptin.
6. Many proteins that are insoluble when refolded in solution can be successfully refolded while immobilized on the Ni-NTA column. Immobilizing one end of the protein during renaturation appears to prevent aggregate formation. Renature using a linear 6 M–1 M urea gradient in 500 mM NaCl, 20% glycerol, Tris-HCl, pH 7.4, containing protease inhibitors. The gradient should be FPLC-mediated to ensure linearity (conventional gradient makers often make nonlinear gradients), and the renaturation should take place over a period of 1.5 h. After renaturation the proteins can be eluted by the addition of 250 mM imidazole. *See also* Holzinger et al. *(9)*.

4. Notes

1. We suggest that expression should be tried in all three media in parallel, and a time-course of expression should be taken after induction. There are often striking differences noted between the level of expression in different media at different times.
2. On minigels, it is usually sufficient to analyze 5-μL samples of each fraction in an equal volume of SDS-PAGE loading buffer, with or without 3% β-mercaptoethanol. Since the fractions that contain GuHCl will precipitate with SDS, they must either be diluted (1:6), dialyzed before analysis, or separated from the guanidinium hydrochloride by TCA precipitation: Dilute samples to 100 μL; add an equal vol of 10% TCA, leave on ice for 20 min, spin for 15 min in a microfuge, wash pellet with 100 μL of ice cold ethanol, dry, and resuspend in sample buffer. If there is any guanidinium hydrochloride present, samples must be loaded immediately after boiling for 7 min at 95°C.
3. The required volume of expression culture is mainly determined by the expression level, cellular location of the protein, and purification conditions. For purification of poorly expressed proteins, the minimum cell-culture volume should be 30 mL. A 50X concentrated cell lysate should be loaded onto the Ni-NTA spin column to increase the amount of 6xHis-tagged protein, the viscosity, and therefore, the yield. For proteins that are expressed at very high levels (>10 mg/L, i.e., equivalent to an expression level of approx 8% of total cellular protein), the cell lysate should be no more than 10–25X concentrated. At an expression level of 10 mg/L, 600 μL of the 10X concentrated cell lysate in buffer B contains approx 60 μg of 6xHis-tagged protein. For lower expression levels (2–5 mg/L), 25X concentrated cell lysates (600 μL cell lysate = 30–75 μg) should be loaded onto the Ni-NTA spin column. For expression levels lower than 1 mg/L, the cell lysate should be concentrated 50–100X.
4. For proteins that are very sensitive to protein degradation, the induction time should be reduced, and a time-course of expression should be determined. In some cases, addition of 0.1–1 mM PMSF after induction is recommended to inhibit PMSF-sensitive proteases. PMSF treatment can result, however, in reduced expression levels.
5. The solution should become translucent when lysis is complete. It is preferable to lyse the cells in buffer B, so that the cell lysate can be analyzed directly by SDS-PAGE. If the cells or the protein do not solubilize in buffer B, then buffer A must be used. *See* **Note 20** for treatment of samples in buffer A prior to SDS-PAGE.
6. The supernatant samples will monitor the expression level of any proteins that have not bound to the resin.

7. It is important not to exceed 2000 rpm (approx 700g) when centrifuging Ni-NTA spin columns. At higher speeds, NTA silica particles become compressed, leading to high flow rates (channeling) and inefficient binding. For low binding kinetics, especially when nondenaturing conditions are used, reloading of the column flowthrough is recommended.
8. Wash the Ni-NTA spin column with buffer C even if buffer A was used initially to solubilize the protein. Most proteins will remain soluble in buffer C. It may not be necessary to repeat the buffer C wash. The number of wash steps required to obtain highly pure protein is determined primarily by the expression level of the 6xHis-tagged protein. When the expression level is high, two wash step are usually sufficient for removal of contaminants. For very low expression levels or highly concentrated lysates, three wash steps may be required to achieve high purity.
9. Most of the 6xHis-tagged protein (>80%) should elute in the first 200 µL eluate, particularly when proteins smaller than 30 kDa are purified. The remainder will elute in the second 200 µL. If dilution of the protein is undesirable, do not combine the eluates or, alternatively, elute in 100–150 µL aliquots in order to increase the protein concentration.
10. Elution can be performed by lowering the pH value of the buffer, either as a continuous or step gradient decreasing from pH 8.0–4.5. Most proteins will be efficiently eluted by wash buffer at pH 4.5; many (particularly monomers) can be eluted at a higher pH.
11. Proteins may be purified on NiNTA resin in either a batch or a column procedure. The batch procedure entails binding the protein to the Ni-NTA resin in solution, and then packing the protein/resin complex into a column for the washing and elution steps. Especially under native conditions, the batch procedure promotes more efficient binding and reduces the amount of debris that is loaded onto the column by decanting the supernatant. In the column procedure, the NiNTA column is packed and washed, and the cell lysate is applied slowly to the column.
12. The purification is performed in 8 M urea, since 6 M guanidinium hydrochloride precipitates in the presence of SDS, making SDS-PAGE analysis of samples difficult. Otherwise, both urea and GuHCl or combinations thereof can be used throughout the whole purification procedure.
13. Alternative elution procedures may be used. Monomers can usually be eluted in buffer D, whereas multimers, aggregates, and proteins with two 6xHis tags will generally elute in buffer E. Elution can also be carried out using a pH 6.5–4.0 gradient in 8 M urea, 0.1 M NaH_2PO_4, 10 mM Tris-HCl. If elution at a higher pH is desired, most proteins can be eluted with buffer C containing 100 mM imidazole or at 100–250 mM imidazole at a constant pH of 7.5.
14. When possible, monitor elution photometrically and collect pools rather than fractions. Discolored or impure reagents may affect optical density readings, and imidazole will absorb light at 280 nm.
15. Do not boil a sample that contains imidazole before SDS-PAGE, since it will hydrolyze acid-labile bonds. Heat the sample for a few minutes at 37°C immediately before loading the gel.
16. If the 6xHis-tagged protein does not bind under these conditions, the concentration of imidazole in the sonication buffer should be reduced to 1–5 mM (*also see* **Notes 17** and **18** for information about "hidden" tags).
17. Many proteins remain soluble during expression and can be purified in their native form under nondenaturing conditions on Ni-NTA resin; others, however, form insoluble precipitates. Since almost all of these proteins are soluble in 6 M guanidinium hydrochloride, NiNTA chromatography and the 6xHis tag provide a universal system for the purification of recombinant proteins. The decision whether to purify the tagged proteins under denatur-

ing or nondenaturing conditions depends on both the solubility and location of the protein, and the accessibility of the 6xHis tag. Proteins that remain soluble in the cytoplasm or are secreted into the periplasmic space can generally be purified under nondenaturing conditions (but note the exception below). If the protein is insoluble or located in inclusion bodies, then it must generally be solubilized by denaturation before it can be purified. Some proteins, however, may be solubilized by the addition of detergents, and it is worth experimenting with different solubilization techniques if it is important to retain the native configuration of the protein. Many proteins that form inclusion bodies are also present at some level in the cytoplasm, and may be efficiently purified in their native form, even at very low levels, on NiNTA resin.

In rare cases, the 6xHis tag is hidden by the tertiary structure of the native protein, so that soluble proteins require denaturation before they can be bound to Ni-NTA resin. If denaturation of the protein is undesirable, the problem can usually be solved by moving the tag to the opposite terminus of the protein. Proteins that have been purified under denaturing conditions can either be used directly or refolded in dilute solution by dialyzing in the presence of reduced and oxidized glutathione. It is also possible to renature proteins on the Ni-NTA column.

18. In the case of insufficient binding of 6xHis to Ni-NTA resin, check for the presence of chelating agents (EDTA/EGTA) and high concentrations of electron-donating groups (NH_4) or components like glycine, histamine, and metals, like zinc, and repeat the binding step. In some cases, for instance when the protein is purified from the medium, it may be necessary to dialyze before binding. If the 6xHis tag is hidden in the native protein structure, improve the exposure by adding small concentrations of urea or detergents to the nondenaturing sample preparation buffer. Slower binding kinetics can be compensated by longer contact times with NTA, preferably under batch binding conditions. Alternatively, try 6xHis at the opposite terminus, or use completely denaturing conditions (buffer A with 10 mM β-mercaptoethanol). Avoid any Ni^{2+} complexing reagents.

19. Background contamination arises from proteins that contain neighboring histidine residues, and thus have some affinity for the resin. Also responsible for contamination are proteins that copurify because they are linked to the 6xHis-tagged protein by disulfide bonds, proteins that associate nonspecifically with the tagged protein, and nucleic acids that associate with the tagged protein. All of these contaminants can be easily removed by washing the resin under the appropriate conditions. Proteins that contain neighboring histidines in the primary structure are almost not common in bacteria, but are quite abundant in mammalian cells. These proteins bind to the resin much more weakly than proteins with a 6xHis tag, and can be easily washed away, even when they are much more abundant than the tagged protein *(2)*. (*See* **Note 4** for additional information.) The addition of 10–20 mM β-mercaptoethanol to the loading buffer will reduce background owing to crosslinked proteins. Do not use more than 1 mM DTT; higher concentrations may reduce the Ni^{2+} ions.

Proteins that are associated with the tagged protein or the resin because of nonspecific interactions and nucleic acids can be removed by washing with low levels of detergent (up to 2% Triton X-100 or 0.5% sarcosyl), increasing the salt concentration up to 1 M NaCl, or including a small amount of ethanol or 30% glycerol to reduce hydrophobic interactions. The optimum levels of any of these reagents should be determined empirically for different proteins.

20. Removal of background proteins and elution of tagged proteins from the column may be achieved by either lowering the pH in order to protonate the histidine residues or by the addition of imidazole, which competes with tagged proteins for binding sites on the

NiNTA resin. Although both methods are equally effective, the imidazole method is somewhat milder and is recommended in cases where the protein would be damaged by a reduction in pH (e.g., tetrameric aldolase) *(10)*.

In bacterial expression systems, it is seldom necessary to wash the bound protein under very stringent conditions, since proteins are expressed to high levels and the background is low. In mammalian systems, however, or under native conditions where many more neighboring histidine residues will be exposed to the resin, it may be necessary to increase the stringency of the washing considerably. This can be done by gradually decreasing the pH of the wash buffer, or by slowly increasing the concentration of imidazole. The pH or imidazole concentration that can be tolerated before elution begins will vary slightly for each protein.

If the concentration of the tagged protein is very low and the background is likely to be high (such as in mammalian expression systems), the 6xHis-tagged protein should be incubated with the resin under conditions that would minimize the degree to which the background proteins compete for the binding sites, i.e., at a slightly lower pH or in the presence of low levels of imidazole. Likewise, the purification process will be optimized if the amount of tagged protein is closely matched to the capacity of the resin used, since the 6xHis-tagged protein has a higher affinity for the NiNTA resin than the background proteins, it can occupy all available binding sites, and very few background proteins will be retained on the resin.

21. Do not use strong reducing agents, such as DTT or DTE. They will reduce the Ni^{2+} ions and cause them to elute from the resin. In most situations, β-mercaptoethanol can be used at concentrations up to 20 mM, but even these small amounts may cause problems if the protein has a strongly reducing nature. Use any reducing agent with care, and if in doubt, test it with a small amount of Ni-NTA resin. Strong chelating agents will chelate the Ni and also cause it to elute from the NTA resin. Care should be taken when using EDTA, EGTA, or any other chelating agents, (**Note**: There are many examples where 1 mM EDTA has been used in the buffers without negative effects.)
22. The composition of the sonication, wash, and elution buffers can be modified to suit the particular application, e.g., by adding low levels of imidazole, 1–2% Tween, 5–20 mM β-mercaptoethanol, 1 mM PMSF, or increased NaCl or glycerol concentrations.
23. If expression levels are too high, one or more of the following may help. Reduce the expression level dramatically (10–15 times) by lowering the IPTG concentration to 0.01 mM. At 0.1 mM IPTG, the transcription rate remains virtually unaffected. Alternatively, the induction time and/or the temperature can be reduced, or cultures of higher cell densities (0.8 A_{600}) can be induced.
24. In case of precipitation during purification, check for aggregates of purified proteins and try Tween or Triton additives (up to 2%), adjust to 10–20 mM β-mercaptoethanol, and check for stabilizing cofactor requirements (e.g., Mg^{2+}). Make sure that the salt concentration is at least 300 mM NaCl. Check room temperature (>20°C) for the denaturing protocol.
25. Background binding can be suppressed by adjusting the amount of Ni-NTA resin according to the 6xHis protein expression level. The binding capacity should not exceed the amount of tagged protein by more than a factor of 2.
26. Do not determine the size of the recombinant product by SDS gel. Adding or even replacing amino acids can shift protein bands, suggesting a molecular weight several kilodaltons different than expected.
27. *See* **refs.** *11–20* for further reading.

References

1. Stüber, D., Matile, H., and Garotta, G. (1990) System for high-level production in *Escherichia coli* and rapid purification of recombinant proteins: application to epitope mapping, preparation of antibodies, and structure-function analysis, in *Immunological Methods*, vol. IV (Lefkovits, I. and Pernis, B., eds.), Academic, New York, pp. 121–152.
2. Janknecht, R., de Martynoff, G., Lou, J., Hipskind, R. A., Nordheim, A., and Stunnenberg, H. G. (1991) Rapid and efficient purification of native histidine-tagged protein expressed by recombinant vaccinia virus. *Proc. Natl. Acad. Sci. USA* **88,** 8972–8976.
3. Lindner, P., Guth, B., Wülfing, C., Krebber, C., Steipe, B., Müller, F., and Plückthun, A. (1992) Purification of native proteins form the cytoplasm and periplasm of *Escherichia coli* using IMAC and histidine tails: A comparison of proteins and protocols. *Methods: A Companion to Methods in Enzymol.* **4(2),** 41–55.
4. Porath, J., Carlsson, J., Olsson, I., and Belfrage, G. (1975) Metal chelate affinity chromatography, a new approach to protein fractionation. *Nature* **258,** 598,599.
5. Sulkowski, E. (1985) Purification of proteins by IMAC. *Trends Biotechnol.* **3,** 17.
6. Hochuli, E., Dobeli, H., and Schacher, A. (1987) New metal chelate adsorbent selective for proteins and peptides containing neighboring histidine residues. *J. Chromatog.* **411,** 177–184.
7. Takacs, B. J. (1979) *Immunological Methods*, vol. 1 (Lefkovits, I. and Pernis, B., eds.), Academic, New York, p. 81.
8. Wingfield, P. T., Palmer, I., and Liang, S.-M. (1995) Folding and purification of insoluble (inclusion-body) proteins from *Escherichia coli*, in *Current Protocols in Protein Science*, vol. 1 (Coligan, J. E., Dunn, B. M., Ploegh, H. L., Speicher, D. W., and Wingfield, P. T. ed. board), Wiley, New York, pp. 6.5.1–6.5.27.
9. Holzinger, A., Phillips, K. S., and Weaver, T. E. (1996) Single-step purification/solubilization of recombinant proteins: application to surfactant protein B. *Biotechniques* **20,** 804–808.
10. Dobeli, H., Trecziak, A., Gillessen, D., Matile, H., Srivastava, I. K., Perrin, L. H., Jakob, P. E., and Certa, U. (1990) Expression, purification, biochemical characterization and inhibition of recombinant Plasmodium falciparum aldolase. *Mol. Biochem. Parasitol.* **41,** 259–268.
11. Hochuli, E. (1990) Purification of recombinant proteins with metal chelate adsorbent, in *Genetic Engineering*, vol. 12 (Setlow, J. K., ed.), Plenum, New York, pp. 87–98.
12. Abate, C., Luk, D., Gentz, R., Rauscher III, F. J., and Curran, T. (1990) Expression and purification of the leucine zipper and DNA binding domains of Fos and Jun: both Fos and Jun contact DNA directly. *Proc. Natl. Acad. Sci. USA* **87,** 1032–1036.
13. Bush, G. L., Tassin, A., Friden, H., and Meyer, D. I. (1991) Purification of a translocation-competent secretory protein precursor using nickel ion affinity chromatography. *J. Biol. Chem.* **266,** 13,811–13,814.
14. Gentz, R., Certa, U., Takacs, B. J., Matile, H., Dobeli, H., Pink, R., Mackay, M., Bone, N., and Scaife, J. G. (1988) Major surface antigen pl90 of Plasmodium falciparum: detection of common epitopes present in a variety of plasmodia isolates. *EMBO J.* **7,** 225–230.
15. Gentz, R., Chen, C., and Rosen, C. A. (1989) Bioassay for trans-activation using purified immunodeficiency virus tat-encoded protein: trans-activation requires mRNA synthesis. *Proc. Natl. Acad. Sci. USA* **86,** 821–824.
16. Gu, J., Stephenson, C. G., and Iadarola, M. J. (1994) Recombinant proteins attached to a Ni-NTA column: use in affinity purification of antibodies. *Biotechniques* **17(2),** 257–262.
17. Hochuli, E., Bannwarth, W., Dobeli, H., Gentz, R., and Stüber, D. (1988) Genetic approach to facilitate purification of recombinant proteins with a novel metal chelate adsorbent. *Biotechnology* **6,** 1321–1325.

18. Le Grice, S. F. J. and Grueninger-Leitch, F. (1990) Rapid purification of homodimer HIV-I reverse transcriptase by metal chelate affinity chromatography. *Eur. J. Biochem.* **187,** 307–314.
19. Stüber, D., Bannwarth, W., Pink, J. R. L., Meloen, R. H., and Matile, H. (1990) New B cell epitopes in the plasmodium falciparum malaria circumsporozoite protein. *Eur. J. Immunol.* **20,** 819–824.
20. Takacs, B. J. and Girard, M.F. (1991) Preparation of clinical grade proteins produced by recombinant DNA technologies. *J. Immunol. Methods* **143,** 231–240.

112

Detection and Immobilization of Proteins Containing the 6xHis Tag

Richard E. Kneusel, Melanie Wulbeck, and Joachim Ribbe

1. Introduction

Many systems have been developed for the heterologous expression of recombinant proteins. They are often based on the fusion of the protein of interest with a naturally occurring protein (glutathione S-transferase [1], maltose binding protein [2], or protein A [3]) and using their natural affinity to substrates (glutathione, amylose, or immunoglobulins) coupled to columns in the purification step. Problems with these systems are that the affinity tag may affect protein structure and function, and that proteins in insoluble fractions cannot be adequately purified.

Immobilized metal ion-affinity chromatography (IMAC) takes advantage of the adsorption of polyhistidine (6xHis) containing proteins to a nickel-nitrilotriacetic acid (Ni-NTA) column (4–6). The highly specific interaction of Ni-NTA to the 6xHis tag does not depend on the three-dimensional structure of the tag or ligand, so it can be exploited even under strongly denaturing conditions necessary for solubilizing inclusion bodies. Bound proteins can be eluted simply by competition with imidazole or by a slight reduction in pH.

A problem often associated with the heterologous expression is the detection and identification of the clone or expressed protein in a complex mixture. Since antibodies are often not available, the sole recourse for initial identification is the functional test, often a complex assay and perhaps made more difficult by required solubilization. Two recent developments promise to alleviate this difficulty: monoclonal antibodies (MAbs) and reporter enzyme Ni-NTA conjugates, both of which can detect histidine-tagged proteins.

The MAb His-1 was prepared by injecting mice with a small his-tagged protein (XDCoH) and isolating hybridomas that in Western blot experiments, recognize exclusively the recombinant protein and not the native XDCoH (7). His-1 is highly specific for the amino acid sequence RGSHHHH and could be used in immunocytochemical studies. Ni-NTA conjugates were made simply by activating alkaline phosphatase with bis(sulfosuccinimidyl) suberate, coupling to NTA, and charging the ligand with nickel sulfate. This conjugated reporter molecule has been used successfully in immunological methods like Western blots and ELISA tests for the direct detection of his-tagged protein.

We report here on the application of the MAb and Ni-NTA conjugate in blotting and screening procedures, and monitoring expression levels. The MAb may also be used for ELISA, immunoprecipitation, immunohistochemistry, immunocytochemistry, and in protein localization and targeting studies.

2. Materials

2.1. Western and Dot Blotting

^{RGS}·His antibody (final concentration, 0.2 mg/mL) and Ni-NTA conjugates (alkaline phosphatase [AP] or horseradish peroxidase [HRP]) are available exclusively from Qiagen (Hilden, Germany).

1. TBS buffer: 150 mM NaCl, 10 mM Tris-HCl, pH 7.4.
2. TBS-Tween/Triton buffer: 500 mM NaCl, 0.05% Tween-20, 20 mM Tris-HCl, pH 7.5; 0.2% Triton X-100.
3. TBS-Tween buffer: 500 mM NaCl, 0.05% Tween-20, 20 mM Tris-HCl, pH 7.5.
4. Blocking buffer: 3% BSA in TBS buffer.

AP staining solutions:

1. Buffer A: 100 mM NaCl, 5 mM MgCl$_2$, 100 mM Tris-HCl, pH 9.5.
2. NBT stock solution: 5% NBT (Nitro blue tetrazolium chloride) in 70% dimethylformamide. Store in aliquots at –20°C.
3. 5-Bromo-4 chloro-3-indolyl phosphate (BCIP) stock solution: 5% BCIP in 100% dimethylformamide. Store in aliquots at –20°C.
4. Staining solution: Add 66 µL NBT and 33 µL BCIP stock solutions to 10 mL buffer A. **NOTE**: Prepare immediately before staining.

HRP staining solutions:

1. Tris-saline: 9% NaCl in 1 M Tris-HCl, pH 8.0.
2. Staining solution: Dissolve 18 mg 4-chloro-1-naphthol in 6 mL methanol, add 24 mL Tris-saline followed by 60 µL H$_2$O$_2$. **NOTE**: The final staining solution is only stable for a short period. Prepare immediately before use.

2.2. Colony Blotting

1. LB agar plates with relevant antibiotics.
2. LB agar plates with relevant antibiotics and 1 mM IPTG.
3. Nitrocellulose filter disks.
4. Blunt-ended forceps.
5. Syringe needle (with India ink—optional).
6. Polystyrene dishes.
7. SDS solution: 10% (w/v) sodium dodecyl sulfate (SDS).
8. Denaturing solution: 0.5 M NaOH, 1.5 M NaCl.
9. Neutralization solution: 1.5 M NaCl, 0.5 M Tris-HCl, pH 7.4 (25°C).
10. 20X SSC: 500 mL: 87.65 NaCl, 50.25 g trisodium citrate · 2 H$_2$O.

2.3. Immobilization

1. HisSorb Strips are available from Qiagen.
2. Buffer A: 10 mM Tris-HCl, pH 7.5, 50 mM NaCl.
3. TBS: 150 mM NaCl, 10 mM Tris-HCl, pH 7.5.

4. PBS: 8 g/L NaCl, 1.16 g/L NaH$_2$PO$_4$, 0.2 g/L KH$_2$PO$_4$, 0.2 g/L KCl, pH 7.5.
5. PBS/Tween: 0.25% Tween-20 in PBS.

2.4. Immunoprecipitation

1. Cell lysis buffer: 50 mM Tris-HCl, pH 8.0, 1% NP-40, 150 mM NaCl.
2. 100 mM glycine buffer, pH 3.0.
3. 1 M Tris-HCl, pH 8.0.
4. Protein A-Sepharose (Pharmacia, Freiburg, Germany).

3. Methods

3.1. ^{RGS}·His Antibody

3.1.1. Blotting

Western and dot-blot procedures take advantage of the specificity of the antibody–antigen interaction to detect and quantify proteins in a complex mixture. Whereas dot blots simply indicate whether the protein of interest is present, Western blot analysis involves separating the protein mixture and can therefore provide additional information, such as size and antibody specificity. In Western blotting, the samples are solubilized with detergents (usually SDS) and reducing reagents (β-mercaptoethanol), separated by SDS-polyacrylamide gel electrophoresis (PAGE), and transferred to a membrane support (nitrocellulose). The membrane is probed with antibodies specific for the protein of interest, and the bound antibody is detected by secondary detection reagents (e.g., anti-IgG enzyme conjugates).

The ^{RGS}·His MAb can be used for the identification of recombinant proteins containing the epitope RGSHHHH (**Fig. 1**). The extract can be prepared using standard techniques under native or nondenaturing conditions. For a detailed protocol, *see* Chapter 111. SDS-PAGE *(8)* and protein transfer to nitrocellulose membranes are carried out according to standard procedures *(9–12)*. Dot blots are best prepared by spotting 1 µL of each sample directly on the nitrocellulose filter. Use a shaker or rocker platform for all wash and incubation steps.

1. Wash the membrane twice for 10 min with TBS buffer.
2. Incubate for 1 h in blocking buffer at room temperature.
3. Wash twice for 10 min in TBS-Tween/Triton buffer.
4. Wash once for 10 min with TBS buffer.
5. Incubate in blocking buffer (1/2 dilution) containing a 1/1000 or 1/2000 dilution of the primary antibody, ^{RGS}·His antibody, at room temperature for 1 h. Membrane can be sealed in plastic bags.
6. Wash twice for 10 min in TBS-Tween/Triton buffer.
7. Wash once for 10 min in TBS buffer.
8. Incubate 1 h in blocking buffer containing secondary antibody (*see* **Note 1**).
9. Wash four times for 10 min in TBS-Tween/Triton buffer.
10. Stain with AP or HRP staining solution until the signal is clearly visible (approx 5–7 min in dot-blotting procedures; up to 15 min in Western blotting procedure) (*see* **Note 2**).
11. Stop the reaction by rinsing the membrane twice with water, and fix the blot in 3% TCA for 5 min.
12. Wash with water to remove TCA.

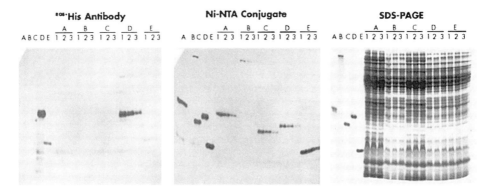

Fig. 1. Detection of 6xHis-tagged proteins in crude cell lysates by Western blotting. Proteins were expressed with N- or C-terminal 6xHis tags in *E. coli* strain M15[pREP4] and purified on Ni-NTA agarose. Purified protein (200 ng, lanes A–E) or tagged proteins in 6–9 μg cell lysate (lane 1, 50 ng; lane 2, 20 ng; lane 3, 10 ng) were separated on 12% denaturing gels and electroblotted to nitrocellulose membranes. Blots were probed with $^{RGS\cdot}$His antibody and AP-conjugated rabbit antimouse IgG or Ni-NTA AP conjugate as indicated. Colorimetric detection with BCIP/NBT was carried out for 5 min or 30 min, respectively. A: (His)$_6$-CAT (45 kDa); B: 2.3 kb ORF-(His)$_6$ (85 kDa); C: DHFR-(His)$_6$ (24 kDa); D: RGS(His)$_6$-DHFRa (26 kDa); E: R(His)$_6$. $^{RGS\cdot}$His antibody displays some affinity towards the sequence R(His)$_6$.

3.1.2. Screening for Positive Clones: Colony Blotting Protocol

Rapid identification of clones correctly expressing a 6xHis-tagged protein is easily achieved using the colony blotting protocol. Colonies are transferred to nitrocellulose filters, protein expression is induced, and protein is liberated from bacterial colonies and bound to nitrocellulose filters *in situ*. The filters are then treated according to the standard Western blot procedure (**Fig. 2**).

1. Plate transformation mix on LB plates containing the relevant antibiotics, and incubate overnight (16 h) at 30°C until colonies are about 1–2 mm in diameter (*see* **Notes 3** and **4**).
2. Remove from incubator, open lids slightly, and allow any condensation to dry for 15–30 min.
3. Place a dry, numbered nitrocellulose filter on the agar surface in contact with the colonies, taking care to avoid introducing air bubbles (*see* **Note 5**).
4. Using a syringe needle, pierce the filter and agar at asymmetric positions to facilitate proper alignment following staining. Grip filter on the sides with a blunt-ended forceps, and peel off in one movement. Each colony should be entirely transferred to the nitrocellulose sheet leaving an apparently empty depressions on the agar plate.
5. Transfer filter (colonies up) to a fresh plate containing antibiotics and IPTG, as described above. Avoid introducing air bubbles.
6. Incubate plates for 4 h at 37°C. Place master plates in 30°C incubator for 8 h to allow colonies to regrow.
7. Prepare a set of polystyrene dishes for colony lysis and binding of DNA to the filters. Each dish should contain a sheet of 3MM paper soaked with one of the following solutions (*see* **Note 6**):
    a. SDS solution;
    b. Denaturing solution;
    c. Neutralization solution;

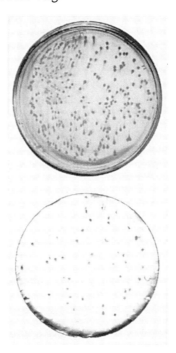

Fig. 2. Detection of positive expression clones with $^{RGS\cdot}$His antibody. Plasmid pQE-40 containing a BamHI, BglII fragment encoding 6xHis-tagged mouse DHFR was digested with BamHI and BglII, ligated, and transformed into E. coli M15[pREP4]. The plated bacteria were grown overnight at 37°C, replica-plated onto a nitrocellulose membrane, and incubated on an LB plate containing 1 mM IPTG for 4 h. Detection: $^{RGS\cdot}$His antibody, AP-conjugated rabbit antimouse IgG, BCIP/NBT.

     d. Neutralization solution;
     e. 2X SSC.
8. Place the nitrocellulose filters (colony side up) on top of the paper in each of these dishes, taking care to exclude air bubbles (colonies above air bubbles will not lyse properly and will generate a higher background in the final staining step).
Incubate at room temperature as follows:

    a. SDS solution   10 s
    b. Denaturing solution   20 s
    c. Neutralization solution 30 s
    d. Neutralization solution 1 min
    e. 2X SSC   2 min
9. Continue with the standard protocol for Western and dot blots using the $^{RGS\cdot}$His antibody (*see* **Note 7**).

3.1.3. Immunoprecipitation and ELISA

Immunoprecipitation utilizes the high specificity of antibodies to form complexes with antigens that can then be purified from other contaminating proteins *(13)*. When this method is used in combination with SDS-PAGE, it is a very sensitive method for the analysis of protein biosynthesis and processing in prokaryotic and eukaryotic cells.

A simple immunoprecipitation consists of the following steps:

1. Cell lysis: The lysis of the cells is often the most crucial step of the immunoprecipitation. It is critical to determine the conditions under which the protein of interest can be solubilized without hindering its immunoreactivity and biological activity. Of the variables that play a role in optimizing lysis conditions are ionic strength and pH of the buffer, type and amount of detergent, and presence of divalent cations and chelating reagents. Many proteins are extremely sensitive to protease degradation, and denatured proteins are much more susceptible than native proteins. Care should be taken to minimize proteolysis, e.g., reduced temperature, protease inhibitors, and so forth.
2. Formation of specific immune complexes.
3. Capture of the antibody–antigen complex.

    The antibody–antigen complexes can be captured using a number of immunoadsorbants:

    Protein A sepharose;

    Protein G sepharose; or

    Anti-IgG antibody that interacts with primary antibody.

3.1.3.1. SAMPLE PROTOCOL

This procedure has been established using eukaryotic cells that were transfected with a 6xHis construct.

1. Suspend cells in cell lysis buffer (1 mL buffer for 10^7 cells).
2. Incubate at room temperature for 5 min.
3. Centrifuge at 14,000g for 5 min.
4. To 100 µL supernatant add 400 µL cell lysis buffer with 2% bovine serum albumin (BSA).
5. Add 10 µL ^{RGS}·His antibody and incubate at room temperature for 1 h.
6. Add 50 µL resuspended protein A-Sepharose, and incubate for 1 h at 4°C.
7. Collect the protein A complexes by centrifugation (12,000g for 20 s at 4°C).
8. Resuspend pellets in SDS-PAGE sample buffer, and heat for 5 min at 95°C prior to SDS-PAGE analysis.
9. Alternatively, resuspend pellet in 100 mM glycine buffer, pH 3.0, centrifuge (12,000g for 20 s at 4°C), and adjust the pH of the supernatant with 1 M Tris-HCl, pH 8.0.

3.1.4. Immunohistochemistry and Immunocytochemistry

Labeled antibodies can be used to demonstrate the presence and subcellular localization antigens in cells and tissues. Cell staining is a highly versatile technique, can detect very small amounts of an antigen if it is highly localized in the cell, and may even be used to determine the relative concentration of the antigen in the cell *(14)*.

There are four steps involved in cell staining procedures:

1. Cell or tissue preparation: Cells used for staining are usually adherent cells grown on a support or suspension cells attached to a solid support. Sections of frozen, paraffin-embedded, and paraformaldehyde-fixed tissue as well as cell smears may also be used.
2. Fixation: Fixation is important to prevent antigen leakage, make the cell permeable for the antibody, preserve antigen conformation so it can be recognized by the antibody, and maintain cell structure. Fixation is most often carried out with organic solvents.
3. Antibody binding: A single labeled antibody may be used or an indirect method employing a labeled secondary reagent. When a labeled primary antibody is used, the signals are

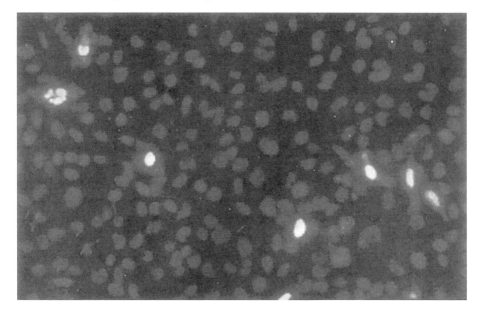

Fig. 3. Immunolocalization of 6xHis-tagged HNF4 in human renal carcinoma cells. Following transfection, tagged protein expressed from the His-HFN4 construct was detected using ^{RGS}·His antibody and a Cy3-conjugated rat antimouse antibody, and visualized by fluorescence microscopy. (Data kindly provided by G. U. Ryffel, University of Essen Medical School, Germany.)

generally cleaner with less background than the indirect method. A labeled secondary antibody often amplifies otherwise weak signals.
4. Detection: Fluorochrome and enzyme-labeled antibodies are the most common detection reagents. Enzyme-conjugated antibodies provide a simple and extremely sensitive detection, however, with a lower resolution. Fluorescent reagents have a much higher resolution; however, their relatively weak emission together with quenching mean that good images can only be obtained if antigens are present at a high concentration.

3.1.4.1. Sample Protocol

Localization of *Xenopus* transcription factor (HNF4) (**Fig. 3**): HNF4 cDNA was combined with the sequence for MRGS(H)$_6$ in the eucaryotic expression vector Rc/CMV (Invitrogen) to yield His-HNF4/CMV. His-HNF4/CMV expression vector was transfected into the human renal carcinoma cell line SK-RC-47 grown on cover slips and incubated for 20 h (*7*).

1. Fix cells in 100% methanol for 15 min at –20°C.
2. Wash cells with PBS.
3. Block with PBS containing 10% goat serum.
4. Incubate with a 1/500 dilution of ^{RGS}·His antibody in cell culture medium for 1 h at 4°C.
5. Wash in three changes of PBS for 5 min each.
6. Incubate with secondary antibody (Cy3-conjugated rat antimouse antibody, Jackson Immuno-Research, code no. 415-166-100, 0.5 mg dissolved in 0.6 mL H$_2$O, diluted 1/200 in PBS with 10% goat serum).
7. Observe labeled antigens using fluorescent microscopy. Cy3 can be excited maximally at 552 nm and emits at 568–574 nm.

Table 1
Development Times of Immunoblots

Quantity of 6xHis-tagged protein	Development time, AP	Development time, HRP
50 ng	3 min	10 s
25 ng	5 min	20 s
10 ng	15 min	30 s
2 ng	20–25 min	1 min
Background	30 min	2 min

3.2. Ni-NTA Conjugate

Ni-NTA conjugates can be used in nearly every application as the ^{RGS}·His MAb. Detection is based on the affinity of Ni-NTA for affinity tags consisting of consecutive histidine residues, the same principle that underlies purification of 6xHis-tagged proteins using Ni-NTA agarose

3.2.1. Blotting

Ni-NTA conjugates can be used for detection of recombinant proteins containing either N or C-terminal histidine tags (**Fig. 1**), independent of the surrounding amino acid composition. They can therefore be used to detect 6xHis-tagged proteins expressed by any construct (*see* **Note 8**). Signals above background (6xHis-free *Escherichia coli* cell lysates) can be detected in 1-µL dots (approx 2 mm in diameter) of *E. coli* cell lysates containing 6xHis-tagged protein when developed with the staining solutions for the times listed in **Table 1**. The protocol uses standard Western or dot-blotting techniques (*see* **Subheading 3.1.1.**). Best results will be obtained if all steps are carried out on a shaker or rocker platform.

1. Wash Western or dot blot membrane twice for 10 min with TBS buffer.
2. Incubate for 1 h in blocking buffer at room temperature.
3. Wash three times for 10 min with TBS-Tween buffer.
4. Incubate the membrane for 1 h at room temperature in TBS-Tween containing a 1/1000 dilution of Ni-NTA conjugate stock solution. Membranes can be sealed in plastic bags.
5. Wash three times for 10 min in TBS-Tween buffer at room temperature.
6. Stain with AP or HRP staining solution until the signal is clearly visible (*see* **Note 2**).
7. Stop the reaction by rinsing the membrane twice in water, and fix the blot in 3% TCA for 5 min.
8. Wash with water to remove TCA.

3.2.2. Screening for Positive Clones: Colony Blotting Protocol

Rapid identification of clones correctly expressing a 6xHis-tagged protein is easily achieved using the colony blotting protocol. Colonies are transferred to nitrocellulose filters, protein expression is induced, and protein is liberated from bacterial colonies and bound to nitrocellulose filters in situ. The filters are then treated according to the standard Western blot procedure.

1. Plate transformation mix on LB plates containing the relevant antibiotics, and incubate overnight (16 h) at 30°C until colonies are about 1–2 mm in diameter (*see* **Notes 3** and **4**).

2. Remove from incubator, open lids slightly, and allow any condensation to dry for 15–30 min.
3. Place a dry, numbered nitrocellulose filter on the agar surface in contact with the colonies, taking care to avoid introducing air bubbles (*see* **Note 5**).
4. Using a syringe needle, pierce the filter and agar at asymmetric positions to facilitate proper alignment following staining. Grip filter on the sides with a blunt-ended forceps, and peel off in one movement. Each colony should be entirely transferred to the nitrocellulose sheet, leaving apparently empty depressions on the agar plate.
5. Transfer filter (colonies up) to a fresh plate containing antibiotics and IPTG, as described above. Avoid introducing air bubbles
6. Incubate plates for 4 h at 37°C. Place master plates in 30°C incubator for 8 h to allow colonies to regrow.
7. Prepare a set of polystyrene dishes for colony lysis and binding of DNA to the filters. Each dish should contain a sheet of 3MM paper soaked with one of the following solutions (*see* **Note 6**):
    a. SDS solution;
    b. Denaturing solution;
    c. Neutralization solution;
    d. Neutralization solution;
    e. 2X SSC.
8. Place the nitrocellulose filters (colony side up) on top of the paper in each of these dishes, taking care to exclude air bubbles (colonies above air bubbles will not lyse properly and will generate a higher background in the final staining step).
    Incubate at room temperature as follows:
    a. SDS solution   10 s
    b. Denaturing solution   20 s
    c. Neutralization solution 30 s
    d. Neutralization solution 1 min
    e. 2X SSC   2 min
9. Continue with the standard protocol for Western and dot blots using the Ni-NTA conjugate (*see* **Note 7**).

3.2.3. Immobilization on Microtiter Plates

Immobilized proteins can be used for a variety of analytical and diagnostic techniques. Bound ligands can be used to screen serum samples as well as for determining the affinity of protein–ligand interactions and monitoring protein–protein interactions. Microtiter plates whose inner surfaces are coated with a spacer bearing an Ni-NTA group can be used for immobilizing proteins or peptides that contain a tag of six consecutive histidine residues (*see* **Notes 9–11**).

These protocols are intended to be used as guidelines. Optimal conditions for each individual protein should be determined.

3.2.3.1. BINDING OF 6XHIS-TAGGED PROTEIN OR PEPTIDE (GENERAL PROTOCOL)

1. Prepare a solution of the 6xHis protein or peptide at a concentration of 50–100 pmol/mL in buffer A (or a cell lysate containing the 6xHis-tagged protein/peptide).
2. Add 200 µL of the protein/peptide solution to each well, and incubate for at least for 2 h at 4°C. When low protein concentrations are used, an increased incubation time (6–12 h) gives better results.

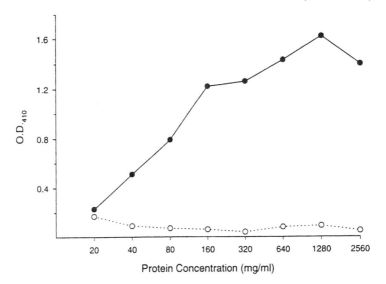

Fig. 4. Detection of 6xHis-tagged TNFα in HisSorb Plates. 6xHis-tagged TNFα (●) or TNFα-D6xHis (○) (generated by protease cleavage of the N-terminal 6xHis tag) was bound to the wells of Ni-NTA HisSorb Strips. Binding was performed overnight. Bound protein was detected using monoclonal antihuman TNFα antibody, peroxidase-labeled goat antimouse IgG, and 30 min of incubation with the substrate ABTS. Absorbances were measured at 410 nm. Each point represents the average of determinations.

3. Wash wells three times with buffer A. When using a cell lysate (nonpurified 6xHis protein) the wash step should be carried out at a slightly reduced pH (e.g., pH 6.3) to wash away untagged proteins.

The wells are now ready for use in the specific application.

3.2.3.2. Binding of a 6xHis-Tagged Protein (6xHis Cytokine) and Detection by ELISA (Sample Protocol)

1. Block wells with PBS/10% fetal calf serum for 1 h. Alternatively, 10% horse serum, 3% BSA, or "Super blocking buffer" from Pierce can be used.
2. Prepare a solution of the 6xHis protein or peptide.
3. Add 200 μL of the protein/peptide to a concentration of 100 pmol/mL in TBS (or a cell lysate containing the 6xHis-tagglution to each well) and incubate overnight at 4°C. For optimal binding, an incubation time of at least 6 h is recommended.
4. Wash five times with PBS-Tween, pH 7.5. Soak wells for at least 1 min/wash, and blot on paper towels after each wash. When using a cell lysate (nonpurified 6xHis protein), the wash step should be carried out at a slightly reduced pH (e.g., pH 6.3) to wash away nontagged proteins.
5. Add primary MAb in blocking buffer, cover plate, and incubate for 6 h or overnight at 4°C.
6. Wash twice with PBS-Tween, pH 7.5.
7. Dilute secondary MAb in PBS containing 3% BSA, add to plate, and incubate at room temperature for 45 min.
8. Wash five times with PBS-Tween, pH 7.5.
9. Add substrate and determine absorbance (**Fig. 4**).

4. Notes

1. Both AP and HRP-conjugated rabbit or goat antimouse IgG may be used. Dilute according to the manufacturer's recommendations.
2. For chemiluminescent detection, CDP-Star from Tropix Inc. can be used with AP-conjugated secondary and with the ECL system from Amersham, it can be used in combination with HRP-conjugated secondary antibodies. Follow the manufacturer's instructions.
3. After spreading the transformation mix, dry the plates inverted with the lids slightly open until small wrinkles develop on the surface of the agar. To prevent streaking, incubation should not be started until all of the suspension has been absorbed into the agar.
4. To avoid expression of toxic proteins without IPTG induction (a result of "leaky" promoters) and to maintain plasmid stability, incubation should be carried out at 30°C. If the expressed protein is not toxic and the plasmids are stable, incubation can be carried out at 37°C, but care should be taken that the colonies do not become too large.
5. Number filters with a water-resistant marking pen. Hold the filter on opposite sides with blunt-ended forceps, and align across the diagonal of the plate. Lower the filter gently onto the agar surface, making contact first along the diagonal and then lowering (but not dropping) the sides.
6. Discard excess fluid so that paper is moist, but not wet. Excess liquid promotes colony swelling and diffusion, which will cause blurred signals.
7. There is often only a slight difference between colonies showing a positive signal and those with background intensity. Different staining times are required with this procedure. Two to 3 min are usually sufficient, but it is very important to monitor color development at this stage. If it is extremely difficult to differentiate between positive clones and background, the cause of the high background should be determined. The following controls should be included:
    a. A plate with the host bacteria without the expression plasmid.
    b. A plate with the host bacteria harboring the expression plasmid without the insert.
    c. A colony blot treated only with the secondary antibody prior to staining.
8. Ni-NTA conjugates also form a complex with a 31-kDa mol-wt standard, bovine carbonic anhydrase (a metalloenzyme with 1 zinc ion/protein molecule *[15]*). Carbonic anhydrase is a component of many SDS-PAGE mol-wt marker sets (e.g., Bio-Rad) and can be used as a positive control.
9. Since the interaction of 6xHis-tagged proteins with immobilized nickel ions is pH-dependent, binding should be carried out around pH 7.5.
10. The binding capacity is approx 20 pmol/well for small peptides (20–30-mers) and approx 3–6 pmol/well for proteins.
11. Binding should be carried out for at least 2 h at room temperature. Binding efficiency can be improved by reducing the concentration of 6xHis-tagged protein or increasing the incubation time (6 h or overnight).

References

1. Smith, D. B. and Johnson, K. S. (1988) Single-step purification of polypeptides expressed in *Escherichia coli* as fusions with glutathione S-transferase. *Gene* **141,** 79–84.
2. Maina, C. V., Riggs, P. D., Grandea, P. D., Slatko, B. E., Moran, L. S., Tagliamonte, J. A., McReynolds, L. A., and Guan, C. (1988) An *Escherichia coli* vector to express and purify foreign proteins by fusion to and separation from maltose binding protein. *Gene* **74,** 365–373.
3. LaVallie, E. R., DiBlasio, E. A., Kovacic, S., Grant, K. L., Schendel, P. F., and McCoy, J. M. (1993) A thioredoxin gene fusion system that circumvents inclusion body formation in the *E. coli* cytoplasm. *Bio/Technology* **11,** 187–193.

4. Porath, J., Carlsson, J., Olsson, I., and Belfrage, G. (1975) Metal chelate affinity chromatography, a new approach to protein fractionation. *Nature* **258,** 598,599.
5. Sulkowski, E. (1985) Purification of proteins by IMAC. *Trends Biotechnol.* **3,** 17.
6. Hochuli, E., Dobeli, H., and Schacher, A. (1987) New metal chelate adsorbent selective for proteins and peptides containing neighboring histidine residues. *J. Chromatog.* **411,** 177–184.
7. Pogge von Strandmann, E., Zoidl, C., Nakhei, H., Holewa, B., Pogge von Strandmann, R., Lorenz, P., Klein-Hitpass, L., and Ryffel, G. U. (1995) *Protein Eng.* **8,** 733–735.
8. Gallagher, S. (1995) One-dimensional gel electrophoresis of proteins, in *Current Protocols in Protein Science*, vol. 1 (Coligan, J. E., Dunn, B. M., Ploegh, H. L., Speicher, D. W., and Wingfield, P. T., ed. board), Wiley, New York, pp. 10.1.1–10.1.29.
9. Knecht, D. A. and Dimond, R. L. (1984) Visualization of antigenic proteins on Western blots. *Anal. Biochem.* **136,** 180–186.
10. Towbin, H., Staehelin, T., and Gordon, J. (1979) Electrophoretic transfer of proteins from polyacrylamide gels to nitrocellulose sheets: Procedure and some applications. *Proc. Nat. Acad. Sci. USA* **76,** 4350–4354.
11. Gallagher, S. (1992) Immunoblotting and immunodetection, in *Current Protocols in Molecular Biology*, vol. 2. (Jannsen, K., ed.), Wiley, New York, pp. 10.8.1–10.8.16.
12. Ursitti, J. A., Mozdzanowski, J., and Speicher, D. W. (1995) Electrophoresis, in *Current Protocols in Protein Science*, vol. 1 (Coligan, J. E., Dunn, B. M., Ploegh, H. L., Speicher, D. W., and Wingfield, P. T., ed. board), Wiley, New York, pp. 10.7.1–10.7.14.
13. Harlow, E. and Lane, D. (1988) *Antibodies. A Laboratory Manual*. Cold Spring Harbor Laboratory Press, Cold Spring Harbor, NY, pp. 421–470.
14. Harlow, E. and Lane, D. (1988) *Antibodies. A Laboratory Manual*. Cold Spring Harbor Laboratory Press, Cold Spring Harbor, NY, pp. 361–420.
15. Tanis, R. J., Ferrell, R. C., and Tashian, R. E. (1974) Amino acid sequence of sheep carbonic anhydrase C. *Biochem. Biophys. Acta* **371,** 534–548.

113

Expression and Purification of Recombinant Proteins Using the pET System

Robert C. Mierendorf, Barbara B. Morris, Beth Hammer, and Robert E. Novy

1. Introduction

The pET System is the most powerful system yet developed for the cloning and expression of recombinant proteins in *Escherichia coli*. Target genes are cloned in pET plasmids under control of strong bacteriophage T7 transcription and (optionally) translation signals; expression is induced by providing a source of T7 RNA polymerase in the host cell *(1–3)*. T7 RNA polymerase is so selective and active that almost all of the cell's resources are converted to target gene expression; the desired product can comprise more than 50% of the total cell protein after a few hours of induction. Another important benefit of this system is its ability to maintain target genes transcriptionally silent in the uninduced state. Target genes are initially cloned using hosts that do not contain the T7 RNA polymerase gene, thus eliminating plasmid instability caused by the production of proteins potentially toxic to the host cell. Once established in a nonexpression host, plasmids are then transferred into expression hosts containing a chromosomal copy of the T7 RNA polymerase gene under *lacUV5* control, and expression is induced by the addition of IPTG. Two types of T7 promoter and several hosts that differ in their stringency of suppressing basal expression levels are available, providing great flexibility and optimizing the expression of a wide variety of target genes. This chapter describes the vectors, hosts, and basic protocols for cloning, expression, and purification of target proteins in the pET System.

1.1. pET Vectors

The pET vectors were originally constructed by Studier and colleagues *(1–3)*. Derivatives of these plasmids have been designed with enhanced features to permit easier subcloning, detection, and purification of target proteins (*see* **Table 1**). Two general categories of vectors are available:

1. Transcription vectors, which are designed for expression of target genes that already carry their own prokaryotic ribosome binding site and ATG start codon.
2. Translation vectors, which contain the highly efficient ribosome binding site from the phage T7 major capsid protein. These are distinguished from the transcription vectors by the addition of a letter suffix following the name, e.g., pET-21a(+), which denotes the

Table 1
pET Vector Classification

Vector	Selection	Promoter	Cloning sites for N-term fusion	Protease cleavage sites	Optional C-term fusion
Transcription[a]					
pET-21(+)[b,c]	Ap	T7lac	BamHI, EcoRI, SacI, SalI, HindIII, EagI, NotI, XhoI		His·Tag®
pET-23(+)	Ap	T7	BamHI, EcoRI, SacI, SalI, HindIII, EagI, NotI, XhoI		His·Tag
pET-24(+)	Kan	T7lac	BamHI, EcoRI, SacI, SalI, HindIII, EagI, NotI, XhoI		His·Tag
Translation[d]					
pET-3a–d[e]	Ap	T7	BamHI		No
pET-5a–c[e]	Ap	T7	BamHI, EcoRI		No
pET-9a–d[e]	Kan	T7	BamHI		No
pET-17b[e]	Ap	T7	HindIII, KpnI, SacI, BamHI, SpeI, BstXI[f], EcoRI, EcoRV, NotI, XhoI		No
pET-23a–d(+)[e]	Ap	T7	BamHI, EcoRI, SacI, SalI, HindIII, EagI, NotI, XhoI		His·Tag
pET-11a–d[e]	Ap	T7lac	BamHI		No
pET-13a–d[e]	Kan	T7lac	BamHI		No
pET-21a–d(+)[e]	Ap	T7lac	BamHI, EcoRI, SacI, SalI, HindIII, EagI, NotI, XhoI		His·Tag
pET-24a–d(+)[e]	Kan	T7lac	BamHI, EcoRI, SacI, SalI, HindIII, EagI, NotI, XhoI		His·Tag
pET-3xa–c[g]	Ap	T7	BamHI		No
pET-17xb[g]	Ap	T7	SacII, HindIII, KpnI, SacI, BamHI, SpeI, BstXI[f], EcoRI, EcoRV, NotI, XhoI		No

Plasmid	Resistance	Promoter	Restriction sites	Protease	Tag
pET-12a–c[i]	Ap	T7	BamHI		No
pET-20b(+)[i]	Ap	T7	NcoI, EcoRV, BamHI, EcoRI, SacI, SalI, HindIII, EagI, NotI, XhoI		His·Tag
pET-22b(+)[i]	Ap	T7lac	MscI, NcoI, BamHI, EcoRI, SacI, SalI, HindIII, EagI, NotI, XhoI		His·Tag
pET-26b(+)[i]	Kan	T7lac	MscI, NcoI, BamHI, EcoRI, SacI, SalI, HindIII, EagI, NotI, XhoI		His·Tag
pET-25b(+)[i]	Ap	T7lac	MscI, NcoI, BamHI, EcoRI, SacI, SalI, HindIII, EagI, NotI, XhoI, NheI		His·Tag + HSV·Tag®
pET-27b(+)[i]	Kan	T7lac	MscI, NcoI, BamHI, EcoRI, SacI, SalI, HindIII, EagI, NotI, XhoI, NheI		His·Tag + HSV·Tag
pET-14b[j]	Ap	T7	NdeI, XhoI, BamHI	Thrombin	No
pET-15b[j]	Ap	T7lac	NdeI, XhoI, BamHI	Thrombin	No
pET-16b[j]	Ap	T7lac	NdeI, XhoI, BamHI	Factor Xa	No
pET-19b[j]	Ap	T7lac	NdeI, XhoI, BamHI	Enterokinase	No
pET-28a–c(+)[k]	Kan	T7lac	His·Tag only: NdeI, NheI His·Tag + T7·Tag®: BamHI, EcoRI, SacI, SalI, HindIII, EagI, NotI, Xho	Thrombin	His·Tag
pET-29a–c(+)[l]	Kan	T7lac	NcoI, EcoRV, BamHI, EcoRI, SacI, SalI, HindIII, EagI, NotI, XhoI	Thrombin	His·Tag
pET-30a–c(+)[m]	Kan	T7lac	NcoI, EcoRV, BamHI, EcoRI, SacI, SalI, HindIII, EagI, NotI, XhoI	Thrombin Enterokinase	His·Tag
pSCREEN™-1b(+)[h,n]	Ap	T7	NcoI, EcoRV, BamHI, EcoRI, SacI, SalI, HindIII, EagI, NotI, XhoI, SmaI, PmeI	Thrombin Enterokinase	His·Tag

(continued)

Table 1 (continued)

Vector	Selection	Promoter	Cloning sites for N-term fusion	Protease cleavage sites	Optional C-term fusion
pET-31b(+)[o]	Ap	T7lac	AlwNI, XhoI, AvaI	Thrombin	His·Tag
pET-32a-c(+)[p]	Ap	T7lac	MscI, BglII, KpnI, NcoI, EcoRV, BamHI, EcoRI, SacI, SalI, HindIII, EagI, NotI, AvaI, XhoI	Enterokinase	His·Tag

[a]For expression from translation initiation signals within a cloned insert.
[b]All vectors having a (+) in their name contain the bacteriophage f1 origin of replication for the production of single-stranded DNA using helper phage.
[c]All vectors with the pET designation carry the pBR322 origin of replication. pSCREEN carries the high-copy number pUC origin.
[d]The cloning sites shown are for N-terminal fusion.
[e]Vectors that allow N-terminal fusion to small (11aa) T7·Tag sequence.
[f]pET-17b and pET-17xb contain dual BstXI sites designed for efficient cloning using non-palindromic linkers (5).
[g]Vectors that allow N-terminal fusion to large (260aa) T7·Tag sequence for stabilization of small target proteins/peptides.
[h]pSCREEN-1b(+) is also the product of Cre-loxP subcloning from the vector λSCREEN-1.
[i]Vectors that allow fusion to N-terminal signal sequence for potential periplasmic localization.
[j]Vectors that allow N-terminal fusion to cleavable His·Tag sequence.
[k]Vectors that allow N-terminal fusion to cleavable His·Tag and/or T7·Tag sequence.
[l]Vectors that allow N-terminal fusion to cleavable S·Tag™ sequence.
[m]Vectors that allow N-terminal fusion to cleavable His·Tag and S·Tag sequences.
[n]Vector that allows N-terminal fusion to cleavable 260aa T7·Tag, His·Tag and S·Tag sequences.
[o]Vector that allows N-terminal fusion to 125aa ketosteroid isomerase sequence for stable high-level expression of small proteins and peptides.
[p]Vectors that allow N-terminal fusion to cleavable thioredoxin, His·Tag and S·Tag sequences.

reading frame relative to the *Bam*HI cloning site. Therefore, all vectors having the suffix a express from the GGA triplet, all vectors having the suffix b express from the GAT triplet, and all vectors having the suffix c express from the ATC triplet of the GGATCC *Bam*HI recognition sequence. Some vectors have a d suffix; these also express from the c frame but contain an upstream *Nco*I cloning site in place of the *Nde*I site in that series for insertion of target genes directly into the ATG start codon.

In general, the translation vectors are used for the expression of target genes derived from eukaryotic sources, whereas the transcription vectors are used for prokaryotic genes (which usually carry compatible ribosome binding sites).

Many strategies can be used for subcloning a protein-coding region of DNA into a pET vector for expression. The pET vectors offer a variety of unique restriction sites suitable for cloning inserts in a defined orientation using standard methods. Another very convenient and efficient method that is useful when cloning polymerase chain reaction (PCR) products is ligation-independent cloning (LIC), which allows directional insertion of target sequences without restriction enzyme digestion or ligation reactions *(4)*. PCR products with appropriate 12–14 base overhangs are created by building defined 5' extensions into the primers. Following amplification the PCR product is purified to remove dNTPs and then treated with T4 DNA polymerase in the presence of one dNTP to generate the specific vector-compatible overhangs. Commercial LIC vector kits are available which contain various pET vectors having compatible ends specifically prepared to accept these products (Novagen, Inc., Madison, WI, www.novagen.com).

The various pET vectors also contain different sequences adjacent to the cloning sites that code for a variety of peptide tags for detection and purification of the target protein. The choice of cloning sites and strategies depends on the type of assay or method of purification used. Options include: the His·Tag oligohistidine domain for convenient, economical purification on any scale and under many conditions, the T7·Tag and HSV·Tag sequences for immunological detection and immunoaffinity purification, the S·Tag peptide, which allows quantification of expressed proteins by a rapid homogeneous assay, detection on Western blots, and affinity purification, the *ompT* and *pelB* signal sequence leaders for potential periplasmic localization, the ketosteroid isomerase fusion for high-level stable expression of small proteins and peptides, and the Trx·Tag™ thioredoxin fusion domain for increased solubility of target proteins in the *E. coli* cytoplasm.

Except for pET-32a–c(+), all of the pET vectors can also express proteins without N-terminal fusions, provided that the appropriate site (*Nde*I or *Nco*I in the translation vectors) is used at the 5' end of the insert coding sequence. Similarly, proteins without vector-encoded C-terminal fusions are obtained by including a translation stop codon in the insert. It should be noted that the expression of desired C-terminal fusions requires the lack of a stop codon in the insert and the proper reading frame.

Except for the pET-5 series, all pET vectors listed in **Table 1** contain translation stop codons in all three reading frames following the cloning and tag regions, as well as a downstream T7 transcription terminator. The terminator is not necessary for the efficient expression of most proteins, but note that many pET plasmids contain the gene for ampicillin resistance (β-lactamase) in the same orientation as the target gene. If the T7 transcription terminator is removed during cloning, IPTG-dependent accumulation of β-lactamase (M_r 31,515) is usually observed along with the target protein, because of efficient read-through transcription by T7 RNA polymerase.

In **Table 1** the translation vectors are organized into functional groups based on their N-terminal fusion sequences. Note that the (+) following the name indicates that the vector contains an f1 origin of replication that allows the production of single-stranded plasmid DNA for mutagenesis and sequencing applications.

1.1.1. Antibiotic Resistance

The selective markers *bla* (β-lactamase, or ampicillin resistance) and kan (kanamycin resistance) are available with the pET vectors and are indicated in **Table 1**. Both types of selection have been widely used, but several simple guidelines are recommended when using vectors carrying the β-lactamase gene (*see* **Subheading 4.1.**). Although ampicillin resistance is commonly used in a variety of cloning vectors, kanamycin may be preferable under certain conditions, including protein expression under GMP standards, and when subcloning target genes from other ampicillin-resistant vectors. Kanamycin may also be preferred over ampicillin in labs that do not need to meet GMP standards for several reasons. Ampicillin selection tends to be lost in cultures because the drug is degraded by the secreted β-lactamase enzyme and by the drop in pH that usually accompanies bacterial fermentation. Other ways of avoiding this loss of drug resistance are replacement with fresh ampicillin-containing media or the use of the related drug, carbenicillin, which is less sensitive to low pH.

Another difference between kan^R and amp^R pETs involves the direction of transcription of the drug-resistance gene. In amp^R pET vectors the β-lactamase promoter is located downstream and in the same orientation as the T7 promoter. With the exception of the pET-5 series, all pET and pSCREEN plasmids have the native T7 transcription terminator located before the β-lactamase promoter. However, this terminator is approx 70% effective. Therefore, T7 RNA polymerase produces a small amount of β-lactamase RNA in addition to the target RNA, resulting in the accumulation of β-lactamase enzyme in induced cultures. In contrast, the kan^R gene is in the opposite orientation from the T7 promoter, so there should not be an increase in kan gene product after induction resulting from read-through transcription from the T7 promoter.

1.1.2. The T7lac Promoter

Even in the absence of IPTG there is some expression of T7 RNA polymerase from the *lacUV5* promoter in λDE3 lysogens. If target gene products are sufficiently toxic to *E. coli*, this basal level can be enough to prevent the establishment of plasmids in λDE3 lysogens. Several solutions to this problem are available with the pET System.

One approach is to use vectors that contain what is termed a T7*lac* promoter (*3,6*; *see* **Table 1**). These plasmids contain a *lac* operator sequence just downstream of the T7 promoter. They also carry the natural promoter and coding sequence for the *lac* repressor (*lac*I), oriented so that the T7*lac* and *lac*I promoters diverge. When this type of vector is used in λDE3 lysogens to express target genes, the *lac* repressor acts both at the *lacUV5* promoter in the host chromosome to repress transcription of the T7 RNA polymerase gene by the host polymerase and at the T7*lac* promoter in the vector to block transcription of the target gene by any T7 RNA polymerase that is made. Only a few target genes have been encountered that are too toxic to be stable in these vectors in BL21(DE3) or HMS174(DE3).

1.2. Bacterial Hosts

1.2.1. Hosts for Cloning

As described above, a powerful feature of the pET system is the ability to clone target genes under conditions of extremely low transcriptional activity, i.e., in the absence of a source of T7 RNA polymerase. Background expression is minimal in the absence of T7 polymerase because the host enzyme does not initiate from T7 promoters and the cloning sites in pET plasmids are in regions weakly transcribed (if at all) by read-through activity of bacterial RNA polymerase. Suitable bacterial hosts for cloning include the *E. coli* K12 strains HMS174, HB101, JM109, DH5α, and NovaBlue. These strains are convenient hosts for initial cloning of target DNA into pET vectors and for maintaining plasmids, since they are *rec*A⁻ and give high transformation efficiencies and good plasmid yields. NovaBlue has the additional advantage of having an F factor that allows helper phage infection and hence the production of single-stranded plasmid DNA for mutagenesis purposes (appropriate only for plasmids carrying the f1 origin of replication).

1.2.2. Hosts for Expression

For protein production, a recombinant plasmid is transferred to host *E. coli* strains containing a chromosomal copy of the gene for T7 RNA polymerase. These hosts are lysogens of bacteriophage DE3, a λ derivative that has the immunity region of phage 21 and carries a DNA fragment containing the *lac*I gene, the *lac*UV5 promoter, and the gene for T7 RNA polymerase *(1)*. This fragment is inserted into the *int* gene, preventing λDE3 from integrating into or excising from the chromosome without a helper phage. Once a λDE3 lysogen is formed, the only promoter known to direct transcription of the T7 RNA polymerase gene is the *lac*UV5 promoter, which is inducible by isopropyl-β-D-thiogalactopyranoside (IPTG). Addition of IPTG to a growing culture of the lysogen induces T7 RNA polymerase, which in turn transcribes the target DNA in the plasmid.

Table 2 shows the genotypes and general applications of some strains commonly used with the pET system. λDE3 lysogens of strain BL21 are the most widely used hosts for target gene expression. As an *E. coli* B strain, BL21 lacks the *lon* protease and the *ompT* outer membrane protease that can degrade proteins during purification *(7,8)*. Thus, at least some target proteins should be more stable in BL21 than in host strains that contain these proteases. Since BL21 is sensitive to rifampicin, it is also possible to inhibit transcription by the host RNA polymerase in cases in which a reduction of background synthesis of host RNA and proteins may be desirable. However, some target genes are more stable in HMS174(DE3) for reasons that are unclear *(3)*.

Also useful are strains B834, B834(DE3), and B834(DE3)pLysS, which are the BL21 parental *met* auxotrophs, and BLR, BLR(DE3), and BLR(DE3)pLysS, which are *rec*A⁻ derivatives of BL21 constructed by Alberto Roca, University of Wisconsin. B834 strains are useful for higher-specific-activity ^{35}S-met labeling and selenomethioninyl labeling for crystallography *(9,10)*. A recent report demonstrates significantly higher production of several target proteins in B834(DE3) than in BL21(DE3), which suggests that there may be other advantages to using the parental strain *(11)*. BLR strains may allow better stabilization of target genes containing repetitive sequences. In addi-

Table 2
Commonly Used pET System Host Strains

Strain	Genotype	Description/application(s)	Antibiotic resistance[a]
NovaBlue	endA1 hsdR17(r_{K12}^- m_{K12}^+) supE44 thi-1 recA1 gyrA96 relA1 lac F'[proA$^+$B$^+$ lacI^{q}ZΔM15::Tn10]	Nonexpression[b] host, general purpose cloning, plasmid preps	Tet
NovaBlue(DE3)	endA1 hsdR17(r_{K12}^- m_{K12}^+) supE44 thi-1 recA1 gyrA96 relA1 lac F'[proA$^+$B$^+$ lacI^{q}ZΔM15::Tn10] (DE3)	recA$^-$ K-12 expression[c] host, recommended for use with NovaTope System	Tet
BL21	F$^-$ ompT hsdS$_B$(r_B^- m_B^-) gal dcm	Control nonexpression host	None
BL21(DE3)	F$^-$ ompT hsdS$_B$(r_B^- m_B^-) gal dcm (DE3)	General purpose expression host	None
BL21(DE3)pLysS	F$^-$ ompT hsdS$_B$(r_B^- m_B^-) gal dcm	High-stringency[d] expression host	Cam
BLR(DE3)	F$^-$ ompT hsdS$_B$(r_B^- m_B^-) gal dcm Δ(srl-recA)306::Tn10 (DE3)	recA$^-$ expression host, recommended for use with tandem repeats	None
BLR(DE3)pLysS	F$^-$ ompT hsdS$_B$(r_B^- m_B^-) gal dcm Δ(srl-recA)306::Tn10 (DE3) pLysS	recA$^-$ high-stringency expression host, recommended for use with tandem repeats	Cam, Tet
HMS174	F$^-$ recA hsdR(r_{K12}^- m_{K12}^+) Rif^R	Control nonexpression host	None
HMS174(DE3)	F$^-$ recA hsdR(r_{K12}^- m_{K12}^+) Rif^R (DE3)	recA$^-$ K-12 expression host	None
HMS174(DE3)pLysS	F$^-$ recA hsdR(r_{K12}^- m_{K12}^+) Rif^R (DE3) pLysS	recA$^-$ K-12 high-stringency expression host	Cam
AD494	Δara-leu7967 ΔlacX74 ΔphoAPvuII phoR ΔmalF3 F'[lac$^+$(lacI^q)pro] trxB::kan	trxB$^-$ nonexpression host; allows disulfide bond formation in E. coli cytoplasm	Kan
AD494(DE3)	Δara-leu7967 ΔlacX74 ΔphoAPvuII phoR ΔmalF3 F'[lac$^+$(lacI^q)pro] trxB::kan (DE3)	trxB$^-$ high-stringency expression disulfide bond formation in E. coli cytoplasm	Kan

Strain	Genotype	Description	Antibiotic
AD494(DE3)pLysS	Δara-leu7967 ΔlacX74 ΔphoAPvuII phoR ΔmalF3 F'[lac+(lacIq)pro] trxB::kan (DE3) pLysS	trxB− high-stringency expression host; allows disulfide bond formation in E. coli cytoplasm	Kan, Cam
B834	F− ompT hsdSB(rB− mB−) gal dcm met	Met auxotroph, parent of BL21, control nonexpression host	None
B834(DE3)	F− ompT hsdSB(rB− mB−) gal dcm met (DE3)	Met auxotroph, parent of BL21, general expression host, ^{35}S-met labeling	
B834(DE3)pLysS	F− ompT hsdSB(rB− mB−) gal dcm met (DE3) pLysS	Met auxotroph, parent of BL21, high-stringency expression host, ^{35}S-met labeling	Cam

[a] Antibiotics normally used during cell growth. The appropriate drug to select for the target plasmid should also be added. Tet = tetracycline (12.5 μg/mL), Cam = chloramphenicol (34 μg/mL), Kan = kanamycin (15 μg/mL).

[b] Here, nonexpression means that the strain does not contain the gene for T7 RNA polymerase and therefore will not express from a T7 promoter. These strains are suitable for expression from E. coli promoters, such as lac, tac, and trp.

[c] Expression means that the strain is a λDE3 lysogen, i.e., it carries the gene for T7 RNA polymerase under lacUV5 control. It is, therefore, suited to expression from T7 promoters.

[d] High stringency means that the strain carries pLysS, a pET-compatible plasmid that produces T7 lysozyme, thereby reducing basal expression of target genes. Even greater stringency is provided by pLysE hosts.

tion, the BLR strains may be more stable as λDE3 lysogens because there is no RecA protein to cleave the λ repressor.

Another λDE3 lysogen that may provide additional advantages for other applications is NovaBlue(DE3). This strain carries the *lac*I^q repressor, which provides tighter control over basal expression than the wild-type repressor in the other strains. This host is especially well-suited for use with the higher copy number pSCREEN vector.

The thioredoxin reductase deficient strain AD494(DE3) is a host that is used to maximize soluble protein expression with the pET vectors. The *trx*B$^-$ cells have been shown to permit disulfide bond formation in the cytoplasm of *E. coli* (*12*). Since the *trx*B mutation in AD494(DE3) is maintained by kanamycin selection, this strain is not appropriate for expression of target genes cloned in kanamycin-resistant pET plasmids.

It should be noted that several popular commercial vectors carry T7 promoters and in principle could be used with the pET expression hosts. However, vectors that carry a T7 promoter without an additional source of *lac* repressor are inappropriate because multiple copies of the operator will titrate repressor and partially induce the gene for T7 RNA polymerase, which is also controlled by *lac* repressor. As a result, basal polymerase activity becomes high enough that many target genes cannot be stably maintained.

1.2.3. pLysS and pLysE Hosts

One way of providing additional stability to target genes is to express them in host strains containing a compatible plasmid that provides a small amount of T7 lysozyme, a natural inhibitor of T7 RNA polymerase (*13,14*). In addition to binding to the polymerase, T7 lysozyme also cleaves a bond in the peptidoglycan layer of the *E. coli* cell wall (*15*). Relatively high levels of T7 lysozyme in the cytoplasm can be tolerated; however, cells will lyse under mild conditions, such as addition of Triton X-100 to 0.1% or freeze/thaw treatment (which can be convenient for the preparation of cell extracts).

pLysS and pLysE contain the T7 lysozyme gene (bp 10,665–11,296 of T7 DNA) cloned in opposite orientations in the *Bam*HI site of pACYC184 (*3,16,17*). In pLysE, the lysozyme gene is expressed at high levels from the *tet* promoter in the plasmid. Cells harboring pLysS accumulate much lower amounts of enzyme, presumably as the result of transcription around the plasmid from the weak T7 φ3.8 promoter immediately following the gene (*3,18*). Both plasmids are maintained by selection with chloramphenicol and contain the p15A origin of replication, which is compatible with the pBR322 and pUC origins in the pET and pSCREEN vectors, respectively.

The presence of either pLysS or pLysE increases the tolerance of λDE3 lysogens for plasmids with toxic inserts: Unstable plasmids become stable, and plasmids that would not otherwise be established can be maintained and expressed. Because pLysE causes slower growth and a tendency toward lysis, its use is somewhat less convenient in most cases. For very toxic genes, the combination of a T7*lac* promoter-containing vector and pLysS is preferable.

1.3. Bacteriophage CE6

An alternative for toxic genes is to introduce the T7 RNA polymerase by infection with bacteriophage CE6, a λ recombinant that carries the cloned polymerase gene under control of the phage p_L and p_I promoters, the *cI857* thermolabile repressor, and the *Sam*7 lysis mutations (*1*). When CE6 infects HMS174, the newly made T7 RNA polymerase transcribes target DNA so actively that normal phage development cannot proceed. Although this method is less convenient than induction of DE3 lysogens, it can be used if target gene products are too toxic to be maintained any other way. No T7 RNA polymerase will be present in the cell before infection, so any target DNA that can be cloned under control of a T7 promoter should be expressible in this way.

2. Materials

2.1. Growth Media

A wide range of growth media is suitable for growth of strains and expression of target DNAs in the pET System. Suitable growth media are M9ZB, LB broth, and TB ("terrific broth").

1. LB: Per liter: 10 g Bacto-tryptone, 5 g yeast extract, 10 g NaCl, pH to 7.5 with 1 N NaOH, autoclave. For plates, add 15 g agar per liter prior to autoclaving.
2. M9ZB *(3)*: Per liter: 10 g N-Z-amine A (Sheffield Products, Norwich, NY), 5 g NaCl, autoclave, cool, add 100 mL 10X M9 salts, 1 mL 1 M MgSO$_4$, 10 mL 40% glucose (from autoclaved stocks).
3. 10X M9 salts: Per liter: 10 g NH$_4$Cl, 30 g KH$_2$PO$_4$, 60 g Na$_2$HPO$_4$·7H$_2$O; autoclave.
4. TB *(19)*: Per liter: 12 g Bacto-tryptone, 24 g yeast extract, 4 mL glycerol; autoclave, cool to 60°C, and add 100 mL sterile potassium phosphate.
5. Potassium phosphate: Per liter: 23.1 g KH$_2$PO$_4$, 125.4 g K$_2$HPO$_4$; autoclave.
6. Autoclaved 80% glycerol, for strain storage.
7. 100 mM IPTG (isopropyl β-D-thiogalactopyranoside): 2.38 g IPTG in 100 mL deionized water. Filter sterilize and store at –20°C.
8. Carbenicillin (disodium salt): 50 mg/mL in deionized water. Store at –20°C. Use at 50 µg/mL.
9. Ampicillin (sodium salt): 25 mg/mL in deionized water. Store at –20°C. Use at 50 µg/mL.
10. Chloramphenicol: 34 mg/mL in ethanol. Store at –20°C. Use at 34 µg/mL.
11. Kanamycin (sulfate): 30 mg/mL in deionized water. Store at –20°C. Use at 30 µg/mL for Kan-resistant plasmids and at 15 µg/mL for AD494 strains.

2.2. Cloning and Screening for Target Genes

1. pET system plasmids, host strains, and vector primers (Novagen, Madison, WI).
2. dNTP mix: 10 mM each dATP, dCTP, dGTP, dTTP, neutralized to pH approx 7.0.
3. 10X *Taq* buffer: 100 mM Tris-HCl, pH 8.8, at 25°C, 500 mM KCl, 1% Triton X-100.
4. 25 mM MgCl$_2$.
5. *Taq* DNA polymerase (Perkin-Elmer, Foster City, CA).
6. TE buffer: 10 mM Tris-HCl, pH 8.0, at 25°C, 1 mM EDTA.
7. Single Tube Protein™ System 3, T7 (Novagen).

2.3. Plasmid Purification

1. Solution I: 50 mM glucose, 25 mM Tris-HCl, pH 8.0, 10 mM EDTA.
2. Solution II: 0.2 N NaOH, 1% sodium dodecyl sulfate (SDS), must be freshly prepared from stocks of 5 N NaOH and 10% SDS.
3. Solution III: 3 M sodium acetate, pH 5.2.
4. TE-buffered phenol:chloroform: isoamyl alcohol (25:24:1).
5. 100% Ethanol.
6. RNase A, 10 mg/mL (DNase-free; *19*).
7. 30% Polyethylene glycol (PEG), 1.5 M NaCl (prepared from autoclaved stocks of 50% (PEG-8000 and 5 M NaCl).

2.4. Cell Fractionation and Metal Chelation Chromatography

1. Cell extract buffer: 50 mM Tris-HCl, pH 8.0, at 25°C, 2 mM EDTA.
2. Chicken egg lysozyme.
3. 1 M MgCl$_2$.
4. 1 mg/mL DNase I.
5. 1% Triton X-100.
6. 2X SDS sample buffer: 250 mM Tris-HCl, pH 6.8, at 25°C, 2% SDS, 20% glycerol, 0.005% bromphenol blue; add 2-mercaptoethanol to 4% immediately before use.
7. His·Bind resin (Novagen) and column.
8. 8X Charge buffer: 400 mM NiSO$_4$.
9. 8X Binding buffer: 40 mM imidazole, 4 M NaCl, 160 mM Tris-HCl pH 7.9.
10. 8X Wash buffer: 480 mM imidazole, 4 M NaCl, 160 mM Tris-HCl, pH 7.9.
11. 4X Elute buffer: 4 M imidazole, 2 M NaCl, 80 mM Tris-HCl, pH 7.9.
12. 4X Strip buffer: 400 mM EDTA, 2 M NaCl, 80 mM Tris-HCl, pH 7.9.

2.5. Site-Specific Protease Cleavage Buffers

1. Factor Xa cleavage buffer: 100 mM NaCl, 50 mM Tris-HCl, pH 8.0, 5 mM CaCl$_2$.
2. Enterokinase cleavage buffer: 50 mM NaCl, 20 mM Tris-HCl, pH 8.0, 2 mM CaCl$_2$.
3. Thrombin cleavage buffer: 20 mM Tris-HCl, pH 8.4, 150 mM NaCl or KCl, and 2.5 mM CaCl$_2$.

3. Methods

3.1. Cloning Target Genes into pET Vectors

3.1.1. Construction of Recombinants

Initial cloning of target genes in pET vectors is straightforward and amenable to standard cloning techniques (19). A wide variety of cloning sites are available that allow fusion with a selection of N-terminal and C-terminal fusion peptide sequences for detection and purification of target proteins. Initial cloning should be done in a *rec*A$^-$ strain, such as NovaBlue, or other similar host that does not carry the T7 RNA polymerase gene. This allows high percentage monomer plasmid yields for examination of the construct sequence, as well as separation of cloning from expression. This separation can be valuable in troubleshooting any difficulties that might arise during later procedures. Once a plasmid is established in a nonexpression host, it is transferred into a λDE3 lysogen for high-level expression. The strains described above for cloning and expression with pET vectors can be prepared for transformation by standard procedures. Expect BL21 and its derivatives to be transformed at about 1/10 the efficiency of the other strains.

3.1.2. Screening

If cloning was successful there are usually many more colonies produced from ligation in the presence of the insert than with the negative control (vector ligated without insert). However, cloning can be successful even though the number of colonies resulting from ligations plus and minus insert are roughly equivalent. Colonies can be screened for inserts without the need for minipreps by direct-colony PCR using vector-specific primers by standard methods; we present one such procedure here.

pET Expression and Purification

In addition, pET constructs can be quickly evaluated for expression of the desired target protein using the Single Tube Protein System 3 (Novagen). For this purpose, PCR-amplified DNA is used directly in a linked in vitro transcription/ translation reaction to produce ^{35}S-labeled or unlabeled protein. The PCR step can be performed with ligations of vector plus insert, or with colonies from the transformation. The colony procedure is especially useful when target sequences are cloned via PCR, since it allows rapid screening for PCR-related introduction of unwanted stop codons in individual clones. To prepare suitable templates for transcription/translation, appropriate primers must be used that allow amplification of the T7 promoter along with the correct orientation of the target insert. The T7 promoter primer is not suitable for this application since T7 RNA polymerase does not transcribe efficiently if the promoter is at the very end of the molecule. The pET upstream primer is an appropriate 5' primer for all pET vectors described here except pET-17b, 17xb, 20b(+), 23(+), 23a-d(+), and pSCREEN. By using an insert-specific 3' primer, the correct orientation of insert is preferentially amplified, which is necessary if the ligation is into a single restriction site (i.e., nondirectional). The T7 terminator primer is appropriate as the 3' primer for nonorientation-specific amplification of target genes in all pET vectors except the pET-5 series. A protocol for postligation PCR is also provided in **Subheading 3.1.2.2.**

3.1.2.1. COLONY PCR

1. Pick a colony from an agar plate using a 200-μL pipet tip or sterile toothpick. Choose colonies that are at least 1 mm in diameter and try to get as many cells as possible. If a "copy" of the colony is desired, touch the pipet tip to a plate before transferring the bulk of the colony to the tube in the next step.
2. Transfer the bacteria to a 1.5-mL tube containing 50 μL of sterile water. Vortex to disperse the pellet.
3. Place the tubes in boiling water or a heat block at 99°C for 5 min to lyse the cells and denature DNases.
4. Centrifuge at 12,000g for 1 min to remove cell debris.
5. Transfer 10 μL of the supernatant to a fresh 0.5-mL tube for PCR. Leave on ice until use.
6. Make a master reaction mix as follows (*see* **Note 1**): Per reaction:
   a. 28.8 μL Sterile water.
   b. 5 μL 10X *Taq* buffer.
   c. 1 μL pET Upstream primer (5 pmol).
   d. 1 μL Downstream primer (5 pmol) (*see* **Note 2**).
   e. 1 μL dNTP mix.
   f. 3 μL 25 m*M* MgCl$_2$.
   g. 0.25 μL (1.25 U) *Taq* DNA polymerase.
   Mix together the above components in a single tube using amounts corresponding to the number of reactions desired.
7. Add 40 μL of the master mix to each sample, mix gently, add 2 drops of mineral oil, cap the tubes, and put the samples in a thermal cycler. Process for 35 cycles for 1 min at 94°C, 1 min at 55°C, and 2 min at 72°C, with a final extension at 72°C for 6 min (*see* **Note 3**).
8. Transfer the reaction from underneath the oil layer to a fresh tube. Inactivate the polymerase by adding 100 μL of chloroform, vortex mixing 30 s, and centrifuging for 1 min. The top aqueous phase (which may appear cloudy) contains the DNA products. If desired, remove a 10-μL sample for gel analysis.
9. Add 2 μL of the reaction directly to a Single Tube Protein System 3 reaction for protein production.

3.1.2.2. POSTLIGATION PCR

1. Assemble the following components for ligation PCR (*see* **Note 1**):
   a. 2 µL Ligation reaction diluted 1:10 in TE (1–2 ng vector plus a 3–10-fold molar excess of insert).
   b. 5 µL 10X *Taq* buffer.
   c. 1 µL pET Upstream primer (5 pmol).
   d. 1 µL Downstream primer (5 pmol) (*see* **Note 2**).
   e. 1 µL 10 mM dNTP mix.
   f. 3 µL 25 mM MgCl$_2$.
   g. 0.25 µL (1.25 U) *Taq* DNA polymerase.
   h. 36.75 µL Sterile deionized water.
   i. 50 µL Total volume.
2. Add the enzyme or DNA last to start the reaction, mix gently, and add 2 drops of mineral oil from a 200-µL pipet tip to prevent evaporation. Optimal results are usually obtained by heating the assembled reaction to 80°C prior to addition of the enzyme or DNA. Insert the tubes in a thermal cycler and carry out the PCR for 30 cycles: 1 min at 94°C, 1 min at the proper annealing temperature (usually 55°C for vector primers), 2 min at 72°C, with a 6 min final extension at 72°C (*see* **Note 3**). Check the products by agarose gel electrophoresis of a 5-µL sample. A strong band should appear that corresponds to the distance between (and including) the primers. Extract the reaction with chloroform as described in **step 8** of **Subheading 3.1.2.1.**
3. Under most conditions, 2 µL of the ligation PCR reaction can be used directly in the transcription/translation reaction without.

3.1.3. Plasmid Purification

After positive clones are identified, the plasmids can be isolated for transformation into expression hosts, restriction mapping, and sequence analysis. The following protocol is a slight modification of one presented in Sambrook et al. *(19)*. Since pET plasmids carry the relatively low-copy pBR322 origin of replication, a 1.5-mL culture yields only 1–2 µg plasmid DNA. This is sufficient for transfer into an expression host, but not enough for other types of analysis. The protocol can be scaled up proportionally if more plasmid is needed (e.g., a 50-mL prep will give 35–75 µg pET plasmid or 300–400 µg pSCREEN plasmids, which have the high-copy pUC origin of replication).

1. Using a sterile loop, toothpick, or pipet tip, transfer a well-isolated colony into 3 mL of LB broth supplemented with the appropriate antibiotic(s), in a Falcon 2059 culture tube. Cap loosely and incubate with shaking at 37°C 6 h to overnight.
2. Transfer 1.5 mL of culture into a 1.5-mL microcentrifuge tube and centrifuge at 12,000g for 1 min.
3. Remove the medium by aspiration, leaving the pellet as dry as possible.
4. Resuspend the cells in 100 µL of ice-cold solution I. Pipet up and down to make sure that the pellet is completely suspended.
5. Add 200 µL of freshly prepared solution II. Mix by inversion and let sit on ice for 3 min.
6. Add 150 µL of ice-cold solution III. Mix by inversion and leave on ice for 5 min.
7. Centrifuge at 12,000g for 5 min. Transfer the clear supernatant to a fresh tube, avoiding the pellet, which tends to break up easily. Spin again if too much particulate matter remains in the supernatant.
8. Add 400 µL TE-buffered phenol:chloroform: isoamyl alcohol, vortex for 30 s, and centrifuge at 12,000g for 1 min at room temperature.

9. Transfer the top aqueous phase to a fresh tube and add 800 μL ethanol. Vortex, leave at room temperature for 2 min, and centrifuge at 4°C, 12,000g for 5 min.
10. Decant the supernatant and add 400 μL ethanol to the pellet. Spin briefly, pour off the ethanol, and allow the pellet to air dry in an inverted position for about 10 min.
11. Resuspend the pellet in 30 μL of TE buffer containing 20 μg/mL RNase A and incubate at 37°C for 15 min.
12. At this point the DNA can be used for transformation of expression hosts. However, it is advisable to check the construct prior to transforming the expression host, either by the Single Tube Protein method described above or by sequencing. Note that RNase must be removed by phenol:chloroform extraction prior to Single Tube Protein System Analysis.

3.1.4. Sequencing

For double-stranded plasmid sequence analysis the preparation must be further processed to remove RNA breakdown products. This can be easily accomplished by precipitation with PEG.

1. Add 10 μL of 30% PEG 1.5 M NaCl, vortex thoroughly, and incubate on ice for 60 min.
2. Centrifuge at 12,000g at 4°C for 10 min. Carefully remove the supernatant, leaving the small transparent DNA pellet behind. Rinse the pellet successively with 70% ethanol and then 100% ethanol, and let air dry.
3. Resuspend the DNA in 20 μL TE. The plasmid is now suitable for alkali denaturation and double-stranded sequencing.

3.1.5. Storage of Strains

Permanent stocks of hosts and pET recombinants are best kept as glycerol stocks. Note that high glycerol concentrations (>10%) may lead to plasmid instability. For the preparation of stock cultures of host strains and pET recombinants, inoculate a single colony into 50 mL media containing appropriate antibiotic(s) in a 250-mL flask. Incubate with vigorous shaking at 37°C during the day until the OD_{600} reaches 0.6–0.8. Remove 0.9 mL to a cryovial, add 0.1 vol of 80% glycerol, mix well, and store at –70°C. Plasmid-bearing strains, particularly those having any tendency toward instability, are titered at the time of freezing to be sure that the vast majority of cells in the culture have the intended host–plasmid combination (*see* **Subheading 3.3.1.**). To inoculate a culture from the frozen stock, a few μL are scraped or melted from the surface, typically with a sterile pipet tip or plastic culture loop, and the remainder is returned to the –70°C freezer without thawing.

3.2. Target Gene Expression

3.2.1. Transformation of Expression Host

For transforming into an expression host, use 1 μL of a 50-fold dilution (approx 1 ng) of plasmid in sterile water or TE and follow any standard transformation procedure *(19)*.

3.2.2. Induction of λDE3 Lysogens

After a target plasmid is established in a λDE3 lysogen, or in one of these strains containing pLysS or pLysE, expression of the target DNA is induced by the addition of IPTG to a growing culture. For pET constructions carrying the "plain" T7 promoter, a concentration of 0.4 mM IPTG is recommended, while 1 mM IPTG is recommended with vectors having the T7*lac* promoter. An example of an induction protocol is presented as follows.

Pick a single colony from a freshly streaked plate and inoculate 50 mL LB containing the appropriate antibiotic in a 250-mL Erlenmeyer flask. For good aeration, add medium up to only 20% of the total flask volume. (NOTE: Include 34 µg chloramphenicol/mL if the cells carry pLysS or pLysE, or 15 µg/mL kanamycin for *trx*B mutants.)

Alternatively, inoculate a single colony or a few microliters from a glycerol stock into 2 mL LB medium containing the appropriate antibiotic. Incubate with shaking at 37°C until the OD_{600} reaches 0.6–1.0. Store the culture at 4°C overnight. The following morning, collect the cells by centrifugation (30 s in a microcentrifuge). Resuspend the cells in 2 mL fresh medium and use this to inoculate 50 mL medium.

1. Incubate with shaking at 37°C until OD_{600} reaches 0.4–1 (0.6 is recommended; about 3 h).
2. Remove samples for the uninduced control and for titering as described in **Subheading 3.3.1.** Add IPTG from a 100 m*M* stock to a concentration of 0.4 m*M* (T7 promoter) or 1 m*M* (T7*lac* promoter) and continue the incubation for 2–3 h. When using AD494 strains, additional shaking at 4°C for 2 or more hours may facilitate the production of properly folded target protein.
3. Place the flasks on ice for 5 min and then harvest the cells by centrifugation at 5000*g* for 5 min at 4°C. Resuspend the cells in 0.25 culture volume of cold cell extract buffer and centrifuge as above. Remove the supernatant and store the cells as a frozen pellet at –70°C or continue with purification (note that inclusion bodies become less soluble on "aging" in the freezer). Also note that the cells will lyse when thawed if they carry pLysS or pLysE.

3.3. Optimizing Expression

3.3.1. Plasmid Stability Test

In most cases, the synthesis of target proteins driven by T7 RNA polymerase compromises the cell's ability to carry out normal functions, thereby slowing its growth. Therefore, in the absence of selection cultures tend to become overgrown with cells that have either lost the target plasmid or the ability to express the target protein. The use of ampicillin for selection requires particular attention because it tends to be broken down rapidly in an actively growing cultures *(3)*.

Immediately before induction, it is recommended to test the culture to determine the fraction of cells that carry the target plasmid, especially when using ampicillin-resistant plasmids. This involves plating on four plates that differ in the additions to the LB-agar used in plating. Plate the cells at a dilution of 10^{-5} on plates that have (a) both IPTG and antibiotic or (b) just IPTG, and at a dilution of 2×10^{-6} on plates that have (c) just antibiotic or (d) nothing added to the agar. For this test, it is convenient prepare IPTG plates by spreading 200 µL of a 100 m*M* sterile solution on the plate containing about 20 mL hardened LB-agar with or without 50 µL/mL ampicillin and allowing it to soak in for 30 min at room temperature before plating the cells. Count the colonies after incubation overnight at 37°C.

A comparison of plates (c) and (d) will give the proportion of cells that contain the plasmid (this should typically be >98% in properly maintained cultures). When the target plasmid has a "plain" T7 promoter and is carried in a λDE3 lysogen, IPTG usually prevents colony formation if the cell is expressing the target gene, but will not do so if the cell has lost the plasmid or the ability to express the target protein. In pLysS hosts, IPTG also inhibits colony formation (except in certain cases, including pET-3 and some vectors that carry the T7*lac* promoter without a cloned insert). In pLysE hosts, IPTG usually does not prevent colony formation unless the target gene product is toxic. Similarly, IPTG occasionally does not suppress colony formation with plasmids carrying the T7*lac* promoter and relatively innocuous target genes, especially in the presence of pLysS.

With the above exceptions, typically <2% of the cells should form colonies on the IPTG plate and <0.01% should form colonies on the IPTG-plus-ampicillin plate. If the target plasmid is unstable, there will be an increase in the number of colonies on the IPTG plates and a corresponding decrease in the number of colonies on the ampicillin-only plate. In most cases instability is caused by loss of the target plasmid rather than to mutations that prevent expression *(3)*.

If appropriate attention is paid to the possibility of plasmid instability, >98% of the cells in the culture will usually contain expressible target plasmid. Cells are usually harvested 2–3 h after induction, enough time for substantial accumulation of target protein but before the culture can be overgrown with cells that have lost plasmid or are otherwise unproductive. However, some target proteins continue to accumulate for much longer times.

3.3.2. Protein Solubility

Since the characteristics of each target gene product are unique, the optimal scheme and time-course for induction can vary. For example, growth at 37°C causes some proteins to accumulate as inclusion bodies, whereas incubation at 30°C leads to soluble, active protein *(20)*. If export of the target protein is desired using the signal sequence leaders present in pET-12a-c, pET-20b(+), pET-22b(+), pET-25b(+), pET-26b(+), or pET-27b(+), growth and induction at 25 or 30°C may be optimal. Inclusion bodies can be an advantage for purification since they are easily isolated by centrifugation to yield highly concentrated and relatively pure protein. With many proteins it has been possible to solubilize the inclusion body protein with urea or guanidine-HCl and then refold by slowly diluting or dialyzing out the denaturant at low protein concentrations *(21)*. Many other refolding schemes have been reported; optimal protocols vary with the target protein and must be determined empirically.

Even when inclusion bodies are formed, some proportion of the target protein is usually soluble within the cell. With the high expression levels of the pET system and the power of the His·Tag purification method, even when most of the target protein mass is in aggregates there may be a significant amount of soluble material as well. It should be noted that solubility does not necessarily indicate that a protein is folded properly; some proteins form soluble aggregates that are inactive. If a significant fraction is soluble and active, the induction can be scaled up in volume and the affinity chromatography step can be used to purify and concentrate the soluble fraction.

Solubility can also be manipulated by the choice of vector, cloning sites, or host cells *(22)*. For example, the use of a pET-32 series vector to produce a thioredoxin fusion protein may increase the yield of soluble product *(23,24)*. The use of AD494 strains may also allow the formation of disulfide bonds in the *E. coli* cytoplasm, which may affect the solubility and/or activity of a given target protein.

3.3.3. Time-Course and Yield

To study the time course of target protein expression after induction, save 500-µL aliquots of the culture at 30- or 60-min intervals following IPTG addition.

1. Collect cells by centrifugation and remove the supernatant.
2. Suspend the cell pellet in 50 µL of 2X SDS sample buffer and freeze at –20°C.
3. When all the time-points are collected, heat to 70°C for 5 min and run 2.5–10 µL/lane on an SDS-polyacrylamide gel using standard methods.

3.3.4. Precautions to Maximize Expression

Simple precautions are advisable to maximize retention of plasmid through the procedures for isolating, maintaining, and expressing target plasmids (*see* **Subheading 3.3.1.**). The use of carbenicillin in place of ampicillin helps to prevent overgrowth of cultures by cells that have lost the plasmid.

Another alternative is to choose a pET vector containing the kanamycin resistance marker instead of the *bla* gene. A more detailed discussion of the potential advantages of kan^R compared to amp^R is presented in **Subheading 1.1.1.** The following isolation protocol usually produces the highest possible fraction of cells containing functional ampicillin-resistant target plasmid.

1. Inoculate a colony from the transformation plate into 2 mL LB + 50 µg/mL carbenicillin and incubate for a few hours, until the culture becomes slightly turbid.
2. Streak a sample on a plate containing carbenicillin to obtain a single colony.
3. As soon as the colony develops (usually overnight at 37°C), inoculate into 2 mL LB + carbenicillin and grow until $OD_{600} = 1.0$.
4. Mix 0.9 mL of culture with 0.1 mL of 80% glycerol in a cryovial and store in a –70°C freezer. If there is any question about the possible stability of the plasmid, perform the plasmid stability test described in **Subheading 3.3.1.** at the time of freezing to determine what fraction of the cells contain functional target plasmid.

The following induction protocol has been used successfully with an extremely toxic gene in pET-22b(+). It involves the use of a high concentration of carbenicillin and replacing the medium twice prior to induction.

1. Inoculate a single colony into 2 mL TB + 200 µg/mL carbenicillin. Grow the cells at 37°C until $OD_{600} = 0.2$–0.6.
2. Collect the cells by centrifugation (30 s in a microcentrifuge), remove the supernatant and resuspend in 2 mL fresh media. Add a 100 µL sample to 8 mL TB + 500 µg/mL carbenicillin and grow the culture at 37°C until $OD_{600} = 0.2$–0.6.
3. Collect the cells by centrifugation at 1000*g* for 5 min and resuspend in fresh TB + 500 µg/mL carbenicillin containing 1 m*M* IPTG. Incubate at 30°C for 2 h before harvest.

3.4. Purification of Target Proteins

3.4.1. Localization and Detection

Many approaches can be used to purify target proteins expressed with the pET System. One advantage of the system is that in many cases the target protein accumulates to such high levels that it constitutes a high percentage of the total cell protein. Therefore, it may be relatively straightforward to isolate the protein in two or three chromatographic steps by conventional methods (ion exchange, gel filtration, and so on). In these cases it is possible to obtain a good idea about relative purity and expression level by SDS-polyacrylamide gel analysis.

However, it is often necessary to follow target proteins through purification procedures using more sensitive and specific assays. Although protein-specific antibodies or

Table 3
Reagents for Detection, Assay, and Purification of pET System Fusion Proteins

Reagent	Applications
T7·Tag monoclonal antibody (MAb)	Western blot, immunoprecipitation, immunofluorescence
T7·Tag antibody alkaline phosphatase conjugate	Western blot
Biotinylated T7·Tag antibody	Western blot, immunofluorescence
T7·Tag antibody agarose	Immunoaffinity purification
HSV·Tag monoclonal antibody	Western blot, immunoprecipitation, immunofluorescence
S·Tag grade S-protein	Rapid homogeneous quantification
Biotinylated S-protein	Western blot
S-protein alkaline phosphatase conjugate	Western blot
S-protein agarose	Affinity purification
Recombinant enterokinase	Removal of fusion peptides
Restriction grade factor Xa	Removal of fusion peptides
Restriction grade thrombin	Removal of fusion peptides
Biotinylated thrombin	Removal of fusion peptides
His·Bind resins	Metal chelation chromatography

activity determinations can be used, it is often more convenient to take advantage of vector-encoded fusion domains for which reagents are available. The pET and pSCREEN plasmids offer a variety of vector-encoded peptide tags, including 11aa and 260aa T7·Tag, N-terminal and C-terminal His·Tag, 15aa S·Tag, 109aa Trx·Tag, and 11aa HSV·Tag sequences, which allow target protein detection, measurement of expression levels and purification by several independent strategies. Reagents available for detection and purification of fusion proteins containing these peptides are listed in **Table 3**.

Prior to purification it is useful to determine where in the cell the target protein is accumulated. Preferential localization to inclusion bodies or the periplasmic space can be advantageous for rapid purification by relatively simple procedures. Several methods for analyzing crude cell fractions are presented below.

3.4.2. Total Cell Protein

For cells induced at around $OD_{600} = 0.6$ and harvested 3 h later, a typical yield is about 0.5–1 mg total protein per milliliter of culture. Uninduced cells have about one-third of this amount of total protein. Since roughly 10 µg of total protein is needed for a gel lane on a typical minigel apparatus (0.75-mm thick, 10-well comb), protein from 10–20 µL of induced cells and 50 µL of uninduced cells should give proper protein band intensities after Coomassie blue staining. For rapid analysis, collect from these amounts of culture by centrifugation ($12,000g$ for 1 min), suspend in 2X SDS sample buffer by vigorous vortex mixing, heat to 70°C, and use for electrophoresis.

Better results are usually obtained when the cells are lysed and DNA sheared prior to addition of sample buffer. Collect 1 mL cells by centrifugation (1 min at $12,000g$) and resuspend in 0.5 mL cold cell extract buffer. Break the cells and shear the DNA by brief sonication with a microtip, or addition of 0.1 mL of 10 mg/mL lysozyme followed by 5 µL each of 1 M $MgCl_2$ and 1 mg/mL DNase I and incubation at 37°C for 15 min. Then, add a sample to an equal volume of 2X SDS sample buffer for gel analysis.

An empirical formula for calculating the volume to load is 180/(concentration factor × OD_{600} at time of harvest), where the concentration factor is the volume harvested/processed volume (typically 0.1–0.05). For Western blots, usually 1/20–1/100 of the protein loaded for Coomassie staining gives a good signal.

3.4.3. Soluble and Insoluble Fractions

As discussed in **Subheading 3.3.2.**, many target proteins are expressed in both soluble and insoluble forms. Crude soluble and insoluble fractions can be prepared by the following protocol. Note that although this procedure is appropriate for analysis by SDS-polyacrylamide gel electrophoresis (PAGE), it is not intended for use with the His·Bind column, since it uses a buffer containing EDTA (*see* **Subheading 3.4.5.** for cell extract preparation for the metal chelation chromatography purification system).

This protocol will work with any of the pET host strains; although the lysozyme addition could be omitted with strains having pLysS or pLysE.

1. Collect induced cells (50 mL culture) by centrifugation at 5000*g* for 5 min. Discard the supernatant and resuspend the cell pellet in 1/10 culture vol (5 mL) of cell extract buffer.
2. Add lysozyme to a concentration of 100 µg/mL; use a 10 mg/mL stock freshly prepared in the buffer used in **step 1**. Then add 1/10 vol (0.5 mL) 1% Triton X-100. Incubate at 30°C for 15 min.
3. Place the tube in an ice bath and sonicate with a microtip (Branson Sonifier model 450, Danbury, CT) to shear the DNA. The solution should lose viscosity after one or two 10-s pulses at a high output setting. Note that over sonication can denature some proteins.
4. Centrifuge at 12,000*g* for 15 min at 4°C. The supernatant contains soluble proteins; add an equal volume of 2X SDS sample buffer to a sample for gel analysis. The pellet is the insoluble fraction; resuspend in 1X SDS sample buffer for gel analysis. An amount of sample corresponding to 30 µL of the original culture volume is usually sufficient for bands to be visualized by Coomassie blue staining.

The insoluble fraction (inclusion bodies) can be further purified by repeated centrifugation and washing steps; however, the product will be contaminated at some level with other proteins and nucleic acids. In many cases purified inclusion bodies are suitable for direct use as antigens for the preparation of antibodies against the target protein.

3.4.4. Periplasmic Fraction

When using vectors having signal sequence leaders, target proteins may be directed to the periplasmic space. The leader is necessary, but not sufficient, for export into the periplasm. Translocation across the cell membrane of *E. coli* is incompletely understood *(25)*. However, it is clear that translocation also can depend on the mature domain of the target protein, which is recognized by SecB, the major chaperone of export. The following osmotic shock protocol *(21)* is a simple method of preparing the periplasmic fraction from λDE3 lysogens. However, this method is not appropriate for use with host strains containing pLysS or pLysE because T7 lysozyme causes disruption of the inner membrane.

1. Centrifuge an induced culture at 7000*g* for 10 min at 4°C. Discard the supernatant.
2. Resuspend cells in 0.4 culture volumes (80 mL/g cells) of 30 m*M* Tris-HCl, pH 8.0, 20% sucrose. Add EDTA to 1 m*M* and incubate 5–10 min at room temperature with shaking or stirring.

3. Centrifuge at 10,000g for 10 min at 4°C. Remove as much supernatant as possible and resuspend the pellet in the same volume used in **step 2** of ice-cold 5 mM MgSO$_4$. Shake or stir for 10 min on ice.
4. Centrifuge as in **step 3**. The supernatant is the periplasmic fraction.

A slightly different procedure has been reported by LaVallie et al. *(24)* for extraction of *trx*A fusion proteins, which can be produced by pET-32 series vectors.

1. Resuspend induced cells in ice-cold 20% sucrose 2.5 mM EDTA, 20 mM Tris-HCl, pH 8.0, to a concentration of 5 OD$_{600}$ U/mL and incubate on ice for 10 min.
2. Centrifuge at 15,000g for 30 s, decant the pellet, and resuspend in the same volume of ice-cold 2.5 mM EDTA, 20 mM Tris-HCl, pH 8.0. Incubate on ice for 10 min.
3. Centrifuge at 15,000g for 10 min. The supernatant is the osmotic shock fraction. Analyze supernatant and pellet by SDS-PAGE.

3.4.5. Rapid Affinity Purification by Metal Chelation Chromatography

Many pET vectors have the advantage of carrying the His·Tag sequence, a stretch of either 6 or 10 consecutive histidine residues that can be expressed at the N-terminal or C-terminal end of the target protein. The His·Tag sequence binds to divalent cations (e.g., Ni^{2+}) immobilized on the His·Bind metal chelation resin. After unbound proteins are washed away, the target protein is recovered by elution with imidazole. Columns can be regenerated and reused many times. This versatile system provides a convenient, economical means of purification without the need to develop new protocols for each protein. It also allows the purification of target proteins under gentle, native conditions for maintaining activity of soluble proteins, as well as under denaturing conditions necessary for solubilization of inclusion bodies. Up to 20 mg of target protein can be purified on a single 2.5-mL column.

This procedure starts with a cell culture that has been induced for target protein production. The target protein should contain a stretch of at least six consecutive histidine residues (His·Tag sequence) located at the N-terminus or C-terminus. The His·Tag sequence can probably also occur elsewhere in the polypeptide and still be used for affinity purification. A detailed discussion of the induction of target proteins is found in **Subheading 3.2.**

The following protocols are for 100 mL of induced culture. At this scale, expect the entire procedure to take about 4–6 h. For proteins that are highly expressed in the pET System, a yield of 20 mg of target protein per 100 mL culture is not unusual. Whereas some target proteins remain soluble in the cytoplasm, other proteins form insoluble aggregates, or inclusion bodies, in *E. coli*. In addition to the target protein, inclusion bodies also contain contaminating bacterial proteins and nucleic acids.

The first protocol described below is for purification of soluble proteins under native conditions. If most or all of the target protein is insoluble, it is possible to solubilize the protein and run the His·Bind purification under denaturing conditions. Modifications required for denaturing conditions are found following the first protocol. Note that the 1X binding, wash, and elute buffers all contain 500 mM NaCl, which reduces nonspecific ionic interactions of cellular proteins with the resin.

It is preferable to prepare the column and cell extract simultaneously, and run the column in the same day. Proteins in the cell extract can be unstable, so it is best to minimize the amount of time the cell extract is left sitting prior to loading it on the column. If desired, after the column has been charged, it can be stored in 1X binding buffer overnight at 4°C. Be careful not to allow the column to dry out.

3.4.5.1. Resin Preparation

1. Dilute the stocks of charge buffer, binding buffer, wash buffer, and elute buffer to 1X with sterile deionized water before use.
2. Gently mix the bottle containing His·Bind resin by inversion until completely suspended. Small polypropylene columns from a number of commercial sources conveniently hold 2.5 mL of settled resin, which can be used to purify up to 20 mg of target protein. Add a few milliliters of sterile deionized water to the dry column and gently push on the column top with a gloved finger to wet the frit and start the column flowing. Using a wide-mouth pipet, transfer the desired amount of slurry to a column. Allow the resin to pack under gravity flow.
3. When the level of storage buffer (20% ethanol) drops to the top of the column bed, use the following sequence of washes to charge and equilibrate the column (where 1 vol is equivalent to the settled bed volume):
   a. 3 Vol sterile deionized water
   b. 5 Vol 1X charge buffer
   c. 3 Vol 1X binding buffer

Note: The resin can also be used in a batchwise fashion for small scale purifications. Settled bed volumes of 50–200 µL can be handled in 1.5-mL microcentrifuge tubes. The His·Bind resin matrix will withstand low-speed centrifugation. For each wash, add the appropriate buffer, invert the tube several times to mix, and spin for 1 min at 400–1000g.

3.4.5.2. Cell Extract Preparation (see Note 4)

1. Harvest the cells by centrifugation at 5000g for 5 min. Decant the supernatant and allow the cell pellet to drain as completely as possible. Resuspend the cells in 4 mL ice-cold binding buffer (the amount of binding buffer can be scaled up or down proportionally; this is for cells from a 100 mL culture). If desired, NP-40 or another nonionic detergent can be added to 0.1% to reduce nonspecific binding. If resuspension is difficult or the extract becomes too viscous, a Dounce homogenizer, blender, or sonicator can be used to break up the cell pellet.
2. With the sample in a tube on ice or in a salt-ice bath, sonicate. The conditions are not specified here because results depend on the type of sonicator probe used, the power setting, and the shape and size of the vessel holding the cells. Avoid long sonication times during which the sample could heat up; instead, break up the sonication into bursts with cooling in between. Sonicate until the sample is no longer viscous. If the DNA is not sheared by sonication, the extract will be so viscous that it will clog up the column. Alternatively, a French press may be used for cell lysis.
3. Centrifuge the lysate at 39,000g for 20 min to remove debris. If soluble proteins are to be run over the affinity column, filter the postcentrifugation supernatant through a 0.45-µ membrane to prevent clogging of the resin (syringe-end filters are convenient for this purpose).

3.4.5.3. Column Chromatography

1. Allow the binding buffer to drain to the top of the column bed and load the column with the prepared extract. A flow rate of about 10 column volume/hour is optimal for efficient purification. If the flow rate is too fast, impurities will contaminate the eluted fraction.
2. Wash the column with 25 mL (10 vol) of 1X binding buffer.

pET Expression and Purification

3. Wash the column with 15 mL (6 vol) of 1X wash buffer (*see* **Note 5a**).
4. Elute the bound protein with 15 mL (6 vol) of 1X elute buffer (*see* **Note 5b**). Alternatively, 10–15 mL of strip buffer may also be used to remove the protein by stripping the column of the Ni^{2+}.
5. *See* **Subheading 3.4.5.7.** for buffer exchange procedures prior to storage.

3.4.5.4. COLUMN REGENERATION

When elution is complete, the column can be regenerated for reuse. This process can be carried out many times. Because some small amount of protein sticks and may not release with EDTA treatment, it is advisable to use a different sample of resin for each protein studied. Following the last elution step the column should be washed with 3 vol of 1X strip buffer. The presence of 100 mM EDTA in the solution will prevent bacterial growth and the column can be recharged as above before use.

When the flow-rate of a column slows noticeably or the resin does not turn a strong blue-green color when charge buffer is added, it is time to clean the resin more thoroughly. Use the following regime for a 2.5-mL column bed:

1. 5 mL 6 M Guanidine-HCl, 0.2 M acetic acid.
2. 5 mL Water.
3. 2.5 mL 2% SDS.
4. 2.5 mL 25% Ethanol.
5. 2.5 mL 50% Ethanol.
6. 2.5 mL 75% Ethanol.
7. 12.5 mL 100% Ethanol.
8. 2.5 mL 75% Ethanol.
9. 2.5 mL 50% Ethanol.
10. 2.5 mL 25% Ethanol.
11. 2.5 mL Water.
12. 12.5 mL 100 mM EDTA, pH 8.0.
13. 7.5 mL Water.
14. 7.5 mL 20% Ethanol. Store at 4°C.

3.4.5.6. PURIFICATION UNDER DENATURING CONDITIONS

For purification under denaturing conditions add solid guanidine HCl or urea (final concentration is 6 M) directly to the concentrated binding buffer, wash buffer, and elute buffer and bring up the volume with deionized water. Adjust the pH to 7.9 with either HCl or NaOH. The columns are prepared as described in **Subheadings 3.4.5.1.** and **3.4.5.3.**, except that binding buffer plus 6 M guanidine HCl or 6 M urea should be used in the final equilibration step. The procedure entails isolating inclusion bodies in regular 1X binding buffer, washing to remove contaminating proteins, and then resuspending them in 1X binding buffer plus either 6 M guanidine HCl or 6 M urea to solubilize the protein.

1. Harvest the cells (100 mL-induced culture) by centrifugation at 5000g for 5 min. Decant the supernatant and allow the cell pellet to drain as completely as possible. Resuspend the cells in 40 mL 1X binding buffer that does not contain either guanidine or urea.
2. Sonicate briefly as described above to resuspend the pellet thoroughly and to shear the DNA.
3. Centrifuge at 20,000g for 15 min to collect the inclusion bodies and cellular debris while leaving other proteins in solution.

4. Remove the supernatant and suspend the pellet in 20 mL 1X binding buffer (without denaturant). Repeat **step 3**. Sonication may be necessary to resuspend the pellet. Sometimes repeating this step several times releases more trapped proteins.
5. Remove the supernatant from the final centrifugation and resuspend the pellet in 5 mL 1X binding buffer containing either 6 M guanidine or 6 M urea. Incubate on ice for 1 h to completely dissolve the protein. Remove any remaining insoluble material by centrifugation at 39,000g for 20 min. Filter the supernatant through a 0.45-µm membrane before loading it on the column.

Column chromatography of target proteins under denaturing conditions is similar to purification under native conditions except that a lower imidazole concentration is used in the wash buffer, since target proteins tend to elute at lower imidazole concentrations in the presence of 6 M guanidine or 6 M urea. All buffers loaded on the column contain guanidine or urea.

1. Allow the binding buffer to drain to the top of the column bed and load the column with the prepared extract.
2. Wash the column with 25 mL (10 vol) of 1X binding buffer.
3. Wash the column with 15 mL of 20 mM imidazole buffer (combine 11 mL of 1X binding buffer with 1 mL of 1X wash buffer).
4. Elute the protein with 1X elute buffer. As just described, a lower imidazole concentration (made by mixing binding and elute buffers) can be used to elute the protein if desired. Most proteins will elute with 300 mM imidazole under denaturing conditions.
5. *See* **Subheading 3.4.5.7.** for buffer exchange procedures prior to storage.
   NOTE: When urea is used, samples may be mixed with sample buffer and loaded directly on an SDS polyacrylamide gel, whereas samples in 6 M guanidine must be diluted 1:5 in water or dialyzed before running on an SDS polyacrylamide gel.

3.4.5.7. Processing the Purified Sample

The buffer of the purified sample may be changed or the sample concentrated by one of several methods. If the protein was purified under denaturing conditions, removal of the denaturing agent may allow refolding of the protein and reconstitution of activity. The salt and imidazole present under elution conditions may interfere with protein activity; thus, it is advisable to remove them prior to conducting an assay. Further, these high levels of imidazole and salt may affect gel mobility.

Three alternative procedures are:

1. Dialyze into the buffer of choice. If refolding is desired, a gradual removal of the 6 M guanidine or urea is recommended (e.g., successive changes into decreasing concentrations of denaturant). After dialysis, the sample may be concentrated by sprinkling solid PEG (15,000–20,000 mol wt) or Sephadex G50 (Pharmacia) on the dialysis tubing. Use dialysis tubing with an exclusion limit of 6000M_r or less, and leave the solid in contact with the tubing until the desired volume is reached, replacing it with fresh solid as necessary.
2. Use plastic disposable micro concentrator units (e.g., Centricon; Amicon) as directed by the manufacturer to both desalt and concentrate the sample by ultra filtration.
3. Desalt the sample by gel on Sephadex (G10, G25, G50; Pharmacia) or Bio-Gel (P6DG, P10, P30; Bio-Rad).

3.4.6. Protease Cleavage

Many pET N-terminal fusion vectors encode a protease cleavage site that allows removal of the vector-derived fusion domain by treatment with one of three different proteases (thrombin, Factor Xa, or enterokinase). Some vector encode two protease sites on the N-terminal side of the insert, which allow the removal of one or more N-terminal fusion sequences. Removal of these sequences is optional; their presence does not appear to affect the biological activity of many target proteins.

Protease treatment may be used as part of the purification protocol to remove residual *E. coli* proteins that bind strongly to the His·Bind resin. Rather than trying to adjust elution conditions, another approach is to cleave off the His·Tag sequence using the appropriate protease and pass the cleavage products over the chromatography resin a second time. Cleaved target protein can be recovered in the flow-through fraction. The short His·Tag leader, any uncut target protein, and *E. coli* proteins that bind the resin should be retained by the column. Another convenient strategy for obtaining pure proteins postdigestion is to use biotinylated thrombin, which can be used exactly as regular thrombin for digestion, but which can then be quantitatively removed from the preparation with streptavidin–agarose.

Each of the site-specific proteases is commercially available and should be used according to the manufacturer's instructions. Typical reaction conditions for Factor Xa and enterokinase are given: Factor Xa: Digest 1 mg recombinant protein at 20–37°C for 8–16 h with 20–50 µg of Factor Xa. Enterokinase: Digest 1 mg recombinant protein at 37°C for 8–24 h with 100 µg of purified enterokinase.

Human thrombin is one of the most active site-specific proteases, which is a cost savings since a very low mass ratio of enzyme to target protein is needed for efficient cleavage. In addition, the enzyme is available from Novagen in biotinylated form to allow removal with streptavidin–agarose after digestion. The unit is defined as that amount of enzyme required to cleave 1 mg of a test protein when incubated in standard digest buffer at 20°C for 16 h. A 1:2000 (w:w) ratio (thrombin:recombinant polypeptide) is generally sufficient for cleavage under the conditions described here. This is equivalent to one unit of thrombin per milligram of recombinant protein. Because each target protein presents the cleavage site somewhat differently, it is recommended that several thrombin concentrations and/or incubation times be tested to optimize cleavage of proteins. Begin with 0.5 U/mg recombinant protein and incubate for 2 h at 20°C in thrombin cleavage buffer. Incubation temperatures ranging from 4–37°C can be used, although we recommend 20°C as the starting point for most proteins. Note that excess thrombin may result in proteolysis at secondary sites. Avoid the presence of serine protease inhibitors during cleavage (for example, thrombin is effectively inhibited by 1 mM PMSF). Cleavage of the fusion peptide may be monitored by SDS-PAGE if the recombinant protein is small enough to see a size shift following its removal. Cleavage of the His·Tag peptide may also be monitored by the failure of the recombinant polypeptide to bind to His·Bind resin as follows (all steps can be carried out at room temperature):

1. Gently shake a tube containing a 1:1 slurry of His·Bind Resin previously charged with NiSO$_4$ to form an even suspension. Transfer 100-µL aliquots into 1.5-mL microcentrifuge tubes (each sample to be assayed requires 100 µL of 1:1 slurry). Centrifuge briefly to

Table 4
Thrombin Activity Remaining After Incubation at 20°C

Sample	After 2 h incubation	After 16 h incubation
Cleavage buffer	100%	95–100%
Cleavage buffer + 1 M urea	100%	93%
Cleavage buffer + 2 M urea	100%	73%
Cleavage buffer + 3 M urea	69%	45%
Cleavage buffer + 0.01% SDS	71%	Not determined

pellet the resin, remove the supernatant, and wash three times with 1 mL of a solution containing 120 mM imidazole, 0.5 M NaCl, and 20 mM Tris-HCl, pH 8.0. Resuspend the equilibrated resin in 50 µL of this buffer.

2. Add 100 µL of digested polypeptide (up to 50 µg protein) to resin. Mix well and incubate for 15 min at room temperature. Invert tubes to mix several times during this period. Include an undigested sample as a positive control for binding to the resin.
3. Centrifuge for 1 min to pellet the resin.
4. The supernatant may be assayed directly for activity of the cleaved protein or protein may instead be visualized by SDS-PAGE.

Controls for this assay can include measuring the activity of digested material that has not been incubated with resin (gives 100% activity) and undigested material that has been mixed with resin (all should bind). Cleavage of the S·Tag peptide can be monitored in a similar way using S-protein agarose.

To enhance cleavage of recombinant proteins, it is possible to carry out thrombin digestion in the presence of protein denaturants, which may expose the cleavage site to the enzyme more effectively. **Table 4** shows the effect of several urea concentrations and 0.01% SDS on thrombin activity after 2 and 16 h of incubation at 20°C.

4. Notes

1. For colony and postligation PCR of multiple samples, it is convenient to prepare a master mix by multiplying the prescribed reagent amounts by $X.5$, where X is the number of reactions, in order to account for pipeting losses.
2. For most pET vectors, appropriate primers for colony PCR without in vitro transcription/translation are the T7 promoter primer and the T7 terminator primer. Exceptions are pET-3xa–c, pET-17xb and pSCREEN-1b(+), in which the T7 gene *10* primer is recommended over the T7 promoter primer, and pET-5a–c, which require the use of the pBR322 *Eco*RI clockwise primer instead of the T7 terminator primer. To analyze the reaction products, first remove the oil overlay by adding 100 µL of chloroform. Add 5 µL of 10X loading dye to the top aqueous phase and load 10–25 µL/lane on a 1% agarose gel containing 0.5 µg/mL ethidium bromide.
3. For colony and postligation PCR, a "hot start" procedure can be used in which the cell lysate samples are prewarmed to 80°C before the addition of the master mix.
4. Cell extract preparation:
   a. It is important that 2-mercaptoethanol, dithiothreitol, and EDTA be avoided in solutions that will be run over the column. The reducing reagents react with Ni^{2+} to form brown precipitates. EDTA will chelate the Ni^{2+} and thereby strip the column of the active affinity group.

If a reducing agent must be used, note that alternative Ni-NTA-based resins manufactured by Qiagen are comparable with up to 20 mM 2-mercaptoethanol.

b. As an alternative to sonication, the DNA can be degraded by brief treatment of the sample with DNase. Add $MgCl_2$ to 10 mM and DNase to 20 µg/mL and incubate at room temperature for 20 min (or until the viscosity is reduced). This treatment may subject the sample to proteolytic activity and might cause some degradation of susceptible target proteins.

c. If necessary, protease inhibitors may be added to the buffers to protect against different degratory enzymes. Because inhibitors are often unnecessary, we recommend proceeding without their addition first. If breakdown is a problem, try adding AESBF or Protease-Inhibitor Cocktail Kit III (without EDTA; Novagen).

5. Column chromatography:

   a. The 1X wash buffer used in **Subheading 3.4.5.3., step 3** contains 60 mM imidazole. A few minor contaminants remain bound to the column after **step 3**; these can be removed with a 100 mM imidazole wash. However, some proteins may partly or entirely elute with this buffer; generally, proteins with a 6 histidine stretch elute at lower imidazole concentrations than those with a 10 histidine stretch. A 100 mM imidazole buffer can be made by combining 13.55 mL 1X binding buffer with 1.45 mL 1X elute buffer. To optimize the purification of specific proteins, buffers having intermediate imidazole concentrations may also be tried. (Note also that some proteolytic breakdown products of the target protein may copurify with it and may appear to be smaller sized contaminants.)

   b. The 1X elute buffer contains 1 M imidazole. If desired, a lower imidazole concentration can be used to elute the target protein (most proteins will elute between 100 and 400 mM imidazole). The binding buffer and elute buffer can be mixed in various ratios to obtain different imidazole concentrations. Alternatively, a linear gradient of imidazole can be run using binding buffer and elute buffer with an appropriate mixing device.

6. General note:

   Although this system has been used to express hundreds of target proteins from many different sources, some constructs may be difficult to express. The following notes discuss some additional factors that may contribute to low expression.

   a. The target protein itself may interfere with gene expression or with the integrity of the cell, or may be unusually susceptible to degradation *(3)*. In these cases it may be possible to express the target protein as insoluble inclusion bodies and refold following purification under denaturing conditions.

   b. Another factor that appears to influence target protein stability is the amino acid immediately following the N-terminal Met (penultimate amino acid). The amino acid at this position determines the removal of N-terminal fMet. Processing is catalyzed by methionyl aminopeptidase and is governed by the following relationship: The degree of removal decreases as the size of the penultimate amino acid side chain increases *(26,27)*. In practice, little or no processing was observed by these authors when the following amino acids occupied the penultimate position: His, Gln, Glu, Phe, Met, Lys, Tyr, Trp, and Arg. Processing ranged from 16 to 97% when the remaining amino acids occupied this position. Tobias et al. *(28)* have determined the relationship between a protein's amino terminal amino acid and its stability in bacteria, i.e., the "N-end rule." They reported protein half-lives of only 2 min when the following amino acids were present at the amino terminus; Arg, Lys, Phe, Leu, Trp, and Tyr. In contrast, all other amino acids conferred half-lives of >10 h when present at the amino terminus in the protein examined. Taken together, these studies suggest that Leu in the penultimate position would be a poor choice since it would likely be exposed by fMet processing and then be targeted for rapid degradation. Therefore, when an *Nde*I site is employed for the production of unfused target proteins from pET vectors, Leu codons in the penultimate position should be avoided. Leu codons in this position are not available when using *Nco*I as the cloning site, since the penultimate codon must begin with G.

c. Occasionally, truncated expression products are observed in addition to full-length target proteins. One obvious explanation is proteolytic degradation; however, secondary site translation initiation is another possibility *(29,30)*. This can occur within an RNA coding sequence when a sequence resembling the ribosome binding site (AAGGAGG) occurs with the appropriate spacing (typically 5–13 nucleotides) upstream of an AUG (Met) codon. These truncated products can be problematic when attempting to purify full-length proteins. One possible solution is to employ pET vectors that allow fusion to affinity tags at both ends of the target protein. The pET-28 and pET-30 series allow His·Tag fusions at both the N- and C-terminus. Full-length proteins would then be expected to elute at higher imidazole concentrations than truncated forms. The pET-29 and pET-30 series allow S·Tag N-terminal fusion and His·Tag C-terminal fusion. In this case the full-length target protein can be isolated by performing sequential affinity purification with immobilized S-protein and His·Bind resin *(31,32)*.

d. Secondary structure in the mRNA transcript can interfere with the AUG translation initiation codon and/or the ribosome binding site *(33–35)*. All pET vectors will generate one of the following transcripts:

   rbs   *Nde*I/*Nco*I
   5'...AAGAAGGAGAUAUACAUAUG...3'
   5'...AAGAAGGAGAUAUACCAUGG...3'

   If poor expression is observed, searching the coding strand of an insert for stretches of complementarity with the above sequences (5-CATATGTATATCTCCTTCTT-3, or 5-CCATGGTATATCTCCTTCTT-3) may reveal whether secondary structure is a potential problem.

e. Excessive rare codon usage in the target gene has also been implicated as a cause for low level expression *(36,37)*. The effect seems to be most severe when multiple rare codons occur near the amino terminus *(38)*. It should be noted, however, that only a subset of codons characterized as rare in highly expressed *E. coli* genes have levels of cognate charged tRNAs that are low enough to present potential problems in translation elongation *(39)*.

f. Unexpected stop codons can be generated by mutation, especially when cloning PCR products. Sequencing can reveal these mutations, but another alternative is to test the construct's ability to produce the target protein by in vitro translation. This test is conveniently performed using the Single Tube Protein System 2 (*see* **Subheading 3.1.2.**).

7. **General Note:** It is recommended to retain the T7 transcription terminator when cloning target genes into the pET vectors (of the vectors described here, only the pET-5 series lacks the terminator). In the absence of the terminator, T7 RNA polymerase transcribes nontarget plasmid sequences with the same efficiency as the target gene. In Ap^R pET plasmids the presence of the terminator appears to increase the expression of some genes, notably those that carry their own ribosome binding site *(3)*. This may be because of the attenuated read-through of T7 RNA polymerase into the *bla* gene, thereby reducing competition of *bla* RNA with target gene transcripts for ribosomes *(3)*.

Acknowledgments

Novagen is grateful to Bill Studier, Alan Rosenberg, and John Dunn of Brookhaven National Laboratories for many helpful discussions. We also thank Alex Hoffmann, Steve Burley, and Jeff Miller for their contributions to the protocols for induction and purification. The vectors, host strains, and many associated reagents herein described are available commercially from Novagen, Inc. (Madison, WI), and updated information is available at www.novagen.com. The following are trademarks of Novagen, Inc.: T7·Tag, His·Tag, S·Tag, Trx·Tag, His·Bind, pSCREEN, and Single Tube Protein.

References

1. Studier, F. W. and Moffatt, B. A. (1986) Use of bacteriophage T7 RNA polymerase to direct selective high-level expression of cloned genes. *J. Mol. Biol.* **189,** 113–130.
2. Rosenberg, A. H., Lade, B. N., Chui, D., Lin, S., Dunn, J. J., and Studier, F. W. (1987) Vectors for selective expression of cloned DNAs by T7 RNA polymerase. *Gene* **56,** 125–135.
3. Studier, F. W., Rosenberg, A. H., Dunn, J. J., and Dubendorff, J. W. (1990) Use of T7 RNA polymerase to direct the expression of cloned genes. *Meth. Enzymol.* **185,** 60–89.
4. Aslanidis, C. and de Jong, P. J. (1990) Ligation-independent cloning of PCR products (LIC-PCR). *Nucleic Acids Res.* **18,** 6069–6074.
5. Seed, B. (1987) An LFA-3 cDNA encodes a phospholipid-linked membrane protein homologous to its receptor CD2. *Nature* **329,** 840–842.
6. Dubendorff, J. W. and Studier, F. W. (1991) Creation of a T7 autogene. Cloning and expression of the gene for bacteriophage T7 RNA polymerase under control of its cognate promoter. *J. Mol. Biol.* **219,** 45–59.
7. Grodberg, J. and Dunn, J. J. (1988) OmpT encodes the *Escherichia coli* outer membrane protease that cleaves T7 RNA polymerase during purification. *J. Bacteriol.* **170,** 1245–1253.
8. White, C. B., Chen, Q., Kenyon, G. L., and Babbitt, P. C. (1995) A novel activity of *omp*T. *J. Biol. Chem.* **270,** 12,990–12,994.
9. Leahy, D. J., Hendrickson, W. A., Aukhil, I., and Erickson, H. P. (1992) Structure of a fibronectin type III domain from tenascin phased by MAD analysis of the selenomethionyl protein. *Science* **258,** 987–991.
10. Wood, W. (1966) Host specificity of DNA produced by *Escherichia coli*: bacterial mutations affecting the restriction and modification of DNA. *J. Mol. Biol.* **16,** 118–133.
11. Doherty, A. J., Ashford, S. R., Brannigan, J. A. and Wigley, D. B. (1995) A superior host strain for the over-expression of cloned genes using the T7 promoter based vectors. *Nucleic Acids Res.* **23,** 2074–2075.
12. Derman, A. I., Prinz, W. A., Belin, D., and Beckwith, J. (1993) Mutations that allow disulfide bond formation in the cytoplasm of *Escherichia coli*. *Science* **262,** 1744–1747.
13. Studier, F. W. (1991) Use of bacteriophage T7 lysozyme to improve an inducible T7 expression system. *J. Mol. Biol.* **219,** 37–44.
14. Moffatt, B. A. and Studier, F. W. (1987) T7 lysozyme inhibits transcription by T7 RNA polymerase. *Cell* **49,** 221–227.
15. Inouye, M., Arnheim, N., and Sternglanz, R. (1973) Bacteriophage T7 lysozyme is an N-acetylmuramyl-L-alanine amidase. *J. Biol. Chem.* **248,** 7247–7252.
16. Chang, A. C. Y. and Cohen, S. N. (1978) Construction and characterization of amplifiable multicopy DNA cloning vehicles derived from the P15A cryptic miniplasmid. *J. Bacteriol.* **134,** 1141–1156.
17. Dunn, J. J. and Studier, F. W. (1983) Complete nucleotide sequence of bacteriophage T7 DNA and the locations of T7 genetic elements. *J. Mol. Biol.* **166,** 477–535 and *erratum* (1984). *J. Mol. Biol.* **175,** 111,112.

18. McAllister, W. T., Morris, C., Rosenberg, A. H., and Studier, F. W. (1981) Utilization of bacteriophage T7 late promoters in recombinant plasmids during infection. *J. Mol. Biol.* **153,** 527–544.
19. Sambrook, J., Fritsch, E. F., and Maniatis, T. (1989) *Molecular Cloning: A Laboratory Manual* (2nd ed.), Cold Spring Harbor Laboratory, Cold Spring Harbor, NY.
20. Schein, C. H. and Noteborn, M. H. M. (1989) Production of soluble recombinant proteins in bacteria. *Bio/Technology* **7,** 1141–1148.
21. Ausubel, F. M., Brent, R., Kingston, R. E., Moore, D. D., Seidman, J. G., Smith, J. A., and Struhl, K. (1989) Expression and purification of maltose binding protein fusions, in *Current Protocols in Molecular Biology* (Riggs, P., ed.), Wiley, New York, pp. 16.6.1–16.6.14.
22. Mierendorf, R., Yaeger, K., and Novy, R. (1994) The pET system: your choice for expression. *Innovations* **1,** 1–3.
23. Novy, R., Berg, J., Yaeger, K., and Mierendorf, R. (1995) pET TRX fusion system for increased solubility of target proteins expressed in *E. coli*. *inNovations* **3,** 7–9.
24. LaVallie, E. R., DiBlasio, E. A., Kovacic, S., Grant, K. L., Schendel, P. F., and McCoy, J. M. (1993) A thioredoxin gene fusion expression system that circumvents inclusion body formation in the *E. coli* cytoplasm. *Bio/Technology* **11,** 187–193.
25. Wickner, W., Driessen, A. J. M., and Hartl, F.-U. (1991) The enzymology of protein translocation across the *Escherichia coli* plasma membrane. *Ann. Rev. Biochem.* **60,** 101–124.
26. Hirel, P.-H., Schmitter, J.-M., Dessen, P., Fayat, G., and Blanquet, S. (1989) Extent of N-terminal methionine excision from *Escherichia coli* proteins is governed by the side-chain length of the penultimate amino acid. *Proc. Natl. Acad. Sci. USA* **86,** 8247–8251.
27. Lathrop, B. K., Burack, W. R., Biltonen, R. L., and Rule, G. S. (1992) Expression of a group II phospholipase A2 from the venom of *Agkistrodon piscivorus* in *Escherichia coli*: recovery and renaturation from bacterial inclusion bodies. *Prot. Exp. Purif.* **3,** 512–517.
28. Tobias, J. W., Shrader, T. E., Rocap, G., and Varchavsky, A. (1991) The N-end rule in bacteria. *Science* **254,** 1374–1377.
29. Preibisch, G., Ishihara, H., Tripier, D., and Leineweber, M. (1988) Translational controls. Unexpected translation initiation within the coding region of eukaryotic genes expressed in *Escherichia coli*. *Gene* **72,** 179–186.
30. Halling, S. M. and Smith, S. (1985) Expression in *Escherichia coli* of multiple products from a chimeric gene fusion: evidence for the presence of procaryotic translational control regions within eucaryotic genes. *Bio/Technology* **3,** 715–720.
31. Kim, J.-S. and Raines, R. T. (1994) Peptide tags for a dual affinity fusion system. *Anal. Biochem.* **219,** 165,166
32. McCormick, M. and Mierendorf, R. (1994) S·Tag: a multipurpose fusion peptide for recombinant proteins. *inNovations* **1,** 4–6.
33. Tessier, L.-H., Sondermeyer, P., Faure, T., Dreyer, D., Benavente, A., Villeval, D., Courtney, M., and Lecocq, J.-P. (1984) The influence of mRNA secondary structure on human IFN-γ gene expression in *E. coli*. *Nucleic Acids Res.* **12,** 7663–7675.
34. Looman, A. C., Bodlaender, J., De Gruyter, M., Vogelaar, A., and Van Knippenberg, P. H. (1986) Secondary structure as a primary determinant of the efficiency of ribosomal binding sites in *Escherichia coli*. *Nucleic Acids Res.* **14,** 5481–5496.
35. Lee, N., Zhang, S.-Q., Cozzitorto, J., Yang, J.-S., and Testa, D. (1987) Modification of mRNA secondary structure and alteration of the expression of human interferon α1 in *Escherichia coli*. *Gene* **58,** 77–86.
36. Zhang, S., Zubay, G., and Goldman, E. (1991) Low-usage codons in *Escherichia coli*, yeast, fruit flies and primates. *Gene* **105,** 61–72.

37. Sorensen, M. A., Kurland, C. G., and Pedersen, S. (1989) Codon usage determines translation rate in *Escherichia coli*. *J. Mol. Biol.* **207,** 365–377.
38. Chen, G.-F. T. and Inouye, M. (1990) Suppression of the negative effect of minor arginine codons on gene expression; preferential usage of minor codons within the first 25 codons of the *Escherichia coli* genes. *Nucleic Acids Res.* **18,** 1465–1473.
39. Ikemura, T. (1985) Codon usage and tRNA content in unicellular and multicellular organisms. *Mol. Biol. Evol.* **2,** 13–34.

X

GENE LOCALIZATION, MAPPING *IN SITU*, AND BIOINFORMATICS

114

Preparation of Tissue Sections and Slides for mRNA Hybridization

Giorgio Terenghi

1. Introduction

The first step for a successful *in situ* hybridization is the fixation of the tissue. This will ensure target nucleic acid retention and preservation of the tissue morphology. Either crosslinking or precipitative fixatives can be used, and a preference for either of the two types of fixative has often been based on the different types of system under investigation *(1–7)*. For hybridization of regulatory peptide mRNA, 4% paraformaldehyde appears to be the most effective, both on tissue blocks and on tissue culture preparations.

When manipulating tissue to be used for mRNA hybridization, it is essential to work in RNase-free conditions. RNase is an ubiquitous and heat-resistant enzyme that degrades any single-stranded RNA molecule very rapidly. Fingertips are particularly rich in ribonuclease, hence clean, disposable gloves should be worn at all times. All equipment and solutions should also be RNase-free. Fixative solutions, as they exert an inhibitory action on this enzyme, are naturally RNase-free.

Fixed tissue is generally processed for cryostat sectioning, but *in situ* hybridization can be equally successful on paraffin embedded tissue *(8,9)*. In any case, it is important always to keep the delay between tissue collection and fixation to a minimum, to avoid nucleic acid degradation, which obviously increases with time delay *(4,10)*. This is not a problem when using experimental animal tissue or cell culture, but it becomes an extremely important point when using surgical or postmortem material. As the degradation curve varies for different mRNAs, there is no fixed rule on an acceptable time limit, but a delay of 15–30 min is considered acceptable in most cases.

The fixed material can be stored in washing buffer only for a limited time (up to 1 mo), but frozen or paraffin tissue blocks can be safely stored for many months or years. Cryostat blocks should be stored at –40°C or below, and room temperature is considered adequate for paraffin blocks. Cryostat sections can be stored dry at –70°C for up to 1 yr, and wax sections keep at room temperature (dewaxed sections can be kept in 70% alcohol at 4°C), given that RNase-free conditions are observed.

Tissue sections should be collected on Vectrabond (Vectra, UK) coated slides to prevent loss of material during the many steps of the hybridization procedure. The best

From: *The Nucleic Acid Protocols Handbook*
Edited by: R. Rapley © Humana Press Inc., Totowa, NJ

tissue adhesion is obtained if the sections are left to dry for at least 4 h (or overnight) at 37°C before use or storage.

2. Materials

Plastic disposable equipment and solutions should be autoclaved before use. RNase inhibitors (i.e., human placental ribonuclease inhibitor, DEPC, heparin, and so on) can be added to solutions containing enzymes, which are heat sensitive.

1. Phosphate buffered saline (PBS): Dissolve in 9 L of distilled water the following: 87.9 g NaCl, 2.72 g KH_2PO_4, 11.35 g anhydrous Na_2HPO_4. Adjust the pH to 7.1–7.2 with HCl before adjusting the total volume to 10 L. This solution can be stored at room temperature ready for use.
2. 4% Paraformaldehyde solution: Dissolve 4 g of paraformaldehyde in 80 mL of 0.01 M PBS with heat, keeping the temperature below 60°C. Stir the slurry until the powder is completely dissolved, and if necessary add few drops of 10 N NaOH to clear the solution. Adjust the volume to 100 mL and leave the solution to cool before using. The fixative should be freshly prepared before use.
3. Vectrabond: Prepare the solution according to the manufacturer instructions, and use on RNase-free coated slides.
4. RNase-free glass slides: Place some slides in metal racks and soak them in 70% ethanol for 10 min before proceeding with the coating according to manufacturer instructions. Always wear gloves during this procedure.
5. Liquid nitrogen.
6. Isopentane.
7. Freezing glue (e.g., Tissue-Tek OCT, Miles, New Haven, CT).
8. Washing buffer: 15% (w/v) sucrose, 0.01% (w/v) sodium azide dissolved in PBS.
9. Plastic disposable equipment and solutions should be autoclaved before use. RNase inhibitors (i.e., human placental ribonuclease inhibitor, diethylpyrocarbonate (DEPC), heparin, and so on) can be added to solutions containing enzymes that are heat sensitive.

3. Methods

3.1. Fixation (see Note 2)

1. Cut tissue into small pieces (approx 1 × 1 × 0.5 cm) using a sterile sharp blade (*see* **Note 2**).
2. Fix the tissue in freshly made 4% paraformaldehyde solution for 6 h at room temperature (*see* **Notes 3** and **4**).
3. After fixation, rinse the tissue blocks in four to five changes of washing buffer (2 h each change or overnight).
4. Store the fixed material in washing buffer at 4°C in labeled containers, ready for cryostat or wax blocking (*see* **Note 5**).

3.2. Preparing the Cryostat Block and Sectioning Tissue

1. Precool a Pyrex or metal beaker by immersion in liquid nitrogen and fill it with isopentane.
2. Immerse the beaker again in liquid nitrogen and freeze the isopentane, then remove the beaker from the flask and allow the isopentane to melt until there is enough liquid to cover the block, but still some solid in the bottom to maintain the temperature as low as possible.
3. Mount the tissue on a cork disk and surround it with special freezing glue (e.g., OCT [Miles]) (*see* **Notes 6** and **7**).

4. Hold the cork disk with forceps and lower it quickly in the melting isopentane to snap-freeze the tissue.
5. Transfer the frozen block in precooled plastic bag for storage in liquid nitrogen or at −40°C.
6. The block should be allowed to warm up to cryostat temperature (−20°C) before cutting is attempted.
7. Mount the block on the cryostat head and cut thin sections (10–30 mm), picking them up onto PLL-coated slides (see **Notes 1**, **5**, and **7**).
8. Dry the sections for at least 4 h (or overnight) at 37°C to obtain maximum tissue adhesion.

4. Notes

1. Batches of coated slides may be prepared in advance and stored in racks, wrapped in the aluminum foil to protect from dust and RNase contamination. Vetrabond-coated slides can be stored at room temperature for up to 1 mo.
2. When mRNA is the target, special care should be taken, and all the specimen handling procedures should be carried out using clean, disposable gloves and sterile instruments, in order to avoid RNase contamination.
3. Animal tissue can be fixed *in situ* by perfusion with 4% paraformaldehyde, followed by immersion fixation of the dissected tissue (1–4 h, depending on the tissue size and fixation obtained with perfusion). This method is strongly recommended if brain or spinal cord tissues are used, as these tissues do not fix well by immersion only, owing to the poor penetration of the fixative in the tissue matrix.
4. Fixative other than paraformaldehyde may be more appropriate when investigating specific target molecules. Some indication of other possible fixatives can be gained from the listed literature. However, it is good practice to test more than one fixative in order to establish which give best hybridization results, as indicated by the highest signal:background noise ratio with optimal preservation of morphology.
5. During storage of the cryostat blocks, do not leave any tissue surface uncovered, as it will dry out and it will become impossible to cut. After cutting, spread a thin layer of OCT glue on the cut surface of the block, and leave at −20°C until frozen. Store the block in a sealed plastic bag or other appropriate container.
6. Tissue blocks should be orientated so that the face you wish to cut is uppermost. Cryostat blocks cut best in the vertical plane, unlike paraffin blocks.
7. Very small pieces of tissue should be mounted on another piece of inert or inappropriate tissue (e.g., liver), suitably trimmed, so that adequate clearance is obtained on cutting.

References

1. Haase, A. T., Brahic, M., and Stowring, L. (1984) Detection of viral nucleic acids by *in situ* hybridization, in *Methods in Virology,* vol. VII (Maramorosch, K. and Koprowski, H., eds.), Academic, New York, pp. 189–226.
2. McAllister, H. A. and Rock, D. L. (1985) Comparative usefulness of tissue fixatives for *in situ* viral nucleic acid hybridization. *J. Histochem. Cytochem.* **33,** 1026–1032.
3. Moench, T. R., Gendelman, H. E., Clements, J. E., Narayan, O., and Griffin, D. E. (1985) Efficiency of *in situ* hybridization as a function of probe size and fixation technique. *J. Virol. Method.* **11,** 119–130.
4. Hofler, H., Childers, H., Montminy, M. R., Lechan, R. M., Goodman, R. H., and Wolfe, H. J. (1986) *In situ* hybridization methods for the detection of somatostatin mRNA in tissue sections using antisense RNA probes. *Histochem. J.* **18,** 597–604.
5. Singer, R. H., Lawrence, J. B., and Villnave, C. (1986) Optimization of *in situ* hybridization using isotopic and non-isotopic detection methods. *BioTechniques* **4,** 230–250.

6. Guitteny, A. F., Fouque, B., Mongin, C., Teoule, R., and Boch, B. (1988) Histological detection of mRNA with biotinylated synthetic oligonucleotide probes. *J. Histochem. Cytochem.* **36,** 563–571.
7. Terenghi, G. and Fallon, R. A. (1990) Techniques and applications of *in situ* hybridization, in *Current Topics in Pathology: Pathology of the Nucleus* (Underwood, J. C. E., ed.), Springer Verlag, Berlin, pp. 290–337.
8. Farquharson, M., Harvie, R., and McNicol, A. M. (1990) Detection of mRNA using a digoxigenin end labelled oligodeoxynucleotide probe. *J. Clin. Pathol.* **43,** 424–428.
9. Unger, E. R., Hammer, M. I., and Chenggis, M. L. (1991) Comparison of 35S and biotin as labels for *in situ* hybridization: use of an HPV model system. *J. Histochem. Cytochem.* **39,** 145–150.
10. Asanuma, M., Ogawa, N., Mizukawa, K., Haba, K., and Mori, A. (1990) A comparison of formaldehyde-preperfused frozen and freshly frozen tissue preparation for the *in situ* hybridization for α-tubulin mRNA in the rat brain. *Res. Comm. Chem. Pathol. Pharmacol.* **70,** 183–192.

115

Use of Digoxigenin-Labeled Probes on Tissue Sections

Giorgio Terenghi

1. Introduction

Nonradioactively labeled probes offer several advantages compared to radioactive ones, as they show long stability, high morphological resolution, and rapid developing time. There are different types of nonradioactive labeling methods available, although digoxigenin-labeled probes *(1)* have become the most widely used for investigation on animal tissue, as they offer the advantage of low background noise and increased sensitivity *(2,3)*. Also, digoxigenin can be used to label either RNA, DNA, or oligonucleotide probes. There have been different opinions on the sensitivity of detection of digoxigenin probes, but it has been shown that the sensitivity of radiolabeled and nonradioactive probes is comparable *(3)*.

The detection of digoxigenin-labeled probes is carried out with immunohistochemical methods, using antidigoxigenin antibodies that are conjugated to either fluorescent or enzymatic reporter molecules *(1)*. However, it has to be remembered that different immunohistochemical detection systems might determine the resolution and detection sensitivity of *in situ* hybridization *(4)*. The variety of detection methods also offers the possibility to carry out double *in situ* hybridization, e.g., using digoxigenin-labeled probes and directly labeled probes, which are then visualized using different immunohistochemical methods *(5)*. Alternatively, digoxigenin and biotin can be used in combination for the identification of two different nucleic acid sequences on the same sections *(6,7)*.

2. Materials

1. 1 M Tris-HCl, pH 8.0: Dissolve 121.1 g Tris-base in 800 mL double-distilled water. Adjust to pH 8.0 with HCl, then adjust volume to 1 L with double-distilled water before autoclaving.
2. 0.5 M EDTA: Add 186.1 g of $Na_2EDTA \cdot 2H_2O$ to approx 600 mL double-distilled water. Stir continuously keeping the solution at 60°C, adding NaOH pellets (approx 20 g) until near pH 8.0. Only then will the EDTA start to dissolve. When completely dissolved, leave the solution to cool down to room temperature, then adjust to pH 8.0 with 10 N NaOH solution. Adjust the volume to 1 L with double-distilled water and autoclave.
3. 1 M Glycine: Dissolve 75 g glycine in 800 mL double-distilled water. When dissolved, adjust volume to 1 L and autoclave.

4. 1 *M* Triethanolamine: Mix 44.5 mL triethanolamine in 200 mL double-distilled water, adjust to pH 8.0 with HCl, then bring to 300 mL volume with double-distilled water before autoclaving.
5. 10X SSC (standard saline citrate): Dissolve 87.65 g NaCl and 44.1 g sodium citrate in 800 mL double-distilled water, adjust to pH 7.0 with 10 *N* NaOH, then bring to 1 L vol. Autoclave a small aliquot to be used for the hybridization buffer. When the solution is used for posthybridization washes, it does not need to be autoclaved (*see* **Note 1**).
6. Deionized formamide: Mix 50 mL of formamide and 5 g of mixed-bed ion-exchange resin (e.g., Bio-Rad AG 501-X8, 20–50 mesh). Stir for 30 min at room temperature, then filter twice through Whatman No.1 filter paper. Store in small aliquots at –20°C.
7. 100X Denhardt's solution: Dissolve 1 g Ficoll (Pharmacia, Piscataway, NJ), 1 g polyvinylpyrrolidone, and 1 g bovine serum albumin (BSA) (Fraction V) in 50 mL of sterile distilled water and autoclave.
8. Herring sperm DNA: Dissolve the DNA (Type XIV sodium salt) in sterile distilled water at a concentration of 10 mg/mL. If necessary, stir the solution on a magnetic stirrer for 2–4 h at 37–40°C to help the DNA to dissolve. Shear the DNA by passing it several times through a sterile 18-g hypodermic needle. Alternatively sonicate on medium-high power for 5 min. Boil the DNA solution for 10 min and store at –20°C in small aliquots. Just before use, heat the DNA for 5 min in a boiling water bath. Chill it quickly in ice-water.
9. 10% Sodium dodecyl sulfate (SDS): Dissolve 100 g SDS in 900 mL of double-distilled water. Heat to 68°C to assist solubilization. Adjust to pH 7.2 by adding a few drops of HCl. Adjust volume to 1 L. This solution does not need autoclaving, as it is an RNase inhibitor.
10. Hybridization buffer: 50% deionized formamide, 5X SSC, 10% dextran sulfate, 5X Denhardt's solution, 2% SDS, and 100 µg/mL denatured sheared herring sperm DNA. Make this solution fresh before use, and store at 50°C.
11. Phosphate buffered saline (PBS): dissolve in 9 L of distilled water the following: 87.9 g NaCl, 2.72 g KH_2PO_4, 11.35 g Na_2HPO_4 anhydrous. Adjust the pH to 7.1–7.2 with HCl before adjusting the total volume to 10 L. Autoclave before use.
12. 0.2% Triton-X100 (v/v) in autoclaved PBS.
13. Proteinase K (stock solution): Dissolve the proteinase K at 0.5 mg/mL concentration in sterile distilled water. Divide into small aliquots and store at –20°C.
14. Permeabilizing solution: Prepare 0.1 *M* Tris-HCl, pH 8.0, 50 m*M* EDTA from stock solutions (1 *M* and 0.5 *M* respectively), diluting 1/10 with sterile distilled water. Just before use (**Subheading 3.1.**, **step 2**), to 100 mL of this solution, prewarmed at 37°C, add 200 µL proteinase K stock solution (final concentration 1 µg/mL).
15. 0.1 *M* Glycine in PBS: Dilute 1 *M* stock solution 1/10 with autoclaved PBS.
16. 4% Paraformaldehyde in PBS: Dissolve 4 g of paraformaldehyde in 80 mL of 0.01 *M* PBS with heat, keeping the temperature below 60°C. Stir the slurry until the powder is completely dissolved, and if necessary add few drops of 10 *N* NaOH to clear the solution. Adjust the volume to 100 mL and leave the solution to cool before using. The fixative should be freshly prepared before use.
17. Acetic anhydride.
18. 0.1 *M* Triethanolamine: freshly prepared 1/10 dilution of 1 *M* stock solution in sterile distilled water.
19. Sterile double-distilled water.
20. Digoxigenin-labeled cRNA probe. The probe should be complementary to the target mRNA. The probe should be ethanol precipitated and dissolved in hybridization buffer immediately before use (**Subheading 3.1.**, **step 8**).

Digoxigenin-Labeled Probes

21. 5X SSC (*see* **Note 1**).
22. 2X SSC, 0.1% SDS (*see* **Note 1**).
23. 0.1X SSC, 0.1% SDS (*see* **Note 1**).
24. 2X SSC (*see* **Note 1**).
25. RNase stock solution: Dissolve pancreatic RNase (RNase A) at a concentration of 10 mg/mL in 10 mM Tris-HCl, pH 7.5, and 15 mM NaCl. Dispense into aliquots and store at −20°C.
26. 10 µg/mL RNase in 2X SSC: Add 100 µL of RNase stock solution to 100 mL 2X SSC prewarmed at 37°C. Prepare freshly before use.
27. Sections on slides.
28. Buffer 1: 0.1 M Tris-HCl, pH 7.5, 0.1 M NaCl, 2 mM MgCl$_2$, 3% BSA. In 800 mL double-distilled water dissolve 12.1 g Tris-base and 5.85 g NaCl. Adjust to pH 7.5, then add 0.2 g MgCl$_2$. Adjust the volume to 1 L and add BSA to 3%.
29. Buffer 2: 0.1 M Tris-HCl, pH 9.5, 0.1 M NaCl, 50 mM MgCl$_2$. In 800 mL double-distilled water dissolve 12.1 g Tris-base and 5.85 g NaCl. Adjust to pH 9.5, then add 4.4 g MgCl$_2$. Adjust the volume to 1 L.
30. Antidigoxigenin antibody/alkaline phosphatase conjugate (Boehringer, Mannheim, Germany): Immediately before use (**Subheading 3.2.**, **step 3**) dilute to 1/500 with buffer 1.
31. Nitroblue tetrazolium chloride (NBT): Dissolve 35 mg NBT in 277 µL 70% dimethylformamide (DMF). Prepare freshly before use.
32. 5-Bromo-4-chloro-3-indolyl-phosphate (BCIP): Dissolve 17 mg BCIP in 222 µL 100% DMF. Prepare freshly before use.
33. Substrate buffer: In 100 mL buffer 2 dissolve 25 mg levamisole (Sigma, St. Louis, MO), then add 277 µL NBT solution and 222 µL BCIP solution just before use.
34. Stop buffer: 20 mM Tris-HCl, pH 7.5, 5 mM EDTA. In 800 mL double-distilled water dissolve 2.42 g Tris-base, then adjust to pH 7.5 with HCl. Add 10 mL 0.5 M EDTA stock solution and adjust volume to 1 L.
35. 5% Pyronin Y: Dissolve 5 g pyronin Y in 100 mL double-distilled water.

3. Methods

All the steps up to hybridization (included) should be carried out in RNase-free conditions. Solutions should be autoclaved or prepared with sterile ingredients using RNase-free equipment. Equipment should be autoclaved, or baked at 250°C for 4 h, as appropriate.

Select the slides, number, and mark them as necessary with pencil (not pen—ink may disappear during the various incubation steps).

3.1. Hybridization

1. Rehydrate the sections by immersion in 0.2% Triton/PBS for 15 min. Wash in PBS twice for 3 min.
2. Carry out the tissue permeabilization by incubating the tissue in permeabilizing solution prewarmed at 37°C, containing 1 µg/mL proteinase K. The normal incubation time is 15–20 min (*see* **Note 2**).
3. Stop the proteinase K activity by immersion in 0.1 M glycine in PBS for 5 min.
4. Immerse the sections in 4% paraformaldehyde for 3 min to postfix the target nucleic acid.
5. Rinse the sections briefly in PBS, twice, to remove the paraformaldehyde.
6. Place the slides in a staining jar containing 0.1 M triethanolamine and, while stirring, add acetic anhydride to 0.25% (v/v) and incubate for 10 min (*see* **Note 3**).

7. Rinse the slides briefly in double-distilled water and dry them at 37–40°C. This takes approx 10 min.
8. Dissolve the probe in hybridization buffer at 50°C to a final concentration of 2.5 ng/μL.
9. Apply 10 μL of diluted probe per section to the dry slides. The volume of diluted probe can be increased for large sections.
10. Using fine forceps, gently place a siliconized cover slip onto the section to spread the probe solution. If there are any air bubbles, remove them by pressing gently on the cover slip with the forceps.
11. Hybridize the section for 16–20 h (see **Note 4**) at suitable temperature in a sealed humid chamber containing 5X SSC. A different hybridization temperature will be needed for various probes according to their T_m (see **Note 5**).
12. Following hybridization, remove the cover slip by immersing the slide in 2X SSC, 0.1% SDS. The cover slips will float off after few minutes soaking.
13. Wash the sections in 2X SSC, 0.1% SDS at room temperature, shaking gently, for four changes of 5 min.
14. Wash the slides in 0.1X SSC, 0.1% SDS at the same temperature used for hybridization, shaking gently, for two changes of 10 min (see **Note 6**).
15. If using cRNA probes, rinse the sections briefly in 2X SSC, twice, then incubate 10 μg/mL RNase A solution in 2X SSC at 37°C for 15 min (see **Note 7**).
16. Rinse briefly in 2X SSC, then PBS before proceeding with the immunohistochemistry detection.

3.2. Immunohistochemistry Detection (see Note 8)

1. Block nonspecific binding by immersing the slides in buffer 1 for 10 min at room temperature.
2. Wipe dry the slides around the tissue but keep the tissue wet.
3. Put on the section a drop of antidigoxigenin antisera conjugated to alkaline phosphatase, diluted 1/500 with buffer 1. Incubate 2 h at room temperature (see **Notes 8** and **9**).
4. Wash in buffer 1 for three changes of 3 min.
5. Equilibrate the sections in buffer 2 for 10 min at room temperature.
6. Immerse the slides in substrate buffer for 10–30 min (or longer if necessary) at room temperature, covering the dish with aluminum foil to keep the reaction in the dark (see **Note 10**).
7. Check the slides under the microscope to assess the development reaction. Stop the reaction or put the slides back in the solution and leave for a longer time as required. The color of the reaction is blue-black.
8. Stop the reaction by immersing in stop buffer for 5 min at room temperature.
9. Counterstain by dipping the slides in 5% pyronin Y solution, for 10–30 s.
10. Rinse well under tap water, approx 5–10 min, until the water is clear.
11. While still wet, mount the slides in aqueous mountant (e.g., Vectashield [Vector Laboratories, Peterborough, UK] or similar).

4. Notes

1. There is no need to autoclave the solutions used for posthybridization washes.
2. The incubation time for proteinase K should be titrated for tissue type, as prolonged proteinase digestion could damage the tissue, with a loss of morphology and of target nucleic acid.
3. The treatment with triethanolamine is carried out in order to acetylate the tissue. This prevents electrostatic interaction between the tissue and the probe, as a result of opposite electrostatic charges, thus reducing the background staining.
4. It has been demonstrated that the hybridization reaction reaches an equilibrium after 4–6 h incubation, when a maximum of hybrid has formed. However, incubation is generally carried out overnight for convenience.

5. The hybridization temperature is dependent on the type of probes that have been used. With hybridization buffer containing 50% formamide, it is suggested that the following range of temperature should be tested initially: 42–48°C for cDNA probes; 42–55°C for cRNA probes; 37–40°C for oligonucleotide probes.
6. High background signal can be removed by prolonged washes of the slides at higher stringencies. Increase the number of washes in 0.1X SSC, 0.1% SDS, also progressively increase the temperature. Take care not to allow the sections to dry out in between any of the washes. Use the same container throughout, changing the solutions quickly.
7. It is essential to remove any trace of SDS from the sections before the incubation with RNase solution, as SDS would inhibit the action of the enzyme. It is desirable to use RNase at this stage of the hybridization procedure if you have been using cRNA probes. RNaseA will degrade the single-stranded cRNA probe that is bound nonspecifically to the section, hence decreasing the background staining. Double-stranded RNA hybrids (cRNA-mRNA) are unaffected by the enzyme.
8. There is a variety of reporter molecules (e.g., alkaline phosphatase, fluorescein isothyocyanate [FITC], and so on) conjugated to antibodies against digoxigenin, available from Boehringer, which may be suitable for different applications. There are also detection kits available from the same supplier, which include all reagents needed for the procedure. It is suggested that the reader refers to the catalog from Boehringer for further details.
9. Antibodies conjugated to other reporter molecules may require different dilutions, which are specified by the supplier.
10. The developing time varies considerably according to the type of tissue and the abundance of the target nucleic acid within the cell. In some cases several hours, or overnight, incubation in substrate buffer is necessary to obtain detectable signal. However, it has to be remembered that prolonged incubation also increases the nonspecific background staining.

References

1. Kessler, C. (1991) The digoxigenin anti-digoxigenin (DIG) technology—A survey on the concept and realization of a novel bioanalytical indicator system. *Mol. Cell. Probes* **5,** 161–205.
2. Morris, R. G., Arends, M. J., Bishop, P. E., Sizer, K., Duvall, E., and Bird, C. C. (1990) Sensitivity of digoxigenin and biotin labeled probes for detection of human papillomavirus by *in situ* hybridization. *J. Clin. Pathol.* **43,** 800–805.
3. Furuta, Y., Shinohara, T., Sano, K., Meguro, M., and Nagashima, K. (1990) *In situ* hybridization with digoxigenin-labeled DNA probes for detection of viral genomes. *J. Clin. Pathol.* **43,** 806–809.
4. Giaid, A., Hamid, Q., Adams, C., Springall, D. R., Terenghi, G., and Polak, J. M. (1989) Non-isotopic RNA probes. Comparison between different labels and detection systems. *Histochemistry* **93,** 191–196.
5. Dirks, R. W., van Gijlswijk, R. P. M., Tullis, R. H., Smit, A. B., van Minnen, J., van der Ploeg, M., and Raap, A. K. (1990) Simultaneous detection of different mRNA sequences coding for neuropeptide hormones by double *in situ* hybridization using FITC- and biotin-labeled oligonucleotides. *J. Histochem. Cytochem.* **38,** 467–473.
6. Herrington, C. S., Burns, J., Graham, A. K., Bhatt, B., and McGee, J. O. D. (1989) Interphase cytogenetics using biotin and digoxigenin labeled probes. II: Simultaneous detection of two nucleic acid species in individual nuclei. *J. Clin. Pathol.* **42,** 601–606.
7. Trask, B. J., Massa, H., Kenwrick, S., and Gitschier, J. (1991) Mapping of human chromosome Xq28 by two colour fluorescence *in situ* hybridization of DNA sequences to interphase cell nuclei. *Am. J. Hum. Genet.* **48,** 1–15.

116

Gene Mapping by FISH

Rafael Espinosa III and Michelle M. Le Beau

1. Introduction

The efforts to localize genes to human chromosomes date back to the early 1970s. Although few techniques were available to map genes, many scientists recognized that the ability to determine the location of genes and DNA sequences on human chromosomes would not only facilitate the identification of disease-related genes, but might also provide important knowledge on the organization of chromosomes and the mechanisms of gene expression. With the introduction of somatic cell genetics and the development of hybrid cell panels in the mid-1970s, investigators were now capable of mapping sequences to whole chromosomes and, in some cases, to specific chromosome regions or bands. Nonetheless, the major breakthrough in mapping efforts was provided by the development of *in situ* hybridization of isotopically labeled probes; this technique provided the first method by which scientists could actually visualize the hybridization of a DNA probe to chromosomes *(1,2)*. Using this technique, genes could be mapped to a few chromosome bands and often to a single band. The disadvantages of this method were the relatively poor spatial resolution owing to scatter of the radioactive emissions, the length of time for the procedure (long autoradiographic exposure times were typically required), and the poor stability of the probes. The introduction of techniques to detect hybridized probes using fluorochromes in the late 1970s circumvented many of these problems *(3,4)*; however, it was not until the end of the next decade that fluorescence *in situ* hybridization (FISH) techniques became widely applicable *(5,6)*.

In the short time since the development of FISH, the technique has had a major impact on efforts to map the human genome because:

1. It is a rapid technique;
2. The efficiency of hybridization and detection are high;
3. The sensitivity and specificity is very high;
4. The spatial resolution is high;
5. Large numbers of cells can be analyzed in a short time;
6. Data can be obtained from nondividing or terminally differentiated cells; and
7. The technique has been adapted for automated systems.

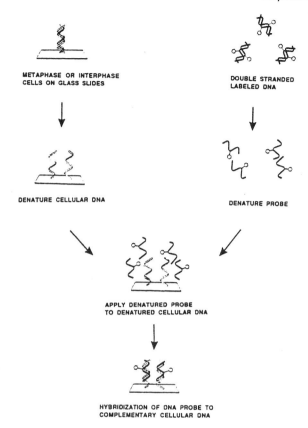

Fig. 1. Schematic diagram of the technique of FISH.

Several authors have addressed the applications of FISH in cytogenetic analysis, cancer diagnosis, and gene mapping (7,8). This chapter examines the role of FISH in gene mapping with an emphasis on detailed techniques.

1.1. Principles of FISH

The technique of FISH is based on the same principle as Southern blot analysis, namely, the ability of single-stranded DNA to anneal to complementary DNA. The essential steps in FISH are described briefly here and are illustrated in **Fig. 1**. As in Southern blot analysis, the target DNA is attached to a substrate; in the case of FISH, the target DNA is the nuclear DNA of interphase cells or the DNA of metaphase chromosomes that are affixed to a glass microscope slide (FISH can also be accomplished with bone marrow or peripheral blood smears, or fixed and sectioned tissue). The test probe is labeled, most commonly by enzymatic incorporation of modified nucleotides (probe labeling systems are described in a subsequent section). To allow hybridization of complementary sequences to occur, the cellular DNA and labeled probe DNA are denatured by heating in a formamide solution to form single-stranded DNA. A solution containing the probe DNA is applied to the microscope slide, the slides are covered with coverslips and sealed, and hybridization is allowed to occur by overnight incubation at 37–40°C. Thereafter, the unbound probe is removed by extensive washes in formamide-SSC, and the slides are processed for probe detection. As described in **Subheading 1.7.1.**, the most suitable probes for FISH analyses are large genomic clones, such as λ phage or cosmid clones. These probes contain ubiquitous repetitive sequences, such as *Alu* and *Kpn*I elements. The crosshybridization of repetitive sequences to nontargeted chromosomes can result in substantial background hybridization signal. This background labeling can be suppressed under appropriate preannealing conditions, using unlabeled total human DNA or the Cot1 DNA fraction, which is enriched in highly repetitive DNA, as a competitor *(9,10)*.

1.2. Probe Labeling Schemes

A variety of schemes have been described for labeling probes with nonradioactive compounds; these include enzymatic incorporation of modified nucleotides and chemical labeling techniques *(11)*. For applications involving *in situ* hybridization, enzymatic incorporation of nucleotides modified with biotin, or digoxigenin by nick translation or polymerase chain reaction (PCR)-labeling techniques results in a high labeling efficiency, and is usually preferred over chemical labeling techniques employing photoreactive compounds, e.g., photobiotin.

1.3. Amplification and Labeling of DNA Using PCR

Although the PCR was introduced initially to amplify single loci in target DNA, it has been used increasingly to amplify multiple loci simultaneously. The major application of "general" amplification of DNA has been the rapid generation of new clones from particular genomic regions. The initial methods were based on amplification of repetitive sequences within the genome, and resulted in the amplification of segments between suitably positioned repeats (interspersed repetitive sequence [IRS-PCR]). Thus, IRS-PCR is applicable only to those species where abundant repeat families have been identified. In humans, the most abundant family of repeats is the *Alu* family, estimated to comprise 900,000 elements in the haploid genome, with an average spacing of 3–4 kb. *Alu*-PCR has been used to create human chromosome- and region-specific libraries, as well as to amplify and label human sequences from somatic cell hybrids or yeast artificial chromosomes (YACs) *(12)*. IRS-PCR or *Alu*-PCR has notable advantages for species-specific amplification in which genomes are mixed, such as in rodent X human hybrids. However, these repeats are not uniformly distributed, a phenomenon that introduces a bias in cloning experiments. This is evidenced by the R-banding pattern that can be generated in hybridizations of probes prepared by *Alu*-PCR of total genomic DNA or chromosome-specific libraries *(13)*. For the purposes of gene mapping, *Alu*-PCR is particularly applicable to the localization of genomic clones with large inserts, such as YAC clones, since the human insert DNA can be amplified preferentially.

In the past few years, several simple PCR techniques have been introduced for the general amplification of DNA. Two major techniques known as degenerate oligonucleotide-primed PCR (DOP-PCR) *(14)* and sequence-independent amplification (SIA) *(15)* employ oligonucleotides of partially degenerate sequence. This feature together with a PCR protocol using a low initial annealing temperature ensures priming from multiple evenly dispersed sites within a given genome. The methods are species-independent and can be used for the efficient amplification of DNA from all species using

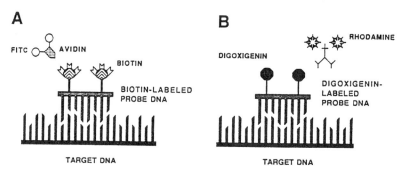

Fig. 2. Schematic diagram of the detection of hybridized probes following FISH. (**A**) FITC-conjugated avidin; (**B**) rhodamine-antidigoxigenin antibody.

the same primers. In the context of FISH applications, these methods are particularly useful for the amplification and labeling of YACs and other genomic clones, the amplification and labeling of microdissected chromosomal material, and the amplification of DNA from small numbers of cells from frozen sections or paraffin-embedded, formalin-fixed specimens for studies such as comparative genomic hybridization analysis of tumors *(16)*.

1.4. Labeling of PCR Products for FISH

The *Alu*-PCR, DOP-PCR, and SIA products from the YACs can be labeled using several different methods. We have used both nick translation (*see* **Subheading 2.3.**) and PCR labeling extensively, and have obtained good results with both methods; however, we have found that PCR labeling generally results in stronger signals. The products can be labeled with biotin in a second PCR by the incorporation of Bio-dUTP (*see* **Subheading 3.6.**).

1.5. Detection of Hybridized Probes

The visualization of hybridized probes can be accomplished in several ways:

1. By fluorochromes that are detected by fluorescence microscopy;
2. By chemiluminescence detected by an emulsion overlay or detected directly by photon counting devices; or
3. By high-density colored precipitate generated by enzymatic assays, e.g., alkaline phosphatase or horseradish peroxidase, or the use of metallic compounds, such as colloidal gold or silver, which are visualized by phase contrast, Nomarski optics, or electron microscopy.

As a result of the high sensitivity of detection, improved spatial resolution, the commercial availability of the reagents, and the greater potential for simultaneous multiprobe analysis, biotin- and digoxigenin-labeling combined with fluorescent detection are currently the most widely used procedures (**Fig. 2**). Biotin-labeled probes are usually detected with fluorescein isothiocyanate (FITC)- or rhodamine-conjugated avidin, whereas digoxigenin-labeled probes can be detected with FITC-labeled or rhodamine-labeled antidigoxigenin antibodies. To enable one to visualize the cellular material, the slides are typically counterstained with DNA-binding fluorochromes, such as propidium iodide or 4,6-diamidino-2-phenylindole (DAPI). DAPI staining induces a chromosomal banding pattern that is identical to G-banding or Q-banding; thus, DAPI is preferred for the analysis of metaphase cells.

Each of the methods described typically uses indirect detection procedures. Recently, direct labeling techniques have been developed in which probes are labeled directly with nucleotides conjugated to fluorochromes *(17)*. This allows microscopic examination immediately after hybridization and eliminates the requirement for the detection steps. The potential applications of FISH are increased significantly by multiple probe hybridization protocols, which allow the delineation of several target sequences simultaneously *(18)*.

1.6. Extended Chromatin Preparations

The determination of the physical locations of genes and DNA segments on individual chromosomes is an important aspect of genome research. Correct orientation and ordering of DNA markers are also critical in the identification of disease-related genes on the basis of chromosome location. During the past 5 yr, FISH has played a major role in this effort; using this technique for the analysis of prometaphase or metaphase chromosomes, it is possible to assign relative positions of genes and DNA segments as close as 1 Mb apart. For higher-resolution mapping, FISH can be applied to interphase nuclei or pronuclei *(19)*. Since the DNA is less condensed in interphase nuclei than in metaphase chromosomes, resolution in the 50–100 kb range can be obtained. However, mapping the distance between two probes in three-dimensional nuclei or compressed two-dimensional nuclei is complex, and requires a large data sampling. A linear correlation between measured interphase distances and kilobase distances is observed up to 500 kb (varies between 500 kb and 1 MB); however, with increasing distances, the measurements become more inaccurate because of chromatin folding.

Recently, a number of techniques for releasing DNA fibers from nuclei have been described, and FISH of extended chromatin fibers has been used for a variety of purposes (**Table 1**). The principle of these methods is that a small region of DNA when extended to the expected length of relaxed duplex DNA spans a distance that is visible through a light microscope *(20–23)*. The resultant preparations differ somewhat for the various methods, and thus, the investigator may wish to select the appropriate method depending on the particular application (**Table 2**). We have used the methods described by Fidlerova et al. *(23)* extensively and have found these procedures to yield optimal preparations routinely. The advantages of these methods to release chromatin fibers are that the DNA is virtually straight, which simplifies length measurements, as opposed to DNA that winds or loops, and the degree of stretching can be controlled by experimental conditions and by the particular method. Highly stretched DNA can be used for high-resolution mapping (NaOH method), whereas DNA stretched to a lesser extent can be used for long-range mapping (YAC mapping with the formamide method). Moreover, fluorescent signal on a stretch of DNA is continuous, or nearly continuous, which may facilitate the detection of small structural differences. Recently, several groups of investigators have described methods to prepare extended chromatin fibers from large genomic clones, such as YACs; these methods are particularly suitable to fine physical mapping by FISH on relatively small genomic regions (100 kb to 2 Mb) *(24)*.

Table 1
Applications of Extended DNA Preparations

Confirm the presence, and determine the position of clones (phage/cosmid/plasmid/cDNA) within larger clones, EX: mapping a cosmid within a YAC
Preparing physical maps:
    Ordering probes using two- or three-color FISH
    Determining the orientation of clones by cohybridizing end clones and entire clones
    Identifying overlapping sequences
    Estimating distances between sequences
Can be combined with linkage mapping to determine whether gaps identified by linkage are genuine and to estimate the physical distance of gaps
Mapping amplified genes, and determining the spacing and arrangement of amplified genes
Identify possible deletions, insertions, inversions, or complex rearrangements in the genome
Generate fingerprint maps of repetitive sequences in the genome
Functional analysis of chromatin:
    Differential packaging of DNA
    Transcriptional activity of DNA
    Metaphase vs interphase DNA
    DNA at different stages of the cell cycle

1.7. Analysis of FISH Experiments

Although FISH is a well-established technique, there has been relatively little standardization in the procedures used for the analysis of slides from FISH studies or in the reporting of the results of these studies in the literature. Nonetheless, there are some general principles that have emerged that should be considered by any investigator performing FISH. First, although the efficiency of hybridization of FISH is very high for most types of probes, it usually does not reach 100%. For cDNA probes, the efficiency may be very low, particularly for short probes that are <600 bp, and it may be necessary to use image acquisition and analysis techniques. Second, the level of resolution of FISH signals is excellent, but the location of the signal may vary slightly from cell to cell, particularly between metaphase cells of varying degrees of condensation. By analyzing multiple cells, including prometaphase or prophase cells, most probes can be mapped to a single chromosomal subband. For these reasons, it is critical that a sufficient number of cells be analyzed both to verify the distribution of signal and to determine the most precise localization of signal.

A third issue relates to the nature of cytogenetic analysis, which relies on the interpretation of chromosomal banding patterns. An inherent feature of all interpretative assays is that they are subject to error. To minimize the chance of errors either in the identification of chromosomes or the designation of bands, we recommend that the analysis of FISH experiments be performed by two well-trained individuals. Similarly, to avoid the possibility of laboratory errors in the case of the localization of new genes, we repeat the probe labeling and hybridizations a second time to confirm that the probe maps to the same site. Additional recommendations are described in **Subheadings 1.7.1.–1.7.3.** for specific applications.

Table 2
Techniques for the Preparation of Extended Chromatin Fibers

Extraction	Comments	Resolution	References
m-AMSA and alkaline extraction	Produces highly elongated, undefined spindle-shaped structures	approx 10 kb	22
Detergent and high salt	Produces halo preparations; however, greater release of DNA is achieved, and DNA is more elongated; looping of DNA limits usefulness for regions of >200 kb	10 kb (upper limit: 200 kb)	21
SDS	SDS dissolves membranes and releases a stream of DNA, which can be spindle-shaped	1–2 kb (upper limit: ≥200 kb)	20
Formamide	Results in a comet-like tail of released chromatin; the borders of disrupted nuclei can be defined	1–5 kb (upper limit: ≥800 kb)	23
NaOH-alcohol	Results in a network of straight chromatin fibers or an irregular network	1 kb (upper limit: ≥800 kb)	23

1.7.1. Localization of Single-Copy Genes/DNA Sequences

The localization of genomic or cDNA clones requires a maximal degree of precision. Given the tremendous efforts to prepare genetic linkage and physical maps of the human genome, accuracy in assigning the location of probes is essential to integrating cytogenetic, genetic linkage, and physical maps, and to the identification of disease-related genes. With respect to genomic clones (λ phage, cosmid, BAC, P1, or YAC clones), we routinely determine the location of signal in 25 DAPI-stained metaphase cells. It is critical that DAPI staining or another method to obtain R or Q/G bands be used. The intensity of the signal obtained for genomic clones is sufficient so that the cells are scored through the microscope, although an imaging system is used to capture and merge images of the banded chromosomes and the hybridization signal for the preparation of photographs for publication. The number of cells containing 0–4 signals on the four chromatids containing the test sequence is determined, as are the number and location of background signals. An additional 10 cells are scored by a second individual who is unaware of the results obtained by the first investigator. The hybridization is repeated, and at least 10 cells are scored to confirm the initial results.

In reporting the results, we describe the number of cells with 0, 1, 2, 3, or 4 chromatids labeled, the distribution of signal in each band (or subband) at the site of specific hybridization, and the location of background signals. The first data allow other scientists to evaluate the specificity of the hybridization. In most hybridizations in which the ratio of probe to placental and Cot1 DNA is optimized, the number of background signals will be very small—on the order of 0–10 in 25 metaphase cells. In those instances where this value is higher, it may be necessary to list only those sites that contain signal doublets, or were labeled more than once. Although the format used to report FISH localizations is variable, a consensus among investigators experienced in gene mapping by FISH is that the minimum data that must be included in a report in the literature are

1. The proportion of cells with signal;
2. The proportion of the total signals at the locus;
3. The location of background signal; and
4. The most precise location of the gene/DNA sequence.

In contrast to genomic clones in which the efficiency of hybridization and detection is high, cDNA clones are often substantially more difficult to localize. The percentage of cells with specific signal may be quite variable, and background signal can be high. In our experience, cDNA probes larger than 1 kb present few problems in mapping, and can be visualized and scored through the microscope; however, clones that are <600 bp can be problematic and may require an imaging system for analysis. Some laboratories with extensive experience at mapping cDNA clones have adopted the following criteria (J. Korenberg, personal communication). An initial analysis is performed on a selection of metaphase cells. For a hybridization to be considered successful, one or both homologs should have signal doublets in >10% of these metaphase cells. Because the incidence of single background signals may be high, only cells with doublet signals should be scored. As a minimum, 20 metaphase cells that have any doublet signals should be scored, and the location of the doublets recorded. At least 50% of the doublets must be located at the same chromosomal site to be considered evidence for determining the map location of the probe. The total number of doublets, the number and percent of doublets at the site of the gene, the location of other doublet signals (>1%), and the most precise assignment should be reported.

1.7.2. Analysis of Cancer-Specific Rearrangements

FISH has played an important role in the molecular-cytogenetic analysis of cancer-specific rearrangements. Specifically, FISH can be used to flank translocation breakpoints with specific DNA probes, to split translocation breakpoints and, hence, facilitate the cloning of the genes affected by these rearrangements *(25)*, and to identify the smallest commonly deleted region of recurring chromosomal deletions *(26,27)*. FISH can also be used in the subsequent stages in the identification of cancer-related genes, such as in the preparation of a genomic contig of a deleted segment or translocation breakpoint region, and to confirm that cDNA clones that are isolated from a specific region are derived from the correct region of the genome.

The analysis of hybridizations to metaphase cells from human tumors is often complicated by the limited number of metaphase cells available, the relatively poor quality of some metaphase cells from tumors, and the presence of normal cells in many samples. To address some of these problems, it may be necessary to hybridize several slides to a single probe (or to use the entire slide, rather than only a 22-mm^2 section), and to hybridize a control probe, such as a centromere-specific probe to assist in the identification of the rearranged chromosome of interest. When material is limited, we frequently cohybridize multiple probes using two-color FISH; we have also found that it is possible to hybridize the same slide multiple times. This is typically accomplished by using a different fluorochrome to detect the second probe, but the same fluorochrome can also be used since the denaturation step removes the initial probe. Subsequent hybridizations are performed as described.

In scoring metaphase cells from tumor samples, some of the same issues that arise in the analysis of single copy probes apply. It is critical that a sufficient number of cells be analyzed, and that the result be confirmed by a second independent observer. Thus, we recommend that a minimum of 10 cells containing the rearrangement be scored by each of two independent observers. Ideally, these are different cells; however, if there is insufficient material, the same cells could be evaluated.

1.7.3. Physical Mapping Using Extended Chromatin Preparations

The applications of FISH to chromatin fibers in genome mapping are numerous, and include the ordering of probes, determination of the physical distance between sequences, determining whether probes overlap, the evaluation of large genomic probes for rearrangements, such as internal deletions, and the mapping and ordering of cDNA probes within larger genomic probes. The stringency of the criteria used for the collection of data varies among these applications. For example, it may be necessary to evaluate only 10 cells to determine if a YAC has an internal deletion, if two probes overlap, or the relative order of probes, whereas estimating the distance between sequences requires a more extensive analysis.

To estimate the distance between probes, one must have a reference or control value; this is generally determined by determining the length of the signal for a probe of known size (in kb) *(20,24)*. For example, the length of the signal for the two test probes, as well as the length of the gap between them is determined. This is accomplished optimally with an image analysis system, The measurements for the gap are normalized to the signal lengths of the two probes of known size. Preliminary studies by several groups of investigators have suggested that the relative length remains constant for DNA stretched to various extents. Maps are determined by averaging the relative lengths from all cells scored; we and other investigators have found that at least 30 cells should be evaluated for these studies *(20)*.

2. Materials

2.1. Culture of Mitogen-Stimulated Lymphocytes

1. Culture medium: 90% RPMI 1640, 10% heat-inactivated fetal bovine serum (HI-FBS) 10 m*M*/L HEPES, 100 U/mL penicillin, 100 mg/mL streptomycin, pH adjusted to 7.2–7.3 with 7.5% sodium bicarbonate. All reagents are from Gibco-BRL.
2. Phytohemagglutinin (PHA, Murex Diagnostics Inc., Dartford, UK, HA-15, reagent-grade).
3. Colcemid (Gibco-BRL, Gaithersburg, MD).

2.2. Preparation of Metaphase Cells

1. Phosphate buffered saline (PBS) (37°C): 9 g/L NaCl, 0.21 g/L KH_2PO_4, 0.726 g/L Na_2HPO_4-7 H_2O.
2. Hypotonic solution: 0.075 *M* KCl (37°C).
3. Fixative solution: 1 part glacial acetic acid:3 parts absolute methanol (room temperature). Prepare immediately before use, and mix solution by rotating container prior to each use, since the components separate.

2.3. Nick Translation

1. 1 µg of probe DNA.
2. 10X Reaction mix: 0.5 *M* Tris-HCl, pH 7.5, 100 m*M* $MgCl_2$.
3. 10X dNTP mixture: 0.3 m*M* dATP, 0.3 m*M* dGTP, 0.3 m*M* dCTP in 50 m*M* Tris-HCl, pH 7.5.
4. Biotin-16-dUTP (Boehringer Mannheim, Indianapolis, IN), 0.3 m*M* (alternatively, other biotin-dUTP derivatives can be used).
5. 10X Digoxigenin-11-dUTP reaction mix—0.3 m*M* digoxigenin-11dUTP (Boehringer Mannheim) in 50 m*M* Tris-HCl.
6. *Escherichia coli* DNA polymerase I (Promega Bio Tech, Madison, WI).
7. DNase I—6.5 mg/mL in 100 m*M* $MgCl_2$ (Sigma, St. Louis, MO). The optimal size of the nick-translated probe should be below 500 bp. The final concentration of DNase added to the reaction must be determined by gel electrophoresis (*see* **step 6**).
8. Dilution buffer: 10 m*M* Tris-HCl, pH 7.5, 1 mg/mL BSA.
9. TE buffer: 10 m*M* Tris-HCl, pH 7.4, 0.1 m*M* EDTA, pH 8.0.
10. Sephadex G-50 spin column (Pharmacia, Washington, DC, Probe Quant G-50, microcolumns).
11. 1X TBE: 45 m*M* Tris-Borate, 1 m*M* EDTA.
12. 0.5 m*M* EDTA.
13. 5% SDS.

2.4. Alu-PCR

1. 10X PCR buffer: 100 m*M* Tris-HCl, pH 8.4, 500 m*M* KCl, 20.0 m*M* $MgCl_2$.
2. 4 m*M* dNTPs: 4.0 m*M* of each dNTP: dATP, dCTP, dGTP, and dTTP.
3. *Alu*-PCR 10 µ*M* primer: CL1, 5'-TCCCAAAGTGCTGGGATTACAG-3' (nt position 23–44 of the consensus *Alu* repeat), CL2, 5'-CTGCACTCCAGCCT GGG-3' (nt position

Gene Mapping by FISH

244–260). Because of the orientation of these primers, *Alu*-PCR products will contain only short *Alu* segments, which may facilitate efficient suppression of *Alu* sequences in subsequent FISH experiments *(12)*.
4. 2.5 U of *Taq* DNA polymerase.
5. DNA template: Genomic DNA can be prepared from normal or tumor cells using standard techniques; for tissue samples, 1 mg of tissue yields approx 3–5 µg of DNA. Purified cloned DNA, genomic yeast clone DNA, or YAC DNA isolated from pulsed-field gels may also be used. DNA can be quantitated using a spectrophotometer, or alternatively, the concentration can be estimated by electrophoresing a small amount of DNA in an agarose gel, and comparing the intensity of the ethidium stained gel under UV light to that of reference DNA.

2.5. DOP-PCR

1. 10X PCR buffer: 100 mM Tris-HCl, pH 8.4, 500 mM KCl, 20.0 mM MgCl$_2$.
2. 2 mM dNTPs: 2.0 mM of each dNTP: dATP, dCTP, dGTP, and dTTP.
3. 10X DOP primer (20 µM) primer 6-MW, 5'-CCGACTCGAGNNNNNNATG TGG-3', with N = A,C,G, or T in approximately equal proportions.
4. 2.5 U of *Taq* DNA polymerase.
5. DNA template (*see* **Subheading 2.4.**): Genomic DNA, purified cloned DNA, or YAC DNA isolated from pulsed-field gels may be used. Procedures for DNA isolation from small amounts of tissue (as little as a few hundred cells) or from paraffin-embedded material have been described *(16)*.

2.6. SIA

1. Buffer A: 40 mM Tris-HCl, pH 7.5, 50 mM NaCl, 10.0 mM MgCl$_2$, 5 mM dithiotreitol, 50 µg/mL BSA, 300 µM of each dNTP, and 1.5 mM Primer A.
2. Buffer B: 6.6 mM Tris-HCl, pH 9.0, 0.25 mM MgCl$_2$, 55 mM KCl, 0.01% (w/v) gelatin, 77 µM of each dNTP, 1.66 mM Primer B.
3. Primer A, 5'-TGGTAGCTCTTGATCANNNNN-3', with N = A,C,G, or T in approximately equal proportions. Primer B, 5'-AGAGTTGGTAGCTCTTGATC-3'.
4. T7 DNA polymerase (Sequenase Version 2.0, USB, Cleveland, OH), 2.5 U of *Taq* DNA polymerase.
5. A variety of DNA templates can be used for SIA as described for *Alu*-PCR and DOP-PCR.

2.7. Labeling of PCR Products for FISH

1. 10X PCR buffer: 100 mM Tris-HCl, pH 8.4, 500 mM KCl, 15.0 mM MgCl$_2$, 0.1% (w/v) gelatin.
2. 1.5 mM dATP, dCTP, and dGTP, 1 mM dUTP, 0.4 mM Bio-16-dUTP.
3. 0.15 mM SIA Primer B, DOP-PCR primer, or each *Alu*-PCR primer.
4. 1.0 U of *Taq* DNA polymerase.

2.8. Hybridization Techniques

1. 20X SSC: 3.0 M NaCl, 0.3 M NaCitrate, pH 7.0.
2. 4X SSC: Dilute 20X SSC 1:5 with dH$_2$O, pH 7.0.
3. 2X SSC: Dilute 20X SSC 1:10 with dH$_2$O, pH 7.0.
4. 3 M potassium acetate, 14.75 g KOAc in 50 mL ddH$_2$O, pH 5.2.
5. Formamide (hybridization buffer): Nucleic-acid-hybridization-grade or molecular-biology-grade deionized with ion-exchange resin. Molecular-biology-grade formamide may be used for posthybridization washes.
6. Carrier DNA: Salmon sperm DNA, 1 µg/µL, sheared mechanically or by sonication.

7. Placental DNA: 1 µg/µL, sheared mechanically or by sonication.
8. Cot1 DNA: 1 µg/µL (Gibco-BRL).
9. Dextran sulfate: Prepare 20% dextran sulfate in 4X SSC solution.
10. Labeled DNA probe: 1 µg/50 µL.
11. 10X RNase: 1 mg/mL in 2X SSC, pH 7.0.
12. Absolute ethanol, pure.

2.9. Detection of Hybridized Probes

1. Blocking solutions: 2% BSA (400 mg BSA/20 mL 4X SSC), 2% goat serum and 1% BSA, or 2% rabbit serum and 1% BSA in 4X SSC. Blocking solutions are used to reduce background signal from nonspecific binding. In general, the best results are obtained by incubating with diluted normal serum obtained from the same species host as the labeled antibody immediately prior to the application of the labeled antibody.
2. Nonionic detergent—Tween-20 or Triton-X 100.
3. DAPI stain—200 ng/mL in 2X SSC.
4. PDD antifade—dissolve 300 mg p-phenylenediamine dihydrochloride in 30 mL, pH 7.0, Dulbecco's PBS, pH adjusted to 8.0 with 0.5 M sodium bicarbonate. Combine 20 mL PDD solution with 80 mL glycerol. Store in the dark at –20°C.
5. Avidin conjugated with a fluorochrome, e.g., FITC, CY3, Texas red, Rhodamine, 1 mg/mL (all are from Vector Laboratories, Burlingame, CA, except for avidin-CY3, which is available from Amersham Life Science, Arlington Heights, IL).
6. 1 mg/mL Antidigoxigenin antibody (sheep) conjugated with a fluorochrome (Boehringer Mannheim).
7. 1 mg/mL Antiavidin antibody (goat) conjugated with a fluorochrome (Vector Laboratories).
8. 1 mg/mL Biotinylated antiavidin antibody (goat) (Vector Laboratories).
9. 1 mg/mL Antisheep antibody (rabbit) conjugated with a fluorochrome (Vector Laboratories).

2.10. Extended Chromatin Preparations

1. 0.07 M NaOH:ethanol (5:2) (alkaline treatment method) or 70% formamide in 2X SSC, pH 7.0 (formamide treatment method).
2. 70, 95, and 100% Ethanol.

3. Methods
3.1. Culture of Mitogen-Stimulated Lymphocytes

1. Peripheral blood—draw 10 mL of peripheral blood aseptically by venipuncture into a heparinized syringe or vacuum tube. Use only preservative-free heparin, since the preservatives in most heparin formulations suppress the growth of cells.
2. Centrifuge at 225g for 8 min, and transfer the buffy coat to a new sterile tube. Determine the number of leukocytes/mL of sample and the total number of cells using a hemacytometer and Unopette test (Becton Dickinson, San Jose, CA, test 5856), or another method of lysing the red blood cells.
3. Initiate two 10-mL cultures in 25-cm^2 flasks using complete culture medium prewarmed to 37°C. The cell aliquot (10^7 leukocytes) should be added in a volume of 1 mL or less; therefore, it may be necessary to concentrate or dilute the sample with medium prior to culture initiation. Bring the volume to 10 mL with culture medium. The final cell density should be 1×10^6 leukocytes/mL. Add PHA to a final concentration of 10 µg/mL.
4. Incubate flasks vertically with caps slightly loose (opened a one-quarter to one-half turn) at 37°C in a humidified 5% CO_2/95% air atmosphere for 72 h.

5. Add Colcemid (final concentration of 0.05 µg/mL), and reincubate for 45–60 min.
6. Process using the protocol for the preparation of metaphase cells.

3.1.1. Preparation of Metaphase Cells

1. Transfer the contents of the culture flask to a conical centrifuge tube.
2. To minimize the loss of cells, rinse the flask with 3–5 mL of PBS. Then add it to the tube prior to centrifugation.
3. Spin the centrifuge tubes for 8 min at 150–225g. Decant the supernatant.
4. Resuspend the pellet in the residual supernatant by gently tapping the tube, and add 10 mL of prewarmed hypotonic KCl. Mix by bubbling air through the solution with a Pasteur pipet.
5. Incubate the tubes for 8 min at 37°C.
6. Centrifuge for 8 min at 150–225g, and decant the supernatant.
7. Resuspend the pellet, and add two to five drops of freshly prepared fixative while gently tapping the tube. Add approx 1–2 mL of fixative in this manner, and then add an additional 8–10 mL of fixative.
8. Centrifuge tubes at 150g, decant the supernatant, resuspend the cells, and add 10 mL of fixative. Repeat this step until the fixative is clear.
9. Store the tube at –20°C, or repeat **step 8** four to eight times with fresh fixative, and prepare slides. Slides prepared immediately are generally of optimal quality. Stored cells should be washed with fresh fixative four to eight times prior to slide preparation.

3.1.2. Slide Preparation

Procedures for preparing slides from cell suspension are numerous and vary from laboratory to laboratory. The following technique works well in our laboratory; however, it may be necessary to modify this technique to obtain optimal results in other laboratories.

1. Clean glass microscope slides by immersing them in 95% ethanol, and wiping with a Kimwipe® or other lint-free cloth.
2. Resuspend the cell pellet in enough fresh fixative (3:1 absolute methanol:glacial acetic acid) to produce a slightly milky cell suspension (the volume of fresh fixative required is approx 10 times the volume of the cell pellet). Take care to dilute the sample correctly, because underdilution can result in an overly dense slide and poor spreading of metaphase cells.
3. Place one drop of fixative on the top of the slide (below the frosted edge); the fixative will spread and cover the surface of the slide. Do not place more than one drop of fixative onto the dry slide, because too much fixative may cause cell nuclei and metaphase cells to accumulate along the edges of the slide.
4. Hold the slide pointing downward at a 60° angle over a steam bath, and drop four to six drops of the cell suspension from a Pasteur pipet held 18–24 in. above the slide.
5. Immediately place the slide over a steam bath created by running hot water (45–50°C) into a pan. Two glass rods, placed 1 1/2 in. apart across the rim of the pan, can be used to hold the slides above the water level in the pan. Remove the slides from the pan when the fixative has dried from the upper surface (2–3 min).
6. Slides may be used immediately or can be stored at –20°C for several months. Slides should be stored in sealed slide boxes containing a desiccant, such as Drierite, and should be thawed slowly at room temperature before use.

3.2. Nick Translation

To nick-translate 1 µg of probe DNA in a 50 µL reaction:

1. Combine 1 µg of probe DNA, 5 µL 10X reaction mix, 5 µL 10X dNTP mix, 2.5 µL Bio-16-dUTP, 20 U DNA polymerase I, and an appropriate dilution of DNase I (1:1000 is a good test dilution). Add the enzymes last.
2. To label a probe with digoxigenin, follow the same steps, but use 5 µL of 10X digoxigenin-11-dUTP reaction mix instead of Bio-16-dUTP.
3. Bring the total volume to 50 µL with filtered dH_2O.
4. Vortex, and then centrifuge briefly.
5. Incubate at 14°C for 2 h.
6. After 2 h, remove tubes and place on ice.
7. Determine probe size by gel electrophoresis (*see* **Note 1**). Take 5 µL of the nick-translated probe, and add 5 µL of gel loading buffer (50% glycerol, 1X TBE, 1% bromophenol blue, 1% xylene cyanol). Load sample and an appropriate marker onto a 1–2% agarose minigel (1X TBE) and run the gel at 60 mA for approx 30 min. The probe size should be between 300–600 bp. Probe size is **critical**; if the probe is too large, add an additional 5 µL of diluted DNase I, and incubate for 30 min. If the probe is too short, repeat the reaction with less DNase I.
8. Stop reaction by adding 1 µL of 0.5 m*M* EDTA, and 1 µL of 5% SDS and heating reaction mix at 65°C for 10 min.
9. Remove unincorporated nucleotides by running the reaction sample through a 50-µL Sephadex G-50 column (Pharmacia).

3.3. Alu-PCR

1. Each PCR reaction contains: 100 ng genomic DNA, yeast clone DNA or YAC DNA isolated from pulsed-field gels, 10 µL 10X PCR buffer, 10 µL 10X dNTPs (final concentration 0.4 m*M*), 2.5 µL *Alu* PCR primer (final concentration 0.25 µ*M*), 2.5 U *Taq* polymerase, and dH_2O to adjust volume to 100 µL.
2. Tubes containing a positive control, such as placental DNA, and a negative control using the same solutions, but without the addition of template DNA should be prepared.
3. The PCR conditions are initial denaturation at 94°C for 5 min, followed by 35 cycles of denaturation at 94°C for 1 min, 30 s at 37°C, and extension at 72°C for 6 min, with a final extension at 72°C for 10 min.
4. The PCR reaction can be evaluated by electrophoresing 5–10 µL in an agarose gel (1% agarose, 1X TBE) at 100 V along with a size marker. A smear of DNA ranging in size from 600 bp to approx 5 kb should be visible in the ethidium bromide-stained gel of amplified human genomic DNA, whereas multiple discrete amplified bands with occasional smears are seen in amplified products of cloned DNA (200–4000 bp), such as YACs; there should be no smear in the negative control.
5. The *Alu*-PCR products can be labeled by nick-translation (*see* **Subheading 2.3.**) or by PCR labeling (*see* **Subheading 3.6.**).

3.4. DOP-PCR

1. Each 50-µL PCR reaction contains: 0.1–10 ng DNA, 5 µL 10X PCR buffer, 5 µL 10X dNTPs (final concentration 0.2 m*M*), 5 µL 10X DOP-primer (final concentration 2 µ*M*), 2.5 U *Taq* polymerase, and dH_2O to adjust volume to 50 µL.
2. Tubes containing a positive control, such as placental DNA, and a negative control using the same solutions, but without the addition of template DNA should be prepared.

3. The PCR conditions are initial denaturation at 94°C for 10 min, followed by five cycles of denaturation at 94°C for 1 min, 30°C for 1.5 min, 3-min transition 30–72°C, and extension at 72°C for 3 min, followed by 35 cycles of denaturation at 94°C for 1 min, annealing at 62°C for 1 min, and extension at 72°C for 3 min with an addition of 1 s/cycle to the extension step, with a final extension at 72°C for 10 min.
4. The PCR reaction can be evaluated by electrophoresing 5–10 µL in an agarose gel (1% agarose, 1X TBE) at 100 V along with a size marker. A smear of DNA ranging in size from 200–2000 bp should be visible in the ethidium bromide-stained gel; there should be no smear in the negative control *(14)*.

3.5. SIA

1. For SIA, 0.5 µL of the diluted YAC DNA (or 10 ng of genomic DNA or cloned DNA) is added to 5 µL of buffer A. The DNA is denatured at 94°C for 2 min, and cooled to 4°C to allow primer A to anneal at random sites. One unit of T7 DNA polymerase (USB: Sequenase version 2.0) is added in 2.5 µL of buffer A, and the temperature is gradually increased to 37°C over an 8-min interval, and kept at 37°C for 8 min, resulting in the synthesis of the first strand of DNA.
2. After denaturation, and annealing, this synthesis step is repeated one more time by adding fresh T7 enzyme in 2.5 µL of buffer A. T7 DNA polymerase is used for this step because it functions well at low temperatures at which random priming complexes are stable, and because it possesses strand displacement capabilities. Strand displacement synthesis enables the enzyme to synthesize long stretches of DNA by displacing other primers that have already annealed to DNA. In the second synthesis step, primer A will also prime on the products from the first round.
3. The products of this second synthesis step are suitable for PCR amplification. The PCR is carried out by adding 90 µL of buffer B, and 2.5 U of *Taq* DNA polymerase. Five low stringency cycles with denaturation at 94°C for 50 s, annealing at 42°C for 5 min, an increase to 72°C for 3 min, and synthesis at 72°C for 3 min were followed by 33 PCR cycles with denaturation at 94°C for 50 s, annealing at 56°C for 1 min, and synthesis at 72°C for 2 min.
4. Tubes containing a positive control such as placental DNA, and a negative control using the same solutions, but without the addition of template DNA should be prepared.
5. The PCR reaction can be evaluated by electrophoresing 5–10 µL in an agarose gel (1% agarose, 1X TBE) at 100 V along with a size marker. A smear of DNA ranging in size from 300–1000 bp should be visible in the ethidium bromide stained gel; there should be no smear in the negative control.

3.6. Labeling of PCR Products for FISH

1. Each 30-µL PCR reaction contains: 1 µL PCR product, 3 µL 10X PCR buffer, 3 µL 10X dNTPs, 1.0 U of *Taq* DNA polymerase, and dH$_2$O to adjust volume to 30 µL.
2. The PCR conditions are 18 cycles of 50 s at 94°C, 1 min at 56°C, and 2 min at 72°C.
3. Tubes containing a positive control, such as placental DNA, and a negative control using the same solutions, but without the addition of template DNA should be prepared.
4. The labeled PCR products (prepared by either biotinylation or labeling with directly labeled nucleotides; *see below*) are ethanol-precipitated and resuspended in 10 µL TE buffer; 8 µL of this DNA is treated with DNase I (200 pg/µL) for 5–10 min at room temperature. After 10 min of heat inactivation at 65°C, the DNA is ethanol-precipitated and resuspended in 10 µL of TE.

5. FISH is performed as described in **Subheading 3.7.** Amplified products from YACs are hybridized at a concentration of 90–120 ng probe/slide, along with 0.63 µg of placental DNA and 1–2 µg Cot1 DNA/slide.

Using the second approach, the amplified products are labeled with the Spectrum Orange™ or Spectrum Green™ fluorophore (Vysis, Inc., Downers Grove, IL) by performing a PCR under the same conditions as described with each dNTP at 150 µM, and the Spectrum Orange-dUTP at 30 µM. Digoxigenin-labeled nucleotides do not yield good results using PCR-labeling methods owing to the poor incorporation of this large labeled nucleotide.

3.7. Hybridization

3.7.1. Probe Preparation

1. Combine 3 µL salmon sperm (approx 1–4 µg), 0.1 µg labeled probe, placental DNA (approx 0.3–1 µg) and Cot1 DNA (between 0 and 1 µg depending on the repetitive elements in the probe DNA (*see* **Note 2**).
2. Ethanol-precipitate by adding 1/20 vol 3M KOAc and 2–2.5 vol of ice-cold pure-grade absolute ethanol. Vortex and then centrifuge for 5 s, chill at –80°C freezer for 30 min, and then centrifuge at 16,000kg at 4°C for 20 min. Decant the supernatant, then blot on paper towels, and air-dry for 30 min or vacuum-dry for 5 min.
3. Add 10 µL of hybridization buffer (5 µL formamide/5 µL 20% dextran sulfate in 4X SSC) to each tube, and vortex 1 h.
4. Denature hybridization mixture at 75°C for 5 min, and preanneal at 37°C for 20–30 min.

3.7.2. Slide Preparation and Denaturation

1. Prepare 200 mL RNase (dilute 10X RNase with 2X SSC pH 7.0, final conc. 100 µg/mL) in a glass staining box and warm to 40°C in a water bath.
2. Warm slides to room temperature, and check under the microscope to locate the metaphase or interphase cells for hybridization. Mark the area with a glass etching pen, and place slides in a glass rack.
3. Place slides in RNase for 1 h.
4. Remove slides and wash 4 × 2 min in 2X SSC and 1 × 2 min each in 70, 80, and 95% ethanol at room temperature.
5. Remove the slides from the glass rack and air-dry.
6. Place the slides in a glass rack, and denature in a solution of 70% formamide/30% 4X SSC (pH 7.0) for 2 min at 75°C. (The solution will cool a few degrees when the slides are added.)
7. After denaturation, dehydrate the slides in a graded alcohol series for 2 min each in 70, 80, and 95% ethanol, air-dry, and then place on a slide warmer at 40°C.
8. Apply 10 µL of denatured hybridization/probe mix to each slide in the premarked area taking care to avoid bubbles, and then cover with a 22 × 22 mm coverslip.
9. When all slides are finished, seal with rubber cement. This can be accomplished easily by applying the rubber cement with a 5–10 mL syringe without the needle attached.
10. Place in a prewarmed humid chamber (A plastic refrigerator box with a lid that seals tightly works well.) Incubate overnight at 37°C.
11. Prepare solutions of 50% formamide/50% 4X SSC (50/50 mix) and 4X SSC (pH 7.0) and warm to 40°C.
12. Remove glue from slides carefully.
13. Soak slides for 2 min in 50/50 mix. The coverslips should slide off gently, but it is sometimes necessary to pull them off very gently.

14. Rinse 3 × 5 min in 50/50 solution agitating gently.
15. Next rinse 4 × 2 min in 4X SSC.
16. Drain slides, and proceed to **Subheading 3.8.**

3.8. Detection of Hybridized Probes

Detection and amplification schemes can be quite complicated and varied. However, detection normally involves blocking of the target DNA with an appropriate serum, followed by detection with an appropriate reporter molecule. Amplification of the signal involves using a fluorochrome-conjugated antibody specific for the reporter molecule. The following procedure works well in our laboratory, but is flexible and can be adjusted to most laboratory conditions.

1. Apply 200 µL of blocking solution, coverslip, and incubate for 30–60 min in a humid chamber at 37°C.
2. Gently remove coverslip, and apply 200 µL of blocking solution containing 5 µg/mL of avidin-fluorochrome (to detect a biotinylated probe) or 5 µg/mL antidigoxigenin-fluorochrome sheep antibody (to detect a digoxigenin-labeled probe). For dual-color detection (simultaneous detection of a biotin-labeled probe and a digoxigenin-labeled probe), apply 200 µL of blocking solution containing 5 µg/µL of avidin-fluorochrome and 5 µg/µL antidigoxigenin antibody conjugated with a different fluorochrome. These detection solutions should be filtered through a 0.2-µm millipore filter before use.
3. Coverslip and incubate for 30–60 min at 37°C in a humid chamber.
4. Remove coverslips and wash 3 × 3 min in 4X SSC/0.1% Triton-X at 39°C. To amplify a biotin-labeled probe, proceed to **step 7**; to amplify a digoxigenin-labeled probe, proceed to step 12. Dual-color amplification can be achieved by amplifying the biotin-labeled probe first (**steps 7–10**) followed by amplification of the digoxigenin-labeled probe (**steps 12–16**).
5. Stain each slide in DAPI stain for 20–60 s. Note: You may need to rinse slide in dH$_2$O before staining in DAPI. Rinse in 2X SSC for 10 s.
6. Coverslip using three to four drops antifade with PDD. Blot the slides on a paper towel, and store in a slide box at 4°C.
7. Apply 200 µL of blocking solution, and coverslip for 0–30 min in a humid chamber at 37°C. Note: Amplification incubation times may vary and should be determined empirically.
8. Gently remove coverslip, and apply 200 µL of blocking solution containing 5 µg/mL of antiavidin antibody conjugated with the appropriate fluorochrome, e.g., amplify avidin-FITC with antiavidin-FITC (*see* **Note 3**).
9. Coverslip and incubate for 30 min at 37°C in a humid chamber.
10. Remove coverslips, and wash 2 × 3 min in 4X SSC/ 0.1% Triton-X at 39°C.
11. If only a biotinylated probe is being amplified, drain and stain slides as described in **steps 5** and **6**. To amplify a digoxigenin probe, proceed to **step 12**.
12. Apply 200 µL of blocking solution, and coverslip for 0–30 min in a 37°C humid chamber.
13. Gently remove coverslip, and apply 200 µL of blocking solution containing 5 µg/mL of antisheep antibody (for sheep antidigoxigenin-fluorochrome) conjugated with the appropriate fluorochrome.
14. Coverslip and incubate for 30 min at 37°C in a humid chamber.
15. Remove coverslips, and wash 2 × 3 min in 4X SSC/0.1% Triton-X at 39°C.
16. Drain and stain slides as described in **steps 5** and **6**.

3.9. Extended Chromatin Preparations

3.9.1. Cultures and Slide Preparation

1. Establish PHA-stimulated lymphocyte cultures as described, and incubate at 37°C for 72 h.
2. The cultures are harvested as described with the exception that the exposure to Colcemid is omitted.
3. Slides are prepared by dropping the cell suspension onto clean moist slides. Before the fix has dried, place slides in PBS for 1 min. Drain slides on a paper towel. Treat immediately with either the alkaline or formamide treatments. The fixed cells can be used immediately to prepare slides or can be stored at −20°C. Slides that are prepared immediately typically yield better preparations and more uniform release of DNA.

3.9.2. Alkaline Treatment

1. Place 100 µL NaOH:ETOH on the top edge of the slide, below the frosted end. Place the edge of the coverslip along the slide (horizontal), and move the coverslip along the slide at a 30° angle. (Alkaline treatment results in the immediate disruption of the nuclei.) Blot residual fluid with a paper towel.
2. Hold the slide horizontally, and rinse with methanol using a pipet. (**Use only small amounts of methanol, and apply dropwise gently.**) Note: A viscous consistency of the fluid dripping off the slide indicates major loss of DNA. Thereafter, the slide can be held vertically and rinsed several more times gently.
3. Air-dry slide, and pass through 70, 95, and 100% ethanol (2 min each).
4. Air-dry, and store at −20°C in the presence of a desiccant.

3.9.3. Formamide Treatment (see **Note 4**)

1. Place 100 µL formamide:SSC on the top edge of the slide, below the frosted end. Place the edge of the coverslip along the slide (horizontal), and move the coverslip along the slide at a 30° angle. Blot residual fluid with a paper towel.
2. Hold the slide horizontally, and rinse with methanol using a pipet. (**Use only small amounts of methanol, and apply dropwise gently.**) Thereafter, the slide can be held vertically and rinsed several more times gently. The nuclei disrupt on contact with methanol, and the released chromatin is fixed simultaneously.
3. Air-dry slides, and pass through 70, 95, and 100% ethanol (2 min each).
4. Air-dry, and store at −20°C in the presence of a desiccant.

4. Notes

1. The size of the nick-translated probe is critical to the outcome of FISH experiments, and should be between 300 and 600 bp. If the fragments are too small, no hybridization or background signal will be observed (a faint dusting of signal may be seen on the chromosomes), since many hybridized fragments will be removed by the posthybridization washes. In contrast, if the probe is too large, nonspecific hybridization can occur, which is visualized as many signals distributed randomly on the chromosomes.
2. Suppression of hybridization signals from ubiquitous repetitive sequences, such as *Alu* or *Kpn*, elements is achieved using total human DNA and the Cot1 fraction of DNA for a reannealing process that is based on rapid reassociation kinetics. The amount of competitor DNA needed varies depending on the number of repetitive elements in any particular genomic sequence. It is critical that the concentration of these competitors be adjusted appropriately, particularly the Cot1 DNA, since too little results in high background and too much results in weakened probe signal. A good starting point is to prepare the probe

with 0.5 µg placental DNA and 0.5 µg Cot1 DNA; the concentration of the Cot1 DNA can then be increased or decreased as needed.
3. Alternatively, biotinylated antiavidin antibody may be substituted for antiavidin fluorochrome. The biotinylated antiavidin must then be followed by additional rounds of washes and detection with avidin-fluorochrome. This method of amplification produces strong specific signal, but may also produce background.
4. Alkaline treatment with NaOH results in the complete disruption of the nuclei, and the slides are covered with a network of straight chromatin fibers (or occasionally an irregular network). With the formamide treatment, the results are a comet-like tail of released chromatin, and the borders of most of the disrupted nuclei can be defined, allowing signals from the same nucleus to be identified. Hybridization signals usually appear as a linear array of dots (approx 1% of signals will appear as a continuous line of fluorescent dots). Signals are longer in NaOH-treated cells owing to the more extended chromatin. For shorter probes, NaOH may be preferable, whereas formamide treatment may be preferable for longer clones, such as YACs. The lower limit of resolution of these methods is 1 kb; the upper limit has not been established yet, but is probably >800 kb.

References

1. Evans, H. J., Buckland, R. A., and Pardue, M. L. (1974) Location of the genes coding for 18 S and 28 S ribosomal RNA in the human genome. *Chromosoma* **48,** 405–426.
2. Harper, M. E. and Saunders, G. F. (1981) Localization of single copy DNA sequences on G-banded chromosomes by *in situ* hybridization. *Chromosoma* **83,** 431–439.
3. Rudkin, G. T. and Stollar, B. D. (1977) High resolution detection of DNA-RNA hybrids *in situ* by indirect immunofluorescence. *Nature* **265,** 472,473.
4. Bauman, J. G., Wiegant, J., Borst, P., and van Duijn, P. (1980) A new method for fluorescence microscopical localization of specific sequences by *in situ* hybridization of fluorochrome-labelled RNA. *Exp. Cell Res.* **128,** 485–490.
5. Lawrence, J. B., Villnave, C. A., and Singer, R. H. (1988) Sensitive, high-resolution chromatin and chromosome mapping *in situ*: presence and orientation of two closely integrated copies of EBV in a lymphoma line. *Cell* **52,** 51–61.
6. Pinkel, D., Landegent, J., Collins, C., Fuscoe, J., Segraves, R., Lucas, J., and Gray, J. W. (1988) Fluorescence *in situ* hybridization with human chromosome-specific libraries: detection of trisomy 21 and translocations of chromosome 4. *Proc. Natl. Acad. Sci. USA* **85,** 9138–9142.
7. Trask, B. J. (1991) Fluorescence *in situ* hybridization: applications in cytogenetics and gene mapping. *Trends Genet.* **7,** 149–154.
8. Le Beau, M. M. (1993) Fluorescence *in situ* Hybridization in Cancer Diagnosis, in Important Advances in Oncology (de Vita, V. T., Jr, Hellman, S., and Rosenberg, S. A., eds.) J. B. Lippincott, Philadelphia, pp. 29–45.
9. Lichter, P., Cremer, T., Borden J., Manuelides, L., and Ward, D. C. (1988) Delineation of individual human chromosomes in metaphase and interphase cells by *in situ* suppression hybridization using recombinant DNA libraries. *Hum. Genet.* **80,** 224–234.
10. Landegent, J. E., Jansen in de Wal, N., Dirks, R. W., Baas, F., and van der Ploeg, M. (1987) Use of whole cosmid cloned genomic sequences for chromosomal localization by non-radioactive *in situ* hybridization. *Hum. Genet.* **77,** 366–370.
11. Lichter, P. and Ward, D. C. (1990) Is non-isotopic *in situ* hybridization finally coming of age? *Nature* **345,** 93–94.
12. Lengauer, C., Green, E. D., and Cremer, T. (1992) Fluorescence *in situ* hybridization of YAC clones after *Alu*-PCR amplification. *Genomics* **13,** 826–828.

13. Baldini, A. and Ward, D. C. (1991) *In situ* hybridization of human chromosomes with *Alu*-PCR products: a simultaneous karyotype for gene mapping studies. *Genomics* **9**, 770–774.
14. Telenius, H., Carter, N. P., Bebb, C., Nordenskjöld, M., Ponder, B. A. J., and Tunnacliffe, A. (1992) Degenerate oligonucleotide-primed PCR: General amplification of target DNA by a single degenerate primer. *Genomics* **13**, 718–724.
15. Bohlander, S. K., Espinosa, R. III, Le Beau, M. M., Rowley, J. D., and Diaz, M. O. (1992) A method for the rapid sequence-independent amplification of microdissected chromosomal material. *Genomics* **13**, 1322–1324.
16. Speicher, M. R., du Manoir, S., Schröck, E., Holtgreve, H., Schoell, B., Lengauer, C., Cremer, T., and Reid, T. (1993) Molecular cytogenetic analysis of formalin-fixed, paraffin-embedded solid tumors by comparative genomic hybridization after universal DNA amplification. *Hum. Mol. Genet.* **11**, 1907–1914.
17. Wiegant, J., Reid, T., Nederlof, P. M., van der Ploeg, M., Tanke, H. J., and Raap, A. K. (1991) *In situ* hybridization with fluoresceinated DNA. *Nucleic Acids Res.* **19**, 3237–3241.
18. Reid, T., Baldini, A., Rand, T., and Ward, D. C. (1992) Simultaneous visualization of seven different DNA probes by *in situ* hybridization using combinatorial fluorescence and digital imaging. *Proc. Natl. Acad. Sci. USA* **89**, 1388–1392.
19. Lawrence, J. B., Carter K. C., and Gerdes, M. J. (1992) Extending the capabilities of interphase chromatin mapping. *Nature Genet.* **2**, 171,172.
20. Parra, I. and Windle, B. (1993) High resolution visual mapping of stretched DNA by fluorescent hybridization. *Nature Genet.* **5**, 17–21.
21. Wiegant, J., Kalle, W., Mullenders, L., Brookes, S., Hoovers, J. M. N., Dauwerse, J. G., van Ommen, G. J. B., and Raap, A. K. (1992) High-resolution *in situ* hybridization using DNA halo preparations. *Hum. Mol. Genet.* **1**, 587–591.
22. Heng, H. H. Q., Squire, J., and Tsui, L.-C. (1992) High-resolution mapping of mammalian genes by *in situ* hybridization to free chromatin. *Proc. Natl. Acad. Sci. USA* **89**, 9509–9513.
23. Fidlerova, H., Senger, G., Kost, M., Sanseau, P., and Sheer, D. (1994) Two simple procedures for releasing chromatin from routinely fixed cells for fluorescence *in situ* hybridization. *Cytogenet. Cell Genet.* **65**, 203–205.
24. Cai, W., Aburatani, H., Stanton, V. P., Housman, D. E., Wang, Y.-K., and Schwartz, D. C. (1995) Ordered restriction endonuclease maps of yeast artificial chromosomes created by optical mapping on surfaces. *Proc. Natl. Acad. Sci. USA* **92**, 5164–5168.
25. Rowley, J. D., Diaz, M. O., Espinosa, R., Patel, Y. D., van Melle, E., Ziemin, S., Taillon-Miller, P., Lichter, P., Evans, G. A., Kersey, J. D., Ward, D. C., Domer, P. H., and Le Beau, M. M. (1990) Mapping chromosome band 11q23 in human acute leukemia with biotinylated probes: Identification of 11q23 translocation breakpoints with a yeast artificial chromosome. *Proc. Natl. Acad. Sci. USA* **89**, 9358–9362.
26. Le Beau, M. M., Espinosa, R. III, Neuman, W. L., Stock, W., Roulston, D., Larson, R. A., Keinanen, M., and Westbrook, C. A. (1993) Cytogenetic and molecular delineation of the smallest commonly deleted region of chromosome 5 in myeloid leukemias. *Proc. Natl. Acad. Sci. USA* **90**, 5484–5488.
27. Boultwood, J., Fidler, C., Lewis, S., Kelly, S., Sheridan, H., Littlewood, T. J., Buckle, V. J., and Wainscoat, J. S. (1994) Molecular mapping of uncharacteristically small 5q deletions in two patients with the 5q-syndrome: Delineation of the critical region on 5q and identification of a 5q-breakpoint. *Genomics* **19**, 425–432.

117

Oligonucleotide PRINS DNA Synthesis

John R. Gosden and Diane Lawson

1. Introduction

The technique for labeling chromosomes by annealing an oligonucleotide DNA primer to the denatured DNA of chromosome preparations on glass slides and extending it enzymatically *in situ* with the incorporation of labeled nucleotides was first described by Koch et al. in 1989 *(1)*. Since then, the technique has been greatly improved in reliability, sensitivity, and resolution, and now provides a viable, rapid alternative to conventional fluorescence *in situ* hybridization (FISH) for many investigations, particularly the identification of chromosome aneuploidy in metastatic tissues and antenatal diagnosis and the analysis of the human chromosome complement of somatic hybrid cell lines *(2–6)*.

2. Materials

2.1. Primed In Situ Synthesis

1. Twin-Frost glass slides and 22 × 40 mm coverslips: The slides must be cleaned by soaking in ethanol to which a few drops of HCl have been added, followed by polishing with a clean piece of muslin, before the cells are deposited on the slide. Coverslips must be cleaned in the same way before use.
2. PRINS buffer (10): 500 mM KCl, 100 mM Tris-HCl, pH 8.3, 15 mM MgCl$_2$, 0.1% BSA.
3. 2'-Deoxyadenosine 5'-triphosphate (dATP): 100-mM solution (Amersham Pharmacia Biotech, St. Albans, UK), diluted 1:10 with sterile distilled H$_2$O.
4. 2'-Deoxycytidine 5'-triphosphate (dCTP): 100-mM solution (Amersham Pharmacia Biotech) diluted 1:10 with sterile distilled H$_2$O.
5. 2'-Deoxyguanosine 5'-triphosphate (dGTP): 100-mM solution (Amersham Pharmacia Biotech) diluted 1:10 with sterile distilled H$_2$O.
6. 2'-Deoxythymidine 5'-triphosphate (dTTP): 100-mM solution (Amersham Pharmacia Biotech) diluted 1:100 with sterile distilled H$_2$O.
7. Biotin-16-2'-deoxyuridine-5'-triphosphate (Bio-16-dUTP) (Roche Diagnostics GmBH, Lewes, Sussex, UK).
8. Digoxigenin-11-deoxyuridine-5'-triphosphate (Dig-11-dUTP) (Roche Diagnostics GmBH).
9. FluoroRed (Amersham International, plc, Buckinghamshire, UK).
10. FluoroGreen (Amersham International).
11. FluoroBlue (Amersham International).

12. Oligonucleotide primer(s) at 250 ng/μL (*see* **Note 1**).
13. *Taq* DNA polymerase (*Taq* [Boehringer], AmpliTaq [Cetus], or Thermoprime[plus] [Advanced Biotechnologies Ltd., Leatherhead, UK]).
14. Rubber cement (vulcanizing solution) (e.g., Tip-Top, Stahlgruber, DS-8011 Poing, Germany) (*see* **Note 2**).
15. Stop buffer (500 mM NaCl, 50 mM EDTA).
16. Flat-bed thermal cycler (*see* **Note 3**).
17. Water bath at 65°C.

2.2. Detection

1. Dried skimmed milk powder.
2. Avidin-DCS-fluorescein isothiocyanate (Av-FITC) (Vector Labs, Burlingame, CA).
3. Avidin-DCS-Texas red (Av-TR) (Vector Labs).
4. Antidigoxigenin-fluorescein (anti-DIG-FITC) (Boehringer Mannheim).
5. Antidigoxigenin-rhodamine (anti-DIG-rhodamine) (Boehringer Mannheim).
6. Propidium iodide (20 μg/mL) (Sigma).
7. 4',6-Diamidino-2-phenylindole 2 HCl (DAPI) (100 μg/mL) (Sigma).
8. VectaShield (Vector Labs).
9. 20X SSC: 3.0 M NaCl, 0.30 M tri-sodium citrate, pH 7.3.
10. Wash buffer: 4X SSC (diluted from stock 20X SSC), 0.05% Triton X-100.
11. Blocking buffer: wash buffer with the addition of 5% skimmed milk powder.
12. Incubator or water bath at 37°C and water bath at 45°C.
13. Microscope equipped for epifluorescence (e.g., Zeiss Axioskop or Leitz Ortholux II with Pleomopak filter system).

3. Methods

3.1. Standard PRINS

1. You will need cells or chromosomes, prepared from peripheral blood lymphocytes (*7*), cultured cells (*8*), or frozen sections (*9*) (*see* **Note 5**).
2. Oligonucleotide primers are prepared on an Applied Biosystems (Foster City, CA) Model 381A DNA synthesizer according to the manufacturer's instructions. Recommendations for some successful chromosome-specific primers are given in **Table 1** (but *see* **Note 4**).
3. The reaction mix is made up as follows: **For each slide,** put 1 μL of each of the diluted nucleotide triphosphates, plus 1 μL of the selected labeled dUTP (biotin, digoxigenin, or a fluorochrome), 5 μL 10X PRINS buffer, and 1 μL of the appropriate oligonucleotide primer (*see* **Note 6**) into a microcentrifuge tube, and add distilled water to 50 μL.
4. Mix thoroughly and add 1 U of your chosen DNA polymerase. Mix carefully and place 40 μL on a clean coverslip.
5. Pick the coverslip up with a slide (this spreads the reaction mix evenly, with the least risk of introducing air bubbles) and seal with rubber cement.
6. Dry the seal (a cold air fan is quick and safe) and transfer the slides to the flat block of a thermal cycler. A suitable basic program for the Hybaid OmniGene™ In Situ, or Hybaid OmniSlide™ is 93°C, 3 min; 60°C, 5–10 min; 72°C, 15 min.
7. On completion of the program, remove the seal (it peels off easily by rubbing one corner) and transfer the slides for 1 min to a Coplin jar containing stop buffer at 65°C. Leave the coverslips in place, unless they come off readily with the seal; they will in any case fall off in the stop buffer. After 1 min, transfer the slides to a stain dish containing wash buffer. They may be held in this solution overnight if convenient (but *see* **Note 7**).

Table 1
Examples of Chromosome-Specific Oligonucleotides and a Primer for All Human Centromeres

F673 (20-mer)	D16Z1, Satellite II	TTCTTTTCATACCGCATTCT
F60 (30-mer)	D17Z1, alphoid	ATTGCACTTCTTTGAGGAGTACCGTAGTAA
G33 (19-mer)	D9Z1, Satellite III	AATCAACCCGAGTGCAATC
I68 (17-mer)	CenP-B Box	CTTCGTTGGAAACGGGA

3.2. Detection

It is important that the slides do not become dry at any time during this process. The following steps apply only to slides in which the PRINS reaction has been labeled with biotin or digoxigenin. Slides in which the reaction used a fluorochrome-dUTP as the label require no detection step, and are simply mounted (*see* **step 6**).

1. Prepare blocking buffer. The milk powder dissolves rapidly if the solution is warmed to 45°C for a few seconds.
2. Put 40 µL blocking buffer on a clean coverslip, shake surplus wash buffer from the slide, and pick up the coverslip containing blocking buffer. Leave (unsealed) at room temperature for 5 min.
3. Dissolve reporter (avidin-fluorochrome or antidigoxigenin-fluorochrome) in blocking buffer. For Av-FITC or Av-TR, 1:500 is a suitable dilution; anti-DIG FITC and anti-DIG rhodamine are better at 1:100 dilution. Make sufficient buffer for a 40 µL/slide. Spin in a microcentrifuge for 5 min. This precipitates any aggregates that may have formed during storage and can cause high and nonspecific background.
4. Remove the coverslip from the slide, shake surplus fluid off both the slide and the coverslip, and add 40 µL of reporter solution to the **same** coverslip. Replace the slide and incubate in a moist chamber (e.g., a sandwich box lined with damp filter paper) at 37°C for 30 min.
5. Warm a reagent bottle containing wash buffer to 45°C in a water bath. Remove coverslips and wash slides 3 × 2 min in 50 mL wash buffer at 45°C.
6. After the final wash, shake off surplus fluid and mount slides in VectaShield containing the appropriate counterstain: For slides labeled with rhodamine or Texas red, this should be DAPI (5 µg/100 µL VectaShield, i.e., 5 µL of DAPI stock/100 µL VectaShield); for slides labeled with FITC, this should be a propidium iodide/DAPI mixture (3.75 µL of each stock/100 µL VectaShield). Use 20–30 µL mountant/slide, blot surplus by covering slide and coverslip with a tissue and pressing gently to expel excess mountant, and seal with rubber cement. Slides may be stored in the dark at 4°C for several months. If the stain shows signs of fading, simply peel off the sealant, soak the slide overnight in 4X SSC, 0.05% Triton X-100 (the coverslip will fall off at this point), and remount as above. **Figure 1** shows some typical results.
7. Multiple sequential PRINS reactions may be performed on the same sample in order to quantify a number of chromosomes. For details of the method, *see* **ref. 6** and **10**.
8. The technique may also be combined with FISH. After the stop buffer, the slides are passed through an ethanol series (70, 90, 100%) and air-dried before performing a normal FISH procedure, **omitting any denaturation of the chromosomal DNA**. Detection of the PRINS product and the hybridized FISH probe is then performed simultaneously *(11)*. This provides a rapid method for identifying the chromosomal target located by the FISH.

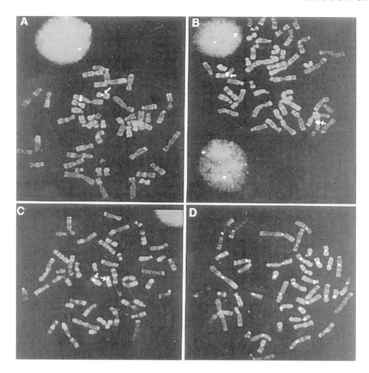

Fig. 1. Examples of PRINS reactions with the primers shown in **Table 1**. All reactions were labeled with biotin-16-dUTP, and the label detected with avidin-FITC. Chromosomes were counterstained with a mixture of DAPI and propidium iodide. (**A**) Chromosome 16. (**B**) Chromosome 9. (**C**) Chromosome 17. (**D**) CenP-B box primer (labels all centromeres).

4. Notes

1. Oligonucleotide primers can be synthesized on an ABI DNA synthesizer and used without further purification other than alcohol precipitation and washing. If this facility is not available, they may be obtained from commercial sources, but purification steps, such as HPLC, are not needed and only increase the cost of the product.
2. The requirement for a suitable seal is that it should be reasonably robust, provide a vapor-tight seal, and be easily and **completely** removed at the end of the procedure. We have found that Tip-Top fulfills all these parameters and is readily available from bicycle repair shops.
3. Thermal cyclers with a flat bed for microscope slides are not yet widely available. Some of the products sold for this purpose are not altogether suitable, since they are *ad hoc* modifications of machines designed for PCR in microtubes, with a plate added to the heated block. Thermal transfer and temperature control in such a system are rarely satisfactory. The procedure can be carried out by transferring slides through a series of water baths at appropriate temperatures, but this too means that temperature control cannot be precise, and the temperature drop during the transfer from water bath to water bath leads to high backgrounds. The most suitable purpose-built products are the OmniGene In Situ and OmniSlide made by Hybaid (Teddington, Middlesex, UK), which hold 4 and 20 slides, respectively.
4. As an alternative, complete systems for chromosome identification by PRINS are becoming available (e.g., Advanced Biotechnologies, Leatherhead, UK).

5. Cell suspensions may be stored in fix (methanol:acetic acid [3:1]) at –20°C for several months. Slides are prepared fresh each week by gently centrifuging the suspension to precipitate the cells, resuspending in fresh fix, repeating this process, and finally resuspending in sufficient fix to give a suitable density and putting one drop on a clean slide, which is allowed to dry at room temperature. The balance of the suspension may then be diluted suitably with fix and returned to –20°C. Using slides more than 1–2 wk old can be successful, but may lead to reduced sensitivity and greater variability.
6. The majority of chromosome-specific alphoid sequences produce adequate signal with a single primer at a concentration of 250 ng/50 µL reaction. In some cases, a clearer signal with less background may be produced with paired primers, at the same concentration, whereas in others, the concentration of primer may be reduced, with a concomitant reduction in crossreaction to related chromosomal sequences.
7. Slides that have been labeled directly with fluorochromes may still be held in this solution overnight if convenient, but should be kept in the dark to prevent bleaching and fading of the label.

References

1. Koch, J. E., Kølvraa, S., Petersen, K. B., Gregersen, N., and Bolund, I. (1989) Oligonucleotide-priming methods for the chromosome-specific labelling of alpha satellite DNA in situ. *Chromosoma* **98**, 259–265.
2. Gosden, J., Hanratty, D., Starling, J., Fantes, J., Mitchell, A., and Porteous, D. (1991) Oligonucleotide primed in situ DNA synthesis (PRINS): a method for chromosome mapping, banding and investigation of sequence organization. *Cytogenet. Cell Genet.* **57**, 100–104.
3. Gosden, J. and Lawson, D. (1994) Rapid chromosome identification by oligonucleotide primed in situ DNA synthesis (PRINS). *Hum. Mol. Genet.* **3**, 931–946.
4. Gosden, J. and Lawson, D. (1995) Instant PRINS: a rapid method for chromosome identification by detecting repeated sequences in situ. *Cytogenet. Cell Genet.* **68**, 57–60.
5. Hindkjaer, J., Koch, J., Terkelsen, C., Brandt, C. A., Kølvraa, S., and Bolund, L. (1994) Fast, sensitive multicolour detection of nucleic acids by primed in situ labelling (PRINS). *Cytogenet. Cell Genet.* **66**, 152–154.
6. Speel, E. J. M., Lawson, D., Hopman, A. H. N., and Gosden, J. (1995) Multi-PRINS: multiple sequential oligonucleotide primed in situ DNA synthesis reactions label specific chromosomes and produce bands. *Hum. Genet.* **95**, 29–33.
7. Spowart, G. (1994) Mitotic metaphase chromosome preparation from peripheral blood for high resolution, in *Methods in Molecular Biology, vol. 29: Chromosome Analysis Protocols* (Gosden, J. R., ed.), Humana, Totowa, NJ, pp. 1–10.
8. Fletcher, J. (1994) Immortalized cells lines: chromosome preparation and binding, in *Methods in Molecular Biology, vol. 29: Chromosome Analysis Protocols* (Gosden, J. R., ed.), Humana, Totowa, NJ, pp. 51–57.
9. Speel, E. J. M., Lawson, D., Ramachers, F. C. S., Gosden, J. R., and Hopman, A. H. N. (1997) Combined immunocytochemistry and PRINS DNA synthesis for simultaneous detection of phenotypic and genomic parameters in cells, in *Methods in Molecular Biology, vol. 71, PRINS and* In Situ *PCR Protocols* (Gosden, J. R., ed.), Humana, Totowa, NJ, pp. 53–59.
10. Multiple sequential oligonucleotide primed *in situ* DNA synthesis (MULTI-PRINS) (1997), in *Methods in Molecular Biology, vol. 71, PRINS and* In Situ *PCR Protocols*, (Gosden, J. R., ed.), Humana, Totowa, NJ, pp. 39–44.
11. Warburton, P. E., Haaf, T., Gosden, J., Lawson, D., and Willard, H. F. (1996) Characterization of a chromosome-specific chimpanzee alpha satellite subset: evolutionary relationship to subsets on human chromosomes. *Genomics* **33**, 220–228.

118

Chromosome-Specific PRINS

Jean-Paul Charlieu and Frank Pellestor

1. Introduction

The identification of individual chromosomes is of great importance in cytogenetics, in order to detect aneuploidies or chromosomal rearrangements associated with genetic diseases. This can be achieved by several techniques based either on the intrinsic staining properties of the chromosomes in producing bands (the banding pattern being specific for each pair of chromosomes) *(1)* or the use of a DNA probe to detect specifically a region of the chromosome by fluorescence *in situ* hybridization (FISH) *(2)*. The use of centromeric α satellite sequences as FISH probes is very popular because of the specificity of these sequences. α Satellite (or alphoid) DNA is a family of tandemly repeated sequences present at the centromere of all human chromosomes *(3)*. Subfamilies, some of them specific for one or a small group of chromosomes, can be identified within alphoid DNA both by the periodic distribution of restriction sites and the nucleotide sequence of the 171-bp basic motif *(4)*. These chromosome-specific subfamilies can therefore be used as FISH probes. This approach is limited, however, since the DNA sequences of some subfamilies are very close to each other, and crosshybridization can occur between the centromeric sequences of several pairs of chromosomes. This is the case with chromosomes 13 and 21, for example, which share 99.7% homology in their alphoid sequences *(5)*. The development of the primed *in situ* (PRINS) technique of labeling DNA *(6–8)* introduced a solution to this problem. The PRINS procedure consists of the use of a small oligonucleotide (usually 18–22 nucleotides) from the sequence of interest as a primer. The primer is annealed to the denatured DNA of a chromosome or cell preparation. An *in situ* DNA synthesis reaction is performed with the incorporation of a labeled precursor (biotin-dUTP or digoxygenin-dUTP), using a thermostable DNA polymerase. A single base mismatch between the target and the probe will produce a less stable hybrid than for a long FISH probe. In addition, if the mismatching nucleotide is located at the 3'-end of the PRINS primer, it will prevent any elongation by the DNA polymerase.

We have developed several chromosome-specific α-satellite primers for PRINS, each of them carrying at least a chromosome-specific nucleotide at its 3'-end, and we describe in this chapter the use of two of them for the detection of human chromosomes 13 and 21. Other primers are available in the literature *(9,10)* or on request, but we are presenting only the conditions of use for the two most difficult, differing only at one position.

2. Material

2.1. Slides

1. Chromosome spreads are prepared from peripheral blood using standard methods (fixation in methanol:acetic acid 3:1).
2. 20X SSC: 3 M NaCl, 0.3 M Na$_3$-citrate (can be stored for several months at room temperature).
3. 70, 90, and 100% ethanol.
4. Formamide (Prolabo, Paris, France): Formamide must be deionized by mixing with Amberlite resin (Sigma, St. Louis, MO), allowing to stand for at least 1 h, and then filtering. Deionized formamide is stored at +4°C.

2.2. PRINS Reaction

1. Primers: Synthetic oligonucleotides are used as primers in the PRINS experiments. Their nucleotide sequences are as follows (*11*):

    13A (chromosome 13): 5'-TGATGTGTGTACCCAGCT-3'
    21A (chromosome 21): 5'-TGATGTGTGTACCCAGCC-3'

    Precipitate the primers by adding 10 vol of 1-butanol, vortex, and centrifuge for 1 min at maximum speed in a bench-top microfuge. Dry the pellets under vacuum, and resuspend in 5 mM Tris-HCl, pH 8.0, to obtain a 50 µM (50 pmol/µL) solution. Store small aliquots (50 µL) at –20°C (*see* **Notes 1** and **2**).
2. 2'-Deoxyadenosine 5'-triphosphate (dATP) (Boehringer Mannheim, Meylan, France): Resuspend in H$_2$O to obtain a 100-mM stock solution (store at –20°C).
3. 2'-Deoxycytosine 5'-triphosphate (dCTP) (Boehringer Mannheim): Resuspend to obtain a 100-mM stock solution (–20°C).
4. 2'-Deoxyguanosine 5'-triphosphate (dGTP) (Boehringer Mannheim): Resuspend to obtain a 100-mM stock solution (–20°C).
5. 2'-Deoxythymidine 5'-triphosphate (dTTP) (Boehringer Mannheim): Resuspend to obtain a 100-mM stock solution (–20°C).
6. Biotin-16-dUTP, 1 mM (Boehringer Mannheim) (–20°C).
7. Glycerol 87% (Prolabo).
8. *Taq* DNA polymerase (Boehringer Mannheim). Store at –20°C.
9. 10X *Taq* buffer (provided with the enzyme) (–20°C).
10. Stop buffer: 500 mM NaCl, 50 mM EDTA, pH 8.0 (can be stored at room temperature for several months).
11. Sterile, deionized, double-distilled water.
12. Water bath at 60°C.
13. Water bath at 72°C.
14. 1.5-mL microcentrifuge tubes (sterilized by autoclaving).
15. Coverslips (22 × 40).
16. Thermal cycling machine equipped with a flat block (e.g., Techne PHC-3).

2.3. Detection

1. Washing solution: 4X SSC, 0.05% Tween 20.
2. Blocking solution: washing buffer plus 5% nonfat dry milk. Make fresh each time.
3. Fluorescein-avidin DCS (FITC-avidin) (Vector Laboratories, Burlingame, CA).
4. Propidium iodide (PI) (Sigma).

Chromosome-Specific PRINS

5. Antifade solution Vectashield (Vector Labs).
6. Staining jars.
7. Microscope equipped for detection of FITC and PI fluorescence.

3. Methods

3.1. Slides

1. Store slides prepared according to standard methods at room temperature for 5 d before use.
2. Just before the PRINS reaction, dehydrate the slides by passage through an ethanol series (70, 90, 100%) at room temperature, 3 min each step, and air-dry.
3. Denature the chromosomal DNA on the slides by immersing them in 70% formamide, 2X SSC, at 72°C for 2 min, dehydrating through an ice-cold ethanol series (70, 90, 100%), and air-drying (*see* **Note 3**).

3.2. PRINS Reaction

1. Prepare a 10X dNTPs mix: Dilute the stock solutions (100 mM) of dATP, dCTP, dGTP, and dTTP 1/10 in sterile, distilled water. In a sterile microcentrifuge tube, mix 10 µL of each diluted dATP, dCTP, and dGTP, 0.25 µL of diluted dTTP, 25 µL of 1 mM biotin-16 dUTP, and 55 µL of glycerol. Mix well and store at –20°C.
2. Prepare the PRINS reaction mix in a sterile 1.5-mL microtube by mixing (for each slide) 4 µL of primer (200 pmol), 5 µL of 10X *Taq* polymerase buffer, 5 µL of 10X dNTPs mix (from step 1), and 0.5 µL of *Taq* polymerase (2.5 U), and add sterile distilled water to a final volume of 50 µL.
3. Preheat the reaction mix at 60°C in a water bath.
4. Place the slide (prepared as in **Subheading 3.1.**) and a coverslip on the plate of the thermal cycler.
5. Set up the program for PRINS: 12 min at the annealing temperature (60°C for primer 13A, 61°C for primer 21A; *see* **Note 4**) and 30 min at 72°C.
6. When starting the program, heat the slide(s) and the coverslip(s) at the annealing temperature for 5 min. Then put the reaction mix onto the slide and cover by the coverslip. Incubate the slide at the annealing temperature for a further 7 min; the temperature is automatically raised to 72°C at the beginning of the elongation step.
7. At the end of the elongation time, transfer the slide to 100 mL of preheated stop buffer (72°C) for 3 min to stop the PRINS reaction and to remove the coverslip. Then transfer the slide to 100 mL of washing solution. The slides can stay in this buffer overnight at 4°C if convenient.

3.3. Detection

1. Wash the slides twice for 3 min at room temperature in washing solution, with gentle agitation.
2. Drain the excess washing solution and apply 100 µL of blocking solution to each slide.
3. Incubate for 10 min at room temperature under a coverslip.
4. Remove the coverslip, drain excess fluid, and apply 100 µL of FITC-avidin diluted to 5 µg/mL in blocking solution to the slide. Cover with a new coverslip and incubate at 37°C for 30 min in a moist chamber.
5. Remove the coverslip and wash the slide three times (5 min each) in washing solution, at room temperature, with gentle agitation.
6. Drain excess fluid and mount the slide (22 × 40 coverslip) with Vectashield antifade solution containing 0.5 µg/mL propidium iodide.
7. Examine the slide by fluorescence microscopy (**Fig. 1**).

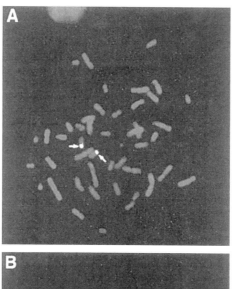
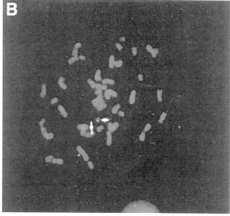

Fig. 1. PRINS detection of chromosomes 13 (**A**) and 21 (**B**). The detection was performed according to the protocol described in the text. The chromosomes were counterstained with propidium iodide. Arrows indicate the chromosome-specific signals.

4. Notes

1. Chromosome-specific primers sometimes differ from each other by only one nucleotide at the 3'-end, as for the primers described here. It is therefore advisable to purify the primers by HPLC to avoid contamination by shorter products arising from premature stops during synthesis. Storage of the primers in small aliquots also prevents degradation of the primers by repeated cycles of freeze-thawing.
2. The concentration of the primers can be determined by using the Beer-Lambert equation:

$$C = A_{260}/\varepsilon_{max} \times L \quad (1)$$

where C is the concentration (M), A_{260} is the absorbance at 260 nm, ε_{max} is the molar extinction coefficient (M^{-1}), and L is the path length (cm) of the spectrophotometer cuvet. The molar extinction coefficient for an oligonucleotide can be determined as follows:

$$\varepsilon_{max} = (\text{number of A} \times 15,200) + (\text{number of C} \times 7050) + \\ (\text{number of G} \times 12,010) + (\text{number of T} \times 8400) \; M^{-1} \quad (2)$$

3. We describe here formamide denaturation, which gave more consistent results in our hands, but it is also possible to denature the chromosomes by heating the slide at 95°C for 3 min as part of the thermal cycle. In this case, omit **step 3** of **Subheading 3.1.**, and run the following program on the PCR machine: 95°C for 3 min, annealing temperature for 7 min, and 72°C for 30 min. The preheated reaction mix is added after the initial denaturation step.
4. The annealing temperature was determined empirically for each primer, and those described here were found to give specific labeling in our hands with our PCR machine and in our laboratory. However, slight adjustments may be necessary if these primers are to be used in other laboratories, since each PCR machine may have a different thermal response curve. The conditions described here must therefore be taken as indications only, and not as absolute rules. When testing new PRINS primers, a good start for the annealing temperature is 5°C under the empirically determined melting temperature (T_m) of the primer: $4°C \times (G + C) + 2°C \times (A + T)$. The annealing temperature is then modified according to the signal and/or the specificity obtained.

References

1. Sumner, A. T. (1994) Chromosome banding and identification: absorption staining, in *Methods in Molecular Biology*, vol. 29, *Chromosome Analysis Protocols* (Gosden, J. R., ed.), Humana, Totowa, NJ, pp. 59–81.
2. Lichter, P. and Ried, T. (1994) Molecular analysis of chromosome aberrations: in situ hybridization, in *Methods in Molecular Biology*, vol. 29, *Chromosome Analysis Protocols* (Gosden, J. R., ed.), Humana, Totowa, NJ, pp. 449–478.
3. Choo, K. H., Vissel, B., Nagy, A., Earle, E., and Kalitsis, P. (1991) A survey of the genomic distribution of alpha satellite DNA on all the human chromosomes, and derivation of a new consensus sequence. *Nucleic Acids Res.* **19,** 1179–1182.
4. Willard, H. F. and Waye, J. S. (1987) Hierarchical order in chromosome-specific human alpha satellite DNA. *Trends Genet.* **3,** 192–198.
5. Jorgensen, A. L., Bostock, C. J., and Bak, A. L. (1987) Homologous subfamilies of human alphoid repetitive DNA on different nucleolus organizing chromosomes. *Proc. Natl. Acad. Sci. USA* **84,** 1075–1079.
6. Koch, J. E., Kølvraa, S., Petersen, K. B., Gregersen, N., and Bolund, L. (1989) Oligonucleotide-priming methods for the chromosome-specific labelling of alpha satellite DNA in situ. *Chromosoma* **98,** 259–265.
7. Gosden, J. and Lawson, D. (1994) Rapid chromosome identification by oligonucleotide-primed in situ DNA synthesis (PRINS). *Hum. Mol. Genet.* **3,** 931–936.
8. Pellestor, F., Girardet, A., Lefort, G., Andréo, B., and Charlieu, J.-P. (1995) Rapid in situ detection of chromosome 21 by PRINS technique. *Am. J. Med. Genet.* **56,** 393–397.
9. Pellestor, F., Girardet, A., Andréo, B., Lefort, G., and Charlieu, J.-P. (1994) The use of PRINS technique for a rapid in situ detection of chromosomes 13, 16, 18, 21, X and Y. *Hum. Genet.* **95,** 12–17.
10. Pellestor, F., Girardet, A., Lefort, G., Andréo, B., and Charlieu, J.-P. (1995) Selection of chromosome specific primers and their use in simple and double PRINS technique for rapid in situ identification of human chromosomes. *Cytogenet. Cell. Genet.* **70,** 138–142.
11. Charlieu, J.-P., Murgue, B., Marçais, B., Bellis, M., and Roizès, G. (1992) Discrimination between alpha satellite DNA sequences from chromosomes 21 and 13 by using polymerase chain reaction. *Genomics* **14,** 515,516.

119

In Situ PCR Amplification of Intracellular mRNA

Raymond H. Chen and Susan V. Fuggle

1. Introduction

The polymerase chain reaction (PCR) is now commonly used in laboratories involved in research studies and clinical diagnostic work *(1,2)*. A major advantage of PCR combined with reverse transcription (RT-PCR) is that it can be used to amplify and detect rare mRNA within a specimen. However, conventional RT-PCR cannot be used either to quantitate the frequency of cells expressing a particular mRNA or determine the cellular origin of the amplified signal. Both of these factors may be relevant in the interpretation of gene expression.

In order to overcome the limitations of conventional RT-PCR, methods have been developed for performing *in situ* RT-PCR (for reviews, *see 3–6*). By performing the reverse transcription and subsequent amplification within the cells fixed onto microscope slides, it is possible to identify the cellular origin of the signal. The technique has an advantage in that it does not require mRNA to be extracted from the sample, and thus, there is no potential for signal loss during the nucleic acid isolation step. Furthermore, unlike conventional PCR, the technique can be used to determine the prevalence of gene expression within a cell population.

The technique is based on the functional hypothesis that enzymes and reagents can freely enter fixed cells, and synthesize and amplify cDNA *in situ*. One important corollary is that the PCR products themselves can also freely enter and egress. Consequently, the success of the technique is dependent on an equilibrium between permeability to reagents and the retention of PCR products. In our experience, we found that carefully controlled fixation and digestion were important in maximizing the *in situ* amplification and retention of signal. The optimization of the equilibrium is critical to the success of the technique, and careful calibration of reaction conditions is mandatory for each primer set used.

In this chapter, we will describe the technique we developed for the detection of granzyme A and perforin mRNA in cytospin preparations of activated human peripheral blood lymphocytes *(7)*. Granzyme A and perforin are functional markers of cytotoxic T-cells, NK, and lymphokine-activated killer cells, and they can be readily detected in adult peripheral lymphocytes after in vitro stimulation *(8)*. Our technique involves directly incorporating biotinylated nucleotides into the PCR product and detecting the labeled product with an antibiotin antibody using a standard immunochemical staining method. This method is not suitable for use on tissue sections, since damaged DNA may serve to prime the reaction, and thus, labeled nucleotides may become incorporated into a nonspecific product *(9–11)*. Methods for *in situ* PCR on tissue sections have been described elsewhere (*see* **refs. *12–22***).

2. Materials

2.1. Glass Slide Preparation

1. Glass slides (Solmedia, Romford, Essex, UK).
2. Decon 90 (Decon Laboratories, Hove, UK).
3. 3-Aminopropyltriethoxysilane (Tespa; Sigma, Poole, Dorset, UK).
4. Acetone (Merck, Poole, Dorset, UK).
5. Diethyl pyrocarbonate (DEPC; Sigma) treated doubly-distilled water (DEPC-ddH$_2$O; *see* **Note 1**).
6. Coverslips (Chance Propper, Smethwick, Warley, UK).
7. 1% Dimethyl dichlorosilane in CCl$_4$ (Merck).

2.2. Cell Preparation

1. Anticoagulated human peripheral blood.
2. RPMI (Imperial Laboratories, Andover, UK) supplemented with 10% fetal calf serum (FCS; Imperial Laboratories), 2 m*M* L-glutamine, 100 U/mL penicillin, and 100 µg/mL streptomycin.
3. Lymphocyte separation medium (Flow Laboratories, Rickmansworth, Herts., UK).
4. Phosphate-buffered saline (PBS; Unipath, Basingstoke, UK).
5. Phorbol-12-myristate-13-acetate (PMA, 500 ng/mL; Sigma).
6. Phytohemagglutinin (PHA, 1 mg/mL; Wellcome, Dartford, UK).
7. Tissue-culture incubator at 37°C.
8. Cytospin centrifuge (Shandon Southern Products, Runcorn, Cheshire, UK).

2.3. Fixation

Starting from this step, it is imperative that all glassware is baked and all reagents are RNase-free (*see* **Note 2**). Chemicals should be reserved for RNA work.

1. 4% Paraformaldehyde (Sigma) in DEPC-PBS: Make fresh; heat suspension to 60°C with constant agitation for at least 1 h to allow paraformaldehyde to dissolve in DEPC-PBS. Cool to room temperature before use.
2. DEPC-3X PBS, DEPC-1X PBS.
3. DEPC-ddH$_2$O.
4. 50% Ethanol in DEPC-ddH$_2$O, 80% ethanol in DEPC-ddH$_2$O, 100% ethanol.

2.4. Proteinase K Digestion

Stock solutions:

1. Proteinase K (Type XI protease [Sigma] 10 mg/mL in 0.1 *M* Tris-HCl, pH 8.0).
2. 1 *M* Tris-HCl, pH 8.0.
3. 500 m*M* EDTA, pH 8.0.

2.5. Hybridization and Reverse Transcription

1. Hybridization solution: 50% formamide (Merck), 10% dextran sulfate (Amersham Pharmacia Biotech, Milton Keynes, Bucks., UK), 300 mM NaCl, 20 mM Tris-HCl, pH 7.6, 5 mM ethylenediaminetetraacetic acid (EDTA), 1X Denhardt's (Sigma), 10 mM dithiothreitol (DTT, Sigma).
2. Antisense primer at 1 µg/mL (*see* **Subheading 2.6.**).
3. Moloney murine leukemia virus reverse transcriptase (RT; Life Technologies, Uxbridge, UK).
4. 5X Reverse transcription buffer (Life Technologies).
5. 100 mM stocks of dATP, dCTP, dGTP, dTTP (Boehringer Mannheim, Lewes, East Sussex, UK).
6. 100 mM DTT (Life Technologies).
7. RNase inhibitor (Promega, Southampton, UK).
8. Bovine serum albumin (BSA [Sigma]) 10 mg/mL in DEPC-H_2O.
9. DEPC-ddH_2O.
10. DEPC-2X SSC.
11. Humidified slide box.
12. Hybridization oven at 42°C.
13. Hybridization oven at 37°C.

2.6. In Situ PCR

1. Granzyme A primers:
   5'-CCA GAA TCT CCA TTG CAC GA
   5'-CTG TAA CTT GAA CAA AAG GT
2. Perforin primers:
   5'-ACA TGG AAA CTG TAG AAG CG
   5'-GGA TTC CAG CTC CAT GGC AG
3. *Taq* polymerase (Promega).
4. 10X *Taq* polymerase buffer (Promega).
5. 100 mM stocks of dATP, dCTP, dGTP, dTTP.
6. 22.5 nmol biotin-11-dUTP (Sigma).
7. 25 mM $MgCl_2$ (Life Technologies).
8. DEPC-ddH_2O.
9. Mineral oil (Sigma).
10. Xylene (Merck).
11. Thermocycler (Hybaid, Teddington, Middlesex, UK).

2.7. Detection of Amplified Products

1. Mouse antibiotin monoclonal antibody (MAb) (Dako, High Wycombe, Bucks., UK).
2. Horseradish peroxidase (HRP)-conjugated, rabbit antimouse Ig antibody (Dako).
3. Human AB serum.
4. BSA (10% stock solution in PBS).
5. Substrate: 3,3-diaminobenzidine tetrahydrochloride (DAB [Sigma]; 0.6 mg/mL in PBS made freshly with the addition of 3 µL/mL 3% H_2O_2 (Thornton and Ross, Huddersfield, UK) immediately before use.
6. Harris' hematoxylin (Sigma).
7. 70, 90, and 100% Ethanol.
8. DPX mountant (Merck).
9. PAP pen (Bayer Diagnostics, Basingstoke, Hants., UK).
10. Humidified staining tray.

3. Methods

3.1. Glass Slide Preparation

1. Glass slides should be thoroughly cleaned by soaking overnight in 10% Decon 90 in double distilled water. Then rinse slides successively with copious amounts of hot tap water, deionized water, and ddH$_2$O, place in racks, dry, wrap in aluminum foil, and bake at 200°C for 4 h to destroy RNase activity. From this point onward, all glassware used should be RNase free (*see* **Note 1**). Then coat slides by incubating in a solution of 2% Tespa in acetone for 2 min. Rinse twice with fresh acetone, twice with DEPC-H$_2$O (*see* **Note 1**), wrap loosely in aluminum foil, and dry at 37°C overnight. Slides may be stored at room temperature before use.
2. Coverslips should be silicon-coated for easy removal during the procedure. Soak the coverslips in 1% dimethyl dichlorosilane in CCl$_4$ for 1 min, rinse with fresh ddH$_2$O, wrap in aluminum foil, bake at 200°C for 4 h, and store at room temperature until required.

3.2. Cell Preparation

1. Dilute human peripheral blood 1:2 with PBS and isolate lymphocytes by centrifugation through lymphocyte-separation medium at 400g at 20°C for 25 min.
2. Collect lymphocytes from the gradient interface and dilute at least 1:2 in PBS; pellet at 400g at 20°C for 7 min.
3. Wash twice further in PBS, centrifuging at 300g at 20°C for 5 min.
4. Resuspend cells in supplemented RPMI medium at a concentration of 1×10^6 cells/mL, and stimulate with 100 ng/mL PMA and 50 µg/mL PHA in RPMI/FCS.
5. Four days later, harvest cells by centrifugation at 300g at 20°C for 5 min.
6. Wash cells in DEPC-PBS buffer and pellet onto Tespa-treated microscope slides by Cytospin centrifugation at 80g at 20°C for 5 min.

3.3. Fixation

1. Place Cytospin preparations on Tespa-treated glass slides in baked slide racks.
2. After allowing the slides to dry for 5 min, fix the cytospin preparations at room temperature according to the following schedule (*see* **Note 2**):
   a. 4% paraformaldehyde, 20 min;
   b. 3X DEPC-PBS, 5 min;
   c. 1X DEPC-PBS, 5 min;
   d. 1X DEPC-PBS, 5 min;
   e. DEPC-ddH$_2$O, 1 min; 50% EtOH in DEPC-ddH$_2$O, 1 min;
   f. 80% EtOH in DEPC-ddH$_2$O, 1 min; and
   g. 100% EtOH, 1 min.
3. After fixation, cytospin preparations may be covered in aluminum foil and stored at –80°C.

3.4. Proteinase K Digestion

1. Place slides in a baked 2-L beaker at 37°C for a 30-min digestion with 10 µg/mL proteinase K in 0.1 M Tris/50 mM EDTA, pH 8.0 (*see* **Note 2**).
2. Subsequently fix slides in 4% paraformaldehyde as in **Subheading 3.3.**

3.5. Hybridization and Reverse Transcription

For all of the manipulations described below, the slides are placed on a bench covered with aluminum foil.

1. Hybridize cytospin preparations with 10 µL of hybridization solution containing 2.5 ng/µL of antisense oligonucleotides to the human granzyme A or perforin genes in a humidified chamber for 2 h at 42°C (see **Note 3**).
2. During this incubation, the preparations are covered with a coverslip, and carefully placed onto the cells using baked forceps.
3. At the end of the incubation, wash slides vigorously in 2X SSC for 5 min to remove coverslips and hybridization buffer.
4. Shake slides vigorously, wipe with tissue to remove excess salt, and then air-dry.
5. Apply 7 µL of reverse transcription mixture to the cell pellet, cover with a coverslip, and incubate for 1 h in a humidified chamber at 37°C.
6. The mixture contains reverse transcriptase (3 U/µL), in a buffer of 75 mM KCl, 10 mM Tris, pH 8.0, 12 mM MgCl$_2$, 2 µg/µL BSA, 10 mM DTT containing 1 mM of each dATP, dGTP, dCTP, dTTP, and 1 U/µL RNase inhibitor.
7. Wash slides extensively in 2X SSC buffer for 5 min, briefly rinse in ddH$_2$O, and air-dry.

3.6. In Situ PCR

Controls are critically important for *in situ* PCR. (Please refer to **Notes 4** and **5** for suggestions.) In order to reduce the quantity of reagents required for the PCR stage, cut coverslips to a size of approx 1 cm^2 using a diamond glass cutter.

1. Add 5 µL of a solution containing *Taq* polymerase at 0.5 U/µL; 1 mM dATP, dGTP, dCTP; 0.9 mM dTTP; 0.1 mM biotin-11-dUTP; 75 mM KCl; 10 mM Tris, pH 8.0; 10 mM MgCl$_2$; and 7 pmol/µL of each 5'- and 3'-oligonucleotide complementary either to the human granzyme A or perforin genes to the slides (see **Note 6**).
2. Place slides on the thermocycler (see **Note 7**), cover the mixture with the small coverslips, and flood the slides with mineral oil to prevent desiccation (see **Note 8**).
3. The amplification proceeds for one cycle at 94°C for 5 min; 30 cycles at 94°C for 1 min, 60°C for 1 min, 72°C for 1 min; one cycle of 72°C for 10 min.
4. After PCR amplification, submerge the slides in xylene for 2 min to remove mineral oil and then leave in the fume hood to allow xylene to evaporate.
5. Spray the slides with 70% alcohol and wipe with tissue to remove any remaining oil (see **Note 9**).
6. When dry, place the slides in a rack and agitate in 2X SSC to remove coverslips. Then wash extensively in 2X SSC, in PBS, and finally air-dry.

3.7. Detection of Amplification Products

1. Draw a circle around the cell pellet with a PAP pen in order to create a barrier to contain reagents for the detection step.
2. The detection step is performed in a humidified chamber.
3. Rinse slides in PBS and incubate for 30 min with 50 µL of mouse antibiotin MAb (1:25 in PBS containing 0.5% BSA).
4. Wash the slides three times in PBS, incubating for 5 min each time.
5. Detect the primary antibody by incubating for 30 min with 50 µL of HRP-conjugated rabbit antimouse Ig antibody (1:50 in PBS with 10% human AB serum and 0.5% BSA).
6. After washing a further three times in PBS, develop the signal with the freshly prepared substrate solution (DAB and H$_2$O$_2$).
7. Counterstain with Harris' hematoxylin.
8. Dehydrate through 70, 90, and 100% ethanol (1 min each), equilibrate in xylene, and mount in DPX medium.

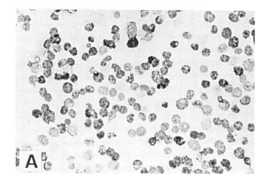 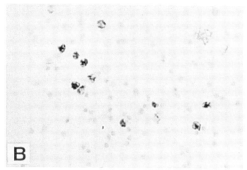

Fig. 1. *In situ* cDNA PCR detection of granzyme A. **(A)** Granzyme A was detected in PHA/PMA stimulated lymphocytes following 30 cycles of PCR amplification. **(B)** PHA/PMA-stimulated lymphocytes were mixed 1:4 with unstimulated, granzyme A-negative, peripheral blood lymphocytes. Large blast cells are granzyme A-positive, whereas the smaller, unstimulated lymphocytes are negative. This demonstrates that, in this system, despite the presence of labeled PCR product in the supernatant, cells negative for granzyme A within a mixed-cell population remain unstained.

Figure 1 illustrates results obtained by using the RT-PCR technique described in this chapter showing granzyme A mRNA in stimulated peripheral blood lymphocytes.

4. Notes

1. It is important to use RNase-free conditions for this technique. Solutions should be treated with 0.1% DEPC for 12 h at 37°C and autoclaved for 30 min before use. Tris buffers cannot be treated directly with DEPC, but should be made with DEPC-treated, autoclaved ddH$_2$O. Glassware should be rendered RNase-free by covering with aluminum foil and baking for 4 h at 200°C. Gloves should be worn at all times. When it is necessary to place slides on the bench, the bench should be covered with aluminum foil and the slides manipulated with baked forceps.
2. Cellular fixation and digestion are critical to the success of this technique. The conditions may vary according to the type of cells used, the size of the PCR product, and the stability of the mRNA. The conditions need to be optimized for each new set of experiments.
3. The duration of hybridization needs to be empirically determined. Although a longer hybridization time should favor oligonucleotide-mRNA binding, labile mRNA may degrade, resulting in a truncated cDNA.
4. Controls are of crucial importance in *in situ* PCR systems. Positive controls using primers specific for a housekeeping gene need to be performed on the test cells to demonstrate the presence of mRNA. Cells known to be positive for the gene of interest should be included to demonstrate that the reaction conditions have been optimized. Negative controls are particularly important in a system such as we describe, where labeled nucleotides are directly incorporated into the PCR product. In this respect, a PCR *in situ* hybridization system has advantages and may be preferred. Some suggested negative control reactions are shown in **Table 1**. A negative result from reaction (a) will demonstrate the absence of endogenous peroxidase in a sample and from (b) the absence of nonspecific binding of the HRP-conjugated secondary antibody. Reactions (c) and (d) will show that there is neither endogenous biotin in the sample nor a signal following RT alone. Reactions (e) and (f) are particularly important controls; a negative result demonstrates that the signal

Table 1
Suggested Negative Controls for *In Situ* cDNA PCR

	a	b	c	d	e	f	g	h
Fixation, digestion	+	+	+	+	+	+	+	+
Hybridization	+	+	+	+	−	+	+	+
Reverse transcription	+	+	+	+	+	+[b]	+	+
PCR	+	+	−	+[a]	+	+	+[c]	+[d]
Antibiotin antibody	−	−	+	+	+	+	+	+
HRP-conjugated secondary antibody[e]	−	+	+	+	+	+	+	+
Substrate	+	+	+	+	+	+	+	+
Expected result	−	−	−	−	−	−	−	−

[a]Omit biotinylated nucleotide.
[b]Omit reverse transcriptase.
[c]Omit primers.
[d]Omit *Taq* polymerase.
[e]HRP—horseradish peroxidase.

results from amplification of the cDNA, and not from priming by nicked DNA or amplification of genomic DNA. Reaction (g) demonstrates the requirement for primers and (h) on *Taq* polymerase to produce a signal.

5. It is possible to perform a Southern blot analysis on the supernatant from the *in situ* PCR reaction *(3)*. This enables the specificity of the amplified product to be confirmed.
6. *In situ* PCR appears to require higher concentrations of reagents than tube PCR ($MgCl_2$, *Taq* polymerase, and nucleotides). Consequently, it may not be possible to transfer conditions optimized for tube PCR directly to an *in situ* PCR system.
7. The heating blocks of traditional PCR thermocyclers, with their discontinuous surface area, do not provide ideal heat conduction for *in situ* PCR. It is preferable to use a machine with specifically designed flat blocks.
8. Desiccation during the PCR process can yield false-positive signals. It is important that the mineral oil completely seal the coverslip.
9. The mineral oil should be completely removed before the immunohistochemistry step. If present, the hydrophobic oil droplets will interfere with the antibody binding and subsequent detection of the PCR product.

References

1. Saiki, R. K., Scharf, S. J., Faloona, F., Mullis, K. B., Horn, G. T., Erlich, H. A., and Arnheim, N. (1985) Enzymatic amplification of beta-globin genomic sequences and restriction site analysis for diagnosis of sickle cell anemia. *Science* **230,** 1350–1354.
2. Rolfs, A., Schuller, I., Finckh, U., and Weber-Rolfs, I. (1992) *PCR: Clinical Diagnosis and Research.* Springer-Verlag, Heidelberg.
3. Nuovo, G. J. (1992) RNA and mRNA reverse transcriptase PCR *in situ* hybridization, in *PCR In Situ Hybridization: Protocols and Applications* (Nuovo, G. J., ed.), Raven, New York, pp. 239–254.
4. Komminoth, P. and Long, A. A. (1993) *In situ* polymerase chain reaction. An overview of methods, applications and limitations of a new molecular technique. *Virchows Archiv. B Cell Pathol.* **64,** 67–73.
5. Chen, R. H. and Fuggle, S. V. (1995) *In situ* cDNA polymerase chain reaction: research and clinical applications in *Progress in Pathology*, vol. 2 (Kirkam, N. and Lemoine, N. R., eds.), Churchill Livingstone, London, UK, pp. 203–218.

6. Teo, I. A. and Shaunak, S. (1995) Polymerase chain reaction *in situ*: an appraisal of an emerging technique. *Histochem. J.* **27,** 647–659.
7. Chen, R. H. and Fuggle, S. V. (1993) *In situ* cDNA polymerase chain reaction: a novel technique of detecting mRNA expression. *Am. J. Pathol.* **143,** 1527–1534.
8. Griffiths, G. M. and Mueller, C. (1991) Expression of perforin and granzymes in vivo: potential diagnostic markers for activated cytotoxic cells. *Immunol. Today* **12,** 415–419.
9. Long, A. A., Komminoth, P., Lee, E., and Wolfe, H. J. (1993) Comparison of indirect and direct *in-situ* polymerase chain reaction in cell preparations and tissue sections. *Histochemistry* **99,** 151–162.
10. Sallstrom, J. F., Zehbe, I., Alemi, M., and Wilander, E. (1993) Pitfalls of *in situ* polymerase chain reaction (PCR) using direct incorporation of labeled nucleotides. *Anticancer Res.* **13,** 1153–1154.
11. Teo, I. A. and Shaunak, S. (1995) PCR *in situ*: aspects which reduce amplification and generate false positive signals. *Histochem. J.* **27,** 660–669.
12. Nuovo, M. A., Nuovo, G. J., MacConnell, P., Forde, A., and Steiner, G. C. (1992) *In situ* analysis of Paget's disease of bone for measles-specific PCR-amplified cDNA. *Diagn. Mol. Pathol.* **1,** 256–265.
13. Nuovo, G. J., Forde, A., MacConnell, P., and Fahrenwald, R. (1993) *In situ* detection of PCR-amplified HIV-1 nucleic acids and tumor necrosis factor cDNA in cervical tissues. *Am. J. Pathol.* **143,** 40–48.
14. Nuovo, G. J., Lidonnici, K., MacConnell, P., and Lane, B. (1993) Intracellular localization of polymerase chain reaction (PCR)-amplified hepatitis C cDNA. *Am. J. Surg. Pathol.* **17,** 683–690.
15. Heniford, B. W., Shum Siu, A., Leonberger, M., and Hendler, F. J. (1993) Variation in cellular EGF receptor mRNA expression demonstrated by *in situ* reverse transcriptase polymerase chain reaction. *Nucleic Acids Res.* **21,** 3159–3166.
16. Nuovo, G. J., Becker, J., Simsir, A., Margiotta, M., Khalife, G., and Shevchuk, M. (1994) HIV-1 nucleic acids localize to the spermatogonia and their progeny. *Am. J. Pathol.* **144,** 1142–1148.
17. Nuovo, G. J., Gallery, F., MacConnell, P., and Braun, A. (1994) *In situ* detection of polymerase chain reaction-amplified HIV-1 nucleic acids and tumor necrosis factor-α RNA in the central nervous system. *Am. J. Pathol.* **144,** 659–666.
18. Patel, V. G., Shum-Siu, A., Heniford, B. W., Wieman, T. J., and Hendler, F. J. (1994) Detection of epidermal growth factor receptor mRNA in tissue sections from biopsy specimens using *in situ* polymerase chain reaction. *Am. J. Pathol.* **144,** 7–14.
19. Staecker, H., Cammer, M., Rubinstein, R., and van de Water, T. R. (1994) A procedure for RT-PCR amplification of mRNAs on histological specimens. *Biotechniques* **16,** 76–80.
20. Lau, G. K., Fang, J. W., Wu, P. C., Davis, G. L., and Lau, J. Y. (1994) Detection of hepatitis C virus genome in formalin-fixed paraffin-embedded liver tissue by *in situ* reverse transcription polymerase chain reaction. *J. Med. Virol.* **44,** 406–409.
21. Mougin, C., Didier, J. M., Bettinger, D., Madoz, L., Coumes-Marquet, S., and Lab, M. (1997) *In situ* PCR to cells and to wax sections: an alternative to *in situ* hybridization for the detection of low copies of human papillomavirus and cytomegalovirus DNA, in *Methods in Molecular Biology, vol. 71, PRINS and In Situ PCR Protocols* (Gosden, J. R., ed.), Humana Press, Totowa, NJ, pp. 77–97.
22. Hendler, F. J., Shum-Siu, A., Heniford, B. W., El Mekki Ahmed, M., and Patel, V. J. (1997) Reverse transcriptase PCR *in situ* on cryopreserved tissue sections, in *Methods in Molecular Biology, vol. 71, PRINS and In Situ PCR Protocols* (Gosden, J. R., ed.), Humana Press, Totowa, NJ, pp. 99–111.

120

An Introduction to Bioinformatics

Henry Brzeski

The purpose of this chapter is to illustrate how to obtain information on DNA and protein sequences from databases. This is most conveniently achieved using a Web browser (Netscape or Internet Explorer). This chapter is loosely based on a course given by the author at the University of Hertfordshire from a set of Web pages that facilitate Internet navigation by using hyperlinks and allow sequences to be copied from the Web page and pasted into the relevant search engine. (The online version of this information can be found at http://www.herts.ac.uk/natsci/Bio/informatics.htm.)

Many simple queries about protein and DNA sequences can now be answered using a Web browser. The purpose of this chapter is to give you a flavor of the sorts of things which are now possible, but first it is necessary to explain some of the jargon. If you know all about the World Wide Web (WWW or 'the Web'), URLs, and hyperlinks then you can bypass the introductory information.

1. Introduction
1.1. The World Wide Web

The World Wide Web and the Internet were not invented by Bill Gates! It was originally put together in the 1960s by, primarily, the U. S. Department of Defense (DOD) to ensure continuity of communication during a war. The DOD relied on the fact that every computer connected to its lines could communicate with any other computer. However, to make the system impregnable to attack, there was not a straightforward connection between each computer. Instead, multiple computers were connected to each other. As a result of this form of connection, there was always more than one way to connect between A and J. It may have been A–B–C–G–J, it may have been A–C–H–I–J, or it may have been, A–G–C–I–J, etc. This provided the resilience to possible attempts at disruption. The original work was done by the Advanced Research Projects Agency, and the network was known as *ARPAnet*. Later on, academic institutions saw this as a wonderful way for researchers to communicate and so they started to connect their computers to this international network or "Internet." This gave rise to the ability to communicate via e-mail and also allowed collaborators to share large amounts of data by transferring the files rapidly via the Internet rather than by mailing a pile of disks or tapes. Now the Internet is being used for advertising and other commercial purposes.

1.2. Navigating on the Web

The files on computers scattered around the world must all have a unique name so that you can access each one specifically. This means that their names sometimes can be rather complex. Initially, it is necessary to tell your Web browser where to go to pick up a specific page. (The address of this page is given a jargon name; it is a "URL," which stands for Universal Resource Locator.) However, navigating (surfing) on the Internet would be tedious if surfers had to type in these names continually. They would soon get bored with typing "http://www.expasy.ch/prosite/," one of the addresses we will used later.) For this reason navigation is accomplished by using "hyperlinks" displayed in the now ubiquitous Web browsers. Hyperlinks can readily be identified on Web pages, because they are usually represented as underlined text in color, or as a button that is 'pressed' by clicking on it with the computer mouse and cursor. If the pointer is positioned over the hyperlink the address or URL it represents will appear in the status bar at the bottom of the browser window. Single click on the hyperlink and the Web browser will load the page at the new address. Hyperlinks can refer to different places in the same document or to totally new addresses. Hyperlinks should be traversed with care, as it is all to easy to follow links without thinking and end up miles from home, both figuratively and literally. It is usually possible to retrace the original path by pressing the **Back** button, but take care, this does not always work. All Web browsers record a history of traversed links, which can be used to connect rapidly to recently visited sites.

1.3. Databases and the Web

Since the early days of DNA and protein sequencing, such information has been deposited in computer databases so that many individuals could access this information. When the World Wide Web greatly expanded the reach of networked computers, it was not long before the Web browser became the interface between a very widely scattered population of researchers and the programs that could sift through the large amounts of data that were being accumulated.

2. DNA Databases

In the following sections, I will introduce the reader to a few of the programs available via the Web for finding and analyzing biochemical information.

2.1. Using Entrez to Search for Relevant Database Entries

Sooner or later a project reaches the point when it is necessary to devise primers to amplify known sequences from cells. This section describes how to obtain sequences for known genes/mRNAs, making it possible to devise primers to characterize genes/mRNAs. The National Center for Biotechnology Information (NCBI) has a very powerful computer with an easy-to-use Web-based interface for accessing sequence information. Follow these instructions to find sequences of particular genes/mRNAs.

1. Start a Web browser (*Netscape* or *Internet Explorer*) by clicking on its icon.
2. Go to **File/Open** in the menubar, enter http://www.ncbi.nlm.nih.gov/Entrez/ in the dialog box and press the **OK** button. This *will not* load the file into a new window.
3. This will load the *Entrez* page which allows users to quiz the databases available at NCBI for textual information connected with any required topic.
4. Click on the **Nucleotide** hyperlink.
5. In the text box enter 'p53' and press the **Search** button.
6. This very simple search will find many records (note the button with **Retrieve N records** at the top right hand side of the page).
7. These matches or 'hits' will consist of database entries containing partial and complete genomic or cDNA sequences from *Homo sapiens* and many other species.
8. Press the **Back** button on the browser.
9. Change the query to p53 & human (the "&" tells the server to find all records which contain both the word "p53" **and** the word "human"—an example of Boolean logic) and press the **Search** button. (You can find out more about Boolean expressions by clicking on **Detailed help** on the *Entrez* page: scroll to the top of the page, find the section labeled **For Experts Only**, and click on **Entering Complex Boolean Expressions**.)
10. This query finds fewer records. You can adjust your query using various required words and Boolean operators. By adding extra keywords, e.g., "complete," "mRNA," etc., you can fine tune your search and hit fewer documents.
11. Once the list is manageable, press the **Retrieve N records** to receive the first summary page of hits.
12. Each hit contains a checkbox, an accession number (the ID of the record), a brief summary of the entry (taken from the file), and various related links.
13. Check the boxes of the relevant hits and then press the **Display** button to retrieve the actual record(s).
14. Each record contains a number of fields that describe the sequence, e.g., the organism, whether the sequence is genomic or derived from mRNA, or relevant published information. Depending on the sequence, the record may contain information on biologically important areas of the sequence, e.g., promoters, start AUG, introns, etc.) followed by the final, and most important, part of the record, the sequence itself.

2.2. Searching for Database Entries That Match a Sequence

The first step of the human genome sequencing project has involved identifying those DNA sequences most important for a cell. These code for proteins synthesized by the cell, which defines the cell's enzymatic complement and therefore its function. For this reason, people have been isolating the mRNAs expressed in cells, converting them to DNA (cDNA), cloning and sequencing them in their thousands. These expressed sequences are given the jargon name of *Expressed Sequence Tag* (EST) and will define the proteins made by a cell. As the sequencing of the human genome progresses, the function of more and more DNA/protein sequences will be identified. It is now routine to generate many ESTs and then to compare them with sequences in the databases to determine their function. A number of such ESTs are given in **Table 1** (2–8) along with one bacterial gene (1). The next procedure illustrates how to compare these sequences against the DNA databanks using *Basic Local Alignment Search Tool* (*BLAST*) to find out what they code for.

1. Identify a sequence to use (this can be an in house sequence or one of those provided in **Table 1**). Copy the sequence to your clipboard.

Table 1

Number	Sequence
1	GGAAAGAAATGCATAAGCTTTTGCCATTC TCACCGGATTCAGTCGTCACTCATGGTGATT TCTCACTTGATAACCTTATTTTTGACGAGG GGAAATTAATAGGTTGTATTGATGTTGGAC GAGTCGGAATCGCAGACCGATACCAGGATC TTGCCATCCTATGGAACTGCCTCGGTGAGT TTTCTCCTTCATTACAGAAACGGCTTTTTCA AAAATATGGTATTGATAATCCTGATATGA ATAAATTGCAGTTTCATTTGATGCTCGATG AGTTTTTCTAATCAGAATTGGTTAATTGGT TGTAACACTGGCAGAGCATTACGCTGACT TGACGGGACGGCGGCTTTGTT
2	TCCTGGNTCTGTTCTTCATCTTCACCTACTTCAAAGTTCCTGAGACTAAA
3	GGCCAAATTTGAAGAGCTCAACATGGATCTGTTCCGGTCTACTATGAAGC
4	GATGTCCAGAAGAATATTCAGGACTTAACGGCTNCAGGNTTTTAACAAAA
5	ATTGGCAGCCACACGGTGCTGGAGCTGCTGGAGGCTGGCTACTTGCCTGT
6	CATCGTGGAGAAGCCCTTCGGGAGGGACCTGCAGAGCTCTGACCGGCTGT
7	GCCCTGTCGAGACACTTGCCTTCTTCACCCAGCTAATCTGTAGGGCTGGA
8	TACATAATGTATTTATATATTTTTTGTATAATCACTATCTTTGTATTTAC

2. Open your Web browser and go to http://www.ncbi.nlm.nih.gov/BLAST/. Click on **Basic *BLAST* Search** to load a page containing the search form.
3. **Paste** the copy on the clipboard into the **search window**.
4. Use the default conditions for the search, i.e., *blastn*.
5. Press the **Search** button and wait while the sequence is compared to the databases and the matches displayed. (There is now a formal queuing system at NCBI, and you will wait for your results as explained on the page.)

The Washington University–Merck collaboration for EST sequencing (http://genome.wustl.edu/est/esthmpg.html) generates a large amount of sequencing information, and pictures of every sequencing gel are available at this site.

The Washington University Medical School Genome Sequencing Centre (http://genome.wustl.edu/gsc/index.shtml) is also involved in sequencing the human genome and information can be found here.

2.2.1. Interpreting the Results

The results from the NCBI *BLAST* server are presented both graphically and textually. The graphical view shows the query sequence as a thick red line with base numbers attached to it. Below this are a series of thin lines which represent matches to the query sequence. The length of the line indicates that part of the query sequence which matches the hit sequence. The color represents the quality of the match.

Below the picture is a list of files which correspond to these matches sorted in match-quality order. The first hyperlink is to the file containing the entire sequence. This is followed by a very brief description of the file. The next number is a numerical score which represents how good the match was. This score is hyperlinked to the actual

Table 2

insert.seq	TGTGAGCGGATAACAATTTCACACAGGAAACAGCTATGACCATGAT TACGAAAGGTGCTTTTGGGGGCCGTCAGGGTCGAGGGTTCCTATTT CCTGGTCTATGGGGTCCCCGGCTTCGGGAAAGATAATGAAAGCCT CATCAGCAGGGAGGAGTTTTTAGGGGGGGTCCGCATGGGGGTCCC CCAAGCGACCGAATTGGCGGCTGAGGCCGTGGTGCTTCATTACAC CGATTTTCGAGCTCGGTACCCGGGGATCCTCTAGAGTCGACCTGC AGGCATGCAAGCTTGGCACTGGCCGTCGTTTTACAACGTCGTGAC TGGGAAAACCCTGGCGTTACCCAACTTAATCGC

match found between your query sequence and the match itself. Finally, the last number gives the statistical significance of the match (the *E* value) and the chances of finding this match by chance.

2.3. Designing PCR Primers

2.3.1. Designing Primers for PCR Using xprimer

xprimer is a Web-based primer design package. Go to http://alces.med.umn.edu/xprimerinfo.html to see a detailed explanation of the various conditions that the primers must fulfil.

The purpose of this exercise is to learn how to design primers using *xprimer* and the sequence in **Table 2** (insert.seq) as the template. This sequence represents an insert in a plasmid plus 50–100 bp of vector sequence on either side.

1. Start a Web browser (*Netscape* or *Internet Explorer*) by clicking on the relevant icon.
2. Go to **File/Open** in the menubar, enter http://alces.med.umn.edu/webprimers.html in the dialog box and press the **OK** button.
3. After connection the Web browser will open the **Primer selection (image)** window.
4. Click in the **Query sequence**: text box (towards the bottom of the page).
5. Paste or type the sequence into this text. *Don't worry about any spaces which might appear.*
6. It is possible to fine tune the search parameters, i.e., primer length or Tm difference, by altering the values in the various list boxes but for the moment use the suggested defaults.
7. Press the **Submit** button.
8. After a few seconds/minutes the results will be returned in the form of a GIF file (it has a '.gif' file extension). This is a format for displaying images on a computer, and it is not possible to copy and paste primer sequences from here! If you want to copy and paste sequences then use the **Text** version of *xprimer* available from the **Primer selection (image)** window.
9. Compare these sequences with those of the M13 forward (GTTTTCCCAGTCACGAC) and reverse (GGAAACAGCTATGACCATG) primers. *Note*: the terms forward and reverse used for M13 primers are not the same as the terms used for forward and reverse primers.
10. Do they match?
11. Can you find these sequences in insert.seq? Remember that these are PCR primer sequences and will be given in a 5' to 3' direction for *each* strand. Remember insert.seq is single stranded and does not include the complementary strand.
12. Do they match?

2.3.2. Checking the Suitability of Your Primers

The final part of primer design is to ensure that the chosen sequences will be specific for the required DNA target. This is achieved by repeating the *BLAST* search performed earlier. However, in this instance the two PCR primers should be used as the queries to ensure that there are no other sequences in the database which might be amplified along with your own sequence. This is obviously not a guarantee against mis-primes, but it will help avoid the more obvious problems, such as unknowingly including regions containing repeated sequences.

2.4. The Human Genome Project

The chromosomal location of completed human DNA sequences can be found at (http://www.ncbi.nlm.nih.gov/genemap/). This Web address contains data on all chromosomes. Selection of a chromosome number will display a figure which contains three parts. First, there are three different ways of displaying the mapping data: two RH (radiation hybrid G3 and GB4) and one genetic map. Second, a drawing of the gene density on this particular chromosome is shown, and, third, the chromosome is drawn as an ideogram. Below this is a wealth of information on what has been sequenced and its relevance, if known. This site is continually updated.

Clicking on a region on the GB4 or G3 map will display the available sequencing information below the mapping data. All this information contains hypertext links to the actual sequencing data itself.

2.4.1. Genes Associated with Human Diseases

The Online Mendelian Inheritance in Man (OMIM) Web site, edited by Dr. Victor A. McKusick and his colleagues at Johns Hopkins University, and elsewhere, contains information on a large number (10,000) of diseases that have been identified as being linked with particular genes. This site can be accessed at http://www3.ncbi.nlm.nih.gov/omim and provides not only a description of the clinical symptoms of the disease but also the genetic lesion that gives rise to it.

It is possible to display the results from a search of this site in two different ways.

1. Searching the **Gene map** (http://www3.ncbi.nlm.nih.gov/Omim/searchmap.html) accesses the database using the name of the disease of interest and will display the result in the order in which the genes are found on the chromosome.
2. Searching the **Morbid map** (http://www3.ncbi.nlm.nih.gov/Omim/searchmorbid.html) will allow a search of the database using the name of the disease of interest (or a general descriptive term such as *anemia*) and will display, in alphabetical order, a list of diseases found that contain the keyword(s). The list contains information on the chromosome location and details of the genetics and clinical symptoms of the disease.

2.5. Sequencing Genomes

The genomes of the following species have been or are being sequenced, and data can be found at the given Web site.

- Human (http://www.ornl.gov/TechResources/Human_Genome/home.html).
- Mouse (http://www.informatics.jax.org).
- *Escherrichia coli* (http://www.genetics.wisc.edu).

- *Haemophilus influenzae* (http://www.tigr.org/tdb/mdb/hidb/hidb.html).
- *Caenorhabditis elegans* (http://www.sanger.ac.uk/Projects/C_elegans).
- *Arabidopsis thaliana* (http://genome-www.stanford.edu/Arabidopsis).
- Rice (http://www.dna.affrc.go.jp:82).
- Yeast (http://genome-www.stanford.edu/Saccharomyces).

A more exhaustive list can be found at http://www.ncbi.nlm.nih.gov/Entrez/Genome/org.html.

3. Protein Databases
3.1. The Databases

There are essentially three databases: the *Protein Information Resource* (*PIR*®; http://pir.georgetown.edu), *SWISS-PROT*® (http://www.expasy.ch/prosite) and *OWL* (http://www.biochem.ucl.ac.uk/bsm/dbbrowser/OWL/OWL.html).

3.2. The Sequence Retrieval System (SRS)

SRSWWW is a World Wide Web interface to the *Sequence Retrieval System* (*SRS*). It can be accessed at a number of different Web sites. *SRSWWW* is widely used because of the simplicity of Web browsers as an interface.

Detailed instructions on how to use *SRS* can be found in the SRS online manual (http://www.expasy.ch/srs5/man/srsman.html). Here I will introduce the basics of the program. In essence, *SRS* will allow the construction of a query that will look for the requested information in a number of databases. This is not as straightforward as it may seem because different databases organize the data into different fields, so it is necessary to construct the query with care. *SRS* ensures that this query construction is as simple as possible.

There are a number of *SRS* Web sites at which users can search various databases. Compare the *SRS* page at Heidelberg (http://www.embl-heidelberg.de/srs5; set up to find nucleic acid and protein database entries) with the version of *SRS* at *SWISS-PROT* (http://www.expasy.ch/srs5).

1. Go to the *SWISS-PROT SRS* page (http://www.expasy.ch/srs5).
2. Press the **Start** button to **Start a new SRS session**.
3. Click on the **TREMBL check box** to deselect it.
4. A detailed explanation of the databases are available by clicking on the hyperlinked database name.
5. Press the **Continue** button.
6. Type oxygen in the first field (leave the default **All text** in the drop-down list box) in the **SRS: Query Form Page**.
7. Press the **Do query** button.
8. This search will find over 1000 entries.
9. Press the **Back** button on your Web browser to return to the **SRS: Query Form Page**.
10. Click on the drop down list box to the left of the first text field and select **Description**.
11. Press the **Do query** button.
12. This search will find about 100 entries.
13. Press the **Back** button to return to the **SRS: Query Form Page**.
14. On the next line change **All text** to **Organism**.

Table 3

1	GDAAKNQLTSNPENTVFDAKRLI
2	EKASGKKIPYKVVARREGDVAACY
3	KLGKSFEMLILGRFIIGVYCGL
4	KGRTFDEIASGFRQGGASQSDKTPEELFHP
5	DDERNGWPVEQVWKEMHKLLPFSPDSVV
6	WRIFTPLLHQIELEKPKPIPYIYGSRG
7	PGAPGGGGGMYPPLIPTRVPTPSNGAPEIP
8	AVFYYSTSIFEKAGVQQPVYATIG

15. Type *Homo sapiens* in the adjacent text box.
16. Press the **Do query** button.
17. This search will find just less than 10 entries.
18. This query is looking for database entries in *SWISS-PROT* that contain only the word "oxygen" in the **Description field** and "*Homo sapien*" in the **Organism field**. (It is possible to change this to an OR search in the drop down list box adjacent to the **Do query** button.)
19. Using this search technique it is possible to find entries from one or many databases using only one set of search parameters.

3.3. Searching for Database Entries That Match a Sequence

The first step of the human genome sequencing project has involved identifying those DNA sequences most important for a cell: the ones coding for proteins synthesized by the cell and, hence, which define the cell's enzymatic complement and thus its function. For this reason, people have been isolating the mRNAs expressed in cells, converting them to DNA (cDNA), cloning and sequencing them in their thousands. These expressed sequences (cDNAs) define those proteins made by a cell and are given the jargon name *Expressed Sequence Tag* (EST). As the sequencing of the human genome progresses, the functions of more and more DNA/protein sequences are identified. It is now routine to generate many ESTs, which are then sequenced and translated into proteins that can be compared with the protein databases to determine their function. You will find a number of such protein sequences derived from ESTs in **Table 3**. Compare these against the protein databanks using *Basic Local Alignment Search Tool* (*BLAST*) to find out what they code for:

1. Choose a sequence to use (this can be an in-house sequence or one of those provided in **Table 3**).
2. Copy the sequence to your clipboard.
3. Now click on **Basic *BLAST* SEARCH** at (http://www.ncbi.nlm.nih.gov/BLAST/) to load a page containing the search form.
4. **Paste** the copy on the clipboard into the search window.
5. Click in the **Program** drop down list box, which at present says *blastn* (*blast nucleic acid*), and choose *blastp* to carry out a blast search on the protein databases. Now press the **Search** button and wait while the chosen sequence is compared to the databases and the matches displayed.

3.3.1. Interpreting the Results

The results from the NCBI *BLAST* server are presented both graphically and textually. The **graphical view** shows the query sequence as a thick red line with base numbers attached to it. Below this are a series of thin lines which represent matches to the query sequence. The length of the line indicates which part of the query sequence matches the hit sequence. The color represents the quality of the match.

Below the picture is a list of files corresponding to these matches and sorted in match-quality order. The first hyperlink (blue and underlined) is to the file containing the entire sequence. This is followed by a very brief description of the file. A numerical score represents how good the match was. This score is hyperlinked to the actual match found between your query sequence and the match **itself**. Finally, the last number gives the statistical significance of the match (the E value) and the chances of finding this match by chance.

Note that the color of the line represents a hit of poor quality. However, clicking on the hyperlinked colored line will display the matching sequence, which will be very similar. The reason for the apparently poor match is that the chance of finding such a short sequence match is high and so the score will be correspondingly low.

3.4. Aligning Protein Sequences Using CINEMA

There are two commonly used programs for sequence alignment: *CLUSTALW* and *pileup*. One of these programs (*CLUSTALW*) can be accessed using a Web browser and a Java Applet called *CINEMA* (Colour Interactive Editor for Multiple Alignment). This applet will access local or database sequences over the Internet and then, once they have been retrieved, align them. The alignments are color coded. If you wish to modify the alignment, it is possible using the *CINEMA* interface.

The purpose of an alignment is to compare two sequences and align the related regions to identify conserved and non-conserved regions. This alignment is built on the underlying assumption that the two sequences being aligned have evolved from a common precursor. If this evolutionary relationship is, in fact, true, then substitution, addition, or deletion of amino acids will be a rare occurrence because of the evolutionary constraints on biological function; the relationship is scored by assigning positive and negative values to matches and mismatches. However, not all amino acid changes are necessarily equally disadvantageous (for instance, the substitution of one hydrophobic amino acid with another is less likely to cause dramatic changes in protein structure than substituting a hydrophobic amino acid with a polar one). Each substitution has a "cost" associated with it and this cost is contained in tables which have such names as PAM-30, PAM-70, BLOSUM-80, and BLOSUM-62. This concept of "cost" is also true for the introduction of gaps into either sequence (addition or deletion of amino acids). The introduction of gaps is an undesirable event and so the introduction of gaps carries penalties in the summation of the final score.

There are penalties to pay for mismatches and introduction of gaps into an alignment. Depending on the specific aims of your own particular alignment, you might want to change these penalties (press the **Advanced button** in the *CLUSTALW* interface window). If you do so, then the results you obtain will probably be different. The default values suggested by the program are a good starting point. Don't be afraid to experiment with these penalties to look for less obvious similarities.

The *CINEMA* home page can be found at http://www.biochem.ucl.ac.uk/bsm/dbbrowser/CINEMA2.1. Alternatively, if this server is proving slow, then try one of the mirror sites at Venus Internet (http://www.venus.co.uk/cinema) or The Weizmann Institute (http://bioinformatics.weizmann.ac.il/CINEMA).

1. Go to the *CINEMA* home page (http://www.biochem.ucl.ac.uk/bsm/dbbrowser/CINEMA2.1) directly into a Web browser.
2. In the top window click on **Applet here** under the *CINEMA* logo.
3. This will load a separate window entitled *CINEMA* which contains multiple color sequences.
4. Select **File/Clear all** in the menu bar to give a clean starting window.
5. Select **Pluglets/Load pluglets** in the menu bar.
6. This opens the **Load Pluglet** window, and you should select **Clustal** and then press the **Load Pluglets** button. **AutoAlign** will appear in the menu bar.
7. Press the **Close** button in the **Load Pluglet** window.
8. Use **SRS** at *SWISS-PROT* to find human globin sequences (**All text** = globin, **Organism** = Homo sapiens) and note the SWISS-PROT file names, e.g., HBA HUMAN.
9. In the *CINEMA* window.
   EITHER
   a. Press the **DB seq** button
   b. In the **Load database sequence** window, delete Enter ID code here and then
   EITHER
     i. Enter the *OWL* code (use the protein ID code, e.g., opsd_sheep, not accession number).
   OR
     i. Change the **Database name** to *SWISSPROT* (or *PIR*) in the drop down list box at the top of the window.
     ii. Enter the *SWISS-PROT* (*PIR*) code, e.g., hba_human (this is *not* case sensitive).
   c. Press the **Get sequence** button.
   d. The requested sequence will be loaded into the *CINEMA* window.
   e. Repeat this procedure until you have accumulated all your sequences.
   OR (if the *OWL* server is down)
   a. Load a database file in the browser window (e.g., from *SWISS-PROT*) by clicking on the relevant hyperlink.
   b. Find the protein sequence at the end of the file.
   c. Drag the mouse across the sequence to select it (you will know it is selected because the sequence is now seen as white text on a black background; don't worry if you have selected numbers.
   d. Choose **Edit/Copy** from the menu bar
   e. Click on the button labeled *CINEMA* in the task bar at the foot of the screen.
   f. Click on the **Seq Editor** button.
   g. Give your sequence a title in the **Accession name** window.
   h. Click in the large text window at the bottom and
   EITHER
     i. Press **Ctrl-V** to paste your sequence from the clipboard
   OR
     i. Right click in the window then select **Paste** from the **Shortcut** menu.
     ii. Press the **Add sequence** button.
     iii. Press the **Clear All** button then repeat this procedure until you have included all the required sequences.
10. Select **AutoAlign/Clustal interface** in the menu bar.

11. In the *CLUSTALW* interface window note that you have a series of dark gray buttons with white text on them which will indicate the progress of the alignment.
12. In the *CLUSTALW* interface window enter a name for your job and then press the **Submit job** button.
    a. You can fine tune your search by pressing the **Advanced** button. This will give you the opportunity of using a different scoring matrix (default is PAM 250) or by altering the criteria used for the alignment, e.g., by changing the penalties for opening or extending a gap.
13. When the alignment has been performed the dark gray buttons will change to light gray and the text will become blue.
14. The alignment will be loaded into the *CINEMA* window so you should clear this first by switching to the *CINEMA* window and choosing **File/Clear all** from the menu bar.
15. Move back to the *CLUSTALW* interface window and view your alignment by pressing the **Load Alignment** button.
16. Note that
    a. The amino acids are color coded according to their properties, e.g., Polar positive—H, K, R (Blue); Polar negative—D, E (Red); Polar neutral—S, T, N, Q (Green); Non-polar aliphatic-A, V, L, I, M (White); Non-polar aromatic-F, Y, W (Purple); P, G (Brown); and C (Yellow). Colors can be viewed and modified by pressing the **COLORS** button.
    b. Where gaps have been introduced this is indicated with a - (dash).
    c. You can move through the aligned sequences using the scroll bars on the *CINEMA* window.

3.5. Comparing 2D Gels in Databases from Different Tissues

Identifying differences between two 2D gels is not always easy. Using the *Flicker* program it is possible to compare two 2D gels on the screen at the same time, and this program will make differences between the two gels more obvious by making the unique spots flicker.

1. Go to the *Flicker* Web site, http://www-lmmb.ncifcrf.gov/flicker.
2. Scroll down the page until you see **C) Lists of 2D PAGE gel images—you pick two from each list to compare** (in Section 1.1), then locate **6.T-lymphocyte phosphoproteins from IL-2/IL-4 dependent cell line 2D gel studies** and click on this. The reason for choosing these gels are that they contain only phosphoproteins; this makes the patterns simpler and so differences are more easily seen.
3. Click on the hyperlink **Select two gels and Flicker Compare them** and then accept the default choices, i.e., **lymphocyte-T_mouse_32P_59g-PPDB - G1-phase** in the upper window and **lymphocyte-T_mouse_32P_59h-PPDB - G2/M-phase** in the lower window.
4. Press the **Go Flicker** button.
5. This may take some time to display the gels but you will eventually see a new page with two 2D gels, one from each stage of the cell cycle, at the bottom of the window.
6. Each gel image will have a set of cross hairs in blue. Look at the two gels and
    a. Decide on a common spot,
    b. Press and hold down the **Control key (Ctl)**, and
    c. Click on the common spot with the mouse in each gel.
7. This will move the gel image and position the cross hairs onto the spot. If you are not happy with the position then repeat this procedure.
8. Scroll back up the screen and click in the ***Flicker*** box to check it. The two images will be viewed in quick succession so that small changes can be easily seen.

3.6. Comparing 2D Gels in House from Different Tissues (Flicker)

This is relatively easy to accomplish as long as you can download your 2D gels onto a Web server that can be accessed using a normal Web addressing system.

1. Log onto the *Flicker* home page, http://www-lmmb.ncifcrf.gov/flicker.
2. Once you have connected with the *Flicker* page you should scroll down the page until you see **4. Flicker compare images from any two URLs**, then scroll further until you get to **Enter two images URLs:**.
3. Click in the **Left image** box and type http://www.herts.ac.uk/natsci/Bio/2Dimages/Image10alt.gif (or select, copy, and paste!).
   EITHER
   a. Click in the **Right image** box and type in the gel address, i.e., http://www.herts.ac.uk/natsci/Bio/2Dimages/Image12alt.gif (or select, copy, and paste!)
   OR
   a. Drag the mouse cursor across the whole of the address you have just typed in **step 3** to highlight it.
   b. Hold down **(Ctrl)**.
   c. Press and release the **C key** to copy the address to the clipboard.
   d. Click in the **Right image** box.
   e. Hold down **Ctrl**.
   f. Press and release the **V key** to paste the address into the box.
   g. Edit the address to give http://www.herts.ac.uk/natsci/Bio/2Dimages/Image12alt.gif (or select, copy, and paste!).
4. Now press the **Go Flicker** button.
5. The relevant gels and the program to view them will now be downloaded. This may take some time but you will eventually see a new page with your two 2D gels. These will be found at the bottom of the window, and if you can't see them then scroll down the window.
6. Each gel image will have a set of cross hairs in blue. Look at the two gels and decide on a common spot, press and hold down **Ctl**, and click on the common spot with the mouse in each gel. This will move the gel image and position the cross hairs on the spot. If you are not happy with the position then repeat this procedure.
7. Now scroll back up the screen and click in the **Flicker box** to check it. The two images will be viewed in quick succession so that small changes can be easily seen.

4. URLs Cited

Chromosomal location of genes, http://www.ncbi.nlm.nih.gov/genemap
Chromosomal location of completed human DNA sequences, http://www.ncbi.nlm.nih.gov/genemap
CINEMA, http://www.biochem.ucl.ac.uk/bsm/dbbrowser/CINEMA2.1
Entrez, NCBI http://www.ncbi.nlm.nih.gov/Entrez
fFlicker (comparing two images), http://www-lmmb.ncifcrf.gov/flicker
Genes associated with human diseases, http://www3.ncbi.nlm.nih.gov/omim
Genomes which have been or are being sequenced:
  Human, http://www.ornl.gov/TechResources/Human Genome/home.html
  Mouse, http://www.informatics.jax.org
  E. coli, http://www.genetics.wisc.edu
  Haemophilus influenzae, http://www.tigr.org/tdb/mdb/hidb/hidb.html
  Caenorhabditis elegans, http://www.sanger.ac.uk/Projects/C elegans
  Arabidopsis thaliana, http://genome-www.stanford.edu/Arabidopsis
  Rice, http://www.dna.affrc.go.jp:82
  Yeast, http://genome-www.stanford.edu/Saccharomyces

A list can be found at http://www.ncbi.nlm.nih.gov/Entrez/Genome/org.html
Human genome, http://www.ornl.gov/TechResources/Human Genome/research.html
Human diseases which have been identified as being linked with particular genes. This site can be accessed at the NCBI home page, http://www3.ncbi.nlm.nih.gov/omim
OWL, http://www.biochem.ucl.ac.uk/bsm/dbbrowser/OWL/OWL.html
PCR primers, *xprimer*, http://alces.med.umn.edu/xprimerinfo.html and http://alces.med.umn.edu/webprimers.html
Positioning ORFs on 2D gels, http://expasy.hcuge.ch/ch2d
ProDom (protein domains), http://protein.toulouse.inra.fr/prodom.html
Protein Information Resource (PIR), http://pir.georgetown.edu
SWISS-PROT, http://www.expasy.ch/sprot/sprot-top.html
Sequence retrieval system (SRS), http://www.expasy.ch/srs5/man/srsman.html or http://www.embl-heidelberg.de/srs5/ c) http://www.expasy.ch/srs5/
Searching the *OMIM* Gene Map, http://www3.ncbi.nlm.nih.gov/Omim/searchmap.html
Searching the *OMIM* Morbid Map, http://www3.ncbi.nlm.nih.gov/Omim/searchmorbid.html
The address of the online version of this chapter, http://www.herts.ac.uk/natsci/Bio/informatics.htm
Washington University–Merck collaboration for EST sequencing, http://genome.wustl.edu/est/esthmpg.html
Washington University Medical School Genome Sequencing Center, http://genome.wustl.edu/gsc/index.shtml

4.1. Other URLs Concerned with Bioinformatics

1. University College London, *A Taste of Bioinformatics*, http://www.biochem.ucl.ac.uk/bsm/dbbrowser/jj. The aim of this tutorial is to provide a gentle introduction to sequence and structure function analysis.
2. DNA Learning Center Cold Spring Harbor Laboratory, *Online DNA Sequence Analysis and Comparison Tutorial* http://vector.cshl.org/SequenceAnlaysisExercise/index1.html.

   This tutorial uses reference human mitochondrial DNA sequences and online resources to:

- Search for like DNA sequences in online databanks
- Locate DNA sequences in genomes
- Compare modern human DNA sequences
- Compare modern human DNA sequences to Neanderthal
- Compare modern human DNA sequences to other organisms

3. EMBnet, DNA analysis tutorial, http://www.ie.embnet.org/other/tut.html. This tutorial considers three popular sets of DNA and protein sequence analysis programs:

- The *Staden Package*, from Rodger Staden et al., MRC Laboratory of Molecular Biology, Cambridge, UK
- *The Wisconsin Package* (*GCG*), from the Genetics Computer Group, Inc., Madison, WI.
- *EGCG* (*Extended GCG*) from a consortium of researchers mostly based in Europe at EMBnet Nodes.

4. University of Adelaide, *A tutorial on sequence analysis: From sequence to structure*, http://www.microbiology.adelaide.edu.au/learn/index.html

   An unusual tutorial by Harry Mangalam subtitled, "one person's cautionary tale of model building."

Index

A

Acrylamide, 73, 648, 697, 750, 773
Actinomycin D, 626
Activation domain library, 761
Adapters, 360
Affinity chromatography, 9
Agarose,
    bead preparation, 83, 91
    gel preparation of 67
    low melting point, 82, 432, 784
    Metaphor™, 510
    nusieve, 725, 861
    plug preparation, 83, 90
    Seaplaque™, 500
Alkaline phosphatase, 138, 145, 155, 157, 334
Alu PCR, *see* PCR
Amberlite™, 527, 749
Amperase™, 575, 577
Amplification, in PCR, *see* PCR
Amplification refractory mutation system, ARMS, 723
AmpliTaq™, *see* Polymerase
AmpliWax™, 576
Anchored PCR, *see* PCR, anchor
Autoradiography,
    direct, 171, 682
    general, 169, 442
    indirect, 171

B

β-Mercaptoethanol, 931, 937
Baculovirus expression, 907
Balked segregant analysis, 675
BCIP, 129, 987
Biodite, 529
Bioinformatics,
    BLAST, 1033
    CINEMA, 1039
    databases, 1032
    Entrez, 1037
    general, 1031
    OMIM, 1036
    Swiss-Prot™, 1037
    World Wide Web (WWW), 1031
    xprimer, 1035
Biotin, 127, 132, 136, 240
Biotin-ON™, 529
BLAST, *see* Bioinformatics, BLAST
Blocking agents, 137
Blood, extraction of DNA from, 4, 6
Blotting,
    Blue dextran, 491
    Bluescript, *see* Vectors
    BluGene™, 336
    capillary transfer, 77
    electroblotting, 79
    fluorography, 172
    from pulsed field gels, 9
    Southern, 77, 84, 155, 163, 681, 917
    vacuum, 79
Buffers, preparation of,
    cacodylate, 136, 734
    denaturation, 77
    GTE, 321
    hybridization , *see* Hybridization, buffer
    kinase, 197, 201, 307, 391
    Klenow, 118
    ligase, 340
    lysis, 208
    MOPS, 38, 178, 252
    MSTET, 489
    nick translation, 123
    PCR, 128, 214, 220, 229, 267, 289, 295, 397, 499, 505, 569, 589
    restriction enzyme, 64, 84
    S1, 202
    sample, 74
    sequenase, 506
    SM, 499, 503
    SSC, 71, 72, 77, 85, 390
    STE, 340
    superscript, 307
    T4 kinase, 119
    TA, 807
    TAE, 101, 340, 348
    TBE, 101, 390, 482, 558, 809
    TENT, 549
    transcription, 251

C

CAT assay, *see* Chloramphenicol acetyl transferase
CAT-ELISA, 795
Caesium chloride, 54
Capillary transfer, *see* Blotting
Capped transcripts, synthesis of, 880
Carbenicillin, 957
Carbohydrates, 17
CEN4 gene, 671
Centricon™, 871, 970
Centri-Sep™, 545

Centri-Spin™, 545
Centrix™, 816
Cerenkov counting, 189, 198, 740
Cetyltriethylammonium bromide (CTEB), 13, 14, 30, 31, 613
CHEF, see Electrophoresis, CHEF
Chemical cleavage, 685
Chemiluminescence, 153
Chloramphenicol acetyl transferase, CAT, 793
Chorulon, 896
Citroclear, 131
Colcemid, 1003
Colony, bacterial,
    purification, 385
    screening, 140, 160, 316, 445, 936
Complementary DNA (cDNA),
    hairpins, 363
    libraries, 261, 289, 305, 355, 373
    library rescue, 311
    preparation, see First strand synthesis
Cosmid library,
    construction, 163, 409
    screening, 408, 411
CPD-Star, 153, 155
CpG islands, 98, 669
Crosslinkers, 320
Cryostat blocks, 982
Cycle sequencing, see DNA sequencing, cycle
Cytochalasin B, 896
Cytospin, 1026

D

DAPI stain, 995
7-Deaza GTP, 506, 510, 530
Denaturation,
    genomic DNA, 177
    PCR DNA, 508
Denaturing gradient gel electrophoresis, DGGE, 705
Denhardts solution, 85, 165, 182, 334, 390, 784, 986
Depurination 77, 96
Diethylamine, 687
Diethylpyrocarbonate (DEPC), 38, 47
Digoxigenin, 129, 132, 985
Differential display, see PCR
Dimethyl sulfate, 554, 555, 818
Dithiotheritol (DTT), 494, 516, 626, 662, 739, 892
DNA amplification fingerprinting (DAF), 675
DNA extraction, see also Extraction of
    bacteria, 29
    biopsy, 53
    blood 4
    colonies, 490, 526
    culture cells 4, 6, 53
    Gram negative bacteria, 29, 53
    Gram positive bacteria, 29
    high molecular weight, 3
    labeling, see Labeling
    mitochondria, 40
    for PCR, 33
    for PFGE, 33
    plant DNA, 13, 53
    plasmid,
        alkaline lysis, 188, 327
        rapid boiling, 327

DNA, extraction of (*cont.*)
    recovery from gels,
        agarose, 69
        polyacrylamide, 516, 558
    solid tissues 3, 4
DNase footprinting, 729
DNA sequencing,
    automated, 541
    cDNA, 499
    chemical, 541, 553
    compressions, 486, 545
    cycle, 525, 533, 535
        degenerate primers in, 533
    direct PCR, 512, 515
        formamide in, 515, 521
        dimethylsulphoxide in, 515
    four-dye labeled primer, 528
    M13, 493
    one dye-labeled primer, 528
    one-step/lane, 557
    partial, 563, 564
    phagemids, 493
    plasmids, 489
    preparation of gels for, 481
    reactions in
        labeling, 516, 537
        termination, 516
    smiling of gels, 485
    solid phase, 523
        biotin, 524, 547
        magnetic beads, 523
        streptavidin, 524
Dot blots, 71
Drying, of gels, 484
Dyad symmetry, 485
Dynabeads, 526

E

ECL, 139, 142, 145
Electroelution, 187
Electrophoresis,
    agarose gels, 67, 251, 507
    buffers, 67
    CHEF, 89, 94
    FIGE (field inversion gel electrophoresis), 100
    formaldehyde, 25, 42, 177, 240
    nondenaturing gels, 50
    PFGE, pulsed field gel, 81,
    polyacrylamide, 73, 648, 697, 750, 773
    SDS-polyacrylamide gels, 886, 925
ELISA, 939, 944
Elongase™, 635
ELOSA, see PCR
Enzyme units, 65
Ethidium bromide, 67, 74, 84, 177,
Ethylmethanesulfonate (EMS), 711
ExAssist, 392, 402
Exon/intron borders, 214, 669
Exon trapping,
    3' terminal, 662
    5' terminal, 653
ExonucleaseIII, 807
Expand™, 635

Index

Expressed sequence tags (EST), 1033
Expression libraries, see Library, expression
Extended chromatin, 997
Extraction of
 bacterial RNA, 47
 DNA from gels, 74
 DNA/RNA, 53
 fungal nucleic acids, 37
 mitochondrial RNA, 23
 mRNA, 9, 195, 213, 219, 239, 261, 289, 305, 308, 355, 645, 875
 plant RNA, 17
 plasmid DNA, see DNA extraction
 total RNA, 41

F

Field strength, 101
FIGE, see Electrophoresis, FIGE
Fill in reactions, see Labeling of DNA
First strand synthesis, 213, 289, 305, 356, 613, 626, 645
Fluorescence in situ hybridization (FISH),
 amplification in, 993
 detection, 994, 1007
 labeling, 993
 principles, 991, 992
 two color, 999
Fluorescein, 139, 149, 526
Fluorescein isothiocyanate, FITC, 989
Fluoricon™, 548
Fluorography, 169
Folligon, 896
Footprinting, 729
Formaldehyde, 25, 177
Formamide, 182, 197, 516, 554, 705, 986
Fusion proteins,
 applications, 367
 expression, 369
 induction, 369
 immunoaffinity purification, 370

G

GC clamping, 709
Gel electrophoresis, see Electrophoresis
Gel retardation, 738, 783, 784
Gel shift assay, 745
GeneClean, 68, 316, 348, 627, 871
Gene families, 595
Gene probes
 biotinylated, 333
 double stranded DNA, 381
 HRP labeled, 158
 labeling, see Labeling
 RNA, 181, 190, 249, 252, 783
 single copy, 96, 161
 single stranded DNA, 204
 specific activity, 190, 879
 specificity, 211
Genescreen, 165
Genomic footprinting, 738
Gultathione-S-transferase, GST, 935
Glycerol, 511
Glycogen, 128
Glycosylation,
 N-linked, 895, 899, 902
 O-linked, 895

Glyoxal, 182
Gradient gels, 486
Guanidinium thiocyanate, 18, 19, 24, 47, 192, 250, 309, 930

H

Hemocytometer, 82, 90
Heparin, 982
Hexamer mix, 124
Hexamine cobolt chloride, 627
High performance liquid chromatography (HPLC),
 applications, 224
 basic components, 106
 of DNA, 105, 523
 PCR products, 106, 721
 quantitation, 107, 108, 112
 retention times, 111
 UV-vis detection, 107
His-Tags, 921, 935
Horseradish peroxidase, 135, 146, 157
Hot start PCR, see PCR, hotstart
Human genome project, 1036
Hybond, 79, 139
Hybridization,
 antibody incubation, 154
 blocking, 154
 buffer, 158, 197, 208
 competition, 163, 164
 cross, 190
 dot blots, 190
 filter, 129
 fluorescein probes, 153
 genomic DNA, 163, 166
 oven, 101
 in pulsed field gels, 97
 subtractive, 305, 312
 washing, 154
Hydrazine, 554
Hydroxylamine, 686

I

Imidazole, 926
Immunocytochemistry, 940
Immunohistochemistry, 940, 988
Insect cell transfection, 914
In situ hybridization, 141, 189, 981
In situ PCR amplification, 1023
Interphase nuclei, 995
Inverse PCR, see PCR, inverse
Island rescue PCR, see PCR, island rescue
Isopropyl β-D thiogalactopyranoside (IPTG), 340
Isotopes, see Radioisotopes

K

Kanamycin, 957
Klenow enzyme, 118, 202

L

Labeling,
 end labeling
  3' DNA 117
  5' DNA 117,
 efficiency of, 128, 151
 fill in reactions 118
 nick translation, 123
 oligonucleotides, 136
 by PCR, of DNA, 127
 random primer labeling, 123, 149
Lambda ZAP®, see Vectors
LA-PCR, 339, 845

Library,
    amplification, 362
    cDNA synthesis, 355
    differential, 314
    expression, 373
    packaging, 362
    PCR screening, 397
    phage display, 449
    plating, 362, 390
    screening, 375, 382
Ligation, 349
Ligation independent cloning (LIC), 951
Long PCR, 339
Loss of heterozygosity, 305
Luciferin, 789
Luria broth medium, 82, 494
Lysostaphin, 30, 34
Lysozyme, 29, 30, 48, 328, 489

M

M13, see Vectors
Maltose binding protein, MBP, 935
Maxam-Gilbert sequencing, see DNA sequencing, chemical
Megaprimer PCR, see Mutagenesis
Metaphase cells, 996, 1000
Methylation interference, 737, 741
Methylation protection, 737
Methylformamide, 559
Microcon-30, 343
Minipreps, see DNA extraction
Minisequencing, see PCR
Minisubmarine gels, 67
Mitochondria, isolation of, 24
Mitogen stimulated lymphocytes, 1000
Molecular weight markers, 67, 82
Monobed resins, 487
MOPS, see Buffers
mRNA, see also Extraction of mRNA
    quantitative determination, 186
Multiple cloning site, 187
Multiplex PCR, see PCR
Mung bean nuclease, 808
Mutagenesis,
    double stranded, 835
    megaprimer PCR, 865
    overlap extension, 857
    primer-directed, 815
    site directed
        phagemids, 827
        uracil containing, 827

N

NCZYM media, 398
Neodymium-iron-boron magnet, 526
Nested deletions, 807
Nick columns, 85, 97
Nick translation, see Labeling
Nitro blue tetrazolium (NBT), 129, 982
Northern blotting, 182, 185, 239, 243, 249, 256
Northwestern assay, 784
Nuclei, preparation of, 209
Nylon membrane, 77, 241, 381
Nytran, 710

O

Oligo (dT) column, preparation of, 10

Oligonucleotides,
    degenerate, 373, 386, 389, 397
    dimer formation, 582
    kinasing, 197
    nested primers, 398, 529
    primer design, see PCR
    probes, see Gene probes
    self complementarity, 583
    stability,
        general, 583
        internal, 584
    SV40, 665
    synthesis, 819
    vectorette sequences, 670
    vector pairs, 502
Oligo™ software, 216, 589, 592, 823
Orange G, 491
Orientation, of inserts, 349
Osmium tetroxide, 687, 689
Overlap extension PCR, see Mutagenesis

P

PAGE, see Electrophoresis, SDS-polyacrylamide gels
Palindromic sequences, 63
Paraffin embedding, 55
Paraformaldehyde, 982
Parallel gradient gels, 712
Phosphate buffered solution (PBS), 220
PCR (Polymerase chain reaction),
    alu PCR, 669, 993
    anchor, 625
    asymmetric, 499, 501
    buffers in, 574
    cloning of, 604
    colony, 959
    contamination of, 570
    continuous, 219
    cycle numbers, 575
    differential display
        nonradioactive, 645
        primers in, 646
    DOP, 1001
    ELOSA coupled, 219
    hotstart, 570, 868
    inverse, 625
    ion cofactors in, 574
    island rescue PCR, 668
    labeling of DNA by, see Labeling, by PCR
    linear, 534
    long range, 633
    multiplex,
        general, 619
        troubleshooting, 621
    plateau effect, 229
    postligation, 960
    preparation of, 215, 569
    primers, design of,
        bioinformatics in, 1035
        degenerate, 591, 595, 596, 597, 599
        digestion efficiencies, 586
        duplex stability, 582
        general, 214, 233, 398, 502, 572, 581, 619
        M13, 825, 846
        for multiplex, 621
        from peptides, 585
        for subcloning, 586
    random amplified polymorphic DNA (RAPDs), 675

PCR (cont.)
  rapid amplification of cDNA ends (RACE),
    general, 613
    3' RACE, 614
    5' RACE, 614, 654
  reamplification, 649
  sequential linear, 499
  single cell, 645
  solid phase minisequencing, 229, 717
  stepdown, 589, 590, 593
  substrate analogues, 574
  touchdown, 589, 590, 593
  troubleshooting, 578
PEG, see Polyethyleneglycol
PFGE, see Electrophoresis, PFGE
Phenylmethylsulfonyl-fluoride (PMSF), 34, 84, 897
Phosphorothioate, 817
Photobiotin, 993
Pileup, 600
Piperidine, 554, 687
Plaque,
  assay, 912
  purification, 376
  screening, 139, 352
Plasmid DNA, see DNA extraction, plasmid
Polyacrylamide, see Electrophoresis, polyacrylamide
Poly (A+) mRNA, 9
Polyethyleneglycol, 48, 49, 494, 829
Polyethylene glycol-6000, 402
Polyhedrin, 916
Polyhistidine (6xHis), see His-Tags
Polymerase,
  AmpliTaq™, 500, 526, 590, 724
  AmpliTaq™Gold, 571
  DNA, 231
  Pyrococcus furiosus (pfu), 324, 633, 862
  RNA, 188
  TaqPlus™, 635
  Thermoprime™, 1012
  Thermus aquaticas (Taq), 525, 558, 633, 719
  Thermus thermophilus (Tth), 573
  Thermosequenase™, 721
  Vent, 324, 862
Polynucleotide kinase, 202, 310, 339
Polysaccharides, spectrophotometry of, 57
Ponceau S, 788
Preflashing, 171
Prep-a-gene™, 325
Primed *in situ* synthesis (PRINS), 1011, 1017
Primer Detective™ software, 216
Primer-directed mutagenesis, see Mutagenesis, primer-directed
Primer extension analysis, 195
Primers, see Oligonucleotides
Prism™, 542
Probes, see Gene probes
Promoters, T3, 526
  RNA, 188
  SP6, 526
  T7, 526
Protease protection, 900
Protein A, 935
Proteinase K, 209, 336
Pulse time, 101
Pyronin Y, 987

Q
Q-banding, 995

QIAEX™, 536
Quantification,
  absolute RNA, 232
  cDNA, 361
  nucleic acids, 57
  relative, 232
  RNA, 42, 229, 254

R
RACE, see PCR, rapid amplification of cDNA ends (RACE)
Radioisotopes,
  characteristics, 170
  uses, 169
RAPDs, see PCR, rapid amplified polymorphic DNA (RAPDs)
Random hexamer labeling, see Labeling
R-banding pattern, 993
Refolding, of protein, 928
Repetitive DNA, 993
Replica plating, 383
Replicative form, 347
Reporter assay, 793
Restriction endonuclease, 63
  buffers used in, see Buffers, restriction enzyme
  digestion in agarose, 91, 93
  inactivation of, 65
  nomenclature, 64
  rare cutting, 93
  star activity, 65
Restriction fragment length polymorphism (RFLP), 679
Reverse transcriptase,
  AMV, 224,
  MoMLV, 615, 626
  Superscript™, 616, 617
Reverse transcription-PCR,
  applications, 54, 213, 220
  quantitative, 226
Riboprobes, see Gene probes, RNA
Ribosomal RNA, 179, 253
Ribosome binding site, 974
RNA extraction, see Extraction of
RNAguard™, 223
RNA-protein interactions, 783
RNA:RNA hybrids, 194
RNase A, 13, 23, 38, 55, 89, 178, 328, 489, 494
RNase H, 627, 665
RNase mapping, 876
RNase P, 875
RNAse protection, 183, 191
RNase T1, 876
RNazol™, 314
RNeasy, 213
RT-PCR, see Reverse transcription-PCR
Run-off transcription, 183, 207

S
S1 mapping, 201
Sample buffer, see Buffers, sample
Screening, see Library, screening
SDS-PAGE, see Electrophoresis, SDS-polyacrylamide gels
Second strand synthesis, 357
Semiquantitative RNA analysis, 213
Sephacryl,
  S-200, 491
  S-400, 342

Sephadex,
    columns, 137
    G25, 137, 197, 198, 863, 970
    G50, 166, 252, 386, 775, 779
Sepharose, 490, 808
Sequenase, 493, 506, 513, 516, 525
Sequence independent amplification, SIA, 993
Sequence tagged sites (STS), 431, 667
Sequencing, see DNA sequencing
Shadowing, UV, 75
Sharks tooth, 481
Signal sequence, 895
Siliconozation, of plates, 484
Silver staining, 648, 696, 700
Single strand confirmation polymorphism, SSCP, 695
Smiling, see DNA sequencing
Solid phase minisequencing, see PCR, solid phase minisequencing
Southern blotting, see Blotting
Southwestern assay,
    considerations, 773
    double replica, 781
    two dimensional, 781
Spectrophotometry,
    DNA/RNA 57
    performing, 58
    polysaccharides, 57
    proteins 57, 965
    scanning, 58
Spectrum Green™, 106
Spectrum Orange™, 1006
Spheroplasts, 100, 419
Splice by overlap extension (SOE), see Mutagenesis, overlap extension
SSC, see Buffers
Star activity, 65, see Restriction endonuclease
STET, 328
Stringency, 138, 154, 155, 159, 167, 241
Sucrose, 208, 901
Synthetic genes, construction of, 609

T
T1 RNase, 183
T4 gene 32 protein, 189, 191
T4 kinase, 119
T/A cloning, 319, 571, 659
TaqTrak™, 503
TEMED, 482, 520, 697, 707, 739
Temperature,
    annealing, 401
    melting 142, 246, 394, 572
Terminal transferase, 615
Thin layer chromatography, 558, 796
Tissue sections, preparation of, 981
Tm, see Temperature, melting
TMAC (Tetramethylammonium chloride), 389
Transcription,
    in vitro, 252, 875
    rates, 207
Transformation,
    E. coli, 313, 351, 383
Translation,
    cell free, 885, 891
    in vitro, 885

Translation (cont.)
    rabbit reticulocyte lysate, 885
    wheat germ, 891
    Xenopus egg extract, 895
Triethanolamine, 986
Triton X100, 84
Trypan blue, 908
Two hybrid screening, 757

U
UDG-cloning, 658
Universal primer, 497
Uracil-N-glycosylase, 570
Urea, 483
UV crosslinking, 79, 785
UV spectrophotometry, see Spectrophotometry

V
Variable number tandem repeat (VNTR), 679
VCS M13, 497
Vectorette, 667
Vectors,
    BacPAK6, 908
    Baculovirus, 910
    lambdaGET, 655
    Lambda ZAP®, 391
    λgt11, 367
    M13, 201, 347
    pACT, 770
    pBluescript, 201, 317, 497
    PCRii, 322
    pET, 947, 948
    pGAD10, 770
    pGEM™, 849, 850
    pHDCAT, 793
    pQE, 922
    pSCREEN, 952
    pTAG4, 921
    pUC18, 323, 836
    pWhitescript, 835,
    pYAC4, 415
    yeast shuttle, 757

W
Western blot, 377, 378, 936
Wisconsin GCG software, 1043

X
X-Gal, 321, 340
XL1 blue cells, 308, 313, 402, 497, 835, 840

Y
Yeast artificial chromosomes,
    filters, 438
    library 415
    screening, 431, 437, 441
    storage, 425
    transport, 425, 428
Yeast chromosomes, separation of, 86, 87
Yeast nitrogen base, 83
Yeast peptone dextrose, 82
Yeast phenotype verfication, 760
Yeast reporter strains, 765
Yeast transformation, 760
YT medium, 391, 494

Z
Zymolase, 82, 435